Hadrosaurs

LIFE OF THE PAST

James O. Farlow, editor

HADROSAURS

Edited by DAVID A. EBERTH and DAVID C. EVANS

Editorial Assistant PATRICIA E. RALRICK

INDIANA UNIVERSITY PRESS Bloomington & Indianapolis

This book is a publication of

Indiana University Press
Office of Scholarly Publishing
Herman B Wells Library 350
1320 East 10th Street
Bloomington, Indiana 47405 USA

iupress.indiana.edu

Telephone 800-842-6796
Fax 812-855-7931

Library of Congress Cataloging-in-Publication Data

Hadrosaurs / edited by David A. Eberth and David C. Evans.
pages cm.–(Life of the past)
Includes bibliographical references and index.
ISBN 978-0-253-01385-9 (cloth)
ISBN 978-0-253-01390-3 (ebook)
1. Hadrosauridae. 2. Hadrosauridae–Anatomy. 3. Hadrosauridae–Geographical distribution. 4. Dinosaurs. I. Eberth, David A. II. Evans, David C. (David Christopher), [date]
QE862.O65H33 2014
567.914–dc23

2014011885

1 2 3 4 5 20 19 18 17 16 15

To David Weishampel and all those, from J. Leidy onward, who have contributed to our knowledge of hadrosaurs.

In particular, we recognize the efforts of Derek J. Main, a tireless promoter of Earth Science education and research. We value his contribution to this volume and mourn his all-too-soon passing.

Those animals of other days will give joy and pleasure to generations yet unborn.

Charles H. Sternberg

Contents

Contributors

Andrey Atuchin, Palaeontological Laboratory of the Institute of Geology and Nature Management, Far East Branch, Russian Academy of Sciences, per. Relochny 1, 675000 Blagoveschensk, Russia

Karl T. Bates, Department of Musculoskeletal Biology II, Institute of Aging and Chronic Disease, University of Liverpool, Sherrington Buildings, Ashton Street, Liverpool, U.K. L69 3GE

Paul M. Barrett, Department of Earth Sciences, Natural History Museum, Cromwell Road, London, U.K. SW7 5BD

Phil R. Bell, School of Environmental and Rural Science, University of New England, Armidale 2351, NSW, Australia

Uwe Bergmann, Stanford Synchrotron Radiation Lightsource, Stanford University, Menlo Park, California 94305

Ivan Y. Bolotsky, Research Center for Paleontology and Stratigraphy, Jilin University, Changchun 130061, China

Yuri L. Bolotsky, Palaeontological Laboratory of the Institute of Geology and Nature Management, Far East Branch, Russian Academy of Sciences, per. Relochny 1, 675000 Blagoveschensk, Russia

Kirstin S. Brink, Department of Ecology and Evolutionary Biology, University of Toronto Mississauga, 3359 Mississauga Road N., Mississauga, Ontario L5L 1C6

Donald B. Brinkman, Royal Tyrrell Museum of Palaeontology, Box 7500, Drumheller, Alberta T0J 0Y0

Caleb M. Brown, Royal Tyrell Museum of Paleontology, Box 7500, Drumheller, Alberta T0J 0Y0

Michael Buckley, Faculty of Life Sciences, University of Manchester, Manchester, U.K. M13 9PL

Michael E. Burns, Department of Biological Sciences, University of Alberta, Edmonton, Alberta T6G 2E9

Nicolás E. Campione, Departments of Earth Science and Organismal Biology, Uppsala University, Box 256, 751 05 Uppsala, Sweden

Rodolfo A. Coria, CONICET–University of Rio Negro–Museo Carmen Funes, Av. Córdoba 55 (8318) Plaza Huincul, Neuquén, Argentina

Philip J. Currie, Department of Biological Sciences, Zoo. 413, Biological Sciences Building, University of Alberta, Edmonton, Alberta T6G 2E9

Fabio M. Dalla Vecchia, Grup de Recerca del Mesozoic, Institut Catalá de Paleontologia Miquel Crusafont (ICP), Escola Industrial 23, E-08201 Sabadell, Spain

Peter Dodson, School of Veterinary Medicine, University of Pennsylvania, 3800 Spruce Street, Philadelphia, Pennsylvania 19104-6045

David A. Eberth, Royal Tyrrell Museum of Palaeontology, Box 7500, Drumheller, Alberta T0J 0Y0

Victoria M. Egerton, School of Earth, Atmospheric and Environmental Sciences, University of Manchester, Manchester, U.K. M13 9PL

Gregory M. Erickson, Department of Biological Science, Florida State University, Tallahassee, Florida, 32306-4295

David C. Evans, Department of Natural History, Royal Ontario Museum, 100 Queen's Park, Toronto, Ontario M5S 2C6

Federico Fanti, Dipartimento di Scienze della Terra e Geologico-Ambientali, Universitá di Bologna, Italy

Andrew A. Farke, Raymond M. Alf Museum of Paleontology, 1175 West Baseline Road, Claremont, California 91711

Rodrigo Gaete, Museu de la Conca Dellá, Carrer del Museu 4, Isona, E-25650, Spain

Àngel Galobart, Grup de Recerca del Mesozoic, Institut Catalá de Paleontologia Miquel Crusafont (ICP), Escola Industrial 23, E-08201 Sabadell, Spain

Albert Garcia Sellés, Grup de Recerca del Mesozoic, Institut Catalá de Paleontologia Miquel Crusafont (ICP), Escola Industrial 23, E-08201 Sabadell, Spain

Terry A. Gates, Department of Biological Sciences, North Carolina State University, Raleigh, NC 27695; North Carolina Museum of Natural Sciences, Raleigh, NC 27601

Michael A. Getty, Denver Museum of Nature and Science, 2001 Colorado Boulevard, Denver, Colorado 80205-5732

Pascal Godefroit, Department of Palaeontology, Royal Belgian Institute of Natural Sciences, rue Vautier 29, 1000 Brussels, Belgium

Merrilee F. Guenther, Department of Biology, Elmhurst College, Elmhurst, Illinois 60126

Rebecca R. Hanna, Museum of the Rockies, Montana State University, Bozeman, Montana 59717

Jason J. Head, Department of Earth and Atmospheric Sciences, University of Nebraska–Lincoln, 228 Bessey Hall, Lincoln, Nebraska 68588

Donald M. Henderson, Royal Tyrrell Museum of Palaeontology, Box 7500, Drumheller, Alberta T0J 0Y0

René Hernández-Rivera, Intituto de Geologiá, Universidad Nacional Autónoma de México, Circuito de Investigación Científica, Cuidad Universitaria, Delegación Coyoacán, 04510 México

Lucia Herrero, The Webb Schools, 1175 West Baseline Road, Claremont, California 91711; Stanford University, 450 Serra Mall, Stanford, California 94305

David W. E. Hone, School of Biological and Chemical Sciences, Queen Mary University of London, Mile End Road, London, U.K. E1 4NS

John R. Horner, Museum of the Rockies, Montana State University, Bozeman, Montana 59717-0040

Frankie D. Jackson, Department of Earth Sciences, Montana State University, Bozeman, Montana 59717

Zubair Jinnah, School of Geosciences, University of the Witwatersrand, Johannesburg, Wits 2050, South Africa

Derek W. Larson, Department of Ecology and Evolutionary Biology, University of Toronto, Toronto, Ontario M5S 2C6

Carolyn Levitt, University of Utah, Department of Geology and Geophysics, Salt Lake City, Utah 84112

Da-Qing Li, Gansu Geological Museum, 6 Tuanjie Road, Chengguan District, Lanzhou, Gansu Province, 730010, China

David W. H. Lloyd, Royal Tyrrell Museum of Palaeontology, Box 7500, Drumheller, Alberta T0J 0Y0

Spencer G. Lucas, New Mexico Museum of Natural History and Science, 1801 Mountain Road NW, Albuquerque, New Mexico 87104

Tyler R. Lyson, Smithsonian Institution, National Museum of Natural History, Washington, D.C. 20560

Susannah C. R. Maidment, Department of Palaeontology, The Natural History Museum, Cromwell Road, London, U.K. SW7 5BD

Derek J. Main (deceased), Department of Earth and Environmental Sciences, University of Texas at Arlington, Arlington, Texas 76019

Phillip L. Manning, School of Earth, Atmospheric and Environmental Sciences, University of Manchester, Manchester, U.K. M13 9PL

Andrew T. McDonald, Department of Earth and Environmental Science, University of Pennsylvania, 254-b Hayden Hall, 240 South 33rd Street, Philadelphia, Pennsylvania 19104

Christopher T. McGarrity, Department of Ecology and Evolutionary Biology, University of Toronto, 25 Willcocks Street, Toronto, Ontario M5S 3B2

Ali Nabavizadeh, Center for Functional Anatomy and Evolution, Johns Hopkins University School of Medicine, Baltimore, Maryland 21287

David B. Norman, Sedgwick Museum and Department of Earth Sciences, University of Cambridge, Downing Street, Cambridge, U.K. CB2 3EQ

Christopher R. Noto, Department of Biological Sciences, University of Wisconsin–Parkside, Kenosha, Wisconsin 53144

Oriol Oms, Universitat Autónoma de Barcelona, Departament de Geologia, Cerdanyola del Vallés, Barcelona, E-08193, Spain

W. Scott Persons IV, Department of Biological Sciences, Zoo. 418, Biological Sciences Building, University of Alberta, Edmonton, Alberta T6G 2E9

Albert Prieto-Márquez, School of Earth Sciences, University of Bristol, Wills Memorial Building, Queens Road, Bristol, U.K. BS8 1RJ

Angel A. Ramírez-Velasco, Posgrado Facultad de Medicina Veterinaria y Zootecnia, Universidad Nacional Autónoma de México, Circuito de Investigación Científica, Cuidad Universitaria, Delegación Coyoacán, 04510 México

Violeta Riera, Universitat Autónoma de Barcelona, Departament de Geologia, Cerdanyola del Vallés, Barcelona, E-08193, Spain

Bruce M. Rothschild, Department of Vertebrate Paleontology, Carnegie Museum of Natural History, 4400 Forbes Ave. Pittsburgh, Pennsylvania 15213

Michael J. Ryan, Department of Vertebrate Paleontology, Cleveland Museum of Natural History, 1 Wade Oval Drive, University Circle, Cleveland, Ohio 44106

James G. Schmitt, Department of Earth Sciences, Montana State University, Bozeman, Montana 59717

William I. Sellers, Faculty of Life Sciences, University of Manchester, Manchester, U.K. M13 9PL

Ricardo Servin-Pichardo, Facultad de Ciencias, Universidad Nacional Autónoma de México, Circuito de Investigación Científica, Cuidad Universitaria, Delegación Coyoacán, 04510 México

Robin Sissons, Philip J. Currie Dinosaur Museum, Pipestone Creek Dinosaur Initiative, County of Grande Prarie No. 1, 10001 84 Ave., Clairmont, Alberta T0H 0W0

Wendy J. Sloboda, Warner, Alberta T0K 2L0

Corwin Sullivan, Key Laboratory of Vertebrate Evolution and Human Origin, Institute of Vertebrate Paleontology and Paleoanthropology, Chinese Academy of Sciences, PO Box 643, Beijing 100044, China

Robert M. Sullivan, New Mexico Museum of Natural History and Science, 1801 Mountain Road NW, Albuquerque, New Mexico 87104

Kohei Tanaka, Department of Geoscience, University of Calgary, 2500 University Dr NW, Calgary, Alberta T2N 1N4

Darren H. Tanke, Royal Tyrrell Museum of Palaeontology, Box 7500, Drumheller, Alberta T0J 0Y0

François Therrien, Royal Tyrrell Museum of Palaeontology, Box 7500, Drumheller, Alberta T0J 0Y0

Khishigjav Tsogtbaatar, Mongolian Paleontological Center, Ulaanbaatar 210351, Mongolia

Bart Van Dongen, Williamson Research Centre for Molecular Environmental Science, University of Manchester, Manchester, U.K. M13 9PL

Bernat Vila, Grupo Aragosaurus-Instituto Universitario de Ciencias Ambientales, Paleontología, Facultad de Ciencias, Universidad de Zaragoza, Pedro Cerbuna 12, E-50009 Zaragoza, Spain

Jonathan R. Wagner, Texas State University–San Marcos, Department of Geography ELA 139, 601 University Drive, San Marcos, Texas 78666-4684

Kebai Wang, Key Laboratory of Vertebrate Evolution and Human Origin, Institute of Vertebrate Paleontology and Paleoanthropology, Chinese Academy of Sciences, PO Box 643, Beijing 100044, China

Mahito Watabe, Hayashibara Museum of Natural Science, 2-3. Shimoishii-1, Okayama 700-0907, Japan

Sam Webb, Stanford Synchrotron Radiation Lightsource, Stanford University, Menlo Park, California 94305

David B. Weishampel, Center for Functional Anatomy and Evolution, Johns Hopkins University School of Medicine, Baltimore, Maryland 21287

Roy A. Wogelius, Williamson Research Centre for Molecular Environmental Science, University of Manchester, Manchester, U.K. M13 9PL

Xing Xu, Key Laboratory of Vertebrate Evolution and Human Origin, Institute of Vertebrate Paleontology and Paleoanthropology, Chinese Academy of Sciences, PO Box 643, Beijing 100044, China

Hai-lu You, Key Laboratory of Evolutionary Systematics of Vertebrates, Institute of Vertebrate Paleontology and Paleoanthropology, Chinese Academy of Sciences, 142 Xizhimenwai Street, Beijing 100044, China

Darla K. Zelenitsky, Department of Geoscience, University of Calgary, 2500 University Dr NW, Calgary, Alberta T2N 1N4

Qi Zhao, Key Laboratory of Vertebrate Evolution and Human Origins, Institute of Vertebrate Paleontology and Paleoanthropology, Chinese Academy of Sciences, PO Box 643, Beijing 100044, China

Reviewers

R

Paul M. Barrett, Department of Earth Sciences, Natural History Museum, Cromwell Road, London, U.K. SW7 5BD

Kirstin S. Brink, Department of Ecology and Evolutionary Biology, University of Toronto Mississauga, 3359 Mississauga Road N., Mississauga, Ontario L5L 1C6

Brooks B. Britt, Department of Geology, Brigham Young University, S-387 ESC, PO Box 24606, Provo, Utah 84602

Caleb M. Brown, Royal Tyrrell Museum of Palaeontology, Box 7500, Drumheller, Alberta T0J 0Y0

Nicolás E. Campione, Departments of Earth Sciences and Organismal Biology, Uppsala University, Box 256, 751 05 Uppsala, Sweden

Katherine Clayton, Paleontology Collections, Natural History Museum of Utah, 301 Wakara Way, Salt Lake City, Utah 84108

Penelope Cruzado-Caballero, Universidad de Zaragoza, Departamento de Ciencias de la Tierra, c/ Pedro Cerbuna, 12 c.p. 50009 Zargoza, Spain

David Dilkes, Department of Biology and Microbiology, University of Wisconsin Oshkosh, 800 Algona Boulevard, Oshkosh, Wisconsin 54901

Peter Dodson, School of Veterinary Medicine, University of Pennsylvania, 3800 Spruce Street, Philadelphia, Pennsylvania 19104-6045

David A. Eberth, Royal Tyrrell Museum of Palaeontology, Box 7500, Drumheller, Alberta T0J 0Y0

David C. Evans, Department of Natural History, Royal Ontario Museum, 100 Queen's Park, Toronto, Ontario M5S 2C6

Andrew A. Farke, Raymond M. Alf Museum of Paleontology, 1175 West Baseline Road, Claremont, California 91711

Denver Fowler, Museum of the Rockies, Paleontology Department, 600 West Kagy Boulevard, Bozeman, Montana 59717

Roland Gangloff, University of California Museum of Paleontology, 1101 Valley Life Sciences Building, Berkeley, California 94720-4780

James Gardner, Royal Tyrrell Museum of Palaeontology, Box 7500, Drumheller, Alberta T0J 0Y0

Terry A. Gates, Department of Biological Sciences, North Carolina State University, Raleigh, NC 27695; North Carolina Museum of Natural Sciences, Raleigh, NC 27601

Pascal Godefroit, Department of Palaeontology, Royal Belgian Institute of Natural Sciences, rue Vautier 29, 1000 Brussels, Belgium

Merrilee F. Guenther, Department of Biology, Elmhurst College, Elmhurst, Illinois 60126

Jason J. Head, Department of Earth and Atmospheric Sciences, University of Nebraska–Lincoln, 228 Bessey Hall, Lincoln, Nebraska 68588

Donald M. Henderson, Royal Tyrrell Museum of Palaeontology, Box 7500, Drumheller, Alberta T0J 0Y0

Casey Holliday, Department of Pathology and Anatomical Sciences, University of Missouri, M318 Medical Sciences Building, Columbia, Missouri 65212

James Kirkland, Utah Geological Survey, PO Box 146100, Salt Lake City, Utah 84114

Derek W. Larson, Department of Ecology and Evolutionary Biology, University of Toronto, Toronto, Ontario M5S 2C6

Jordan Mallon, Canadian Museum of Nature, Palaeobiology Department, PO Box 3443 Stn. "D," Ottawa, Ontario K1P 6P4

Richard McCrea, Peace Region Palaeontology Research Centre, 255 Murray Drive, Box 1540, Tumbler Ridge, British Columbia V0C 2W0

Andrew T. McDonald, Department of Earth and Environmental Science, University of Pennsylvania, 254-b Hayden Hall, 240 South 33rd Street, Philadelphia, Pennsylvania 19104

David B. Norman, Sedgwick Museum and Department of Earth Sciences, University of Cambridge, Downing Street, Cambridge, U.K. CB2 3EQ

Albert Prieto-Márquez, School of Earth Sciences, University of Bristol, Wills Memorial Building, Queens Road, Bristol, U.K. BS8 1RJ

Patricia E. Ralrick, Royal Tyrrell Museum of Palaeontology, Box 7500, Drumheller, Alberta, T0J 0Y0

Raymond R. Rogers, Geology Department, Macalester College, 1600 Grand Avenue, Saint Paul, Minnesota 55105

Eric Snively, Department of Biology, University of Wisconsin-La Crosse, 1725 State Street, La Crosse, Wisconsin 54601

David Spalding, 1105 Ogden Road, Pender Island, British Columbia V0N 2M1

Daisuke Suzuki, Sapporo Medical University, South 1 West 17, Chuo-ku, Sapporo 060-8556 Japan

François Therrien, Royal Tyrrell Museum of Palaeontology, Box 7500, Drumheller, Alberta T0J 0Y0

Khishigjav Tsogtbaatar, Mongolian Paleontological Center, Ulaanbaatar 210351, Mongolia

David B. Weishampel, Center for Functional Anatomy and Evolution, Johns Hopkins University School of Medicine, Baltimore, Maryland 21287

Lawrence Witmer, Department of Biomedical Sciences, College of Osteopathic Medicine, Ohio University, Life Sciences Building, Room 123, Athens, Ohio 45701

Holly Woodward, Oklahoma State University Center for Health Sciences, 1111 West 17th Street, Tulsa, Oklahoma 74107

Hai-lu You, Key Laboratory of Evolutionary Systematics of Vertebrates, Institute of Vertebrate Paleontology and Paleoanthropology, Chinese Academy of Sciences, 142 Xizhimenwai Street, Beijing 100044, China

Preface

P

HADROSAURS – ALSO KNOWN AS DUCK-BILLED DINOSAURS – are one of the best-known groups within Dinosauria due to their abundance in the fossil record, notable diversity, and near global distribution in the Late Cretaceous. Their success was likely driven by a combination of factors that included, most importantly, anatomically-unique and functionally-complex jaws and dentitions that processed plants more efficiently than those of any "reptile" before or since. Ultimately, the ubiquity of hadrosaurs in the Cretaceous fossil record has allowed us to learn more about dinosaurian paleobiology and paleoecology than we have from any other group.

In recent years, a number of dinosaur groups have been the subject of renewed scientific interest. In 2005, sauropod studies experienced a scientific renaissance with the benchmark publications *The Sauropods: Evolution and Paleobiology* and Indiana University Press's *Thunder Lizards: The Sauropodomorph Dinosaurs*. In 2010, after a decade-long surge of interest in horned dinosaurs, that group received similar treatment in Indiana University Press's *New Perspectives on Horned Dinosaurs*. During the last five years it has been the hadrosaurs' time in the spotlight. Due to the rapidly growing fossil record as well as widespread international collaborations, researchers from around the world are now studying new specimens and taxa of hadrosaurs to clarify their origins, patterns of evolution, function, paleobiology, paleobiogeography, and preservation.

It was with this perspective that we (Davids 1 and 2) convened the International Hadrosaur Symposium (September 22–23, 2011). A collaboration between the Royal Tyrrell Museum and the Royal Ontario Museum, the goal of the IHS was to bring together an international slate of scientists and enthusiasts to share their research on and passion for duck-billed dinosaurs. Hosting the event at the Royal Tyrrell Museum made perfect sense to us; after all, few places in the world can boast the abundance and quality of hadrosaur fossils as are found in the classic Upper Cretaceous nonmarine strata of southern Alberta, and the Tyrrell's collections.

Fifty-plus presentations by an international roster of dinosaur specialists and up-and-coming students rounded out two days of hardcore hadrophilia (apologies to Peter Dodson for blatantly ripping off his terminology). The IHS was also an opportunity for all of us to honor the contributions of David Weishampel (David 3).

Setting the international tone were our five keynote presenters: Rodolfo Coria (Argentina), Pascal Godefroit (Belgium), Jack Horner (U.S.A.), Khishigjav Tsogtbaatar (Mongolia), and our honored guest, David Weishampel (U.S.A.). The watershed nature of the meeting was recognized by all attendees and, collectively, we managed to overcome unanticipated obstacles such as an impending strike by Air Canada employees, which resulted in last minute rerouting of flights and late appearances by some attendees. Be it known that we truly appreciate the efforts everyone made to attend the symposium.

This volume comprises most of the content from the symposium, and more. Because we believe this volume and its contents to be a uniquely comprehensive treatment of hadrosaurs, we chose simply to call it *Hadrosaurs*. The scope of the volume encompasses not only the well-known hadrosaurids proper, but also Hadrosauroidea, which allows the former group to be evaluated in a broader perspective.

The volume's 36 chapters are organized into the following six parts, followed by an afterword by Jack Horner:

Overview includes only one chapter, written by David Weishampel. David has spent a large part of his career studying dinosaurian paleobiology and, arguably, his most significant contributions are hadrosaurian. He has conducted pioneering work on hadrosaurian parental care, feeding mechanisms, functional morphology of bizarre structures, and phylogeny. In this chapter he uses data from the second edition of *The Dinosauria* to document patterns of research on ornithopods over the past two centuries, and uses his wisdom to surmise where researchers may be focusing in the future.

New Insights Into Hadrosaur Origins includes six chapters that document new and historical materials that shed light on the evolution and diversity of hadrosauroids before the origins of true hadrosaurids. David Norman (David 4) reviews taxa that have been implicated in the origin of Hadrosauroidea, and presents some provocative ideas about the evolution of ornithopods leading up to hadrosaurids. A standout chapter by Tsogtbataar et al. describes an exciting new taxon from the Djadokhta Formation of Mongolia, important for understanding the origin of hadrosaurids (it is rendered beautifully on the cover of the book by Julius Csotonyi). McDonald et al., You et al., and Barrett et al. provide new information about known specimens, and help sort out some long-standing questions about these specimens. Similarly, Main et al. and Larson et al. remind us of the importance of the North American hadrosauroid record for understanding the origins of Hadrosauridae.

Hadrosaurid Anatomy and Variation includes contributions by Gates, Bell, Farke and Herrero, Evans, Campione, Brink, and colleagues, and focuses on the anatomy of a variety of hadrosaurid taxa from western North America. Gates, Bell, Farke and Herrero, and colleagues describe new specimens from stratigraphic units that are rapidly proving to be important sources of new information about hadrosaur diversity and distributions, whereas the contributions by Evans, Campione, Brink, and colleagues provide in-depth descriptions and interpretations of known taxa and specimens. The morphological details provided here will lead undoubtedly to improved comparative studies.

Biogeography and Biostratigraphy documents the distribution of hadrosaurids in time and space. Here, chapters by Ramírez-Velasco et al. and Dalla Vecchia and colleagues stand out as exceptionally detailed overviews of hadrosaur occurrences in Mexico and Europe, respectively. Similarly, contributions by Bolotsky et al., Coria, and Sullivan and Lucas go a long way to help improve our understanding of hadrosaurian diversity in eastern Asia, South America, and the Southwestern U.S.A. The contribution by Tanke and Evans underscores the importance of properly documenting locality data for specimens.

Function and Growth includes seven contributions that address function, growth, and life habits. Studies of hadrosaur morphology, locomotion, and function by Maidment et al., Persons and Currie, and Henderson employ evolving techniques in computer modeling and engineering that we hope will spark discussion and renewed interest in this topic. Nabavizadeh revisits the all-important question of jaw kinetics via predentary morphology, and Guenther's comparison of postcrania is a step toward identifying different developmental pathways in hadrosaurs. Erickson and Zelenitsky describe ontogenetic changes in tooth morphology/histology in *Hypacrosaurus stebingeri* that reflect dietary changes during development. Lastly, Brinkman's size-distribution data are the basis of conclusions that challenge conventional wisdom related to growth rates in hadrosaurs.

Preservation, Tracks, and Traces is the last part of the volume and includes eight chapters, including contributions by Manning et al., Prieto-Márquez and Wagner, and Bell on skin and skin traces. Of particular note is the chapter on the origins of the classic *Maiasaura* bonebed by Schmitt et al., which many of us have awaited for years (no pressure anymore, Jim!). Contributions by Eberth et al. and Hone et al. present more evidence that some hadrosaurs lived in large, segregated herds, perhaps rivaling in size those of centrosaurian ceratopsians. Back in Alberta, Therrien et al. provide the first evidence of hadrosaur tracks from the Oldman Formation of Alberta, and Tanke and Rothschild provide an exhaustive survey of paleo-osteopathologies in hadrosaurs from Dinosaur Provincial Park.

Nomenclature note – Unlike most forms of science, taxonomy can be quite democratic. Over the last two decades, numerous clade names and definitions have been proposed for the hadrosaurian part of the ornithopod family tree. Rather than imposing a particular taxonomic scheme on the book's contributors, we chose to allow contributors to employ their preferred taxonomy. Not surprisingly, the book reveals little consensus. In particular, readers may find differential use of the terms Hadrosauridae, Hadrosaurinae, and Saurolophinae across the book's chapters a bit confusing. In order to address this, and other similar confusions, we asked authors to cite their taxonomic sources where necessary.

In summary, we have tried our best to present a group of well-balanced and consistently edited manuscripts, while allowing the authors to express their individual styles. We hope that you all enjoy the volume and find it useful for years to come.

David A. Eberth
David C. Evans

Acknowledgments

WE THANK ALL OF THE PARTICIPANTS WHO ATTENDED THE International Hadrosaur Symposium in 2011 and helped us realize that this volume would be a successful venture. We thank the authors for helping provide a cohesive and coherent product, and especially for being so patient as the clock kept ticking. Special thanks to the reviewers who did their jobs in a timely manner and were often willing to look at manuscripts more than once, thus ensuring that contributions were of high quality both scientifically and editorially.

Our sincere gratitude goes to Bob Sloan and Jim Farlow, and the rest of the great team at Indiana University Press for all their help with this project.

We thank our respective home organizations who gave freely of their time, physical resources, and manpower. Special thanks to the Royal Tyrrell Museum for providing the FTP site, printing services, physical layout space, and so much else that was critical to the compilation of this volume. DAE thanks J. Gardner, D. Brinkman, F. Therrien, and D. Henderson for advice on numerous scientific and editorial items, and W. Taylor and D. Braman for editorial assistance. DCE thanks N. Campione and D. Larson for editorial and scientific assistance.

We are particularly grateful to A. Keibel and the Royal Tyrrell Museum Cooperating Society for financial and administrative support during the symposium and throughout this project.

We thank J. Csotonyi for the exceptional cover artwork of *Plesiohadros djadokhtaensis* and the wonderful (and first) artistic rendering of Djadokhtan paleoenvironments during a wet climatic phase. We also thank D. Dufault and L. Panzarin for their original artistic contributions to the volume.

Last, but certainly not least, we recognize P. Ralrick for her colossal contributions to this project. Patty served as our technical editor and editorial assistant, helped review manuscripts, listened to the occasional rant, and also indexed the volume. Without her attention to detail, this project would have taken twice as long and would not have been done half as well. Thanks, Patty, we hope that now you are very satisfied.

1 Overview

A History of the Study of Ornithopods: Where Have We Been? Where Are We Now? and Where Are We Going?

1

David B. Weishampel

ABSTRACT

Where ornithopod studies have been and where they are going is fascinating. I try to provide answers for the history of the study of ornithopod dinosaurs by collecting bibliographic data from the second edition of *The Dinosauria*. The resulting publication curves were examined for 10 intrinsic factors, nearly all of which increase through the first decade of the twenty-first century. These measures are used to take stock of present-day ornithopod studies and, finally, to try to predict our future as ornithopod researchers in this historically contingent world.

INTRODUCTION

From a historical perspective, knowledge about a taxonomic group can be judged by its publication rate. A zero rate may indicate a momentarily stalled interest in the group or a cessation of interest in it altogether (e.g., Kalodontidae Nopcsa, 1901), while a low rate suggests less than vigorous or meager research activity focused on the group (say, during a war or when there are few publishing scientists). Finally, a high publication rate may have many reasons, including new discoveries and new taxonomic recognition, and evolutionary controversy, to name a few.

Compilations of taxa are not new to studies of dinosaurs, or even tetrapods or invertebrates (Sepkoski et al., 1981; Benton, 1985, 1998; Dodson, 1990; Weishampel, 1996; Sepkoski, 2002; Fastovsky et al., 2004; Wang and Dodson, 2004). However, this present compilation and survey differs from previous varieties in that it focuses on the number of papers published and the research areas those papers address.

For Ornithopoda – the most abundant and diverse of which are hadrosaurids – the record of publication begins in 1825 with the publication of Mantell's *Iguanodon*, and finishes with the numerous papers, some being issued via conventional journals as well as online-only journals, with no hard copies, of the present day. What this record looks like is presented in Figure 1.1. How it was obtained and how it is interpreted are the subjects of this chapter.

Caveat: although this volume is the product of a symposium dedicated predominantly to hadrosaurs, which includes hadrosaurids proper as well as hadrosauroids, it has been extended by the organizers to include iguanodontians as well. By stretching it slightly more to include iguanodontians, we are practically down to the base of Ornithopoda. Hence, this chapter is about hadrosaurs – and more.

MATERIALS AND METHODS

In order to evaluate the rate of publication of papers dealing with ornithopod dinosaurs, the number of papers was tabulated on a per-decade basis from 1820–2010 from the bibliography of *The Dinosauria*, second edition (Weishampel et al., 2004). Containing 90 published pages of references on all dinosaurian taxa, this book is likely to be comprehensive enough for our current purposes. Because the decade of 2000–2010 was incomplete in that volume, the remainder of this decade was filled in proportionally based on the approximate representation during the first three and one-half years of the decade. That is, the 2000–2010 decadal numbers are projections based on tabulations from the first three and one-half years. Total papers and papers for each research category (see below) were adjusted by multiplying the raw totals for the first three and one-half years of the 2000–2010 decade by a factor of 2.86 to yield a total proportionally equivalent to other decades. This kind of correction was judged preferable to changing data sources (e.g., Web of Science), which would have resulted in an under-sampling of the more obscure literature.

In addition to the total curve, I have attempted to characterize the papers that went into this total by identifying nine categories of research (Table 1.1). I provide general description of these categories, denoted in boldface text, below. These categories were usually assessed by title alone, but occasionally it was necessary to consult the paper itself to determine to which category it belonged. I made no account of footprints and eggshell papers, because it was often impossible to assess affinities of the tracks or shell beyond Dinosauria from the title of the paper.

Table 1.1. Categories of Ornithopod Research Identified in This Survey

General taxonomy
Functional morphology
Phylogeny
Biostratigraphy and taphonomy
Biogeography
Paleoecology
Soft tissue
Growth
Faunistics

General taxonomy refers to those publications announcing new specific or generic taxa, or new taxonomic revisions that do not come under the heading of phylogeny (see below). For example, Gilmore's (1913) announcement of *Thescelosaurus neglectus* is here considered a work of general taxonomy.

Functional morphology is the category for papers involving a biomechanical or functional interpretation of an ornithopod anatomical system. An example of a functional morphology study is Alexander's (1985) work on stance and gait in ornithopods among other dinosaurs.

Phylogeny refers to those studies that attempt to portray the evolutionary history, or phylogeny, of the group. In recent years, these studies have emphasized cladistics in phylogenetic reconstruction (e.g., Prieto-Márquez, 2010), but also include a number of pre-Hennigian analyses (e.g., Galton, 1972).

Biostratigraphy and taphonomy papers involve the geologic disposition of ornithopod specimens, whether within or among rock units. Rogers (1990) provided an example of how bonebed taphonomy can provide evidence for drought-related mortality in dinosaurs that include hadrosaurs.

Biogeography includes studies that examine the geographic distribution of ornithopods either from a dispersal or vicariant perspective, or both. For example, Casanovas et al. (1999) examined the global distribution of lambeosaurine hadrosaurids, whereas Upchurch et al. (2002) considered the full spectrum of controls on dinosaur diversity, including that of ornithopods, as a function of biogeography and biostratigraphy.

Paleoecology papers include those of Carrano et al. (1999) on convergence—or lack thereof—among ornithopods and ungulate mammals, and Varricchio and Horner (1993) on the significance of bonebeds in paleoecological interpretations, and are intended to address the reconstruction of particular taxonomically bound or free ecosystems of the past.

Soft tissue studies have been generally limited to skin impressions. Examples include Osborn (1912) on the "mummy" of *Edmontosaurus annectens* in the American Museum of Natural History.

Growth includes papers associated with aspects of ontogenetic development. The impact of growth on ornithopod studies is relatively recent. Here I note Dodson (1975) on the taxonomic significance of growth in *Lambeosaurus* and *Corythosaurus*, as well as various studies by Horner and colleagues (e.g., Horner et al., 1999, 2000) focused on the cellular basis of bone growth.

Faunistics includes papers whose principal purpose is to establish or review fossil assemblages that include ornithopods. For example, Lapparent (1960) reviewed the dinosaurs, including many ornithopods, from the "Continental intercalaire" of northern Africa.

Usually contributions were entered once in a category. However, a study can contribute here to several categories. For instance, Ostrom (1961) included discussion of general taxonomy, functional morphology, phylogeny, and other subjects in his major review of North American hadrosaurs, and so it was added to each of these categories.

WHERE HAVE WE BEEN?

Where we have been can be determined by looking at the total curve of ornithopod publications (Fig. 1.1A). Beginning in the 1820s, the number of papers published per decade rises to a high of 15 in the 1870s. It then declines to 4 in the 1890s, and increases again, to 24, in the 1920s. The 1940s see a drop to 7, followed by a persistent, long-term increase to the decade of the 2000s, which is characterized by nearly 200 papers, amounting to almost 2 papers per month!

Before turning to several intrinsic factors, I want to examine three kinds of extrinsic events that may have influenced these numbers and patterns. For possible influences due to world events, the European revolutions of 1848, the American Civil War, World War I, the Russian Revolution, the fall of communism, and the combined Iraq and Afghan wars appear to have no substantial influence on rate of publication, whereas the 1929 stock market crash and the subsequent worldwide financial depression followed by World War II are likely factors in the decline of publication rates in the 1930s and 1940s. Regarding technological influences, there are no great fluctuations in rate of publication for technological events, except for the last two events. It is probably safe to say that the invention of personal computers, particularly laptops (1970s), in combination with the development of the World Wide Web and internet (1990s) made a huge impact on the rate of ornithopod publications. With the initiation of web publishing, this trend is certain to continue. Finally, scientific influences probably account for smaller perturbations in the total curve. For example, the discovery of the *Iguanodon* assemblage from Bernissart probably accounts for the rise in ornithopod publications during the 1870s and 1880s. The

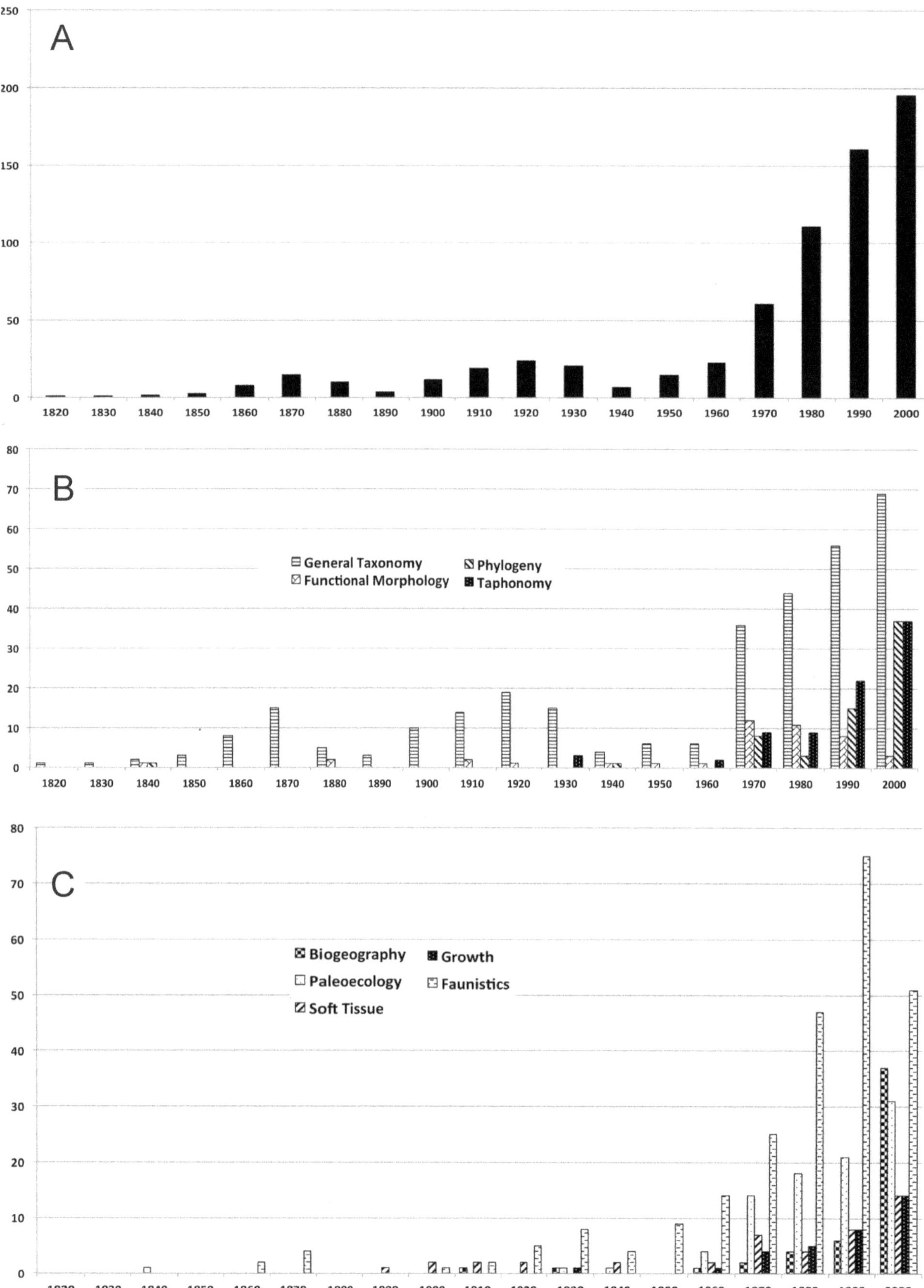

1.1. Publication trends on ornithopod dinosaurs. (A) Total publication record of ornithopod dinosaurs from 1820 to 2000 tabulated by decade; (B) Total publications of general taxonomy, functional morphology, phylogeny, and biostratigraphy and taphonomy, tabulated by decade; (C) Total publications in biogeography, paleoecology, soft tissue, growth, and faunistics, tabulated by decade.

rise in publication rates during the 1910s, 1920s, and 1930s can certainly be attributed to the Great Canadian Dinosaur Rush in Alberta. Finally, as a personal homage, I consider John H. Ostrom's first monographic publication–his 1961 treatment of the hadrosaurids of North America–to signal the beginning of what has turned out to be a plethora of ornithopod publications to the present day.

Intrinsic factors, on the other hand, are some of the subjects that I am interested in, which also have given Ornithopoda pride of place in the world of dinosaur publishing. General taxonomy and faunistics are the largest contributors to the total sample, whereas the rest have relatively low influence.

General taxonomy (Fig. 1.1B) has as long a history, beginning with the first publication on *Iguanodon* by Mantell (1825) and early on encompassing the first publication on *Hadrosaurus* by Leidy (1858). Furthermore, it mirrors fairly well the total publication curve, with a high point of 69 publications during the decade of 2000–2010.

Functional morphology (Fig. 1.1B) has a long, but patchy history, beginning with the publication of Mantell (1848) on the teeth and jaws of *Iguanodon*. It has never been common, but increases significantly in the 1970s and 1980s, with renewed interest in ornithopod jaw mechanics. Functional morphology has been in decline since this time.

Phylogeny (Fig. 1.1B) also has a long and equally patchy history, beginning with Owen's (1842) christening of Dinosauria. Thereafter, there is a long hiatus until the 1970s, when we see an irregular publication record reflecting the large impact of cladistics on phylogeny estimates. The 1990s and 2000s indicate an important increase in cladistic studies, peaking near 40 publications.

Biostratigraphy and taphonomy (Fig. 1.1B) have a relatively short history, confined to the period of the 1930s to the present, and within this span only relatively abundant since the 1970s, with the publications of Dodson (1971), Rogers (1990), and Varricchio and Horner (1993). There is a steady increase in biostratigraphic and taphonomic publications from the 1980s to the 2000s, indicative of increased interest in the sedimentological aspects of ornithopod fossils.

Biogeography (Fig. 1.1C) is in its infancy, with its concentration of publications only evident from the 1960s onward. This is roughly the same time as the scientific ascendancy of plate tectonics and phylogenetic systematics, and thus, may be a direct product of these two revolutions in the natural sciences (Sereno, 1997, 1999a, 1999b; Upchurch et al., 2002). Biogeography reaches its zenith in the decade of 2000; in all likelihood it will continue to increase.

Paleoecology (Fig. 1.1C) has a relatively short history. With a few notable exceptions (Mantell, 1844; Nopcsa, 1934), the history of paleoecology papers really began in the 1960s. There has been a steady increase in the number of paleoecology publications since then, to a high of more than 30 publications in the decade of 2000–2010.

Soft tissue (Fig. 1.1C), consisting almost entirely of the study of integumentary impressions, has a reasonable steady and long history, increasing steadily since the 1970s. It is presently on a very large upswing, in large part because of the discovery of exceptionally well preserved specimens (particularly in northeastern China) and a more focused evaluation of variation in integumentary patterns (Bell, this volume).

Growth (Fig. 1.1C) has a very modest history. It has been common only since the 1970s, and appears to be on a steep upswing to nearly a dozen papers for the decade of 2000–2010. This increase probably represents the rise in fossil bone histology studies in ornithopods (e.g., Chinsamy, 1995; Horner et al., 2000).

Finally, **faunistics** (Fig. 1.1C) has a long history, approximately paralleling general taxonomy and the total curve, at least since the 1860s. Faunistics seems to drop off during the decade of the 2000s, but this downturn should be treated with skepticism because it is almost certainly an artifact of sampling extrapolation. Examples taken from the 1990s and 2000s include Csiki (1997), Ryan and Russell (2001), López-Martinez et al. (2001), and Zhou et al. (2003).

WHERE ARE WE NOW?

Before we all assembled for the International Hadrosaur Symposium, we all probably thought we knew where our science was. At a minimum, that was what we came to Drumheller to report on. It was hadrosaur taxonomy, North American, Asian, South American, and European hadrosaurs, and ornithopod brains. It was also hadrosaur gigantism and age, hadrosaur jaws and herbivory, locomotor mechanics, taphonomy, integument, tracks, and various aspects of development. This was where we thought our discipline was as we began the symposium.

Eighty-eight percent of the symposium talks (n = 34 talks, 16 posters) fall within the categories discussed here (Braman et al., 2011). Most are taxonomic, phylogenetic, or biogeographic in scope. Another half-dozen or more pertain to functional morphology, growth, and taphonomy–a good sampling of the categories examined here (an acclaim delivered independently twice over–the organizers and I both got it right!).

Symposium percentages are all the same order of magnitude compared to those obtained for the decade of 2000–2010, but there are several differences. General taxonomic presentations at the symposium were nearly 25% fewer than from 2000–2010, phylogeny was 19% fewer, taphonomy was 15% fewer, biogeography was 28% fewer, paleoecology was

19% fewer, and faunistics was 13% fewer. Soft tissue remained approximately the same. Interestingly, functional morphology was 14% more and growth was 6% more than from the decade of 2000–2010. While it is tempting to assign significance to individual percentages, they are probably no more than sampling errors when comparing a very small number of symposium talks with the projected breakdown of categories for an entire decade.

WHERE ARE WE GOING?

I am certainly no prognosticator, even about my own research field. Like all historical sciences, our ability to predict the future is fraught with the kinds of unpredictability that derives from historical contingency. There is little inevitability that guides us in the progress of our science – just as there is little that links the hand-cranked ice-cream maker (1840s) to the electron microscope (1930s), a transition that happened in only nine decades. What about going from the invention of the Band-Aid (1930s) to the home computer in five decades? Who would have predicted these changes?

But the contents of this volume give an inkling of where we are headed, at least in the short run. I see continued fieldwork, the wellspring of our science. Its direct consequences – new species and taxonomic revisions – are likely to be accompanied by a healthy continuance of studies focused on comparative anatomy, both bony and inferred soft tissue. To do so requires a healthy dose of phylogenetic systematics, which now should be part of everyone's toolkit. In functional morphology, finite element analyses and tooth-wear studies have appeared on the horizon and I hope these will be coupled with cladistic analyses to produce even more outstanding work. Finally, growth studies are very likely to continue in the future: the small bit of bone given up for a thin-section is bound to yield disproportionately much more subtle and profound information than if it were left with the rest of the bone.

Still, things do not always work out that way. Contingency makes history messy. Things come out of left field and WHAM! Someone discovers the most amazing specimen or means by which colors can be inferred from skin impressions. All of a sudden, with no way of predicting, we are all scrambling to do research on the melanosomes of what could turn out to be red-, green-, and yellow-striped ornithopods!

ACKNOWLEDGMENTS

I thank David Eberth and David Evans for their kind invitation to join them at their fantastic first International Hadrosaur Symposium in Drumheller, Alberta, Canada. Their generosity and that of François Therrien and the other hosts at the Royal Tyrrell Museum of Palaeontology are most commendable. And to throw in a bronze plaque of *Corythosaurus intermedius* (ROM 845); well, I sure had a good time! I also thank Ali Nabavizadeh and Cat Sartin for their help on and reading of this manuscript and Jack Horner, Cora Jianu, and Pilar Yagüe for their own individual inspirations.

LITERATURE CITED

Alexander, R. McN. 1985. Mechanics of posture and gait of some large dinosaurs. Zoological Journal of the Linnean Society 83:1–25.

Bell, P. 2014. A review of hadrosaurid skin impressions; Chapter 25 in D. A. Eberth, and D. C. Evans (eds.), Hadrosaurs. Indiana University Press, Bloomington, Indiana.

Benton, M. J. 1985. Mass extinction among non-marine tetrapods. Nature 316:811–814.

Benton, M. J. 1998. The quality of the fossil record of the vertebrates; pp. 269–303 in S. K. Donovan and C. R. C. Paul (eds.), The Adequacy of the Fossil Record. John Wiley & Sons, New York.

Braman, D., D. A. Eberth, D. C. Evans, and W. Taylor (compilers). 2011. International Hadrosaur Symposium Abstract Volume. Royal Tyrrell Museum. Drumheller, Alberta. 171 pp.

Carrano, M. T., C. M. Janis, and J. J. Sepkoski. 1999. Hadrosaurs as ungulate parallels: lost lifestyles and deficient data. Acta Palaeontologia Polonica 44:237–261.

Casanovas, M. L., X. Pereda-Suberbiola, and D. B. Weishampel. 1999. First lambeosaurine hadrosaurid from Europe: palaeobiogeographical implications. Geological Magazine 136:205–211.

Chinsamy, A. 1995. Ontogenetic changes in the bone histology of the Late Jurassic ornithopod *Dryosaurus lettowvorbecki*. Journal of Vertebrate Paleontology 15:96–104.

Csiki, Z. 1997. Legături paleobiogeografice ale faunei de vertebrate continentale Maastrichtian superioare din Bazinul Haţeg. Nymphaea 23–25:45–68.

Dodson, P. 1971. Sedimentology and taphonomy of the Oldman Formation (Campanian), Dinosaur Provincial Park, Alberta (Canada). Palaeogeography, Palaeoclimatology, Palaeoecology 10:21–74.

Dodson, P. 1975. Taxonomic implications of relative growth in lambeosaurine hadrosaurs. Systematic Zoology 24:37–44.

Dodson, P. 1990. China reaches the top. American Paleontologist 17: online supplement. Available at www.museumoftheearth.org/files/pubtext/supplements/suppl_557c.pdf. Accessed summer 2012.

Fastovsky, D. E., Y. Huang, J. Hsu, J. Martin-McNaughton, P. M. Sheehan, and D. B. Weishampel. 2004. Shape of Mesozoic dinosaur richness. Geology 32:877–880.

Galton, P. M. 1972. Classification and evolution of ornithopod dinosaurs. Nature 239:464–466.

Gilmore, C. W. 1913. A new dinosaur from the Lance Formation of Wyoming. Smithsonian Miscellaneous Collections 61:1–5.

Horner, J. R., A. de Ricqlès, and K. Padian. 1999. Variation in skeletochronological indicators of the hadrosaurid dinosaur *Hypacrosaurus:* implications for age assessment of dinosaurs. Paleobiology 25:295–304.

Horner, J. R., A. de Ricqlès, and K. Padian. 2000. The bone histology of the hadrosaurid dinosaur *Maiasaura peeblesorum:* growth dynamics and physiology based on an ontogenetic series of skeletal elements. Journal of Vertebrate Paleontology 20:109–123.

de Lapparent, A. F. 1960. Les dinosauriens du "Continental intercalaire" du Sahara central. Memoire de la Societé geologique de France 88A:1–57.

Leidy, J. 1858. *Hadrosaurus foulkii,* a new saurian from the Cretaceous of New Jersey. Proceedings of the Academy of Natural Science of Philadelphia 1859:215–218.

López-Martinez, N., J. I. Canudo, L. Ardèvol, X. Pereda-Suberbiola, X. Orue-Etxebarria, G. Cuenca-Bescós, J. I. Ruiz-Omeñaca, X. Murilaga, and M. Feist. 2001. New dinosaur sites correlated with upper Maastrichtian pelagic deposits in the Spanish Pyrenees: implications for the dinosaur extinction pattern in Europe. Cretaceous Research 22:41–61.

Mantell, G. A. 1825. Notice on the *Iguanodon,* a newly discovered fossil reptile, from the sandstone of the Tilgate Forest, in Sussex. Philosophical Transactions of the Royal Society of London 115:179–186.

Mantell, G. A. 1844. Medals of Creation. H. G. Bohn, London, U.K., 289 pp.

Mantell, G. A. 1848. On the structure of the jaws and teeth of the *Iguanodon.* Philosophical Transactions of the Royal Society of London 138:183–202.

Nopcsa, F. 1901. Synopsis und Abstammung der Dinosaurier. Földtani Közlöny 31:247–288.

Nopcsa, F. 1934. The influence of geological and climatological factors on the distribution of non-marine fossil reptiles and Stegocephalia. Quarterly Journal of the Geological Society of London 90:76–140.

Osborn, H. F. 1912. Integument of the iguanodont dinosaur *Trachodon.* Memoirs of the American Museum of Natural History 1:33–54.

Ostrom, J. H. 1961. Cranial morphology of the hadrosaurian dinosaurs of North America. Bulletin of the American Museum of Natural History 122:33–186.

Owen, R. 1842. Report on British fossil reptiles. Part II. Report of the British Association for the Advancement of Science 1841:60–204.

Prieto-Márquez, A. 2010. Global phylogeny of Hadrosauridae (Dinosauria: Ornithischia) using parsimony and Bayesian methods. Zoological Journal of the Linnean Society 159:435–502.

Rogers, R. R. 1990. Taphonomy of three dinosaur bone beds in the Upper Cretaceous Two Medicine Formation, northwestern Montana: evidence for drought-related mortality. Palaios 5:394–413.

Ryan, M. J., and A. P. Russell. 2001. Dinosaurs of Alberta (exclusive of Aves); pp. 279–297 in D. H. Tanke, and K. Carpenter (eds.), Mesozoic Vertebrate Life. Indiana University Press, Bloomington, Indiana.

Sepkoski, J. J., Jr. 2002. A compendium of fossil marine animal genera; pp. 1–560 in D. Jablonski, and M. Foote (eds.), Bulletins of American Paleontology, No. 363. Paleontological Research Institution, Ithaca, New York.

Sepkoski, J. J., Jr., R. K. Bambach, D. M. Raup, and J. W. Valentine. 1981. Phanerozoic marine diversity and the fossil record. Nature 293:435–437.

Sereno, P. C. 1997. The origin and evolution of dinosaurs. Annual Review of Earth and Planetary Sciences 25:435–489.

Sereno, P. C. 1999a. The evolution of dinosaurs. Science 284:2137–2147.

Sereno, P. C. 1999b. Dinosaurian biogeography: vicariance, dispersal and regional extinction. National Science Museum of Tokyo Monographs 15:249–257.

Upchurch, P., C. A. Hunn, and D. B. Norman. 2002. An analysis of dinosaurian biogeography: evidence for the existence of vicariance and dispersal patterns caused by geological events. Proceedings of the Royal Society of London B269:613–621.

Varricchio, D. J., and J. R. Horner. 1993. Hadrosaurid and lambeosaurid bone beds from the Upper Cretaceous Two Medicine Formation of Montana: taphonomic and biological implications. Canadian Journal of Earth Sciences 30:997–1006.

Wang, S. C., and P. Dodson. 2004. Estimating the diversity of dinosaurs. Proceedings of the National Academy of Science 103:13601–13605.

Weishampel, D. B. 1996. Fossils, phylogeny, and discovery: a cladistic study of the history of tree topologies and ghost lineage durations. Journal of Vertebrate Paleontology 16:191–197.

Weishampel, D. B., P. Dodson, and H. Osmólska (eds.). 2004. The Dinosauria, Second Edition. University of California, Berkeley, California, 861 pp.

Zhou Z., P. R. Barrett, and J. Hilton. 2003. An exceptionally preserved Lower Cretaceous ecosystem. Nature 421:807–814.

2

New Insights into Hadrosaur Origins

Iguanodonts from the Wealden of England: Do They Contribute to the Discussion Concerning Hadrosaur Origins?

2

David B. Norman

ABSTRACT

The earliest known hadrosaur-like ornithopod is represented by a tooth from the early Cenomanian (Cambridge Greensand) of England. The Wealden outcrops (late Berriasian–early Aptian) of England include a range and variety of iguanodonts that, in many anatomical respects, presage the structures seen in a succession of Albian–Maastrichtian, hadrosaur-like neoiguanodontians, hadrosauromorphans, and euhadrosaurians. The anatomy and taxonomic assignments applicable to known Wealden iguanodonts are reviewed, albeit briefly, and the recently published proposition that Wealden taxonomic diversity was far higher than previously supposed is regarded as unfounded. A new systematic analysis has generated a consistent topological framework that provides the basis for a consideration of the general pattern of assembly of anatomical features, within the neoiguanodontian lineage, that culminated in the appearance of true hadrosaurs (euhadrosaurians) during the Late Cretaceous. The general topology generated by the present analysis largely conforms to previous analyses. However, the primary region of inconsistency is located across a range of taxa that appear to form a plexus of late Early and early Late Cretaceous age; they are widely distributed geographically, vary in their degrees of preservation, and have been described mostly in the last two decades. A revised classification is proposed, based upon the new topology, and generalized phylogenetic inferences have also been drawn from the successional pattern and its associated character distributions. The systematic pattern and therefore the phylogenetic (evolutionary) origin of euhadrosaurians from within the plexus of derived neoiguanodontians is potentially tractable. However, questions focused upon the geographic (area of) origin of hadrosaurs are unlikely to be resolved satisfactorily because of definitional instability (an inherent problem of fossil-based systematic analyses), compounded by the more or less constant flow of new discoveries.

INTRODUCTION

The zenith of ornithopod evolution is represented by the Late Cretaceous duck-billed, or hadrosaurian dinosaurs (e.g., Lull and Wright, 1942; Ostrom, 1961; Horner et al., 2004; Prieto-Márquez, 2010), which were highly speciose, geographically widespread, and anatomically (and probably behaviorally) complex herbivores. However, the details governing the evolutionary transition from derived (neoiguanodontian) to the definitive hadrosaurian state, although understood in general terms, have proved elusive. Initially (encompassing the time between the 1870s and 1970s) the fossil record was comparatively mute on the subject: "middle" Cretaceous (Albian–Cenomanian) ornithopods were extremely rare and poorly described, as well as unreliably dated. As a consequence, evolutionary hypotheses were necessarily speculative (e.g., Gilmore, 1933; Ostrom, 1961; Rozhdestvensky, 1966; Taquet, 1975). The closing decades of the twentieth century and the opening decade of the twenty-first century mark a turning point during which a considerable number of new ornithopod taxa have been identified from both the older and established fossil hunting grounds as well as many new geographic locations. However, it seems that the abundant new data has increased ambiguity, rather than creating the expected resolution or increasing levels of consensus concerning the ancestry of the clade referred to herein as Euhadrosauria (Weishampel, Norman, and Grigorescu, 1993 [= Hadrosauridae sensu lato, e.g., Lull and Wright, 1942; Horner et al., 2004; Prieto-Márquez, 2010]).

Hadrosaur origins can be explored through a number of independent, yet correlated, lines of investigation: the chrono-geographical evidence suggestive of their first appearance in the fossil record; the study of ornithopod taxa that are positioned adjacent to the clade Euhadrosauria; the construction of parsimony-based and Bayesian likelihood trees (Evans, 2010; Prieto-Márquez, 2010); and the evaluation of the anatomical transformations (and phylogenetics) implied by the topology of such trees. In combination, these approaches should be able to reveal when, where, and how hadrosaurian anatomy was assembled (and, by implication,

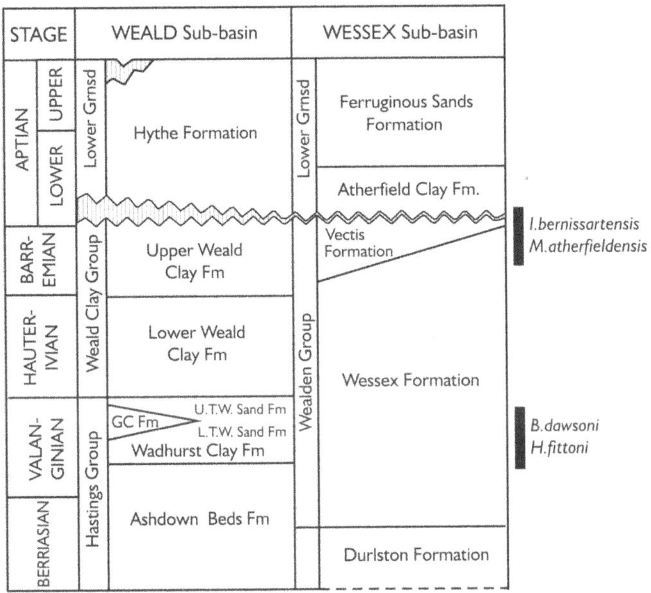

2.1. Stratigraphy of the Wealden of southern England. Abbreviations: Fm, Formation; GC Fm, Grinstead Clay Formation; L.T.W. Sand Fm, Lower Tunbridge Wells Sand Formation; U.T.W. Sand Fm, Upper Tunbridge Wells Sand Formation; Lower Grnsd, Lower Greensand. Derived from Batten (2011). The vertical bars on the right-hand side indicate the approximate stratigraphic distribution of the four principal Wealden neoiguanodontian taxa.

utilized by these animals in a biological sense) in the lineage(s) ancestral to the first diagnosable members of the clade Euhadrosauria.

This contribution concerns itself with updating our current understanding of the anatomy, taxonomy and systematics of an Early Cretaceous group of neoiguanodontians from the Wealden of northwest Europe (Fig. 2.1). Given their older chronostratigraphic occurence relative to hadrosaurids, these taxa contribute to an analysis of taxa that are considered topologically basal to hadrosaurids. This review probes our understanding of an important phase in ornithopod evolution, and highlights areas where more research is needed.

Institutional abbreviations MIWG, Museum of the Isle of Wight Geology, Sandown, Isle of Wight, U.K.; NHMUK, Natural History Museum, London, U.K.; RBINS [formerly IRSNB], Royal Belgian Institute of Natural Sciences, Brussels, Belgium.

A REVIEW OF WEALDEN IGUANODONTIANS

A large number of names have become associated with medium-to-large-bodied Wealden-aged *Iguanodon*-like ornithopods, or iguanodonts (Fig. 2.2); these include historical names such as *Vectisaurus valdensis* Hulke, 1879 (Norman, 1990); *Sphenospondylus gracilis* Lydekker, 1888; and *Iguanodon seelyi* Hulke, 1882. However, renewed scientific interest in the Wealden since 2008 has resulted in a proliferation of

WEALDEN TAXA (Norman/McDonald)	WEALDEN TAXA (Paul, Carpenter & Ishida, Hulke, Lydekker)
Barremian/Aptian	**Barremian/Aptian**
Iguanodon bernissartensis Boulenger, 1881 [v]	*I. bernissartensis* [v]
	I. seelyi Hulke, 1882 [jss]
	Dollodon seelyi (Carpenter & Ishida, 2010) [nd-jss]
Mantellisaurus atherfieldensis (Hooley, 1925) [v]	*M. atherfieldensis* [v]
	Vectisaurus valdensis Hulke, 1879 [nd]
	Sphenospondylus gracilis (Lydekker, 1888) [nd]
	Proplanicoxa galtoni Carpenter & Ishida, 2010 [nd-jss]
	Dollodon bampingi Paul, 2008 [nd-jss]
	**Mantellodon carpenteri* Paul, 2012 [nd-jss]
Valanginian	**Valanginian**
Barilium dawsoni (Lydekker, 1888) [v]	*B.dawsoni* [v]
Iguanodon anglicus Holl, 1829 [servo statua]	*Kukufeldia tilgatensis* McDonald, Barrett & Chapman, 2010 [nd-jss]
	Torilion dawsoni Carpenter & Ishida, 2010 [jos]
	Sellacoxa pauli Carpenter & Ishida, 2010 [nd-jss]
Hypselospinus fittoni (Lydekker, 1889) [v]	*H. fittoni* [v]
	Wadhurstia fittoni Carpenter & Ishida, 2010 [jos]
	Huxleysaurus hollingtoniensis Paul, 2012 [nd-jss]
	Darwinsaurus evolutionis Paul, 2012 [nd-jss]

2.2. The taxonomy of Wealden iguanodontian dinosaurs tabulated according to (in the left column) the interpretation of Norman (2010, 2011a, 2011b, 2012, 2013, in press; McDonald, pers. comm., 2012), compared with the taxonomy (in the right column) introduced by Paul (2008); Carpenter and Ishida (2010); and McDonald, Barrett, and Chapman (2010). Asterisk indicates non-Wealden taxon. Abbreviations: nd, nomen dubium; jos, junior objective synonym; jss, junior subjective synonym; v, valid taxon.

additional taxa that have been recognized on the morphological variation present in the sample. Recently proposed taxa include *Dollodon bampingi* (Paul, 2008); *Barilium dawsoni* (Norman, 2010); *Hypselospinus fittoni* (Norman, 2010); *Kukufeldia tilgatensis* (McDonald, Barrett, and Chapman, 2010); *Torilion dawsoni* (Carpenter and Ishida, 2010); *Wadhurstia fittoni* (Carpenter and Ishida, 2010); *Sellacoxa pauli* (Carpenter and Ishida, 2010); *Proplanicoxa galtoni* (Carpenter and Ishida, 2010); *Dollodon seelyi* (Carpenter and Ishida, 2010); *Huxleysaurus hollingtoniensis* (Paul, 2012); *Darwinsaurus evolutionis* (Paul, 2012); and *Mantellodon carpenteri* (Paul, 2012).

This proliferation of Wealden taxa suggests that there was considerable taxonomic diversity among these animals during Wealden time (Fig. 2.2). Recognition of this diversity is significant, because phylogenetic analyses suggest these taxa provide insights into the morphological changes that occurred in the evolutionary transition to early hadrosauroids from more primitive iguanodonts. However, the validity of a number of these recently named taxa has been questioned (Norman, 2013). The taxa erected by Paul, and Carpenter and Ishida are poorly–or incorrectly–diagnosed. Based on detailed study of the original material, many of these new taxa have been considered nomina dubia that are either unambiguously or subjectively synonymous with one of just four osteologically distinct Wealden-aged iguanodont taxa: *Barilium*, *Hypselospinus*, *Iguanodon*, and *Mantellisaurus* (Norman, 2011a, 2011b, 2012, 2013, in press; McDonald, 2012a). In addition, it appears that these novel, but dubious, taxa were proposed (and, in part, justified) on the basis of a fundamental lack of understanding of the stratigraphy of the Wealden and the provenance of the taxa that have been collected from Wealden exposures (Norman, 2013). The anatomy and spatio-temporal distribution of the iguanodonts from the Weald is reviewed below.

SYSTEMATIC PALEONTOLOGY

ORNITHISCHIA Seeley, 1887
ORNITHOPODA Marsh, 1881
DRYOMORPHA Sereno, 1986
ANKYLOPOLLEXIA Sereno, 1986
BARILIUM Norman, 2010
BARILIUM DAWSONI (Lydekker, 1888)

Lydekker (1888) described the partial skeleton of a large iguanodont that had been recovered during quarrying through ferruginous sandstones at Shornden, an open-cast quarry on an area of land about 1.5 km north of Hastings town center (Norman, 2011a). The material included dorsal and caudal vertebrae, portions of the pelvis, and parts of the hindlimb. The ilium was regarded as sufficiently distinct from anything previously described as pertaining to *Iguanodon* that it merited the establishment of a new taxonomic name, *I. dawsoni*. Norman (2010, 2011a) redescribed the type material, presented a formal diagnosis, and proposed the new combination *Barilium dawsoni* (Lydekker, 1888).

Taxonomic Discussion

McDonald et al. (2010) proposed that an isolated, nearly complete dentary of large size with two embedded dentary crowns (NHMUK OR28660), collected from one of the quarries at Whiteman's Green, Cuckfield, can be diagnosed as a new taxon of Valanginian neoiguanodontian: *Kukufeldia tilgatensis* McDonald, Barrett, and Chapman, 2010. This specimen is of considerable historical interest, because it was first studied and described by Mantell (1848) and later by Owen (1855). Their diagnosis currently rests upon one character: an apparently unique pattern of vascular foramina on the outer surface near the anterior tip of the jaw, and this is supported by some subsidiary evidence concerning the comparative straightness of the anterior part of the dentary ramus. The distribution of vascular foramina on the external surface of the dentary is a character of dubious validity, given the variation in the pattern of vascular openings that may be seen between the left and right jaws of single individuals, let alone that which may be seen in different individuals (pers. obs.).

Apart from the pattern of dentary foramina, the distinction concerning the straightness of the dentary ramus relies upon an alleged association of another partial skeleton (NHMUK R1834) to the taxon *B. dawsoni*. The latter includes the anterior portion of an eroded dentary that appears to be arched, rather than straight. Unfortunately, NHMUK R1834 was incorrectly assigned to the taxon *B. dawsoni* by McDonald et al. (2010); it can be referred, quite unambiguously (Norman, 2010, in press), to the Valanginian taxon *Hypselospinus fittoni* (Lydekker, 1889) on the basis of detailed shared similarities between the ilium of the holotype of *H. fittoni* and that of NHMUK R1834 (Norman, in press). At present, *K. tilgatensis* comprises just the dentary of a large neoiguanodontian that is considered to be a nomen dubium and to be potentially referable to the contemporary large, robust neoiguanodontian taxon *Barilium dawsoni*. Furthermore, the dentary teeth are very similar in form to those of NHMUK R2358, which have been referred to *B. dawsoni*. Carpenter and Ishida (2010) proposed, in October of that year, a new taxonomic combination *Torilion dawsoni* (Fig. 2.2) for the holotype material named *Iguanodon dawsoni*; this proposal can be suppressed, because it is a junior objective synonym of *Barilium dawsoni* (Lydekker, 1888). Furthermore, Carpenter and Ishida proposed a new genus and species (*Sellacoxa pauli*) on the basis of a photograph of the right side of a large partial skeleton (NHMUK R3788) collected by Charles Dawson from Old Roar Quarry, near Hastings (Norman, 2011a). Naish and Martill (2008), using appropriately cautious remarks, suggested that its anatomy was unusual and perhaps indicative of a new species. The description by Carpenter and Ishida (2010) is erroneous (Norman, 2011a, 2011b) because these authors had evidently not examined the specimen closely and therefore failed to recognize preservational anomalies, missing pieces, or additional anatomical features visible on the other (left) side of the specimen (Norman, 2011a). *Sellacoxa pauli* (NHMUK R3788) is considered to be a nomen dubium (its diagnosis is incorrect), and this articulated partial skeleton is considered to be referable to the hypodigm of *B. dawsoni* (as originally argued by Norman, 1977, 2010, 2011a; Blows, 1998). The taxon *Sellacoxa pauli* Carpenter and Ishida, 2010, has been proposed to be a nomen dubium and that it can

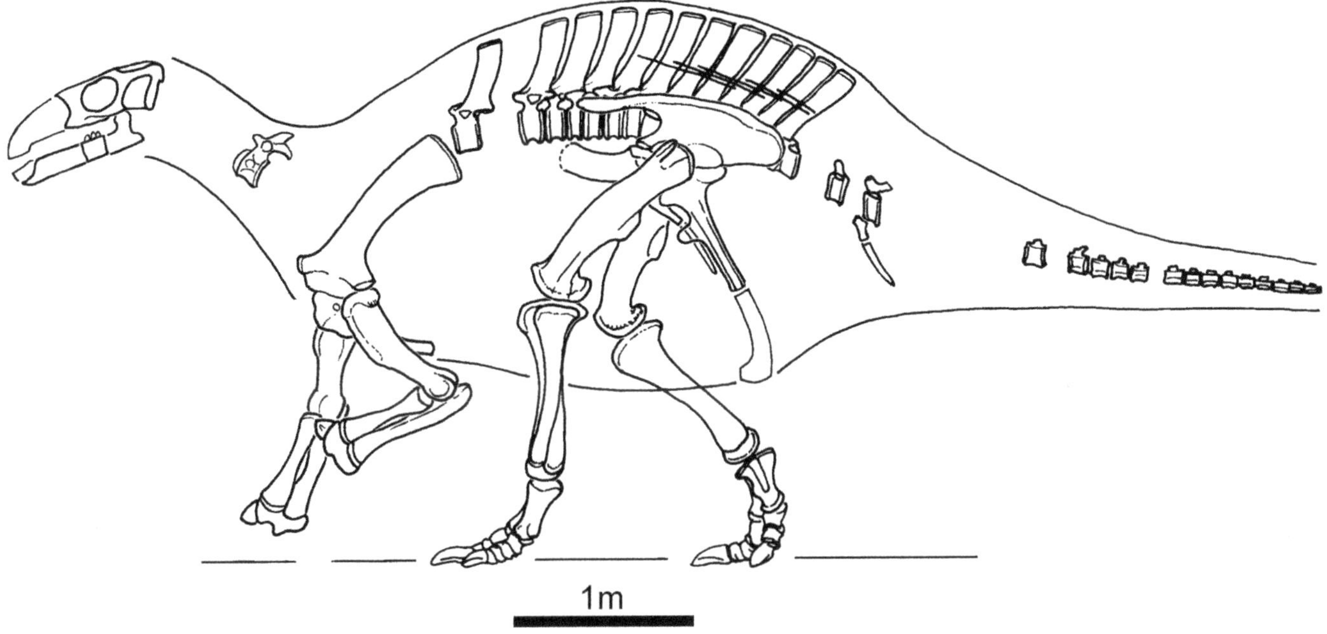

2.3. *Barilium dawsoni*. Preliminary skeletal reconstruction based upon the holotype and referred material (from Norman, 2011a).

be relegated into synonym with *Barilium dawsoni* (Norman, 2011a; Fig. 2.2).

Description

Craniodental Anatomy The dentary is robust and has parallel upper and lower edges and an elevated coronoid process that arises from a shelf lateral to the most posterior alveoli. The large, and visibly crushed, replacement dentary tooth crown preserved in NHMUK OR28660 generally resembles those seen in another referred specimen (NHMUK R2358) that comprises part of a robust dentary with three embedded teeth; these are of additional interest because they resemble the morphology seen in the lectotype tooth (*I. anglicus:* Norman, 2011b:fig. 27.23A).

Vertebrae Dorsal vertebrae are notable for having very tall, deep, and slightly inclined spines; anterior dorsals have slightly waisted, cylindrical vertebral centra; posterior dorsals become more axially compressed and develop everted edges. Sacrals are very poorly known, while the caudals are distinctive: those nearest the sacrum are squat, subrectangular in axial view, and somewhat inclined forward (Norman, 2011a:fig. 6); these are succeeded by deeper-bodied, hexagonal (more typical iguanodont) caudals, whereas the caudals toward the tip of the tail tend to have very angular sides and their articular faces tend to be deeply concave (Norman, 2011a:fig. 7).

Girdles and Limbs The pectoral (shoulder) girdle and forelimb are robust. The scapula (based upon the referred specimen NHMUK R2848) is long, curved, and expands towards its upper end. The coracoid is notably broad and dished, and has a prominent and completely enclosed coracoid foramen near the suture with the scapula (Norman, 2011a:fig. 17A). One specimen (NHMUK R2357) includes the "handle" portion of a hatchet-shaped sternal bone (Norman, 2011a:fig. 17B). The principal forearm bones (radius and ulna) are very robust; the carpals and metacarpal I cap the ends of the ulna and radius and are fused into a solid block that supports a fused, squat pollex. The form of the remaining bones of the hand is unknown. The hip (pelvic) bones include a very distinctive ilium, which has a long, robust, preacetabular process that is twisted along its length and bears a large rib facet near its base. The main body of the ilium is slab sided, thick along its dorsal edge with minimal lateral swelling and an inflection along its upper edge (posterodorsal to the ischiadic peduncle). The postacetabular process is deep and rounded in profile, and does not develop a ventrolateral ridge that delimits a vaulted brevis fossa; a well-developed brevis fossa is present in all other Wealden iguanodonts. The shape of the shaft of the ischium is unknown, but proximally the external surface of the shaft adjacent to the obturator process displays a pronounced vertical ridge that runs along the ischial shaft (NHMUK R2357) rather than forming a flat, rugose facet seen typically in this area in specimens attributable to *H. fittoni;* and the pubis appears to develop a thick, deep, and slightly upwardly curved prepubic process and the dorsoventrally compressed (strap-like) pubic shaft is unlikely to have extended to the end of the ischial shaft. The hindlimb is poorly known (Norman, 2011a).

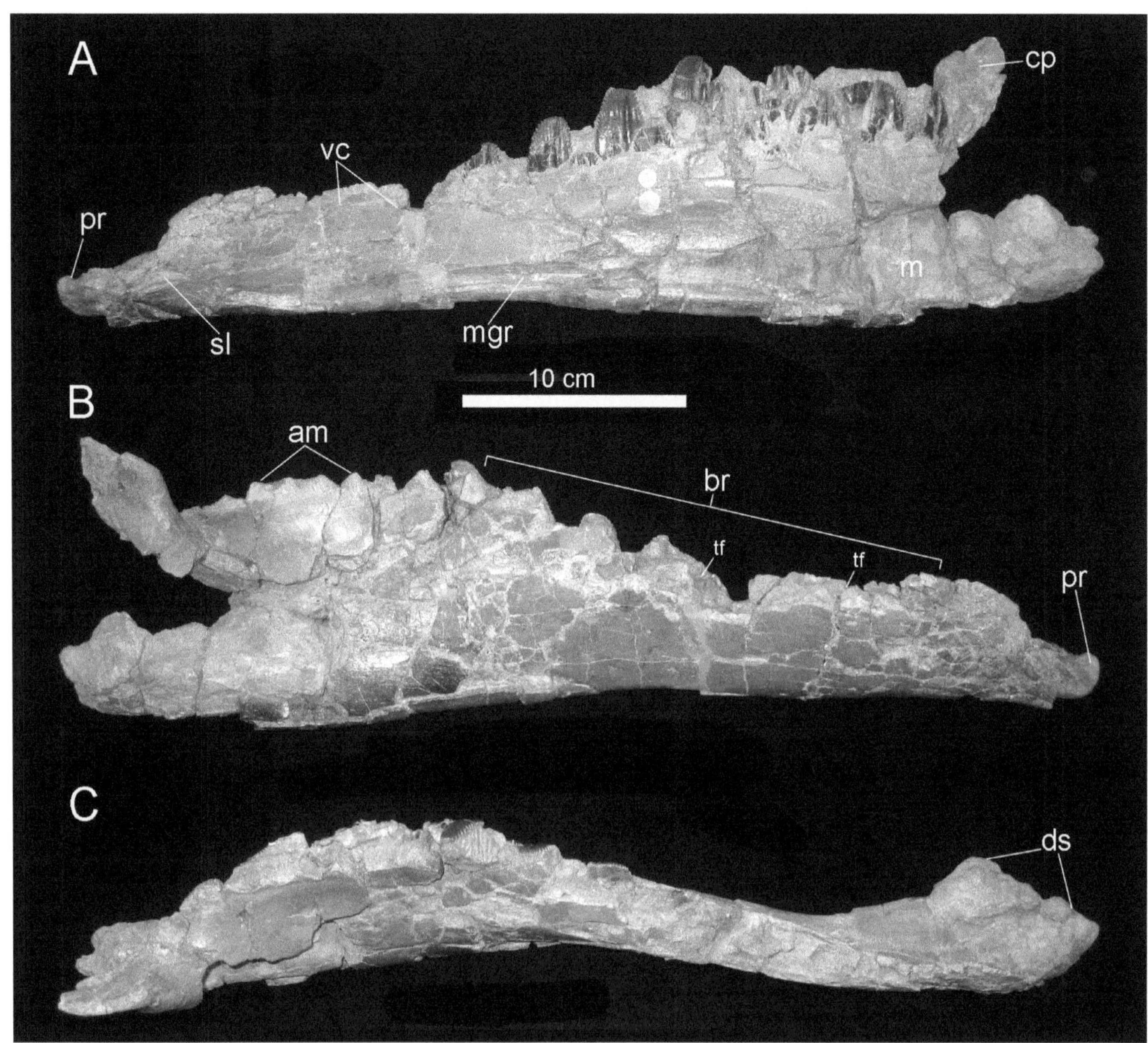

2.4. *Hypselospinus* cf. *fittoni,* NHMUK R1831. Dentary (right) with teeth preserved in situ. (A) medial; (B) lateral; (C) dorsal views. Abbreviations: am, alveolar margin; br, badly broken portion of the dentary; cp, coronoid process; ds, dentary symphysis; m, matrix; mgr, Meckelian groove; pr, anterior lateral process of the dentary; sl, "slot-and-lip" portion of the dentary symphysis; tf, tooth fragments in alveolar bone; vc, vascular channel. Scale bar equals 10 cm (from Norman, in press).

Reconstruction of Barilium

The composite reconstruction of the skeleton (Fig. 2.3) suggests that this dinosaur was likely to have been at least facultatively quadrupedal, and it may in fact have been an obligate quadruped.

HYPSELOSPINUS Norman, 2010
HYPSELOSPINUS FITTONI (Lydekker, 1889)

Taxonomic Discussion

Lydekker (1889) described some portions of a skeleton recovered from the same quarry near Hastings that produced the type material of *Barilium dawsoni*. The type material was redescribed by Norman (2010, 2011b, in press) and on the basis of distinctive features of the holotype ilium, which was supplemented by better-preserved referred material – including specimens that were previously attributed to *Iguanodon hollingtoniensis* Lydekker, 1889. *I. hollingtoniensis* is now regarded as a junior subjective synonym of *Hypselospinus fittoni* (Norman,

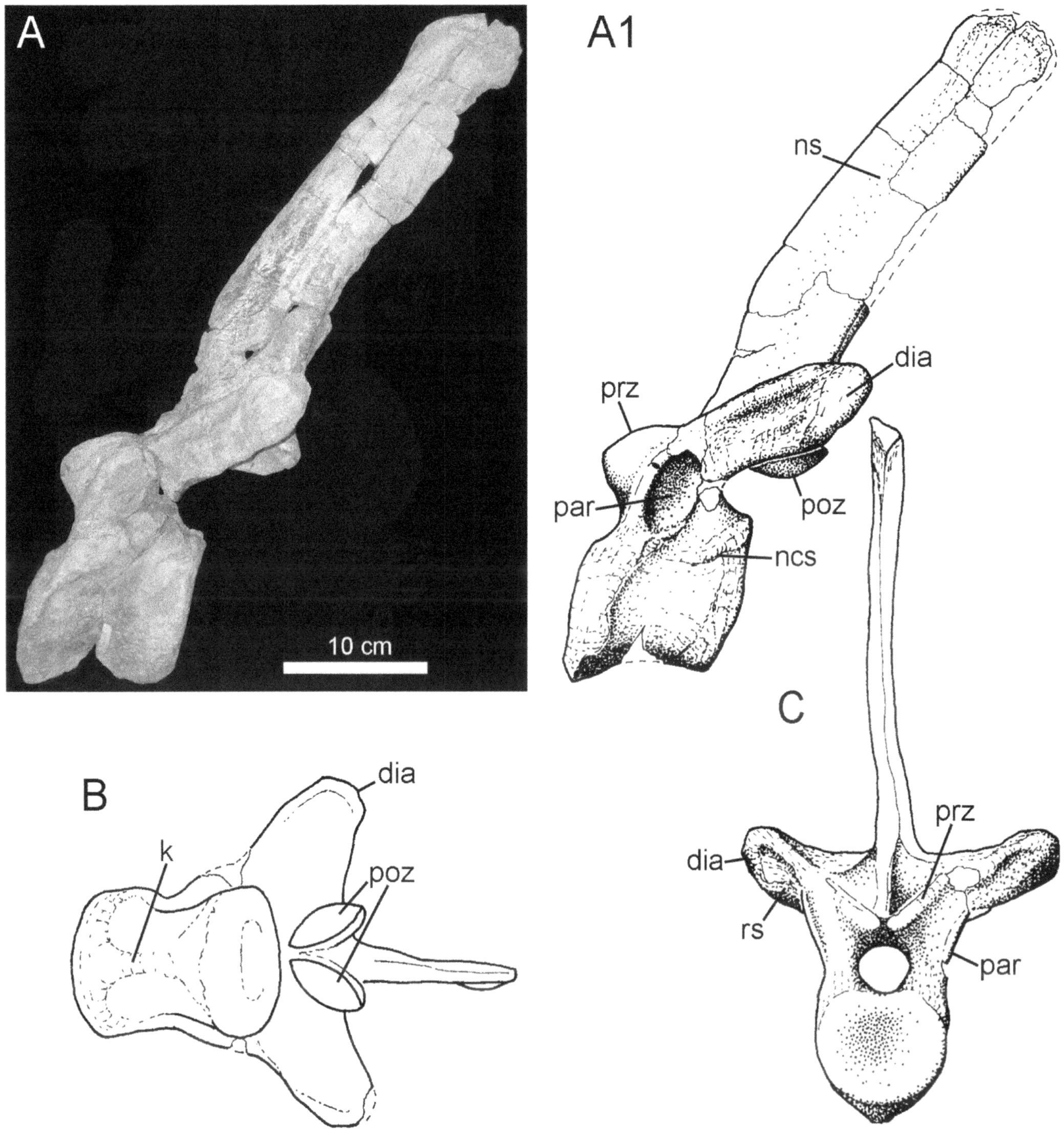

2.5. *Hypselospinus* cf. *fittoni,* NHMUK R604. Third dorsal. (A, A1) Lateral ([A] is a reversed image of the right side); (B) ventral; (C) anterior. Abbreviations: dia, diapophysis; k, midline keel; ncs, neurocentral suture; ns, neural spine; par, parapophysis; poz, posterior zygapophysis; prz, anterior zygapophysis; rs, rugose surface for ligamentous attachment of the neck of the rib. Scale bar equals 10 cm (from Norman, in press).

2010). *Iguanodon fittoni* was rediagnosed and renamed as a new nomenclatural combination: *Hypselospinus fittoni.* This iguanodont appears to be generally somewhat smaller (body length ~6 m) and less robustly built than specimens typical of *B. dawsoni.* A number of unsupportable claims concerning the osteology, taxonomic status, and affinities of material referable to this renamed taxon have been made by Paul (2008), as outlined in Norman (2010). Carpenter and Ishida (2010: October), subsequent to Norman (2010: May), published an alternative name for the type material of *I. fittoni*: *Wadhurstia fittoni.* The latter can safely be suppressed because it is a junior objective synonym of *Hypselospinus fittoni.*

Description

This taxon has a body form that is distinct from its sympatric contemporary *B. dawsoni:* it is smaller in body length overall, less robust, and has a much more lightly built vertebral column with slender, tall neural spines along the dorsal, sacral, and caudal regions.

Craniodental Anatomy A crushed and distorted, yet almost complete, dentary (NHMUK R1831; Fig. 2.4) demonstrates that this bone was more slender than the specimen (NHMUK OR28660) that can be attributed to *B. dawsoni,* and another referred specimen (NHMUK R1834; Norman, in press) indicates that the anterior part of the dentary is deflected ventrally (contra McDonald et al., 2010). Several tooth crowns are preserved in NHMUK R1831 (Norman, 2010, 2011b, in press) and are quite distinct in surface details from those referred to *B. dawsoni.* Whereas the enameled face of the crown is shield shaped and fringed with mammilate, ledge-shaped denticles (as in *B. dawsoni*), the ridge pattern seen on the lingual face of the crown differs considerably. There is a prominent primary ridge running the length of the crown distal (posterior) to the midline, and mesial (anterior) to this is a series of strand-like minor ridges that extend down the remainder of the crown for varying distances.

Vertebrae Most characteristically, the dorsals develop remarkably long, backwardly inclined, narrow spines that in life should have given the appearance of a tall midline ridge (Fig. 2.5). The dorsals differ markedly from those of *B. dawsoni.* The caudals do not exhibit the low (squat), inclined angular form seen in *B. dawsoni;* in contrast, they appear to have compressed, tall centra and support equally elongate, narrow spines. Middle and posterior caudals display the gradual loss of the caudal rib and become lower and more apparently elongate, eventually developing angular (hexagonal) cross sections; they do not seem to show the deeply concave articular surfaces seen in *B. dawsoni.*

Girdles and Limbs The pectoral girdle differs only in size from that of *B. dawsoni:* most of the elements appear similar in general shape and the sternals are hatchet shaped. The forelimb resembles that of *B. dawsoni* in the shape of each element of upper and lower arm, but there is a clear difference in the shape of the characteristic pollex spine (thumb-spike). Whereas in *B. dawsoni* the pollex is short, laterally compressed and bluntly truncated, that of *H. fittoni* appears to be tall, laterally compressed and pointed (triangular in lateral aspect; Fig. 2.6); this thumb-spike resembles Mantell's (1827) classic "nasal horn." The manus (Fig. 2.6) resembles that described in *Iguanodon bernissartensis* (Norman, 1980) in the relative shape and proportions of each digit, although overall it seems to have rather shorter digits than might have been expected, judged by the dimensions of the associated radius and ulna.

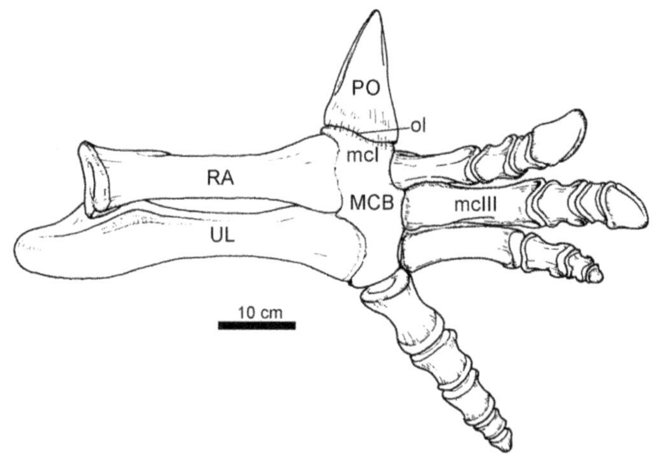

2.6. *Hypselospinus* cf. *fittoni,* NHMUK R1831 (R1832/R1833). Reconstructed antebrachium and manus in lateral view. Abbreviations: mcl/mcIII, metacarpals; MCB, metacarpo-carpal block; ol, ossified ligaments; PO, pollex ungual; RA, radius; UL, ulna. Scale bar equals 10 cm (from Norman, in press).

The pelvis comprises an ilium with a narrow, untwisted, elongate preacetabular process with a low, curved medial ridge; the deep, central portion of the ilium is flat and has a relatively compressed dorsal edge (with, at most, a slight lateral expansion on its dorsal margin above and behind the ischiadic peduncle); the postacetabular process tapers (as upper and lower borders converge) to form a blunt, rounded transverse bar; below the latter is a low-vaulted brevis fossa, demarcated laterally by the presence of a ridge (Norman, 2010:fig. 5). This iliac morphology is clearly distinct from that seen in *B. dawsoni* (Norman, 2011b:fig. 8), which has a transversely thick and axially twisted preacetabular process, a broad dorsal edge to the main blade, a deep postacetabular process that is inflected medially toward its ventral edge but has no lateral ridge, and a brevis shelf that is either absent or very reduced in extent (NHMUK R3788). The pubis (Fig. 2.7A) is incomplete but has a deep and slightly upwardly curved, parallel-sided, prepubic process with an anterior tip that appears to be moderately dorsoventrally expanded. In contrast to the pubis of *B. dawsoni,* the pubic shaft is cylindrical. The ischium (Fig. 2.7B) has a robust, curved (J-shaped) shaft that appears to be twisted along its length and ends in an enlarged anteriorly expanded "boot"; the proximal external surface of the shaft bears a flattened, scarred facet adjacent to the flap-like obturator process (obt) is positioned close to the proximal end of the shaft and offered mechanical support to the pubic shaft.

The hindlimb, as in the case of *Barilium,* is not known from good-quality articulated material; it differs little in morphology from what is known in *B. dawsoni* (Fig. 2.8).

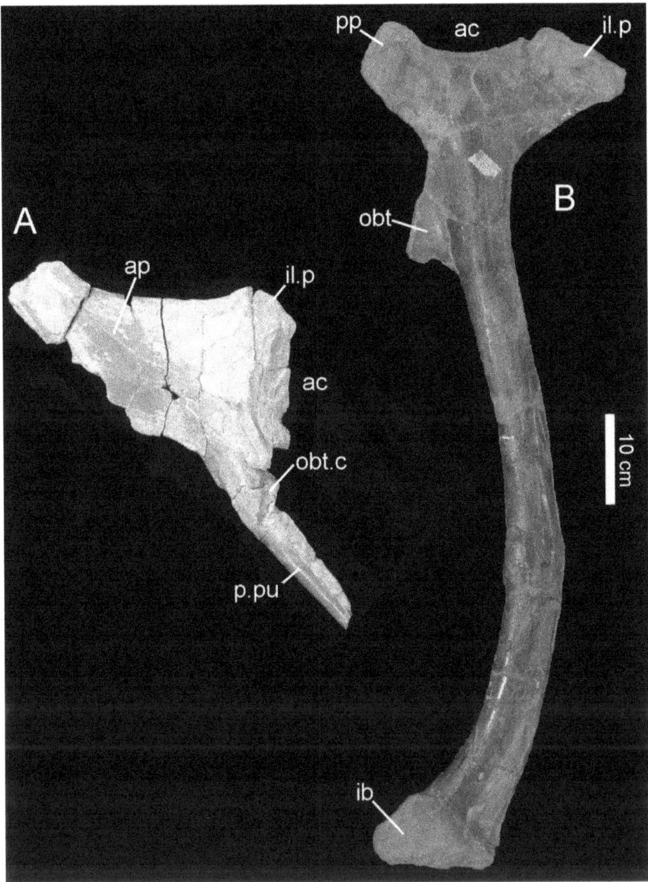

2.7. *Hypselospinus* cf. *fittoni,* NHMUK R811. (A) pubis partial (right, this is a reversed image) in lateral view; (B) ischium complete (left) in lateral view. Abbreviations: ac, acetabular margin; ap, anterior blade of the pubis; ib, ischial "boot"; il.p, iliac peduncle; obt, obturator process; obt.c, obturator channel; pp, pubic peduncle; p.pu, posterior ramus of the pubis. Scale bar equals 10 cm (from Norman, in press).

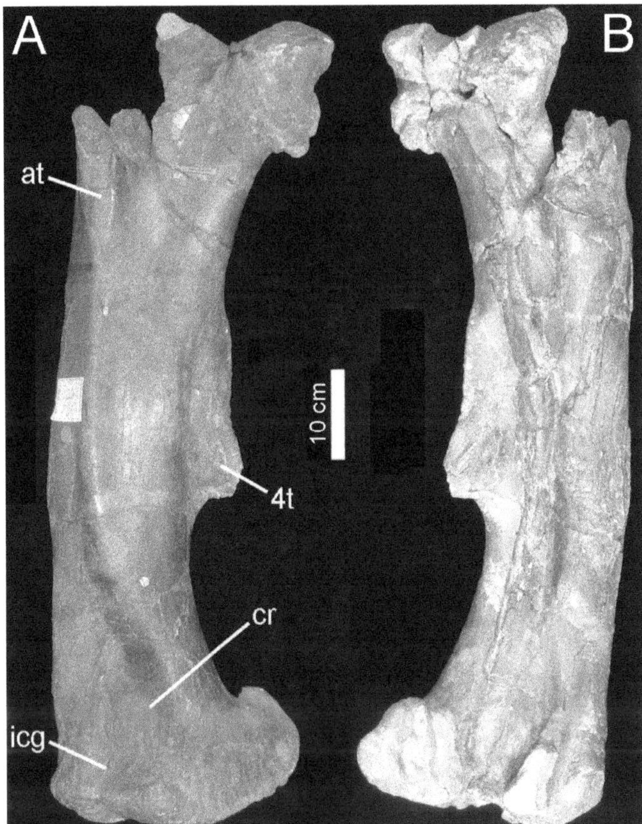

2.8. *Hypselospinus* cf. *fittoni,* holotype of *Iguanodon hollingtoniensis,* NHMUK R1148. (A, B) femur, right, the original specimen as preserved (May 2011) in dorsal and ventral views respectively; the ventral view reveals the extent of longitudinal crushing. Abbreviations: 4t, fourth trochanter; at, anterior (lesser) trochanter; cr, crushing of the dorsal part of the medial condyle; icg, anterior intercondylar groove. Scale bar equals 10 cm (modified from Norman, in press).

Hindlimb material of this taxon was first illustrated by Lydekker (1889). There is a prominent crested fourth trochanter that probably terminated in a marginally pendent tip that does not resemble that seen in camptosaurs (contra Lydekker, 1889).

Reconstruction of Hypselospinus

A preliminary reconstruction (Fig. 2.9) of *H. fittoni* based on the type and referred material has been described in detail (Norman, in press). The skull was probably more slender and elongate than that of *B. dawsoni* based on the morphology of the lower jaw. The vertebral column is notable for the comparatively small proportions of dorsal centra and the attenuation of the neural spines, which form a sail-like structure reminiscent of the even taller "sail" seen in the gracile neoiguanodontian *Ouranosaurus* (Taquet, 1976). This reconstruction is tentative because it is a composite based on a number of skeletons of individuals of differing size and there are uncertainties about the relative proportions of the fore and hindlimbs (as well as within-limb proportions).

MANTELLISAURUS Paul, 2007
MANTELLISAURUS ATHERFIELDENSIS (Hooley, 1925)

The posthumous work by Hooley (1925) based on a nearly complete skeleton recovered (in 1914) from broken blocks of shale following a cliff collapse near Atherfield Point, Isle of Wight, provided the first detailed anatomical description of any Wealden-aged *Iguanodon*-like ornithopod 100 years after the first *Iguanodon* teeth were described by Mantell. This paper founded a new species: *Iguanodon atherfieldensis* Hooley, 1925. The importance of this discovery and its description cannot be overemphasized, given the previous century of attempts to identify and name new species using material that was often inadequate and compounded by the startling failure to provide detailed descriptions when material was, in fact, available.

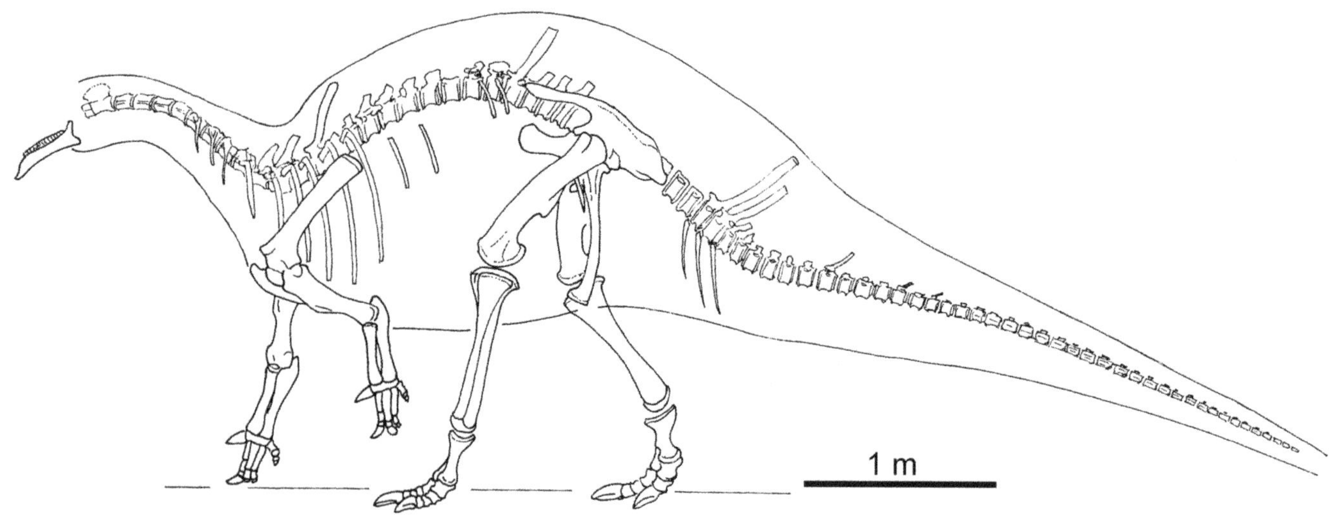

2.9. *Hypselospinus fittoni.* Preliminary skeletal reconstruction based upon the holotypes of *Iguanodon fittoni* Lydekker, 1889, and *I. hollingtoniensis* Lydekker, 1889, supplemented by information from several additional referred partial skeletons (from Norman, in press).

Noteworthy, in the latter respect, is the remarkable fully articulated skeletal material collected between 1878 and 1881 from Bernissart in Belgium, which was described only superficially by Louis Dollo (Norman, 1980, 1986, 1987). What became increasingly obvious, with the benefit of hindsight, was that material (notably that collected from the Isle of Wight) described variously under the names *Vectisaurus*, *Sphenospondylus*, or *Iguanodon mantelli* – the latter name usually considered synonymous with the "Mantel-piece" collected from Maidstone in 1834 (Norman, 1993) – would eventually be referred to *I. atherfieldensis* (Norman, 1986, 1990, 2004).

Taxonomic Discussion

Recently, the taxon *Iguanodon atherfieldensis* has been subjected to revision. Paul (2007) proposed *Mantellisaurus* as a new generic name for *I. atherfieldensis.* The reasoning for this change relied on osteological differences originally regarded as sufficient to distinguish these forms as "osteological species" (Norman, 1986:327). Having proposed the generic name *Mantellisaurus*, Paul (2008) then extended his taxonomic revision of Wealden iguanodonts by creating an entirely new taxon, *Dollodon bampingi* (Fig. 2.2) for the gracile skeleton (RBINS R57 [formerly "IRSNB 1551"]; see Norman, 1986) and referred to, historically, as "*Iguanodon mantelli*" (e.g., Dollo, 1882; Casier, 1960). The first monograph on this specimen (Norman, 1986) referred it to *Iguanodon atherfieldensis.* The case for erecting the new binomial *Dollodon bampingi* was supported by a list of diagnostic characters derived from some simplistic outline drawings, some "fleshed-out" restorations of the heads of these animals and the interpretation of photographs of mounted specimens (named "technical restorations" by Paul, 2008:202). Norman (2012) evaluated the diagnostic characters proposed by Paul and demonstrated that none could be considered to be valid, and that on that basis alone, the new name should be considered a nomen dubium: a very similar conclusion was reached independently by McDonald (2012a).

Description

Mantellisaurus atherfieldensis attained a probable adult body length of about 7 m. The type material (NHMUK R5764) represents a disarticulated partial skull and skeleton collected from the Isle of Wight, that is ontogenetically immature and has an estimated body length of approximately 5.5 m; the referred skeleton from Bernissart (RBINS R57) shows some residual features associated with immaturity, and is approximately 6.5 m long; and the length of the "Mantel-piece" individual (Norman, 1993) from Maidstone (NHMUK OR3741) is estimated (based on femoral length) at probably a little in excess of 7 m. Some material collected recently from the Isle of Wight exhibits very interesting anatomical variation (Martill and Naish, 2001:MIWG 6344).

Craniodental Anatomy The skull (Fig. 2.10) of this species is known in considerable detail (Norman, 1986). The lower jaw is elongate and its lower margin is gently arched towards its anterior tip; the coronoid process is comparatively short, vertical and slightly expanded anteriorly at its apex. The posterior end of the lower jaw is marked by a large surangular with a distinct surangular foramen and the angular is visible in lateral aspect. Dentary teeth are comparatively simple in construction with primary and secondary ridges alone on the lingual enameled surface resembling the

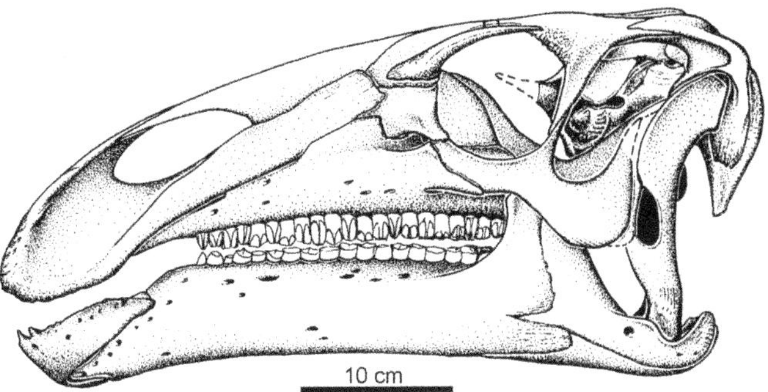

2.10. *Mantellisaurus atherfieldensis.* Skull restoration based upon the "Chase skull," NHMUK R11521.

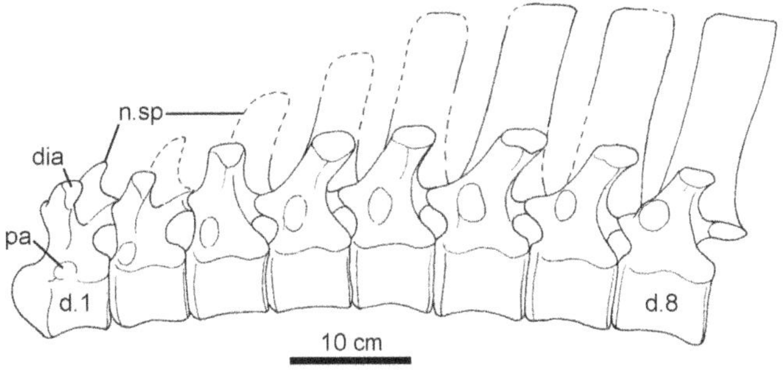

2.11. *Mantellisaurus atherfieldensis.* Anterior dorsal vertebrae, reconstruction in lateral view based upon examination of the original material of RBINS R57 and NHMUK R5764 (the holotype of *I. atherfieldensis*). Abbreviations: d.1–d.8, dorsals numbered in sequence; dia, diapophysis; n.sp, neural spine; pa, parapophysis (after Norman, 1986:fig. 29B).

2.12. *Mantellisaurus atherfieldensis,* holotype of *I. atherfieldensis* Hooley, 1925, NHMUK R5764. Articulated sequence of mid-dorsal vertebrae as preserved (same as Norman, 2011b:fig. 27.42B).

pattern seen in examples of *B. dawsoni*. Maxillary teeth have narrower crowns than dentary teeth and have an extremely prominent distally offset primary ridge.

Vertebrae Cervical vertebrae exhibit the following characteristics: strongly opisthocoelous; low cylinders with ventral keels and a mid-height ridge that is expanded near the anterior condylar margin to form a parapophysis; neural arch develops a small midline spine lateral to which are prominent, stout diapophyses for the attachment of ribs; prezygapophyses are widely spaced and do not project beyond the articular margin of the centrum, whereas the postzygapophyses are long, arched, and divergent (and overlap the succeeding centrum). The general form of cervical vertebrae is seen in the first dorsal vertebra reconstructed in Figure 2.11.

Mid-dorsal vertebrae develop elongate spines in the articulated skeleton RBINS R57 (Fig. 2.11), but preservation is usually not nearly so good in Wealden specimens: all are broken in the holotype skeleton (Fig. 2.12). Ossified tendons are distributed in the form of a layered lattice across the taller neural spines. The centra are spool shaped and bear a modest ventral keel. The articular faces, which bear remnant opisthocoely across the cervicodorsal transition, have

2.13. The "Saull Sacrum" illustrated in ventral view, NHMUK OR37685. Specimen referred to *Mantellisaurus* cf. *atherfieldensis.* Scanned from the original lithograph in Owen (1855:pl. 3). This specimen was one of the key specimens that Richard Owen used in order to diagnose his new "sub-order" Dinosauria (Owen, 1842).

predominately amphiplatyan faces. Posterior dorsals develop centra that are broader and deeper than anterior members of the series, and also become slightly opisthocoelous in the region adjacent to the sacrum.

Sacral Vertebrae One specimen (Fig. 2.13) comprises a nearly complete sacrum (lacking the sixth true sacral) with portions of an attached ilium (NHMUK OR37685), which is attributable to this species. The sacrum comprises seven fused vertebrae in mature specimens (fusion is incomplete in immature individuals) and involves the incorporation of a posterior dorsal with a free (non-sacralized) rib. There is a narrow keel present, unlike *I. bernissartensis*, which exhibits a broad, longitudinal midline sulcus.

Girdles and Limbs The pectoral girdle and forelimb bones differ little from those described for previous taxa (and the contemporary *I. bernissartensis*) except that they tend to be smaller and less robust. The blade of the scapula tends to have a narrower shaft and the blade flares distally to a greater extent than in *I. bernissartensis.* The coracoid also exhibits a discrete foramen (cf) externally, which is different from the coracoid "notch" seen in *I. bernissartensis.*

The sternal bone (Fig. 2.14C) has the classic "styracosternan" hatchet-like shape. The humerus (Fig. 2.14A) is sinuous. The ulna and radius (Fig. 2.14B) are comparatively slender and bowed, thus suggesting the possibility of some axial rotation between these elements. The wrist and hand are worth mentioning because they are distinctive (Fig. 2.15). The carpals are sutured together, but they are neither as massive nor as tenaciously bound by ossified ligaments as is the case in previous examples (above; Norman, 2011a, in press) or in *I. bernissartensis* (below; Norman, 1980). The first metacarpal is fused to the carpals and forms an oblique, roller-like structure for articulation with the base of the pollex; the latter is relatively diminutive and, unlike the *Barilium* and *Hypselospinus*, is genuinely conical rather than transversely compressed or truncated. In its general shape the pollex of *M. atherfieldensis* echoes, on a smaller scale, the conical pollex of *I. bernissartensis.* The central bones of the hand (metacarpals II–IV) are slender and more elongate than those known in either *Hypselospinus* or *Iguanodon.*

The ilium (Fig. 2.16) has a long, slender, preacetabular process (prp) that is buttressed by a curved medial ridge. The main body of the iliac blade is vertical, but the dorsal edge is thickened and everted so that it overhangs the lateral surface. Farther posteriorly, the dorsal edge thickens and becomes more everted, forming a beveled structure (boss) posterodorsal to the ischiadic peduncle. The dorsal edge of the postacetabular process beyond the iliac boss is inflected downward before terminating in a short transverse bar. Beneath this bar there is a narrow, vaulted brevis fossa (br.f). In overall shape the ilium resembles that of *Hypselospinus fittoni* from the Valanginian of the Weald Sub-basin; however, the preacetabular process is more slender and transversely thicker, whereas the equivalent portion of *H. fittoni* is more strongly compressed laterally and considerably deeper; the central portion of the iliac blade is shallower than in *H. fittoni;* and the postacetabular process differs also in having a far less pronounced brevis fossa than in *H. fittoni* and, as a direct consequence, the posterior bar is also much narrower.

The pubis (Fig. 2.16) has a thin, deep, prepubic process that expands distally, whereas the pubic shaft is narrow and short; there is a massive iliac peduncle, and beneath this a broad cup-shaped depression forms the anterior part of the acetabulum. The proximal part of the pubic shaft has a finger-like dorsal process that nearly encircles the obturator foramen (obt.f); its posterior surface forms a flattened vertical surface for attachment of the adjacent part of the ischium

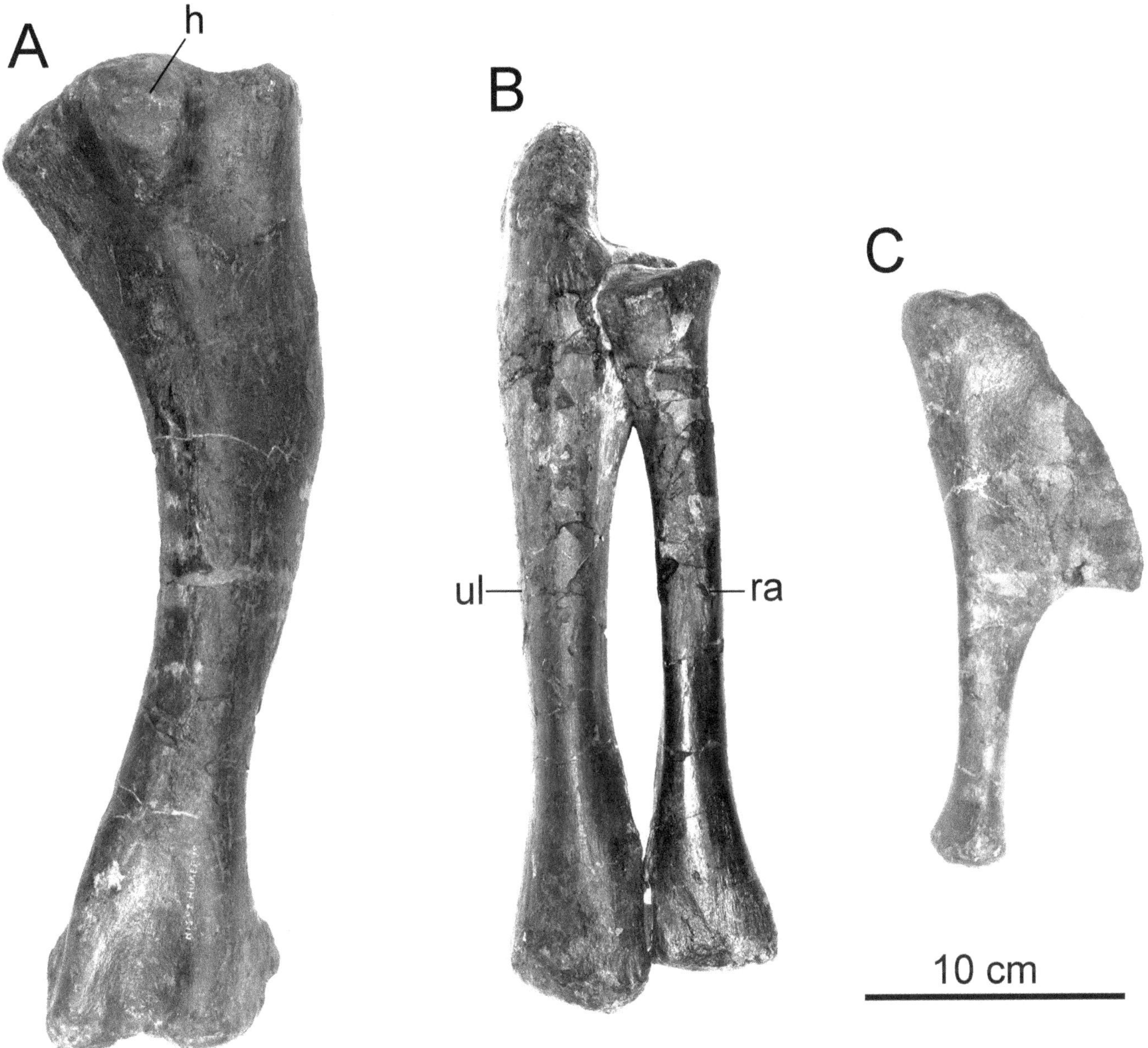

2.14. *Mantellisaurus atherfieldensis,* holotype of *I. atherfieldensis* Hooley, 1925, NHMUK R5764. (A) humerus, right in dorsal view; (B) radius and ulna, right lateral view; (C) right sternal bone in ventral view. Abbreviations: h, articular head of the humerus; ra, radius; ul, ulna (same as Norman, 2011b:fig. 27.43C–E).

and, when articulated, the obturator foramen is completely enclosed. The shaft of the ischium is long, slender, and only slightly arched along its length (the arching is perhaps exaggerated in Figure 2.16, and the distal boot is too large) and has a modest anterodistal expansion.

The hindlimb (Fig. 2.17) is not particularly distinctive, as is true of most similar-sized iguanodonts. The femur (Fig. 2.17A, B) has a shaft that is more slender, less angular-sided, and less curved along its length than that seen in *B. dawsoni* and *H. fittoni* from the Weald Sub-basin (any remaining curvature of the shaft is present only below the fourth trochanter [4t]); and the anterior trochanter (at) is narrower, less robust, more laterally compressed, and more closely appressed to the lateral surface of the greater trochanter, when compared to the latter taxa. The lower leg elements (Fig. 2.16C, D) are not distinctive, except insofar as they are more slender and lightly built than in the contemporaneous taxon *I. bernissartensis*, and the proximal tarsals are firmly attached (but not fused) to the crus (Fig. 2.17: ast, cal).

The pes (Fig. 2.18A, B) is slender and functionally three toed. Neither the holotype (NHMUK R5764) nor the referred specimen (RBINS R57) have metatarsal I preserved. A well-preserved and articulated pes that is commensurate and that is the same stratigraphic age has been referred to

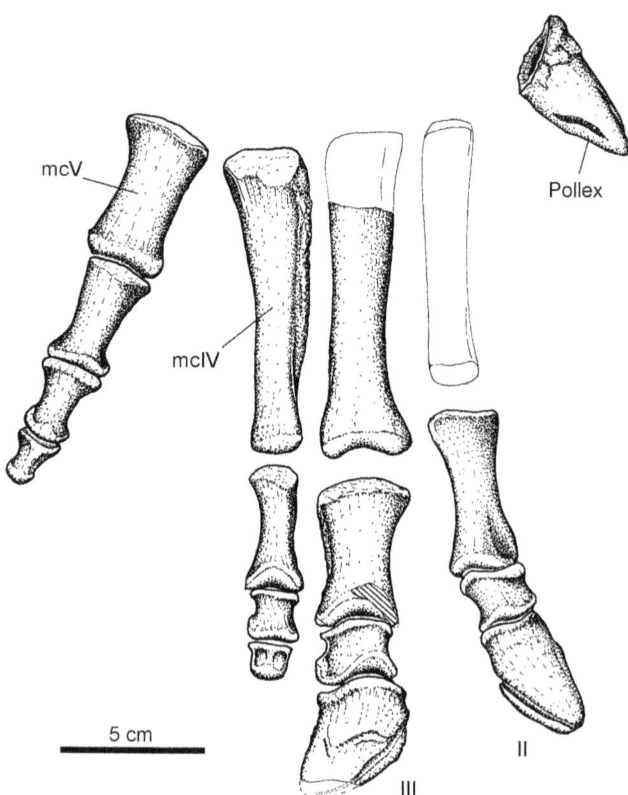

2.15. *Mantellisaurus atherfieldensis,* holotype of *I. atherfieldensis* Hooley, 1925, NHMUK R5764. The associated elements of the right manus in dorsal view. Abbreviation: mc, metacarpal (from Norman, 1977).

2.16. *Mantellisaurus.* Attempted reconstruction of the articulated pelvic bones in left lateral view. The curvature of the ischial shaft is exaggerated; the shaft is typically straighter, more angular-sided proximally, and has a smaller distal "boot" (from Norman, 1977, 2011b). Abbreviations: ac, acetabulum; boss, faceted dorsal margin of the iliac blade; br.f, brevis fossa; il, ilium; is, ischium; obt, obturator process; obt.f, obturator foramen; prp, preacetabular process of the ilium; pu, pubis; sac, supra-acetabular crest; step, beveled external surface of the ischiadic peduncle.

M. atherfieldensis (Norman, 1986; NHMUK R1829) exhibits a narrow, splint-like metatarsal I. The sympatric contemporary *I. bernissartensis* has a small, laterally compressed metatarsal I (Norman, 1980).

Reconstruction of Mantellisaurus

The reconstruction in a bipedal pose (Fig. 2.19) is based primarily upon the proportions of the holotype skeleton (NHMUK R5764) and that of the referred skeleton (RBINS R57). The pectoral girdle and forelimb are notably less robust than those seen in either of the Valanginian taxa.

IGUANODON Mantell, 1825
IGUANODON BERNISSARTENSIS Boulenger (in Beneden, 1881)

Although extremely well known in mainland Europe, where more than 30 complete and partial skeletons have been recovered in Belgium, Germany, France, and Spain, *Iguanodon bernissartensis* is comparatively rare in Britain. The first occurrence of this morphotype was a hindlimb, pelvis, and some caudal vertebrae of large size collected in 1870 by John Whitaker Hulke at Brook Chine, Isle of Wight (late Barremian: NHMUK R2501-R2514; associated bones of an almost complete ilium and hindlimb that were given separate registered numbers). This material was eventually described and illustrated as *Iguanodon seelyi* Hulke, 1882 (Norman, 2012). Rather interestingly, from a purely historical perspective, this paper appeared shortly after Hulke had spent time studying the newly excavated Bernissart material in Brussels. Hulke was informed (as he quite candidly reported in this article) that formal description of the Belgian remains was not expected for a number of years; however, he must have been aware of the strong anatomical similarity between the material that he had discovered earlier and that of the large iguanodont skeletons from Bernissart. Hulke's specimen consisted of the ilium, portions of both hindlimbs (including an almost complete pes), a major portion of a humerus, and a small number of caudal vertebrae (Norman, 2012); it should be noted there are also a number of additional complementary skeletal elements that are commensurate with the holotype that were also recovered from Brook Chine by Hulke and Rev. William Fox. Although these specimens were recovered at about the same time, they were not recorded in the same series of accession numbers. Dollo (1882) synonymized

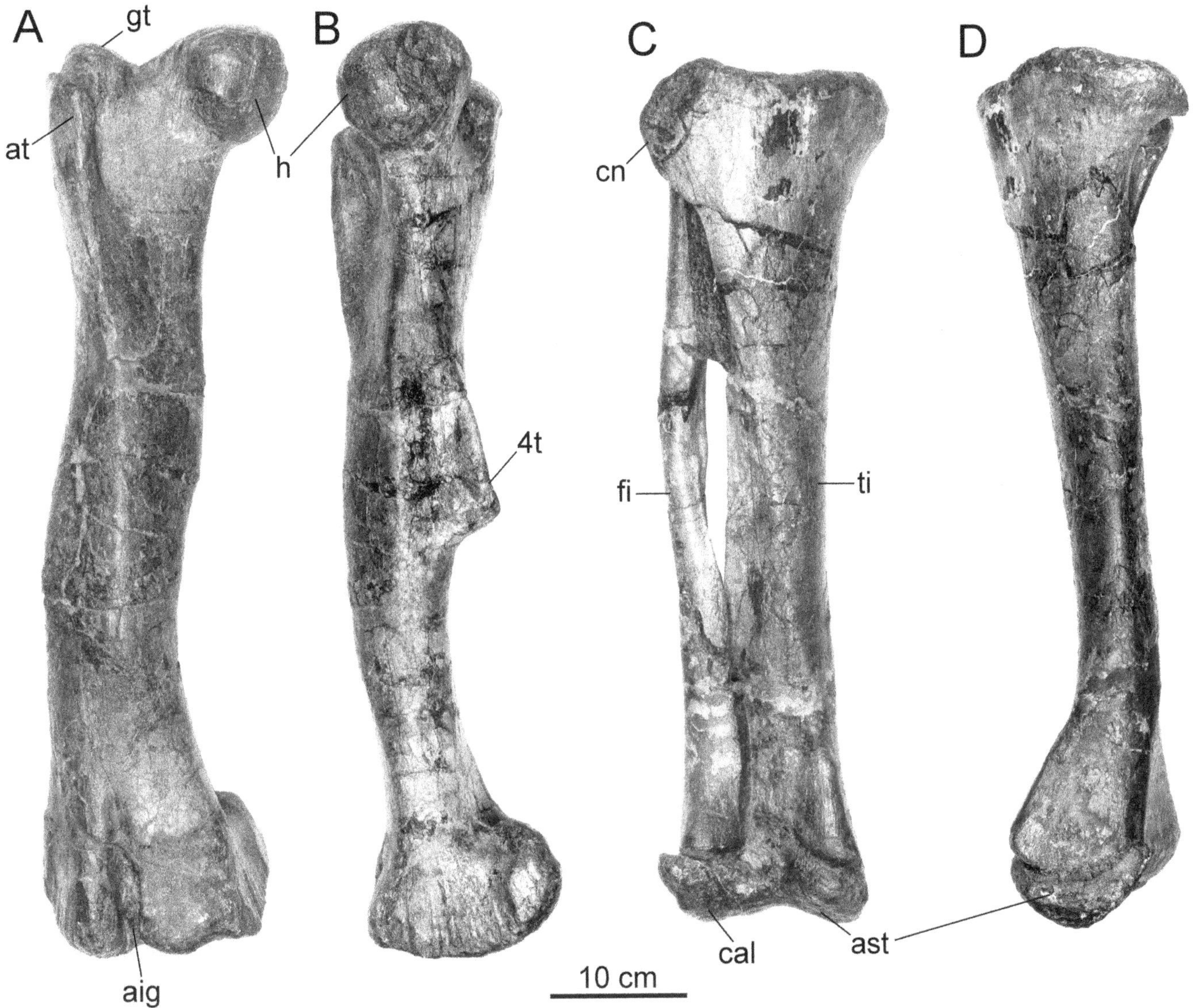

2.17. *Mantellisaurus atherfieldensis,* holotype of *I. atherfieldensis* Hooley, 1925, NHMUK R5764. The principal bones of the hindlimb. (A, B) femur in anterior (dorsal) and medial views, respectively; (C, D) the crus in anterior (dorsal) and medial views, respectively. Abbreviations: 4t, fourth trochanter; aig, anterior intercondylar groove; ast, astragalus; at, anterior (lesser) trochanter; cal, calcaneum; cn, cnemial crest; fi, fibula; gt, greater trochanter; h, femoral articular head; ti, tibia (same as Norman, 2011b:fig.27.46).

Hulke's material with *Iguanodon bernissartensis* Boulenger, which had been named within the text of a critical report by P.-J. van Beneden (1881).

Taxonomic Discussion

Despite some lingering concerns about the status of *Iguanodon seelyi*, the iconic taxon *I. bernissartensis* has remained largely unaffected by recent taxonomic revisions. However, an entirely unnecessary confusion concerns a new taxonomic combination, *Dollodon seelyi* (Fig. 2.2). Carpenter and Ishida (2010:148) consider the ilia of RBINS R57 that were renamed *Dollodon bampingi* by Paul (2008) to be "practically indistinguishable from that of *Iguanodon seelyi* Hulke, 1882 and therefore the species *seelyi* has priority over *bampingi*." However, the two ilia are distinguishable. *Dollodon* is an invalid taxon (see above, *D. bampingi*) and referring another specimen to this genus by creating a new species without reference to diagnostic characters is logically and taxonomically unacceptable (McDonald, 2012a).

Description

Mature specimens of *I. bernissartensis* attained a body length in the range 10–13 m (*I. seelyi* being at the top end of the range) and, by virtue of their size and robustness, these remains are

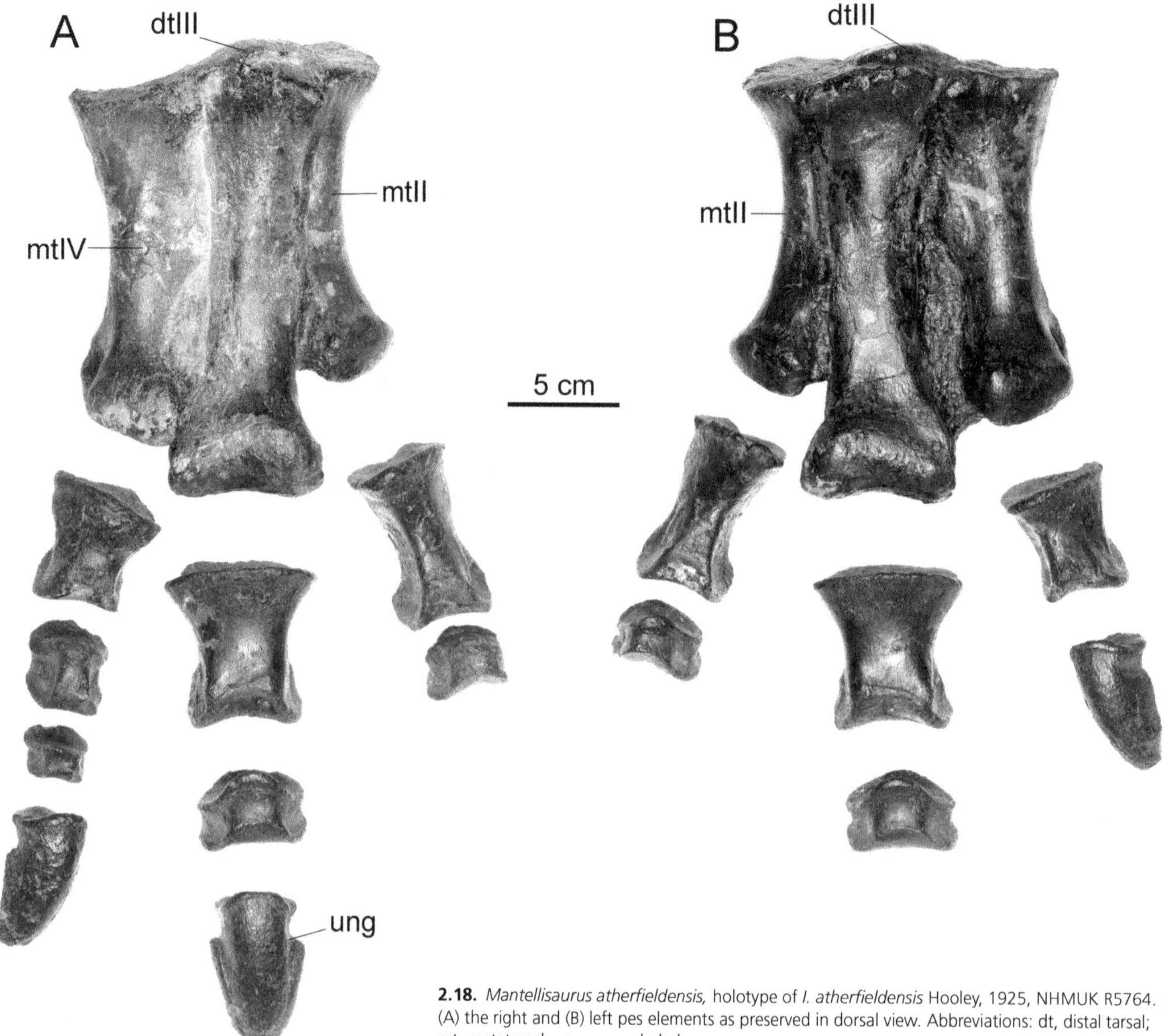

2.18. *Mantellisaurus atherfieldensis,* holotype of *I. atherfieldensis* Hooley, 1925, NHMUK R5764. (A) the right and (B) left pes elements as preserved in dorsal view. Abbreviations: dt, distal tarsal; mt, metatarsal; ung, ungual phalanx.

readily distinguished from those of the contemporaneous taxon *Mantellisaurus.* Subadult remains overlap the size range of *Mantellisaurus* (Norman, 1980), but can generally be distinguished anatomically without too much difficulty.

Craniodental Anatomy The skull of *I. bernissartensis* (Fig. 2.20) has been described in detail (Norman, 1980) and is distinctive in both its proportions and osteology. Compared to that of *Mantellisaurus* (Fig. 2.10) the skull is taller and less elongate rostrally. The lower jaw is deep and robust, parallel sided, and is less arched along its ventral margin. The other notable feature of the skull of *I. bernissartensis* compared to that of *M. atherfieldensis* is the double palpebral. Apart from their generally larger size, maxillary and dentary teeth are very similar in appearance to those described for *Mantellisaurus.*

Vertebrae The large size of the cervical and dorsal vertebrae (Fig. 2.21) are distinctive, and the spines of the dorsals are not as tall, relative to centrum height, as in *Mantellisaurus;* the form of the dorsal centra also shows an exaggerated change in shape from anterior to posterior along the series. Anterior dorsals tend to have comparatively narrow, tall centra, whereas posterior dorsals have centra that are broad and short with everted articular margins; the posterior articular surfaces (initially flat with a slight depression at the center) also become increasingly concave nearer the sacrum. The sacrum typically comprises eight fused vertebrae (including the sacrodorsal) and examples with nine fused caudals (by incorporation of the first caudal) are known (Norman, 1980), and there is typically a broad ventral sulcus on the posterior sacrals. Caudal vertebrae form slightly taller than broad subrectangular bodies

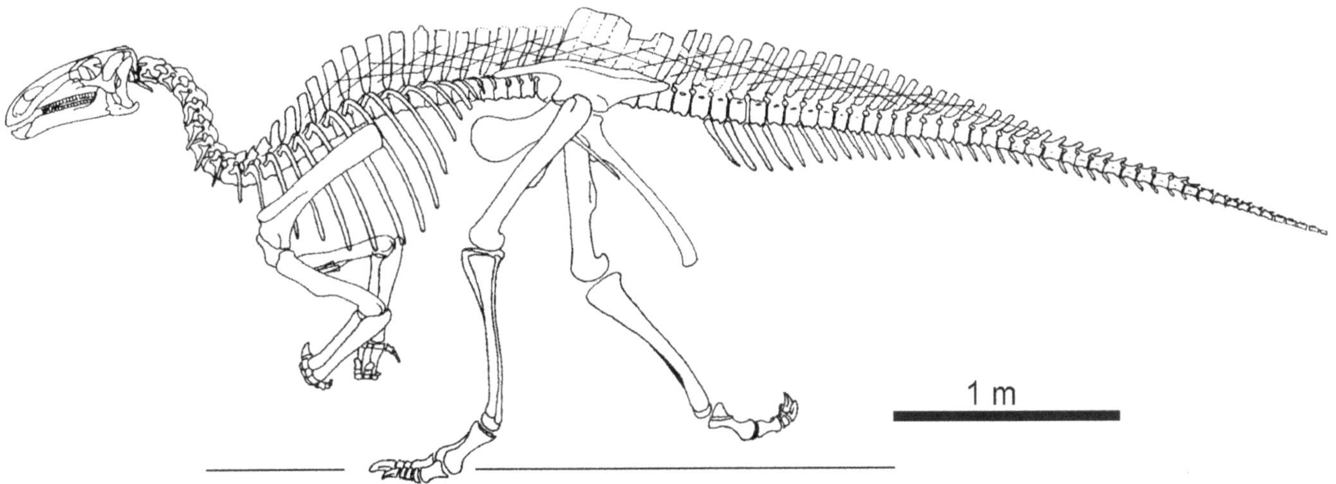

2.19. *Mantellisaurus.* Skeletal reconstruction based primarily upon the articulated, albeit crushed and distorted, skeleton from Bernissart, RBINS R57 (formerly IRSNB 1551 [after Norman, 1986]).

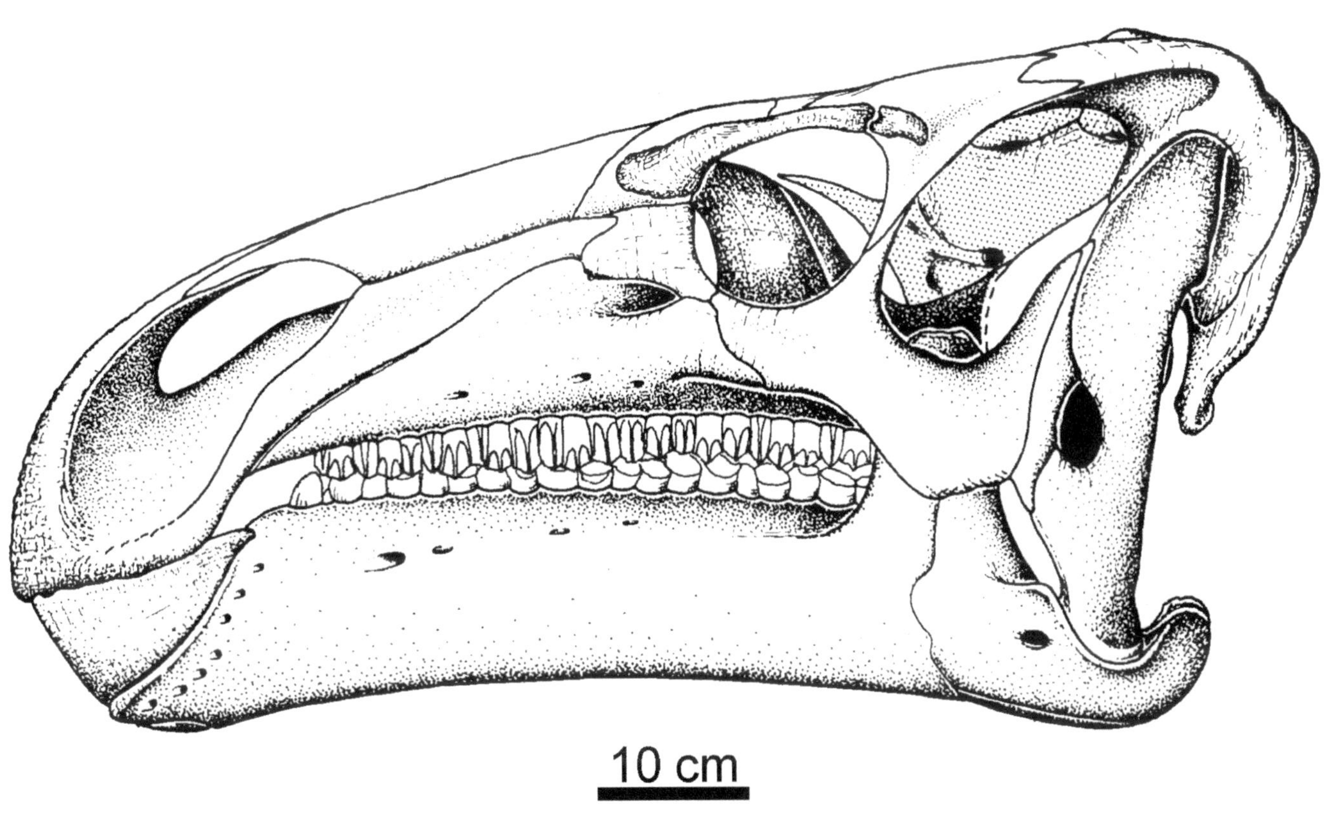

2.20. *Iguanodon bernissartensis* Boulenger, 1881. Skull reconstruction, in lateral view, based on several of the Bernissart specimens (from Norman, 1980:fig. 2).

anteriorly, with prominent horizontal caudal ribs and large chevron facets. More posteriorly in the caudal series, the loss of the caudal ribs and diminution of size often result in a change of centrum morphology from hexagonal cylinders to elongate, round cylinders. Ossified tendons are present as a latticelike array along the sides of the neural spines of the entire dorsal series, across the sacrum, and along the anterior third of the tail; they may also form what appear to be collapsed bundles lying in the longitudinal recess formed between the base of the neural spines and the adjacent transverse processes, either as preservational artifact or reflecting their involvement in different parts of the epaxial musculature.

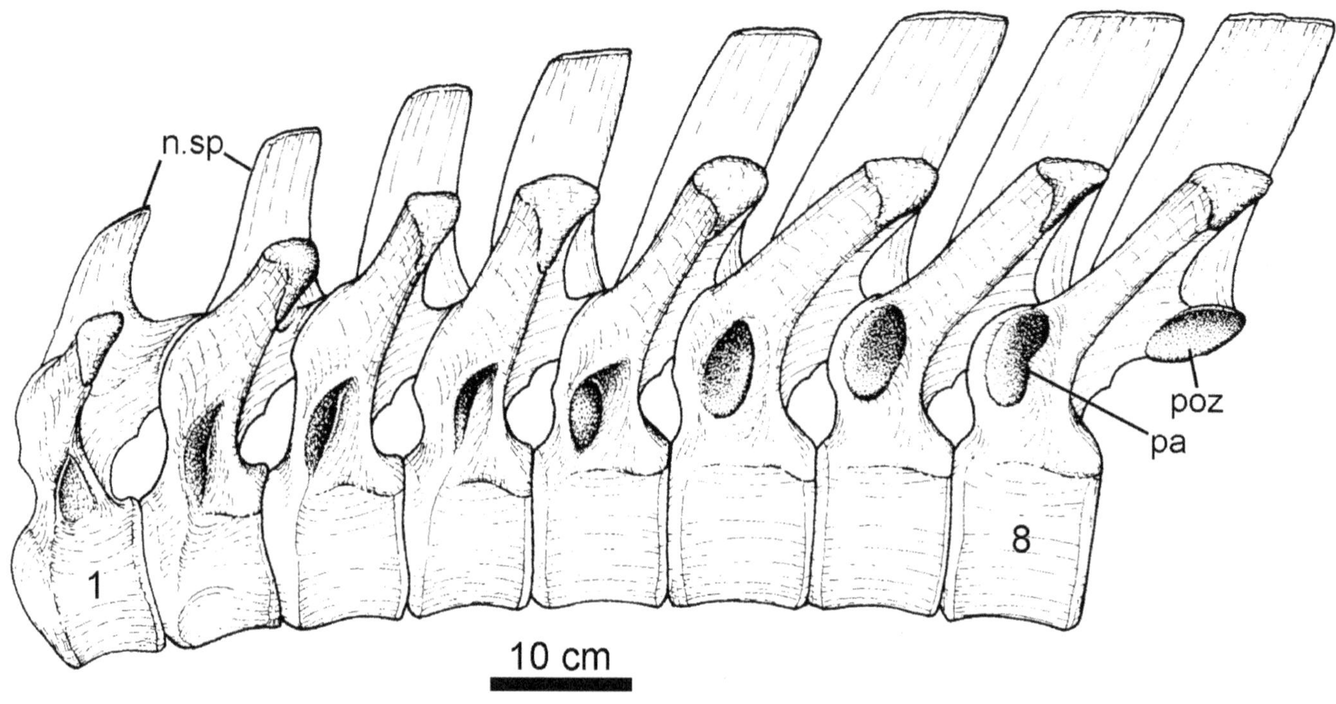

2.21. *Iguanodon bernissartensis.* Anterior dorsal series based upon an articulated series of vertebrae preserved in the Conservatoire Collections of the RBINS [individual "S"]. Abbreviations: 1–8, serial arrangement of dorsals; n.sp, neural spine; pa, parapophysis; poz, posterior zygapophysis (after Norman, 1980).

Girdles and Limbs Size and robustness are key features that distinguish these elements from those of the contemporary *Mantellisaurus.* The scapula tends to be generally less curved along its length and less expanded towards the distal end of the blade than in *Mantellisaurus,* although there is variation in both curvature and distal expansion among the individuals collected at Bernissart (pers. obs., 2009). The coracoid has a well-developed coracoid notch (co.n) rather than the discrete coracoid foramen seen in the much smaller *Mantellisaurus.* The sternal bones are comparatively very large and the handle of the hatchet tends to be more curved; and associated with these in some individuals there is also an unusual, somewhat irregular mass of bony material located in the center of the chest and referred to as an "intersternal ossification" (Norman, 1980:fig. 56). The humerus is very robust and nearly straight rather than strongly sigmoid (obscured by crushing), and has a massively thickened deltopectoral crest. The forearm bones are equally massive and parallel, with very little gap between the shafts of the two bones, which supports the contention that this dinosaur used its forelimbs for walking and body-weight support. The carpals and metacarpal I are fused into a large block (Norman, 1980:fig. 59); the metacarpal has a roller-like articular surface for articulation of the large, conical, and slightly curved pollex. Unlike the condition seen in Hastings Group taxa (*B. dawsoni* and *H. fittoni*), the pollex does not become fused to the carpometacarpal block in mature specimens and remained freely mobile at its base. A small, flattened phalanx is occasionally seen lodged in the base of the pollex ungual. The central metacarpals are more massive, and in proportion shorter, than those of *Mantellisaurus* (Fig. 2.15).

The pelvis (Fig. 2.22) is distinct from that seen in *Mantellisaurus* (Fig. 2.16). The ilium has a very robust, thick, preacetabular process, which is supported by an enlarged medial ridge (note the shape of its cross section in silhouette; Fig. 2.22). The main part of the iliac blade is vertical, but the upper edge is thick and posteriorly it becomes more so–so that it forms a somewhat everted and curved ledge that overhangs the ischiadic peduncle; there is no abrupt inflection along the upper margin of the postacetabular process that characterizes the ilium of *Mantellisaurus.* The dorsal margin of the posterior ilium is elongate and pointed in profile and its ventral surface forms a broad, shallowly vaulted brevis fossa (br.f) bounded laterally by a prominent ridge (Fig. 2.22). The ischiadic peduncle does not exhibit the prominent lateral and stepped expansion typical of all other Wealden iguanodontians. The prepubic process forms an elongate anterior blade that is transversely thick (Fig. 2.22, silhouette), but dorsoventrally narrow along much of its length, before expanding distally; this is distinct from the thinner and deeper blade that is typical of *Mantellisaurus* (Fig. 2.16). The pubic shaft forms a tapering rod that is much shorter than the shaft of the ischium. The ischium has a shaft that is elongate and stout, with a generally rounded, rather than angular, cross section; the shaft is curved along its length (J-shaped) and ends in a prominent, anteriorly expanded "boot." This structure is distinct from the narrower, angular-sided, and far straighter ischial shaft with a small boot that characterizes *Mantellisaurus*.

In the hindlimb the femur is large and very stout compared to that of *Mantellisaurus*; the fourth trochanter is very elongate and forms a very thick crested blade along the postero-interior edge of the mid-shaft. The shin or lower half of the limb is similar in overall shape in the two taxa, but the difference in robustness is notable; this is especially so in the massive construction of the pes. The first metatarsal of the foot in *I. bernissartensis* forms a small, oblique, flattened spatula-shaped splint bone that is distinct from the thin, pencil-like, metatarsal that lies parallel to the shaft of metatarsal II in one articulated example of *Mantellisaurus* from the Isle of Wight.

Reconstruction of Iguanodon bernissartensis

The reconstruction of *I. bernissartensis* presented here is modified from an often-published version created originally by the artist Gregory Paul (e.g., Brett-Surman, 1997; Fig. 2.23). Changes introduced are as follows: the joint between the forelimb and manus has been straightened; mammal-style scapular rotation against the ribcage has been removed (this was physically impossible judged by the anatomy of the pectoral girdle); and a lattice of ossified tendons has been added. The overall impression of an *Iguanodon* in a hurry depicted in the earlier reconstruction is potentially misleading in an animal that attained a body length of 12 m and probably weighed in excess of 5 metric tons.

SYSTEMATIC ANALYSIS

The brief anatomical survey of these Wealden taxa provides an opportunity to align them, systematically, with other reasonably well-known taxa from younger stratigraphic stages, with the intention of exploring their phylogenetic relationships generally, as well as exploring the origin of euhadrosaurs. Numerous phylogenetic hypotheses have been proposed (e.g., Norman 2002, 2004, in press; Weishampel et al., 2003; Horner et al., 2004; Prieto-Márquez, Gaete, et al., 2006; Dalla Vecchia, 2009; McDonald et al., 2010; Prieto-Márquez, 2010; McDonald, 2012b). These analyses have either directly or indirectly addressed the relationships of those animals that are proximate to the clade most commonly referred to as Hadrosauridae (= Euhadrosauria, Weishampel et al. 1993), but have resulted in little general consensus on the relationships of individual taxa.

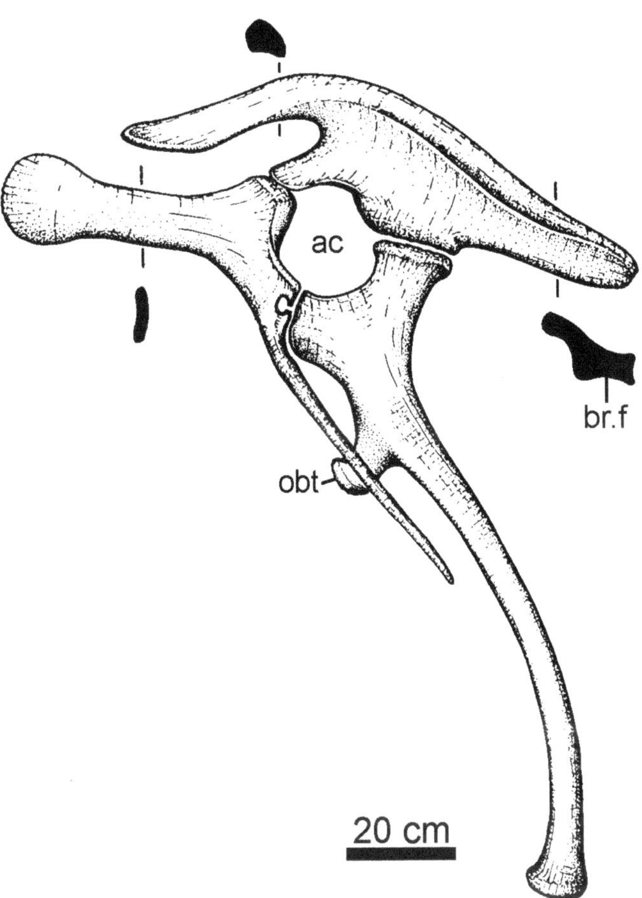

2.22. *Iguanodon bernissartensis* Boulenger. Holotype pelvis partially restored, RBINS R51. Abbreviations: ac, acetabulum; br.f, brevis fossa; obt, obturator process. Cross-sectional views shown as solid black (after Norman, 1980).

Data and Methodological Framework

A new systematic analysis is presented here for the explicit purpose of exploring the nature of putative phylogenetic relationships between a range of more derived non-hadrosaurian ornithopods. Twenty-four taxa (Appendix 2.1) were selected for this analysis because they are known from generally well preserved skulls and/or skeletons, and have been reasonably well described. A significant number of additional taxa have been named in very recent years but these are, on the whole, more fragmentary, and their addition to the analyses materially affects the resolution and stability of tree topology. These have been removed a priori. More comprehensive analyses incorporating these additional taxa, as well as more basal ornithopods, are being considered in more detail elsewhere (Norman, in press). *Lesothosaurus* (based on the descriptions of Thulborn [1970, 1972], and the supplementary information from Sereno [1991]) is used as the outgroup taxon (Butler et al., 2008) in order to polarize character-states. The character-states and their codings are presented in Appendix 2.2, and these comprise a suite of 92 characters that have

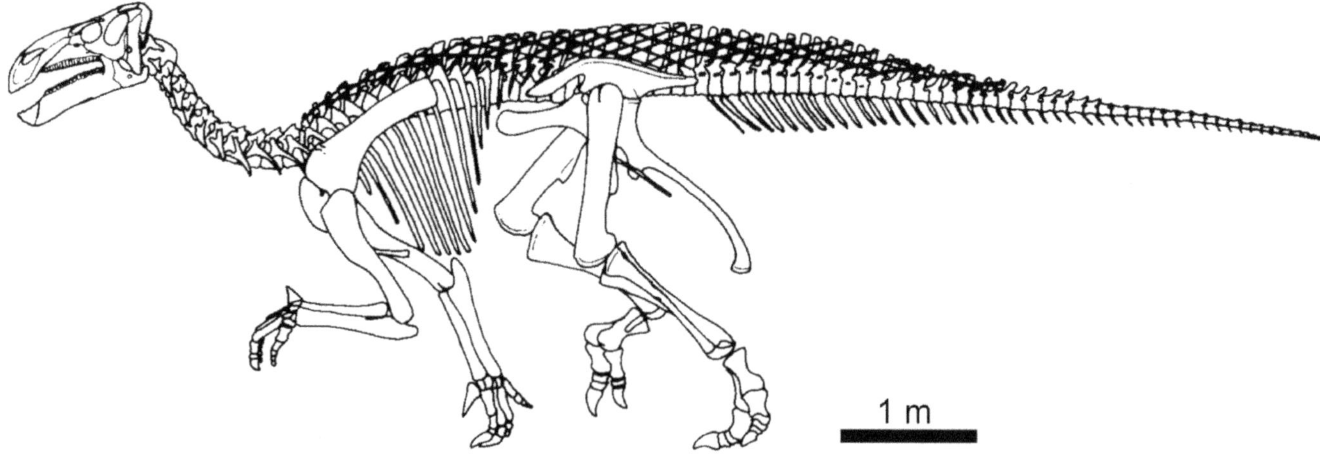

2.23. *Iguanodon bernissartensis.* Reconstruction of the skeleton, modified from an earlier original drawing by Gregory S. Paul (from Brett-Surman, 1997).

been generated after reassessment and revision of previously published character lists (e.g., Norman, 2004; McDonald et al., 2010; Prieto-Márquez, 2010; Wu and Godefroit, 2012). Unusually, compared to the general trend in the literature, the number of characters used is fewer rather than greater. The data matrix was constructed in MacClade 4.06 (Maddison and Maddison, 2003), and the analysis was undertaken using PAUP* 4.0b10 (Swofford, 2002). All characters were equally weighted, and were first analyzed as "unordered" using the Branch and Bound search option (and the analysis was run twice using ACCTRAN and DELTRAN optimizations). Because a substantial number of characters used in this analysis are multistate in nature, a second run was undertaken using the "ordered" character option. The matrix was again analysed using Branch and Bound search option and was run under both ACCTRAN and DELTRAN optimization protocols.

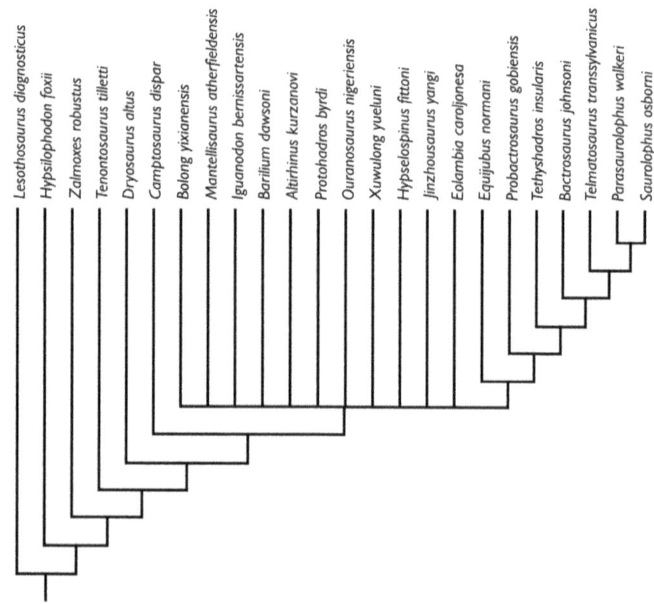

2.24. The strict consensus tree (based on 12 MPTs) generated by running the data matrix with characters unweighted and coded as unordered.

Results

Running the data matrix with characters unordered produced a fairly well resolved topology in the strict consensus tree. Twelve most-parsimonious trees (MPTs) were produced, each with a tree length of 262 steps (Consistency Index [CI]: 0.56, Retention Index [RI]: 0.74). The strict consensus tree based on these analyses is represented in Figure 2.24. The lack of resolution revolves around the character distributions seen in the 11 taxa that fall across the center of the cladogram. The 50% majority rule tree (Fig. 2.25, top) and Adams consensus tree (Fig. 2.25, bottom) derived from the 12 MPTs produce a measure of resolution (as expected) within this taxon plexus. A second search of the data matrix using the same protocols, but with the characters run as "ordered," generated a well-resolved topology represented by three MPTs of 267 steps, a CI of 0.55, and a RI of 0.75. The strict consensus tree shown in Figure 2.26 is identical to the 50% majority rule tree (Fig. 2.25, top) generated from the first analysis. The strict consensus tree reveals an unresolved trichotomy between *Barilium, Iguanodon,* and *Mantellisaurus* within a clade (= Iguanodontidae) containing the additional taxa *Bolong* and *Jinzhousaurus.*

CHARACTER DISTRIBUTION AND PHYLOGENETIC IMPLICATIONS

The degree of resolution seen in the tree generated through this analysis (Fig. 2.26) provides a topology that is available

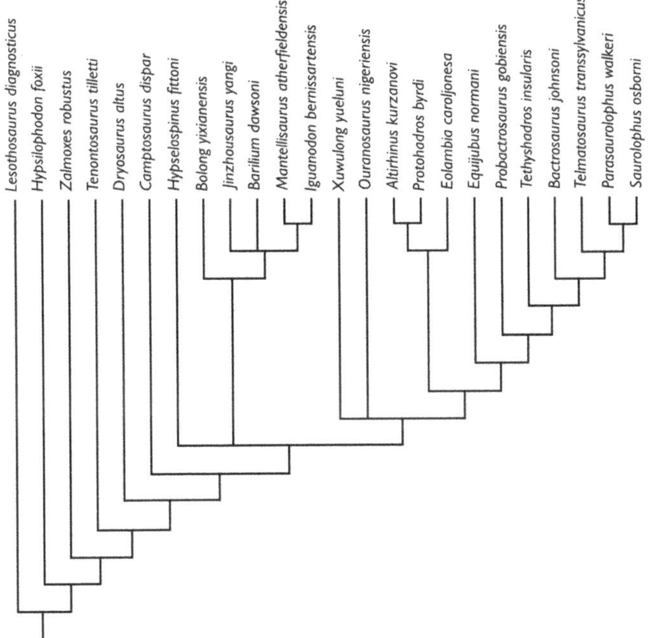

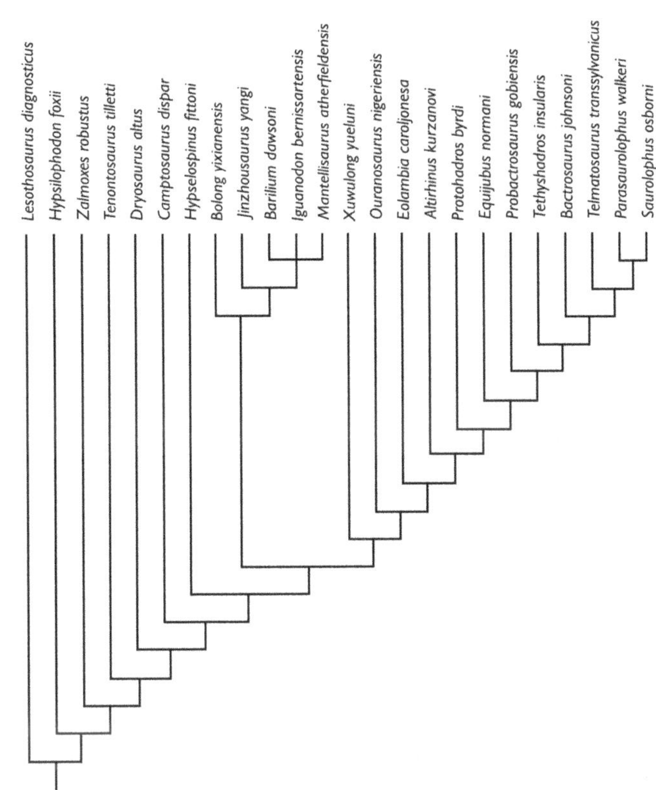

2.25. (Left) Adams consensus tree based upon 12 MPTs generated using characters coded as unordered; (right) The 50% majority rule tree resulting from analysis of the same 12 MPTs.

for morpho-phylogenetic interpretation within this taxonomic subset of the ornithopod lineage, and forms a framework that highlights the anatomical acquisitions that resulted in the attainment of true hadosaurian morphology (represented by the clade Euhadrosauria [sensu Weishampel et al., 1993]).

Basal Ornithopod Taxa

Basal, or "hypsilophodontian-grade" ornithopods sensu lato (Norman et al., 2004; see also Butler et al., 2008), are generally small–medium sized (1–3 m long), bipedal cursors with simple leaf-shaped teeth in both upper and lower jaws; their dental morphology is similar in morphology to that described by Thulborn (1970) and is typical of all basal ornithischians. At the base of the tree, the outgroup is polarized (in this analysis, crudely) against what appear to be more derived taxa in the absence of a consideration of the substantially greater diversity of hypsilophodontian-grade ornithopods (see Butler et al., 2008).

Clypeodonta clade nov.

Phylogenetic Definition The stem-based definition of Clypeodonta ("shield-tooths") is *Parasaurolophus walkeri* and all taxa positioned more closely to *P. walkeri* than to *Thescelosaurus neglectus*. However, the topology of more basal stem taxa falls outside the scope of this account, and Clypeodonta in this analysis occurs at the node that includes *Hypsilophodon foxii*, *Parasaurolophus walkeri*, their common ancestor, and all of its descendants.

Character Acquisition at This Node

ACCTRAN: 19, 20, 45, 47, 50, 52, 53, 54, 55, 56, 59, 60, 65, 79, 80, 85.

DELTRAN: 20, 45, 47, 50, 52, 53, 54, 55, 56, 59, 60, 79, 80.

Condensed Diagnosis

1. Dentary crowns laterally compressed, asymmetrical, and shield shaped in lingual aspect only. The lingual surface of the crown is demarcated from the root by an oblique cingulum.
2. Dentary teeth are curved apicobasally along their length and describe an arc (a convexity lingually) as they emerge from the alveolus during growth.
3. Crowns of dentary and maxillary teeth display an asymmetrical distribution of enamel (dentary crowns have thicker enamel lingually, and maxillary crowns have thicker enamel labially).
4. Dentary crowns bear a prominent primary ridge on the lingual surface that is flanked by a variable number of less prominent subsidiary (accessory, or tertiary) ridges.

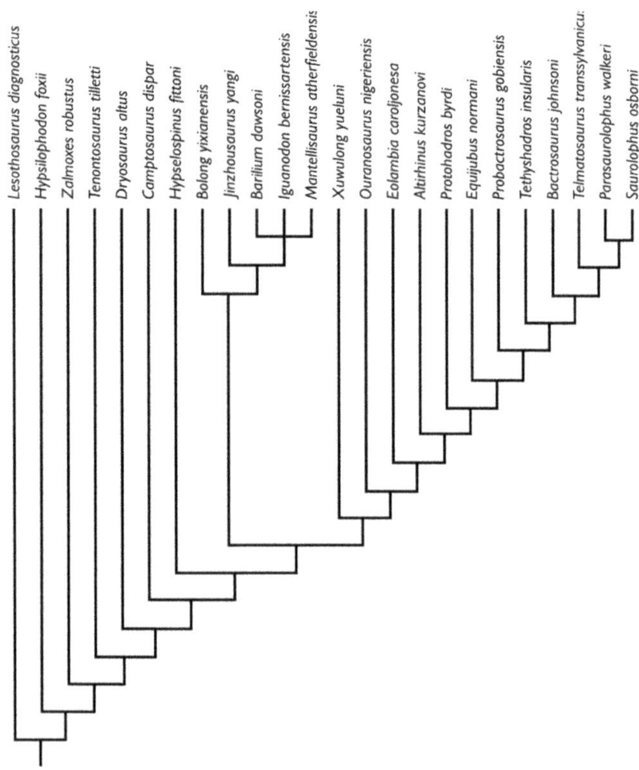

2.26. The strict consensus tree (based upon three MPTs) generated by running the data matrix with characters unweighted and coded as ordered.

5. Maxillary crowns are transversely compressed, asymmetrical and shield shaped, but their thickly enameled labial surface bears a variable number of subsidiary ridges, and a prominent primary ridge is not present.

Comments Clypeodontans (and more basal ornithopods)–*Hypsilophodon*, rhabdodontids (*Zalmoxes* spp., *Rhabdodon, Mochlodon*), and tenontosaurs–are skeletally conservative. However, the specialized modifications seen in clypeodontan teeth exclude a substantial diversity of basal ornithopod taxa–including *Jeholosaurus, Othnielia, Gasparinisaura, Orodromeus, Parksosaurus, Thescelosaurus,* and *Bugenasaura* (Butler et al., 2008). Tenontosaurs exhibit larger size and some graviportal adaptations that converge upon those seen in more derived taxa such as *Camptosaurus*, but the latter taxon and more derived forms exhibit a fundamentally different dental morphology.

Clypeodontan dentary crowns are laterally compressed, are inclined labially, and have denticulate margins; the crowns sit upon curved roots and, as a consequence, the teeth move along a lingual-to-labial arc as they emerge from the dental alveolus, instead of rising vertically from the alveolus as is the case in basal ornithopods, and basal ornithischians more generally. When viewed lingually, dentary crowns are broad and exhibit a clearly defined ("shield-like" surface), and are more thickly enameled than on their labial surfaces (Norman et al., 2004:fig. 18.3). The lingual crown surface exhibits incomplete, oblique ledges (sometimes referred to as a "cingulum") that define a V-shaped junction between crown and root. The enameled crown surface bears a prominent primary ridge flanked by a variable number of much less prominent subsidiary ridges. Maxillary teeth are also transversely compressed, but it is the labial surface of the crown that is more thickly enameled and traversed (apicobasally) by a variable number of low ridges; a primary ridge is not present. The roots of maxillary teeth do not appear to display the lengthwise curvature seen in dentary teeth.

It may also be noted that the singular combination of a prominent primary ridge on the dentary crown and no primary ridge on the maxillary crown may contribute to character combinations that define a more restrictively defined clypeodontan clade–which includes *Rhabdodon, Zalmoxes* spp., *Mochlodon*, and *Tenontosaurus* spp.–and that this clade represents a sister clade to the Dryomorpha. In addition, the unusually specialized basal ornithischians known as heterodontosaurids (Butler et al., 2008) and most notably *Heterodontosaurus tucki* (Norman et al., 2011) homoplasticly exhibit specialized, superficially shield-like dentary and maxillary tooth crown morphologies, with asymmetrical enamel distribution and prominent enamel ridges. However, these teeth are straight rooted, and the detailed structure of these teeth is quite distinct from that described in clypeodontans. Conventional usage of the name Iguanodontia (sensu Sereno, 1986, 2005) is becoming increasingly problematic because it recognizes a clade that includes *Tenontosaurus tilletti* and all taxa positioned closer to hadrosaurs (e.g., *Parasaurolophus walkeri*) and yet excludes *Hypsilophodon* and *Thescelosaurus*. The conventional node Iguanodontia (sensu Sereno, 2005) occurs immediately above the node here named Clypeodonta (see also discussion in Norman, in press).

Dryomorpha

Phylogenetic Definition This node-based clade is defined as *Parasaurolophus walkeri, Dryosaurus altus*, their common ancestor, and all of its descendants.

Character Acquisition at This Node

ACCTRAN: 4, 20, 53, 54, 58, 59, 61, 65, 68, 81, 85, 89, 91.

DELTRAN: 12, 20 33, 53, 54, 58, 59, 82, 83, 85.

Condensed Diagnosis

1. Maxillary tooth crowns are apicobasally elongate and characterized by the possession of a narrow, elevated primary ridge on the labial surface of the crown that is positioned slightly distal to the midline.
2. Dentary tooth crowns exhibit a clearly defined, but low, primary ridge that is offset distally to the midline.
3. In lingual view a shoulder-like edge is formed between the denticle that marks the tip of the primary ridge and the point of inflection that marks the start of the subvertical mesial margin of the crown.
4. The quadrate bears a well-defined semicircular notch on the central portion of the anterior margin of the jugal wing that is spanned anteriorly by the quadratojugal. With the quadratojugal in articulation a fully enclosed quadrate (paraquadratic) foramen is formed.
5. Predentary with divergent, ventrolateral processes.
6. The obturator process of the ischium is positioned proximally (separated by an embayment from the pubic peduncle of the ischium) on the anteromedial edge of the shaft of the ischium.
7. The ischial shaft has a subcircular cross section along almost its entire length and there is a small anterior expansion of the distal tip of the ischial shaft that forms a "boot."
8. The femoral shaft displays an anterior intercondylar groove that is defined by a pair of ridges that form on the adjacent edges of the dorsal part of the distal articular condyles.

Comments There is a marked contrast between the style of morphological differentiation of the dentition in basal clypeodontans and that in dryomorphans, which suggests that a new morpho-functional trajectory (probably linked to oral food processing) had emerged. Additional characters, such as the development of a bilobate posteroventral processes on the predentary, probably served to reinforce a structurally weak dentary symphysis and may reflect alterations to jaw function. A ventrally bilobate predentary also appears sporadically in basal clypeodonts (rhabdodontids), but this is interpreted as an example of convergence (homoplasy) in the larger-bodied tenontosaurs and the very late appearing (Maastrichtian) rhabdodontids. The ischial characters (such as the positioning of the obturator process and curvature of the ischial shaft) also appear sporadically and probably convergently in Maastrichtian rhabdodontids.

Ankylopollexia

Phylogenetic Definition This node-based clade (Fig. 2.26) can be defined as *Parasaurolophus walkeri*, *Camptosaurus dispar*, their common ancestor, and all of its descendants.

Character Acquisition at This Node

ACCTRAN: 2, 15, 46, 51, 57, 69, 72.

DELTRAN: 2, 4, 15, 30, 45, 51, 57, 64, 68, 69, 72.

Condensed Diagnosis

1. Conical pollex ungual.
2. Short, block-shaped metacarpal I is sutured at an oblique angle against the radiale and carpal 2.
3. Carpo-metacarpal block formed by suturing of the individual elements.
4. Manus digit IV bears an ungual phalanx that is small and shows no obvious grooves for attachment and growth of a claw or hoof.
5. Occlusal plane of the premaxilla and predentary deflected ventrally.
6. Ventral margin of the jugal is arched and forms a prominently angled projection posteriorly.
7. Acromion process of the scapula curved anteriorly at its proximal end.

Comment It is unfortunate that the structure of the carpus and manus is presently unknown among dryosaurs, since this might have a significant bearing on the status of the ankylopollexian clade.

Neoiguanodontia ("New Iguanodonts")

Phylogenetic Definition The node-based clade can be defined as *Parasaurolophus walkeri*, *Hypselospinus fittoni*, their common ancestor, and all of its descendants.

Character Acquisition at This Node

ACCTRAN: 9, 13, 27, 34, 40, 48, 61, 62, 66, 67, 70, 73, 80, 81, 87, 88, 89, 90, 92.

DELTRAN: 40, 48, 62, 66, 67, 70, 73, 80, 81, 87, 88, 89, 90, 92.

Condensed Diagnosis

1. The marginal denticles on the mesial and distal margins of maxillary and dentary tooth crowns form curved ledges that are ornamented with mammillae/papillae.

2. The posterodistal corner of dentary crowns, when viewed lingually, exhibit inrolling of the marginal denticulate edge to form an oblique ledge-like structure (sometimes referred to as a cingulum).
3. Posterior dentary dentition extends medial to the coronoid process, and from which the dentition is separated by a narrow, horizontal ledge that represents a posterior extension of the lateral cheek recess.
4. Coronoid process elevated and its axis lies perpendicular to the long axis of the dentary ramus.
5. Antorbital fenestra (and fossa) reduced in size and forms an oblique channel between the maxilla and lacrimal.
6. The articulation between the jugal and maxilla comprises an oblique, finger-like posterolateral projection from the maxilla that fits into a complementary elongate slot on the anteroventral surface of the jugal.
7. The supraoccipital is excluded from the posterodorsal margin of the foramen magnum by a shelf-like structure formed by dorsomedial processes of the exoccipitals that meet in the midline.
8. Sternal bones develop an oblique, posterolateral, rod-like extension that ends in an articular boss.
9. Metacarpals II–IV of the manus are elongate, bundled together and held in place by development of collateral ligaments that are sometimes ossified.
10. Manus unguals II and III are dorsoventrally flattened, asymmetrical and generally hoof-like.
11. Posterior pubic ramus is slender and significantly shorter than the shaft of the ischium (this character is homoplastic in rhabdodontids and tenontosaurids).
12. Pedal ungual phalanges are dorsoventrally flattened and bluntly truncated distally, but retain well-developed claw grooves bilaterally.

Comments The position of *Hypselospinus* with respect to the sister clade comprising *Bolong, Jinzhousaurus, Barilium, Iguanodon,* and *Mantellisaurus* (= Iguanodontidae) and Hadrosauriodea needs to be more accurately determined (Norman, in press). It is also clear, following the systematic review above that the anatomies of the contemporary taxa *Bolong* and *Jinzhousaurus* are very similar. A range (plexus, as in Figure 2.24) of neoiguanodontian taxa forms a cluster between *Hypselospinus* and Hadrosauromorpha. These taxa have proved difficult to arrange consistently in any published phylogeny, and this may well reflect anatomical incompleteness and/or a phase of comparatively rapid evolution (and possibly disparate character acquisition) among and between ornithopods during the early Late Cretaceous.

Hadrosauromorpha clade nov.

Phylogenetic Definition This stem-based clade is defined as *Parasaurolophus walkeri* and all taxa positioned more closely to *P. walkeri* than to *Probactrosaurus gobiensis.*

Character Acquisition between *Probactrosaurus* and the Succeeding Node

ACCTRAN: 1, 4, 16, 20, 41, 43, 48, 54, 55, 56, 60, 68, 69, 71, 72, 77, 78, 87, 88.

DELTRAN: 1, 9, 10, 19, 20, 41, 43, 48, 54, 55, 68, 69, 71, 72, 77, 87.

Condensed Diagnosis

1. Dentary dentition with small, diamond-shaped crowns integrated into a closely packed dental magazine.
2. Dentary and maxillary crowns bear a single median primary ridge (carina) on their labial or lingual surfaces, respectively.
3. Surangular foramen absent.
4. A shallow embayment in the jugal wing of the quadrate to accommodate a disc-shaped quadratojugal that completely occludes the quadrate (paraquadratic) foramen.
5. Modification of the acromion process to form a pendant promontory orientated along the long-axis of the scapular blade.
6. Carpus represented by no more than two small discoidal elements.
7. Phalanges of digit I of the manus absent.
8. Metacarpal I absent.
9. Ilium has a region of the dorsal margin, posterodorsal to the ischiadic peduncle, that forms a discrete everted lip (pendule, see below) that overhangs the ischiadic peduncle of the ilium.
10. Postacetabular ramus of the ilium forms a laterally flattened bar-like structure.
11. Pedal ungual phalanges are strongly dorsoventrally compressed, short proximodistally, and very broad with a rounded anterior margin; distinct claw grooves on the lateral margins are absent, and are therefore truly hoof-shaped when compared to those seen in more basal taxa.

Note here that the term "pendule" is suggested for the everted and downturned tongue-shaped structure that projects lateroventrally from the dorsal margin of the ilium. This word is proposed for two reasons. First, it obviates the need for repeated complex descriptive phraseology to describe the feature. Second, this structure has been mistakenly referred to as an "antitrochanter" (e.g., Lull and Wright, 1942). The true antitrochanter is an articular facet located on the posterior margin of the acetabulum and is not topographically similar or anatomically homologous to the development of the dorsal margin of the ilium. Similarly, the term "supra-acetabular crest" has been proposed more recently for this structure (Prieto-Márquez, 2011); unfortunately this latter usage creates a homonym with the anatomically distinct supra-acetabular crest (a specialized feature derived from the supra-acetabular buttress of tetrapods [as per Romer, 1956]) that forms a prominent ridge located on the acetabular margin of the ilium in derived archosaurs (e.g., Charig, 1972).

Comments The sister taxon to this clade, *Probactrosaurus gobiensis*, displays important anatomical differences that distinguish this and all non-hadrosauromorphan taxa from hadrosauromorphans: dentary crowns are comparatively small, and almost diamond-shaped, but they retain an asymmetrical aspect when viewed lingually, and bear accessory ridges running parallel to the distally offset primary ridge. In addition, a surangular foramen is present, and the quadrate has a semicircular embayment in the jugal wing rather than the shallow embayment seen in *Tethyshadros* (Dalla Vecchia, 2009) and more derived hadrosauromorphans. The acromion process of the scapula curves forward (rather than forming an overhanging promontory-like structure that lies parallel to the main axis of the scapular blade as in hadrosauromorphans).

In the forelimb, the radius, ulna, and metacarpals are elongate and slender in *Probactrosaurus* and hadrosauromorphans, which is suggestive of a general trend toward gracility in this part of the skeleton. However, *Probactrosaurus* retains a small, spike-like pollex, which *implies* the presence of an at least partially competent and ossified carpus (Norman, 2002). The ilium of *Probactrosaurus* has a modestly everted dorsal margin (Norman, 2002), but there is no evidence of either strong eversion or development of a tongue-shaped pendule, as described in *Tethyshadros*. The unguals of the pes of *Probactrosaurus* are comparatively elongate and truncated at their tips (Norman, 2002).

Bactrosaurus and *Telmatosaurus* retain a consistent topology in many different analyses (Prieto-Márquez, 2010, 2011; Wang et al., 2010; McDonald, 2012b; Wu and Godefroit, 2012), as, respectively, successive outgroup taxa to the well-established node-based clade (Euhadrosauria) represented in this analysis by the sister taxa *Parasaurolophus* and *Saurolophus*.

Euhadrosauria

By tradition, these latter taxa are placed within the clade Hadrosauridae. The clade Euhadrosauria was originally proposed by Weishampel et al. (1993) and was diagnosed at that time on the basis of five unambiguous and two ambiguous synapomorphies. This node-based clade may be defined using phylogenetic nomenclature protocol as *Saurolophus* and *Parasaurolophus*, their most common ancestor, and all of its descendants.

Abbreviated Diagnosis

1. Loss of all but the primary ridge on dentary and maxillary tooth crowns.
2. Ilium with its posterodorsal margin everted and ventrally deflected forming a pendule positioned dorsal to the posteroventral margin of the lateral expansion of the ischiadic peduncle.
3. Pendule on the ilium anteroposteriorly shorter than deep.
4. No lateral expansion of the ischiadic peduncle, so that this area is flush with the lateral wall of the ilium.
5. See Prieto-Márquez (2010:457–461) for an extended consideration of the character states that may be used to diagnose Hadrosauridae (= Euhadrosauria, this account).

Comments Prieto-Márquez (2010:456) argued that the clade Euhadrosauria was not "defined" (the implication being that it lacked a phylogenetic definition, although it should be recognized that such definitions did not come into practice until later than 1993) and he thus considered it to be ambiguous; he also objected to its usage because he claimed that it violated recommendations in the ICZN (1999) concerning the naming of "family group" taxa by not having as its root the name of a nominal taxon. Invoking the ICZN "family group" concept to derived iguanodontians implies a degree of stability of taxon relationships that is not consistent with the type of accumulative science practiced by paleontologists. The recent history of systematic evaluations of hadrosaurians and their near relatives typifies this problem. Consistency is not the general rule, even though it is an obvious aspiration (cf., Horner, 1985, 1990, 1992; Sereno, 1986, 1998; Weishampel and Horner, 1990; Weishampel et al., 1993; Head, 1998; Norman, 2002, 2004; Horner et al., 2004; Prieto-Márquez, Weishampel, and Horner, 2006; Gates and Sampson, 2007; Evans and Reisz, 2007; Godefroit et al., 2008; Dalla Vecchia, 2009;

Evans, 2010; Prieto-Márquez, 2010, 2011; Wang et al., 2010; McDonald, 2012b; Wu and Godefroit, 2012).

Prieto-Márquez's suggestion that the stem of the family group name (Euhadrosauria) does not derive from the nominal taxon name *Hadrosaurus*, is at best specious: the origin of the name is self-evident and the use of the prefix serves a specific role by polarizing the nomenclature in order to reflect one aspect of tree topology. It can also be observed that a logical extension of Prieto-Márquez's concerns about the validity of the family group name Hadrosauridae is that the nominal taxon Hadrosauridae is based upon a taxon that has in the past been considered a nomen dubium (Prieto-Márquez, Weishampel, and Horner, 2006), whose position within any topology is probabilistic (Prieto-Márquez, 2010), and could be considered sedis mutabilis–especially considering the fragmentary and incomplete nature of its remains.

More recently Prieto-Márquez (2011) has contradicted previous work by proposing that *Hadrosaurus foulkii* is indeed a valid taxon, which he assigns (on its own) to the subfamily Hadrosaurinae (as in Prieto-Márquez, 2010). In addition he now defines the node-based clade Hadrosauridae as "the clade stemming from the the most recent common ancestor of *Hadrosaurus foulkii* and *Parasaurolophus walkeri*" (Prieto-Márquez, 2011:67). His clade Saurolophidae (= Euhadrosauria in this account) is given a node-based definition: "the last common ancestor of *Saurolophus osborni, Lamebosaurus lambei*, and all its descendants" (Prieto-Márquez, 2011:67; it is not made explicit why he has abandoned the use of *Parasaurolophus* in favour of *Lambeosaurus* as one specifier in this instance) and he concludes with the proposition that this latter clade (Saurolophidae = Euhadrosauria in this account) includes "the two major hadrosaurid clades: Saurolophinae and Lambeosaurinae" (Prieto-Márquez, 2011:67). Phylogenetic definition of the taxonomic scheme proposed by Weishampel et al. (1993) with respect to Hadrosauridae (the least inclusive, node-based clade containing *Telmatosaurus transsylvanicus* and *Parasaurolophus walkeri*) and Euhadrosauria (see above) solves a number of problems put forward by Prieto-Márquez (2011), and necessitates only adopting Saurolophinae for the clade traditionally recognized as Hadrosaurinae. However, if future phylogenetic analyses recover *Hadrosaurus foulkii* as a member of this clade, Hadrosaurinae would have precedence.

A DISCUSSION: THE ORIGIN OF HADROSAURS

The topology generated by this analysis provides an interpretative framework for exploring the morphological transition from derived neoiguanodontian ornithopod to euhadrosaurian and can be compared directly to those that have been produced more recently.

Topological Variation

Prieto-Márquez (2010) offered a detailed analysis of the systematics of hadrosaurians and included a consideration of a few of the more basal forms considered herein (Fig. 2.27). The stem-based clade Hadrosauroidea was redefined as comprising *Hadrosaurus foulkii* and all taxa more closely related to it than to *Iguanodon bernissartensis*, including Hadrosaurinae (represented solely by *H. foulkii*) and Saurolophidae (Prieto-Márquez, 2011:67). *Equijubus* and *Probactrosaurus* occupy significantly more basal positions in this topology than in the resolved tree generated by the latest analyses (compare Figs. 2.26 and 2.30). However, the relative positions of *Eolambia, Protohadros*, and *Bactrosaurus* are topologically more consistent.

McDonald (2012b) produced an alternative tree depicting ornithopod relationships (Fig. 2.28), which incorporated a greater proportion of the taxa considered herein and is therefore more truly comparable to the one generated in this account (Figs. 2.26, 2.30). The topology shows substantial similarity to that which has been proposed in this account. Basal taxa appear with the same topology as seen in Figure 2.30, and all the topological variation is to be found within a range of what might be termed "intermediate" taxa leading to *Tethyshadros* (in this particular instance *Altirhinus, Equijubus, Xuwulong, Eolambia, Probactrosaurus*, and *Protohadros*). Above *Tethyshadros*, the topology of the two trees is again identical. The node-group names follow conventional usage and amplify those outlined by Prieto-Márquez (2010, 2011). It should also be noted that Hadrosauroidea has a different composition to that suggested by Prieto-Márquez (compare Fig. 2.27).

Wu and Godefroit (2012) provided another analysis that is pertinent to this review (Fig. 2.29). Basal relationships are consistent with those shown in Figure 2.26. *Iguanodon* and *Mantellisaurus* plus *Ouranosaurus* are recognized as a clade (Iguanodontidae), although this node is weakly supported. More derived taxa show differences in relative position, the most notable of which is the comparatively basal positions of *Equijubus* and *Probactrosaurus*. Clade names were placed at nodes, and it is notable that Hadrosauridae incorporates *Bactrosaurus* as well as *Tethyshadros, Telmatosaurus*, and Euhadrosauria, and that the latter conforms to the concept of this clade created by Weishampel et al. (1993).

The latest contribution to this on-going debate focused on hadrosaur origins is presented in summary form in Figure 2.30. Although this topology was resolved from three MPTs, the topology is not entirely unambiguous. The clade

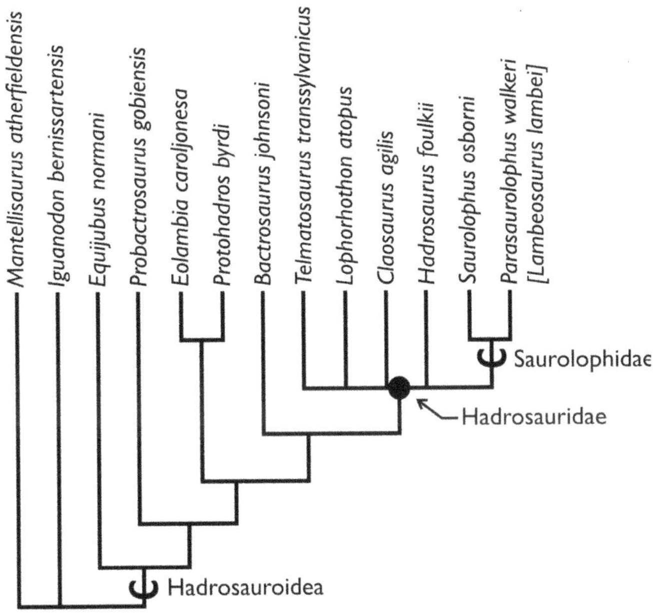

2.27. Cladogram generated by the analysis of Prieto-Márquez (2010, 2011). Modified in order to present the topology that reflects (with a few exceptions) the taxa considered in this account. Clade names and positions indicated as in the original version.

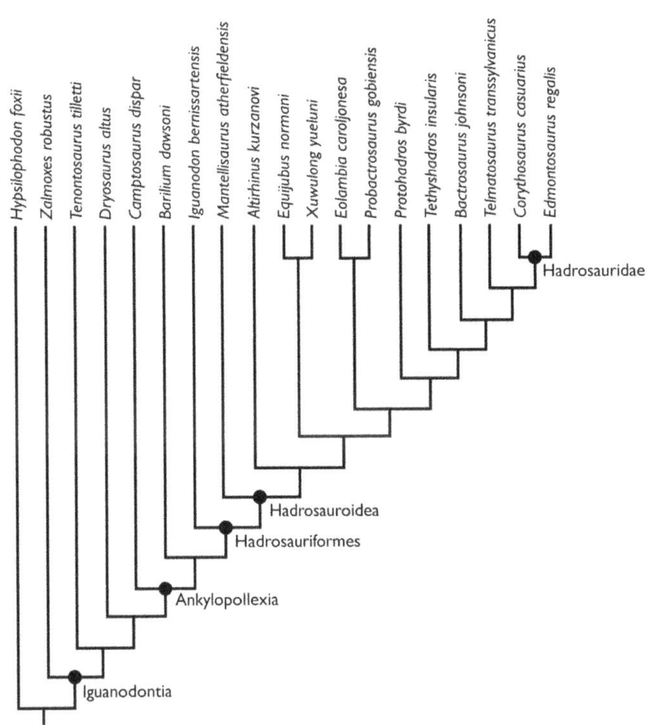

2.28. Cladogram generated by the analysis of McDonald (2012b). Modified in order to present the topology that reflects the taxa considered in this account. Clade names and positions indicated as in the original version.

Clypeodonta is recognized in order to differentiate those ornithopods that specialize their teeth in order to create an integrated dentition of distinctively shield-shaped teeth, from those that retain a simpler set of non-imbricating, straight, leaf-shaped teeth that do not develop interdentally continuous wear surfaces. Basal clypeodonts (hypsilophodonts, rhabdodonts, and tenontosaurs) have a distinct set of tooth morphologies that may eventually prove to unite them into a sister clade to more derived dryomorphans, although that is not recovered here. The node-based clade Neoiguanodontia recognizes a range of taxa that demonstrate the acquisition of the definitive *Iguanodon*-like tooth morphology (notably mammilate denticles along mesial and distal crown margins; dentary crowns with a shouldered crown edge, inrolled enameled ledges, and crown subdivision by primary and secondary ridges) as well as a subsidiary set of postcranial features. Five taxa (*Bolong, Jinzhousaurus, Barilium, Iguanodon,* and *Mantellisaurus*) form a relatively poorly supported clade (= Iguanodontidae), but this is not formally defined at this stage. Succeeding taxa demonstrate the gradual acquisition of hadrosaur-like anatomical features, and this inconsistency reflects the wide geographic spread of these taxa and the difficulties inherent in trying to resolve phylogenetic relationships among groups of taxa that were evidently evolving relatively rapidly during the early Late Cretaceous. This evolutionary "plexus" (see Fig. 2.24) is succeeded by more derived taxa such as *Probactrosaurus* that exhibit a number of hadrosaur-like characteristics. Here, the term Hadrosauromorpha encompasses taxa that are more derived than the pollex-spike-bearing *Probactrosaurus.* Hadrosauromorpha comprise *Tethyshadros, Bactrosaurus,* and *Telmatosaurus,* as well as the potential non-euhadrosaurians (non-saurolophids sensu Prieto-Márquez, 2010, 2011) such as *Hadrosaurus foulkii, Claosaurus agilis,* and *Lophorhothon atopus.*

An Evolutionary Narrative of Hadrosaur Origins

Clypeodonta Clypeodontans exhibit a unique combination of features involved in the differentiation of the dentary and maxillary dentitions (partially integrated opposing occlusal surfaces and the creation of incipient dental magazines of smaller teeth). Dryomorphans link these dental modifications with others that stabilize the anterior end of the lower jaw (the ventrally bilobed predentary). The restructuring of the pelvis, notably through the development of a structurally dominant, J-shaped ischium (functionally replacing the elongate, but slender, posterior pubic ramus) suggests the need to support a more massive and by implication more complex gut, which reflects the increasing dietary sophistication implied from the structure of the jaws and teeth. Furthermore, the specialization of the knee joint, through the development of a defined anterior intercondylar groove, may

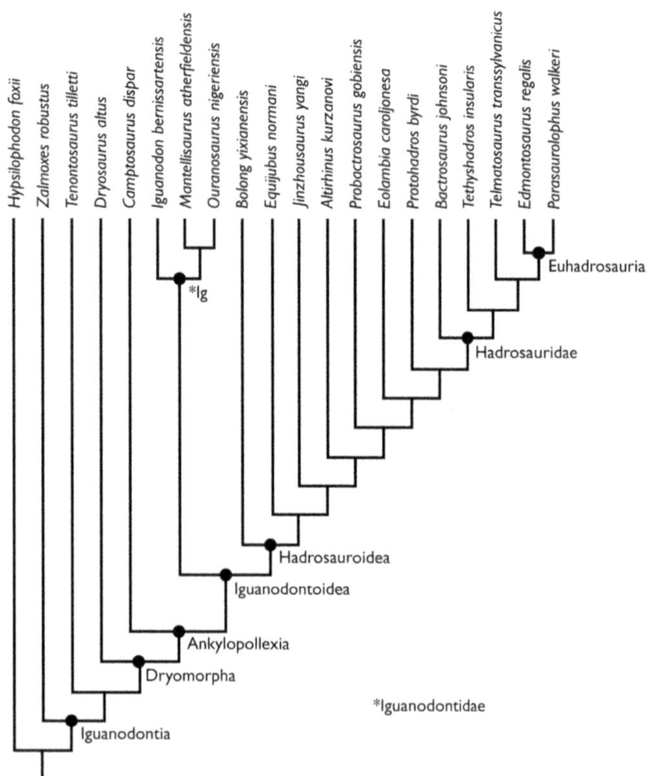

2.29. Cladogram generated by the analysis of Wu and Godefroit (2012). Modified in order to present the topology that reflects the taxa considered in this account. Clade names and positions indicated as in the original version.

be correlated with increasing strength of the joint (to cope with the more massive gut) and at the same time maintaining or even improving joint mechanics and locomotor ability.

Neoiguanodontia Neoiguanodontians are notable for the development of larger and longer (jaw-dominated) skulls with larger and increasingly structurally sophisticated teeth. Teeth form incipient dental magazines that show increasing signs of integration into mutually supportive arrangements, creating the equivalent of single functional megatooth in each jaw. The leverage exerted by the jaw muscles is augmented by the development of a tall, perpendicular coronoid process, and there are a number of lineages that appear to experiment with the functional separation and specialization of food gathering (the premaxillary and predentary beak) and food processing (cheek teeth) through the development of a diastema. It is very notable that some of the more derived taxa tend to develop a broader, down-turned snout (*Altirhinus, Protohadros, Probactrosaurus*), which anticipates a feature seen among euhadrosaurs and suggests that these animals were functionally modifying their food gathering abilities. Whether this trend can be linked to vegetational changes is uncertain at present. The skeletons of neoiguanodontians are generally large (6–12 m in body length), their guts are clearly very large and heavy, and their limbs become increasingly stout and graviportally adapted. Linked to the generally pedestrian build of their bodies, these forms develop a specialized spike-like pollex that is anchored to an enlarged and fused, or partially fused, set of carpometacarpals. These anatomical features correlate with stoutly constructed pectoral and forelimb skeletons that were evidently adapted for weight support and locomotion. The pollex may have had multiple functions, but as a potential stilleto-like weapon of defense, wielded at close quarters, this may well have proved advantageous in animals that lacked obvious cursorial abilities.

Hadrosauromorpha Rather unexpectedly, the taxa that form successional sister taxa to the clade Euhadrosauria exhibit relatively conservative anatomies, insofar as their skulls are concerned. Within their jaws, the fully integrated and interdentally cemented dentary magazine (megatooth) with interlocked, diamond-shaped dentary teeth and multiple successional crowns is firmly established. The coronoid process is both tall and expanded apically, suggesting that both the volume of musculature that can be recruited to jaw closure and its lever-arm mechanics have been augmented. The dental magazine also begins to migrate distally along the jaw, medial to the coronoid process. Counterintuitively, given the recognition (above) of a diffuse trend focusing upon elaboration of the shape and proportions of the lower jaw in derived neoiguanodontians, there is little evidence for elongation, deepening, or curvature of the dentary. The diastema is comparatively abbreviated and the specialized functional separation between the cropping (beak) and food processing (dental battery) regions in the jaw that had become apparent in (for example) *Mantellisaurus, Ouranosaurus, Altirhinus,* and *Protohadros* is far less evident.

The most derived non-hadrosauromorphan neoiguanodontian in this analysis (*Probactrosaurus*) exhibits the onset of a trend that culminates in the loss of both the substantial carpometacarpus and manus digit I (the latter taxon being still characterized by its offset, small conical pollex ungual). As a functionally linked consequence, hadrosauromorphans show a reduction in robustness of the antebrachium and manus, which had previously been associated with the specialized locomotor and weight-supporting structures, consistently seen in more basally positioned neoiguanodontians.

No hadrosauromorphan exhibits a pollex spine or the heavily co-ossified carpus (even though modest-sized hoof-like unguals are present on digits II and III of the manus), and the forelimb and manus when considered as a whole present comparatively gracile proportions combining a short, sinuous humerus with elongation and slenderness in the more distal elements (antebrachium and metacarpals). The pectoral girdle shows a reduction in the size of the coracoid and modification to the structure of the scapular

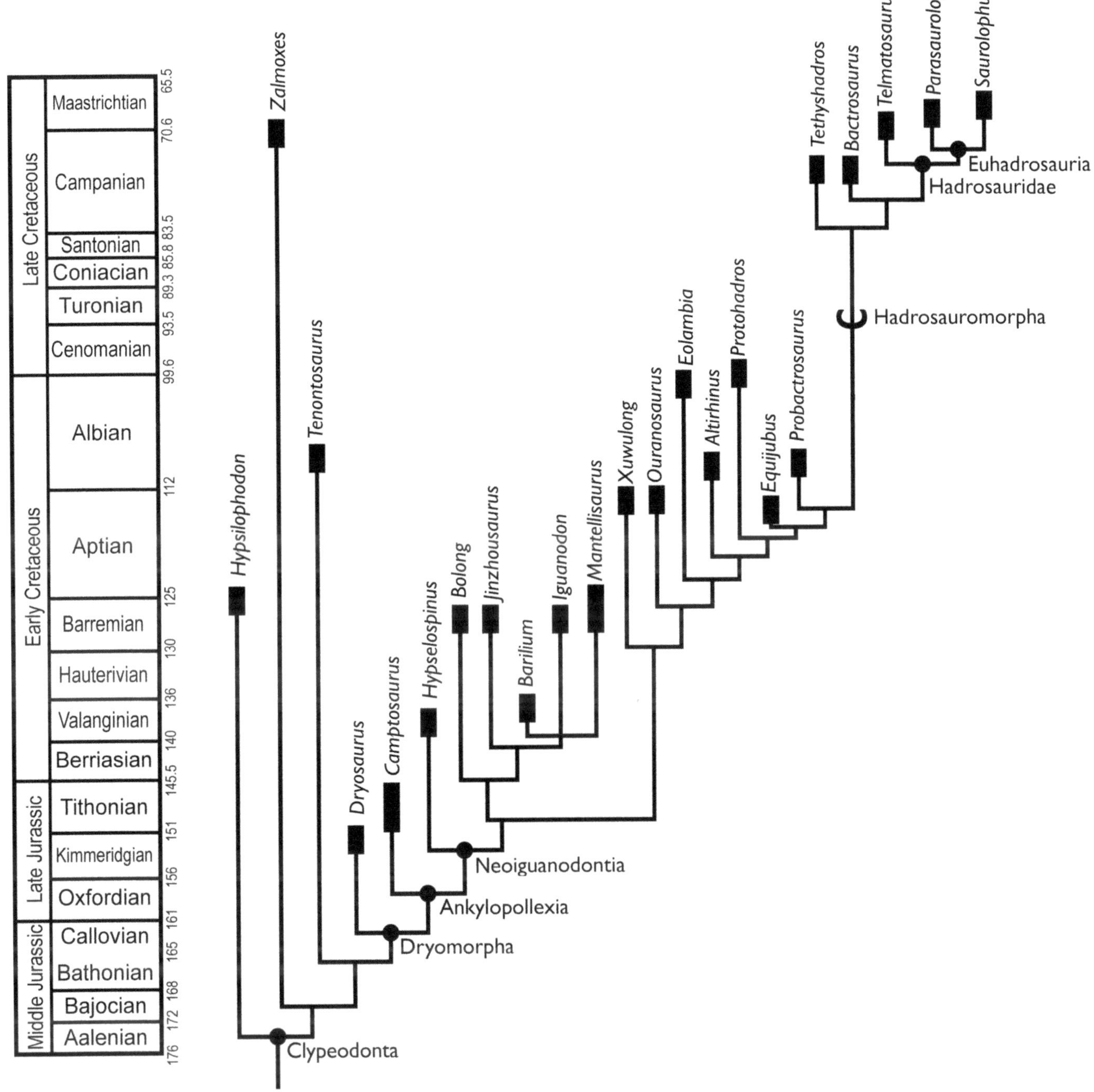

2.30. Summary cladogram calibrated against the geological timescale (Ma ages listed along right side of timescale). New clade names discussed in the text.

acromion that appears to correlate with the gracility of the forelimb. The pelvis is modified primarily by the elaboration of the dorsal margin of the ilium to create a discrete pendule and a postacetabular process that is bar-like and laterally compressed. Since the dorsal margin of the ilium and the brevis shelf of the postacetabular process are areas for the origin of significant hindlimb musculature (Norman, 1986; Maidment et al., this volume), these anatomical changes are suggestive of changes in hindlimb functionality. The hindlimb is characterized by a columnar (straight) femoral shaft, a globular femoral head (which lacks the femoral head notch seen in more basal forms), and a completely enclosed (tunnel-like) anterior intercondylar groove. The pedal unguals become short and remarkably hoof shaped, rather than elongate with truncated tips and claw grooves as seen in more basally positioned neoiguanodontians.

Euhadrosaurians Derived hadrosauromorphans (euhadrosaurians) represent a culmination of the general trends seen through more primitive neoiguanodontians and hadrosauromorphans. The elongation of upper and lower jaws creates a prominent diastema and an undoubted functional separation between the cropping and food processing components of jaw action. Linked to this functional differentiation, the muzzle/beak (and its ensheathing rhamphotheca) can be seen to become increasingly diverse in morphology (ranging from a droop-tipped and comparatively narrow morphology common among lambeosaurines to extreme transverse expansions bordered by upturned margins seen in some hadrosaurines/saurolophines). These variations in muzzle form are suggestive of ecological separation (niche partitioning) with respect to feeding guilds within euhadrosaurians (Carrano et al., 1999). The dental magazines are highly integrated and migrate distally (posteriorly) along the jaw so that they come to lie partly behind the elevated and anteriorly curved coronoid process; the effect of these changes is to increase the lever-arm mechanics (and hence efficiency) of the jaw muscles by promoting the development of a "bent first-order lever" from the traditional diapsid low-efficiency third-order lever mechanics that are associated with most non-mammalian vertebrate jaws. In addition to the jaw anatomy changes, the architecture of the roof of the skull becomes modified in a variety of ways: the temporal region becomes robust in order to withstand the stresses imposed upon the skull roof by the large and complex jaw adductors, and modifications of the nasal vestibule and dorsally projecting crests have been linked to a variety of biological functions. These include olfaction, sound production, and visual recognition (Ostrom, 1961; Hopson, 1975; Weishampel, 1981) that are suggestive of increasingly complex social interactions between euhadrosaurs. The postcranial skeleton differs little from that seen in hadrosauromorphans, except that the sacrum incorporates many more sacrals and produces an extremely strong region to support the stresses generated through weight-support using the hind limbs almost exclusively. It is also worth noting that some euhadrosaurs are the largest of all ornithopods (15+ m long) and yet retained a facultatively bipedal locomotor strategy. As noted above, the forelimbs are comparatively slender and distally elongate, but the unguals of digits II and III are hoof shaped. It is presumed their forelimbs were used for support while feeding upon low browse or when moving slowly and cautiously when feeding or moving in, for example, crowded colonial nesting sites, or indulging in nest building and related activities. So it appears that the mechanical efficiency of their hindlimb support and locomotor system was considerable. In contrast it is the case that among more basal neoiguanodontians an upper size range (~11 m long) is accompanied by a consistent tendency to become specialized by becoming secondarily obligate quadrupeds (Norman, 1980).

The evidence based upon times of occurrence in the fossil record (Prieto-Márquez, 2010:fig. 10) suggests that the pattern of diversification of euhadrosaurians displays a significant lag phase during the Coniacian–Santonian before a log-phase diversification in the Campanian and an equilibration during the Maastrichtian. Whether this pattern is an artifact of preservation in the fossil record, or represents some element of "bottle-necking" associated with the process of assembly of euhadrosaurian anatomy (and implicit biology), cannot yet be resolved satisfactorily.

ACKNOWLEDGMENTS AND DEDICATION

I would like to thank a very old friend, Dave Eberth, for inviting me to attend this symposium. Circumstances dictated that his co-organizer and co-editor, David Evans, had the unenviable task of presenting my talk because a family bereavement prevented me from attending the meeting. I am also indebted to David Weishampel for savagely criticizing another draft manuscript that is of relevance to this contribution (and hence thank him for indirectly contributing to this article). David Evans and Paul Barrett, my formal reviewers, were unstinting in their critical commentaries and made a number of valuable observations and cogent points that helped to turn a far too rapidly produced first draft into a better-structured, and more cogent, revised version. None of the above can take any blame for the remaining errors; the latter will, perforce, fall upon my shoulders.

This chapter is, of course, dedicated to a dear friend whom I first met when he visited me in London and Oxford University in the early 1980s (our having corresponded for some time before over matters relating to ornithopod jaw mechanisms and their implications). Suffice it to say we saw eye to eye rather than tooth to claw pretty much immediately and became firm friends (we are, after all, both named David Bruce and were born in the same year); and, despite a few ups and downs along the way, we have never lost that core friendship. So it is with pleasure (and a little amusement) that I contribute to a volume on hadrosaurs in Dave's honor when I know so little about such exotic beasts, especially given that old canard that Dave W. has undoubtedly forgotten more about hadrosaurs than Dave N. will ever know.

LITERATURE CITED

Batten, D. J. 2011. Wealden geology; pp. 7–14 in D. J. Batten (ed.), English Wealden Fossils. The Palaeontological Association, London, U.K.

Beneden, P. J. van. 1881. Sur l'arc pelvien chez les dinosauriens de Bernissart. Bulletin de l'Académie des Sciences, Belge 1:600–608.

Blows, W. T. 1998. A review of Lower and Middle Cretaceous dinosaurs of England. New Mexico Museum of Natural History and Science Bulletin 14:29–38.

Brett-Surman, M. K. 1997. Ornithopods; pp. 330–346 in J. O. Farlow, and M. K. Brett-Surman (eds.), The Complete Dinosaur. Indiana University Press, Bloomington, Indiana.

Butler, R. J., P. Upchurch, and D. B. Norman. 2008. The phylogeny of the ornithischian dinosaurs. Journal of Systematic Palaeontology 6:1–40.

Carpenter, K., and Y. Ishida. 2010. Early and "middle" Cretaceous Iguanodonts in time and space. Journal of Iberian Geology 36:145–164.

Carrano, M. T., C. M. Janis, and J. J. Sepkoski, Jr. 1999. Hadrosaurs as ungulate parallels: lost lifestyles and deficient data. Acta Palaeontologica Polonica 44:237–261.

Casier, E. 1960. Les Iguanodons de Bernissart: a la mémoire de Louis Dollo (1857–1931). Editions du patrimoine du l'Institut royal des Sciences naturelles de Belgique, Bruxelles, Belgium, 134 pp.

Charig, A. J. 1972. The evolution of the archosaur pelvis and hind limb: an explanation in functional terms; pp. 121–155 in K. A. Joysey, and T. S. Kemp (eds.), Studies in Vertebrate Evolution. Oliver and Boyd, Edinburgh, U.K.

Cope, E. D. 1869. Synopsis of the extinct Batrachia, Reptilia, and Aves of North America. Transactions of the American Philosophical Society 14:1–252.

Dalla Vecchia, F. M. 2009. *Tethyshadros insularis,* a new hadrosauroid dinosaur (Ornithischia) from the Upper Cretaceous of Italy. Journal of Vertebrate Paleontology 29:1100–1116.

Dollo, L. 1882. Première note sur les dinosauriens de Bernissart. Bulletin de la Musée royale d'Histoire naturelle de Belgique 1:55–80.

Evans, D. C. 2010. Cranial anatomy and systematics of *Hypacrosaurus altispinus,* and comparative analysis of skull growth in lambeosaurine hadrosaurids (Dinosauria: Ornithischia). Zoological Journal of the Linnean Society 159:398–434.

Evans, D. C., and R. R. Reisz. 2007. Anatomy and relationships of *Lambeosaurus magnicristatus,* a crested hadrosaurid dinosaur (Ornithischia) from the Dinosaur Park Formation, Alberta. Journal of Vertebrate Paleontology 27:373–393.

Gates, T. A., and S. D. Sampson. 2007. A new species of *Gryposaurus* (Dinosauria: Hadrosauridae) from the Late Campanian Kaiparowits Formation, southern Utah, USA. Zoological Journal of the Linnean Society 151:351–376.

Gilmore, C. W. 1933. On the dinosaurian fauna of the Iren Dabasu Formation. Bulletin of the American Museum of Natural History 67:23–78.

Godefroit, P., S. Hai, T. Yu, and P. Lauters. 2008. New hadrosaurid dinosaurs from the uppermost Cretaceous of northeastern China. Acta Palaeontologica Polonica 53:47–74.

Head, J. J. 1998. A new species of basal hadrosaurid (Dinosauria: Ornithischia) from the Cenomanian of Texas. Journal of Vertebrate Paleontology 18:718–738.

Holl, F. 1829. Handbuch der Petrefactenkunde Pt 1. Quedlinberg, Leipzig, pp. 232.

Hooley, R. W. 1925. On the skeleton of *Iguanodon atherfieldensis* sp. nov., from the Wealden shales of Atherfield (Isle of Wight). Quarterly Journal of the Geological Society of London 81:1–61.

Hopson, J. A. 1975. The evolution of cranial display structures in hadrosaurian dinosaurs. Paleobiology 1:21–43.

Horner, J. R. 1985. Evidence for polyphyletic origination of the Hadrosauridae (Reptilia: Ornithischia). Proceedings of the Pacific Division of the American Association for the Advancement of Science 4:31–32.

Horner, J. R. 1990. Evidence of diphyletic origination of the hadrosaurian (Reptilia: Ornithischia) dinosaurs; pp. 179–187 in K. Carpenter, and P. J. Currie (eds.), Dinosaur Systematics: Approaches and Perspectives. Cambridge University Press, Cambridge, U.K.

Horner, J. R. 1992. Cranial morphology of *Prosaurolophus* (Ornithischia: Hadrosauridae). With descriptions of two new hadrosaurid species and an evaluation of hadrosaurid phylogenetic relationships. Museum of the Rockies Occasional Paper 2:1–119.

Horner, J. R., D. B. Weishampel, and C. Forster. 2004. Hadrosauridae; pp. 438–463 in D. B. Weishampel, P. Dodson, and H. Osmólska (eds.), The Dinosauria, Second Edition. University of California Press, Berkeley, California.

Hulke, J. W. 1879. *Vectisaurus valdensis,* a new Wealden dinosaur. Quarterly Journal of the Geological Society of London 35:421–424.

Hulke, J. W. 1882. Description of some *Iguanodon* remains indicating a new species, *I. seelyi.* Quarterly Journal of the Geological Society of London 38:135–144.

ICZN. 1999. International Code of Zoological Nomenclature, Fourth Edition. International Trust for Zoological Nomenclature, Natural History Museum, London, U.K., 306 pp.

Lull, R. S., and N. E. Wright. 1942. Hadrosaurian Dinosaurs of North America. Geological Society of America Special Papers 40. 242 pp.

Lydekker, R. 1888. Note on a new Wealden iguanodont and other dinosaurs. Quarterly Journal of the Geological Society of London 44:46–61.

Lydekker, R. 1889. Notes on new and other dinosaur remains. Geological Magazine VI (Decade III):352–356.

Maddison, D. R., and W. P. Maddison. 2003. MacClade: Analysis of Phylogeny and Character Evolution. Version 4.06. Sinauer Associates, Sunderland, Massachusetts.

Maidment, S. C. R., K. T. Bates, and P. M. Barrett. 2014. Three-dimensional computational modeling of pelvic locomotor muscle moment arms in *Edmontosaurus* (Dinosauria, Hadrosauridae) and comparisons with other archosaurs; chapter 25 in D. A. Eberth, and D. C. Evans (eds.), Hadrosaurs. Indiana University Press, Bloomington, Indiana.

Mantell, G. A. 1825. Notice on the *Iguanodon,* a newly discovered fossil reptile, from the sandstone of Tilgate forest, in Sussex. Philosophical Transactions of the Royal Society of London CXV: 179–186.

Mantell, G. A. 1827. Illustrations of the Geology of Sussex: With Figures and Descriptions of the Fossils of Tilgate. Lupton Relfe, London, U.K., 92 pp.

Mantell, G. A. 1848. On the structure of the jaws and teeth of the *Iguanodon.* Philosophical Transactions of the Royal Society of London 138: 183–202.

Marsh, O. C. 1881. Classification of the Dinosauria. American Journal of Science 7: 81–86.

Martill, D. M., and D. Naish (eds.). 2001. Dinosaurs of the Isle of Wight. Palaeontological Association, London, U.K., 433 pp.

McDonald, A. T. 2012a. The status of *Dollodon* and other basal iguanodonts (Dinosauria: Ornithischia) from the Lower Cretaceous of Europe. Cretaceous Research 33:1–6.

McDonald, A. T. 2012b. Phylogeny of basal iguanodonts (Dinosauria: Ornithischia): an update. PLoS ONE 7(5):e36745.

McDonald, A. T., P. M. Barrett, and S. D. Chapman. 2010. A new basal iguanodont (Dinosauria: Ornithischia) from the Wealden (Lower Cretaceous) of England. Zootaxa 2569:1–43.

Naish, D., and D. M. Martill. 2008. Dinosaurs of Great Britain and the role of the Geological Society of London in their discovery: Ornithischia. Journal of the Geological Society of London 165:613–623.

Norman, D. B. 1977. On the anatomy of the ornithischian dinosaur *Iguanodon.* Ph.D. dissertation, King's College London, London, U.K., 631 pp.

Norman, D. B. 1980. On the ornithischian dinosaur *Iguanodon bernissartensis* from Belgium. Mémoires de l'Institut royal des Sciences naturelles de Belgique 178:1–105.

Norman, D. B. 1986. On the anatomy of *Iguanodon atherfieldensis* (Ornithischia: Ornithopoda). Bulletin de l'Institut royal des Sciences naturelles de Belgique 56:281–372.

Norman, D. B. 1987. On the discovery of fossils at Bernissart (1878–1921) Belgium. Archives of Natural History 13:131–147.

Norman, D. B. 1990. A review of *Vectisaurus valdensis,* with comments on the family Iguanodontidae; pp. 147–162 in K. Carpenter, and P. J. Currie (eds.), Dinosaur Systematics: Approaches and Perspectives. Cambridge University Press, Cambridge, U.K.

Norman, D. B. 1993. Gideon Mantell's "Mantel-piece": the earliest well-preserved ornithischian dinosaur. Modern Geology 18:225–245.

Norman, D. B. 2002. On Asian ornithopods (Dinosauria: Ornithischia). 4. Redescription of *Probactrosaurus gobiensis* Rozhdestvensky, 1966. Zoological Journal of the Linnean Society (London) 136:113–144.

Norman, D. B. 2004. Basal Iguanodontia; pp. 413–437 in D. B. Weishampel, P. Dodson, and H. Osmólska (eds.), The Dinosauria, Second Edition. University of California Press, Berkeley, California.

Norman, D. B. 2010. A taxonomy of iguanodontians (Dinosauria: Ornithopoda) from the lower Wealden Group (Valanginian) of southern England. Zootaxa 2489:47–66.

Norman, D. B. 2011a. On the osteology of the lower Wealden Group (Valanginian) ornithopod *Barilium dawsoni* (Iguanodontia: Styracosterna). Special Papers in Palaeontology 86:165–194.

Norman, D. B. 2011b. Ornithopod dinosaurs; pp. 407–475 in D. J. Batten (ed.), Field Guide to the Wealden of England. The Palaeontological Association, Oxford, U.K.

Norman, D. B. 2012. Iguanodontian taxa (Dinosauria: Ornithischia) from the Lower Cretaceous of Britain and Belgium; pp. 174–212 in P. Godefroit (ed.), Bernissart Dinosaurs and Early Cretaceous Terrestrial Ecosystems. Indiana University Press, Bloomington, Indiana.

Norman, D. B. 2013. On the taxonomy and diversity of Wealden euiguanodontian dinosaurs (Ornithischia: Ornithopoda); pp. 385–404 in T. Malvesy (ed.), Fourth Georges Cuvier Symposium. Revue de Paléobiologie, Genève, Switzerland.

Norman, D. B. in press. On the osteology, comparative morphology and systematics of *Hypselospinus fittoni* (Ornithopoda: Euiguanodontia) and its bearing on euiguanodontian phylogeny. Zoological Journal of the Linnean Society.

Norman, D. B., A. W. Crompton, R. J. Butler, L. B. Porro, and A. J. Charig. 2011. The Lower Jurassic ornithischian dinosaur *Heterodontosaurus tucki* Crompton and Charig, 1962: cranial anatomy, functional morphology, taxonomy and relationships. Zoological Journal of the Linnean Society 163:182–276.

Norman, D. B., H.-D. Sues, L. Witmer, and R. A. Coria. 2004. Basal Ornithopoda; pp. 393–412 in D. B. Weishampel, P. Dodson, and H. Osmólska (eds.), The Dinosauria, Second Edition. University of California Press, Berkeley, California.

Ostrom, J. H. 1961. Cranial morphology of the hadrosaurian dinosaurs of North America. Bulletin of the American Museum of Natural History 122:33–186.

Owen, R. 1842. Report on British Fossil Reptiles. Part 2. Report of the British Association for the Advancement of Science (Plymouth) 11(1841):60–204.

Owen, R. 1855. Monograph of the Fossil Reptilia of the Wealden and Purbeck Formations. Part II. Dinosauria (*Iguanodon*). Palaeontographical Society Monographs 8:1–54.

Paul, G. S. 2007. Turning the old into the new: a separate genus for the gracile iguanodont from the Wealden of England; pp. 69–77 in K. Carpenter (ed.), Horns and Beaks. Indiana University Press, Bloomington, Indiana.

Paul, G. S. 2008. A revised taxonomy of the iguanodont dinosaur genera and species. Cretaceous Research 29:192–216.

Paul, G. S. 2012. Notes on the rising diversity of Iguanodont taxa, and Iguanodonts named after Darwin, Huxley, and evolutionary science; pp. 123–133 in P. H. Hurtado, F. T. Fernandez-Baldro, and J. I. C. Sanagustin (eds.), Actas de V Jornadas Internacionales sobre Paleontologia de Dinosaurios y su Entorno. Colectivo Arqueologico y Paleontologico de Salas, C.A.S, Salas de Los Infantes, Burgos, Spain.

Prieto-Márquez, A. 2010. Global phylogeny of Hadrosauridae (Dinosauria: Ornithopoda) using parsimony and Bayesian methods. Zoological Journal of the Linnean Society 159:435–502.

Prieto-Márquez, A. 2011. Revised diagnosis of *Hadrosaurus foulkii* Leidy, 1858 (the type genus and species of Hadrosauridae Cope, 1869) and *Claosaurus agiiis* Marsh, 1872 (Dinosauria: Ornithopoda) from the Late Cretaceous of North America. Zootaxa 2765:61–68.

Prieto-Márquez, A., R. Gaete, G. Rivas, A. Galobart, and M. Boada. 2006. Hadrosauroid dinosaurs from the Late Cretaceous of Spain: *Pararhabdodon isonensis* revisited and *Koutalisaurus kohlerorum*. Journal of Vertebrate Paleontology 26:929–943.

Prieto-Márquez, A., D. B. Weishampel, and J. R. Horner. 2006. The dinosaur *Hadrosaurus foulkii,* from the Campanian of the East Coast of North America, with a re-evaluation of the genus. Acta Palaeontologica Polonica 51:77–98.

Romer, A. S. 1956. Osteology of the Reptiles. University of Chicago Press, Chicago, Illinois, 772 pp.

Rozhdestvensky, A. K. 1966. [New iguanodonts from Central Asia. Phylogenetic and taxonomic interrelationships of late Iguanodontidae and early Hadrosauridae]. Palaeontologicheskii Zhurnal 1966:103–116. [Russian]

Seeley, H. G. 1887. On the classification of the fossil animals commonly named Dinosauria. Proceedings of the Royal Society of London 43:165–171.

Sereno, P. C. 1986. Phylogeny of the bird-hipped dinosaurs. National Geographic Research 2:234–256.

Sereno, P. C. 1991. *Lesothosaurus,* "Fabrosaurids" and the early evolution of Ornithischia. Journal of Vertebrate Paleontology 11:234–256.

Sereno, P. C. 1998. A rationale for phylogenetic definitions, with application to the higher-level taxonomy of Dinosauria. Neues Jahrbuch für Geologie und Paläontologie Abhandlungen 109:41–83.

Sereno, P. C. 2005. Stem Archosauria version 1.0 website. Available at www.taxonsearch.org /Archive/stem-archosauria-1.0.php. Accessed summer 2012.

Swofford, D. L. 2002. PAUP*, Phylogenetic Analysis Using Parsimony (*and other methods). Version 4.0b10. Sinauer Associates, Sunderland, Massachusetts.

Taquet, P. 1975. Remarques sur l'évolution des iguanodontidés et l'origine des hadrosauridés. [Comments on the evolution of Iguantodontids and the origin of Hadrosaurids]: problèmes actuels de paléontologie-évolution des vertebras [current problems in paleontology–vertebrate evolution]. Paris, Colloque international, Centre national de la Recherche scientifique Paris 218:1–8. [French].

Taquet, P. 1976. Géologie et paléontologie du gisement de Gadoufaoua (Aptien du Niger). Cahiers de Paléontologie. [Geology and paleontology of the Gadoufaouna Locality (Aptian of Niger)]. Centre national de la Recherche scientifique Paris:1–191. [French]

Thulborn, R. A. 1970. The skull of *Fabrosaurus australis,* a Triassic ornithischian dinosaur. Palaeontology 13:414–432.

Thulborn, R. A. 1972. The postcranial skeleton of the Triassic ornithischian dinosaur *Fabrosaurus australis.* Palaeontology 15:29–60.

Wang, X.-L., R. Pan, R. J. Butler, and P. M. Barrett. 2010. The postcranial skeleton of the iguanodontian ornithopod Jinzhousaurus yangi from the Lower Cretaceous Yixian Formation of western Liaoning, China. Earth and Environmental Science Transactions of the Royal Society of Edinburgh 101:135–159.

Weishampel, D. B. 1981. The nasal cavity of lambeosaurine hadrosaurids (Reptilia: Ornithischia): comparative anatomy and homologies. Journal of Paleontology 55:1046–1057.

Weishampel, D. B., and J. R. Horner. 1990. Hadrosauridae; pp. 534–561 in D. B. Weishampel, P. Dodson, and H. Osmólska (eds.), The Dinosauria. University of California Press, Berkeley, California.

Weishampel, D. B., C. M. Jianu, Z. Csiki, and D. B. Norman. 2003. Osteology and phylogeny of *Zalmoxes* (n.g.), an unusual euornithopod dinosaur from the latest Cretaceous of Romania. Journal of Systematic Palaeontology 1:65–123.

Weishampel, D. B., D. B. Norman, and D. Grigorescu. 1993. *Telmatosaurus transsylvanicus* from the Late Cretaceous of Romania: the most basal hadrosaurid dinosaur. Palaeontology 36:361–385.

Wu, W.-H., and P. Godefroit. 2012. Anatomy and relationships of *Bolong yixianensis,* an Early Cretaceous iguanodontoid dinosaur from western Liaoning, China; pp. 293–333 in P. Godefroit (ed.), Bernissart Dinosaurs and Early Cretaceous Terrestrial Ecosystems. Indiana University Press, Bloomington, Indiana.

Appendix 2.1. Taxon-Character Matrix

Lesothosaurus diagnosticus	0000000000 0000000000 0000000000 0000000000 0000000000 0000000000 0000000000 0000000000 0000000000 00
Hypsilophodon foxii	0000000000 0000000111 0000000000 0000000000 0000101001 0111110021 0000100000 0000000011 0000100000 00
Zalmoxes robustus	0010000010 00000?1011 00?10?0001 0112101000 0100111111 1111111021 101??00??? ????100111 1111000010 ?0
Tenontosaurus tilletti	1010101110 0???0?0101 100100001? 1100101000 0000101111 1111110021 1011100000 0001000312 1002101010 00
Dryosaurus altus	101?102110 110000000 0000100001 0101010000 0000010111 1112211011 1000000?0 0?00?000112 0111200020 10
Camptosaurus dispar	1111101010 0?00010001? 0001000011 1110101000 0000111111 2122111111 0011000110 0101000112 0111200010 00
Mantellisaurus atherfieldensis	1111102110 1110100000 0001011011 1110201111 0100111211 2122111111 1111011111 0111101114 1021201131 11
Iguanodon bernissartensis	1111102110 1110100000 1001011111 1110101111 0100111211 2122111111 1111011111 0111112113 1111201131 11
Ouranosaurus nigeriensis	3111111110 0211100000 1001001?11 111?202001 0000111211 2122111111 1121011111 0111001114 11112?1131 ??
Altirhinus kurzanovi	211110112? 0211100000 100100??1? 1111211111 0001212211 2122111111 ????11111 0111001113 101?2?11?? ?1
Eolambia caroljonesa	2111011?? ??11000000 10?10?10?1 ??1?202101 0001112211 2123112111 ??1??11111 011?101113 ?111211131 ?1
Jinzhousaurus yangi	211?102121 1?101?0000 ??0111??1? 1111010?0 000011?2?1 2122111111 1111?11111 0111?01112 ?0112?1131 ??
Hypselospinus fittoni	?1???????? ?????????? ?????????? ????111001 ???0112211 21221?1111 ?121?11111 011?000113 1111201131 ?1
Barilium dawsoni	?????????? ?????????? ?????????? ????101111 ???0112211 21221?1111 ?111?11111 01??100013 111?201131 ?1
Equijubus normani	211?101010 1?101?1000 10?100??1? 1111011?1 0101102211 1122111111 11???????1 ???0011?? 11???????? ??
Protohadros byrdi	211010???? ??10100010 10?100???? 1111212111 0101212211 1123111111 ?????????? ?????????? ?????????? ??
Probactrosaurus gobiensis	2110101 0?? 1?1?101??0 00?10??01? 1111101111 0101212221 2122112111 ?11??12112 011?001113 1111201131 21
Bactrosaurus johnsoni	111101?21 ?211110 0?1 0011001011 1211112211 1111222321 2123222111 111??12?22 12??002113 1111212231 22
Tethyshadros insularis	1111101021 1?200?1011 000100??1? 1111?01111 1111?1?321 21232?2112 1111012222 12?1003213 11012?2??1 21
Parasaurololphus walker	2111100021 0222110011 0112111111 1211211212 1112222321 3113222112 1111022222 1211013214 1111212241 22
Saurolophus osborni	3111110021 1221111011 1112011111 121?201212 1112222321 3113222112 1111022222 1211013214 1000212241 22
Telmatosaurus transsylvanicus	1111101121 1?211????1 10?100101? ??1?101112 1102222321 3113222112 111??2???? ?????????? ?????12141 ??
Xuwulong yueluni	111101?21 0??11????0 0?0100??1? 111?101101 ??001?1??? 2122111111 ?11??1???? ???001113 11112????? ??
Bolong yixianensis	1111011?0 1?0?????0 ??0????11 111?10?001 ???0111211 2122111111 1?1??1?111 1111022?? ?1112?11?1 ?1

Appendix 2.2. Character Choice and State Definitions for the Phylogenetic Analysis Conducted in This Study

1. Premaxillary rostrum: dorsal aspect. Margins converge to a blunt tip (0), modest rounded expansion (1), occlusal margin is broad and rounded in dorsal view such that its overall width approaches that of the skull roof (2), flared occlusal margins that form a "spoon-bill" structure in dorsal view (3).
2. Premaxilla: level of occlusal margin relative to that of the maxillary tooth row. Not at all (or slightly) ventrally offset from alveolar margin of the maxilla (0), strongly ventrally offset (1).
3. Premaxillary teeth: present (0), absent (1).
4. Premaxillary denticles: absent (0), present (1).
5. External naris: confined to area above oral margin of premaxilla (0), posterior margin extended posteriorly to lie above the maxilla (1).
6. Premaxilla: anterolateral margin of the nasal fossa, above the occlusal edge of the premaxilla is reflected dorsally to form a distinct rim: absent (0), present (1).
7. Premaxilla-lacrimal contact: absent (0), present (1) posterolateral premaxillary process extends posterodorsally to also contact/overlap the prefrontal (2).
8. Premaxillary posterolateral process: tapers to point (0), posterior tip is bluntly truncated (1).
9. Antorbital fenestra shape. Opening when viewed laterally: large and subtriangular (0), small and subcircular (1), absent (2).
10. Antorbital fenestra location: between lacrimal and maxilla (0), on anterodorsal margin (premaxillary suture) of maxilla and not visible in lateral view of the articulated skull (1).
11. Lacrimal-nasal contact: present (0), absent (1).
12. Maxilla: dorsal process morphology. Flattened mound-like structure (0), narrow, finger-like process (1), laterally flattened subtriangular plate (2).
13. Jugal: anterior process. Tapering (0), expanded (1), expanded and truncated anteriorly (2).
14. Jugal-maxilla suture: scarf joint (0), "finger-in-recess" (oblique finger-like process of the maxilla fits into a slot formed in the medioventral surface of the anterior jugal ramus) (1), butt-jointed against a broad facet on the lateral surface of the ascending process of the maxilla (2).
15. Jugal and its free ventral margin: generally strap-like with little undulation along the ventral edge (0), marked sinuous ventral edge with marked ventral deflection posteriorly (1).
16. Jugal-ectopterygoid contact: present (0), absent (1). *Difficult to score in many instances.*
17. Jugal contribution to the margin of the infratemporal fenestra: jugal forms part of the margin (0), jugal forms the entire margin by excluding the quadratojugal (1).
18. Quadratojugal foramen: absent (0), present (1). *Very limited distribution.*
19. Quadrate (paraquadratic) foramen: between quadratojugal and quadrate. Present (0), absent (1).
20. Quadrate: embayment on anterolateral wing. Relatively small and semicircular in outline (0), broad embayment whose edges are marked by a scarf suture for a close-fitting quadratojugal (1).
21. Quadrate: posterior margin of the shaft. Bowed anteriorly and the dorsal portion tilted posteriorly (0), straight (1). *Often difficult to score because of postmortem distortion.*
22. Quadrate-articular condyle: transversely expanded (0), laterally compressed so that it forms (almost) a simple rounded condyle (1).
23. Palpebral (supraorbital) bone(s): present (0), absent or fused to orbital margin (1). *Given the looseness of attachment of this bone to the orbital margin (and therefore the lack of a reliable osteological marker) evidence of absence is subjective.*
24. Frontal: shape. Arched and narrow embayed laterally so that the orbital cavity is exposed dorsally (0), flat and transversely broad plate, roofing the orbital cavity dorsally (1), anteroposteriorly abbreviated (2).
25. Frontal: forms part of the dorsal margin of the orbital cavity. Present (0), absent (1).
26. Postorbital-squamosal contact: postorbital forms a tapering finger-like squamosal process the overlaps the squamosal (0), this process develops a bifurcate tip (1). *McDonald et al. (2010) proposed that this process might be coded in three ways: "blunt," "pointed," or "bifurcate." However, there is variation in this structure as it is illustrated in skull reconstructions, which makes the distinction between "blunt" and "pointed" one that is potentially subjective and risks misinterpretation.*
27. Foramen magnum dorsal margin: supraoccipital in dorsal margin (0), supraoccipital excluded from dorsal margin by exoccipitals (1).
28. Foramen magnum ventral margin: basioccipital in ventral margin (0), basioccipital excluded from ventral margin by the exoccipitals (1). *This character, though commonly used and perhaps of more value at a coarser scale (e.g., across higher-level taxonomic groupings), is often difficult to assess reliably (within individual taxa), and may vary ontogenetically and be subject by taphonomic influence.*
29. Paroccipital wing shape: horizontal and dorsoventrally expanded distally (0), pendant distal tip (1).
30. Basipterygoid process orientation: anteroventral (0), posterolateral (1).
31. Predentary occlusal margin: smooth edged (0), denticulate (1).
32. Predentary shape (in plan view): subtriangular (0), arcuate (1), broad and subrectangular (2).
33. Predentary ventral lobe: median tab sometimes notched in the midline (0), deeply incised in the midline and with strongly bifurcate lobes (1).
34. Predentary rostral surface: smooth (0); bearing a pair of oblique grooves, one on either side of midline (1), midline groove (2).
35. Mandibular diastema (the gap between the posterior end of the predentary and the first dentary alveolus): absent (0), present (1), greater than three crown widths (2). *Difficult to assess in instances in which the predentary is not articulated with the dentary and/or when the anterior part of the dentary is not well preserved.*
36. Dentary ramus shape in lateral view: straight (0), arched along its ventral edge (1).
37. Dentary ramus (tooth-bearing portion) shape: tapers anteriorly (0), parallel dorsal and ventral borders (1), deepens anteriorly (2).
38. Dentary coronoid process profile: axis of coronoid oblique (posterodorsal orientation) (0), perpendicular (1), anteriorly inclined (2).
39. Dentary coronoid process shape: dorsal tip unexpanded (0), expanded (1).
40. Dentary coronoid process position: laterally offset and dentition [alveoli] curves laterally into its base (0), posterior dentition extends medial to the middle of the coronoid process (1), posterior dentition extends posterior to the coronoid process (2).
41. Surangular foramen: present (0), absent (1).
42. Surangular-angular suture: obliquely inclined (0), horizontal (1). *Difficult to score faithfully because this feature may be based upon skull reconstructions which may, or may not, be accurate in this area.*
43. Angular: lateral exposure. Visible laterally (0), not visible (contact with surangular is positioned ventrally or medially) (1).
44. Replacement crowns present: one (0), two (1), three or more (2).
45. Wear facet distribution on dentary and maxillary crowns: irregular and discontinuous distribution on individual crowns (0), wear facets continuous across adjacent crowns, producing a uniformly narrow cutting/grinding surface (1), the oldest and successional crowns contribute to the wear surface to varying degrees to produce a transversely broader cutting/grinding occlusal (2).
46. Relative crown width: maxillary crowns equal in width to dentary crowns (0), narrower than dentary crowns (1), equal in width to dentary crowns, but "miniaturized" (2).
47. Enamel surface distribution on tooth crowns: equally distributed on both sides of crown (0), asymmetrical distribution, thicker on one surface of the crown (1), enamel is restricted to one side of the crown (2).
48. Marginal denticle shape: simple cones (0), tongue-shaped (1), curved ledges with mammillae (2), absent, or reduced to small irregular papillae (3).
49. Tooth roots: tapering cylinders (0), longitudinally grooved to accommodate adjacent, closely packed teeth (1), highly angular sided "prismatic" roots (2).
50. Dentary tooth curvature of long axis: the root–crown axis of the tooth is straight (0), the long axis of the tooth is bowed lingually so that the occlusal portion of the crown becomes directed labiodorsally (1).
51. Dentary teeth, crown shape in lingual view: simple, symmetrical leaf-shaped profile (0), broad, shield-like (1), in unworn examples occlusal margin forms a distinct shoulder mesially (2), mesiodistally compressed and diamond shaped (3).
52. Dentary teeth, presence of oblique, thickened ledges at the base of the enamelled lingual face of the crown: absent (0), present (1).

Appendix 2.2. (continued)

53. Dentary teeth, primary ridge: absent (0), median position (1), distally offset (2).
54. Dentary teeth, ridge pattern: simple median swelling (0), prominent primary ridge with variable number of parallel subsidiary ridges (1), parallel primary and secondary ridge divide crown face into three zones (2), primary ridge alone (3).
55. Dentary teeth, relative size in mandible: large and shield shaped (0), miniaturized (1).
56. Dentary, lateral alveolar wall shape: shaped by dentary crowns (0), narrow, parallel-sided grooves (1).
57. Maxillary teeth shape 1: equal width to dentary crowns (0), narrower and more lanceolate than opposing dentary crowns (1), lanceolate but equal in width to dentary crowns (2).
58. Maxillary teeth shape 2: Root–crown long axis of the tooth straight (0); long axis bowed, convex labially (1).
59. Maxillary teeth, labial ridges 1: simple median swelling (0), very prominent primary ridge (1), several subsidiary ridges and no obvious primary ridge (2).
60. Maxillary teeth, labial ridges 2: simple medial swelling (0), array of primary, secondary, and subsidiary ridges (1), single median primary ridge with no other ridges present (2).
61. Axis vertebra. Neural spine shape: low and sloping (0), tall and expanded (1).
62. Cervical vertebrae. Centrum articular surfaces: amphiplatyan (0), opisthocoelous (1).
63. Dorsal vertebrae. Neural spine shape: low and rectangular (0), tall and narrow (1). Extremely tall, relative to the anteroposterior dimension of spine (2). Dubious. *Biomechanical: small animals have short neural spines. Only two taxa,* Ouranosaurus *and* Hypselospinus, *have extremely elongate neural spines—i.e., probable homoplasy.*
64. Epaxial ossified tendons: arranged in linear bundles (0), form a layered lattice against the neural spines (1). Dubious: *reflects the accommodation space available: taller spines enable a lattice-like arrange of tendons to exist.*
65. Ossified tendons form a posterior caudal sheath: absent (0), present (1). *Rarely preserved, and where a sheath of tendons is known, its occurrence seems to be restricted to basal clypeodontans and may be plesiomorphic within cerapodans.*
66. Scapular acromion: prominent on the proximodorsal margin of scapula (0), occupying a median position on the proximal shaft and curved toward the dorsal edge of the proximal scapular blade (1), developed into a promontory that overhangs the proximal end of the scapula and is not curved toward the dorsal border (2).
67. Sternal shape: reniform (0), hatchet shaped with a stout, short "handle" (1), pronounced elongation of the "handle" of the hatchet relative to the "blade" (2).
68. Carpals: fully ossified (0), fused to form a carpometacarpal I block (1), reduced to no more than two small ossicles (2).
69. Metacarpal I: elongate dumbbell shaped (0), short, block-like and fused against carpals (1), absent (2).
70. Metacarpals II–IV: spreading (0), robust, closely appressed (1), slender and elongate (2).
71. Manus digit I: present (0), absent (1).
72. Ungual of manus digit I: claw-like (0), subconical (1), lost (2).
73. Unguals of manus digits II and III: claw-like (0), flattened, twisted and hoof-like (1).
74. Manus digit III: four phalanges (0), three phalanges (1).
75. Ilium, preacetabular process: long, laterally compressed (0), axially twisted so that lateral surface faces dorsolaterally (1).
76. Ilium: profile of dorsal edge. Horizontal to slightly arched, no significant notch in its profile posterodorsal to the ischiadic peduncle (0), sinuous profile created by the presence of a broad saddle-like notch (1).
77. Ilium: dorsal margin development: no transverse thickening of the dorsal edge in the region posterodorsal to the ischial peduncle (0), transversely thickened, beveled edge (1), thickened edge developed into a thick rolled edge (2), prominently everted, with a downturned, flap-shaped pendule that overhangs the ischiadic peduncle (3).
78. Ilium: postacetabular process in profile. Vertical plate with rounded edge (0), generally triangular, tapering posteriorly (1), laterally compressed and relatively narrow, rectangular bar (2), upturned plate (3).
79. Pubis: anterior ramus form: short (0), elongate (1).
80. Pubis: anterior ramus shape. Short, deep and blunt (0), rod-shaped (1), laterally compressed parallel-sided blade (2), expanded distally (3), strongly transversely compressed and deeply expanded distal portion (4).
81. Pubis: posterior ramus. Terminates bluntly adjacent to distal end of ischium (0), shorter than ischium, tapers to a point (1).
82. Ischium: shaft morphology 1. Straight (0), bowed (1).
83. Ischium: shaft morphology 2. Compressed and blade-like along length of shaft (0), sub-cylindrical shaft (1), narrow, angular-sided shaft (2).
84. Ischium: shaft morphology 3. Distal end unexpanded (0), distal end expanded into "boot" (1), distal end laterally expanded, rather than expanded anteroposteriorly (2).
85. Ischium: obturator process. Absent (0), positioned near mid-shaft (1), positioned close to pubic peduncle, from which it is separated by an embayment (2).
86. Femoral head grooved posteriorly: present (0), absent (1).
87. Femur: curvature of shaft. Femoral shaft bowed along its length (0), proximal half of the shaft straight, distal half of shaft curved caudally (1), femoral shaft straight (2).
88. Femur: fourth trochanter. Pendant (0), large, triangular (1), curved, laterally compressed eminence (2).
89. Femur: distal extensor groove. Absent (0), very broad trough (1), U-shaped trough (2), partially enclosed by expansion of adjacent anterior condyles (3), edges of trough meet to form a fully-enclosed tunnel (4).
90. Femur: distal condyles. Moderately expanded anteroposteriorly (0), strongly expanded and partly occluding flexor channel (1). *Most probably a size-related mechanical feature.*
91. Metatarsal I: well developed and articulates with proximal phalanx (0), slender and splint-like (1), absent (2). *Difficult to assess in many instances. As in the case of a palpebral bone on the orbital margin, is absence actual evidence of absence, or a biostratinomic artifact?*
92. Pedal phalanges: shape. Dorsoventrally flattened but elongate and pointed (0), elongate, bluntly truncated tip with prominent claw grooves retained (1), anterior margin broadly rounded in dorsal view, lateral claw grooves absent (2).

Osteology of the Basal Hadrosauroid *Equijubus normani* (Dinosauria, Ornithopoda) from the Early Cretaceous of China

3

Andrew T. McDonald, Susannah C. R. Maidment, Paul M. Barrett, Hai-lu You, and Peter Dodson

ABSTRACT

The basal hadrosauroid *Equijubus normani* is rediagnosed and fully described based upon the holotype and only known specimen, IVPP V 12534, from the Lower Cretaceous Xinminpu Group of Gansu Province, China. *Equijubus* can be diagnosed by a suite of cranial (finger-like process on the maxillary process of the jugal, rostrally elongate lacrimal) and vertebral (epipophyses present on third cervical vertebra, hyposphene present on at least dorsal vertebrae 6–8 and 11–15) autapomorphies, and a unique combination of characters. Comprehensive description of the skull and partial postcranium of IVPP V 12534 allows *Equijubus* to be compared to other iguanodontians from China and elsewhere. A better understanding of the anatomy of *Equijubus* will be useful for assessing the diversity of Asian iguanodonts, and allow it to be more confidently placed in a phylogenetic context.

INTRODUCTION

During much of the nineteenth and twentieth centuries, Europe boasted the richest fossil record of basal (i.e., non-hadrosaurid) iguanodontian dinosaurs in the world (Buffetaut and Le Loeuff, 1991; Weishampel et al., 2003; Norman, 2010, 2011a, this volume; Barrett et al., 2011). However, recent discoveries have revealed an abundance of new taxa from Asia, particularly China; recently named taxa include *Nanyangosaurus zhugeii* (Xu et al., 2000), *Jinzhousaurus yangi* (Wang and Xu, 2001; Barrett et al., 2009; Wang et al., 2010), *Shuangmiaosaurus gilmorei* (You, Ji, et al., 2003), *Equijubus normani* (You, Luo, et al., 2003), *Lanzhousaurus magnidens* (You et al., 2005), *Penelopognathus weishampeli* (Godefroit et al., 2005), *Jintasaurus meniscus* (You and Li, 2009), *Bolong yixianensis* (Wu et al., 2010), and *Xuwulong yueluni* (You et al., 2011). Although it is represented by one of the most complete skeletons known for these taxa, *Equijubus* has received only a brief description (You, Luo, et al., 2003).

We present herein a complete osteological description of the holotype and only known specimen of *Equijubus*, including elements that were not described or figured in the original description, and we provide a revised diagnosis for the taxon. The additional information presented on *Equijubus* in this description will facilitate comparison with other basal iguanodonts and allow its phylogenetic affinities to be more adequately tested. As is evident from the long list of taxa above, new Chinese iguanodonts are being discovered at a rapid pace, and it is essential when diagnosing new taxa that full comparisons with other iguanodonts are possible. Detailed descriptions of existing Chinese taxa, such as *Equijubus* and *Jinzhousaurus* (Barrett et al., 2009; Wang et al., 2010), will aid such comparative study.

Institutional Abbreviations AMNH, American Museum of Natural History, New York; CEUM, College of Eastern Utah Prehistoric Museum, Price, Utah; CM, Carnegie Museum of Natural History, Pittsburgh, Pennsylvania; IVPP, Institute of Vertebrate Paleontology and Paleoanthropology, Beijing, China; MB, Museum für Naturkunde, Berlin, Germany; MNHN, Muséum national d'Histoire naturelle, Paris, France; NHMUK, Natural History Museum, London, U.K.; SDSM, South Dakota School of Mines and Technology, Rapid City, South Dakota; USNM, National Museum of Natural History, Smithsonian Institution, Washington, D.C.

SYSTEMATIC PALEONTOLOGY

DINOSAURIA Owen, 1842
ORNITHISCHIA Seeley, 1887
ORNITHOPODA Marsh, 1881
IGUANODONTIA Dollo, 1888, sensu Sereno, 2005
ANKYLOPOLLEXIA Sereno, 1986, sensu Sereno, 2005
STYRACOSTERNA Sereno, 1986, sensu Sereno, 2005
HADROSAURIFORMES Sereno, 1997, sensu Sereno, 1998
HADROSAUROIDEA Cope, 1870, sensu Sereno, 2005

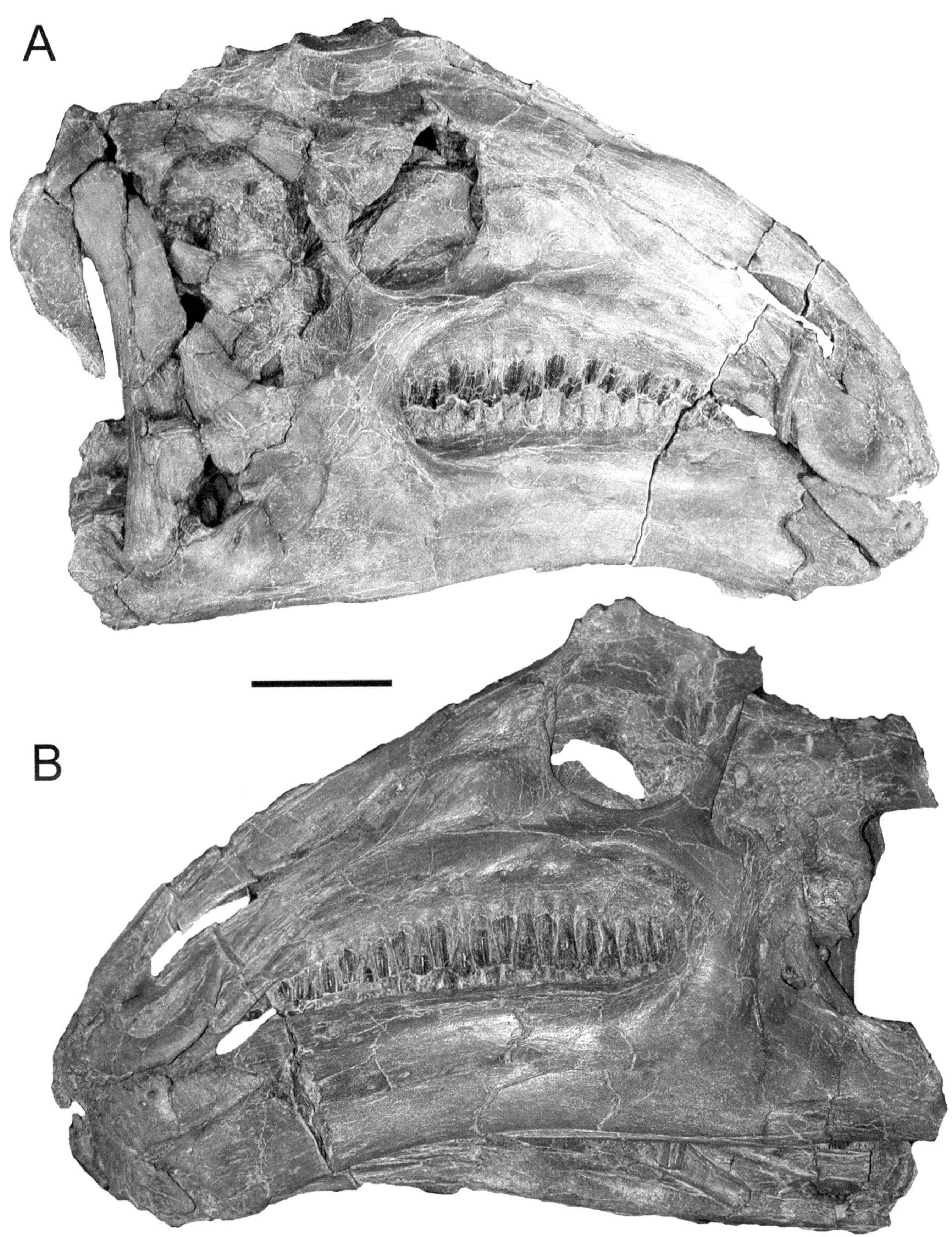

3.1. Fully assembled skull of IVPP V 12534, holotype of *Equijubus normani,* in (A) right lateral and (B) left lateral views. Scale bar equals 10 cm.

EQUIJUBUS NORMANI You, Luo, Shubin, Witmer, Tang, and Tang, 2003 (Figs. 3.1–3.20)

Holotype IVPP V 12534, complete skull and partial postcranium of a single individual.

Diagnosis For genus and species by monotypy. Characters derived from the original diagnosis of You, Luo, et al. (2003) are marked with an asterisk (*). Basal hadrosauroid characterized by four autapomorphies: (1), rostrodorsally curved finger-like process that arises from the maxillary process of the jugal at the jugal-lacrimal contact*; (2), elongate rostral ramus of lacrimal that extends along the dorsal margin of the maxilla and terminates rostral to the apex of the ascending process of the maxilla*; (3), epipophyses present on third cervical vertebra; and (4), hyposphene present on at least dorsal vertebrae 6–8 and 11–15. Also distinguished by the following unique combination of characters: rostral ramus of lacrimal tapers to a point, ventrolateral process of premaxilla contacts prefrontal (also noted by Paul, 2008:201, as lack of contact between the lacrimal and nasal); quadrate gradually curves caudally along its entire length in lateral view (also noted by Paul, 2008:201, as "shaft nearly straight"); and convex dorsal margin of ilium (also noted by Carpenter and Ishida, 2010).

Horizon Middle Grey Unit, Xinminpu Group; Albian, Early Cretaceous (Tang et al., 2001).

Type Locality Gongpoquan Basin, Mazongshan area, Gansu Province, China.

DESCRIPTION

Cranium and Mandible

Although it is nearly complete, the skull of IVPP V 12534 is extremely compressed transversely, obscuring the medial surfaces of most of the cranial elements (Fig. 3.1); the dorsal surface of the skull roof is visible in only right lateral view (Fig. 3.1A). Some pieces, including the predentary, rostral ramus of the right dentary, rostral ends of both premaxillae, the articulated right postorbital and squamosal, and the left paroccipital process can be separated from the rest of the skull and described in greater detail. See You, Luo, et al. (2003:348–350, fig. 1C) for a reconstruction and measurements of the fully assembled skull. See Appendix 3.1 for additional cranial and postcranial measurements of IVPP V 12534.

Premaxilla and Nasal The oral margin of the premaxilla is gently convex and edentulous (Fig. 3.2A, B). The rostral end of the premaxilla is ventrally inflected relative to the ventral margin of the maxilla, such that the oral margin is positioned well below the maxilla (Fig. 3.2A, B). The caudolateral corner of the oral margin curves caudodorsally towards the rostroventral process of the maxilla (Fig. 3.2A). The oral margin of each premaxilla expands laterally to form a transversely broad snout (Fig. 3.2C, D). Each premaxilla bears two rostrocaudally elongate denticles on its oral margin (Fig. 3.2D), as in *Mantellisaurus* (NHMUK R5764), *Ouranosaurus* (cast of MNHN GDF 300), and *Eolambia* (CEUM 35635). Caudal to the denticles, the ventral surface of the premaxilla is flat and somewhat rugose along the interpremaxillary suture (Fig. 3.2D). Dorsal to the oral margin, the lateral surface of the premaxilla is concave, forming the rostral end of the narial fossa (Fig. 3.2A, B).

Caudodorsal to the narial fossa, the premaxilla splits into two processes, the dorsomedial and ventrolateral processes; the point at which these processes diverge forms the rostral margin of the external naris (Fig. 3.2A, B). The dorsomedial process is overlapped laterally by the premaxillary process of the nasal, which forms the dorsal margin of the external naris (Figs. 3.2A, B, 3.3, 3.4). The dorsomedial processes of the left and right premaxillae meet along the interpremaxillary suture (Fig. 3.2C). The ventrolateral process forms the ventral margin of the external naris and contacts the maxilla along its ventral margin, curving over the rostroventral process of the maxilla and extending caudally to contact the lacrimal and prefrontal (Figs. 3.2A, 3.3, 3.4). The ventrolateral process also contacts the prefrontal in *Hippodraco* (McDonald, Kirkland, et al., 2010), *Theiophytalia* (Brill and Carpenter, 2006), *Dakotadon* (Weishampel and Bjork, 1989), *Iguanodon bernissartensis* (Norman, 1980), *Mantellisaurus* (Norman, 1986), and *Jinzhousaurus* (Barrett et al., 2009), but does not in *Altirhinus* (Norman, 1998) or *Xuwulong* (You et al., 2011). The ventrolateral process tapers towards its caudal end, as in *Hippodraco* (McDonald, Kirkland, et al., 2010), *Dakotadon* (Weishampel and Bjork, 1989), *Iguanodon* (Norman, 1980), and *Mantellisaurus* (Norman, 1986) in contrast to the dorsoventrally expanded ventrolateral processes of more derived hadrosauroids such as *Eolambia* (Kirkland, 1998; Head, 2001) and *Probactrosaurus gobiensis* (Norman, 2002) (Figs. 3.3, 3.4). The ventrolateral process contacts the nasal along its dorsal margin (Figs. 3.3, 3.4). The ventrolateral process of the premaxilla and the nasal contribute equally to the caudal margin of the external naris.

The nasal contacts the premaxilla rostrally and ventrally, the prefrontal ventrally, the frontal caudally, and its counterpart medially (Figs. 3.3, 3.4). Rostral to its contact with the ventrolateral process of the premaxilla, the ventral margin of the nasal curves rostrodorsally toward the dorsomedial process of the premaxilla, forming the dorsal margin and part of the caudal margin of the external naris. The sutures

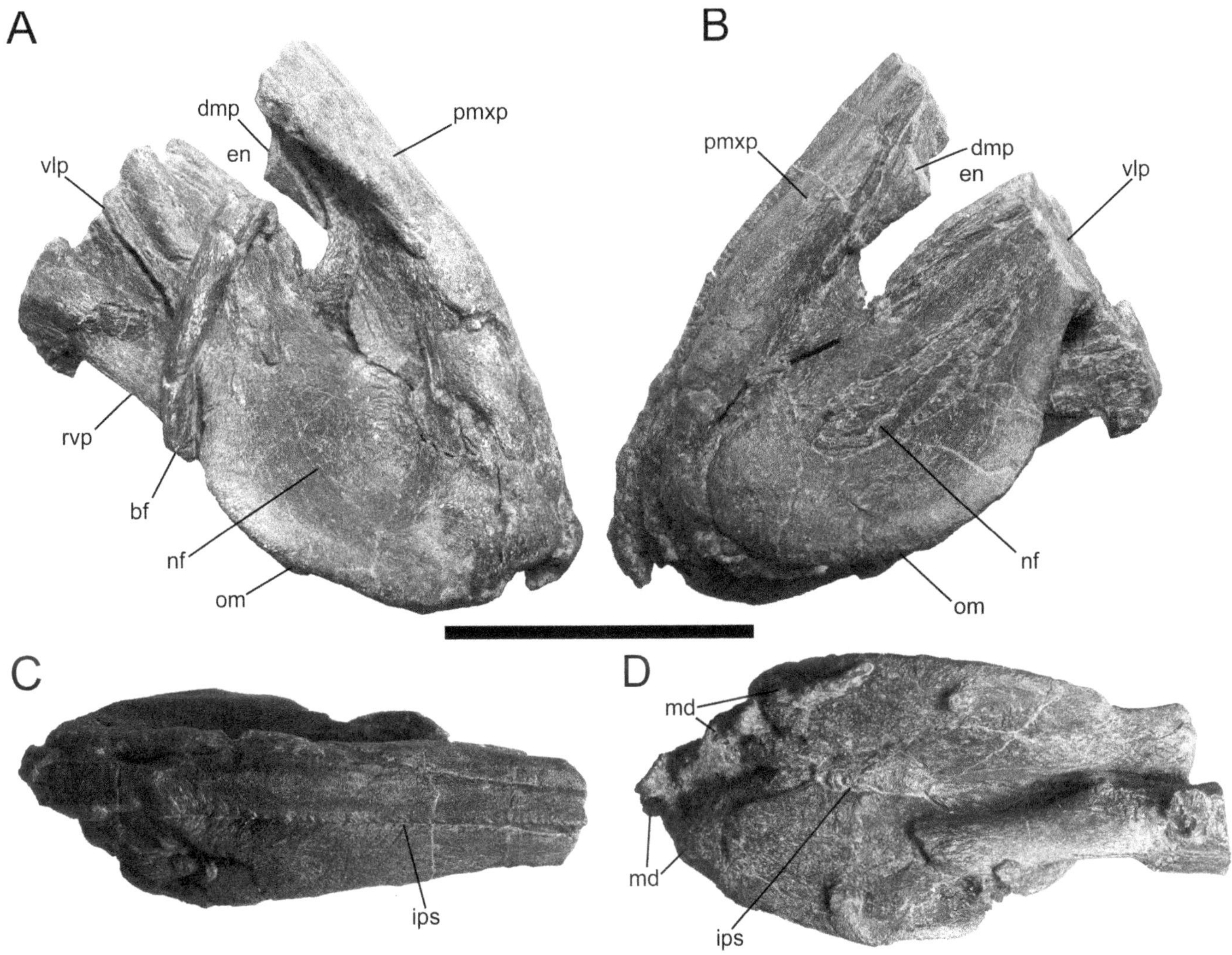

3.2. Articulated premaxillae of IVPP V 12534, holotype of *Equijubus normani*. (A) right premaxilla in lateral view; (B) left premaxilla in lateral view; (C) premaxillae in dorsal view; (D) same in ventral view. Abbreviations: bf, bone fragment; dmp, dorsomedial process; en, external naris; ips, interpremaxillary suture; md, marginal denticles; nf, narial fossa; om, oral margin; pmxp, premaxillary process of nasal; rvp, rostroventral process of right maxilla; vlp, ventrolateral process. Scale bar equals 10 cm.

between the nasals and frontals are difficult to discern due to damage and compression of the skull; their shape is not certain (Figs. 3.3, 3.4).

Maxilla, Palatine, and Pterygoid The rostroventral process of the maxilla curves rostroventrally to contact the ventrolateral process of the premaxilla (Figs. 3.2A, 3.3, 3.4). There is a short diastema between the rostral end of the rostroventral process and the first alveolus. The presence of a rostrodorsal process cannot be ascertained due to the articulation of the premaxillae with the maxillae of IVPP V 12534. The ventral margin of the maxilla is gently concave in lateral view (Figs. 3.3, 3.4). The lateral surface of the maxilla is pierced by several deep, irregularly distributed neurovascular foramina dorsal to the tooth row and near the dorsal margin of the maxilla (Figs. 3.3A, 3.4A), as in *Dakotadon* (SDSM 8656; Weishampel and Bjork, 1989).

The ascending process of the maxilla is rostrocaudally broad and triangular, and contacts the ventrolateral process of the premaxilla and the lacrimal along its rostral margin, and the maxillary process of the jugal along its caudal margin (Figs. 3.3, 3.4). The antorbital fossa is not visible in lateral view. The jugal process of the maxilla is finger-like and projects caudolaterally to lock into a recess on the medial surface of the maxillary process of the jugal (Fig. 3.5), as in *Iguanodon bernissartensis* (Norman, 1980), *Mantellisaurus* (Norman, 1986), *Altirhinus* (Norman, 1998), *Probactrosaurus gobiensis* (Norman, 2002), and *Eolambia* (Kirkland, 1998; CEUM 34356).

The right palatine and right and left pterygoids are partially visible through the orbits and infratemporal fenestrae of IVPP V 12534 (Figs. 3.3, 3.4). However, due to damage and the overlap of other cranial elements, little can be determined

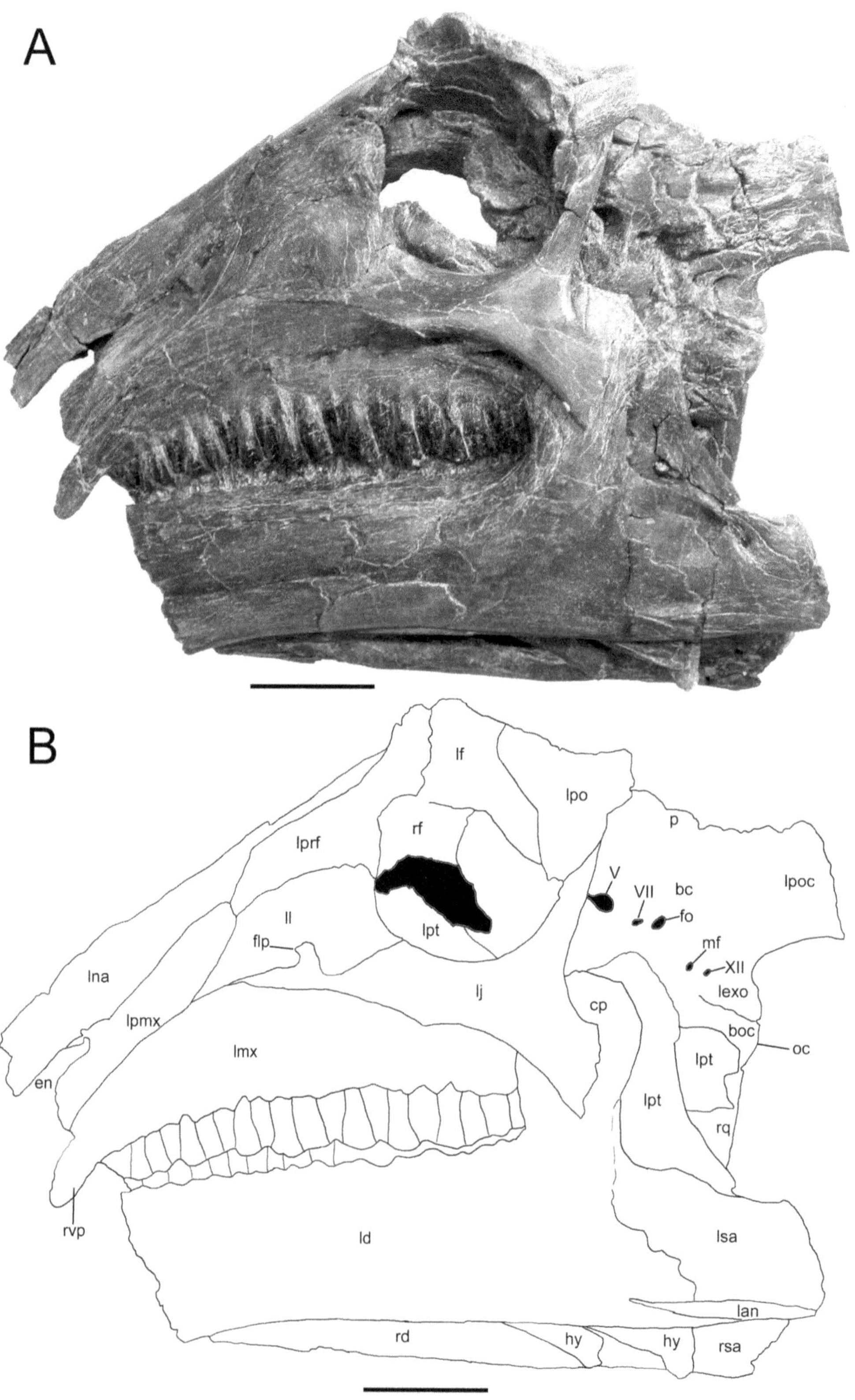

3.3. Skull of IVPP V 12534, holotype of *Equijubus normani.* (A) skull in left lateral view; (B) tracing of the skull in left lateral view. Solid black lines indicate sutures and contacts between cranial bones; dashed lines indicate uncertain sutures and contacts. Predentary, rostral ramus of right dentary, rostral ends of the premaxillae, articulated right postorbital and squamosal, and the left paroccipital process have been removed. Abbreviations: bc, braincase; boc, basioccipital; cp, coronoid process of the left dentary; en, external naris; flp, finger-like process of the maxillary process of the left jugal; fo, fenestra ovalis; hy, hyoid; lan, left angular; ld, left dentary; lexo, left exoccipital; lf, left frontal; lj, left jugal; ll, left lacrimal; lmx, left maxilla; lna, left nasal; lpmx, left premaxilla; lpo, left postorbital; lpoc, left paroccipital process; lprf, left prefrontal; lpt, left pterygoid; lsa, left surangular; mf, metotic foramen; oc, occipital condyle; p, parietal; rd, right dentary; rf, right frontal; rq, right quadrate; rsa, right surangular; rvp, rostroventral process of left maxilla; V, VII, and XII, foramina for cranial nerves V, VII, and XII. Scale bars equal 10 cm.

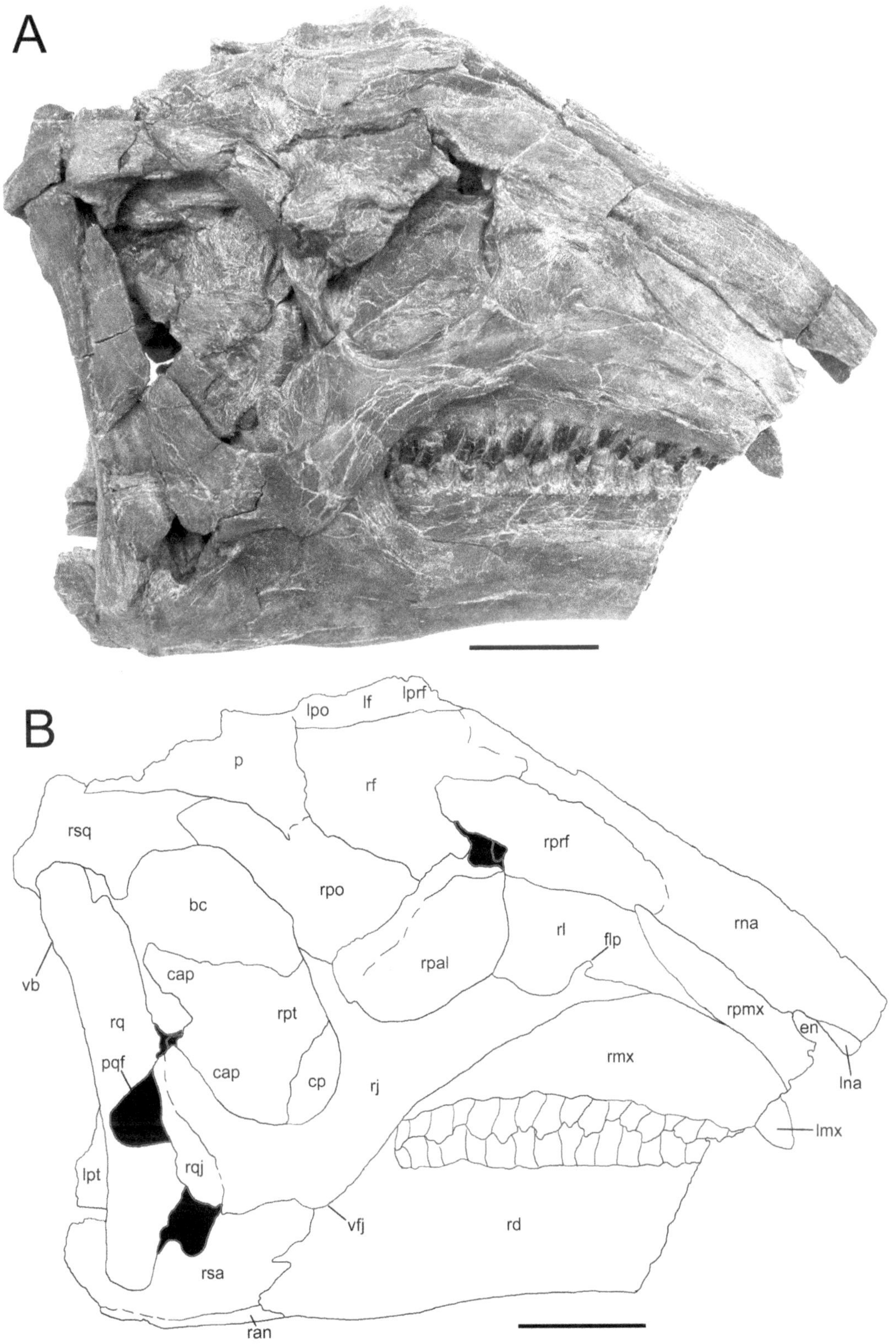

3.4. Skull of IVPP V 12534, holotype of *Equijubus normani.* (A) skull in right lateral view; (B) tracing of the skull in right lateral view. Solid black lines indicate sutures and contacts between cranial bones; dashed lines indicate uncertain sutures and contacts. Predentary, rostral ramus of right dentary, rostral ends of the premaxillae, and the left paroccipital process have been removed. The articulated right postorbital and squamosal were left in place. Abbreviations: bc, braincase; cap, caudal alar process of the right pterygoid; cp, coronoid process of the right dentary; en, external naris; flp, finger-like process of the maxillary process of the right jugal; lf, left frontal; lmx, left maxilla (rostroventral process); lna, left nasal; lpo, left postorbital; lprf, left prefrontal; lpt, left pterygoid; p, parietal; pqf, paraquadrate foramen; ran, right angular; rd, right dentary; rf, right frontal; rj, right jugal; rl, right lacrimal; rmx, right maxilla; rna, right nasal; rpal, right palatine; rpmx, right premaxilla (ventrolateral process); rpo, right postorbital; rprf, right prefrontal; rpt, right pterygoid; rq, right quadrate; rqj, right quadratojugal; rsa, right surangular; rsq, right squamosal; vb, vertical buttress; vfj, ventral flange of right jugal. Scale bars equal 10 cm.

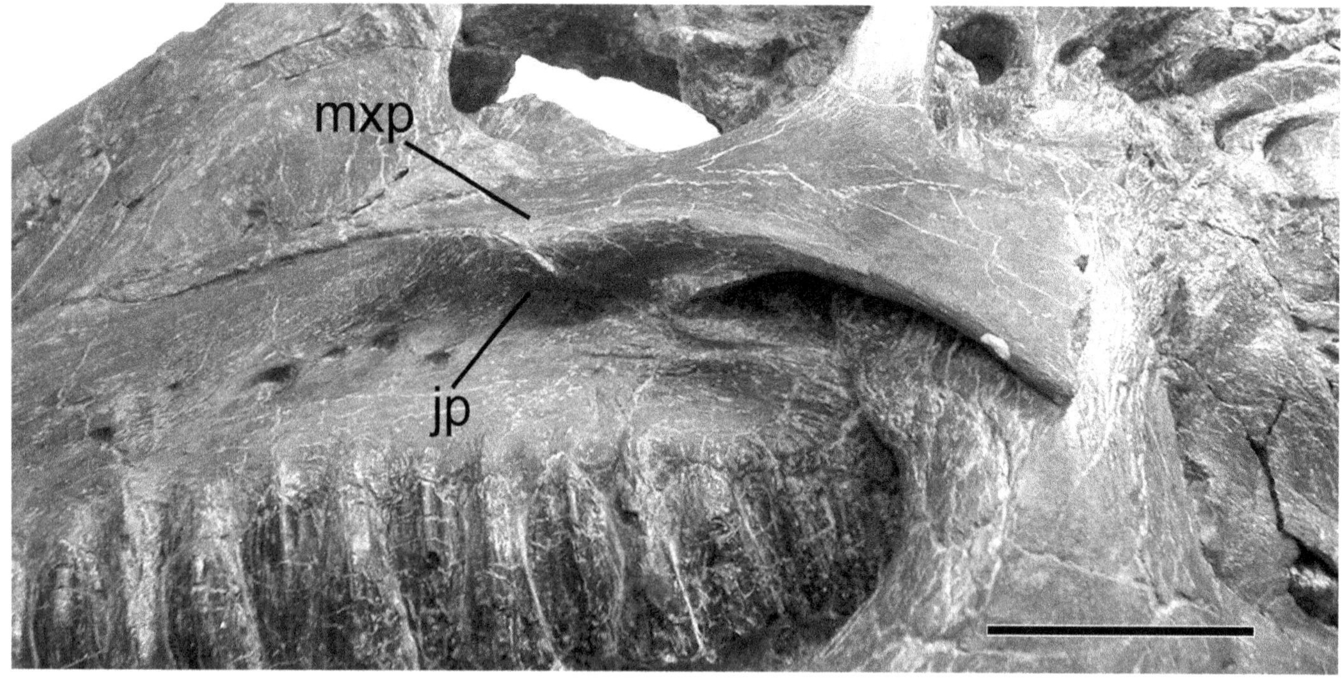

3.5. Left maxilla-jugal articulation of IVPP V 12534, holotype of *Equijubus normani*. Abbreviations: jp, jugal process of maxilla; mxp, maxillary process of jugal. Scale bar equals 10 cm.

regarding their morphology. The caudal alar processes of the right pterygoid can be identified and are in contact with the braincase and the medial wing of the right quadrate (Fig. 3.4).

Jugal The maxillary process of the jugal tapers to a point and extends rostrally to meet the apex of the ascending process of the maxilla (Figs. 3.3, 3.4). The contact between the maxillary process and the lacrimal is unusually complex; a finger-like process arises from the maxillary process and appears to overlap the lateral surface of the lacrimal (Figs. 3.3, 3.4). This feature is present on both sides of the skull and has not been observed in any other iguanodontian, and therefore is an autapomorphy of *Equijubus normani*. Many other basal iguanodonts exhibit a tapering maxillary process, such as *Mantellisaurus* (NHMUK R11521), *Altirhinus* (Norman, 1998), *Jinzhousaurus* (Barrett et al., 2009), and *Xuwulong* (You et al., 2011), but the dorsal margin of the maxillary process is smooth and lacks the finger-like process of *Equijubus*. The maxillary process of the jugal widens abruptly caudal to the autapomorphic finger-like process and forms the ventral margin of the orbit. The dorsal and ventral margins of the maxillary process are parallel between the rostroventral margin of the orbit and the base of the postorbital process (Fig. 3.5). The postorbital process of the jugal projects caudodorsally and meets the stouter jugal process of the postorbital, forming the caudal margin of the orbit (Figs. 3.3, 3.4). Caudoventral to the base of the postorbital process, the ventral margin of the jugal is sinuous with a rounded, ventrally directed flange ventral to the infratemporal fenestra (Fig. 3.4). Caudal to this flange, the jugal abruptly expands dorsoventrally, giving off a dorsally directed process that forms part of the caudal margin of the infratemporal fenestra and excludes the quadratojugal from the fenestra (Fig. 3.4).

Lacrimal and Prefrontal The rostral ramus of the lacrimal tapers to a point, wedging between the maxilla and the ventrolateral process of the premaxilla (Figs. 3.3, 3.4); the rostral ramus also tapers in *Dakotadon* (SDSM 8656) and *Xuwulong* (You et al., 2011). The rostral ramus of the lacrimal is elongated in IVPP V 12534, extending along the dorsal margin of the maxilla to terminate rostral to the apex of the ascending process of the maxilla. This morphology has not been observed in other iguanodonts and is an autapomorphy of *Equijubus;* in other iguanodonts–such as *Iguanodon* (Norman, 1980), *Altirhinus* (Norman, 1998), *Shuangmiaosaurus* (You, Ji, et al., 2003), *Jinzhousaurus* (Barrett et al., 2009), and *Xuwulong* (You et al., 2011)–the rostral ramus of the lacrimal does not extend along the dorsal surface of the maxilla rostral to the apex of the ascending process. The ventral margin of the lacrimal is very slightly concave dorsally between its rostral end and the aforementioned finger-like process of the jugal. Caudal to the finger-like process of the jugal, the ventral margin of the lacrimal is convex ventrally (Figs. 3.3, 3.4). The caudal margin of the lacrimal is concave to form the rostral margin of the orbit. The caudal extension of the

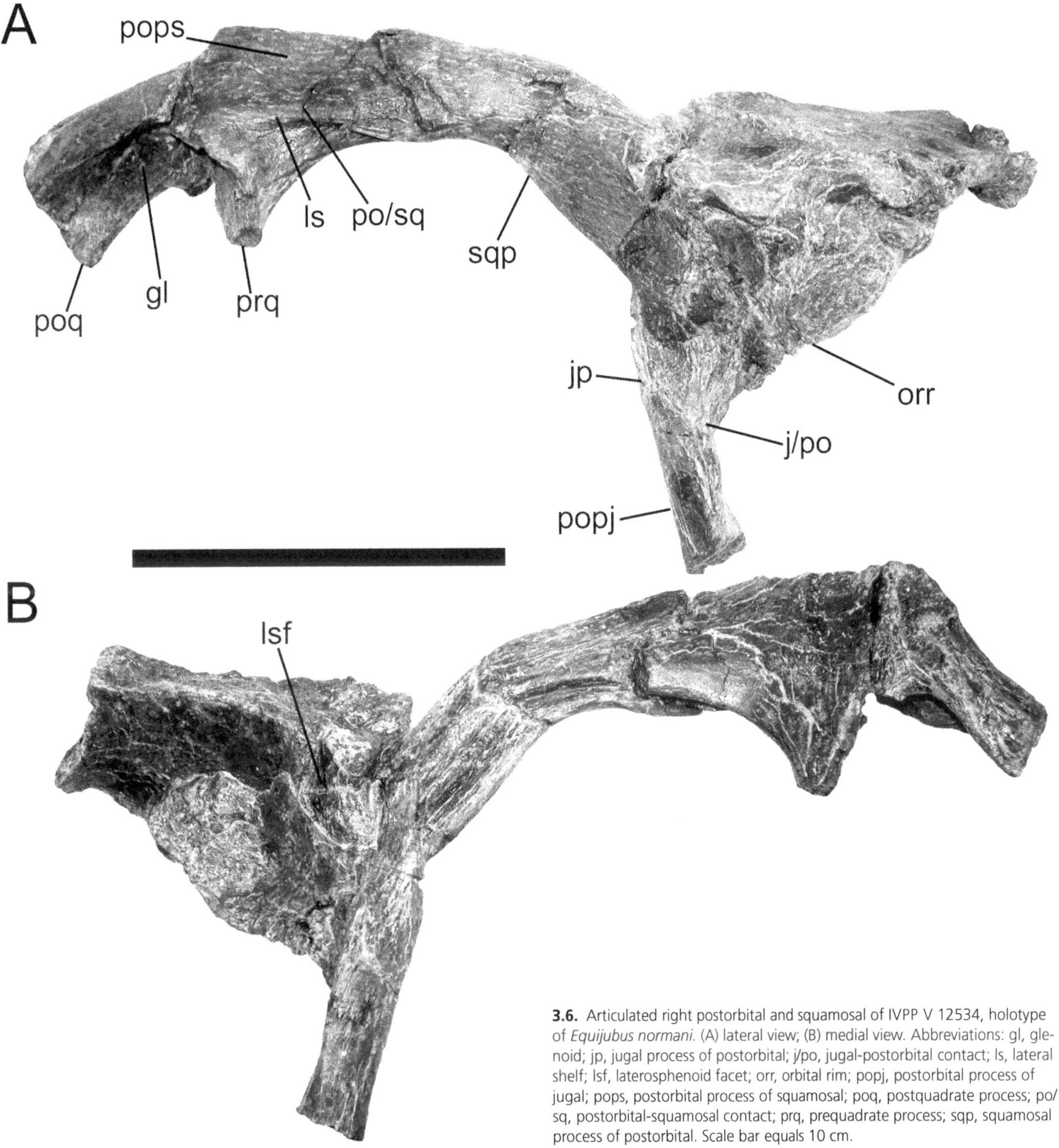

3.6. Articulated right postorbital and squamosal of IVPP V 12534, holotype of *Equijubus normani*. (A) lateral view; (B) medial view. Abbreviations: gl, glenoid; jp, jugal process of postorbital; j/po, jugal-postorbital contact; ls, lateral shelf; lsf, laterosphenoid facet; orr, orbital rim; popj, postorbital process of jugal; pops, postorbital process of squamosal; poq, postquadrate process; po/sq, postorbital-squamosal contact; prq, prequadrate process; sqp, squamosal process of postorbital. Scale bar equals 10 cm.

ventrolateral process of the premaxilla excludes the lacrimal from contact with the nasal.

Due to the articulated nature of the skull, damage to the right prefrontal, and severe deformation of the left prefrontal, little can be determined regarding the anatomy of this element. The prefrontal comprises the rostrodorsal margin of the orbit. Rostral to the orbital rim, the nasal process of the prefrontal is dorsoventrally expanded (Figs. 3.3, 3.4).

Postorbital and Squamosal The postorbital forms the caudodorsal margin of the orbit (Figs. 3.3, 3.4). The orbital rim is rugose (Fig. 3.6A). Rostromedial to the orbital rim, the postorbital meets the frontal along an almost straight suture (Figs. 3.3, 3.4). The jugal process of the postorbital projects rostroventrally to contact the postorbital process of the jugal (Figs. 3.3, 3.4, 3.6A). The squamosal process arches dorsally and projects caudally to meet the postorbital process of the

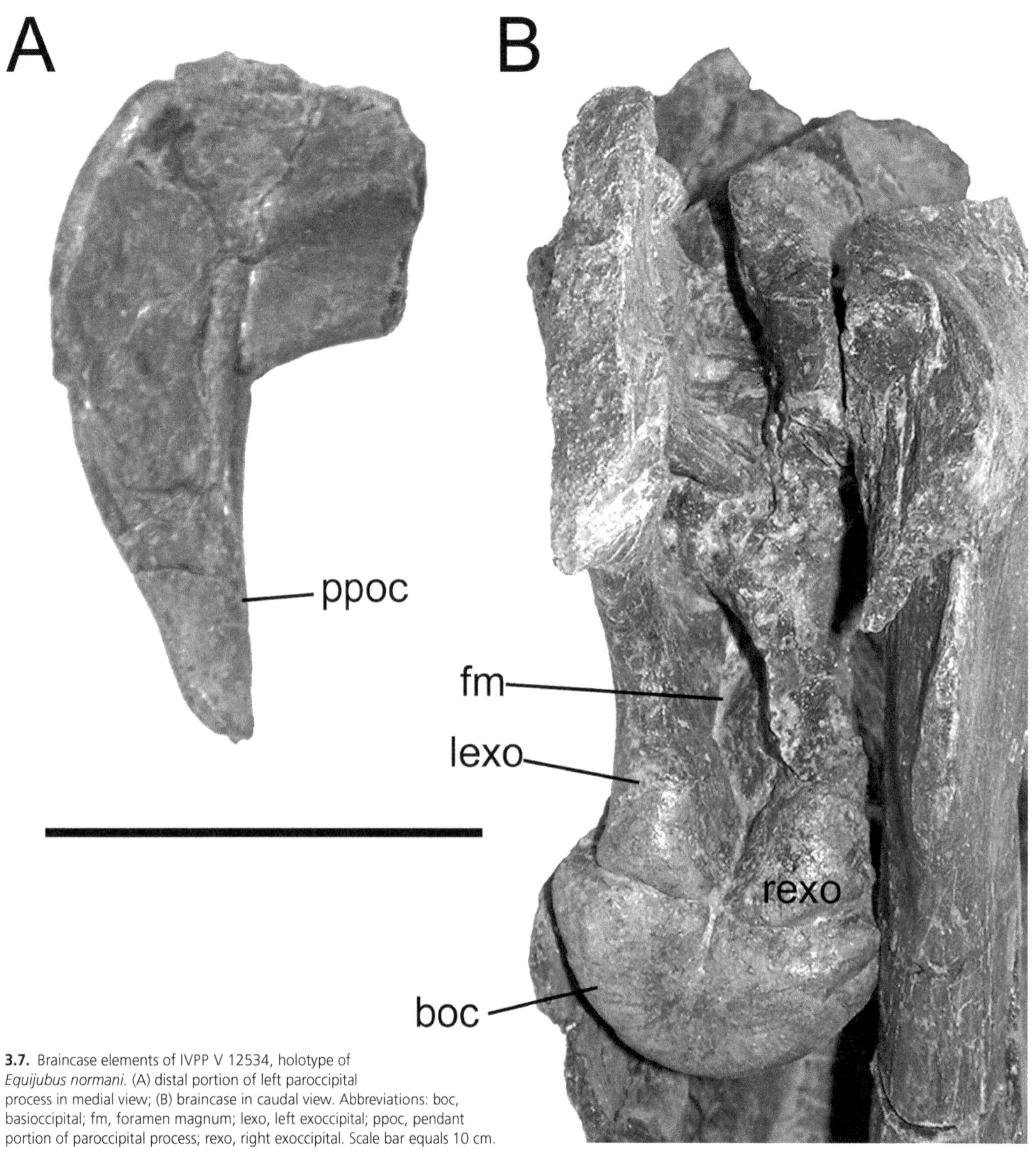

3.7. Braincase elements of IVPP V 12534, holotype of *Equijubus normani.* (A) distal portion of left paroccipital process in medial view; (B) braincase in caudal view. Abbreviations: boc, basioccipital; fm, foramen magnum; lexo, left exoccipital; ppoc, pendant portion of paroccipital process; rexo, right exoccipital. Scale bar equals 10 cm.

squamosal (Figs. 3.4, 3.6A). The caudal end of the squamosal process is rounded and overlaps the lateral surface of the squamosal. The medial surface of the postorbital bears a shallow facet that would articulate with the rostral end of the laterosphenoid (Fig. 3.6B).

The postorbital process of the squamosal bears a rounded shelf on its lateral surface, forming the origin site for M. adductor mandibulae externus superficialis (Ostrom, 1961; Fig. 3.6A). Dorsal to this shelf, the lateral surface of the postorbital process is gently convex dorsally and slopes dorsomedially. Caudal to the postorbital process, two prongs, the prequadrate and postquadrate processes, project ventrally rostral and caudal to the glenoid that would receive the dorsal condyle of the quadrate (Figs. 3.4, 3.6A). The

3.8. Predentary of IVPP V 12534, holotype of *Equijubus normani.* (A) predentary in dorsal view; (B) same in ventral view; (C) same in left lateral view; (D) same in right lateral view. Abbreviations: bdm, base of dorsomedial process; brl, broken base of right lobe of ventromedial process; dg, groove for contact with dentary; ll, left lobe of ventromedial process; md, marginal denticle; mmd, median marginal denticle; rg, groove on rostral surface. Scale bar equals 10 cm.

morphology of the caudomedial process of the squamosal cannot be ascertained due to crushing.

Quadratojugal and Quadrate The quadratojugal is a narrow sliver of bone wedged between the caudal end of the jugal and the lateral wing of the quadrate (Fig. 3.4). The quadratojugal does not overlap the notch in the lateral wing of the quadrate but instead forms the rostral margin of an open paraquadrate foramen.

The quadrate gradually curves caudally along its entire length in lateral view (Fig. 3.4). The ventral condyle of the quadrate is mediolaterally broad and rostrocaudally narrow, with an enlarged lateral condylar surface compared to the medial condyle surface. Approximately one-third of the way up of the quadrate from the ventral condyle is the semicircular quadratojugal notch in the lateral wing of the quadrate (Fig. 3.4); as stated above, the paraquadrate foramen is open, as in *Iguanodon bernissartensis* (Norman, 1980), *Mantellisaurus* (Norman, 1986), and *Altirhinus* (Norman, 1998). The medial wing of the quadrate is obscured by the lateral wing. Immediately ventral to the dorsal condyle of the quadrate is a distinct vertical buttress on the caudal margin of the quadrate (Fig. 3.4).

Neurocranium Due to the compression and articulation of the skull and – in the case of the lateral surface of the braincase – surface damage to the bone, the morphology and sutural relationships of the neurocranium are difficult to ascertain. The midline interfrontal suture is visible on the skull roof, as are the left and right frontoparietal sutures perpendicular to it (Fig. 3.4). The frontal participates in the dorsal rim of the orbit. The only suture discernible on the lateral surface of the braincase is that between the left exoccipital and the basioccipital (Fig. 3.3). The lateral wall of the braincase is pierced by five foramina: the fenestra ovalis, the metotic foramen, and the foramina for cranial nerves V, VII, and XII.

The left paroccipital process is preserved, and its distal portion can be detached from the rest of the skull (Fig. 3.3). The distal portion of the paroccipital process is pendant, with the straight pendant portion directed ventrally (Fig. 3.7A). The caudal aspect of the braincase is visible, although compression of the skull means that the sutures between the supraoccipital and the exoccipitals cannot be discerned. The ventral margin of the foramen magnum is formed exclusively by the left and right exoccipitals; the basioccipital is excluded

(Fig. 3.7B). The occipital condyle is directed caudoventrally. The basioccipital bears a rostrocaudally oriented groove on its ventral surface.

Predentary Like the rest of the skull and mandible, the predentary has suffered marked transverse compression, rendering its true overall shape ambiguous (Fig. 3.8A, B); both of the lateral processes of the predentary appear to project caudally, but it cannot be determined whether they were divergent or parallel to each other. The lateral surfaces of the lateral processes are flat, whereas the medial surfaces are convex (Fig. 3.8C, D). The lateral processes taper at their caudal ends in all views (Fig. 3.8A–D). The ventral surface of each lateral process bears an elongated groove that narrows caudally, forming the contact surface for the rostral ramus of the dentary (Fig. 3.8B).

The dorsal margin of each lateral process bears a row of at least four marginal denticles; four intact denticles are present on the left lateral process, whereas all the denticles on the right lateral process are broken at their bases (Fig. 3.8A, C, D). The median denticle is not broken and is the tallest denticle on the predentary. The median denticle and the intact denticles on the left lateral process are rostrocaudally compressed, subtriangular prongs (Fig. 3.8C), as in *Altirhinus* (Norman, 1998) and *Probactrosaurus gobiensis* (Norman, 2002). Ventral to the denticles on each lateral process is a row of deep neurovascular foramina that extends along the lateral surface of the lateral process parallel to the denticle row and penetrates to the medial surface of the predentary (Fig. 3.8A, C, D). On the rostral surface of the predentary, ventrolateral to the median denticle, are two grooves that extend dorsomedially to ventrolaterally (Fig. 3.8C).

Although part of the ventromedial process is missing, it is clear that it was bifurcated into two ventrolaterally directed lobes, as indicated by the unbroken edges of the preserved left lobe (Fig. 3.8C); this is akin to the predentaries of *Altirhinus* (Norman, 2002) and *Xuwulong* (You et al., 2011), but differs from the non-bifurcated ventromedial process of *Jinzhousaurus* (Barrett et al., 2009). The left lobe of the ventromedial process is triangular with a rounded apex (Fig. 3.8C). The broken base of the dorsomedial process is visible between the lateral processes (Fig. 3.8A).

Dentary The left dentary of IVPP V 12534 is missing its rostral ramus, but the right dentary is complete, and its rostral ramus can be removed for closer inspection (Fig. 3.9A–C). At its rostral end, the dentary exhibits a shallow groove for articulation with the right lateral process of the predentary (Fig. 3.9A). There is a short and mediolaterally compressed diastema caudal to this predentary groove, forming a sharp margin between the predentary groove and the first alveolus (Fig. 3.9C). Medial to the predentary groove, the symphysis projects medially to form a rugose horizontal shelf (Fig. 3.9B, C). The symphysis is oriented rostrolaterally to caudomedially relative to the lateral surface of the dentary in dorsal view (Fig. 3.9C). Caudal to the symphysis is a shallow furrow, the Meckelian groove, which deepens caudally (Fig. 3.9B). The lateral surface of the dentary is pierced by several large neurovascular foramina, with two foramina on the ventrolateral surface of the dentary caudal to the symphysis and a row of irregularly spaced foramina that extends parallel to the tooth row (Fig. 3.9A, D).

The dorsal and ventral margins of the dentary are parallel in lateral view, with a ventrally inflected ventral margin that curves towards the symphyseal region (Fig. 3.9D), as in *Iguanodon bernissartensis* (Norman, 1980), *Mantellisaurus* (Norman, 1986), *Altirhinus* (Norman, 1998), *Probactrosaurus gobiensis* (Norman, 2002), and *Eolambia* (Kirkland, 1998). This differs from the straight dentaries of *Penelopognathus* (Godefroit et al., 2005), *Lanzhousaurus* (You et al., 2005), *Jinzhousaurus* (Barrett et al., 2009), and *Xuwulong* (You et al., 2011). The ventral margin also curves ventrally near the caudal end of the dentary, beginning approximately at the 11th dentary tooth position, and forms a ventrally convex bulge ventral to the base of the coronoid process (Fig. 3.9D). The coronoid process itself arises from another, laterally convex bulge on the lateral surface of the dentary, which begins approximately at the 13th tooth position (Fig. 3.9D). The coronoid process of the left dentary is better exposed than that of the right dentary due to breakage of the left jugal. The coronoid process projects vertically and is rostrocaudally expanded near its apex; the coronoid process attains its greatest rostrocaudal width ventral to its apex (Fig. 3.3).

Surangular and Angular From the coronoid process, the dorsal margin of the surangular slopes caudoventrally toward the glenoid fossa in which the ventral condyle of the quadrate sits (Fig. 3.9E). Rostroventral to the glenoid fossa, the lateral surface of the surangular is pierced by a small surangular foramen (Fig. 3.9E). Caudal to the glenoid fossa, the surangular curves caudodorsally to form the articular process. The angular is visible in lateral view and contacts the ventral margin of the surangular along a horizontal suture (Fig. 3.3).

Hyoids The left and right hyoids are preserved, but only slivers of them can be observed due to compression of the skull.

Dentition The dentary tooth crowns are mesiodistally broad and oblong, as in more basal iguanodontians such as *Altirhinus* (Norman, 1998), *Iguanodon* (Norman, 1980), and *Kukufeldia* (McDonald, Barrett, and Chapman, 2010). The dentary teeth bear numerous ridges on their lingual surfaces, with a distally offset primary ridge, a similarly prominent secondary ridge mesial to the primary, and multiple less prominent accessory ridges arising from the marginal denticles

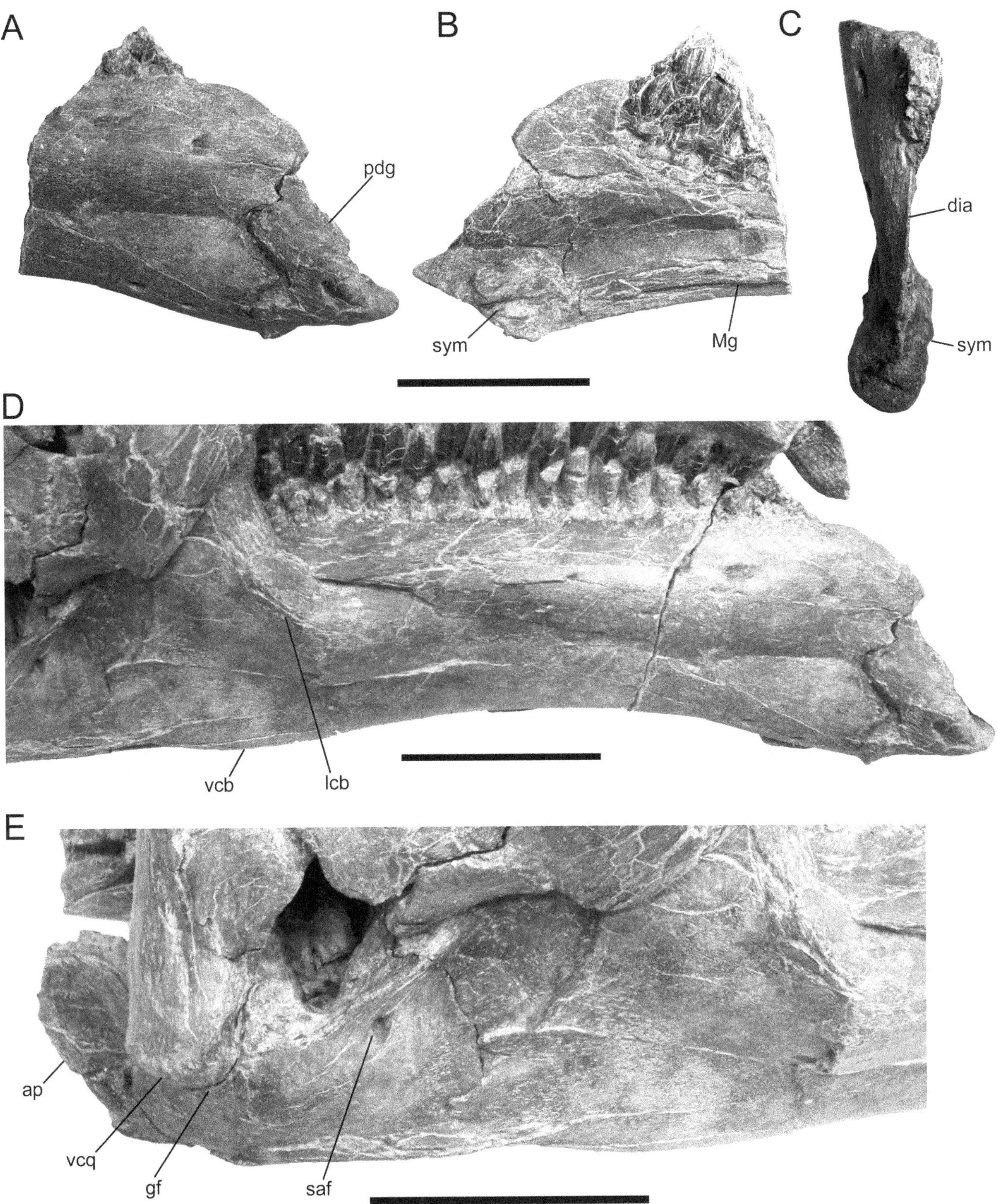

3.9. Right dentary and surangular of IVPP V 12534, holotype of *Equijubus normani*. (A) rostral ramus of right dentary in lateral view; (B) same in medial view; (C) same in dorsal view; (D) right dentary with rostral ramus attached in lateral view; (E) right surangular in lateral view. Abbreviations: ap, articular process; dia, diastema; gf, glenoid fossa; lcb, bulge at base of coronoid process on lateral surface of dentary; Mg, Meckelian groove; pdg, groove for predentary; saf, surangular foramen; sym, symphysis; vcb, bulge ventral to base of coronoid process along ventral margin of dentary; vcq, ventral condyle of quadrate. Scale bars equal 10 cm.

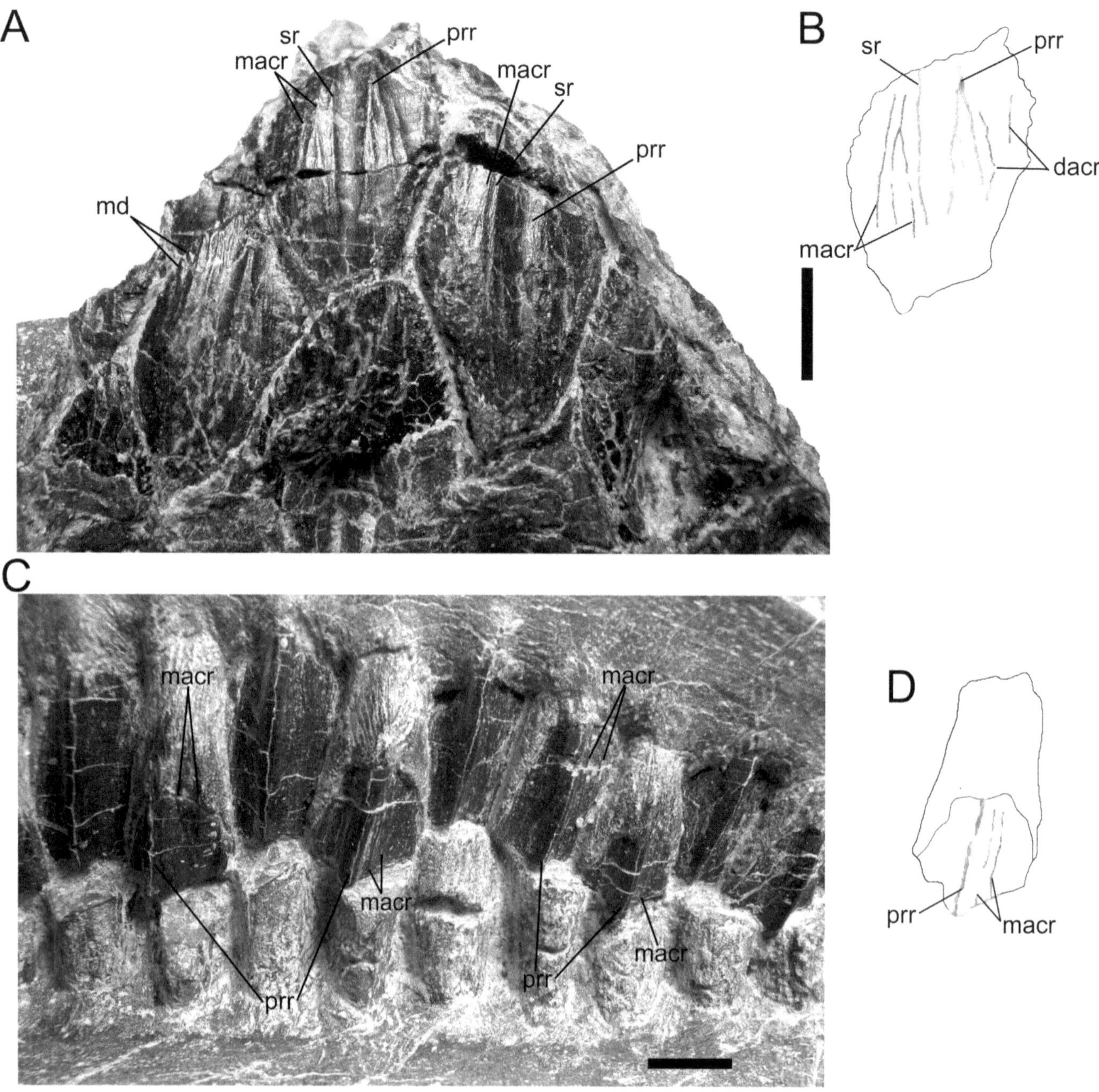

3.10. Dentition of IVPP V 12534, holotype of *Equijubus normani.* (A) dentary teeth in lingual view; (B) tracing of best-preserved dentary tooth (topmost tooth in [A]) in lingual view; (C) maxillary teeth in labial view; (D) tracing of representative maxillary tooth (fourth from left in [C]) in labial view. Abbreviations: dacr, distal accessory ridge; macr, mesial accessory ridge; md, marginal denticle; prr, primary ridge; sr, secondary ridge. Scale bars equal 1 cm.

mesial and distal to the primary and secondary ridges (Fig. 3.10A, B). The marginal denticles are tongue shaped with mammilated edges, and steadily increase in size from the apex of the crown to its base (Fig. 3.10A). There is only one replacement tooth per alveolus, and only one active tooth per alveolus to contribute to the occlusal plane.

Examples of unworn maxillary teeth are not visible in IVPP V 12534; however, the worn crowns visible in labial view are lozenge shaped (Fig. 3.10C). The maxillary teeth bear multiple ridges on their labial surfaces, with a distally offset primary ridge. There are several faint accessory ridges mesial to the primary ridge (Fig. 3.10C, D). As is the case for the dentary teeth, only one maxillary tooth per alveolus forms part of the occlusal plane. The morphology of the marginal denticles cannot be discerned.

The numbers of tooth positions in the dentaries and maxillae cannot be ascertained due to overlap of the caudal ends of these elements by the coronoid processes and jugals

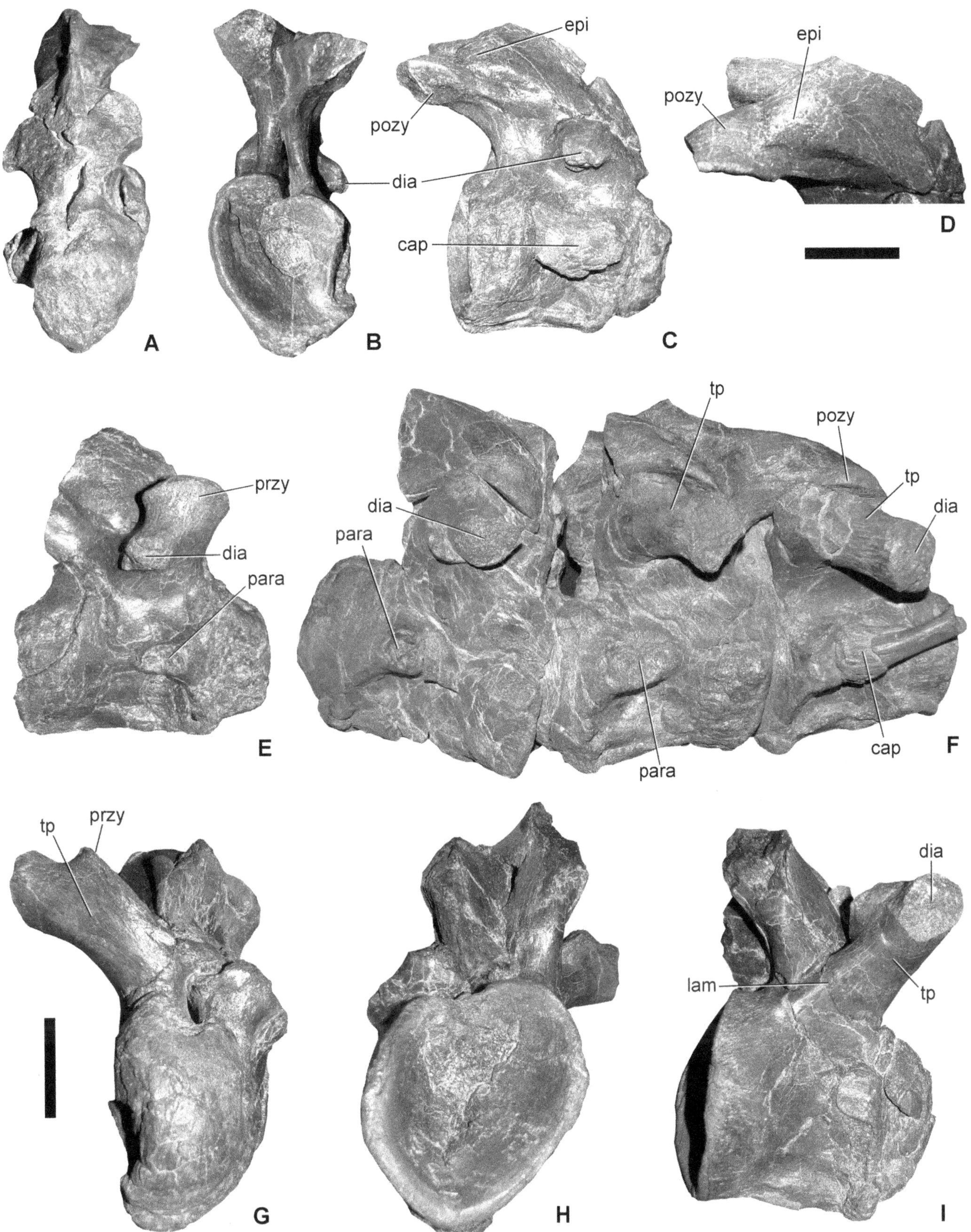

3.11. Representative cervical vertebrae of IVPP V 12534, holotype of *Equijubus normani*. (A–D) cervical 3 in (A) cranial view, (B) caudal view, (C) right lateral view, and (D) detail of epipophyses and postzygapophyses in oblique dorsal–right lateral view; (E) cervical 4 in right lateral view; (F) cervicals 7 to 9 in left lateral view; (G–I), cervical 10 in (G) cranial, (H) caudal and (I) right lateral view. Abbreviations: cap, capitulum of cervical rib; dia, diapophysis; epi, epipophysis; lam, lamina; para, parapophysis; pozy, postzygapophysis; przy, prezygapophysis; tp, transverse process. Scale bars equal 5 cm.

(Figs. 3.3, 3.4). There are 18 active dentary teeth exposed in the right dentary; the active teeth of the left dentary are largely hidden under the left active maxillary teeth. There are 18 active teeth visible in the right maxilla, and 20 in the left maxilla.

Axial Skeleton

Nine cervical vertebrae, 17 dorsal vertebrae, and 6 sacral vertebrae are preserved. The atlas and axis are not present, indicating that there were originally 11 cervical vertebrae in *Equijubus*. The first two dorsal vertebrae are transitional, bearing features of both cervical and dorsal vertebrae, but in keeping with You, Luo, et al. (2003) and other descriptions of basal iguanodonts (e.g., Norman, 2004) they are considered part of the dorsal column herein. Eleven cervicals and 17 dorsals are known in other basal hadrosauriforms such as *Mantellisaurus*, *Iguanodon*, and *Ouranosaurus* (Taquet, 1976; Norman, 1980, 1986). *Jinzhousaurus* also has 11 cervicals (Wang et al., 2010).

Cervical Vertebrae All of the cervicals are strongly transversely crushed, obscuring details such as the shape of the cranial and caudal articular facets. All cervicals are opisthocoelous, with a ball-like, convex cranial articular facet and a strongly concave caudal articular facet. Neural spines are not preserved on any of the vertebrae. Representative cervical vertebrae are shown in Figure 3.11.

The centrum of cervical 3 is strongly transversely crushed (Fig. 3.11A–D). The lateral surfaces of the centrum converge ventrally to form a strong keel extending craniocaudally from the cranial to the caudal articular facet; the prominence of this feature may have been accentuated by crushing. The parapophysis is located cranially and dorsally on the lateral surface of the centrum just caudal to the ball-like cranial articular facet. It is raised relative to the sides of the centrum, and the cervical rib capitulum is preserved in articulation on both sides. In lateral view the neural arch is the same length as the centrum. Diapophyses are positioned on the distal ends of short, stout processes that extend laterally from the neural arch dorsal and slightly caudal to the parapophyses. Prezygapophyses are not preserved. Postzygapophyses are elongate and extend caudal to the caudal articular facet. Although they are crushed together, there is a deep groove extending between the postzygapophyses indicating that they were originally separated. The dorsal surfaces of the postzygapophyses bear a groove that separates distinct epipophyses from the surfaces of the postzygapophyses. Epipophyses are known in a variety of saurischian dinosaurs, some basal ornithischians, and the basal iguanodontian *Tenontosaurus* (e.g., Ostrom, 1970; Butler et al., 2008), but to our knowledge have not previously been reported in any other dryomorph ornithopod. For example, they are not present in *Iguanodon* (Norman, 1980), *Mantellisaurus* (Norman, 1986), or *Jinzhousaurus* (Wang et al., 2010). The presence of epipophyses on cervical 3 therefore represents an autapomorphy of *Equijubus* within Dryomorpha.

Cervical 4 is similar in many respects to cervical 3, although ribs are not articulated with the parapophyses, allowing their morphology to be described (Fig. 3.11E). Parapophyses, again present on the cranial part of the lateral centrum, are oval with the long axis horizontal. They are held on small, raised tubercles and angled caudolaterally. A shallow ridge extends horizontally from the caudal part of the parapophyses caudally toward the caudal articular facet. The right prezygapophysis of cervical 4 is also preserved. It projects laterally level with the dorsal margin of the neural canal and its articular surface faces dorsomedially. It is oval in dorsal view with the long axis trending craniomedially. The diapophysis is located on the caudoventral surface of the prezygapophysis on a raised tubercle and is angled caudoventrally. Postzygapophyses are not preserved.

Cervicals 5 and 6 are similar to cervicals 3 and 4, except that in both cases, the double-headed cervical ribs are preserved in articulation on the left side. Diapophyses on these vertebrae are larger, circular, face laterally, and are supported on a tubercle. Postzygapophyses, preserved in cervical 5, are similar to those of cervical 3, although there is no evidence of epipophyses on these or subsequent vertebrae.

Cervicals 7 to 9 are preserved in articulation with each other (Fig. 3.11F). Cervical ribs articulate with the parapophyses of all three vertebrae on the right-hand side and cervical 9 on the left. In general their morphology is similar to the other cervicals, although the diapophyses extend caudolaterally and are positioned on distinct transverse processes that are circular in cross section.

In cervicals 10 and 11 (Fig. 3.11G–I) the transverse processes are more elongate and robust. They project dorsolaterally and are supported by a lamina that extends from the caudal articular facet craniodorsally and extends along the ventral margin of the transverse process in a position equivalent to the posterior centrodiapophyseal lamina of saurischians (Wilson, 1999). Prezygapophyses are located on the transverse process and do not extend farther cranially than the cranial margin of the process. The lateral surface of the centrum of cervical 11 bears several nutrient foramina.

With the exception of the presence of epipophyses on cervical 3, the cervical column is similar in most respects to those of other basal hadrosauriforms (e.g., *Iguanodon*, Norman, 1980; *Mantellisaurus*, Norman, 1986; *Jinzhousaurus*, Wang et al., 2010).

Cervical Ribs A small portion of a Y-shaped left cervical rib is preserved (Fig. 3.12). It has a relatively long capitulum and a much shorter tuberculum. Both processes are transversely compressed and slightly concave on their medial surfaces. Both articular facets are round in cross section. A ridge extends from between the tuberculum and capitulum caudally and disappears farther down the rib shaft. The dorsal margin of the shaft of the rib is gently convex upward. Medially the shaft is concave; it is broken distally. The rib is very similar in morphology to the eighth cervical rib of *Iguanodon bernissartensis* (Norman, 1980:fig. 32). Several other rib fragments are also preserved (see above), but offer little additional anatomical information.

Dorsal Vertebrae Dorsals 1 and 2 are preserved in articulation with each other and show several features indicating that they are transitional in morphology between cervicals and dorsals (Fig. 3.13A–C). Two dorsal vertebrae exhibiting transitional features are also present in the basal hadrosauriforms *Mantellisaurus* (Norman, 1986) and *Iguanodon* (Norman, 1980), and the hadrosauroid *Probactrosaurus* (Norman, 2002). Centra are similar in morphology to the cervicals: they are transversely crushed and bear a prominent ventral keel that has probably been accentuated by crushing. Both vertebrae are opisthocoelous. This contrasts with the condition in *Mantellisaurus*, in which the cranial articular facets are not as strongly convex as they are in the cervicals (Norman, 1986), but is similar to the condition in *Iguanodon bernissartensis*, in which the transitional vertebrae remain opisthocoelous (Norman, 1980). Parapophyses are oval in shape, with the long axis angled cranioventrally, and are situated partially on the centrum and partially on the neural arch, bridging the neurocentral suture. The neurocentral suture itself is visible on both vertebrae, and on the left side of dorsal two a well-developed concavity is present on the lateral surface of the centrum just ventral to it. This appears to be preservational, however, because it is not present on the right side. Transverse processes extend from a point dorsal to the parapophyses and are broken and crushed, as is much of the neural arch, and few details of the prezygapophyses and postzygapophyses can be determined.

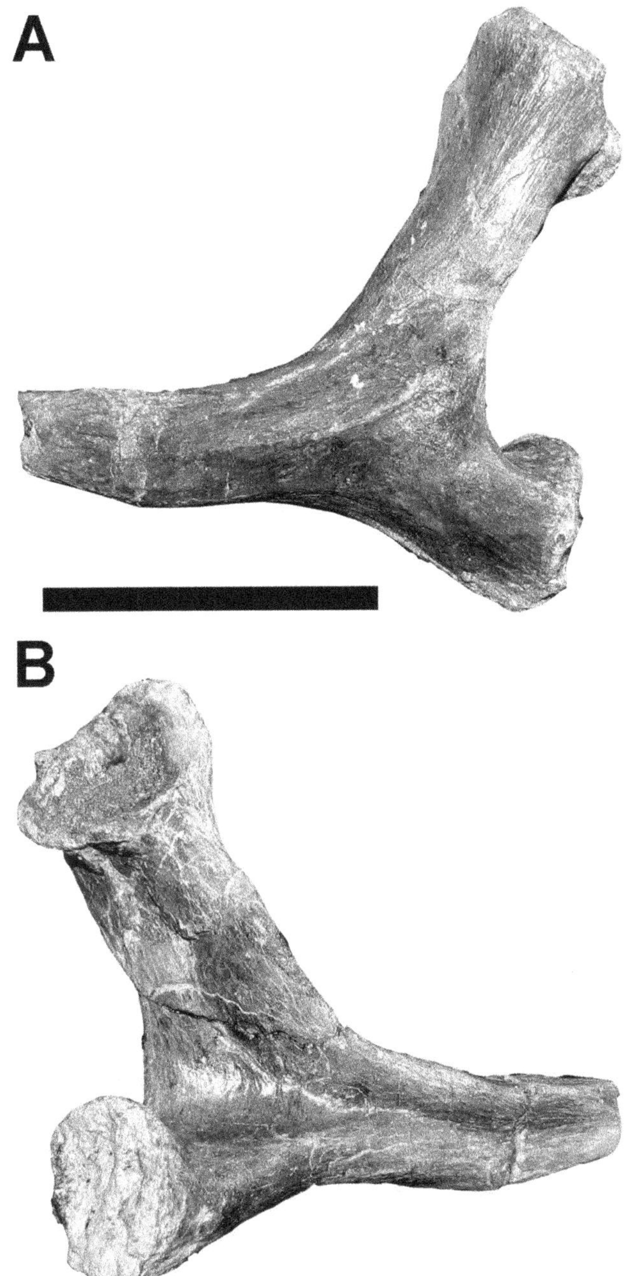

3.12. Cervical rib of IVPP V 12534, holotype of *Equijubus normani*. (A) lateral view; (B) medial view. Scale bar equals 5 cm.

In dorsals 3 and 4, only centra are preserved. These are more amphiplatyan, with very slightly convex cranial articular facets and very slightly concave caudal articular facets. Laterally the sides of the centra are rugose and textured proximal to the articular facets, perhaps due to soft tissue that would have bound the vertebrae together. The sides of the centra are pierced by irregular small foramina, as in *Jinzhousaurus* (Wang et al., 2010), and a ventral keel is present extending craniocaudally.

The centrum of dorsal 5 is similar to those of 3 and 4, except that the cranial and caudal articular facets are flat (Fig. 3.13D–F). The neural arch is partially preserved and the neurocentral suture is visible on both sides. The parapophyses are located caudal to the prezygapophyses on the neural arch, and are large, rounded, and angled caudolaterally. On the left side, the rib capitulum is preserved in articulation. The right transverse process is broken, but the left is preserved, although it is crushed and has rotated to project caudally. A lamina extends from the prezygapophysis along the cranial margin of the transverse process (in the same position as the prezygodiapophyseal lamina of saurischians [Wilson, 1999]), and the bone surface in this area is rugose,

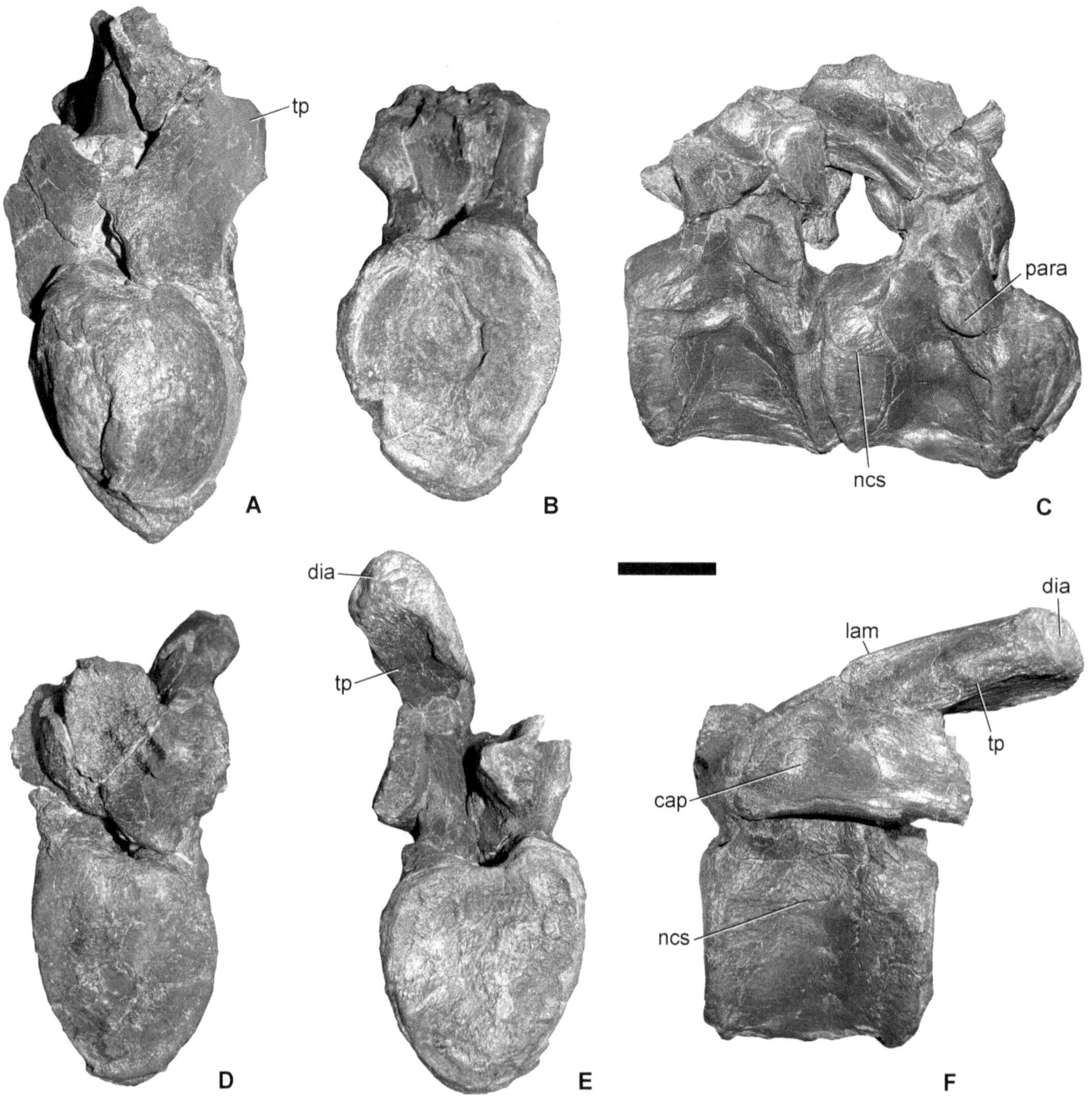

3.13. Representative cranial dorsal vertebrae of IVPP V 12534, holotype of *Equijubus normani.* (A) dorsal 1 in cranial view; (B) dorsal 2 in caudal view; (C) dorsals 1 and 2 in right lateral view; (D–F) dorsal 5 in (D) cranial, (E) caudal and (F) left lateral view. Abbreviations: cap, capitulum of dorsal rib; dia, diapophysis; lam, lamina; ncs, neutrocentral suture; para, parapophysis; tp, transverse process. Scale bar equals 5 cm.

suggestive of soft tissue attachment. The transverse process is roughly triangular in longitudinal cross section with the apex pointing cranioventrally, as in *Jinzhousaurus* (Wang et al., 2010) and *Iguanodon* (Norman, 1980). The diapophysis is oval with the long axis trending horizontally and is slightly convex. The postzygapophyses of dorsal 4 are preserved in articulation with the prezygapophyses of dorsal 5, obscuring details of their anatomy.

Dorsals 6 and 7 are very similar in morphology to dorsal 5, and the left transverse processes on both are preserved but rotated to project caudally as in the latter. The parapophyses are located slightly dorsal to the prezygapophyses on the neural arch. The postzygapophyses of dorsal 6 are broken distally, but a thin, transversely compressed plate of bone extends from dorsal to the neural canal caudally between the postzygapophyses. This feature appears to be

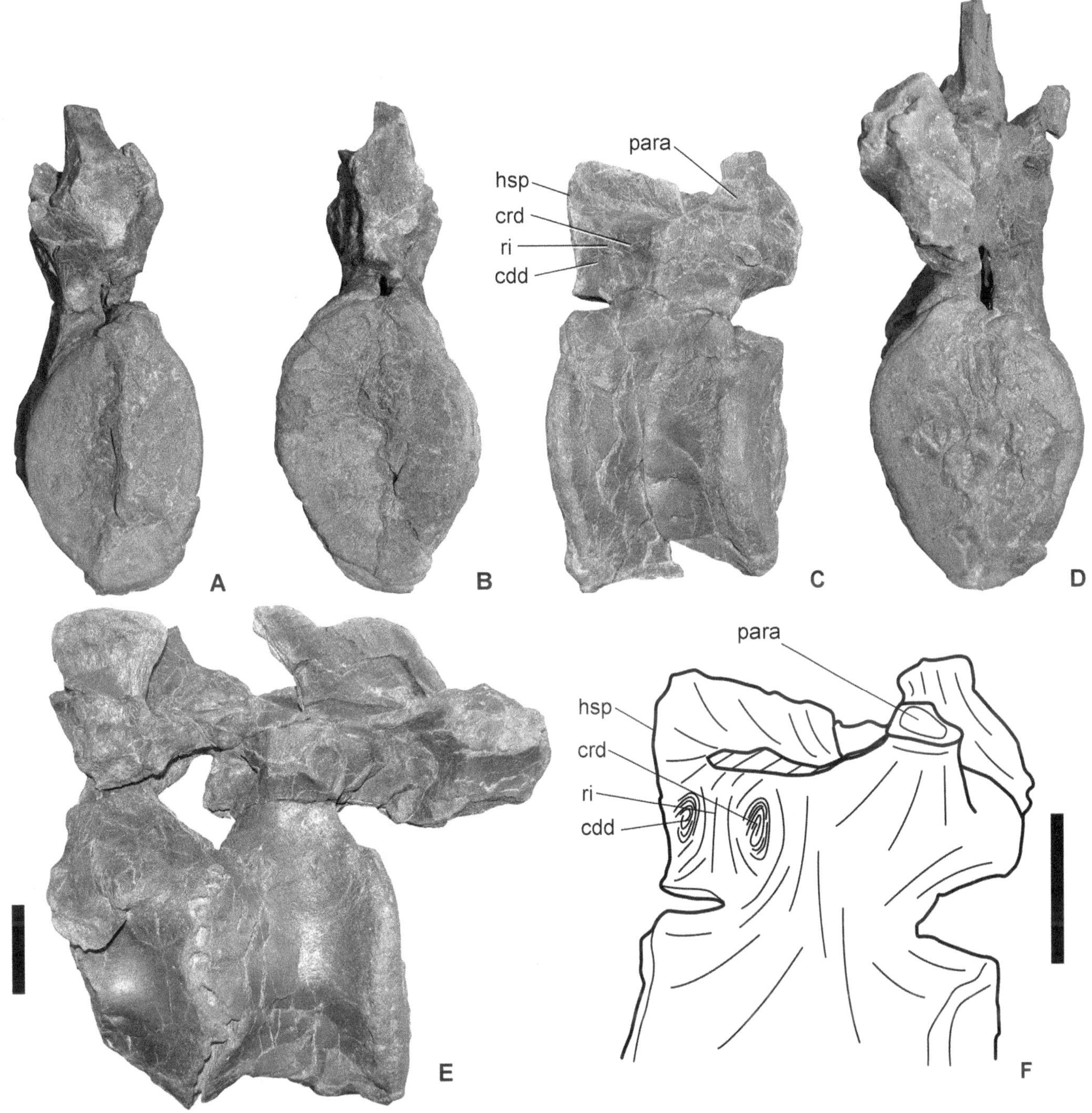

3.14. Representative caudal dorsal vertebrate of IVPP V 12534, holotype of *Equijubus normani*. (A) dorsal 13 in cranial view; (B) dorsal 13 in caudal view; (C) dorsal 13 in right lateral view; (D) dorsal 16 in cranial view; (E) dorsals 16 and 17 in right lateral view; (F) detail of neural spine of dorsal 13 in right lateral view showing hyposphene morphology. Abbreviations: cdd, caudal depression on hyposphene; crd, cranial depression on hyposphene; hsp, hyposphene; para, parapophysis; ri, vertical ridge on hyposphene. Scale bars equal 5 cm.

a hyposphene (Fig. 3.14), a feature known in many archosauromorphs and saurischians (Apesteguía, 2005), but that has not–to our knowledge–previously been reported in any ornithischian, and therefore represents an autapomorphy of *Equijubus*. The capitulum of the rib of dorsal 7 is preserved in articulation on the left side, obscuring some details of the anatomy of the neural arch, but a plate-like hyposphene appears to be present.

Only the centrum and lower part of the neural arch are preserved on dorsal 8. The centrum is unchanged from the morphology of the cranial dorsals, and the neurocentral suture is clearly visible. Once again the rib capitulum adheres

3.15. Sacral vertebrae with associated ilia of IVPP V 12534, holotype of *Equijubus normani*, in (A) dorsal and (B) ventral view. Abbreviations: gr, groove; li, left ilium; ns, neural spine; ot, ossified tendons; ri, right ilium. Scale bar equals 10 cm.

to the left parapophysis, located caudodorsal to the prezygapophyses, which are broken. The hyposphene is again present, although the postzygapophyses are broken. Only the centra are preserved for dorsals 9 and 10, and the centrum and ventral part of the neural arch in dorsal 11. The morphology of these centra does not differ from that of the other dorsals.

The postzygapophyses of dorsal 11 are preserved in articulation with the prezygapophyses of dorsal 12. Immediately dorsal to the postzygapophyses, the base of the neural spine of dorsal 11 is preserved. It is rounded in cross section and transversely broad caudally, but extends cranially as a thin, plate-like sheet and is similar in morphology to that of *Mantellisaurus* (Norman, 1986) and *Iguanodon bernissartensis* (Norman, 1980). The prezygapophyses of dorsal 12 are separated by a groove ventrally, presumably for the hyposphene of dorsal 11. The parapophyses are oval and situated at the base of the transverse processes, which are broken and not preserved. The long axis of the parapophysis extends cranioventrally, and a ridge arises from the cranial margin and extends cranially to form the craniolateral margin of the neural canal in cranial view. The postzygapophyses of dorsal 12 are preserved in articulation with the prezygapophyses of dorsal 13. As in dorsal 11, the neural spine of dorsal 12 arises immediately dorsal to the postzygapophyses and is transversely relatively thickened in this area. Cranially, the neural spine becomes transversely thinner and plate-like. Cranial to the neural spine a deep prespinal fossa is present caudal to the prezygapophyses. A hyposphene is present.

Dorsal 13 is similar in morphology to dorsal 12 but the hyposphene is better preserved, allowing more details of the anatomy to be observed (Fig. 3.14). The hyposphene comprises a medially situated, transversely compressed plate of bone that arises dorsal to the neural canal and extends dorsally to the ventromedial margin of the postzygapophyses. Laterally, a depression on the hyposphene defines the caudal margin of the neural arch. A dorsoventral ridge extends up the hyposphene and separates this cranial depression from a second, caudal depression on its transverse surface. The transverse processes of dorsal 13 are broken, but at their bases they appear dorsoventrally compressed. The parapophysis, clearly preserved on the right, is located between the prezygapophysis and transverse process.

The centra of dorsals 14–17 become increasingly larger, moving caudally along the dorsal vertebral column; all bear irregular nutrient foramina on their lateral surfaces. Dorsal 14 is similar to dorsal 13 in morphology. Once again, the region of the hyposphene is well preserved, and features such as the two depressions separated by a dorsoventral ridge can also be observed on this vertebra. The centrum of dorsal 17 is preserved independently of its neural arch, which is articulated with the neural arches of dorsals 16 and 17. The postzygapophyses and hyposphene of dorsal 15 are well preserved and similar in morphology to those of dorsals 12 and 13. A portion of the lower part of the neural spine is preserved. It extends from dorsal to the postzygapophyses cranially, is angled slightly caudally, and is transversely compressed. Transverse processes are broken on this vertebra but they appear to be craniocaudally elongate and dorsoventrally compressed, at least at their bases.

Dorsals 16 and 17 are preserved in articulation, and only a portion of the centrum of dorsal 17 is present (Fig. 3.14E). The parapophyses are still located on the neural arch, and have not migrated onto the transverse process. This contrasts with the condition in *Iguanodon bernissartensis, Mantellisaurus*, and *Jinzhousaurus*, in which the parapophysis migrates along the transverse process, eventually forming a conjoined facet with the diapophysis in dorsal 17 (Norman, 1980, 1986; Wang et al., 2010). The transverse process of dorsal 17 is small, and has been crushed so that it now projects dorsally. It is relatively short and dorsoventrally flattened, being unsupported by buttresses or laminae, in contrast to

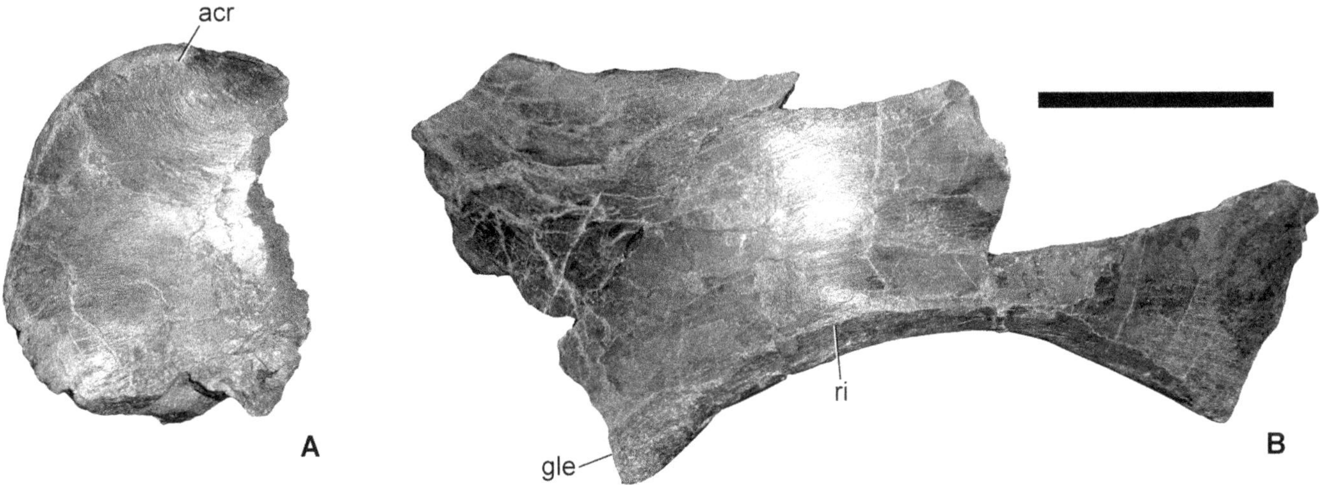

3.16. Fragments of left scapula of IVPP V 12534, holotype of *Equijubus normani,* in lateral view. (A) proximal plate; (B) blade. Abbreviations: acr, acromion process; gle, glenoid; ri, ridge. Scale bar equals 10 cm.

more cranial transverse processes. A change in shape of the transverse processes along the dorsal series is also seen in *Iguanodon bernissartensis* (Norman, 1980) and *Mantellisaurus* (Norman, 1986). Small portions of the bases of the neural spines are preserved; these are strongly transversely compressed and angled slightly caudally.

Sacral Vertebrae Six vertebrae are fused together to form the sacrum. Eight vertebrae usually comprise the sacrum in *Iguanodon bernissartensis* (Norman, 1980), while seven co-ossified vertebrae form the sacrum in *Mantellisaurus* (Norman, 1986).

The sacrum is preserved with both ilia in natural articulation, and has been crushed transversely, so that details of the articulation between the sacrum and ilia are obscured (Fig. 3.15). The ventral portions of the first three sacrals and the caudal half of the sixth sacral have broken and are not preserved. Small foramina pierce the lateral surface of the centrum of sacral 4. A deep ventral groove is present on sacral 5, but was not present on sacral 4. The more caudal sacrals of *Iguanodon bernissartensis* also bear a groove ventrally (Norman, 1980), but this is not the case in *Mantellisaurus* (Norman, 1986) or *Probactrosaurus* (Norman, 2002). The neural spines, which are broken dorsal to the ilia, are craniocaudally broad and transversely compressed. Ossified tendons extend along the bases of the neural spines and are better preserved on the left-hand side. The total length of the sacral rod, as preserved, is 460 mm.

Appendicular Skeleton

The appendicular skeleton is in general very fragmentary and poorly preserved.

Scapula Two fragments represent the partial proximal plate and middle section of the left scapula blade, but could not be fitted together (Fig. 3.16). The partial plate includes the acromion process and dorsal part of the articulation for the coracoid, but lacks the glenoid. The lateral surface of the plate is strongly concave and is bounded dorsally by a low, rounded, striated swelling, which represents the acromion process. In the basal hadrosauroids *Bactrosaurus* (AMNH 6553), *Gilmoreosaurus* (AMNH 30725, 30727), and *Jinzhousaurus* (Wang et al., 2010), and in hadrosaurids, the acromion process projects laterally and the proximal plate is relatively small. *Equijubus* exhibits a condition more similar to that of more basal iguanodontians, in which the acromion process forms a distinct tubercle but is not folded strongly laterally (e.g., *Camptosaurus dispar* [USNM 4282, 5473]; *Uteodon aphanoecetes* [CM 11337]; *Hypselospinus* [NHMUK R1629; Norman, 2010]; *Barilium* [NHMUK R2848; Norman, 2011b]; *Iguanodon bernissartensis* [Norman, 1980]). The coracoid articulation is subtriangular in proximal view, with the apex of the triangle pointed medially. It is strongly rugose. The ventral part of the medial surface has a broad longitudinal ridge, which lies at the same level as the broadest part of the coracoid articular surface. The area dorsal to this is shallowly concave.

The central section of the scapula has a strongly concave ventral margin in lateral view and it appears that the proximal plate would have been expanded farther ventrally than dorsally with respect to the blade midline. In the majority of iguanodontians–such as *Camptosaurus dispar* (USNM 4282, 5473), *Iguanodon bernissartensis* (Norman, 1980), and *Gilmoreosaurus* (AMNH 30725, 30727)–the ventral margin of the scapula is dorsally concave; however, in *Hypselospinus* (NHMUK R1629) and *Probactrosaurus* (Norman, 2002)

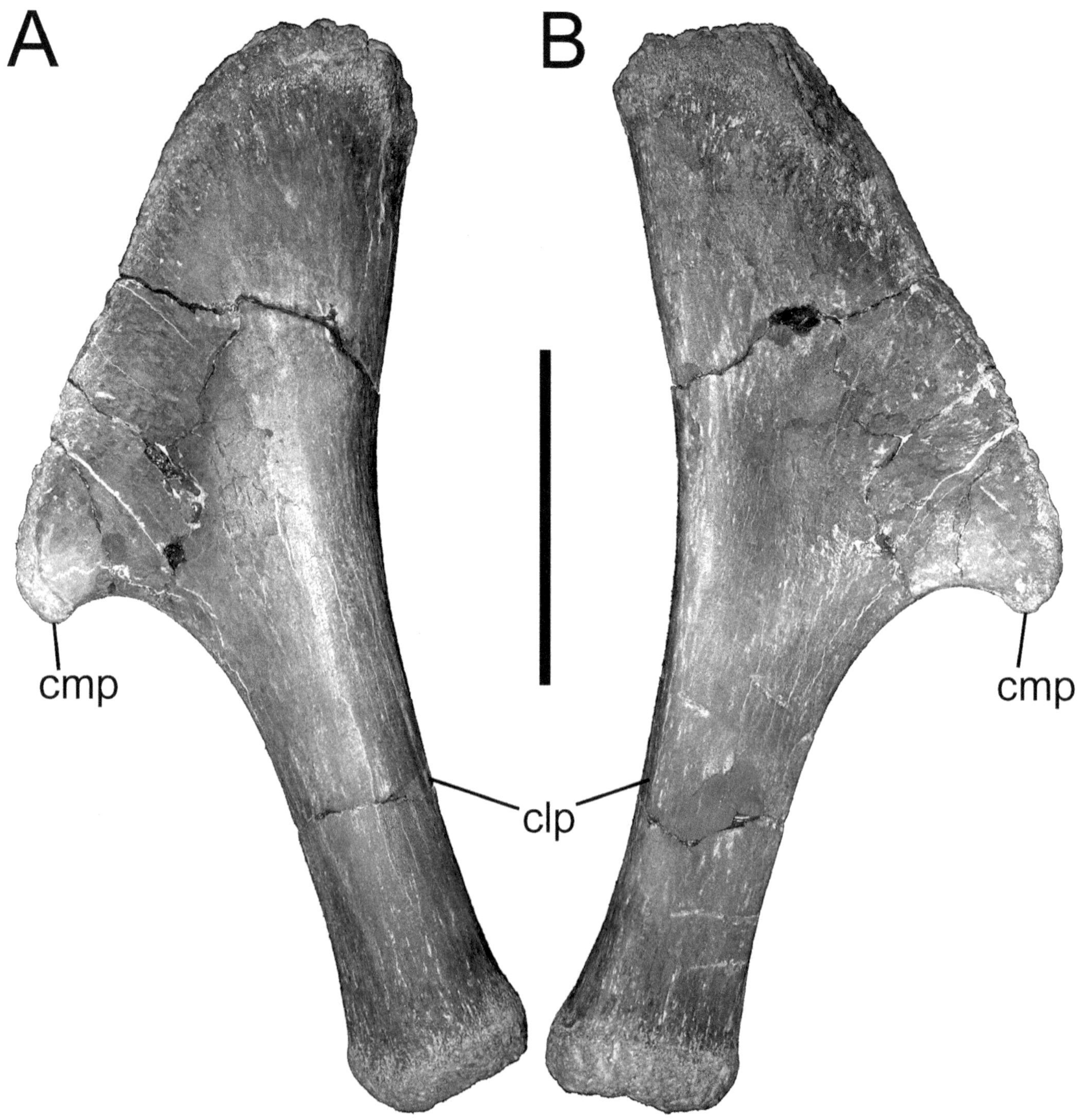

3.17. Left sternal of IVPP V 12534, holotype of *Equijubus normani*. (A) ventral view; (B) dorsal view. Abbreviations: clp, caudolateral process; cmp, caudomedial process. Scale bar equals 10 cm.

the ventral margin of the scapula blade is straight, allowing *Equijubus* to be distinguished from these taxa. A low ridge extends parallel to the ventral margin along the most strongly concave part of the ventral margin, forming a shallow ventrolaterally facing surface. The rest of the lateral blade surface is laterally bowed. The blade becomes more transversely compressed toward its distal end. Although the dorsal margin is incomplete, the curvature of the ventral margin suggests the presence of a distal expansion, as is present in the majority of iguanodontians (e.g., *Dysalotosaurus*, MB R.1707; *Camptosaurus dispar*, USNM 4282, 5473; *Barilium*, NHMUK R2848; *Bactrosaurus*, AMNH 6553), although the scapular blade is not distally expanded in *Probactrosaurus* (Norman, 2002) or in *Iguanodon bernissartensis* (Norman, 1980). In medial view,

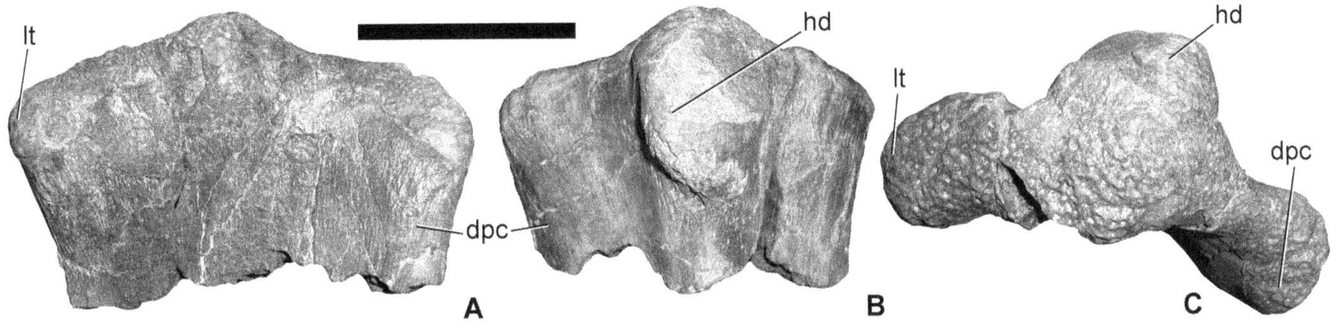

3.18. Proximal end of left humerus of IVPP V 12534, holotype of *Equijubus normani*, in (A) cranial, (B) caudal and (C) proximal view. Abbreviations: dpc, deltopectoral crest; hd, humeral head; lt, lesser tuberosity. Scale bar equals 10 cm.

the blade is generally concave, with few surface features. Toward the distal end, there are a large number of prominent longitudinal striations. Again, a low ridge or swelling is present along the concave ventral margin, which fades out caudally and causes the ventral margin to be thicker than the dorsal margin.

Sternals The left sternal is complete, whereas the otherwise well-preserved right sternal is broken along its caudomedial margin. At the craniolateral corner of the sternal, a concave facet, perhaps for a sternal rib, is present. The sternal is dorsoventrally thickened along its cranial margin, and the bone surface is rough and unfinished for the coracosternal cartilage. The main body of the sternal is slightly convex along its medial margin and straight along its lateral margin (Fig. 3.17). The rod-like caudolateral process of the sternal is directed caudolaterally, is oval in cross section, and is fringed by striations distally. The distal end of the caudolateral process is roughened. A small, hook-like, caudally directed caudomedial process is present (Fig. 3.17). The sternal of *Equijubus* is overall similar to that of *Jinzhousaurus*, except that the caudomedial process of the latter is more prominent and is directed caudomedially (Wang et al., 2010:fig. 6c, d). The sternal of *Lanzhousaurus* lacks the hook-like caudomedial process of *Equijubus* (You et al., 2005:fig. 3b).

Humerus Only the proximal portion of the left humerus is preserved (Fig. 3.18). In proximal view, the cranial surface is concave, with this concavity interrupted by a small swelling marking the cranial expansion of the humeral head. By contrast, the caudal margin of the humerus is divided into two concavities by the humeral head, which is strongly expanded caudally. Extending laterally from the humeral head, the proximal surface of the humerus expands slightly craniocaudally, before terminating in a narrow, rounded apex. Medial to the humeral head, the lesser tuberosity maintains the same width for its entire length, ending in a bluntly rounded process. In proximal view, the deltopectoral crest and tubercle are separated by an angle of approximately 120 degrees. In cranial view, the proximal margin of the humerus slopes slightly dorsally from the medial tubercle to the head, where the angle changes to slope ventrally to the top of the deltopectoral crest. In caudal view, a robust ridge extends ventrally from the ventral surface of the humeral head, dividing the caudal surface into two concave areas. The fragmentary nature of the humerus precludes detailed comparisons with other taxa, although the preserved portion appears similar to the proximal humerus in *Iguanodon bernissartensis* (Norman, 1980), *Mantellisaurus* (Norman, 1986), and *Jinzhousaurus* (Wang et al., 2010).

Ilium Parts of both ilia are preserved in natural articulation with the sacrum (Fig. 3.19). They have been transversely crushed against the sacral vertebrae, and as a result a series of gentle undulations on the lateral surface of the better-preserved left ilium correspond to the anatomy of the underlying vertebrae. The preacetabular process is not preserved on either side. The postacetabular process is entirely broken on the right and is not preserved; on the left it is broken at its distal end.

The dorsal margin of the ilium is gently convex, in contrast to the condition in most other hadrosauriforms, in which it is generally rather straight (e.g., *Mantellisaurus*, NHMUK R11521; *Jinzhousaurus*, Wang et al., 2010; *Probactrosaurus*, Norman, 2002), although it is also strongly convex in *Barilium* (NHMUK R802) and in NHMUK R3741, the famous "Mantel-piece" (Carpenter and Ishida, 2010). The dorsal margin is transversely thickened relative to the body of the ilium and fringed in a series of prominent striations. Just caudal to the ischial peduncle, the dorsal margin of the ilium becomes deeper dorsoventrally, thicker transversely, and is folded over into a laterally everted rim, possibly an incipient version of the supra-acetabular process of more derived iguanodontians; the striations continue under this surface. The development of the laterally everted rim appears to be comparable to that of *Mantellisaurus* (NHMUK R11521) and *Probactrosaurus* (Norman, 2002). The caudal

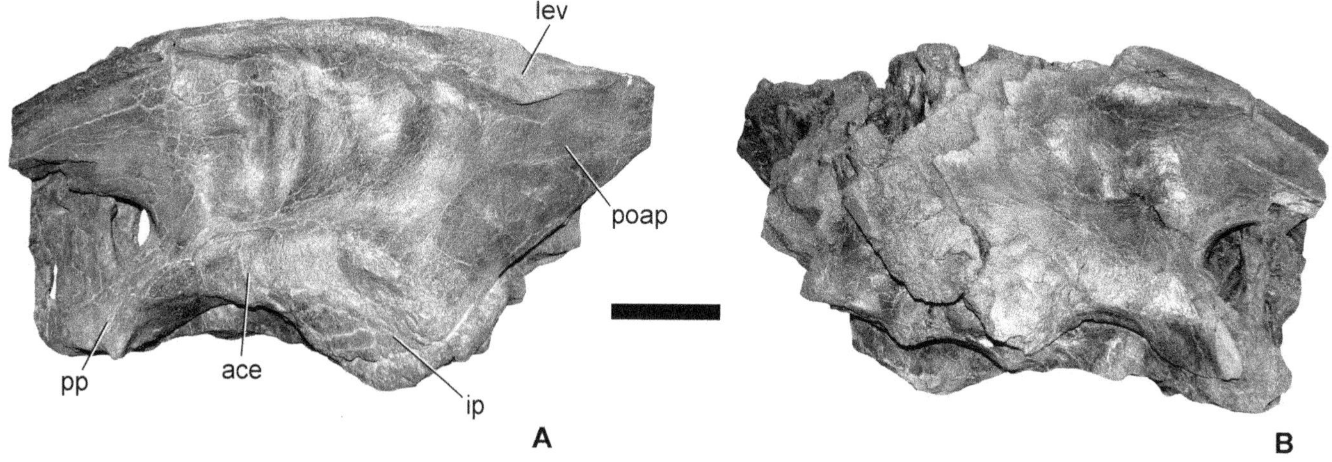

3.19. Ilia of IVPP V 12534, holotype of *Equijubus normani,* in lateral view. (A) left ilium; (B) right ilium. Abbreviations: ace, acetabulum; ip, ischial peduncle; lev, laterally everted rim; poap, postacetabular process; pp, pubic peduncle. Scale bar equals 10 cm.

end of the postacetabular process is broken on the left ilium, but the ventral margin of the process extends at a steep angle cranioventrally to the ischial peduncle, suggesting that the postacetabular process was relatively short and dorsoventrally deep. No brevis shelf is developed in *Equijubus*, similar to *Probactrosaurus* (Norman, 2002), but in contrast to the conditions in *Iguanodon bernissartensis* (Norman, 1980) – in which there is a relatively broad brevis shelf – and in *Mantellisaurus* (NHMUK R11521) and *Jinzhousaurus* (Wang et al., 2010).

As in most neornithischians (Butler et al., 2008) the ischial peduncle is large, transversely thickened, and angled ventrolaterally. This contrasts with the situation in more basal iguanodontians (e.g., *Camptosaurus dispar*, USNM 4282) in which the ischial peduncle faces ventrally, but is similar to the condition in other hadrosauriforms (e.g., *Mantellisaurus*, NHMUK R11521; *Probactrosaurus*, Norman, 2002; *Gilmoreosaurus*, AMNH 30735). The pubic peduncle is broken distally, but is roughly triangular in cross section, with the flattened base of the triangle forming the internal surface of the acetabulum. This surface is prominently striated. The acetabulum is strongly dorsally concave in lateral view, similar to that of *Mantellisaurus* (NHMUK R11521) but in contrast to the condition in *Probactrosaurus* (Norman, 2002), *Hypselospinus* (NHMUK R1635), and *Gilmoreosaurus* (AMNH 30735), in which the concavity described by the acetabular margin is much weaker.

Femur Fragments of both femora are preserved in three separate pieces (Fig. 3.20). The proximal end of the right femur preserves the lesser trochanter and a small part of the greater trochanter. The greater trochanter is better preserved on the left femur. The lesser trochanter forms a finger-like process that terminates ventral to the greater trochanter and is separated from it by a distinct cleft. The dorsal surface of the greater trochanter is rounded in lateral view. The femoral head is not preserved on either femur, but the preserved portion of the greater trochanter suggests it was separated from the head by a saddle-shaped sulcus in dorsal view. The mid-shaft region is present on the left femur, but extremely badly crushed and eroded. The fourth trochanter is visible and projects caudally. It takes the form of an elongate crest that is concave medially and fringed with striations. The distal condyles from the right side are preserved, although they have become offset from each other by craniocaudal crushing and much of the lateral condyle is eroded on its caudal surface. A deep intercondylar extensor groove is present cranially, and was probably almost entirely enclosed by expansion of the lateral and medial condyles. The ventral surface of the medial condyle is strongly convex ventrally, and it curves strongly caudally in medial view. Detailed comparisons with other taxa are precluded by the poor state of preservation; however, features of the greater, lesser, and fourth trochanters and the distal condyles do not appear to differ significantly from the femora of *Iguanodon* (Norman, 1980) or *Probactrosaurus* (Norman, 2002).

Other Appendicular Material Three large conjoined bone fragments appear to be parts of the appendicular skeleton, but are too poorly preserved for a reliable identification or any features of their anatomy to be seen.

DISCUSSION

Ontogenetic Age of IVPP V 12534

All the neural arches of the preserved vertebrae are fused to their respective centra, suggesting that the holotype of *Equijubus normani*, IVPP V 12534, is a skeletally mature individual

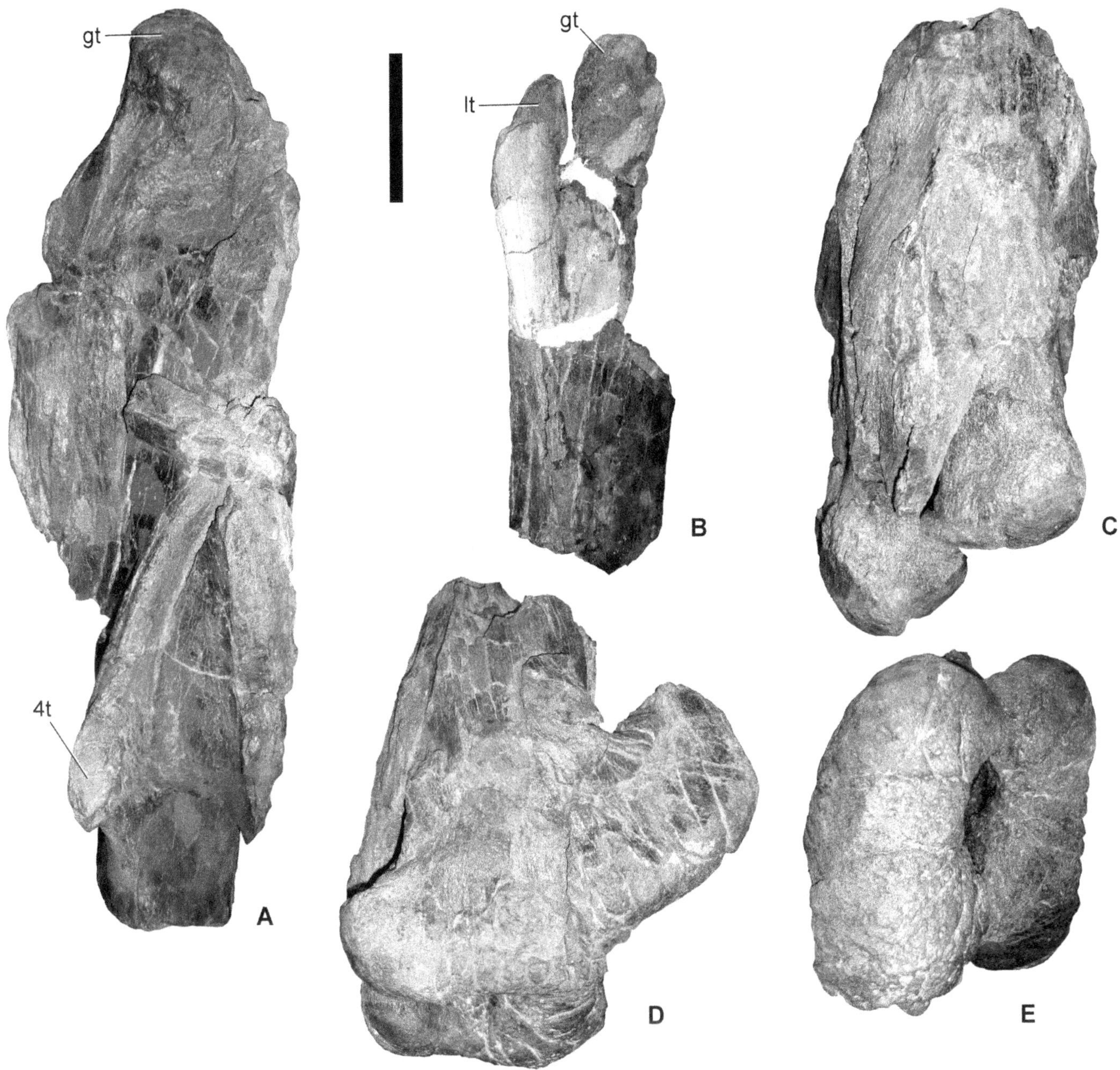

3.20. Fragments of femora of IVPP V 12534, holotype of *Equijubus normani.* (A) left femur in caudal view. (B–E) right femur: (B) proximal end in cranial view; (C–E) distal end in (C) cranial, (D) medial and (E) distal view. Abbreviations: 4t, fouth trochanter; gt, greater trochanter; lt, lesser trochanter. Scale bar equals 10 cm.

(Brochu, 1996). There are only six vertebrae in the sacrum, a low number compared to other basal iguanodonts (see above), but the first and sixth preserved sacrals are incomplete; thus, the low sacral count might be due to incomplete preservation of the sacrum rather than lack of fusion between sacral vertebrae.

Changes Made to Previous Diagnoses of Equijubus

Two characters have been removed from the original diagnosis of You, Luo, et al. (2003):

1. "Very large lower temporal fenestra," which is later described as being "twice the size of the orbit" (You, Luo, et al., 2003:349). The lower temporal (= infratemporal) fenestra is complete

only on the right side of the skull. The size of the right infratemporal fenestra relative to the right orbit is exaggerated by the mediolateral compression of the skull, which has crushed the right frontal and postorbital rostroventrally into the orbit (Fig. 3.4). The shape and relative size of the right infratemporal fenestra of IVPP V 12534 are not dissimilar to those of *Iguanodon* (Norman, 1980:fig. 2), *Altirhinus* (Norman, 1998:fig. 3), and possibly *Xuwulong* (the only skull of this taxon has been crushed caudoventrally, as indicated by the displacement of the jugal relative to the quadrate [You et al., 2011:fig. 3]).

2. "Lacking the median primary ridge on the crown of the dentary teeth" (You, Luo, et al., 2003:349). Reinterpretation of the dentary teeth of IVPP V 12534 indicates that there is actually a distally offset primary ridge (Fig. 3.10A, B).

The emended diagnosis of *Equijubus normani* proposed by Paul (2008) consists of two proportional characters and a number of qualitative statements. Paul's diagnosis includes the autapomorphic finger-like process of the jugal originally noted by You, Luo, et al. (2003) and included in the revised diagnosis herein (autapomorphy 1), as well as the distinctive morphology of the lacrimal noted by You, Luo, et al. (2003) and included herein (autapomorphy 2); Paul phrased his description of the lacrimal somewhat differently ("lacrimal long," "anterior process wedges between premaxilla and maxilla," "ventral edge at level of dorsal edge of maxilla" [Paul, 2008:201]). Paul (2008) also noted the lack of contact between the lacrimal and the nasal, which is related to the contact between the premaxilla and prefrontal that we included above in the unique combination of characters. Paul (2008:201) also mentioned that the shaft of the quadrate is "nearly straight," similar to our observation that the quadrate gently curves caudally along its entire length.

However, the same general issues noted by Barrett et al. (2009) in Paul's (2008) emended diagnosis of *Jinzhousaurus yangi* and by McDonald (2012a) in the diagnosis of "*Dollodon bampingi*" (= *Mantellisaurus atherfieldensis*) apply to that of *Equijubus normani*. The remaining characters used by Paul (2008:201) in the emended diagnosis of *E. normani* are ambiguous subjective statements that would be better treated as quantitative characters (e.g., premaxilla "maxillary process is shallow," "dorsal apex of maxilla sited anteriorly," "quadratojugal tall," "quadrate tall," quadrate "lateral foramen set moderately low," quadrate "dorso-posterior buttress small," "diastema long," "main body of ilium deep"), possible taphonomic consequences ("primary palpebral absent"), or widely distributed among basal iguanodonts (e.g., "premaxilla projects well below level of tooth rows" [Barrett et al., 2009], tooth "battery tightly packed" [Norman, 2004]). Paul (2008:201) also noted that the antorbital fossa is "small"; however, the antorbital fossa is not visible in lateral view (Figs. 3.3, 3.4). The proportional character "premaxillary tip to anterior orbital rim/latter to paraoccipital process tip length ratio ~1.0" (Paul, 2008:201) should be employed with caution until a morphometric analysis is carried out to assess intraspecific variation in basal iguanodont cranial proportions using taxa for which multiple skulls are known (e.g., *Iguanodon bernissartensis* [Norman, 1980]), as has been carried out for several hadrosaurids (Dodson, 1975; Evans, 2010; Campione and Evans, 2011). Paul (2008) also mentioned that 23 tooth positions are present in the maxilla; however, the caudal end of the right maxilla is obscured by the right jugal and the coronoid process of the right dentary, and the caudal end of the left maxilla is obscured by the coronoid process of the left dentary, rendering an accurate count of alveoli impossible. Finally, the character "dentary pre-coronoid process length/minimum depth ratio under 4" is also problematic; McDonald (2012a) demonstrated that the proportions of basal iguanodont dentaries can exhibit a considerable degree of intraspecific variation, so it is entirely possible that an individual of *Equijubus normani* will be discovered in which this ratio is greater than four.

Phylogenetic Relationships of Equijubus

The new information on the anatomy of *Equijubus* presented here allows its phylogenetic placement to be more confidently tested. Previous phylogenetic analyses have recovered *Equijubus* in a variety of positions. The analysis by You, Luo, et al. (2003) in the original description of *Equijubus* found it to be the most basal hadrosauroid, more derived than *Iguanodon, Ouranosaurus, Altirhinus,* and *Jinzhousaurus,* but more basal than *Probactrosaurus.* In contrast, the analysis of Norman (2002) placed *Equijubus* (referred to as "Mazongshan sp.") in a much more basal position–basal to *Iguanodon, Mantellisaurus, Ouranosaurus,* and *Altirhinus,* and in a polytomy with *Lurdusaurus.* However, Norman (2002) did not code *Equijubus* for any postcranial characters; You, Luo, et al. (2003) did code *Equijubus* for characters related to the cervical vertebrae, sacrum, and sternal (characters 41–43 and 45), but not for characters related to the ilium and femur. A subsequent analysis of Norman (2004) placed *Equijubus* in the same basal position as that of Norman (2002).

Wang et al. (2010) reanalyzed the datasets of You, Luo, et al. (2003) and Norman (2004) with new information on *Jinzhousaurus,* and again found *Equijubus* to be a basal hadrosauroid and a more basal iguanodontian outside Hadrosauroidea, respectively. The analyses of Sues and Averianov

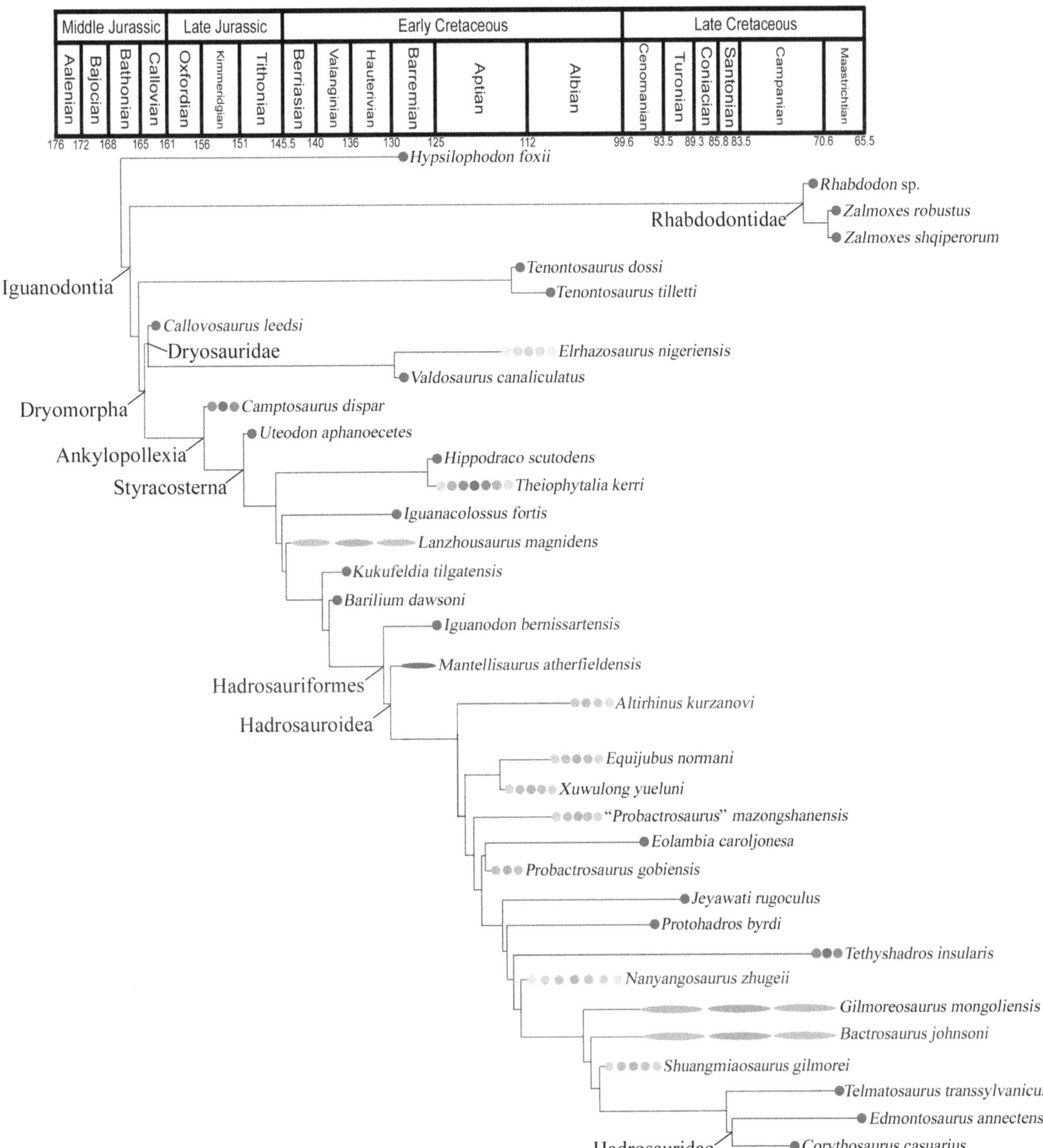

3.21. Phylogenetic relationships of *Equijubus normani*. Time-calibrated phylogeny of basal iguanodonts using the maximum agreement subtree of McDonald (2012b). Timescale based upon Walker and Geissman (2009); numerical ages are in millions of years. Uncertainty in taxon ages indicated by lighter circles or ellipses. Modified from McDonald (2012b).

(2009) and Prieto-Márquez (2010) also included *Equijubus*, and both found it to be a basal hadrosauroid close to *Probactrosaurus gobiensis* as did You, Luo, et al. (2003); however, both of these analyses focused primarily on derived hadrosauroids and hadrosaurids and incorporated only a limited sample of more basal members of Iguanodontia. Finally, the analysis of McDonald, Kirkland, et al. (2010) found *Equijubus* to be a basal hadrosauroid in a polytomy with *Altirhinus*; a clade of *Jinzhousaurus* plus *Penelopognathus*, and a clade composed of *Eolambia*, *Probactrosaurus gobiensis*, and all more derived iguanodontians. Although this analysis included an extensive sample of basal iguanodonts, *Equijubus* was not coded for any characters related to the pectoral girdle or femur.

Equijubus was recently included in the global phylogenetic analysis of basal iguanodonts by McDonald (2012b). The coding of *Equijubus* included both cranial and postcranial features described by You, Luo, et al. (2003) and those described for the first time herein. For example, *Equijubus* was coded for characters related to the scapula (101^1), ilium (111^1, 112^3, and 114^0), and femur (123^1, 125^0, 126^1, 127^3, 128^1, and 133^0) in the analysis of McDonald (2012b); this analysis was the first to code *Equijubus* for all preserved postcranial characters. The maximum agreement subtree of McDonald (2012b) placed *Equijubus* as the sister taxon of *Xuwulong* (You et al., 2011), a basal hadrosauroid from the Xinminpu Group of the neighboring Yujingzi Basin (Fig. 3.21). *Equijubus* and *Xuwulong* were united by a single ambiguous synapomorphy (75^1: presence of a rostrocaudally directed groove on the ventral surface of the basioccipital). The clade of *Equijubus* plus *Xuwulong* was more derived than *Iguanodon bernissartensis*, *Mantellisaurus*, and *Altirhinus*, but more basal than "*Probactrosaurus*" *mazongshanensis* (= *Gongpoquansaurus mazongshanensis* [You, Li, and Dodson, this volume]), *Probactrosaurus gobiensis*, and *Eolambia*.

CONCLUSIONS

Equijubus normani is diagnosed by a suite of cranial and vertebral autapomorphies and a unique combination of characters. An enhanced diagnosis and complete description of *Equijubus* allow it to be readily distinguished from other basal iguanodonts, including the rapidly growing record of basal iguanodont taxa from China. The new anatomical data presented here can be used in comparisons with other iguanodonts, and incorporated into future phylogenetic analyses. Detailed osteological descriptions of existing taxa are crucial to distinguishing additional taxa and, ultimately, in fully assessing diversity of basal iguanodonts in Asia and globally.

ACKNOWLEDGMENTS

This chapter is primarily the work of the first three authors; ATM wrote the cranial description, and SCRM and PMB wrote the postcranial sections. ATM thanks M. Lamanna (CM) for access to the skull of IVPP V 12534 while it was on loan; and J. Bartlett and J. Bird (CEUM), R. Allain (MNHN), S. Chapman and L. Steel (NHMUK), and S. Shelton (SDSM) for access to comparative material. SCRM and PMB thank X. Xu, F. Zheng, and C. Sullivan (IVPP); C. Mehling (AMNH); A. Henrici (CM); D. Schwarz-Wings (MB); and M. Carrano (USNM) for access to specimens in their care; and P. Mannion (University College London) for assistance during data collection in China. We are grateful to D. Evans and D. Norman for reviews that improved the manuscript. ATM's research was funded by grants from the Jurassic Foundation, Evolving Earth Foundation, University of Pennsylvania Paleobiology Summer Stipend, and Utah Friends of Paleontology. SCRM's visit to China was funded by a grant from the Jurassic Foundation; visits to collections housing comparative material were funded by Natural Environment Research Council grant number NE/G001898/1 awarded to PMB. HY and PD's work on Chinese iguanodonts is supported by National Science Foundation grant NSF EAR 1024671. HY's work on Chinese iguanodonts is also supported by the National Natural Science Foundation of China.

LITERATURE CITED

Apesteguía, S. 2005. Evolution of the hyposphene-hypantrum complex within Sauropoda; pp. 248–267 in V. Tidwell and K. Carpenter (eds.), Thunder-Lizards: The Sauropodomorph Dinosaurs. Indiana University Press, Bloomington, Indiana.

Barrett, P. M., R. J. Butler, R. J. Twitchett, and S. Hutt. 2011. New material of *Valdosaurus canaliculatus* (Ornithischia: Ornithopoda) from the Lower Cretaceous of southern England. Special Papers in Palaeontology 86:131–163.

Barrett, P. M., R. J. Butler, X.-L. Wang, and X. Xu. 2009. Cranial anatomy of the iguanodontoid ornithopod *Jinzhousaurus yangi* from the Lower Cretaceous Yixian Formation of China. Acta Palaeontologica Polonica 54:35–48.

Brill, K., and K. Carpenter. 2006. A description of a new ornithopod from the Lytle Member of the Purgatoire Formation (Lower Cretaceous) and a reassessment of the skull of *Camptosaurus;* pp. 49–67 in K. Carpenter (ed.), Horns and Beaks: Ceratopsian and Ornithopod Dinosaurs. Indiana University Press, Bloomington, Indiana.

Brochu, C. A. 1996. Closure of neurocentral sutures during crocodilian ontogeny: implications for maturity assessment in fossil archosaurs. Journal of Vertebrate Paleontology 16:49–62.

Buffetaut, E., and J. Le Loeuff. 1991. Une nouvelle espèce de *Rhabdodon* (Dinosauria, Ornithischia) du Crétacé supérieur de l'Hérault (Sud de la France). Comptes rendus de l'Académie des sciences, Série II, 312:943–948.

Butler, R. J., P. Upchurch, and D. B. Norman. 2008. The phylogeny of the ornithischian dinosaurs. Journal of Systematic Palaeontology 6:1–40.
Campione, N. E., and D. C. Evans. 2011. Cranial growth and variation in edmontosaurs (Dinosauria: Hadrosauridae): implications for latest Cretaceous megaherbivore diversity in North America. PLoS ONE 6(9):e25186.
Carpenter, K., and Y. Ishida. 2010. Early and "middle" Cretaceous iguanodonts in time and space. Journal of Iberian Geology 36:145–164.
Cope, E. D. 1870. Synopsis of the extinct Batrachia, Reptilia, and Aves of North America. Transactions of the American Philosophical Society 14:1–252.
Dodson, P. 1975. Taxonomic implications of relative growth in lambeosaurine hadrosaurs. Systematic Zoology 24:37–54.
Dollo, L. 1888. Iguanodontidae et Camptonotidae. Comptes rendus de l'Académie des Sciences 106:775–777.
Evans, D. C. 2010. Cranial anatomy and systematic of *Hypacrosaurus altispinus,* and a comparative analysis of skull growth in lambeosaurine hadrosaurids (Dinosauria: Ornithischia). Zoological Journal of the Linnean Society 159:398–434.
Godefroit, P., H. Li, and C.-Y. Shang. 2005. A new primitive hadrosauroid dinosaur from the Early Cretaceous of Inner Mongolia (P. R. China). Comptes rendus palevol 4:697–705.
Head, J. J. 2001. A reanalysis of the phylogenetic position of *Eolambia caroljonesa* (Dinosauria, Iguanodontia). Journal of Vertebrate Paleontology 21:392–396.
Kirkland, J. I. 1998. A new hadrosaurid from the upper Cedar Mountain Formation (Albian–Cenomanian: Cretaceous) of eastern Utah: the oldest known hadrosaurid (Lambeosaurine?); pp. 283–295 in S. G. Lucas, J. I. Kirkland, and J. W. Estep (eds.), Lower and Middle Cretaceous Terrestrial Ecosystems. New Mexico Museum of Natural History and Science Bulletin 14.
Marsh, O. C. 1881. Principal characters of the American Jurassic dinosaurs, part IV. American Journal of Science, Series 3, 21:417–423.
McDonald, A. T. 2012a. The status of *Dollodon* and other basal iguanodonts (Dinosauria: Ornithischia) from the Lower Cretaceous of Europe. Cretaceous Research 33:1–6.
McDonald, A. T. 2012b. Phylogeny of basal iguanodonts (Dinosauria: Ornithischia): an update. PLoS ONE 7(5):e36745.
McDonald, A. T., P. M. Barrett, and S. D. Chapman. 2010. A new basal iguanodont (Dinosauria: Ornithischia) from the Wealden (Lower Cretaceous) of England. Zootaxa 2569:1–43.
McDonald, A. T., J. I. Kirkland, D. D. DeBlieux, S. K. Madsen, J. Cavin, A. R. C. Milner, and L. Panzarin. 2010. New basal iguanodonts from the Cedar Mountain Formation of Utah and the evolution of thumb-spiked dinosaurs. PLoS ONE 5(11):e14075.
Norman, D. B. 1980. On the ornithischian dinosaur *Iguanodon bernissartensis* from the Lower Cretaceous of Bernissart (Belgium). Mémoire de l'Institut royal des Sciences naturelles de Belgique 178:1–103.
Norman, D. B. 1986. On the anatomy of *Iguanodon atherfieldensis* (Ornithischia: Ornithopoda). Bulletin de l'Institut royal des Sciences naturelles de Belgique: Sciences de la Terre 56:281–372.
Norman, D. B. 1998. On Asian ornithopods (Dinosauria: Ornithischia). 3. A new species of iguanodontid dinosaur. Zoological Journal of the Linnean Society 122:291–348.
Norman, D. B. 2002. On Asian ornithopods (Dinosauria: Ornithischia). 4. *Probactrosaurus* Rozhdestvensky, 1966. Zoological Journal of the Linnean Society 136:113–144.
Norman, D. B. 2004. Basal Iguanodontia; pp. 413–437 in D. B. Weishampel, P. Dodson, and H. Osmólska (eds.), The Dinosauria, Second Edition. University of California Press, Berkeley, California.
Norman, D. B. 2010. A taxonomy of iguanodontians (Dinosauria: Ornithopoda) from the lower Wealden Group (Cretaceous: Valanginian) of southern England. Zootaxa 2489:47–66.
Norman, D. B. 2011a. Ornithopod dinosaurs; pp. 407–475 in D. J. Batten (ed.), English Wealden Fossils. Palaeontological Association Field Guides to Fossils Number 14. The Palaeontological Association, London.
Norman, D. B. 2011b. On the osteology of the lower Wealden (Valanginian) ornithopod *Barilium dawsoni* (Iguanodontia: Styracosterna). Special Papers in Palaeontology 86:165–194.
Norman, D. B. 2014. Iguanodonts from the Wealdean of England: do they contribute to the discussion concerning hadrosaur origins? chapter 2 in D. A. Eberth and D. C. Evans (eds.), Hadrosaurs. Indiana University Press, Bloomington, Indiana.
Ostrom, J. H. 1961. Cranial morphology of the hadrosaurian dinosaurs of North America. Bulletin of the American Museum of Natural History 122:33–186.
Ostrom, J. H. 1970. Stratigraphy and palaeontology of the Cloverly Formation (Lower Cretaceous) of the Bighorn Basin area, Wyoming and Montana. Bulletin of the Peabody Museum of Natural History, Yale University 35:1–234.
Owen, R. 1842. Report on British fossil reptiles, part II. Reports of the British Association for the Advancement of Science 11:60–204.
Paul, G. S. 2008. A revised taxonomy of the iguanodont dinosaur genera and species. Cretaceous Research 29:192–216.
Prieto-Márquez, A. 2010. Global phylogeny of hadrosauridae (Dinosauria: Ornithopoda) using parsimony and Bayesian methods. Zoological Journal of the Linnean Society 159:435–502.
Seeley, H. G. 1887. On the classification of the fossil animals commonly named Dinosauria. Proceedings of the Royal Society of London 43:165–171.
Sereno, P. C. 1986. Phylogeny of the bird-hipped dinosaurs (Order Ornithischia). National Geographic Research 2:234–256.
Sereno, P. C. 1997. The origin and evolution of dinosaurs. Annual Review of Earth and Planetary Sciences 25:435–489.
Sereno, P. C. 1998. A rationale for phylogenetic definitions, with application to the higher-level taxonomy of Dinosauria. Neues Jahrbuch für Geologie und Paläontologie, Abhandlungen 210:41–83.
Sereno, P. C. 2005. Stem Archosauria Version 1.0. TaxonSearch. Available at www.taxonsearch.org/Archive/stem-archosauria-1.0.php. Accessed November 16, 2010.
Sues, H.-D., and A. Averianov. 2009. A new basal hadrosauroid dinosaur from the Late Cretaceous of Uzbekistan and the early radiation of duck-billed dinosaurs. Proceedings of the Royal Society B 276:2549–2555.
Tang, F., Z.-X. Luo, Z.-H. Zhou, H.-L. You, J. A. Georgi, Z.-L. Tang, and X.-Z. Wang. 2001. Biostratigraphy and palaeoenvironment of the dinosaur-bearing sediments in Lower Cretaceous of Mazongshan area, Gansu Province, China. Cretaceous Research 22:115–129.
Taquet, P. 1976. Ostéologie d'*Ouranosaurus nigeriensis,* Iguanodontide du Crétacé Inférieur du Niger [Osteology of *Ouranosaurus nigeriensis,* Iguanodontid from the Lower Cretaceous of Niger]; pp. 57–168 in Géologie et Paléontologie du Gisement de Gadoufaoua (Aptien du Niger) [Geology and Paleontology of the Gadoufaoua deposit (Aptian of Niger)]. Available at www.paleoglot.org/files/Taquet_76.pdf.
Walker, J. D., and J. W. Geissman. 2009. 2009 GSA geologic time scale. GSA Today 19:60–61.
Wang, X., and X. Xu. 2001. A new iguanodontid (*Jinzhousaurus yangi* gen. et sp. nov.) from the Yixian Formation of western Liaoning China. Chinese Science Bulletin 46:1669–1672.
Wang, X.-L., R. Pan, R. J. Butler, and P. M. Barrett. 2010. The postcranial skeleton of the iguanodontian ornithopod *Jinzhousaurus yangi* from the Lower Cretaceous Yixian Formation of western Liaoning, China. Earth and Environmental Transactions of the Royal Society of Edinburgh 101:135–159.
Weishampel, D. B., and P. R. Bjork. 1989. The first indisputable remains of *Iguanodon* (Ornithischia: Ornithopoda) from North America: *Iguanodon lakotaensis,* sp. nov. Journal of Vertebrate Paleontology 9:56–66.
Weishampel, D. B., C.-M. Jianu, Z. Csiki, and D. B. Norman. 2003. Osteology and phylogeny of *Zalmoxes* (n. g.), an unusual euornithopod dinosaur from the latest Cretaceous of Romania. Journal of Systematic Palaeontology 1:65–123.
Wilson, J. A. 1999. A nomenclature for vertebral laminae in sauropods and other saurischian dinosaurs. Journal of Vertebrate Paleontology 19:639–653.
Wu, W., P. Godefroit, and D. Hu. 2010. *Bolong yixianensis* gen. et sp. nov.: a new iguanodontoid dinosaur from the Yixian Formation of western Liaoning, China. Geology and Resources 19:127–133.
Xu, X., X.-J. Zhao, J.-C. Lu, W.-B. Huang, Z.-Y. Li, and Z.-M. Dong. 2000. A new iguanodontian from Sangping Formation of Neixiang, Henan and its stratigraphical implications. Vertebrata PalAsiatica 38:176–191.
You, H.-L., and D.-Q. Li. 2009. A new basal hadrosauriform dinosaur (Ornithischia: Iguanodontia) from the Early Cretaceous of northwestern China. Canadian Journal of Earth Sciences 46:949–957.
You, H.-L., Q. Ji, and D.-Q. Li. 2005. *Lanzhousaurus magnidens* gen. et sp. nov. from Gansu Province, China: the largest-toothed herbivorous dinosaur in the world. Geological Bulletin of China 24:785–794.
You, H.-L., D.-Q. Li, and W.-C. Liu. 2011. A new hadrosauriform dinosaur from the Early Cretaceous of Gansu Province, China. Acta Geologica Sinica 85:51–57. [English Edition]
You, H.-L., Q. Ji, J.-L. Li, and Y.-X. Li. 2003. A new hadrosauroid dinosaur from the mid-Cretaceous

of Liaoning, China. Acta Geologica Sinica 77:148–154. [English Edition]

You, H.-L., Z.-X. Luo, N. H. Shubin, L. M. Witmer, Z.-L. Tang, and F. Tang. 2003. The earliest-known duck-billed dinosaur from deposits of late Early Cretaceous age in northwest China and hadrosaur evolution. Cretaceous Research 24:347–355.

You, H.-L., D.-Q. Li, and P. Dodson. 2014. *Gongpoquansaurus mazongshanensis* (Lü, 1997) comb. nov. (Ornithischia: Hadrosauroidea) from the Early Cretaceous of Gansu Province, Northwestern China; chapter 4 in D. A. Eberth and D. C. Evans (eds.), Hadrosaurs. Indiana University Press, Bloomington, Indiana.

Appendix 3.1. Measurements of IVPP V 12534, the Holotype of *Equijubus normani*

Anatomical features	Measurements (cm)
Right dentary, length of symphysis	5.6
Right dentary, total length along ventral margin	38.7
Right dentary, length of diastema, from predentary groove to rostral margin of first alveolus	8.1
Right dentary, depth at midpoint	9.4
Left dentary, height of coronoid process, from shelf at base of the process to the apex	11.8
Left dentary, greatest rostrocaudal width of coronoid process	6.7
Right jugal, dorsoventral depth of ventral flange	5.4
Right postorbital, rostrocaudal length of squamosal process along dorsal margin	12.3
Right quadrate, height along rostral margin	27.4
Right quadrate, greatest mediolateral width of ventral condyle	7.2
Right quadrate, greatest rostrocaudal width of ventral condyle	3.3
Cervical and dorsal vertebrae, centra lengths:	
C3	11.0
C4	10.1
C5	11.1
C6	12.0
C7	12.9
C8	11.2
C9	10.5
C10	12.5
C11	12.2
D1	11.8
D2	10.8
D3	9.6
D4	9.8
D5	10.1
D6	10.0
D7	10.0
D8	9.8
D9	11.5
D10	10.0
D11	9.6
D12	9.5
D13	9.1
D14	8.8
D15	9.0
D16	9.5
D17	Incomplete
Sacrum, craniocaudal length of preserved portion	46.0
Left sternal, craniocaudal length	30.5

Gongpoquansaurus mazongshanensis (Lü, 1997) comb. nov. (Ornithischia: Hadrosauroidea) from the Early Cretaceous of Gansu Province, Northwestern China

4

Hai-lu You, Da-Qing Li, and Peter Dodson

ABSTRACT

This chapter proposes a new generic name "*Gongpoquansaurus*" for "*Probactrosaurus*" *mazongshanensis* Lü, 1997. *Gongpoquansaurus mazongshanensis* comb. nov. is a basal hadrosauroid distinguishable from other taxa by the unique combination of these cranial features: transversely elongated supratemporal fenestra, broad and flat dorsal surface of the parietals with a deep median groove, and large and straight nuchal crest. Recognizing its own generic status, rather than as a species of *Probactrosaurus*, increases basal hadrosauroids to four genera in the Early Cretaceous Mazongshan area of northwestern Gansu Province, China.

INTRODUCTION

Recent years have seen a surge of new basal hadrosauroid dinosaurs (the most inclusive taxon containing *Parasaurolophus walkeri* Parks, 1922, but not *Iguanodon bernissartensis* Boulenger [in Beneden, 1881]) in China. Among them, three genera (*Equijubus normani* You et al., 2003, *Jintasaurus meniscus* You and Li, 2009, and *Xuwulong yueluni* You et al., 2011) and one species (*Probactrosaurus mazongshanensis* Lü, 1997) were named based on material from the Early Cretaceous Xinminpu Group in the Mazongshan (also called Beishan) area of Gansu Province, northwestern China (Fig. 4.1), making it one of the richest hadrosauroid-yielding areas in the world. However, the assignment of the "*mazongshanensis*" species to "*Probactrosaurus*" has been questioned, and this taxon probably represents a new genus (Norman, 2002, 2004; Carpenter and Ishida, 2010; McDonald et al., 2010; Buffetaut and Suteethorn, 2011).

In this chapter, based on a review of previous comments on the status of *Probactrosaurus mazongshanensis* Lü, 1997 and new comparisons to its related taxa, a new generic name *Gongpoquansaurus* is proposed for "*Probactrosaurus*" *mazongshanensis* mainly based on its unique cranial features.

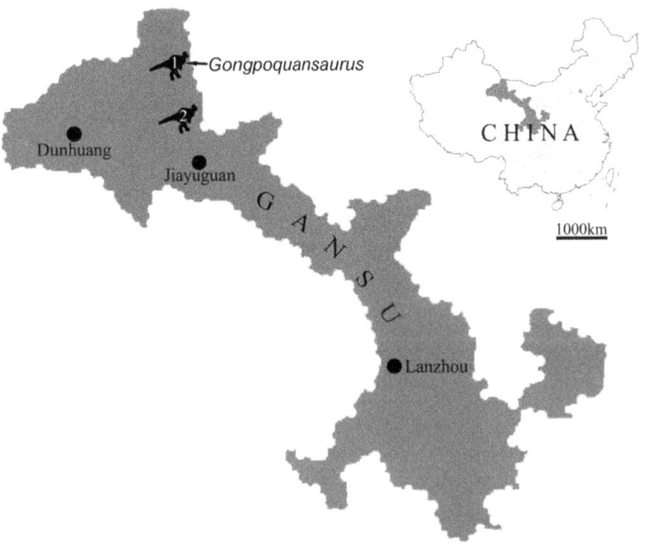

4.1. Localities of basal hadrosauriform dinosaurs from the Early Cretaceous Zhonggou Formation (Xinminpu Group) in the Mazongshan area of Gansu Province, northwestern China. (1) locality of *Gongpoquansaurus mazongshanensis* (Lü, 1997) comb. nov. and *Equijubus normani* You et al., 2003 in the Gongpoquan Basin; (2) locality of *Jintasaurus meniscus* You and Li, 2009, and *Xuwulong yueluni* You et al., 2011, in the Yujingzi Basin.

Institutional Abbreviations IVPP, Institute of Vertebrate Paleontology and Paleoanthropology, Chinese Academy of Sciences, Beijing, China.

SYSTEMATIC PALEONTOLOGY

DINOSAURIA Owen, 1842
ORNITHISCHIA Seeley, 1887
ORNITHOPODA Marsh, 1881
IGUANODONTIA Dollo, 1888
(sensu Sereno, 2005)
ANKYLOPOLLEXIA Sereno, 1986
(sensu Sereno, 2005)
STYRACOSTERNA Sereno, 1986
(sensu Sereno, 2005)

HADROSAURIFORMES Sereno, 1997
(sensu Sereno, 1998)
HADROSAUROIDEA Cope, 1870
(sensu Sereno, 2005)
GONGPOQUANSAURUS gen. nov.

Gongpoquansaurus mazongshanensis
(Lü, 1997) comb. nov.
Probactrosaurus mazongshanensis Lü, 1997
(original description).
?Probactrosaurus mazongshanensis
(Norman, 2002, 2004).
"Probactrosaurus" mazongshanensis
(You and Li, 2009).
"Probactrosaurus" mazongshanensis
(McDonald et al., 2010).
"Probactrosaurus" mazongshanensis
(Carpenter and Ishida, 2010).
"Probactrosaurus" mazongshanensis
(Buffetaut and Suteethorn, 2011).

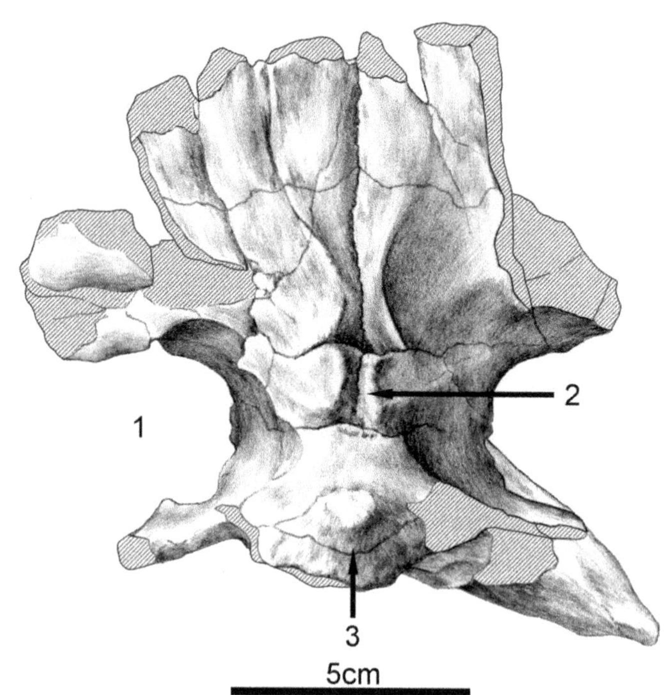

4.2. Holotype (IVPP V. 11333) of *Gongpoquansaurus mazongshanensis* comb. nov. Caudal part of the skull in dorsal view, showing its unique combination of features: (1) transversely elongated supratemporal fenestra; (2) a deep median groove on the broad and flat dorsal surface of the parietals; and (3) large and straight nuchal crest.

Type Species *Gongpoquansaurus mazongshanensis* (Lü, 1997) comb. nov.

Etymology Gongpoquan is the name of the basin where the material of the type species was recovered.

Diagnosis As for the type and only known species.

Holotype IVPP V. 11333 (Field Number 9208–21): Caudal part of an incomplete skull.

Referred Material IVPP V. 11334–16: part of the left quadrate; IVPP V. 11334–10, 11: two nearly complete maxillary teeth; IVPP V. 11334–12, 13, 14, 15: four isolated dentary teeth; IVPP V. 11334–1: four cervical vertebrae (including one nearly complete tenth cervical vertebra); IVPP V. 11334–2: second dorsal vertebra; IVPP V. 11334–3: nearly complete sacrum; IVPP V. 11334–4: two series of caudal vertebrae (some with neural spines and chevrons); IVPP V. 11334–17: complete left scapula; IVPP V. 11334–18: part of the left humerus; IVPP V. 11334–8: part of the preacetabular process of the left ilium; IVPP V. 11334–9: part of the postacetabular process of the right ilium; IVPP V. 11334–6, 7: parts of the right and left pubes; IVPP V. 11334–5: right femur missing the proximal part.

Locality and Horizon Gongpoquan Basin, Town of Mazongshan, Subei Mongol Autonomous County, Jiuquan City, Gansu Province. middle gray unit of Zhonggou Formation, Xinminpu Group, Lower Cretaceous (Albian). This fossil locality is now included in the Subei Gongpoquan Dinosaur Geopark of Gansu Province established in 2006, which covers an area of 2288 km^2.

Amended Diagnosis *Gongpoquansaurus mazongshanensis* is a basal hadrosauroid distinguishable from other taxa by the following unique combination of features: transversely elongated supratemporal fenestra, broad and flat dorsal surface of the parietals with a deep median groove, and large and straight nuchal crest (Fig. 4.2).

Comments and Comparisons All the specimens were collected from the southeastern part of the Gongpoquan Basin in the summer of 1992 during the Sino-Japanese Silk Road Dinosaur Expedition (Dong, 1997). The locality of the holotype is shown in figure 1 (Dong, 1997:7). All the referred specimens were probably recovered from the same quarry as the holotype (Dong, Z.-M., pers. comm.). Because it is common for more than one iguanodontian taxon to occur in the same horizon and locality, and another iguanodontian (*Equijubus normani*) has already been recovered from this horizon in the same (southeastern) part of the Gongpoquan Basin, there is a possibility that not all the referred specimens belong to *Gongpoquansaurus mazongshanensis*. Therefore, we prefer to restrict the diagnosis of this taxon, basing it on the holotype only.

Based on dentary crowns, Norman (2002) considered *"Probactrosaurus" mazongshanensis* to be more closely related to *Altirhinus* than to *Probactrosaurus gobiensis*. Buffetaut and Suteethorn (2011) found that the maxillary teeth of *Siamodon nimngami* are very similar to those of *"Probactrosaurus" mazongshanensis*, with a prominent median primary

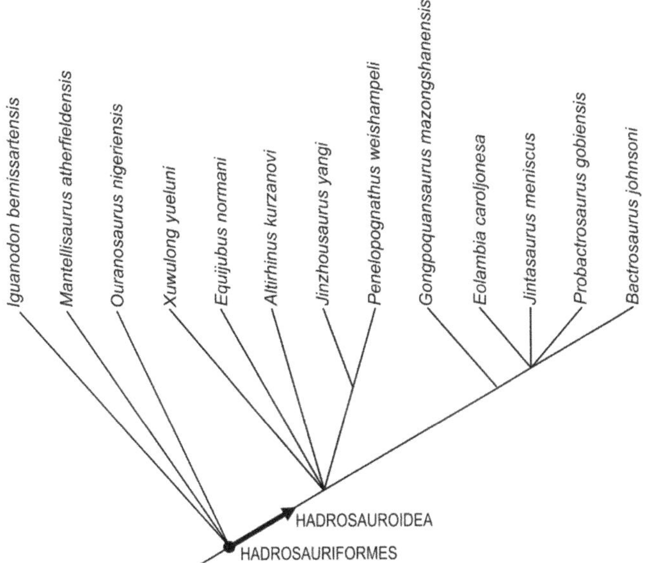

4.3. Cladogram showing the phylogenetic relationships of *Gongpoquansaurus mazongshanensis* comb. nov. This is based on the results of cladistic analyses of McDonald et al. (2010) and McDonald (2011), with the addition of *Xuwulong yueluni* based on You et al. (2011).

ridge, and strongly denticulate margins. As suggested above, the association between the referred teeth and the holotype needs to be confirmed.

In McDonald et al. (2010)'s cladistic analysis, which included *"Probactrosaurus" mazongshanensis, Equijubus normani* and *Jintasaurus meniscus*, *"Probactrosaurus" mazongshanensis* is more derived than *Equijubus* and less derived than *Jintasaurus*, with several intermediate taxa in between (Fig. 4.3). This supports the elevation of *"Probactrosaurus" mazongshanensis* as a new genus.

Carpenter and Ishida (2010) noticed that *"Probactrosaurus" mazongshanensis* is an immature individual as evidenced by the incomplete fusion of the neural arch with the centrum, and found substantial differences between it and *Probactrosaurus gobiensis*. In addition to the diagnostic cranial and cervical features identified here, they also noted the following differences: teeth with large marginal crenulations (small, less developed in *P. gobiensis*); dentary teeth significantly larger than maxillary teeth (nearly same size in *P. gobiensis*); sacrum rounded ventrally (keeled in *P. gobiensis*); and scapula expanded distally (nearly parallel-sided in *P. gobiensis*).

Jintasaurus, a slightly more derived hadrosauroid from the nearby Yujingzi Basin, shows clear differences with *Gongpoquansaurus:* the supratemporal fenestra is rostrolateral-caudomedially directed in *Jintasaurus* but transversely elongated in *Gongpoquansaurus*, and the parietals of *Jintasaurus* do not possess a median deep groove.

DISCUSSION

Four basal hadrosauroids have been recovered from two basins in the Mazongshan area of northwestern Gansu Province (Fig. 4.1). The northern Gongpoquan Basin yielded *Gongpoquansaurus mazongshanensis* (Lü, 1997) comb. nov. and *Equijubus normani* You et al., 2003, whereas the southern Yujingzi Basin produced *Jintasaurus meniscus* You and Li, 2009 and *Xuwulong yueluni* You et al., 2011. Although all four are basal hadrosauroids, the Gongpoquan and Yujingzi basins each yield a relatively derived hadrosauroid (*Gongpoquansaurus* and *Jintasaurus*, respectively) and a relatively more basal form (*Equijubus* and *Xuwulong*, respectively) (Fig. 4.3).

Both *Gongpoquansaurus* and *Equijubus* are from the middle gray unit of the Xinminpu (= Xinminbao) Group in the Gongpoquan Basin, and the age of the Xinminpu Group in the Mazongshan area is considered as Aptian–Albian in age based on analysis of spores and pollens (Tang et al., 2001; You et al., 2003). Both *Jintasaurus* and *Xuwulong* are from the middle gray unit in the Yujingzi Basin, and the dinosaur-bearing rocks here are considered as belonging to the Xinminpu Group as well, and can be generally correlated to those in the Gongpoquan Basin with an age assignment of Aptian–Albian (You and Li, 2009; Li et al., 2010; Makovicky et al., 2010; You et al., 2010; You et al., 2011).

The Xinminpu Group includes two formations: the lower Xiagou and the upper Zhonggou formations. Our 2010 and 2011 field work confirms that the dinosaur-bearing Xinminpu Group in the Mazongshan area belongs to Zhonggou Formation (Liu, Y.-Q., pers. comm.). In the Hanxia Section about 200 km south of the Gongpoquan Basin, the Zhonggou and Xiagou formations are separated by andesites (Niu, 1987) dated at 113.7 Ma (Liu, Y.-Q., pers. comm.). This volcanic activity, which occurred at about the Aptian-Albian boundary (112 Ma), is expressed regionally across several localities and extended northeastward to Tebch in Inner Mongolia (110 Ma; Eberth et al., 1993). In this context the Zhonggou Formation is most likely of Albian age. An Albian age interpretation for the Zhonggou Formation is further supported by the palynological analysis of the lower red unit in the Gongpoquan Basin (Tang et al., 2001).

ACKNOWLEDGMENTS

We are grateful to the crew of the former Fossil Research and Development Center of the Third Geology and Mineral Resources Exploration Academy of the Gansu Provincial Bureau of Geo-Exploration and Mineral Development for discovering, excavating, and preparing most of the Gansu dinosaur specimens, especially *Lanzhousaurus magnidens*.

Funding was provided by the National Natural Science Foundation of China (41072019 and 40672007) and the IVPP Key Projects to HLY, the National Science Foundation of the United States (1024671) to PD, and Gansu Department of Land and Resources to QLD.

LITERATURE CITED

Beneden, P. J. van. 1881. [On the pelvic arch in the dinosaurs of Bernissart]. Bulletins de L'Académie royal de Belgique 3ème série 1:600–608. [French]

Buffetaut, E., and V. Suteethorn. 2011. A new iguanodontian dinosaur from the Khok Kruat Formation (Early Cretaceous, Aptian) of northeastern Thailand. Annales de paleontologie 97:51–62.

Carpenter, K., and Y. Ishida. 2010. Early and "middle" Cretaceous iguanodonts in time and space. Journal of Iberian Geology 36:145–164.

Cope, E. D. 1870. Synopsis of the extinct Batrachia, Reptilia, and Aves of North America. Transactions of the American Philosophical Society 14:1–252.

Dollo, L. 1888. Iguanodontidae et Camptonotidae. Comptes rendus de l'Académie des sciences, Paris 106:775–777.

Dong, Z. (ed.). 1997. Sino-Japanese Silk Road Dinosaur Expedition. China Ocean Press, Beijing, 114 pp.

Eberth, D. A., D. A. Russell, D. R. Braman, and A. L. Deino. 1993. The age of the dinosaur-bearing sediments at Tebch, Inner Mongolia, People's Republic of China. Canadian Journal of Earth Sciences 30:2101–2106.

Li, D.-Q., M. A. Norell, K.-Q. Gao, N. D. Smith, and P. J. Makovicky. 2010. A longirostrine tyrannosauroid from the Early Cretaceous of China. Proceedings of the Royal Society Biological Sciences Series B 277:183–190.

Lü, J.-C. 1997. A new Iguanodontidae (*Probactrosaurus mazongshanensis* sp. nov.) from Mazongshan area, Gansu Province, China; pp. 27–47 in Z. Dong (ed.), Sino-Japanese Silk Road Dinosaur Expedition. China Ocean Press, Beijing, China.

Makovicky, P. J., D.-Q. Li, K.-Q. Gao, M. Lewin, G. M. Erickson, and M. A. Norell. 2010. A giant ornithomimosaur from the Early Cretaceous of China. Proceedings of the Royal Society Biological Sciences Series B 277:191–198.

Marsh, O. C. 1881. Principal characters of American Jurassic dinosaurs. Part V. American Journal of Science, Third Series, 21:417–423.

McDonald, A. T. 2011. The phylogeny, taxonomy, and biogeography of basal Iguanodonts (Dinosauria: Ornithischia); pp. 101–104 in D. Braman, D. A. Eberth, D. C. Evans, and W. Taylor (compilers), Hadrosaur Symposium Abstract Volume. Royal Tyrrell Museum of Palaeontology, Drumheller, Alberta.

McDonald, A. T., P. M. Barrett, and S. D. Chapman. 2010. A new basal iguanodont (Dinosauria: Ornithischia) from the Wealden (Lower Cretaceous) of England. Zootaxa 2569:1–43.

Niu, S. W. 1987. Late Mesozoic strata of Jiuquan in Gansu. Journal of Stratigraphy 11:1–22.

Norman, D. B. 2002. On Asian ornithopods (Dinosauria: Ornithischia). 4. *Probactrosaurus* Rozhdestvensky, 1966. Zoological Journal of the Linnean Society 136:113–144.

Norman, D. B. 2004. Basal Iguanodontia; pp. 413–437 in D. B. Weishampel, P. Dodson, and H. Osmólska (eds.), The Dinosauria, Second Edition. University of California Press, Berkeley, California.

Parks, W. A. 1922. *Parasaurolophus walkeri,* a new genus and species of crested trachodont dinosaur. University of Toronto Studies, Geological Series 13:1–32.

Owen, R. 1842. Report on British fossil reptiles. Report of the British Association of Advanced Sciences 9:60–204.

Seeley, H. G. 1887. On the classification of the fossil animals commonly named Dinosauria. Proceedings of the Royal Society of London 43:165–171.

Sereno, P. C. 1986. Phylogeny of the bird-hipped dinosaurs (Order Ornithischia). National Geographic Research 2:234–256.

Sereno, P. C. 1997. The origin and evolution of dinosaurs. Annual Review of Earth and Planetary Sciences 25:435–489.

Sereno, P. C. 1998. A rationale for phylogenetic definitions, with application to the higher-level taxonomy of Dinosauria. Neues Jahrbuch für Geologie und Paläontologie Abhandlungen 210:41–83.

Sereno, P. C. 2005. Stem Archosauria Version 1.0. TaxonSearch. Available at www.taxonsearch.org/Archive/stem-archosauria-1.0.php. Accessed January 15, 2012.

Tang, F., Z.-X. Luo, Z.-H. Zhou, H.-L. You, J. A. Georgi, Z.-L. Tang, and X.-Z. Wang. 2001. Biostratigraphy and palaeoenvironment of the dinosaur-bearing sediments in Lower Cretaceous of Mazongshan area, Gansu Province, China. Cretaceous Research 22:115–129.

You, H.-L., and D.-Q. Li. 2009. A new basal hadrosauriform dinosaur (Ornithischia: Iguanodontia) from the Early Cretaceous of northwestern China. Canadian Journal of Earth Sciences 46:949–957.

You, H.-L., D.-Q. Li, and W.-C. Liu. 2011. A new hadrosauriform dinosaur from the Early Cretaceous of Gansu Province, China. Acta Geologica Sinica 85:51–57. [English Edition]

You, H.-L., Z.-X. Luo, N. H. Shubin, L. M. Witmer, Z.-L. Tang, and F. Tang. 2003. The earliest-known duck-billed dinosaur from deposits of late Early Cretaceous age in northwest China and hadrosaur evolution. Cretaceous Research 24:347–355.

You, H.-L., K. Tanoue, and P. Dodson. 2010. A new species of *Archaeoceratops* (Dinosauria: Neoceratopsia) from the Early Cretaceous of the Mazongshan Area, northwestern China; pp. 59–67 in M. J. Ryan, B. J. Chinnery-Allgeier, and D. A. Eberth (eds.), New Perspectives on Horned Dinosaurs. Indiana University Press, Bloomington, Indiana.

Postcranial Anatomy of a Basal Hadrosauroid (Dinosauria: Ornithopoda) from the Cretaceous (Cenomanian) Woodbine Formation of North Texas

5

Derek J. Main, Christopher R. Noto, and David B. Weishampel

ABSTRACT

New fossil material recovered from the Arlington Archosaur Site (AAS; North Arlington, Tarrant County, Texas) includes the most complete hadrosauroid (Dinosauria: Ornithopoda) recovered from the Woodbine Formation to date, and allows for a detailed study of the postcranial anatomy of a basal hadrosauroid. The depositional setting of the AAS was a delta plain, and the host facies comprise fine-grained mudstones and claystones overlying a dense peat bed. The ornithopod fossil-bearing horizon occurs within a paleosol: a clay-rich, heavily rooted histic gleysol with numerous calcareous concretions. Hadrosauroid fossils recovered include cervical, dorsal, and caudal vertebrae; ribs, a scapula and coracoid; an ilium, ischium, and pubis; and a partial juvenile femur. Most of the remains represent a single adult individual, which experienced a complex biostratinomic and fossil-diagenetic history following death. The predominance of relatively dense bones from the axial column and limbs, and disarticulation are indicative of hydraulic sorting and limited transport of the remains. The new hadrosauroid material from the Woodbine Formation demonstrates a unique mix of plesiomorphic hadrosaurid characters and derived iguanodontian features. Whereas its exact taxonomic affinity cannot be determined presently, it is expected that this specimen will prove informative in understanding hadrosauroid evolution.

INTRODUCTION

The Arlington Archosaur Site (AAS) is located in the Dallas–Fort Worth Metroplex of North Texas, within the suburban city of Arlington, in Tarrant County (Fig. 5.1). It was discovered independently in 2003 by University of Texas, Arlington, students Phil Kirchhoff and Bill Walker, and local fossil collector Art Sahlstein. Among the first fossils found and reported were crocodyliform and ornithopod dinosaur remains, leading one of us (DJM) to propose the site's name. Unfortunately, the landowner at the time of discovery refused to grant land access. From 2003 to 2007, the 2200 acres of fossil-rich land in northern Arlington was unavailable for paleontological exploration and scientific study. It was not until a new landowner, the Huffines family, purchased the land in 2007 that formal exploration and excavation began.

Amateur fossil hunters have collected scattered dinosaur remains in the Woodbine Formation for many years, but little of this material is described in the scientific literature. Isolated skeletal remains of theropods, nodosaurid ankylosaurs, and iguanodontian ornithopods are reported by Lee (1997a) from multiple localities in the greater Dallas–Fort Worth metroplex. A proximal left femur, complete tibia, and fibula are described from a locality in the Lewisville Member of the Woodbine Formation, and a humerus and teeth have been described from the overlying Arlington Member. All of this material is assigned to indeterminate hadrosaurids (Lee, 1997a). Ornithopod tracks from the Woodbine are described by Lee (1997b), and assigned to the ichnotaxon *Caririchnium protohadrosaurichnos*. The best known ornithopod from the Woodbine Formation, *Protohadros byrdi*, was discovered at Flower Mound in 1994, and was later named in honor of its discoverer, Dallas Paleontological Society member Gary Byrd (Head, 1998). *Protohadros byrdi* is based on a nearly complete skull and scant postcrania from the Arlington Member of the Woodbine Formation (Head, 1998), and whereas it was originally described as the basal-most member of Hadrosauridae, subsequent analyses have recovered it as a non-hadrosaurid hadrosauroid (Norman, 2002; Horner et al., 2004; Prieto-Márquez, 2010a).

Given the proximity of fossiliferous strata of the Woodbine Formation to a major and expanding metropolitan area, it is surprising that more significant fossil material has not been recovered from the unit. Such material would likely provide valuable insight into the turnover of taxa and the origin of major clades (e.g., hadrosaurids) that occurred during the middle Cretaceous, for which there is little fossil evidence from North America (Jacobs and Winkler, 1998).

This contribution describes the stratigraphy, depositional setting, taphonomy, and postcranial anatomy of a basal hadrosauroid from the Woodbine Formation, and thus, further

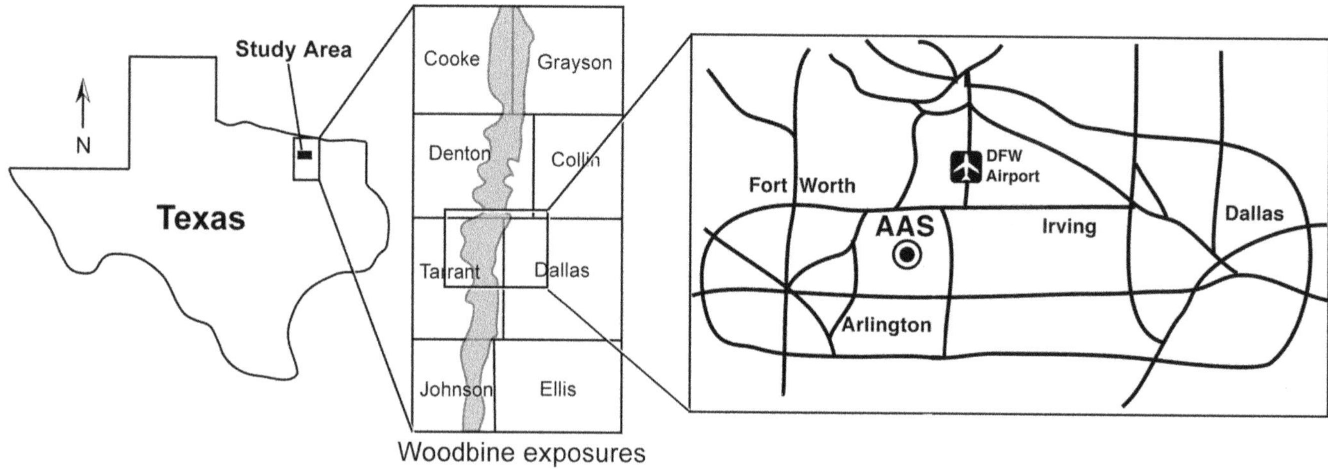

5.1. Location of the Arlington Archosaur Site (AAS) on a Texas map, with outlines of counties in which Woodbine Formation exposures occur. The AAS study area is in the city of Arlington, in northeastern Tarrant County.

elaborates the nature of the formation's hadrosauroid fauna. The new ornithopod material consists of specimens UTA-AASO-2003 and UTA-AASO-125 (University of Texas, Arlington–Arlington Archosaur Site field numbers), now stored in the collections of the Perot Museum of Nature and Science, Dallas, Texas. (Field numbers provided by C. R. Noto, August 2013. At the time of publication all material described here has been received by the Perot Museum of Nature and Science, Dallas, Texas, where it will be accessioned and assigned specimen numbers [A. R. Fiorillo, pers. comm., August 2013]). All specimens described here were recovered from the AAS, intermittently, from 2003 to 2008. The adult material includes a femur; scapula; coracoid; ilium; ischium; pubis; ribs; an axis; and cervical, dorsal, and caudal vertebrae. Although an assemblage of hadrosauroid postcrania such as this is potentially important for assessing phylogenetic relationships (Brett-Surman, 1979; Davies, 1983; Maryańska and Osmólska, 1984; Casanovas et al., 1999; Prieto-Márquez et al., 2006; Brett-Surman and Wagner, 2007; Poole, 2008; McDonald et al., 2010; Campione, this volume) we are not yet able to determine whether the AAS ornithopod material represents a known taxon, such as *Protohadros byrdi*, or a new taxon. Such conclusions must await further discoveries and research.

Institutional Abbreviations SMU, Shuler Museum of Paleontology, Southern Methodist University, Dallas, Texas; UTA-AASO, University of Texas, Arlington–Arlington Archosaur Site, Arlington, Texas.

Anatomical Abbreviations **ac**, acetabulum; **ap**, pseudoacromion process; **C**, coracoid; **ccv**, cranial convexity; **cf**, coracoid foramen; **cp**, capitulum; **cr**, coracoid ridge; **crp**, cranial pubic process; **cs**, coracoid suture; **ctr**, base of cranial trochanter; **dia**, diapophysis; **dr**, deltoid ridge; **eb**, expanded boot; **fn**, femoral neck; **ftr**, fourth trochanter; **gl**, glenoid; **gtr**, greater trochanter; **haf**, hemal arch facet; **IL**, ilium; **IS**, ischium; **ilp**, iliac peduncle; **isp**, ischial peduncle; **obp**, base of obturator process; **od**, odontoid process; **P**, pubis; **pap**, preacetabular process; **par**, parapophysis; **pop**, postacetabular process; **poz**, postzygapophysis; **prz**, prezygapophysis; **pup**, pubic peduncle; **R**, rib; **S**, scapula; **sac**, suprailiac crest; **scl**, scapula labrum; **scs**, scapula suture; **svf**, facets for sacral vertebrae; **tb**, tuberculum; **V**, vertebra; **vl**, ventral lip.

GEOLOGIC SETTING

Woodbine Formation and Members

R. T. Hill was the first geologist to map and describe the Cretaceous of Texas, which extends from North Texas to the Big Bend region (Hill, 1901; Alexander, 1976), and named the Woodbine Formation of North Texas for the village of Woodbine in Cooke County. Woodbine exposures extend as an irregular and narrow, north–south-oriented band from Cooke to Johnson counties (Johnson, 1974; Fig. 5.1). Woodbine sediments were sourced from the Ouachita Mountains in southern Oklahoma and were deposited in fluvial, deltaic, and shelf environments in coastal plain and shoreline settings of the subsiding East Texas Basin (Dodge, 1952; Oliver, 1971; Trudel, 1994). The Woodbine Formation unconformably overlies the Grayson Marl of the Washita Group and, in turn, is overlain conformably by the Eagle Ford Group (Dodge, 1952, Oliver, 1971; Fig. 5.2). A minimum age for the Woodbine Formation has been determined using the ammonite biostratigraphy of the overlying Eagle Ford Group,

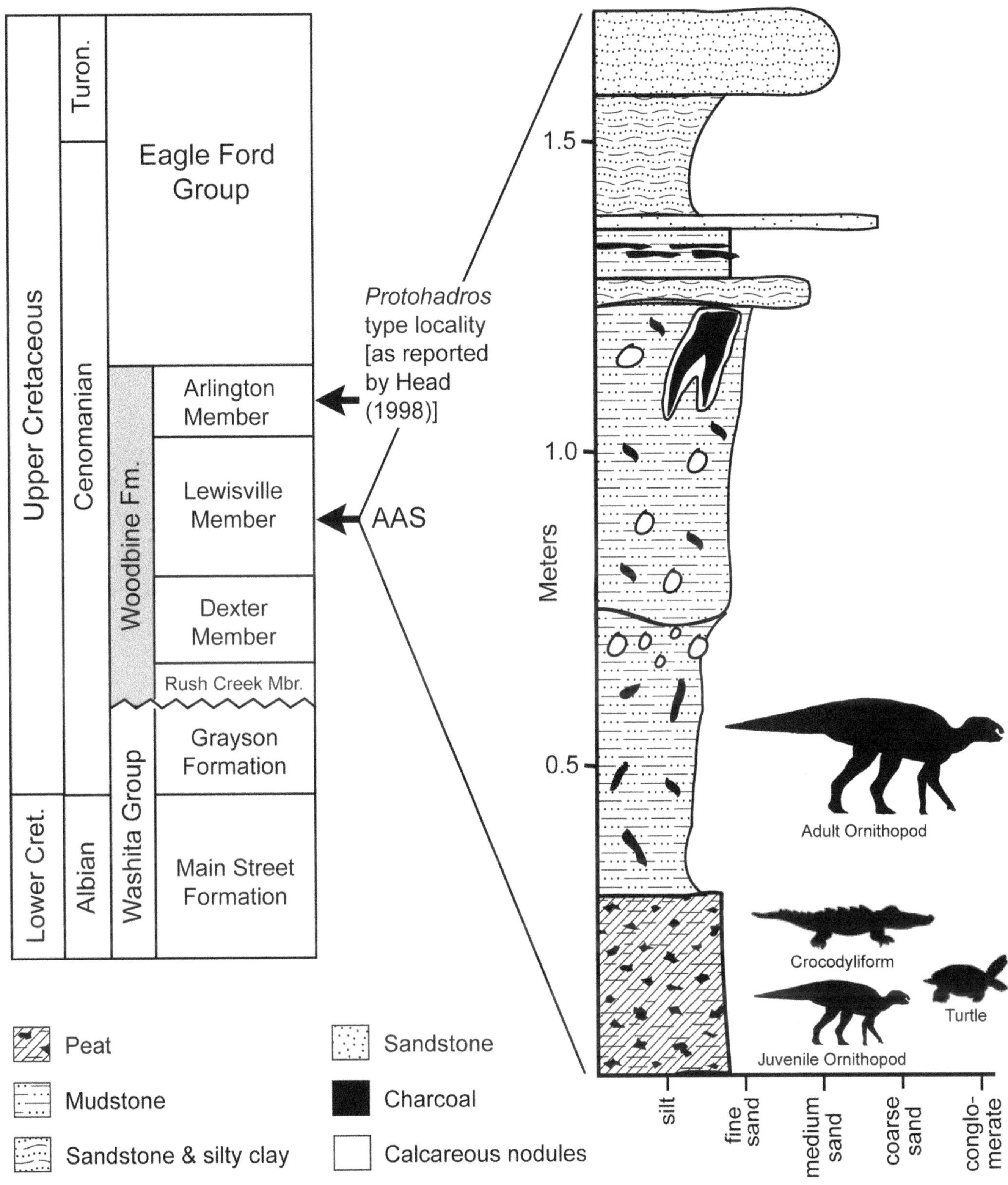

5.2. North Texas Cretaceous stratigraphy, with the Woodbine Formation highlighted in gray. The Arlington Archosaur Site lies in the lower to middle Lewisville Member of the Woodbine Formation. The site consists of ~1.7 m of sediments representing a delta-plain environment. The adult ornithopod material discussed here was discovered in a paleosol overlying a carbonaceous peat bed containing crocodyliform, turtle, and juvenile ornithopod remains (not discussed herein). North Texas stratigraphic column modified after Jacobs and Winkler (1998).

which lies within the *Conlinoceras tarrantense* zone at the base of the middle Cenomanian. This approach indicates that the Woodbine is no younger than 95 Ma (Kennedy and Cobban, 1990; Head, 1998).

The Woodbine Formation comprises four members. In ascending order these are Rush Creek, Dexter, Lewisville, and Arlington (Dodge, 1968, 1969; Fig. 5.2). The Rush Creek Member consists of a southern province distinguished by offshore bars and lagoonal deposits, and a northern province distinguished by lower shoreface mudstones and sandstones, and estuarine sandstones (Bergquist, 1949; Johnson, 1974). The Dexter Member consists of coastal-plain distributary and meander belt facies (Oliver, 1971). The Lewisville Member consists of deltaic sandstones and mudstones, and marine mudstones and shales deposited adjacent to active deltas (Oliver, 1971; Trudel, 1994; Main, 2005). The Lewisville Member is highly fossiliferous, containing mollusks, ammonites, and foraminifers (Trudel, 1994), and the material described here. A basal unit of lenticular, cross-bedded sandstones with localized lignite beds was designated as the Red Branch Member by Bergquist (1949), but subsequently was included as part of the Lewisville Member (Beall, 1964; Dodge, 1969; Trudel, 1994). The Arlington Member comprises nonmarine delta plain deposits dominated by fluvial sandstones (Murlin, 1975). The type locality of *Protohadros* occurs within this member (Head, 1998).

Stratigraphy and Paleoenvironments of the AAS

The AAS occurs within the Lewisville Member of the Woodbine Formation based on lithostratigraphic and biostratigraphic criteria. The presence of the invertebrates *Anchura* and *Gyrodes* place the AAS in the lower to middle strata of the Lewisville Member (Stephenson, 1952). The palynomorphs *Dichastopollenites* and *Stellatopollis*, recovered from the AAS, affirm a mid-Cenomanian age (May, 1975).

The sediments of the AAS suggest sedimentation in a low-lying coastal plain setting. All beds dip slightly to the east. The lowermost part of the section consists of rippled shallow marine sandstones that are heavily marked by traces of *Diplocraterion*, *Thalassinoides*, and *Arenicolites* (*Skolithos* ichnofacies [Seilacher, 2007]). Overlying these nearshore sandstones is a thin (50–60 cm thick) peat that has locally high claystone, sulphur, and pyrite content (Main, 2009). The peat is laterally continuous, cropping out along the base of adjacent hillsides. It is also highly fossiliferous, containing the mixed remains of numerous vertebrates, and carbonized wood, branches, twigs, and fallen logs (Main, 2009: Main et al., in press). Overlying the peat is a claystone-rich, heavily rooted histic gleysol with numerous calcareous concretions (Main, 2009; Fig. 5.2). The uppermost part of the AAS section is distinguished by a lag deposit overlain by heterolithic mudstones and, subsequently, finely laminated and rippled sandstones.

We interpret the AAS section as recording fine-grained, organic-rich sediments that were deposited successively on a flood-inundated coastal plain. Within the peat, claystones, and mudstones, numerous coprolites and fossil wood have been recovered that can be used to study the ecosystem of the region. The large amount of fossil plant and wood at the site indicates a plant-rich, coastal wetland, such as a fen or mangrove forest. Several large (>3 m) and numerous partial (<2 m) logs were excavated from the peat bed immediately underlying the primary dinosaur bonebed. The ornithopod fossil-bearing horizon occurs within the paleosol, in association with calcareous concretions. Within the concretions are charcoal fragments and several large (28–36 cm long) charcoal tree stumps and roots (Fig. 5.2). We propose that concretion formation indicates seasonal dryness, and that the charcoal tree stumps and roots are possible evidence of wildfires. The charcoal (wildfire) horizons occur below, within, and above the vertebrate fossil horizons. The charcoal-rich conglomerates contain numerous charcoal and fossil wood fragments (1–4 cm long) bound in an iron-rich sandstone. We speculate that the charcoal and other clasts were transported and deposited during distributary-channel flooding.

Taphonomy

The ornithopod fossil remains were recovered from an approximately 6 m^2 surface, through ~0.6 m of section within the paleosol. Most of the specimens were originally recovered from the hillside by amateur fossil collectors and then, later, donated to the collection once formal excavation and mapping began. Thus, the precise positions of all the remains are uncertain, limiting our taphonomic interpretations. A simple map of the largest specimens has been reconstructed using anecdotal information provided by the collectors, and their photographs (Fig. 5.3). Elements were associated but disarticulated, thus appearing to belong to a single individual. An important exception is the femur (UTA-AASO-125), which, in addition to its smaller size (discussed below), also lacks many of the surface weathering features seen on the other specimens. This suggests that it was collected from the same bed, but from a different location along the hillside.

Evidence suggests that the remains experienced a complex biostratinomic and fossil-diagenetic history following the death of the individual. All specimens vary in color between dark brown to light orange. The predominance of relatively dense bones from the axial column and limbs, and disarticulation suggest some degree of hydraulic sorting and,

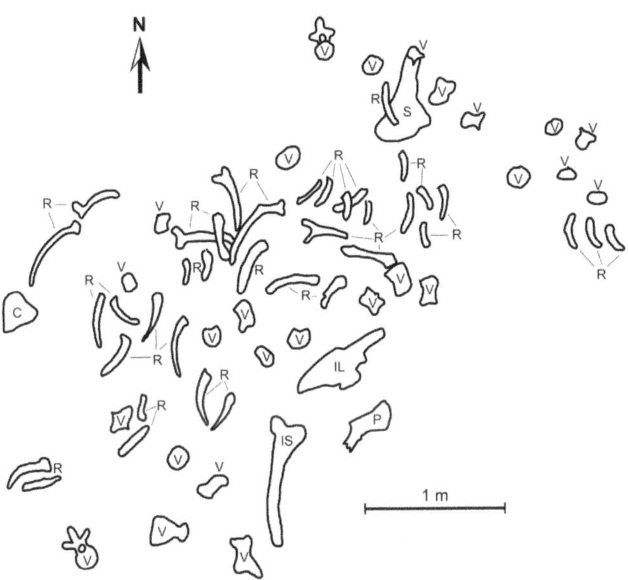

5.3. Reconstructed map of Arlington Archosaur Site ornithopod quarry based on discussions with original excavators. Bone locations are approximate, as is the scale.

possibly, transport of the remains, although some elements may be missing due to the activity of amateur fossil collectors and earth-moving equipment prior to formal excavation.

All the bones discussed here, except the femur, demonstrate weathering stages 2–4 (Behrensmeyer, 1978), including flaking, cortical exfoliation, small mosaic cracking, and deeper longitudinal cracking, indicating prolonged subaerial exposure prior to final burial. In contrast, the femur exhibits weathering stages 1–2.

The remains underwent postburial fossil-diagenetic alteration in the form of plastic and brittle deformation. Several bones, most notably a cervical vertebra (described below) and caudal vertebrae, show plastic deformation of the neural spine and/or centrum. Plastic deformation likely occurred early in diagenesis due to clay shrink-swell cycles in the paleosol. Slickensides are present in this horizon, with one preserved in situ along the posterior surface of a dorsal vertebra. Brittle deformation due to lithostatic compression is present in the dorsal margin of the ilium and the coracoid. In both cases these elements were flattened leaving large, cemented cracks. Several bones have undergone more recent post-fossilization weathering and now exhibit a simple few to a network of non-cemented cracks from the action of roots, moisture, and temperature extremes that characterize the modern seasonal climate of North Texas. This weathering is most intense close to the surface and lessens with depth. This overall taphonomic signature is similar to that described by González Riga and Astini (2007) in early to mid Late Cretaceous overbank deposits containing titanosaur sauropods in the Neuquén Basin of Argentina. There, sauropod remains apparently underwent prolonged subaerial exposure as well as disarticulation and sorting from sporadic, low-energy ancient hydraulic activity.

Despite lacking a knowledge of precise bone positions, the presence of these remains in an active delta-plain environment, their disarticulated and sorted nature, and the overall taphonomic similarity with material in an Argentinean sauropod bonebed strongly suggest that the AAS ornithopod material is parautochthonous. Upon death the individual likely experienced prolonged episodes of subaerial exposure and carcass rotting that alternated with episodes of short-distance transport and sorting due to sporadic, low-energy flooding from a nearby distributary channel. Upon final burial in muds, climatic seasonality and soil-forming processes resulted in extensive plastic deformation of the elements. Brittle deformation of elements resulted later from lithostatic compression by overlying strata. Finally, recent actions by humans and natural erosion exposed remaining bones to extensive post-fossilization weathering and wear.

SYSTEMATIC PALEONTOLOGY

DINOSAURIA Owen, 1842
ORNITHISCHIA Seeley, 1888
ORNITHOPODA Marsh, 1882
HADROSAUROIDEA Cope, 1869
HADROSAUROIDEA indet.

Material Hadrosauroid represented by at least two individuals. UTA-AASO-2003 appears to represent an adult individual, and includes a scapula, coracoid, ilium, ischium, pubis, cervical and dorsal ribs, an axis, and cervical, dorsal, and caudal vertebrae. UTA-AASO-125 is a partial femur from a subadult or juvenile individual.

Locality and Horizon All specimens were recovered from UTA Location 50 (the Arlington Archosaur Site; AAS), northern Arlington, Tarrant County, TX. GPS coordinates for the location, maps of the AAS excavation, and collections database with all relevant site data are on file with the Perot Museum of Nature and Science, Dallas, Texas. The specimens are from a single bone-bearing horizon in a clay-rich paleosol within the Lewisville Member of the Woodbine Formation (Cenomanian). UTA-AASO-125 is likely from the same horizon, but the exact location is unknown.

Comments Hadrosauroid remains designated UTA-AASO-2003 were recovered as disarticulated but associated elements from an ~6 m^2 area through ~0.6 m of section. Most of the material was collected by UT Arlington students Phil Kirchhoff and Bill Walker (who also discovered the site) in collaboration with Art Sahlstein. Initial excavation

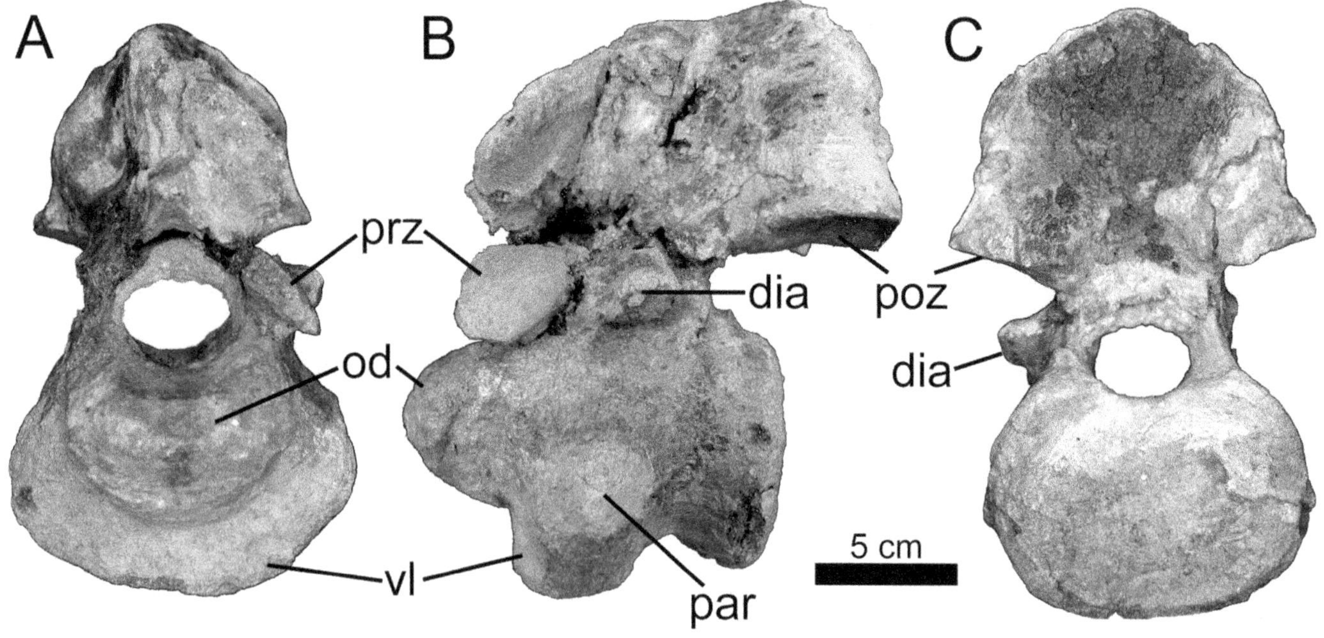

5.4. Axis of UTA-AASO-2003. (A) anterior view; (B) left lateral view; (C) posterior view.

of the site occurred prior to the authors' involvement with the project, and before large-scale excavation and study began. Maps of the initially recovered material were generated by interviewing Kirchhoff and Sahlstein, and by reviewing photos of the initial dig. All known AAS hadrosauroid material was prepared at the Earth and Environmental Sciences Department, UT Arlington. The material was prepared by DJM and paid student interns under his supervision. All prepared fossils were later donated to the Perot Museum of Nature and Science, Dallas, Texas.

DESCRIPTION

Cervical Vertebrae The axis is relatively complete, with minor deformation to the neural spine. The centrum is longer than it is tall and shorter than the large neural spine; its maximum length is 96 mm and it is 92 mm tall. The centrum is strongly opisthocoelous with a pronounced odontoid process fused to the anterior articular surface and buttressed by a broad ventral lip (Fig. 5.4). This lip terminates in a subrectangular parapophysis on each side of the centrum. The odontoid process is similar in form to that of *Zalmoxes*, *Camptosaurus*, *Bactrosaurus*, and *Telmatosaurus*, although it is proportionally larger than in *Camptosaurus* and it lacks the dorsal deflection of *Bactrosaurus* (Weishampel et al., 1993, 2003; Godefroit et al., 1998; Dalla Vecchia, 2006; Carpenter and Wilson, 2008). The fused odontoid process and ventral buttress are features typical of iguanodontians, and are absent in most hadrosaurids (although present in *Telmatosaurus* [Weishampel et al., 1993; Horner et al., 2004]). The prezygapophyses are oblong structures that sit flush with the sides of the neural spine. Small, rounded transverse processes occur caudal to the prezygapophyses; each terminates in a diapophysis. The neural arch and spine are robust, extending dorsally 156 mm above the centrum. The vertebral canal is wider than it is tall. The craniodorsal margin of the neural spine is blade like in lateral view and narrows dorsally. It extends cranially over the centrum as a pointed process but does not extend beyond the odontoid process. In lateral view the neural spine has a gentle slope (~60°) that rises caudally. Caudally, the neural arch bifurcates into separate, broad postzygapophyses with ventrally flattened articular facets. A plate from each postzygapophysis extends dorsally and meet at the caudal apex of the neural spine, creating an arch surrounded by a deep, caudally oriented fossa. The caudal articular surface of the centrum is deeply concave and slightly reniform in shape. In caudal view, the outline of the axis is roughly hourglass-shaped. The overall morphology of the axis is similar to the iguanodontians *Iguanodon bernissartensis*, *Camptosaurus*, and *Bactrosaurus* with some affinities to the primitive hadrosaur *Telmatosaurus* (Norman, 1980; Weishampel et al., 1993; Godefroit et al., 1998; Dalla Vecchia, 2006; Carpenter and Wilson, 2008).

The postaxial cervical vertebrae are roughly equal in length with strongly opisthocoelous centra, a character typical of hadrosauroids (Horner et al., 2004). The centra are

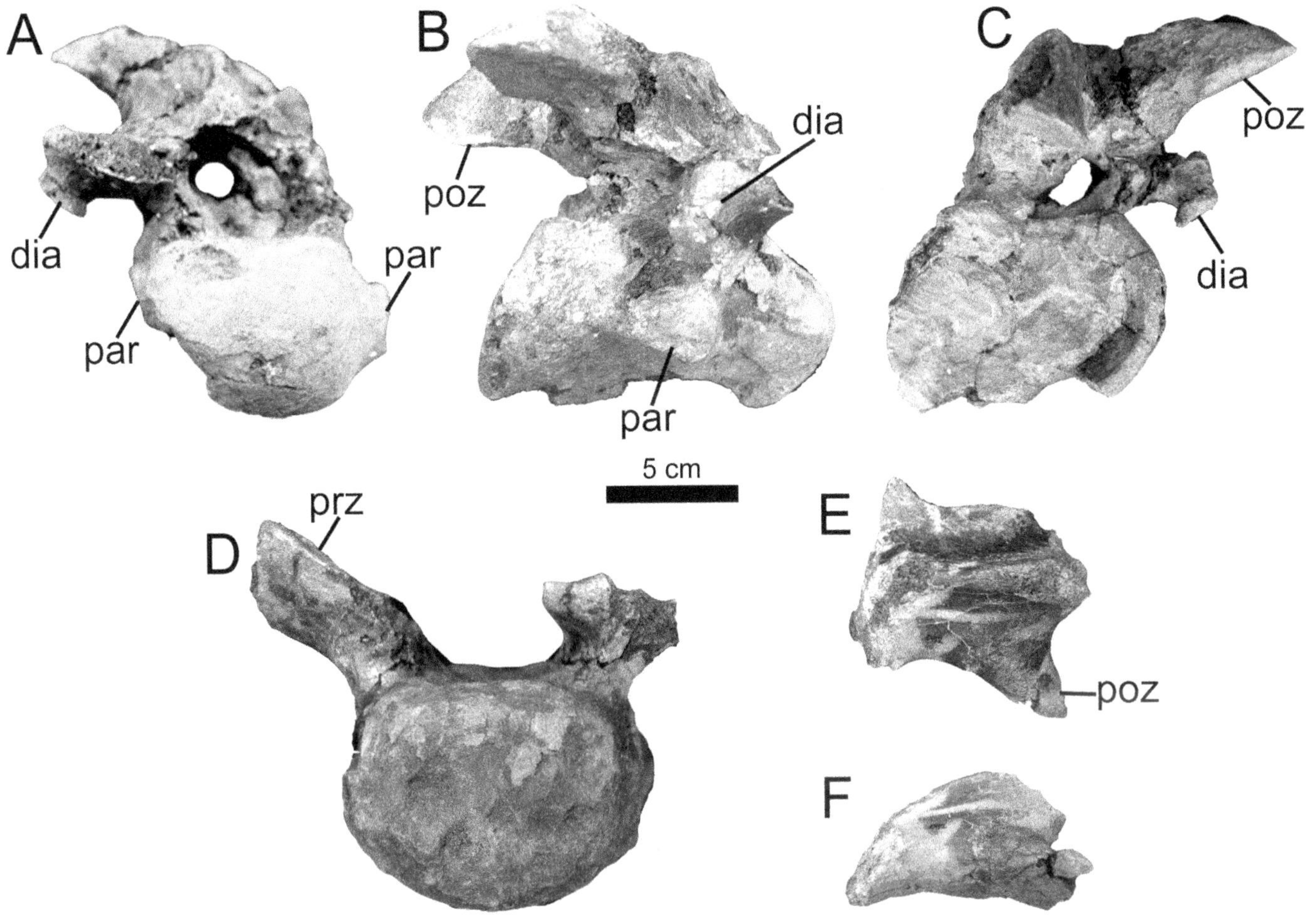

5.5. Postaxial cervical vertebrae of UTA-AASO-2003. (A) cranial cervical vertebra in anterior view; (B) right lateral view of same; (C) posterior view of same; (D) caudal cervical vertebra in anterior view; (E) isolated cervical neural arch in dorsal view; (F) left lateral view of same. Note the cranial cervical in (A–C) is deformed, with the neural arch deflecting laterally to the right.

broad mediolaterally and elongate axially, with a slight constriction in the middle of the centrum. Both in cranial and caudal view, the centra appear reniform and possess a ventral keel (Fig. 5.5A–C). The neural arch is vertically short and gently slopes caudodorsally. The neural spine is weakly developed or absent on the more cranial cervicals (Fig. 5.5A–C, F). The neural arch and spine broadens caudally approaching the dorsals (although the most caudal cervicals are incomplete). The prezygapophysis is elongate laterally and extends cranially over the centrum with an ovoid articular facet (Fig. 5.5D). The diapophysis is a rounded process that is deflected ventrocaudally from the transverse process. The postzygapophyses are elongate, wedge-shaped, and sweep caudolaterally at a wide angle. The postzygapophyses possess broad, ellipsoidal articular surfaces that are flattened and extend sharply outward.

Dorsal Vertebrae The dorsal vertebrae consist of at least eight relatively complete specimens – more if miscellaneous fragments are included. Due to the fragmentary and disarticulated nature of the vertebrae, the complete number of vertebrae in the dorsal series is unknown. The cervicodorsal transition is seen in the proximal most dorsals, as they are low with small centra and dorsolaterally directed transverse processes (Fig. 5.6A–C). The neural arch and spines thicken along the dorsal series, with the neural spines increasing in height and straightening dorsoventrally. In one well-preserved cranial dorsal vertebra, the centrum is slightly opisthocoelous and subrounded, broadening posteriorly in lateral view. Above the neurocentral suture, two thin ridges of bone form an arch rising dorsally to the transverse process; within this arch is the parapophysis, which is small and sits at mid-height laterally beneath the transverse process. The transverse process rises dorsolaterally from the neural arch at a sharp angle. It is rectangular in lateral view, ending in a thick, subrounded diapophysis. The vertically arched transverse process is typical of the cranial dorsal series in hadrosauroids. A partial vertebra

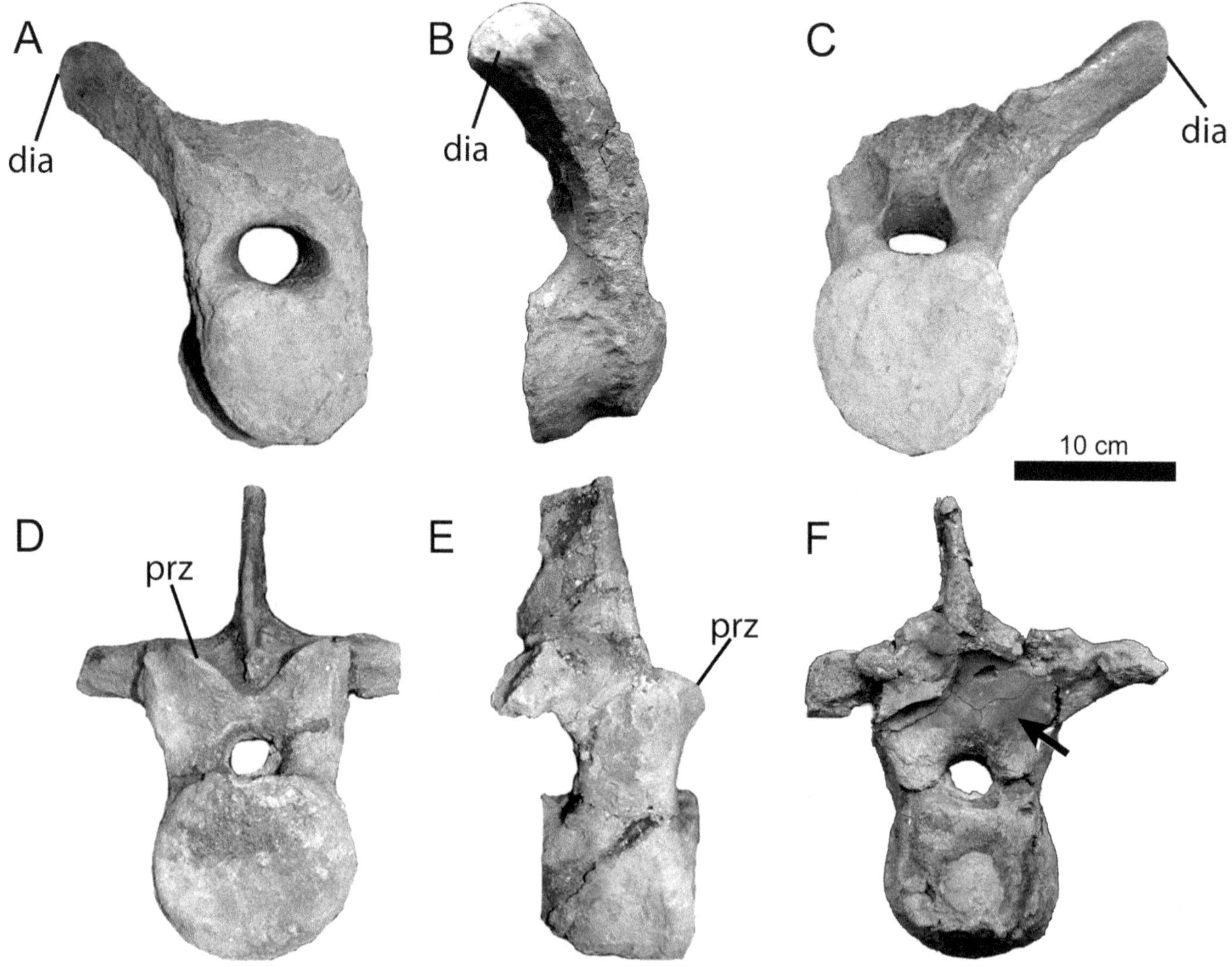

5.6. Dorsal vertebrae of UTA-AASO-2003. (A) cranial dorsal vertebra in anterior view; (B) right lateral view of same; (C) posterior view of same; (D) caudal dorsal vertebra in anterior view; (E) right lateral view of same; (F) posterior view of same. Arrow notes slickensides on surface of bone.

from the middle dorsal series is composed of a complete centrum with neural arch, a partial neural spine and partial transverse processes (Fig. 5.6D–F). The centrum is amphiplatyan, rounded in cranial and caudal view, and broadens caudally. The transverse processes project laterally at a near 90° angle to the neural spines, a feature that is typical of the middle to sacral dorsal vertebrae. The exact lengths of the transverse processes are unknown as the specimen is incomplete. The neural spine is rectangular in lateral view and projects dorsally. The prezygapophyses are subrounded with lateroventrally facing ellipsoidal facets and project dorsomedially from the base of the neural arch, but do not extend far over the centra. The postzygapophyses are not preserved. The remainder of the dorsal vertebrae are represented by centra only. All specimens are amphiplatyan and relatively rounded in cranial and caudal view.

Dorsal Ribs The ribs are represented by partial, disarticulated elements from the middle to the caudal end of the series (Fig. 5.7). The ribs are similar to those of *Tethyshadros* and *Iguanodon*, with the most diagnostic characters residing in the tuberculum and capitulum (Norman, 1980; Dalla Vecchia, 2009). The tuberculum is shorter than the capitulum and the shaft of the most caudal ribs is long and ventrally, rather than caudolaterally, situated. The capitulum is rod like; however, the tuberculum is not preserved. The mid-dorsal ribs (Fig. 5.7D, E) are curved ventrolaterally and appear triangular in cross section. The capitulum and tuberculum are much larger and better defined in the mid-dorsal series. The tuberculum is a small, rounded process that projects vertically away from the surface of the rib. The capitulum is a thin, rectangular projection that extends away from the tuberculum at a 30° angle, ending in a flattened articular surface.

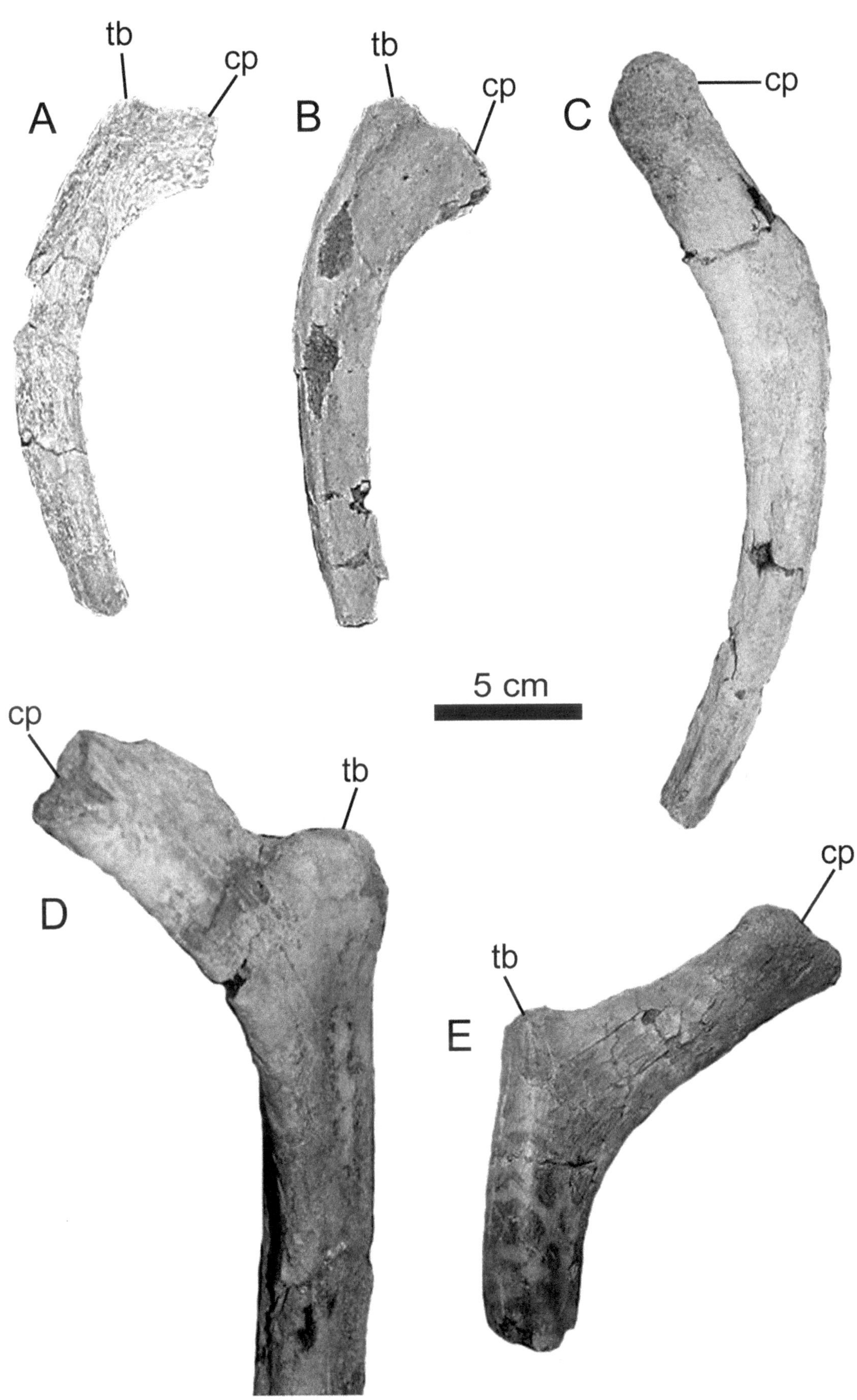

5.7. Dorsal ribs of UTA-AASO-2003. (A) cranial dorsal rib; (B) cranial dorsal rib; (C) caudal dorsal rib; (D) middle dorsal rib; (E) middle dorsal rib.

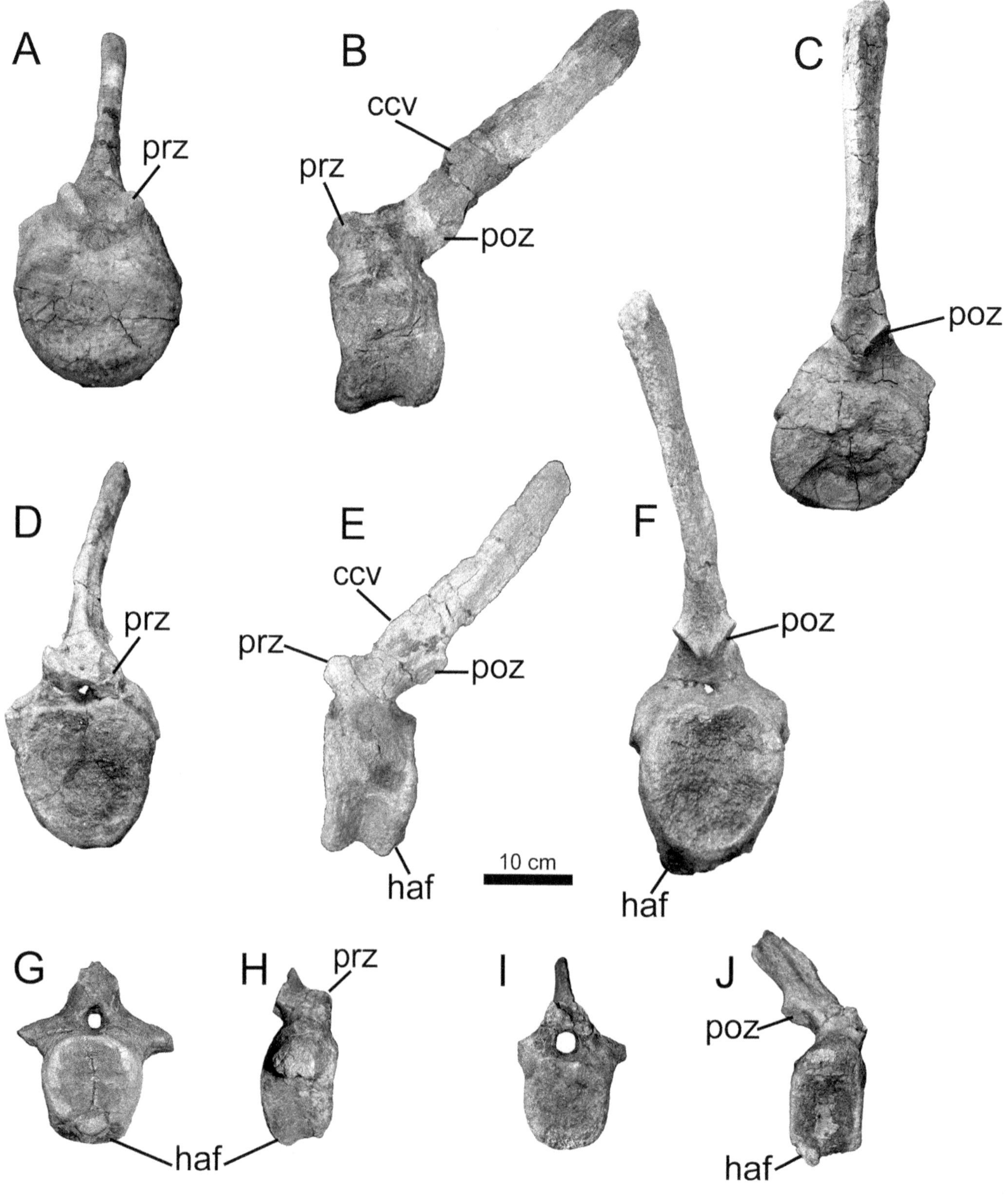

5.8. Caudal vertebrae of UTA-AASO-2003. (A) first or second caudal vertebra in anterior view; (B) left lateral view of same; (C) posterior view of same; (D) anterior view; (E) proximal caudal vertebra in left lateral view; (F) posterior view of same; (G) middle caudal vertebra in posterior view; (H) right lateral view of same; (I) middle to distal caudal vertebra in anterior view; (J) right lateral view of same.

Caudal Vertebrae There are six caudal vertebrae: two proximal caudals represent the first, and second or third caudals; several middle to distal caudals are present; there are multiple fragmentary centra. In the proximal caudals the centra are round, amphiplatyan, and longitudinally compressed (Fig. 5.8A, B, D, E). The dorsal surface is defined by broad, triangular neural arches. The prezygapophyses are small, ovoid facets that project obliquely while the postzygapophyses are ventrolaterally directed, oblong facets attached to the posterior edge of the neural spine. The neural spines are rectangular in lateral view, tall (>30 cm long), relatively narrow, and deflected distally by ~45° to 60° (Fig. 5.8B, E). The neural spine is three times the length of the centrum and unexpanded distally. Its cranial edge is relatively straight, except for a slight convexity, similar to *Eolambia* (McDonald et al., 2012). The transverse processes are broken, but are attached at the base of the neural arch. In the middle caudal series, the centrum is heart shaped and the ventral surface is rounded, with a pronounced ventrally deflected hemapophysis for a chevron (hemal arch; Fig. 5.8G, I). Prezygapophyses are prominent while the postzygapophyses project ventrolaterally from the neural arch, with ovoid articular facets visible in lateral view. The neural spine is deflected distally, but at a greater angle (~60°) than proximal caudals. The opening for the neural canal is small and circular. Transverse processes are present and extend laterally with an oval cross section. The remaining middle to distal caudals are incomplete; all have heart-shaped centra and lack neural spines and zygapophyses. Transverse processes become less prominent in more distal caudals. The distal-most caudals have more elongate centra in lateral view.

Ossified Tendons Ossified tendon fragments were recovered but found isolated away from the caudal vertebrae. None were recovered on or associated with the caudal series. They are thin, cylindrical elements, pencil like in appearance. They are thicker along the proximal surface and thin distally.

Scapula The left scapula is a broad, robust element 356 mm in length from the ventral margin to the mid-shaft (Fig. 5.9A–C). The blade of the specimen was damaged during excavation, and is incomplete. The partial blade is straight in lateral aspect and oval in cross section, with no proximal constriction and a mild medial curvature to accommodate the ribcage. The cranial edge of the blade rises gently into the pseudoacromion process, which is a convex, rounded ridge that thins cranially, similar to that of *Eolambia* (McDonald et al., 2012) and it is gently inclined laterally toward the coracoid suture, as in *Probactrosaurus* (Norman, 2002). The pseudoacromion process curves ventrally into the coracoid suture. A low deltoid ridge is present, extending caudally as a low prominence from the pseudoacromion process. The coracoid suture is convex and bi-faceted into two thick, prominent surfaces in cranioventral view, similar to *Camptosaurus* and *Mantellisaurus* (Paul, 2007; Carpenter and Wilson, 2008). The coracoid suture is thick (94 mm) and curves into the scapular part of the glenoid. The proximal blade expands ventrally to produce a deep, pronounced (60 mm wide × 150 mm long) arched depression that terminates in a broad laterally projected buttress. It is similar in form to the iguanodontians *Camptosaurus* and *Iguanodon* and the hadrosauroid *Probactrosaurus* (Norman, 1986, 2002, 2004; Carpenter and Wilson, 2008). The angle of ascent of the scapula labrum away from the shaft is steep and similar in form to *Eolambia* (McDonald et al., 2012).

Coracoid The left coracoid is 200 mm long, slightly convex in lateral view, and concave in medial view (Fig. 5.9C–E). A large, elliptical coracoid foramen occurs near the caudal margin at the scapular contact. The glenoid surface is found along the caudoventral margin and is roughly equal in size to its counterpart on the scapula. The scapular suture is elongate and rugose at the glenoid contact, and tapers into a thin and blade-like ridge dorsocranially. The coracoid ridge is present on the cranial margin, forming a low, rounded prominence. Along the cranioventral margin is the base of the sternal process, which is broken and thus its exact shape cannot be determined. Overall the morphology of the coracoid strongly resembles that of *Eolambia* and *Probactrosaurus* (Norman, 2002; McDonald et al., 2012).

Ilium The left ilium has a vertical height of 212 mm and a total length of 604 mm, as preserved. The preacetabular process is incomplete missing its distal end (Fig. 5.10). The preserved portion of the preacetabular process is stout and rounded dorsally with a thin ventral margin, giving the process a nearly triangular cross section. It is gently deflected ventrolaterally and contains a well-developed medial shelf, similar to *Eolambia* and *Probactrosaurus*, (Norman, 2002; McDonald et al., 2012). The preacetabular notch is open. The pubic peduncle is short (36 mm) and subrounded with a slight inclination cranially, similar to *Bactrosaurus*, *Probactrosaurus*, *Tethyshadros*, and *Eolambia* (Godefroit et al., 1998; Norman, 2002; Dalla Vecchia, 2009; McDonald et al., 2012). The broad ischial peduncle is 176 mm in length and rectangular. The acetabular margin is 93 mm in length and relatively shallow.

The dorsal edge of the ilium underwent some postburial distortion, displacing the suprailiac crest dorsally (Fig. 5.10A, C). Its dorsal margin caudal to the pubic peduncle was likely straight or slightly convex in lateral view, and sigmoidal in dorsal view. The suprailiac crest begins as a narrow ridge just caudal to the pubic peduncle and it widens caudally, producing a thick (20 mm) curved ridge that deflects caudoventrally, and ends just caudal to the ischial peduncle,

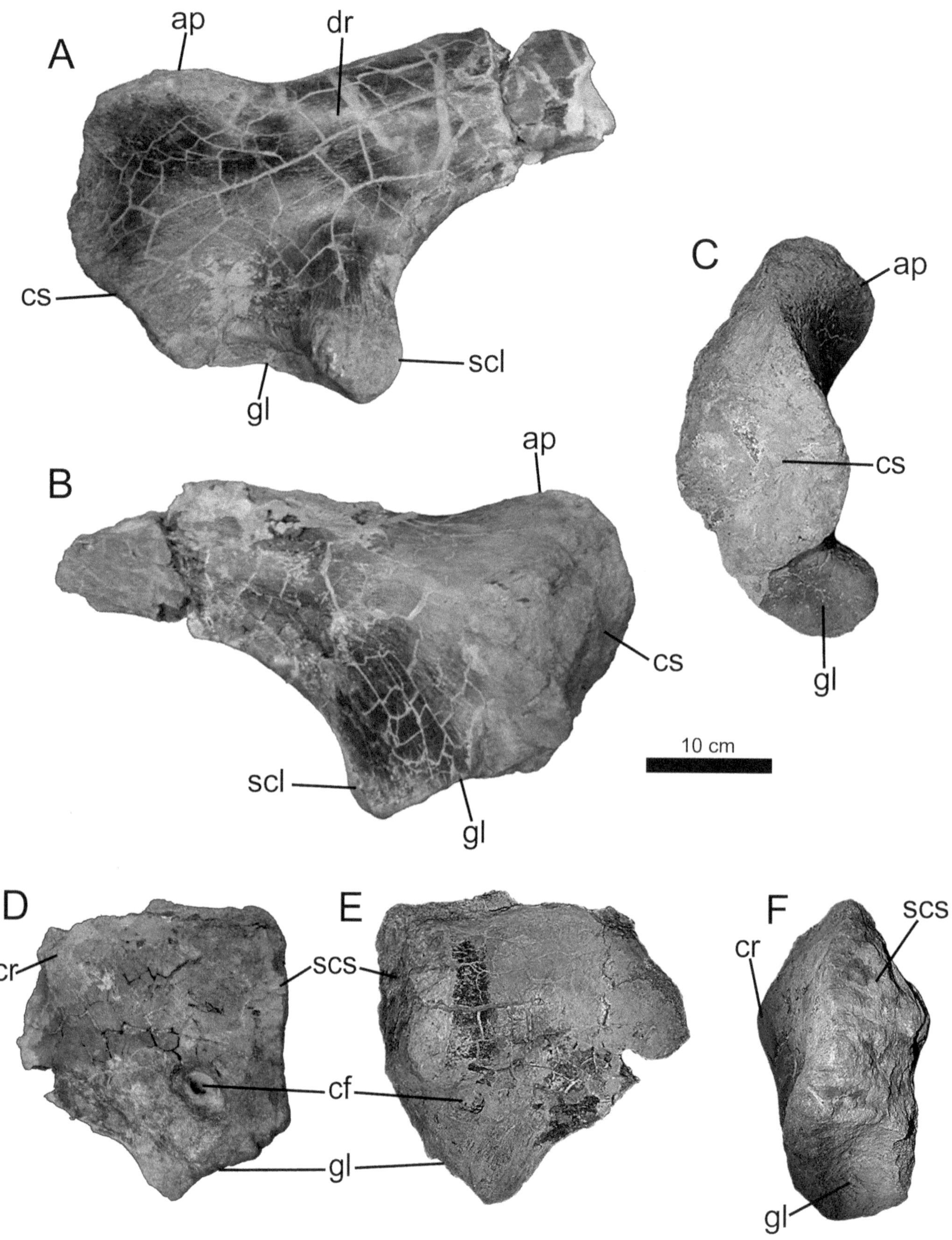

5.9. Pectoral girdle of UTA-AASO-2003. (A) left scapula in lateral view; (B) medial view of same; (C) anterior view of same; (D) left coracoid in lateral view; (E) medial view of same; (F) posterior view of same.

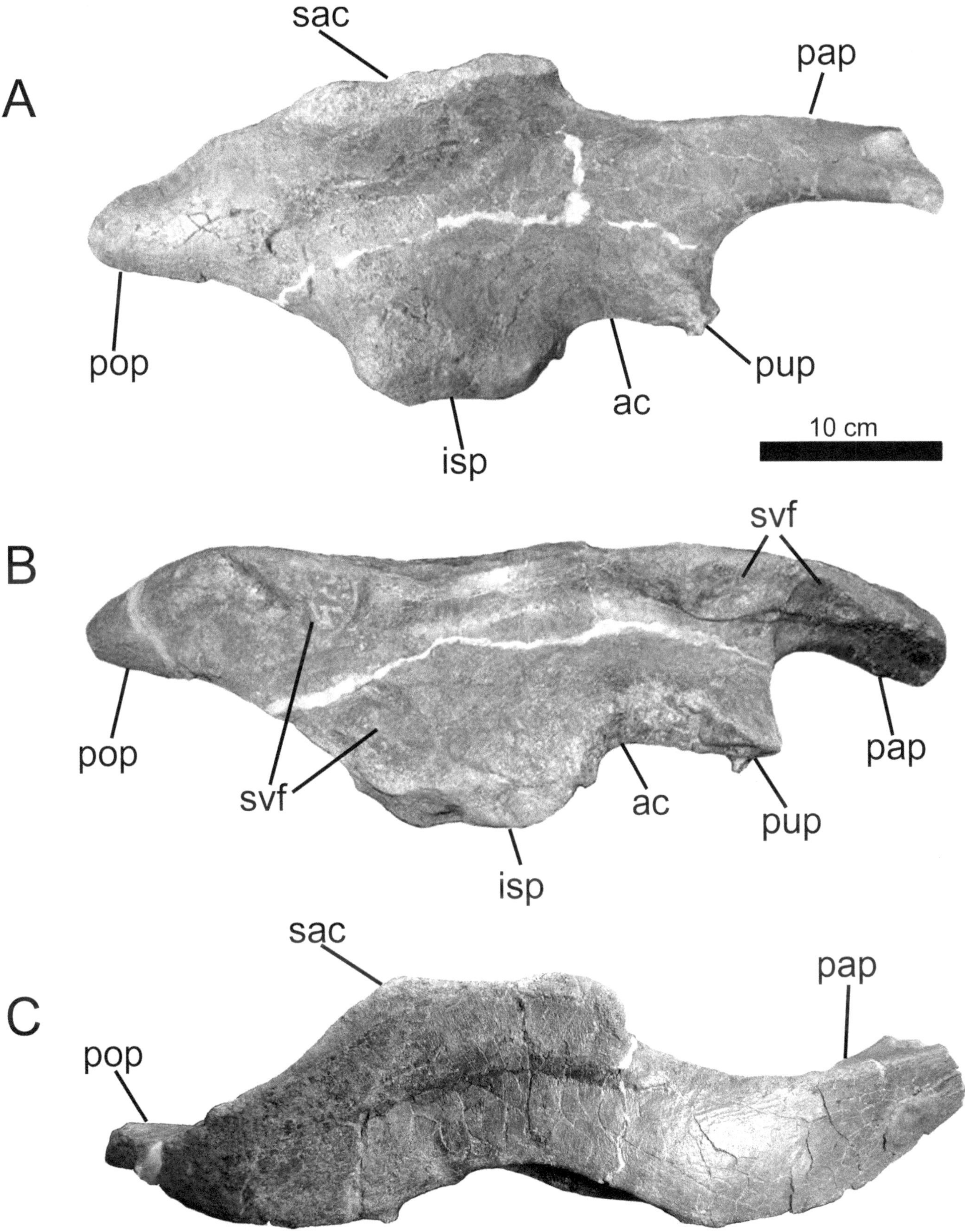

5.10. Left ilium of UTA-AASO-2003. (A) lateral view; (B) medial view; (C) dorsal view.

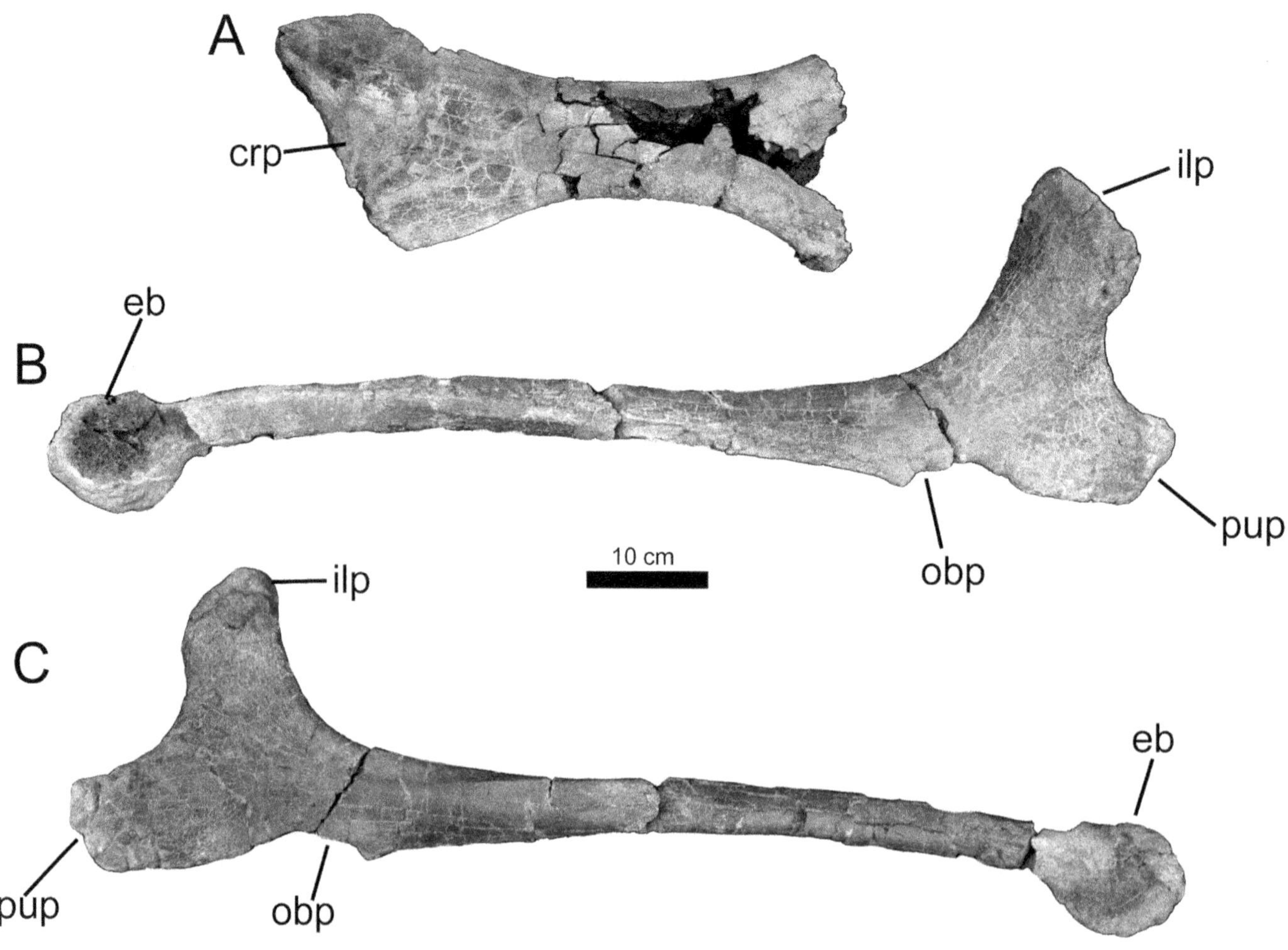

5.11. Pelvic elements of UTA-AASO-2003. (A) left pubis in lateral view; (B) left ischium in medial view; (C) lateral view of same.

where it grades into the postacetabular process. This more cranial position of the suprailiac crest ("antitrochanter") is distinctive of early hadrosaurids (Brett-Surman and Wagner, 2007). The postacetabular process is broad and triangular in lateral view, as is typical of derived iguanodontids and basal hadrosauroids (Norman, 2004; Brett-Surman and Wagner, 2007). A medial brevis shelf is absent. The medial surface of the ilium is concave, with the deepest portion overlying the ischial peduncle. Multiple facets for sacral rib attachment are visible (Fig. 5.10B). The ilium fits well into the first iliac morphotype recognized by Chapman and Brett-Surman (1990) and bears a strong resemblance to the ilium of *Claosaurus* (Prieto-Márquez, 2011).

Ischium The left ischium is nearly complete, demonstrating the typical triradiate ornithopod form, although it shows multiple fractures along the shaft (Fig. 5.11). The ischium is 720 mm long, 256 mm from the pubic peduncle to the iliac peduncle, and ranges in width from 76 to 52 mm from the obturator process to the boot. The acetabulum is denoted by a gentle sloping, shallow concavity. The pubic peduncle is smaller than the iliac peduncle, similar to the ratios seen in *Probactrosaurus*, *Bactrosaurus*, and *Eolambia* (Godefroit et al., 1998; Kirkland, 1998; Norman, 2002; McDonald et al., 2012). The iliac peduncle is a broad (98 mm cranial width) and subrectangular process. The pubic peduncle is small (50 mm cranial width), narrow and rectangular. The craniomedial part of the proximal blade is weathered and the shape of the obturator process cannot be determined. A ridge extends dorsolaterally away from the obturator process down the length of the shaft, flattening distally. This tapering ridge is similar to the ischial ridge noted in *Probactrosaurus*, although more sublime (Norman, 2002). The shaft is straight for most of its length, with a gentle ventral deflection distally, and terminating in an ischial knob, or boot. The ischium fits within the Type 1 *Gilmoreosaurus* classification used by Brett-Surman and Wagner (2007).

Pubis A partial left pubis, 420 mm long as preserved, is primarily represented by the cranial pubic process, with

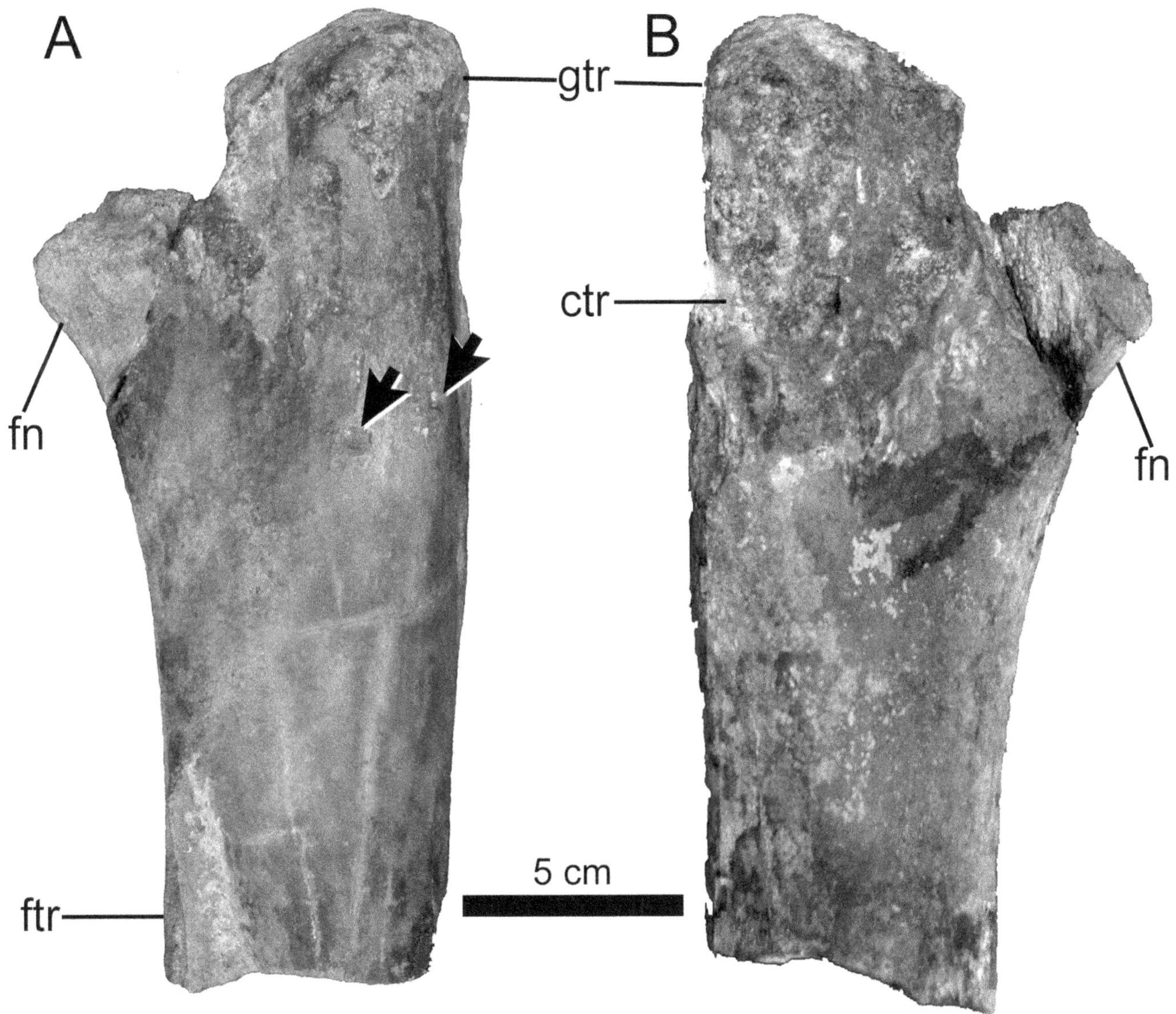

5.12. Right femur UTA-AASO-125. (A) posterior view; (B) anterior view. Arrows denote crocodyliform bite marks (see Noto et al., 2012).

the remainder shattered along the caudal shaft and acetabular margin (Fig. 5.11). The cranial pubic process is shallowly convex along the ventral margin and concave along the dorsal margin. The cranial pubic process is 180 mm in vertical height, twice that of the minimum constriction of the neck. It is laterally compressed and thins cranially with an axe-like cranial margin, similar to *Eolambia* (Kirkland, 1998; McDonald et al., 2012), however the margin is not completely preserved. The prepubic neck is relatively long, 12 cm and thin, and similar in form to that of *Tethyshadros* (Dalla Vecchia, 2009). The dorsal margin of the neck is thick and rounded, while the ventral margin is thin. Both the iliac and ischial peduncles are broken and missing, as is the acetabular margin. The pubis is morphologically similar to that of *Tethyshadros* and *Eolambia* (Kirkland, 1998; Dalla Vecchia, 2009; McDonald et al., 2012).

Femur The femur is represented by a partial right element, but its smaller size relative to the other pelvic elements suggests it likely belongs to a different individual (UTA-AASO-125). The femur preserves the proximal third of the bone, including a partial femoral neck, the greater and cranial trochanters, and a small dorsal section of the fourth trochanter (Fig. 5.12). The medial surface is gently bowed, while the lateral surface is straight. The femoral head is missing, with only a fragmentary base indicating the neck was narrow and formed a saddle-shaped depression between the head and greater trochanter. The greater trochanter is craniocaudally elongate and mediolaterally compressed, with a convex dorsal surface, forming the majority of the proximolateral end of the femur. The greater and cranial trochanters were separated by a narrow vertical cleft. The cranial trochanter is weathered but shows it was small and occupied a craniolateral position on the femur. The fourth trochanter is missing, and its ventral extent truncated by a break in the shaft, so its exact shape cannot be determined. What is preserved shows it was positioned along the caudomedial surface of the shaft about a third of the way down, as opposed to about half way down the shaft as is seen in many hadrosauroids. The fourth trochanter was proximally narrow, probably forming a ridge as in *Eolambia*, *Telmatosaurus*, and *Bactrosaurus* (Weishampel et al., 1993; Godefroit et al., 1998; McDonald et al., 2012). Of note, this element preserves evidence of predatory behavior via two pit marks along the caudal shaft (Fig. 5.12A). The pit marks are round impressions in the bone surface attributed to the feeding behavior of a large crocodyliform, discussed in detail in Noto et al. (2012).

DISCUSSION

Taxonomic Affinity

The discovery of the AAS postcranial material adds significantly to what is known about the anatomy of basal hadrosauroids from the Woodbine Formation. This new material represents a non-hadrosaurid hadrosauroid of uncertain affinity pending further analysis. It can confidently be assigned to the Hadrosauroidea on the basis of an ischium with a shaft that is nearly straight (McDonald et al., 2010:character 64). It is excluded from Hadrosauridae by possession of a large, dish-shaped coracoid, dorsocranially oriented pseudoacromion process of the scapula, triangular postacetabular process, and odontoid fused to the body of the axis (Horner et al., 2004; Brett-Surman and Wagner, 2007). The remaining mixture of basal and derived characters in the AAS taxon highlights the mosaic nature of hadrosauroid evolution prior to the emergence of Hadrosauridae, as noted by Dalla Vecchia (2009).

The gentle slope of the axial neural spine is similar to the iguanodontians *Iguanodon*, *Tenontosaurus*, and *Camptosaurus*, and is a feature not seen in hadrosaurids. The axis of hadrosaurids typically has a rounded, crescent, or blade-like neural spine that expands outwards dorsally, as seen in *Kritosaurus*, *Gryposaurus*, and *Edmontosaurus* (Parks, 1920; Lull and Wright, 1942; Davies, 1983). In the dorsal ribs, the length of separation between the capitulum and tuberculum, extended rectangular shape, and gentle descending slope to the rib shaft are morphologically similar to *Iguanodon bernissartensis*, and not seen in hadrosaurids such as *Edmontosaurus* (Lambe, 1920; Lull and Wright, 1942; Norman, 1980; Campione, this volume). However, heart-shaped centra of proximal and middle caudal vertebrae is a feature seen in the Hadrosauridae (Horner et al., 2004).

In the appendicular skeleton, the scapula and coracoid are primitively iguanodontian, and are similar in form to *Camptosaurus* and *Iguanodon* and to the basal hadrosauroids *Eolambia* and *Probactrosaurus* (Norman, 1980, 1986, 2002, 2004; Carpenter and Wilson, 2008; McDonald et al., 2012). The ilium contains an unusual combination of traits, but is most morphologically similar to *Bactrosaurus*, *Probactrosaurus*, *Cedrorestes*, and *Eolambia* (Godefroit et al., 1998; Norman, 2002; Gilpin et al., 2007; McDonald et al., 2012). The postacetabular process of the ilium retains a primitive triangular shape similar to *Camptosaurus*, *Cedrorestes*, *Eolambia*, and *Mantellisaurus* (Gilpin et al., 2007; Paul, 2007; Carpenter and Wilson, 2008; McDonald et al., 2012), whereas the pubic peduncle is extremely small – a feature usually found in hadrosaurids (Horner et al., 2004). Another derived feature is the more cranial position of the suprailiac crest, positioned between the acetabulum and ischial peduncle, a trait UTA-AASO-2003 shares with *Claosaurus* and hadrosaurids (Brett-Surman and Wagner, 2007; Prieto-Márquez, 2011). In the femur the more proximal origin of the fourth trochanter is unlike many hadrosauroids and a condition also found in *Telmatosaurus* and *Bactrosaurus*, although this may be ontogenetic (Weishampel et al., 1993; Godefroit et al., 1998). UTA-AASO-2003 appears to be more derived than *Probactrosaurus* due to possessing a short pubic peduncle, development of an incipient "antitrochanter" on the suprailiac crest, and overall cranial orientation of the crest. Likewise it appears more primitive than *Telmatosaurus* in possessing an iguanodontian-grade pectoral girdle. Further work, including a comprehensive cladistic analysis, is necessary to test the generality of this hypothesis.

Comparison with Woodbine Ornithopods

The discovery of the AAS specimens widens the scope of knowledge on the Woodbine Formation ornithopods and corresponds well to previously discovered material; however, direct comparison with previously collected material is hampered by their fragmentary nature and lack of overlapping elements with the AAS specimen. The partial femur described by Lee (1997a), which was also found in the Lewisville Member, shares several features in common with the AAS femur (UTA-AASO-125) including a convex, laterally compressed greater trochanter, a cranial trochanter separated from the latter by a narrow groove, and a proximal origin for the fourth trochanter. This suggests that the femur (SMU 73062) and associated tibia and fibula may belong to the same taxon as the AAS specimen.

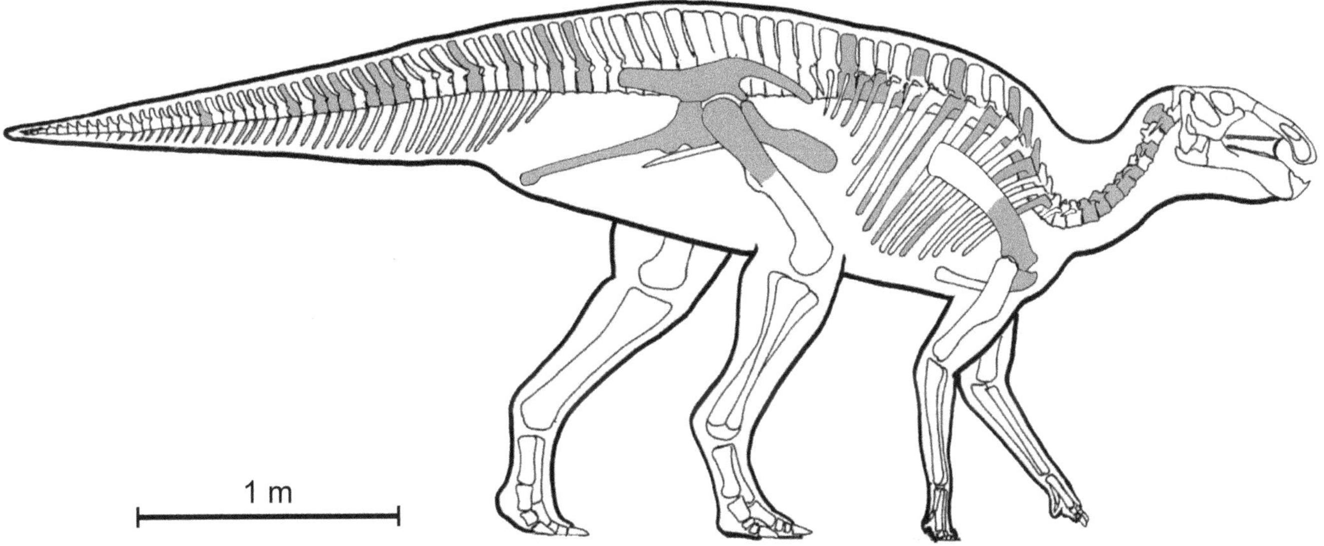

5.13. Reconstruction of AAS ornithopod using a generalized hadrosauroid skeleton as a model. Recovered fossils discussed in text are shaded gray. Artwork by Tracy Ford.

Because UTA-AASO-2003 is composed primarily of postcranial elements (Fig. 5.13) and *Protohadros* is based upon crania (Head, 1998), it is currently difficult to verify whether they represent independent taxa, or if the AAS specimen represents the postcranial skeleton of *Protohadros*. Both the AAS specimen and *Protohadros* are basal hadrosauroids of comparable grade. Although basal hadrosauroid taxa are widespread throughout the Cretaceous, most taxa do not overlap in time or space (Prieto-Márquez, 2010b). The AAS occurs in the Lewisville Member of the Woodbine Formation, whereas the type locality of *Protohadros* was reported by Head (1998) as occurring in the overlying, younger Arlington Member. However, the stratigraphic distribution of *Protohadros* is unknown, and the widespread presence of hadrosauroid material throughout the Lewisville and Arlington Members in the area suggests this material may all belong to the same taxon (i.e., *Protohadros*), which implies it was a widespread and common component of the Woodbine fauna during the early to middle Cenomanian. More cranial material and additional stratigraphic work are needed before this question can be sufficiently answered.

CONCLUSIONS

The Arlington Archosaur Site represents an important addition to our knowledge of middle Cretaceous terrestrial communities, and includes new remains of a non-hadrosaurid hadrosauroid dinosaur. UTA-AASO-2003 has a complex taphonomic history, that consisted of periods of prolonged aerial exposure and low-energy hydraulic transport, early and late diagenetic distortion, followed by recent, post-fossilization weathering. Bite marks on the femur (UTA-AASO-125) suggest that individuals of this taxon were consumed by resident AAS crocodyliforms (Noto et al., 2012). This new specimen is distinguished by its mosaic combination of basal and derived traits, and will contribute to our understanding of basal hadrosauroid anatomy and evolution. Currently there are insufficient skeletal and stratigraphic data to determine whether UTA-AASO-2003 and UTA-AASO-125 belong to *Protohadros*, or represents a separate taxon. Because the AAS is an ongoing excavation, any additionally recovered material will further clarify the nature of the AAS specimens as well as the diversity of the Woodbine ornithopod fauna.

ACKNOWLEDGMENTS

We thank the Huffines family and Robert Kimball of HC LOBF for granting land access to conduct work at the Arlington Archosaur Site. P. Kirchhoff and B. Walker, A. Sahlstein and B. Carter all donated the specimens discussed herein. The PaleoMap Project, the Earthwatch Institute, the Arlington Tomorrow Foundation, and the Jurassic Foundation provided research funds; and the Dallas Paleontological Society and the Western Interior Paleontological Society provided scholarships to DJM. We heartily thank Arlington Archosaur Site project volunteers A. Sahlstein, R. Fry, A. Miramontes, D. Sumerfelt, B. Carter, A. Osen, R. Colvin, R. Zack, P. Scoggins, J. Beeck, N. Campbell, C. K. Bingman, N. Van Vranken, and T. Dalbey, along with members of the Dallas Paleontological Society and UTA Dinosaurs class students for their assistance excavating the Dinosaur Quarry at the AAS. We include here a special note of thanks to A. Miramontes for his preparation of the specimens discussed herein; his patience and dedication in the lab made much of this study possible. F. M. Dalla Vecchia and J. Kirkland provided important literature references, comments, and advice. Many thanks go to A. McDonald for viewing images of AAS fossils and offering his insight, and for providing data and literature on *Eolambia;* K. Poole for discussing iguanodontian cladistics,;and D. Tanke for reviewing AAS fossils and providing comments. T. Ford for provided the excellent paleo-artwork. This work would not have been possible without L. Jacobs and D. Winkler at the Shuler Museum of Paleontology at Southern Methodist University for providing access to the *Protohadros* type specimen. We would also like to thank the editors and one anonymous reviewer for their detailed critique of the original manuscript. Their efforts significantly improved this work.

LITERATURE CITED

Alexander, N. 1976. Father of Texas Geology, Robert T. Hill. Bicentennial Series in American Studies 4. Southern Methodist University Press, Dallas, Texas, 136 pp.

Beall, A. D. 1964. Fabric and mineralogy of Woodbine Sediments, East Central Texas. Ph.D. dissertation, Stanford University, Stanford, California, 171 pp.

Behrensmeyer, A. K. 1978. Taphonomic and ecologic information from bone weathering. Paleobiology 4:150–162.

Bergquist, H. R. 1949. The Woodbine Formation of Cooke, Grayson and Fanin Counties, Texas. USGS Oil and Gas Inventory. Preliminary Map 98.

Brett-Surman, M. K. 1979. Phylogeny and paleobiogeography of the hadrosaurian dinosaurs. Nature 277:560–562.

Brett-Surman, M. K., and J. R. Wagner. 2007. Discussion of character analysis of the appendicular anatomy in Campanian and Maastrichtian North American hadrosaurids: variation and ontogeny; pp. 135–169 in K. Carpenter (ed.), Horns and Beaks: Ceratopsian and Ornithopod Dinosaurs. Indiana University Press, Bloomington, Indiana.

Campione, N. E. 2014. Postcranial anatomy of *Edmontosaurus regalis* (Hadrosauridae) from the Horseshoe Canyon Formation, Alberta, Canada; chapter 13 in D. A. Eberth and D. C. Evans (eds.), Hadrosaurs. Indiana University Press, Bloomington, Indiana.

Carpenter, K., and Y. Wilson. 2008. A new species of *Camptosaurus* (Ornithopoda: Dinosauria) from the Morrison Formation (Upper Jurassic) of Dinosaur National Monument, Utah and a biomechanical analysis of its forelimbs. Annals of the Carnegie Museum 76:227–263.

Casanovas, M. L., X. P. Suberbiola, J. V. Santafe, and D. B. Weishampel. 1999. First lambeosaurine hadrosaurid from Europe: paleobiological implications. Geological Magazine 136:205–211.

Chapman, R. E., and M. K. Brett-Surman. 1990. Morphometric observations on hadrosaurid dinosaurs; pp. 163–177 in K. Carpenter and P. J. Currie (eds.), Dinosaur Systematics: Approaches and Perspectives. Cambridge University Press, Cambridge, U.K.

Cope, E. D. 1869. Synopsis of the extinct Batrachia, Reptilia and Aves of North America. Transactions of the American Philosophical Society 14:1–252.

Dalla Vecchia, F. M. 2006. *Telmatosaurus* and the other hadrosaurids of the Cretaceous European Archipelago: an overview. Natura Nascosta 32:1–55.

Dalla Vecchia, F. M. 2009. *Tethyshadros insularis,* a new hadrosauroid dinosaur (Ornithischia) from the Upper Cretaceous of Italy. Journal of Vertebrate Paleontology 29:1100–1116.

Davies, K. L. 1983. Hadrosaurian dinosaurs of Big Bend National Park, Brewster County, Texas. M.S. thesis, University of Texas at Austin, Austin, Texas, 235 pp.

Dodge, C. F. 1952. Stratigraphy of the Woodbine Formation in the Arlington area, Tarrant Co., Texas. Field and Laboratory 20:66–78.

Dodge, C. F. 1968. Stratigraphic nomenclature of the Woodbine Formation Tarrant County, Texas; pp. 107–125 in C. F. Dodge (ed.), Stratigraphy of the Woodbine Formation, Tarrant County, Texas. Geological Society of America Field Trip Guidebook, South Central Section.

Dodge, C. F. 1969. Stratigraphic nomenclature of the Woodbine Formation Tarrant County, Texas. Texas Journal of Science 21:43–62.

Gilpin, D., T. DiCroce, and K. Carpenter. 2007. A possible new basal hadrosaur from the Lower Cretaceous Cedar Mountain Formation of Eastern Utah; pp. 79–89 in K. Carpenter (ed.), Horns and Beaks: Ceratopsian and Ornithopod Dinosaurs. Indiana University Press, Bloomington, Indiana.

Godefroit, P., Z.-M. Dong, L. Bultynck, H. Li, and L. Feng. 1998. Cretaceous dinosaurs and mammals from Inner Mongolia. New *Bactrosaurus* (Dinosauria: Hadrosauridae) material from Iren Dabasu (Inner Mongolia, P. R. China). Bulletin de l'Institut royal des Sciences naturelles de Belgique Supplement 68:3–70.

González Riga, B. J., and R. A. Astini. 2007. Preservation of large titanosaur sauropods in overbank fluvial facies: a case study in the Cretaceous of Argentina. Journal of South American Earth Sciences 23:290–303.

Hill, R. T. 1901. Geography and geology of the Black and Grand Prairies, Texas. U.S. Geological Survey, 21st Annual Report, 1899–1900, pt. 7, 666 pp.

Head, J. 1998. A new species of basal hadrosaurid (Dinosauria, Ornithischia) from the Cenomanian of Texas. Journal of Vertebrate Paleontology 18:718–738.

Horner, J. R., D. B. Weishampel, and C. A. Forster. 2004. Hadrosauridae; pp. 438–463 in D. B. Weishampel, P. Dodson, and H. Osmólska (eds.), The Dinosauria, Second Edition. University of California Press, Berkeley, California.

Jacobs, L. L., and D. A. Winkler. 1998. Mammals, archosaurs and the Early to Late Cretaceous transition in North-Central Texas; pp. 253–280 in Y. Tomida, L. J. Flynn, and L. L. Jacobs (eds.), Advances in Vertebrate Paleontology and Geochronology. National Science Museum Monographs 14, National Science Museum, Tokyo, Japan.

Johnson, R. O. 1974. Lithofacies and depositional environments of the Rush Creek Member of the Woodbine Formation (Gulfian) of North Central Texas. M.S. thesis, University of Texas at Arlington, Arlington, Texas, 158 pp.

Kennedy, W. J., and W. A. Cobban. 1990. Cenomanian ammonite faunas from the Woodbine Formation and lower part of the Eagle Ford Group, Texas. Palaeontology 33:75–154.

Kirkland, J. 1998. A new hadrosaurid from the Upper Cedar Mountain Formation (Albian–Cenomanian: Cretaceous) of Eastern Utah: the oldest known hadrosaurid (Lambeosaurine?); pp. 283–302 in S. G. Lucas, J. I. Kirkland, and J. W. Estep (eds.), Lower and Middle Cretaceous Terrestrial Ecosystems. New Mexico Museum of Natural History and Science Bulletin 14.

Lambe, L. M. 1920. The hadrosaur *Edmontosaurus* from the Upper Cretaceous of Alberta. Geological Survey Memoir 120:1–73.

Lee, Y. N. 1997a. The Archosauria from the Woodbine Formation (Cenomanian) in Texas. Journal of Paleontology 71:1147–1156.

Lee, Y. N. 1997b. Bird and dinosaur footprints in the Woodbine Formation (Cenomanian), Texas. Cretaceous Research 18:849–864.

Lull, R. S., and N. E. Wright. 1942. Hadrosaurian Dinosaurs of North America. Geological Society of America Special Papers 40. 242 pp.

Main, D. J. 2005. Paleoenvironments and paleoecology of the Cenomanian Woodbine Formation of Texas, Paleobiogeography of the hadrosaurs (Dinosauria: Ornithischia). M.S. thesis, University of Texas at Arlington, Arlington, Texas, 304 pp.

Main, D. J. 2009. Delta plain environments and ecology of the Woodbine Formation (Cenomanian) at the Arlington Archosaur Site, North Texas. Geological Society of America Abstracts with Programs 41:103.

Main, D. J., D. Parris, B. Grandstaff, and B. Carter. In press. A new lungfish (Dipnoi: Ceratodontidae) from the Cretaceous Woodbine Formation, Arlington Archosaur Site, North Texas. Texas Journal of Science.

Marsh, O. C. 1882. Classification of the Dinosauria. American Journal of Science, Third Series, 23:81–96.

Maryańska, T., and H. Osmólska. 1984. Postcranial anatomy of *Saurolophus angustirostris* with

comments on other hadrosaurs. Paleontologia Polonica 46:119–141.
May, F. 1975. *Dichastopollenites reticulatus,* gen. et sp. nov.: potential Cenomanian guide fossil from southern Utah and northeastern Arizona. Journal of Paleontology 49:528–533.
McDonald, A. T., D. G. Wolfe, and J. I. Kirkland. 2010. A new basal hadrosauroid (Dinosauria: Ornithopoda) from the Turonian of New Mexico. Journal of Vertebrate Paleontology 30:799–812.
McDonald, A. T., J. Bird, J. I. Kirkland, and P. Dodson. 2012. Osteology of the basal hadrosaouroid *Eolambia caroljonesa* (Dinosauria: Ornithopoda) from the Cedar Mountain Formation of Utah. PloS ONE 7(10):e45712.
Murlin, J. R. 1975. Stratigraphy and depositional environments of the Arlington Member, Woodbine Formation (Upper Cretaceous), Northeast Texas. M.S. thesis, University of Texas at Arlington, Arlington, Texas, 214 pp.
Norman, D. B. 1980. On the ornithischian dinosaur *Iguanodon bernissartensis* from the Lower Cretaceous of Bernissart Belgium. Institut royal des Sciences naturelles de Belgique Mémoire 178:1–99.
Norman, D. B. 1986. On the anatomy of *Iguanodon atherfieldensis* (Onithischia: Ornithopoda). Bulletin of the Institute of Science and Nature, Belgium. Sciences de la Terre 56:281–372.
Norman, D. B. 2002. On Asian ornithopods (Dinosauria: Ornithischia). 4. *Probactrosaurus* Rozhdestvensky, 1966. Archosaurian anatomy and paleontology; pp. 113–144 in D. B. Norman, and D. J. Glower (eds.), Essays in Memory of Alick D. Walker. Zoological Journal of the Linnean Society 36:113–144.
Norman, D. B. 2004. Basal Iguanodontia; pp. 413–437 in D. B. Weishampel, P. Dodson, and H. Osmólska (eds.), The Dinosauria, Second Edition. University of California Press, Berkeley, California.
Noto, C. R., D. J. Main, and S. K. Drumheller. 2012. Feeding traces and paleobiology of a Cretaceous (Cenomanian) crocodyliform: example from the Woodbine Formation of Texas. Palaios 27:105–115.
Oliver, W. B. 1971. Depositional systems within the Woodbine Formation (Upper Cretaceous), Northeast Texas: reports of investigations. Bureau of Economic Geology, University of Texas at Austin 73:1–28.
Owen, R. 1842. Report on British fossil reptiles. Part II. Report on the eleventh meeting of the British Association for the Advancement of Science. July 1841:66–204.
Parks, W. A. 1920. The osteology of the trachodont dinosaur *Kritosaurus incurvimanus.* University of Toronto Geology Series 11:1–72.
Paul, G. S. 2007. Turning the old into the new: a separate genus for the gracile iguanodont from the Wealden of England; pp. 69–77 in K. Carpenter (ed.), Horns and Beaks: Ceratopsian and Ornithopod Dinosaurs. Indiana University Press, Bloomington, Indiana.
Poole, K. E. 2008. A new specimen of iguanodontian dinosaur from the Cedar Mountain Formation, Grand County, Utah. M.S. thesis, Washington University, Pullman, Washington, 62 pp.
Prieto-Márquez, A. 2010a. Global phylogeny of Hadrosauridae (Dinosauria: Ornithopoda) using parsimony and Bayesian methods. Zoological Journal of the Linnean Society 159:435–502.
Prieto-Márquez, A. 2010b. Global historical biogeography of hadrosaurid dinosaurs. Zoological Journal of the Linnean Society 159:503–525.
Prieto-Márquez, A. 2011. Revised diagnoses of *Hadrosaurus foulkii* Leidy, 1858 (the type genus and species of Hadrosauridae Cope, 1869) and *Claosaurus agilis* Marsh, 1872 (Dinosauria: Ornithopoda) from the Late Cretaceous of North America. Zootaxa 2765:61–68.
Prieto-Márquez, A., R. Gaete, G. Rivas, A. Galobart, and M. Boada. 2006. Hadrosauroid dinosaurs from the Late Cretaceous of Spain: *Pararhabdodon isonensis* revisited and *Koutalisaurus kohlerorum,* gen. et sp. nov. Journal of Vertebrate Paleontology 26:929–943.
Seeley, H. G. 1888. On the classification of the fossil animals commonly named Dinosauria. Proceedings of the Royal Society of London 43:165–171.
Seilacher, A. 2007. Trace Fossil Analysis. Springer Press, New York, 226 pp.
Stephenson, L. W. 1952. Larger invertebrate fossils of the Woodbine Formation (Cenomanian) of Texas. U. S. Geological Survey Professional Paper 242:1–226.
Trudel, P. 1994. Stratigraphic sequences and facies architecture of the Woodbine–Eagle Ford interval, Upper Cretaceous, North Central Texas. M.S. thesis, Tarleton State University, Stephenville, Texas, 105 pp.
Weishampel, D. B., C. M. Jianu, Z. Csiki, and D. B. Norman. 2003. Osteology and phylogeny of *Zalmoxes* (n. g.), an unusual euornithopod dinosaur from the latest Cretaceous of Romania. Journal of Systematic Paleontology 1:65–123.
Weishampel, D. B., D. Norman, and D. Grigorescu. 1993. *Telmatosaurus transsylvanicus* from the Late Cretaceous of Romania: the most basal hadrosaurid dinosaur. Paleontology 36:361–385.

A Re-evaluation of Purported Hadrosaurid Dinosaur Specimens from the "Middle" Cretaceous of England

Paul M. Barrett, David C. Evans, and Jason J. Head

ABSTRACT

"Trachodon cantabrigiensis" and *'Iguanodon' hillii* are poorly known ornithopod taxa from the "middle" Cretaceous of England that were both named on the basis of a single isolated tooth. The holotype of *"T." cantabrigiensis* was collected from the Cambridge Greensand Member of the Melbury Marly Chalk Formation (Cenomanian) and was reworked from late Albian deposits, while that of *'I.' hillii* is from the slightly younger Totternhoe Stone Member of the Zig Zag Chalk Formation (middle Cenomanian). Both taxa have been suggested to represent hadrosaurids, and if correctly identified would form the earliest-known records of this predominantly latest Cretaceous dinosaur clade. Here we provide redescriptions of these teeth, and evaluate their morphology with reference to the phylogenetic distributions of hadrosauroid dental characters. In addition, we incorporate *"T. cantabrigiensis"* into a morphometric analysis of iguanodontian tooth crown shape. Results indicate that although both taxa share many features in common with hadrosaurids, neither can be referred to this clade. Nevertheless, both taxa appear to nest in a grade of hadrosauroid taxa, including *Bactrosaurus* and *Telmatosaurus*, which are among the closest relatives of hadrosaurids. Consequently, these isolated European records need to be considered when discussing the paleobiogeography of hadrosaurid origins.

INTRODUCTION

The Lower Cretaceous portion of the Purbeck Limestone Formation (Tithonian–Berriasian) and the Wealden Group (upper Berriasian–lowermost Aptian) of central and southern England have yielded historically important dinosaur faunas that contain diverse ornithopod dinosaurs (see recent reviews in Naish and Martill [2008], Galton [2009], and Norman [2011a]). These faunas include basal ornithopods (*Hypsilophodon* [Galton, 1974, 2009]) and numerous iguanodontians, including dryosaurids (*Valdosaurus* and other indeterminate material [Galton, 2009; Barrett et al., 2011]) and many ankylopollexians, such as *Barilium, Hypselospinus, Iguanodon, Kukufeldia, Mantellisaurus,* and *Owenodon* (Paul, 2007, 2008; Galton, 2009; Carpenter and Ishida, 2010; McDonald, Barrett, and Chapman, 2010; Norman, 2010, 2011b; McDonald, 2012). By contrast only a small amount of ornithopod material has been recovered from post-Wealden deposits, namely the Lower Greensand Group (Aptian–lower Albian), which has yielded a partial skeleton of *Mantellisaurus* (the historically important "Mantel-piece": Norman, 1993) and other fragmentary material (e.g., Keeping, 1883), the Cambridge Greensand Member of the West Melbury Marly Chalk Formation (Seeley, 1879; Lydekker, 1888a; see below), and the Totternhoe Stone Member of the Zig Zag Chalk Formation (Newton, 1892; see below). Although material from these units is rare it is potentially important paleobiogeographically and phylogenetically, as specimens from contemporary units in North America and eastern Asia indicate that this interval witnessed the beginning of the hadrosauroid radiation (e.g., Head, 1998; Kirkland, 1998; Norman, 1998, 2002; You et al., 2003; Wang et al., 2011). Few other European localities yield material of this age (Weishampel et al., 2004), so specimens from these formations have the potential to provide some insight into the distribution and species richness of "middle" Cretaceous hadrosauroids.

Dating the Cambridge Greensand Member has been contentious, but there is now a general consensus that it is of early Cenomanian age, although many of its macrofossils were probably reworked from the upper part of the underlying late Albian Gault Formation (see Pereda-Suberbiola and Barrett [1999] and Unwin [2001] for reviews of the sedimentology, vertebrate paleontology, and dating of this unit). The Cambridge Greensand vertebrate fauna includes numerous chelonians, fish, ichthyosaurs, dinosaurs, plesiosaurs, and pterosaurs, and rare birds, crocodilians, and squamates. Ankylosaur specimens comprise the majority of the dinosaur remains recovered, but sauropod and ornithopod specimens also occur (see brief summary in Pereda-Suberbiola and Barrett, 1999). Several putative ornithopod taxa have been named on the basis of these finds, including *"Anoplosaurus major"* (partim), *"Eucercosaurus tanyspondylus," "Syngonosaurus macrocercus,"* and *"Trachodon cantabrigiensis"* (Seeley, 1869, 1879; Lydekker, 1888a), all of which are currently considered to be nomina dubia and either iguanodontian

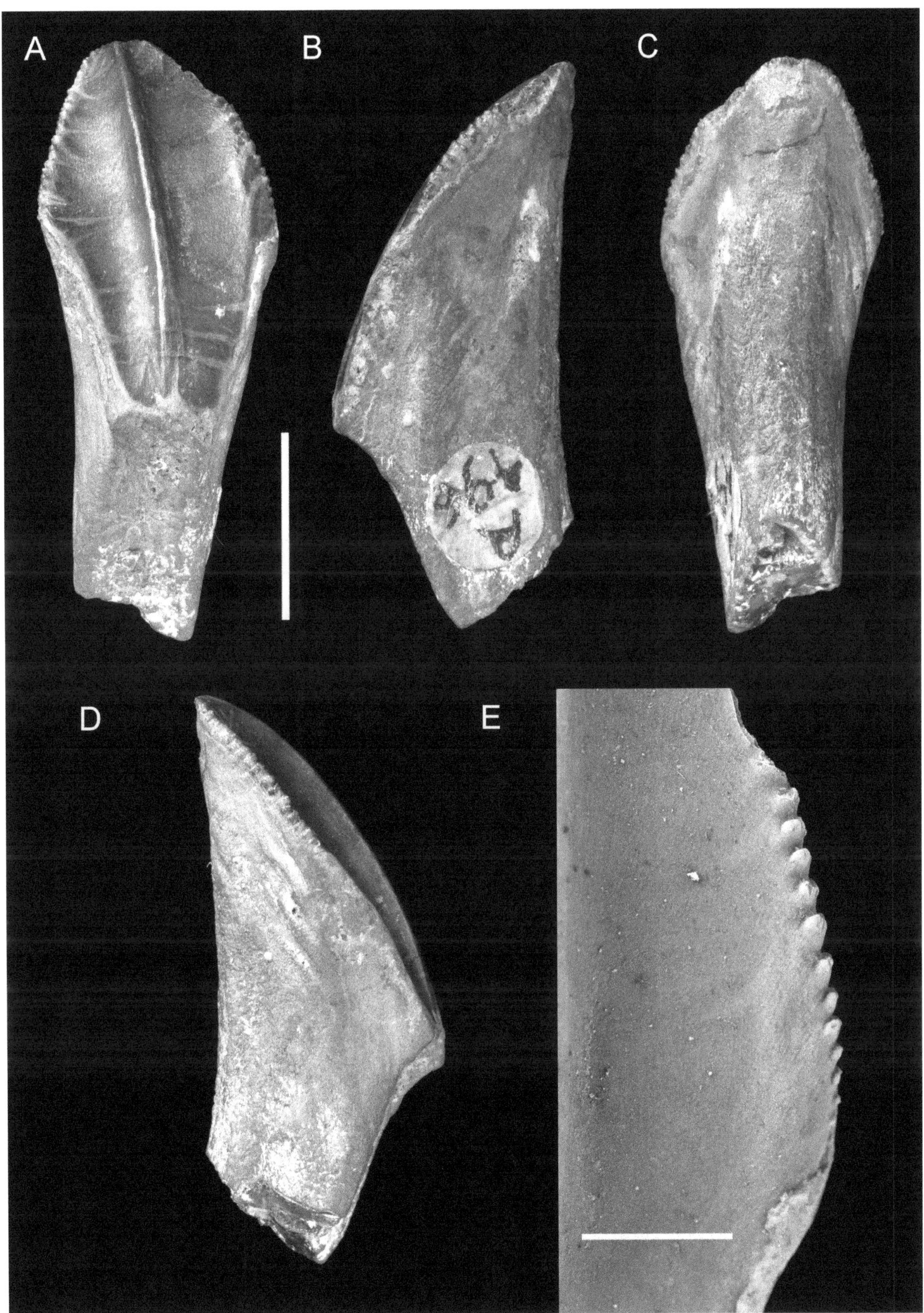

6.1. Holotype left dentary tooth of "*Trachodon cantabrigiensis*" (NHMUK R496). (A) lingual view; (B) mesial view; (C) labial view; (D) distal view; (E) close-up on medial margin showing denticles with a single row of mammillae. Scale bars equal 10 mm (A–D) and 3 mm (E).

ornithopods, or chimeras of iguanodontian and ankylosaur material (Pereda-Suberbiola and Barrett, 1999; Horner et al., 2004; Norman, 2004; Vickaryous et al., 2004). Of these, "*Trachodon cantabrigiensis*" Lydekker, 1888a, which is represented by the isolated holotype tooth (NHMUK R496; Fig. 6.1) and a set of referred material (Lydekker, 1888b; Nopcsa, 1923), is of special interest as, historically, it has been regarded as an early member of the Hadrosauridae (or a "trachodontid" in earlier works).

Owen (1861:30, pl. 7, figs. 15–16) was the first person to describe NHMUK R496 – as a "young *Iguanodon*" – but this identification was questioned by Leidy (1865:86–87) who noted its resemblance to teeth of *Hadrosaurus* and "*Trachodon.*" The "trachodontid" affinity of the tooth was upheld by Lydekker (1888a), who erected the name "*Trachodon cantabrigiensis*" for its reception, and referred additional unassociated specimens (all pedal elements) from the Cambridge Greensand to the species (Lydekker, 1888b). Nopcsa (1923) also referred more specimens to the taxon (an isolated maxilla and pedal phalanges) and commented upon the similarities between all of these elements and those of hadrosaurids. The hadrosaurid identity of the holotype tooth was subsequently accepted by many authors, usually in the context of systematic lists of taxa or paleobiogeographic discussions (e.g., Steel, 1969; Kirkland, 1998; Horner et al., 2004; Dalla Vecchia, 2006; Benton et al., 2010). Head (1998) was the first to cast doubt on this identification, although the specimen was not formally removed from Hadrosauridae. Rather, a combination of reasons, including the absence of other definitive hadrosaurid material from the English dinosaur record (and the European pre-Campanian record more generally), the geographic isolation of England from the Aptian onward, and the possibility of homoplasy in the distribution of various iguanodontian dental characters were cited as confounding factors in accepting the hadrosaurid identification of the tooth (Head, 1998:733). Sues and Averianov (2009) also questioned the possible hadrosaurid affinities of "*T. cantabrigiensis*," regarding it as a more basal hadrosauroid, although they provided no evidence in support of this conclusion.

A second putative hadrosaurid, '*Iguanodon*' *hillii* Newton, 1892, was recovered from the Totternhoe Stone Member of the Zig Zag Chalk Formation (middle Cenomanian). It is known from a single partial tooth crown (GSM 1966; Fig. 6.2) and no other ornithopod material has been recovered from any other Late Cretaceous British fossil reptile locality (Benton and Spencer, 1995; Weishampel et al., 2004). '*Iguanodon*' *hillii* has received much less attention in the literature than "*Trachodon cantabrigiensis*," but it is has also been regarded as an indeterminate hadrosaurid (Horner et al., 2004; Dalla Vecchia, 2006) or an indeterminate iguanodontian (Head, 1998), in both cases without discussion. Naish and Martill (2008:620) questioned Head's (1998) conclusion stating that "the tooth is highly derived," but offered no alternative identification or other remarks on this specimen.

Here, we provide reappraisals of the putative hadrosaurid specimens from the English "middle" Cretaceous and discuss their possible significance. Definitions for clade names used herein follow Sereno (1998): Hadrosauroidea is a stem-based name defined as all hadrosauriforms closer to *Parasaurolophus* than to *Iguanodon* (*Iguanodon* is used in a strict sense herein, referring only to the genotype species *I. bernissartensis*), whereas Hadrosauridae is a node-based name defined as *Saurolophus*, *Parasaurolophus*, their most recent common ancestor, and all of its descendants. For convenience, non-hadrosaurid hadrosauroids are termed "basal hadrosauroids" herein.

Institutional Abbreviations CAMSM, Sedgwick Museum of Earth Sciences, University of Cambridge, Cambridge, U.K.; CEUM, College of Eastern Utah, Price, Utah; FMNH, Field Museum of Natural History, Chicago, Illinois; GSM (formerly BGS), British Geological Survey, Keyworth, U. K.; IVPP, Institute of Vertebrate Paleontology and Paleoanthropology, Beijing, China; NHMUK, The Natural History Museum, London, U. K.; ROM, Royal Ontario Museum, Toronto, Ontario; SMU, Shuler Museum of Paleontology, Southern Methodist University, Dallas, Texas.

DESCRIPTIONS

"Trachodon cantabrigiensis" *Lydekker, 1888a*

Holotype The holotype (NHMUK R496) is an isolated, well-preserved, and unworn left dentary tooth, lacking only the basal part of the root. In lingual view (Fig. 6.1A), the tooth crown is apicobasally elongate, mesiodistally expanded with respect to the root, and asymmetrically diamond shaped in outline, with the apex inclined slightly distally. The medial crown margin describes an almost continuous smooth curve from base to apex, while the distal margin is divided into apical and basal portions by a distinct break in slope that occurs at mid-crown height. The apical portion of the distal margin is straight, while the basal portion is gently concave. The crown apex is bluntly rounded and the crown basal margin is gently convex in an apical direction, forming a shallow notch for the reception of a replacement tooth. Only the lingual surface of the crown bears enamel. The crown is 23.5 mm long and 12mm wide, giving a height-to-width ratio of ~1.96. In mesial or distal view (Fig. 6.1B, D), the crown surface is labially inclined, so that the enameled lingual surface faces dorsolingually. This also gives the tooth a very gently bowed appearance in mesial or distal view. In apical view, the crown has a subtriangular cross section, with the apex of this triangle directed labially.

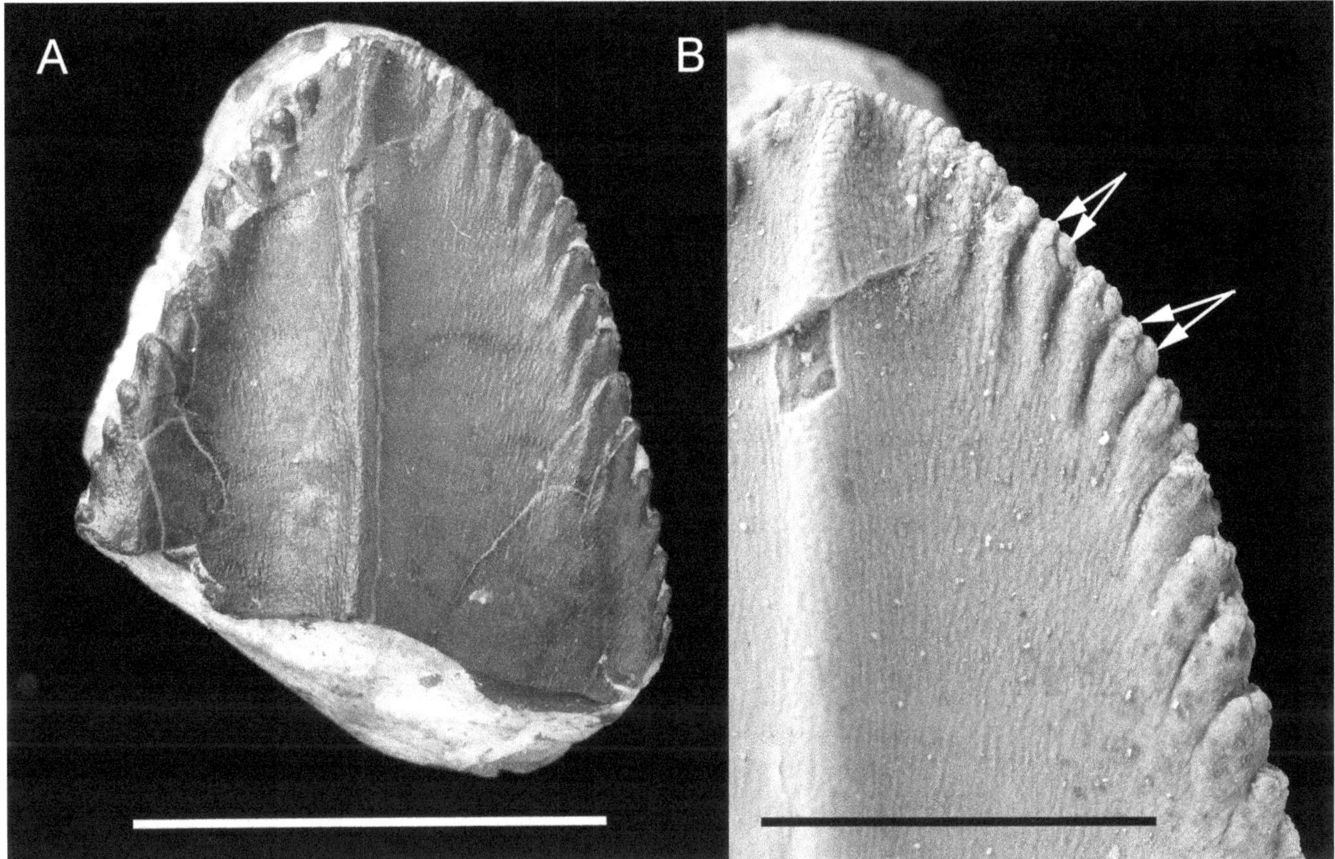

6.2. Holotype tooth of *'Iguanodon' hillii* (GSM 1966). (A) ?lingual view; (B) detail of marginal denticles, coated in ammonium chloride, with arrows highlighting the bifurcated rows of mammillae. Scale bars equal 10 mm (A) and 5 mm (B).

A sharp, very slightly distally positioned, and prominent primary ridge divides the lingual surface of the crown into two subequal areas (Fig. 6.1A). There are no secondary ridges and the lingual crown surface is otherwise smooth and unornamented, although the enamel has a fine granular texture. Small denticles are present on the apical halves of both the mesial and distal crown margins, with approximately 22 present along the distal margin and a minimum of 19 along the mesial margin (Fig. 6.1A, E). (The true number along the mesial margin is likely to have been substantially higher than that present distally, as approximately one-quarter of the denticulate part of the mesial margin is slightly damaged, which has removed the denticles from this area). The majority of the denticles are subtriangular in outline in lingual view, bear mammillae, and are oriented at approximately 45° to the apicobasal axis of the crown. However, those denticles adjacent to the crown apex are reduced to small rounded nubbins of enamel, project apically, and lack mammillae.

The broken base of the root has an arch-shaped cross section, which has a concave lingual margin, straight mesial and distal margins, and a convex labial margin. The lingual concavity is created by a pronounced groove that would have housed a replacement tooth. This groove extends along the lingual surface of the root, terminating at the base of the crown where it creates the notched basal margin of the crown (see above). In labial or lingual view (Fig. 6.1A, C), the root and unenameled part of the crown merge into each other and exhibit two well-defined vertical facets, which would have accommodated the margins of adjacent tooth crowns. The lingual-most of these is small and triangular in outline and covers the area positioned immediately labial to the lingual groove. A second larger facet, which covers most of the unenameled crown/root surface, is situated labial to the first and is divided from the lingually positioned facet by a distinct change in slope, which forms a low apicobasally extending ridge. The labial boundary of this second facet is formed by the mesiodistally convex labial surface of the tooth. These facets demonstrate that the unerupted tooth would have been in contact with at least five other teeth, set within a complex dental battery.

Referred Material Lydekker (1888b:245) provisionally referred other specimens from the Cambridge Greensand to *"Trachodon cantabrigiensis"*: a ?pedal first phalanx (NHMUK OR33884); a more distally positioned phalanx (mentioned as a second or third phalanx, not referred to either the manus or pes: NHMUK OR33885); and two pedal ungual phalanges (NHMUK OR33886 [suggested to be from digit 3]; NHMUK OR33887). Lydekker (1888b) provided no justification for these referrals, was unable to present exact locality information, and gave no indication that they were associated with either each other or with the holotype tooth. Nopcsa (1923:193–194, pl. 7, fig. 3) also referred two unassociated, poorly provenanced phalanges to this taxon, an ungual (CAMSM B55744) and non-ungual phalanx (CAMSM B55398). All of these elements are partially encrusted in matrix and/or heavily abraded, obscuring many salient anatomical features. For convenience, all will be described as though the foot was held in a plantigrade stance.

The first pedal phalanx (NHMUK OR33884) is poorly preserved and it cannot be determined which digit it belonged to, or to which side of the body. In dorsal view, the proximal and distal ends of the phalanx are obviously expanded with respect to the shaft, but are so abraded that the true extent of this expansion cannot be ascertained. The shaft has a subtriangular cross section at midlength, as one surface of the phalanx (here termed lateral, for convenience) is anteroposteriorly expanded relative to the medial surface. In dorsal view, the surface of the shaft is bounded laterally by a low, prominent ridge (representing the dorsal margin of the expanded lateral surface), and the rest of the surface is either gently concave mediolaterally (proximal to shaft midlength), or gently convex (distal to shaft midlength). In lateral view, the dorsal margin of the phalanx is almost straight and slopes ventrally from the proximal to the distal ends; in medial view, the dorsal margin is gently concave. In either lateral or medial view, the ventral margin of the phalanx is arched dorsally. A prominent boss of bone is present on the ventral surface, positioned just distal to the proximal articular surface, and is also visible in both lateral and medial views. In proximal view, the phalanx has a subtrapezoidal outline. The articular surfaces of both the proximal and distal ends are too poorly preserved to warrant description.

The other non-ungual pedal phalanges (NHMUK OR33885, CAMSM B55398) are anteroposteriorly shorter than they are wide mediolaterally, with length-to-width ratios of approximately 1.9 and 2.3, respectively. In dorsal view, the proximal and distal margins of the phalanges are subequal in mediolateral width and there is no distinct shaft. The proximal articular surfaces are concave, both dorsoventrally and mediolaterally, whereas the distal articular surfaces are mediolaterally concave and dorsoventrally convex (saddle shaped). In both phalanges, the ventral surface is mediolaterally and anteroposteriorly concave, while the dorsal surface is mediolaterally convex and anteroposteriorly flat, sloping slightly ventrally from the proximal to the distal end. Collateral ligament pits are absent, and the distal articulations are not developed into distinct ginglymi. All of the articular surfaces are wider than tall.

Two of the ungual phalanges (NHMUK OR33886, CAMSM B55744) are hoof shaped in dorsal view. The proximal articular surface supports a short "neck" that expands mediolaterally to form the main body of the ungual, which is offset from the proximal neck by a distinct notch both laterally and medially, and has a broad, subtriangular outline that tapers to a bluntly rounded distal margin. The dorsal surface of the ungual is mediolaterally and proximodistally convex and shallow attachment grooves extend from each of the notches for a short distance. In proximal view, the articular surface is wider than it is tall, with a subelliptical outline. It is concave mediolaterally and dorsoventrally and is divided into subequal lateral and medial portions by a low ridge that extends dorsoventrally along the approximate midline of the articular surface. In lateral view, the dorsal margins of the unguals are convex, while the ventral margins are concave. A subtle midline ridge extends along the ventral surface of CAMSM B55744, which is absent in NHMUK OR33886. The third ungual phalanx (NHMUK OR33887) is similar to the other two, but lacks the development of the neck and notches separating the proximal articular surface from the rest of the phalanx, and so has an elongate, sub-triangular overall outline in dorsal view.

Other Cambridge Greensand "Trachodontids" A partial abraded maxilla (listed as an indeterminate dinosaur by Seeley [1869:23]: CAMSM B55527) was regarded as a "Trachodontid" by Nopcsa (1923:193). It is broken into two parts, which comprise a short section from the anterior portion of a maxilla, with eight, narrow parallel dental grooves. These grooves are closely packed and lack the impressions of individual tooth crowns. The tooth row is medially inset, creating a rounded buccal recess. No teeth are preserved, and no other anatomical details can be determined.

Several other specimens in the collections of the Sedgwick Museum are cataloged as "trachodontids" following identifications made by Nopcsa during his visit to the collections, including a heavily abraded ungual phalanx (CAMSM B55434), two non-ungual phalanges (CAMSM B55397, B55399), and the proximal portion of a large dorsal rib (CAMSM B55390), although Nopcsa did not provide formal referrals for these specimens.

'Iguanodon' hillii *Newton, 1892*

Unfortunately, only the apical part of the holotype tooth crown is preserved (GSM 1966: Fig. 6.2) and it is too incomplete to determine whether it was a maxillary or dentary tooth. For convenience, the tooth is described as a left dentary tooth in order to standardize terminology and facilitate comparisons. The labial crown surface is partially embedded within matrix.

The preserved portion of the crown has a subtriangular outline in lingual view, tapering apically (Fig. 6.2A). The unenameled labial surface of the crown is strongly convex mesiodistally, giving the tooth a D-shaped cross section in basal view. The lingual surface of the tooth is enamelled and bears a strong, slightly distally offset primary ridge that divides the crown surface into a large mesial and smaller distal portion. There are no secondary ridges. The enamel has a fine granular texture and faint parallel wrinkles extend horizontally across the crown surface mesiodistally (although these are not as prominent as in the figure provided by Newton [1892:fig. 1A]). The crown apex is unworn and the mesial and distal crown margins bear numerous denticles.

The mesial crown margin bears 16 large denticles, which are separated from each other by deep sulci. Basally, the denticles are oriented at approximately 45° to the apicobasal axis of the crown, but this angle changes progressively towards the crown apex, with more apically positioned denticles oriented at 80–90°. These denticles are subequal in size and are separated from the tooth apex by a group of 6 very small apically extending denticles, which are effectively small nubbins of enamel that are more like mammillae than true denticles. The largest, basal-most denticles each bear a row of mammillae (with around 4–5 mammillae per denticle). However, progressing apically around the crown margin the tips of each denticle bifurcate, to produce a forked structure: the apically positioned part of this fork bears a row of mammalliae (consisting of 2–4 mammillae), whereas the basally positioned part is comprised of a single, tongue-like process (Fig. 6.2B). This bifurcation of the denticle tips is subtle in lingual view, but more apparent in mesial view.

The denticles on the distal surface of the tooth are poorly preserved, but appear to follow similar trends in terms of size and orientation (12 large denticles are preserved, with a minimum of 2–3 small nubbin-like denticles just distal to the crown apex). Unfortunately, the poorer preservation of the apical-most distal denticles precludes examination of their tips, and it cannot be determined if they were bifurcated. However, the basal-most denticles possess single rows of mammillae that are identical to those on the basally positioned mesial denticles.

DISCUSSION

Morphological Comparisons

Holotype of "*Trachodon cantabrigiensis*" NHMUK R496 possesses a melange of dental features seen in hadrosaurids and basal hadrosauroids. The proportions of the "*Trachodon cantabrigiensis*" tooth are similar to those of *Eolambia* and *Protohadros* dentary teeth (with crown height-to-width ratios of approximately 2.1 and 1.8, respectively; Head, 1998; Kirkland, 1998), but are less elongate of those than *Levnesovia*, *Probactrosaurus*, *Telmatosaurus*, or hadrosaurids, which usually have crown height-to-width ratios >3.0 (e.g., Norman, 2002; Dalla Vecchia, 2006; Sues and Averianov, 2009). Distal recurvature of the tooth apex is present in NHMUK R496, strongly in *Telmatosaurus* (Weishampel et al., 1993; Dalla Vecchia, 2006), and some teeth of *Bactrosaurus* (Godefroit et al., 1998), but is absent in *Probactrosaurus* (Norman, 2002), *Protohadros* (Head, 1998), and hadrosaurids (e.g., Lull and Wright, 1942; Ostrom, 1961; Horner et al., 2004; Sues and Averianov, 2009).

A prominent primary ridge (or carina) is ubiquitous among basal hadrosauroids and hadrosaurids (e.g., Lull and Wright, 1942; Ostrom, 1961; Head, 1998; Kirkland, 1998; Norman, 2002; Horner et al., 2004; Prieto-Márquez, 2010a). Secondary ridges are present on the lingual surfaces of the tooth crowns in *Bactrosaurus* (Godefroit et al., 1998), *Levnesovia* (Sues and Averianov, 2009), *Probactrosaurus* (Norman, 2002), *Protohadros* (Head, 1998), *Telmatosaurus* (Weishampel et al., 1993), and subadult hadrosaurids (Hall, 1993). These features are absent or weakly developed in adult hadrosaurids (e.g., Lull and Wright, 1942; Ostrom, 1961) and NHMUK R496. The slightly distal offset of the primary ridge in NHMUK R496 is similar to that seen in the majority of basal hadrosauroids (Weishampel et al., 1993; Godefroit et al., 1998; Head, 1998; Norman, 2002; Sues and Averianov, 2009), differing from the more centrally positioned carinae of hadrosaurids (Prieto-Márquez, 2010a). In "*T. cantabrigiensis*" the carina lacks the sinuous curve present in some lambeosaurine hadrosaurids (Prieto-Márquez, 2010a).

The retention of small marginal denticles positioned around the apical half of the crown margins is also common among basal hadrosauroids and hadrosaurids (e.g., Weishampel et al., 1993; Head, 1998; Kirkland, 1998; Norman, 2002; Horner et al., 2004; Dalla Vecchia, 2006; Sues and Averianov, 2009), although the denticles are lost in some deeply nested hadrosaurid lineages (Prieto-Márquez, 2010a). Mammillae are present on the marginal denticles of "*T. cantabrigiensis*" and "basal hadrosauroids" such as *Probactosaurus* (Norman, 2002), *Protohadros* (Head, 1998), and *Telmatosaurus* (Dalla Vecchia, 2006), but are generally lost in hadrosaurids (Norman, 2002; Prieto-Márquez, 2010a).

Referred Material of "*Trachodon cantabrigiensis*" The narrow, closely packed, and straight alveolar grooves of the poorly preserved maxillary fragment (CAMSM B55527) do resemble those of hadrosaurids (e.g., Ostrom, 1961; Horner et al., 2004), but are also similar to those of a variety of other basal hadrosauroids and basal iguanodontians, including *Mantellisaurus* (Norman, 1986), *Probactrosaurus* (Norman, 2002), and *Telmatosaurus* (Weishampel et al., 1993; Dalla Vecchia, 2006).

All of the isolated ungual phalanges referred to "*T. cantabrigiensis*" lack the distinctive spade-shaped outline seen in *Bactrosaurus* (Godefroit et al., 1998) and hadrosaurids (Lull and Wright, 1942; Horner et al., 2004). Three of these (CAMSM B55744, NHMUK OR33886–7) are similar to those of non-hadrosaurid ornithopods such as *Iguanodon* (Norman, 1980), *Mantellisaurus* (Norman, 1986), and *Probactrosaurus* (Norman, 2002). The remaining two (CAMSM B55400, B55430) are strongly abraded and could be referable to either a non-hadrosaurid iguanodontian or an ankylosaur, whose pedal phalanges are somewhat similar (see Pereda-Suberbiola and Barrett [1999] for descriptions of ankylosaur unguals from the Cambridge Greensand Member). The isolated dorsal rib (CAMSM B55390) and non-ungual phalanges (NHMUK OR33884–5, CAMSM B55397–99) bear no distinctive features that allow them to referred to any particular dinosaur clade with confidence and these should all be regarded as Dinosauria indet.

'Iguanodon' hillii The fragmentary nature of the holotype specimen of *'Iguanodon' hillii* precludes detailed comparison. Many features of the tooth are shared with the majority of basal hadrosauroids and hadrosaurids, including a strong primary ridge, the presence of mammillate marginal denticles, and the absence of secondary ridges (see comparative comments on "*Trachodon cantabrigiensis*," above). Although the tooth crown is incomplete basally, the absence of secondary ridges in *'I.' hillii* is probably genuine, as in those basal hadrosauroids that possess secondary ridges these normally extend onto the apical part of the tooth (e.g., Head, 1998; Norman, 2002). In addition, this clearly prevents referral of the specimen to *Iguanodon* or other similar iguanodontian taxa, such as *Mantellisaurus*. The distal offset of the primary ridge distinguishes the tooth of *'I.' hillii* from those of hadrosaurids (e.g., Prieto-Márquez, 2010a), although the lack of distal recurvature is more similar to the hadrosaurid condition than that seen in "*T. cantabrigiensis*" (see above) or *Telmatosaurus* (Weishampel et al., 1993; Dalla Vecchia, 2006). Bifurcated rows of mammilae similar to those of *'I.' hillii* are also present in some specimens of *Telmatosaurus* (Dalla Vecchia, 2006).

Character Analysis and Phylogenetic Positions of "Trachodon cantabrigiensis" *and* 'Iguanodon' hillii

In order to assess the systematic position of "*Trachodon cantabrigiensis*" and *'Iguanodon' hillii*, we scored the holotype teeth for characters listed in three recent phylogenetic analyses of hadrosauriformes: Norman (2002), Sues and Averianov (2009, which includes most of the characters presented in Horner et al. [2004]), and Prieto-Márquez (2010a). Dental characters comprise approximately 6% (21 of 370 characters in Prieto-Márquez, 2010a) to 18% (12 of 67 characters in Norman, 2002) of the total number of characters in these and other hadrosauroid phylogenetic analyses. However, only a subset of the dental characters can be scored for either "*T. cantabrigiensis*" or *'I.' hillii*, and therefore they cannot be given formal phylogenetic treatments here. Nevertheless, given the morphological variation and distribution of these dental characters with Hadrosauroidea, even isolated teeth can be systematically placed within the clade with a reasonable degree of confidence in reference to specific phylogenies, as below.

Using the matrix of Norman (2002), NHMUK R496 can be scored for six characters (29–32, 35, and 38), half of which represent the plesiomorphic state (characters 29, 35, and 38). Of the derived character states, the restricted distribution of the dentary tooth enamel (30[1]) occurs in the least inclusive clade formed by *Protohadros* and hadrosaurids. NHMUK R496 has mammillated papillae 31[1]) that occur in non-hadrosaurid hadrosauroids, but lacks the reduced papillae of hadrosaurids (31[2]). The angular morphology of the tooth root suggests a derived condition of tooth emplacement (32[2]), which occurs in *Protohadros, Bactrosaurus, Probactrosaurus*, and hadrosaurids. GSM 1966 can only be scored for four characters (29–31 and 35) and is identical to "*T. cantabrigiensis*" in all scores. It should be noted that characters 29 and 35 of Norman (2002) are somewhat problematic, as they combine several morphological features into one character (e.g., presence or absence of secondary ridges and crown shape in the case of character 29) and there are no metrics (such as a ratio for crown height vs. length) to distinguish between states for broad and narrow crowns. We have tried to make our criteria for scoring characters compatible with those used by Norman (2002), but acknowledge that others might operationalize these character descriptions differently.

"*Trachodon cantabrigiensis*" can be scored for four characters in the matrix of Sues and Averianov (2009). Of these, three characters, 97(0), 98(0), and 100(0), exhibit the plesiomorphic state and are phylogenetically uninformative. A single derived state is present (character 99), in which the tooth crown is dominated by a single primary ridge. This character optimizes as an unambiguous, unreversed synapomorphy of the least inclusive clade formed by (*Bactrosaurus* + *Levnesovia*) and Hadrosauridae in the four most-parsimonious trees (MPTs) recovered by this analysis. Importantly, NHMUK R496 lacks the more symmetrical (98[0]) dentary crowns that are characteristic of more derived hadrosaurids. *'Iguanodon' hillii* has identical character scores to "*T. cantabrigiensis*" for this analysis.

In terms of the analysis presented by Prieto-Márquez (2010a), NHMUK R496 can be scored for characters 4–11 and 14, and some of the component character states optimize unambiguously on to the consensus tree presented in this study. NHMUK R496 possesses the following character states (those with unambiguous distributions are marked with an asterisk): 4(1)*, 5(3)*, 6(1)*, 7(0), 8(2), 9(1), 10(0), 11(1)*, and 14(1). Character 4(1) suggests placement in an *Eolambia+Protohadros* clade and 6(1) indicates inclusion in a *Lophorhothon* + Hadrosauridae clade. Character 5(3) optimizes as a hadrosaurid synapomorphy in this analysis and 11(1) suggests similarity to a clade formed by (*Corythosaurus* + *Lambeosaurus*) + *Hypacrosaurus*. The character scores of NHMUK R496 for characters 8, 9, and 10 imply that that it cannot be referred to a variety of derived hadrosaurid clades. The absence of a sinuous primary ridge (character 7[0]) indicates that NHMUK R496 cannot be assigned to Saurolophinae (although a secondary reversal to this primitive state does occur among derived saurolophines). '*Iguanodon*' *hillii* can be scored for characters 5–7, 9–11, and 14 and is identical in all states to "*T. cantabrigiensis*," except in retaining the primitive state for character 11.

Character analysis of NHMUK R496 and GSM 1966 in the context of the phylogenetic analyses presented by Norman (2002) and Sues and Averianov (2009) suggests that both "*Trachodon cantabrigiensis*" and '*Iguanodon*' *hillii* are basal hadrosauroids that lie just outside Hadrosauridae. The characters of the teeth suggest that they are both more derived than *Altirhinus*, *Eolambia*, and *Probactrosaurus*, and are consistent with them being *Protohadros-Bactrosaurus-Telmatosaurus* grade taxa. The analysis of Prieto-Márquez (2010a) supports the inclusion of both taxa in Hadrosauroidea and provides some evidence in favor of hadrosaurid affinities in both cases. The latter placements depend almost exclusively on the absence of secondary ridges in these specimens, which optimizes as a hadrosaurid synapomorphy. However, the utility of this character is slightly problematic, given that the presence or absence of faint secondary ridges can vary ontogenetically and positionally within hadrosauroid dentitions. For example, the presence or absence of secondary ridges is variable on the maxillary teeth of *Levnesovia* (Sues and Averianov, 2009) and *Jayewati* (McDonald, Kirkland, et al., 2010), suggesting that the loss of secondary ridges is likely convergent in "*T. cantabrigiensis*" and derived hadrosaurids. As "*T. cantabrigiensis*" lacks many of the features optimizing as hadrosaurid dental characters in all three analyses, we consider that the available evidence is more consistent with a basal hadrosauroid position for this taxon.

Table 6.1. Taxa Used in Morphometric Analysis and Their Specimen Number or Literature Reference

Taxon	Reference
Brachylophosaurus canadensis	FMNH PR 682
*Bactrosaurus johnsoni**	Godefroit et al. (1998)
Charonosaurus jiayinensis	Godefroit et al. (2001)
Corythosaurus casuarius	ROM 1947
Edmontosaurus annectens	NHMUK unnumbered
*Eolambia caroljonesa**	CEUM 9758
*Equijubus normani**	IVPP V12534
Gryposaurus notablis	ROM 873
Levnesovia transoxiana	Sues and Averianov (2009)
Mantellisaurus atherfieldensis	NHMUK R5190 and unnumbered
*Owenodon hoggii**	NHMUK R2998
Probactrosaurus gobiensis	Norman (2002)
Protohadros byrdi	SMU 74582
Telmatosaurus transylvanicus	NHMUK R3386
"*Trachodon cantabrigiensis*"	NHMUK R496

Note: Asterisks indicate taxa whose tooth morphology was estimated using a composite shape from two teeth.

Morphometric Analysis

Many of the discrete characters used in phylogenetic analysis of iguanodontians are crude approximations of shape, so to examine the phylogenetic significance of precise, quantified tooth shape in determining the phylogenetic affinities of "*T. cantabrigensis*," we quantified tooth morphology in select iguanodontian taxa (Table 6.1). Unfortunately, the tooth of '*Iguanodon*' *hillii* is too incomplete to include in this analysis.

Methods We translated the overall shape of the enamel surface of dentary teeth into 2-D coordinate systems and performed eigenshape analysis, using the resultant shape variables to determine the similarity of tooth shape in "*T. cantabrigensis*" to other taxa. Eigenshape analysis determines covariance between shape outlines by comparing the net angular deviation (phi function) of coordinate points describing those outlines, between shapes (Lohmann, 1983; MacLeod, 1999). Singular value decomposition of the covariance matrix of phi functions for multiple shapes provides the principal components, eigenvalues, and shape variables used for comparisons (MacLeod, 1999; Krieger, 2010). The method requires a single homologous landmark starting point, and we used the anteroventral margin of the enamel surface as the landmark coordinate after orienting all specimen images to correspond to the left side of the dentary. All images were digitized using tpsDIG2 (Rohlf, 2007) and outlines were resampled to 100 coordinate points. For denticulate enamel margins, outlines points were digitized at the apex of individual denticles. Several taxa did not preserve a complete enamel surface for any individual tooth, and shape

Table 6.2. Eigenvalues and % Variance Explained for Jolliffe Cut-off of 75%

Eigenvalue	Value	% Variance
1	0.419	35.9
2	0.259	22.2
3	0.143	12.2
4	0.099	8.5

was estimated by creating a composite tooth from adjacent elements (Table 6.1).

Eigenshape analysis was performed using PAST v.2.14 (Hammer et al., 2001), and the significant principle components were determined using the Jolliffe cut-off, which retains those components whose eigenvalues are greater than the average of all eigenvalues (Jolliffe, 1986; see also Mutsvangwa and Douglas, 2007). Component scores for all taxa were used in a cluster analysis to create a phenetic tree based on tooth shape for comparison with phylogenetic trees based on larger datasets (e.g., McDonald, Barrett, and Chapman, 2010; McDonald, Kirkland, et al., 2010). We created a pairwise distance matrix by reducing the coordinate distances for the first four components to single dimensional Euclidean distances, and performed a UPGMA cluster analysis (Fig. 6.3C).

Results Analysis resulted in four significant components with eigenvalues accounting for 78.8% for all shape variance within the sample (Table 6.2). Shape space defined by the first three component scores is shown in Figure 6.3. For all significant components, "*T. cantabrigensis*" is excluded from the space defined by the sample of examined hadrosaurid taxa (Fig. 6.3A, B). Instead, the taxon occupies a position close to *Equijubus*, *Protohadros*, and the British *Iguanodon*-like taxa, despite possessing discrete apomorphies of more derived taxa. Among other taxa, *Telmatosaurus* occupies an extreme position in coordinate space, reflecting its apomorphically narrow, elongate, and recurved dentary crown surface. Hadrosaurids occupy a narrow range of coordinate space. Although this range represents only a small sample of hadrosaurid species-richness, it includes representatives of most of the primary divisions within the clade and we do not expect that increased taxonomic sampling would result in a greatly expanded shape space.

Comparison of the UPGMA cluster analysis topology with phylogenetic hypotheses of iguanodontians reveals some similarities in the proximal grouping of terminal taxa, but tooth morphometrics shows an overall poor fit to phylogenetic hypotheses. Hadrosaurids cluster together with the exception of *Charonosaurus*, and hadrosaurines cluster within Hadrosauridae. *Levnesovia* and *Bactrosaurus* are closely clustered, which reflects their overall anatomical similarity (e.g., Godefroit et al., 1998; Sues and Averianov, 2009). Other aspects of the cluster topology are not consistent with published phylogenies. *Mantellisaurus* clusters with *Probactrosaurus*, *Bactrosaurus*, and *Levnesovia*, despite cranial and postcranial data that suggest it is a basal iguanodontian relative to those taxa (McDonald, Barrett, and Chapman, 2010; McDonald, Kirkland, et al., 2010). The highly apomorphic dentition of *Telmatosaurus* results in it clustering with the group consisting of all other taxa, as opposed to it being recognized as the sister taxon to Hadrosauridae (Weishampel et al., 1993). *Eolambia* clusters with the majority of hadrosaurids despite being recognized as a more basal iguanodontian (Head, 2001).

"*Trachodon cantabrigiensis*" clusters with *Protohadros*, *Equijubus*, and *Owenodon*, taxa that constitute a paraphyletic assemblage within Iguanodontia (McDonald, Barrett, Chapman, 2010; McDonald, Kirkland, et al., 2010). This grouping does not strongly support a precise hypothesis of relationships for "*T. cantabrigiensis*," but, in combination with the exclusion of the taxon from Hadrosauridae based on morphometry, suggests a position among basal hadrosauroids, consistent with the distribution of discrete characters. The conflict between morphometric tooth data and discrete dental characters, and between dental and osteological characters, may in part be due to dental ecomorphology. Dietary specification is correlated to tooth morphology in herbivorous mammals, but the potential for diet to influence tooth size and shape in ornithopods has yet to be examined.

Validity and Affinities of English "Hadrosaurids"

"*Trachodon cantabrigiensis*" possesses a unique character combination among basal hadrosauroids and hadrosaurids: the presence of a distally offset carina, the absence of secondary ridges, and the relatively short, wide tooth crown. However, the latter character is subject to variation depending on the position of the tooth within the tooth row: more mesially and distally positioned teeth in hadrosaurids often have lower crown height-to-width ratios than those in the middle of the tooth row (Gallimore and Evans, 2009). As the exact position of the holotype tooth within the tooth row cannot be established, we adopt a conservative approach and refrain from proposing a new generic name for this very limited material and regard the taxon as a nomen dubium (see also Horner et al., 2004). Nevertheless, it is likely that this single tooth, and possibly other isolated indeterminate material from the Cambridge Greensand Member (such as the isolated maxillary fragment described above), represents an otherwise unknown basal hadrosauroid taxon, given its geographic and stratigraphic provenance and its anatomical distinctiveness from other European hadrosauroids and more basally positioned iguanodontians.

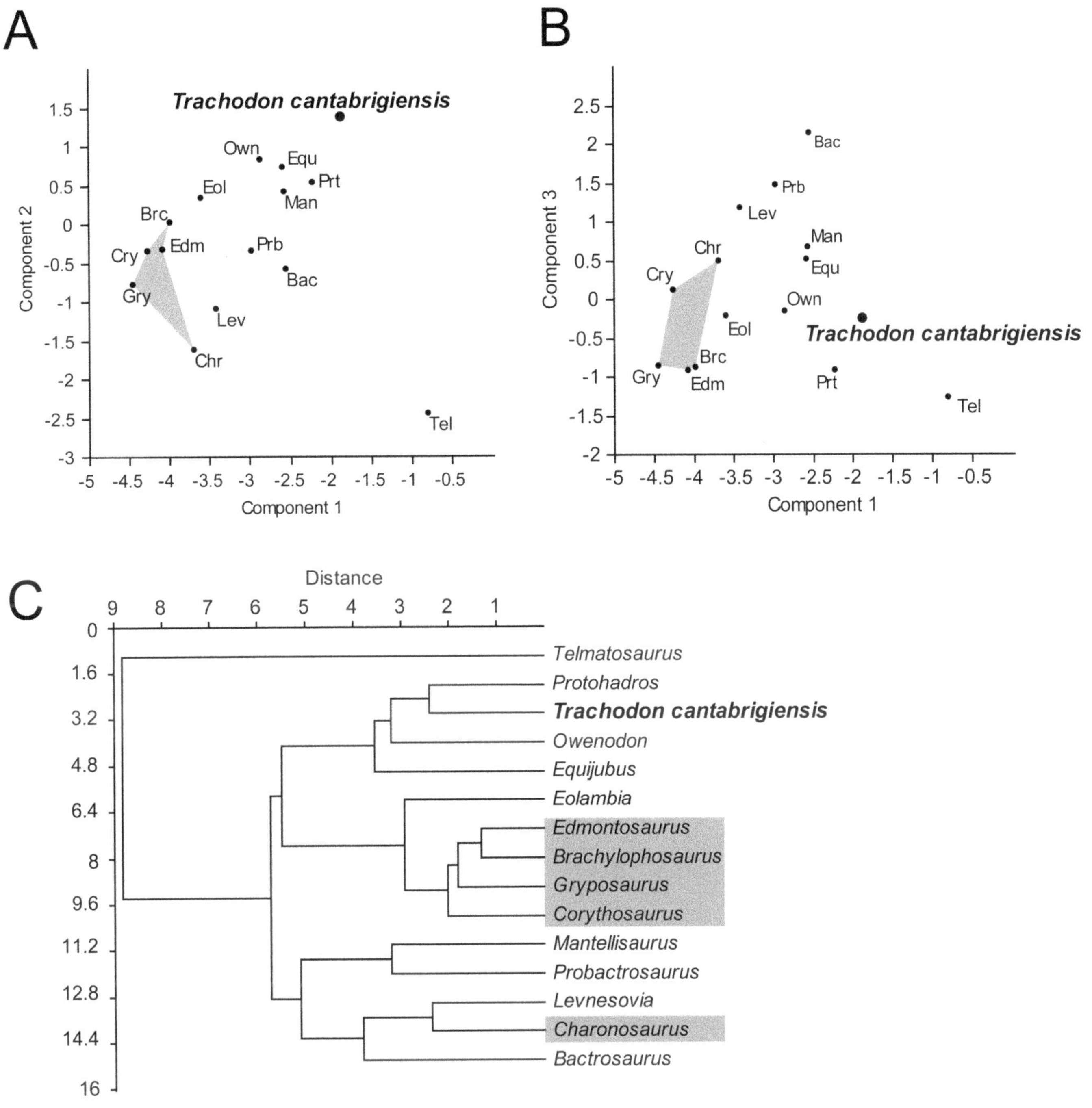

6.3. Morphometric analysis of hadrosauroid tooth shape. (A) Principal component space defined by components 1 and 2; (B) principal component space defined by components 1 and 3; (C) UPGMA cluster analysis of hadrosauroid shape distances for the first four eigenvalue scores. Gray areas indicate shape spaces defined by Hadrosauridae.

'Iguanodon' hillii can be distinguished from *"Trachodon cantabrigiensis"* on the basis of its denticle morphology, suggesting that it, too, represents an otherwise unknown basal hadrosauroid taxon from the "middle" Cretaceous of England. However, the tooth possesses no autapomorphic features and as far as can be determined is similar in morphology to teeth of *Telmatosaurus*. We also regard this taxon as a nomen dubium (see also Horner et al., 2004).

CONCLUSIONS

Morphometric analysis and consideration of dental character distributions indicate that neither *"Trachodon cantabrigiensis"* nor *'Iguanodon' hillii* can be referred to Hadrosauridae, contrary to previous accounts (e.g., Horner et al., 2004). Nevertheless, these specimens represent the most derived non-hadrosaurid hadrosauroids to have been recovered from western Europe and are taxa that, on the basis of the limited available evidence, are potentially very close to the origin of Hadrosauridae. All other basal hadrosauroids are known from the "middle" to early Late Cretaceous of eastern Asia (e.g., *Bactrosaurus, Equijubus, Levnesovia, Probactrosaurus*) and North America (*Eolambia, Protohadros*), along with the relictual taxon *Telmatosaurus* from the latest Cretaceous of Romania (Norman, 2004; Sues and Averianov, 2009). The "middle" Cretaceous English taxa therefore extend significantly the already broad Laurasian geographic distribution of basal hadrosauroids.

Frustratingly, the apparent phylogenetic proximity of "*T. cantabrigiensis*" and '*I.*' *hillii* to *Bactrosaurus, Protohadros, Telmatosaurus*, and hadrosaurids, along with the western European provenance of the former taxa, clouds further the paleobiogeography of hadrosaurid origins, by offering another geographic region that yields plausible hadrosaurid sister taxa (see Horner et al. [2004] and Prieto-Márquez [2010b] for recent discussions on this issue). However, the early (late Albian) occurrence of "*T. cantabrigiensis*" is noteworthy. Although it is not the earliest-known hadrosauroid (contra Sues and Averianov [2009]; various Asian taxa are potentially of Aptian age [e.g., Norman, 1998; You et al., 2003]), the dental morphology of "*T. cantabrigiensis*" exhibits a set of derived character states relative to those present in other late Early Cretaceous hadrosauroids (e.g., *Altirhinus, Equijubus, Probactrosaurus*) that place it much closer to hadrosaurids than any of these near-contemporaries. In addition, it occurs earlier than all of the other most-proximate hadrosaurid outgroups, which are Cenomanian in age or younger (Head, 1998; Sues and Averianov, 2009; Prieto-Márquez, 2010b). If our inferred phylogenetic position for "*T. cantabrigiensis*" is correct, this could imply that hadrosaurid origin occurred close to the Early–Late Cretaceous boundary, rather than in the Santonian as previously suggested (see Sues and Averianov, 2009; Prieto-Márquez, 2010a, b).

ACKNOWLEDGMENTS

This chapter is dedicated to David Weishampel, the doyen of hadrosaurid workers. He was one of PMB's Ph.D. examiners and was influential in JJH's graduate training. We thank M. Riley (CAMSM) and P. Shepherd (GSM) for arranging the loan of material and P. Hurst (NHMUK, Photographic Unit) for his excellent photographs of the holotype teeth. A. McDonald and J. Kirkland provided photographs of the teeth of *Equijubus* and *Eolambia*, respectively; D. Norman is thanked for numerous interesting discussions on the Cambridge Greensand dinosaur fauna; and D. Eberth and Kh. Tsogtbaatar provided helpful reviews of the manuscript.

LITERATURE CITED

Barrett, P. M., R. J. Butler, R. J. Twitchett, and S. Hutt. 2011. New material of *Valdosaurus canaliculatus* (Ornithischia: Ornithopoda) from the Lower Cretaceous of southern England. Special Papers in Palaeontology 86:131–163.

Benton, M. J., and P. S. Spencer. 1995. Fossil Reptiles of Great Britain. Chapman and Hall, London, 386 pp.

Benton, M. J., Z. Csiki, D. Grigorescu, R. Redelstorff, P. M. Sander, K. Stein, and D. B. Weishampel. 2010. Dinosaurs and the island rule: the dwarfed dinosaurs from Hateg Island. Palaeogeography, Palaeoclimatology, Palaeoecology 293:438–454.

Carpenter, K., and Y. Ishida. 2010. Early and "middle" Cretaceous iguanodonts in time and space. Journal of Iberian Geology 36:145–164.

Dalla Vecchia, F. M. 2006. *Telmatosaurus* and the other hadrosaurids of the Cretaceous European archipelago. An overview. Natura Nascosta 32:1–55.

Gallimore, G., and D. C. Evans. 2009. Morphometric analysis of hadrosaurid dental battery variation. Journal of Vertebrate Paleontology, Program and Abstracts 2009:101A.

Galton, P. M. 1974. The ornithischian dinosaur *Hypsilophodon* from the Wealden of the Isle of Wight. Bulletin of the British Museum (Natural History), Geology 25:1–152.

Galton, P. M. 2009. Notes on Neocomian (Lower Cretaceous) ornithopod dinosaurs from England–*Hypsilophodon, Valdosaurus, "Camptosaurus," "Iguanodon"*–and referred specimens from Romania and elsewhere. Revue de Paléobiologie 28:211–273.

Godefroit, P., Z.-M. Dong, P. Bultynck, H. Li, and L. Feng. 1998. Sino-Belgian Cooperation Program, "Cretaceous dinosaurs and mammals from Inner Mongolia." 1. New *Bactrosaurus* material (Dinosauria: Hadrosauroidea) material from Iren Dabasu (Inner Mongolia, P. R. China). Bulletin de l'Institut royal des Sciences naturelles de Belgique, Sciences de la Terre 68(Supplement):3–70.

Hall, J. P. 1993. A juvenile hadrosaurid from New Mexico. Journal of Vertebrate Paleontology 13:367–369.

Hammer, Ø., D. A. T. Harper, and P. D. Ryan. 2001. PAST: Paleontological Statistics Software Package for Education and Data Analysis. Palaeontologia Electronica 4(1):Article 4, 9 pp.

Head, J. J. 1998. A new species of basal hadrosaurid (Dinosauria, Ornithischia) from the Cenomanian of Texas. Journal of Vertebrate Paleontology 18:718–738.

Head, J. J. 2001. A reassessment of the phylogenetic position of *Eolambia caroljonesa* (Dinosauria: Iguanodontia). Journal of Vertebrate Paleontology 21:392–396.

Horner, J. R., D. B. Weishampel, and C. A. Forster. 2004. Hadrosauridae; pp. 438–463 in D. B. Weishampel, P. Dodson, and H. Osmólska (eds.), The Dinosauria, Second Edition. University of California Press, Berkeley, California.

Jolliffe, I. T. 1986. Principal Component Analysis. Springer-Verlag, New York, 487 pp.

Keeping, W. 1883. The fossils and palaeontological affinities of the Neocomian deposits of Upware and Brickhill (Cambridgeshire and Bedfordshire). Deighton, Bell, and Company, Cambridge, U.K., xi + 167 pp. + 8 pls.

Kirkland, J. I. 1998. A new hadrosaurid from the upper Cedar Mountain Formation (Albian–Cenomanian: Cretaceous) of eastern Utah–the oldest known hadrosaurid (Lambeosaurinae?); pp. 283–295 in S. G. Lucas, J. I. Kirkland, and J. W. Estep (eds.), Lower and Middle Cretaceous Terrestrial Ecosystems. New Mexico Museum of Natural History and Science Bulletin 14.

Krieger, J. D. 2010. Controlling for curvature in the quantification of leaf form; pp. 27–72 in A. M. T. Elewa (ed.), Morphometrics for Nonmorphometricians. Springer-Verlag, Berlin.

Leidy, J. 1865. Cretaceous reptiles of the United States. Smithsonian Contributions to Knowledge 14:v + 1–135 + pls. 1–20.

Lohmann, G. P. 1983. Eigenshape analysis of microfossils: a general morphometric method for describing changes in shape. Mathematical Geology 15:659–672.

Lull, R. S., and N. E. Wright. 1942. Hadrosaurian Dinosaurs of North America. Geological Society of America Special Papers 40. 242 pp.

Lydekker, R. 1888a. A new Wealden iguanodont and other dinosaurs. Quarterly Journal of the Geological Society of London 44:46–61 + pl. 3.

Lydekker, R. 1888b. Catalogue of the fossil Reptilia and Amphibia in the British Museum (Natural History), Part 1: Containing the orders Ornithosauria, Crocodilia, Dinosauria, Squamata, Rhynchocephalia, and Proterosauria. British Museum (Natural History), London, xxviii + 309 pp.

MacLeod, N. 1999. Generalizing and extending the eigenshape method of shape space visualization and analysis. Paleobiology 25:107–138.

McDonald, A. T. 2012. The status of *Dollodon* and other basal iguanodonts (Dinosauria: Ornithischia) from the Lower Cretaceous of Europe. Cretaceous Research 33:1–6.

McDonald, A. T., P. M. Barrett, and S. D. Chapman. 2010. A new basal iguanodont (Dinosauria, Ornithischia) from the Wealden (Lower Cretaceous) of England. Zootaxa 2569:1–43.

McDonald, A. T., J. I. Kirkland, D. D. DeBlieux, S. K. Madsen, J. Cavin, A. R. C. Milner, and L. Panzarin. 2010. New basal iguanodonts from the Cedar Mountain Formation of Utah and the evolution of thumb-spiked dinosaurs. PLoS ONE 5(11):e14075.

Mutsvangwa, T., and T. S. Douglas. 2007. Morphometric analysis of facial landmark data to characterize the facial phenotype associated with fetal alcohol syndrome. Journal of Anatomy 210:209–220.

Naish, D., and D. M. Martill. 2008. Dinosaurs of Great Britain and the role of the Geological Society of London in their discovery: Ornithischia. Journal of the Geological Society, London 165:613–623.

Newton, E. T. 1892. Note on an iguanodont tooth from the Lower Chalk ("Totternhoe Stone"), near Hitchin. Geological Magazine (New Series, Decade 3) 9:49–50.

Nopcsa, F. 1923. Notes on British dinosaurs. Part VI: *Acanthopholis*. Geological Magazine 60:193–199 + pls. 7–8.

Norman, D. B. 1980. On the ornithischian dinosaur *Iguanodon bernissartensis* of Bernissart (Belgium). Mémoire de l'Institut royal des Sciences naturelles de Belgique 178:1–104.

Norman, D. B. 1986. On the anatomy of *Iguanodon atherfieldensis* (Ornithischia: Ornithopoda). Bulletin de l'Institut royal des Sciences naturelles de Belgique 56:281–372.

Norman, D. B. 1993. Gideon Mantell's "mantel-piece": the earliest well-preserved ornithischian dinosaur. Modern Geology 18:225–245.

Norman, D. B. 1998. On Asian ornithopods (Dinosauria: Ornithischia). 3. A new species of iguanodontid dinosaur. Zoological Journal of the Linnean Society 122:291–348.

Norman, D. B. 2002. On Asian ornithopods (Dinosauria: Ornithischia). 4. *Probactrosaurus* Rozhdestvensky, 1966. Zoological Journal of the Linnean Society 136:113–144.

Norman, D. B. 2004. Basal Iguanodontia; pp. 413–437 in D. B. Weishampel, P. Dodson, and H. Osmólska (eds.), The Dinosauria, Second Edition. University of California Press, Berkeley, California.

Norman, D. B. 2010. A taxonomy of iguanodontians (Dinosauria: Ornithopoda) from the lower Wealden Group (Cretaceous: Valanginian) of southern England. Zootaxa 2489:47–66.

Norman, D. B. 2011a. Ornithopod dinosaurs; pp. 407–475 in D. J. Batten (ed.), English Wealden fossils (Palaeontological Association Field Guides to Fossils Number 14). The Palaeontological Association, London, U.K.

Norman, D. B. 2011b. On the osteology of the lower Wealden (Valanginian) ornithopod *Barilium dawsoni* (Iguanodontia: Styracosterna). Special Papers in Palaeontology 86:165–194.

Ostrom, J. H. 1961. Cranial morphology of the hadrosaurian dinosaurs of North America. Bulletin of the American Museum of Natural History 122:33–186.

Owen, R. 1861. The fossil Reptilia of the Cretaceous formations. Supplement 2. Dinosauria (*Iguanodon*). Palaeontographical Society Monographs 12(Number 52):27–30 + pl. 7.

Paul, G. S. 2007. Turning the old into the new: a separate genus for the gracile iguanodont from the Wealden of England; pp. 69–77 in K. Carpenter (ed.), Horns and Beaks: Ceratopsian and Ornithopod Dinosaurs. Indiana University Press, Bloomington, Indiana.

Paul, G. S. 2008. A revised taxonomy of the iguanodont dinosaur genera and species. Cretaceous Research 29:192–216.

Pereda-Suberbiola, X., and P. M. Barrett. 1999. A systematic review of ankylosaurian dinosaur remains from the Albian–Cenomanian of England. Special Papers in Palaeontology 60:177–208.

Prieto-Márquez, A. 2010a. Global phylogeny of Hadrosauridae (Dinosauria: Ornithopoda) using parsimony and Bayesian methods. Zoological Journal of the Linnean Society 159:435–502.

Prieto-Márquez, A. 2010b. Global historical biogeography of hadrosaurid dinosaurs. Zoological Journal of the Linnean Society 159:503–525.

Rohlf, F. J. 2007. TpsDig2. Department of Ecology and Evolution, State University of New York at Stony Brook. Available at life.bio.sunysb.edu/morph/soft-dataacq.html. Accessed fall 2012.

Seeley, H. G. 1869. Index to the fossil remains of Aves, Ornithosauria, and Reptilia from the secondary system of strata arranged in the Woodwardian Museum of the University of Cambridge. Deighton, Bell, and Co., Cambridge, U.K., xxiii + 143 pp.

Seeley, H. G. 1879. On the Dinosauria of the Cambridge Greensand. Quarterly Journal of the Geological Society of London 35:591–636 + pls. 34–35.

Sereno, P. C. 1998. A rationale for phylogenetic definitions, with application to the higher-level taxonomy of Dinosauria. Neues Jahrbuch für Geologie und Paläontologie, Abhandlungen 210:41–83.

Steel, R. 1969. Ornithischia. Handbuch der Paläoherpetology, Teil 15. Gustav Fischer Verlag, Stuttgart, Germany, 84 pp.

Sues, H.-D., and A. Averianov. 2009. A new basal hadrosauroid dinosaur from the Late Cretaceous of Uzbekistan and the early radiation of duck-billed dinosaurs. Proceedings of the Royal Society B 276:2549–2555.

Unwin, D. M. 2001. An overview of the pterosaur assemblage from the Cambridge Greensand (Cretaceous) of Eastern England. Mitteilungen aus dem Museum fur Naturkunde in Berlin, Geowissenschaftliche Reihe 4:189–221.

Vickaryous, M. K., T. Maryańska, and D. B. Weishampel. 2004. Ankylosauria; pp. 363–392 in D. B. Weishampel, P. Dodson, and H. Osmólska (eds.), The Dinosauria, Second Edition. University of California Press, Berkeley, California.

Wang, X.-L., R. Pan, R. J. Butler, and P. M. Barrett. 2011. The postcranial skeleton of the iguanodontian ornithopod *Jinzhousaurus yangi* from the Lower Cretaceous Yixian Formation of western Liaoning, China. Earth and Environmental Science Transactions of the Royal Society of Edinburgh 101:135–159.

Weishampel, D. B., D. B. Norman, and D. Grigorescu. 1993. *Telmatosaurus transsylvanicus* from the Late Cretaceous of Romania: the most basal hadrosaurid dinosaur. Palaeontology 36:361–385.

Weishampel, D. B., P. M. Barrett, R. A. Coria, J. Le Loeuff, X. Xu, X.-J. Zhao, A. Sahni, E. M. P. Gomani, and C. R. Noto. 2004. Dinosaur Distribution; pp. 517–606 in D.B. Weishampel, P. Dodson, and H. Osmólska (eds.), The Dinosauria, Second Edition. University of California Press, Berkeley, California.

You, H.-L., Z.-X. Luo, N. H. Shubin, L. M. Witmer, Z.-L. Tang, and F. Tang. 2003. The earliest-known duck-billed dinosaur from deposits of late Early Cretaceous age in northwest China and hadrosaur evolution. Cretaceous Research 24:347–355.

A New Hadrosauroid (*Plesiohadros djadokhtaensis*) from the Late Cretaceous Djadokhtan Fauna of Southern Mongolia

7

Khishigjav Tsogtbaatar, David B. Weishampel, David C. Evans, and Mahito Watabe

ABSTRACT

The Djadokhta Formation (Campanian) has been intensely surveyed and sampled for its fossil vertebrate fauna for almost a century. Its dinosaur fauna is historically defined as being dominated by the ornithischians *Protoceratops* and *Pinacosaurus*, as well as oviraptorid theropods, but recent collecting efforts continue to yield new taxa. Here we describe and name the first ornithopod from this unit. The material consists of an almost complete skull and associated limb material collected from the Alag Teeg locality. *Plesiohadros djadokhtaensis* nov. gen. et nov. sp. is characterized by a unique combination of plesiomorphic and derived characters within Hadrosauroidea, and an autapomorphically rugose, raised rim around the rostrodoral margin of the orbit. In order to place *P. djadokhtaensis* into its paleoecological and stratigraphic context within these Upper Cretaceous rocks of Mongolia, we also provide a review of the geological setting of the Djadokhta Formation in the Gobi Desert. *Plesiohadros* is the largest member of the Djadokhta fauna, and the only known hadrosauroid from the Campanian of Mongolia. An updated phylogenetic analysis indicates that *P. djadokhtaensis* is a derived hadrosauroid proximate to Hadrosauridae (sensu Sereno). *Plesiohadros* is posited as the sister taxon to a clade that includes *Lophorhothon* and Hadrosauridae, and is more closely related to Hadrosauridae than to *Bactrosaurus*, *Telmatosaurus*, or *Tethyhadros*. The complete replacement of non-hadrosaurid hadrosauroids with true hadrosaurids in the Maastrichtian of Mongolia highlights the complex pattern of faunal interchange between Asian and North America in the latest Cretaceous.

INTRODUCTION

After nearly 100 years of scientific exploitation, the Upper Cretaceous rocks of the Gobi Desert in southern Mongolia continue to yield new and unusual dinosaur taxa. In the last decade alone over a dozen new taxa have been named–including the theropods *Tsaagan*, *Mahakala*, and *Kol*; and the ceratopsians *Bainoceratops*, *Yamaceratops*, and *Gobiceratops* (Norell et al., 2006; Turner et al., 2007, 2009; Tereschenko and Alifanov, 2003; Makovicky and Norell, 2006; Alifanov, 2008, respectively). Ornithopods are considerably less diverse in these deposits than in contemporaneous Upper Cretaceous dinosaur-bearing units of western North America and Europe (Horner et al., 2004; Weishampel and Jainu, 2011; Dalla Vecchia, this volume), which are, in general, dominated by hadrosaurids. Only one new ornithopod taxon, *Haya griva*, has been identified from the Late Cretaceous of Mongolia in the last three decades (Makovicky et al., 2011).

In this chapter, we describe a new genus and species of hadrosauroid ornithopod, *Plesiohadros djadokhtaensis*, from the Alag Teeg locality, southern Gobi Desert, Mongolia. *Plesiohadros* is the second record of an ornithopod from the Djadokhta Formation (Barsbold and Perle, 1983), and the only named hadrosauroid taxon from Campanian-age rocks in Mongolia. In order to place *Plesiohadros djadokhtaensis* into its paleoecological and stratigraphic context within these Upper Cretaceous rocks of Mongolia, we outline the geological setting of the Djadokhta Formation at Alag Teeg and elsewhere in the Gobi Desert, review the different vertebrate assemblages of Djadokhta age, and assess the chronostratigraphy of the Djadokhta Formation. *P. djadokhtaensis* is then described, and its systematic relationships within Hadrosauroidea are assessed. Finally, we comment on the biostratrigraphy and biogeographic implications of this new form within the context of our updated hadrosauroid phylogeny.

Institutional Abbreviations MPC, Mongolian Paleontology Centre, Ulan Baatar, Mongolia; ROM, Royal Ontario Museum, Toronto, Ontario; MNHN GDF, Muséum national d'Histoire naturelle, Paris, France.

GEOLOGICAL SETTING

The Djadokhta Formation was named by Berkey and Morris (1927), based on the fossiliferous sandstone-dominant beds at Bayan Zak (= Flaming Cliffs) in the central part of the Gobi Desert. It is this unit that has yielded a rich dinosaur fauna, mammals, and dinosaur eggs within nest structures. Although its depositional environments were originally considered lacustrine (Verzilin, 1979, 1982), it is now clear that they mainly represent eolian environments (Eberth, 1993; Fastovsky et al., 1997; Loope et al., 1998). Other Campanian-age localities that consist of eolian beds rich in dinosaur fossils are also known in the Gobi Desert of both Mongolia and China, and have been correlated with the Djadokhta Formation on the basis of similarities in vertebrate faunal composition and depositional environments (Eberth, 1993; Jerzykiewicz et al., 1993). Below, we briefly review the major Djadokhta localities of Mongolia.

Bayn Dzak Bayn Dzak is situated 15 km east-northeast of Bulgan Sum in the South Gobi Aimag (Fig. 7.1). Known as the "Flaming Cliffs," this locality was originally described by Roy Chapman Andrews of the Central Asiatic Expedition, American Museum of Natural History, in the 1920s and has subsequently been described by other researchers (see Dashzeveg et al., 2005). The fossiliferous beds at this locality consist of eolian sandstones, fluvial-eolian sandstones with paleosol layers, and red-colored mudstone strata of lacustrine origin. Although many specimens, including *Protoceratops andrewsi*, *Pinacosaurus grangeri*, *Oviraptor philoceratops*, turtles, and lizards, as well as dinosaur eggs and nests (elongatoolithid form), have been found in these beds, no ornithopod specimens have been reported from this locality to date (Dashzeveg et al., 2005). This locality has also produced the holotypes of *Velociraptor mongoliensis* and *Saurornithoides mongoliensis* (Osborn, 1924). The fauna was recently summarized by Dashzeveg et al. (2005), who cited *Oviraptor* (= *Rinchenia*) *mongoliensis* as present at this locality. However, *Rinchenia mongoliensis* is known only from the Maastrichtian Nemegt Formation (Osmólska et al., 2004), whereas the holotype of *Oviraptor philoceratops* was collected at Bayn Dzak (Osborn, 1924).

Tögrögiin Shiree Locality Tögrögiin Shiree (Tugrikin-Shireh of Fastovsky et al., 1997), discovered by the joint Polish-Mongolian Expeditions in the late 1960s and early 1970s, is situated 26 km northwest of Bulgan Sum of the South Gobi Aimag (Fig. 7.1). The fossiliferous beds at this locality consist mainly of fine- to medium-grained sandstones with large-scale cross stratification (foresets inclined more than 30°). Trace fossils (simple and branching vertical sand pipes) and dinosaur footprints are abundant on the bedding planes of these beds. These sediments have been interpreted as originating as eolian dune sands (Fastovsky et al., 1997). The combined thickness of the eolian sandstone beds is 50 m (Watabe and Suzuki, 2000). Small, presumed infant individuals of a hadrosauroid have been found in the eolian beds at this locality (Barsbold and Perle, 1983). These specimens will be described elsewhere. An abundance of

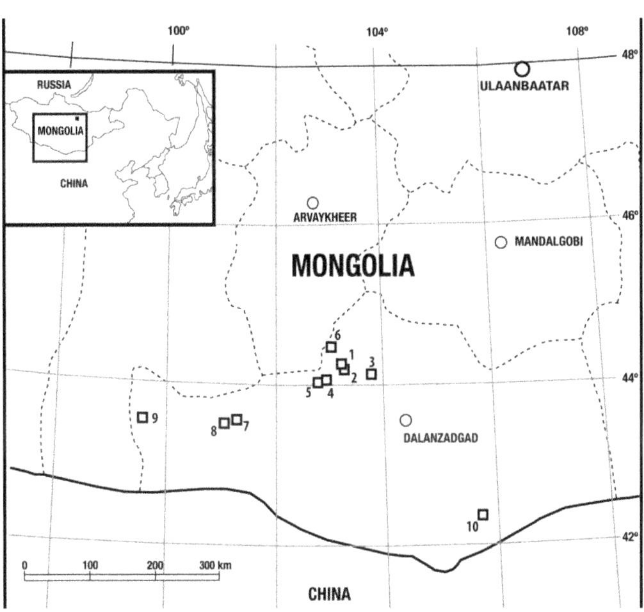

7.1. Map of Mongolia showing the major Djadokhta Formation localities in the Gobi Desert. Numbers correspond to the following localities: (1) Alag Teeg; (2) Tögrögiin Shiree; (3) Bayn Dzak; (4) Üüdin Sair; (5) Dzamyn Khond; (6) Abdrant Nuru; (7) Ukhaa Tolgod; (8) Khoolson; (9) Hermiin Tsav-II; (10) Shilüüt Uul.

vertebrate fossils have been found from the eolian beds, including *Velociraptor*, *Protoceratops*, *Shuvuuia* (an alvarezsaurid), lizards, multituberculate mammals, an enantiornithine bird, and dinosaur eggs and nests (Mikhailov, 1991) (elongatoolithid and *Protoceratopsidovum* forms [Mikhailov et al., 1994]). The fauna was summarized by Dashzeveg et al. (2005), who did not recognize any ornithischians from the locality. However, skeletal remains of *Protoceratops andrewsi* are common at this locality (Osmólska, 1993; Fastovsky et al., 1997; Fastovsky et al., 2011)

Alag Teeg The specimens of *Plesiohadros djadokhtaensis* nov. gen. et sp. reported herein were found at Alag Teeg (Fig. 7.1), situated 3 km north of Tögrögiin Shiree (Tverdokhlebov and Tsybin, 1974). Alluvial and paludal facies dominate at this locality, thus differing from most other correlative localities (D. Eberth, pers. comm., 2012). Eolian sandstone beds with large-scale cross stratification and invertebrate trace fossils overlie the fluvial beds and consist of white and red sandstone, and red and brown mudstone layers. Total thickness of the fluvial part of the section at Alag Teeg is 33 m (Suzuki and Watabe, 2000; Watabe and Suzuki, 2000).

The cranial and postcranial bones of *P. djadokhtaensis* were discovered in mudstone beds in the lowermost part of the geological section at the locality. An articulated pes referred to this species was discovered at a different site. The pes was buried in a mudstone layer with its dorsoventral long axis vertical, and thus perpendicular to the bedding plane, which suggests that the animal was killed in a "mud trap" (Watabe and Tsogtbaatar, 2007).

Üüdin Sair Üüdin Sair is situated 35 km southwest of Tögrögiin Shiree. The section consists of a lower portion of large-scale inclined strata comprising alternating sandstone and mudstone layers, and an upper portion of fine-grained reddish sandstone with large-scale cross stratification. The lower portion is interpreted as the remains of large-scale point bar deposits and the upper portion is interpreted as consisting of eolian dune deposits. Although there are many vertebrate fossils at the locality, including protoceratopsids (a large form, *Udanoceratops tschizhovi*, and several smaller forms), *Pinacosaurus*, mammals, lizards, turtles, and dinosaur eggs and nests (elongatoolithid form), no ornithopod specimens have so far been recognized in these collections.

Dzamyn Khond Dzamyn Khond, 9 km southwest of Üüdin Sair (Fig. 7.1), is a small outcrop consisting of fine-grained red sandstones with cross stratification. The beds are considered to be eolian dune deposits. An unnamed oviraptorine (represented by the well-preserved skeleton IGM 100/42, Clark et al., 2002), *Pinacosaurus*, lizards, and mammals have been found from a small outcrop here; and different kinds of dinosaur eggs (at least three forms, belonging to the elongatoolithid form) and nests have also been found.

Ukhaa Tolgod Ukhaa Tolgod, located 220 km southwest of Bayn Dzak (Fig. 7.1), provides one of the most diverse of all Djadokhta-age assemblages (Dingus et al., 2008). The fossiliferous strata consist of fine-grained, cross-stratified red sandstones, considered to be eolian dune deposits. Worked by the Mongolian Academy of Sciences–American Museum of Natural History expeditions, this locality constitutes one of the most recently discovered of the Djadokhta-age sites to be exploited, and has already yielded upward of a dozen new species within the last two decades. *Khaan*, *Tsaagan*, and numerous other theropods; *Pinacosaurus*; *Protoceratops*; lizards; turtles; mammals; as well as several kinds of dinosaur eggs and nests have been recovered at Ukhaa Tolgod, as summarized recently by Dingus et al. (2008).

Sedimentological Environments of the Djadokhta Formation

Fastovsky et al. (1997) demonstrated the widespread existence of eolian depositional environments in the Djadokhta Formation at Tögrögiin Shiree, more specifically an erg (= sand sea) with interconnected eolian dunes. Fluvial environments occurred in patches between the dunes (at Alag Teeg and Bayn Dzak). Based on the rich discoveries of vertebrate fossils including dinosaurs (and their eggs, nests, and infant individuals) in Djadokhta eolian beds, it suggests the possibility that the eolian (arid) environments of the Djadokhta age were inhabited by dinosaurs and other vertebrates. Ornithopod specimens found at Alag Teeg are the only records of this taxonomic group in the Djadokhta Formation, making *Plesiohadros djadokhtaensis* an apparently minor component of the dinosaur fauna in this overall arid setting. However, the eolian depositional environments and lithology seem to preferentially preserve the remains of small vertebrates such as small-sized theropods, lizards, and mammals. But the occurrence of aquatic vertebrates, such as turtles and crocodiles, is rare. The dominance of eolian deposits at these vertebrate fossil bearing localities suggests that the faunal composition is preservationally biased in favor of small body sizes.

Age of the Djadokhta Formation

The time interval represented by the Djadokhta Formation is difficult to estimate. Based on patterns of mammalian biostratigraphy, exposures of the Djadokhta Formation throughout the region have been suggested to be Campanian in age (Kielan-Jaworowska, 1974; ~80 Ma, Jerzykiewicz and Russell, 1991; Jerzykiewicz et al., 1993). However, because the lower and upper boundaries of the Djadokhta Formation cannot be observed in outcrops at the type and other localities, it is unknown if a Campanian age applies to the entire formation throughout the region. The thick eolian redbeds that dominate these localities do not lend themselves easily to age assessments using paleomagnetostratigraphic means (J. Hicks, pers. comm.); however, recent paleomagnetostratigraphic work by Dashzeveg et al. (2005) suggests that the Byan Dzak sediments were deposited during the rapid succession of polarity changes during the Late Campanian (~71–75 Ma).

Djadokhta-like outcrops are also present at Khulsan, Nemegt, and Khermeen Tsav in the Nemegt Basin. The upper boundary with the Nemegt Formation can be observed at Nemegt. At Khermeen Tsav, the upper and lower boundaries with fluvial beds are observed. The upper boundary of the formation is transitional with the overlying Nemegt Formation, where Djadokhta-like redbeds are referred to as the Baruungoyot Formation (Eberth et al., 2009).

SYSTEMATIC PALEONTOLOGY

DINOSAURIA Owen, 1842
ORNITHISCHIA Seeley, 1887
ORNITHOPOD Marsh, 1881
IGUANODONTIA Sereno, 1986
HADROSAUROIDEA sensu Sereno, 2005
PLESIOHADROS, gen. nov.

Type Species *PLESIOHADROS DJADOKHTAENSIS* sp. nov.

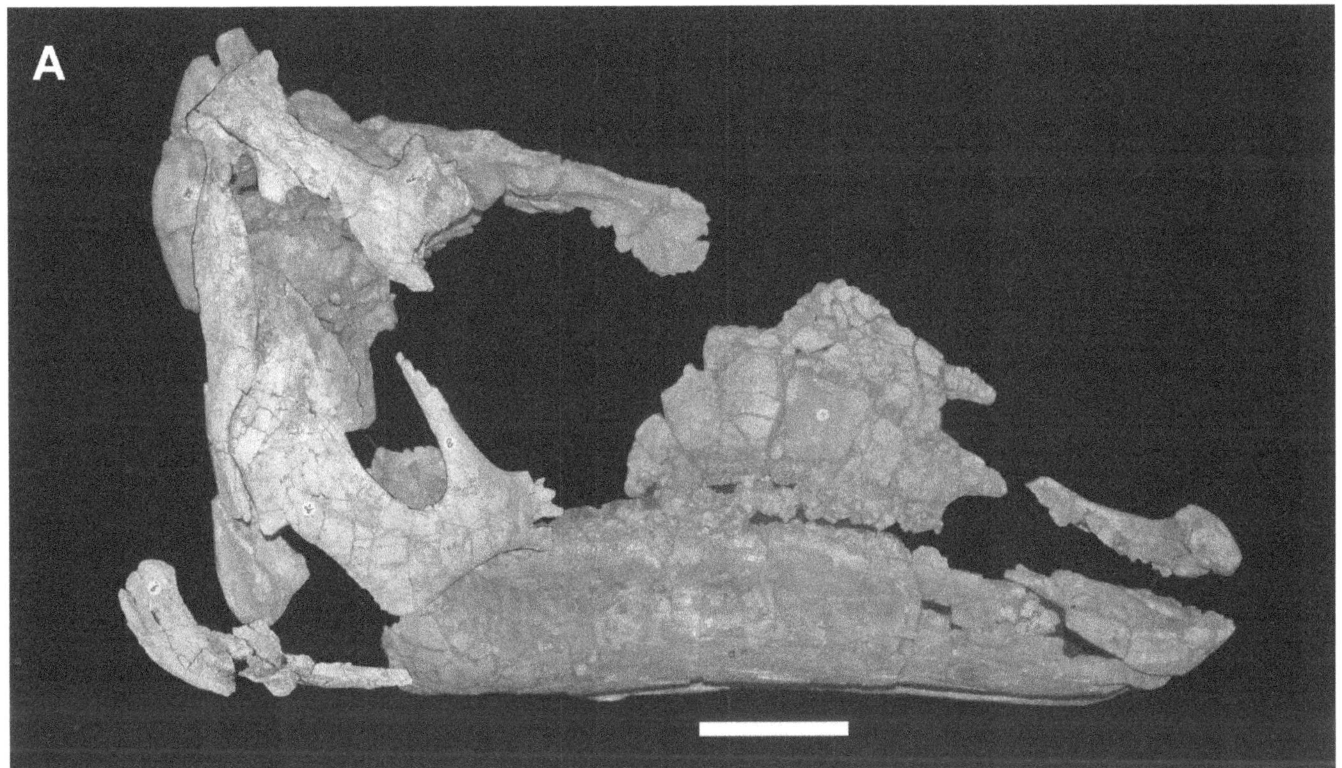

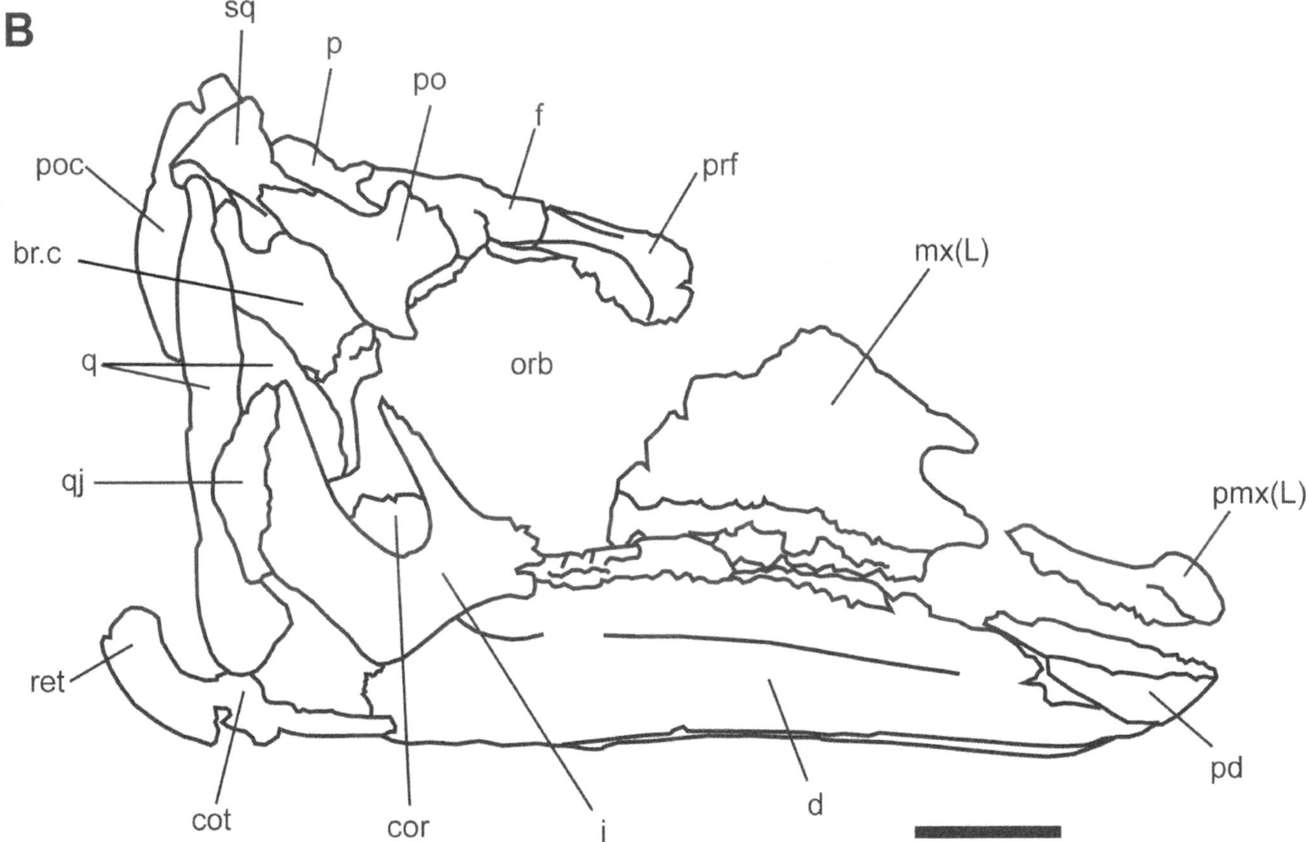

7.2. Photomosaic schematic of the reassembled skull of *Plesiohadros djadokhtaensis* (MPC-D100/745). (A) in right lateral view; (B) interpretive line drawing of schematic. Abbreviations: br.c, braincase; cor, coronoid process; cot, glenoid for the lateral quadrate condyle; d, dentary; f, frontal; j, jugal; mx(L), maxilla (left); orb, orbit; p, parietal; pd, predentary; pmx(L), premaxilla (left); po, postorbital; poc, paroccipital process; prf, prefrontal; q, quadrate; qj, quadratojugal; ret, retroarticular process; sq, squamosal. Scale bar equals 10 cm.

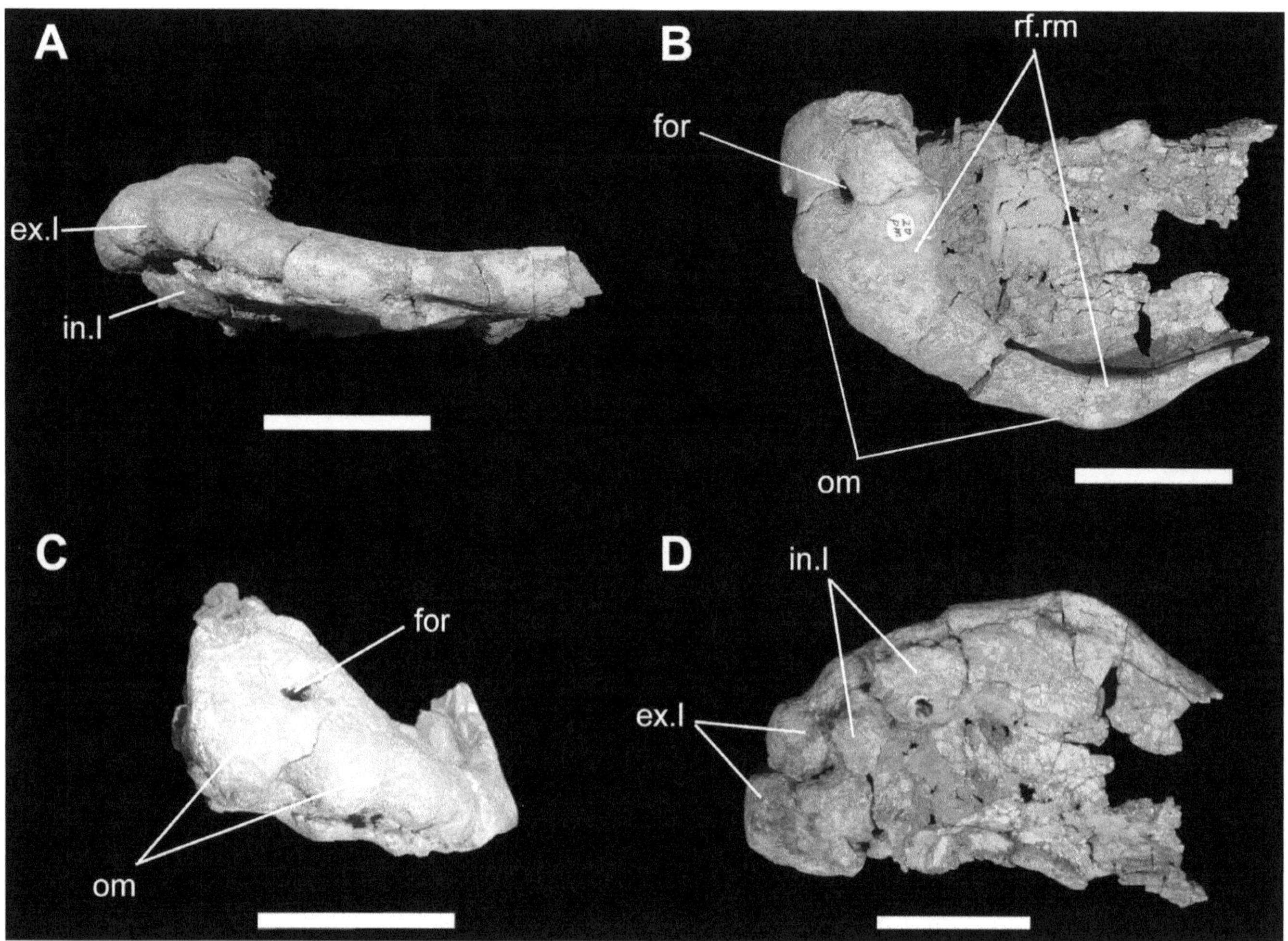

7.3. Partial left premaxilla of *Plesiohadros djadokhtaensis* (MPC-D100/745) in (A) lateral; (B) dorsal; (C) rostral; and (D) ventral views. Scale bars equal 5 cm. Abbreviations: ex.l, external lip; for, foramen; in.l, internal lip; om, oral margin; rf.rm, reflected rim.

Etymology *Plesios* ("near"); *hadros* ("hadrosaurid"); in reference to its systematic position as a proximate sister taxon to Hadrosauridae.

Diagnosis As per the type and only species.

Plesiohadros djadokhtaensis sp. nov.

Etymology From the Djadokhta Formation.

Holotype MPC-D100/745: partial skull, lower jaw, cervical vertebra, hyoid, metacarpals, and phalanges.

Referred Material MPC-D100/751: tibia, fibula, astragalus, metatarsals, phalanges.

Locality and Horizon Djadokhta Formation, ?Upper Campanian, Upper Cretaceous. MPC-D100/745 was collected in sediments exposed in the northern hills of Alag Teeg, Gobi Desert, South Gobi Aimag. MPC-D100/751 was collected in exposures of the southern main cliff of the same general locality.

Diagnosis Large non-hadrosaurid hadrosauroid characterized by the following unique combination of characters: ornamentation on lingual surface of the dentary teeth includes only one or two accessory ridges located mesial and distal to primary carina; length of diastema between first dentary tooth and predentary on the dentary extremely long, equal to approximately one-third of tooth row; caudal extent of dentary tooth row does not extend beyond the apex of the coronoid process; reflected rim around oral margin of the premaxilla; premaxillary bill constricted abruptly behind the oral margin; poorly defined quadratojugal embayment of the quadrate; acute angle between postorbital bar and jugular bar on the infratemporal fenestra; and proximal end of metacarpal III aligned with that of metacarpals II and IV. *Plesiohadros* is also characterized by one autapomorphy: prefrontal flares dorsolaterally to form a rugose, everted, wing-like rim around rostrodorsal orbital margin.

DESCRIPTION

Based on our reconstruction (Fig. 7.2), the length of the skull of *Plesiohadros djadokhtaensis* is approximately 820 mm, with

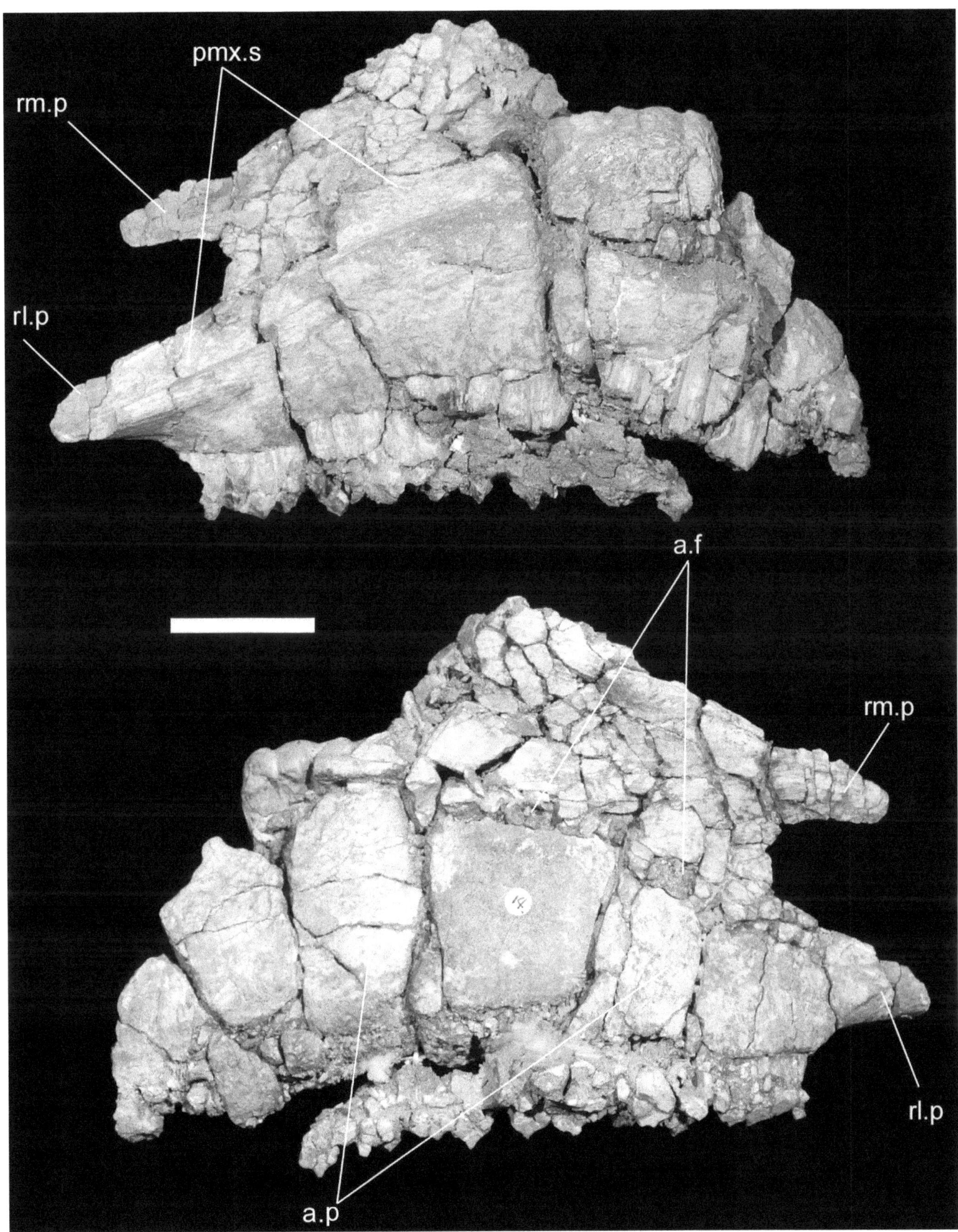

7.4. Rostral half of left maxilla of *Plesiohadros djadokhtaensis* (MPC-D100/745) in lateral (top) and medial (bottom) views. Scale bars equal 5 cm. Abbreviations: a.f, alveolar foramina; a.p, alveolar parapet; pmx.s, sutural surface for premaxilla; rm.p, rostromedial process; rl.p, rostrolateral process.

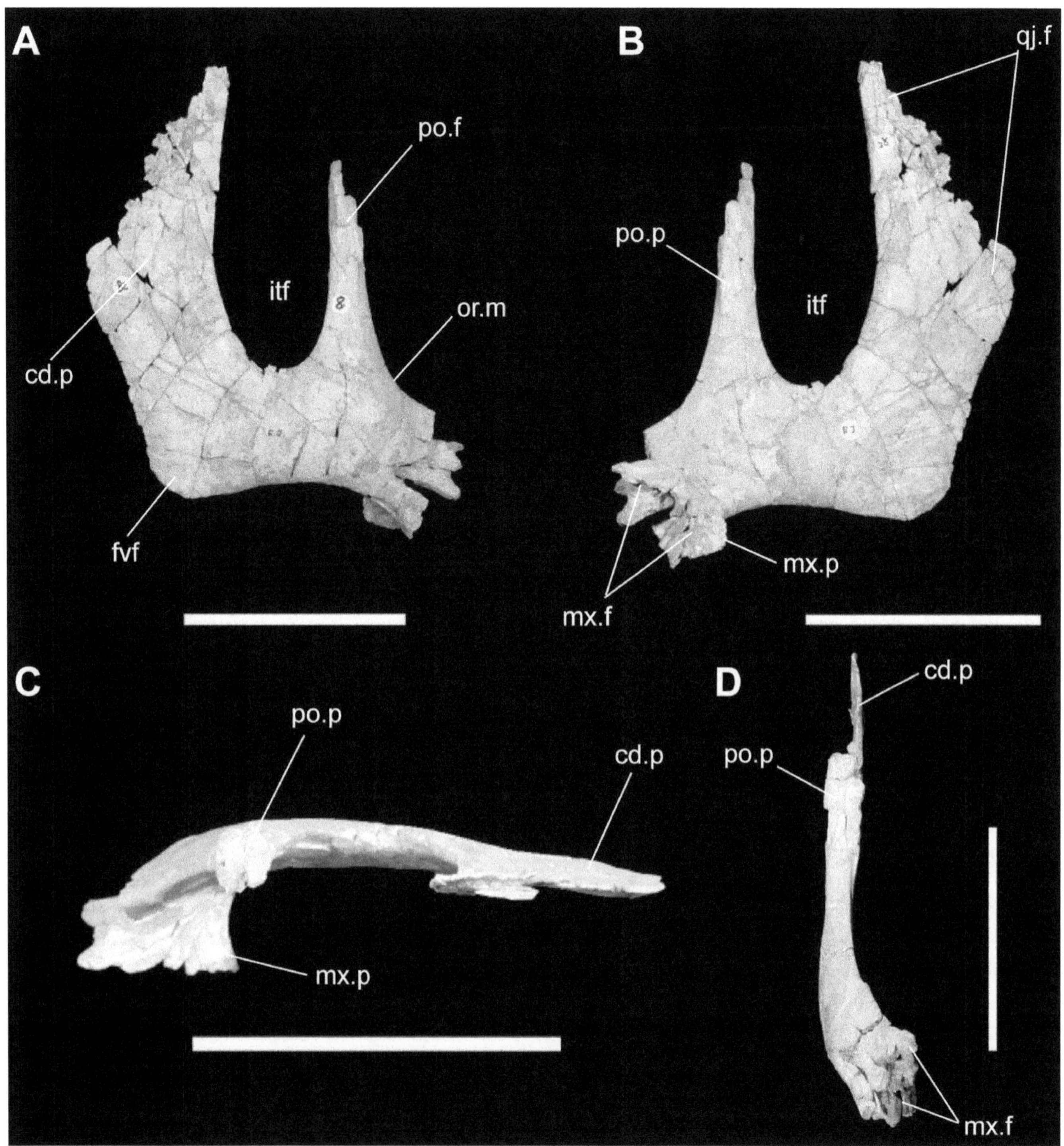

7.5. Right jugal of *Plesiohadros djadokhtaensis* (MPC-D100/745) in (A) lateral; (B) medial; (C) dorsal; and (D) rostral views. Scale bars equal 10 cm. Abbreviations: cd.p, caudal process; fvf, free ventral flange on jugal; itf, infratemporal fenestra; mx.f, maxillary facet; mx.p, maxillary process; or.m, orbital margin; po.f, postorbital facet; po.p, postorbital process; qj.f, facet for the rostral process of the quadratojugal.

a height through the quadrates of 420 mm and a width across the paroccipital processes of 180 mm. These are approximate dimensions given that the specimen is highly weathered and fractured. The region of the external naris, the antorbital fenestrae and fossa, and the orbit are not preserved. The shape of the supratemporal fenestra is unknown, but it is at least 120 mm long. The infratemporal fenestra is much shorter than tall and ovate in lateral view, judging from the shape of the postorbital process of the jugal (see below). The occiput is triangular and narrow dorsally. The ventral part of

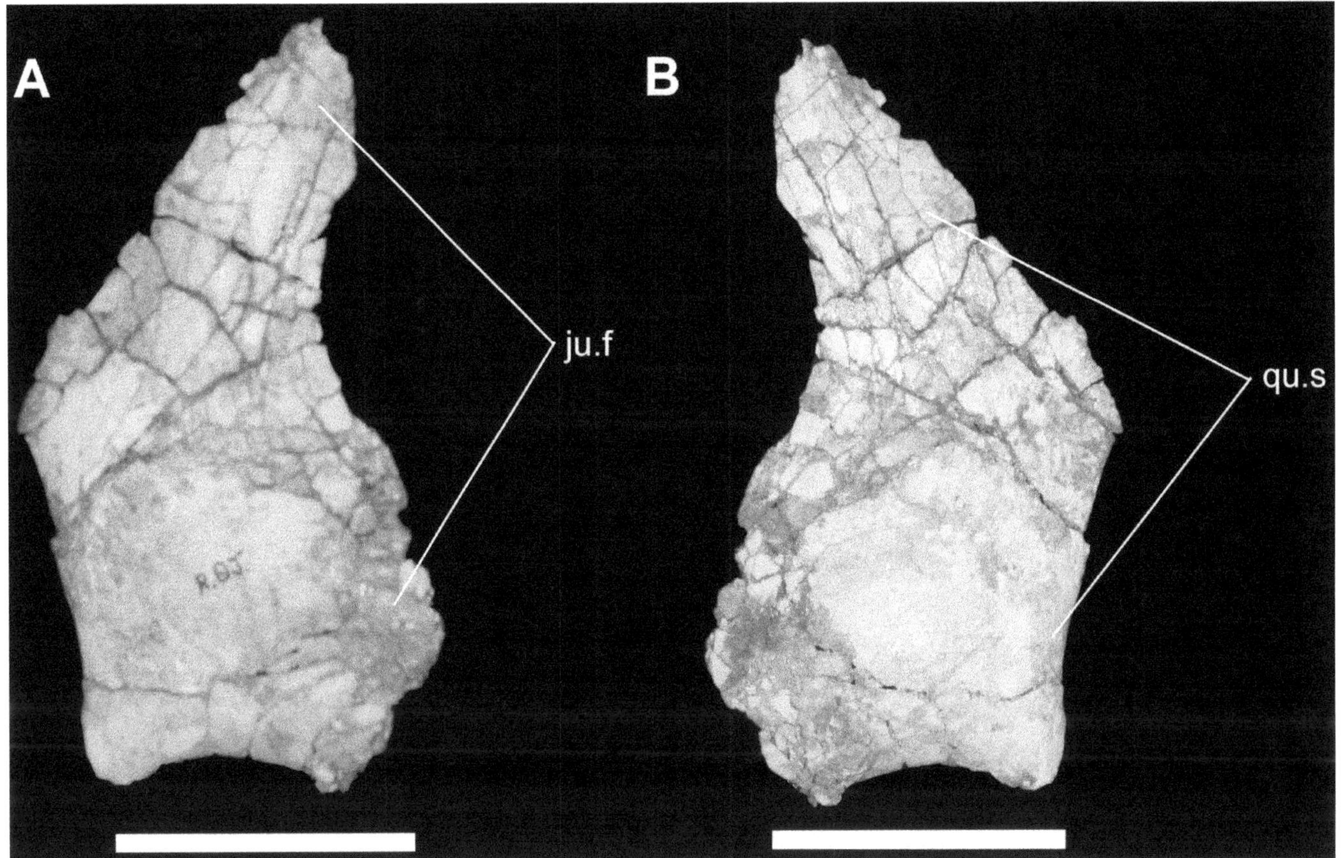

7.6. Right quadratojugal of *Plesiohadros djadokhtaensis* (MPC-D100/745) in (A) lateral; (B) medial views. Scale bars equal 5 cm. Abbreviations: ju.f, facet for the caudal process of jugal; qu.s, sutural surface for quadrate.

the quadrates splay distinctly laterally. The foramen magnum is 36 mm tall and 53 mm wide.

Although most of the skull of *Plesiohadros djadokhtaensis* is known, thus far, no nasal, palatine, ectopterygoid, palpebral, sclerotic elements, and stapes have been recovered. Most of the axial skeleton, forelimb (with the exception of the manus), and the entire pelvic girdle and femur are not preserved.

Premaxilla Only the most rostral portion of the left premaxilla is preserved (Fig. 7.3). The oral margin is reflected to form a distinct prenarial fossa. The degree of reflection is more similar to that seen in the saurolophines *Gryposaurus notabilis* (ROM 873) and *Prosaurolophus maximus* (ROM 787) than to the weakly reflected oral margin of non-hadrosaurid hadrosauroids including *Bactrosaurus johnsoni* (Godefroit et al., 1998), *Protohadros byrdi* (Head, 1998), *Jinzhousaurus yangi* (Barrett et al., 2009), and *Tethyshadros insularis* (Dalla Vecchia, 2009). In dorsal view, the rostral margin is broadly arcuate and constricted abruptly behind the oral margin in dorsal view. There is no obvious primary or accessory premaxillary foramina, as are present in saurolophines (Horner et al., 2004). The oral margin has a double-layer morphology consisting of an external denticle-bearing layer that can be seen externally and an internal palatal layer of thickened bone that is set back slightly from the oral margin and separated from the denticulate layer by a deep sulcus bearing neurovascular foramina. This morphology is present in all saurolophines and lambeosaurines, as well as *Bactrosaurus johnsoni* (Prieto-Márquez, 2011a; Campione et al., 2013) and *Ouranosaurus nigeriensis* (MNHN GDF 300). The nasal process and caudolateral process are not preserved.

Maxilla The left maxilla is highly fractured but virtually complete (Fig. 7.4). It has a roughly isosceles shape, being half as high as it is long. The rostromedial process is separated from the somewhat larger rostrolateral process by a deep embayment. Behind this region is a flat articular surface leading to the dorsal process that accommodated the caudolateral premaxillary process. The ventral margin of the jugal articulation is somewhat sigmoidal. The ectopterygoid shelf is relatively prominent, though highly fractured, and is relatively short and caudoventrally oriented. The length of the ectopterygoid shelf relative to the total length of the maxilla is similar to that in *Gilmoreosaurus mongoliensis*

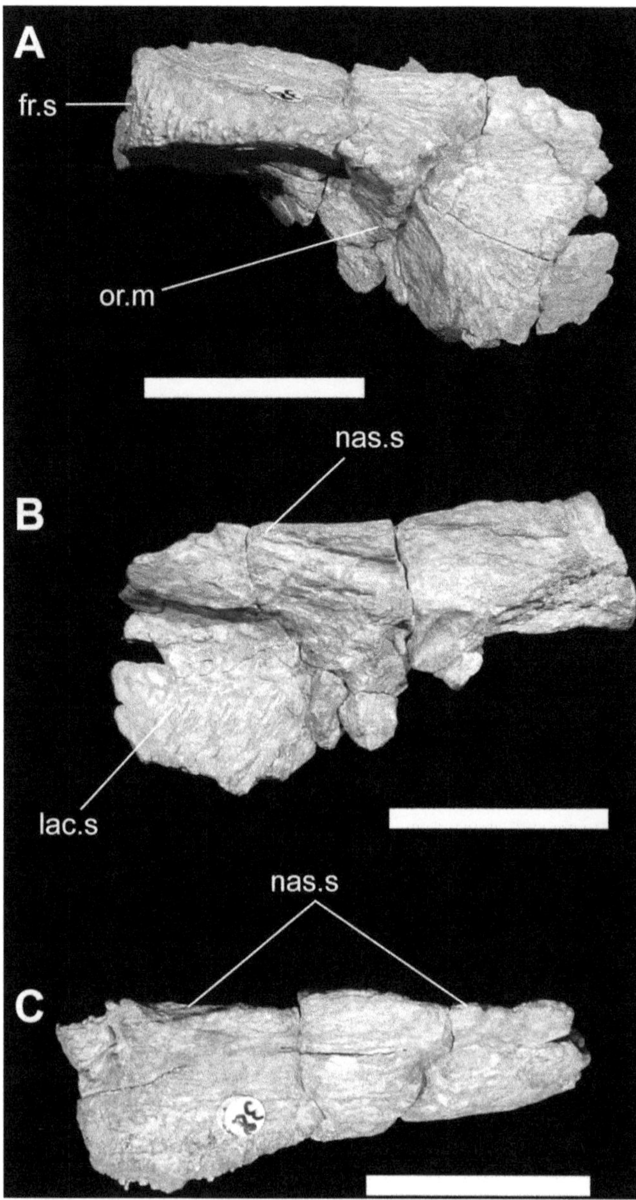

7.7. Right prefrontal of *Plesiohadros djadokhtaensis* (MPC-D100/745) in (A) lateral; (B) medial; and (C) dorsal views. Scale bars equal 5 cm. Abbreviations: fr.s, sutural surface for frontal; lac.s, sutural surface for lacrimal; nas.s, sutural surface for nasal; or.m, orbital margin.

(Prieto-Márquez and Norell, 2010), *Bactrosaurus johnsoni* (Godefroit et al., 1998), and *Levnesovia transoxiana* (Sues and Averianov, 2009), but is considerably shorter than that of hadrosaurids. An arcuate row of alveolar foramina, connected by a neurovascular groove, is on the medial maxillary surface, which otherwise is flat. There are at least 18 tooth positions (5–6 tooth positions/5 cm) arranged in a linear fashion (although this may be due to the lateral crushing of the element). This is fewer than typically present in lambeosaurines and saurolophines of equivalent tooth row length (Horner et al., 2004).

Jugal The right jugal is shallowly concave medially, and slightly higher than long as preserved (the rostral process incomplete; Fig. 7.5). Only the caudal margin of the maxillary process is preserved. Contact with the ectopterygoid in this area appears to be absent, while articulation with the palatine is enlarged. This condition characterizes saurolophines and lambeosaurines, as well as *Bactrosaurus johnsoni* (Godefroit et al., 1998), *Levnesovia transoxiana* (Sues and Averianov, 2009), and *Telmatosaurus transsylvanicus* (Weishampel et al., 1993), whereas the ectopterygoid contacts the jugal in *Protohadros* (Head, 1998), *Eolambia caroljonesa* (McDonald et al., 2012), and non-hadrosauroid ornithopods. The postorbital process extends 90 mm above the base of the orbit and gives this margin a strongly obtuse angle. The ventral flange of the jugal is pronounced beneath the infratemporal fenestra, as in saurolophines and lambeosaurines, but unlike the smoothly rounded ventral profile of the jugal in *Tethyshadros* (Dalla Vecchia, 2009), *Protohadros* (Head, 1998), *Equijubus normani* (You et al., 2003), and *Fukuisaurus* (Kobayashi and Azuma, 2003). The ventral margin of the caudal blade is shallowly concave. The caudal process is relatively narrow, as in *Tethyshadros* (Dalla Vecchia, 2009) and *Bactrosaurus johnsoni* (Godefroit et al., 1998), as well as the saurolophines *Maiasaura peeblesorum* (ROM 44770), *Brachylophosaurus canadensis* (Cuthbertson and Holmes, 2010), and *Gryposaurus notabilis* (ROM 873). The jugal contributes extensively to the margin of the infratemporal fenestra, where an acute angle marks the junction between the postorbital process and the body of the jugal, as in hadrosaurids. A shallow depression on the medial surface of the caudal jugal process marks the articulation with the quadratojugal.

Quadratojugal The thin right quadratojugal of *Plesiohadros* is roughly pentagonal in lateral view (Fig. 7.6). Externally it is marked by the facet for the jugal and internally by a long sinuous articulation, which receives the scalloped rostral margin of the quadrate.

Prefrontal The right prefrontal, where it forms the rostrodorsal margin of the orbit, is flat, triangular, and more than three times as long as wide in dorsal view (Fig. 7.7). Articulation with the lacrimal (the latter bone is not preserved) occurs within a crescentic fossa on the medial surface of the prefrontal. The orbital rim bears numerous vertical ridges that may correspond to fusion with one or more supraorbital elements. Contact with the frontal is along the complex medial and ventral surface. The most unusual aspect of the prefrontal is that the rostrodorsal margin of the orbit flares dorsolaterally to form a rugose, everted, wing-like margin, which is unusual in hadrosauroids.

Frontal The right frontal is more completely preserved than the left. As preserved, the somewhat triangular right frontal is 50% longer than wide (Fig. 7.8). In dorsal view,

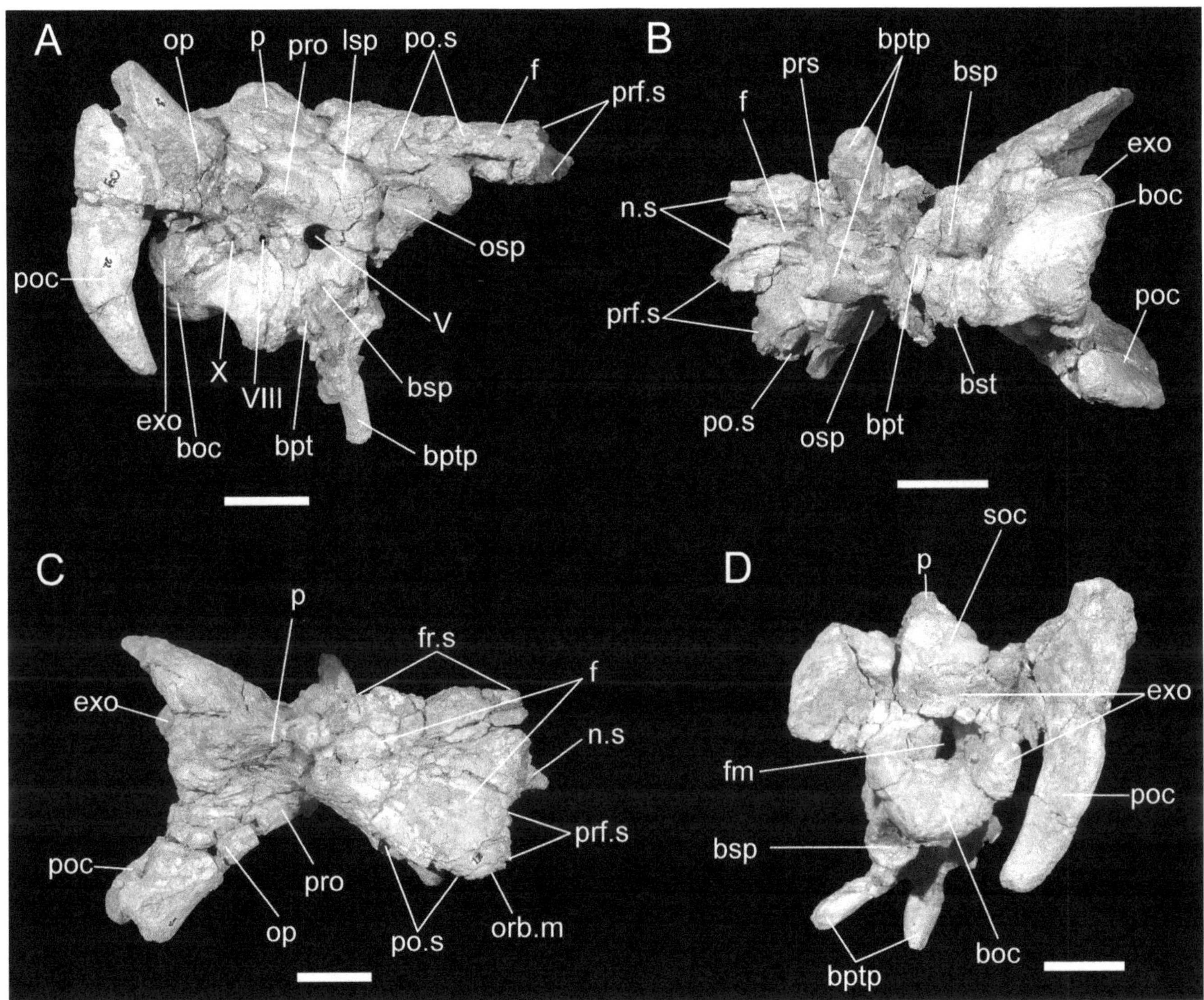

7.8. Braincase of *Plesiohadros djadokhtaensis* (MPC-D100/745) in (A) right lateral; (B) ventral; (C) dorsal; and (D) caudal views. Scale bars equal 10 cm. Abbreviations: boc, basioccipital; bpt, body of basisphenoid; bptp, basipterygoid process; bsp, basisphenoid; bst, basal tubera; exo, exoccipital; f, frontal; fm, foramen magnum; fr.s, interfrontal suture; lsp, laterosphenoid; n.s, suture for nasal; op, opisthotic; orb.m, orbital margin; osp, orbitosphenoid; p, parietal; poc, paroccipital process; po.s, suture for postorbital; prf.s, suture for prefrontal; pro, prootic; prs, parasphenoid; soc, supraoccipital; V, VIII, X, foramina for cranial nerves.

there are two facets for the nasal (medially) and prefrontal (laterally) at the rostral end of the otherwise flat frontal. Laterally, the frontal forms a small part of the orbital margin, between its prefrontal and postorbital contact surfaces. Contact with the parietal is difficult to interpret due to poor preservation, but appears to consist of a somewhat sinuous, transverse butt joint. The interfrontal suture is undulatory and more complex caudally than rostrally. The head of the laterosphenoid fits into a small, well-defined fossa that spans the undersurface of frontal-postorbital suture. The impression of the olfactory bulbs and tracts (cranial nerve [c.n.] I) are found on the ventral surface of the frontal. In overall morphology, the frontal is similar to that of *Bactrosaurus johnsoni* (Godefroit et al., 1998), *Levnesovia* (Sues and Averianov, 2009), and *Jintasaurus* (You and Li, 2009), although it is somewhat narrower relative to its total length.

Postorbital The tri-radiate postorbital has a long caudal (squamosal) process that overlaps the reciprocal process of the squamosal in a scarf joint (Fig. 7.9). The ventral (jugal) process appears short, but is incomplete distally. The caudal margin of the ventral process is nearly co-linear with the ventral margin of the caudal process. The caudal process appears to be relatively long, as in *Jintasaurus* (You and Li, 2009), *Bactrosaurus johnsoni* (Godefroit et al., 1998), *Levnesovia* (Sues and Averianov, 2009), and most hadrosaurids. The distal end of the caudal process is bifurcated at its contact with the squamosal. The stout medial process makes an extensive, highly interdigitated joint with the frontal. The orbital

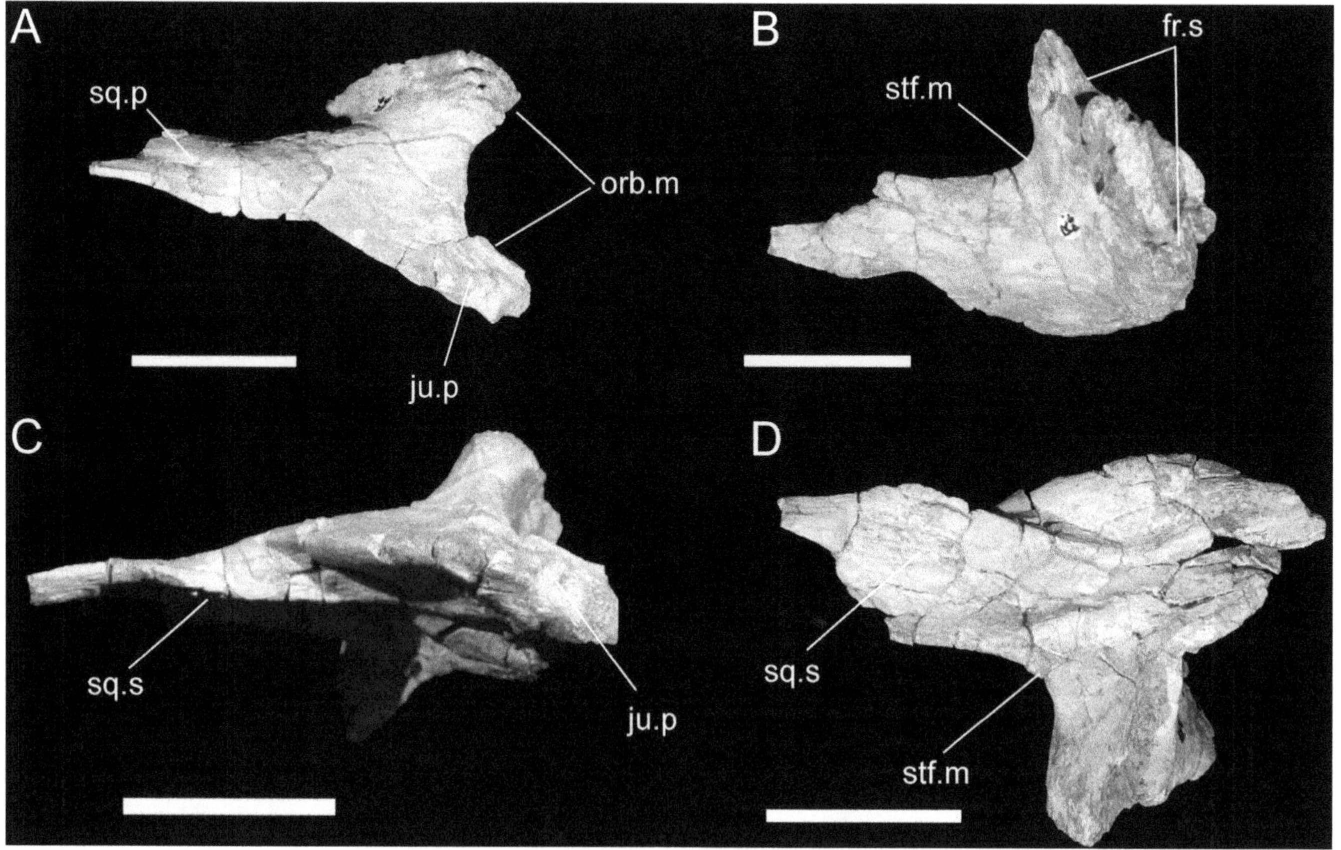

7.9. Right postorbital of *Plesiohadros djadokhtaensis* (MPC-D100/745) in (A) lateral; (B) dorsal; (C) ventral; and (D) medial views. Scale bars equal 5 cm. Abbreviations: fr.s, sutural surface for frontal; ju.p, jugal process; orb.m, orbital margin; sq.p, squamosal process; sq.s, sutural surface for squamosal; stf.m, margin of supratemporal fenestra.

margin is relatively smooth compared to that of the hadrosauroids *Jeyawati*, *Eolambia*, and *Protohadros* (McDonald et al., 2010).

Squamosal The rostral (postorbital) process of the squamosal is relatively long and straight in lateral view (Fig. 7.10). The rostral process is dorsoventrally tall and makes a unilobed contact with the caudal process of the postorbital. The medial process extends toward the midline, but it is not known how much of the caudal aspect of the parietal it concealed on the occiput. The cotylus for articulation with the head of the quadrate is well developed and located along the ventral margin of the squamosal body. It is framed by a short, robust prequadrate process and a pendant postquadrate process that contacts the paroccipital process caudally.

Quadrate The right quadrate of *P. djadokhtaensis* is tall, approximately 50% of the skull length (Fig. 7.11). The dorsal condyle of the quadrate is incomplete on its caudal aspect; therefore it is unknown whether this surface bore a strong buttress that met the postquadrate process of the squamosal. The middle of the paraquadratic notch, which is reduced to a poorly defined embayment, occurs 65% down the shaft of the quadrate. In ventral view, the ventral condyle of the quadrate narrows slightly in the middle, giving a bicondylar morphology to this joint. The lateral condyle is expanded rostrocaudally so that the condyles appear subtriangular in ventral view; furthermore the lateral condyle is rostrocaudally longer than the medial, as in *Bactrosaurus johnsoni* (Godefroit et al., 1998), *Gilmoreosaurus mongoliensis* (Prieto-Márquez and Norell, 2010), and all hadrosaurids. However, the lateral condyle lacks the distinctive enlarged, globular morphology characteristic of hadrosaurids (Horner et al., 2004). The pterygoid ala, roughly triangular in shape, extends rostromedially to where it would contact the quadrate ala of the pterygoid.

Pterygoid Only a partial right pterygoid is known for *P. djadokhtaensis* (Fig. 7.12). On the medial surface of the quadrate ramus is a prominent, nearly linear buttress whose distal limit articulated with the medial side of the body of the quadrate. The U-shaped articulation for the basipterygoid is visible at the rostral terminus of this buttress. Articulations with the ectopterygoid and maxilla are found on the rostral and ventral margins of the pterygoid.

Parietal The parietal is a single, median, hourglass-shaped element. It contacts the frontals rostrally, the postorbital rostrolaterally, the caudolateral braincase wall ventrally, and the supraoccipital caudoventrally. Caudally, the parietal narrows abruptly to form the long and prominent sagittal crest, which is only slightly down-warped along its length, unlike the strongly downwarped morphology of lambeosaurines (Horner et al., 2004). The length of the sagittal crest is more than half the length of the supratemporal fenestra.

Supraoccipital The supraoccipital is triangular in caudal view (Fig. 7.8). It occupies the midline between the overlying parietal and underlying opisthotic/exoccipital complex ventrally. The caudal surface of the supraoccipital is nearly vertical. The ventral margin of the supraoccipital is horizontal and developed as a strong ridge along its suture with the opisthotic/exoccipital complex.

Basioccipital In caudal view, the basioccipital is reniform where it forms the occipital condyle (Fig. 7.8). Here it contributes approximately one third to the lower margin of the foramen magnum. Large exoccipital condyloids articulate with the lateral aspect of the occipital condyle. There is only a short collum, approximately the length of the occipital condyle in ventral view. Rostral to the collum, the basioccipital broadly contributes to the caudal half of the basal tubera, where it complexly joins with the basisphenoid.

Basisphenoid The basisphenoid forms the rostral half of the basal tubera, making a complex transverse suture with the basioccipital (Fig. 7.8). Between the tubera is the continuation of the groove separating the basioccipital portion of the tubera, which spans the basisphenoid-basioccipital joint. The basipterygoid processes project approximately 5 cm (average of right and left) in a slightly ventrolateral direction. The basipterygoid processes appear to relatively longer, diverge less strongly, and are directed more ventrally than in *Levnesovia* (Sues and Averianov, 2009) and *Bactrosaurus johnsoni* (Godefroit et al., 1998), but these features could be accentuated by mediolateral crushing on the holotype braincase. The synovial facet for the pterygoid is on the rostrolateral terminus of the basipterygoid process. The basisphenoid contributes to the ventral margin of the trigeminal foramen. A horizontal groove marks the passage of the ophthalmic division of the trigeminal nerve (c.n. V_1), while a vertical groove marks the bony wall of the maxillary and mandibular divisions of the trigeminal nerve (c.n. $V_{2,3}$). Immediately caudal to and paralleling the channel for the maxillary and mandibular divisions of the trigeminal nerve is a groove for hyomandibular branch of facial nerve (c.n. VII). Only a small part of the parasphenoid is preserved ahead of the body of the basisphenoid.

Lateral Wall of Braincase The orbitosphenoid is fragmentary in MPC-D100/745, which provides no information except the location of what appears to be the optic canal (c.n. II). The rostral contact with the presphenoid is not preserved and the presphenoid is missing. However, its caudal contact with the laterosphenoid is visible as an undulatory joint. The head of the laterosphenoid is broad as it contacts the undersurface of the frontal and postorbital (Fig. 7.8). Otherwise, it is gently curved where it forms the lateral wall of the braincase. It contributes the rostrodorsal margin of the large trigeminal foramen, approximately where the laterosphenoid meets the basisphenoid along their horizontal contact.

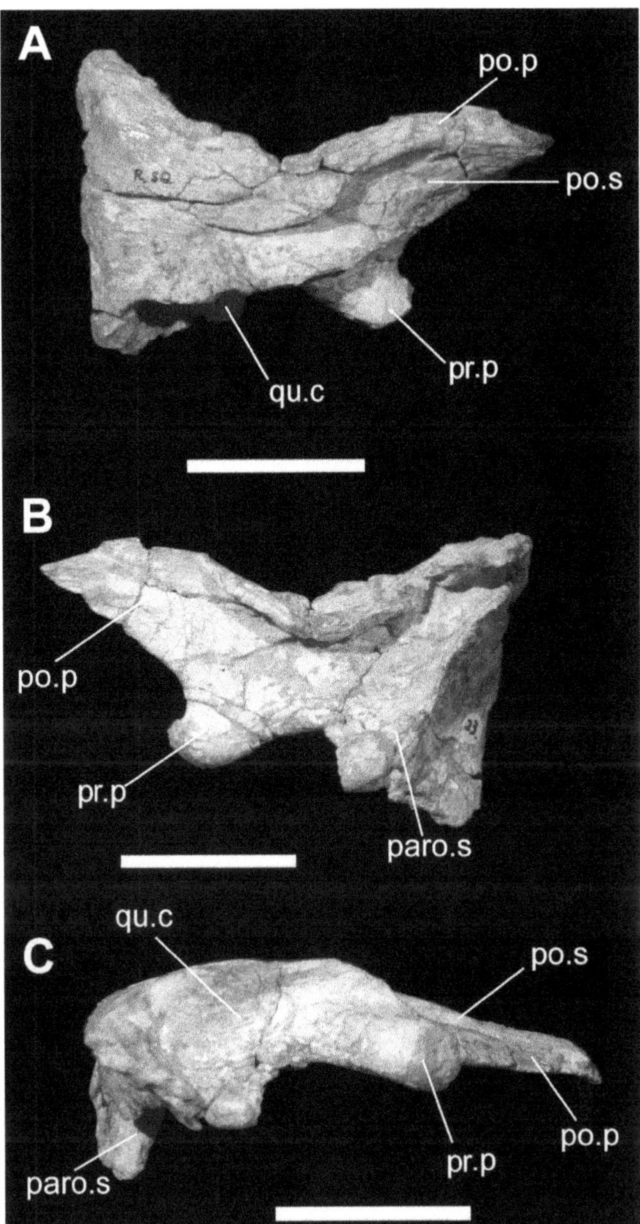

7.10. Right squamosal of *Plesiohadros djadokhtaensis* (MPC-D100/745) in (A) lateral; (B) medial; and (C) ventral views. Scale bars equal 5 cm. Abbreviations: paro.s, sutural surface for paroccipital process of exoccipital/opisthotic complex; po.p, postorbital process; po.s, sutural surface for postorbital; pr.p, prequadratic process; qu.c, quadrate cotylus.

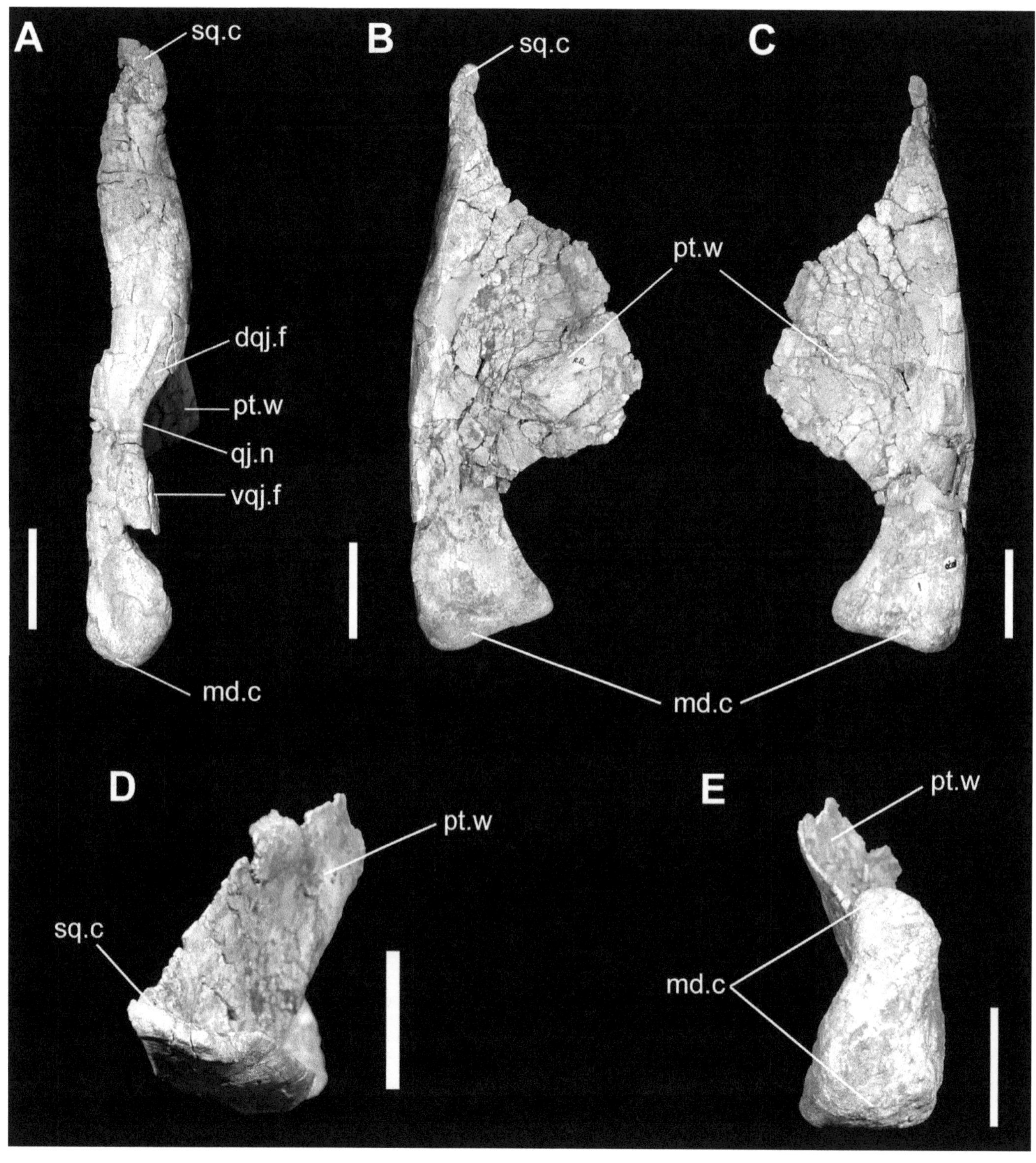

7.11. Right quadrate of *Plesiohadros djadokhtaensis* (MPC-D100/745) in (A) lateral; (B) rostral; (C) caudal; (D) dorsal; and (E) ventral views. Scale bars equal 5 cm. Abbreviations: dqj.f, dorsal quadratojugal facet; md.c, mandibular condyle; pt.w, pterygoid wing; sq.c, squamosal condyle; qj.n, quadratojugal notch; vqj.f, ventral quadratojugal facet.

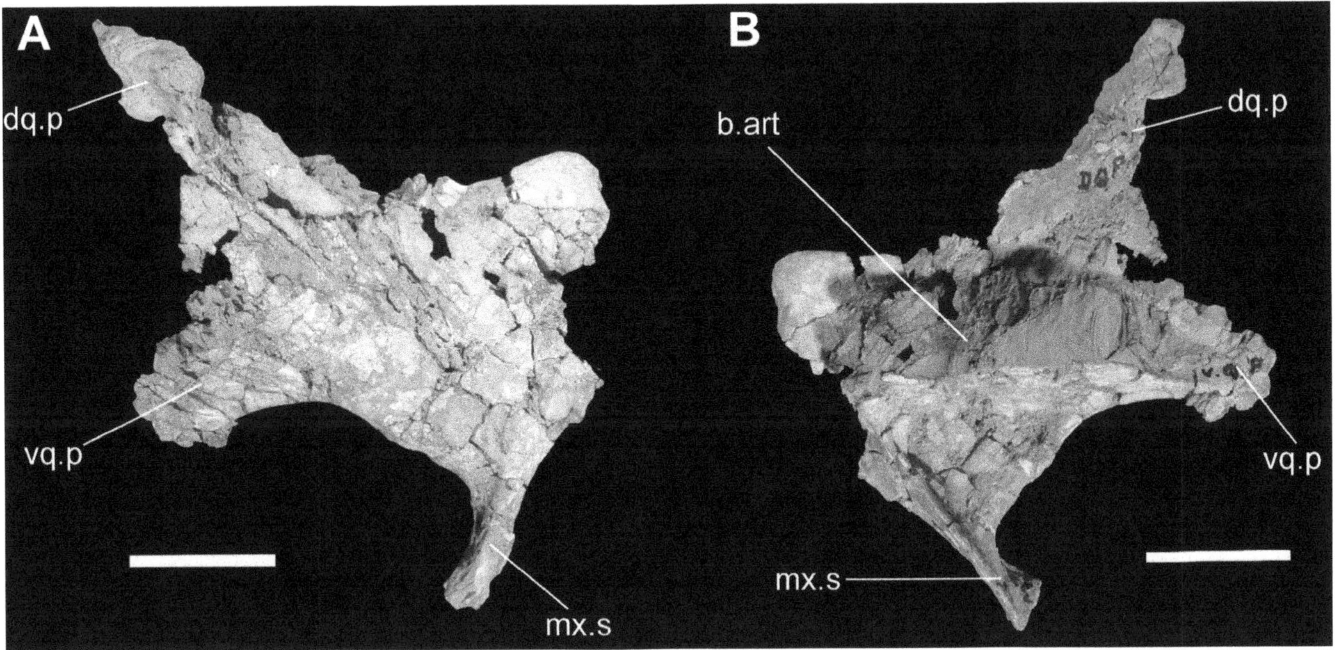

7.12. Right pterygoid of *Plesiohadros djadokhtaensis* (MPC-D100/745) in (A) lateral; and (B) caudomedial views. Scale bars equal 5 cm. Abbreviations: dq.p, dorsal quadrate process; b.art, basicranial articulation; mx.s, sutural surface for maxilla; vq.p, ventral quadrate process.

The prootic makes an oblique scarf joint over the opisthotic as it forms the middle element of the lateral side of the braincase. Rostrally, it contacts the laterosphenoid in a simple butt joint, where it also contributes the dorsocaudal margin of the large trigeminal foramen (c.n. V). Its lateral surface continues as the crista prootica from the opisthotic. At the base of the prootic, near its junction with the opisthotic, is the foramen ovale, which is overlain by the footplate of the stapes (not preserved).

The fused opisthotic and exoccipital form the paroccipital process, which is long, ventrally pendant, and slightly rostrally curved. In caudal view, the presumed exoccipital portion of the complex forms relatively large condyloids that abut either side of the occipital condyle. Foramina for the glossopharyngeal (c.n. IX), vagus (c.n. X), spinal accessory (c.n. XI), and hypoglossal (c.n. XII) nerves are obscured by crushing in this region.

Lower Jaw

Predentary In dorsal view, the gracile predentary is shovel shaped with a gently rounded rostral margin (Fig. 7.13). In ventral view, the predentary is more rectangular than in *Protohadros* (Head, 1998), *Gilmoreosaurus* (Prieto-Márquez and Norell, 2010) and *Bactrosaurus johnsoni* (Godefroit et al., 1998), although it is not as pronounced in this respect as most hadrosaurids. There are numerous denticles and neurovascular foramina across the entire rostral margin. Denticles are organized into doublets: pairs of relatively small, rounded, and triangular denticles project dorsally from a common base along the oral surface of the predentary, and would have fit into a continuous transverse slot on an underside of the premaxilla. The median denticle doublet is the largest of the series, and there are five more doublet sets on left side, where the complete series is preserved. The ventral process is robust and likely bilobed, but only the left side is preserved. There is a short median dorsal process that is triangular in dorsal view. Large lateral processes contact the external surface of the dentary rostral to the tooth row.

Dentary The left dentary is long and moderately high (Fig. 7.14). The symphyseal process turns medially to end in a linear and roughly horizontal symphysis. The diastema between the predentary articulation and the first dentary tooth is extremely long, approximately one-third the length of the tooth row. The oral margin of the dentary rostral to the tooth row is moderately down-turned, and the body of the dentary is deep below the middle of the tooth row. The coronoid process is stout, oriented approximately 100° from the long axis of the dentary. The terminus of the coronoid process is taller than wide, and has distinct rostral and caudal expansions (Fig. 7.14A). On the caudal aspect of the coronoid process, there is no facet for the coronoid bone. The most caudoventral aspect of the dentary ends behind the apex of the coronoid process; here and on the caudal aspect of the coronoid process, the dentary articulates with the surangular. The tooth row is generally poorly preserved, but the caudal

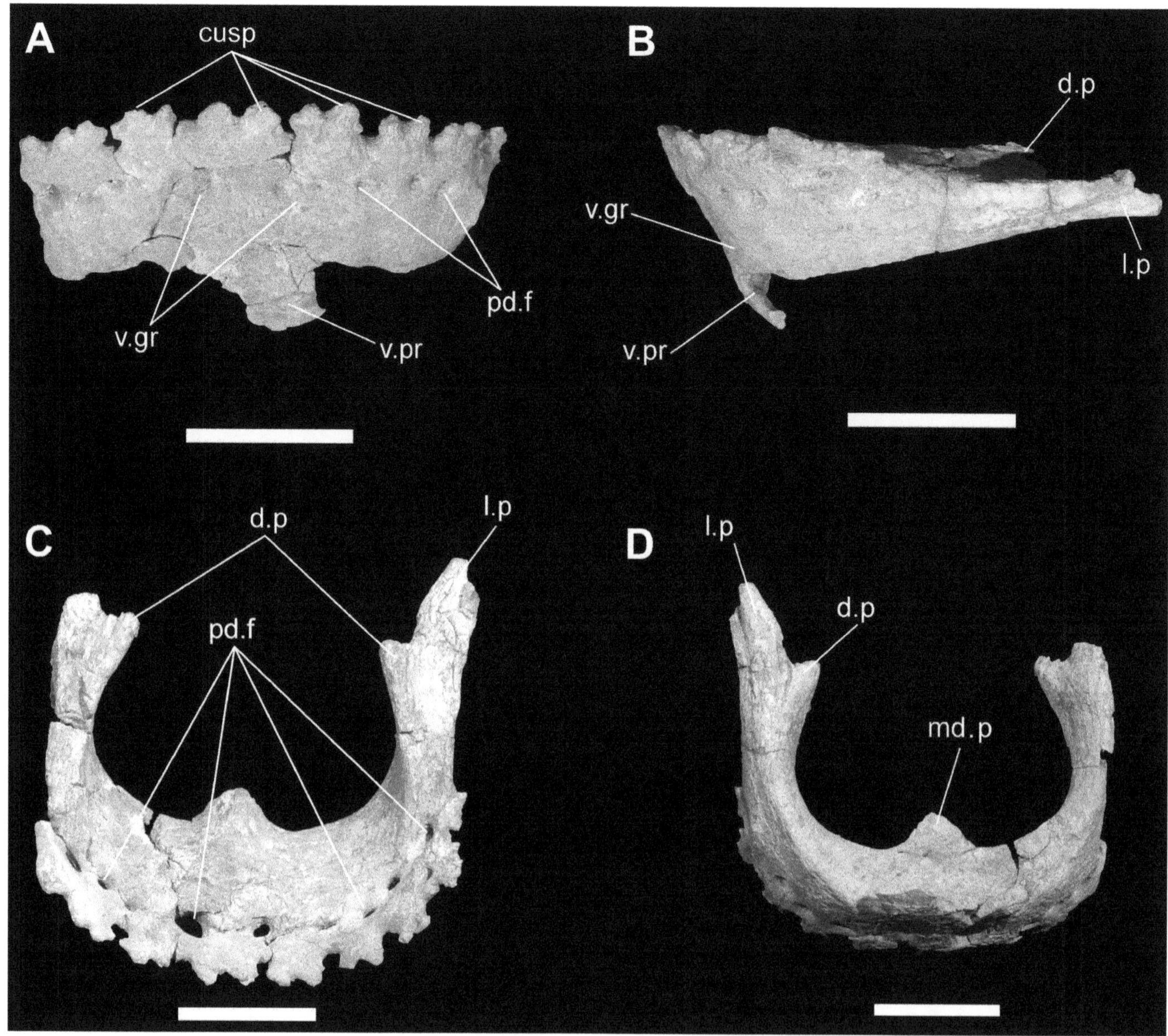

7.13. Predentary of *Plesiohadros djadokhtaensis* (MPC-D100/745) in (A) rostral; (B) left lateral; (C) dorsal; and (D) ventral views. Scale bars equal 5 cm. Abbreviations: cusp, denticulations on oral margin; d.p, dorsal process; l.p, lateral process; md.p, median dorsal process; pd.f, predentary foramina; v.gr, vascular groove; v.pr, ventral process.

end of the tooth row terminates beyond the rostral margin of the coronoid process, as in *Jeyawati rugoculus* (McDonald et al., 2010) and *Bactrosaurus johnsoni* (Godefroit et al., 1998). Although it is difficult to determine the exact number of individual tooth files, there appear to be more than 32 in the dentary.

Surangular The surangular is poorly preserved (Fig. 7.15). Only a small part of the thin rostral process that contacts the coronoid process is preserved. The robust central region of the surangular forms most of the mandibular glenoid, but is also fragmented and incomplete. It is unknown whether or not a surangular foramen on the lateral body of the bone was present. Caudally, the surangular forms the lateral portion of the stout, up-turned, and medially curved retroarticular process. The retroarticular process preserves articulation facets with the small articular distally, the splenial medially, and the angular ventrally. Given the ventrally positioned contact surface for the angular, it appears that this bone was largely obscured by the surangular when the jaw is viewed laterally, as in derived hadrosauroids, including hadrosaurids (Horner et al., 2004).

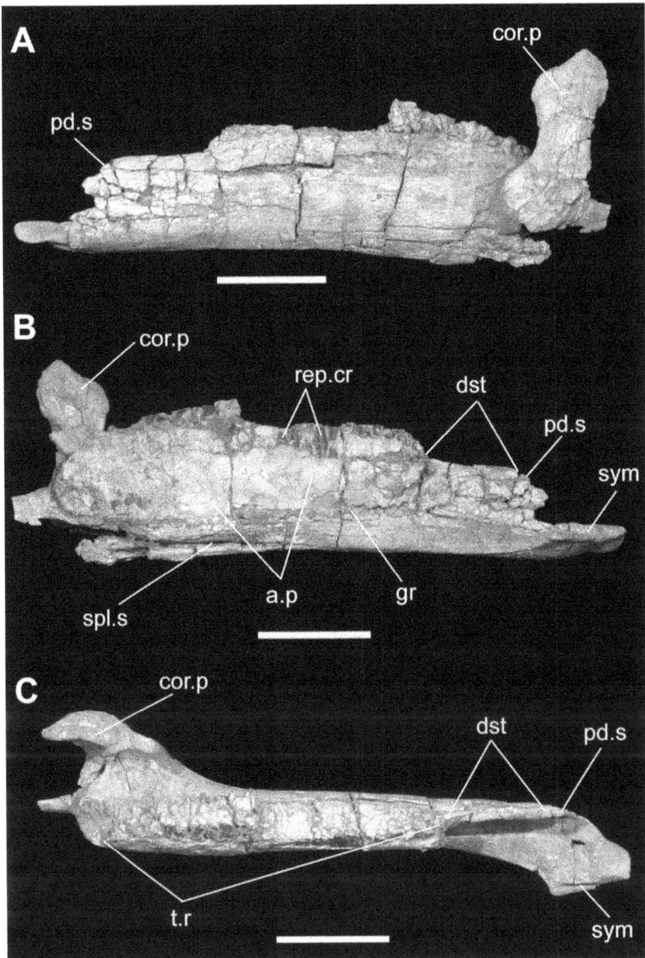

7.14. Left dentary of *Plesiohadros djadokhtaensis* (MPC-D100/745) in (A) lateral; (B) medial; and (C) dorsal views. Scale bars equal 10 cm. Abbreviations: a.p, alveolar parapet; cor.p, coronoid process; dst, diastema between first dentary tooth and predentary; gr, groove extending across alveolar foramina; pd.s, sutural surface for predentary; rep.cr, replacement dentary crowns; spl.s, sutural surface for splenial; sym, dentary symphysis; t.r, tooth row.

7.15. Right surangular of *Plesiohadros djadokhtaensis* (MPC-D100/745) in (A) lateral; (B) dorsal; and (C) medial views. Scale bars equal 5 cm. Abbreviations: an.f, angular facet; ar.f, articular facet; cot, mandibular cotyle; ret, retroarticular process; spl.f, splenial facet.

Dentition

Maxillary teeth in the holotype of *P. djadokhtaensis* are not well preserved, but they are preserved in place and provide important information on the anatomy of individual teeth (Fig. 7.16). Unfortunately, the caudal end of the tooth row is not preserved and the total number of tooth positions is therefore unknown; however, the number of tooth positions in the maxillary tooth row is at least 20, with a single functional tooth and one replacement tooth per position. Each tooth is straight and symmetrical and has a well-developed primary carina on the buccal surface. Very slight secondary ridges are present on mesial and distal to the primary ridge on some, but not all, teeth.

The dentary tooth row, consisting of more than 22 tooth positions, is straight in occlusal view (Fig. 7.17). The dentary crowns are relatively broad and slightly asymmetrical in lingual view, and the mid-dentary tooth crown in Fig. 7.17 is approximately twice as tall as it is wide. The apex of each tooth is slightly offset mesially. Ornamentation on the enamel face of the crown consists of a primary ridge, and one or two accessory ridges mesial and distal to the primary ridge. The mesial secondary ridge occurs on most crowns, and is more prominent than the distal secondary ridge, when present, on the same tooth. Small denticles are present along the slightly raised mesial and distal margins of the tooth crown. At least one tooth is functional and there are at least two replacement teeth per tooth position, but precise counts

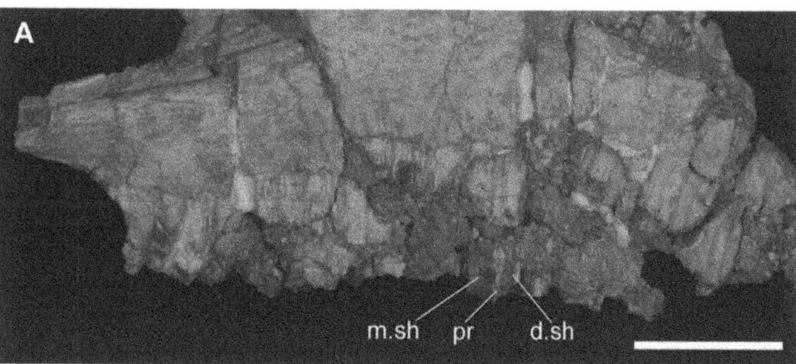

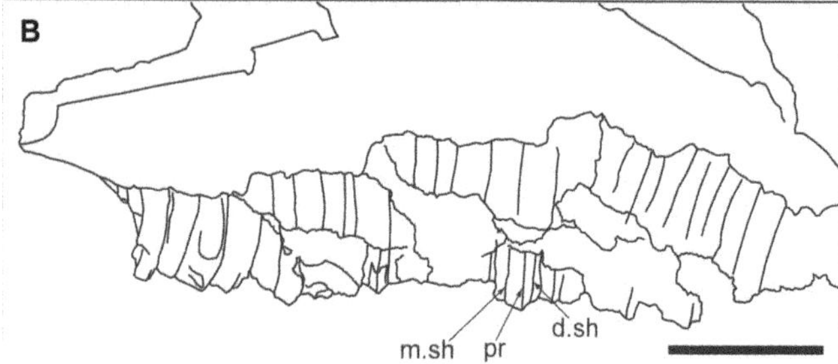

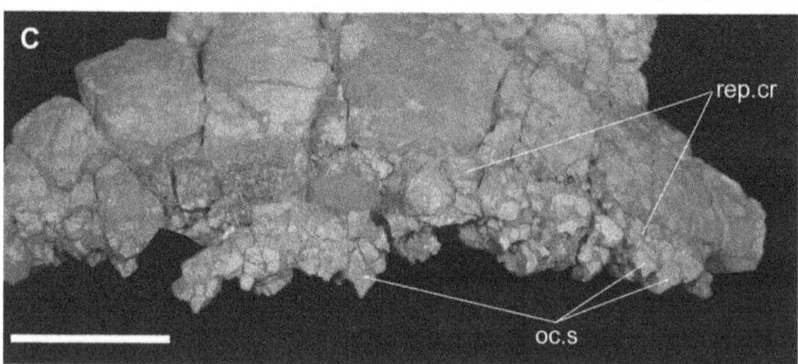

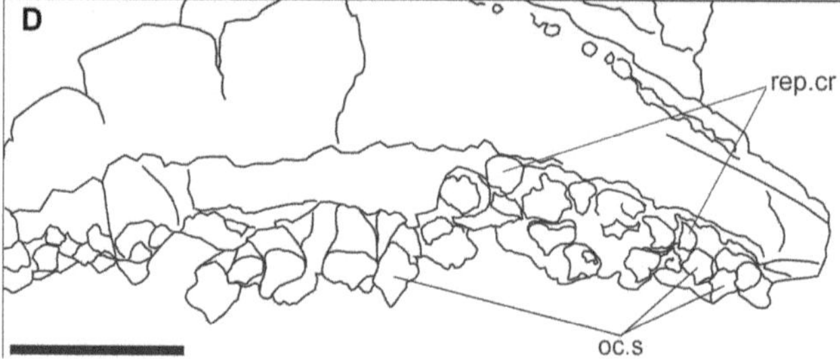

7.16. Left maxillary teeth of *Plesiohadros djadokhtaensis* (MPC-D100/745) in (A, B) lateral; and (C, D) medial views. (A, C) photographs and (B, D). interpretive line drawings. Scale bars equal 5 cm. Abbreviations: d.sh, distal shoulder; m.sh, mesial shoulder; oc.s, occlusal surfaces; pr, primary ridge; rep.cr, replacement maxillary crowns.

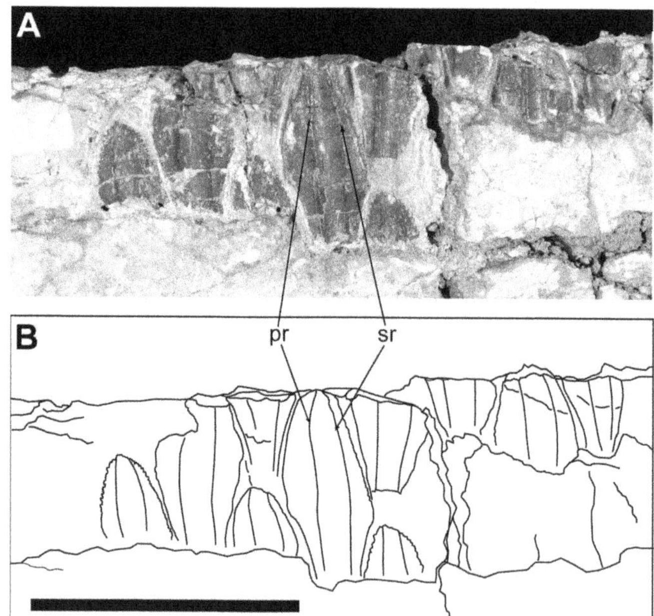

7.17. Left dentary teeth of *Plesiohadros djadokhtaensis* (MPC-D100/745) in medial view. (A) photograph and (B) interpretive line drawing. Scale bar equals 5 cm. Abbreviations: pr, primary ridge; sr, secondary ridge.

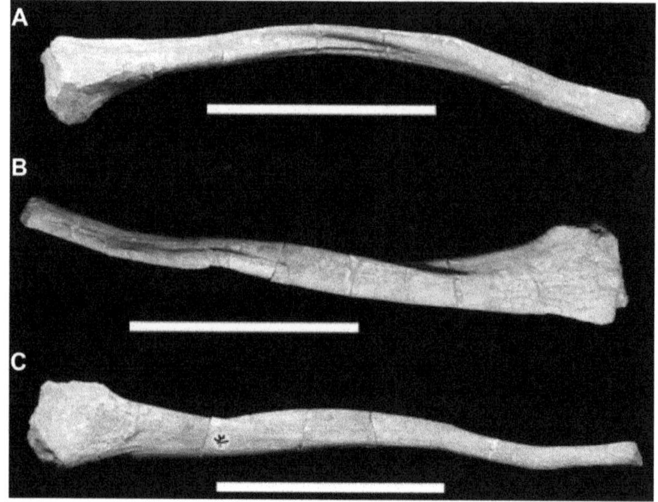

7.18. Right hyoid of *Plesiohadros djadokhtaensis* (MPC-D100/745) in (A) dorsal; (B) lateral; and (C) medial views. Scale bars equal 10 cm.

of functional and replacement teeth are difficult determine due to poor preservation. The dentary teeth closely resemble those of *Bactrosaurus johnsoni* in terms of their overall proportions and ornamentation of the enamel face (Godefroit et al., 1998).

Hyoid The long slender right hyoid is slightly curved and twisted along its length (Fig. 7.18). Rostrally blunt, it gently flares into a plate-like caudal end.

Postcranium

The atlas and the manus were found in direct association with the holotype skull, and are therefore interpreted as part of that individual. The articulated ankle and foot (MPC-D100/751) was found in that same formation at the same locality as the holotype, and it pertains to a similarly sized large hadrosauroid. MPC-D100/751 is referred to *Plesiohadros*

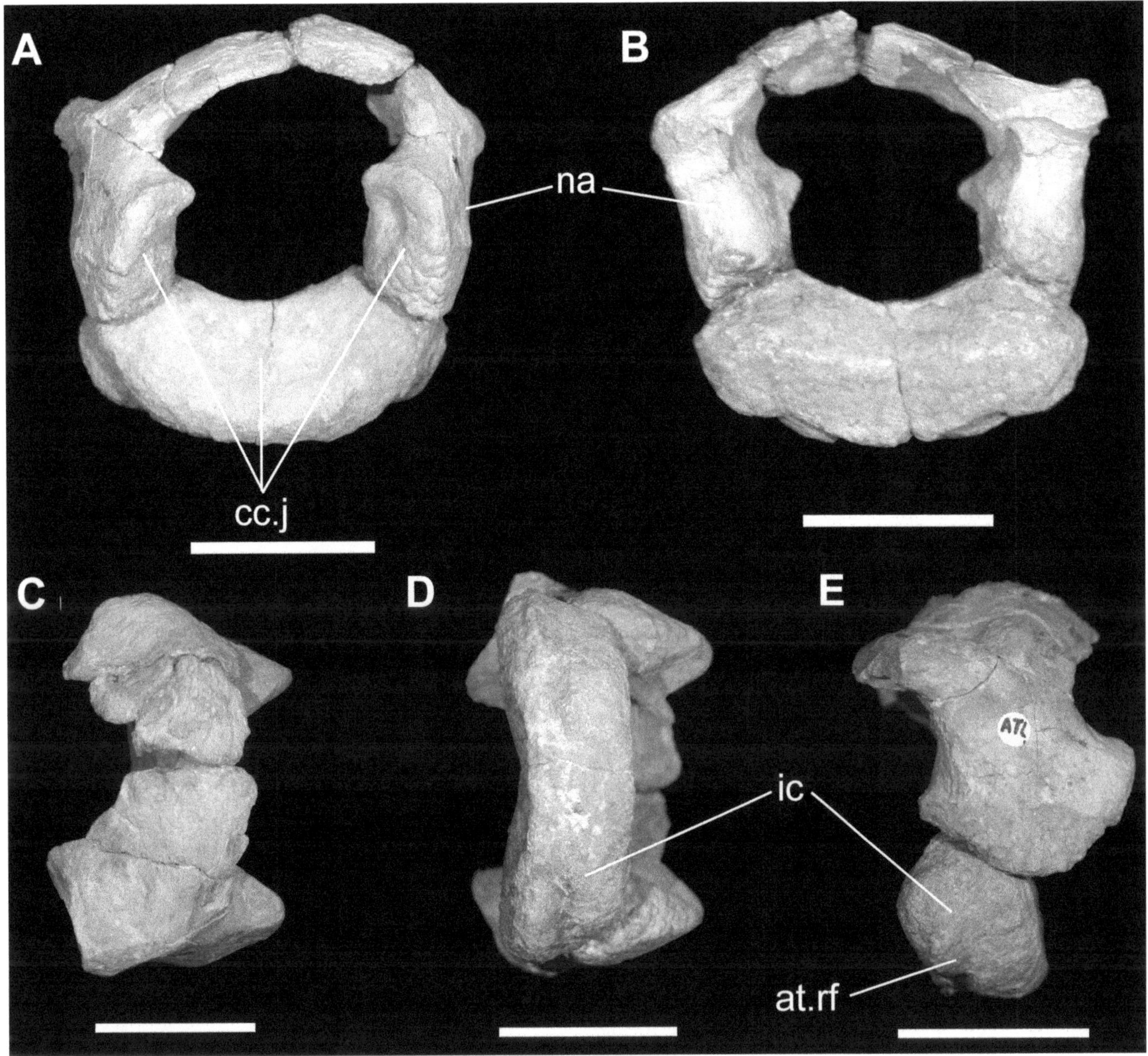

7.19. Atlas of *Plesiohadros djadokhtaensis* (MPC-D100/745) in (A) cranial; (B) caudal; (C) dorsal; (D) ventral; and (E) right lateral views. Scale bars equal 5 cm. Abbreviations: at.rf, atlantal rib facet; cc.j, craniocervical joint surface; ic, intercentrum; na, neural arch.

djadokhtaensis on the basis of stratigraphic and geographic proximity to the holotype.

Atlas The atlas is complete, consisting of paired unfused pleurocentra and a single neurocentrum (Fig. 7.19). The pleurocentra form the lateral and dorsal margins of the neural canal; they meet, but do not fuse on the dorsal midline. The internal surface of the ventrally positioned neurocentrum marks where the dens of the axis articulates with the atlas. The atlantal rib (Fig. 7.20) widens proximally to articulate with the diapophysis of the atlas.

Manus The left manus of *P. djadokhtaensis* is preserved (Fig. 7.21). Digit I, including its metacarpal, is not known. Metacarpals II–V are slender and long (midshaft width < 0.15 length) and metacarpals II, III, and IV align with each other proximally. Metacarpal V is conical and appears to bear two phalanges. The penultimate phalanx of digit III is wedge shaped, and digit IV terminates in a hoof-shaped ungual.

Tibia All that is known of the tibia of *P. djadokhtaensis* is its distal quarter from the left side (Fig. 7.22). Here the medial malleolus and ventral surface articulate with the astragalus.

7.20. Atlantal rib of *Plesiohadros djadokhtaensis* (MPC-D100/745) in (A) medial; and (B) lateral views. Scale bars equal 5 cm. Abbreviations: ct, merged capitulum and tuberculum.

The remainder of the cranial surface of the tibia forms the facet for the fibula.

Fibula Only the distal fibula is preserved. Medially, it forms an articular facet for the lateral malleolus of the tibia (Fig. 7.22). The distal end of the fibula appears to be only moderately expanded, as seen in most hadrosauroids, and does not exhibit the enlarged distal end as described for *Parasaurolophus cyrtocristatus* (Brett-Surman and Wagner, 2007).

Astragalus The astragalus was found in articulation with the distal tibia (Fig. 7.22). It is typically hadrosauroid in morphology, cupped proximally for the tibia and very broadly convex where it articulates with the metatarsus. The cranial ascending process is short and shifted to the medial side, as in hadrosaurids (Horner et al., 2004; Brett-Surman and Wagner, 2007).

Pes Three metatarsals and digits are preserved in the referred specimen of *P. djadokhtaensis* (Fig. 7.23). Metatarsal II is the most gracile; its proximal end wraps around that of metatarsal III. This distal end of metatarsal II is slightly twisted laterally and bears a facet to articulate with the first phalanx of digit I. Metatarsal III is the longest; its greatest width is at its proximal and distal ends; the latter forms an almost bicondylar condition for articulation with the first phalanx of digit III. Metatarsal IV is slim, has an axially long contact with metatarsal III, and forms a unicondylar contact with the first phalanx of digit IV. As in metatarsal II, the long axis of metatarsal IV is slightly divergent distally from that of digit III.

The pedal digital formula of *P. djadokhtaensis* is 0-3-4-4-0. The first phalanx of each digit is the largest, followed by relatively thin, wedge-shaped phalanges that terminate in a hoof-like ungual.

PHYLOGENETIC ANALYSIS

Methods

The phylogeny of hadrosauroids has undergone major revisions in the last decade, as numerous new taxa have been discovered and described. A number of large-scale, taxonomically inclusive phylogenetic analyses have also been conducted (e.g., You et al., 2003; Horner et al., 2004; Norman, 2004; Evans and Reisz, 2007; Sues and Averianov, 2009; Prieto-Márquez, 2010b; Godefroit, Bolotsky, and Lauters, 2012; Godefroit, Escuillié, et al., 2012; McDonald, 2012; Wu and Godefroit, 2012). In order to provide a preliminary hypothesis with regard to the phylogenetic position of *Plesiohadros*, the type and referred material were scored into a modified version of the data matrix published by Sues and Averianov (2009). The analysis of Prieto-Márquez (2010b) and modifications thereof were not considered because most iterations have focused on more derived hadrosauroids. The Sues and Averianov (2009) matrix was expanded to include the recently described taxa *Jinzhousaurus yangi* and *Tethyshadros insularis*, which were scored for all characters in the matrix using descriptions in the literature (Barrett et al., 2009; Dalla Vecchia, 2009; Wang et al., 2011). The majority of character scorings were left unchanged (although some scorings for specific taxa were modified, see data matrix in Appendix 7.1). All characters were considered unordered. The character matrix for the analysis includes 42 taxa and 138 characters. The data matrix was analyzed using TNT (Goloboff et al., 2008) under the default setting TBR branch-swapping algorithm with 10,000 random addition sequences. Bremer Decay and bootstrap analyses were conducted to assess the robustness of the results.

Results

The analysis resulted in 10 most-parsimonious trees (MPTs), each with a tree length of 309 steps, a Consistency Index (CI) of 0.505, and a Retention Index (RI) of 0.869. The strict consensus cladogram is shown in Figure 7.24. *Plesiohadros djadokhtaensis* is recovered as the sister taxon to the least inclusive clade containing *Lophorhothon atopus* and Hadrosauridae (sensu Sereno, 2005 and Horner et al., 2004). This position is supported by eight unambiguous synapomorphies, including: a broadly arcuate premaxilla that constricts abruptly behind oral margin (character 3: state 1); the presence of a free ventral flange on the jugal that is dorsoventrally constricted beneath infratemporal fenestra (character 60: state 1); free ventral flange is rounded or lobate (character 61: state 1); free ventral flange is small, with a ratio of depth of jugal at constriction below infratemporal fenestra to length

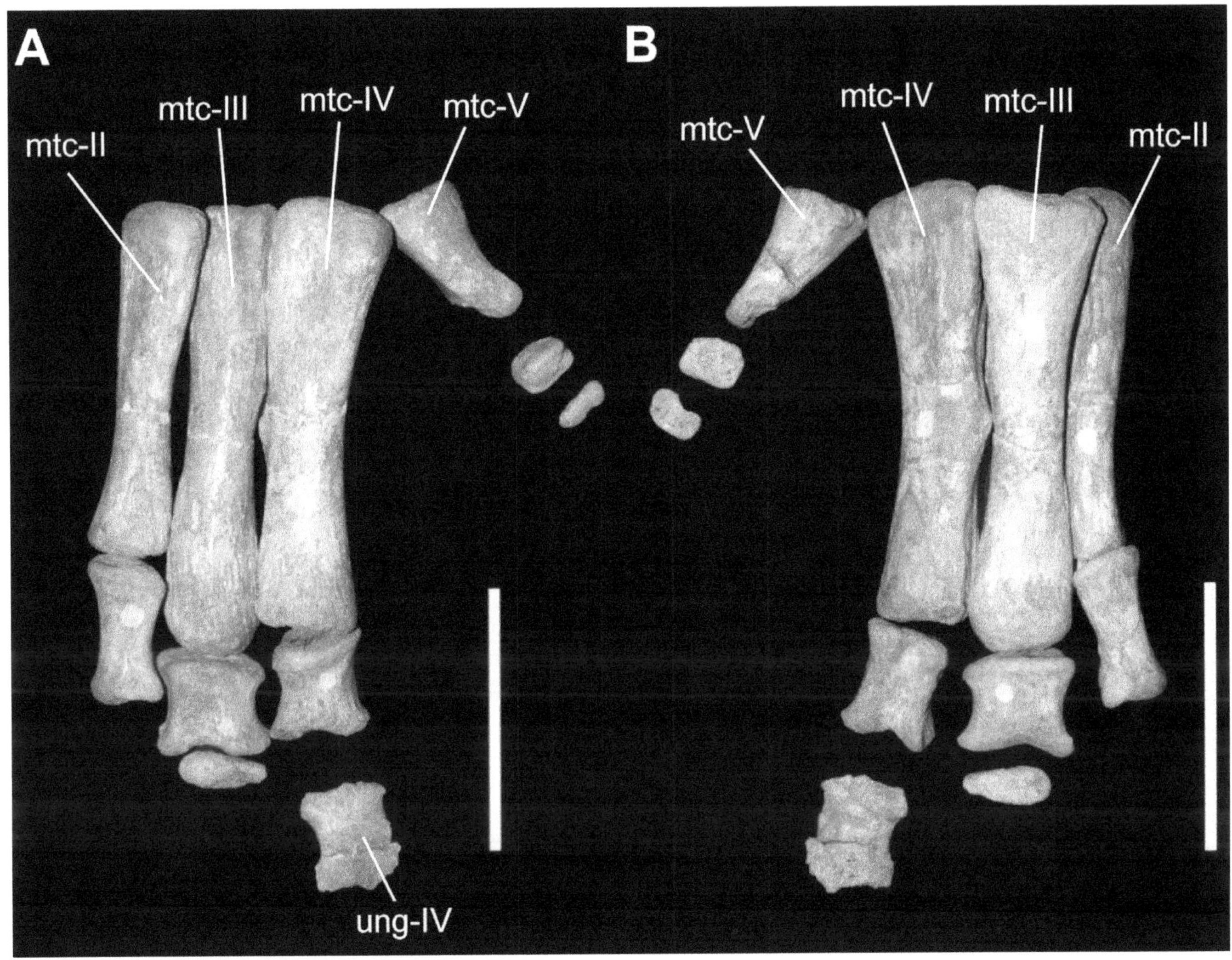

7.21. Left manus of *Plesiohadros djadokhtaensis* (MPC-D100/745) in (A) caudal; and (B) cranial views. Scale bars equal 10 cm. Abbreviations: mtc-II, metacarpal II; mtc-III, metacarpal III; mtc-IV, metacarpal IV; mtc-V, metacarpal V; ung-IV, ungual of digit IV.

of free ventral flange between 0.7 to 0.9 (character 62: state 1); angle between postorbital bar and jugular bar is acute (character 63: state 1); mandibular condyle of quadrate is subtriangular in distal view with a lateral condyle expanded anteroposteriorly relative to the medial condyle (character 67: state 1); a long diastema length of the dentary, greater than one-fifth the length of the tooth row (character 80: state 1); and the coronoid process is formed almost entirely by the dentary with the surangular reduced to thin sliver along posterior margin and does not reach distal end of coronoid process (character 85: state 1). An interesting result of the analysis is that *P. djadokhtaensis* does not belong to Hadrosauridae (sensu Horner et al., 2004), despite its probable Late Campanian age, but is a derived non-hadrosaurid hadrosauroid.

The analysis also supports the hypothesis of Weishampel and Jainu (2011) that *Bactrosaurus* and *Levnesovia* are part of the least inclusive clade containing *Telmatosaurus* and hadrosaurids (Weishampel et al., 1993). This clade (=Hadrosauridae sensu Weishampel et al., 1993; and Godefroit, Escuillié, et al., 2012) on the cladogram is supported by three unambiguous synapomorphies: the apex of the maxilla (in lateral view) is at or rostral to center (character 52: state 1); there is no contact between the ectopterygoid and the jugal (character 56: state 1); and the absence of a surangular foramen (character 89: state 1). The newly recovered clade formed by *Telmatosaurus*, *Bactrosaurus*, *Levnesovia*, and *Tanius* is also supported by three unambiguous synapomorphies: the presence of a premaxillary foramen (character 4: state 1); a lunate shape to the posterior margin of the external nares (character 15: state 0); and a tooth crown that is dominated by the presence of one primary ridge and faint secondary ridges (character 99: state 1). The recovery of this "bactrosaur" clade supports the hypothesis of Sues and Averianov (2009) of a

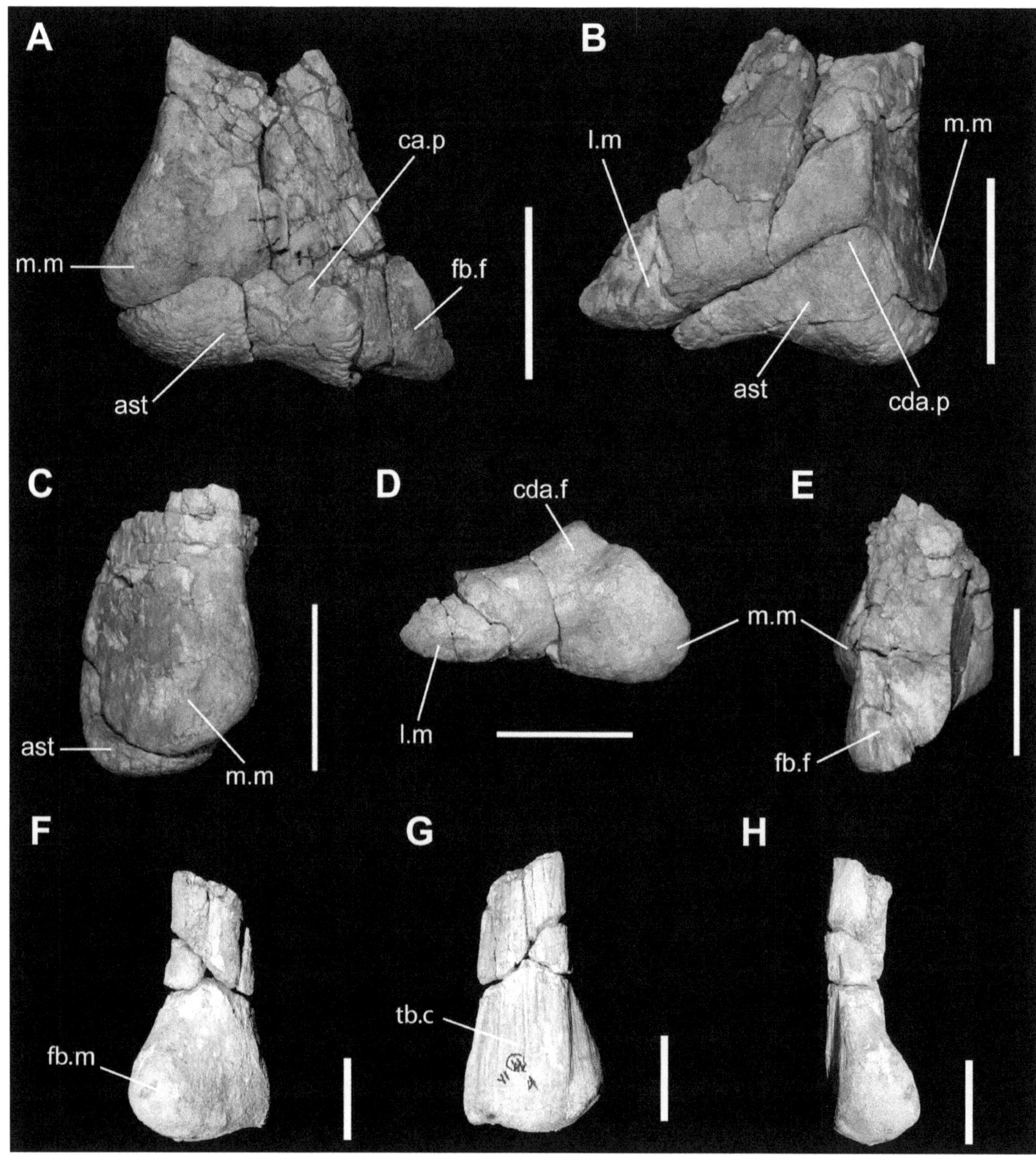

7.22. Left distal tibia and fibula of *Plesiohadros djadokhtaensis* (MPC-D100/751). Left distal tibia in (A) cranial; (B) caudal; (C) medial; (D) ventral; and (E) lateral views. Scale bars equal 10 cm. Left distal fibula in (F) lateral; (G) medial; and (H) cranial views. Scale bars equal 5 cm. Abbreviations: ast, astragalus; ca.p, cranial ascending process of the astragalus; cda.f, facet for caudal ascending process of astragalus; cda.p, caudal ascending process of the astragalus; fb.f, fibula facet; fb.m, fibula maleolus; l.m, lateral maleolus of tibia; m.m, medial maleolus of tibia; tb.c, tibia contact.

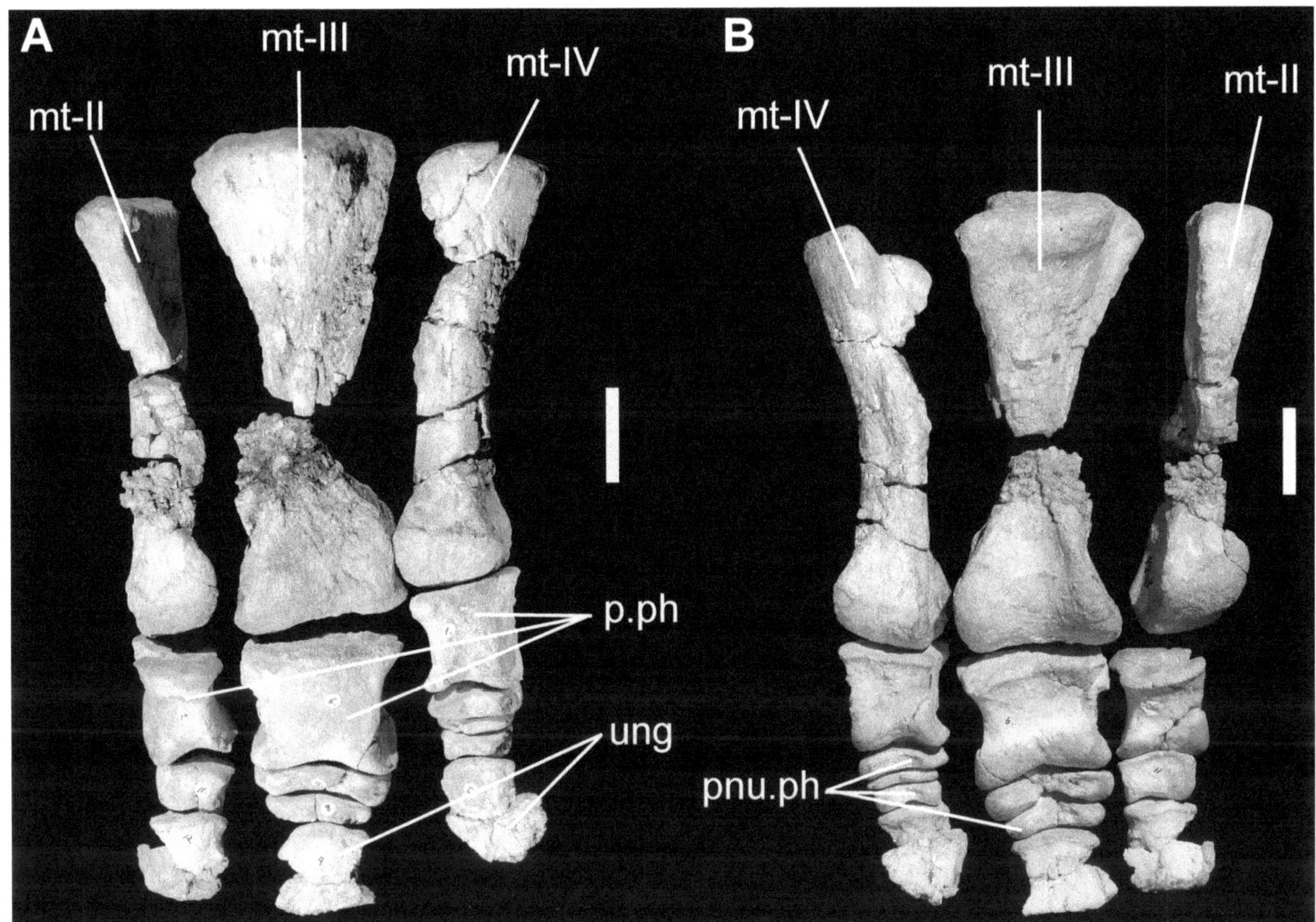

7.23. Left pes of *Plesiohadros djadokhtaensis* (MPC-D100/751) in (A) dorsal; and (B) ventral views. Scale bars equal 5 cm. Abbreviations: mt II–IV, metatarsals of the second through fourth digits; p.ph, proximal pedal phalanges; pnu.ph, penultimate pedal phalanges; ung, unguals.

Late Cretaceous radiation of derived non-hadrosaurid (sensu Horner et al., 2004) hadrosauroids in Asia.

DISCUSSION

Plesiohadros djadokhtaensis is characterized by a unique combination of plesiomorphic and derived characters relative to more basal iguanodontians and hadrosaurids, as well as an autapomorphic prefrontal that flares dorsolaterally to form a rugose, everted, wing-like rim around the rostrodorsal orbital margin, which may have even supported a small horn-like keratinous projection in life. Similar circumorbital rugosity has been described on the postorbital of the hadrosauroid *Jeyawati* from the Turonian of western North America (McDonald et al., 2010).

Plesiohadros djadokhtaensis is the first hadrosauroid identified from the fossiliferous Djadokhta Formation in Mongolia, and the first new member of this clade named from the Late Cretaceous of Mongolia in three decades. Remarkably, this is only the second report of remains referable to Hadrosauroidea from these deposits, despite extensive sampling and research (Dashzeveg et al., 2005; Dingus et al., 2008). The only previous report of hadrosauroid material from this formation was the brief note on numerous articulated infant hadrosauroid skeletons from the Tögrögiin Shiree locality by Barsbold and Perle (1983). Together, these represent the only Campanian occurrences hadrosauroids in Mongolia. In the Maastrichtian, two species of hadrosaurids are recognized: the common *Saurolophus angustirostris* and the rare *Barsboldia sicinskii*, both from the Nemegt Formation (Bell, 2011; Prieto-Márquez, 2011b). However, hadrosaurids are both taxonomically diverse and numerically abundant in the Campanian–Maastricthian of North America (Horner et al., 2004; Campione and Evans, 2011), as well as in the Amur region of Eastern Russia and adjacent China (e.g., Godefroit, Bolotsky, and Lauters, 2012; Bolotsky et al., this volume) and possibly the European archipelago (Weishampel and Jainu, 2011; Dalla Vecchia, this volume). It thus appears that

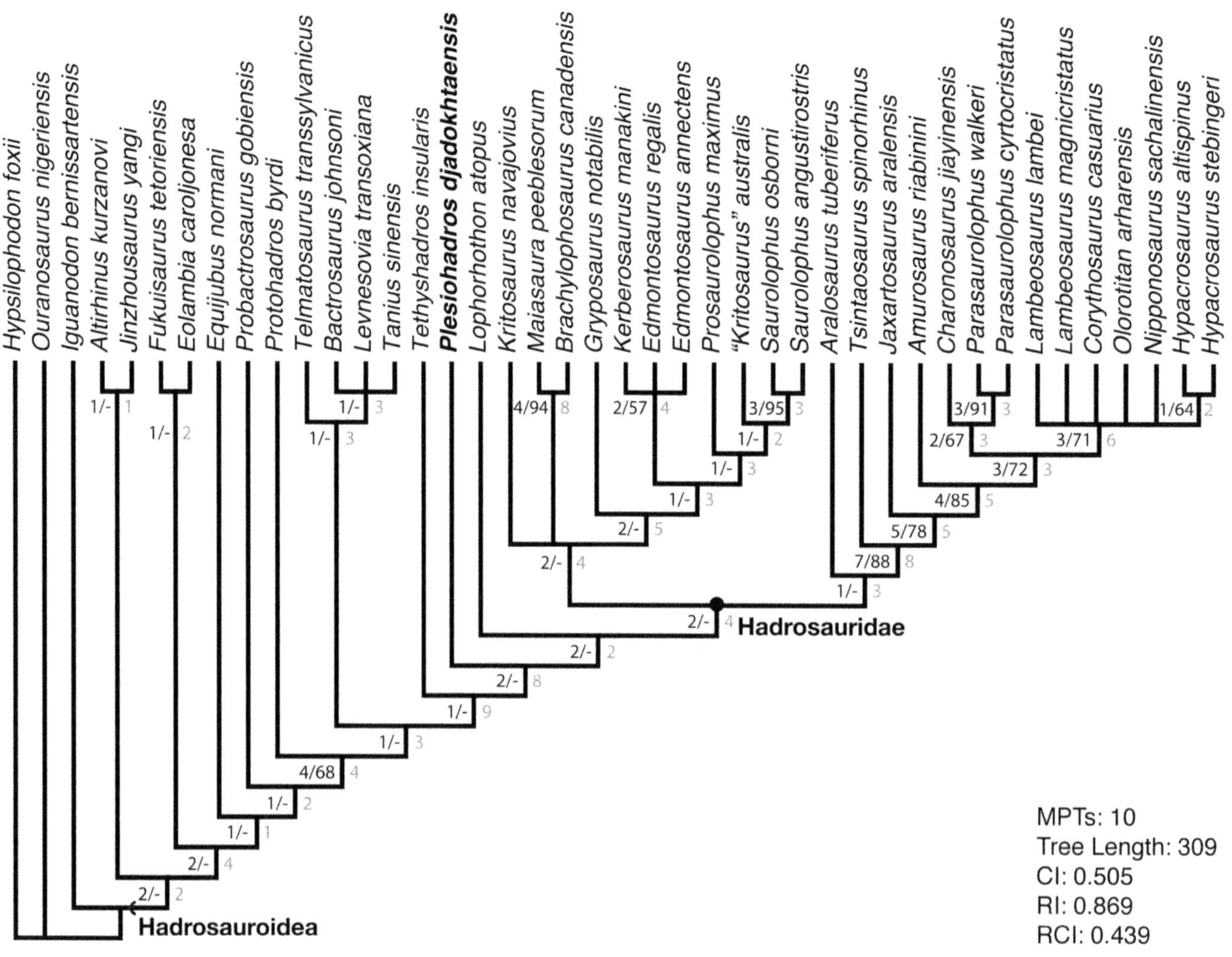

7.24. Strict consensus tree of ten most-parsimonious trees (tree length = 309). See text for explanation of phylogenetic analysis and tree statistics. Numbers in black text to the left of the nodes are Bremer Decay (left) and bootstrap (right) values, whereas numbers in gray text to the right of the nodes indicate the number of synapomorphies that support the node. Only bootstrap values in excess of 50% are reported.

hadrosauroids were less diverse and less abundant in the latest Cretaceous of Mongolia than in many other parts of Laurasia during the Campanian and Maastrichtian.

Plesiohadros is the largest member of the Djadokhta fauna. With a reconstructed skull length of approximately 820 mm, *Plesiohadros* likely attained the body size of a typical hadrosaurid, such as *Hypacrosaurus* (Evans, 2010) or *Brachylophosaurus* (Cuthbertson and Holmes, 2010), which have comparable skull lengths and reach total body lengths in excess of 10 m and estimated body masses of at least 2 tons (Paul, 1997). The Djadokhta fauna is characterized by the presence of the small-bodied ceratopsian *Protoceratops andrewsi*, as well as a number of small theropod taxa including *Oviraptor* and *Velociraptor*; apart from *Plesiohadros*, the largest members of the fauna are the ankylosaur *Pinacosaurus* and an unnamed tyrannosaur (Longrich, 2010).

The probable late Campanian age of *Plesiohadros djadokhtaensis* is significant, as this new taxon represents the youngest occurrence on a non-hadrosaurid hadrosauroid from Mongolia, after which it appears that non-hadrosaurid hadrosauroids were replaced by the immigration of true hadrosaurids into the area, including *Saurolophus* and *Barsboldia*, which likely originated after a dispersal event from western North America to Asia via Beringia (Prieto-Márquez, 2010a). At present, there are no dinosaur fossil–bearing localities in either China or Mongolia where true hadrosaurids and non-hadrosaurid hadrosauroids co-occur.

The Nemegt Formation is considered to represent a time when relatively more humid climates prevailed in the Gobi region of southern Mongolia and northern China (e.g., Jerzykiewicz and Russell, 1991). In contrast, the Djadokhta and Baruungoyot formations were deposited when more arid climatic conditions prevailed in the same area. Accordingly, the disappearance of derived non-hadrosaurid hadrosauroids in Mongolian Late Cretaceous (post–Djadokhta age) might have been a response to the shift from arid to more mesic climatic conditions. There is no fossil record of hadrosauroids from the Baruungoyot Formation; the pattern of turnover of ornithopod assemblages from hadrosauroid to hadrosaurid is not clearly shown in the Gobi desert. Likewise, the immigration of derived hadrosaurids from North America to Mongolia (Central Asia) may be related to this regional shift from more arid to more mesic climates.

As in Mongolia, there are no definitive occurrences of non-hadrosaurid hadrosauroids in the hadrosaur-dominated assemblages of the Maastrichtian of North America (Campione and Evans, 2011; Evans et al., 2011; Campione et al., 2013), suggesting that non-hadrosaurid hadrosauroids were completely replaced by hadrosaurids in Laurasia by the Maastrichtian, except where they (e.g., *Telmatosaurus transsylvanicus* and *Tethyshadros insularis*) persisted as island relicts in the European archipelago (Weishampel et al., 1993; Dalla Vecchia, 2009; Weishampel and Jainu, 2011). The other exception to this might be in the Iren Dabasu Formation in Inner Mongolia (China). The stratigraphy and sedimentology of the Iren Dabasu Formation, which hosts the derived non-hadrosaurid hadrosauroids *Gilmoreosaurus mongoliensis* and *Bactrosaurus johnsoni* (Gilmore and Granger, 1933; Prieto-Márquez and Norell, 2010; Prieto-Márquez, 2011a), has recently been restudied by Van Itterbeeck et al. (2005). Although originally considered Senonian in age (Currie and Eberth, 1993), Van Itterbeeck et al. (2005) concluded that the Iren Dabasu Formation is most probably latest Campanian–early Maastrichtian in age, based on their interpretation of the microfossil assemblage, particularly four species of charophytes and eight species of ostracods. If correct, these would be the youngest occurrence of non-hadrosaurid hadrosauroids in Asia, although Sues and Averianov (2009) argued that the data presented by Van Itterbeeck et al. (2005) suggesting a Maastrichtian age for the Iren Dabasu Formation is equivocal.

Although the age and occurrence of *Plesiohadros* suggest that hadrosaurids replaced derived hadrosauroids in Mongolia in the early Maastrichtian, they do not reveal whether Hadrosauridae originated in North America (Prieto-Márquez, 2010a) or Asia (Godefroit et al., 2008). The occurrence and phylogenetic position of *P. djadokhtaensis*, together with other hadrosauroids in the context of this phylogenetic analysis, supports the hypothesis that Eurasia played a key role in hadrosaurid origins (Russell, 1993). However, the new data provided by *P. djadokhtaensis* suggests that the dispersal event took place in the common ancestor of the enigmatic North American form *Lophorhothon atopus* and Hadrosauridae (using Fitch optimization of continental areas on the phylogeny in Fig. 7.24). This predates the origin of Hadrosauridae proper, at least under most definitions proposed in the last two decades (Godefroit et al., 1998; Horner et al., 2004; Sereno, 2005; Sues and Averianov, 2009), but because *Hadrosaurus foulkii* is not included in this analysis, it cannot be evaluated for the definition of Prieto-Márquez (2010b). Furthermore, there are hadrosaurids in North America (e.g., *Gryposaurus latidens*) and Asia (e.g., *Aralosaurus*, *Jaxartosaurus*) that are older than the late Campanian age of *Plesiohadros* (Prieto-Márquez, 2010a). However, this study does support the hypothesis that some Maastrichtian hadrosaurids of Asia likely represent immigration events from North America (e.g., Prieto-Márquez, 2010a). It is important to note that the phylogeny presented here is not complete in terms of taxon sampling, which is likely to affect the general biogeographic conclusions presented here. However, it is becoming increasingly clear that the Late Cretaceous faunal interchange between Asia and North America may be more complex than previously considered (Xu et al., 2010; Prieto-Márquez, 2010a; Evans et al., 2013), and more phylogenetic work on the Mongolian dinosaurs will help to clarify general patterns of Late Cretaceous biogeography.

ACKNOWLEDGMENTS

The authors thank Dr. R. Barsbold for his support of this project and scientific discussions. We thank Nicolás Campione for discussions and assistance with the manuscript, and Derek Larson and Patty Ralrick for editorial assistance. DCE was supported by a Discovery Grant from the Natural Sciences and Engineering Research Council of Canada. We also thank Andrew McDonald and David Eberth for reviews and editing that improved the chapter.

LITERATURE CITED

Alifanov, V. R. 2008. The tiny horned dinosaur *Gobiceratops minutus* gen. et sp. nov. (Bagaceratopidae, Neoceratopsia) from the Upper Cretaceous of Mongolia. Paleontological Journal 42:621–633.

Barrett, P. M., R. J. Butler, W. Xiao-Lin, and X. Xing. 2009. Cranial anatomy of the iguanodontoid ornithopod *Jinzhousaurus yangi* from the Lower Cretaceous Yixian Formation of China. Acta Palaeontologica Polonica 54:35–48.

Barsbold, R., and A. Perle. 1983. [On taphonomy of joint burial of juvenile dinosaurs and some aspects of their ecology]. Transactions of the Joint Soviet-Mongolian Paleontological Expedition 24:121–125. [Russian]

Bell, P. R. 2011. Cranial osteology and ontogeny of *Saurolophus angustirostris* from the Late Cretaceous of Mongolia with comment on *Saurolophus osborni* from Canada. Acta Palaeontologica Polonica 56:703–722.

Berkey, C. P., and F. K. Morris. 1927. Geology of Mongolia: Natural History of Central Asia, Volume II. American Museum of Natural History, New York, 475 pp.

Bolotsky, Y. L., P. Godefroit, I. Y. Bolotsky, and A. Atuchin. 2014. Hadrosaurs from the Far East: historical perspective and new *Amurosaurus* material from Blagoveschensk (Amur region, Russia); chapter 17 in D. A. Eberth and D. C. Evans (eds.), Hadrosaurs. Indiana University Press, Bloomington, Indiana.

Brett-Surman, M. K., and J. R. Wagner. 2007. Discussion of character analysis of the appendicular anatomy in Campanian and Maastrichtian North American hadrosaurids: variation and ontogeny; pp. 135–169 in K. Carpenter (ed.), Horns and Beaks: Ceratopsian and Ornithopod Dinosaurs. Indiana University Press, Bloomington, Indiana.

Campione, N. E., and D. C. Evans. 2011. Cranial growth and variation in edmontosaurs (Dinosauria: Hadrosauridae): implications for latest Cretaceous megaherbivore diversity in North America. PLoS ONE 6(9):e25186.

Campione, N. E., K. S. Brink, E. A. Freedman, C. T. McGarrity, and D. C. Evans. 2013. '*Glishades ericksoni*,' an indeterminate juvenile hadrosaurid from the Two Medicine Formation of Montana: implications for the diversity of non-hadrosauroids in the Late Campanian-Maastrichtian of North America. Palaeobiodiversity and Palaeoenvironments 93:65–75.

Clark, J. M., M. A. Norell, and T. Rowe. 2002. Cranial anatomy of *Citipati osmolskae* (Theropoda, Oviraptorosauria), and a reinterpretation of the holotype of *Oviraptor philoceratops*. American Museum Novitates 3364:1–24.

Currie, P. J., and D. A. Eberth. 1993. Palaeontology, sedimentology and palaeoecology of the Iren Dabasu Formation (Upper Cretaceous), Inner Mongolia, People's Republic of China. Cretaceous Research 14:127–144.

Cuthbertson, R. S., and R. B. Holmes. 2010. The first complete description of the holotype of *Brachylophosaurus canadensis* Sternberg, 1953 (Dinosauria: Hadrosauridae) with comment on intraspecific variation. Zoological Journal of the Linnean Society 159:373–397.

Dalla Vecchia, F. M. 2009. *Tethyshadros insularis*, a new hadrosauroid dinosaur (Ornithischia) from the Upper Cretaceous of Italy. Journal of Vertebrate Paleontology 29:1110–1116.

Dalla Vecchia, F. M. 2014. An overview of the latest Cretaceous hadrosauroid record in Europe; chapter 15 in D. A. Eberth and D. C. Evans (eds.), Hadrosaurs. Indiana University Press, Bloomington, Indiana.

Dashzeveg, D., L. Dingus, D. B. Loope, C. C. Swisher III, T. Dulam, and M. R. Sweeney. 2005. New stratigraphic subdivision, depositional environment, and age estimate for the Upper Cretaceous Djadokhta Formation, southern Ulan Nur Basin, Mongolia. American Museum Novitates 3498:1–31.

Dingus, L., D. B. Loope, D. Dashzeveg, C. C. Swisher III, C. Minjin, M. J. Novacek, and M. A. Norell. 2008. The geology of Ukhaa Tolgod (Djadokhta Formation, Upper Cretaceous, Nemegt Basin, Mongolia. American Museum Novitates 3616:1–40.

Eberth, D. A. 1993. Depositional environments and facies transitions of dinosaur-bearing Upper Cretaceous redbeds at Bayan Mandahu (Inner Mongolia, People's Republic of China). Canadian Journal of Earth Sciences 30:2196–2213.

Eberth, D. A., Y. Kobayashi, Y.-N. Lee, O. Mateus, F. Therrien, D. K. Zelenitsky, and M. A. Norell. 2009. Assignment of *Yamaceratops dorngobiensis* and associated redbeds at Shine Us Khudag (eastern Gobi, Dorngobi Province, Mongolia) to the restricted Javkhlant Formation (Upper Cretaceous). Journal of Vertebrate Paleontology 29:295–302.

Evans, D. C. 2010. Cranial anatomy and systematics of *Hypacrosaurus altispinus*, and a comparative analysis of skull growth in lambeosaurine hadrosaurids (Dinosauria: Ornithischia). Zoological Journal of the Linnean Society 159:398–434.

Evans, D. C., and R. R. Reisz. 2007. Anatomy and relationships of *Lambeosaurus magnicristatus*, a crested hadrosaurid dinosaur (Ornithischia) from the Dinosaur Park Formation, Alberta. Journal of Vertebrate Paleontology 27:373–393.

Evans, D. C., P. M. Barrett, and K. L. Seymour. 2011. Revised identification of a reported *Iguanodon*-grade ornithopod tooth from the Scollard Formation, Alberta, Canada. Cretaceous Research 33:11–14.

Evans, D. C., D. W. Larson, and P. J. Currie. 2013. A new dromaeosaurid (Dinosauria: Theropoda) with Asian affinities from the Latest Cretaceous of North America. Naturwissenschaften 100(11):1041–1049.

Fastovsky, D. E., D. Badamgarav, H. Ishimoto, M. Watabe, and D. B. Weishampel. 1997. The paleoenvironments of Tugrikin-Shireh (Gobi Desert, Mongolia) and aspects of the taphonomy and paleoecology of *Protoceratops* (Dinosauria: Ornithishichia). Palaios 12:59–70.

Fastovsky, D., D. Weishampel, M. Watabe, R. Barsbold, K. Tsogtbaatar, and P. Narmandakh. 2011. A nest of *Protoceratops andrewsi* (Dinosauria, Ornithischia). Journal of Paleontology 85:1035–1041.

Gilmore, C. W., and W. Granger. 1933. Two new dinosaurian reptiles from Mongolia: with notes on some fragmentary specimens. American Museum Novitates 679:1–20.

Godefroit, P., Y. L. Bolotsky, and P. Lauters. 2012. A New Saurolophine Dinosaur from the Latest Cretaceous of Far Eastern Russia. PLoS ONE 7(5):e36849.

Godefroit, P., F. Escuillié, Y. L. Bolotsky, and P. Lauters. 2012. A new basal hadrosauroid dinosaur from the Upper Cretaceous of Kazakhstan; pp. 335–358 in P. Godefroit (ed.), Bernissart Dinosaurs and Early Cretaceous Terrestrial Ecosystems. Indiana University Press, Bloomington, Indiana.

Godefroit, P., S. Hai, T. Yu, and P. Lauters. 2008. New hadrosaurid dinosaurs from the uppermost Cretaceous of northeastern China. Acta Palaeontologica Polonica 53:47–74.

Godefroit, P., Z.-M. Dong, P. Bultynck, H. Li, and L. Feng. 1998. Sino-Belgian Cooperation Program "Cretaceous dinosaurs and mammals from Inner Mongolia" 1. New Bactrosaurus (Dinosauria: Hadrosauridae) material from Iren Dabasu (Inner Mongolia, P. R. China). Bulletin de l'Institut royal des Sciences naturelles de Belgique 68(Supplement):3–70.

Goloboff, P. A., J. S. Farris, and K. C. Nixon. 2008. TNT, a free program for phylogenetic analysis. Cladistics 24:774–786.

Head, J. J. 1998. A new species of basal hadrosaurid (Dinosauria, Ornithischia) from the Cenomanian of Texas. Journal of Vertebrate Paleontology 18:718–738.

Horner, J. R., D. B. Weishampel, and C. A. Forster. 2004. Hadrosauridae; pp. 438–463 in D. B. Weishampel, P. Dodson, and H. Osmólska (eds.), The Dinosauria, Second Edition. University of California Press, Berkeley, California.

Jerzykiewicz, T., and D. A. Russell. 1991. Late Mesozoic stratigraphy and vertebrates of the Gobi Basin. Cretaceous Research 12:345–377.

Jerzykiewicz, T., P. J. Currie, D. A. Eberth, P. A. Johnson, E. H. Koster, and J.-J. Zheng. 1993. Djadokhta Formation correlative strata in Chinese Inner Mongolia: an overview of the stratigraphy, sedimentary geology, and paleontology and comparisons with the type locality in the pre-Altai Gobi. Canadian Journal of Earth Sciences 30:2180–2195.

Kielan-Jaworowska, Z. 1974. Multituberculate succession in the Late Cretaceous of the Gobi Desert (Mongolia). Palaeontologia Polonica 30:23–44.

Kobayashi, Y., and Y. Azuma. 2003. A new iguanodontian (Dinosauria: Ornithopoda) from the Lower Cretaceous Kitadani Formation of Fukui Prefecture, Japan. Journal of Vertebrate Paleontology 23:166–175.

Longrich, N. R. 2010. The function of large eyes in *Protoceratops:* a nocturnal ceratopsian? pp. 308–327 in M. J. Ryan, B. J. Chinnery-Allgeier, and D. A. Eberth (eds.), New Perspectives on Horned Dinosaurs. Indiana University Press, Bloomington, Indiana.

Loope, D. B., L. Dingus, C. C. Swisher III, and C. Minjin. 1998. Life and death in a late Cretaceous dune field, Nemegt basin, Mongolia. Geology 26:27–30.

Makovicky, P. J., and M. A. Norell. 2006. *Yamaceratops dorngobiensis,* a new primitive ceratopsian (Dinosauria: Ornithischia) from the Cretaceous of Mongolia. American Museum Novitates 3530:1–42.

Makovicky, P. J., B. M. Kilbourne, R. W. Sadleir, and M. A. Norell. 2011. A new basal ornithopod (Dinosauria, Ornithischia) from the Late Cretaceous of Mongolia. Journal of Vertebrate Paleontology 31:626–640.

Marsh, O. C. 1881. Classification of the Dinosauria. American Journal of Science 7:81–86.

McDonald, A. T. 2012. Phylogeny of basal iguanodonts (Dinosauria: Ornithischia): an update. PLoS ONE 7(5):e36745.

McDonald, A. T., D. G. Wolfe, and J. I. Kirkland. 2010. A new basal hadrosauroid (Dinosauria: Ornithopoda) from the Turonian of New Mexico. Journal of Vertebrate Paleontology 30:799–812.

McDonald, A.T., J. Bird, J. I. Kirkland, and P. Dodson. 2012. Osteology of the basal hadrosauroid *Eolambia caroljonesa* (Dinosauria: Ornithopoda) from the Cedar Mountain Formation of Utah. PLoS ONE 7(10):e45712.

Mikhailov, K. E. 1991. Classification of fossil eggshells of amniotic vertebrates. Acta Palaeontologica Polonica 36:193–238.

Mikhailov, K. E., K. Sabath, and S. Kurzanov. 1994. Eggs and nests from the Cretaceous of Mongolia; pp. 88–115 in K. Carpenter, K. F. Hirsch, and J. R. Horner (eds.), Dinosaur Eggs and Babies. Cambridge University Press,Cambridge, U.K.

Norell, M. A., J. M. Clark, A. H. Turner, P. J. Makovicky, R. Barsbold, and T. Rowe. 2006. A new dromaeosaurid theropod from Ukhaa Tolgod (Ömnögov, Mongolia). American Museum Novitates 3545:1–51

Norman, D. B. 2004. Basal Iguanodontia; pp. 413–437 in D. B. Weishampel, P. Dodson, and H. Osmólska (eds.), The Dinosauria, Second Edition. University of California Press, Berkeley, California.

Osborn, H. F. 1924. Three new Theropoda, *Protoceratops* zone, central Mongolia. American Museum Novitates 144:1–12.

Osmólska, H. 1993. Were the Mongolian "fighting dinosaurs" really fighting? Revue de Paleïobiologie volume spÈcial 7:161–162.

Osmólska, H., P. J. Currie, and R. Barsbold. 2004. Oviraptorosauria; pp. 165–183 in D. B. Weishampel, P. Dodson, and H. Osmólska, (eds.), The Dinosauria, Second Edition. University of California Press, Berkeley, California.

Owen, R. 1842. Report on British fossil reptiles, part 2. Report of the British Association for the Advancement of Science 60–204.

Paul, G. 1997. Dinosaur models: the good, the bad, and using them to estimate the mass of dinosaurs; pp. 39–45 in D. L. Wolberg, E. Stump, and G. D. Rosenberg (eds.), Dinofest International: Proceedings of a Symposium held at Arizona State University. Academy of Natural Sciences, Philadelphia, Pennsylvania.

Prieto-Márquez, A. 2010a. Global historical biogeography of hadrosaurid dinosaurs. Zoological Journal of the Linnean Society 159:503–525.

Prieto-Márquez, A. 2010b. Global phylogeny of Hadrosauridae (Dinosauria: Ornithopoda) using parsimony and Bayesian methods. Zoological Journal of the Linnean Society 159:435–502.

Prieto-Márquez, A. 2011a. Cranial and appendicular ontogeny of *Bactrosaurus johnsoni,* a hadrosauroid dinosaur from the Late Cretaceous of Northern China. Palaeontology 54:773–792.

Prieto-Márquez, A. 2011b. A reappraisal of *Barsboldia sicinskii* (Dinosauria: Hadrosauridae) from the Late Cretaceous of Mongolia. Journal of Paleontology 85:468–477.

Prieto-Márquez, A., and M. A. Norell. 2010. Anatomy and relationships of *Gilmoreosaurus mongolensis* (Dinosauria: Hadrosauridae) from the Late Cretaceous of Central Asia. American Museum Novitates 3694:1–49.

Russell, D. A. 1993. The role of Central Asia in dinosaurian biogeography. Canadian Journal of Earth Sciences 30:2002–2012.

Seeley, H. G. 1887. The classification of the Dinosauria. Report to the British Association for the Advancement of Science 57:698–699.

Sereno, P. C. 1986. Phylogeny of the bird-hipped dinosaurs (Order Ornithischia). National Geographic Research 2:234–256.

Sereno, P. C. 2005. Stem Archosauria 1.0. TaxonSearch. Available at www.taxonsearch.org /Archive/stem-archosauria-1.0.php. Accessed January 10, 2013.

Sues, H.-D., and A. Averianov. 2009. A new basal hadrosauroid dinosaur from the Late Cretaceous of Uzbekistan and the early radiation of duck-billed dinosaurs. Proceedings of the Royal Society B 276:2549–2555.

Suzuki, S., and M. Watabe. 2000. Report on the Japan-Mongolia Joint Paleontological Expedition to the Gobi Desert, 1995. Hayashibara Museum of Natural Sciences Research Bulletin 1:45–57.

Tereschenko, V. S., and V. R. Alifanov. 2003. *Bainoceratops efremovi,* a new protoceratopid dinosaur (Protoceratopidae, Neoceratopsia) from the Bain-Dzak locality (south Mongolia). Paleontological Journal 37:293–302.

Turner, A. H., S. J. Nesbitt, and M. A. Norell. 2009. A large alvarezsaurid from the Cretaceous of Mongolia. American Museum Novitates 3648:1–14.

Turner, A. H., D. Pol, J. A. Clarke, G. M. Erickson, and M. A. Norell. 2007. A basal dromaeosaurid and size evolution preceding avian flight. Science 317:1378–1381.

Tverdokhlebov, V. P., and I. U. Tsybin. 1974. Origin of Upper Cretaceous sites of dinosaurs of Tugrikin-Us and Alag Teg; pp. 314–319 in Mesozoic and Cenozoic Faunas and Biostratigraphy of Mongolia. Transactions of the Joint Soviet-Mongolian Paleontological Expedition 1. Nauka, Moscow. [Russian]

Van Itterbeeck, J. V., D. J. Horne, P. Bultynck, and N. Vandenberghe. 2005. Stratigraphy and palaeonvironment of the dinosaur-bearing Upper Cretaceous Iren Dabasu Formation, Inner Mongolia, People's Republic of China. Cretaceous Research 26:699–725.

Verzilin, N. N. 1979. Melovye otlozheniya Khara-Khutula i problema saynshandskoy svity. [The Cretaceous deposits of Khara Khutul and the problem of the Sayn Shandin svita]. Moskovskoye Obshchestvo Ispytateley Prirody Bulletin, Otdel Geologicheskii 54, 123. [Russian]

Verzilin, N. N. 1982. Paleolimnologicheskoe znachenie teksturnych osobiennostiey verchnemelovych otlozhenij Mongolii. [Paleolimnologic significance of the textural features of the Upper Cretaceous deposits of Mongolia]; pp. 81–100 in G. G. Martinson (ed.), Mezozoiskie ozernye basseiny Mongoli [Mesozoic lacustrine basins of Mongolia]. Nauka, Leningrad. [Russian]

Wang, X.-L., R. Pan, R. J. Butler, and P. M. Barrett. 2011. The postcranial skeleton of the iguanodontian ornithopod *Jinzhousaurus yangi* from the Lower Cretaceous Yixian Formation of western Liaoning, China. Earth and Environmental Science Transactions of the Royal Society of Edinburgh 101:135–159.

Watabe, M., and S. Suzuki. 2000. Report on the Japan-Mongolia Joint Paleontological expedition to the Gobi Desert, 1997. Hayashibara Museum of Natural Sciences Research Bulletin 1:69–82.

Watabe, M., and K. Tsogtbaatar. 2007. Unique environments for dinosaurs in the Late Cretaceous of Mongolia: Djadokhta Formation. Palaeontological Society of Japan 156th Regular Meeting Abstract Volume, Tokushima, p. 18. [Japanese]

Weishampel, D. B., and C.-M. Jainu. 2011. Transylvanian Dinosaurs. Johns Hopkins University Press, Baltimore, Maryland, 301 pp.

Weishampel, D. B., D. B. Norman, and D. Grigorescu. 1993. *Telmatosaurus transsylvanicus* from the Late Cretaceous of Romania: the most basal hadrosaurid dinosaur. Palaeontology 36:361–385.

Wu, W., and P. Godefroit. 2012. Anatomy and relationships of *Bolong yixianensis,* an Early Cretaceous iguanodontoid dinosaur from Western Liaoning, China; pp. 293–333 in P. Godefroit (ed.), Bernissart Dinosaurs and Early Cretaceous Terrestrial Ecosystems. Indiana University Press, Bloomington, Indiana.

Xu, X., K. Wang, X. Zhao, C. Sullivan, and S. Chen. 2010. A new leptoceratopsid (Ornithischia: Ceratopsia) from the Upper Cretaceous of Shandong, China and its implications for Neoceratopsian evolution. PLoS ONE 5(11):e13835.

You, H.-L., and D. Q. L. Li. 2009. A new basal hadrosauriform dinosaur (Ornithischia: Iguanodontia) from the Early Cretaceous of northwestern China. Canadian Journal of Earth Sciences 46:949–957.

You, H.-L., Z.-X. Luo, N. H. Shubin, L. M. Witmer, Z.-L. Tang, and F. Tang. 2003. The earliest-known duck-billed dinosaur from deposits of late Early Cretaceous age in northwest China and hadrosaur evolution. Cretaceous Research 24:347–355.

Appendix 7.1. Character Data Matrix Used in Phylogenetic Analysis

Hypsilophodon foxii	0001100000 0000100000 0000000000 0000001100 0000000000 0100000000 0000010000 0000000000 0000000000 0000000000 0000000000 ?000000000 0000000010 00000000
Iguanodon bernissartensis	0000000000 0000100000 0000000000 0000000000 0000000000 0000001000 0000000000 0000000000 0000000001 0000000000 0000000000 0000000100 0000000000 00000000
Ouranosaurus nigeriensis	0000?00000 0001100000 0000000000 000000110? 0?0?0001?0 0000000000 00000000?? ??100?0001 000?00?000 0000?00?0? 0001000000 000?000100 0000000?00 00000?00
Fukuisaurus tetoriensis	000100???? ?????????? ?????????? ?????????? ????10??0 100000?000 000?001??? ??????0000 0??000?000 ?000000000 ?????????? ?????????? ?????????? ????????
Altirhinus kurzanovi	000000000 1000100000 0000010000 00000000?? ??????1000 0000000000 00000000?? ????00?001 00?0000000 0001000000 0?????0000 00000000110 000000001? 0????00?
Eolambia caroljonesa	000100000? 000??0???? ????00000 000000?101 ??0?0100?? 1100?0?000 000??000?? ??011???00 0000000000 0000010000 ????0????? ?0010??0?? 0000???00 000???0?
Probactrosaurus gobiensis	000000000? 0000????00 0???000000 0000?00101 0?0?0100?0 000010?000 0000000??? ?0011?0000 001000?001 0001010000 0010000000 00010?111? 0000010000 000??000
Equijubus normani	00???0000? 0000100000 0000?00000 ?000000??? ??0??10?00 0????01000 0000000??? ?0?11?0001 0??000?001 0000000000 01100????? ?????????? 0000?????? ????????
Jinzhousaurus yangi	00000?0000 0000100000 0000000000 010000??00 0000?10??0 0000??1000 001?00???? ??0???0000 ???1?0??0? 00?0?00??? 0??0000000 1000000110 000?0???10 0???????
Protohadros byrdi	101000000? 0000?00000 00???00000 ?000100??? ?????100?1 100010?000 000?110??? ???11?0011 0010001001 0011?10000 ?????????? ?????????? ?????????? ???????0
Telmatosaurus transsylvanicus	0001000000 0000000000 0000000000 0000100000 0000010000 1101111000 0000110100 0?0110??00 1111000011 0010010010 ?110?10?00 ?001?????? ?????????? 1??????0
Tethyshadros insularis	00000?000? 0000100000 0000?00000 ?00?000?00 ?00?011??0 110???1000 0000?10??? ?0????0000 ???100?11? 1??1?0010? 1?10?000?? 0101011201 0110100112 ????111?
Bactrosaurus johnsoni	0001010001 0000000000 0000?0000? 0000001001 0000010001 1100111000 0000111110 01011?0000 001000?011 0011010010 ?110010000 00010?1??? 0000010110 110??010
Plesiohadros djadokhtaensis	?1100100?0 00???????? ????00000 0000?0??01 ?00?0??0?? 11001????1 111?111110 ?1?11?0001 011?10???? ?010?1000? ?????????? ??????1211 ?????????? ????????
Levnesovia transoxiana	?????????? ?????????? ????100000 0000001001 00000?0?01 110011?00? ??00111110 01011?0000 0010000011 0011010010 ?1100??000 ?001??1??0 00??0101?? 110??010
Tanius sinensis	?????????? ?????????? ????0????? 0000?01001 00000????? ?????1?00? ??00?1?110 01011????? ?????????? ?????????? 1111?11??? ?0010????? 00000????? 110???10
Aralosaurus tuberiferus	????????00 000??????? 0?11110000 0000101101 1100010100 1101101?1 110011?110 010??????? ?????????? ?010111110 ?????????? ?001?????? ?????????? 1?????1?
Tsintaosaurus spinorhinus	100001?001 001????0?? ?1?1000010 0101101100 100011011? 21011????? 1???101100 10011?1111 101111??11 1111111111 1011110111 1111??1?0? 1110100?11 110??110
Jaxartosaurus aralensis	?????????0 10???????? ?000000110 0111101100 10000????? ?????????? ???????110 110?1????? ?????????? ?????????? ?????????? ?????????? ?????????? ????????
Nipponosaurus sachalinensis	???????11? 111??????? ?????????? ?????010?0 ?????1???? 2??1111111 1210????1? ?1???????? 1??111??11 1?11111110 2111?11111 ???1?11211 111????11 11??1110
Amurosaurus riabinini	?????????? 1????????? ?????00111 0111101110 21101?111? 211111?111 1110111110 110111??11 111111?11? 1111111111 ?110?10111 11011?1??? 111?110111 110??110
Lambeosaurus lambei	1010010112 1111?00021 1000000111 1111101110 2111111111 2111111111 1110111110 1111111111 1111111111 1111111111 2111111111 1111111211 1111111111 11011110
Lambeosaurus magnicristatus	1010010112 1111?00021 1000000111 1111101110 2111111111 2111111111 1110111110 1111111111 1111111111 1111111111 2111111111 1111111211 1111111111 11011110
Corythosaurus casuarius	1010010112 1111?00022 2000000111 1111101110 2111111111 2111111111 1110111110 1111111111 1111111111 1111111111 2111111111 1111111211 1111111111 11011110
Hypacrosaurus altispinus	1010010112 1111?00022 2000000111 1111101110 2111111111 2111111111 2110111110 1111111111 1111111111 1111111111 2111111111 111111?211 1111111111 11011110
Hypacrosaurus stebingeri	1010010112 1111?00022 2000000111 1111101110 2111111111 2111111111 2110111110 1111111111 1111111111 1111111111 2111111111 111111?211 1111111111 11011110
Olorotitan arharensis	10?00?0112 1?11?00021 ?00000011? 1111101110 2?1?111111 2?111?1111 11111?11?0 ?1????1111 11??11?11? 1111???11 2111111111 111?11?2?? 111????11 1?0????0
Parasaurolophus walkeri	1010010102 1011?00010 0000000111 2121111110 2111111110 2111111101 1110111110 1111111101 1111111111 1111111111 1111111111 1110011211 1111111111 11111?10

Appendix 7.1. (continued)

Parasaurolophus cyrtocristatus	1010010102 1011?00010 0000000111 2121111110 2111111110 2111111101 1110111110 1111111101 1111111111 1111111111 ?111111111 1110011211 1111111111 11111?10
Charonosaurus jiayinensis	?????????? 1????????? ????????1? 2121111110 ?1111?111? 211111?111 1110111110 11?11???11 1?1111?11? 1111??1111 ?111110111 111?1?1??? 11111?1??1 111??110
Lophorhothon atopus	?????????1 000??11100 0121100000 0000?010?? 00001?0??0 1????1?00? ???01?1?10 110?1????? ?????????? ???1010010 ?????????? ?????????? ???????11? ????1?1?
Kritosaurus navajovius	1111011000 000???100 0121?00000 0110101001 1000111000 1101110101 111011111? 111???1101 1?11111111 1111110110 ?110??0?11 ?101?????? 1110110??? 1??1111?
Maiasaura peeblesorum	1111111000 0000000000 0112001000 2010101001 100?111000 1101110101 1211111110 111111110 1111111111 11111110110 1010110111 01011112?1 1111110112 11011111
Brachylophosaurus canadensis	1111111000 0001000000 0111001000 2000101001 1000111001 1101110101 1211111110 1111111101 1111111111 1111110110 1010111111 01011112?1 1110110112 11011111
Gryposaurus notabilis	1211111000 0001111100 0121010000 0000101001 1000111000 1101110101 1210111111 1111111101 1111111111 1111110110 1110110111 0101011201 1111110112 11011110
Kerberosaurus manakini	?????????? ???1?12100 0000?10000 01001??001 ?00??1100? 110111?101 111?1111?1 ???1?????? ?????????? ?1???????0 ????????11 ?????????? ?????????? ???????0
Saurolophus osborni	1200111002 0?01112200 0122001000 0110101001 2001111000 1101111101 1111111101 1111111101 1111111111 1111110110 1110110111 0101011211 1110100112 11011110
Saurolophus angustirostris	1200111002 0?01112200 0122001000 0110101001 2001111000 1101111101 1111111101 1111111101 1111111111 1111110110 1110110111 0101011211 1110100112 11011110
Prosaurolophus maximus	1201111001 0001112200 0121010000 0100101001 1001111000 1101110101 1111111111 1111111101 1111111111 1111110110 1110111111 0101011 2?1 1111110112 11011110
Edmontosaurus regalis	1211111001 0001012100 0000010000 0000101001 100111100? 1101110101 1111111111 1111111101 1111111111 1111110110 1110111111 01010112?1 1111110112 11011110
Edmontosaurus annectens	1211111001 0001012100 0000010000 0000101001 100111100? 1101110101 1111111111 1111111101 1111111111 1111110110 1110111111 01010112?1 1111110112 11011110
"*Kritosaurus*" *australis*	?????1???? 00??????00 0????00??? 0100101001 ?00??1??0? 11???????? ?????????? 1?1?1?0?10 ?????????? ?1111????? ?????10111 0????????? 1111100110 1???????

Note: Characters correspond with those published in Sues and Averianov (2009).

Hadrosauroid Material from the Santonian Milk River Formation of Southern Alberta, Canada

8

Derek W. Larson, Nicolás E. Campione, Caleb M. Brown, David C. Evans, and Michael J. Ryan

ABSTRACT

The first hadrosauroid remains from the Santonian Milk River Formation are described here and include a right frontal, partial left jugal, right surangular, fragmentary right quadrate, dentary, teeth, and postcranial elements. When considered within a phylogenetic context, this material indicates the presence in this formation of at least one non-lambeosaurine hadrosaurid, possibly with close affinities to the approximately contemporaneous *Lophorhothon* from southeastern North America. The material described here represents the only geographic occurrence of Santonian hadrosaurids from the northern part of the Western Interior Basin. Given the putative occurrence of latitudinally arrayed northern and southern dinosaur "provinces" on Laramidia during the late Campanian, the Milk River Formation fauna has implications for understanding the origin of the faunal differentiation that occurred later in the Cretaceous.

INTRODUCTION

The Santonian Milk River Formation preserves some of the oldest terrestrial fossils from the Late Cretaceous of Alberta, Canada (Russell, 1935; Fox, 1968; Payenberg et al., 2002; Larson, 2008). These are part of the poorly sampled Coniacian through early Campanian global Late Cretaceous terrestrial fossil record (Fastovsky et al., 2004; Weishampel et al., 2004; Barrett et al., 2009). As one of the few relatively fossiliferous Santonian units preserving a terrestrial vertebrate assemblage in North America, the Milk River Formation provides a key dataset for helping to understand the origins of Hadrosauridae, Ceratopsidae, and Tyrannosauridae.

The fauna of the Milk River Formation was first described by Russell (1935) based on material collected from localities in Black Coulee (formerly Deadhorse Coulee), near the eastern edge of the exposed unit (Fig. 8.1A), and approximately 40 km southwest and landward of the paleoshoreline as estimated from shoreface deposits in core (Payenberg, 2002). Russell (1935) described mollusc, amphibian, turtle, and dinosaur material, including the occurrence of the oldest hadrosaurid material known at the time. Subsequent study of vertebrate microfossil bonebeds and isolated material from Verdigris and Black coulees (Fig. 8.1B) documented the occurrence of fish, amphibians, lizards, snakes, champsosaurs, turtles, dinosaurs, and mammals in the formation (Fox, 1968, 1972, 1975; Fox and Naylor, 1982; Wilson et al., 1992; Wu and Brinkman, 1993; Gao and Fox, 1996; Brinkman, 2003; Larson, 2008). The dinosaur teeth have been systematically studied by Baszio (1997) and Larson (2008).

Descriptions of dinosaurian macrofossils from the Milk River Formation are limited to the original study by Russell (1935) and the description of pachycephalosaurid domes (Sullivan, 2003; Evans et al., 2013). The rarity of vertebrate macrofossil remains in the formation is surprising given the great abundance of material preserved in later-occurring Campanian and Maastrichtian deposits (Ryan and Russell, 2001; Currie and Koppelhus, 2005; Ryan and Evans, 2005; Bell, 2010; Campione and Evans, 2011; Eberth et al. 2013).

Fieldwork conducted from 2007 to 2009 in the Milk River Formation by the Royal Ontario Museum and Cleveland Museum of Natural History has resulted in the collection of significant new vertebrate macrofossil material (Larson et al., 2012; Ryan et al., 2012). Additionally, undocumented dinosaur fossils in the collections of the Royal Ontario Museum (collected in 1949 and 1950) and the University of Alberta (collected in 1970 and 1977) have recently been cataloged and prepared. The majority of the large dinosaur material collected from the Milk River Formation to date can be confidently assigned to Hadrosauroidea, with possible euhadrosaurian affinities. For the purposes of this study, we use the traditional taxonomic definitions of Hadrosauroidea, Hadrosauridae, Euhadrosauria, Hadrosaurinae, and Lambeosaurinae (sensu Weishampel et al., 1993; Sereno, 1998). The Milk River material includes the first cranial remains of this group recovered from the unit, as well as numerous, phylogenetically informative parts of the postcranium. Here, we describe this material for the first time, assess its possible systematic position, and discuss its implications on the evolutionary and biogeographic history of Hadrosauroidea.

Institutional Abbreviations CMN, Canadian Museum of Nature, Ottawa, Ontario; NMMNH, New Mexico Museum of Natural History and Science, Albuquerque, New Mexico;

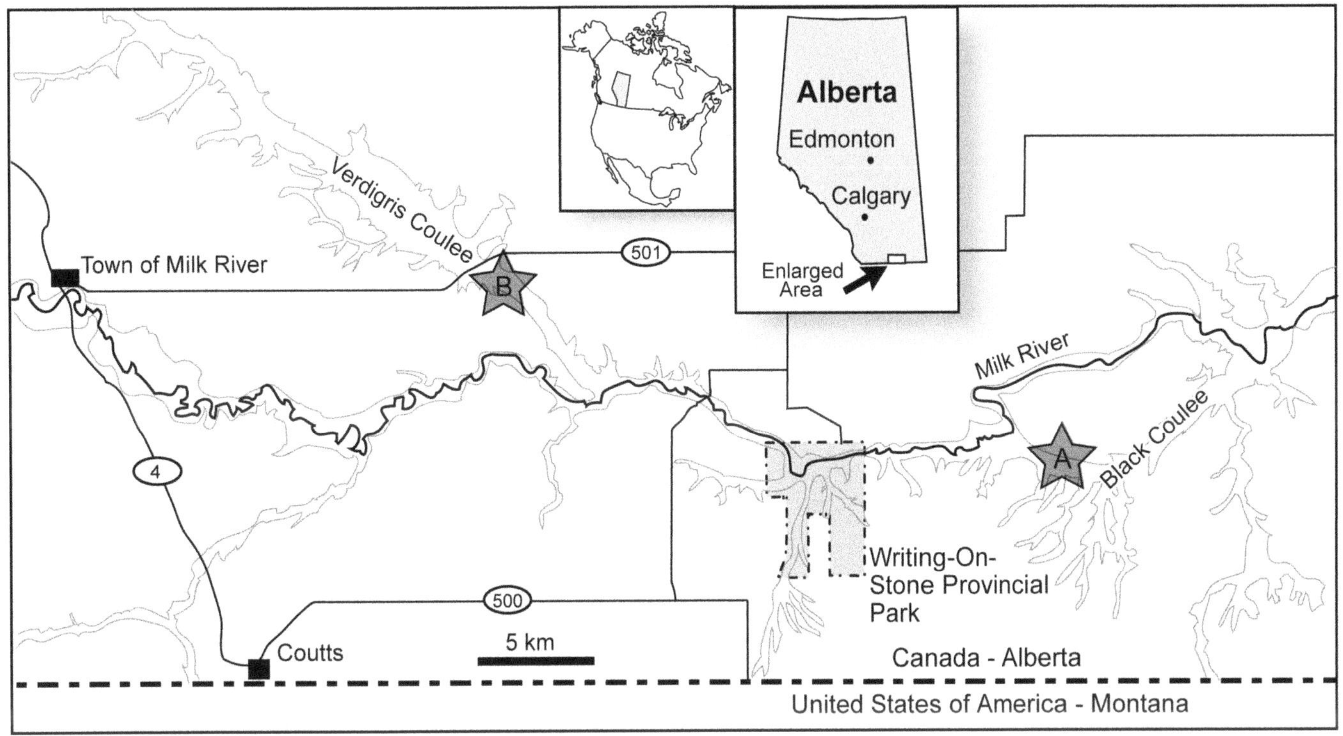

8.1. The two regions in southern Alberta where the Milk River Formation material described here was collected. (A) Black Coulee, where a partial dentary (TMP 1983.118.0001) and dentary fragments and teeth (TMP 2007.035.0006) were collected, as well as Locality 10 described by Russell (1935); and (B) Verdigris Coulee, including localities where the frontal (TMP 2007.035.0012), jugal (UALVP49350), surangular (UALVP49351), and tibia (TMP 2007.035.0134) were collected. Map of southern Alberta modified from Braman (2001).

ROM, Royal Ontario Museum, Toronto, Ontario; TMP, Royal Tyrrell Museum of Palaeontology, Drumheller, Alberta; UALVP, University of Alberta Laboratory for Vertebrate Paleontology, Edmonton, Alberta.

GEOLOGICAL SETTING

The Milk River Formation is a marine to non-marine siliciclastic wedge in the Western Interior Basin (WIB), partially equivalent to the Telegraph Creek and Eagle formations of Montana (Payenberg et al., 2002). Named by Dowling (1916), the unit was formally subdivided by Meijer Drees and Myhr (1981) into the Telegraph Creek, Virgelle, and Deadhorse Coulee members (Fig. 8.2). The Milk River Formation overlies the Colorado/Alberta Group marine shales and is overlain by the marine Pakowki Formation (Payenberg et al., 2002). The formation is laterally extensive throughout southern Alberta; however, outcrops are located only in southernmost Alberta, east of the town of Milk River. Of the three members, only the nonmarine Deadhorse Coulee Member preserves vertebrate fossil remains. At the time the first fossil assemblage from the unit was thoroughly described by Russell (1935, 1936), the member was referred to as the "upper" Milk River Formation. The Deadhorse Coulee Member consists of muddy siltstone floodplain deposits with laterally discontinuous centimeter- to decimeter-thick coals, interbedded with fluvial sandstone beds. Although the member is not completely exposed in outcrop, additional data from well-logs and outcrop studies suggest that it is 52 to 75 m thick in southern Alberta and thins southward (Leahy and Lerbekmo, 1995; Payenberg et al., 2002).

The absolute age of the Milk River Formation is reasonably well constrained (Payenberg et al., 2002). The base of the Telegraph Creek Formation of Montana (which is stratigraphically equivalent to the Telegraph Creek Member of the Milk River Formation) corresponds to the base of the *Desmoscaphites bassleri* ammonite zone. This biostratigraphic zone marks the time interval between the latest Santonian to the Santonian-Campanian boundary in the WIB (Landman and Cobban, 2007). The oldest known age of this biostratigraphic zone is 84.09 ± 0.40 Ma, and its base is estimated at 83.99 Ma (Ogg and Smith, 2004). No radiometric dates have been reported from the Milk River Formation itself. However, Leahy and Lerbekmo (1995) identified the 33r-34n magnetochron boundary (i.e., the end of the Long Cretaceous Normal) at 38.0 m above the base of the Deadhorse Coulee Member. The age of this magnetochron boundary has been assessed at ~84 Ma (Ogg and Smith,

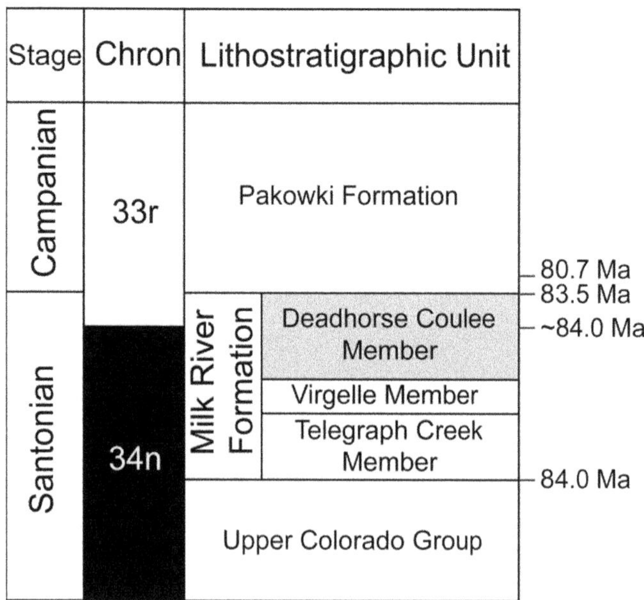

8.2. Lithostratigraphic column of southern Alberta at the Santonian–Campanian transition. All known vertebrate remains from the Milk River Formation are recovered from the Deadhorse Coulee Member. Abbreviation: Ma, mega-annum (million years).

2004). Braman (2001) also did not identify a palynological transition to a Campanian flora above this magnetochron boundary in the Deadhorse Coulee Member, leaving the top of the unit unconstrained but without any Campanian affinities in the flora. Whether the uppermost part of the member is Campanian or not, all of the known vertebrate remains have been recovered below this magnetochron boundary and can be confidently assigned to the latest Santonian.

SYSTEMATIC PALEONTOLOGY

DINOSAURIA Owen, 1842
ORNITHISCHIA Seeley, 1887
ORNITHOPODA Marsh, 1881
HADROSAUROIDEA Sereno, 1986

Material The cranial material consists of a mostly complete right frontal (TMP 2007.035.0012; Fig. 8.3), partial left jugal (UALVP 49350; Fig. 8.4), partial right quadrate (TMP 2007.035.0004; Fig. 8.5H, I), partial left dentary (TMP 1983.118.0001; Fig. 8.6), fragments of left dentary (TMP 2007.035.0083; Fig. 8.5F, G), fragment of other dentaries (TMP 1994.377.0017, TMP 2007.035.0016), complete right surangular (UALVP 49351; Fig. 8.7), as well as isolated and associated teeth (CMN 8729, TMP 2007.035.0006 [associated with dentary fragments], TMP 2007.035.0111; Fig. 8.5A–C). The postcranial material consists of numerous vertebrae (ROM 59705, TMP 1994.377.0005, TMP 2007.035.0009, TMP 2007.035.0044; Fig. 8.5J, K), left humerus (ROM 59702; Fig. 8.8), metacarpal V (TMP 2007.035.0007; Fig. 8.5L), manal phalanx IV-1 (TMP 2007.035.0111; Fig. 8.5M), one right and one left femur (CMN 8732 and ROM 59700; Fig. 8.5N, O), four tibiae (CMN 8733 [two elements], TMP 2007.035.0134, TMP 2007.035.0135; Fig. 8.5P), a fibula (ROM 59701; Fig. 8.5Q), a left astragalus (ROM 56636), a large left metatarsal IV (ROM 59704; Fig. 8.5R), a small partial right metatarsal IV (ROM 59703; Fig. 8.5S), and three pedal phalanges, including II-1 (TMP 2007.035.0008; Fig. 8.5T), III-2 (TMP 2007.035.0013; Fig. 8.5U) and pedal ungual III (TMP 2007.035.0001; Fig. 8.5V).

Locality and Horizon With the exception of some isolated teeth, all of the known hadrosauroid material from the Milk River Formation was recovered from a stratigraphic level around 26.0 m above the base of the Deadhorse Coulee Member, although uncollected elements potentially identified as hadrosauroid in the field are noted to occur 4.5 m above the base of the Deadhorse Coulee Member, illustrating the potential of material to occur anywhere in the lower 26.0 m of the unit, and all more than 12.0 m below the 33r-34n magnetochron boundary noted by Leahy and Lerbekmo (1995). All material was collected as isolated elements from Black Coulee and Verdigris Coulee, except CMN 8729, 8732, 8733, and probably ROM 59700, 59701, 59702, 59703, 59704, and 59705, which were recovered from a single bonebed in Black Coulee (Locality 10, Russell, 1935). This site has been relocated with the help of the Gilchrist family of the Deer Creek Ranch, Alberta, and detailed locality data are on file at the ROM.

Comments Although all of the material is identified as belonging to Hadrosauroidea, the surangular (UALVP 49351) and humerus (ROM 59702) display characters unequivocally identifying them as euhadrosaurian. However, ROM 59702 was likely recovered from Russell's (1935) Locality 10. Locality 10 is a small bonebed of limited extent that preserves disarticulated ceratopsian elements, crocodylian and theropod teeth, as well as hadrosaurid teeth. Also, associated hadrosauroid postcrania were collected by Russell from this site on probably two separate occasions. The first material was collected in 1934 while Russell was employed by the Geological Survey of Canada and included a left femur (CMN 8732), and right and left tibiae (CMN 8733). A second collection, likely also from this site, made by Russell in 1950 while he was employed by the ROM, contains a right femur (ROM 59700), fibula (ROM 59701), humerus (ROM 59702), two metatarsal IV (ROM 59703 and 59704), and vertebra (ROM 59705). The lack of overlapping material between the two groups and their consistent sizes (except for ROM 59704) suggests the possible presence of an associated individual at this site. It is also possible that the isolated teeth (CMN 8729) recovered from this site are also associated with this material, as they are not as greatly worn as those commonly found in vertebrate microfossil localities. Therefore, it is probable that much of the indeterminate hadrosauroid material recovered from this site, if from the same individual, belongs to a single euhadrosaurian taxon.

DESCRIPTION

Cranial Bones

Frontal An almost complete right frontal (TMP 2007.035.0012) is preserved, with the exception of its medial and anterior margins, and parts of the postorbital and parietal contact surfaces (Fig. 8.3). The dorsal surface of the frontal is smooth with a slight central depression, which results in a slightly raised, dome-like appearance posteriorly. A similar morphology is present in *Gryposaurus* (Prieto-Márquez, 2010b), *Naashoibitosaurus* (NMMNH P-16106), and juvenile *Edmontosaurus* (ROM 53502), and is unlike the pronounced dorsal doming that is present in lambeosaurines (Ostrom, 1961). The anterior margin exhibits several deep pockets and accompanying processes indicate complex sutures with the nasal and prefrontal, but the relative contributions of these elements is unclear given the incomplete nature of this region. The lateral margin comprises the contact surface for a large postorbital. The postorbital did not contact the prefrontal anteriorly, as the frontal makes a small (approximately 10 mm) contribution to the rostrolateral margin of external orbital rim. The small contribution of the frontal to the orbital margin is most similar to taxa such as *Lophorhothon* (Langston, 1960), *Bactrosaurus* (Godefroit et al., 1998), *Levnesovia* (Sues and Averianov, 2009), and *Acristavus* (Gates et al., 2011). The size of the contribution is much smaller in *Gryposaurus* (TMP 1980.022.0001), much larger in *Edmontosaurus* (Campione and Evans, 2011), and absent altogether in *Naashoibitosaurus* (NMMNH P-16106), *Prosaurolophus* (CMN 2870), *Saurolophus* (Bell, 2010), and lambeosaurines. The frontal contacts its counterpart along the midline of the skull, but the medial margin of TMP 2007.035.0012 is incomplete. The broken surface in this region reveals multiple large (3–5 mm in diameter) vascular canals.

The ventral surface of the bone contains the orbital and endocranial fossa. The lateral two-thirds are marked by a broad, shallow concavity that forms the dorsal border of the orbit, while the medial third bears a shallow, longitudinally oriented depression that conforms to the dorsal aspect of the olfactory tract. The contact surfaces for the laterosphenoid and the orbitosphenoid separate the laterally situated orbital depression from the medial endocranial fossa. Posteriorly, the ventromedial surface of the frontal preserves the cerebral fossa as a prominent round concavity. The frontal, as preserved, measures at least 72 mm in width and 99 mm in length, making it comparable in size to that found in a hadrosaurid skull approximately 600–1000 mm in length.

Although the frontal is not complete, a conservative estimate of the frontal length-to-width ratio is less than one, representing the plesiomorphic state for hadrosauroids (Godefroit et al., 2003). As a result, the morphology of the frontal is consistent with that seen in the non-hadrosaurid hadrosauroid *Bactrosaurus* (Godefroit et al., 1998), in basal hadrosaurids such as *Lophorhothon* (Langston, 1960), *Aralosaurus* and *Jaxartosaurus* (Rozhdestvensky, 1968; Godefroit et al., 2004), as well as some hadrosaurines including *Naashoibitosaurus* (NMMNH P-16106) and *Gryposaurus* (CMN 2877)–and unlike derived lambeosaurines, which have a very short and broad frontal (Godefroit et al., 2003; Evans, 2010). Notably, some derived hadrosaurines such as *Edmontosaurus* (ROM 53502) and *Prosaurolophus* (CMN 185) have frontals nearly two times longer than their width.

Jugal The anterior portion of the left jugal (UALVP 49350; Fig. 8.4) is preserved from the anterior-most margin to the cranioventral corner of the lower temporal fenestra. The jugal does not exhibit a dorsoventral expansion of the anterior process, typical of hadrosaurids. Rather, the jugal exhibits a slight rounded dorsoventral expansion, similar to that of *Lophorhothon* (Langston, 1960), *Bactrosaurus* (Godefroit et al., 1998), and *Levnesovia* (Sues and Averianov, 2009). The dorsal margin of the anterior expansion is thick transversely, relative to the ventral portion, and its dorsal surface bears diagonally oriented ridges, as well as a small trough that extends longitudinally along the dorsal margin of the anterior process. Together with the lacrimal, this trough would have formed a foramen. The base of the postorbital process is complete, but it is broken dorsal to its contact with the postorbital. The medial portion of the anterior expansion is dominated by a wide and flaring contact for the maxilla. The medial surface of the jugal exhibits several closely spaced circular pits just ventral to the postorbital process, which may indicate a pathology or taphonomic alteration. The jugal is relatively small (preserved length = 88 mm, height of anterior expansion = 39 mm, anterior margin to base of postorbital process = 68 mm), and it may pertain to a subadult individual due to its relatively small size.

Quadrate A fragment of a right quadrate (TMP 2007.035.0004) is preserved (Fig. 8.5H, I). The preserved portion represents the region just dorsal to the constriction associated with the quadratojugal. The lateral surface is smooth. The anterior and ventral margins are not complete, but the lateral surface preserves a longitudinal ridge that represents the base of the pterygoid flange, the majority of which is not preserved. The size of the fragment suggests

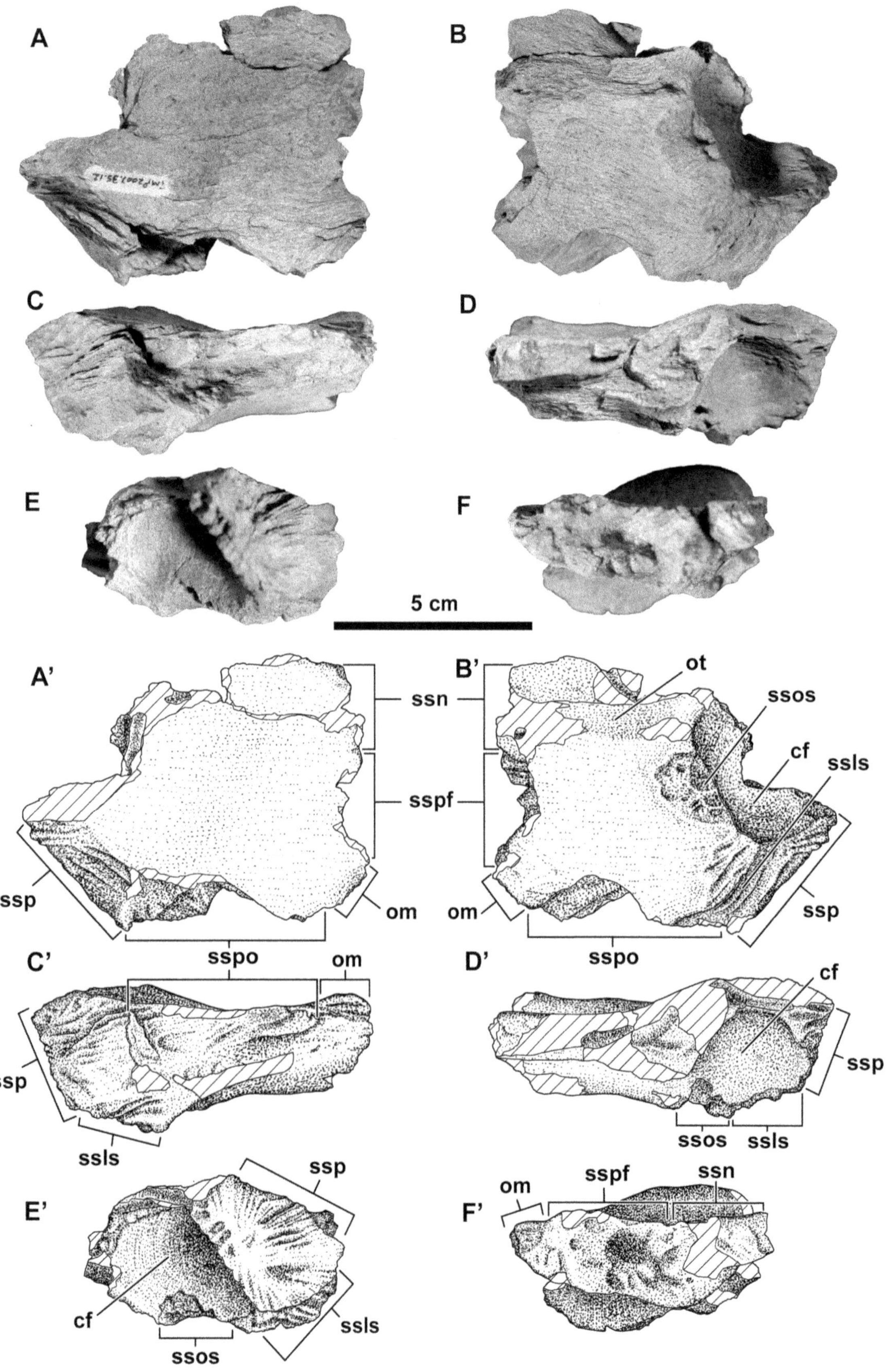

8.3. (A–F) photographs and (A′–F′) interpretive drawings of hadrosaurid frontal (TMP 2007.035.0012) in (A) dorsal; (B) ventral; (C) lateral; (D) medial; (E) posterior; and (F) anterior views. Scale bar equals 5 cm. Abbreviations: cf, cerebral fossa; om, orbital margin; ot, olfactory tract; ssls, sutural surface for laterosphenoid; ssn, sutural surface for nasal; ssos, sutural surface for orbitosphenoid; sspf, sutural surface for prefrontal; ssp, sutural surface for parietal; sspo, sutural surface for postorbital.

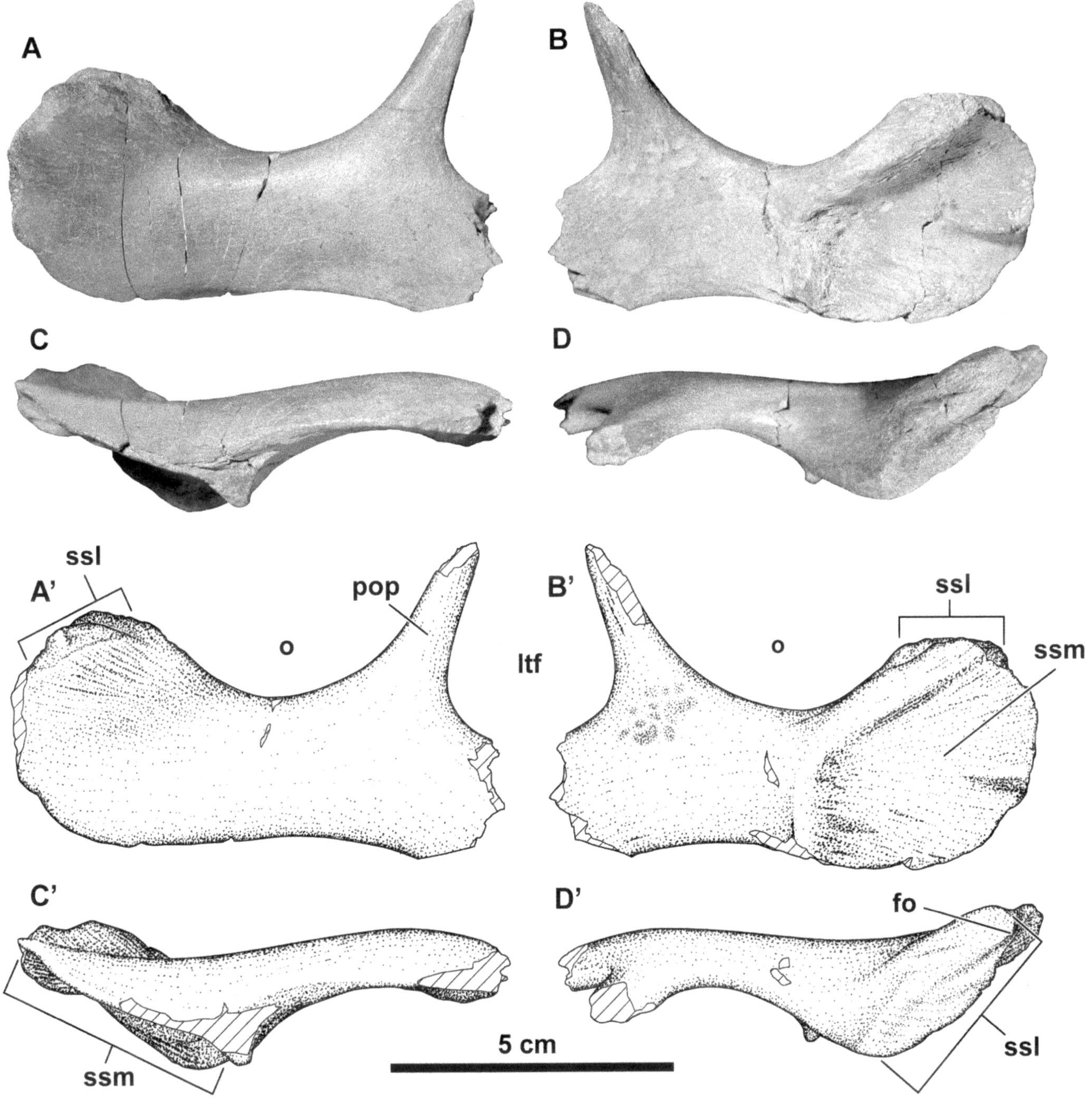

8.4. (A–D) photographs and (A'–D') interpretive drawings of hadrosaurid jugal (UALVP49350) in (A) lateral; (B) medial; (C) ventral; (D) dorsal views. Scale bar equals 5 cm. Abbreviations: fo; foramen; ltf, lateral temporal fenestra; o, orbit; pop, postorbital process of jugal; ssl, sutural surface for lacrimal; ssm, sutural surface for maxilla.

that it belonged to a relatively small individual (skull length less than 700 mm).

Dentary The most complete dentary (TMP 1983.118.0001; Fig. 8.6), a left, is missing the anterior margin and the portion posterior to the base of the coronoid process. It measures 270 mm as preserved, roughly 330 mm if complete, and has a mid-tooth-row height of 60 mm. The tooth row, although not quite complete anteriorly, preserves 31 alveoli and measures 228 mm in length. No teeth are preserved in the jaw. The width of the alveoli range from 6.5 to 7.7 mm, with a slight decrease in width anteriorly. The tooth row is straight in occlusal view, without distinct bowing, and the posterior end of the tooth row extends posterior to the apex of the coronoid. A distinct diastema is present anteriorly, but its length cannot be determined due to the incompleteness of the anterior end of the tooth row and anterior-most dentary. An associated

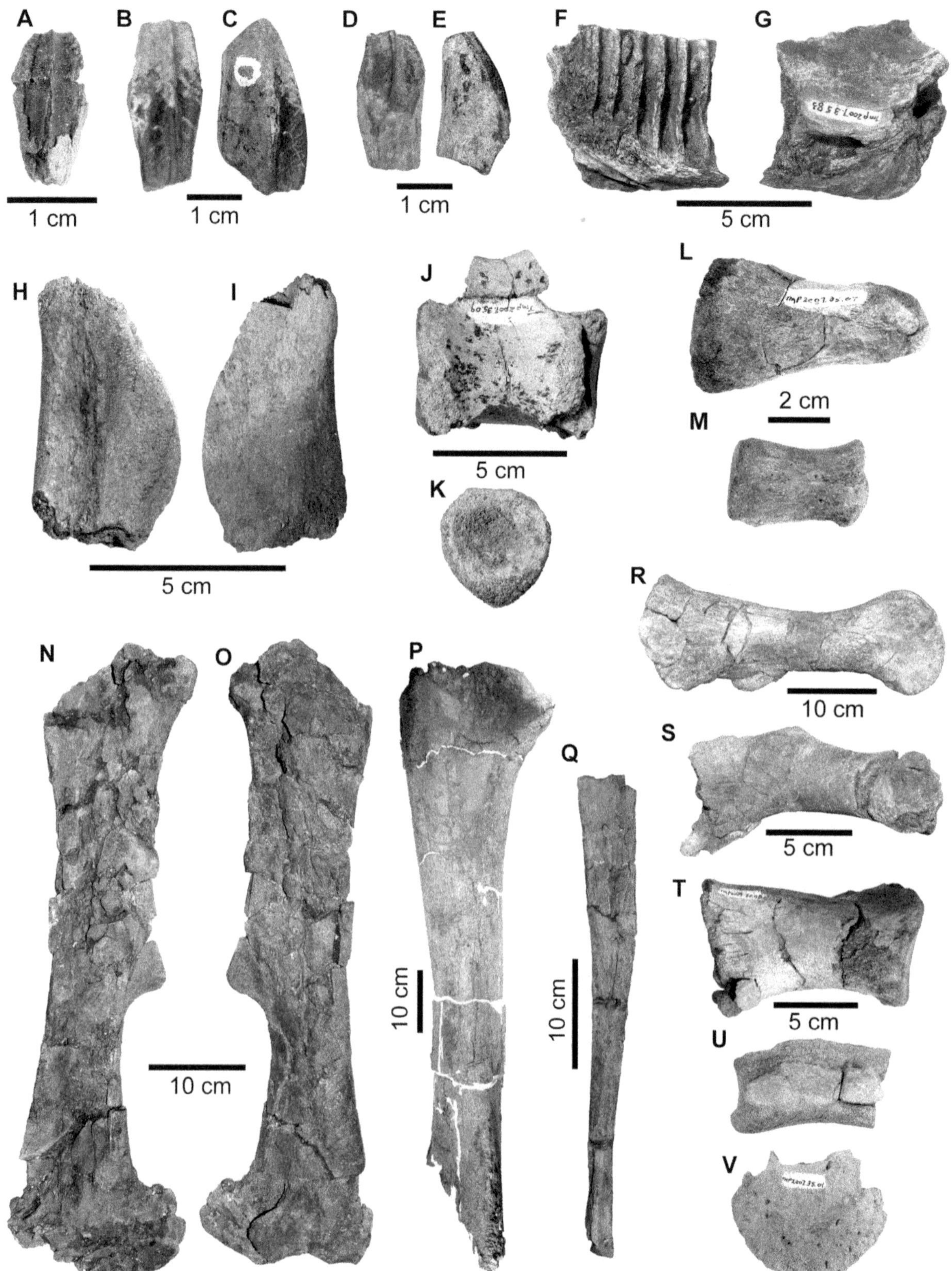

8.5. Photographs of isolated and associated hadrosauroid material from the Milk River Formation: (A) isolated tooth (TMP 2007.035.0111) in medial view; (B–E) isolated teeth (CMN 8729) in (B, D) medial and (C, E) anterior or posterior views; (F, G) fragment of left dentary (TMP 2007.035.0083) in (F) medial and (G) anterior/posterior views; (H, I) fragment of right quadrate (TMP 2007.035.0004) in (H) medial and (I) lateral views; (J) Isolated caudal vertebra (TMP 2007.035.0009) in lateral view; (K) fragment of vertebral centrum (TMP 2007.035.0044) in anterior or posterior view; (L) metacarpal V (TMP 2007.035.0007); (M) manal phalanx IV-1 (TMP 2007.035.0111) in lateral view; (N, O) left femur (ROM 59700) in (N) medial and (O) lateral views; (P) left tibia (TMP 2007.035.0134) in lateral view; (Q) fibula (ROM 59701) in lateral view; (R) large right metatarsal II (ROM 59704) in medial view; (S) small partial right metatarsal IV (ROM 59703) in dorsal view; (T) pedal phalanx II-1 (TMP 2007.035.0008) in dorsal view; (U) pedal phalanx III-2 (TMP 2007.035.0013) in dorsal view; and (V) pedal ungual III (TMP 2007.035.0001) in dorsal view. Scale bars as indicated.

8.6. Photographs of left hadrosauroid dentary (TMP 1983.118.0001) in (A) lateral; (B) dorsal; and (C) medial views. Scale bar equals 10 cm.

coronoid process and anterior symphysis are also preserved, but they make no direct contact with the preserved portion of the main body of the dentary. The anteroposterior length of the anterior symphysis is 49 mm. The dorsal portion of the coronoid does not appear to have an anteriorly expanded head, but the surangular does not make a large contribution to the coronoid.

Several dentary fragments (TMP 1994.377.0017, TMP 2007.035.006, TMP 2007.035.0016, and TMP 2007.035.0083) are known, but none preserve teeth. The largest dentary fragment (TMP 2007.035.0083; Fig. 8.5F, G) is from an area immediately anterior to the coronoid process and preserves seven tooth row positions averaging 8.3 mm in anteroposterior length and becoming smaller posteriorly. TMP 1994.377.0017 is a small section of the middle portion of the tooth row from a small animal. The preserved section is 27 mm long, with a height of 23 mm, and bears 6 alveoli. The alveoli average 4.5 mm in width. The section of dentary is just anterior to the coronoid base, as indicated by the distinct laterally thickening and widening of the Meckelian groove medially. The other dentary fragments are highly worn and show no discernible features other than a few alveolar grooves.

Surangular A nearly complete right surangular (UALVP 49351; Fig. 8.7) is relatively small compared to most hadrosaurids, being 94 mm long and 38 mm wide. As is typical of hadrosauroids (Weishampel et al., 1993), it is dorsoventrally flattened, but expanded mediolaterally in the middle to accommodate the quadrate condyle. A small foramen is located on the articular surface. The lateral surface is well preserved, and it lacks a surangular foramen, the derived condition observed in hadrosaurids (Horner et al., 2004). Posteriorly, the surangular tapers to form the retroarticular process.

Dentition The teeth assigned by Russell (1935) to cf. *Kritosaurus* sp. (CMN 8729; Fig. 8.5B–E) and other relatively complete teeth recovered subsequently (TMP 2007.035.0006, TMP 2007.035.0111; Fig. 8.5A) all appear to be from the

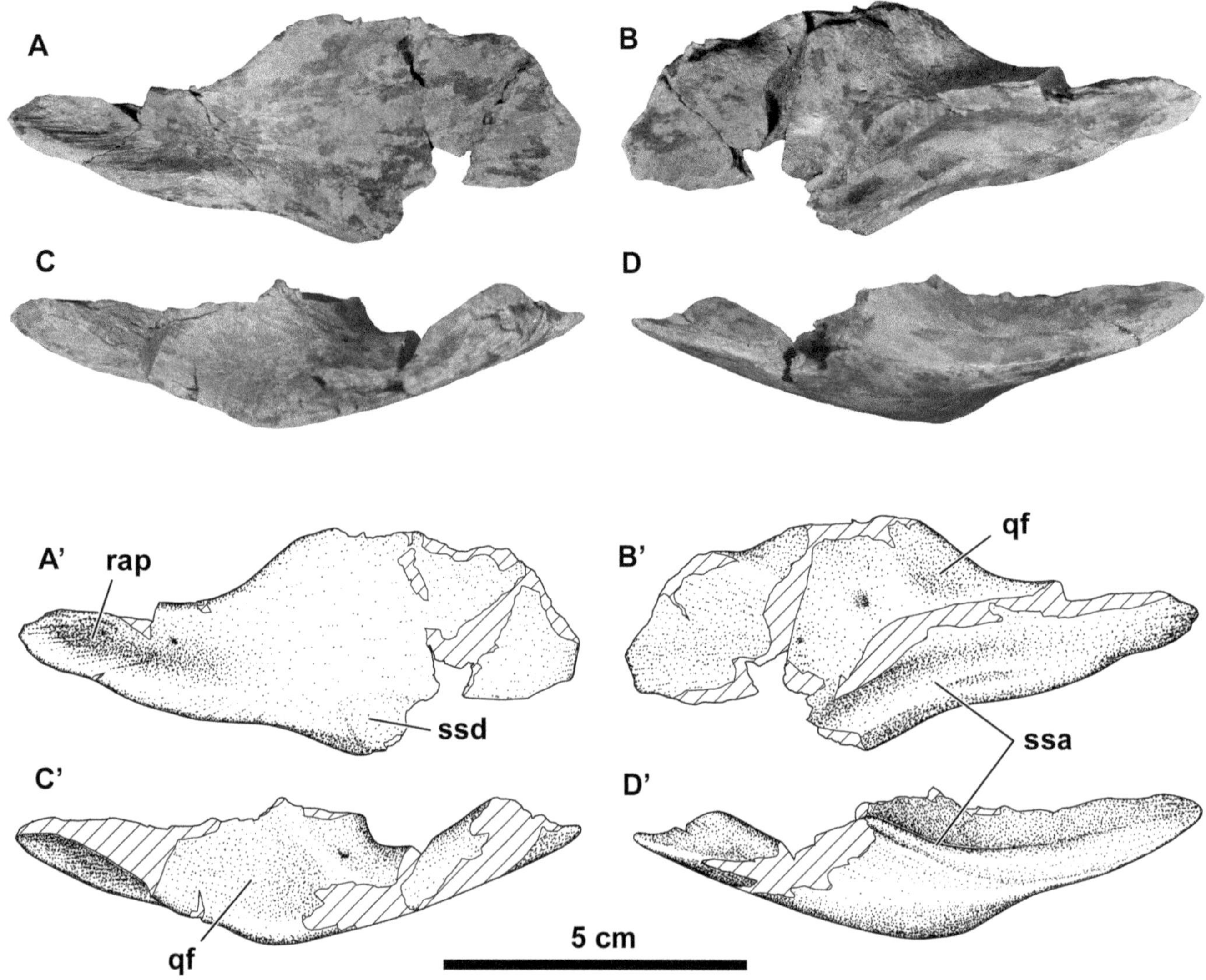

8.7. (A–D) photographs and (A'–D') interpretive drawings of hadrosaurid surangular (UALVP49351) in (A) ventral; (B) dorsal; (C) lateral; (D) medial views. Scale bar equals 5cm. Abbreviations: qf; quadrate fossa; rap, retroarticular process; ssa, sutural surface for angular; ssd, sutural surface for dentary.

dentary based on the flattened enameled surface of the tooth crown. These teeth are characterized by large distinct papillae along their mesial and distal margins, and the lack of secondary ridging on the enamel-covered surface of the crown. One tooth from CMN 8729 differs from the others. It is slightly asymmetrical with curved mesial and distal margins (Fig. 8.5B–E) whereas the other specimens have symmetrical enameled faces. The size of the marginal papillae is consistent with the morphology of non-euhadrosaurian hadrosaurids and lambeosaurines, but is unlike that in most hadrosaurines, which lack strong papillae (Head, 1998; Horner et al., 2004; Sues and Averianov, 2009). However, most of the other hadrosauroid tooth crowns recovered from the Milk River Formation do not preserve the mesiodistal margins, and are identifiable only as an indeterminate derived hadrosauroid lacking secondary ridging. These worn and fragmentary teeth are indistinguishable from those of hadrosaurids from vertebrate microfossil localities in younger units in the WIB (Baszio, 1997), which are best referred to Hadrosauroidea indet. based on synapomorphies. However, the lack of clearly non-hadrosaurid material suggests that identification of these teeth as Hadrosauridae indet. may be advisable in Santonian through Maastrichtian strata of western North America.

Postcranial Bones

Vertebrae Multiple isolated vertebrae are preserved (ROM 59705, TMP 2007.035.0009, and TMP 2007.035.0044). The largest (TMP 2007.035.0009; Fig. 8.5J), a caudal, is amphiplatyan to slightly amphicoelous, with the centrum longer than its height, and higher than wide. The centrum is hexagonal in cross section, with distinct facets for the chevrons ventrally. The neural arch is preserved, and fully fused to the centrum, but the zygapophyses and neural spine are missing. The centrum is 63 mm long, 48 mm tall, and 38 mm wide. A second, smaller dorsal centrum (TMP 2007.035.0044; Fig. 8.5K) is less complete, preserving only one end. It shows distinctly open neurocentral sutures, and is ovoid with a ventral keen in cross section. This centrum is 41 mm tall and 39 mm wide. A fragmentary neural arch (ROM 59705) was recovered from Locality 10 (Russell, 1935), but it lacks any diagnostic features.

Humerus The left humerus, (ROM 59702; Fig. 8.8) is small and incomplete with a preserved length of 306 mm and a minimum diaphyseal circumference of 98 mm. The deltopectoral crest extends over half the total length of the element, as is typical for euhadrosaurians. However, unlike hadrosaurids, the deltopectoral crest is not robust and the lateroventral corner of the deltopectoral crest is rounded and not sharply angled (Prieto-Márquez, 2011a). The proximal end is incomplete; however, based on the preserved length of the deltopectoral crest it is likely that only the distal-most region is missing and a total length can be estimated at 370 mm. The distal end of the humerus has condyles for articulation with the ulna medially and the radius laterally. The mediolateral expansion of the distal end relative to the humeral shaft is typical of most hadrosaurids, but it is not as strongly expanded as in *Parasaurolophus* (Horner et al., 2004).

Metacarpal V The preserved metacarpal V (TMP 2007.035.0007; Fig. 8.5L). is abraded and missing a portion of its proximal articulation surface. It is 79 mm long and 46 mm wide (proximally) and 92 mm in diaphyseal circumference. It is cylindrical, but with a greatly decreasing thickness distally–giving it an appearance of a truncated cone. The proximal articular surface is flat and the distal articular surface is greatly convex.

Manal Phalanges Phalanx IV-1 (TMP 2007.035.0111) is complete and measures 46 mm in length and 61 mm in circumference (Fig. 8.5M). It is flattened so as to be oval in cross section, with the proximal articular facet flat and the distal articular surface saddle shaped.

Femur Two femora (CMN 8732 and ROM 59700), both approximately 590 mm in length, were recovered from Russell's Locality 10. CMN 8732 was described and figured by Russell (1935). ROM 59700 (Fig. 8.5N, O) is relatively complete, but not as well preserved as CMN 8732 due to extensive crushing. Both exhibit a short, mediolaterally flat and proximodistally broad fourth trochanter as is typical in hadrosauroids. The distal end is not well preserved and the development of the extensor tunnel is uncertain due to crushing and incompleteness in this region.

Tibia In addition to those documented by Russell in 1935 (CMN 8733), two additional hadrosauroid tibiae have been collected, including one large (TMP 2007.035.0134; Fig. 8.5P) and one small specimen (TMP 2007.035.0135). All are typical of hadrosauroids in having well-developed condyles, strong cnemial crests proximally, and a mediolaterally broad distal head with medial and lateral malleoli. TMP 2007.035.0135 is 506 mm long and 166 mm in minimum diaphyseal circumference. TMP 2007.035.0134 has a minimum preserved length of 910 mm and a proximal width of 260 mm.

Fibula A partial fibula (ROM 59701) is missing its proximal and distal ends (Fig. 8.5Q). Its length as preserved is 407 mm, suggesting that it is from an individual comparable in size to CMN 8732 and CMN 8733. The proximal end is flattened mediolaterally and is anteroposteriorly broader than the preserved distal end, as in all hadrosauroids (Horner et al., 2004).

Astragalus An isolated left astragalus (ROM 56636) is semilunate in shape with a weak ascending process, and is tallest (27 mm) in the area of the articulation with the calcaneum. It measures 95 mm in width and 70 mm in maximum length. Part of the ascending process and portions of the caudodorsal surface are broken. It is indistinguishable from other known hadrosauroid astragali.

Metatarsal IV Two metatarsals IV are preserved. ROM 59704 (Fig. 8.5R) measures 324 mm in length, and 204 mm in minimum diaphyseal circumference. ROM 59703 (Fig. 8.5S) is very small compared to ROM 59704 and is likely from a juvenile. It is 156 mm in preserved length and 106 mm in circumference. Both ends are broken and a total length cannot be determined with confidence; however, its size is consistent with those of other elements recovered from Locality 10. The morphology of these elements is typical of hadrosauroids (Prieto-Márquez, 2011b; Zheng et al., 2011; Campione, this volume). They are robust with anteroposteriorly broad proximal ends. They are distinctly angled in anterior and posterior views, which is the result of the flat, medial articulation for metatarsal III and the lateral projection of the distal portion of metatarsal IV.

Pedal Phalanges Several pedal phalanges are known. Phalanx II-1 (TMP 2007.035.0008 [right]; Fig. 8.5T) is complete except for fragments missing from the dorsal and proximo-dorsal margin. It is 118 mm long, 68 mm wide (distally), 49 mm tall (distally), and 165 mm in circumference. Phalanx III-2 (TMP 2007.035.0013; Fig. 8.5U) is poorly preserved, but shows the flattened and shortened morphology. It is 46 mm long and at least 97 mm wide. A pedal ungual III (TMP 2007.035.0001; Fig. 8.5V) is 89 mm in width and 27 mm in height. It is distinctly spade shaped in dorsal view, as it is constricted proximally, and flares mediolaterally as it tapers distally. The flared distal margin is nearly semicircular in

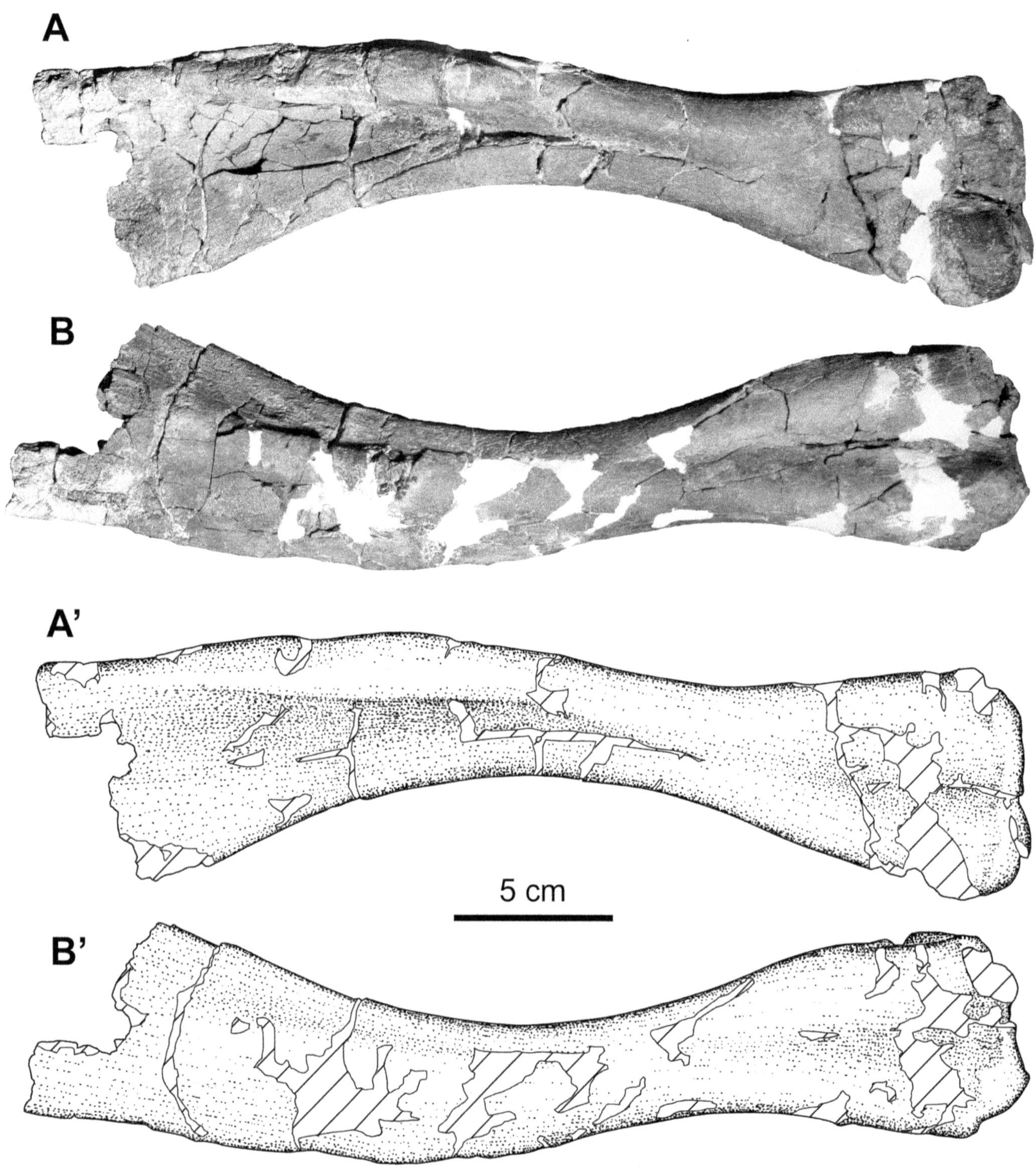

8.8. (A–B) photographs and (A′–B′) interpretive drawings of hadrosaurid humerus (ROM 59702) in (A) medial; (B) lateral views. Scale bar equals 5 cm.

shape. It is also exhibits numerous nutritive foramina on the plantar and distal margins. This ungual morphology is typical for hadrosaurids (Horner et al., 2004) and is also seen in *Bactrosaurus* (Godefroit et al., 1998).

PHYLOGENETIC ANALYSIS

Methods

In order to assess the systematic position of the Milk River Formation hadrosauroid material we conducted two different analyses. Both are based on the phylogenetic character matrix of Sues and Averianov (2009). First, we coded the specimens for the characters in the matrix (Table 8.1). In total, the material can be coded for 14 cranial and 6 postcranial characters (14% of the total characters). Given that all of the hadrosauroid material was collected from microsite sampling or from a bonebed, it is impossible to determine whether the material pertains to one or more species. As a result, in the first analysis, we treat each derived character from Table 8.1 separately by mapping each onto the strict consensus phylogeny of Sues and Averianov (2009). Only unambiguous changes were examined, and autapomorphic reversals were ignored unless they have the potential to affect the relative placements of the material within the phylogenetic tree.

In the second analysis, we treated all of the hadrosauroid material coded in Table 8.1 as pertaining to a single taxon. Although this cannot be assumed outright, it is defensible for two reasons: the coding of these elements (as evident from the first analysis) do not conflict with each other; and all the materials are derived from a relatively restricted stratigraphic horizon and would have been from penecontemporaneous individuals. The phylogenetic analysis was conducted using an heuristic search with 10,000 random addition sequence replicates using tree-bisection reconnection in PAUP v. 4.0b10 (Swofford, 2002). Robustness of the most-parsimonious topology was assessed in PAUP using Bremer support and bootstrap values (100 replicate random addition sequence heuristic, 10,000 replicates–limited to a maximum of 500 trees saved per replicate). Given the geologic age and lack of autapomorphies of the Milk River hadrosauroid remains, we evaluated the biogeographic history of hadrosaurids based on the phylogenetic tree recovered in this study. Ancestral biogeographic nodes were reconstructed using the Fitch parsimony optimization criterion (Ronquist, 1994) in MacClade (Maddison and Maddison, 2005).

Results

The synapomorphy-based assessment of derived characters (Table 8.1) exhibited by the recovered elements (Fig. 8.9A) reveal that only the humerus (ROM 59702) and the surangular (UALVP 49351) can be identified as pertaining to Euhadrosauria (Fig. 8.9A:clade 5). The teeth and postcranial elements exhibit the derived state at the node of *Bactrosaurus* plus all hadrosaurids (Fig. 8.9A:clade 3). The dentary (TMP 1983.118.0001) indicates a hadrosaurid of at least *Telmatosaurus* grade (Fig. 8.9A:clade 4). Interestingly, the frontal (TMP 2007.035.0012) and jugal (UALVP 49350) exhibit no derived characters in the matrix of Sues and Averianov (2009), but can be excluded from groups with the derived states associated with a shortened frontal (i.e., Lambeosaurinae).

The phylogenetic-based analysis recovered 64 most-parsimonious trees of 299 steps (CI = 0.535, RI = 0.855, RC = 0.457; Fig. 8.9B). The Milk River operational taxonomic unit (OTU) is recovered as the sister taxon to *Lophorhothon*. This sister pair forms the sister taxon to all other hadrosaurines. The sister relationship between these taxa is supported by one unambiguous synapomorphy (character 58), an unexpanded anterior jugal, which is a reversal within this topology, with the derived condition present in all other euhadrosaurians as well as *Aralosaurus*. Bootstrap and Bremer support values for the most-parsimonious topologies (Fig. 8.9B) are weak with most bootstrap values being less than 70% and most Bremer support values being one. The topology recovered here is otherwise identical to that reported by Sues and Averianov (2009).

The results of the biogeographic optimization (Fig. 8.10) hypothesize an Asian origin for Hadrosauridae and Euhadrosauria, but a North American radiation of hadrosaurines.

DISCUSSION

Affinities and Implications of the Milk River Hadrosauroid Material

The North American hadrosauroid fossil record spans much of the Cretaceous. Older formations (Barremian–Turonian age) preserve non-hadrosaurid iguanodontian taxa (Head, 1998, 2001; Brill and Carpenter, 2007; McDonald, Kirkland, et al., 2010; McDonald, Wolfe, et al., 2010), whereas younger formations (Campanian–Maastrichtian) preserve hadrosaurids to the exclusion of non-hadrosaurids (Lull and Wright, 1942; Horner et al., 2004; Evans et al., 2012; Campione et al., 2013). A significant gap in the fossil record of hadrosauroids exists from the start of the Coniacian through the Santonian. Therefore, it is important to assess evolutionary patterns in hadrosauroids of this age in order to evaluate the origin of hadrosaurids. To date, Coniacian–Santonian-aged terrestrial deposits in North America have yielded only fragmentary dinosaur remains (Weishampel et

Table 8.1. Characters That Could Be Scored in the Sues and Averianov (2009) Matrix and the Condition Exhibited by an Exemplar Specimen from the Milk River Formation

Character	Condition	Exemplar	State
31	Frontal has elongate ectocranial surface	TMP 2007.035.0012	0
32	Frontal participates in orbital margin	TMP 2007.035.0012	0
33	Frontal absent	TMP 2007.035.0012	0
34	Frontal upward doming	TMP 2007.035.0012	0
58	Jugal with dorsoventrally narrow anterior expansion	UALVP 49350	0
79	Dentary anterior to tooth row (lateral view) approximately straight	TMP 1983.118.0001	0
81	Dentary tooth row terminates posterior to apex of coronoid process	TMP 1983.118.0001	1
82	Dentary tooth row straight in occlusal view	TMP 1983.118.0001	1
83	Dentary with narrow parallel-sided tooth troughs	TMP 1983.118.0001	1
85	Coronoid process apex only slightly expanded anteriorly	TMP 1983.118.0001	0
88	Surangular with more ventrally exposed surface	UALVP 49351	1
89	Surangular lacks surangular foramen	UALVP 49351	1
99	Dentary teeth lack secondary ridges	CMN 8729	1
100	Dentary teeth have straight medial carina	CMN 8729	0
112	Humerus has long deltopectoral crest	ROM 59702	1
113	Humerus has low deltopectoral crest	ROM 59702	0
114	Humerus has mediolaterally compressed condyles	ROM 59702	1
132	Cnemial crest of tibia extends onto diaphysis	TMP 2007.035.0134	1
137	Pedal ungual III is flared	TMP 2007.035.0001	1
138	Pedal ungual III lacks a plantar keel	TMP 2007.035.0001	0

al., 2004; Prieto-Márquez, 2010a). With the exception of isolated teeth, definitive hadrosauroid material of this age has been previously documented as present only in the southern part of North America (Langston, 1960; Kaye and Russell, 1973; Hunt and Lucas, 1993; Carpenter et al., 1995; Williamson et al., 2000; Ramirez-Velasco et al., 2012; Table 8.2). The material described here from the Milk River Formation of southern Alberta represents the only skeletal concentration of Santonian-age hadrosaurid material currently known from northern North America, and represents some of the oldest specimens that can be unambiguously referred to Hadrosauridae.

The presence of a long deltopectoral crest on the humerus (character 112: state 1) and the lack of a surangular foramen (character 89: state 1), indicates that at least one euhadrosaurian taxon is present in the Milk River assemblage (Fig. 8.9A). Because none of the material can be referred to Lambeosaurinae, the Milk River hadrosauroid assemblage was apparently dominated by at least one euhadrosaurian of possible hadrosaurine affinities, although other non-lambeosaurine hadrosauroid taxa may have been present. Previous work conducted on the hadrosauroid material from the Milk River Formation suggested the possible presence of *Kritosaurus* based on the unspecified morphology of isolated teeth (Russell, 1935). However, the teeth of hadrosauroids are known to be highly variable in crown shape along the tooth row (Horner et al., 2004; Gallimore and Evans, 2009; Prieto-Márquez, 2010a), and until a comprehensive and quantitative analysis of tooth variation in this clade is conducted, the incompleteness and isolated nature of the Milk River Formation hadrosauroid teeth preclude any confident generic or specific assignment. The small-size, high-aspect ratio of the dentary crowns (CMN 8729), and the lack of secondary ridges on the teeth suggest that they can be regarded as pertaining to Hadrosauroidea of at least *Bactrosaurus* grade, and the broad enamel face of the dentary teeth appear to preclude their assignment to Lambeosaurinae. Most derived characters of the elements recovered from the formation, including teeth, are consistent in their appearance on the topology of Sues and Averianov (2009) between *Bactrosaurus* and Hadrosaurinae. The frontal and jugal, although they exhibit no derived characters in this matrix, closely resemble those taxa close to the origins of Euhadrosauria and Hadrosauridae.

Our phylogenetic results (Figs. 8.9B, 8.10) suggest that the potentially composite Milk River Formation OTU has affinities with *Lophorhothon* from the Selma Formation (now Mooreville Chalk) of Alabama (Langston, 1960). This hypothesis is notable because the lower unnamed member of the Mooreville Chalk Formation from which *Lophorhothon* was recovered is dated from middle Santonian to early Campanian (Mancini et al., 2008), and may suggest that this grade of hadrosauroid was widespread in North America during this time. However, this cannot be confidently concluded given the limited material from the Milk River Formation, as well as the inconsistent phylogenetic position of *Lophorhothon* (Horner et al., 2004; Sues and Averianov, 2009; Prieto-Márquez, 2010a).

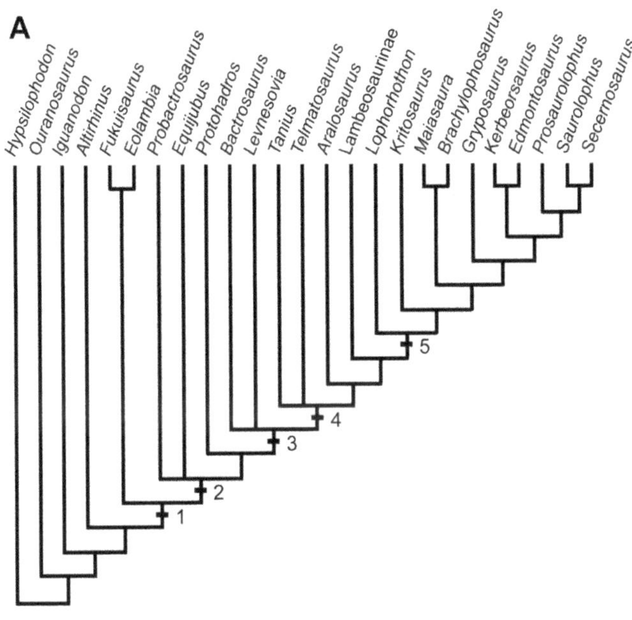

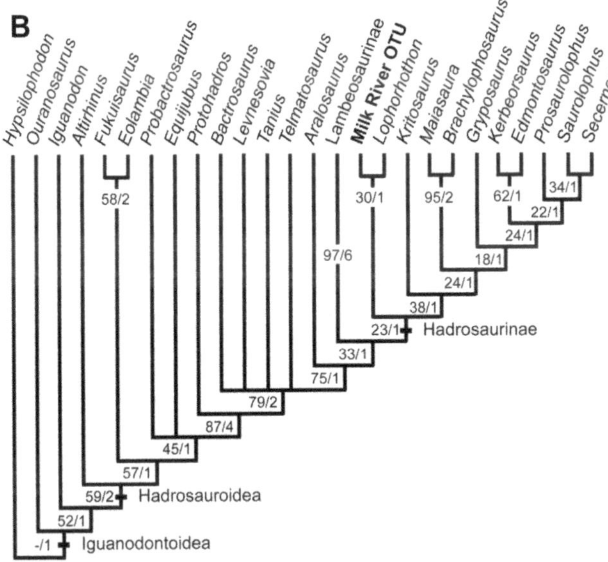

8.9. (A) The strict consensus phylogeny of Sues and Averianov (2009) with the position of derived characters present in the Milk River hadrosauroid added. (1) Character 114: state 1; (2) character 83: state 1 (equivocal for *Equijubus;* (3) character 89: state 1, character 99: state 1, character 132: state 1 (equivocal for *Equijubus* and *Protohadros*), character 137: state 1 (equivocal for *Equijubus* and *Protohadros*); (4) character 81: state 1 (equivocal for *Tanius*), character 81: state 1 (equivocal for *Tanius*); and (5) character 88: state 1 (equivocal for *Aralosaurus* and *Tanius*), character 112: state 1. (B) Strict consensus of 64 most-parsimonious trees with a CI = 0.535, RI = 0.855, RC = 0.457, based on the matrix of Sues and Averianov (2009) including the Milk River material as an OTU.

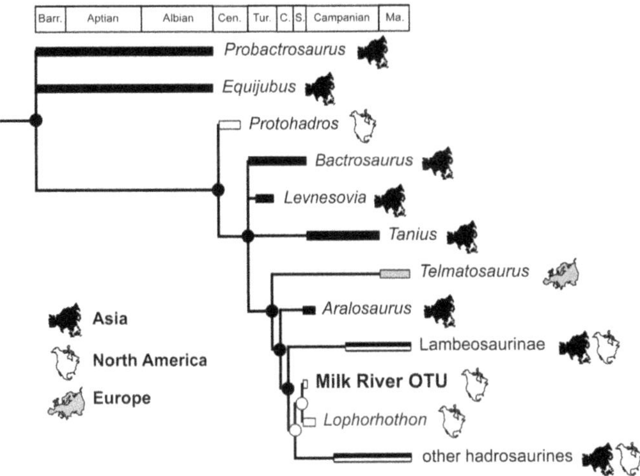

8.10. Time-calibrated phylogenetic tree of subset of taxa based on the Adams consensus of the 64 most-parsimonious trees. The biogeographic reconstruction of each node is indicated by the shaded circles. Black indicates Asia, white indicates North America, and gray indicates Europe.

The Milk River hadrosauroid material provides conclusive evidence of a hadrosaurid of Santonian age in western North America, prior to the emergence of the highly diverse hadrosaurid faunas of the Campanian. Furthermore, along with *Lophorhothon*, it suggests the widespread occurrence of hadrosaurines in North America during this time, although this is tentative given the fragmentary nature of the available fossils. The preliminary phylogenetic assessment of the Milk River material as an early hadrosaurine, if valid, represents one of the oldest occurrences of Hadrosaurinae anywhere in the world, and as a result, indicates Euhadrosauria and Hadrosaurinae originated no later than the Santonian.

Biogeographic Implications

The biogeographic optimization presented here suggests that Euhadrosauria originated in Asia, and hadrosaurines dispersed to North America prior to the end of the Santonian (Fig. 8.10). In Asia, *Aralosaurus*, widely regarded as the most primitive lambeosaurine (Horner, 1992; Godefroit et al., 2004; Evans and Reisz, 2007; Godefroit et al., 2008; Sues and Averianov, 2009), also occurs in the Santonian (Kordikova et al., 2001). This, along with the presence of close Euhadrosaurian outgroup taxa in the Turonian–Santonian of Asia, suggests that the timeframe of the origin and early radiation of Euhadrosauria is becoming better understood. Future discoveries and more in-depth comparisons of Santonian material (including *Aralosaurus*, *Lophorhothon*, and the Milk River fossils) will help elucidate the pattern of euhadrosaurian evolution and dispersal.

Perhaps more importantly, the Milk River Formation preserves the only record of Santonian hadrosauroids from the northern part of the Western Interior Basin, whereas all other occurrences are from Utah, Kansas, and more southerly localities (Langston, 1960; Carpenter et al., 1995; Williamson et al., 2000; Ramirez-Velasco et al., 2012). It has long been recognized that dinosaur faunas of the Campanian differ in

Table 8.2. Occurences of Santonian Hadrosauroid Material in North America

Identification	Previous ID	Stage	Formation	Location	References
Hadrosauroidea indet.	Hadrosauridae indet.	l. Santonian–l. Campanian	Menefee	New Mexico, United States	(Hunt and Lucas, 1993; Williamson et al., 2000)
Claosaurus agilis	*Claosaurus agilis*	m. Santonian–l. Campanian	Smoky Hill Chalk Mem. Niobrara	Kansas, United States	(Marsh, 1872; Carpenter et al., 1995)
Lophorhothon atopus	*Lophorhothon atopus*	m. Santonian–l. Campanian	Mooreville Chalk	Alabama, United States	(Langston, 1960; Mancini et al., 2008)
Hadrosaurinae indet.	cf. *Kritosaurus* and hadrosaurid?	l. Santonian	Milk River	Alberta, Canada	(Russell, 1935; Payenberg et al., 2002)
Huehuecanauhtlus tiquichensis	*Huehuecanauhtlus tiquichensis*	Santonian	unnamed	Michoacán State, Mexico	(Ramirez-Velasco et al., 2012)
N/A	In press	Santonian	Straight Cliffs	Utah, United States	(Gates et al., 2013)
Hadrosauroidea indet.	Hadrosaurinae indet.	m. Coniacian–m. Santonian	Eutaw	Mississippi, United States	(Kaye and Russell, 1973)
Hadrosauroidea indet.	Hadrosaurinae indet.	l. Santonian	Crevasse Canyon	New Mexico, United States	(Hunt and Lucas, 1993; Williamson et al., 2000)

their species composition in the northern and southern regions of the WIB, which has led to the hypothesis that marked ecologically driven dinosaur provinciality characterizes the Late Cretaceous of North America (Lehman, 1997, 2001; Gates et al., 2010; Sampson et al., 2010; but see Vavrek and Larsson, 2010). The circumstances surrounding the origins and extent of these putative faunal provinces are unclear, but these provinces likely predate the Campanian, the timeframe in which they have been documented (Sampson et al., 2010). Currently, the Milk River Formation provides the only Santonian-age data point within the northern faunal province. Therefore, future discoveries of hadrosaurs and other dinosaurs are likely to be important in testing the ideas about the nature and origins of putative faunal provinciality (Lehman, 2001; Sampson et al., 2010).

ACKNOWLEDGMENTS

We thank the diligent efforts of the 2007–2009 Southern Alberta Dinosaur Research Group, the skilled preparation done by E. LeBlanc, C. McGarrity, I. Morrison (ROM), and G. Jackson, the late S. Green, and the volunteers in the Department of Vertebrate Paleontology, Cleveland Museum of Natural History. Assistance with material was provided by M. Currie (CMN), G. Housego (TMP), and K. Seymour (ROM). Collection access was granted by K. Sheppard and M. Currie (CMN), K. Seymour (ROM), B. Strilisky (TMP), and P. Currie (UALVP). Special thanks to M. van Zetten for her help editing figures. Funding for this research was provided by the Jurassic Foundation and the Dinosaur Research Institute (to DWL) and the Natural Sciences and Engineering Research Council of Canada (to DCE).

LITERATURE CITED

Barrett, P. M., A. J. McGowan, and V. Page. 2009. Dinosaur diversity and the rock record. Proceedings of the Royal Society of London, Series B, 276:2667–2674.

Baszio, S. 1997. Systematic palaeontology of isolated dinosaur teeth from the latest Cretaceous of South Alberta, Canada. Courier Forschungstitut Senckenberg 196:33–77.

Bell, P. R. 2010. Redescription of the skull of *Saurolophus osborni* Brown 1912 (Ornithischia: Hadrosauridae). Cretaceous Research 32:30–44.

Braman, D. R. 2001. Terrestrial palynomorphs of the Upper Santonian–?Lowest Campanian Milk River Formation, southern Alberta, Canada. Palynology 25:57–107.

Brill, K., and K. Carpenter. 2007. A description of a new ornithopod from the Lytle Member of the Purgatoire Formation (Lower Cretaceous) and a reassessment of the skull of *Camptosaurus;* pp. 49–67 in K. Carpenter (ed.), Horns and Beaks: Ceratopsian and Ornithopod Dinosaurs. Indiana University Press, Bloomington, Indiana.

Brinkman, D. B. 2003. A review of nonmarine turtles for the Late Cretaceous of Alberta. Canadian Journal of Earth Sciences 40:557–571.

Campione, N. E. 2014. Postcranial anatomy of *Edmontosaurus regalis* (Hadrosauridae) from the Horseshoe Canyon Formation, Alberta, Canada: description and photographic atlas; chapter 13 in D. A. Eberth and D. C. Evans (eds.), Hadrosaurs. Indiana University Press, Bloomington, Indiana.

Campione, N. E., and D. C. Evans. 2011. Cranial growth and variation in edmontosaurs (Dinosauria: Hadrosauridae): implications for latest Cretaceous megaherbivore diversity in North America. PLoS ONE 6(9):e25186.

Campione, N. E., K. S. Brink, E. A. Freedman, C. T. McGarrity, and D. C. Evans. 2013. '*Glishades ericksoni,*' an indeterminate juvenile hadrosaurid from the Two Medicine Formation of Montana: implications for the diversity of non-hadrosauroids in the Late Campanian-Maastrichtian of North America. Palaeobiodiversity and Palaeoenvironments 93:65–75.

Carpenter, K., D. Dilkes, and D. B. Weishampel. 1995. The dinosaurs of the Niobrara Chalk Formation (Upper Cretaceous, Kansas). Journal of Vertebrate Paleontology 15:275–297.

Currie, P. J., and E. B. Koppelhus (eds.). 2005. Dinosaur Provincial Park: A Spectacular Ancient Ecosystem Revealed. Indiana University Press, Bloomington, Indiana, 648 pp.

Dowling, D. B. 1916. Water supply, southeastern Alberta. Geological Survey of Canada Summary Report 1914:102–110.

Eberth, D. A., D. C. Evans, D. B. Brinkman, F. Therrien, D. Tanke, and L. S. Russell. 2013. Dinosaur biostratigraphy of the Edmonton Group (Upper Cretaceous), Alberta, Canada: Evidence for climate influence. Canadian Journal of Earth Sciences 50(7):701–726.

Evans, D. C. 2010. Cranial anatomy and systematics of *Hypacrosaurus altispinus,* and a comparative analysis of skull growth in lambeosaurine hadrosaurids (Dinosauria: Ornithischia). Zoological Journal of the Linnean Society 159:398–434.

Evans, D. C., and R. R. Reisz. 2007. Anatomy and relationships of *Lambeosaurus magnicristatus,* a crested hadrosaurid dinosaur (Ornithischia) from the Dinosaur Park Formation, Alberta. Journal of Vertebrate Paleontology 27:373–393.

Evans, D. C., P. M. Barrett, and K. L. Seymour. 2012. Revised identification of a reported *Iguanodon*-grade ornithopod tooth from the Scollard Formation, Alberta, Canada. Cretaceous Research 33:11–14.

Evans, D. C., R. K. Schott, D. W. Larson, C. M. Brown, and M. J. Ryan. 2013. The oldest North American pachycephalosaurid and the hidden diversity of small-bodied ornischian dinosaurs. Nature Communications 4:1828.

Fastovsky, D. E., Y. Huang, J. Hsu, J. Martin-McNaughton, P. M. Sheehan, and D. B. Weishampel. 2004. Shape of Mesozoic dinosaur richness. Geology 32:877–880.

Fox, R. C. 1968. Early Campanian (Late Cretaceous) mammals from Alberta, Canada. Nature 220:1046.

Fox, R. C. 1972. A primitive therian mammal from the Upper Cretaceous of Alberta. Canadian Journal of Earth Sciences 9:1479–1494.

Fox, R. C. 1975. Fossil snakes from the Upper Milk River Formation (Upper Cretaceous), Alberta. Canadian Journal of Earth Sciences 12:1557–1563.

Fox, R. C., and B. G. Naylor. 1982. A reconsideration of the relationships of the fossil amphibian *Albanerpeton.* Canadian Journal of Earth Sciences 19:118–128.

Gallimore, G., and D. C. Evans. 2009. Morphometric analysis of hadrosaurid dental battery variation. Journal of Vertebrate Paleontology, Program and Abstracts 2009:101A.

Gao, K., and R. C. Fox. 1996. Taxonomy and evolution of Late Cretaceous lizards (Reptilia: Squamata) from western Canada. Bulletin of the Carnegie Museum of Natural History 33:1–107.

Gates, T. A., J. R. Horner, R. R. Hanna, and C. R. Nelson. 2011. New unadorned hadrosaurine hadrosaurid (Dinosauria, Ornithopoda) from the Campanian of North America. Journal of Vertebrate Paleontology 31:798–811.

Gates, T. A., E. K. Lund, C. A. Boyd, D. D. DeBlieux, A. L. Titus, D. C. Evans, M. A. Getty, J. I. Kirkland, and J. G. Eaton. 2013. Ornithopod dinosaurs from the Grand Staircase–Escalante National Monument region, Utah, and their role in paleobiogeographic and macroevolutionary studies; pp. 463–481 in A. L. Titus and M. A. Loewen (eds.), At the Top of the Grand Staircase: The Late Cretaceous of Southern Utah. Indiana University Press, Bloomington, Indiana.

Gates, T. A., S. D. Sampson, L. E. Zanno, E. M. Roberts, J. G. Eaton, R. L. Nydam, J. H. Hutchison, J. A. Smith, M. A. Loewen, and M. A. Getty. 2010. Biogeography of terrestrial and freshwater vertebrates from the Late Cretaceous (Campanian) Western Interior of North America. Palaeogeography, Palaeoclimatology, Palaeoecology 291:371–387.

Godefroit, P., V. Alifanov, and Y. Bolotsky. 2004. A re-appraisal of *Aralosaurus tuberiferus* (Dinosauria, Hadrosauridae) from the Late Cretaceous of Kazakhstan. Bulletin de l'Institut royal des Sciences naturelles de Belgique 74, Supplement:139–154.

Godefroit, P., Y. L. Bolotsky, and V. Alifanov. 2003. A remarkable hollow-crested hadrosaur from Russia: an Asian origin for lambeosaurines. Comptes rendus palevol 2:143–151.

Godefroit, P., Z.-M. Dong, P. Bultynck, H. Li, and L. Feng. 1998. Sino-Belgian Cooperation Program "Cretaceous dinosaurs and mammals from Inner Mongolia" 1. New Bactrosaurus (Dinosauria: Hadrosauroidea) material from Iren Dabasu (Inner Mongolia, P. R. China). Bulletin de l'Institut royal des Sciences naturelles de Belgique 68, Supplement:3–70.

Godefroit, P., S. Hai, T. Yu, and P. Lauters. 2008. New hadrosaurid dinosaurs from the uppermost Cretaceous of northeastern China. Acta Palaeontologica Polonica 53:47–74.

Head, J. J. 1998. A new species of basal hadrosaurid (Dinosauria, Ornithischia) from the Cenomanian of Texas. Journal of Vertebrate Paleontology 18:718–738.

Head, J. J. 2001. A reanalysis of the phylogenetic position of *Eolambia caroljonesa* (Dinosauria, Iguanodontia). Journal of Vertebrate Paleontology 21:392–396.

Horner, J. R. 1992. Cranial morphology of *Prosaurolophus* (Ornithischia: Hadrosauridae) with descriptions of two new hadrosaurid species and an evaluation of hadrosaurid phylogenetic relationships. Museum of the Rockies Occasional Paper 2:1–119.

Horner, J. R., D. B. Weishampel, and C. A. Forster. 2004. Hadrosauridae; pp. 438–463 in D. B. Weishampel, P. Dodson, and H. Osmólska (eds.), The Dinosauria, Second Edition. University of California Press, Berkeley, California.

Hunt, A. P., and S. G. Lucas. 1993. Cretaceous vertebrates of New Mexico. New Mexico Museum of Natural History and Science Bulletin 2:77–92.

Kaye, J. M., and D. A. Russell. 1973. The oldest record of hadrosaurian dinosaurs in North America. Journal of Paleontology 47:91–93.

Kordikova, E. G., P. D. Polly, V. A. Alifanov, Z. Roček, G. F. Gunnell, and A. O. Averianov. 2001. Small vertebrates from the Late Cretaceous and early Tertiary of the northeastern Aral Sea Region, Kazakhstan. Journal of Paleontology 75:390–400.

Landman, N. H., and W. A. Cobban. 2007. Redescription of the Late Cretaceous (late Santonian) ammonite *Desmoscaphites bassleri* Reeside, 1927, from the Western Interior of North America. Rocky Mountain Geology 42:67–94.

Langston, W., Jr. 1960. The vertebrate fauna of the Selma Formation of Alabama Part VI: The Dinosaurs. Fieldiana: Geology Memoirs 3:317–361.

Larson, D. W. 2008. Diversity and variation of theropod dinosaur teeth from the uppermost Santonian Milk River Formation (Upper Cretaceous), Alberta: a quantitative method supporting identification of the oldest dinosaur tooth assemblage in Canada. Canadian Journal of Earth Sciences 45:1455–1468.

Larson, D. W., N. R. Longrich, D. C. Evans, and M. J. Ryan. 2012. A new species of *Neurankylus* from the Milk River Formation (Cretaceous: Santonian), Alberta, Canada, and a revision of the type species *N. eximius;* pp. 389–405 in D. B. Brinkman, P. A. Holroyd, and J. D. Gardner (eds.), Morphology and Evolution of Turtles. Springer, Dordrecht, Netherlands.

Leahy, G. D., and J. F. Lerbekmo. 1995. Macrofossil magnetobiostratigraphy for the upper Santonian–lower Campanian interval in the Western Interior of North America: comparisons with European stage boundaries and planktonic foraminiferal zonal boundaries. Canadian Journal of Earth Sciences 32:247–260.

Lehman, T. M. 1997. Late Campanian dinosaur biogeography in the western interior of North America; pp. 223–240 in D. L. Wolberg, E. Stump, and G. D. Rosenburg (eds.), Dinofest International: Proceedings of a Symposium held at Arizona State University. Academy of Natural Sciences, Philadelphia, Pennsylvania.

Lehman, T. M. 2001. Late Cretaceous dinosaur provinciality; pp. 310–328 in D. H. Tanke and K. Carpenter (eds.), Mesozoic Vertebrate Life. Indiana University Press, Bloomington, Indiana.

Lull, R. S., and N. E. Wright. 1942. Hadrosaurian Dinosaurs of North America. Geological Society of America Special Papers 40. 242 pp.

Maddison, W. P., and D. R. Maddison. 2005. MacClade 4.08 Sinauer Associates, Sunderland, Massachusetts.

Mancini, E. A., J. Obid, M. Badali, K. Liu, and W. C. Parcell. 2008. Sequence-stratigraphic analysis of Jurassic and Cretaceous strata and petroleum exploration in the central and eastern Gulf coastal plain, United States. American Association of Petroleum Geologists Bulletin 92:1655–1689.

Marsh, O. C. 1872. Notice of a new species of *Hadrosaurus.* American Journal of Science, Series 3:301.

Marsh, O. C. 1881. Classification of the Dinosauria. American Journal of Science 7:81–86.

McDonald, A. T., D. G. Wolfe, and J. I. Kirkland. 2010. A new basal hadrosauroid (Dinosauria: Ornithopoda) from the Turonian of New Mexico. Journal of Vertebrate Paleontology 30:799–812.

McDonald, A. T., J. I. Kirkland, D. D. DeBlieux, S. K. Madsen, J. Cavin, A. R. Milner, and L. Panzarin. 2010. New basal iguanodonts from the Cedar Mountain Formation of Utah and the evolution of thumb-spikes dinosaurs. PLoS ONE 5(11):e14075.

Meijer Drees, N. C., and D. W. Myhr. 1981. The Upper Cretaceous Milk River and Lea Park formations in southeastern Alberta. Bulletin of Canadian Petroleum Geology 29:42.

Ogg, J. G., and A. G. Smith. 2004. The geomagnetic polarity time scale; pp. 63–86 in F. M. Gradstein, J. G. Ogg, and A. G. Smith (eds.), A Geologic Time Scale 2004. Cambridge University Press, Cambridge, U.K.

Ostrom, J. H. 1961. Cranial morphology of the hadrosaurian dinosaurs of North America. Bulletin of the American Museum of Natural History 122:33–186.

Owen, R. 1842. Report on British fossil reptiles, part 2. Report of the British Association for the Advancement of Science 60–204.

Payenberg, T. H. D. 2002. Litho-, chrono- and allostratigraphy of the Santonian to Campanian Milk River and Eagle formations in southern Alberta and north-central Montana: implications for differential subsidence in the Western Interior Foreland Basin. Ph.D. dissertation, Univeristy of Toronto, Toronto, Ontario, 221 pp.

Payenberg, T. H. D., D. R. Braman, D. W. Davis, and A. D. Miall. 2002. Litho- and chronostratigraphic relationships of the Santonian Campanian Milk River Formation in southern Alberta and Eagle Formation in Montana utilising stratigraphy, U-Pb geochronology, and palynology. Canadian Journal of Earth Sciences 39:1553–1577.

Prieto-Márquez, A. 2010a. Global phylogeny of hadrosauridae (Dinosauria: Ornithopoda) using parsimony and Bayesian methods. Zoological Journal of the Linnean Society 159:435–502.

Prieto-Márquez, A. 2010b. The braincase and skull roof of *Gryposaurus notabilis* (Dinosauria, Hadrosauridae), with a taxonomic revision of the genus. Journal of Vertebrate Paleontology 30:838–854.

Prieto-Márquez, A. 2011a. Revised diagnoses of *Hadrosaurus foulkii* Leidy, 1858 (the type genus and species of Hadrosauridae Cope, 1869) and *Claosaurus agilis* Marsh, 1872 (Dinosauria: Ornithopoda) from the Late Cretaceous of North America. Zootaxa 2765:61–68.

Prieto-Márquez, A. 2011b. Cranial and appendicular ontogeny of *Bactrosaurus johnsoni,* a hadrosauroid dinosaur from the Late Cretaceous of Northern China. Palaeontology 54:773–792.

Ramirez-Velasco, A. A., M. Benammi, A. Prieto-Márquez, J. A. Ortega, and R. Hernandez-Rivera. 2012. *Huehuecanauhtlus tiquichensis,* a new hadrosauroid dinosaur (Ornithischia: Ornithopoda) from the Santonian (Late Cretaceous) of Michoacan, Mexico. Canadian Journal of Earth Sciences 49:379–395.

Ronquist, F. 1994. Ancestral areas and parsimony. Systematic Biology 43:267.

Rozhdestvensky, A. K. 1968. Hadrosaurs of Kazakhstan; pp. 97–141 in [Upper Paleozoic and Mesozoic Amphibians and Reptiles]. Akademia Nauk S.S.S.R., Moscow, Russia. [Russian]

Russell, L. S. 1935. Fauna of the upper Milk River beds, Southern Alberta. Transactions of the Royal Society of Canada IV:115–128.

Russell, L. S. 1936. Oil and gas possibilities along Milk River, Southeastern Alberta. Canada Department of Mines Bureau of Economic Geology, Geological Survey of Canada Paper 36–12:115–137.

Ryan, M. J., and A. P. Russell. 2001. Dinosaurs of Alberta (exclusive of Aves); pp. 279–297 in D. H. Tanke and K. Carpenter (eds.), Mesozoic Vertebrate Life. Indiana University Press, Bloomington, Indiana.

Ryan, M. J., and D. C. Evans. 2005. Ornithischian dinosaurs; pp. 312–348 in P. J. Currie and E. B. Koppelhus (eds.), Dinosaur Provincial Park: A Spectacular Ancient Ecosystem Revealed. Indiana Univeristy Press, Bloomington, Indiana.

Ryan, M. J., D. C. Evans, P. J. Currie, C. M. Brown, and D. B. Brinkman. 2012. New leptoceratopsids from the Upper Cretaceous of Alberta, Canada. Cretaceous Research 35:69–80.

Sampson, S. D., M. A. Loewen, A. A. Farke, E. M. Roberts, C. A. Forster, J. A. Smith, and A. L. Titus. 2010. New horned dinosaurs from Utah provide evidence for intracontinental dinosaur endemism. PLoS ONE 5(9):e12292.

Seeley, H. G. 1887. On the classification of the fossil animals commonly named Dinosauria. Proceedings of the Royal Society of London 43:165–171.

Sereno, P. C. 1986. Phylogeny of the bird-hipped dinosaurs (Order Ornithischia). National Geographic Research 2:234–256.

Sereno, P. C. 1998. A rationale for phylogenetic definitions, with application to the higher-level taxonomy of Dinosauria. Neues Jahrbuch für Geologie und Paläontologie, Abhandlungen 210:41–83.

Sues, H.-D., and A. Averianov. 2009. A new basal hadrosauroid dinosaur from the Late Cretaceous of Uzbekistan and the early radiation of duck-billed dinosaurs. Proceedings of the Royal Society B 276:2549–2555.

Sullivan, R. M. 2003. Revision of the dinosaur *Stegoceras* Lambe (Ornithischia, Pachycephalosauridae). Journal of Vertebrate Paleontology 23:181–207.

Swofford, D. L. 2002. PAUP*. Phylogenetic Analysis Using Parsimony (*and Other Methods). Version 4.0b10. Sinauer Associates, Sunderland, Massachusetts.

Vavrek, M. J., and H. C. E. Larsson. 2010. Low beta diversity of Maastrichtian dinosaurs from North America. Proceedings of the National Academy of Science 107:8265–8268.

Weishampel, D. B., D. B. Norman, and D. Grigorescu. 1993. *Telmatosaurus transsylvanicus* from the Late Cretaceous of Romania: the most basal hadrosaurid dinosaur. Palaeontology 36:361–385.

Weishampel, D. B., P. M. Barrett, R. A. Coria, J. Le Loeuff, X. Xu, X. Zhao, A. Sahni, E. M. P. Gomani, and C. R. Noto. 2004. Dinosaur distribution; pp. 517–608 in D. B. Weishampel, P. Dodson, and H. Osmólska (eds.), The Dinosauria, Second Edition. University of California Press, Berkeley, California.

Williamson, T. E., S. G. Lucas, and A. B. Heckert. 2000. Review of Hadrosauridae (Dinosauria, Ornithischia) from the San Juan Basin, New Mexico. New Mexico Museum of Natural History Bulletin 17:191–213.

Wilson, M. V. H., D. B. Brinkman, and A. G. Neuman. 1992. Cretaceous Esocoidei (Teleostei): early radiation of the pikes in North American fresh waters. Journal of Paleontology 66:839–846.

Wu, X. C., and D. B. Brinkman. 1993. A new crocodylomorph of "mesosuchian" grade from the Upper Cretaceous Upper Milk River Formation, southern Alberta. Journal of Vertebrate Paleontology 13:153–160.

Zheng, R., A. A. Farke, and G.-S. Kim. 2011. A photographic atlas of the pes from a hadrosaurine hadrosaurid dinosaur. PalArch's Journal of Vertebrate Paleontology 8:12.

Hadrosaurid Anatomy and Variation

3

New Hadrosaurid (Dinosauria, Ornithopoda) Specimens from the Lower–Middle Campanian Wahweap Formation of Southern Utah

9

Terry A. Gates, Zubair Jinnah, Carolyn Levitt, and Michael A. Getty

ABSTRACT

Hadrosaurid ornithopods from the early to middle Campanian are rare in North America, but the Wahweap Formation of southern Utah yields specimens that are increasing the known diversity of hadrosaurids from this poorly understood time period. A new genus and species of lambeosaurine hadrosaurid, *Adelolophus hutchisoni*, is described on the basis of an isolated maxilla. This single element is distinct from all other known lambeosaurine hadrosaurid maxillae in the large medial wall, raised palatine process, and other features associated with the raised medial wall. Another locality in the middle of the Wahweap Formation yielded two individuals of presumably the same species, an adult and a juvenile. The specimens show autapomorphic morphology of the caudal vertebrae neural spines and centra, along with a unique suite of iguanodontian characters. Given that the only diagnostic skull elements belong to the juvenile individual, the phylogenetic position of this species is unclear, although the postcranial skeleton possesses a mix of primitive and derived traits. The site represents a swampy environment containing autochthonous specimens; among these, long bones exhibit slightly preferred orientations.

INTRODUCTION

In contrast to the long history of Cretaceous dinosaur collecting in Montana, United States, and Alberta, Canada (Currie, 2005), southern Utah, United States, has been intensively prospected for dinosaurs only since 2000. As in other North American terrestrial formations, hadrosaurid dinosaurs are some of the most common fossils discovered in southern Utah, and are providing important data on biogeographic patterns and paleoecology of Campanian dinosaurs (Gates and Evans, 2005; Gates et al., 2010). The hadrosaurids from the Kaiparowits Formation (late Campanian) are becoming well known, with multiple specimens of the two known species of the saurolophine *Gryposaurus*, *G. notabilis*? (Gates et al., 2013) and *G. monumentensis* (Gates and Sampson, 2007). The only lambeosaurine known from the formation is an undetermined species of *Parasaurolophus* (Gates et al., 2013).

The Wahweap Formation underlies the Kaiparowits Formation within the Grand Staircase–Escalante National Monument of southern Utah, and it is divided into several members (Lower, Middle, Upper, Capping Sandstone; Fig. 9.1) (Jinnah and Roberts, 2011). Deposition of this unit took place between ~81–77 Ma, based on a dated volcanic ash bed, detrital zircons, and estimated sedimentation rates (Jinnah et al., 2009). Although it is slower to produce fossil material, this formation is providing important information on the hadrosaurid dinosaurs that lived in southern Utah during the early and middle Campanian (Gates et al., 2013). One of the most significant taxa is a new saurolophine named *Acristavus gagslarsoni*, which is currently the earliest member of a small clade that contains this taxon and the hadrosaurids *Maiasaura peeblesorum* and *Brachylophosaurus canadensis* (Gates et al., 2011). Several other hadrosaurid specimens have been located within the Wahweap in addition to the referred specimen of *Acristavus* (UMNH VP 16607). This chapter describes specimens from two of these sites, both briefly mentioned in Gates et al. (2013) but not described in detail or discussed in reference to the taxonomy of Prieto-Márquez (2010a). An isolated lambeosaurine maxilla is described here as the second new hadrosaurid species from the Wahweap Formation. A bonebed that preserves two individuals of a possible new species is also documented. These specimens increase the known hadrosaurid diversity in the southern region of the Western Interior Basin during the middle Campanian.

Institutional Abbreviations AMNH, American Museum of Natural History, New York; CPC, Colección Paleontológica de Coahuila (Paleontological Collection of Coahuila) Saltillo, Coahuila, Mexico; MOR, Museum of the Rockies, Bozeman, Montana; NCSM, North Carolina Museum of Natural Sciences, Raleigh, North Carolina; RAM, Raymond M. Alf Museum of Paleontology, Claremont, California; ROM, Royal Ontario Museum, Toronto, Ontario; TCMI, Children's Museum of Indianapolis, Indianapolis, Indiana; TMP, Royal Tyrrell Museum of Palaeontology, Drumheller, Alberta;

UCMP, University of California Museum of Paleontology, Berkeley, California; UMNH, Natural History Museum of Utah, Salt Lake City, Utah.

SYSTEMATIC PALEONTOLOGY

DINOSAURIA Owen, 1842
ORNITHISCHIA Seeley, 1888
ORNITHOPODA Marsh, 1882
HADROSAURIDAE Cope, 1869
LAMBEOSAURINAE Parks, 1923
ADELOLOPHUS, gen. nov.

Etymology *Adelo* (= "unknown," Greek) + *lophus* (= "crest," latinized from the Greek *lophos*); meaning "unknown crest," in reference to unknown morphology of the skull and crest of this taxon.

Diagnosis As for type and only species.

ADELOLOPHUS HUTCHISONI, sp. nov.

Etymology Named in honor of Dr. Howard Hutchison, discoverer of the type specimen and longtime proponent of southern Utah vertebrate paleontology.

Holotype UCMP 152028, a partial maxilla.

Locality and Horizon Wahweap Formation (?Upper Member), Grand Staircase–Escalante National Monument, Garfield Co., Utah, United States.

Diagnosis Lambeosaurine hadrosaurid characterized by the following autapomorphies: maxilla with medial wall of the anterior process taller than in other lamebosaurines; maxillary medial wall of posterior process taller than in all known lambeosaurines, extending up to the dorsal process and sloping posteroventrally (causally placing the palatine process well above the position found in other lambeosaurines); thickened ridge extending from ventral articulation of jugal with maxilla posterodorsal to the palatine process.

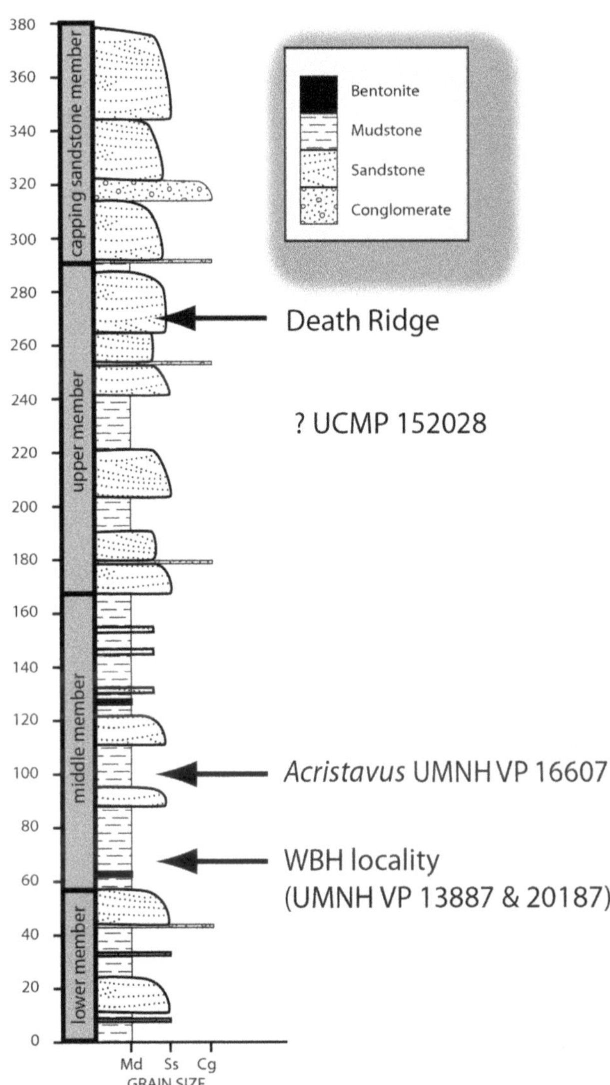

9.1. Composite stratigraphic section of the Wahweap Formation. Hadrosaurid specimens discussed in text are labeled reflecting their stratigraphic occurrences. The locality of UCMP 152028 is unknown, but based on an examination of the host matrix it is likely to have originated from the Upper Member. Death Ridge is a locality that has produced at least one hadrosaurid (not mentioned in the text) and several other important fossils.

Description

The isolated maxilla UCMP 152028 (Fig. 9.2) is incomplete, with anterior-most and posterior-most regions broken and missing, along with the most of the ectopterygoid shelf. Iron oxide–rich sediment covers other portions of the maxilla, especially the area around the dorsal process and teeth. Nonetheless, UCMP 152028 clearly exhibits the diagnostic lambeosaurine characteristics of a tall and narrow dorsal process, a well-developed premaxillary shelf, and an overall triangular shape. In general aspect, it is more similar to *Parasaurolophus* spp. maxillae than to lambeosaurin maxillae, although it is distinctive from the former as detailed below.

The anterior-most region of the maxilla was broken prior to fossilization as evidenced by the presence of iron oxide–rich sediment on the broken surface. The remaining premaxillary shelf predominates the anterodorsal maxilla as is in other lambeosaurine hadrosaurids. The lateral edge of the anterior half of the premaxillary shelf is contiguous with the lateral face of the maxilla, but, more posteriorly, a small lateral ridge rises and continues onto the dorsal process, similar to the condition in taxa such as *Corythosaurus* (Sternberg, 1935), *Hypacrosaurus altispinus* (Evans, 2010), and *Parasaurolophus* sp. (UMNH VP 16666.1). However, the medial wall of the premaxillary shelf rises to a higher level

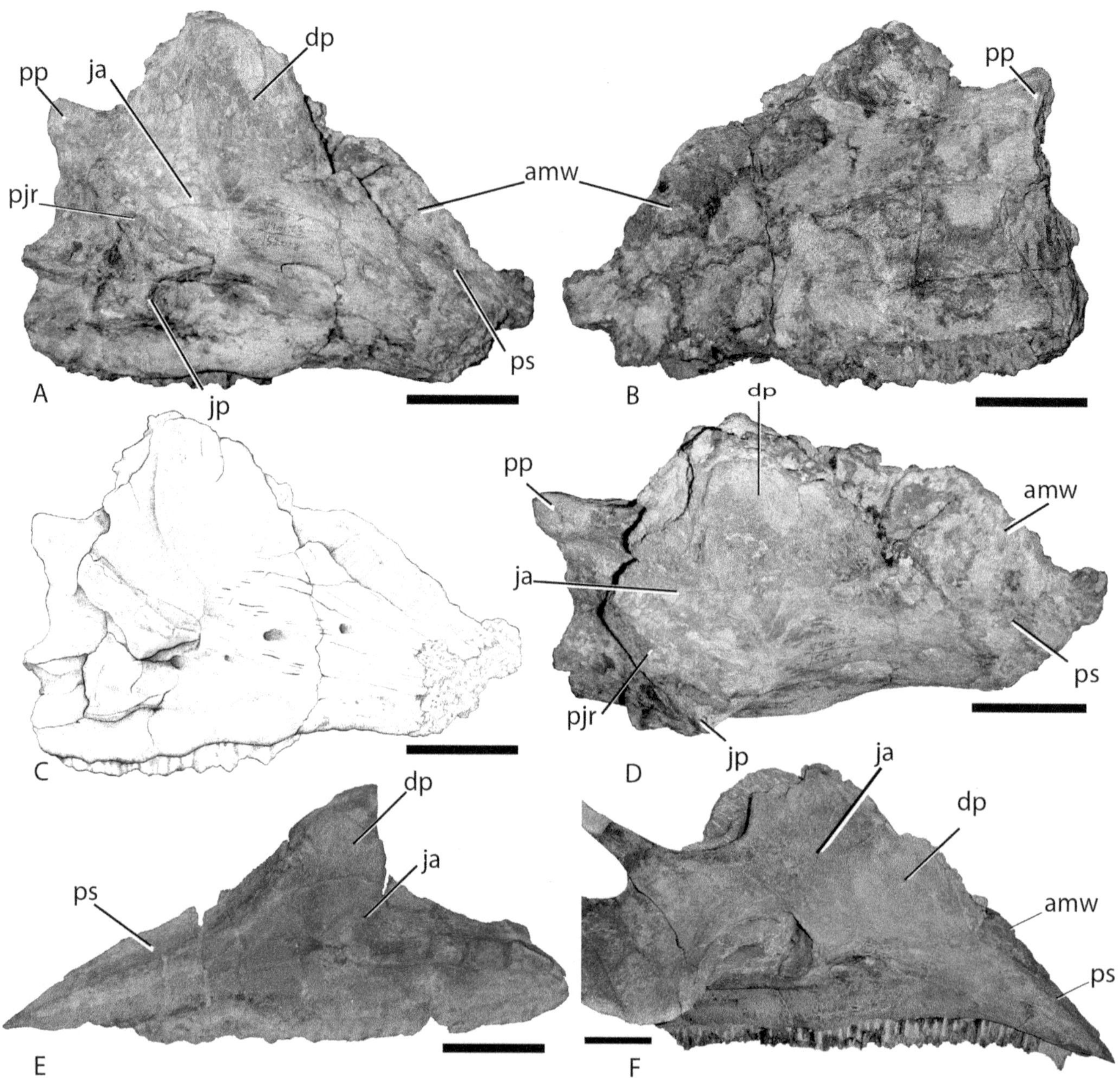

9.2. Type specimen *Adelolophus hutchisoni,* UCMP 152028. (A) lateral view; (B) medial view; (C) illustration of lateral view; (D) dorsolateral view; (E) left maxilla of CPC-59 *Velafrons coahuilensis;* and (F) UMNH VP 16666.1 maxilla and articulated jugal of *Parasaurolophus* sp. Note the bi-level structure between the medial wall and dorsal process and unique raised palatine-jugal ridge in (D). Abbreviations: amw, anterior medial wall; dp, dorsal process; ja, jugal articulation; jp, jugal process; pjr, palatojugal ridge; pp, palatine process; ps, premaxillary shelf. Scale bars equal 5 cm.

than is seen on every other lambeosaurine except *Aralosaurus tubiferus* (Godefroit, Alifanov, and Bolotsky, 2004). This species possesses both medial and lateral walls bounding the premaxillary shelf. Lambeosaurins, on the other hand, have a much shorter medial wall that, in most cases, rises only a short distance from the premaxillary shelf and then turns medially (Sternberg, 1935; Godefroit, Bolotsky, and Van Itterbeeck, 2004; Gates et al., 2007; Evans, 2010; Prieto-Márquez et al., 2012). *Olorotitan* is a stark contrast from the other lambeosaurins with a medial wall that rises in a smooth arch, but terminates well below the dorsal border of the maxillary dorsal process (Godefroit et al., 2012). Parasaurolophs, such as UMNH VP 16666.1, have medial walls that are intermediate between *Adelolophus* and typical lambeosaurins. In UCMP 152028, the well-developed medial wall is broken about the mid-height of the dorsal process. Therefore, the morphology of the leading edge is unclear for much of its length between the lower broken edge and a small, presumably complete section at the same elevation as the dorsal margin of the dorsal process. This feature turns medially to a small degree, but much less so than in *Parasaurolophus* sp. (UMNH VP 16666.1). On the dorsal margin of the medial wall, a narrow space is present between it and the dorsal process that likely would have clasped the anteroventral margin of the lacrimal.

The dorsal process of the maxilla is the prominent central feature of UCMP 152028, and as in other lambeosaurines, is dorsally elongated (Horner et al., 2004). It measures 50 mm anteroposteriorly at its midpoint, and measures 127 mm from the maxillary tooth line to the dorsal-most tip of the process. The morphology of the dorsal region of the dorsal process has not been discussed in the literature, but is markedly different between lambeosaurins and parasaurolophs, with UCMP 152028 having the general morphology of the latter group. In short, lambeosaurins have a sharply tapering dorsal process (Sternberg, 1935; Godefroit, Bolotsky, and Van Itterbeeck, 2004; Prieto-Márquez et al., 2006; Evans, 2010; Godefroit et al., 2012), whereas *Parasaurolophus walkeri* (ROM 768) has an elongated dorsal process body but a rounded dorsal profile, and UMNH VP 16666.1 has a shortened, but sub-squared to rounded, dorsal process. The shape of the *Adelolophus* dorsal process is quite similar to that of *P. walkeri*.

An impression of the jugal articulation facet reveals the morphology of the ventral portion of the anterior process of the jugal as dorsoventrally tall and anteroposteriorly narrow as in *Hypacrosaurus* (Evans, 2010) and *Parasaurolophus* (Sullivan and Williamson, 1999). Based on the comparable morphology of the narrow ventral region of the anterior jugal process and the rounded as opposed to sharply tapering dorsal process, we speculate that *Adelolophus* has a jugal morphology similar to *Parasaurolophus*, where the asymmetrical anterior jugal process extends a small sharp tapered process anteriorly over the maxillary dorsal process. Lambeosaurins have a dorsoventrally symmetrical anterior jugal process (Evans and Reisz, 2007; Evans, 2010; Prieto-Márquez, 2010a).

Posterior to the jugal articulation facet, the palatine process rises as a prominent triangular feature, larger in relative size compared to other lambeosaurines (Sternberg, 1935; Godefroit, Bolotsky, and Van Itterbeeck, 2004; Gates et al., 2007; Evans, 2010; Prieto-Márquez et al., 2012; Godefroit et al., 2012). The size of the palatine process is unknown in parasaurolophs. In addition to its large relative size, the palatine process is unique among lambeosaurines in that it originates from the palatine ridge at approximately 65% the height of the dorsal process, and terminates at approximately 80% the height of the dorsal process (as measured from the base of the jugal process to the tip of the dorsal process, 5.6 cm). For the lambeosaurins *Hypacrosaurus altispinus* (Evans, 2010), *Olorotitan* (Godefroit et al., 2012), and *Amurosaurus* (Godefroit, Bolotsky, and Van Itterbeeck, 2004), the palatine process reaches a maximum of approximately 50% of the height of the dorsal process, based on published specimens such as *H. stebingeri* (MOR 549 6-19-8-9). *Aralosaurus* has a relatively low palatine process (Godefroit, Alifanov, and Bolotsky, 2004), although precise ratios could not be determined from the literature. The elevated palatine process in UCMP 152028 appears to be related to the high medial wall that continues from the anterior end of the maxilla. No other hadrosaurid possesses such an elevated posteromedial wall.

Another unique feature of the UCMP 152028 is a large thickened ridge that runs posterodorsally from the base of the jugal articulation facet straight to the palatine process. The medial wall descends posteriorly to the broken posterior end of the maxilla. Eroded bone prevents examination of the most of the posterior region of the bone, including the ectopterygoid shelf.

The base of the maxilla, excluding the teeth, is slightly rugose with an inflated, well-rounded contour and an accessory protrusion ventral to the dorsal process. This condition differs drastically from that of North American lambeosaurins, which tend to have smooth, tapering bases toward the tooth rows (e.g., *Hypacrosaurus stebingeri* MOR 549 6-19-8-9; *Corythosaurus intermedius* ROM 777; *Velafrons coahuilensis* CPC-59), and to a lesser degree in at least one specimen of *Parasaurolophus* (UMNH VP 16666.1). The morphology of UCMP 152028 is rounded to a greater degree than in UMNH VP 16666.1, and the position of the accessory inflation underlies the jugal in the latter specimen. The incompletely preserved tooth row consists of approximately 24 or 25 tooth positions, and measures 19.25 cm in length.

Medially, UCMP 152028 does not possess any distinguishing characteristics beyond those related to the features discussed above. The medial wall dominates the dorsal half of the element, being mostly dorsoventrally flat except for the anterior eversion along the premaxillary shelf. The row of nutrient foramina form an arch in the dorsal half of the element, as in other lambeosaurines.

SAUROLOPHINAE Brown, 1914 (sensu Prieto-Márquez, 2010a) Gen. et. sp. indet.

Material UMNH VP 20187, partial juvenile skeleton, including skull and appendicular elements; elements from the right side include maxilla, jugal, postorbital, surangular, scapula, humerus, radius, pubis, ischium, tibia, metatarsal II, metatarsal IV; elements from the left side include dentary, coracoid, ilium, ischium, femur. UMNH VP 13881, partial adult skeleton comprising cervical vertebrae, dorsal vertebrae, sacral vertebrae, and anterior, mid-, and distal caudal vertebrae, ribs, and chevrons; elements from the right side

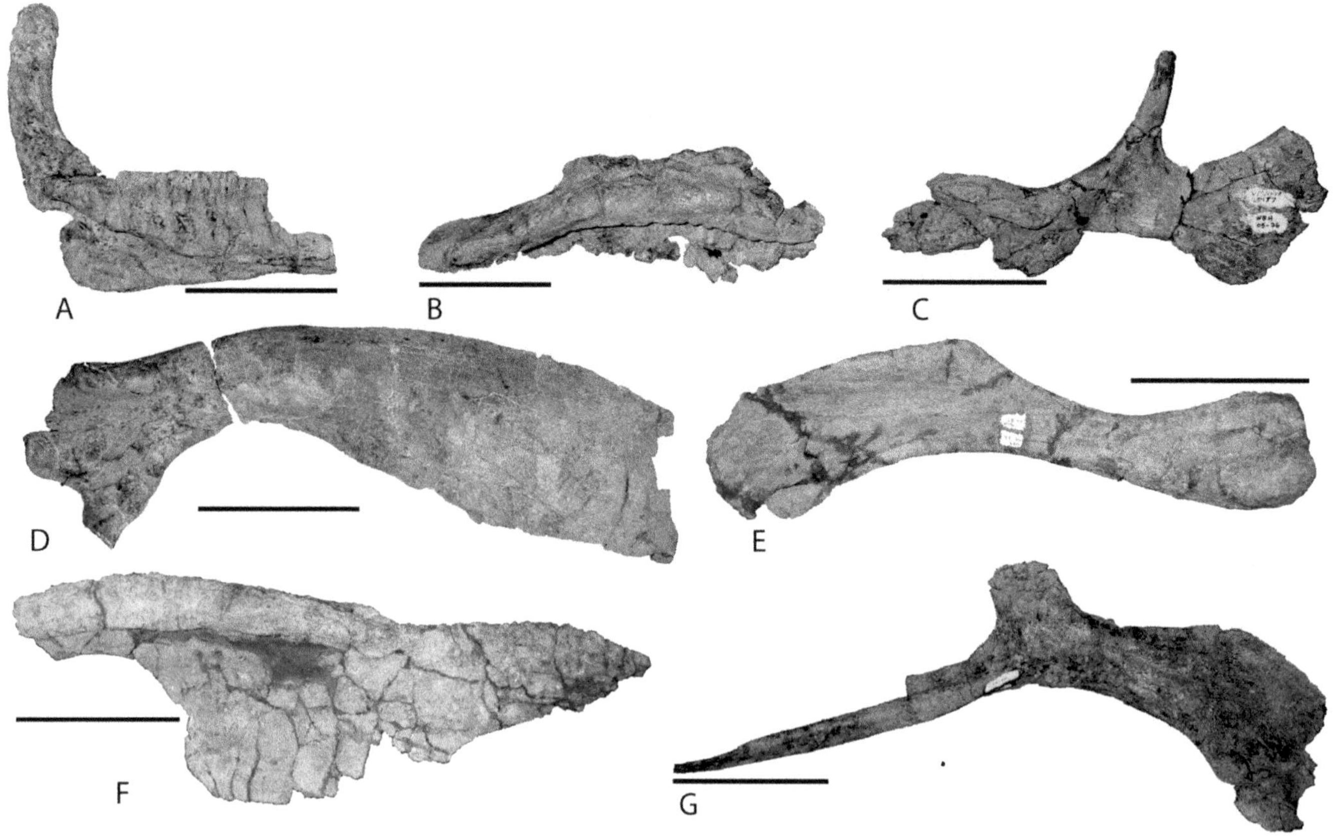

9.3. Skull and appendicular elements from UMNH VP 20187. (A) partial left dentary in medial view; (B) right maxilla in lateral view; (C) right jugal in medial view; (D) left scapula in lateral view; (E) right humerus in posterior view; (F) partial left ilium in lateral view; (G) right pubis in lateral view. Scale bars for (A–C) equal 5 cm. Scale bars for (D–G) equal 10 cm.

include postorbital, sternal, ulna, ilium, pubis, tibia, fibula, metatarsal II, metatarsal IV, phalanges IV-2, 3?; elements from the left side include coracoid, sternal, metacarpals III and IV, pubis, ischium, femur, calcaneum, astragalus, phalanges II-2, IV-1, IV-3?; phalanx III-3 from unknown side.

Locality and Horizon WBH Locality (UMNH VP Locality 324), Middle Member, Wahweap Formation, Grand Staircase–Escalante National Monument, Garfield Co., Utah, United States. The site is located 9 m above an altered volcanic ash dated to 80.1 ± 0.03 Ma (Jinnah et al., 2009).

Comments Two individual saurolophine specimens were recovered from the WBH site (UMNH VP Locality 324), a juvenile (UMNH VP 20187) and an adult (UMNH VP 13881) (see below for discussion of age determination). Skeletal elements that overlap between the two specimens will be discussed in tandem, with specific reference to characteristics not observed on elements of both ages. All skull elements except for a partial postorbital are associated with the younger individual; therefore, characteristics described are subject to varying degrees of ontogenetic change.

Description

Maxilla The partial juvenile right maxilla does not reveal substantial taxonomic information (Fig. 9.3B). It is evident that the low-angled dorsal process is positioned toward the anterior end of the maxilla, and that the palatine process is large and separated from the dorsal process by a relatively wide gap. The ectopterygoid shelf is poorly developed compared to that of maxillae from adult specimens in other saurolophine species such as *Gryposaurus* spp. (e.g., ROM 873, RAM 6797), *Prosaurolophus maximus* (e.g.., TMP 1984.001.0001) and *Edmontosaurus annectens* (NCSM 23119), although preservation could be a causal factor in the divergent morphology. In lateral view, a dorsally directed, shallow, and elongated arch defines the ventral ramus of the maxilla bordering the tooth row.

Jugal The mostly complete right jugal reveals important anatomical information (Fig. 9.3C). The anterior process is elongated, as seen in more primitive iguanodontians (Norman, 2004) and in the saurolophines *Acristavus* and *Maiasaura* (Gates et al., 2011), and the medial side of the process has a long, straight articular facet. The posterior portion of the process constricts to become the jugal neck. The postorbital process extends posterodorsally behind the jugal neck. The inclination of this process is subtle, but based on the fact that this is a juvenile with a presumably shorter face than those of adults (Campione and Evans, 2011), it is likely that its degree of posterior inclination increases ontogenetically. And even though inclination of the postorbital process is an important phylogenetic character, we do not feel compelled to speculate on phylogenetic significance of the trait in this taxon until suitable adult-sized specimens are recovered. The posterior process of the jugal is wide and expanded, and in this respect is more similar to species of *Gryposaurus* (Gates and Sampson, 2007) than to brachylophosaurins (Gates et al., 2011). Additionally, the posteroventral flange is not well developed, as observed in the latter clade. This element possesses a unique suite of characters when compared to currently known hadrosaurid taxa. The only hadrosaurids known with a similar anterior process, *Acristavus* and *Maiasaura* (Gates et al., 2011), have a dorsally projecting postorbital process that is not inclined, and a well-developed posteroventral flange with a narrow posterior process. Species of *Gryposaurus* have a slightly inclined postorbital process and a wider posterior process with a similarly developed posteroventral flange, but lack the long narrow morphology of the anterior process (Gates and Sampson, 2007). Saurolophines closely related to *Prosaurolophus* (Gates and Sampson, 2007; Prieto-Márquez, 2010a; McGarrity, 2011) do not have any of these characteristics.

Postorbital Poor preservation in the juvenile and the adult specimens makes accurate description difficult and precludes meaningful comparisons with other taxa.

Dentary The incomplete juvenile dentary (Fig. 9.3A) consists of the posterior region of the bone because the anterior portion was broken prior to burial. The few preserved tooth rows do not extend beyond the coronoid process, which rises slightly posterodorsally from the main ramus of the dentary and then curves anteriorly towards its dorsal terminus. This exact morphology is not known from other hadrosaurids, but again, since this specimen is from a juvenile, its distinct morphology may represent ontogenetic variation, rather than an apomorphy. It is evident that the surangular did not contribute to much of the coronoid process because the articular facet does not rise to the apex of the process. The base of the dentary is straight when viewed laterally.

Surangular The body of the right surangular (UMNH VP 20187) is present, but all processes are broken and the element is incomplete. A surangular foramen is clearly absent, and it appears that the articular would not have been visible laterally because there is no obvious laterally facing articular surface for the element.

Cervical Vertebrae Several disarticulated cervical vertebrae are known from the adult skeleton, but most are not in good enough condition to warrant detailed description. One specimen is well preserved (Fig. 9.4A, B), however, and has a unique morphology not seen in other hadrosaurids. The centrum is nearly heart shaped, with transverse processes that extends horizontally from the centrum. The lamina connecting the neural arch and postzygapophyses is nearly straight, slopes posterodosally, but does not curve posteriorly as in other hadrosaurid taxa. In general, it is more similar to that of *Iguanodon atherfieldensis* (Norman, 1986) and *Bactrosaurus* (Godefroit et al., 1998). Likewise, the postzygapophyses do not curl posteriorly as in other hadrosaurid taxa (Lambe, 1920; Lull and Wright, 1942), giving the entire neural arch complex a flattened appearance as in both *I. atherfieldensis* (Norman, 1986) and *Bactrosaurus* (Godefroit et al., 1998). It is unclear if the angulation of the aforementioned lamina maybe skewed dorsally, but there is no evidence of crushing or plastic deformation of the lamina and postzygapophyses. The neural spine is anteroposteriorly long but dorsoventrally short.

Dorsal Vertebrae No complete dorsal vertebrae were observed during this study. Pieces of neural spine are anteroposteriorly broad, as expected for a hadrosaurid.

Caudal Vertebrae Anterior caudals have heart-shaped amphiplatyan centra in anterior view. A strong anteroposterior keel occurs on the ventral surface of each centrum. The neural spines of all caudal vertebrae, at least midway through the series, have a wide raised ridge with sharp edges that runs the entire length of the lateral sides of each spine, creating anterior and posterior keels. Additionally, the neural spines are all inclined posteriorly at the base but gradually curve dorsally (Fig. 9.4C). Spines also increase in anteroposterior length toward their distal ends, giving them a club-like appearance as in *Bactrosaurus* (Godefroit et al., 1998). Midcaudals (Fig. 9.4D) look similar to the anterior caudals, except that the centrum increases in length and the neural spine is inclined posteriorly even more so than in *Bactrosaurus* (Godefroit et al., 1998).

Distal caudal vertebrae have a distinctly hexagonal shape in anterior view with marked corners. Each side of these vertebrae is slightly indented leaving a raised ridge connecting two corners of the anterior and posterior hexagonal epiphyses. The most unique feature of these vertebrae are paired foramina on the base of each centrum (Fig. 9.4E, F). Foramina are common on the bases of hadrosaurid caudal vertebrae, but the unique characteristic of those in UMNH VP 13881 is that the foramina are always paired and nearly equal in size. One of us (TAG) has observed large variation in the size and presence or absence of foramina in the distal caudal vertebrae of every other hadrosaurid species examined,

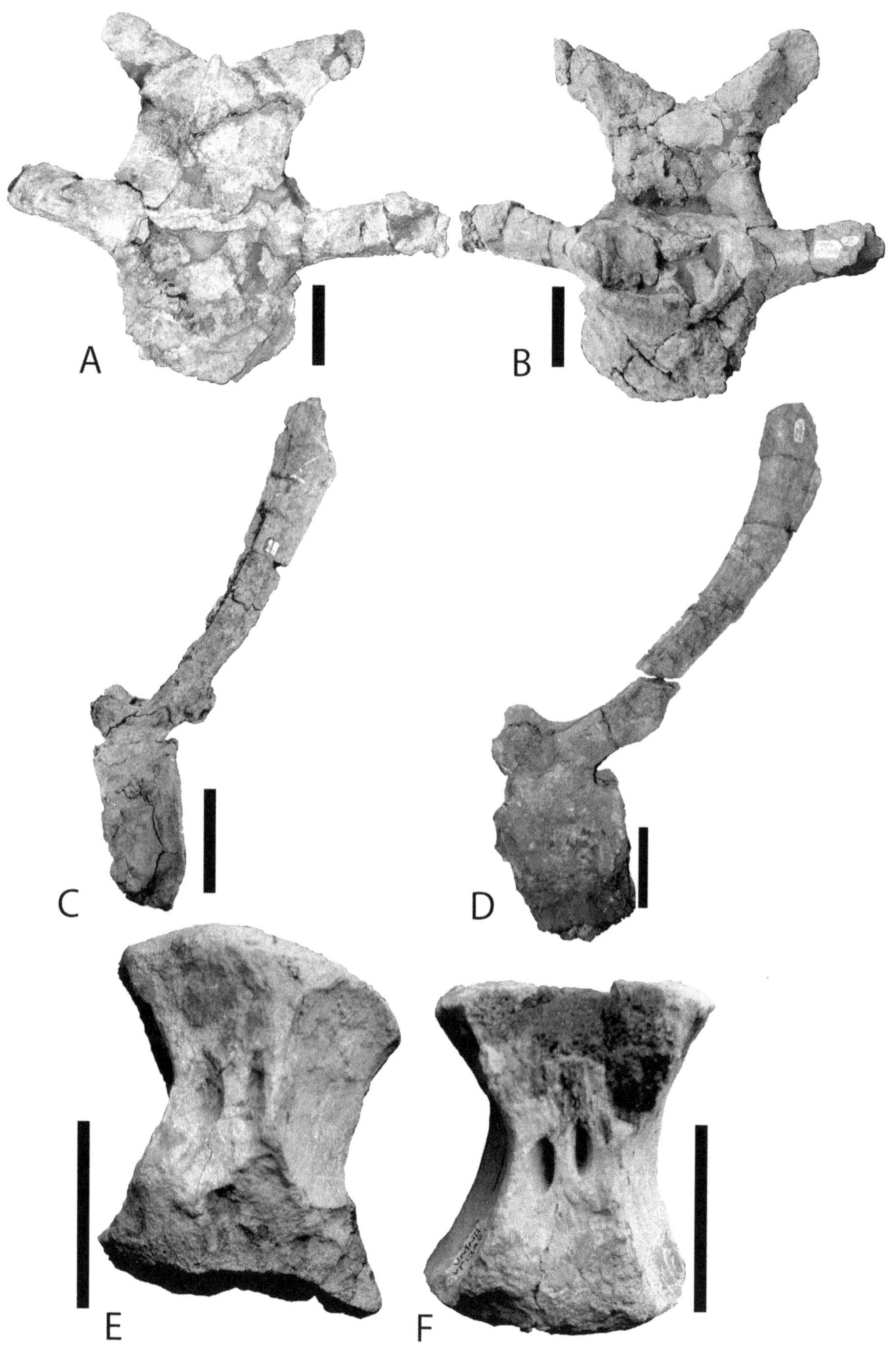

9.4. Vertebrae from UMNH VP 13887. (A) cervical vertebra in anterior view; (B) cervical vertebra in A seen in posteroventral view; (C) anterior caudal vertebra in lateral view; (D) anteromedial caudal in lateral view; (E) distal caudal in ventral view; (F) another distal caudal in ventral view. Scale bars equal 5 cm.

except one caudal vertebra in an MOR specimen identified as *Maiasaura* (specimen number unrecorded).

Pectoral Girdle UMNH VP 13881 includes a partial scapula. The scapula is fairly robust with a triangular acromion process that is similar in morphology to those of most other saurolophines, but differs from *Gryposaurus monumentensis*, which has a rounded acromion process (Gates et al., 2013). The juvenile skeleton UMNH VP 20187 preserves a mostly complete scapula (Fig. 9.3D). In this specimen, the acromion process is the same shape and same relative robustness as the adult from the same site. Also, the angulation of the glenoid is also comparable. The scapular blade of the UMNH VP 20187 has a narrow neck and a broad posterior blade. It is unclear from the preserved adult scapula if the condition seen in the juvenile is retained into adulthood.

A coracoid associated with UMNH VP 13881 is poorly preserved (Fig. 9.5A). It contains the coracoid foramen and a broad contact surface for the scapula. Based on the preserved material, it is difficult to determine the angulation of the glenoid fossa. The ventral process does not extend beyond the main body of the coracoid as in hadrosaurids (Prieto-Márquez, 2010a).

Articulated sternals were found associated with UMNH VP 13887 (Fig. 9.6). They look similar to those reported for *Brachylophosaurus* (Prieto-Márquez, 2007) except that the shafts are shorter. These elements are quite different from those of *Gryposaurus latidens* in that the head is rounded instead of elongated in the latter taxon (Prieto-Márquez, 2012).

Forelimb Only the juvenile skeleton (UMNH VP 20187) preserves a humerus (Fig. 9.3E), which has a characteristic hadrosaurid shape. Although the humeral head is not preserved, it is largely complete. The deltopectoral crest extends approximately 40% of the length of the humerus. The distal condyles lack much definition, but the ulnar condyle is distinctly larger than the radial condyle.

A single ulna is associated with UMNH VP 13881 (Fig. 9.5C). This element possesses a concave proximal articulation surface, with the wide rounded olecranon process that ascends for a short distance. The triangular proximal region tapers into a nearly cylindrical morphology in cross section at its distal articulation surface. Articulation with the radius occurs on the lateral side of the ulna in shallowly concave grooves found on the ulnar head and distal end of the element.

The single right radius (not figured) found at the site is from the juvenile skeleton (UMNH VP 20187). It is a near featureless cylindrical element with a lateral bend. The proximal articulation facet is concave, whereas the distal is convex.

Left metacarpals (MC) III and IV are preserved in UMNH VP 13881 (Fig. 9.5B). metacarpal III is the longest of the two bones, and has convex proximal and distal ends. The lateral side is flattened for articulation with metacarpal IV. A scar two-fifths of the way down the anterior shaft marks the articulation with metacarpal IV. A small scar on the medial proximal shaft marks the articulation with metacarpal II. On metacarpal IV, the proximal end is flattened to slightly convex and laterally expanded. The medial side is flat where it meets metacarpal III. Its distal end is mediolaterally compressed and convex. Overall, the element tapers distally throughout its length.

Ilium A complete adult right ilium (Fig. 9.5H) and a partial juvenile left ilium (Fig. 9.3F) were collected from the WBH site. The adult specimen shows all characteristics of hadrosaurids, a long preacetabular process, two lobed ischial peduncles, and a postacetabular process with no brevis shelf. The supra-acetabular process is broken. The juvenile ilium consists of most of the central plate with broken preacetabular and postacetabular processes, and an ischial peduncle. The preserved portion of the preacetabular process (Fig. 9.5H) is typically hadrosaurid in exhibiting an elongate, anteroventrally projecting process (Lull and Wright, 1942; Horner et al., 2004). The adult preacetabular process descends below the level of the peduncles, continuing to curve posteriorly, which differs from *Brachylophosaurus canadensis* (e.g., MOR 794) that possesses a straighter process (Prieto-Márquez, 2007). The dorsal surface of the preacetabular process does not swell as in other hadrosaurids such as *Gryposaurus notabilis* (ROM 764) or *Saurolophus osborni* (AMNH 5220). A large shelf overhangs the lateral side of the preacetabular process in the juvenile specimen, closely resembling that found on *G. monumentensis* (UMNH VP 12265), but which is not present on the adult specimen. The central plate is flat and relatively narrow. The supra-acetabular process was not well developed in this individual or was poorly preserved. Posteriorly, the postacetabular process of the adult ilium has subparallel dorsal and ventral margins with a slightly pointed terminal margin; in contrast, that of the juvenile has a more primitive triangular morphology, again more similar to *G. monumentensis* (UMNH VP 12265) than to other saurolophines including *G. notabilis* (e.g. ROM 764).

Pubis A left and right pubis is known from UMNH VP 13881 (Fig. 9.5I), and a right pubis is present in UMNH VP 20187 (Fig. 9.3G). The morphology is generally typical for saurolophines: pubic blade with a short neck and an elongated anterior process (blade) that is dorsoventrally expanded. As in *Gryposaurus latidens* (Prieto-Márquez, 2012) and *Gryposaurus notabilis* (ROM 764), the blade of the adult has subparallel dorsal and ventral margins, with the dorsal margin being expanded and the ventral margin connecting seamlessly to the neck of the blade. The juvenile and the adult pubes differ drastically in morphology of the pubic

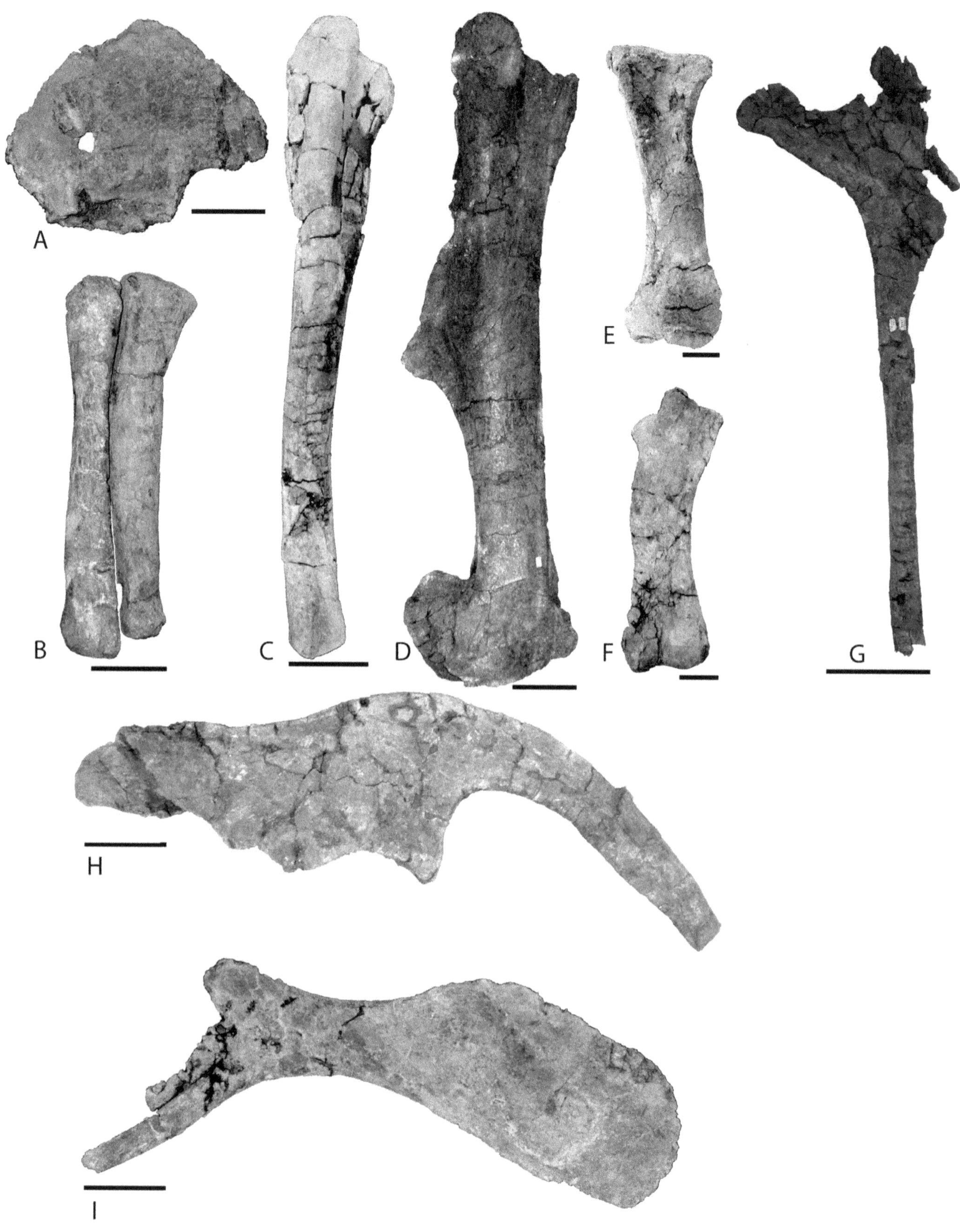

9.5. Appendicular elements from UMNH VP 13887. (A) left coracoid in medial view; (B) left metacarpals II and III in anterior view; (C) right ulna in posterior view; (D) left femur in lateral view; (E) right metatarsal II in lateral view; (F) right metatarsal IV in anterior view; (G) left ischium in lateral view; (H) right ilium in lateral view; (I) left pubis in medial view. Scale bars for (A), (B), (E), and (F) equal 5 cm. Scale bars for (C) (D), (G), (H), and (I) equal 10 cm.

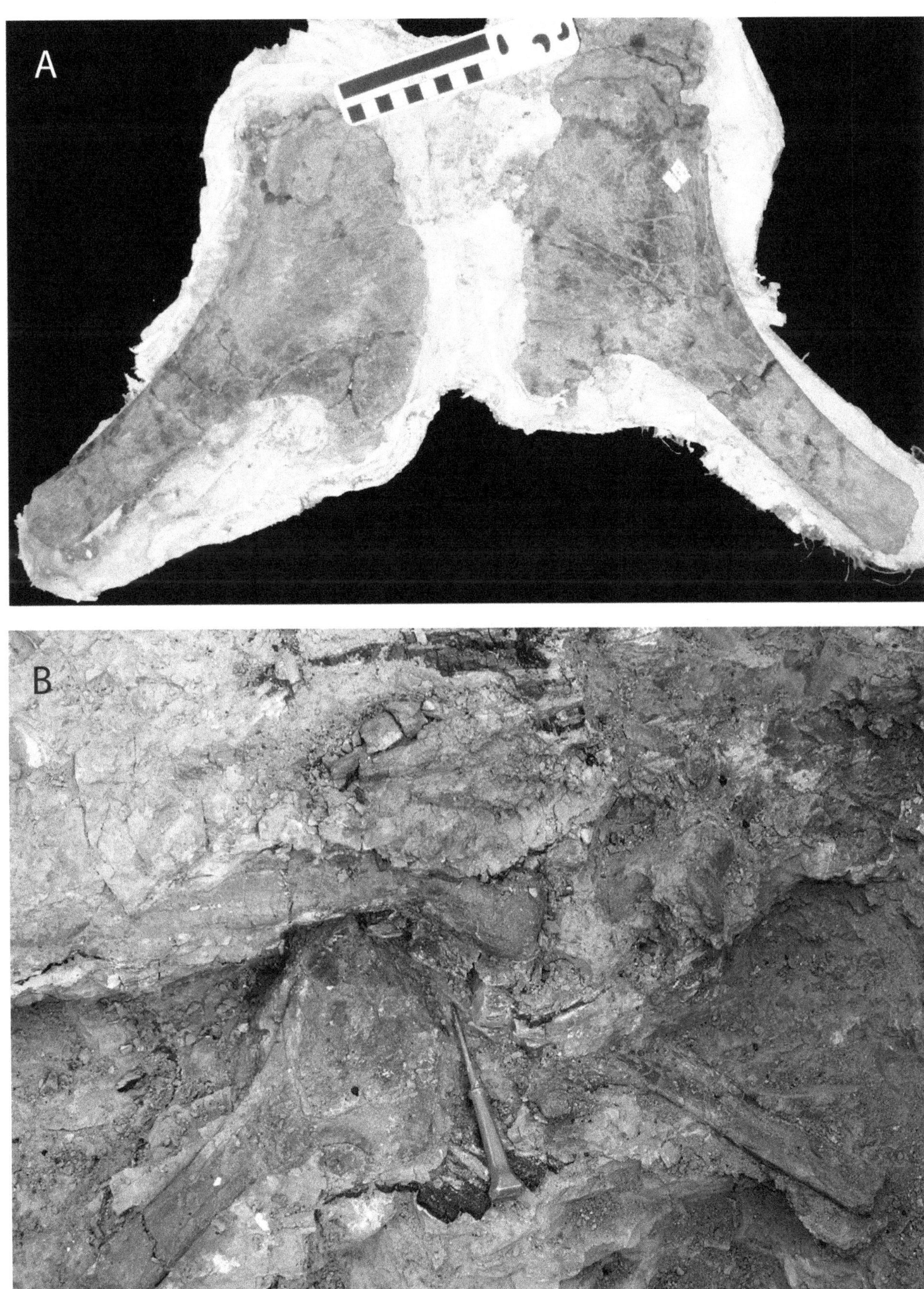

9.6. Sternals from UMNH VP 13887. (A) sternals prepared in jacket as found in quarry; (B) sternals pictured in quarry shown in original articulation. Scale bar equals 10 cm.

blade. In UMNH VP 20187 the ventral margin–and therefore the distal end of the blade–is deflected ventrally, and the dorsal expansion of the blade is more subtle and spread out along the dorsal margin. A morphology almost identical to this is seen in *G. monumentensis* (UMNH VP 12265). The blade morphologies demonstrated on specimens from WBH are not comparable to other hadrosaurs outside of the genus *Gryposaurus*.

Ischium UMNH VP 13887 includes a nearly complete left ischium (Fig. 9.5G), and UMNH VP 20187 includes a partial right and left ischium. Little can be said about the juvenile specimens because of poor preservation, so this discussion is based on the adult specimen (UMNH VP 13887). The iliac peduncle is anteroposteriorly compressed slightly, whereas the pubic peduncle is not well enough preserved to warrant description. The obturator process is broken, so it is unknown if this feature is closed as in more primitive iguanodontians or open as in hadrosaurids. A straight ischial shaft swells slightly in the center, but does not preserve the distal tubercle. This element does not differ significantly from other saurolophine ischia (e.g., *Prosaurolophus maximus* ROM 787; *Gryposaurus notabilis* ROM 764).

Hindlimb Both UMNH VP 13887 (Fig. 9.5D) and 20187 have an associated left femur. The adult specimen has a greater trochanter that is well offset from the femur. The fourth trochanter in the same specimen has a distinctly triangular shape. Its apex is set to ventral end of the trochanter. Also, the base of the trochanter is connected to the femur within a depression on the femoral shaft. Brett-Surman and Wagner (2007) suggested that there was a large amount of variation between individual hadrosaurid femora, and we do not make further comparisons here.

Both UMNH VP 13887 and 20187 preserve a right tibia, and the latter preserves a right fibula. These elements are not identifiably different from the corresponding elements of other hadrosaurids and are not discussed further.

Pes The proximal end of a right metatarsal II from UMNH VP 13887 (Fig. 9.5E) is more anteroposteriorly elongated than wide, and its medial side is concave. A sharp ridge is present on the anterior surface near the articulation with metatarsal III. Its distal end is wide. A right metatarsal IV (Fig. 9.5F) from the same specimen shows a proximal articulation surface that is concave. Again, the medial surface of the proximal end is concave to receive metatarsal III, as is the surface at the distal end. Two-fifths of the way down the medial surface there is a large, flattened protuberance resembling the condition described in *Brachylophosaurus* (Prieto-Márquez, 2007). The general characteristics of these metatarsals are also present in other hadrosaurids (Parks, 1920; Lull and Wright, 1942; Horner et al., 2004; Brett-Surman and Wagner, 2007; Prieto-Márquez, 2007). Numerous manal and pedal phalanges have been recovered from both the juvenile and adult skeletons. These lack any diagnostic traits and are not discussed further. Preserved unguals are hoof-like, but lack a ventral keel (sensu Prieto-Márquez, 2010a).

Femoral Histology and Ontogenetic Status of UMNH VP 20187

The left femur of UMNH VP 20187 was thin-sectioned and analyzed to assess the ontogenetic age of the individual. Poor preservation did not allow preparation of histological samples from the larger WBH specimen. The femur length (FL) of the smaller individual (UMNH VP 20187) is 64.7 cm. Based on size, this suggests that it is a late juvenile (FL = 50 cm) or subadult (FL = 68 cm) if we assume this taxon has a size/ontogenetic stages relationship comparable to *Maiasaura peeblesorum* (Horner et al., 2000); however, size does not always correlate precisely with age, so size-independent criteria such as bone histology are best for assessing ontogenetic stages (Johnson, 1977). We sectioned the femur at midshaft (circumference, 21.6 cm), 23 cm from the distal end. The medullary cavity is 23 mm in diameter. There is very little preserved cancellous bone (Fig. 9.7), but the preserved trabeculae are composed of unordered primary bone. Visible in cross-polarized light, woven bone dominates the inner and middle cortex. The tissue structure in the inner cortex nearest to the medullary cavity can best be described as unordered. The vascular canals are small and longitudinal. The vascular canals are very dense throughout the inner cortex of the bone and oriented slightly radially. Both of these features are evidence of fast growth (deMarerie et al., 2002, 2004. The bone tissue is dominated by primary osteons that are very dense throughout the inner cortex (Fig. 9.7), whereas secondary osteons are rare, being limited to a remodeled area of biomechanical stress located ventral to the fourth trochanter, which is the *m. caudofemoralis* attachment. This limited presence of secondary osteons is typical of the late juvenile growth stage in the saurolophine *Maiasaura* (Horner et al., 2000). Nearer to the periosteum, the tissue becomes more ordered, with dense but reticular vascular canals.

Lines of arrested growth (LAGs) do not appear regularly in hadrosaur long bones until they are half grown (Cooper et al., 2008). UMNH VP 20187 does not preserve any LAGs or identifiable "rest lines." Thus, there are two alternative interpretations for the histology preserved in this specimen: it is possible that this animal was growing so fast that it did not deposit LAGs; or the individual may have been no more than a year old when it died. It is difficult to discern between the two alternatives using histological data. The bone also lacks any evidence of an external fundamental system, a feature that indicates the near cessation of skeletal growth in adult animals (Horner et al., 1999). Lastly, there is no evidence in the external cortex of lamellar-zonal bone, indicative of slowing growth, as found in adult *Maiasaura* (Horner et al., 2000). In summary, although this animal lacks LAGs that would allow us to assign a minimum age, we conclude that this animal was growing rapidly when it died on the basis of the dense radial and reticular vascular canals, and tissue dominated by primary bone.

When compared with the femur histology of an adult specimen of *Hypacrosaurus stebingeri* (MOR 549, FL = 120 cm; Horner et al., 1999), UMNH VP 20187 does not preserve any adult characters. Secondary osteons dominate the inner and outer cortex (Horner et al., 1999), contrary to UMNH VP 20187, with a cortex that is dominated by primary osteons. Furthermore, the femur of MOR 549 preserves at least eight LAGs and an external fundamental system (Horner et al., 1999). Because UMNH VP 20187 has a considerably smaller femur length (64.7 cm), no LAGs, a lack of an external fundamental system, and vascular evidence for sustained continued rapid growth, it does not represent an adult individual. The femur length compared with those from *Maiasaura peeblesorum* (Horner et al., 2000) suggests that the ontogenetic stage of this animal was somewhere between late juvenile and subadult. However, a comparison of bone microstructures between UMNH VP 20187 and *M. peeblesorum* indicates that UMNH VP 20187 shares more histological similarities to the juvenile stage of *M. peeblesorum*. Thus, we hypothesize that this animal was young but almost certainly more than one year old, despite the lack of growth lines. Of course, this age estimate is based on an assumption that the WBH OTU grew at the same rate as *Maiasaura*. Such an assumption is speculative given the differences observed between the Horner et al. (2000) analysis and histological results from UMNH VP 20187.

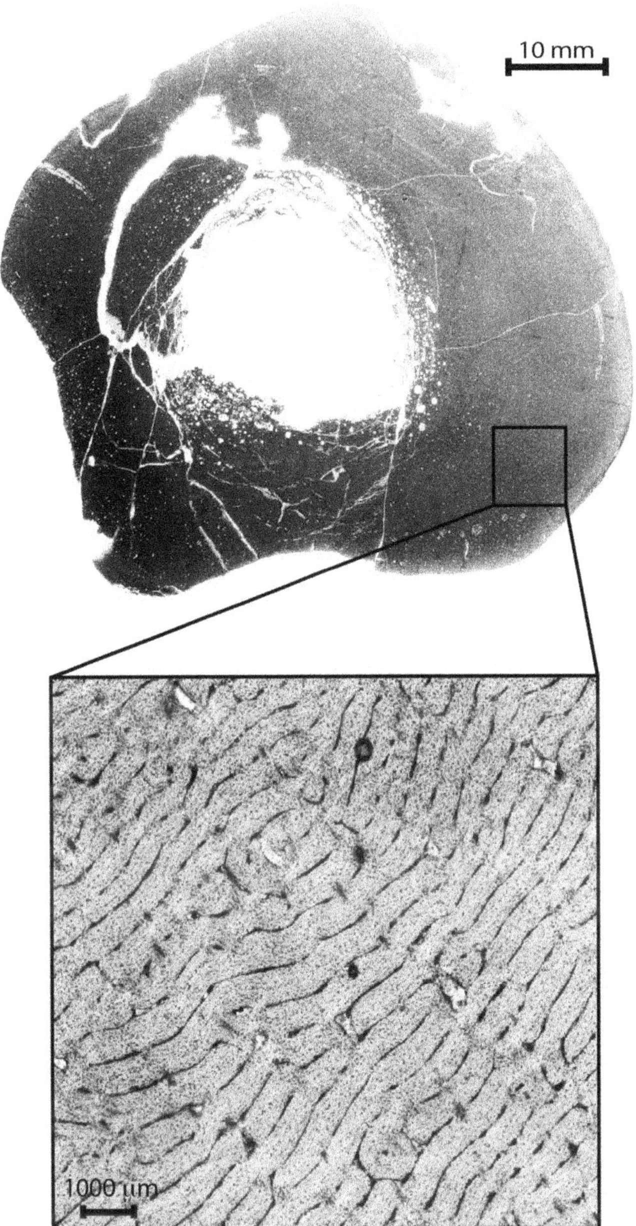

9.7. Thin-section of UMNH VP 20187 taken at the midshaft of the femur, with magnification of inner cortex.

Taxonomic Identification

The juvenile (UMNH VP 20187) and adult (UMNH VP 13887) within the WBH site are tentatively referred to the same taxon, but this assumption is complicated by differences in the morphology of the pubis and ilium. The shape of these bones differs markedly between the two individuals, and thus could be evidence for a species-level distinction. However, all other shared elements exhibit a typical hadrosaurid morphology, thus providing no additional evidence for these being different taxa.

If these skeletons are assumed to be from the same species, which seems likely given that they are from the same site, then the postcranial skeletons found at the WBH site can be diagnosed by a unique combination of characters and possible autapomorphies: straight ventral margin of jugal anterior process; moderately expanded jugal posteroventral flange; cervical vertebrae with posterodorsally and mediolaterally expanded neural lamina; anterior and mid-caudals with neural spines that are posterodorsally curved and broader dorsally than at the base; and distal caudal vertebrae with paired subequal sized foramina on their ventral surface. Despite this suite of characters, we do not feel compelled to name a new taxon in the absence of adult cranial material and the aforementioned pubic differences.

Most of the traits present in the juvenile skull, including the triangular anterior jugal process and the moderate size of the posteroventral flange would likely have been retained into adulthood; similar regions are ontogenetically stable in *Hypacrosaurus stebingeri* and *Maiasaura peeblesorum* (TAG, pers. obs.). Both of these traits are primitive among iguanodontians (Norman, 2004), with the former also found in the hadrosaurids *Acristavus* (Gates et al., 2011) and *Maiasaura* (Horner, 1983; Gates et al., 2011). However, unlike the latter two taxa that demonstrate a well-offset posteroventral flange, the flange in UMNH VP 20187 shows a relatively conservative morphology. The morphology of the maxilla likely changed through ontogeny; and given the fact that the small element does not preserve many informative characters, we do not use it in our taxonomic assessment. Another useful trait is the presence of a long ectopterygoid shelf, which is found in Hadrosauridae (sensu Prieto-Márquez, 2010a).

The partial ilium of the juvenile is also difficult to interpret, because the morphology of this element likely changed through ontogeny. However, this bone is quite similar to that in *Gryposaurus monumentensis*, which has autapomorphic traits that include a shortened, non-tapering, and robust preacetabular process, as well as a markedly triangular postacetabular process (Gates et al., 2013). Although not necessarily unique, the dorsal margin of the iliac plate is flat between the anterior and posterior processes and is, thus, unlike other iguanodontians (hadrosaurids included) that have either a swelling or depressions in this area (Prieto-Márquez, 2010a; Gates et al., 2013).

The adult ilium is more similar to those of other hadrosaurids, and aside from subtle differences in the degree of dorsal expansion of the preacetabular process and depth of the iliac plate, is not diagnostic to the species level within Hadrosauridae. An ilium is not known for either the Utah or Montana specimens of *Acristavus*.

The pubis of UMNH VP 20187 is nearly identical to the pubis of *G. monumentensis* (Gates et al., 2013) whereas UMNH VP 13887 is similar to *G. notabilis* (ROM 764; Parks, 1920) and *Maiasaura* (TCMI 2008.89.2). The preserved pubis on the type specimen of *Acristavus* (MOR 1155) is not complete, but the proximal portion of the prepubic process is similar to that of the latter taxa. No other iguanodontians possesses the combination of morphological traits described here, including *Acristavus gagslarsoni*, which is currently the only named saurolophine from the Wahweap Formation.

The possible autapomorphic traits present on the caudal vertebrae of UMNH VP 13887 are here tentatively accepted as such. That said, the shape of the anterior and middle caudal neural spines (Fig. 9.4C, D), and the nature and distribution of foramina on the ventral surface of the distal caudals (Fig. 9.4E, F) are not well understood for other iguanodontians (see above) because comparatively little emphasis has been placed on the morphology of caudal vertebrae in general. So, although the foramina exhibit unique morphologies per the specimens and taxa examined by Gates, there is a dearth of comparative information about this feature in the literature (cf., Godefroit et al., 1998). Accordingly, more research is needed before the morphology of UMNH VP 13887 is accepted as unique.

Phylogenetic Analyses

In order to inform the taxonomic assignment of the hadrosauroid material described above, two phylogenetic analyses were run on the WBH material, one analysis of only the adult material from UMNH VP 13887 (Fig. 9.8A) and the second with all material from WBH (Fig. 9.8C). The character set of Gates et al. (2011) was used as the basis of the analyses, with the inclusion of WBH material (UMNH VP 13887, 16 of 116 characters scored; UMNH VP 20187, 41 of 116 characters scored; for scorings see Appendix 9.1). The final data matrix consisted of 116 characters and 16 taxa. TnT v.1.1 (Goloboff et al., 2003) was used to analyze the data matrix. *Iguanodon* was considered the outgroup taxon, and all characters were unordered.

The analysis resulted in the recovery of two most parsimonious trees (204 tree score) when only the adult material was considered. In one of the most parsimonious trees, the WBH taxon is placed one node above *Iguanodon* at the base of the tree (Fig. 9.8A). The second most parsimonious tree placed the WBH OTU within the *Prosaurolophus-Saurolophus* clade (Fig. 9.8B). The large disparity in phylogenetic hypotheses demonstrates that the recovered skeletal material of the adult specimen is a mélange of both primitive and derived traits, as might be expected in an Early to Middle Campanian hadrosauroid.

With both the juvenile and adult skeletal material scored as a single OTU, a single most-parsimonious tree (210 tree score) was obtained. In this case, the WBH OTU is placed above *Telmatosaurus* and below *Corythosaurus*, thereby forming the sister taxon of Hadrosauridae (senso Prieto-Márquez [2010a]; Fig. 9.8C). Given the unclear ontogenetic trajectory of the skull elements used to produce the phylogenetic tree in Figure 9.8C, the placement of the WBH OTU remains unknown within the greater context of hadrosauroids. This analysis remains difficult to interpret because juvenile individuals of a given taxon will often be recovered in a more basal systematic position than their corresponding adult (e.g., Tsuihiji et al., 2011). UMNH VP 13887 does possess two ambiguous synapomorphies of Hadrosauridae as defined by Prieto-Márquez (2010a): coracoid reduced in size relative to the scapula (205[1]), and recurved and caudoventrally directed ventral process of the coracoid (210[1]). The coracoid shows the only synapomorphies of Hadrosauridae, but there are other characters in which this bone is too poorly preserved for us to accurately gauge ratios or make interpretations of relative size. Elements such as the large scapula are too poorly preserved to allow us to confidently identify other synapomorphies, and the presumed juvenile elements from UMNH VP 20187 are not considered here.

Taphonomy of the WBH Locality

The WBH site, UMNH VP Locality 324, is one of the most significant sites yet discovered in the Wahweap Formation because it contains the most complete hadrosaurid skeletons known from this unit, comprising two individuals of different ontogenetic age. The geology and taphonomy of the WBH site was studied in order to understand the factors that resulted in the deposition of this unique assemblage.

The middle member of the Wahweap Formation consists largely of fine-grained (claystone-siltstone) floodplain deposits, which can be interpreted as the deposits of swamps, oxbow lakes, and waterlogged soils. Isolated channel-sandstone bodies form a subordinate component of the middle member (Jinnah and Roberts, 2011). The WBH site occurs in a laterally extensive bed of mudrock, interpreted as a distal-floodplain deposit resting approximately 9 m above the 80.1 Ma ash bed of Jinnah et al. (2009). Horizontal laminations in the mudrock–together with crab claws, fish scales, plant hash, and coal stringers–indicate subaqueous deposition in a low-energy floodplain pond. The abundant presence of taxodiaceous leaves further supports this interpretation. In addition to these features, close to one dozen coalified trees were observed crisscrossing the quarry deposit, with bones lying only millimeters above them. Directional data gathered from elongate bones (Fig. 9.9) were recorded and analyzed using one-sample directional statistics in PAST v. 2.13 (Hammer et al., 2001). The mean orientation of these elements is 44° relative to north. The resultant Rayleigh statistic of 0.112 suggests probability of random distribution at 0.245, well outside the 95% confidence interval, and resulting in an interpretation of uniform distribution. However, a χ^2 test of the directional data calculated from eight bins shows the probability of uniform distribution at 0.038, a result consistent with a slightly preferred directionality for the WBH long bones. According to Hammer and Harper (2006) an instance of failing to reject the null hypothesis in a Rayleigh test but rejecting the null hypothesis in a χ^2 test should be interpreted as an indication of multiple preferred directions within a dataset, although not uniform distribution. Visual inspection of the rose diagrams shows a bimodal distribution of elements. Elongate bones at the site are generally aligned exhibiting either a northeast-southwest trend or, less commonly, a northwest-southeast trend. Even though bimodal

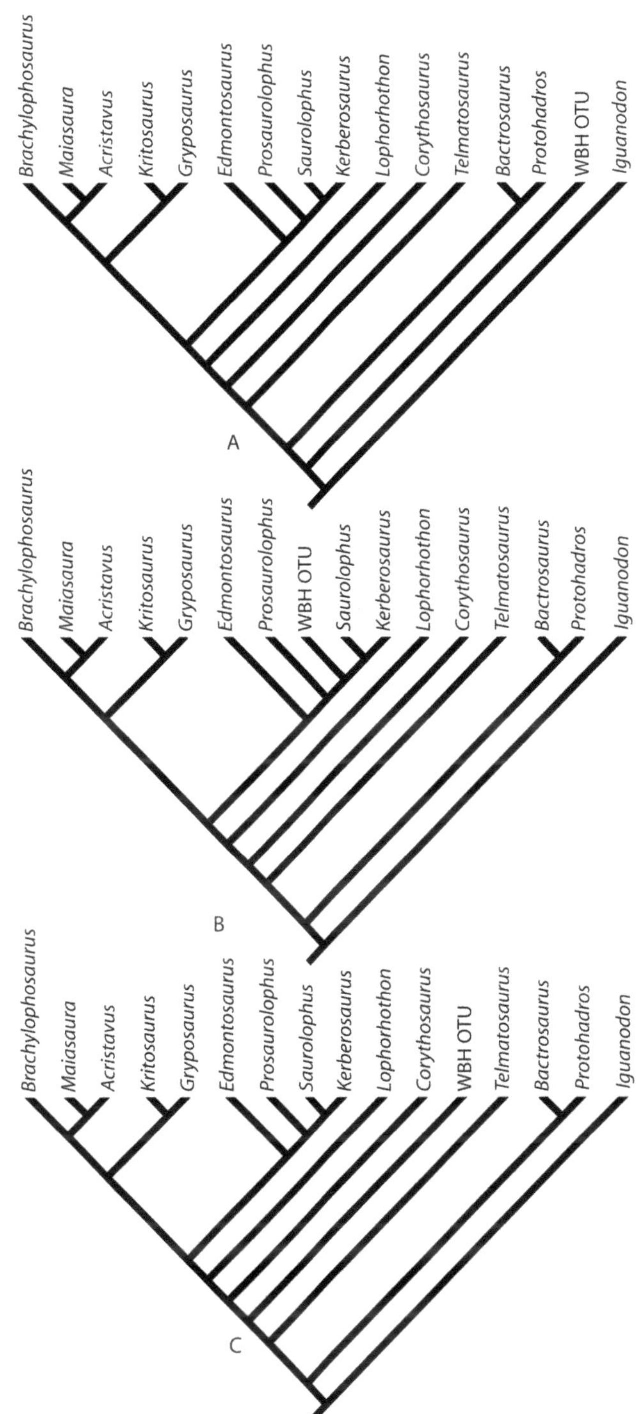

9.8. Phylogenetic trees showing relationships of hadrosaurids discovered in WBH quarry. (A) first of two most-parsimonious trees produced from matrix of only adult skeletal material (tree length 204); (B) second of two most-parsimonious trees produced from only adult WBH material (tree length 204); (C) tree produced from matrix of juvenile and adult material (one tree produced; tree length 210). Matrices are based on that published by Gates et al. (2011). See text for details.

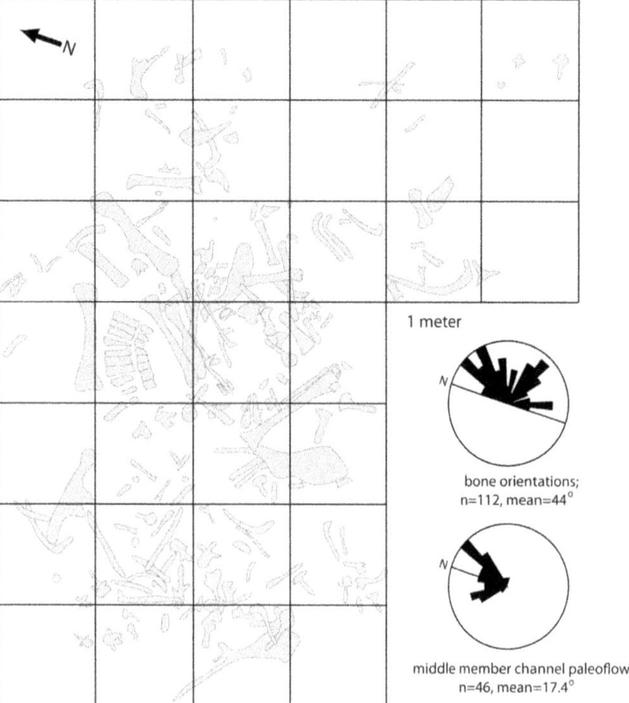

9.9. Quarry map from WBH, UMNH VP Locality 324. Left side marks edge of quarry facing hill. Remaining sides are open faces. Upper rose diagram shows unidirectional orientations (east) of all long bones found in the WBH locality. Lower rose diagram shows paleocurrent directions derived from cross-bedded sandstones in the Middle Member of the Wahweap Formation.

distributions (such as this) are credited to shallowly flowing water, there is minimal evidence at the site for such flow, so this causal agent must be considered only a tentative possibility. Nonetheless, a northeast flow is consistent with the overall paleocurrent direction for the Wahweap Formation based on paleocurrent data collected from trough cross-bedding in channels (Fig. 9.9). Just above the fossiliferous horizon there is at least 10 m of floodplain mudrock, suggesting that a relatively long-lived floodplain was established at the site at the time of deposition.

Preburial condition of bones seems relatively good, with only minor predepositional breakage (40% of ribs and 30% of vertebrae), minimal evidence for subaqueous abrasion, and no evidence of tooth marks. Based on the fine-grained nature of the matrix, the lack of major current alignment, and the lack of subaqueous abrasion on the bones, the WBH fossil assemblage is interpreted as autochthonous, with the two hadrosaurid individuals likely dying at the site and, subsequently, disarticulating via decomposition, current action in calm to slightly flowing water, and, possibly, scavenging.

DISCUSSION

Significance of Adelolophus

Albeit a single element, UCMP 152028 clearly is differentiable from all other known hadrosaurid maxillae. It is interpreted, therefore, as representing a new taxon. *Adelolophus hutchisoni* is not the first new lambeosaurine to be recognized on the basis of a single, isolated maxilla. Wagner and Lehman (2009) described *Angulomastacator daviesi* on material of similar completeness. Distinguishing between species of closely related taxa may be a concern when establishing a new taxon using a single element (such as a maxilla). However, at least two species of *Parasaurolophus* (*P. walkeri* ROM 768 and an unidentified species from the Kaiparowits Formation) have quite distinct maxilla morphology (TAG, pers. obs.). Thus, we conclude that the combination of unique characteristics present in UCMP 152028 and the known species-level variation in maxillary morphology observed within other lambeosaurine taxa warrants the establishment of A. *hutchisoni*.

The precise stratigraphic provenance of UCMP 152028 is not known for certain. However, assuming that it came from the Upper Member of the Wahweap Formation, which is the stratigraphically highest fossiliferous member of the formation, we may still conclude that this maxilla represents the earliest definitive lambeosaurine skeletal material yet known from North America (~78 Ma based on dates in Jinnah et al. [2009]).

The morphology of UCMP 152028 shows that all of the definitive morphological features that characterize lambeosaurine maxillae–as well as all other skull modifications correlated with the maxillary characters–were present by the middle Campanian. This evidence does not support or refute the proposition that Lambeosaurinae and Saurolophinae diverged from a common ancestor during the Santonian, as recently set forth by Prieto-Márquez (2010b). Yet if evidence is ever found of lambeosaurine skeletal material of such a derived grade as in UCMP 152028 within the lower portions of the Wahweap Formation (close to 81 Ma [Jinnah et al., 2009]), then closer scrutiny will have to be paid to the timing and geography of the phylogenetic split between Saurolophinae and Lambeosaurinae.

WBH Locality Behavioral Implications

Does the discovery of two hadrosaurid individuals of different sizes and ontogenetic ages at one WBH site (containing no other macrovertebrate remains) constitute evidence for intraspecies social behaviors (parent-juvenile interactions or herding) in hadrosaurids? We conclude that such intraspecies social behaviors cannot be inferred from this assemblage for the following reasons. First, it remains unclear whether the two specimens are from the same species. With further discoveries of more complete remains, this problem may be remedied. Second, and assuming that the specimens are from the same species, all social behaviors–such as herding, parent-offspring bonding, etc.–that could possibly be elicited to explain the presence of these two hadrosaurid individuals in a fossil site, require evidence of those behaviors (not preserved), and some form of evidence that the two individuals at least died at the same time (and thus lived at about the same time). In fact, direct evidence of such behaviors and time of death are rarely preserved at fossil localities. Despite previous attempts to reconstruct hadrosaurid sociality using taphonomic evidence (Horner and Makela, 1979; Fiorillo, 1987; Horner, 1987; Varricchio and Horner, 1993), we feel that the WBH site does not provide sufficient evidence to infer hadrosaurid social interactions.

Wahweap Hadrosaurid Diversity

Based on the present report and Gates et al. (2011, 2013), there are a minimum of four hadrosaurid taxa known from the Wahweap Formation. These include *Acristavus gagslarsoni* (Gates et al., 2011), *Adelolophus hutchisoni*, the WBH OTU, and a new species based on a partial skeleton from the Upper Member (UMNH VP 9548; Gates et al., 2013). The latter specimen is similar to *Brachylophosaurus canadensis* but distinct from *Acristavus* and the WBH OTU. Even though at least three of the specimens are known to be stratigraphically well separated from one another (see Fig. 9.1), too few specimens are known from the Wahweap Formation to provide meaningful stratigraphic ranges for the taxa they represent. Nonetheless, the notion that these hadrosaurids do not overlap in time may have some validity based on biostratigraphic comparisons with dinosaur-rich sections of similar age in Montana. Specifically, *Acristavus* in southern Utah occurs in strata that are chronostratigraphically equivalent to those that host the type specimen of this same species in Montana. In both cases, these strata are stratigraphically distinct from those containing *Brachylophosaurus*. Similarly, a saurolophine maxilla, UMNH VP 9548, occurs at a level similar in age to those strata that host *Brachylophosaurus* in Montana. Such data, limited though they may be, suggest that the preliminary biostratigraphic pattern emerging from the Wahweap Formation does not differ from that documented in the better-sampled sections in the northern part of the Western Interior Basin.

ACKNOWLEDGMENTS

We thank the following people: Howard Hutchison for his work collecting UCMP 152028 and allowing us to study this specimen; Pat Holroyd for allowing the specimen on loan to the UMNH; Grand Staircase–Escalante National Monument staff, especially Alan Titus, for funding, permitting, logistical, excavation assistance, and the helicopter expedition that led to the discovery of the WBH site; Scott Foss and Laurie Bryant from the Utah BLM for permitting; Scott Sampson for the creation and preservation of the Kaiparowits Basin Project; Eric Lund, Utah Friends of Paleontology, UMNH volunteers, and all others who helped excavate and prepare the skeletal elements from the WBH site; Randy Irmis for permission to histosection WBH material. Albert Prieto-Márquez provided comparative images and publications. Virginia Greene provided the illustration of UCMP 152028. Helpful reviews provided by David Evans and Derek Larson greatly increased the strength of the chapter. Funding was provided by the National Science Foundation, the Bureau of Land Management, Geological Society of America Research Grant (CL), Arthur J. Boucot Award–Paleontological Society Student Research Grant (CL), and the Grand Staircase Institute of Science Student Research Grant (CL).

LITERATURE CITED

Brett-Surman, M. K., and J. R. Wagner. 2007. Discussion of character analysis of the appendicular anatomy in Campanian and Maastrichtian North American hadrosaurids: variation and ontogeny; pp. 135–170 in K. Carpenter (ed.), Horns and Beaks: Ceratopsian and Ornithopod Dinosaurs. Indiana University Press, Bloomington, Indiana.

Brown, B. 1914. *Corythosaurus casuarius*, a new crested dinosaur from the Belly River Cretaceous, with provisional classification of the family Trachodontidae. Bulletin of the American Museum of Natural History 33:559–565.

Campione, N. E., and D. C. Evans. 2011. Cranial growth and variation in edmontosaurs (Dinosauria: Hadrosauridae): implications for latest Cretaceous megaherbivore diversity in North America. PloS ONE 6(9):e25186.

Cooper, L. N., A. H. Lee, M. L. Taper, and J. R. Horner. 2008. Relative growth rates of predator and prey dinosaurs reflect effects of predation. Proceedings of the Royal Society B 275:2609–2615.

Cope, E. D. 1869. Synopsis of the extinct Batrachia, Reptilia, and Aves of North America. Transactions of the American Philosophical Society 14:1–252.

Currie, P. J. 2005. History of research; pp. 3–33 in P. J. Currie, and E. B. Koppelhus (eds.), Dinosaur Provincial Park: A Spectacular Ancient Ecosystem Revealed. Indiana University Press, Bloomington, Indiana.

deMarerie, E., J. Cubo, and J. Castanet. 2002. Bone typology and growth rates: testing and quantifying 'Amprino's rule' in the mallard (*Anas platyrhynchos*). Current Research in Biology 325:221–230.

deMarerie, E., J.-P. Robin, D. Verrier, J. Cubo, R. Groscolas, and J. Castanet. 2004. Assessing a relationship between bone microstructure and growth rate: a fluorescent labeling study in the king penguin chick (*Aptenodytes patagonicus*). Journal of Experimental Biology 207:869–879.

Evans, D. C. 2010. Cranial anatomy and systematics of *Hypacrosaurus altispinus*, and a comparative analysis of skull growth in lambeosaurine hadrosaurids (Dinosauria: Ornithischia). Zoological Journal of the Linnean Society 159:393–434.

Evans, D. C., and R. R. Reisz. 2007. Anatomy and relationships of *Lambeosaurus magnicristatus*, a crested hadrosaurid dinosaur (Ornithischia) from the Dinosaur Park Formation, Alberta. Journal of Vertebrate Paleontology 27:373–393.

Fiorillo, A. R. 1987. Significance of juvenile dinosaurs from Careless Creek Quarry (Judith River Formation), Wheatland County, Montana; pp. 88–95 in P. J. Currie, and E. H. Koster (eds.), Fourth Symposium on Mesozoic Terrestrial Ecosystems, Short Papers. Tyrrell Museum of Palaeontology, Drumheller, Alberta.

Gates, T. A., and D. C. Evans. 2005. Biogeography of Campanian hadrosaurid dinosaurs from western North America; pp. 33–39 in D. R. Braman, F. Therrien, E. B. Koppelhus, and W. Taylor (eds.), Dinosaur Park Symposium Short Papers, Abstracts, and Programs. Royal Tyrrell Museum of Palaeontology, Drumheller, Alberta.

Gates, T. A., and S. D. Sampson. 2007. A new species of *Gryposaurus* (Dinosauria: Hadrosauridae) from the Late Campanian Kaiparowits Formation. Zoological Journal of the Linnean Society 151:351–376.

Gates, T. A., J. R. Horner, R. R. Hanna, and C. R. Nelson. 2011. New unadorned hadrosaurine hadrosaurid (Ornithopoda: Dinosauria) from the Campanian of North America. Journal of Vertebrate Paleontology 31:798–811.

Gates, T. A., S. D. Sampson, C. R. Delgado-Jesus, L. E. Zanno, D. A. Eberth, R. Hernandez-Rivera, M. C. A. Martinez, and J. I. Kirkland. 2007. *Velafrons coahuilensis*, a new lambeosaurine hadrosaurid (Dinosauria: Ornithopoda) from the late Campanian Cerro del Pueblo Formation, Coahuila, Mexico. Journal of Vertebrate Paleontology 27:917–930.

Gates, T. A., E. K. Lund, C. A. Boyd, D. D. DeBlieux, A. L. Titus, D. C. Evans, M. A. Getty, J. I. Kirkland, and J. G. Eaton. 2013. Ornithopod dinosaurs from the Grand Staircase–Escalante National Monument region, Utah, and their role in paleobiogeographic and macroevolutionary studies; pp. 463–481 in A. L. Titus and M. A. Loewen (eds.), At the Top of the Grand Staircase: The Late Cretaceous of Southern Utah. Indiana University Press, Bloomington, Indiana.

Gates, T. A., S. D. Sampson, L. E. Zanno, E. M. Roberts, J. G. Eaton, R. L. Nydam, J. H. Hutchison, J. A. Smith, M. A. Loewen, and M. A. Getty. 2010. Biogeography of terrestrial and freshwater vertebrates from the late Cretaceous (Campanian) Western Interior of North America. Palaeogeography, Palaeoclimatology, Palaeoecology 291:371–387.

Godefroit, P., V. Alifanov, and Y. L. Bolotsky. 2004. A re-appraisal of *Aralosaurus tubiferus* (Dinosauria, Hadrosauridae) from the Late Cretaceous of Kazakhstan. Bulletin de l'Institut royal des Sciences naturelles de Belgique 74:139–154.

Godefroit, P., Y. L. Bolotsky, and J. Van Itterbeeck. 2004. The lambeosaurine dinosaur *Amurosaurus riabinini*, from the Maastrichtian of Far Eastern Russia. Acta Palaeontologica Polonica 49:585–618.

Godefroit, P., Z. Dong, P. Bultynck, H. Li, and L. Feng. 1998. Sino-Belgian Cooperation Program: "Cretaceous dinosaurs and mammals from Inner Mongolia" 1: New *Bactrosaurus* (Dinosauria: Hadrosauroidea) material from Iren Dabasu (Inner Mongolia, P.R. China). Bulletin de l'Institut royal des Sciences naturelles de Belgique 68 (Supplement):3–70.

Godefroit, P., Y. L. Bolotsky, and I. Y. Bolotsky. 2012. Osteology and relationships of *Olorotitan arharensis*, a hollow-crested hadrosaurid dinosaur from the latest Cretaceous of Far Eastern Russia. Acta Palaeontologica Polonica 57:527–560.

Goloboff, P., J. Farris, and K. Nixon. 2003. T.N.T.: Tree analysis using new technology. Available at www.cladistics.com. Accessed February 2012.

Hammer, Ø., and D. Harper. 2006. Paleontological Data Analysis. Blackwell, Malden, Massachusetts, 351 pp.

Hammer, Ø., D. A. T. Harper, and P. D. Ryan. 2001. PAST: Paleontological statistics software package for education and data analysis. Palaeontologia Electronica 4:1–9.

Horner, J. R. 1983. Cranial osteology and morphology of the type specimen of *Maiasaura peeblesorum* (Ornithischia: Hadrosauridae), with discussion of its phylogenetic position. Journal of Vertebrate Paleontology 3:29–38.

Horner, J. R. 1987. Ecologic and behavioral implications derived from a dinosaur nesting site; pp. 50–63 in S. J. Czerkas, and E. C. Olson (eds.), Dinosaurs Past and Present, Volume 2. Natural History Museum of Los Angeles County, Los Angeles, California.

Horner, J. R., and R. Makela. 1979. Nest of juveniles provides evidence of family structure among dinosaurs. Nature 282:296–298.

Horner, J. R., A. de Ricqlès, and K. Padian. 1999. Variation in dinosaur skeletochronology indicators: implications for age assessment and physiology. Paleobiology 25:295–304.

Horner, J. R., A. de Ricqlès, and K. Padian. 2000. Long bone histology of the hadrosaurid dinosaur *Maiasaura peeblesorum:* growth dynamics and physiology based on an ontogenetic series of skeletal elements. Journal of Vertebrate Paleontology 20:115–129.

Horner, J. R., D. B. Weishampel, and C. Forster. 2004. Hadrosauridae; pp. 438–463 in D. B. Weishampel, P. Dodson, and H. Osmólska (eds.), The Dinosauria, Second Edition. University of California Press, Berkeley, California.

Jinnah, Z. A., and E. M. Roberts. 2011. Facies associations, paleoenvironment, and base-level changes in the Upper Cretaceous Wahweap Formation, Utah, U.S.A. Journal of Sedimentary Research 81:266–283.

Jinnah, Z. A., E. M. Roberts, A. L. Deino, J. S. Larsen, P. K. Link, and C. M. Fanning. 2009. New ^{40}Ar/^{39}Ar and detrital zircon U-Pb ages for the Upper Cretaceous Wahweap and Kaiparowits formations on the Kaiparowits Plateau, Utah: implications for regional correlation, provenance, and biostratigraphy. Cretaceous Research 30:287–299.

Johnson, R. 1977. Size independent criteria for estimating relative age and the relationships among growth parameters in a group of fossil reptiles (Reptilia: Ichthyosauria). Canadian Journal of Earth Sciences 14:1916–1924.

Lambe, L. M. 1920. The hadrosaur *Edmontosaurus* from the Upper Cretaceous of Alberta. Canada Department of Mines Memoir 120:1–79.

Lull, R. S., and N. E. Wright. 1942. Hadrosaurian Dinosaurs of North America. Geological Society of America Special Papers 40. 242 pp.

Marsh, O. C. 1882. Classification of the Dinosauria. American Journal of Science (Third Series) 23:81–96.

McGarrity, C. T. 2011. Cranial anatomy and variation of *Prosaurolophus maximus* (Dinosauria: Hadrosauridae). M.Sc. thesis, University of Toronto, Toronto, Ontario, 159 pp.

Norman, D. B. 1986. On the anatomy of *Iguanodon atherfieldensis* (Ornithischia: Ornithopoda). Bulletin de l'Institut royal des Sciences naturelles de Belgique: Sciences de la Terre 56:281–372.

Norman, D. B. 2004. Basal Iguanodontia; pp. 413–437 in D. B. Weishampel, P. Dodson, and H. Osmólska (eds.), The Dinosauria, Second Edition. University of California Press, Berkeley, California.

Owen, R. 1842. Report on British fossil reptiles. Report of the British Association of Advanced Sciences 9:60–204.

Parks, W. A. 1920. The osteology of the trachodont dinosaur *Kritosaurus incurvimanus*. University of Toronto Studies, Geology Series 11:1–75.

Parks, W. A. 1923. *Corythosaurus intermedius*, a new species of trachodont dinosaur. University of Toronto Studies, Geological Series 15:5–57.

Prieto-Márquez, A. 2007. Postcranial osteology of the hadrosaurid dinosaur *Brachylophosaurus canadensis* from the Late Cretaceous of Montana; pp. 91–116 in K. Carpenter (ed.), Horns and Beaks: Ceratopsian and Ornithopod Dinosaurs. Indiana University Press, Bloomington, Indiana.

Prieto-Márquez, A. 2010a. Global phylogeny of hadrosauridae (Dinosauria: Ornithopoda) using parsimony and Bayesian methods. Zoological Journal of the Linnean Society 159:435–502.

Prieto-Márquez, A. 2010b. Global historical biogeography of hadrosaurid dinosaurs. Zoological Journal of the Linnean Society 159:503–525.

Prieto-Márquez, A. 2012. The skull and appendicular skeleton of *Gryposaurus latidens*, a saurolophine hadrosaurid (Dinosauria: Ornithopoda) from the Early Campanian (Cretaceous) of Montana, USA. Canadian Journal of Earth Sciences 49:510–532.

Prieto-Márquez, A., L. M. Chiappe, and S. H. Joshi. 2012. The lambeosaurine dinosaur *Magnapaulia laticaudus* from the Late Cretaceous of Baja California, northwestern Mexico. PloS ONE 7(6):e38207.

Prieto-Márquez, A., R. Gaete, G. Rivas, A. Galobart, and M. Boada. 2006. Hadrosauroid dinosaurs from the Late Cretaceous of Spain: *Pararhabdodon isonensis* revisited and *Koutalisaurus kohlerorum*, gen. et sp. nov. Journal of Vertebrate Paleontology 26:929–943.

Seeley, H. G. 1888. On *Thecospondylus daviesi* (Seeley), with some remarks on the classification of the Dinosauria. Quarterly Journal of the Geological Society of London 44:79–86.

Sternberg, C. M. 1935. Hooded hadrosaurs of the Belly River Series of the Upper Cretaceous: a comparison, with descriptions of new species. Bulletin of the Geological Survey of Canada 77:1–37.

Sullivan, R. M., and T. E. Williamson. 1999. A new skull of *Parasaurolophus* (Dinosauria: Hadrosauridae) from the Kirtland Formation of New Mexico and a revision of the genus. New Mexico

Museum of Natural History and Science Bulletin 15:1–52.

Tsuihiji, T., M. Watabe, K. Tsogtbaatar, T. Tsubamoto, R. Barsbold, S. U. Suzuki, A. H. Lee, R. C. Ridgely, Y. Kawahara, and L. M. Witmer. 2011. Cranial osteology of a juvenile specimen of *Tarbosaurus bataar* (Theropoda, Tyrannosauridae) from the Nemegt Formation (Upper Cretaceous) of Bugin Tsav, Mongolia. Journal of Vertebrate Paleontology 31:1–21.

Varricchio, D. J., and J. R. Horner. 1993. Hadrosaurid and lambeosaurid bone beds from the Upper Cretaceous Two Medicine Formation of Montana: taphonomic and biologic implications. Canadian Journal of Earth Sciences 30:997–1006.

Wagner, J. R., and T. Lehman. 2009. An enigmatic new lambeosaurine hadrosaur (Reptilia: Dinosauria) from the Upper Shale Member of the Campanian Aguja Formation of Trans-Pecos Texas. Journal of Vertebate Paleontology 29:605–611.

Appendix 9.1. Character Codings for the Skeletal Material Recovered from the WBH Site

Adult only (UMNH VP 13881)	?????????? ?????????? ?????????? ?????????? ?????????? ?????????? ?????????? ?????????? ???11??1?0 ???????11 ??1101111? ?0??11
All elements (UMNH VP 20187 and UMNH VP 13881)	?????????? 0???????10 ?????????? ?????????? 1?1210??10 000110001? ?????????? ?????????? ???11??1?0 11??10??11 ?011011111? 10??11

Note: Codings are based on the phylogenetic matrix of Gates et al. (2011).

New Saurolophine Material from the Upper Campanian–Lower Maastrichtian Wapiti Formation, West-Central Alberta

10

Phil R. Bell, Robin Sissons, Michael E. Burns, Federico Fanti, and Philip J. Currie

ABSTRACT

A saurolophine hadrosaurid from the Upper Cretaceous Wapiti Formation of west-central Alberta is described on the basis of an incomplete skull, which was found in association with a partial postcranial skeleton. The new specimen comprises the first semi-complete cranial material of a hadrosaur from the Wapiti Formation, and includes the right prefrontal, left maxilla, both jugals, right quadratojugal, left quadrate, left dentary, and ?left angular. The skull has similarities to those of similar-sized individuals of *Edmontosaurus;* however, the short dentary diastema in the Wapiti saurolophine is unlike that of *Edmontosaurus*. Histological analysis indicates the individual was likely a subadult. The new skeleton comes from the Red Willow Coal Zone at the top of Unit 4 of the Wapiti Formation, placing it on or close to the Campanian-Maastrichtian boundary. A phylogenetic analysis recovers the Wapiti hadrosaur in a polytomy with other saurolophines. Although it could not be assigned to a particular genus, the new specimen underscores the potential of this formation to fill gaps in the terrestrial fossil record of southern Alberta.

INTRODUCTION

The Upper Cretaceous deposits of the Wapiti Formation in west-central Alberta preserve evidence of a high-latitude vertebrate fauna representative of exclusively nonmarine environments. An increasing number of fossil localities in the Grande Prairie region document a typical upper Campanian–lower Maastrichtian dinosaur fauna, which to date includes ankylosaurids, ceratopsids, hadrosaurids, tyrannosaurids, *Troodon*, and saurornitholestine theropods (Currie et al., 2008; Fanti and Miyashita, 2009). However, with the exception of two ceratopsian bonebeds (Fanti and Currie, 2007; Currie et al., 2008), bones are typically fragmentary and isolated from both a geographic and stratigraphic perspective. Furthermore, the only dinosaur taxon named on the basis of specimens collected from the Wapiti deposits is the ceratopsid *Pachyrhinosaurus lakustai* (Currie et al., 2008). As a consequence, taxonomic resolution within the Wapiti Formation remains limited.

The Wapiti Formation ranges in age from 79 to 67 Ma (Fanti and Catuneanu, 2010), and includes the Campanian-Maastrichtian boundary (70.6 Ma; Ogg et al., 2004). The number of terrestrial formations spanning this interval is limited in North America, and faunal successions across this boundary are correspondingly poorly understood. The contemporaneous faunas of the Horseshoe Canyon Formation in southern Alberta are among the best known from this interval, although few studies exist on faunal succession and patterns of diversity from this formation (Russell and Chamney, 1967; Straight, 2003; Wu et al., 2007; Larson et al., 2010; Eberth et al., 2013). Increased taxonomic resolution of the Wapiti Formation will provide important information on the palaeobiogeography, faunal succession, and diversity of Late Cretaceous high-latitude terrestrial faunas in North America.

TMP 1989.092.0001 is an incomplete hadrosaurid specimen comprising a partial, disarticulated skull and postcrania (Figs. 10.1, 10.2). It represents, to date, the most complete hadrosaurid skeleton from the Wapiti Formation and the first definitive saurolophine from this unit. Initial collections in 1989 recovered parts of the tail, pelvic girdle, and hindlimb, along with ossified tendon fragments and skin impressions. Further excavations in 2003 recovered parts of the skull, ribs, and forelimb, as well as several manal and pedal elements. The articulated series of cervical and dorsal vertebrae, and pelvic girdle elements, indicate that the animal came to rest on its left side before the skeleton disarticulated further (Fig. 10.1). This chapter describes the partial skull of TMP 1989.092.0001 and attempts to resolve its phylogenetic position. The importance of this specimen is discussed in relation to its evolutionary and stratigraphic context.

Institutional Abbreviations AMNH, American Museum of Natural History, New York, New York; BMNH, British Museum of Natural History, London, England; CMN, Canadian Museum of Nature, Ottawa, Ontario; MOR, Museum of the

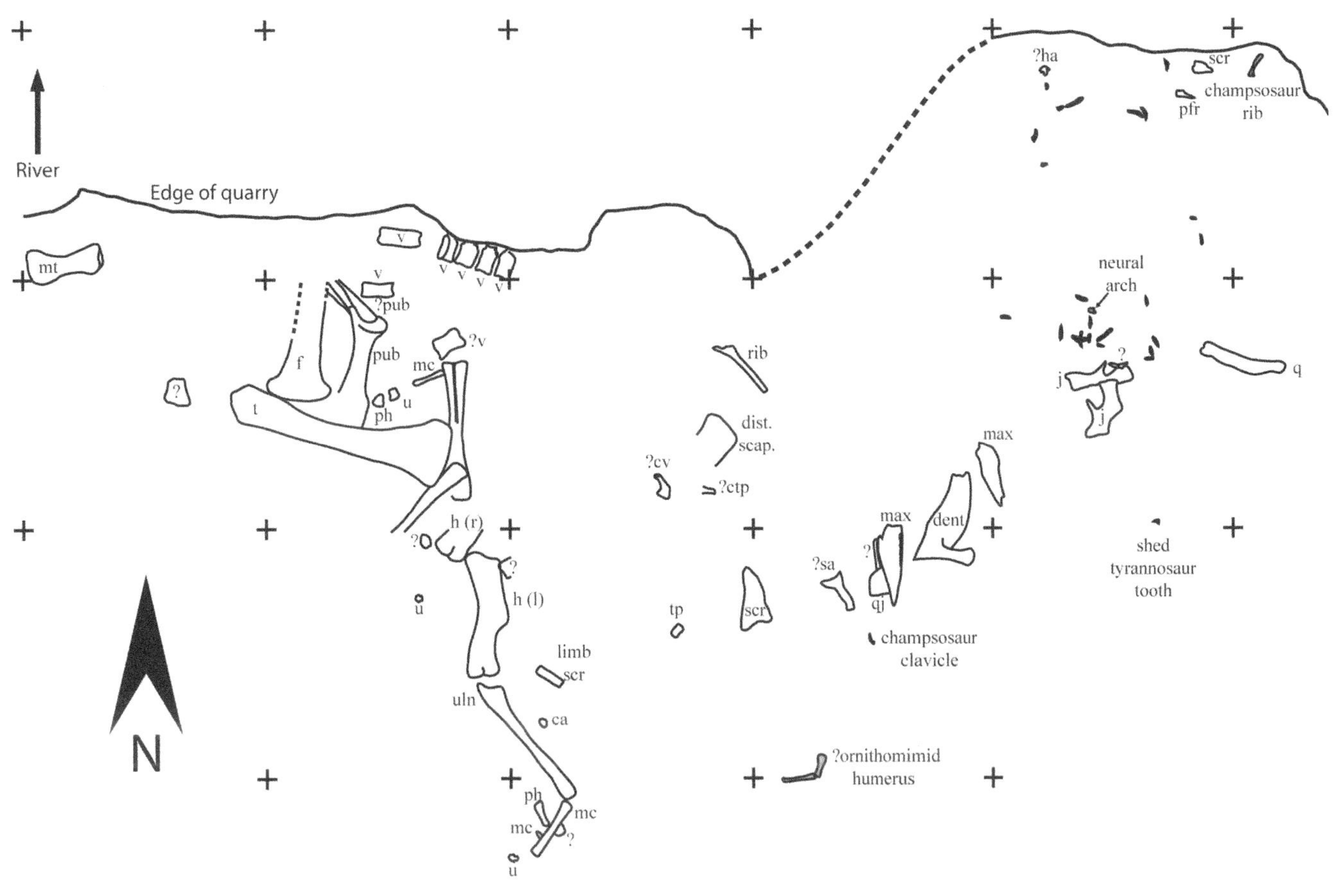

10.1. Quarry map of TMP 1989.092.0001. Black elements are hadrosaur teeth. Non-hadrosaurid elements are in gray. Grids are 1 m squares. Abbreviations: ca, carpal; ctp, cervical transverse process; cv, cervical vertebra; dent, dentary; f, femur; h, humerus; ha, haemal arch; j, jugal; max, maxilla; mc, metacarpal; mt, metatarsal; ph, phalanx; pfr, prefrontal; pub, pubis; q, quadrate; qj, quadratojugal; sa, surangular; scap., scapula; scr, unidentified scrap; t, tibia; tp, transverse process; u, ungual; uln, ulna; v, vertebra.

Rockies, Bozeman, Montana; ROM, Royal Ontario Museum, Toronto, Ontario; SM, Senckenberg Museum, Frankfurt, Germany; TMP, Royal Tyrrell Museum of Palaeontology, Drumheller, Alberta.

LOCALITY AND GEOLOGICAL SETTING

TMP 1989.092.0001 was collected from the south bank of the Red Willow River approximately 7 km northeast of Elmworth, directly across from the Red Willow Park campground (N55°04′88″; W119°31′65″; Fig. 10.3). The Wapiti Formation crops out extensively along this primarily east-west oriented drainage system and has been the object of intense prospecting during the last twenty years. A number of isolated hadrosaur ichnites, postcranial elements, and well-preserved deciduous leaves have been documented in this formation across a distance of ~30 km along the Red Willow River in Alberta and British Columbia.

Associated but undescribed ceratopsian bones (TMP 1990.124.0001) were also collected approximately 100 m downstream from the hadrosaur skeleton quarry. Tanke (2004) reported *Champsosaurus* elements and turtle shell from the quarry of TMP 1989.092.0001. Original quarry maps also indicate that an ornithomimid ulna (putatively the first record of this taxon from the Wapiti Formation) was collected from this site; however, it remains unprepared. Recent stratigraphic studies on the Wapiti sedimentary successions allow exposures along this section of the Red Willow River to be referred to the upper part of Unit 4 of the Wapiti Formation and, in particular, to the Red Willow Coal Zone (sensu Fanti and Catuneanu, 2009, 2010). The Red Willow coals are up to 1.5 m thick, extend laterally for a maximum of 5 km, and are commonly interbedded with dark gray, carbonaceous mudstone. Thus the Red Willow Coal Zone is believed to represent deposition on a low gradient alluvial plain under high-accommodation conditions associated with reduction in sediment supply, high freshwater table, and abundant vegetation (Fanti and Catuneanu, 2009). A robust stratigraphic framework based on palynological, magnetostratigraphic, and sedimentological data document that the

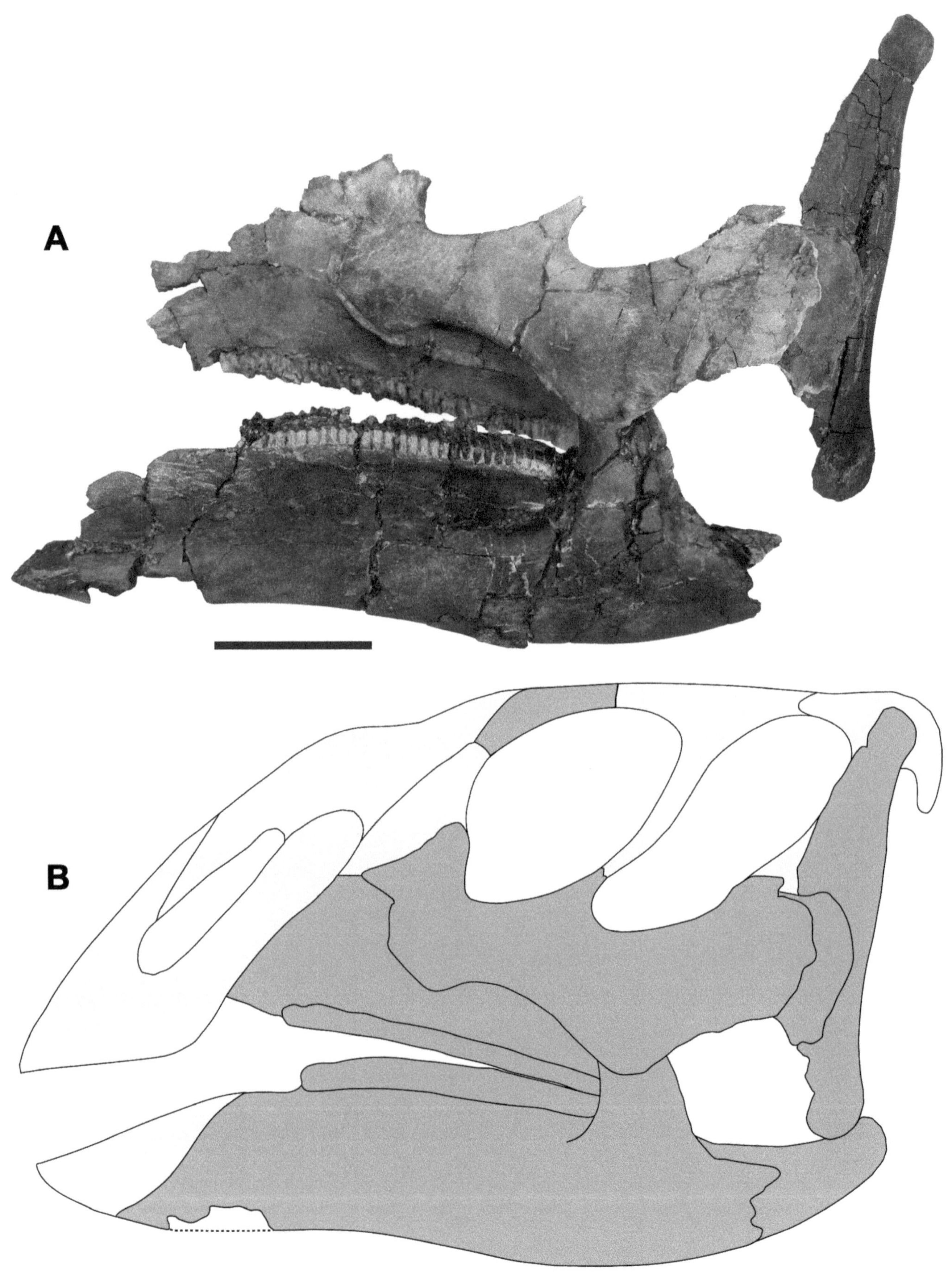

10.2. Skull of TMP 1989.092.0001. (A) Composite photograph and (B) illustration showing reconstructed areas in light grey. Scale bar equals 10 cm.

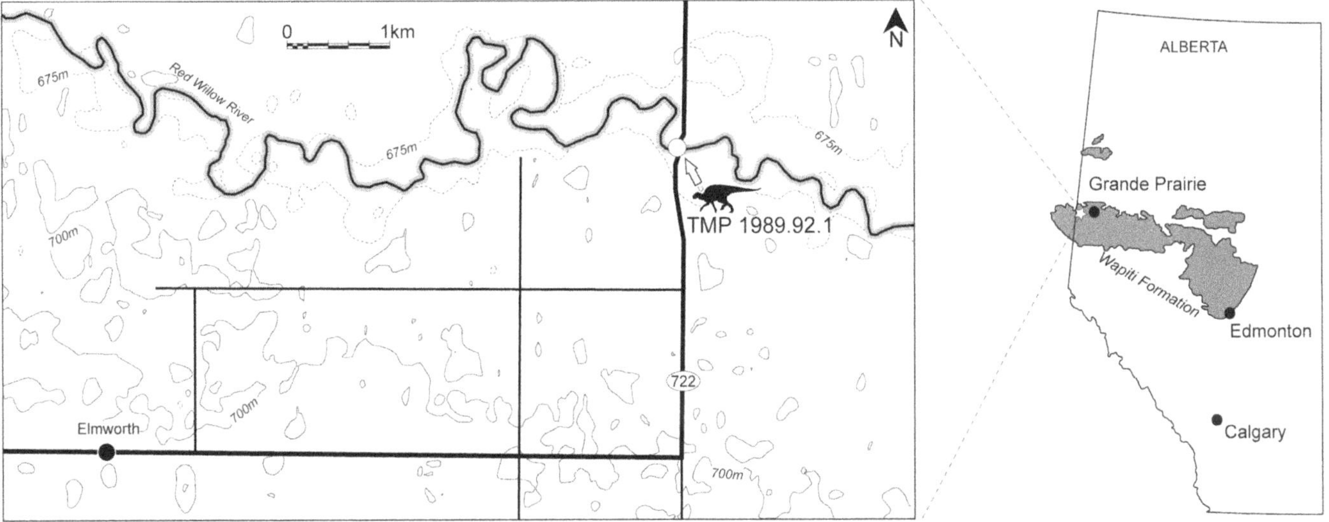

10.3. Locality showing extent of the Wapiti Formation in central-western Alberta (right side) and the discovery site of TMP 1989.092.0001 on the Red Willow River.

Campanian-Maastrichtian boundary (70.6 Ma; Ogg et al., 2004) lies within the Red Willow Coal Zone (Srivastava, 1970; Jerzykiewicz and Sweet, 1988; Dawson and Kalkreuth, 1989; Sweet et al., 1989; Dawson et al., 1994; Fanti and Catuneanu, 2009, 2010). Therefore, it is possible to attribute a latest Campanian to earliest Maastrichtian age to TMP 1989.092.0001 (Fig. 10.4). Consequently, these beds are also equivalent in age to the transition between the Morrin/Tolman members of the Horseshoe Canyon Formation (Eberth and Braman, 2012), and also to the Drumheller Marine Tongue, which has been dated at 70.6 Ma (Lerbekmo and Braman, 2002; Eberth and Deino, 2005; Wu et al., 2007; Eberth and Braman, 2012). The skeleton is preserved in a sandy channel lag characterized by abundant plant remains. Skeletal remains lie just a few centimeters below a large erosive cut of glacial deposits.

SYSTEMATIC PALEONTOLOGY

DINOSAURIA Owen, 1842
ORNITHISCHIA Seeley, 1888
HADROSAURIDAE Cope, 1869
SAUROLOPHINAE sensu Prieto-Márquez, 2010
SAUROLOPHINAE indet.
(Figs. 10.5–10.10)

Material TMP 1989.092.0001, incomplete articulated to associated skull and skeleton. Skull comprises the right prefrontal, both maxillae, both jugals, right quadratojugal, left quadrate, left dentary, and ?left angular along with numerous teeth. The postcranium consists of an associated forelimb and manus, articulated hind limb and pelvis along with multiple dorsal and caudal vertebrae, all of which remain unprepared.

Locality South bank of the Red Willow River directly across from Red Willow Campground (N55°04′88″; W119°31′65″), Wapiti Formation, County of Grande Prairie, Alberta, Canada.

Horizon Uppermost Unit 4, Wapiti Formation (uppermost Campanian or lowermost Maastrichtian; Fanti and Catuneanu, 2009).

Description

Jugal Both left and right jugals were recovered from TMP 1989.092.0001; however, the right one is better preserved (Fig. 10.5; Table 10.1). Constriction of the dorsal and ventral margins of the posterior process give the posterior process a hatchet shape in lateral view. The jugal flange is moderately developed as in *Edmontosaurus* (CMN 8509) and *Gryposaurus notabilis* (ROM 873) but not exaggerated as it is in *Maiasaura peeblesorum* (ROM 44770) and *Brachylophosaurus canadensis* (CMN 8893). The jugal flange is triangular and unlike the more rounded condition in both *Prosaurolophus maximus* and *Saurolophus osborni* (Bell, 2010). The ventral margin of the jugal (between the posterior process and the ventral margin of the anterior process) is concave as in most saurolophines except *P. maximus* (CMN 2277, ROM 787, ROM 1928), where it is comparatively straight (Fig. 10.6). The anterior process is strongly asymmetrical as is typical of Saurolophinae (Horner et al., 2004) but unlike the triangular process seen in *M. peeblesorum* or *B. canadensis* (Cuthbertson and Holmes, 2010); the ventral margin of this process is sigmoidal

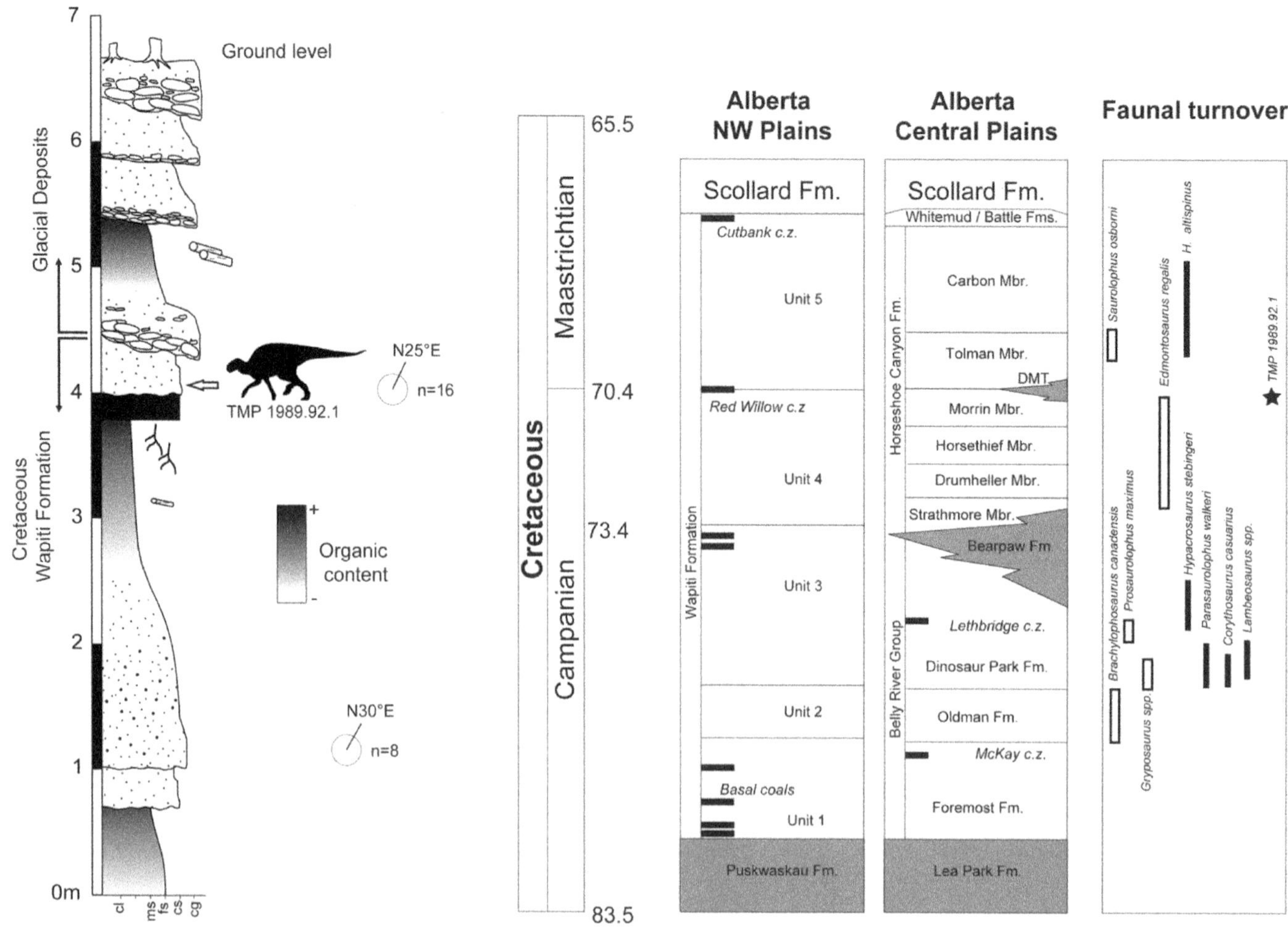

10.4. Lithostratigraphy of the TMP 1989.092.0001 quarry (far left) and stratigraphic and biostratigraphic correlations. Other taxa listed occur in the Belly River and Edmonton Groups. White bars are saurolophines; black bars are lambeosaurines.

whereas the dorsal margin is concave. A sigmoidal ventral margin of the anterior process is seen in species of *Gryposaurus* (Gates and Sampson, 2007), *P. maximus* (ROM 787, ROM 1928), *Saurolophus angustirostris* (Bell, 2011), and immature *E. annectens* (ROM 53519, CMN 8509; Fig. 10.6). In other saurolophines, the ventral margin of the anterior process is relatively straight (*M. peeblesorum*) or convex (*S. osborni*; AMNH 5220, CMN 8796).

In TMP 1989.092.0001, *B. canadensis*, *S. angustirostris*, and species of *Gryposaurus*, a sigmoidal ventral margin of the anterior process is related to a prominent triangular spur at the anterior limit of the jugal. This spur is prominent in juvenile *E. annectens* (CMN 8509; ROM 53519) but is reduced in larger individuals. Therefore the presence of this feature in TMP 1989.092.0001 may be related to its relatively small size and immaturity (see histological discussion). The medial surface of the anterior process contacts the maxilla ventrally and lacrimal dorsally. This contact is truncated posteriorly by an anterodorsally inclined ridge on the medial surface of the jugal. The distal end of the postorbital process is missing. The postorbital process is angled slightly posteriorly as in *P. maximus* and *S. osborni* (Bell, 2010). However, in *E. annectens*, posteriad deflection of the postorbital process is linked to development of the postorbital "pocket" in large, presumably adult individuals (N. Campione, pers. comm., 2012). Consequently, orientation of the postorbital process in TMP 1989.092.0001 may also be a function of absolute size and/or age.

Maxilla Both maxillae from TMP 1989.092.0001 were recovered but are missing the anterior-most part of that element (anteroventral process of Horner, 1992) and the anterodorsal process (Fig. 10.7; Table 10.1). The anterior and posterior edges of the dorsal process are posterodorally-inclined and subvertical, respectively, forming a right triangle as in *P. maximus* (Horner, 1992) and *S. osborni* (AMNH 5221). In *B. canadensis* (Prieto-Márquez, 2005; Cuthbertson and Holmes, 2010), the dorsal process is shaped like an isosceles triangle. This feature is variable in *Edmontosaurus* (ROM

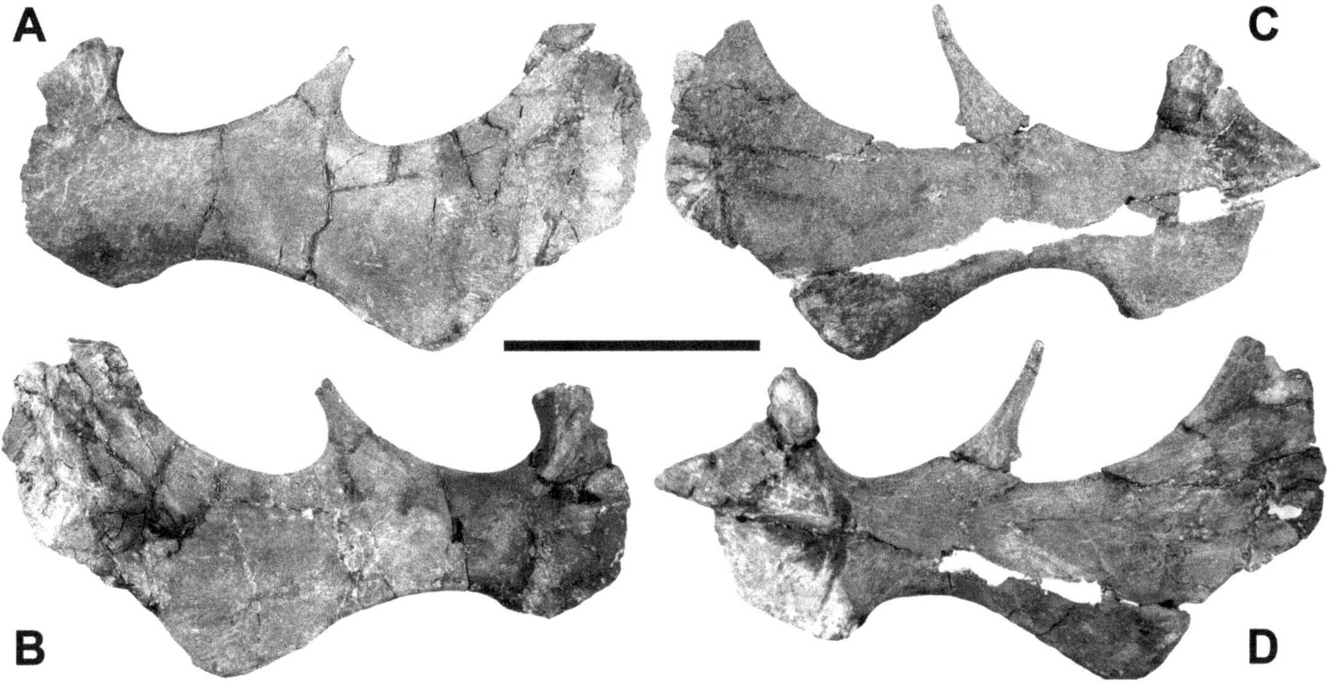

10.5. Jugals of TMP 1989.092.0001. Left jugal in (A) lateral and (B) medial views. Right jugal in (C) lateral and (D) medial views. Scale bar equals 10 cm.

53527, BMNH 8937). The apex of the dorsal process is anterior of the midpoint of the maxilla, whereas in all other saurolophines the apex falls at the midpoint. The lateral surface of the dorsal process formed a lap joint with the jugal. The anteroventral margin of the jugal facet is broadly arcuate as for most saurolophines except *B. canadensis*, in which it is almost straight and parallel with the ectopterygoid shelf (Wagner and Lehman, 2009). Medially, at the base of the dorsal process, there are two elliptical foramina that form a horizontal line. The posterior foramen is larger and is continuous with the posteroventral edge of the dorsal process, opening onto the lateral surface of the maxilla. In lateral view, the dorsal and ventral margins of the posterior process taper posteriorly. The ectopterygoid process is mediolaterally narrow, posterodorsally deflected, and does not extend beyond the posterior margin of the maxilla. The ectopterygoid process is separated ventrally from the rounded posterior margin of the maxilla by a cleft. The ectopterygoid shelf extends along the lateral surface of the posterior process of the maxilla. This shelf is horizontal posteriorly, curving dorsally more anteriorly. In cross section, the entire dorsolateral half of the posterior process is dorsomedially inclined.

The anterior half of the maxilla is triangular in lateral view. The premaxillary facet of the maxilla attains a maximum angle of 45° (although this value may differ in more complete specimens) from the horizontal in lateral view (Fig. 10.7) and is steeper than in all other saurolophines

Table 10.1. List of Select Cranial Measurements for TMP 1989.092.0001

Element	Measurement	Value (cm)
Jugal	Max. length	27.3 (R); 25.1 (L)
	Max. height of ant. process	11.4 (R); 9.9 (L)
Maxilla (L)	Length	31.2
	Max. height	14.3
Quadrate (L)	Height	30.0
Dentary (L)	Total length	49.7
	Tooth row length	12.8

Note: L, left; R, right.

except *Gryposaurus notabilis* (53° in ROM 873). However, this feature is likely influenced by snout length, which has been shown to be a function of ontogeny in species of *Edmontosaurus* (Campione and Evans, 2011). At least six maxillary foramina pierce the anterolateral surface of the maxilla in a rough line angled anterodorsally. The medial surface of the maxilla is featureless. A convex row of nutrient foramina extend the length of the element. Each foramen corresponds to the base of an alveolus. Between the nutrient foramina and the tooth row is a convex (in medial view) groove that extends the length of the maxilla. The gently curving alveolar margin in TMP 1989.092.0001 is similar to most saurolophines except species of *Gryposaurus*, in which it is sigmoidal (Gates and Sampson, 2007). The dental lamina obscures the height of the alveoli medially so the number of teeth at each position cannot be determined. There are 40 tooth positions within

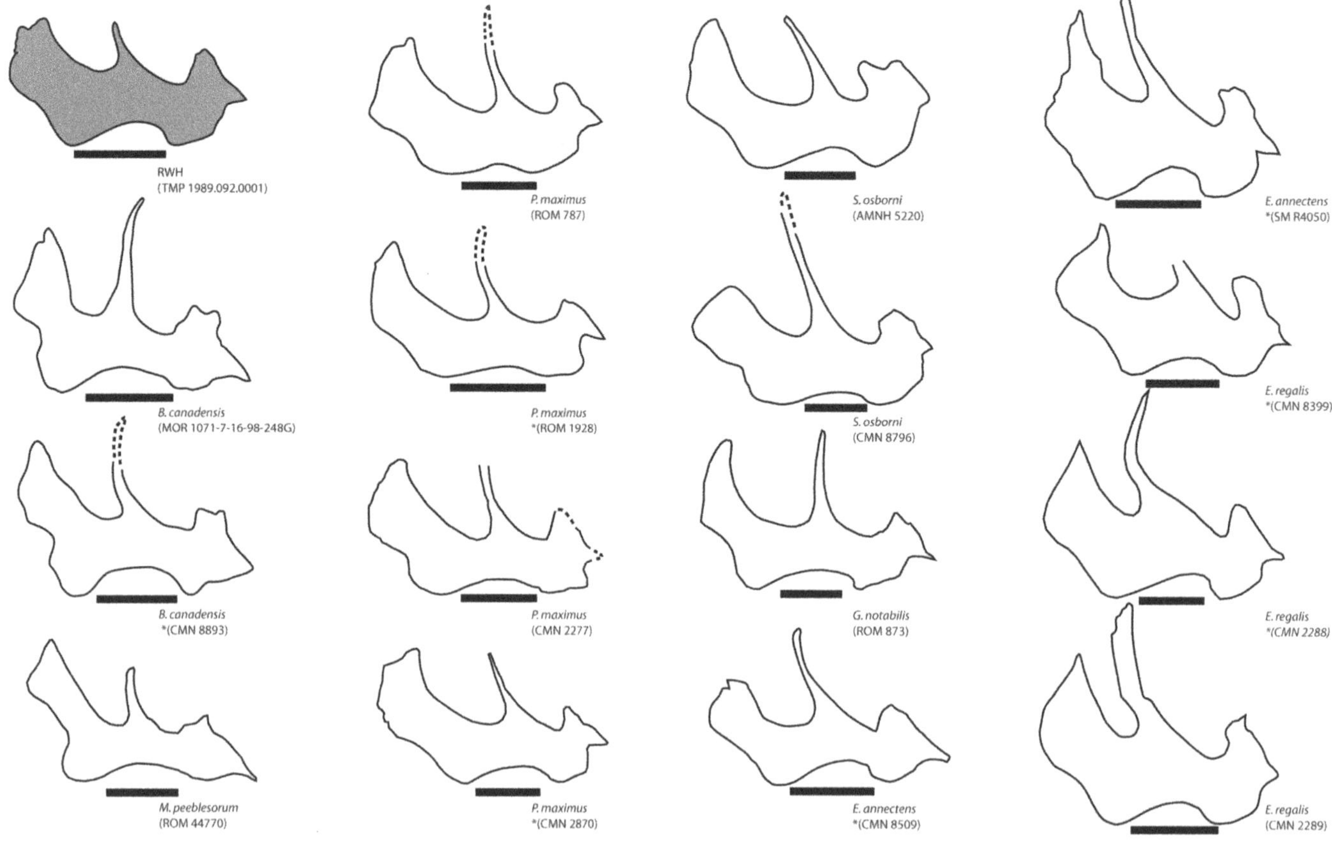

10.6. Comparison of saurolophine jugals in right lateral view. Specimens denoted by an asterisk indicate reversed images. RWH = Red Willow Hadrosaur. Scale bar equals 10 cm.

the maxilla, and up to 3 teeth may be exposed on the occlusal surface at each tooth position.

Prefrontal An incomplete bone recovered with the skull is interpreted as part the right prefrontal (Fig. 10.8). Although it was examined by one of us (RS), it was lost before it could be prepared fully. The fragment, which corresponds to the orbital margin, is anteroposteriorly elongate but is broken medially. Dorsally, the surface is concave due to the dorsally flared lateral margin, similar to *P. maximus* (Prieto-Márquez, 2010). Flaring of the lateral margin does not approach the exaggerated condition in *S. osborni* (AMNH 5220) and there is no indication of a posterodorsal process (Bell, 2010). The lateral margin is weakly crenulated as in *P. maximus* (Horner, 1992).

Quadratojugal The right quadratojugal is incompletely preserved. The posterior margin is C-shaped and the posteroventral margin forms an acute angle with anteroventral margin (Fig. 10.9). Medially, the posterior margin is bevelled where it forms a lap joint with the quadratojugal notch on the quadrate. Most of the dorsal half of the quadratojugal is missing and the anterior margin is also damaged.

Quadrate The left quadrate measures 30 cm dorsoventrally. The posterior margin is nearly straight in lateral view, as is typical of most saurolophines (Horner et al., 2004) except *S. angustirostris* (Bell, 2011; Fig. 10.10). The dorsal head is squared off where it would have inserted into the corresponding cotylus on the squamosal. The head of the quadrate is mediolaterally compressed. In lateral view, the dorsal half of the quadrate tapers dorsally more strongly than in *S. osborni* (Bell, 2010), species of *Gryposaurus* (ROM 873; Gates and Sampson, 2007), *P. maximus* (ROM 787, ROM 1928, CMN 2870), and *B. canadensis* (Prieto-Márquez, 2005; Cuthbertson and Holmes, 2010). The quadrate buttress (sensu Gates and Sampson, 2007) is weakly expressed, as in *S. osborni* (Bell, 2010). The quadratojugal notch occupies most of the ventral half of the anterior edge of the quadrate. This embayment is roughly triangular as in *Gryposaurus* (Gates and Sampson, 2007) but unlike the semicircular notch in *S. osborni* (AMNH 5221; Bell, 2010). A narrow ridge is present on the ventral half of the quadratojugal notch. The ridge extends roughly parallel to the posterior margin of the quadrate on the bevel of the quadratojugal notch. Medially, the pterygoid process is broken and only its base is still intact. The base of the pterygoid process extends nearly the entire height of the element, terminating short of the dorsal

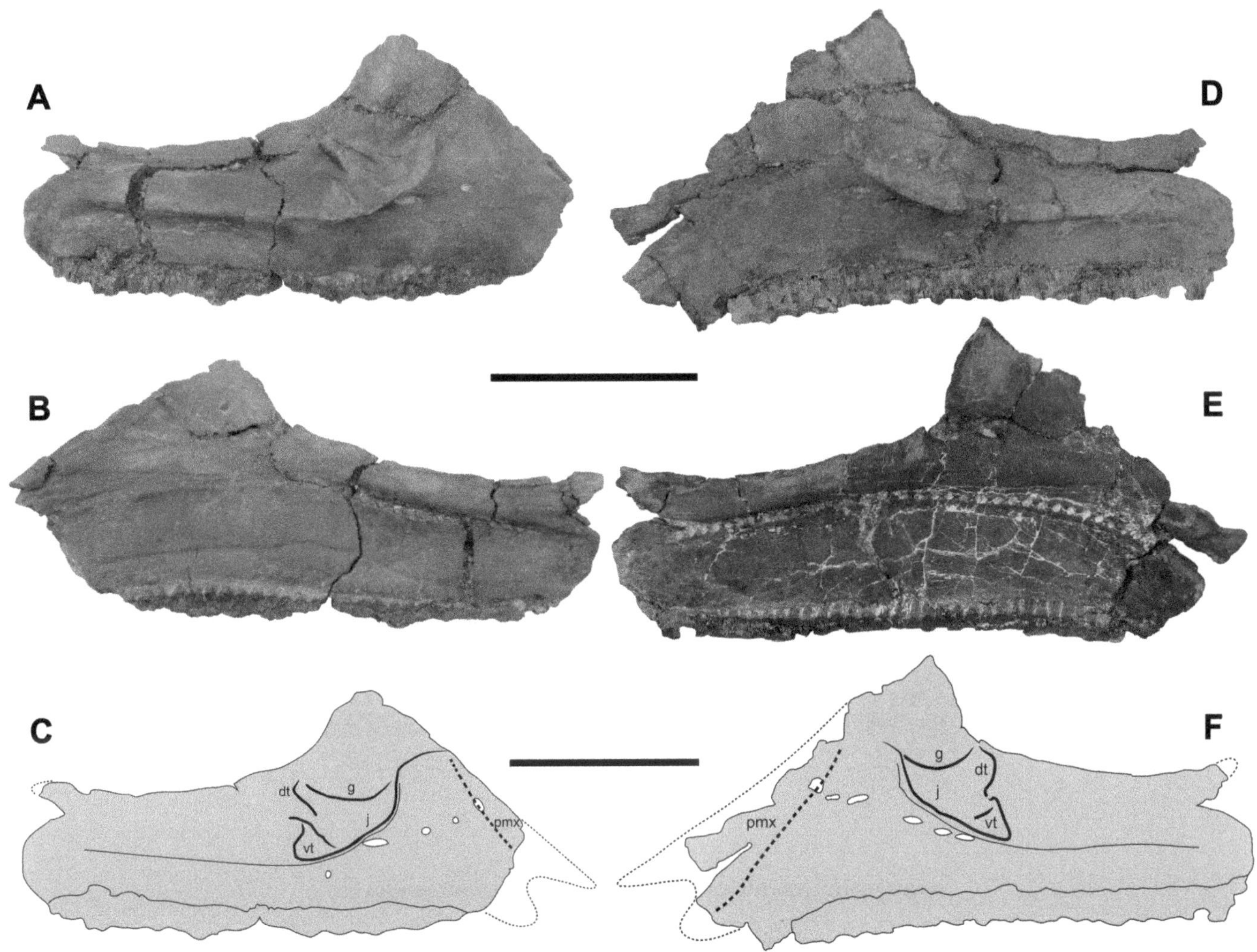

10.7. Maxillae of TMP 1989.092.0001. Left maxilla in (A, C) lateral and (B) medial views. Right maxilla in (D, F) lateral and (E) medial views. Illustrations (C, F) showing sutural contacts. Abbreviations: dt, dorsal tubercle; g, groove; j, jugal facet; pmx, premaxillary contact; vt, ventral tubercle. Scale bar equals 10 cm.

quadrate head. The quadrate is roughly triangular in ventral view where it forms the condylar surface for articulation with the lower jaw. The small medial condyle is indistinct from the larger and more ventrally placed lateral condyle, as is typical for Hadrosauridae (Horner et al., 2004).

Lower Jaw The left dentary is virtually complete although the margins of the anterior edentulous portion are damaged (Fig. 10.11). As preserved, the dentary is 48 cm long (Table 10.1). The dental shelf (sensu Sullivan and Williamson, 1999) is shorter relative to that of any other saurolophine; the ratio between the length of the dental shelf and the length between the anterior-most tooth and the posterior edge of the coronoid process is 0.15. Within Hadrosauridae, only the lambeosaurines *Charonosaurus* and *Parasaurolophus* have relatively shorter dental shelves (Sullivan and Williamson, 1999; Prieto-Márquez, 2010:character 33 and supplementary material therein). Anteriorly, the dental shelf deflects ventrally to form the contact surface for the predentary. The entire edentulous portion of the dentary (i.e., dental shelf plus the dentary-predentary contact surface) is equivalent to about one-third of the length of the tooth row. The length of the edentulous part of the dentary compared to its overall length is shorter than in other saurolophines but is within the range of interspecific variation observed for other members of that clade (Fig. 10.12). Medial and ventral offsetting of the anterior edentulous portion of the dentary is minimal as in *Edmontosaurus* (ROM 867, ROM 801, CMN 8509) and *P. maximus* (ROM 1928, CMN 2870), and less pronounced than in species of *Gryposaurus* (ROM 873; Gates and Sampson, 2007). It is not as straight, however, as it is in *S. osborni* (AMNH 5220; Bell, 2010). The laterally offset coronoid process is anterodorsally inclined as is usual for Hadrosauridae (Prieto-Márquez, 2010). The dorsal terminus of the coronoid process is mediolaterally flattened and subcircular in lateral aspect.

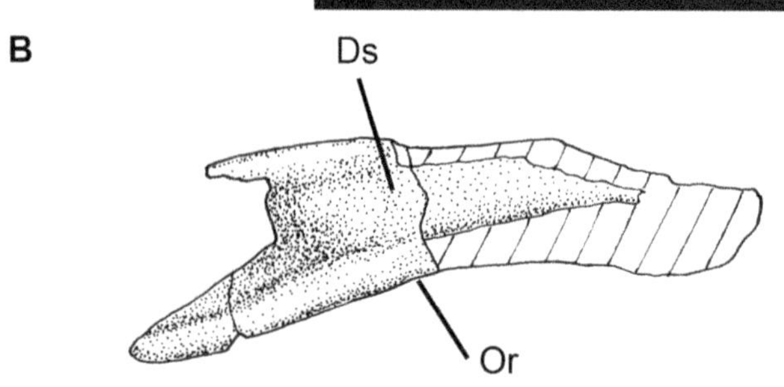

10.8. Unprepared right prefrontal (lost) of TMP 1989.092.0001 in dorsal oblique view. (A) photograph and (B) interpretive illustration. Abbreviations: Ds, dorsal sulcus; Or, orbital margin. Scale bar equals 5 cm.

Medially, much of the thin dental lamina that covered the dental battery is missing, revealing at least 43 tooth positions. Each alveolus supports up to 4 teeth, of which 2 are usually exposed on the occlusal surface. The enameled lingual surface of each tooth is diamond shaped with the ratio of average height to mesiodistal length being 2.7 for teeth from the middle region of the tooth row, and 1.8 for more mesially placed teeth (Fig. 10.11C). The single median carina is straight, typical of other saurolophines (Horner et al., 2004). All but the most mesial teeth are non-papillate. At the base of each alveolus is a circular foramen that, when taken together form a concave-up arc in medial aspect. Posterior to the dental battery is the triangular splenial process. The medial surface of the splenial process is flattened to form a lap joint with the splenial.

An incomplete fragment of the ?left angular measures 11.3 cm long. It is mediolaterally flattened and tapers dorsoventrally at one end (Fig. 10.11D–E). In lateral aspect, the angular is gently bowed to conform to the curvature of the surangular to which it was attached. No other post-dentary bones were recovered.

Skin Impressions Several small patches of skin impression were recovered with TMP 1989.092.0001. Unfortunately, their locations with respect to the skeleton are unknown. In all cases, scales are polygonal to irregular in outline and

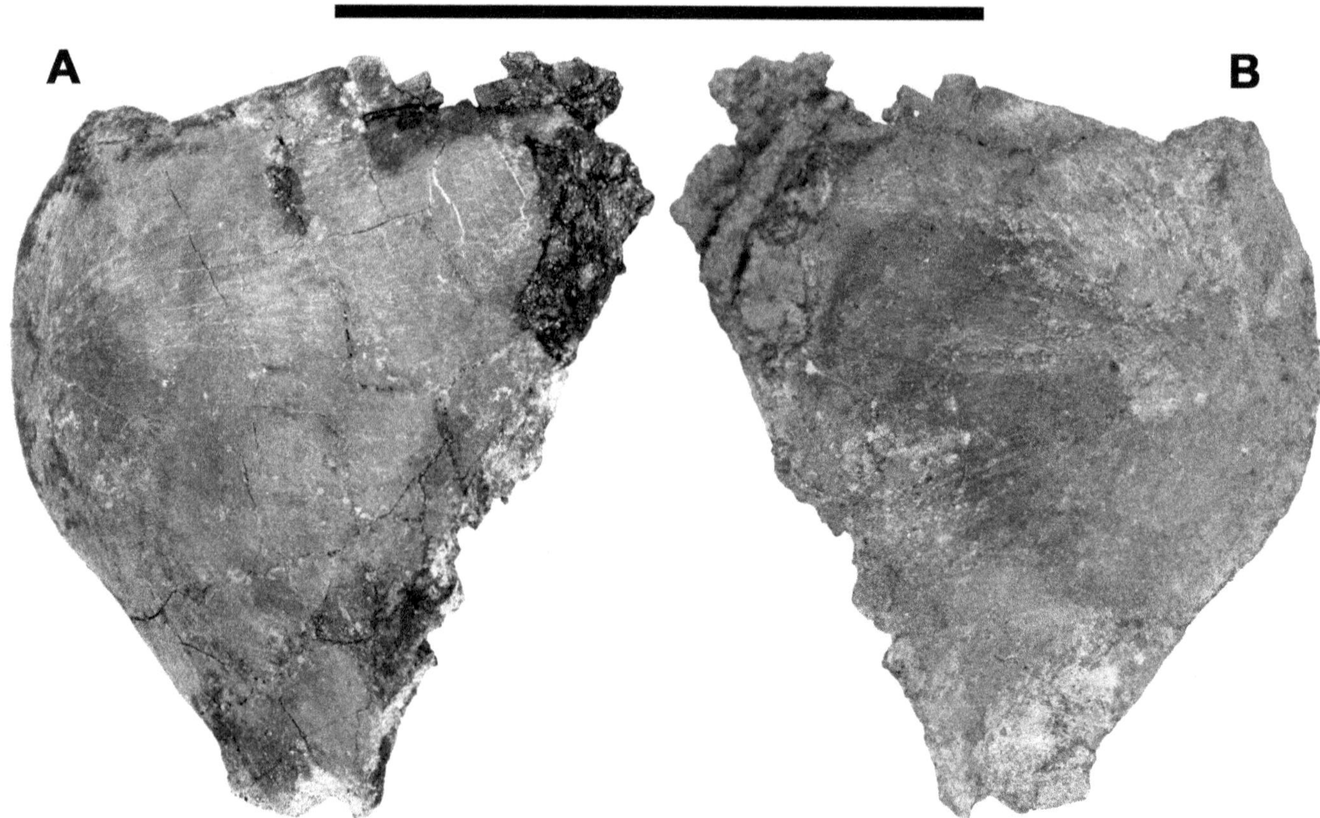

10.9. Right quadratojugal of TMP 1989.092.0001 in (A) lateral and (B) medial views. Scale bar equals 10 cm.

do not exceed 4 mm in diameter (Fig. 10.13). In one fragment, the irregular scales also have radial corrugations on the margins of the scales. Irregular, corrugated scales are known in *Saurolophus* (Bell, 2012, this volume) and several unidentified hadrosaurids from New Mexico (Anderson et al., 1998) and Utah (Anderson et al., 1999), including a specimen tentatively referred to *Gryposaurus* sp. (Clayton et al., 2011). Scales are regularly arranged and there is no evidence of "cluster areas" (sensu Osborn, 1912) or larger feature-scales (Bell, 2012).

HISTOLOGY

Methods

The left metatarsal IV (TMP 1989.092.0001) was selected for paleohistological analysis. Elements more typically employed for histological analysis (such as the femur) were not available for sectioning. Prior to sectioning, the metatarsal was measured and photographed with a Nikon D100 digital camera with an AF Micro-Nikkor 60 mm lens. A sample was removed from near mid-shaft and stabilized via resin impregnation using Buehler EpoThin Low Viscosity Resin and Hardener. Thin-sections were cut at the Department of Biological Sciences, University of Alberta, with a diamond powder disc on a precision saw, grinding, and polishing with CeO_2 powder after the desired optical transmission was reached (i.e., not based on thickness).

Results

The thin section consists of an 18.6 mm thick sample of the cortex from the posterior surface of the diaphyseal midshaft. This represents 53% of the full diaphyseal radius (based on the circumference). It is not possible to ascertain the real size of the medullary sinuses in the marrow cavity because the cancellous bone structure has been taphonomically collapsed. The cortex is characterized by woven-fibered azonal primary fibrolamellar bone. Primary osteons are embedded throughout the matrix and no secondary osteons are present in the section (Fig. 10.14). No external fundamental system (EFS) of subperiosteal lamellar bone was observed. The specimen also lacks any discernible annuli or lines of arrested growth (LAGs). This may only mean that a LAG was not present in the outer portion of the cortex sampled in this study. The vascular canals increase in diameter nearer to the periosteal surface.

10.10. Left quadrate of TMP 1989.092.0001 in (A) medial and (B) lateral views. Scale bar equals 10 cm.

PHYLOGENETIC ANALYSIS

Methods

Three hundred and seventy characters (84 parsimony informative, determined using the cstatus command in PAUP* 4.0b10 for 32-bit Microsoft Windows; Swofford, 2002) as defined by and corresponding to those of Prieto-Márquez (2010) were used to assess the phylogenetic position of TMP 1989.092.0001 within Hadrosauroidea (Appendix 10.1). The

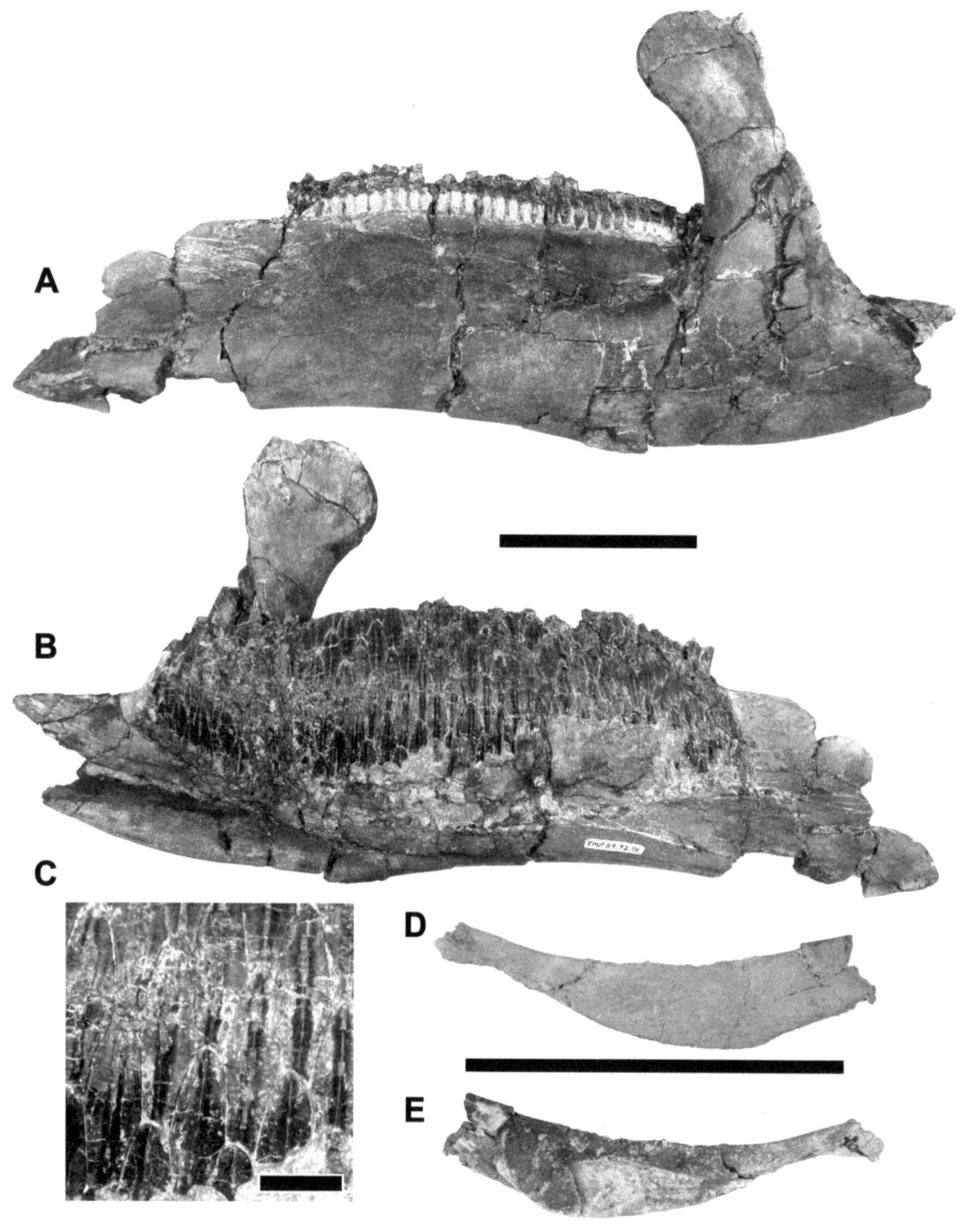

10.11. Mandibular elements of TMP 1989.092.0001. Left dentary in (A) lateral and (B) medial views; (C) close up of teeth from mid tooth row; ?left angular in (D) medial and (E) lateral aspect. Scale bars equal 10 cm except that for (C), which equals 2 cm.

characters used by Prieto-Márquez (2010) were selected because they represent the most recent and comprehensive list of characters available. The analysis included 12 outgroup taxa and 42 ingroup taxa (41 taxa plus TMP 1989.092.0001). The Nexus file was edited in Mesquite 2.75 Build 566 (Maddison and Maddison, 2011). Although the matrix had 34% missing data, TAXEQ3 (Wilkinson, 2003) returned no safely removable taxa. All characters were assigned equal weight and, unlike the analysis by Prieto-Márquez (2010), treated as unordered.

Using TNT 1.1 (Goloboff et al., 2008), a heuristic parsimony search of 1000 replicates using random addition sequences followed by tree-bisection-reconnection (TBR) branch swapping retaining ten trees per replicate. Bootstrap values were calculated via 1000 replicates using heuristic searches, each search using random additional sequences with TBR and 25 replicates. Parsimony reconstruction of ancestral character states was done in Mesquite 2.75 Build 566 (Maddison and Maddison, 2011).

Results

Phylogenetic analysis resulted in 180 equally most parsimonious trees (best score hit 54 times), each with a length of 1005 steps, consistency index of 0.48, and retention index of 0.73. The strict consensus tree is shown in Figure 10.15. TMP 1989.092.0001 is recovered in a broad polytomy within Saurolophinae (sensu Horner et al., 2004), therefore its precise phylogenetic position and relationship to taxa within that clade is unclear. Due to the unresolved positions of taxa within Saurolophinae, the present analysis differs from most other phylogenetic analyses; however, the remainder of the tree retains the topology presented by Prieto-Márquez (2010) on whose character matrix this study was based.

DISCUSSION

Ontogenetic Status of TMP 1989.092.0001

As suggested by Klein and Sander (2008), any description of a new taxon should be accompanied by a description of its bone histology, preferentially leading to a more accurate determination of its ontogenetic status than that provided by morphology alone. This would help avoid controversy surrounding specimens or individual elements that are morphologically similar but differ in absolute size. In this study we examined the histology of a single bone (the left metatarsal IV from TMP 1989.092.0001). We assumed – given the amount of variation previously documented in other dinosaur taxa (Amprino, 1967; Ricqlès, 1976; Reid, 1990; Chinsamy et al., 1998; Horner et al., 2000) – that the histology described here is typical for other bones in the skeleton or conspecifics. In addition, metatarsals are less frequently employed in histological studies on dinosaurs than other limb bones (most often the femur), so few comparisons can be made with homologous bones in other hadrosaur taxa. Therefore, our histological results are treated with caution in this study.

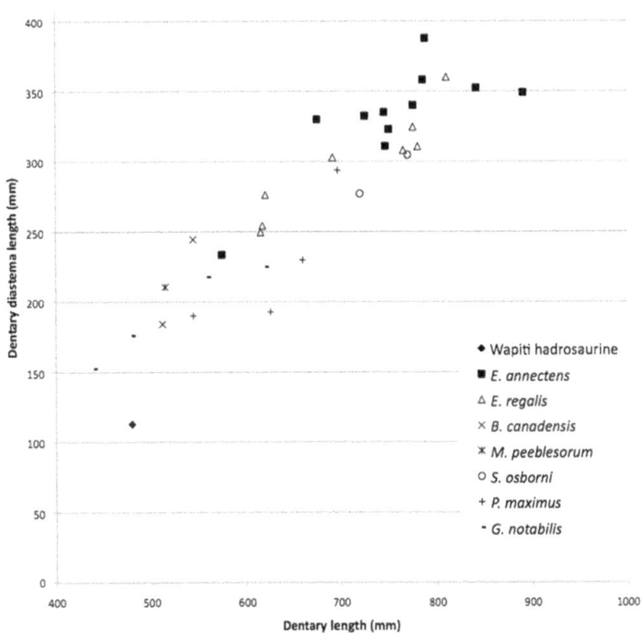

10.12. Allometry of dentary diastema length versus total dentary length in a variety of saurolophines. Diastema is taken as the entire edentulous portion of the jaw anterior of the first tooth.

Laminar (i.e., fibrolamellar) bone is not predominant until the subadult stage in *Maiasaura* (Horner et al., 2000); however, the dense external cortex is still basically formed of primary fibrolamellar tissues in the adult. Therefore, the fact that fibrolamellar bone forms a majority of the primary cortex in TMP 1989.092.0001 is not indicative of any particular age group. The number and occurrence of LAGs vary among taxa and even among different bones of the same individual (e.g., Chinsamy et al., 1998; Horner et al., 2000), so a simple count of LAGs in this specimen (even if they were present) would not necessarily indicate a specific age for the individual animal. Although the degree of vascularity is indicative of relative age in some saurischian taxa (Chinsamy, 1993), this value would be uninformative here in the absence of data from other individuals of the same species. The lack of an EFS indicates that the bone has not yet stopped or substantially slowed its growth. Furthermore, the lack of any secondary osteons implies that extensive remodelling of the bone had not yet begun. Therefore, based on metatarsal IV histology, TMP 1989.092.0001 most closely matches the subadult relative age identified by Horner et al. (2000) for *Maiasaura*.

10.13. Skin impressions of TMP 1989.092.0001 showing (A–E) polygonal arrangement of matrix scales. The location of these impressions with regard to the body is unknown. Scale bar equals 5 cm.

Evolutionary and Biogeographic Implications

Until recently, paleontological investigation of the Wapiti Formation has been limited due to patchy and difficult-to-access exposures. At this stage, only a single dinosaur taxon (the centrosaurine *Pachyrhinosaurus lakustai*) has been positively identified to species level within the Wapiti Formation. *Pachyrhinosaurus lakustai* comes from rocks approximately three million years older than TMP 1989.092.0001 at the base of Unit 4. It is unique to the Wapiti Formation, differing from other pachyrhinosaurs in Alaska (Fiorillo and Tykoski, 2012) and southern Alberta (Currie et al., 2008; Ryan et al., 2010).

TMP 1989.092.0001 shares a combination of morphological characters with immature *Edmontosaurus*, most notably the jugal (prominent anterior spur; rounded and "moderately" developed jugal flange). However, as noted, some of these characters also occur in other saurolophines such as *Gryposaurus*. Purported differences (steeply oriented postorbital process of the jugal and steep premaxillary facet on the maxilla of TMP 1989.092.0001) between *Edmontosaurus* and TMP 1989.092.0001 are likely influenced by ontogenetic status. The short dental shelf on the dentary, however, differs significantly from the elongate shelf in *Edmontosaurus* (including immature individuals) and raises the possibility that TMP 1989.092.0001 may in fact represent a distinct taxon. Given the incompleteness of the skull and the inherent problems associated with ontogenetic variables (which are still relatively poorly understood in saurolophines), we regard TMP 1989.092.0001 as Saurolophinae indet.

Although TMP 1989.092.0001 cannot be assigned to a particular genus, it underscores the potential of the Wapiti Formation to yield associated and articulated dinosaur remains. Previously documented dinosaur localities within the Wapiti Formation identify dense (but disarticulated) monodominant ceratopsian bonebeds (Currie et al., 2008) and rare microvertebrate sites (Fanti and Miyashita, 2009). In the only published report of hadrosaurid material from the Wapiti Formation, Fanti and Miyashita (2009) identified

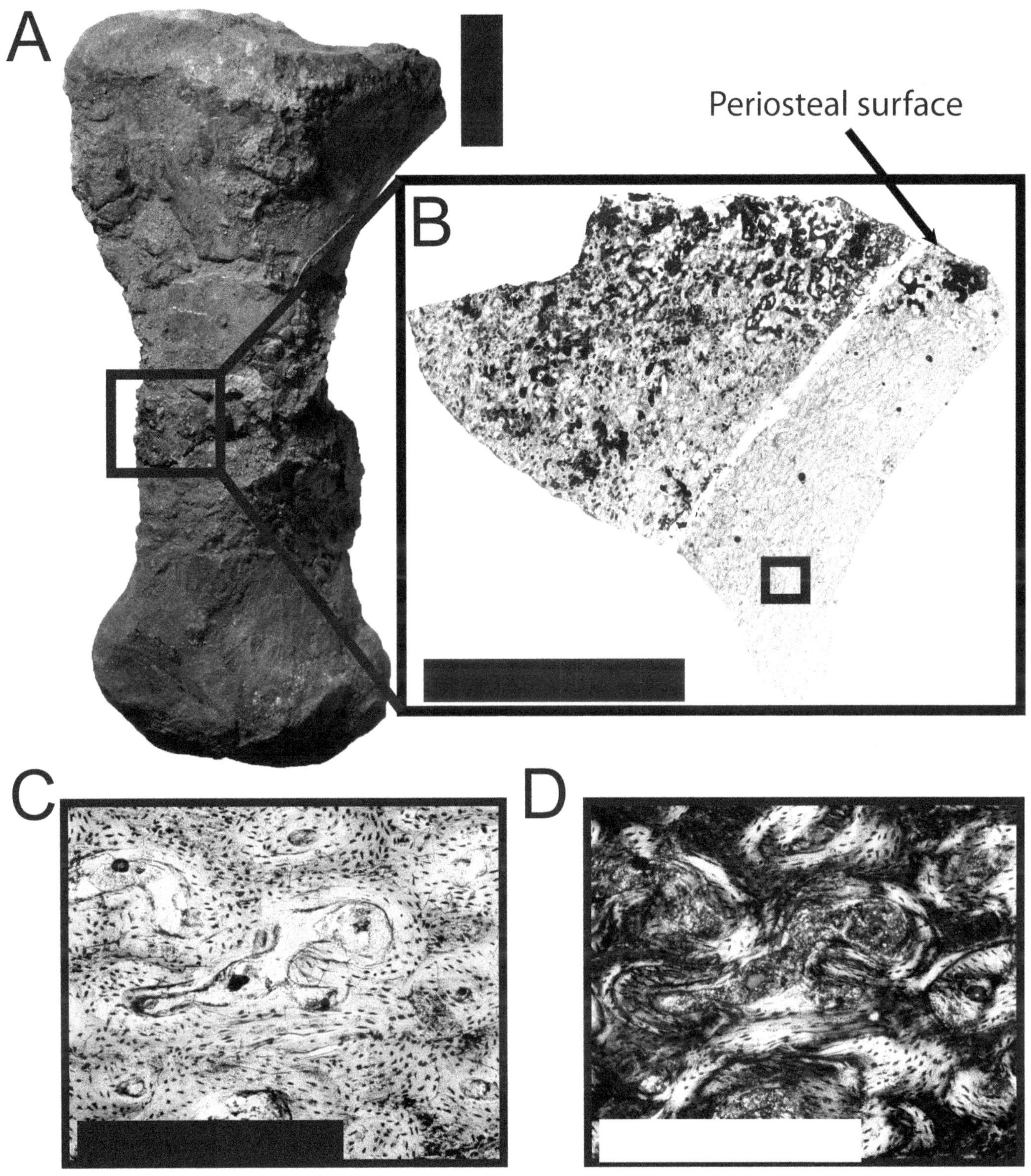

10.14. Long bone histology of TMP 1989.092.0001. (A) The thin-section was cut through the mid-diaphysis of left metatarsal IV and (B) shows primary fibrolamellar bone. Square in B corresponds to (C, D) which shows higher magnification images in (C) plane-polarized and (D) cross-polarized light. Scale bars equal 50 mm in (A), 10 mm in (B) and 0.50 mm in (C) and (D).

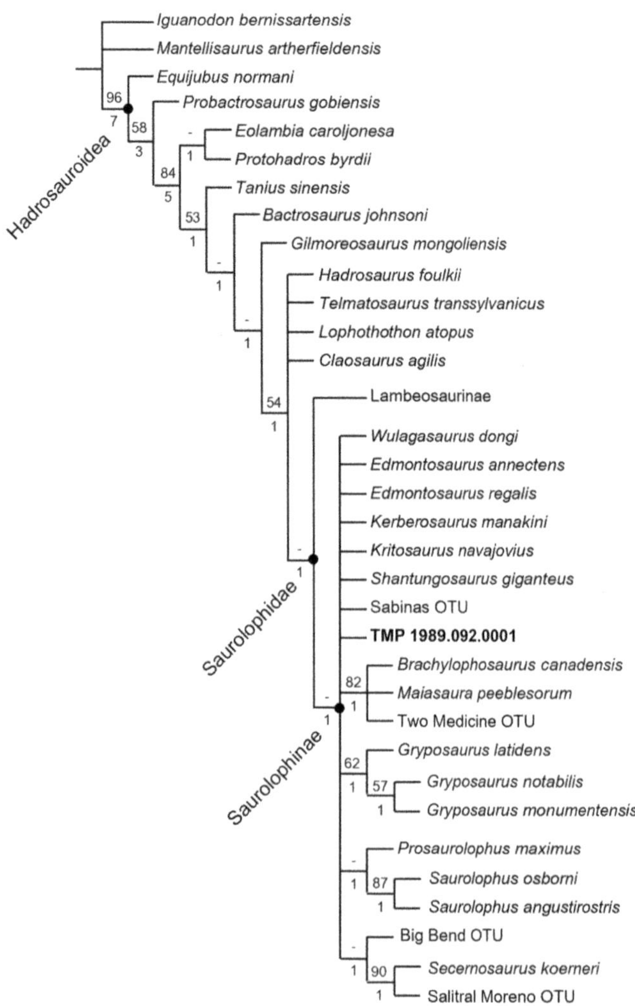

10.15. Phylogenetic relationships of Hadrosauroidea showing the inferred position of TMP 1989.092.0001within the Saurolophinae (higher taxonomy follows Horner et al., 2004). Based on a heuristic analysis of 54 taxa and 370 characters from Prieto-Márquez (2010) using TNT 1.1 (Goloboff et al., 2008). This strict consensus tree was generated with a consistency index of 0.48 and retention index of 0.73. One hundred and eight trees retained. Best score tree length 1005. At each node, the numbers above a branch represent bootstrap percentages; those below represent Bremer supports. Bootstrap percentages below 50 are indicated by a hyphen.

isolated teeth and postcranial elements from hatchling-sized individuals as possible lambeosaurines. The presence of a saurolophine in the latest Campanian Wapiti Formation is not surprising, considering hadrosaurs are known from penecontemporaneous ecosystems at considerably higher latitudes (Gangloff and Fiorillo, 2010; Evans et al., 2012). Nevertheless, the present study joins a growing list of publications documenting the high-latitude vertebrate and invertebrate fauna from this formation (Currie et al., 2008; Fanti and Miyashita, 2009; Nydam et al., 2010; Bell et al., in press) and highlights the need for continued exploration in this region. Moreover, the Wapiti Formation occupies an important geographic position between the polar fauna of Alaska and well-documented faunas from southern Alberta.

ACKNOWLEDGMENTS

TMP 1989.092.0001 was discovered by S. Marsland and collected by the TMP with assistance from Grande Prairie Regional College, especially B. Hunt. The Paslawski family is warmly thanked for their hospitality and for providing access to TMP 1989.092.0001, which was found on their land. We thank D. Tanke, D. Macleod, B. Bavington-Sanchez, and C. Capobianco (TMP) for skillful preparation; and D. Brinkman for facilitating preparation of TMP 1989.092.0001. Thanks to R. Stockey (University of Alberta) for use of thin-sectioning equipment. A. Prieto-Márquez kindly made available the character matrix used in this study and N. Campione provided comparative skull measurements of numerous saurolophines. D. Evans (ROM), C. Mehling (AMNH), and B. Strilisky (TMP) are thanked for providing access to specimens in their care. We also thank D. Evans and D. Eberth for the invitation to contribute to this volume. N. Campione (Univ. of Toronto), D. Evans, and P. Ralrick (editorial assistant) provided detailed and insightful reviews that greatly benefitted this contribution.

LITERATURE CITED

Amprino, R. 1967. Bone histophysiology. Guys Hospital Report 116:51–69.

Anderson, B. G., R. E. Barrick, M. L. Droser, and K. L. Stadtman. 1999. Hadrosaur skin impressions from the Upper Cretaceous Nelsen Formation, Book Cliffs, Utah: morphology and paleoenvironmental context; pp. 295–301 in D. D. Gillette (ed.), Vertebrate Fossils of Utah. Miscellaneous Publication 99-1. Utah Geological Survey, Salt Lake City, Utah.

Anderson, B. G., S. G. Lucas, R. E. Barrick, A. B. Heckert, and G. T. Basabilvaso. 1998. Dinosaur skin impressions and associated skeletal remains from the Upper Cretaceous of southwestern New Mexico: new data on the integument morphology of hadrosaurs. Journal of Vertebrate Paleontology 18:739–745.

Bell, P. R. 2010. Redescription of the skull of *Saurolophus osborni* Brown 1912 (Ornithischia: Hadrosauridae). Cretaceous Research 32:30–44.

Bell, P. R. 2011. Cranial osteology and ontogeny of *Saurolophus angustirostris* from Mongolia with comments on *S. osborni* from Canada. Acta Palaeontologica Polonica 56:703–722.

Bell, P. R. 2012. Standardised terminology and potential taxonomic utility for hadrosaurid skin impressions: a case study for *Saurolophus* from Canada and Mongolia. PLoS ONE 7(2):e31295.

Bell, P. R. 2014. A review of hadrosaurid skin impressions; chapter 34 in D. A. Eberth and D. C. Evans (eds.), Hadrosaurs. Indiana University Press, Bloomington, Indiana.

Bell, P. R., F. Fanti, J. Acorn, and R. L. Sissons. 2013. Fossil mayfly larvae (Ephemeroptera, cf. Heptageniidae) from the Late Cretaceous Wapiti Formation, Alberta, Canada. Journal of Paleontology 87(1):147–150.

Campione, N. E., and D. C. Evans. 2011. Cranial growth and variation in edmontosaurs (Dinosauria: Hadrosauridae): implications for latest Cretaceous megaherbivore diversity in North America. PLoS ONE 6(9):e25186.

Chinsamy, A. 1993. Image analysis and the physiological implications of the vascularisation of femora in archosaurs. Modern Geology 19:101–108.

Chinsamy, A., T. Rich, and P. Vickers-Rich. 1998. Polar dinosaur bone histology. Journal of Vertebrate Paleontology 18:385–390.

Clayton, C. E., R. B. Irmis, M. A. Getty, E. K. Lund, W. J. Nicholls, and M. A. Loewen. 2011.

Non-osseous dermal scutes and integument impressions from an exceptionally preserved hadrosaurid dinosaur skeleton, Upper Cretaceous Kaiparowits Formation of Utah; pp. 23–27 in D. R. Braman, D. A. Eberth, D. C. Evans, and W. Taylor (compilers), Hadrosaur Symposium Abstract Volume. Royal Tyrrell Museum of Palaeontology, Drumheller, Alberta.

Cope, E. D. 1869. Synopsis of the extinct Batrachia, Reptilia and Aves of North America. Transactions of the American Philosophical Society 14:1–252.

Currie, P. J., W. Langston, Jr., and D. H. Tanke. 2008. A new species of *Pachyrhinosaurus* (Dinosauria, Ceratopsidae) from the Upper Cretaceous of Alberta; pp. 1–108 in P. J. Currie, W. Langston, Jr., and D. H. Tanke, A New Horned Dinosaur from an Upper Cretaceous Bone Bed in Alberta. NRC Research Press, Ottawa, Ontario.

Cuthbertson, R. S., and R. B. Holmes. 2010. The first complete description of the holotype of *Brachylophosaurus canadensis* Sternberg, 1953 (Dinosauria: Hadrosauridae) with comments on intraspecific variation. Zoological Journal of the Linnean Society 159:373–397.

Dawson, F., and W. Kalkreuth. 1989. Preliminary results of a continuing study of the stratigraphic context, distribution and characteristics of coals in the Upper Cretaceous to Paleocene Wapiti Formation, northwestern Alberta. Contribution to Canadian Coal Geosciences, Geological Survey of Canada, Paper 89–8:43–48.

Dawson, F., W. Kalkreuth, and A. Sweet. 1994. Stratigraphy and coal resource potential of the Upper Cretaceous to Tertiary strata of northwestern Alberta. Geological Survey of Canada Bulletin 466:1–60.

Eberth, D. A., and D. R. Braman. 2012. A revised stratigraphy and depositional history for the Horseshoe Canyon Formation (Upper Cretaceous) southern Alberta plains. Canadian Journal of Earth Sciences 49:1053–1086.

Eberth, D. A., and A. Deino. 2005. New $^{40}Ar/^{39}Ar$ ages from three bentonites in the Bearpaw, Horseshoe Canyon, and Scollard formations (Upper Cretaceous-Paleocene) of southern Alberta; pp. 23–24 in D. Braman, F. Therrien, E. Koppelhus, and W. Taylor, (compilers), Dinosaur Park Symposium: Short Papers, Abstracts and Program. Royal Tyrrell Museum of Palaeontology, Drumheller, Alberta.

Eberth, D. A., D. C. Evans, D. B. Brinkman, F. Therrien, D. H. Tanke, and L. Russell. 2013. Dinosaur biostratigraphy of the Edmonton Group (Upper Cretaceous), Alberta, Canada: evidence for climate influence. Canadian Journal of Earth Sciences 50:701–726.

Evans, D. C., M. J. Vavrek, D. R. Braman, N. E. Campione, T. A. Dececchi, and G. D. Zazula. 2012. Vertebrate fossils (Dinosauria) from the Bonnet Plume Formation, Yukon Territory, Canada. Canadian Journal of Earth Sciences 49:396–411.

Fanti, F., and O. Catuneanu. 2009. Stratigraphy of the Upper Cretaceous Wapiti Formation, west-central Alberta, Canada. Canadian Journal of Earth Sciences 46:263–286.

Fanti, F., and O. Catuneanu. 2010. Fluvial sequence stratigraphy: the Wapiti Formation, west-central Alberta, Canada. Journal of Sedimentary Research 80:320–338.

Fanti, F., and P. J. Currie. 2007. A new *Pachyrhinosaurus* bonebed from the Late Cretaceous Wapiti Formation; pp. 39–43 in D. Braman (compiler), Ceratopsian Symposium: Short Papers, Abstracts and Program. Royal Tyrrell Museum of Palaeontology, Drumheller, Alberta.

Fanti, F., and T. Miyashita. 2009. A high latitude vertebrate fossil assemblage from the Late Cretaceous of west-central Alberta, Canada: evidence for dinosaur nesting and vertebrate latitudinal gradient. Palaeogeography, Palaeoclimatology, Palaeoecology 275:37–53.

Fiorillo, A. R., and R. S. Tykoski. 2012. A new species of the centrosaurine ceratopsid *Pachyrhinosaurus* from the North Slope (Prince Creek Formation: Maastrichtian) of Alaska. Acta Palaeontologica Polonica 57:561–573.

Gangloff, R. A., and A. R. Fiorillo. 2010. Taphonomy and paleoecology of a bonebed from the Prince Creek Formation, North Slope, Alaska. Palaios 25:299–317.

Gates, T., and S. Sampson. 2007. A new species of *Gryposaurus* (Dinosauria: Hadrosauridae) from the late Campanian Kaiparowits Formation, southern Utah, USA. Zoological Journal of the Linnean Society 151:351–376.

Goloboff, P., J. Farris, and K. Nixon. 2008. TNT, a free program for phylogenetic analysis. Cladistics 24:774–786.

Horner, J. R. 1992. Cranial morphology of *Prosaurolophus* (Ornithischia: Hadrosauridae) with descriptions of two new hadrosaurid species and an evaluation of hadrosaurid phylogenetic relationships. Museum of the Rockies Occasional Paper 2:1–119.

Horner, J. R., A. de Ricqlès, and K. Padian. 2000. Long bone histology of the hadrosaurid dinosaur *Maiasaura peeblesorum:* growth dynamics and physiology based on an ontogenetic series of skeletal elements. Journal of Vertebrate Paleontology 20:115–129.

Horner, J. R., D. Weishampel, and C. Forster. 2004. Hadrosauridae; pp. 438–463 in D. Weishampel, P. Dodson, and H. Osmòlska (eds.), The Dinosauria, Second Edition. University of California Press, Berkeley, California.

Jerzykiewicz, T., and A. Sweet. 1988. Sedimentological and palynological evidence of regional climatic changes in Campanian to Paleocene sediments of the Rocky Mountain Foothill, Canada. Sedimentary Geology 59:29–76.

Klein, N., and M. Sander. 2008. Ontogenetic stages in the long bone histology of sauropod dinosaurs. Paleobiology 34:247–263.

Larson, D. W., D. B. Brinkman, and P. R. Bell. 2010. Faunal assemblages from the upper Horseshoe Canyon Formation, an early Maastrichtian cool-climate assemblage from Alberta, with special reference to the *Albertosaurus* bonebed. Canadian Journal of Earth Sciences 47:1159–1181.

Lerbekmo, J., and D. Braman. 2002. Magnetostratigraphic and biostratigraphic correlation of late Campanian and Maastrichtian marine and continental strata from the Red Deer Valley to the Cypress Hill, Alberta, Canada. Canadian Journal of Earth Sciences 39:539–557.

Maddison, W. P., and D. R. Maddison. 2011. Mesquite: a modular system for evolutionary analysis. Version 2.75. Available at mesquiteproject .org. Accessed fall 2012.

Nydam, R. L., M. W. Caldwell, and F. Fanti. 2010. Boreoteiioidean lizard skulls from Kleskun Hill (Wapiti Formation; upper Campanian), west-central Alberta, Canada. Journal of Vertebrate Paleontology 30:1090–1099.

Ogg, J., F. Agterberg, and F. Gradstein. 2004. The Cretaceous Period; pp. 344–383 in F. Gradstein, and J. Ogg (eds.), A Geologic Time Scale. Cambridge University Press, Cambridge, U.K.

Osborn, H. F. 1912. Integument of the iguanodont dinosaur *Trachodon*. American Museum of Natural History Memoir 2:33–54.

Owen, R. 1842. Report on British fossil reptiles, part 2. Report on the eleventh meeting of the British Association for the Advancement of Science July 1841:66–204.

Prieto-Márquez, A. 2005. New information of the cranium of *Brachylophosaurus canadensis* (Dinosauria, Hadrosauridae), with a revision of its phylogenetic position. Journal of Vertebrate Paleontology 25:144–156.

Prieto-Márquez, A. 2010. Global phylogeny of the Hadrosauridae (Dinosauria: Ornithopoda) using parsimony and Bayesian methods. Zoological Journal of the Linnean Society 159:435–502.

Reid, R. E. H. 1990. Zonal "growth rings" in dinosaurs. Modern Geology 15:19–48.

Ricqlès, A. de. 1976. On bone histology of fossil and living reptiles, with comments on its functional and evolutionary significance; pp. 123–150 in A. d'A. Bellairs and C. B. Cox (eds.), Morphology and Biology of Reptiles. Academic Press, London, U.K.

Russell, D. A, and T. P. Chamney. 1967. Notes on the biostratigraphy of dinosaurian microfossil faunas in the Edmonton Formation (Cretaceous), Alberta. National Museum of Canada Natural History Papers 35:1–2.

Ryan, M. J., D. A. Eberth, D. B. Brinkman, P. J. Currie, and D. H. Tanke. 2010. A new *Pachyrhinosaurus*-like ceratopsid from the Upper Dinosaur Park Formation (late Campanian) of southern Alberta, Canada; pp. 141–155 in M. J. Ryan, B. J. Chinnery-Allgeier, and D. A. Eberth (eds.), New Perspectives on Horned Dinosaurs. Indiana University Press, Bloomington, Indiana.

Seeley, H. G. 1888. On the classification of the fossil animals commonly named Dinosauria. Proceedings of the Royal Society of London 43:165–171.

Srivastava, S. 1970. Pollen biostratigraphy and paleoecology of the Edmonton Formation (Maestrichtian), Alberta, Canada. Palaeogeography, Palaeoclimatology, Palaeoecology 7:221–276.

Straight, W. H. 2003. Stratigraphic distribution, taphonomy, and isotope paleoecology of the dinosaurian fauna in the latest Campanian lower Horseshoe Canyon Formation, Alberta, Canada. Ph.D. dissertation, North Carolina State University, Raleigh, North Carolina, 181 pp.

Sullivan, R. M., and T. E. Williamson. 1999. A new skull of *Parasaurolophus* (Dinosauria: Hadrosauridae) from the Kirtland Formation of New Mexico and a revision of the genus. New Mexico Museum of Natural History and Science Bulletin 15:1–52.

Sweet, A., B. Ricketts, A. Cameron, and D. Norris. 1989. An integrated analysis of the Brackett Coal Basin, Northwest Territories. Geological Survey of Canada, Paper 89-1:85–89.

Swofford, D. L. 2002. Phylogenetic analysis using parsimony (and other methods). Version 4.0b10.

Sinauer Associates, Sunderland, Massachusetts, 40 pp.

Tanke, D. H. 2004. Mosquitoes and mud: the 2003 Royal Tyrrell Museum of Palaeontology expedition to the Grande Prairie region (north-western Alberta, Canada). Alberta Paleontological Society Bulletin 19:3–31.

Wagner, J. R., and T. M. Lehman. 2009. An enigmatic new lambeosaurine hadrosaur (Reptilia: Dinosauria) from the Upper Sale Member of the Campanian Aguja Formation of Trans-Pecos Texas. Journal of Vertebrate Paleontology 29:605–611.

Wilkinson, M. 2003. Missing entries and multiple trees: instability, relationships, and support in parsimony analysis. Journal of Vertebrate Paleontology 23:311–323.

Wu, X., D. B. Brinkman, D. A. Eberth, and D. R. Braman. 2007. A new ceratopsid dinosaur (Ornithischia) from the uppermost Horseshoe Canyon Formation (upper Maastrichtian), Alberta, Canada and the succession of ceratopsians through the Edmonton Group; pp. 168–172 in D. R. Braman (compiler), Ceratopsian Symposium: Short Papers, Abstracts and Program. Royal Tyrrell Museum of Palaeontology, Drumheller, Alberta.

Appendix 10.1. Character Codings for the Cranial Material from the Red Willow Hadrosaurine

RW_Hadrosaur
1121311?2 0112121101 ?????????? ?011010?03 210111121? ?????100?? ?????????? ?????????? ??????1010 211212321? ?110011123 1111011201 001?0????? ?????????

Variation in the Skull Roof of the Hadrosaur *Gryposaurus* Illustrated by a New Specimen from the Kaiparowits Formation (late Campanian) of Southern Utah

11

Andrew A. Farke and Lucia Herrero

ABSTRACT

The Kaiparowits Formation (late Campanian) of Grand Staircase–Escalante National Monument, southern Utah, has yielded two species of *Gryposaurus*. A partial skull, including the braincase and associated skull roof, was recently collected within the upper part of the middle unit of the formation. Because a prong of the nasals inserts between the frontals, the specimen is confirmed as *Gryposaurus*, but a species assignment cannot be made without the rest of the skull. Nevertheless, the morphology of the postorbital differs from that in most other *Gryposaurus* specimens. In the holotype specimen of *Gryposaurus monumentensis*, the posterior portion of the postorbital is sharply angled dorsally, elevating the squamosal and providing a kinked profile to the skull roof in lateral view. In the new partial skull, the dorsal surface of the postorbital, and thus the lateral profile of the skull roof, is horizontal. *Gryposaurus notabilis* shows intraspecific variability in this trait, with smaller and presumably ontogenetically younger specimens possessing a straight postorbital, contrasting with a kinked postorbital in most larger and presumably older specimens. Statistical analysis indicates a correlation between the angulation of the postorbital and skull width. Thus, we suggest that the profile of the postorbital changed allometrically and/or ontogenetically in *Gryposaurus*, and consequently this feature is not phylogenetically informative within the genus. The stratigraphic position, morphology, and small size of the specimen relative to the holotype of *G. monumentensis* suggest that the new fossil may possibly represent a small (and potentially ontogenetically younger) *G. monumentensis*.

INTRODUCTION

Gryposaurus is a widespread genus of hadrosaurine (sensu Horner et al., 2004) hadrosaurid in Campanian-aged strata of western North America, with a number of specimens of varying completeness known from Alberta, Montana, and Utah (see Prieto-Márquez [2010] for a comprehensive review). Three species are currently recognized (Prieto-Márquez, 2010), including *G. notabilis* (described by Lambe [1914]; incorporating its junior synonym *G. incurvimanus*, originally described by Parks [1919]; Fig. 11.1C, D), *G. monumentensis* (described by Gates and Sampson [2007]; Fig. 11.1B), and *G. latidens* (described by Horner [1992]). Skulls of *Gryposaurus* are relatively unornamented, with the most notable feature being a strongly arched nasal. Species of *Gryposaurus* also are united by a shared unique configuration of the nasal-frontal contact, among other synapomorphies (Gates and Sampson, 2007; Prieto-Márquez, 2010).

Gryposaurus is an apparently common component of the faunal assemblage in the Kaiparowits Formation of southern Utah, United States. The Kaiparowits Formation, which crops out primarily in Grand Staircase–Escalante National Monument, preserves a terrestrial ecosystem dating from approximately 76.6–74.5 Ma (Roberts et al., 2005; 2013). Fieldwork by several institutions has recovered a variety of non-avian dinosaurs, with hadrosaurids represented by *Parasaurolophus* sp. and two species of *Gryposaurus* (Weishampel and Jensen, 1979; Gates et al., 2013; Gates and Sampson, 2007). Specimens of the taxonomically distinct *Gryposaurus* sp. have been briefly noted and illustrated (Gates et al., in press), but are otherwise undescribed. *Gryposaurus monumentensis* was named on the basis of a nearly complete skull, RAM 6797 (Fig. 11.1B), with several other referred specimens (Gates and Sampson, 2007). The two taxa are apparently stratigraphically separated, with *Gryposaurus* sp. in the lower unit of the Kaiparowits Formation and *G. monumentensis* only in the middle unit (Gates et al., 2013).

In 2007, paleontologist Donald Lofgren discovered the skull roof and associated braincase of a *Gryposaurus*, slightly above the stratigraphic position of the *G. monumentensis* holotype. The specimen (RAM 12065; Figs. 11.1A, 11.2) was excavated in 2008 by a team from the Raymond M. Alf Museum of Paleontology and the Webb Schools, and after preparation tentatively identified as *Gryposaurus*. Several differences between this skull and the *G. monumentensis* holotype spurred an investigation into whether RAM 12065 was a new taxon, a

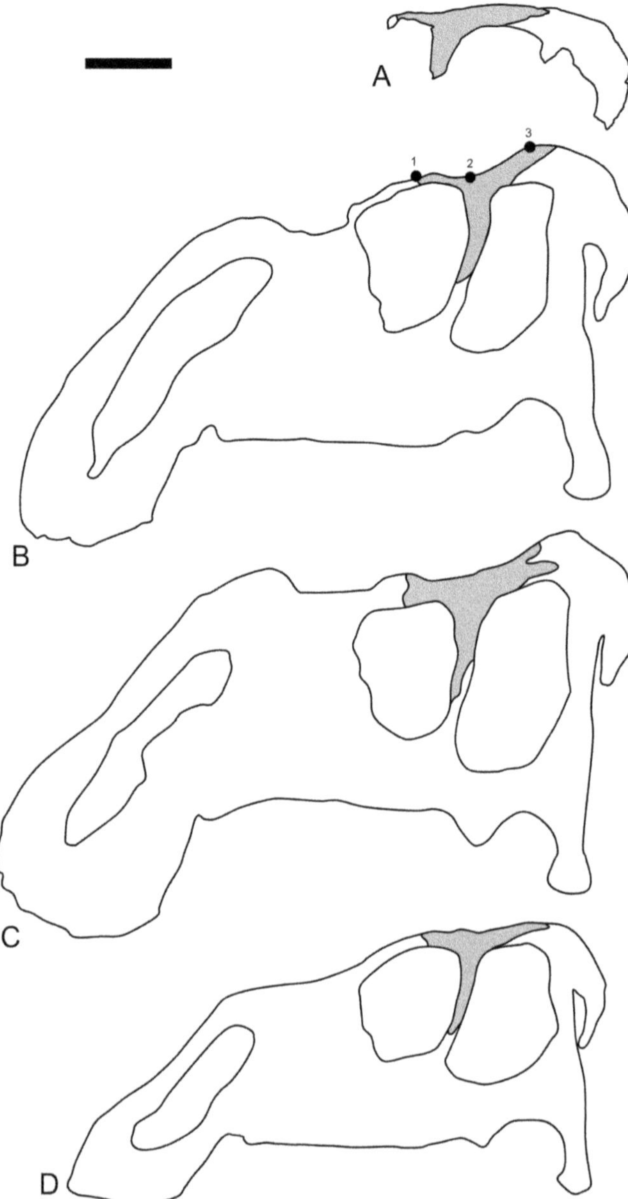

11.1. Schematics of representative skulls of *Gryposaurus* in left lateral view. Note in particular the variation in the flexion of the postorbital (highlighted in gray) and associated elevation of the squamosal. (A) *Gryposaurus* sp., RAM 12065; (B) *Gryposaurus monumentensis,* RAM 6797; (C) *Gryposaurus notabilis,* ROM 873; (D) *Gryposaurus notabilis,* TMP 1980.022.0001. (A) and (B) are from the Kaiparowits Formation of Utah; (C) and (D) are from the Dinosaur Park Formation of Alberta. In (B) the points used to measure the angle of flexion of the postorbital are marked; (1) is the frontal/postorbital suture, (2) is the point of maximum flexion, and (3) is the posterior extent of the postorbital. Skulls are oriented so that the maxillary tooth row is roughly horizontal. Scale bar equals 10 cm.

non-diagnostic specimen, or an individual or ontogenetic variant of *G. monumentensis.*

Gates and Sampson (2007) hypothesized that the postorbital/squamosal configuration of *G. monumentensis* was unique. In the holotype, RAM 6797, the posterior process of the postorbital has a strong dorsal inclination and the squamosal is consequently elevated far dorsal to the rest of the skull roof (Fig. 11.1B). Prieto-Márquez (2010) noted without elaboration that this feature occurs to some degree in specimens of *G. notabilis* (Fig. 11.1C). In the specimen RAM 12065, the postorbital is relatively horizontal in lateral view (Figs. 11.1B, 11.2A, B), and the specimen is 10–20% smaller than RAM 6797 (depending upon the feature under comparison). We thus hypothesized that these differences in postorbital morphology, as well as other characteristics, were allometric and/or ontogenetic in nature. Work on cranial ontogeny in hadrosaurs historically has focused on lambeosaurine taxa (e.g., Dodson, 1975; Evans, 2010), although some recent studies have considered allometry and ontogeny in hadrosaurine crania (e.g., Prieto-Márquez, 2005; Bell, 2011; Campione and Evans, 2011; McGarrity et al., 2013). Within this context, renewed attention to *Gryposaurus* allows an additional point of comparison for allometric trends in hadrosaurine skulls.

Given recent attention to the taxonomy of *Gryposaurus*, as well as the biostratigraphy of the Kaiparowits Formation, it is particularly important to better understand variation within this taxon. In this chapter, we describe RAM 12065 and compare it with other *Gryposaurus* specimens. We also analyze trends in postorbital shape relative to size and address the utility of this character in the taxonomy of *Gryposaurus*.

Institutional Abbreviations AMNH, American Museum of Natural History, New York, New York; CMN, Canadian Museum of Nature, Ottawa, Ontario; MSMN, Museo Civico di Storia Naturale di Milano, Milan, Italy; UMNH, Natural History Museum of Utah, Salt Lake City, Utah; RAM, Raymond M. Alf Museum of Paleontology, Claremont, California; ROM, Royal Ontario Museum, Toronto, Ontario; TMP, Royal Tyrrell Museum of Palaeontology, Drumheller, Alberta.

Anatomical Abbreviations **ap**, alar process; **dtf**, dorsotemporal fenestra; **f**, frontal; **fm**, foramen magnum; **fv**, fenestra vestibuli; **I**, foramen for olfactory tract; **II**, foramen for optic nerve; **III**, foramen for oculumotor nerve; **IV**, foramen for trochlear nerve; **ica**, foramen for internal carotid artery; **lat**, laterosphenoid; **mtf**, metotic foramen; **nss**, nasal sutural surface; **oe**, opisthotic-exoccipital; **p**, parietal; **pcf**, precotylar fossa; **pf**, pituitary fossa; **pfss**, prefrontal sutural surface; **po**, postorbital; **pr**, prootic; **qc**, quadrate cotylus; **sac**, sphenoid artery canal; **so**, supraoccipital; **sq**, squamosal; **V**, foramen for trigeminal nerve; $\mathbf{V_1}$, sulcus for ophthalmic branch of trigeminal nerve; $\mathbf{V_{2\text{–}3}}$, sulcus for maxillary and mandibular branches of trigeminal nerve; **VI**, foramen for abducens nerve; **VII**, foramen for the facial nerve; **XII**, foramen for hypoglossal nerve. For sulci, fossae, and foramina indicating soft-tissue structures, note that the shorthand for cranial nerves necessarily omits other minor structures (nerves, veins, lymphatics, etc.) that may accompany the major cranial neurovasculature.

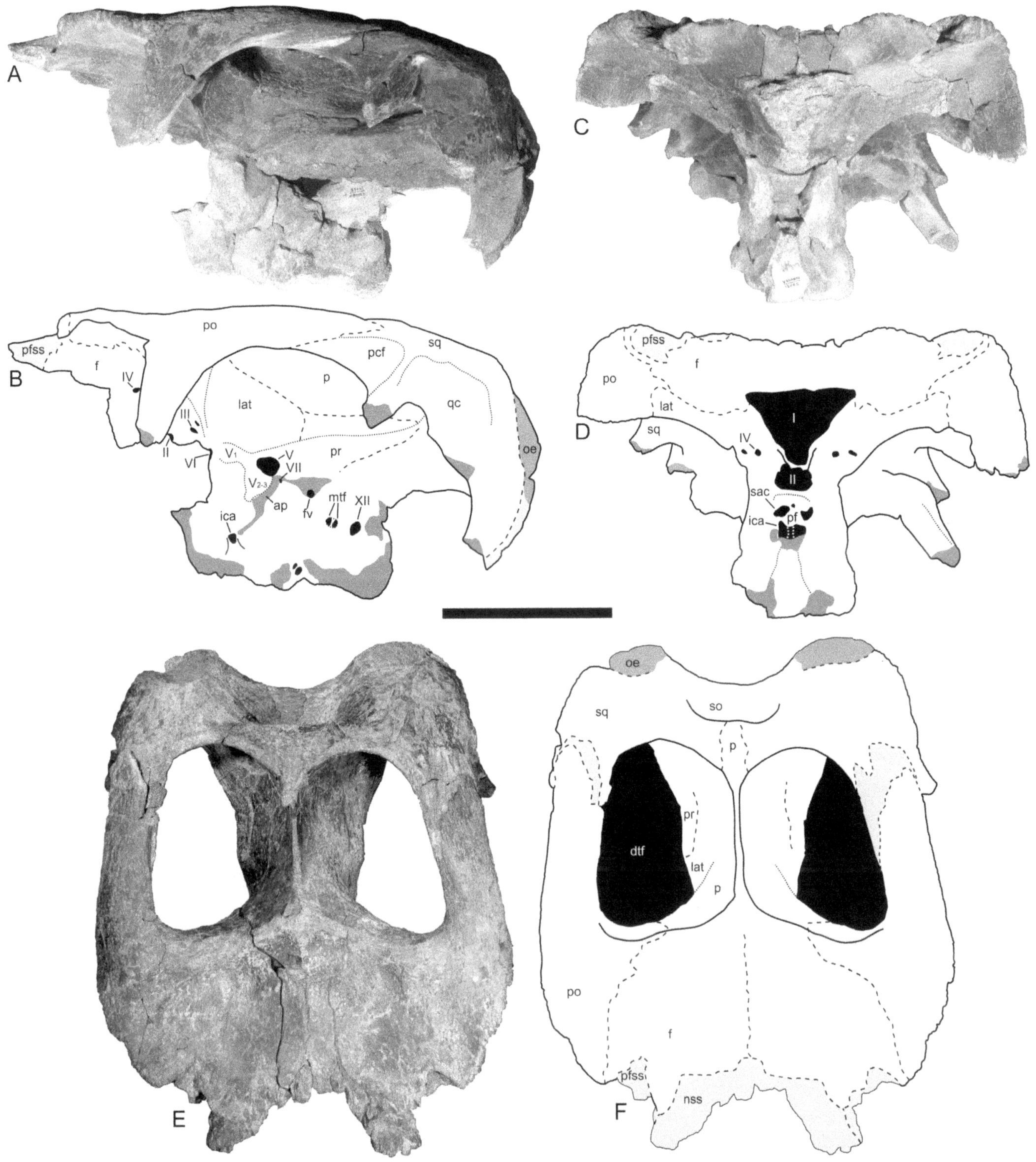

11.2. *Gryposaurus* sp., RAM 12065, photographs and interpretive drawings of partial skull. (A–B) lateral view; (C–D) anterior view; (E–F) dorsal view. Dashed lines indicate visible sutures, light gray areas indicate sutural surfaces, and dark gray areas indicate broken bone surfaces. Scale bar equals 10 cm.

SYSTEMATIC PALEONTOLOGY

DINOSAURIA Owen, 1842
ORNITHISCHIA Seeley, 1888
HADROSAURIDAE Cope, 1869
HADROSAURINAE Cope, 1869
GRYPOSAURUS Lambe, 1914
GRYPOSAURUS sp.
(Figs. 11.1A, 11.2, 11.3)

Referred Specimen RAM 12065, a partial skull including frontals, postorbitals, parietals, squamosals, and basicranium.

Locality RAM V200810, Grand Staircase–Escalante National Monument, Garfield County, Utah, United States. Detailed locality data are on file at the RAM and available to qualified researchers upon request.

Stratigraphic Horizon and Lithology Upper portion of the middle unit of the Kaiparowits Formation (sensu Roberts et al., 2005; Roberts, 2007), Upper Campanian. Within the local section, the estimated position of RAM V200810 is approximately 150 m above the locality for the holotype skull of *Gryposaurus monumentensis* (RAM 6797), which is situated 10 m below a bentonite tentatively correlated with bentonite KBC-109 sensu Roberts et al. (2005; Roberts et al., 2013; Roberts, pers. comm., 2012). The age of this ash bed was recently recalculated as 75.51 ± 0.15 Ma (Roberts et al., 2013). Furthermore, RAM V200810 is above a bentonite correlated with KBC-144 sensu Roberts et al. (2005), placing the site within the uppermost portion of the middle unit or potentially the base of the upper unit. The isolated specimen was preserved dorsum-down within a thick, cross-bedded channel sandstone, with no other associated cranial or postcranial elements.

Description

Most aspects of the skull and braincase of *Gryposaurus* spp. have been described in considerable detail elsewhere (Lambe, 1914; Parks, 1920; Ostrom, 1961; Waldman, 1969; Gates and Sampson, 2007; Prieto-Márquez, 2010), and so the following text primarily focuses on areas that are of particular significance for identifying RAM 12065, or that differ from previously described specimens. RAM 12065 was abraded prior to burial, presumably due to fluvial action, and so portions of the exoccipitals and postorbitals were damaged. The ventral-most portions of the squamosals, postorbital, and exoccipital were lost to erosion prior to discovery, and portions of the braincase itself were similarly lost or damaged (Fig. 11.2). The maximum width of the skull across the widest point of the postorbitals is 260 mm, and the length along the midline is 256 mm. The skull is 213 mm wide at the posterior ends of the postorbitals. RAM 12065 is referred to *Gryposaurus* sp. based on the morphology of the nasofrontal suture (see below), but cannot be assigned to species without the rest of the skull (following the diagnoses of Prieto-Márquez [2010]).

Frontal In dorsal view, the paired frontals form much of the skull roof anterior to the dorsotemporal fenestrae (Fig. 11.2E, F). The sutural surface for the nasals is broad and thin, depressed only 1–2 mm below the dorsum of the skull. The posterior margin of the suture can be roughly divided into three zones (Fig. 11.2F). The first and most lateral section is angled towards the midline in dorsal view; the anterolateral edge slopes strongly towards the prefrontal sutural surface. The next section is oriented mediolaterally and its surface is relatively horizontal. The medial-most section extends posteriorly into the frontals, forming the nasofrontal sutural configuration that is diagnostic of *Gryposaurus* (Gates and Sampson, 2007; Prieto-Márquez, 2010). The prefrontal sutural surface of each frontal is divided into a medial and lateral region by a vertical ridge. As seen in *Gryposaurus* as well as some other hadrosaurines (e.g., *Brachylophosaurus canadensis; Maiasaura peeblesorum* [Prieto-Márquez, 2005]), the frontal extends to the orbital margin but is compressed to a narrow process between the postorbital and prefrontal. The dorsal surface of each frontal has a shallow depression just anterior to the dorsotemporal fenestrae (Fig. 11.2E); this feature also occurs in some *G. notabilis* (ROM 873, TMP 1980.022.0001) but not in others (TMP 1991.081.0001). The condition cannot be determined in RAM 6797 due to incomplete preservation.

Postorbital The postorbital in RAM 12065 contrasts with many *Gryposaurus* specimens (RAM 6797, ROM 873; Fig. 11.1A, B) in that the posterior process is unelevated relative to the anterior portion; similar anatomy is seen in TMP 1980.022.0001 and other specimens (Table 11.1). Although a portion of the posterior process was lost due to pre-burial breakage, the preserved portion of the suture with the squamosal indicates that the process was mostly undivided (except for a small spur on the left side; Figure 11.2E, F). This differs from the prominently forked postorbitals of AMNH 5350 and TMP 1991.081.0001, but is similar to RAM 6797. Contrasting with RAM 6797, AMNH 5350, and TMP 1980.022.0001, but similar to ROM 873 and TMP 1991.081.0001, the postorbital forms approximately half of the dorsal margin of the infratemporal fenestra in lateral view. The width of the fenestra is absolutely greater in RAM 12065 (90 mm) than in RAM 6797 (80 mm).

In dorsal view, the lateral margin of the posterior process of the postorbital in RAM 12065 is straight, compared to the concave curvature seen in *G. notabilis* and RAM 6797. The latter morphology imparts a strongly pinched appearance to the skull roof, so that the squamosals appear to bulge laterally (e.g., AMNH 5350; Prieto-Márquez, 2010:fig. 3). In RAM 12065,

Table 11.1. Measurements of Specimens of *Gryposaurus* spp., Sorted by Increasing Skull Width

Specimen	Taxon	Skull Width (mm)	Postorbital Angle
CMN 8784	*Gryposaurus* sp.	132	176°
TMP 1980.022.0001	*Gryposaurus notabilis*	200	170°
TMP 1991.081.0001	*Gryposaurus notabilis*	205	171°
ROM 764	*Gryposaurus notabilis*	210	163°
RAM 12065	*Gryposaurus* sp.	255	173°
RAM 6797	*Gryposaurus monumentensis*	310	153°
MSMN V345	*Gryposaurus notabilis*	310	156°
AMNH 5350	*Gryposaurus notabilis*	313	161°
CMN 2278	*Gryposaurus notabilis*	321	168°
ROM 873	*Gryposaurus notabilis*	324	160°

Note: See the text and Figure 11.1B for a description of the measurements.

the skull roof tapers smoothly in the medial direction from anterior to posterior.

Parietal The parietal forms a low crest separating the left and right dorsotemporal fenestrae, but cannot be distinguished from the frontal at the area of contact between the two bones. Unlike the *G. notabilis* specimens AMNH 5350 and TMP 1980.022.0001 but similar to CMN 2278, the posterior-most portion of the parietal is slightly mediolaterally expanded at its terminus between the two squamosals (Fig. 11.2F), rather than matching the width of the parietal crest. This region is not preserved in the *G. monumentensis* holotype.

Squamosal In lateral view, the precotyloid fossa is triangular, with the apex directed posterodorsally (Fig. 11.2A, B). In the *G. monumentensis* holotype RAM 6797 as well as *G. notabilis* specimens TMP 1980.022.0001 and AMNH 5350, the apex of this fossa is slightly hooked ventrally, but this does not occur as markedly in RAM 12065, UMNH VP 16669 (*Gryposaurus* sp.), or *G. notabilis* (CMN 2278 and TMP 1991.081.0001). The cotylus for the quadrate is squared off dorsally, with a prominent overhang delineating that margin. In RAM 6797, the cotylus is tucked slightly ventrally relative to the precotyloid fossa. Less overlap between these structures occurs in RAM 12065, due in part to the relatively lower position of the squamosal in the skull roof in the latter specimen.

Dorsotemporal Fenestrae The dorsotemporal fenestrae in RAM 12065 are roughly quadrilateral and anteroposteriorly short in comparison to most other *Gryposaurus* specimens, including RAM 6797, most closely resembling the condition in AMNH 5350. The lateral border of the fenestra is longer than the medial border, and the anterior end is wider than the posterior end (Fig. 11.2E, F). In RAM 12065, the right fenestra is 114 mm in maximum anteroposterior length and 79 mm in maximum width, with corresponding dimensions of 109 mm and 78 mm on the left side.

Braincase Most of the sutures between adjacent elements in the braincase are obliterated by fusion, so little can be said of the precise relationships between bones (Fig. 11.3). Positions and morphologies of the cranial nerve foramina (Fig. 11.2A–D) are consistent with those noted for other *Gryposaurus* specimens (Gates and Sampson, 2007; Prieto-Márquez, 2010). There is no evidence for a ventral extension of the presphenoid as noted by Prieto-Márquez (2010) for *Gryposaurus notabilis* specimen AMNH 5350, suggesting that the feature is variably ossified within *Gryposaurus*. The laterosphenoid closely matches the condition in the *Gryposaurus monumentensis* holotype (RAM 6797), although RAM 12065 differs in that the region anteroventral to the depression for the mandibular branch of the trigeminal nerve is much less markedly rugose. The metotic foramen is definitively split by two laminae (most clearly preserved on the right side), similar to AMNH 5350. The equivalent region on the external surface of the braincase in RAM 6797 is not completely prepared, so the morphology of the metotic foramen cannot be determined in that specimen. Aspects of the foramen vestibuli are damaged, so the occurrence of a bony spur projecting into the foramen as seen in RAM 6797 but lacking in AMNH 5350 cannot be evaluated.

A low-resolution CT scan of RAM 12065 (data archived at RAM) does not offer information on the neurovascular patterns beyond that seen in the specimen itself, but allows an estimation of endocast volume at 307 cm^3. The angle of flexure between the cerebrum and cerebellum is approximately 150°, contrasting with the angulation of approximately 165° for the endocast of AMNH 5350 illustrated by Ostrom (1961:fig. 56).

STATISTICAL ANALYSIS

Methods

Based on visual inspection of specimens of *Gryposaurus* spp. (Fig. 11.1), we hypothesized that flexion of the postorbital in lateral view was negatively correlated with cranial size. In other words, smaller skulls should have a relatively unflexed

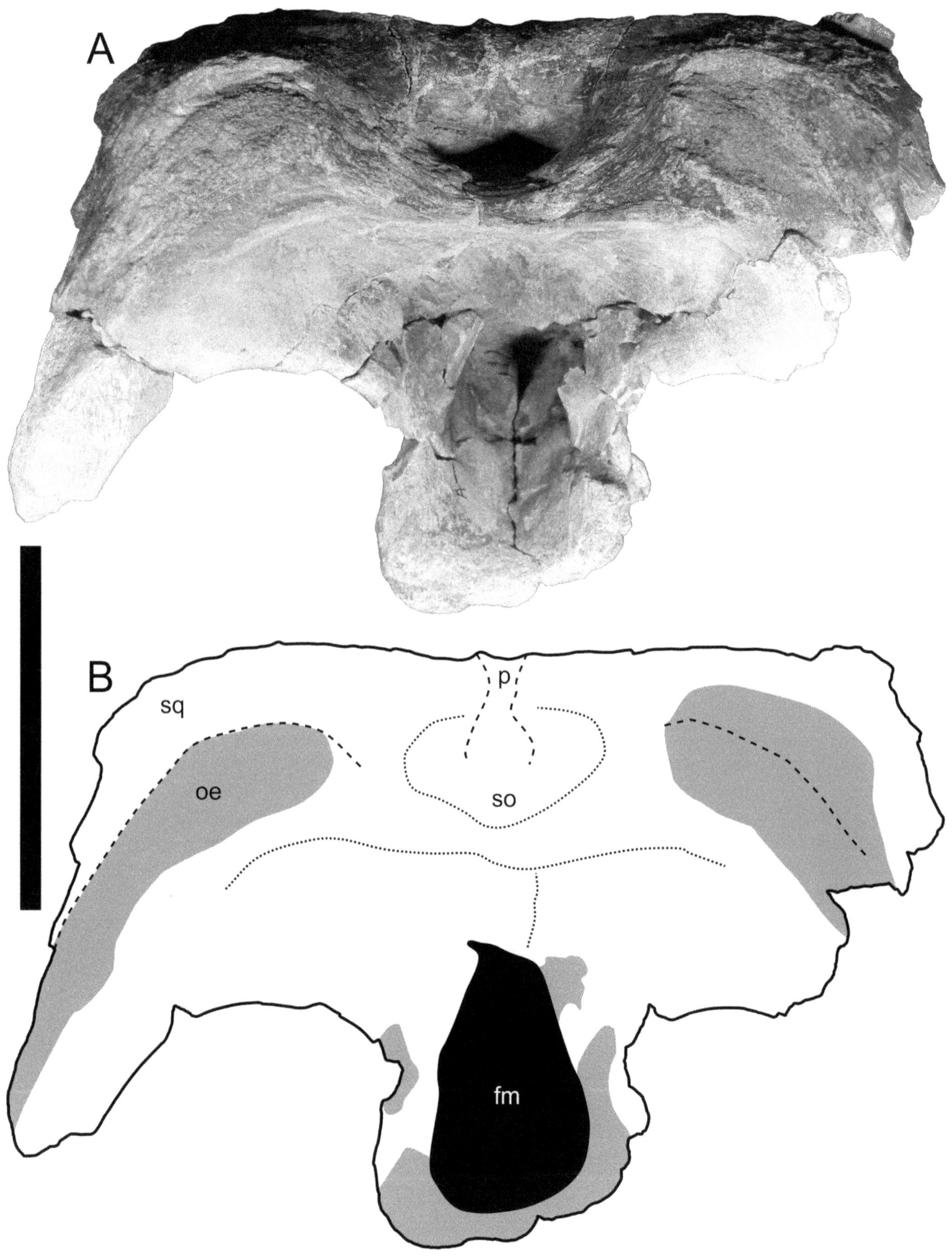

11.3. *Gryposaurus* sp., RAM 12065, photograph and interpretive drawing of partial skull. (A–B) posterior view. Dashed lines indicate visible sutures, dotted lines indicate topographic landmarks, and dark gray areas indicate broken bone surfaces. Scale bar equals 10 cm.

postorbital (and thus an angle of flexion approximating 180°), whereas larger skulls would have more flexed postorbitals (and thus an angle of flexion less than 180°). In order to test this hypothesis, we assembled measurements from 10 different skulls referred to *Gryposaurus* (Table 11.1). Because total skull length could not be measured on many specimens, we used maximum width across the orbits at the postorbitals as a proxy for overall skull size. For specimens that were incomplete on one side, we doubled the measurement from the midline. Specimens were measured directly or from casts whenever possible, although several measurements had to be estimated from photographs or data points in a previously published graph (Prieto-Márquez, 2010:fig. 12).

Flexion of the postorbital in lateral view was measured using three points: (1) the highest point of the postorbital above the orbit (usually coinciding with the postorbital-frontal suture); (2) the vertex at the lowest point on the postorbital between points 1 and 3; and (3) the highest point of the postorbital posteriorly (Fig. 11.1B). This angle was measured to the nearest degree on digital photographs using the program ImageJ. The posterior portion of the postorbital was missing in the smallest specimen in the sample, CMN 8784, so the angle was estimated using the posterior-most preserved point.

A mixed-species sample was utilized in order to increase sample size. This assumes that species of *Gryposaurus* went through similar ontogenetic trajectories; given the gross similarity between currently known specimens, this is considered a reasonable assumption.

The relationship between skull width and postorbital flexion was described using Spearman's r, a measure of correlation. This non-parametric test was chosen due to the non-normal distribution of angular measurements (Sokal and Rohlf, 1995). Spearman's r for the sample was calculated in PAST 2.06 (Hammer et al., 2001).

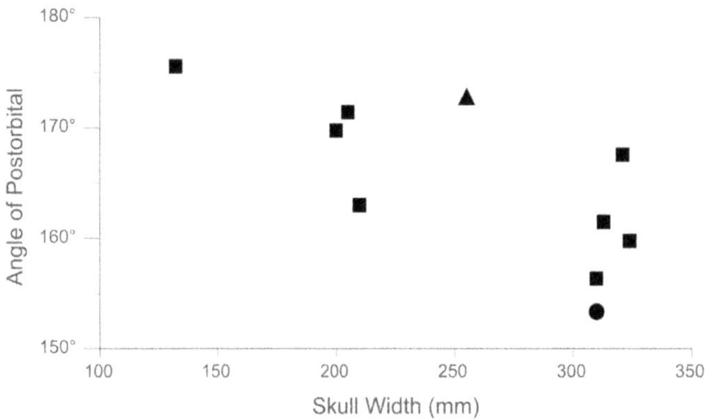

11.4. Plot of skull width versus postorbital flexion angle in *Gryposaurus*, from data in Table 11.1. Specimens from the Dinosaur Park Formation are indicated by squares, RAM 12065 is indicated by a triangle, and RAM 6797 (*G. monumentensis* holotype) is indicated by a circle.

Results

Maximum skull width at the postorbitals was negatively correlated with the angle of postorbital flexion in our sample (Fig. 11.4; Spearman's $r = -0.657$; $p = 0.039$; $n = 10$). In other words, narrower skulls were associated with "flatter" postorbitals, and wider skulls were associated with more flexed postorbitals.

DISCUSSION

The statistical results show that postorbital flexion (and consequently elevation of the squamosal relative to the rest of the skull) increases with increasing skull width and presumably also increasing overall skull size. Although skull size cannot exclusively be used as a proxy for ontogenetic stage, it seems likely that the change in postorbital flexion was at least partly ontogenetic, and not just random individual variation. This is bolstered by the unflexed postorbital in obviously juvenile specimens such as CMN 8784. RAM 15053, an isolated and incomplete juvenile hadrosaurine postorbital (potentially *Gryposaurus*), is also entirely unflexed (not figured). The specimen was recovered from approximately the same stratigraphic level as RAM 12065, and the preserved anterior portion of the posterior ramus is 27 mm wide (75 percent smaller than the equivalent measure in RAM 12065). Sexual dimorphism is another possible factor, but the sample size is too small to evaluate the hypothesis. Other possible ontogenetic changes, in addition to those affecting the postorbital, include an increased prominence in the nasal arch as well as a posterior migration of the apex of the arch (Prieto-Márquez, 2010). In any case, postorbital flexion and associated position of the squamosal on the skull roof should not be considered reliable taxonomic characters within *Gryposaurus*. Consequently, the diagnoses of *G. notabilis* and *G. monumentensis* should be amended to remove this character (Gates and Sampson, 2007; Prieto-Márquez, 2010).

In addition to the variation in postorbital flexion, other minor aspects of variation were documented here. These include the shape of the precotyloid fossa, constriction of the posterior process of the postorbitals in dorsal view, breadth of the posterior portion of the parietal, and shape of the dorsotemporal fenestrae. In all cases, no clear interspecific, ontogenetic, or geographic patterns are evident, although a larger sample size may provide clearer demonstration of any trends.

A narrow infratemporal fenestra has been suggested as a feature distinguishing *G. monumentensis* from *G. notabilis* (Gates and Sampson, 2007; Prieto-Márquez, 2010). RAM

12065 has a comparatively broad fenestra; if this specimen pertains to *G. monumentensis* then it would suggest that the feature is variable within the species. Unfortunately, the rest of the infratemporal fenestra is not preserved in RAM 12065, and neither are other species-diagnostic features. Thus, more material is needed to evaluate the degree of individual variation for this trait within *G. monumentensis*.

Bell (2011) documented allometric changes in the postorbital of *Saurolophus angustirostris* that parallel the changes seen in *Gryposaurus* spp. In *S. angustirostris*, the dorsal margin of the postorbital is relatively straight in small individuals (~35% maximum skull length) but is strongly kinked in skulls that are ~80% of maximum skull size and larger. Although superficially similar to the changes in *Gryposaurus*, there are some important qualitative differences. Specifically, when the maxillary tooth row is used as a datum for horizontal, only the posterior process of the postorbital is elevated in *Gryposaurus* (Fig. 11.1), whereas both the caudal and orbital processes are elevated in *S. angustirostris*. This apparent difference may be due at least in part to the occurrence of a prominent nasal crest in *Saurolophus*, which is also associated with modification of the profiles of the prefrontal and frontal relative to the primitive condition.

Interestingly, no consistent allometric trends in postorbital flexion are evident in *Edmontosaurus* spp., judging by the sample illustrated by Campione and Evans (2011:fig. 2). Large and small specimens alike may show considerable flexion or no flexion at all. Similar to the pattern in *Saurolophus*, both the posterior and orbital processes are elevated.

The functional effects of these ontogenetic changes in skull roof morphology of *Gryposaurus* were probably minor, yet should be considered. Of all cranial muscles, the adductor mandibulae muscles that presumably attached to the margins of the dorsotemporal fenestrae would have been most affected by changes to the bones in this region. Flexion of the postorbital (and associated elevation of the squamosal) would add some length to the overall muscle body. Yet cross-sectional area rather than length is most important for overall force production. More relevant, then, is that these changes in osteology may have shifted the whole muscle mass slightly anteriorly. In terms of lever mechanics, this would decrease the horizontal component and increase the vertical component of force applied to the coronoid process relative to the unflexed morphology, resulting in a slightly increased vertical bite force.

This study is in broad agreement with past investigations of hadrosaurs that emphasized the importance of allometry and ontogeny in distinguishing species (e.g., Dodson, 1975; Evans, 2010; Campione and Evans, 2011; McGarrity et al., 2013). The morphology of RAM 12065 certainly differs from that of the holotype for *Gryposaurus monumentensis*, and the former specimen is stratigraphically higher than any other known specimen of *G. monumentensis*. Thus, a weak case could be made for erecting a new species. However, because other taxonomically informative nasals are missing and most variation reasonably can be attributed to allometry and/or ontogeny, we conservatively assign RAM 12065 to *Gryposaurus* sp., and posit that it possibly (but not definitively) represents *G. monumentensis*. Overall, the variation represented in the sample from the Kaiparowits Formation mirrors the variation in the sample from the Dinosaur Park Formation, particularly with regard to postorbital flexion. Ongoing work in the Kaiparowits Formation and documentation of currently undescribed specimens will undoubtedly add to the understanding of variation within *Gryposaurus*.

ACKNOWLEDGMENTS

We thank the United States Bureau of Land Management and Grand Staircase–Escalante National Monument, especially S. Foss and A. Titus, for permitting and other logistical assistance during field work; all fossils were collected under permit UT06-001S. Numerous volunteers from the Alf Museum and the Webb Schools assisted in the field. Particular thanks are due to D. Lofgren for his leadership in the field and for discovering RAM 12065. RAM 15053 was discovered by W. Allan. Helpful discussion, comparative photographs, and measurements were kindly provided by D. Evans, T. Gates, J. Horner, and A. Prieto-Márquez. E. Roberts provided information on stratigraphic correlations. Additionally, comments from the editors and an anonymous reviewer improved the manuscript. We thank B. Stirlisky (TMP) for assistance in collections and A. Fragomeni (RAM) for curatorial assistance. Pomona Valley Health Center and J. Lightfoote facilitated access to a CT scanner. The David B. Jones Foundation and Mary Stuart Rogers Foundation funded portions of this work.

LITERATURE CITED

Bell, P. R. 2011. Cranial osteology and ontogeny of *Saurolophus angustirostris* from the Late Cretaceous of Mongolia with comments on *Saurolophus osborni* from Canada. Acta Palaeontologica Polonica 56:703–722.

Campione, N. E., and D. C. Evans. 2011. Cranial growth and variation in edmontosaurs (Dinosauria: Hadrosauridae): implications for latest Cretaceous megaherbivore diversity in North America. PLoS ONE 6(9):e25186.

Cope, E. D. 1869. Synopsis of the extinct Batrachia, Reptilia and Aves of North America. Transactions of the American Philosophical Society 14:1–252.

Dodson, P. 1975. Taxonomic implications of relative growth in lambeosaurine hadrosaurs. Systematic Zoology 24:37–54.

Evans, D. C. 2010. Cranial anatomy and ontogeny of *Hypacrosaurus altispinus,* and a comparative analysis of skull growth in lambeosaurines (Ornithischia: Hadrosauridae). Zoological Journal of the Linnean Society 159:398–434.

Gates, T. A., and S. D. Sampson. 2007. A new species of *Gryposaurus* (Dinosauria: Hadrosauridae) from the late Campanian Kaiparowits Formation, southern Utah, USA. Zoological Journal of the Linnean Society 151:351–376.

Gates, T. A., E. K. Lund, C. A. Boyd, D. D. DeBlieux, A. L. Titus, D. C. Evans, M. A. Getty, J. I. Kirkland, and J. G. Eaton. 2013. Ornithopod dinosaurs from the Grand Staircase–Escalante National Monument region, Utah, and their role in paleobiogeographic and macroevolutionary studies; pp. 463–481 in A. L. Titus and M. A. Loewen (eds.), At the Top of the Grand Staircase: The Late Cretaceous of Southern Utah. Indiana University Press, Bloomington, Indiana.

Hammer, Ø., D. A. T. Harper, and P. D. Ryan. 2001. PAST: Paleontological statistics software package for education and data analysis. Palaeontologia Electronica 4(1):1–9.

Horner, J. R. 1992. Cranial morphology of *Prosaurolophus* (Ornithischia: Hadrosauridae) with descriptions of two new hadrosaurid species and an evaluation of hadrosaurid phylogenetic relationships. Museum of the Rockies, Occasional Paper 2:1–119.

Horner, J. R., D. Weishampel, and C. Forster. 2004. Hadrosauridae; pp. 438–463 in D. Weishampel, P. Dodson, and H. Osmólska (eds.), The Dinosauria, Second Edition. University of California Press, Berkeley, California.

Lambe, L. M. 1914. On *Gryposaurus notabilis,* a new genus and species of trachodont dinosaur from the Belly River Formation of Alberta, with a description of the skull of *Chasmosaurus belli.* Ottawa Naturalist 27:145–155.

McGarrity, C. T., N. E. Campione, and D. C. Evans. 2013. Cranial anatomy and variation of *Prosaurolophus maximus* (Dinosauria: Hadrosauridae). Zoological Journal of the Linnean Society 167:531–568.

Ostrom, J. H. 1961. Cranial morphology of the hadrosaurian dinosaurs of North America. Bulletin of the American Museum of Natural History 122:33–186.

Owen, R. 1842. Report on British fossil reptiles, part 2. Report of the Eleventh Meeting of the British Association for the Advancement of Science July 1841:66–204.

Parks, W. A. 1919. Preliminary description of a new species of trachodont dinosaur of the genus *Kritosaurus, Kritosaurus incurvimanus.* Transactions of the Royal Society of Canada, Series 3, 13:51–59.

Parks, W. A. 1920. The osteology of the trachodont dinosaur *Kritosaurus incurvimanus.* University of Toronto Studies, Geology Series 11:1–75.

Prieto-Márquez, A. 2005. New information on the cranium of *Brachylophosaurus canadensis* (Dinosauria, Hadrosauridae), with a revision of its phylogenetic position. Journal of Vertebrate Paleontology 25:144–156.

Prieto-Márquez, A. 2010. The braincase and skull roof of *Gryposaurus notabilis* (Dinosauria, Hadrosauridae), with a taxonomic revision of the genus. Journal of Vertebrate Paleontology 30:838–854.

Roberts, E. M. 2007. Facies architecture and depositional environments of the Upper Cretaceous Kaiparowits Formation, southern Utah. Sedimentary Geology 197:207–233.

Roberts, E. M., A. L. Deino, and M. A. Chan. 2005. $^{40}Ar/^{39}Ar$ age of the Kaiparowits Formation, southern Utah, and correlation of contemporaneous Campanian strata and vertebrate faunas along the margin of the Western Interior Basin. Cretaceous Research 26:307–318.

Roberts, E. M., S. D. Sampson, A. L. Deino, S. A. Bowring, and R. Buchwaldt. 2013. The Kaiparowits Formation: a remarkable record of Late Cretaceous terrestrial environments, ecosystems, and evolution in Western North America; pp. 85–106 in A. L. Titus and M. A. Loewen (eds.), At the Top of the Grand Staircase: The Late Cretaceous of Southern Utah. Indiana University Press, Bloomington, Indiana.

Seeley, H. G. 1888. On the classification of the fossil animals commonly named Dinosauria. Proceedings of the Royal Society of London 43:165–171.

Sokal, R. R., and F. J. Rohlf. 1995. Biometry: The Principles and Practices of Statistics in Biological Research, Third Edition. W. H. Freeman, New York, New York, 880 pp.

Waldman, M. 1969. On an immature specimen of *Kritosaurus notabilis* (Lambe), (Ornithischia: Hadrosauridae) from the Upper Cretaceous of Alberta, Canada. Canadian Journal of Earth Sciences 6:569–576.

Weishampel, D. B., and J. A. Jensen. 1979. *Parasaurolophus* (Reptilia: Hadrosauridae) from Utah. Journal of Paleontology 53:1422–1427.

A Skull of *Prosaurolophus maximus* from Southeastern Alberta and the Spatiotemporal Distribution of Faunal Zones in the Dinosaur Park Formation

12

David C. Evans, Christopher T. McGarrity, and Michael J. Ryan

ABSTRACT

A skull of *Prosaurolophus maximus* from the Dinosaur Park Formation of the Sage Creek Grazing reserve, southern Alberta, represents a southeasterly range extension of this taxon within this unit, approximately 130 km southeast of the type locality at Dinosaur Provincial Park, Alberta. The specimen was found with a juvenile specimen of *Styracosaurus albertensis* in the same quarry. The direct association of *Prosaurolophus maximus* and *Styracosaurus albertensis* confirms that the faunal associations used to establish faunal zones in the Dinosaur Provincial Park area are manifested on a larger, regional geographic scale in the northern Western Interior Basin.

INTRODUCTION

The ornithischian assemblage of the Dinosaur Park Formation (DPF) at Dinosaur Provincial Park (DPP) has a nonrandom biostratigraphic distribution within the approximately 75 m thick DPF (Currie and Russell, 2005; Eberth and Getty, 2005; Ryan and Evans, 2005). Ryan and Evans (2005) proposed three stratigraphically successive biozones within the DPF. In ascending order these are Zone 1, characterized by an association of the centrosaurine *Centrosaurus* and the lambeosaurine *Corythosaurus* in the lowest 30 m of the DPF; Zone 2, characterized by *Styracosaurus* and *Lambeosaurus lambei* from 30 m–50 m above the base of the DPF; and Zone 3, a poorly sampled interval high in section (~50 m+) characterized by occurrences of *Vagaceratops irvinensis* (known from DPP and southeastern Alberta), an indeterminate pachyrhinosaur (Ryan et al., 2010) and *Lambeosaurus magnicristatus*. Ryan et al. (2012) subsequently designated these zones as Dinosaur Park Faunal Zones (DPFZ) 1 to 3. Mallon et al. (2012) tested this biozonation scheme using more sophisticated quantitative methods, and found strong support for the hypothesis. Based on their quantitative results, Mallon et al. (2012) identified two major faunal zones, which they identified as Megaherbivore Assemblage Zones (MAZ 1 and 2), each of which appear to be divisible into two subzones. MAZ-1 corresponds almost exactly to the *Centrosaurus-Corythosaurus* zone proposed by Ryan and Evans (2005), whereas the subzones of MAZ-2 correspond to faunal zones 2 and 3 of Ryan and Evans (2005). Ryan et al. (2012) designated three Dinosaur Park Formation Faunal Zones (DPFZ 1–3) and argued that this arrangement (and that of Ryan and Evans, 2005) is preferred over that of Mallon et al. (2012) because their zones (1) are limited to the Dinosaur Park Formation and time-equivalent strata; (2) include a taxonomically distinct, uppermost zone not included by Mallon et al. (2012); and (3) are not limited to being defined or characterized by the inclusion of megavertebrates or herbivores only.

Two saurolophine hadrosaurid (sensu Prieto-Márquez, 2010b) taxa are recognized from the DPF, *Gryposaurus notabilis* and *Prosaurolophus maximus* (Ryan and Evans, 2005; Prieto-Márquez, 2010a). Specimens of the former are stratigraphically restricted to Megaherbivore Assemblage Zone 1 (Dinosaur Park Formation Faunal Zone 1), whereas the biostratigraphic range of *Prosaurolophus*, the most common saurolophine in the formation, spans Megaherbivore Assemblage Zone 2 (DPFZs 2 and 3). With an occurrence at approximately 52 m above the Oldman–Dinosaur Park Formation contact, *Prosaurolophus* is the stratigraphically highest identifiable hadrosaurid in the DPF section at DPP (Mallon et al., 2012).

The biostratigraphic succession of closely related ornithischian taxa within the Dinosaur Park Formation at DPP can be attributed to either the replacement of one species with another closely related taxon (immigration/emigration), or phyletic change within a sedentary population through time, or a combination of these. Non-overlapping temporal ranges and close phylogenetic relationships of taxa do not reject a phyletic interpretation in some cases. From bottom to top, the DPF records a westward-transgressing shoreline that shifted from southwestern Saskatchewan into southeastern Alberta over a span of at least 1.5 million years (Eberth, 2005). Accordingly, the documented replacement patterns could result from the sampling of different paleoenvironments and their associated communities at successive times. Application of a habitat-tracking model derived from the Dinosaur Park locality dataset can be used to make a number of predictions regarding the distribution of individual genera and species. For example, taxa in the upper part of the formation are hypothesized as more abundant in lower coastal plain and shoreline-proximal settings, whereas taxa found low in the formation are interpreted as more abundant farther updip of the shoreline in association with upper coastal plain settings.

These hypotheses are testable through field collection and documentation of specimens from geographically and paleogeographically disparate localities across southern Alberta and Saskatchewan, and by using previously published chronostratigraphic data. Due to the historically poor vertebrate sampling of regions of Alberta outside of DPP, further fieldwork is required to determine the nature of the faunal shifts within the Belly River Group (Dinosaur Park and the older Oldman and Foremost formations). Of the hadrosaurid species known from the DPF at DPP, only the lambeosaurines *Lambeosaurus magnicristatus* (Evans and Reisz, 2007) and *Corythosaurus casuarius* (Evans, 2002) have been documented outside of that area.

In 1989, a juvenile specimen of *Styracosaurus albertensis* (TMP 1989.097.0001) was discovered in a small area of badlands on the Sage Creek Grazing Reserve, approximately 10 km north of the Onefour Research Station in the extreme southeastern corner of Alberta (Fig. 12.1). During the excavation of this specimen, one of us (MJR) uncovered a partial skull of *Prosaurolophus maximus* associated with the *Styracosaurus* material. In this chapter, we describe the anatomy of this specimen in order to establish its precise taxonomic identification and provide more data on the anatomy and variation in *Prosaurolophus*. In addition, its direct association with *S. albertensis* confirms that the faunal associations used to establish faunal zones in the Dinosaur Provincial Park region are manifested on a larger geographic scale in the northern Western Interior Basin.

Institutional Abbreviations AMNH, American Museum of Natural History, New York, New York; CMN, Canadian Museum of Nature, Ottawa, Ontario; MOR, Museum of the Rockies, Bozeman, Montana; ROM, Royal Ontario Museum, Toronto, Ontario; TMP, Royal Tyrrell Museum of Palaeontology, Drumheller, Alberta; USNM, United States National Museum, Smithsonian Institution, Washington, D.C.

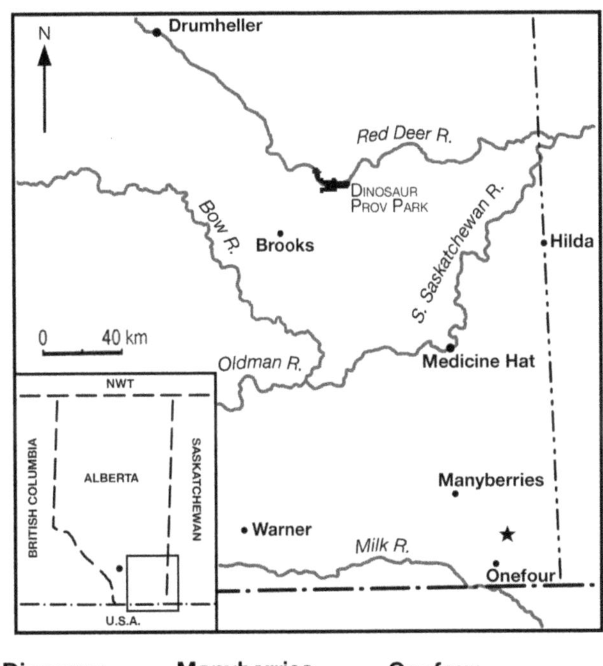

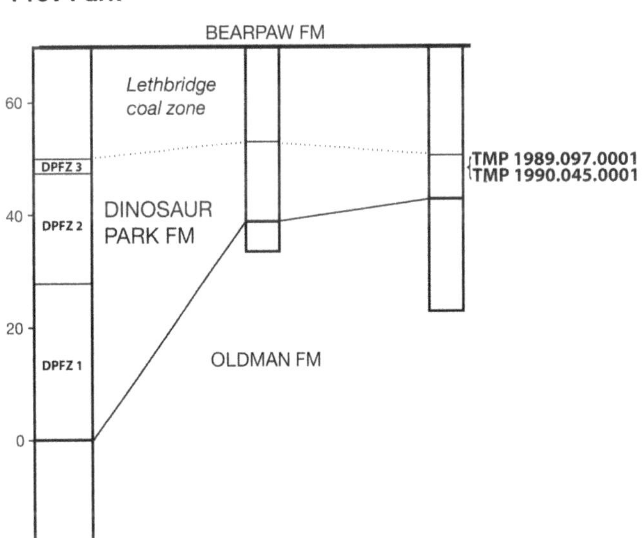

12.1. (Top) locality map. The *Prosaurolophus maximus* skull described herein (TMP 1990.045.0001) and a juvenile *Styracosaurus albertensis* (TMP 1989.097.0001) were collected in the same quarry on the Sage Creek grazing reserve (★). (Bottom) generalized stratigraphic correlations of the Oldman and Dinosaur Park formations in southern Alberta. DPFZ# = Dinosaur Park Faunal Zone. DPFZ 1 is characterized by the presence of *Centrosaurus apertus* and *Corythosaurus casuarius;* DPFZ 2 is characterized by the presence of *Styracosaurus albertensis* and *Lambeosaurus lambei;* DPFZ 3 is characterized by the presence of *Vagaceratops irvinensis* and a Pachyrostra-grade centrosaurine (TMP 2002.076.0001).

SYSTEMATIC PALEONTOLOGY

ORNITHISCHIA Seeley, 1887
ORNITHOPODA Marsh, 1881
HADROSAURIDAE Cope, 1869
SAUROLOPHINAE sensu Prieto-Márquez, 2010b
PROSAUROLOPHUS Brown, 1916
PROSAUROLOPHUS MAXIMUS Brown, 1916

12.2. Quarry of *Styracosaurus albertensis,* TMP 1989.097.0001, and *Prosaurolophus maximus,* TMP 1990.045.0001. People are sitting in the quarry of TMP 1989.097.0001; overburden in the foreground covers the location of TMP 1990.045.0001 (gray outline), which was collected from the trench around the parietal of TMP 1989.097.0001 (inset). Inset photograph by Darren Tanke.

Referred Specimen TMP 1990.045.0001 consists of the nasal, prefrontal, frontal, parietal, postorbital, and squamosal bones, which are described below.

Locality and Horizon The skull was collected from Sage Creek Grazing Reserve, near Onefour, southeastern Alberta (12U, 545,414 mE; 5,451,340 mN [WGS 84]), approximately 130 km southeast of the holotype locality at Dinosaur Provincial Park, Alberta (Fig. 12.2). The specimen was collected from the DPF, less than 5 m below the base of the Lethbridge Coal Zone (i.e., the lowest coal bed). The specimen was preserved in fine-grained, friable mudstone with the bones coated in thin layer of ironstone.

Comments TMP 1990.045.0001 was preserved dorsum up in very close proximity to an associated skull and articulated skeleton of a subadult *Styracosaurus albertensis* (TMP 1989.097.0001; Ryan et al., 2007:fig. 13A), and within the same mudstone layer. The bone is poorly permineralized and soft, making preparation difficult.

DESCRIPTION

TMP 1990.045.0001 is a complete but crushed skull roof assignable to *Prosaurolophus* based on the morphology of the nasal crest (Fig. 12.3). The size and robustness of the diagnostic nasal crest suggest that this specimen is a large adult, similar in size to AMNH 5386 and USNM 12712. TMP 1990.045.0001 consists of the nasal, prefrontal, frontal, parietal, postorbital, and squamosal bones, which are described below. It conforms in overall morphology to other *Prosaurolophus* specimens (AMNH 5386, CMN 2277, ROM 1928, USNM 12712, MOR 553M-8-27-9-538A), except where noted.

Nasal Both nasals are nearly complete in TMP 1990.045.0001 (Fig. 12.3), except for the terminal end of the supranarial process of the left nasal, which is not preserved. The nasal crest is complete, except for the tip of its apex, but mediolateral crushing has distorted the nasals and displaced the apex of the crest 102 mm from the midline of the skull roof. The nasal contacts the premaxilla rostromedially and laterally, the prefrontal and lacrimal caudolaterally, the frontal caudoventrally, and its complement medially along most of its length. The supranarial process is straight and forms the entire dorsal margin of the naris. The nasal is laterally convex and robust above the bony external naris, and also forms the V-shaped caudal margin of the bony external naris. Whereas the premaxilla is not preserved in TMP 1990.045.0001, its typical caudolateral shape is confirmed by the morphology of the overlapping contact on the nasal.

Caudal to the external naris, the nasal thins mediolaterally and extends caudodorsally to form the solid nasal crest. The crest is excavated caudomedially and circumscribes the

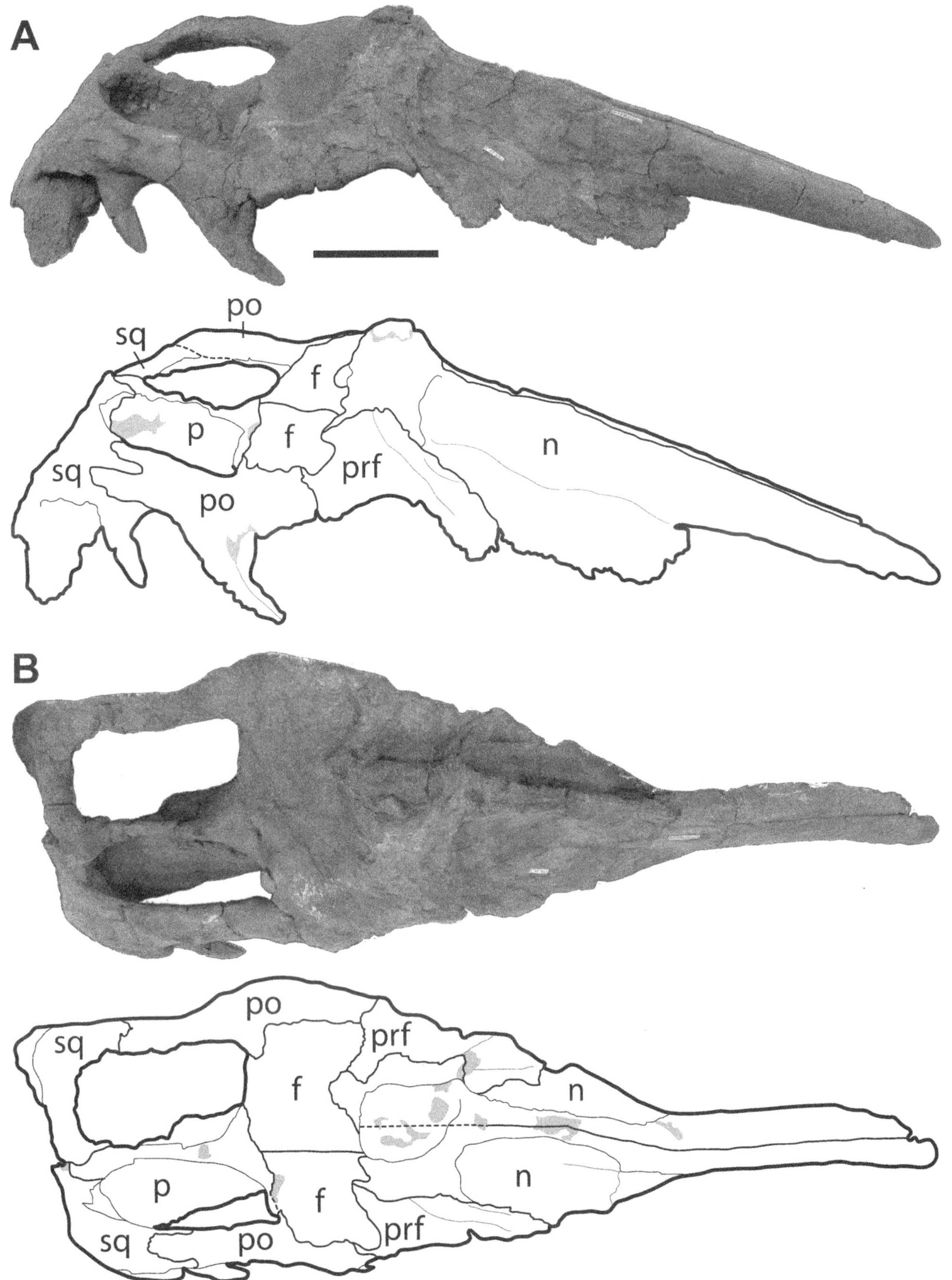

12.3. Incomplete skull of *Prosaurolophus maximus,* TMP 1990.045.0001 in (A) lateral, and (B) dorsal views. Abbreviations: f, frontal; n, nasal; p, parietal; prf, prefrontal; po, postorbital; sq, squamosal. Scale bar equals 10 cm.

caudal margin of the circumnarial depression. The crest of TMP 1990.045.0001 is not as deeply excavated as most specimens of *P. maximus*, which is most likely the result of mediolateral crushing. The nasal crest is rostrocaudally short and is formed entirely by the nasals. In lateral view, the crest extends a short distance beyond the skull roof, above the rostrodorsal margin of the orbit. The orientation of the crest apex, known to be variable in *Prosaurolophus* (McGarrity et al., 2013), is unclear in TMP 1990.045.0001 due to crushing and breakage. There is a slight dorsal bowing of the nasals rostral to the crest, although it is less pronounced in other specimens (see McGarrity et al., 2013). The crest extends laterally and rostrally below its apex, forming a pocket-like caudodorsal limit of the circumnarial depression.

Caudal to the crest, the nasal tapers to a dorsoventrally compressed, triangular process where it contacts the frontal. In dorsal view, the conjoined nasals bifurcate caudally and meet the frontals in a W-shaped joint above the midpoint of the orbit (e.g., Horner, 1992). This frontal process of the nasal is dorsoventrally thick and overlaps the nasal process of the frontal, which acts as the base of the crest.

Prefrontal The prefrontal (Fig. 12.3) is an L-shaped bone that forms the rostrodorsal margin of the orbit. It is nearly complete in TMP 1990.045.0001, except for the rostral tip of the right prefrontal. The prefrontal contacts the lacrimal (not preserved) rostroventrally, the nasal dorsomedially, the frontal caudomedially, and the postorbital caudolaterally. The prefrontal extends caudodorsally and broadens dorsoventrally, where it forms the dorsal and rostral margins of the orbit. There is a small lacrimal depression on the lateral surface of the prefrontal that is separated from the circumnarial depression by a slight ridge (as defined by McGarrity et al., 2013). This depression is rostroventrally oriented, but at a slightly more ventral angle to the ventral margin of the circumnarial depression. The depth of this depression appears to be variable in specimens of *Prosaurolophus maximus*, but is clearly present in TMP 1990.045.0001, and most closely resembles the lacrimal depression exhibited by ROM 1928.

Along the orbit, the prefrontal is dorsoventrally thick, with a large lateral protrusion that transitions into the smooth, rounded dorsal surface of the postorbital. While this protrusion is present in TMP 1990.045.0001, it is far less pronounced than in other specimens (e.g., TMP 1984.001.0001), and continues onto the postorbital. The contact with the postorbital occurs at mid-length along the dorsal orbital margin, and this contact precludes the frontal from contributing to the dorsal portion of the orbital margin.

The preservation of the caudal and dorsal margins of the prefrontal is poor in TMP 1990.045.0001; however, its morphology is consistent with other specimens of *Prosaurolophus maximus* (McGarrity et al., 2013). The prefrontal extends dorsomedially toward the crest above the orbit, and thins mediolaterally to a thin flange that overlaps the nasal medially. Between its contact with the nasal and the postorbital, a rostrally directed V-shaped process in the caudomedial margin of the prefrontal contacts the rostral tip of the frontal in an interdigitating suture.

Frontal The frontal forms the skull roof immediately caudal to the nasal crest. It contacts the nasal rostrodorsally, the prefrontal rostrolaterally, its complement medially, the postorbital laterally, and the parietal caudally. The triangular frontal process of the nasal rests dorsally on the nasal process of the frontal in a lap joint. The frontals do not contribute to the nasal crest, but instead form a base for the crest to rest on, consistent with previous studies (e.g., Horner, 1992; McGarrity et al., 2013).

The paired frontals extend rostrally along the midline of the skull and insert a small triangular process between the dorsally raised, W-shaped frontal processes of the nasal. There is slight variation in the morphology of the right frontal in this specimen. The lateral rostrally directed process of the right frontal exhibits a second V-shaped bifurcation at its contact with the frontal process of the nasal. Additionally, both this process and the rostrally directed lateral process of the left frontal extend beyond the back of the crest, which is farther than it extends in other specimens (McGarrity et al., 2013). However, this anomaly may be due to the extreme mediolateral crushing exhibited by this specimen. The frontals are flat dorsally, and do not exhibit an upward doming over the braincase; however, in TMP 1990.045.0001 there is a short dorsally extended sagittal ridge along the midline of the skull roof with shallow concave depressions on either side, similar to CMN 2870. Lateral to the contact with the nasal, the frontal extends rostrolaterally to contact the prefrontal and postorbital in interdigitated sutures. The union of the postorbital and prefrontal excludes the frontal from contributing to the orbital rim. The caudal contact of the frontal with the parietal is also highly interdigitated.

Parietal The parietal (Fig. 12.3) is an elongate, prism-shaped bone along the center of the skull roof that forms the medial margin of both supratemporal fenestrae. It is poorly preserved in TMP 1990.045.0001, with the exception of the sagittal crest. The parietal contacts the frontal rostrally, the postorbital rostrolaterally, and the squamosal caudolaterally. The rostral end of the bone has broad, transverse contacts with each of the frontals that are highly interdigitated. Rostrolateral processes extend to contact the postorbitals, and thereby exclude the frontals from the rostral margin of the supratemporal fenestrae. The most prominent feature of the parietal is the long sagittal crest, which consists of a triangular caudally oriented plateau that tapers to a mediolaterally thin ridge and extends along the midline from the frontals to the squamosals. The surface of this plateau is weakly depressed along the midline, and is less pronounced than it is in other specimens of *Prosaurolophus maximus* (McGarrity et al., 2013). The sagittal crest separates the supratemporal fenestrae, and forms the medial margin of each fenestra for most of its length. From the sagittal crest, the parietal extends ventrally and laterally to contact the sphenoid bones that form the lateral wall of the braincase. Caudal to the sagittal crest, the parietal is weakly expanded to meet the medially directed parietal processes of the squamosals. The parietal prevents the squamosals from contacting one another except for a short portion on the caudal-most portion of the skull roof in TMP 1990.045.0001, consistent with other *Prosaurolophus* specimens (AMNH 5386, ROM 1928, TMP 1984.001.0001).

Postorbital The triradiate postorbital (Fig. 12.3) is complete in TMP 1990.045.0001. It forms the part of the lateral margin of the skull in the orbital region of the skull, and contributes to the margins of the orbit, supratemporal fenestra, and infratemporal fenestra. The postorbital contacts the prefrontal rostrally and rostromedially, the frontal rostromedially, the parietal medially, the jugal ventrocaudally, and the squamosal caudomedially. The medial contact between the postorbital and the laterosphenoid cannot be observed due to poor preservation. The postorbital contacts the prefrontal near the midpoint of the dorsal margin of the orbit. As in the prefrontal, the postorbital is dorsoventrally thick, rugose, and striated along its contribution to the orbit. Due to poor preservation along the orbital margin, the foramina described by Horner (1992) that penetrate the prefrontal ramus above the orbit cannot be observed in TMP 1990.045.0001. Caudomedial to the prefrontal contact, the postorbital contacts the frontal in an interdigitating joint. The postorbital also has a short contact with the parietal caudal to the frontal joint, where it forms the rostrolateral margin of the supratemporal fenestra.

The postorbital has a long and distally tapered descending jugal process caudal to the orbit that extends rostroventrally to unite with the postorbital process of the jugal along a long scarf joint to form the postorbital bar. The jugal process of the postorbital is triangular in cross section. The rostral surface bears a shallow pocket reminiscent of that of *Edmontosaurus* along the caudodorsal margin of the orbit, similar to AMNH 5386 and USNM 12712. Caudal to the jugal process, the postorbital gently curves and extends caudally to form the lateral half of the intertemporal bar that separates the supratemporal fenestra from the infratemporal fenestra. The cylindrical squamosal process of the postorbital contacts the squamosal medially along most of its length. The caudal portion of the squamosal process of the postorbital thins mediolaterally and bifurcates caudally where it receives the squamosal dorsal to the prequadratic process of the squamosal.

Squamosal The squamosal (Fig. 12.3) forms the caudolateral corner of the skull roof. The right squamosal is almost complete in TMP 1990.045.0001, but the ventral portion of the postquadratic process is not preserved. The squamosal contacts the postorbital rostrolaterally, the parietal caudomedially. In other specimens of this genus, it also contacts the exoccipital caudoventrally and the quadrate ventrally, but these contact surfaces are not preserved in TMP 1990.045.0001. The rostral process of the squamosal contacts the bifurcated squamosal process of the postorbital laterally in a scarf joint to form the intertemporal bar, which separates the supratemporal fenestra from the infratemporal fenestra.

The caudal region of the squamosal extends medially towards the midline of the skull, where it hooks rostrally to meet the parietal in a tall butt joint. The thin sagittal crest of the parietal restricts most of the contact between the squamosals, except for their caudal-most portion along the caudal margin of the skull, where they meet in a short contact. Between the contact with the postorbital and the parietal, the squamosal contributes the lateral and caudal margins of the supratemporal fenestra. The most prominent feature of the squamosal is the pocket-like quadrate cotylus, which is bounded between two prominent descending processes that receive the head of the quadrate, which is not preserved in this specimen. The short, triangular prequadratic process extends rostroventrally at the same angle as that of the jugal process of the postorbital. Caudal to the quadrate cotylus, the postquadratic process extends ventrally, although the postquadratic process is not complete here due to breakage. In other specimens of *Prosaurolophus maximus*, the caudal margin of the postquadratic process contacts the paroccipital process of the exoccipital along its entire length (McGarrity et al., 2013), but the exoccipital is not preserved in TMP 1990.045.0001.

DISCUSSION

Prosaurolophus maximus is known from over a dozen articulated skulls from the DPF at DPP, and from associated and isolated disarticulated skull elements from the younger Two Medicine Formation of Montana (McGarrity et al., 2013). The Sage Creek specimen, TMP 1990.045.0001, can be assigned to *Prosaurolophus maximus* on the basis of several derived characters, including the presence of a small, laterally pocketed nasal crest located above the anterior margin of the orbit (Horner, 1992). Furthermore, the crest is more similar to that seen in specimens collected from Dinosaur Provincial Park in its relatively short length, and differs from a specimen collected from the Two Medicine Formation formally referred to *P. blackfeetensis* (Horner, 1992) in the absence of significant dorsal bowing of the nasal anterior to the crest; however, these features do overlap in variation between specimens collected from these formations (Prieto-Márquez, 2010b; McGarrity et al., 2013).

TMP 1990.045.0001 was collected approximately 130 km southeast of Dinosaur Provincial Park (Fig. 12.1), and over 100 km east of the younger Montana occurrences of this taxon, and therefore represents the most eastward occurrence of this species recorded to date. Correlation of the host stratum of TMP 1990.045.0001 with the DPP section provides the opportunity to test the palaeogeographic robustness of the ornithischian biostratigraphic zonation proposed by Ryan and Evans (2005) and verified statistically by Mallon et al. (2012). TMP 1990.045.0001 was recovered from a small outcrop of the Belly River Group approximately 10 km north of the Onefour Research Station, southeastern Alberta (Fig. 12.1). Despite the isolated nature and low relief of the outcrop, the exposed section records the characteristic, relatively sandy upper Dinosaur Park Formation and the base of the Lethbridge Coal Zone (LCZ), which is visible in section in the immediate area of the quarry (Fig. 12.2). The specimen was collected less than 5 m below the base of the LCZ, within the Dinosaur Park Formation. The DPF is approximately 28 m thick in this area and is litho- and chronostratigraphically equivalent to the upper part of the formation at DPP (Eberth and Hamblin, 1993). Within this package, the total thickness of the LCZ is approximately 20 m whether measured at Onefour, Manyberries, or DPP. The base of the LCZ represents a chronostratigraphic datum across southeastern Alberta (Eberth, 2005). A bentonite layer just below the basal coal of the LCZ near Onefour has been dated at 74.9 ± 0.1 Ma, but recent recalibration of this date is likely to push it back several hundred thousand years (Eberth and Hamblin, 1993; D. A. Eberth, pers. comm., 2012). Based on this information, the Sage Creek *Prosaurolophus* is interpreted as occurring within the chronostratigraphic ranges of *P. maximus* from Dinosaur Provincial Park and, more broadly, Megaherbivore Assemblage Zone 2 of Mallon et al. (2012).

The discovery of TMP 1990.045.0001 in the same quarry as an almost complete skeleton of a subadult *Styracosaurus albertensis* (TMP 1989.097.0001) represents a rare occurrence in which specimens of a large hadrosaurid and ceratopsid that can be identified to the species level have been recovered from the same host stratum within the DPF. TMP 1989.097.0001 was described by Ryan et al. (2007) and referred to *Styracosaurus albertensis* based on the spike-like ornamentation at epiparietal loci 3 and 4, and the caudolateral orientation of the P3 process. This unique occurrence documents the direct association of these taxa outside of the immediate vicinity of the Dinosaur Park locality for the first time. Correlation of these Sage Creek *P. maximus* and *S. albertensis* specimens with the well-known DPF stratigraphic section at Dinosaur Provincial Park suggests that the association of these taxa is consistent across southern Alberta within a relatively restricted temporal interval near the end of "Dinosaur Park time," approximately 75.5–75.0 Ma.

Ornithischian specimens from the Dinosaur Park Formation identifiable to the species level are rare beyond the Red Deer River valley, and the spatiotemporal distribution of taxa that characterize the proposed faunal zones at the DPP locality remains poorly documented (Evans and Reisz, 2007). Although a large number of specimens have been collected along the South Saskatchewan River by the University of Alberta and the Royal Tyrrell Museum in the last half century, few have been formally described and placed into detailed chronostratigraphic context in the literature. Recently, Eberth et al. (2010) documented the distribution of *Centrosaurus apertus* bonebeds along the South Saskatchewan River as far south as the Hilda area. Evans (2002) described the skull of a subadult *Corythosaurus casuarius* (TMP 1997.012.0232) from the same set of outcrops. TMP 1997.012.0232 was collected from the lowermost 10 m of the DPF, and all of the *Centrosaurus* bonebeds occur within the lower 25 m of the DPF in the Hilda sections. There the DPF has a total thickness approximately equal to that found in Dinosaur Provincial Park, and the lower 25 m of the DPF appears to be chronostratigraphically equivalent to the section that hosts *Corythosaurus* and *Centrosaurus* at DPP (MAZ 1 of Mallon et al., 2012), yet predates all of the specimens collected from the DPF in the areas of Onefour and Manyberries, which correlate to the upper half of the Dinosaur Park Formation at the DPP locality (Eberth and Hamblin, 1993) and MAZ 2 (Mallon et al., 2012). Holmes et al. (2001) described specimens of *Vagaceratops* [= *Chasmosaurus*] *irvinensis* from Dinosaur Provincial Park, Irvine, and Onefour localities, all from the Dinosaur Park Formation within 20 m of the base of the Lethbridge Coal Zone. Evans and Reisz (2007) showed that the two known specimens of *Lambeosaurus magnicristatus* are from roughly time-correlative, uppermost DPF strata at Dinosaur Provincial Park and the Manyberries locality of the referred specimen, approximately 120 km to the southeast.

Together with the reported association of *Prosaurolophus* and *Styracosaurus* from Sage Creek documented herein, these occurrences suggest that the faunal zones proposed by Ryan and Evans (2005) and Mallon et al. (2012) are consistent within the Dinosaur Park Formation on a regional scale across southern Alberta, and reinforces their paleoecological and biostratigraphic significance. Furthermore, it appears that the faunal replacements proposed by Ryan and Evans (2005), Evans and Reisz (2007), and Ryan et al. (2010, 2012), and validated by Mallon et al. (2012) likely occurred across the geographic extent of the Dinosaur Park Formation; however, further fieldwork across Alberta, Saskatchewan and Montana is required to determine the precise nature of these faunal shifts in the late Campanian.

ACKNOWLEDGMENTS

We thank Brandon Strilisky and Jim Gardner for access to the TMP collections. Donna Sloan and Darren Tanke provided information on the excavation of TMP 1990.045.0001. We thank Don Brinkman, Caleb Brown, Philip Currie, Dave Eberth, and Jordan Mallon for discussions, and we are indebted to Nicolás Campione for assistance with the manuscript. Jordan Mallon and Dave Eberth provided reviews that improved the manuscript. Funding was provided by an NSERC Discovery Grant to DCE and a University of Toronto Fellowship to CTM.

LITERATURE CITED

Brown, B. 1916. A new crested trachodont dinosaur *Prosaurolophus maximus.* Bulletin of the American Museum of Natural History 35:701–708.

Cope, E. D. 1869. Synopsis of the extinct Batrachia, Reptilia and Aves of North America. Transactions of the American Philosophical Society 14:1–252.

Currie, P. J., and D. A. Russell. 2005. The geographic and stratigraphic distribution of articulated and associated dinosaur remains; pp. 537–567 in P. J. Currie and E. B. Koppelhus (eds.), Dinosaur Provincial Park: A Spectacular Ancient Ecosystem Revealed. Indiana University Press, Bloomington, Indiana.

Eberth, D. A. 2005. The geology; pp. 54–82 in P. J. Currie and E. B. Koppelhus (eds.), Dinosaur Provincial Park: A Spectacular Ancient Ecosystem Revealed. Indiana University Press, Bloomington, Indiana.

Eberth, D. A., and M. A. Getty. 2005. Ceratopsian bonebeds: occurrence, origins, and significance; pp. 501–536 in P. J. Currie and E. B. Koppelhus (eds.), Dinosaur Provincial Park: A Spectacular Ancient Ecosystem Revealed. Indiana University Press, Bloomington, Indiana.

Eberth, D. A., and A. P. Hamblin. 1993. Tectonic, stratigraphic, and sedimentologic significance of a regional discontinuity in the upper Judith River Group (Belly River wedge) of southern Alberta, Saskatchewan, and northern Montana. Canadian Journal of Earth Sciences 30:174–200.

Eberth, D. A., D. B. Brinkman, and V. Barkas. 2010. A centrosaurine mega-bonebed from the Upper Cretaceous of southern Alberta: implications for behaviour and death events; pp. 495–508 in M. J. Ryan, B. J. Chinnery-Allgeier, and D. A. Eberth (eds.), New Perspectives on Horned Dinosaurs. Indiana University Press, Bloomington, Indiana.

Evans, D. C. 2002. A juvenile *Corythosaurus* skull (Ornithischia: Hadrosauridae) from the Dinosaur Park Formation, southeastern Alberta; pp. 11–14 in B. J. Dougherty and A. D. McCracken (eds.), Canadian Paleontology Conference Program and Abstracts 12. Geological Association of Canada, Paleontology Division, St. John's, Newfoundland.

Evans, D. C., and R. R. Reisz. 2007. Anatomy and relationships of *Lambeosaurus magnicristatus,* a crested hadrosaurid dinosaur (Ornithischia) from the Dinosaur Park Formation, Alberta. Journal of Vertebrate Paleontology 27:373–393.

Holmes, R. B., C. Forster, M. Ryan, and K. M. Shepherd. 2001. A new species of *Chasmosaurus* (Dinosauria: Ceratopsia) from the Dinosaur Park Formation of southern Alberta. Canadian Journal of Earth Sciences 38:1423–1438.

Horner, J. R. 1992. Cranial morphology of *Prosaurolophus* (Ornithischia: Hadrosauridae) with descriptions of two new hadrosaurid species and an evaluation of hadrosaurid phylogenetic relationships. Museum of the Rockies Occasional Paper 2:1–119.

Marsh, O. C. 1881. Classification of the Dinosauria. American Journal of Science 7:81–86.

Mallon, J. C., D. C. Evans, M. J. Ryan, and J. S. Anderson. 2012. Megaherbivorous dinosaur turnover in the Dinosaur Park Formation (upper Campanian) of Alberta, Canada. Palaeogeography, Palaeoclimatology, Palaeoecology 350–352:124–138.

McGarrity, C. T., N. E. Campione, and D. C. Evans. 2013. Cranial anatomy and variation of *Prosaurolophus maximus* (Dinosauria: Hadrosauridae). Zoological Journal of the Linnean Society 167:531–568.

Prieto-Márquez, A. 2010a. The braincase and skull roof of *Gryposaurus notabilis* (Dinosauria, Hadrosauridae), with a taxonomic revision of the genus. Journal of Vertebrate Paleontology 30:838–854.

Prieto-Márquez, A. 2010b. Global phylogeny of Hadrosauridae (Dinosauria: Ornithopoda) using parsimony and Bayesian methods. Zoological Journal of the Linnean Society 159:435–502.

Ryan, M. J., and D. C. Evans. 2005. Ornithischian Dinosaurs; pp. 312–348 in P. J. Currie and E. B. Koppelhus (eds.), Dinosaur Provincial Park: A Spectacular Ancient Ecosystem Revealed. Indiana Univeristy Press, Bloomington, Indiana.

Ryan, M. J., R. Holmes, and A. P. Russell. 2007. A revision of the late Campanian centrosaurine ceratopsid genus *Styracosaurus* from the Western Interior of North America. Journal of Vertebrate Paleontology 27:944–962.

Ryan, M. J., D. A. Eberth, D. B. Brinkman, P. J. Currie, and D. H. Tanke. 2010. A new *Pachyrhinosaurus*-like ceratopsid from the upper Dinosaur park formation (Late Campanian) of Southern Alberta, Canada; pp. 141–155 in M. J. Ryan, B. J. Chinnery-Allgeier, and D. A. Eberth (eds.), New Perspectives on Horned Dinosaurs. Indiana University Press, Bloomington, Indiana.

Ryan, M. J., D. C. Evans, P. J. Currie, C. M. Brown, and D. B. Brinkman. 2012. New leptoceratopsids from the Upper Cretaceous of Alberta, Canada. Cretaceous Research 35:69–80.

Seeley, H. G. 1887. On the classification of the fossil animals commonly named Dinosauria. Proceedings of the Royal Society of London 43:165–171.

Postcranial Anatomy of *Edmontosaurus regalis* (Hadrosauridae) from the Horseshoe Canyon Formation, Alberta, Canada

13

Nicolás E. Campione

ABSTRACT

Edmontosaurus regalis is common in the latest Campanian deposits of Alberta, Canada, and is known from numerous articulated cranial specimens and isolated bonebed material. Despite its large sample size and long research history, only the cranial skeleton has been described in detail with little reference to postcrania, the majority of which remained unprepared at the time of its original description. This study presents the first full description of the postcrania of *E. regalis* based on CMN 2289, the paratype of the species. In addition, this study presents the most completely figured postcranial skeleton of a hadrosaurid to date. CMN 2289 preserves an almost complete dorsal vertebral column, along with a complete pelvis and almost complete set of limbs, which allows for comparisons with other hadrosaurids. Quantitative analyses of dorsal vertebral morphology reveal important variations within a single individual, with implications for phylogenetic character construction. There is an overall increase in the centrum height and width along the dorsal series, compared to an overall decrease in centrum length. A high amount of variation is noted in spine height along the dorsal series. Limb proportions in CMN 2289 between the distal and proximal limb segments, as well as between the forelimb and hindlimb, are similar to other hadrosaurids and suggest little ontogenetic variation. A revised body mass estimate of *E. regalis* is presented at 7963 kg, twice that proposed by previous studies and similar in size to its sister taxon, *E. annectens*.

INTRODUCTION

Hadrosaurids are among the most common dinosaurs found in the Upper Cretaceous deposits of North America and Asia, but remains are also known from several other continents, suggesting that by the end of the Cretaceous they had reached a cosmopolitan distribution (Bonaparte et al., 1984; Case et al., 2000; Weishampel et al., 2004; Prieto-Márquez, 2010a). Hadrosaurids are known from multiple complete and articulated specimens, large bonebeds, and from complete ontogenetic growth series (Parks, 1922; Gilmore, 1924b; Lull and Wright, 1942; Horner and Currie, 1994; Evans, 2010). As a result, much more is known about hadrosaurid dinosaurs than about many other dinosaurian groups. However, anatomical descriptions of hadrosaurids tend to focus on cranial material, and often ignore the postcrania, largely due to the high proportion of cranial phylogenetic characters compared to postcranial characters (e.g., 69% cranial characters in Prieto-Márquez [2010b]; 68% in Horner et al. [2004]; and 77% in Godefroit et al. [2008]). Only a few complete holotype and paratype postcranial skeletons of hadrosaurids have been described and figured in detail, several of which were published almost a century ago, including *Brachylophosaurus canadensis* (Cuthbertson and Holmes, 2010), *Saurolophus angustirostris* (Maryańska and Osmólska, 1984), *Parasaurolophus walkeri* (Parks, 1922), *Gryposaurus notabilis* (Parks, 1920), and *Hypacrosaurus altispinus* (Brown, 1913).

The genus *Edmontosaurus* was named by Lambe in 1917 based on the type species *Edmontosaurus regalis* from Alberta, Canada. However, material that closely resembles this species had long been known from the Lance and Hell Creek formations in Montana, Wyoming, and North and South Dakota (Cope, 1883; Marsh, 1892). As a result, material attributed to *Edmontosaurus* represents some of the first hadrosaurid material to be discovered in North America. Given its long history, the systematics of *Edmontosaurus* have varied significantly (Lull and Wright, 1942; Brett-Surman, 1979; Horner et al., 2004); however, most recent studies – including a quantitative analysis of cranial variation in this taxon – agree on the validity of only two species, *E. annectens* (Marsh) and *E. regalis* Lambe (Prieto-Márquez, 2010b; Campione and Evans, 2011).

Edmontosaurus regalis (Fig. 13.1) was first described in detail by Lambe (1920). The description includes a detailed evaluation of the cranial anatomy from both the articulated holotype cranium (CMN 2288) and the disarticulated cranium of the paratype (CMN 2289). The holotype and paratype specimens were both discovered along the Red Deer

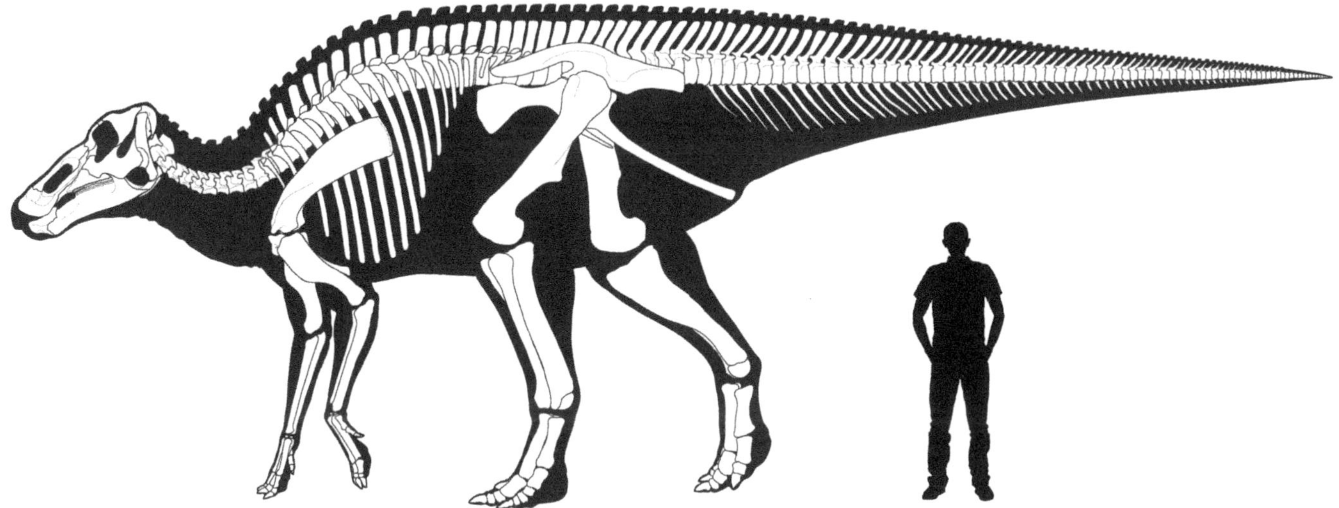

13.1. Reconstructed skeleton of *Edmontosaurus regalis* based on referred specimen ROM 801. Black represents the reconstructed soft tissue, and the human for scale is 180 cm tall.

River Valley, Alberta, in the lower Edmonton Formation, which is now known as the Horseshoe Canyon Formation (Gibson, 1977). Both are known from the lower beds of the formation, traditionally referred to Unit 1 by Eberth (2004; Eberth and Currie, 2010), but most recently ascribed to the Horsethief Member, latest Campanian in age (Eberth and Braman, 2012; Eberth et al., 2013). Levi Sternberg discovered the holotype in 1912 at the south end of the Horsethief Canyon opposite the mouth of the Three Hills Creek and approximately 60 m above the Red Deer River, below the number 9 coal seam. George F. Sternberg collected the paratype during a later expedition in 1916 on the west side of the Red Deer River, approximately 30 m above river level (also in the Horsethief Member), 11 km northwest of the town of Morrin and just north of the Morrin Bridge, in Section 16, Township 31, Range 21. As the holotype postcranial skeleton was only partially preserved and fragmentary, Lambe (1920) based the postcranial description on the paratype. However, the complete preparation of the postcranial skeleton was not completed until several years later (exact date unknown) and, therefore, most elements have never been figured or described. These include all but three of the presacral vertebrae, the sacrum, the hindlimbs, and both the pectoral and pelvic girdles. The latter are perhaps the most important given that these elements present diagnostic characters at higher taxonomic levels, and often account for 15–20% of phylogenetic characters (e.g., Horner et al., 2004; Prieto-Márquez, 2010b). In addition, Lambe (1920) provided only rough outline drawings of the postcranial skeleton. To date, the postcranial skeleton of *Edmontosaurus regalis* has not been properly photographed and described, especially within the context of current knowledge on hadrosaurid anatomy.

This chapter presents a full description of the postcranial anatomy of CMN 2289, the paratype specimen of *Edmontosaurus regalis*, along with a complete photographic atlas of the postcranial skeleton. Variation will be discussed within the context of intraspecific and interspecific characters, which may provide useful information for future hadrosaurid phylogenetic and systematic analyses.

Institutional Abbreviations AMNH, American Museum of Natural History, New York, New York; CMN, Canadian Museum of Nature, Ottawa, Ontario; DMNH, Denver Museum of Nature and Science, Denver, Colorado; ROM, Royal Ontario Museum, Toronto, Ontario; USNM, National Museum of Natural History, Smithsonian Institution, Washington D.C.; YPM, Yale Peabody Museum, New Haven, Connecticut.

Anatomical Abbreviations **4tr**, fourth trochanter; **ac**, acetabulum; **ap**, acromion process (Fig. 13.13); **ap**, ascending process (Fig. 13.22); **bg**, blood groove; **bt**, biceps tubercle; **cap**, capitulum; **cc**, cnemial crest; **cf**, coracoid foramen; **cfl**, attachment site for the *m. caudofemoralis longus*; **coa**, coracoid articulation; **cp**, cranial process; **ct**, cranial trochanter; **df**, deltoid fossa; **dp**, deltopectoral crest; **dr**, deltoid ridge; **eng**, entepicondylar groove; **et**, extensor tunnel; **gl**, glenoid; **gt**, greater trochanter; **ha**, hemal arch articulation site; **hh**, humeral head; **ic**, inner condyle; **il**, ilium; **ila**, iliac attachment site; **ilp**, iliac peduncle; **is**, ischium; **isp**, ischial peduncle; **it**, attachment site for the *m. iliotibialis*; **itr**, attachment site for the *m. ischiotrochantericus*; **lc**, lateral condyle; **lcr**, last cervical rib; **ld**, attachment site for the *m. latissimus dorsi*;

lm, lateral malleolus; **lp**, lateral process; **mc**, medial condyle; **mf**, medial flange; **mm**, medial malleolus; **mr**, medial ridge; **mvg**, medial ventral groove; **ns**, neural spine; **oc**, outer condyle; **of**, obturator foramen; **op**, obturator process (Fig. 13.19); **op**, olecranon process (Fig. 13.15); **par**, parapophysis (Figs. 13.2, 13.4–13.7); **par**, postacetabular ridge (Fig. 13.17); **pifi**, attachment site for the *m. puboischiofemoralis internus*; **poap**, postacetabular process; **pop**, postpubic process; **poz**, postzygapophysis; **pp**, pubic peduncle; **prap**, preacetabular process; **prp**, prepubic process; **prz**, prezygapophysis; **przs**, prezygapophyseal septum; **pu**, pubis; **r**, rib; **sap**, supra-acetabular process; **sca**, scapula articulation; **tf**, transverse foramen; **tp**, transverse process; **tpf**, transverse process fossa; **tub**, tuberculum; **vf**, vertebral foramen; **vg**, ventral groove; **vp**, ventral process.

SYSTEMATIC PALEONTOLOGY

ORNITHOPODA Marsh, 1881
HADROSAUROIDEA Sereno, 1986
HADROSAURIDAE Cope, 1869
HADROSAURINAE Lambe, 1918
EDMONTOSAURUS REGALIS Lambe, 1917

Holotype CMN 2288.

Holotype Postcranial Skeleton Numerous partial cervical, dorsal, and caudal vertebrae; numerous rib fragments; partial proximal end of the right scapula; partial prepubic process of the left pubis.

Paratype CMN 2289.

Paratype Postcranial Skeleton Three cervical vertebrae, 15 dorsal vertebrae, 5 caudal vertebrae, several dorsal ribs, right and left scapulae, right coracoid, right and left sternal plates, right and left humeri, right and left ulnae, left radius, left metacarpal II, two isolated manal distal phalanges, right and left pubis, right and left ischium, left ilium, right and left femur, right and left tibia, complete right fibula and a partial left fibula, right astragalus, complete right metatarsals II and IV, and a partial right metatarsal III.

Diagnosis See Campione and Evans (2011).

Comments The material described here includes all complete and prepared type material of *Edmontosaurus regalis*. The postcranial skeleton of CMN 2288 remains unprepared and cannot be described at this time. However, as described by Lambe (1920), the postcranial skeleton of CMN 2288 is not nearly as complete as CMN 2289 and for this reason only the latter is described here. For a complete list of specimens referred to *E. regalis*, see Campione and Evans (2011).

DESCRIPTION

Axial Skeleton

Cervical Vertebrae Only a single cervical vertebra was described by Lambe (1920), which the author suggested belonged to the middle of the cervical series. Since preparation was completed, three cervical vertebrae are completely preserved in CMN 2289. Although a complete cervical count cannot be determined for this specimen, CMN 8399, a smaller individual of *E. regalis*, has 13 cervical vertebrae. Comparisons made with CMN 8399 and the complete series of *Brachylophosaurus canadensis* (Cuthbertson and Holmes, 2010) indicate that the cervical vertebrae of CMN 2289 are from the cranial or middle region of the cervical series (Fig. 13.2A–E; the one described by Lambe), the middle to caudal portion of the cervical series (Fig. 13.2F–J), and the caudal-most region, perhaps representing the last cervical vertebra based on the location of the parapophysis on the lateral surface of the centrum (Fig. 13.2K–O) and the length of its associated rib (described later). All three are strongly opisthocoelous, as is common for the cervical vertebrae of all hadrosaurids and most ornithopods (Horner et al., 2004; Boyd et al., 2009). All have well-developed postzygapophyses that project caudolaterally beyond the caudal aspect of the centrum to articulate with the prezygapophyses of the succeeding vertebra. The postzygapophyses are approximately 2.5 times the craniocaudal length of the neural arch in the cranial-most cervical vertebra, over three times in the middle cervical vertebra, and approximately two times in the last cervical vertebra. The prezygapophyses face dorsally and slightly medially, whereas the opposing postzygapophyses face ventrally and slightly laterally.

Despite the presence of only three complete cervical vertebrae in CMN 2289, overall changes in vertebral morphology along the cervical series can still be described. In the cranial-most cervical vertebra, the cranial and caudal faces are approximately in line with each other in lateral view. In comparison, their position relative to each other is offset in the caudal cervical vertebrae, so that the caudal face sits above the cranial face. This shift is related to the curvature of the neck in hadrosaurids (Fig. 13.1), in which cranial cervical vertebrae are in line with one another, whereas caudal cervical vertebrae are oriented caudodorsally relative to preceding vertebrae as they rise up to meet the dorsal series. The ventral surfaces of the cervical vertebrae are flat, and marked by a craniocaudal groove, which is strongly developed in the cranial cervical vertebra (Fig. 13.2E), but less pronounced in the more caudal vertebrae. The centrum of the cranial cervical vertebra is deeply excavated laterally, a condition that disappears in the more caudal cervical and dorsal vertebrae.

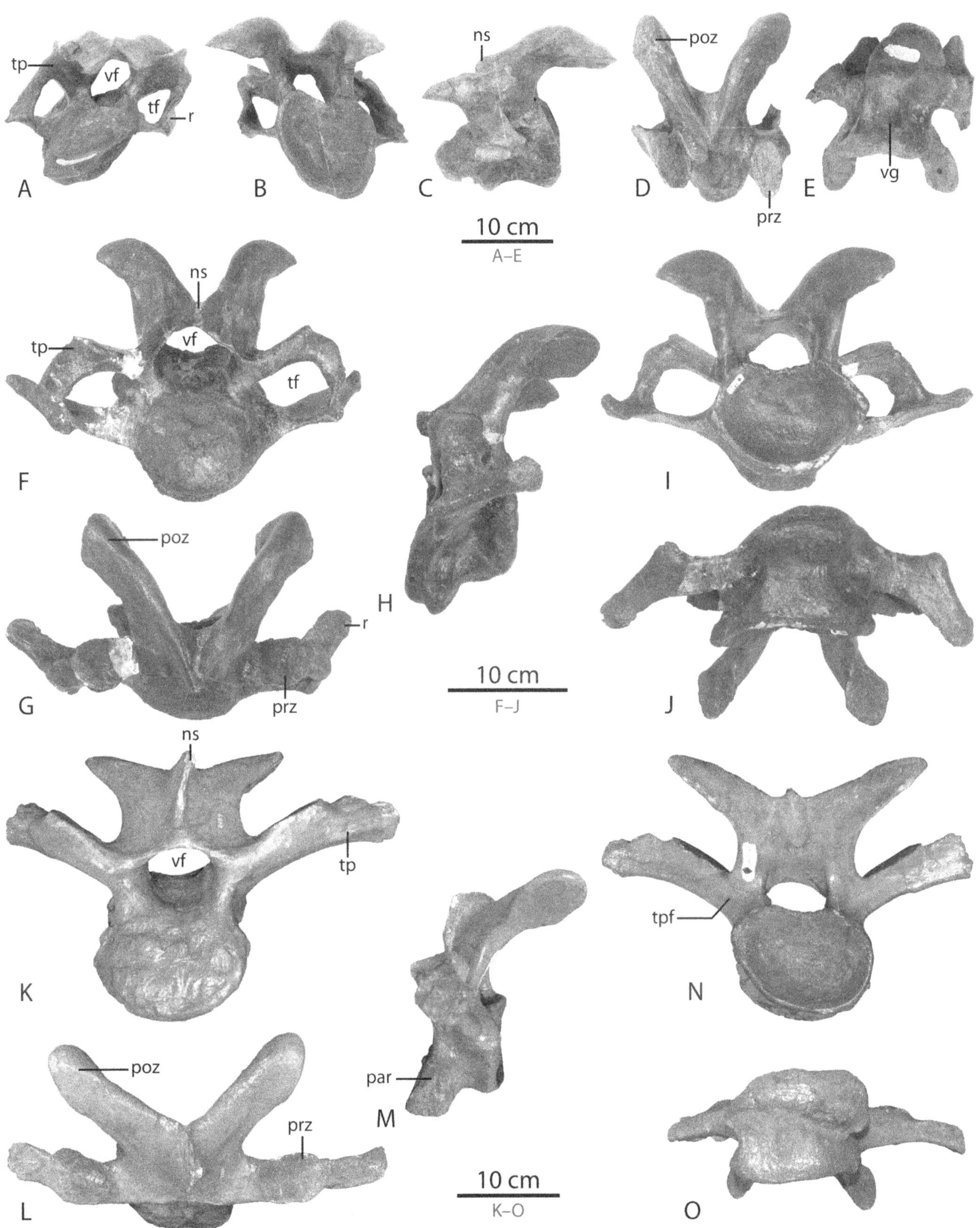

13.2. Cervical vertebrae of CMN 2289. (A–E) cranial-most cervical vertebra in (A) cranial; (B) caudal; (C) lateral; (D) dorsal; (E) ventral views. (F–J): middle to caudal cervical vertebra in (F) cranial; (G) dorsal; (H) lateral; (I) caudal; (J) ventral views. (K–O) the last cervical vertebra, interpreted on the basis of the location of the parapophysis on the central body and the length of the rib that articulates well with it, in (K) cranial; (L) dorsal; (M) lateral; (N) caudal; (O) ventral views.

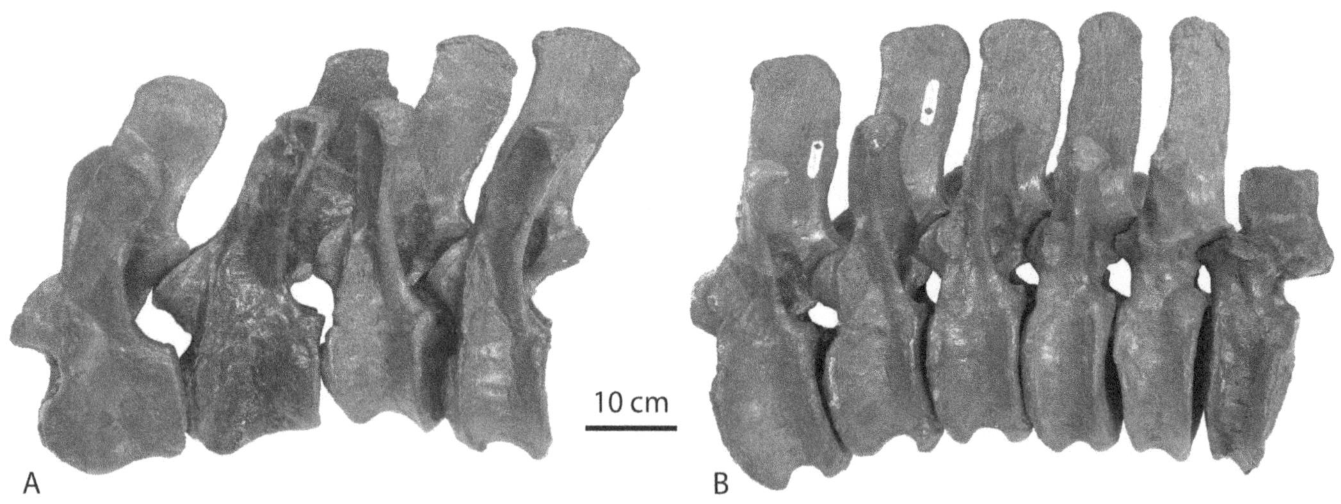

13.3. Dorsal vertebrae, in lateral view, that appear to have either articulated or at least been in close approximation. (A) four dorsal vertebrae that are from the cranial portion of the dorsal series, and (B) six dorsal vertebrae that are from the mid-region of the dorsal series.

In the cranial cervical vertebra the transverse processes are short and most of their dorsal surface consists of the prezygapophyses. The diapophyses, for the tubercular head of the cervical rib, form a small tubercle that extends laterally from each prezygapophysis. The transverse processes are longer in the more caudal cervical vertebrae, and the diapophyses are elongated relative to the prezygapophyses, which remain on the dorsal surface of the transverse processes. The transverse processes of the two cranial-most vertebrae are circular in cross section, whereas the caudal cervical vertebra has well-developed fossae on the caudal surfaces of the transverse processes (Fig. 13.2N). This is similar to the morphology seen in the succeeding dorsal vertebrae. The shape of the articular surface of the zygapophyses changes from craniocaudally elongate in the cranial cervical vertebra, to a circular shape in the middle vertebra, and a mediolaterally elongate zygapophysis in the caudal-most cervical vertebra. The vertebral foramen, or neural canal, is larger in the caudal cervical vertebra (height = 58.9 mm; width = 43.6 mm) than in the cranial vertebra (30.8 mm; 37.1 mm), suggesting a slight, progressive enlargement of the neural canal through the cervical series; a morphological shift not exhibited by *Brachylophosaurus canadensis* (Cuthbertson and Holmes, 2010). In addition, the neural canals in the cranial-most dorsal vertebrae are similar in size to the cranial cervical vertebra (See Table A1b in Appendix 13.1), indicating that the canal is largest at the cervicodorsal transition, and then decreases in size caudally. The neural spines of the cranial-most two cervical vertebrae are small tubercle-like projections that extend cranially with no significant dorsal projection. In comparison, the caudal-most cervical vertebra has a more typical, dorsally projecting neural spine with no cranial extension.

Dorsal Vertebrae The dorsal vertebral series, along with the sacrum, is the best-preserved portion of the vertebral column in CMN 2289. Fifteen dorsal vertebrae are preserved, suggesting that only a few vertebrae are missing; hadrosaurids are known to have anywhere from 16 to 20 dorsal vertebrae (Horner et al., 2004). Since these vertebrae are not in articulation, their exact position along the vertebral column cannot be determined; however, based on their size and morphology, a relative position can be determined and the variation along the dorsal series can be described. Two vertebral segments were assembled, one comprising 4 cranial dorsal vertebrae (Fig. 13.3A) and the other comprising 6 mid-dorsal vertebrae (Fig. 13.3B). These segments represent vertebrae that either articulated with each other or were very closely associated. The remaining 5 vertebrae are caudal to the two segments, and their relative order was approximated based on morphological patterns of variation noted between these vertebrae that were consistent with those noted in more cranial vertebrae. The cranial-most dorsal vertebra is not the first in the dorsal series, as it does not articulate well with what is interpreted as the last cervical vertebra. However, based on the angle to the neural spine relative to the centrum (35°), and the strongly opisthocoelous centrum, this vertebra is perhaps the third or fourth dorsal vertebra.

In comparison to the cervical vertebrae, the dorsal vertebrae are only opisthocoelous in the cranial-most vertebrae (Fig. 13.4). Caudally, they take on a more amphiplatyan to platycoelous appearance. In cranial or caudal view, the centrum is heart shaped in the more cranial vertebrae (Figs. 13.4, 13.5) – a morphology that becomes more circular caudally (Figs. 13.6–13.8). The prezygapophyses are not associated with the transverse processes, as in the cervical series, but are

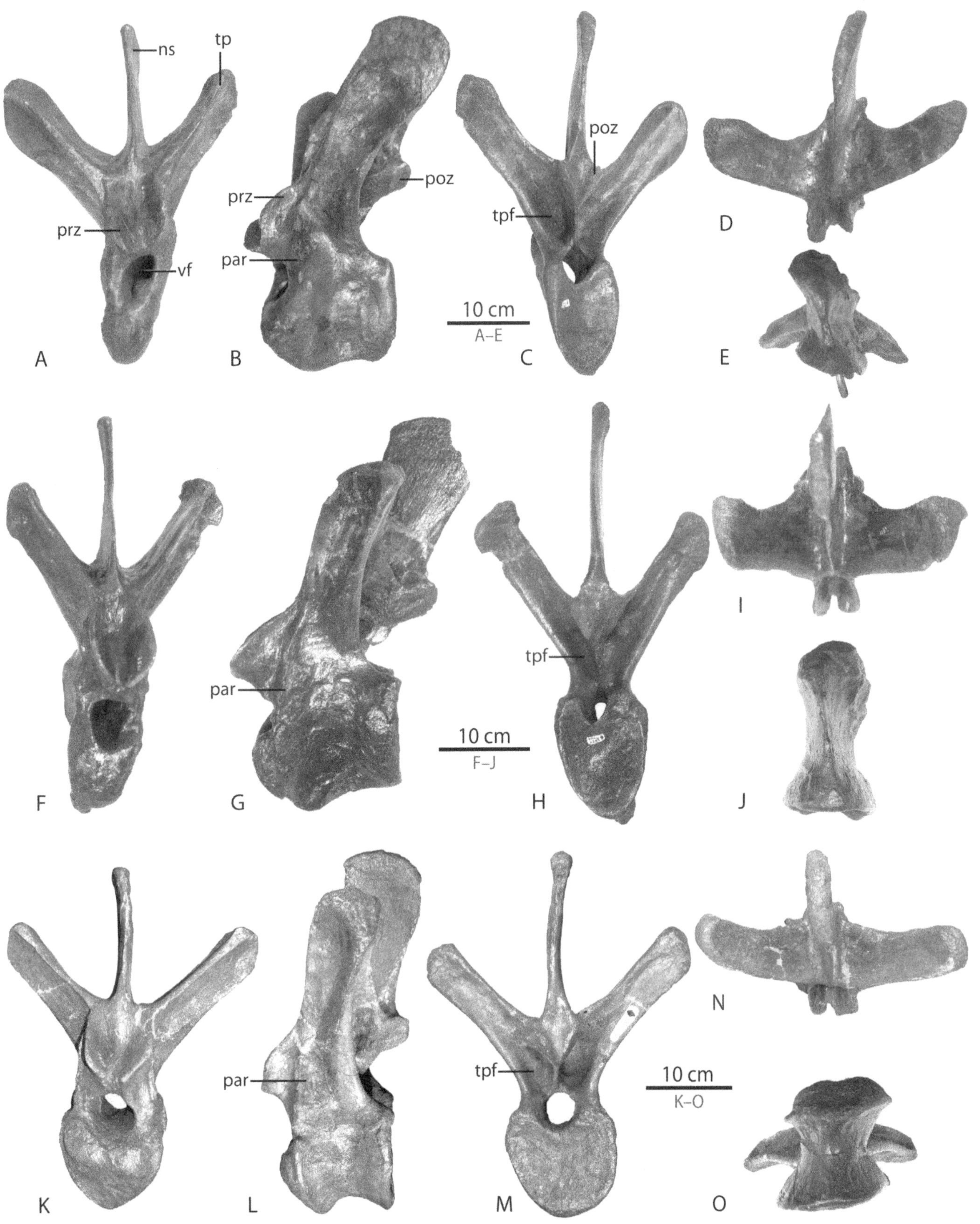

13.4. The three cranial-most dorsal vertebrae preserved in CMN 2289 in (A, F, K) cranial; (B, G, L) lateral; (C, H, M) caudal; (D, I, N) dorsal; (E, J, O) ventral views.

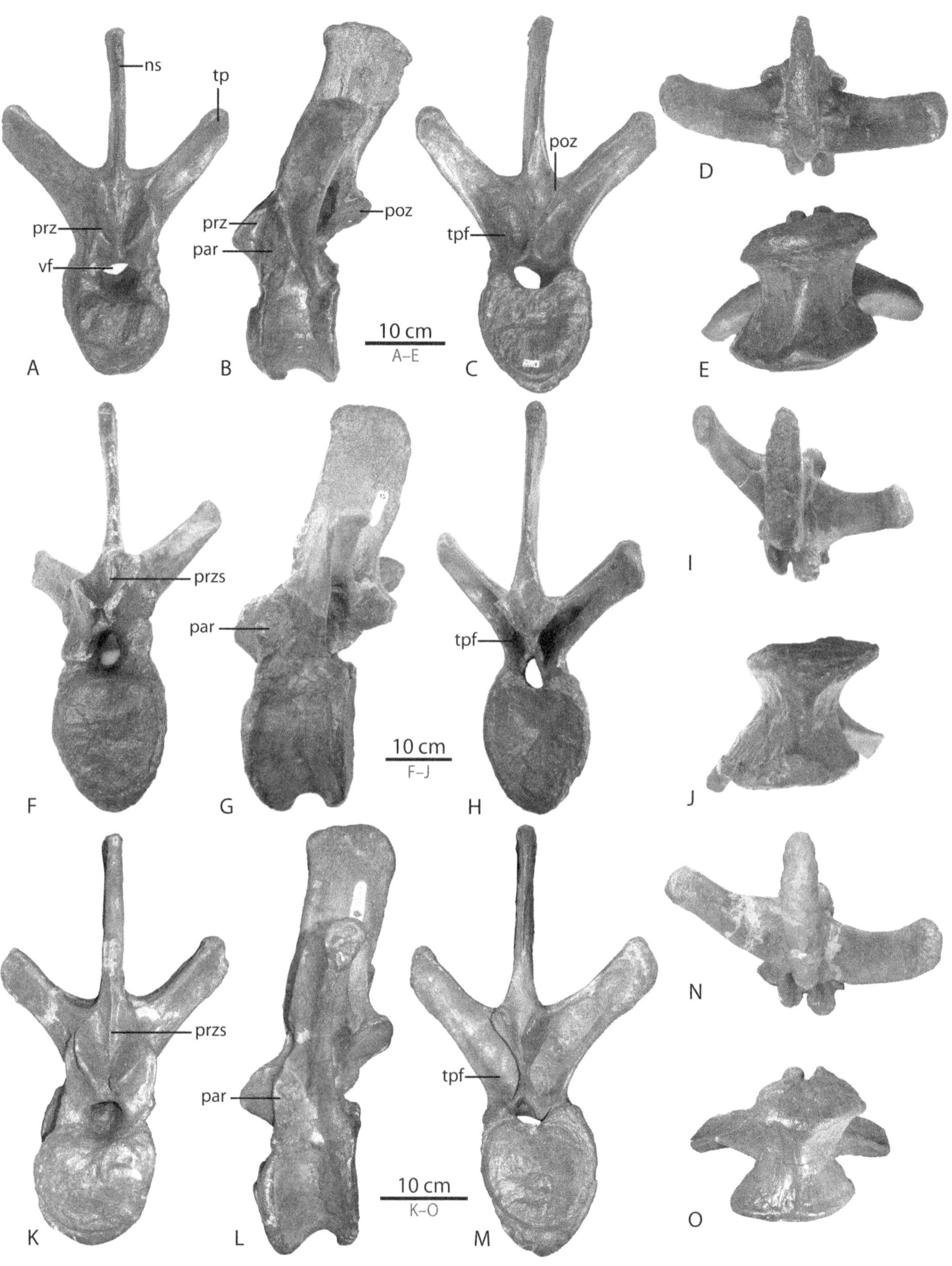

13.5. Dorsal vertebrae ranked four to six along the dorsal series in (A, F, K) cranial; (B, G, L) lateral; (C, H, M) caudal; (D, I, N) dorsal; (E, J, O) ventral views.

located at the base of the neural arch immediately cranial to the neural spine. Both the pre- and postzygapophyses are strongly angled inward and outward, respectively, in cranial vertebrae. These become progressively horizontal caudally, indicating that the main axis of movement allowed by the vertebrae changed progressively down the vertebral column from dorsoventral to mediolateral. Between the right and left prezygapophyses there is a vertical thin septum dividing the articulation surfaces. This septum is present caudal to the mid-dorsal vertebrae (Figs. 13.5–13.8) and is absent in the cranial dorsal vertebrae (Fig. 13.4).

The transverse processes, like those of the last cervical vertebra, are long with a fossa on the caudal surface; this fossa opens caudally in cranial dorsal vertebrae and shifts caudoventrally along the dorsal series. Transverse processes are angled dorsolaterally in the cranial dorsal vertebrae. This angle progressively increases, relative to the sagittal plane, along the dorsal series, so that in the caudal-most vertebrae the transverse processes project at a right angle from the neural spine. The diapophyses remain at the distal end of the transverse processes, whereas the parapophyses are at the base of the neural arch. The parapophyses in the cranial dorsal vertebrae are dorsoventrally broad and extend to the neurocentral suture (Fig. 13.4). Caudally, the parapophyses becomes progressively more symmetrical in shape and lie adjacent to the prezygapophyses. In the caudal-most dorsal vertebrae (Fig. 13.7B), the parapophyses are smaller, circular, and lie at the base of the transverse process. The shape of the capitulum for the dorsal ribs mirrors these changes in morphology. In comparison to the cervical vertebrae, the dorsal vertebrae have well-developed neural spines that are between 1.5 to 2.5 times taller than their respective centra. The neural spines are angled caudally in cranial dorsal vertebrae and gradually become more vertical caudally along the dorsal series. The cranial-most dorsal vertebrae are the most angled and both the middle and caudal-most dorsal vertebrae have nearly vertical neural spines.

Sacrum The sacrum is complete in CMN 2289 and is comprised of nine fused vertebrae (Fig. 13.9). The centra are rounded; however, the three caudal-most sacral vertebrae are flattened, and slightly concave ventrally (Fig. 13.9F) – a morphology generally referred to as the median ventral groove, typical of hadrosaurine dinosaurs, and distinct from lambeosaurines, which typically possess a ventral ridge (Horner et al., 2004; Prieto-Márquez, 2011). The neural spines are approximately as tall as their respective centra. All nine vertebrae have well-developed transverse processes that extend laterally at a right angle to the neural spine, similar to the caudal-most dorsal vertebrae. The transverse processes of the caudal two sacral vertebrae may have contacted the medial surface of the ilium, but the two caudal-most sacral vertebrae do not possess robust sacral ribs characteristic of S1–S7. The cranial seven sacral vertebrae possess laterally projecting transverse processes that contact the ilium near its medial margin. Unlike the caudal-most sacral vertebrae, the transverse processes are attached to the vertebral bodies via a craniocaudally thin lamina of bone that leads to the sacral ribs. The ribs extend laterally to approximately the same level as the transverse processes. The ribs expand craniocaudally at their distal-most end and fuse with one another to form a craniocaudally long attachment surface with the medial surface of the ilium (Fig. 13.9).

Caudal Vertebrae Five caudal vertebrae are completely preserved, along with an isolated neural arch and two hemal spines (Figs. 13.10, 13.11). The three largest caudal vertebrae do not possess articulation surfaces for hemal arches and are interpreted as proximal caudal vertebrae (Fig. 13.10). The centra are amphiplatyan, and although the cranial and caudal faces are circular, the centra have flattened ventral surfaces. The transverse processes are narrow, compared to the sacral and presacral vertebrae, and extend laterally. The neural spines are club shaped, with a highly rugose distal end, and are strongly angled caudally. The angle, from vertical, is smallest in the cranial-most caudal (15°) and become progressively larger down the caudal series. The caudal-most caudal has an angle of 57° from vertical. The orientation of the pre- and postzygapophyseal articulation is strongly angled inward and outward, respectively, similar to the cranial dorsal vertebrae, thus suggesting limited mediolateral movement at the base of the tail.

The two remaining caudal vertebrae (Fig. 13.11A–J) have well-developed double facets for the hemal arches on their cranial and caudal ventral margins. Between the right and left hemal facets is a deep fossa that runs the craniocaudal length of the centrum. This fossa is ellipsoidal in shape in the more cranial vertebra (Fig. 13.11E) and becomes craniocaudally elongate in the caudal vertebra (Fig. 13.11J). The transverse processes become gradually shorter through the caudal series, and the caudal-most vertebra has no transverse processes.

The hemal spines are typically Y-shaped with expanded proximal heads. The heads converge on each other distally to form a long mediolaterally thin spine. The two preserved spines are different in length, suggesting that one has a more cranial position along the caudal series than the other (Fig. 13.11L, M). However, the differences in length are restricted to the spine; both heads are similar in length.

Cervical Ribs The two cranial-most cervical vertebrae possess dicephalic cervical ribs, which are fused to their respective vertebrae (Fig. 13.2). The tuberculum fuses with the diapophysis on the transverse process and the capitulum with the lateral surface of the centrum; together they form a large, circular foramen (the transverse foramen) between

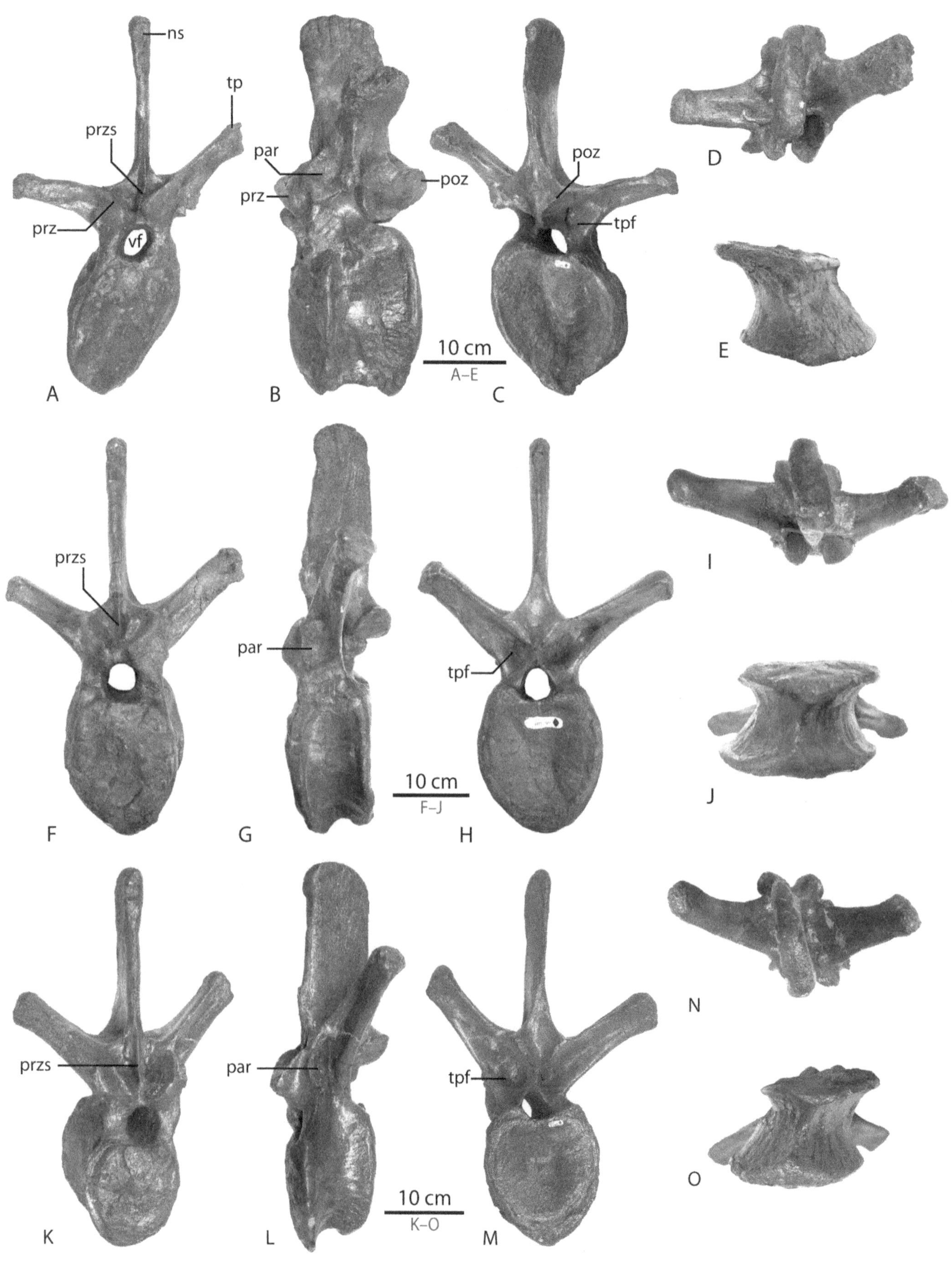

13.6. Dorsal vertebrae ranked seven to nine along the dorsal series in (A, F, K) cranial; (B, G, L) lateral; (C, H, M) caudal; (D, I, N) dorsal; (E, J, O) ventral views.

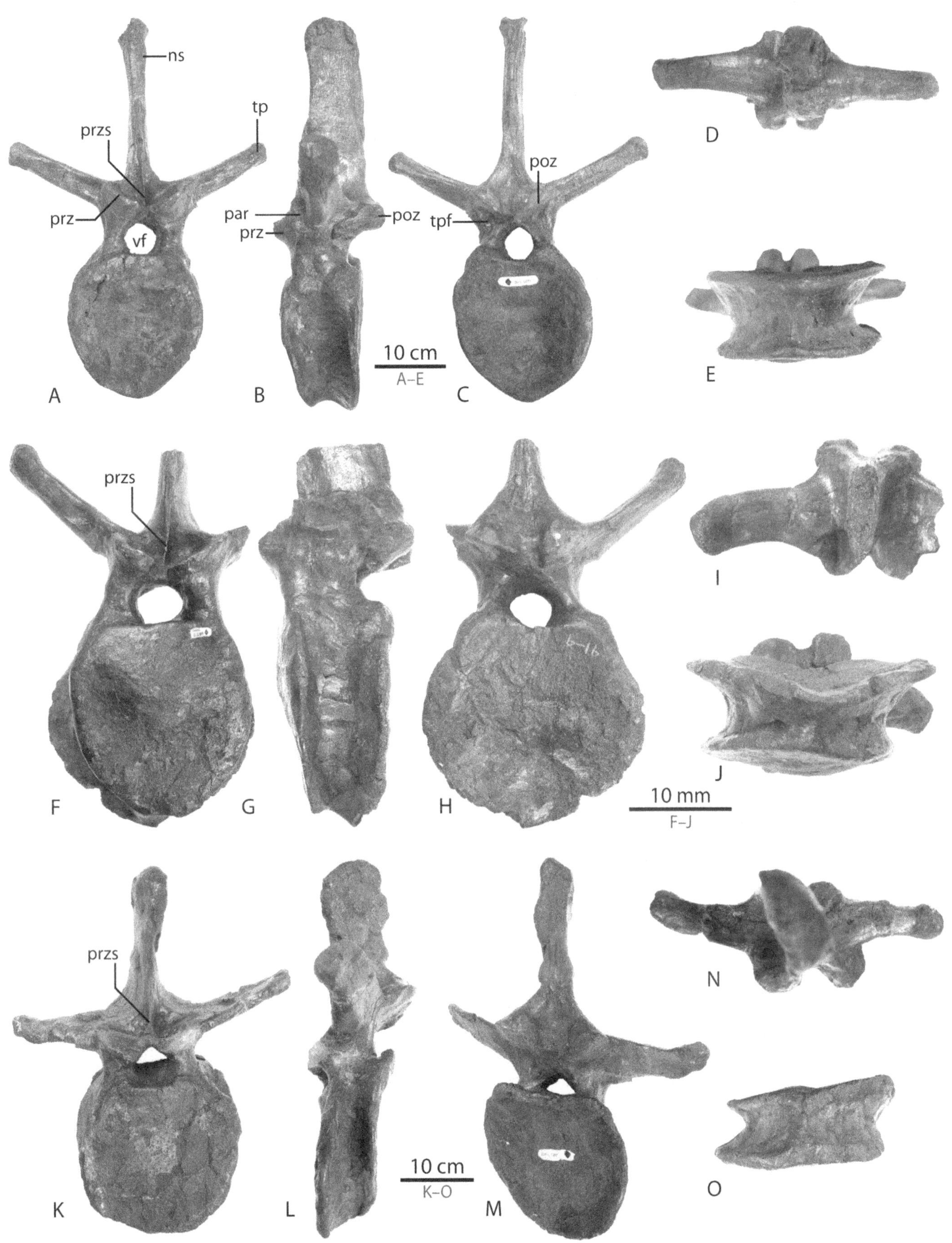

13.7. Dorsal vertebrae ranked 10 to 12 along the dorsal series in (A, F, K) cranial; (B, G, L) lateral; (C, H, M) caudal; (D, I, N) dorsal; (E, J, O) ventral views.

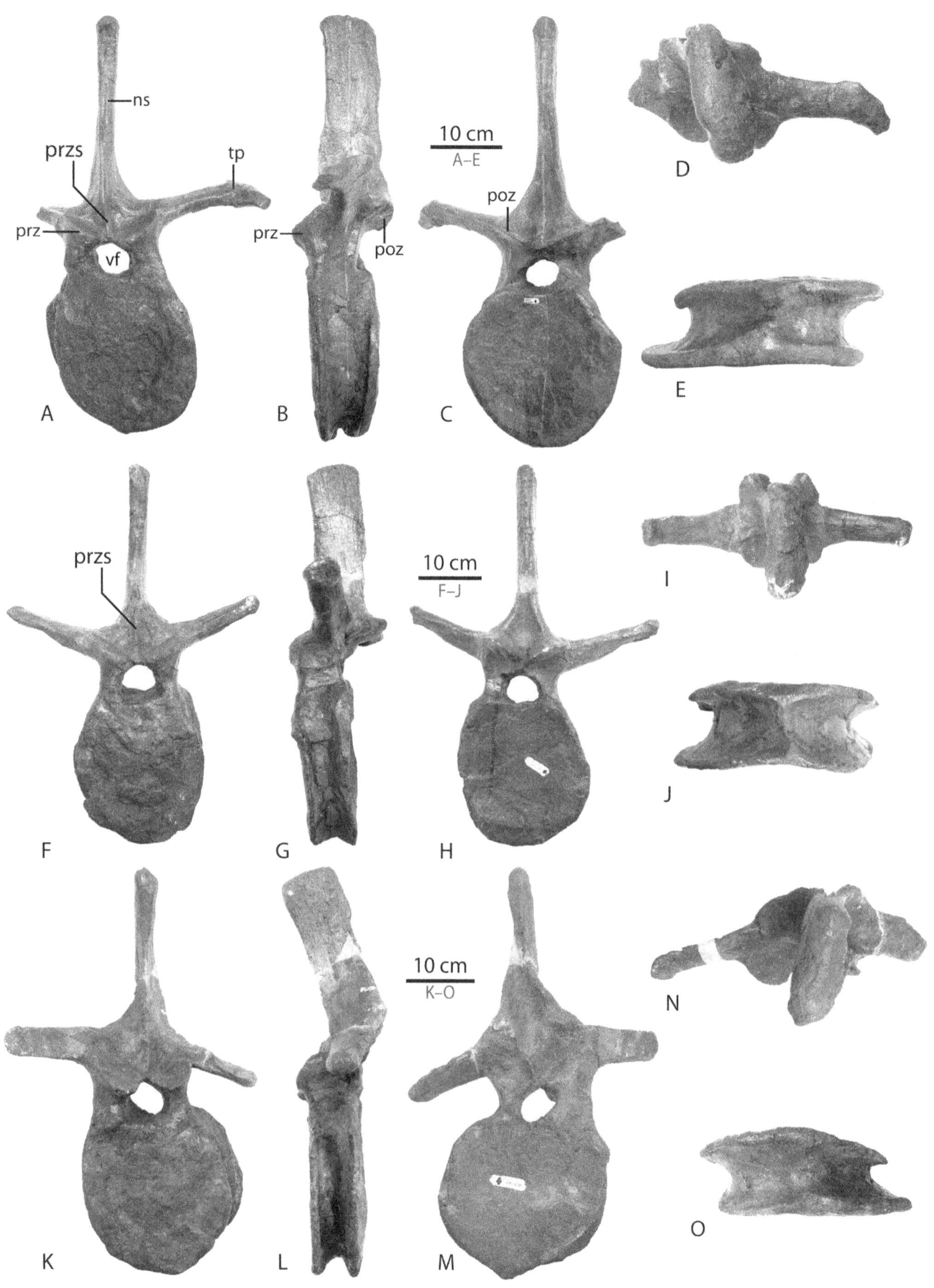

13.8. Dorsal vertebrae ranked 13 to 15 along the dorsal series in (A, F, K) cranial; (B, G, L) lateral; (C, H, M) caudal; (D, I, N) dorsal; (E, J, O) ventral views.

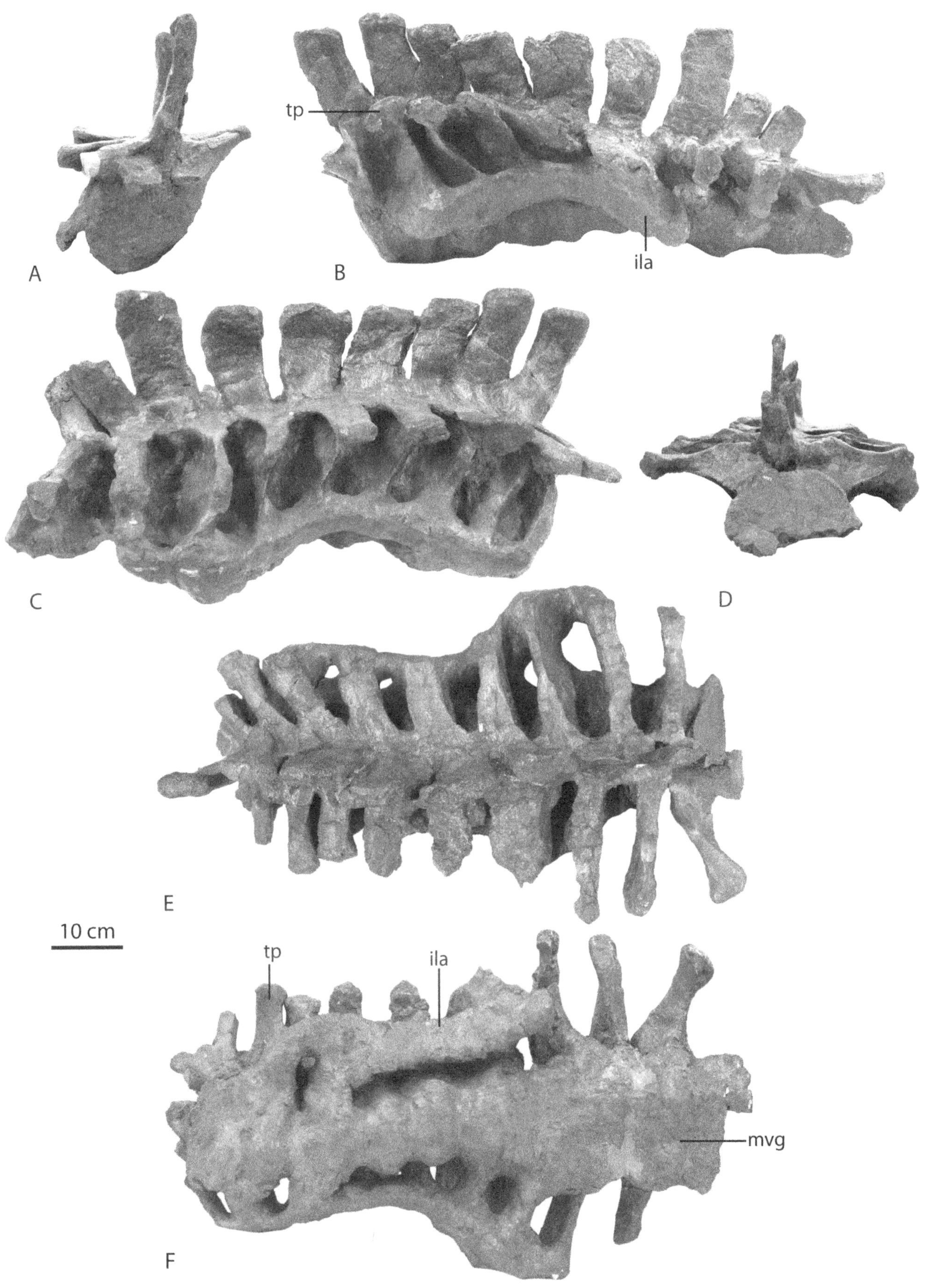

13.9. Fused sacrum of CMN 2289 in (A) cranial; (B) right lateral; (C) left lateral; (D) caudal; (E) dorsal; and (F) ventral views.

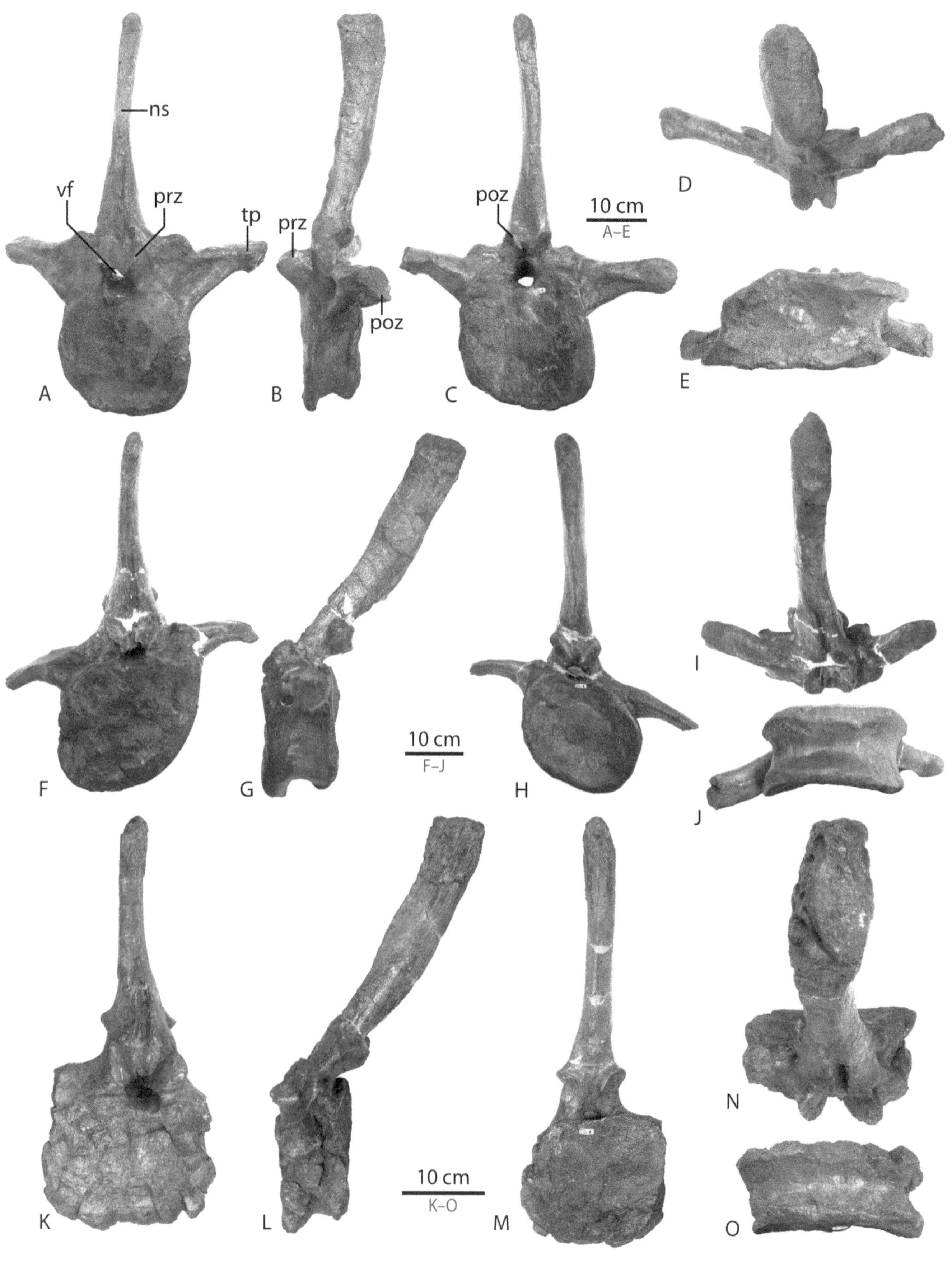

13.10. Cranial-most caudal vertebrae with no evidence of attachment site for hemal spines. (A, F, K) cranial; (B, G, L) lateral; (C, H, M) caudal; (D, I, N) dorsal; (E, J, O) ventral views.

the rib and the centrum, similar in size to the neural canal. A long rib articulates with the putative last cervical vertebra (Fig. 13.12), but it is not long enough to have reached the sternal plates. Unlike the long dorsal ribs, the shaft of the last cervical rib is strongly bowed caudally and the capitulum, at the proximal-most end, is angled slightly dorsally. When the rib is articulated with the last vertebra, the dorsal curvature of the capitulum results in a mediolaterally elongate transverse foramen. As a result, this foramen appears to progressively change from a circular to a mediolaterally elongate shape down the cervical series and is completely lost in the dorsal vertebrae; it likely served major neurovasculature structures leading to and from the cranium. The tuberculum is reduced in the last cervical rib, compared to the more cranial cervical ribs, and more closely resembles the tuberculum on the dorsal ribs.

Dorsal Ribs The majority of dorsal ribs are preserved in CMN 2289, although most are not complete (Fig. 13.12). Most have a long capitular head that extends at almost a right angle to the rib shaft. The capitulum is dorsoventrally broad with a dorsoventrally elongate articulation surface. This corresponds with the dorsoventrally elongate parapophysis on the cranial dorsal vertebrae. In accordance with morphological changes that occur to the parapophyses along the dorsal series, the capitulum becomes progressively dorsoventrally narrower, which results in a circular articulation facet in the caudal dorsal ribs. The tuberculum is highly reduced compared to the capitulum and is sometimes limited to the outer surface of the rib, at the corner between the capitulum and the rib shaft. Cranially, the tubercula are mediolaterally elongate, but they become progressively narrower caudally. The cranial dorsal ribs are the longest, between 935 and 1270 mm long with a 220 to 240 mm long capitulum. The largest ribs have a pronounced craniocaudally expanded flange that extends along the proximal two-thirds of the shaft. Caudal dorsal ribs are much shorter than the cranial dorsal ribs. The only complete rib was approximately 207 mm long, with a 136 mm long capitulum.

Pectoral Girdle

Scapula Both scapulae are preserved in CMN 2289. The scapula is a long, strongly arched bone approximately 1050 mm long (Fig. 13.13A–E). The proximal surface comprises two concave regions that, in medial view, form an obtuse angle with one another (approximately 140°). The coracoid articulates with the cranial concave region, whereas the caudal region represents the scapular portion of the glenoid fossa. In proximal view, the coracoid articulation surface is circular in shape (Fig. 13.13B) and faces slightly medially; it is large compared to the glenoid fossa. The scapular contribution to the glenoid fossa is teardrop in shape and faces caudoventrally; it forms two-thirds of the total glenoid fossa (formed by the scapula and coracoid). The acromion process is prominent and extends laterally and cranially from the scapular blade. Distally, the acromion process grades into the deltoid ridge that extends laterally across the proximal one-third of the scapula. The deltoid ridge results in a deep deltoid fossa in lateral view. Distally, the deltoid ridge gradually becomes less prominent laterally as it extends towards the ventral margin of the scapula, at which point it can no longer be defined. Distal to the deltoid ridge, the scapula is flat mediolaterally and gradually becomes broader dorsoventrally. The distal end is the dorsoventrally widest portion of the scapular blade, approximately 50% larger than the dorsoventrally narrowest portion of the scapula, immediately distal to the glenoid.

Coracoid The coracoid is a small, robust bone that is much smaller in size compared to the scapula (Fig. 13.13F–H). It forms the ventral third of the glenoid fossa, and together with the scapula forms a figure eight–shaped glenoid fossa. The glenoid fossa of the coracoid is short; the length of the coraco-scapular articulation is 1.36 times the length of the former. Both surfaces form an angle of approximately 120° to each other. On the lateral surface, the coracoid foramen pierces the element ventral to the coraco-scapular articulation and cranial to the glenoid. On the medial surface the foramen pierces the element at the level of the glenoid margin, between the glenoid and the coraco-scapular articulation. The prominent biceps tubercle is present along the cranial margin. It is 49 mm in craniocaudal width and extends laterally approximately 30 mm from the body of the coracoid. Distally, the coracoid is curved and extends caudoventrally to form the ventral process. The ventral process is well developed, approximately 70% of the dorsoventral height at the base of the ventral process.

Sternal Plate Both the right and left sternal plates are preserved, but the right is better preserved and complete (Fig. 13.13I–L). The sternal plate is comprised of a long shaft that is triangular in cross section proximally and forms a robust proximal head with a highly rugose texture. Distally, the shaft becomes progressively flat dorsoventrally and only slightly mediolaterally expanded at the distal-most end, which also exhibits a rugose texture. Extending medially from the proximal end is a dorsoventrally flattened flange, which together with the left sternal plate forms the sternum. The flange is shorter in length than the distal shaft.

Forelimb

Humerus Both right and left humeri are preserved, although the left is partially reconstructed. The humerus of CMN 2289 (Fig. 13.14) is typical of other hadrosaurids.

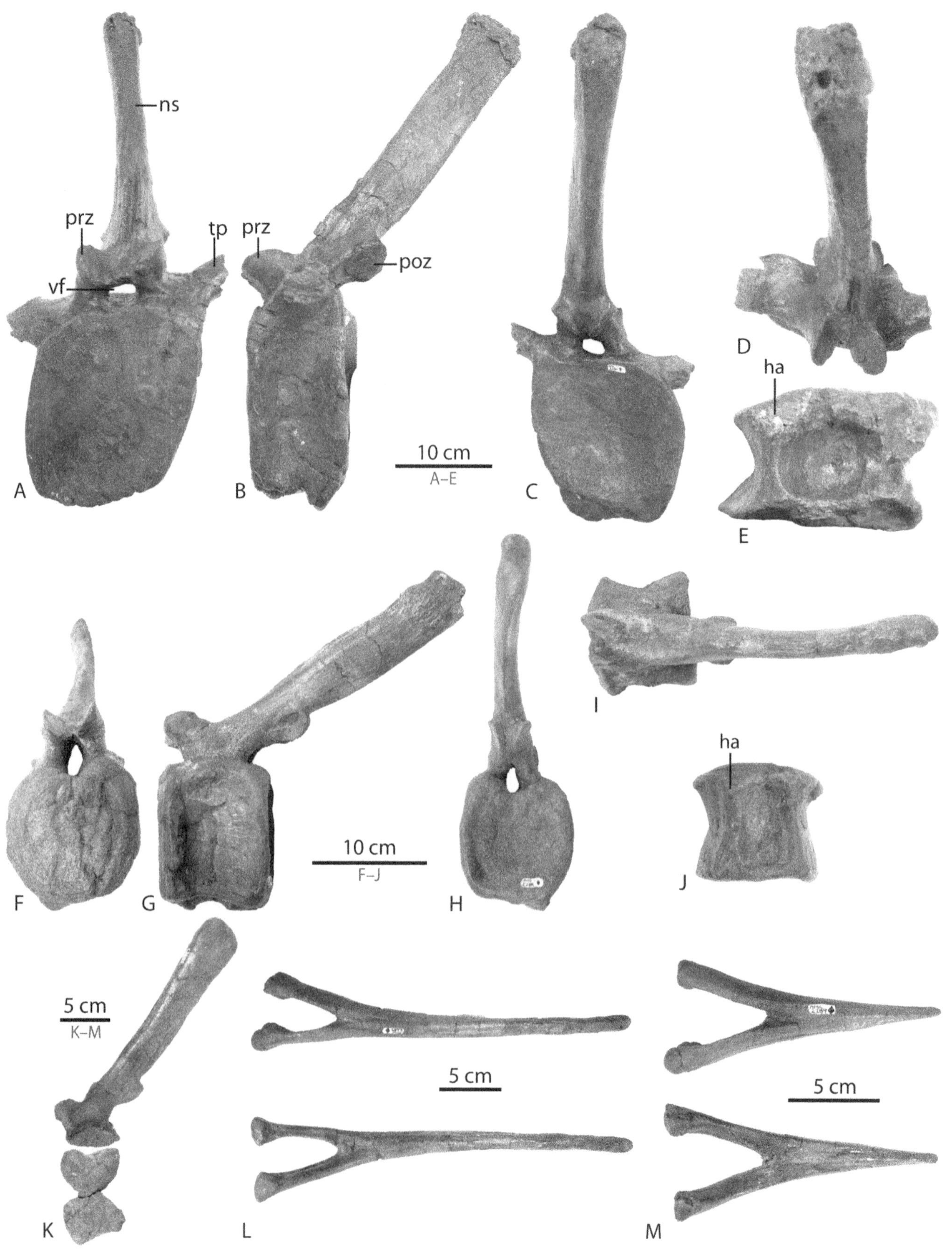

13.11. (A–E) middle caudal vertebra; (F–J) distal caudal vertebrae; (K) isolated caudal neural arch; (L) middle to proximal hemal spine; and (M) middle to distal hemal spine. Caudal vertebrae are in (A, F) cranial; (B, G) lateral; (C, H) caudal; (D, I) dorsal; (E, J) ventral views. Isolated spine in lateral (top) and articular (bottom) views. Hemal spines in cranial (top) and caudal (bottom) views.

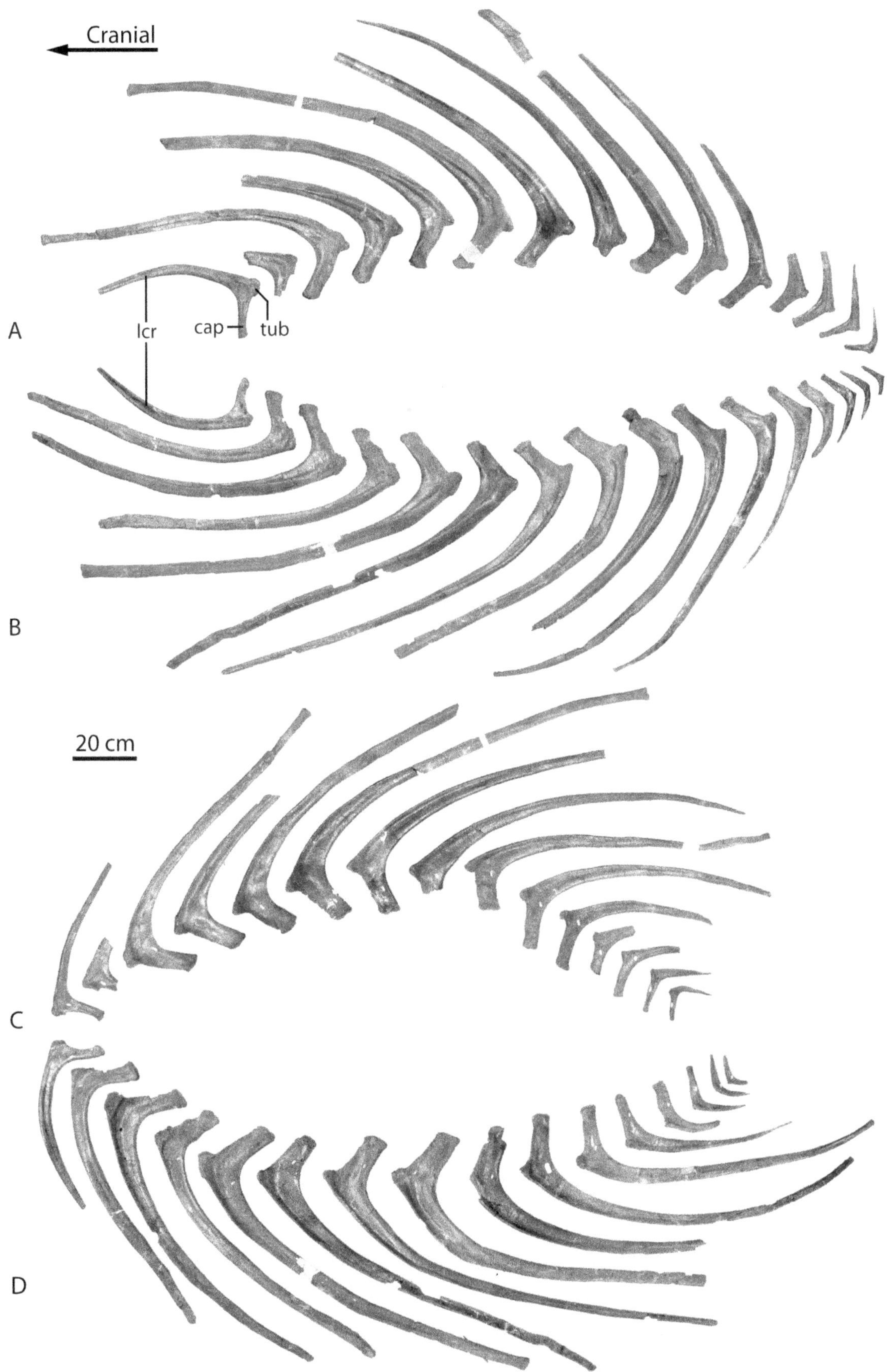

13.12. Last cervical rib and dorsal ribs of CMN 2289, organized from cranial (left) to caudal (right) based on the angle of the head, relative to the rib shaft and the size of the tuberculum. (A, B) ribs from the right and left sides, respectively, in cranial view; (C, D) caudal view.

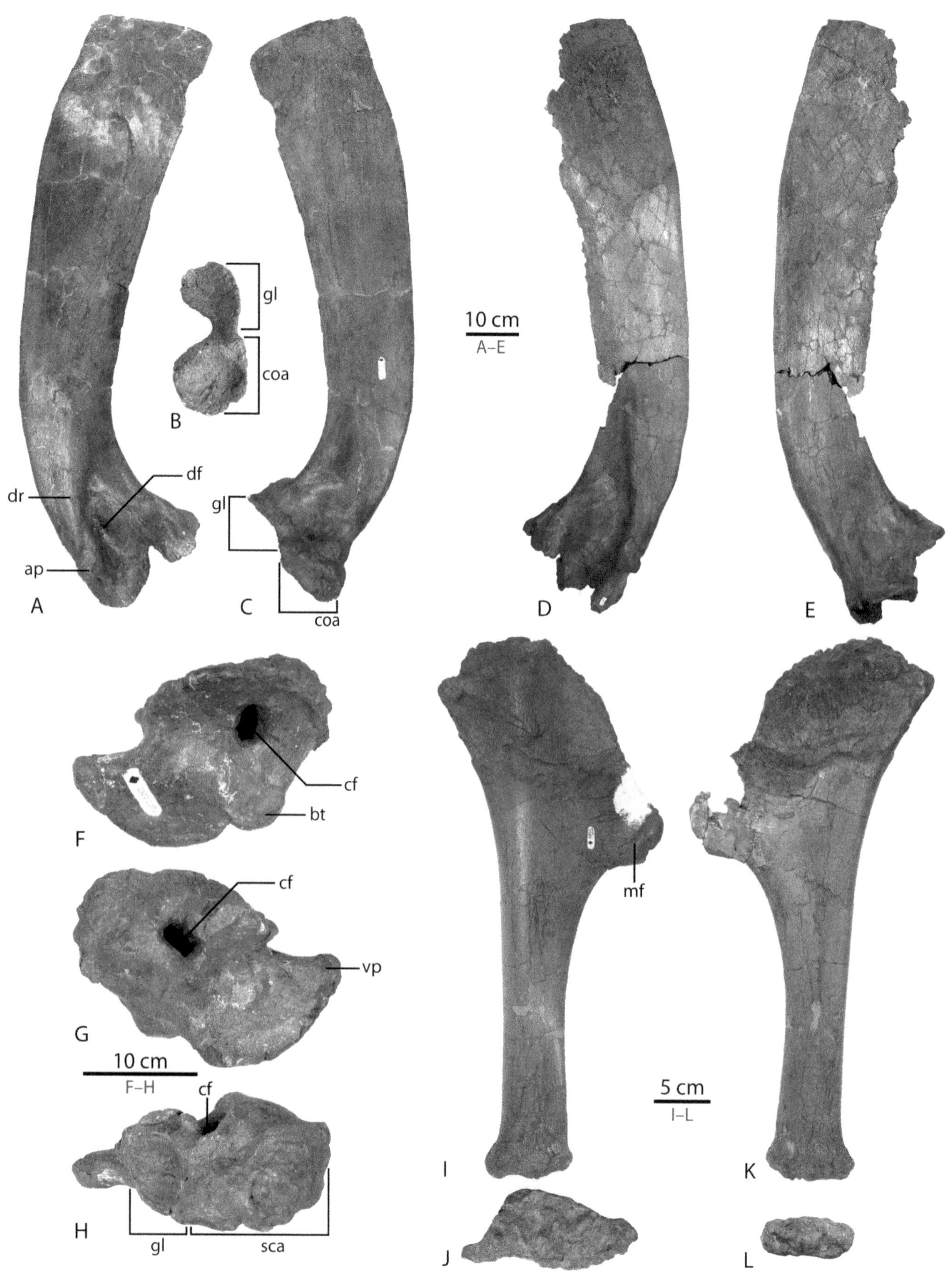

13.13. Pectoral girdle of CMN 2289. (A–E) left and right scapulae in (A) left lateral; (B) left proximal; (C) left medial; (D) right lateral; and (E) right medial. Right coracoid in (F) lateral; (G) medial; and (H) proximal views. Right sternal plate in (I) ventral; (J) proximal; (K) dorsal; (L) distal views.

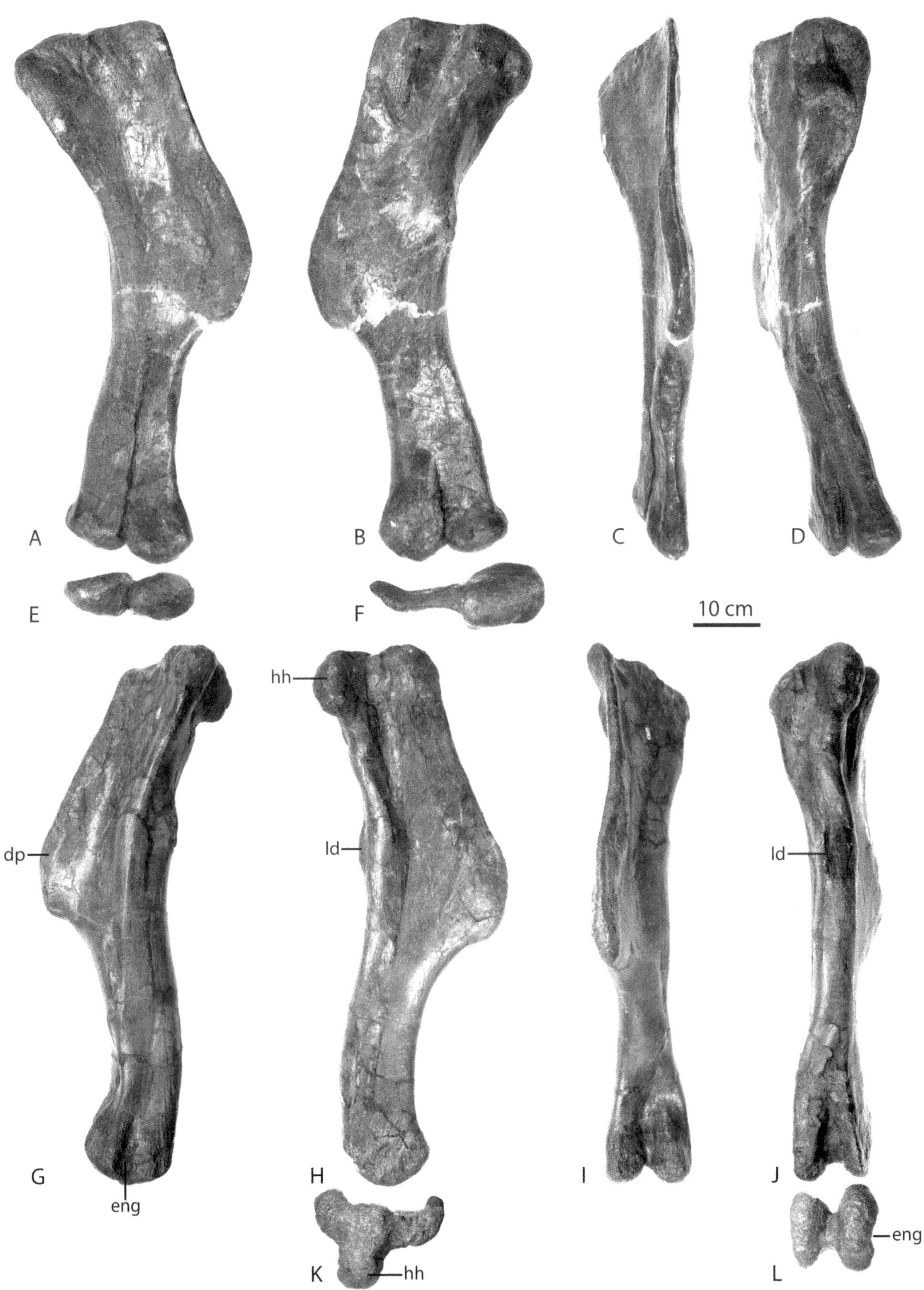

13.14. Left and right humerus of CMN 2289. Left humerus in (A) cranial; (B) caudal; (C) lateral; (D) medial; (E) distal; and (F) proximal views. Right humerus in (G) medial; (H) lateral; (I) cranial; (J) caudal; (K) proximal; and (L) distal views.

It exhibits a well-developed deltopectoral crest that spans over half the length of the entire bone, and is almost double the craniocaudal width of the humeral diaphysis (see Table A1d in Appendix 13.1). The proximal head is mediolaterally broad with a ball-shaped condyle in the middle that makes up about one-third of the proximal head. The condyle faces caudodorsally to articulate with the glenoid fossa. The deltopectoral crest is flattened and extends laterally. The lateral corner of the deltopectoral crest is strongly angled and exhibits a rugose texture for attachment of the pectoral musculature. There is an elongate tuberosity along the caudal portion of the humerus (92.7 mm long; approximately 10% of the humeral length), located opposite the deltopectoral crest, and may represent the attachment site for the *m. latissimus dorsi* (Maidment and Barrett, 2011; Fig. 13.14H, J). Distal to the deltopectoral crest the shaft is circular in cross section. A ridge on the cranial surface of the humerus extends from the medial corner of the proximal head to the medial condyle of the distal head. Both the medial (ulnar) and lateral (radial) condyles are well developed and subequal in size, with a pronounced intercondylar groove separating them. The medial condyle exhibits a shallow longitudinal entepicondylar groove on the medial side (Fig. 13.14G). There is no evidence of an ectepicondylar groove on the lateral surface of the lateral condyle.

Ulna Both right and left ulnae are preserved in CMN 2289 (Fig. 13.15). The ulna is a long, thin bone (length is 14 times the craniocaudal midshaft width) that is longer than the humerus (total ulnar length = 775 mm). However, based on the length of the radius (673 mm) and the length of the prearticular region of the ulna (720 mm), the zeugopodium is approximately the same length as, or only slightly longer than the stylopodium. The olecranon process is short (approximately 61 mm). Distal to the olecranon, the ulna possesses two prominent processes. The cranial process is the most prominent and its proximal border is at a right angle to the olecranon process. The lateral process is smaller than the cranial process. They are separated by a deep trough that extends past the middle of the shaft and represents the region of proximal articulation between the ulna and radius. Distally, the ulna is thinner mediolaterally, and there are longitudinal striations that extend to the distal head, which is lunate in distal view.

Radius The radius forms the lateral portion of the zeugopodium. Only the left is preserved in CMN 2289 (Fig. 13.16A–F). Both the trochlear and distal ends are slightly expanded relative to the radial shaft. The trochlear head is slightly concave. The radial shaft is straight and uniform in shape and ends at the slightly expanded distal head.

Manus Metacarpal II (Fig. 13.16G–L), as well as two isolated distal phalanges (Fig. 13.16M–X), are the only bones of the manus that are complete enough to be identified and described. Metacarpal II is a long and medial laterally thin bone. It is craniocaudally broad compared to its mediolateral width and it narrows distally, and then expands at the distal end. It is flattened laterally where it articulates with metacarpal III. Both the proximal and distal ends are slightly expanded craniocaudally, ellipsoidal in shape in proximal and distal view, and convex in medial view.

The distal phalanges are typical of hadrosaurids. They have a well-developed proximal articular surface and thin distally to a point in lateral view. They are flattened ventrally and convex dorsally. Based on the elongate shape of one of the distal phalanges (Fig. 13.16S–X), it likely represents the ungual for digit two. It is better preserved than the other distal phalanx and on its dorsal surface has two well-developed sulci that extend from the lateral margins, between the proximal end and the lateral expansion of the ungual, to a large foramen on the dorsal surface that leads into the bone (Fig. 13.16S). This foramen lies proximal of the distal margin of the ungual. The distal margin is highly rugose and grooved. The other distal phalanx (Fig. 13.16M–R) is approximately equal in width and length and likely pertained to digit three. This ungual possesses a less pronounced blood groove on its dorsal surface that does not appear to end at a foramen. The distal margin is not as rugose; it has a few notable grooves along with a deep notch at the midline, visible in dorsal view (Fig. 13.16M).

Pelvic Girdle

Ilium Only the left ilium is present in CMN 2289 (Fig. 13.17). It is a robust bone, approximately 1260 mm long. The preacetabular process is the thinnest region of the ilium (dorsoventrally) and projects cranially and curves ventrally to form an angle of approximately 150° with the main body of the ilium. Cranially, it is slightly expanded dorsoventrally, and swollen mediolaterally along its dorsal and ventral margins for attachment of the *m. iliotibialis* and *m. puboischiofemoralis internus*, respectively (Maidment and Barrett, 2011). The lateral surface is also marked by strong longitudinal striations and rugosities. These cover most of the lateral surface of the preacetabular process, but are mostly concentrated along the dorsal and ventral margins. In medial view, a strong medial ridge extends from the base of the preacetabular process over the acetabular region along the dorsal margin of the ilium. This ridge represents the region of attachment between the ilium and the transverse processes of the sacral vertebrae. The region above the acetabulum is the tallest dorsoventrally and is the most robust area of the ilium. The acetabulum is concave in lateral view with well-developed cranial and caudal peduncles for the pubis and ischium, respectively;

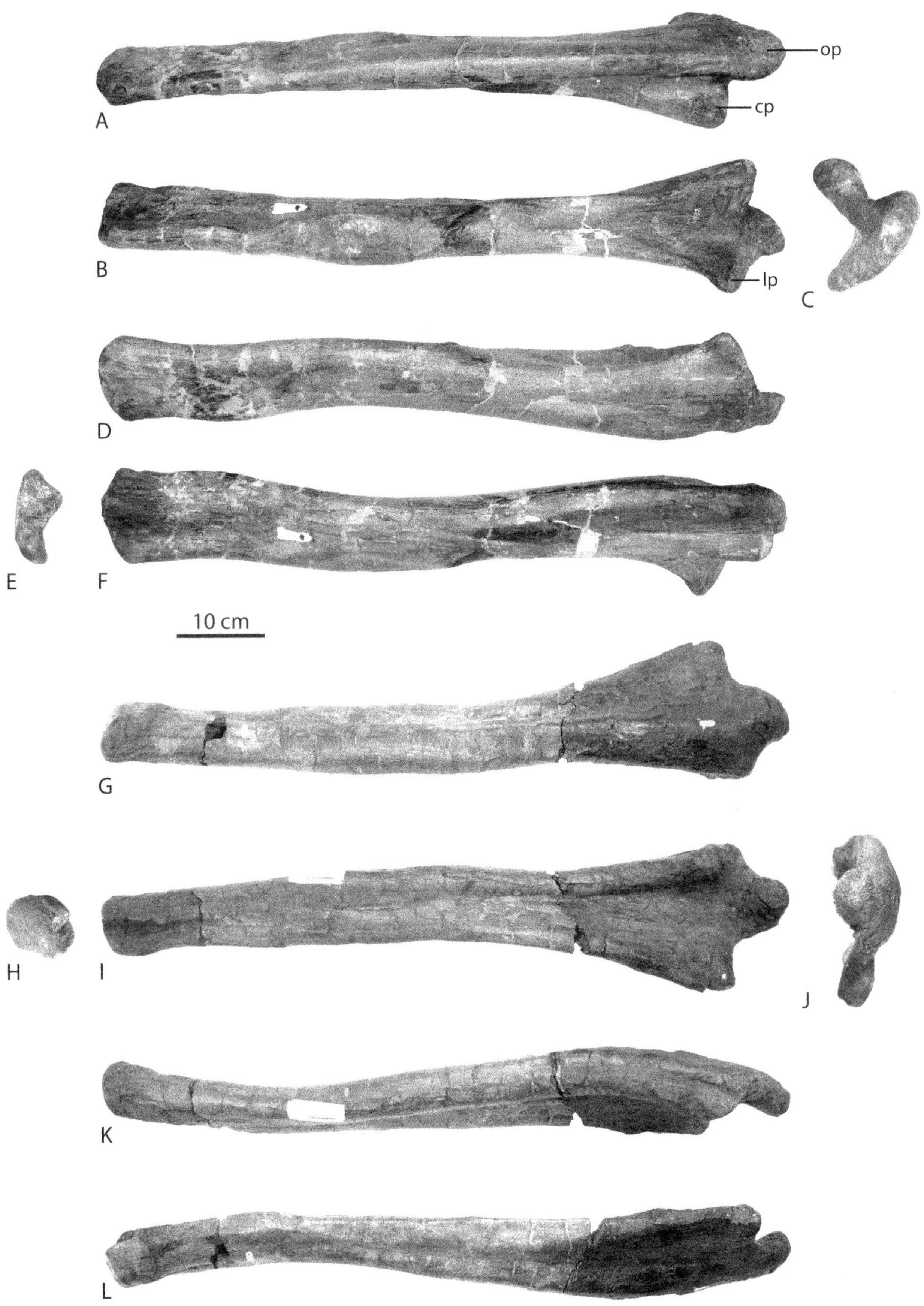

13.15. Left and right ulnae of CMN 2289. (A–F) left and, (G–L) right ulnae in (A, G) caudal; (B, I,) cranial; (C, J) proximal; (D, K) lateral; and (E, H) distal; and (F, L) medial views.

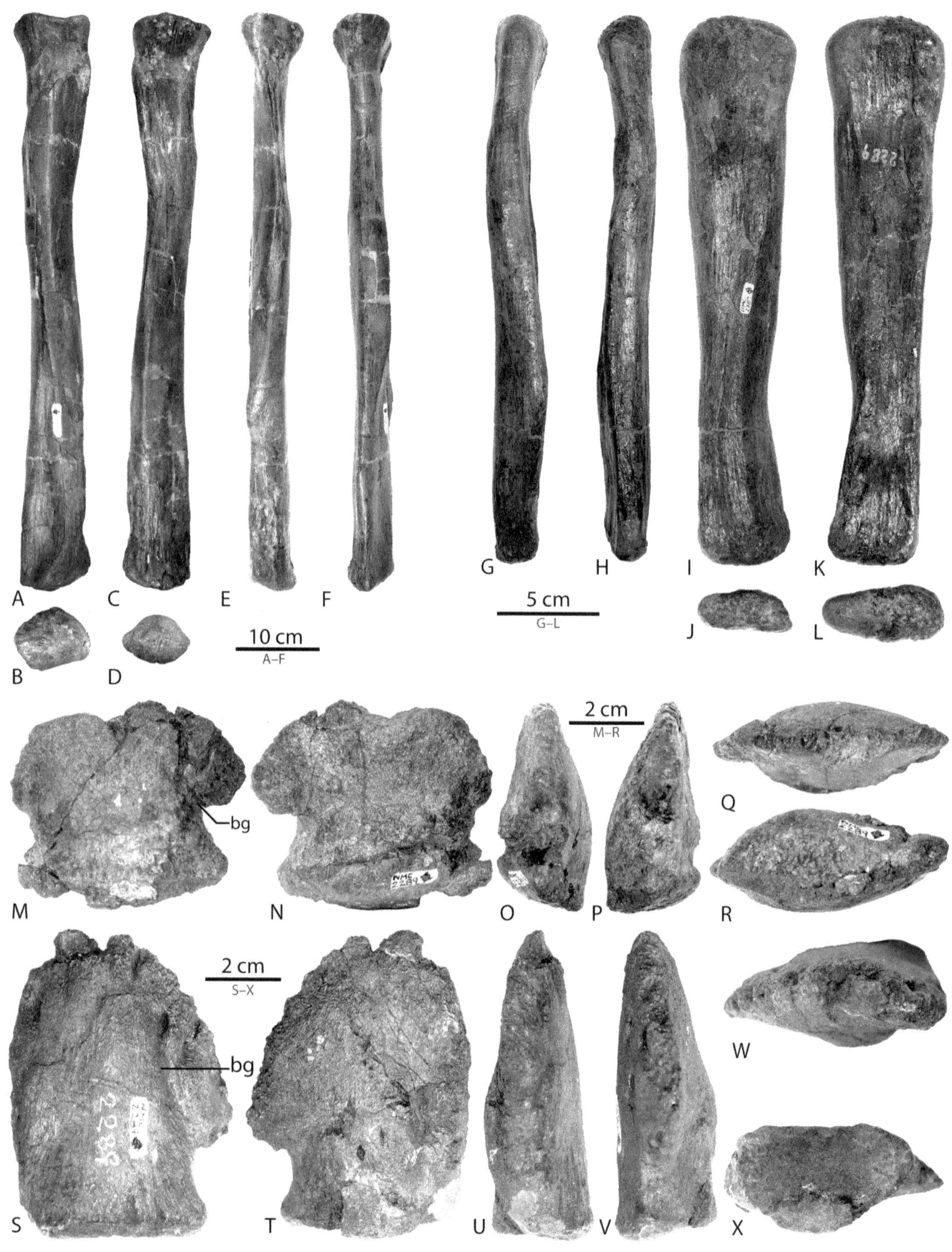

13.16. Forelimb zeugopodial and autopodial elements of CMN 2289. Right radius in (A) medial; (B) proximal; (C) lateral; (D) distal; (E) caudal; and (F) cranial views. Left metacarpal II in (G) cranial; (H) caudal; (I) medial; (J) distal; (K) lateral; and (L) proximal views. (M–R) distal phalanx of digit three and, (S–X) distal phalanx of digit two in (M, S) dorsal; (N, T) ventral; (O, U) left aspect; (P, V) right aspect; (Q, W) distal; and (W, X) proximal views.

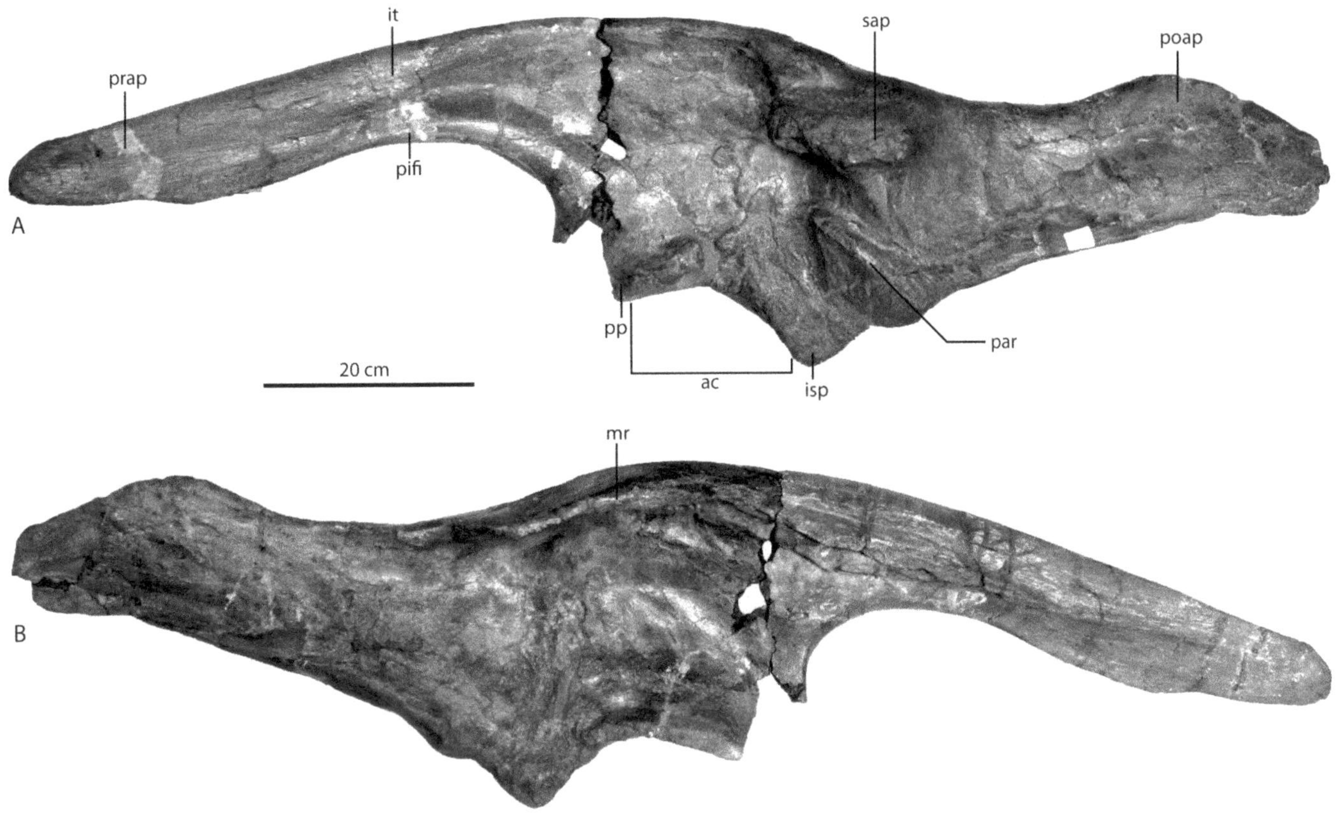

13.17. Left ilium of CMN 2289 in (A) lateral; and (B) medial views.

unfortunately, the pubic peduncle is not complete, and its size relative to the ischial peduncle cannot be determined. The supra-acetabular process is present above the caudal extent of the acetabulum. It is well developed in CMN 2289 and projects laterally and ventrally, almost to the level of the postacetabular ridge to which it is connected to by a thin lamina of bone. The ventral extent of the supra-acetabular process extends slightly caudally and is located craniodorsally to the postacetabular ridge. The development of the supra-acetabular process results in a marked embayment on the dorsal surface of the ilium in lateral and medial views. The postacetabular process extends caudally and slight dorsally. It is approximately equal in length to that of the central blade (measured from the cranial margin of the pubic peduncle to the caudal extent of the postacetabular ridge). The postacetabular process is dorsoventrally tall compared to the preacetabular process, and is not longitudinally symmetrical in lateral view; it exhibits a slight ventral curve at its caudal-most end.

Pubis Both right and left pubes are preserved in CMN 2289 (Fig. 13.18). In both, the postpubic process and a portion of the ischial peduncle are missing. The prepubic process is slightly sigmoidal in lateral view; it is flattened mediolaterally and tall dorsoventrally. The medial surface is slightly convex, whereas the lateral surface is flat. In lateral view, the prepubic process is thinnest cranial to the acetabulum and is symmetrical in this region. Cranially, the prepubic process expands dorsoventrally and is over two times the dorsoventral height of the prepubic construction. As is typical in hadrosaurids, the dorsoventrally expanded region of the pubis is highly asymmetrical – the cranioventral corner projects ventrally, relative to the long axis of the process. However, it is not distinctly rectangular as in *Gryposaurus* (Parks, 1920) and *Saurolophus* (Maryańska and Osmólska, 1984). The acetabular surface faces caudolaterally, and the upper and lower processes represent the peduncles for the ilium and ischium, respectively. The iliac peduncle is much more robust than the ischial peduncle, and its border is rugose.

Ischium Right and left ischia are preserved; however, the left is missing the distal half (Fig. 13.19). At 1255 mm, the ischium is as long as the femur. It is comprised of a robust proximal end that is flattened mediolaterally, and a long, thin distal shaft that is less than 5% of the length of the shaft and is rounded in cross section. The ischium is especially gracile in comparison to lambeosaurines, such as *Hypacrosaurus* (Gilmore, 1924a), and although the distal

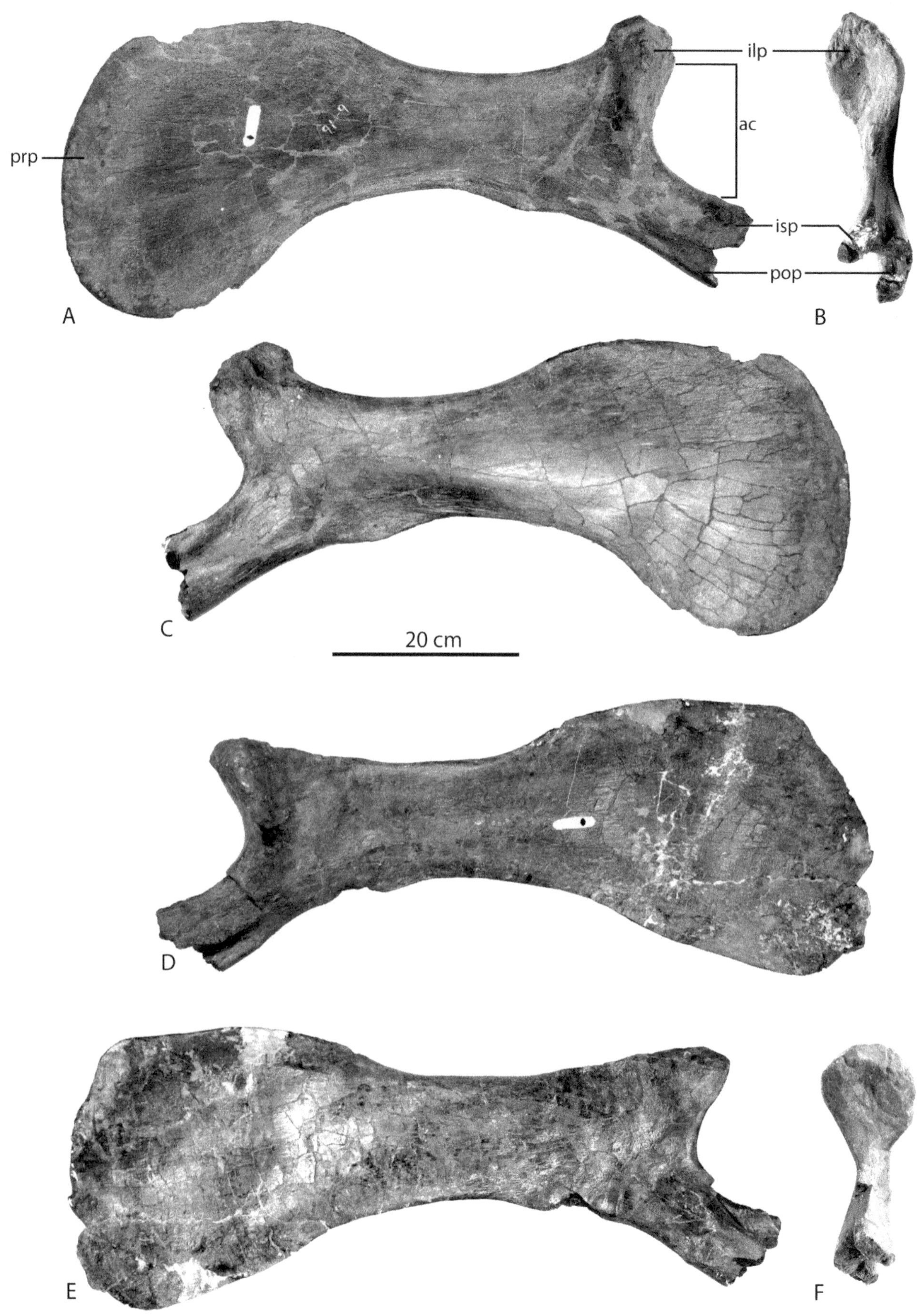

13.18. (A–C) left and, (D–E) right pubes of CMN 2289 in (A, D) lateral; (B, F) articular (acetabular); and (C, E) medial views.

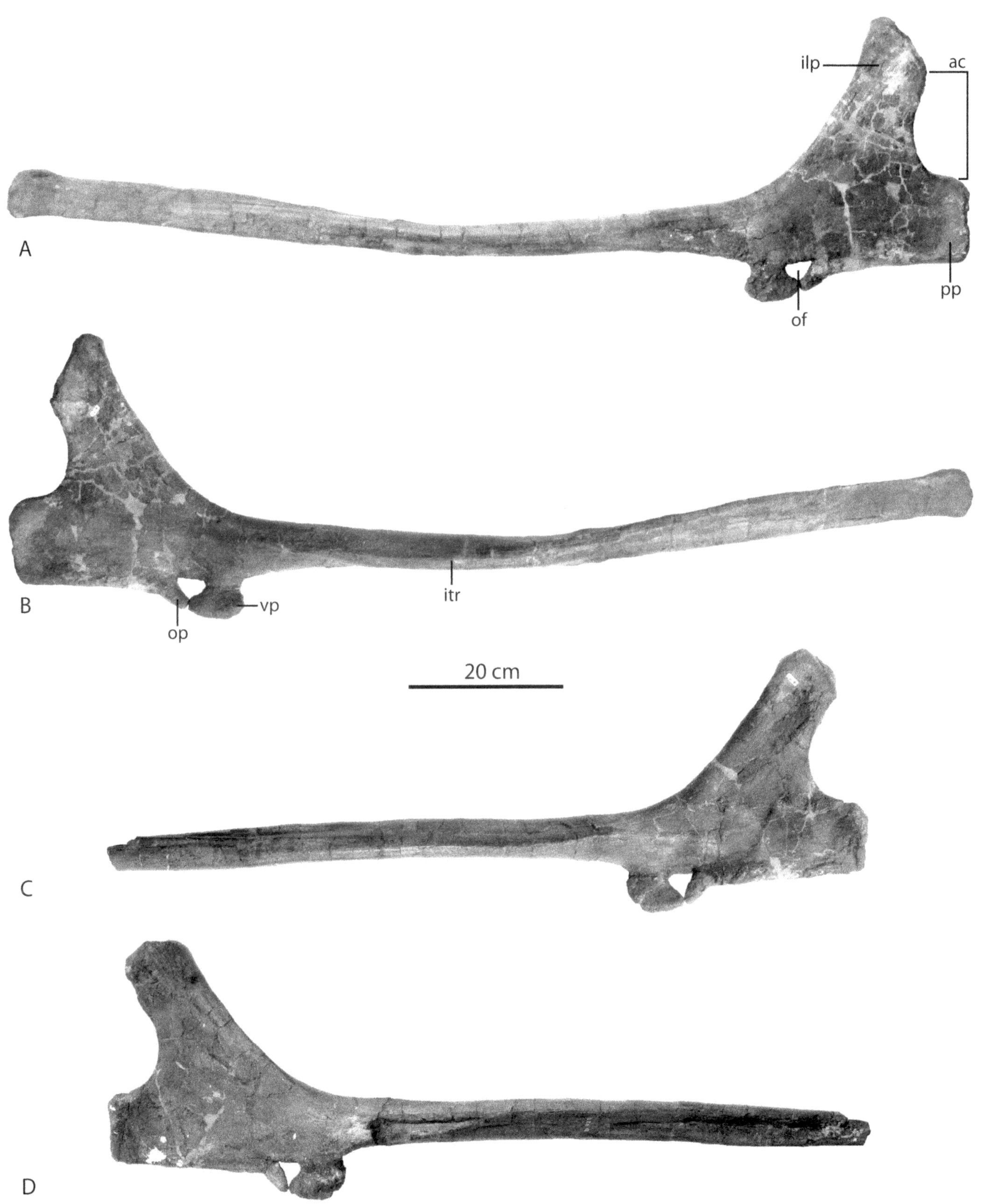

13.19. Right and left ischia of CMN 2289. Right in (A) lateral; and (B) medial views. Left in (C) lateral; and (D) medial views.

end is slightly expanded, it is not "booted" as is common in lambeosaurines (Horner et al., 2004). The proximal end is comprised of two squared processes that are connected by a strongly concave margin that forms the ischial contribution to the acetabulum. The iliac process is longer than the pubic process; it projects dorsally, but the articular face for the ilium is angled cranially, which results in an angle of 120° between the articulation surface and the long axis of the iliac process. The pubic process extends cranially for approximately 70 mm in line with the long caudal shaft; it is 90 mm tall. The obturator process extends caudally along the ventral margin of the ischial shaft. Together, the obturator process of the ischium and the postpubic process of the pubis form the obturator foramen. Caudal to the obturator process is a ventral process that at its distal end expands craniocaudally and almost encloses another foramen, sometimes also referred to as the obturator foramen (Lull and Wright, 1942). The foramen is not fully enclosed in CMN 2289 as in other hadrosaurids (Gilmore, 1924a; Prieto-Márquez, 2007). The lack of fusion is consistent on both the right and left ischia and does not appear to be pathological. The caudal shaft is long and slender, with a medial ridge that extends the length of the shaft and is visible in medial view. The ridge likely represents the site of attachment from the *m. ischiotrochantericus* (Maidment and Barrett, 2011; Fig. 13.19B).

Hindlimb

Femur The right and left femora are present in CMN 2289; however, the left is much better preserved (Fig. 13.20). The proximal end of the femur comprises the femoral head medially, as well as the greater and cranial trochanters laterally. The femoral head extends medially to fit in the acetabulum of the pelvis. The head is broad craniocaudally, with a slightly flattened dorsal surface and well-developed striations that extend from the medial-most region of the head down to the diaphysis. The greater trochanter is a broad, paddle-like structure lateral to the femoral head and extends dorsally above the level of the head. It is broad craniocaudally and flattened mediolaterally, and is the site of attachment for the *m. puboischiofemoralis externus* (Maidment and Barrett, 2011). The cranial trochanter is located distal to the greater trochanter and is separated from the latter by a small gap. The cranial trochanter projects cranially relative to the femoral diaphysis and is the site of attachment for the *m. iliofemoralis* (Maidment and Barrett, 2011). The diaphysis is thickest proximal to the fourth trochanter. The fourth trochanter lies at approximately the middle of the diaphysis; it is triangular in shape, flattened mediolaterally, and extends caudally from the diaphysis. Two distinct muscle scars are found on the medial surface of the fourth trochanter (Fig. 13.20B, H). The largest lies nearest the caudal end of the trochanter and is elongated in a proximodistal direction; it represents the attachment site for the *m. caudofemoralis longus* (Maidment and Barrett, 2011). The second scar lies above that of the *m. caudofemoralis longus*, and it is smaller and not as elongate; it is likely the attachment site for the *m. puboischiofemoralis internus* (Maidment and Barrett, 2011). The diaphysis is straight, circular, and narrowest distal to the fourth trochanter. The distal head of the femur comprises the lateral and medial condyles. These are approximately equal in size and are expanded craniocaudally with the caudal end extending farther than its cranial counterpart. There are well-developed intercondylar depressions between the condyles both cranially and caudally. The caudal depression is the deepest; however, the cranial depression is almost enclosed cranially by an inward deflection of the lateral and medial condyles, thus forming the extensor tunnel (Fig. 13.20C, E).

Tibia Both tibiae are preserved in CMN 2289 (Fig. 13.21). The tibia is slightly shorter than the femur. Both proximal and distal heads are expanded and are oriented perpendicular to each other. The proximal head is expanded craniocaudally and is the more robust of the two, comprising two condyles and the cnemial crest. The outer and inner condyles are on the medial side of the proximal head. They expand caudally, resulting in a deep depression between the condyles. The cnemial crest is lateral to the condyles, projects caudolaterally, and extends down the proximal third of the tibial shaft. The shaft of the tibia is straight and circular in cross section. The distal head is expanded mediolaterally with lateral and medial malleoli. The lateral malleolus extends below the level of the medial malleolus and is thin craniocaudally compared to the medial malleolus, which is expanded craniocaudally for reception of the astragalus. Both malleoli have rugose distal surfaces and are separated from each other by a shallow groove, visible in cranial and caudal views.

Fibula The right fibula is complete in CMN 2239 (Fig. 13.22A–F), whereas only the proximal end of the left fibula is preserved. The fibula is long and slender, with both proximal and distal heads expanded craniocaudally. The proximal head is the largest, with the cranial portion of the head extending cranially from the fibular shaft. The head is lunate in proximal view, and the proximal surface is rugose. A trough, on the medial surface of the fibula, extends from the proximal head to the middle of the shaft. Distal to the midshaft, the fibula is circular in cross section, but then gradually expands craniocaudally and flattens mediolaterally. The distal head is not as expanded craniocaudally as the proximal head and also extends more cranially than caudally. In distal view, the distal head of the fibula is teardrop shaped.

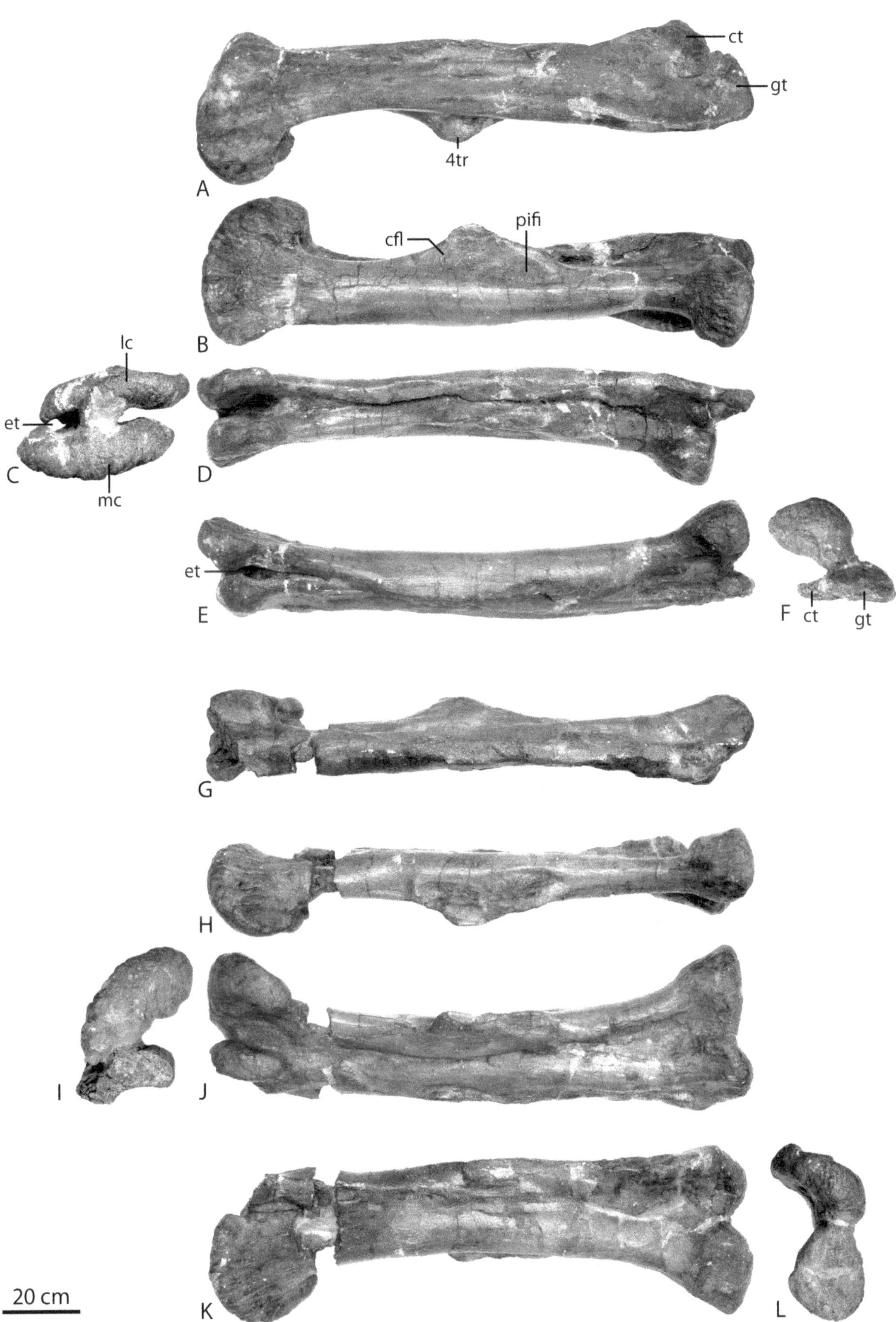

13.20. Left and right femora of CMN 2289 in (A, G) lateral; (B, H) medial; (C, I) distal; (D, J) caudal; (E, K) cranial; and (F, L) proximal views.

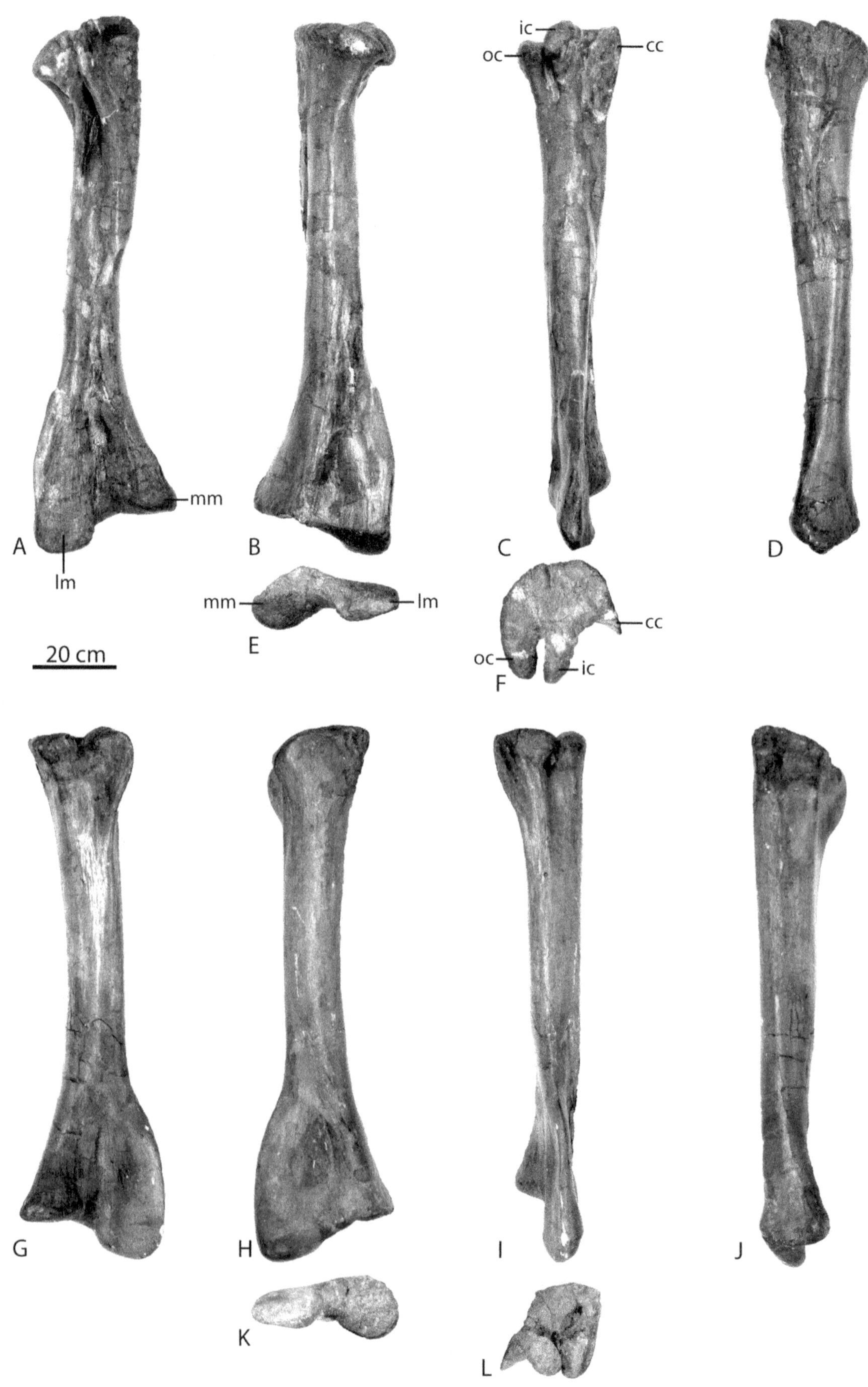

13.21. Right and left tibiae of CMN 2289. (A–F) right; and (G–L), left tibiae in (A) (G) cranial; (B, H), caudal; (C, I) lateral; (D, J) medial; (E, K) distal; and (F, L) proximal views.

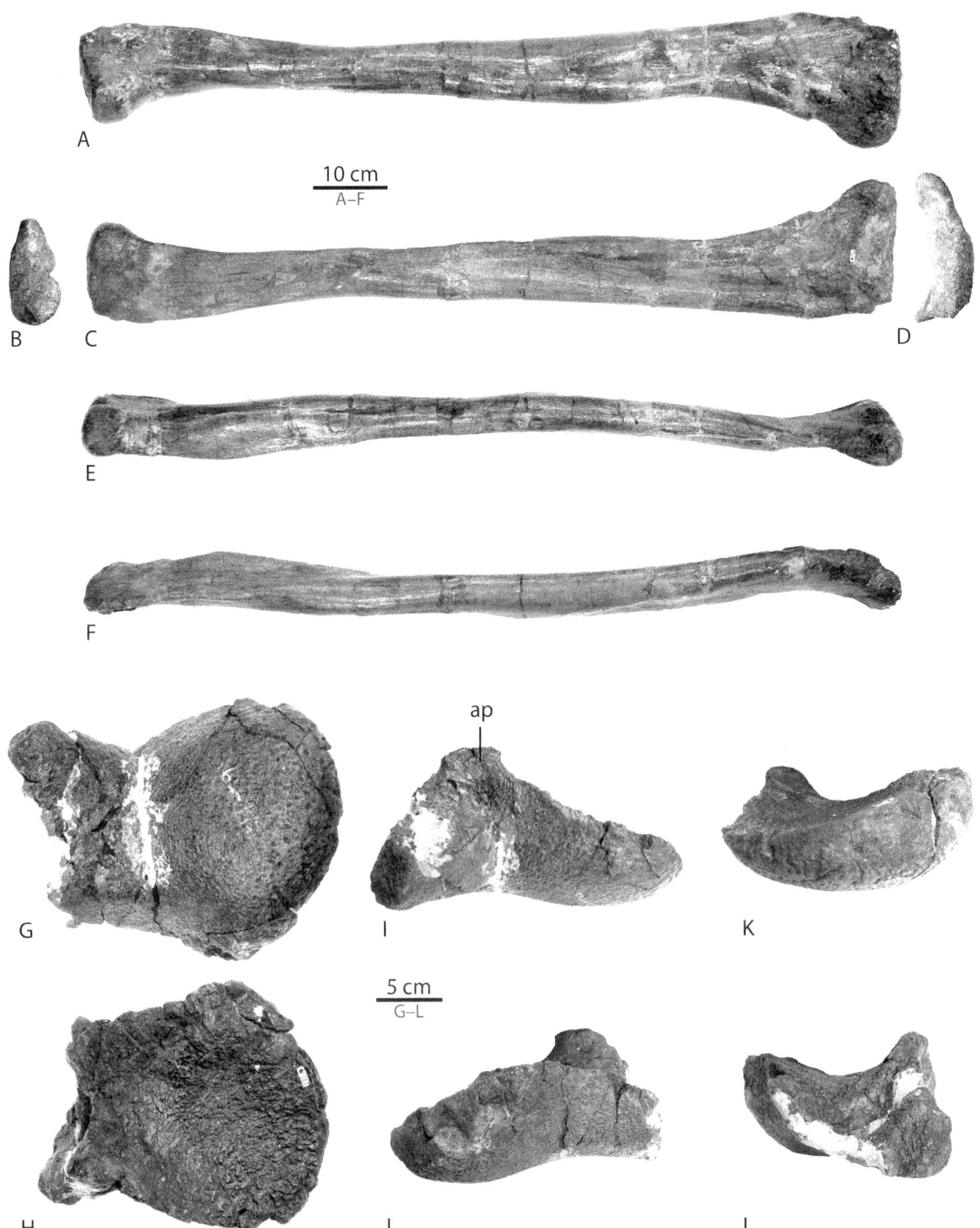

13.22. Right fibula of CMN 2289 in (A) lateral; (B) distal; (C) medial; (D) proximal; (E) cranial; and (F) caudal views. Right astragalus in (G) distal; (H) proximal; (I) cranial; (J) caudal; (K) medial; and (L) lateral views.

Astragalus Only the right astragalus is preserved in CMN 2289 (Fig. 13.22G–L). In both proximal and distal views, the astragalus is roughly hourglass shaped; it is craniocaudally expanded both laterally and medially (greater in the latter), and narrows in the middle. The proximal surface is concave medially, and articulates with the inner malleolus of the tibia; this surface is highly rugose. Lateral to this concavity, the astragalus is raised proximally and fits between the tibial malleoli. Viewed proximally, the lateral-most extent of the astragalus is flattened and contacts the calcaneum (not preserved). The proximal surface is strongly convex, highly rugose, and forms the medial portion of the ankle joint.

Pes Only the right (Zheng et al., 2011) metatarsals are preserved in CMN 2289. The third is only partially complete (Fig. 13.23). Metatarsal II is the narrowest (mediolaterally) of the three metatarsals (Fig. 13.23A–F). The proximal end is fan shaped in lateral and medial views, proximally flattened in comparison to the distal end, and slightly expanded mediolaterally. The lateral surface of metatarsal II is slightly convex, whereas the medial surface is flattened where it would have adjoined metatarsal III. In lateral view, metatarsal II is narrowest at its midpoint and expands craniocaudally at its distal end. A small medial process is present distal to the midpoint along the cranial margin. The distal end is expanded both craniocaudally and mediolaterally and projects slightly laterally relative to the shaft. It is highly convex distally where it would articulate with the tarsus. In distal view, the head is parallelogram shaped with a rugose border.

Metatarsal III preserves only a portion of the shaft and the distal head (Fig. 13.23G–K). It is much more robust than metatarsal II. The shaft is slightly sigmoidal, most obviously seen in caudal view. Also in caudal view, there is a longitudinal ridge that extends from the broken proximal end to the middle of the shaft. The shaft is rectangular is cross section. The distal head has depressions on both lateral sides, though the right side is more pronounced. The distal surface is convex in lateral view and slightly concave in cranial or caudal view, giving it a sellate appearance. In distal view, the head has a rugose surface texture and an hourglass shape.

Metatarsal IV is robust like metatarsal III and is notable for its characteristic angled appearance (Fig. 13.23L–Q). The proximal head is triangular in proximal view, flattened in lateral view, and not expanded relative to the shaft. The proximal medial surface of metatarsal IV is slightly concave where it would have adjoined the lateral surface of metatarsal III. A prominent eminence extends medially distal to the concavity and gives metatarsal IV its characteristic shape relative to the other metatarsals. Distal to the eminence, the medial surface is rounded and craniocaudally thin relative to the expanded distal head. The lateral surface is rounded for the entire length of the shaft. The distal head is rounded, with depressions on both lateral and medial sides. In distal view, the head is parallelogram shaped with a well-developed ridge that extends from the caudomedial border along the caudal margin of the medial depression.

DISCUSSION

Pelvic Morphology

The majority of the postcranial skeleton of CMN 2289, including the diagnostic pelvic girdle (Fig. 13.24), was not prepared at the time of the original description (Lambe, 1920). Its overall morphology is typical of other hadrosaurines (= Saurolophinae of Prieto-Marquez, 2010b). The ilium has a well-developed supra-acetabular process caudal to the acetabulum and a slightly dorsally projecting postacetabular process that is mediolaterally flattened and dorsoventrally broader than the preacetabular process. Although these characters are common to hadrosaurids, some also vary ontogenetically. For instance, the supra-acetabular process is poorly developed in a smaller specimen of *E. regalis*, CMN 8399, suggesting that the development of this process is ontogenetic. The extent of the curvature (or angle) of the preacetabular process relative to the iliac body is a commonly cited phylogenetic character (e.g., Suzuki et al., 2004; Prieto-Márquez, 2010b); it is not strongly curved in CMN 2289 compared to *Brachylophosaurus canadensis* (Prieto-Márquez, 2007) and *Saurolophus angustirostris* (Maryańska and Osmólska, 1984). However, the extent of the ventral curvature may also vary ontogenetically in hadrosaurids, in which juveniles tend to have less of a curvature than more mature individuals (Brett-Surman and Wagner, 2007), making it a difficult character to interpret phylogenetically and taxonomically.

Edmontosaurus regalis has a typical hadrosaurine ischium, as indicated by the slender ischial shaft without a dorsoventrally expanded distal end. This morphology is most notable when compared to the very robustly built and "booted" ischium of *Hypacrosaurus altispinus* (Gilmore, 1924a). A large foramen is present along the ventral margin of the ischium. It was previously identified by Lull and Wright (1942) as the obturator foramen, despite its location on the ischium, and not the pubis. Notably, the ventral margin of this foramen is not fused in CMN 2289, whereas it is completely fused in other hadrosaurines, such as *Saurolophus angustirostris* (Maryańska and Osmólska, 1984) and *Brachylophosaurus canadensis* (Prieto-Márquez, 2007), and in the lambeosaurines *Hypacrosaurus altispinus* (Gilmore, 1924a) and *Corythosaurus intermedius* (ROM 845). Variation in this foramen has not been included in previous phylogenetic analyses despite the fact that it is also unfused in non-hadrosaurid iguanodontians (e.g., Norman, 1980, 1986; Godefroit

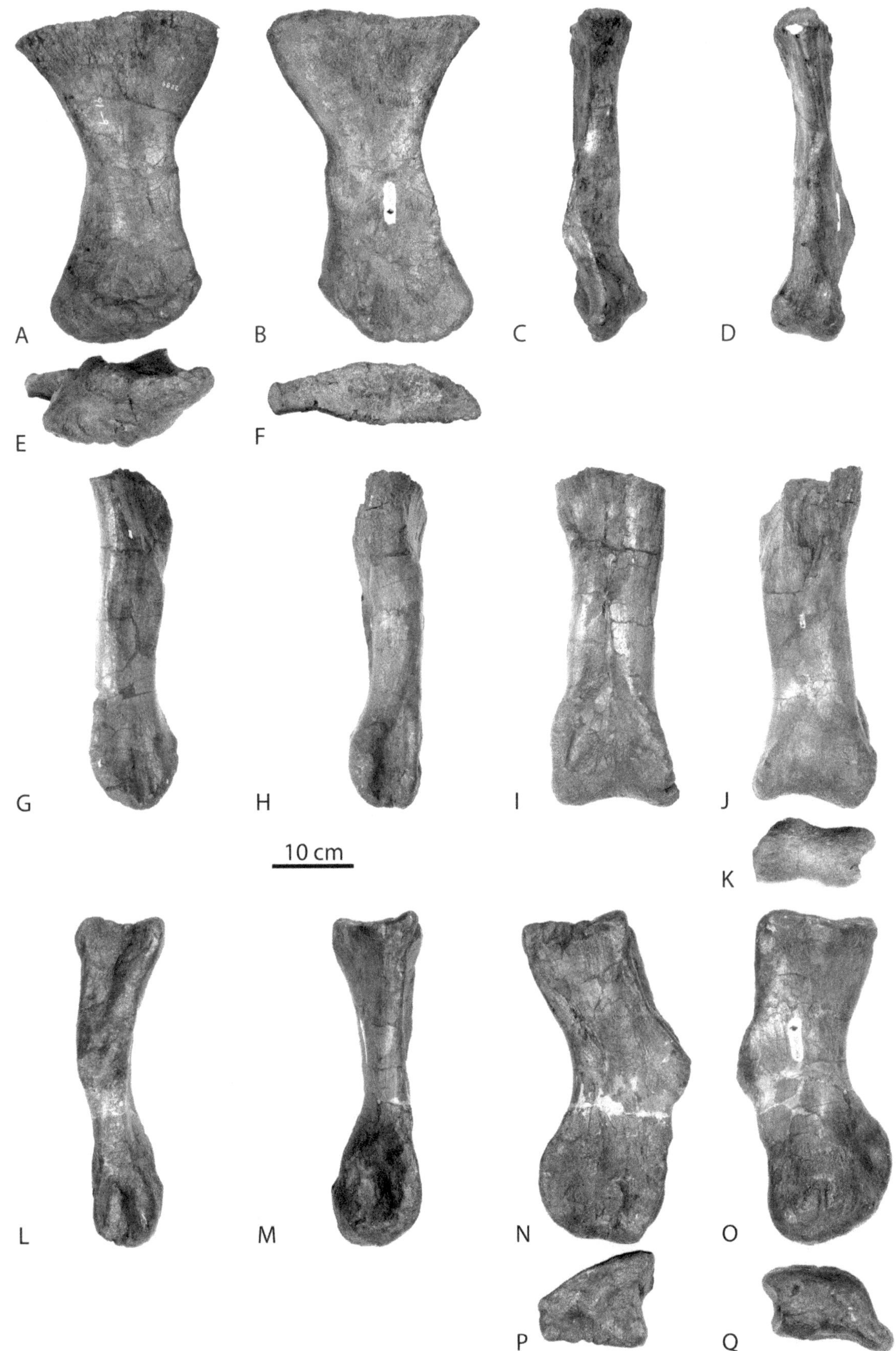

13.23. Proximal portion of the right pes of CMN 2289. Metatarsal II in (A) medial; (B) lateral; (C) cranial; (D) caudal; (E) distal; and (F) proximal views. Metatarsal III (proximally incomplete) in (G) medial; (H) lateral; (I) cranial; (J) caudal; and (K) distal views. Metatarsal IV in (L) medial; (M) lateral; (N) cranial; (O) caudal; (P) proximal; and (Q) distal views.

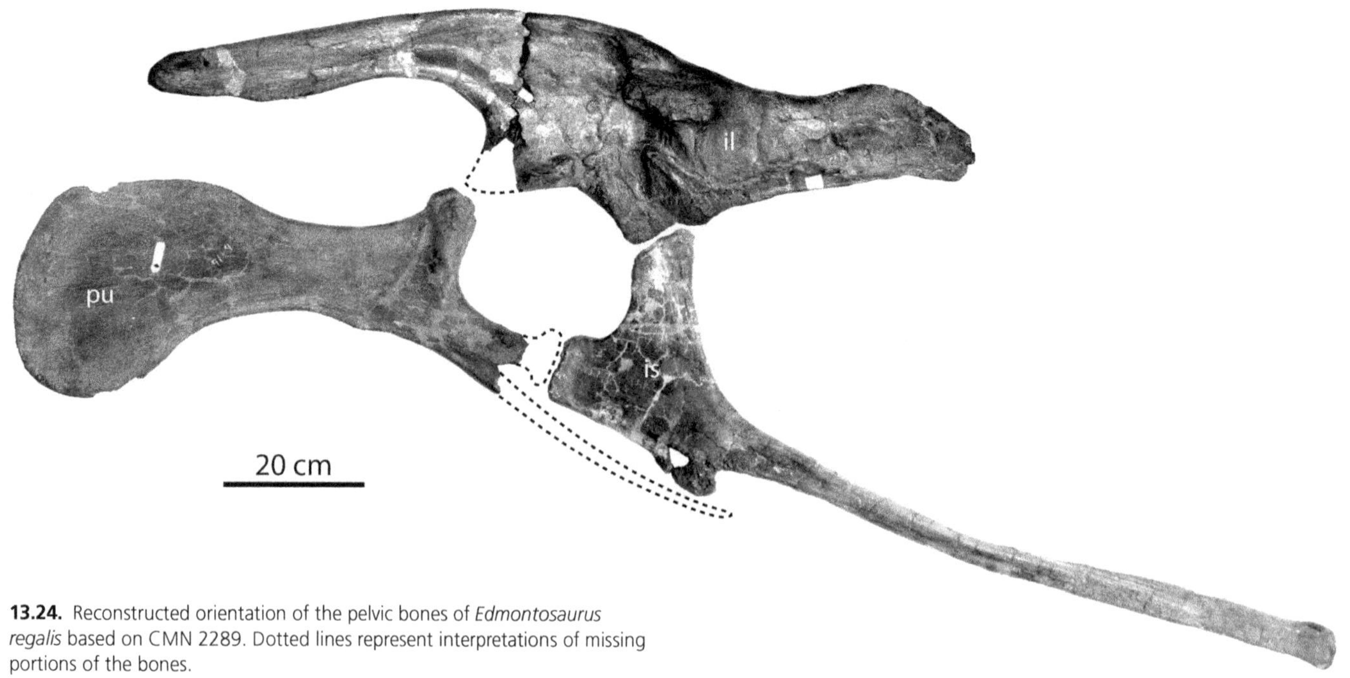

13.24. Reconstructed orientation of the pelvic bones of *Edmontosaurus regalis* based on CMN 2289. Dotted lines represent interpretations of missing portions of the bones.

et al., 1998) and in other hadrosaurids *Shantungosaurus giganteus* (Hu, 1973) and *Tsintaosaurus spinorhinus* (Young, 1958). The absence of fusion of this foramen in non-hadrosaurids suggests that an open foramen may be plesiomorphic for hadrosaurids; however, at this time, the significance of this variation still hinges on further study of ontogenetic variation and interspecific comparisons that are outside the scope of this study.

Based on the observed variation in the pelvis of hadrosaurids, as well as those previously presented (e.g., Chapman and Brett-Surman, 1990; Brett-Surman and Wagner, 2007), it is apparent that the nature of variation in the pelvis of hadrosaurids remains poorly understood. Future research, especially incorporating quantitative morphometric techniques (using both landmark and outline analyses), will provide important insights into the evolution of this complex structure.

Vertebral Variation

CMN 2289 preserves most of its dorsal vertebrae, allowing for quantification of variation along the series. Only the relative positions of the dorsal vertebrae along the series could be determined, and their rank order is used in all analyses (see Table A1b in Appendix 13.1). Centrum length decreases caudally along the dorsal series, whereas the width and height (measured along the caudal surface) increase (Fig. 13.25A). As a result, the dorsal vertebrae of *E. regalis* are flatter and broader caudally relative to the cranial dorsal vertebrae. This contrasts with other hadrosaurines such as *Brachylophosaurus canadensis* (known from the seven cranial-most dorsal vertebrae; Fig. 13.25C; Cuthbertson and Holmes, 2010) and *Saurolophus angustirostris* (Maryańska and Osmólska, 1984), whose dorsal vertebrae increase in length, width, and height caudally along the series.

In hadrosaurids, a gradual increase in the height of the dorsal vertebral spines has been previously noted (Horner et al., 2004). Most importantly, the length of the spines of the caudal dorsal vertebrae relative to their respective centra is commonly used as a phylogenetic character that differentiates lambeosaurines, which have tall dorsal spines, from the short neural spines on the dorsal vertebrae of hadrosaurines (e.g., Horner et al., 2004; Prieto-Márquez, 2010b, 2011). In *E. regalis*, the craniocaudal width (measured at the base of the spine) decreases along the dorsal series, whereas spine height (measured from the vertebral foramen to the tip of the spine) shows no apparent trend and is highly variable (Fig. 13.25B). Dorsal neural spines exhibit an initial increase in height caudally, but show variable heights at approximately the middle of the dorsal series, and then decrease in height leading up to the sacrum. The first sacral vertebra has a spine that is shorter (221 mm) than any of the dorsal vertebrae.

The results presented here (Fig. 13.25) indicate that hadrosaurids exhibit important variation along the dorsal vertebral series, and that this variation may itself differ between hadrosaurid species (e.g., *E. regalis* and *B. canadensis*) and hence be phylogenetically significant. Interspecific variation of the dorsal vertebrae in hadrosaurids has been previously recognized and used in phylogenetic analyses (e.g., Horner et al., 2004; Godefroit et al., 2008; Prieto-Márquez, 2010b). However, the characters are often vague about the specific vertebra used for character coding, citing an undefined broad region (e.g., caudal dorsal to sacral vertebrae). Given the variation noted in *E. regalis* (Fig. 13.25A, B), it is evident that in order to make relevant phylogenetic comparisons, it is important to specify or code homologous vertebrae in order to avoid potentially significant errors in character coding. However, at this time, further quantitative studies incorporating more taxa are needed in order to gain a more comprehensive understanding of the variation along the dorsal series and determine its phylogenetic significance.

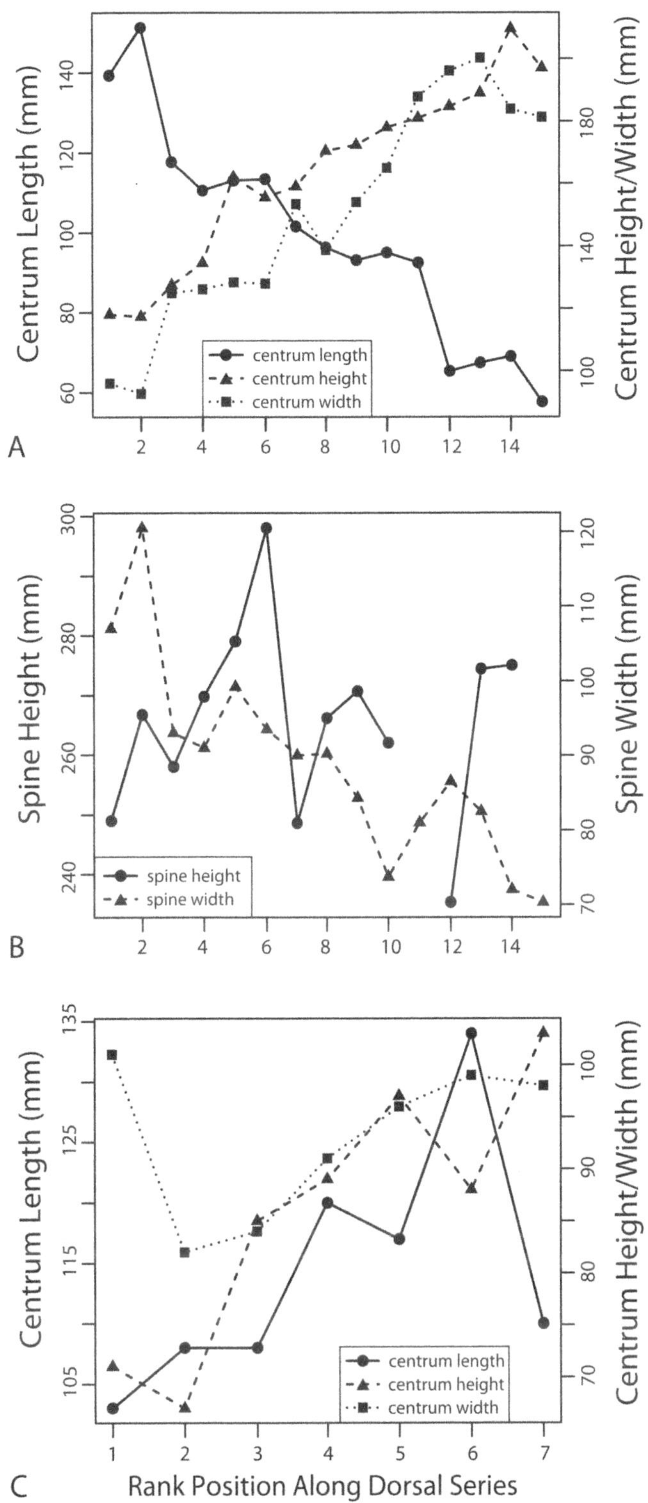

13.25. Plots showing the sequence of morphological changes that occur along the dorsal vertebral series from cranial to caudal. (A) and (B) represent data from CMN 2289 (see Appendix 13.1), and (C) represents data on *Brachylophosaurus canadensis* (Cuthbertson and Holmes, 2010). Since actual position along the dorsal series could not be obtained, the vertebrae of CMN 2289 were ranked according to their interpreted position relative to each other.

Limb Proportions and Body Mass Estimates of Edmontosaurus regalis

CMN 2289 preserves an almost complete set of limbs, missing only portions of the autopodium. As a result, it permits the re-evaluation of certain limb proportions in *Edmontosaurus regalis* in comparison to other hadrosaurids and provides a new estimate of body mass for the specimen, and approximate average body mass for the species.

The total length of the forelimb of CMN 2289 is 1.62 m (sum of the lengths of the humerus, radius, and metacarpal II; see Table A1d in Appendix 13.1), of which the zeugo- and autopodium comprise 57% of the length. The hind limb is 2.8 m long (sum of the lengths of the femur, tibia, and metatarsal II) of which the zeugo- and autopodium comprise 54%. Therefore, the relative proportions of the distal limb relative to the proximal limb are consistent between the fore- and hindlimbs in this specimen. Similar proportions are found in the fore- and hindlimbs of other hadrosaurids (e.g., *Gryposaurus notabilis:* forelimb 55%, hindlimb 56%, ROM 764; *Saurolophus osborni:* 61% and 56%, AMNH 5220; *Corythosaurus casuarius*, 63% and 56%, AMNH 5338; and *Lambeosaurus clavinitialis:* 63% and 56%, CMN 8703 [all measurements, except for CMN 2289, were taken from Lull and Wright, 1942]). In addition to not being particularly variable interspecifically, zeugo- and autopodium to stylopodium proportions do not appear to vary greatly ontogenetically (juvenile *Lambeosaurus lambei:* 61% and 56%, AMNH 5340). Comparisons of total fore- and hindlimbs lengths to each other indicate that CMN 2289 has a forelimb that is 58% the length of the hindlimb, a value similar to that in other hadrosaurids, both adult and juvenile (e.g., *S. osborni:* 60%; *C. casuarius:* 60%). These comparisons, however, represent a preliminary sample, and future allometric analyses based on a more comprehensive sample, which are outside the scope of this study, will more adequately test these intra- and interspecific patterns. In particular, comparisons between

Table 13.1. Body Mass Estimate (in Kilograms) for *Edmontosaurus* Species

Taxon	Study	Exemplar	Mass
E. regalis	Anderson et al. (1985)[RE]	CMN 2289	3,800
E. regalis	Campbell and Marcus (1992)[RE]	CMN 2289	2,822
E. regalis	Paul (1997)[VR]	CMN 8399	2,900
E. regalis	Henderson (this volume)[VR]		10,000
E. annectens	Colbert (1962)[VR]	AMNH 5730	3,071
E. annectens	Anderson et al. (1985)[RE]	AMNH 5730	4,000
E. annectens	Campbell and Marcus (1992)[RE]	AMNH 5730	2,931
E. annectens	Paul (1997)[VR]	DMNH 1493	2,400
E. annectens	Paul (1997)[VR]	AMNH 5730	3,200
E. annectens	Seebacher (2001)[VR]	YPM 1190; USNM 2414	3,990.8
E. annectens	Seebacher (2001)[VR]	AMNH 5730	7,594.4
E. regalis	This study[RE]	CMN 2289	5973–9954*
E. annectens	This study[RE]	AMNH 5730	5203–8671*

Notes:
[RE] Estimates based on a scaling regression equation of extant taxa.
[VR] Estimates based on a volumetric reconstruction (scale or computational).
* Estimated body mass range, given equation 2, and a 25% prediction error presented by Campione and Evans (2012).

fore- and hindlimbs, and between proximal and distal limb bones, will be important for addressing locomotor performance in hadrosaurids (Norman, 1980, 1986; Maidment et al., 2012) and testing hypotheses of long-range migratory behavior in this group (Bell and Snively, 2008; Fricke et al., 2008) within a functional and ontogenetic context.

The body mass of *Edmontosaurus regalis* was first estimated by Anderson et al. (1985) at 3800 kg, based on the specimen described here (CMN 2289) and their bipedal equation. The body mass was subsequently estimated by Campbell and Marcus (1992) at 2822 kg, based on a regression of femoral circumference to body mass in extant avians. Finally, Paul (1997) obtained a body mass of 2900 kg based on a volumetric scale model; however, the reconstruction was based on a smaller individual of *E. regalis* (CMN 8399). Comparatively more estimates have been provided for the sister-species of *E. regalis*, *E. annectens*, which range between 2400 kg and 7600 kg (Table 13.1).

Here I present a revised estimate of body mass for *E. regalis* (based on CMN 2289), using the combined humeral and femoral circumferences, which have been shown to be strong predictors of body mass in extant terrestrial quadrupeds (Campione and Evans, 2012). Based on the phylogenetically corrected equation presented in Campione and Evans (2012), associated with a 25% prediction error, the body mass of *E. regalis* is estimated at 7,936 kg ± 1,991 kg. Unlike the estimate provided by Anderson et al. (1985) and Campbell and Marcus (1992), the body mass estimate presented here assumes quadrupedality in *E. regalis* instead of bipedality, and hence incorporates data from both the humerus and femur. Quadrupedality in hadrosaurids has been the subject of debate (e.g., Ostrom, 1964; Galton, 1970; Norman, 1980; Dilkes, 2001; Sellers et al., 2009); however, the common occurrence of manal footprints in hadrosaurid trackways (Currie, 1983; Lockley and Wright, 2001; Horner et al., 2004) as well as similarities noted in the pelvis between obligate quadrupedal ornithischians and hadrosaurids (Maidment et al., this volume) suggest that hadrosaurids were probably quadrupedal most of the time and, therefore, body mass estimates of hadrosaurids based on limb measurements should include data from both the fore- and hindlimbs. Given the inclusion of data from both limbs, it is not surprising that the new estimate is much higher than the original bipedally based estimate of Anderson et al. (1985); if both humeral and femoral circumferences are included in the quadrupedal formula of the latter, a body mass of 6614 kg is obtained, which is much closer to that presented here.

The body mass estimate of 7936 kg is appreciably heavier than previous estimates of *E. regalis*, as well as most estimates of *E. annectens* (Table 13.1). However, it is similar to that obtained by Henderson (this volume) for *Edmontosaurus*, based on a virtual volumetric reconstruction technique (Henderson, 1999). Given its apparent size (Fig. 13.1), it is perhaps not surprising that the body mass of *E. regalis* is similar to that of extant elephants (*Loxodonta africana:* 6500 kg; Christiansen, 2004). Based on the same equation (Campione and Evans, 2012), the body mass of *E. annectens*, 6937 kg ± 1734 kg (AMNH 5730; humerus circumference: 250.5 mm; femur circumference: 512.3 mm), is not significantly different from that of *E. regalis*, suggesting a similar size for both sister species. Finally, recent mass estimates of hadrosaurids from the Dinosaur Park Formation (Brown et al., 2012) presented a range from 2900 kg in *Prosaurolophus maximus* to 4500 kg in *Parasaurolophus walkeri*, suggesting that *E. regalis* was one of the largest, if not the largest, hadrosaurid to have lived during the Late Cretaceous of western North America.

ACKNOWLEDGMENTS

I am indebted to E. Dawkins for her help in constructing the numerous plates needed for this study, and to D. Dufault for her reconstruction of the skeleton of *Edmontosaurus regalis* in Figure 13.1. Many thanks also to D. Evans, D. Eberth, K. Brink, C. Brown, D. Larson, C. VanBuren and M. Vavrek for numerous valuable discussions, in particular K. Brink and D. Larson for editing and commenting on the manuscript, and C. VanBuren for assistance with siding distal limb elements. Thanks to K. Shepherd for allowing access to the CMN collection and to both M. Currie and C. Kennedy for their help handling the numerous bones belonging to CMN 2289 as well as for their aid in setting up photographs. Special thanks to an anonymous reviewer and D. Eberth for their comments and reviews, which greatly improved the quality of this chapter. Finally, I thank D. Eberth and D. Evans for organizing and inviting me to participate in the International Hadrosaur Symposium 2011.

LITERATURE CITED

Anderson, J. F., A. Hall-Martin, and D. A. Russell. 1985. Long-bone circumference and weight in mammals, birds and dinosaurs. Journal of the Zoological Society of London A 207:53–61.

Bell, P. R., and E. Snively. 2008. Polar dinosaurs on parade: a review of dinosaur migration. Alcheringa 32:271–284.

Bonaparte, J. F., M. R. Franchi, J. R. Powell, and E. G. Sepulveda. 1984. La formación Los Alamitos (Campaniano-Maastrichtiano) del sudeste de Río Negro, con descripción de *Kritosaurus australis* n. sp. (Hadrosauridae). Significado paleogeografico de los vertebrado. Revista de la Asociación Geologica de Argentina 39:284–299.

Boyd, C. A., C. M. Brown, R. D. Scheetz, and J. A. Clarke. 2009. Taxonomic revision of the basal neornithischian taxa *Thescelosaurus* and *Bugenasaura*. Journal of Vertebrate Paleontology 29:758–770.

Brett-Surman, M. K. 1979. Phylogeny and palaeobiogeography of hadrosaurian dinosaurs. Nature 277:560–562.

Brett-Surman, M. K., and J. R. Wagner. 2007. Discussion of character analysis of the appendicular anatomy in Campanian and Maastrichtian North American hadrosaurids: variation and ontogeny; pp. 135–169 in K. Carpenter (ed.), Horns and Beaks: Ceratopsian and Ornithopod Dinosaurs. Indiana University Press, Bloomington, Indiana.

Brown, B. 1913. A new trachodont dinosaur, *Hypacrosaurus*, from the Edmonton Cretaceous of Alberta. Bulletin of the American Museum of Natural History 32:395–406.

Brown, C. M., D. C. Evans, N. E. Campione, L. J. O'Brien, and D. A. Eberth. 2012. Evidence for taphonomic size bias in the Dinosaur Park Formation (Campanian, Alberta), a model Mesozoic terrestrial alluvial-paralic system. Palaeogeography, Palaeoclimatology, Palaeoecology 372:108–122.

Campbell, K. E., and L. Marcus. 1992. The relationships of hindlimb bone dimensions to body weight in birds. Natural History Museum of Los Angeles County Science Series 36:395–412.

Campione, N. E., and D. C. Evans. 2011. Cranial growth and variation in edmontosaurs (Dinosauria: Hadrosauridae): implications for latest Cretaceous megaherbivore diversity in North America. PLoS ONE 6(9):e25186.

Campione, N. E., and D. C. Evans. 2012. A universal scaling relationship between body mass and proximal limb bone dimensions in quadrupedal terrestrial tetrapods. BMC Biology 10:60.

Case, J. A., J. E. Martin, D. S. Chaney, M. Reguero, S. A. Marenssi, S. M. Santillana, and M. O. Woodburne. 2000. The first duck-billed dinosaur (Family Hadrosauridae) from Antarctica. Journal of Vertebrate Paleontology 20:612–614.

Chapman, R. E., and M. K. Brett-Surman. 1990. Morphometric observations on hadrosaurid ornithopods; pp. 163–177 in K. Carpenter and P. J. Currie (eds.), Dinosaur Systematics: Approaches and Perspectives. Cambridge University Press, Cambridge, U.K.

Christiansen, P. 2004. Body size in proboscideans, with notes on elephant metabolism. Zoological Journal of the Linnean Society 140:523–549.

Colbert, E. H. 1962. The weights of dinosaurs. American Museum Novitates 2076:1–16.

Cope, E. D. 1869. Synopsis of the extinct Batrachia, Reptilia and Aves of North America. Transactions of the American Philosophical Society 14:1–252.

Cope, E. D. 1883. On the characters of the skull in the Hadrosauridae. Proceedings of the Academy of Natural Sciences of Philadelphia 35:97–107.

Currie, P. J. 1983. Hadrosaur trackways from the Lower Cretaceous of Canada. Acta Palaeontologica Polonica 28:63–73.

Cuthbertson, R. S., and R. B. Holmes. 2010. The first complete description of the holotype of *Brachylophosaurus canadensis* Sternberg, 1953 (Dinosauria: Hadrosauridae) with comments on intraspecific variation. Zoological Journal of the Linnean Society 159:373–397.

Dilkes, D. W. 2001. An ontogenetic perspective on locomotion in the Late Cretaceous dinosaur *Maiasaura peeblesorum* (Ornithischia: Hadrosauridae). Canadian Journal of Earth Sciences 38:1205–1227.

Eberth, D. A. 2004. Revising the Edmonton Group: a framework for assessing biostratigraphy and climate change; pp. 7–10 in H. Allen (ed.), Alberta Palaeontological Society, Eighth Annual Symposium, Abstracts. Mount Royal College, Calgary, Alberta.

Eberth, D. A., and D. R. Braman. 2012. A revised stratigraphy and depositional history for the Horseshoe Canyon Formation (Upper Cretaceous), southern Alberta plains. Canadian Journal of Earth Sciences 49:1053–1086.

Eberth, D. A., and P. J. Currie. 2010. Stratigraphy, sedimentology, and taphonomy of the *Albertosaurus* bonebed (upper Horseshoe Canyon Formation; Maastrichtian), southern Alberta, Canada. Canadian Journal of Earth Sciences 47:1119–1143.

Eberth, D. A., D. C. Evans, D. B. Brinkman, F. Therrien, D. Tanke, and L. S. Russell. 2013. Dinosaur biostratigraphy of the Edmonton Group (Upper Cretaceous), Alberta, Canada: Evidence for climate influence. Canadian Journal of Earth Sciences 50(7):701–726. Available at doi:10.1139/cjes-2012-0185. Accessed January 2013.

Evans, D. C. 2010. Cranial anatomy and systematics of *Hypacrosaurus altispinus*, and a comparative analysis of skull growth in lambeosaurine hadrosaurids (Dinosauria: Ornithischia). Zoological Journal of the Linnean Society 159:398–434.

Fricke, H. C., R. R. Rogers, and T. A. Gates. 2008. Hadrosaurid migration: inferences based on stable isotope comparisons among Late Cretaceous dinosaur localities. Paleobiology 35:270–288.

Galton, P. M. 1970. The posture of hadrosaurian dinosaurs. Journal of Paleontology 44:464–473.

Gibson, D. W. 1977. Upper Cretaceous and Tertiary coal-bearing strata in the Drumheller-Ardley region, Red Deer River Valley, Alberta. Geological Survey of Canada, Paper 76–35:1–41.

Gilmore, C. W. 1924a. On the skull and skeleton of *Hypacrosaurus*, a helmet-crested dinosaur from the Edmonton Cretaceous of Alberta. Canada Department of Mines Bulletin, Geological Series 38:49–64.

Gilmore, C. W. 1924b. A new species of hadrosaurian dinosaur from the Edmonton Formation (Cretaceous) of Alberta. Canada Department of Mines Bulletin, Geological Series 43:13–26.

Godefroit, P., S. Hai, T. Yu, and P. Lauters. 2008. New hadrosaurid dinosaurs from the uppermost Cretaceous of northeastern China. Acta Palaeontologica Polonica 53:47–74.

Godefroit, P., Z.-M. Dong, P. Bultynck, H. Li, and L. Feng. 1998. Sino-Belgian Cooperation Program "Cretaceous dinosaurs and mammals from Inner Mongolia" 1: New Bactrosaurus (Dinosauria: Hadrosauroidea) material from Iren Dabasu (Inner Mongolia, P. R. China). Bulletin de l'Institut royal des Sciences naturelles de Belgique 68, Supplement:3–70.

Henderson, D. M. 1999. Estimating the masses and centers of mass of extinct animals by 3-D mathematical slicing. Paleobiology 25:88–106.

Henderson, D. M. 2014. Duck soup: the floating fates of hadrosaurs and ceratopsians at Dinosaur Provincial Park; chapter 27 in D. A. Eberth and D. C. Evans (eds.), Hadrosaurs: Proceedings of the International Hadrosaur Symposium at the Royal Tyrrell Museum. Indiana University Press, Bloomington, Indiana.

Horner, J. R., and P. J. Currie. 1994. Embryonic and neonatal morphology and ontogeny of a new species of *Hypacrosaurus* (Ornithischia, Lambeosauridae) from Montana and Alberta; pp. 312–336 in K. Carpenter, K. F. Hirsch, and J. R. Horner (eds.), Dinosaur Eggs and Babies. Cambridge University Press, New York.

Horner, J. R., D. B. Weishampel, and C. A. Forster. 2004. Hadrosauridae; pp. 438–463 in D. B. Weishampel, P. Dodson, and H. Osmólska (eds.), The Dinosauria, Second Edition. University of California Press, Berkeley, California.

Hu, C. 1973. A new hadrosaur from the Cretaceous of Chucheng, Shantung. Acta Geologica Sinica 2:179–206.

Lambe, L. M. 1917. A new genus and species of crestless hadrosaur from the Edmonton Formation of Alberta. Ottawa Naturalist 31:65–73.

Lambe, L. M. 1918. On the genus *Trachodon* of Leidy. Ottawa Naturalist 31:135–139.

Lambe, L. M. 1920. The hadrosaur *Edmontosaurus* from the Upper Cretaceous of Alberta. Memoirs of the Canada Department of Mines, Geological Survey 120:1–79.

Lockley, M., and J. L. Wright. 2001. Trackways of large quadrupedal ornithopods from the Cretaceous: a review; pp. 428–442 in D. H. Tanke and K. Carpenter (eds.), Mesozoic Vertebrate Life. Indiana University Press, Bloomington, Indiana.

Lull, R. S., and N. E. Wright. 1942. Hadrosaurian Dinosaurs of North America. Geological Society of America Special Papers 40. 242 pp.

Maidment, S. C. R., and P. M. Barrett. 2011. The locomotor musculature of basal ornithischian dinosaurs. Journal of Vertebrate Paleontology 31:1265–1291.

Maidment, S. C. R., D. H. Linton, P. Upchurch, and P. M. Barrett. 2012. Limb-bone scaling indicates diverse stance and gait in quadrupedal ornithischian dinosaurs. PLoS ONE 7(5):e36904.

Maidment, S. C. R., K. T. Bates, and P. M. Barrett. 2014. Three-dimensional computational modeling of pelvic locomotor muscle moment arms in *Edmontosaurus* (Dinosauria, Hadrosauridae) and comparisons with other archosaurs; chapter 25 in D. A. Eberth and D. C. Evans (eds.), Hadrosaurs. Indiana University Press, Bloomington, Indiana.

Marsh, O. C. 1881. Classification of the Dinosauria. American Journal of Science 7:81–86.

Marsh, O. C. 1892. Notice of new reptiles from the Laramie Formation. American Journal of Science, Series 3, 43:449–453.

Maryańska, T., and H. Osmólska. 1984. Postcranial anatomy of *Saurolophus angustirostris* with comments on other hadrosaurs. Palaeontologia Polonica 46:119–141.

Norman, D. B. 1980. On the ornithischian dinosaur *Iguanodon bernissartensis* from the Lower Cretaceous of Bernissart (Belgium). Bulletin de l'Institut royal des Sciences naturelles de Belgique 178:1–103.

Norman, D. B. 1986. On the anatomy of *Iguanodon atherfieldensis* (Ornithischia: Ornithopoda). Bulletin de l'Institut royal des Sciences naturelles de Belgique 56:281–372.

Ostrom, J. H. 1964. A reconsideration of the paleoecology of hadrosaurian dinosaurs. American Journal of Science 262:975–997.

Parks, W. A. 1920. The osteology of the trachodont dinosaur *Kritosaurus incurvimanus*. University of Toronto Studies, Geological Series 11:1–76.

Parks, W. A. 1922. *Parasaurolophus walkeri,* a new genus and species of crested trachodont dinosaur. University of Toronto Studies, Geological Series 13:1–32.

Paul, G. 1997. Dinosaur models: the good, the bad, and using them to estimate the mass of dinosaurs; pp. 39–45 in D. L. Wolberg, E. Stump, and G. D. Rosenberg (eds.), Dinofest International: Proceedings of a Symposium held at Arizona State University. Academy of Natural Sciences, Philadelphia, Pennsylvania.

Prieto-Márquez, A. 2007. Postcranial osteology of the hadrosaurid dinosaur *Brachylophosaurus canadensis* from the Late Cretaceous of Montana; pp. 91–115 in K. Carpenter (ed.), Horns and Beaks: Ceratopsian and Ornithopod Dinosaurs. Indiana University Press, Bloomington, Indiana.

Prieto-Márquez, A. 2010a. Global historical biogeography of hadrosaurid dinosaurs. Zoological Journal of the Linnean Society 159:503–525.

Prieto-Márquez, A. 2010b. Global phylogeny of Hadrosauridae (Dinosauria: Ornithopoda) using parsimony and Bayesian methods. Zoological Journal of the Linnean Society 159:435–502.

Prieto-Márquez, A. 2011. A reappraisal of *Barsboldia sicinskii* (Dinosauria: Hadrosauridae) from the Late Cretaceous of Mongolia. Journal of Paleontology 85:468–477.

Seebacher, F. 2001. A new method to calculate allometric length-mass relationships of dinosaurs. Journal of Vertebrate Paleontology 21:51–60.

Sellers, W. I., P. L. Manning, T. Lyson, K. Stevens, and L. Margetts. 2009. Virtual palaeontology: gait reconstruction of extinct vertebrates using high performance computing. Palaeontologia Electronica 12.3.13A:1–26.

Sereno, P. C. 1986. Phylogeny of the bird-hipped dinosaurs (Order Ornithischia). National Geographic Research 2:234–256.

Suzuki, D., D. B. Weishampel, and N. Minoura. 2004. *Nipponosaurus sachalinensis* (Dinosauria: Ornithopoda): anatomy and systematic position within Hadrosauridae. Journal of Vertebrate Paleontology 24:145–164.

Weishampel, D. B., P. M. Barrett, R. A. Coria, J. Le Loeuff, X. Xu, X. Zhao, A. Sahni, E. M. P. Gomani, and C. R. Noto. 2004. Dinosaur distribution; pp. 517–608 in D. B. Weishampel, P. Dodson, and H. Osmólska (eds.), The Dinosauria, Second Edition. University of California Press, Berkeley, California.

Young, C. C. 1958. The dinosaurian remains of Laiyang, Shantung. Palaeontologia Sinica, New Series C 16:1–138.

Zheng, R., A. A. Farke, and G.-S. Kim. 2011. A photographic atlas of the pes from a hadrosaurine hadrosaurid dinosaur. PalArch's Journal of Vertebrate Paleontology 8:1–12.

Appendix 13.1. Measurements of the Postcranial Skeleton of CMN 2289

Table A1a. Cervical Vertebrae Measurements in Millimeters

	TH	PzL	PdL	PPL	CL	IntPzL	SpH	CH1	CH2	CW1	CW2	ScH	ScW
Cervical 1	149.8	137.7	63.4	162.2	119	50.5	21	67.3	87.7		94.7	37.1	30.8
Cervical 2	174.3	153.2	54.4	137.2	83.9	46.7	20.9	75.7	87.1	129	89.2	47	56.3
Cervical 3	216.6	169.7	46.5	132.4	90	89.5	60	76.8	86.3	127.8	122.6	43.6	58.9

Abbreviations: **CH1**, centrum height along the cranial aspect; **CH2**, centrum height along caudal aspect; **CL**, centrum length; **CW1**, centrum width along cranial aspect; **CW2**, centrum width along caudal aspect; **IntPzL**, width between postzygapophyses; **PdL**, pedicle craniocaudal length; **PPL**, pre and postzygapophyseal length; **PzL**, post-zygapophysis length (from neural spine); **ScH**, spinal canal height at cranial opening; **ScW**, spinal canal width at cranial opening; **SpH**, spine height; **TH**, total height.

Table A1b. Dorsal Vertebrae Measurements in Millimeters

	TH	PdL	PPL	CL	SW1	SW2	SH1	SH2	CH1	CH2	CW1	CW2	ScH	ScW
Dorsal 1	365	84.4	185.2	139.3	107	93	249	202.3	92.3	118	67	96	32.2	33
Dorsal 2	375	103	219.8	151.3	120.5		266.8	213.5	106	117.3	68.7	92.7	37	31.2
Dorsal 3	380	83.5	161.2	117.7	93	87	258	217	94	127.5	114	125	47.4	44.3
Dorsal 4	420	75.5	164	110.5	91		269.8	231	99.5	134.6	114.7	126.2	44.5	40.2
Dorsal 5	460	79	180.8	113	99.2	102.8	279	251	139.6	162	116	128.4	43.2	33.3
Dorsal 6	465	76.5	175.3	113.3	93.5	103	298	252.6	121	155.5	128	128	38.6	33.7
Dorsal 7	440	60	160.1	101.5	90	94	248.6	236	149.7	159		153.4	34.9	30
Dorsal 8	460	66.8	160	96.3	90.2	93.4	266.2	241.5	145	170.5	118.3	138.7	40.7	39.4
Dorsal 9	470	57.8	133	93.1	84.3	67.4	270.7	246.6	157.7	172.4	134	154	47.3	34.5
Dorsal 10	475	51.5	128	95	73.7	65	262	254	162.5	178	153.2	165	40	42
Dorsal 11		64.7	131.3	92.5	81				159.3	181	148.5	187.8	32.3	41.9
Dorsal 12	445	64.7	154	65.3	86.5	86.4	235.3	222.2	165.7	184.7	183.7	196.1	36.6	51.5
Dorsal 13	505	57.5	123.5	67.4	82.5	72.6	274.4	265.4	194.6	189	178.5	200.2	34.3	53.8
Dorsal 14	512	58.9	138.5	69	72	79.9	275	256.5	192.5	209.7	171.3	183.9	40.8	49.1
Dorsal 15		61.9	137.1	57.6	70.3				182.5	197	201.7	181.3	36.4	52.1

Abbreviations: **CH1**, centrum height along the cranial aspect; **CH2**, centrum height along caudal aspect; **CL**, centrum length; **CW1**, centrum width along cranial aspect; **CW2**, centrum width along caudal aspect; **PdL**, pedicle craniocaudal length; **PPL**, pre and postzygapophyseal length; **ScH**, spinal canal height; **ScW**, spinal canal width; **SH1**, spine height (from spinal canal); **SH2**, spine height (from the postzygapophyses); **SW1**, spine craniocaudal width (at base); **SW2**, spine craniocaudal width (distally); **TH**, total height.

Table A1c. Caudal Vertebrae Measurements in Millimeters

	TH	PdL	PPL	CL	SpW1	SpW2	SpH1	SpH2	CH1	CH2	CW1	CW2
Caudal 1	540	65.9	121.3	77.4	51.8	72.3	345	332	158.7	186.4	185.9	200
Caudal 2	485	64	122.2	81.5	49.3	66		310	161	165	174.3	168
Caudal 3	465	58	121.7	73.2	46.1	60	340	315	131.4	148.2	178.3	166
Caudal 4	467	54.2	123.5	82.1	45.7	63	310	287.1	151.4	162		
Caudal 5	285	49.8	108	81.7	33.6	43.5	205	179.7	102	103.6	98.5	95.7

Abbreviations: **CH1**, centrum height along the cranial aspect; **CH2**, centrum height along caudal aspect; **CL**, centrum length; **CW1**, centrum width along cranial aspect; **CW2**, centrum width along caudal aspect; **PdL**, pedicle craniocaudal length; **PPL**, pre and postzygapophyseal length; **SpH1**, spine height (from spinal canal); **SpH2**, spine height (from the postzygapophyses); **SpW1**, spine craniocaudal width (at base); **SpW2**, spine craniocaudal width (distally); **TH**, total height.

Table A1d. Appendicular Skeleton Measurements in Millimeters

Measurement	Right	Left
Scapula		
Length	1070	1045
Width at constriction	154.4	150.6
Width at distal end		222.3
Width at coracoid	129	138.1
Glenoid length	136	132.4
Coracoid		
Glenoid length	75.5	
Length at scapula	102.7	
Width of biceps tubercle	49	
Length of biceps tubercle	30	
Sternal plate		
Distal head width	77.7	
Minimum shaft width	51.8	
Cranial blade length	187	
Cranial blade maximum width	150	
Humerus		
Length	690	
Minimum shaft circumference	275	
Deltopectoral crest length	382	
Deltopectoral crest width	189.4	
Proximal head width	165.5	
Mediolateral width of condyle	56	
Distal head width	118	
Minimum shaft width (craniocaudal)	96	
Ulna		
Length	776	775
Prearticular length	720	720
Minimum circumference	220	186
Minimum shaft width (mediolateral)		67
Minimum shaft width (craniocaudal)		41.2
Mid-shaft width (mediolateral)		84.6
Mid-shaft width (craniocaudal)		53.3
Radius		
Length		673
Mid-shaft width (mediolateral)		57.3
Mid-shaft width (craniocaudal)		43.3
Proximal head width (mediolateral)		99.3
Proximal head width (craniocaudal)		81.7
Distal head width (mediolateral)		85.5
Distal head width (craniocaudal)		59.2
Minimum circumference		167
Metacarpal II		
Length	257.5	
Minimum width (mediolateral)	33.6	
Minimum width (craniocaudal)	20	
Proximal head width	57.7	
Distal head width	46	

Measurement	Right	Left
Ilium		
Ilium length	1260	
Preacetabular length	585	
Postacetabular length	540	
Width at supra-acetabular process	281.5	
Length between peduncles	286.7	
Width above acetabulum	244	
Maximum width of preacetabular process	101.5	
Length of supra-acetabular process	141.1	
Pubis		
Prepubic process length	645	625
Width at constriction	119	120.8
Width at expansion	272.5	282.8
Ischium		
Total length		1255
Postacetabular length		1190
Pubic peduncle length	67	75
Pubic peduncle height	87.3	113.6
Obturator foramen length	32.3	37
Distal width		74.2
Minimum shaft width	45	40.1
Iliac peduncle width	86.2	111.4
Femur		
Length		1280
Minimum circumference		532
Minimum shaft width (mediolateral)		147
Minimum shaft width (craniocaudal)		146.5
Medial condyle length		340
Lateral condyle length		355
Fourth trochanter to distal head length		415
Fourth trochanter to proximal head length		675
Tibia		
Length		1174
Minimum circumference		405
Proximal head width		245
Distal head width		343
Fibula		
Length	1070	
Minimum circumference	193	
Proximal head width	173.1	199
Distal head width	128.6	
Pes		
Metatarsal II length	345	
Metatarsal II minimum width	101	
Metatarsal II proximal width	232	
Metatarsal III minimum width	96.6	
Metatarsal III distal width	140.7	
Metatarsal IV length	345	

Cranial Morphology and Variation in *Hypacrosaurus stebingeri* (Ornithischia: Hadrosauridae)

14

Kirstin S. Brink, Darla K. Zelenitsky, David C. Evans, John R. Horner, and François Therrien

ABSTRACT

Hypacrosaurus stebingeri is known from one of the most complete ontogenetic series of dinosaurs, yet the adult skeletal material has never been fully described. Here, the first detailed description of all well-known *H. stebingeri* crania is presented, accompanied by the first multivariate statistical analysis for lambeosaurine crania that includes *H. stebingeri*. Based on our anatomical review of the taxon, we find a new autapomorphy: the elongate lingulate shape of the caudolateral process of the premaxilla that comprises the majority of the cranial crest in lateral view. The statistical analysis indicates that the characters differentiating species of lambeosaurines are the crest-snout angle and the position of the apex of the crest. The shape and size of the cranial crest of *H. stebingeri* is most similar to the crest of *Hypacrosaurus altispinus* and *Corythosaurus* in morphospace, a pattern that supports recent phylogenetic and ontogenetic hypotheses of lambeosaurine dinosaurs.

INTRODUCTION

Hypacrosaurus stebingeri is represented by numerous referred embryonic, juvenile, subadult, and adult specimens, collected from the Upper Cretaceous (Campanian) upper Two Medicine Formation of Montana and the lithologically and stratigraphically equivalent Oldman Formation of Alberta (Horner and Currie, 1994). In the original description of *H. stebingeri*, emphasis was placed on morphological variation in the abundant embryonic and nestling cranial and postcranial elements (Horner and Currie, 1994). Since the original description, a subadult skull of *H. stebingeri* was described and compared to skulls of other lambeosaurine species at a similar growth stage (Brink et al., 2011), but the largest known (presumably adult) material has yet to be described in detail.

According to Horner and Currie (1994) *H. stebingeri* lacks autapomorphies and is defined by a unique suite of characters also present in *Corythosaurus* (Dinosaur Park Formation, Alberta), *Hypacrosaurus altispinus* (Horseshoe Canyon Formation, Alberta; St. Mary River Formation, Alberta and Montana), and *Lambeosaurus* spp. (Dinosaur Park Formation, Alberta). They placed *H. stebingeri* in the genus *Hypacrosaurus* because of morphological similarities with *Lambeosaurus*, including (1) a restricted external naris, (2) a wide narial crest, and (3) restricted dorsal centra with very tall neural spines. Therefore, *H. stebingeri* was proposed as a candidate metaspecies between *Lambeosaurus* and *H. altispinus* in an anagenetic model of evolution within this lineage, as were other putatively transitional dinosaur species in other clades from the upper Two Medicine Formation (Horner et al., 1992). Subsequent studies of lambeosaurine interrelationships conflict on whether *H. stebingeri* is a metaspecies or whether it can be identified as a distinct evolutionary entity (Suzuki et al., 2004; Prieto-Márquez et al., 2006; Evans and Reisz, 2007; Gates et al., 2007; Evans, 2010; Prieto-Márquez, 2010; Prieto-Márquez et al., 2012). Several analyses (Evans and Reisz, 2007; Gates et al., 2007; Evans, 2010) of the lambeosaurine in-group relationships recover a monophyletic *Lambeosaurus* and a monophyletic *Hypacrosaurus*, although *Hypacrosaurus* was unresolved with respect to *Corythosaurus* and *Olorotitan arharensis*. These analyses fail to reject *H. stebingeri* as a transitional metaspecies ancestral to *Hypacrosaurus altispinus*. Other analyses (Suzuki et al., 2004; Prieto-Márquez et al., 2006; Prieto-Márquez, 2010; Prieto-Márquez et al., 2012) result in *Hypacrosaurus* being paraphyletic, with *H. stebingeri* more closely related to *Corythosaurus* and *Lambeosaurus*, and *H. altispinus* more closely related to *Velafrons coahuilensis* or *O. arharensis*.

The abundance of available lambeosaurine material at multiple growth stages has allowed for comparative analyses of growth and allometry in lambeosaurine dinosaurs (Dodson, 1975; Evans, 2010). These analyses suggest that patterns of facial growth do not vary greatly within lambeosaurin lambeosaurines, whereas patterns of crest growth are highly allometric and are species specific. The lambeosaurine crest is an important morphologic complex for species identification and character data critical to resolving the evolutionary relationships within this clade. Here, we provide the first detailed description of the anatomy and variation in morphologically mature specimens of *H. stebingeri*, based on three skulls including the complete skull of the holotype. In particular, description of the phylogenetically informative crest results in the identification of another possible autapomorphy in *H. stebingeri*. A multivariate morphometric analysis, the first to include *H. stebingeri*, documents the range of crest shape variation in this taxon and other lambeosaurine dinosaurs, and affirms its overall phenetic similarity to *H. altispinus* and *Corythosaurus*.

Institutional Abbreviations AMNH, American Museum of Natural History, New York; CMN, Canadian Museum of Nature, Ottawa, Ontario; FMNH, Field Museum of Natural History, Chicago, Illinois; MOR, Museum of the Rockies, Bozeman, Montana; NSM-PV, National Museum of Nature and Science, Tokyo, Japan; ROM, Royal Ontario Museum, Toronto, Ontario; TCMI, The Children's Museum of Indianapolis, Indianapolis, Indiana; TMP, Royal Tyrrell Museum of Palaeontology, Drumheller, Alberta; UALVP, University of Alberta Laboratory for Vertebrate Paleontology, Edmonton, Alberta; USNM, Department of Paleobiology, U.S. National Museum of Natural History, Smithsonian Institution, Washington, D.C.; YPM, Peabody Museum of Natural History, New Haven, Connecticut.

Anatomical Abbreviations **a**, angular; **bo**, basioccipital; **bpt**, basipterygoid process; **bsp**, basisphenoid-parasphenoid complex; **cp**, cultriform process; **d**, dentary; **ec**, ectopterygoid; **ex**, exoccipital-opisthotic complex; **f**, frontal; **fm**, foramen magnum; **fo**, fenestra ovalis; **j**, jugal; **l**, lacrimal; **lsp**, laterosphenoid; **m**, maxilla; **n**, nasal; **opi**, opisthotic; **p**, parietal; **pl**, palatine; **pmd**, caudodorsal process of premaxilla; **pml**, caudolateral process of premaxilla; **po**, postorbital; **pp**, paroccipital process; **pr**, prootic; **pre**, presphenoid; **prf**, prefrontal; **pt**, pterygoid; **q**, quadrate; **qj**, quadratojugal; **sa**, surangular; **so**, supraoccipital; **sp**, splenial; **sq**, squamosal.

MATERIALS AND METHODS

Three articulated skulls of *Hypacrosaurus stebingeri* form the basis of this description. MOR 549, the holotype, is nearly complete, lacking only the ceratobranchials and vomer (Figs. 14.1, 14.2). The right dentary, postdentary bones, predentary, and maxilla are present but not in articulation. The skull is crushed laterally, so the width of all elements appears slightly smaller than natural. This skull was collected with a complete postcranial skeleton near Badger Creek, Glacier County, Two Medicine Formation (TMF) of Montana, about 80 m below the contact between the TMF and the Bearpaw Formation (Horner and Currie, 1994).

MOR 455 (Fig. 14.3) consists of an articulated crest and braincase, snout, disarticulated right jugal and right quadrate. It is the largest of the three adult specimens and is not crushed. The dorsal surface of the crest is eroded and much of the bone surface is not preserved. Matrix still covers a majority of the specimen. The exact provenance of this specimen is unknown, but it is thought to be from the Landslide Butte area (JRH, pers. obs., 2009).

MOR 553s (Figs. 14.4–14.6) is an articulated braincase lacking the nasal, and is very well preserved. This braincase was collected from a bonebed amongst many other disarticulated yet well-preserved cranial elements of *H. stebingeri* and *Prosaurolophus* (Varricchio, 1995).

All material of *H. stebingeri* was analyzed, measured, and photographed for descriptive and comparative purposes. Comparative materials of *Corythosaurus*, *Hypacrosaurus*, and *Lambeosaurus* were examined at other museums, and some skulls were used for morphometric analysis.

In order to augment qualitative comparisons, we performed a multivariate analysis of skull shape in lambeosaurine hadrosaurs using Principal Component Analysis (PCA). Because adult specimens with fully developed cranial crests are the most useful for species identification (Dodson, 1975; Evans, 2010), only adult skulls were used here (skull length greater than 85% of the maximum observed skull length in each taxon; Evans, 2010). Results of this analysis are compared to previous bivariate and multivariate analyses (Dodson, 1975; Evans, 2010) to test whether species can be differentiated based on linear measurements of the cranium in multidimensional space. The dataset used in this analysis (Appendix 14.1) is a compilation of measurements from Dodson (1975), Evans (2010), and newly collected data, which includes 39 adult skulls from 6 species of lambeosaurine dinosaurs (2 *H. stebingeri*, 4 *H. altispinus*, 15 *Corythosaurus* [7 *C. intermedius*, 8 *C. casuarius*] and 17 *Lambeosaurus* [2 *L. magnicristatus*, 10 *L. lambei*, 5 *L. clavinitalis*]). Fifteen cranial variables were measured (Evans, 2010). Six variables (1–6) correspond to the snout and temporal region of the skull, and nine (7–15) correspond to the cranial crest. The metrics for these measurements are illustrated in the supplemental material of Evans (2010). Variable 15, crest area, was measured from photographs using Image J software (Rasband, 2008).

As the dataset contains measurements from skulls that were imperfectly preserved, some data are missing (19%; Appendix 14.1). Frequently, specimens missing important measurements are excluded from analyses, but recent testing of missing data estimation methods has demonstrated that the use of Bayesian Principal Components Analysis (BPCA; Oba et al., 2003) is the most robust method for use in morphometric multivariate analyses (Campione and Evans, 2011; Brown et al., 2012). This method uses an expected maximization approach combined with a Bayesian estimation method to calculate the likelihood of the value of an estimated data point (Oba et al., 2003; Stacklies et al., 2007). In order to preserve a large dataset, missing data are estimated here using

14.1. *Hypacrosaurus stebingeri,* MOR 549, left lateral view.

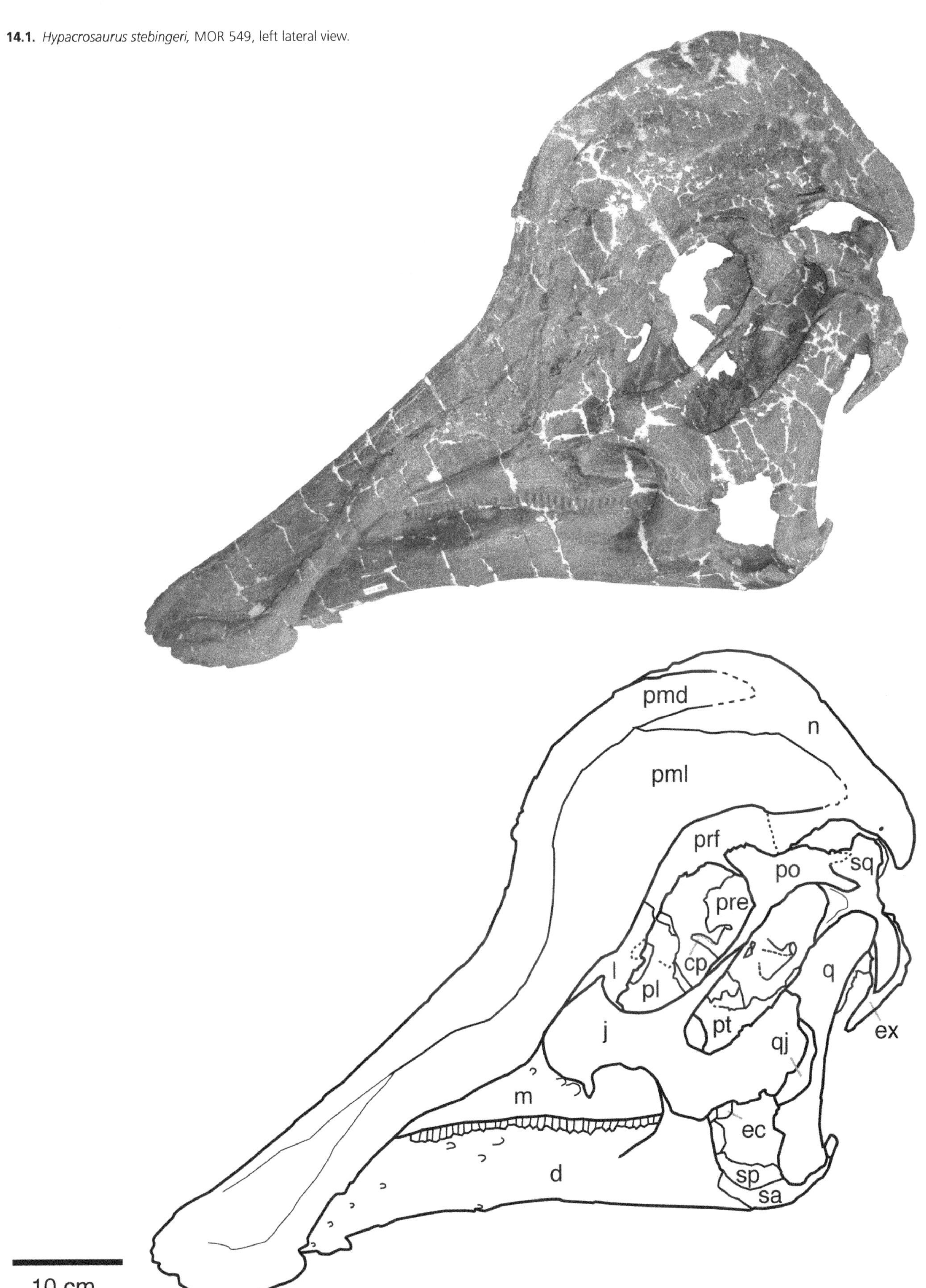

BPCA (Appendix 14.1). All measured and estimated measurements were log transformed, and the PCA was performed using a correlation matrix to account for any high variance in the dataset. An ANOVA test and Tukey-Kramer Honestly Significant Difference (HSD) were performed to analyze significant differences among the results of the PCA. Analyses were completed using the programs R (www.R-project.org); including the packages vegan (Oksanen et al., 2010) and pcaMethods (Stacklies et al., 2007); and JMP (Version 9. SAS Institute Inc., Cary, NC, 1989–2007).

SYSTEMATIC PALEONTOLOGY

DINOSAURIA Owen, 1842
ORNITHISCHIA Seeley, 1888
HADROSAURIDAE Cope, 1869
LAMBEOSAURINAE Parks, 1923
HYPACROSAURUS Brown, 1913

Type Species *Hypacrosaurus altispinus* (Brown, 1913).

Amended Diagnosis (after Evans, 2010) Lambeosaurine hadrosaurid that has the following autapomorphies: (1) elongate neural spines in the cranial region of the dorsal vertebral column (dorsal vertebrae 6–10) that are greater than five times taller than their respective centra, and (2) complete closure of the premaxilla-nasal fontanelle (PNF) in adult individuals. *Hypacrosaurus altispinus* shares with *H. stebingeri* elongation and enlargement of the dorsal vertebra neural spines in the shoulder region and complete closure of the premaxilla-nasal fontanelle in adults (all individuals with skull lengths ≥ 85% of the length of the largest known skull of that species). Complete closure of the PNF may also occur convergently in *Parasaurolophus*.

HYPACROSAURUS STEBINGERI Horner and Currie, 1994

Holotype MOR 549 (Figs. 14.1, 14.2): Complete adult skull and post-cranial skeleton.

Locality and Horizon Badger Creek, Glacier County, Montana, Two Medicine Formation, about 80 m below the contact between the Two Medicine Formation and the Bearpaw Formation (Horner and Currie, 1994)

Revised Diagnosis Lambeosaurine hadrosaurid dinosaur characterized by a cranial crest comprised predominantly of the caudolateral process of the premaxilla in lateral view that is elongate and lingulate in shape. The caudolateral process contribution to the crest is longer than tall, with a slightly convex dorsal margin. It differs from *Corythosaurus* in that the caudolateral process of premaxilla is not divided into distinct caudal and rostral lobes. It differs from *Lambeosaurus* in the following features: sloping facial profile with apex of crest above orbit; caudal margin of crest formed predominantly by nasal; nasal with rostral bifurcation; caudodorsal process of premaxilla without accessory rostroventral flange overlapping the lateral surface of nasal in rostral region of crest; incomplete enclosure of ophthalmic canal. It differs from *H. altispinus* in the following features: external nares elongate; postorbital bifurcate; jugal with rounded ventral flange; nasal bifurcation is more pronounced in juveniles; caudolateral process of premaxilla lingulate, forming majority of cranial crest in lateral view in adult specimens.

Referred Specimens NSM-PV 20377-20378, ROM 53592-94, ROM 61784-86, TCMI 2001.96.01-03, TMP 1994.385.0001, TMP 2007.10.0001-0003, and all specimens listed in Horner and Currie (1994). All of the referred specimens were collected from Glacier County, in the upper Two Medicine Formation of Montana. Detailed locality data can be found on file at their respective institutions.

DESCRIPTION

Hypacrosaurus stebingeri is similar in overall skull shape to previously described lambeosaurine hadrosaurids. Therefore, only significant and informative characters will be described here.

Maxillary Complex

Premaxilla The premaxillae are complete in MOR 549, and mostly complete in MOR 455 (Figs. 14.1–14.3). The paired premaxillae are the largest bones of the skull, forming the rostral portion of the cranial crest, the facial profile, and the oral margin. The premaxillae of MOR 455 are disarticulated along the unfused interpremaxillary suture, exposing a groove present on the medial surface of each element, which Horner and Currie (1994) suggest carried vessels and nerves to the rostral end of the snout. The external nares of *H. stebingeri* are similar to *Lambeosaurus* and *Corythosaurus*, being elongate and open along the rostral end of the snout, whereas the naris of *H. altispinus* is small and lacriform (ROM 702, CMN 8501).

The premaxilla is divided into a caudodorsal and caudolateral process, the latter of which composes the majority of the cranial crest in lateral view (Figs. 14.1–14.3). The caudolateral process displays the most variation between crests of *H. stebingeri*, and also in the crests of other lambeosaurines. The caudolateral process is divided into caudal and rostral lobes, which are most distinct in *Corythosaurus* and *Lambeosaurus* (Horner et al., 2004), and are more difficult to identify in *Hypacrosaurus* (Fig. 14.7). In *H. altispinus*, the caudolateral process ranges from lancet shaped in some specimens to

14.2. *Hypacrosaurus stebingeri,* MOR 549, right lateral view.

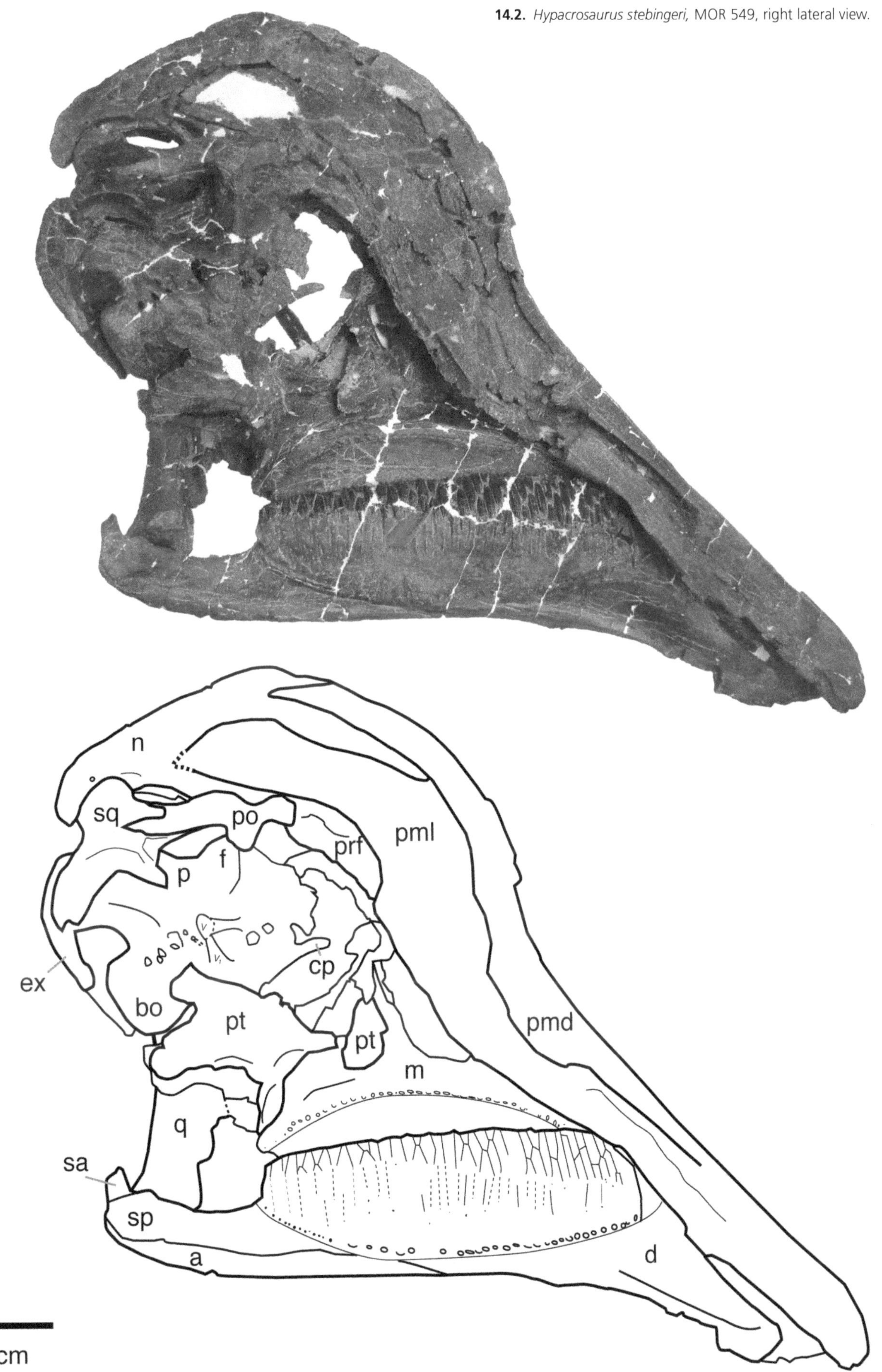

14.3. *Hypacrosaurus stebingeri,* MOR 455, right lateral view.

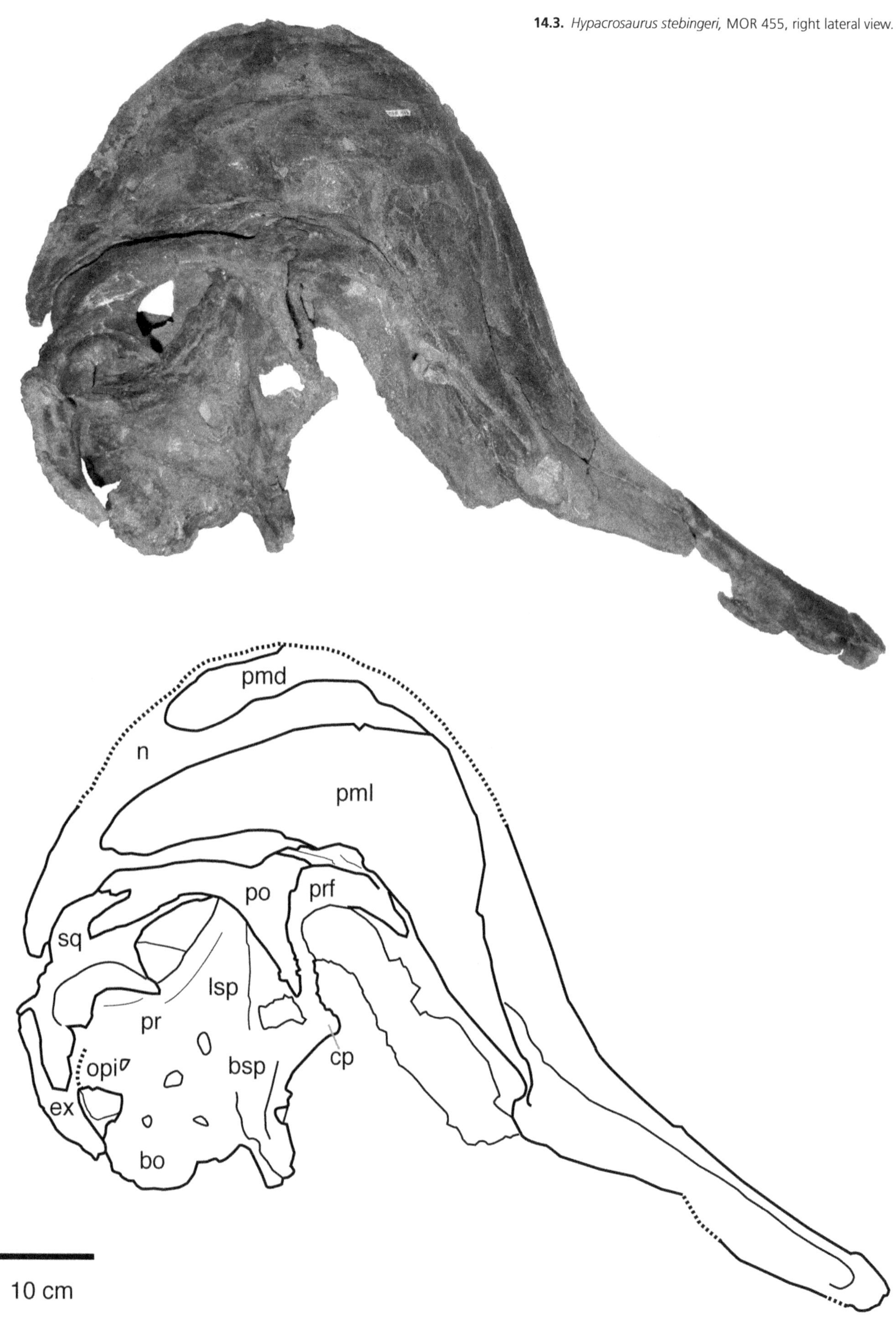

somewhat circular (Evans, 2010), and approaches the shape of the process as observed in *Lambeosaurus*, where the rostral lobe is dorsoventrally expanded. In *H. stebingeri*, the caudolateral process is elongate and lancet shaped in both crested specimens, similar to the caudal lobe of the process in *C. intermedius* (e.g., ROM 845) and *C. casuarius* (e.g., ROM 841; Fig. 14.7). The process of the crest of MOR 455 is expanded rostrally and approaches the shape of the process in *Lambeosaurus*, being inclined slightly caudodorsally (Fig. 14.3). The process of the holotype has a straight to slightly convex dorsal margin where it contacts the nasal, and extends caudally beyond the apex of the crest (Figs. 14.1, 14.2).

Despite intraspecific variation, the shape and proportion of the contribution of the caudolateral process to the crest is autapomorphic for *H. stebingeri*. The contribution of the caudolateral process of the premaxilla and the nasal to the lateral surface area of the crest is species specific. The contribution of the caudolateral process of the premaxilla to lateral crest area is ~5% of total crest area for *H. stebingeri*, but only ~12% of total crest area for *C. casuarius* and *C. intermedius*, ~26% for *H. altispinus* , and ~31% for *L. lambei* (Fig. 14.7).

The caudodorsal process is bifurcated, the lateral lobe visible on the lateral crest surface, and the medial lobe extending along the caudal margin of the crest, separating the nasals dorsally. The extension of the caudodorsal process is also seen in *Lambeosaurus* (e.g., CMN 2869, ROM 1218), in which it evolved convergently (Evans and Reisz, 2007). The caudodorsal process of the premaxilla does not have an accessory rostroventral flange that overlaps the lateral surface of the nasal in the rostral region of the crest, also an autapomorphy of *H. altispinus* (Evans, 2007). Similar to *Corythosaurus* (e.g., ROM 870, 871), the nasal overlaps the caudodorsal process of the premaxilla, forming a complex W-shaped suture. The overlap is not as extensive as that seen in *H. altispinus* (e.g., CMN 8501)

The PNF (Maryańska and Osmólska, 1979) is closed in both adult specimens. This is similar to all known skulls of *H. altispinus* and adult skulls of *Lambeosaurus magnicristatus*, but differs from *L. lambei* and *Corythosaurus*, where the PNF remains open in the largest skulls. The closure of the PNF is an ontogenetic event, as it remains open in subadult *H. stebingeri* skulls (Brink et al., 2011).

Nasal The nasal forms the caudodorsal margin of the crest, united along the caudal and caudoventral margin only, about 2 cm from the caudal tip of the crest (Figs. 14.1–14.3). The nasals are articulated caudally with the frontals, caudolaterally with the prefrontals, and rostrally with the premaxillae. The paired nasals are separated cranially along the midline by the caudodorsal processes of the premaxillae, and are united caudally at the articulation with the frontals. The caudoventral margin of the nasals in adult specimens of *H. stebingeri* curves to follow the shape of the squamosals (Figs. 14.1–14.3), similarly to some *Corythosaurus casuarius* (AMNH 5240; Brown, 1914) and *Lambeosaurus magnicristatus* skulls (e.g., CMN 8705), but differing from the straighter ventral margin present in *H. altispinus* adults (e.g., CMN 8501, ROM 702). A small lateral foramen is situated on the caudolateral margin of the nasals.

Rostrally, the nasal is divided into a lateral and medial process, the lateral process is covered by the caudolateral process of the premaxilla, and the medial process overlaps the caudodorsal process of the premaxilla and forms the dorsal margin of the PNF in juveniles and subadults. The medial process is bifurcated into rostral and caudal branches forming a W-shaped suture (Figs. 14.1–14.3). This bifurcation is most visible in juvenile and subadult specimens (Brink et al., 2011; Evans, 2010). The bifurcation and extent of the overlap is significantly different from that of *H. altispinus*, which has significantly reduced caudal branch and an elongate rostral branch in adults (e.g., CMN 8501), forming an S-shaped suture rather than a W-shaped suture (Evans, 2010). The nasal's contribution to the caudal margin of the crest in lateral view in adult specimens of *H. stebingeri* is significantly less than in *C. casuarius* (e.g., ROM 871) and *Velafrons* (Gates et al., 2007), in which the majority of the crest is formed by the nasal.

A thin sheath of bone descends ventrally from the medial margin of the dome, separating the narial cavities from each other, and forms the caudomedial margin of the lateral diverticulum (Weishampel, 1981). Caudal to the diverticulum, the caudal end of the nasal is square for articulation with the enlarged anterior shelf of the frontal.

Jugal The paired jugals articulate rostrally with the premaxilla, rostromedially with the lacrimal and maxilla, caudally with the quadratojugal and quadrate, and dorsally with the postorbital via the ascending process of the jugal (Fig. 14.1). The maxillary process is dorsoventrally expanded into a large, symmetrically curved plate. In medial view, the tallest extent of the maxillary process is thickened to articulate with the lateral margin of the dorsal process of the maxilla. The ascending process of the jugal is elongate, and forms the caudal margin of the orbit and the rostral margin of the large infratemporal fenestra. The contact of the caudal blade of the jugal and the quadratojugal forms the ventral margin of the infratemporal fenestra, which tends to form an acute, subrectangular arch similar to that of *H. altispinus* (e.g., TMP 2006.015.0001, TMP 1982.010.0001). The jugal is dorsoventrally constricted beneath the infratemporal fenestra, offsetting the subtemportal blade (Fig. 14.1). The rounded ventral flange of the subtemporal blade is similar in morphology to that of *Corythosaurus* (e.g., ROM 871, CMN 34825) and *Lambeosaurus* (e.g., ROM 888, ROM 758), and differs from the autapomorphic angular flange present in *H. altispinus* (e.g.,

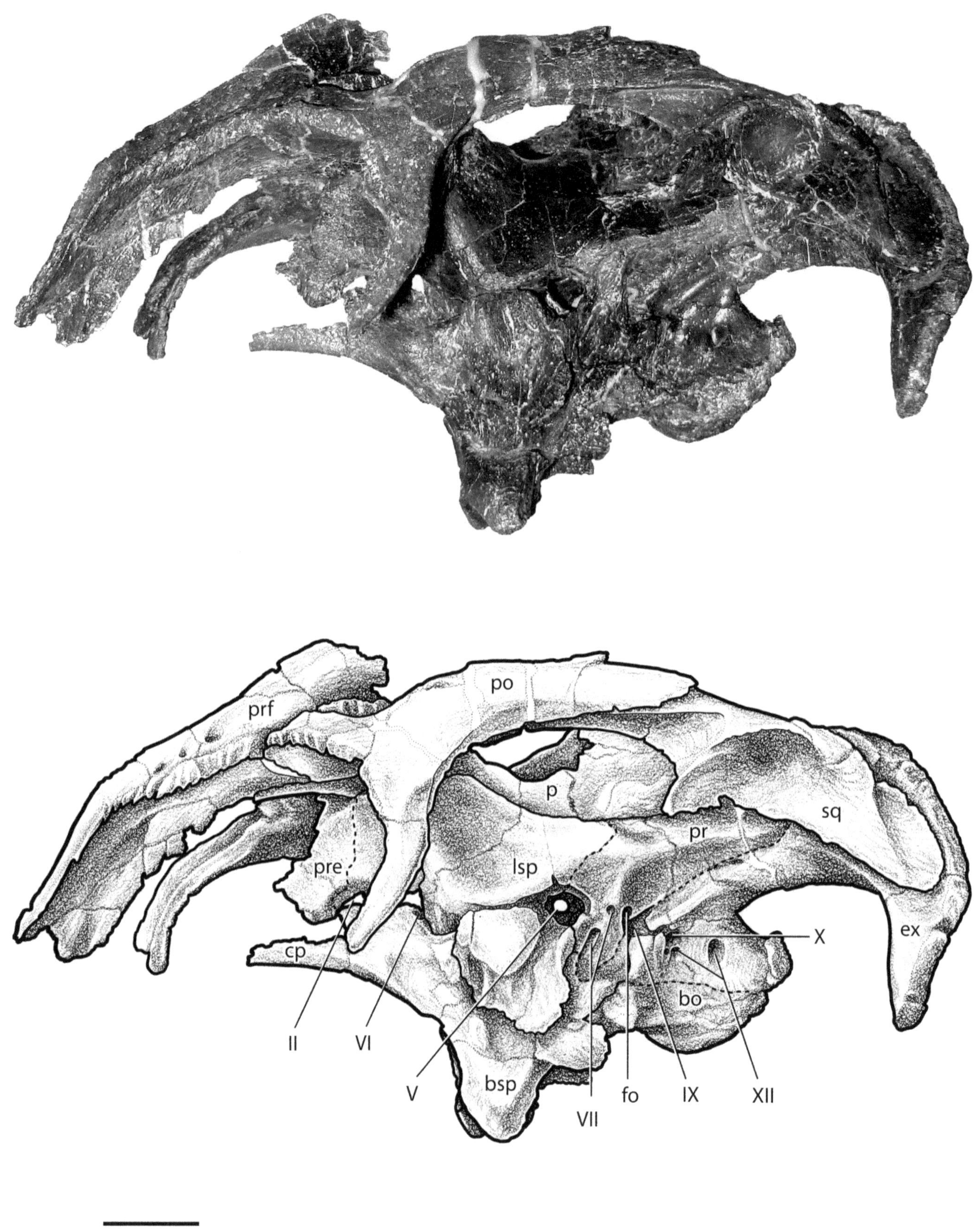

14.4. *Hypacrosaurus stebingeri,* MOR 553s, left lateral view. Roman numerals indicate cranial nerves.

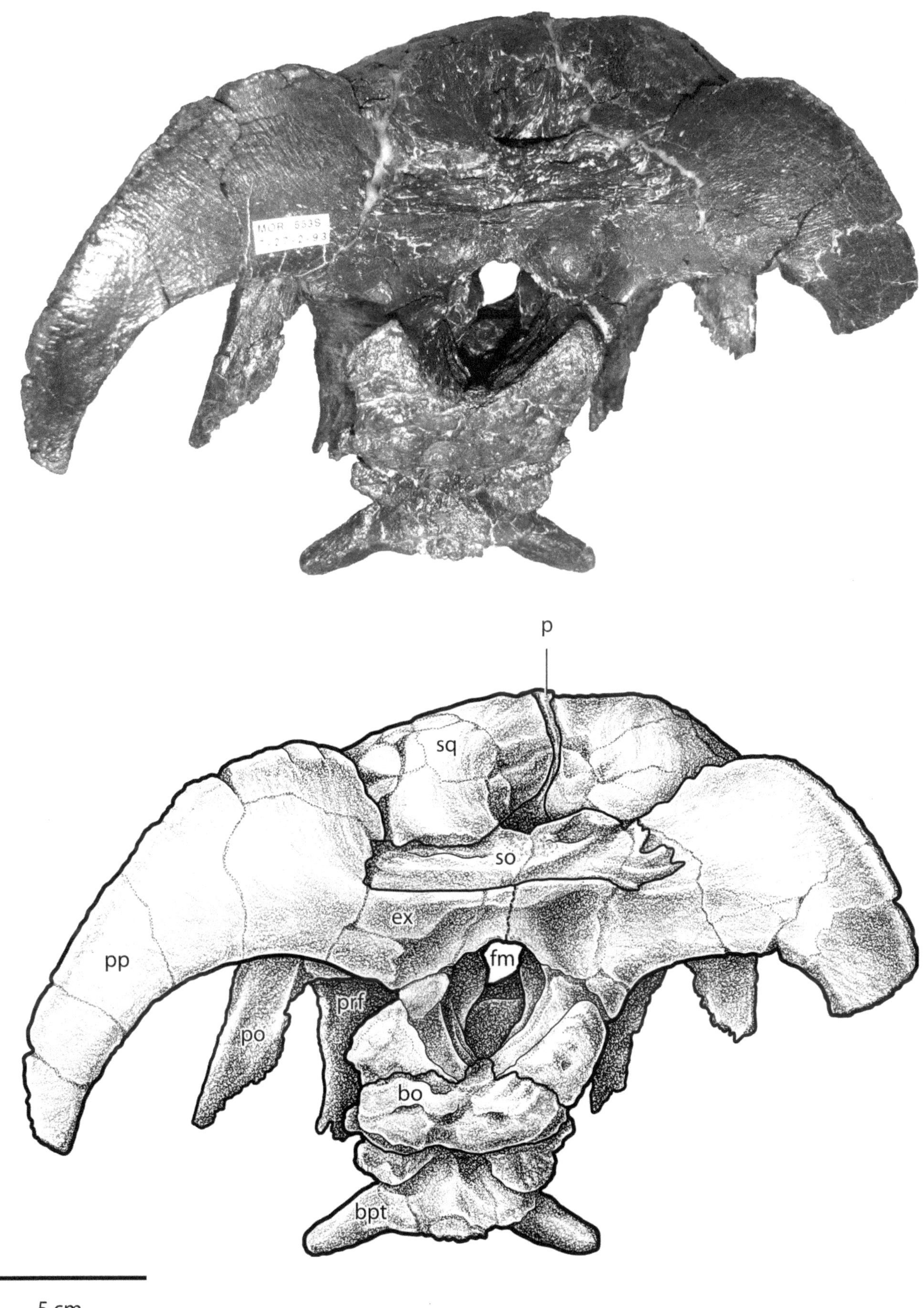

14.5. *Hypacrosaurus stebingeri,* MOR 553s, occipital view.

CMN 8501, TMP 1982.010.0001, TMP 2006.015.0001; Evans, 2010).

Squamosal The squamosal articulates ventrally with the supraoccipital and exoccipital-opisthotic complex, rostromedially with the parietal, rostrodorsally with the postorbital, and medially to its complement (Figs. 14.1–14.6). The ventrolateral cotylus is deep and broad as in other lambeosaurines (Gilmore, 1937), and articulates with the dorsal margin of the quadrate. The squamosal forms a scarf joint with the postorbital rostrally, creating the dorsal margin of the supratemporal fenestra. In caudal view, the squamosal has an arched lateroventral surface that forms a joint with the paraoccipital processes of the exoccipital-opisthotic complex (Fig. 14.5). The squamosals are rounded caudally, and come into contact with each other dorsal to the supraoccipital notch to form a relatively flat caudal margin of the skull. The paired bones meet medially dorsal to the supraoccipital notch, and are separated caudally by an expansion of the parietal sagittal crest. Only one adult (MOR 553s; Figs. 14.5, 14.6) and one subadult (USNM 11893) show this caudal separation of the squamosals, suggesting that the extent of the parietal sagittal crest is a source of intraspecific and ontogenetic variation for *H. stebingeri* (Brink et al., 2011). This separation is present in *Velafrons*, but is absent in *H. altispinus* and *Corythosaurus* (Evans and Reisz, 2007; Gates et al., 2007; Evans, 2010).

Postorbital The paired postorbitals articulate rostrally with the prefrontal, medially with the frontal, parietal, and laterosphenoid, ventrally with the ascending process of the jugal, and caudally in a scarf joint with the squamosal (Figs. 14.1–14.4, 14.6). As in other lambeosaurines (Gilmore, 1937; Ostrom, 1961; Evans and Reisz, 2007), the prefrontal and postorbital join in an interdigitate suture, and form a small convexity on the dorsal margin of the orbit. The caudal process of the postorbital is bifurcated into a long lateral blade and a relatively shorter medial blade (Figs. 14.1–14.4, 14.6). In specimens where this bifurcation is not preserved, the articular surface for the missing blade can be traced on the dorsal surface of the squamosal (MOR 549). This bifurcation is similar to that of *Corythosaurus* (e.g., CMN 34825, ROM 871), *Lambeosaurus* (e.g., ROM 758, CMN 2869), and *Velafrons* (Gates et al., 2007), but differs from that of *H. altispinus* (e.g., TMP 1982.010.0001, CMN 8501), which is unbranched (Evans, 2010).

Neurocranial Complex

Frontal The frontal contacts the nasal rostrally, the parietal caudally, and is thickened laterally where it meets the prefrontal and the postorbital in an interdigitate suture that excludes the frontal from the orbital rim (Fig. 14.6). The frontals unite along their medial surface dorsally to form an elevated dome, and articulate with the orbitosphenoid-presphenoid complex ventrally to house the telencephalon (Evans, 2010). Rostrally, the distinct sutural surface for articulation with the nasals is visible in disarticulated braincases, and is square along the rostral margin of the frontal, similar to *Velafrons* (Gates et al., 2007). The frontals are notched caudally at the midline for articulation with the parietals. The ratio of the interfrontal suture length (length of suture visible in dorsal view on articulated skulls) to width (maximum width of articulated frontals in dorsal view) is 0.52 (41 mm/79 mm) for MOR 553s. The shortening of the frontal is apparently not as reduced as in *H. altispinus* , in which the adults have a ratio of 0.46–0.49 (Evans, 2010).

Parietal The parietal contacts the frontal rostrally, the supraoccipital caudoventrally, the prootic and laterosphenoid ventrally, and the squamosals caudally at all growth stages (Fig. 14.6). This element forms the caudodorsal margin of the telencephalon. Rostrolateral extensions of the parietal extend to meet the postorbitals in an interdigitate suture, but the parietal remains separated from the prefrontal by the frontals (Fig. 14.6). The parietal forms the medial and rostromedial margins of the supratemporal fenestrae. The caudal half of the bone has a robust median crest that rises to the level of the squamosal border. This crest separates the squamosals dorsally in *Velafrons* (Gates et al., 2007), and in MOR 553s, but does not in MOR 549 or MOR 455 (Figs. 14.5, 14.6). A small foramen is apparent on each side of the parietal crest in dorsal view. The rostral margin of the parietals is widely expanded and rounded, and the interfrontal process that separates the frontals caudally is indistinct and can appear larger when viewed endocranially as opposed to dorsally (Evans, 2010). The parietals show no midline suture between the left and right sides, which suggests that the bones form from a single ossification center (Horner and Currie, 1994). As with the frontal, the parietal is rostrocaudally compact (53 mm length/215 mm width, MOR 553s; 55 mm length/183 mm width, MOR 455). This differs from the parietals of adult *H. altispinus* and *Corythosaurus*, in which the parietals are nearly as long as they are wide (Evans, 2010).

Supraoccipital The supraoccipital contacts the squamosals dorsolaterally, the parietal rostrodorsally, the exoccipitals ventrolaterally, and the prootic rostrally. The bone is subtriangular in shape in caudal view, widest at its ventrolateral contact with the exoccipitals, and tapers dorsally until it forms the base of the nuchal notch, a small opening bordered dorsally by the paired squamosals (Fig. 14.5). The nuchal notch closes as the animal matures (Horner and Currie, 1994) and the supraoccipital appears to extend dorsally to close the notch in adult skulls (e.g., MOR 553s). As in other lambeosaurines, the supraoccipital is excluded from the foramen magnum by the medial contact of the exoccipitals

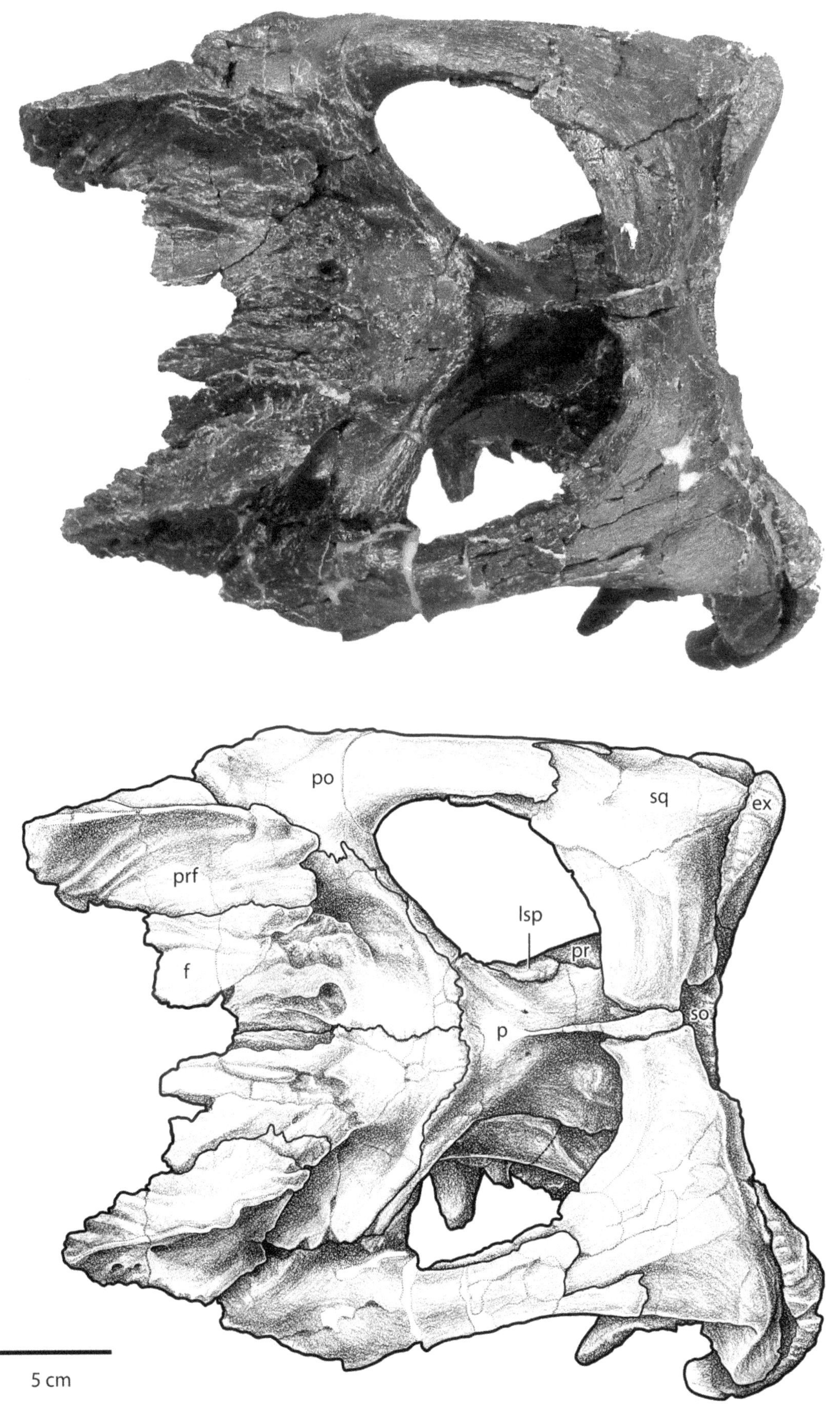

14.6. *Hypacrosaurus stebingeri,* MOR 553s, dorsal view.

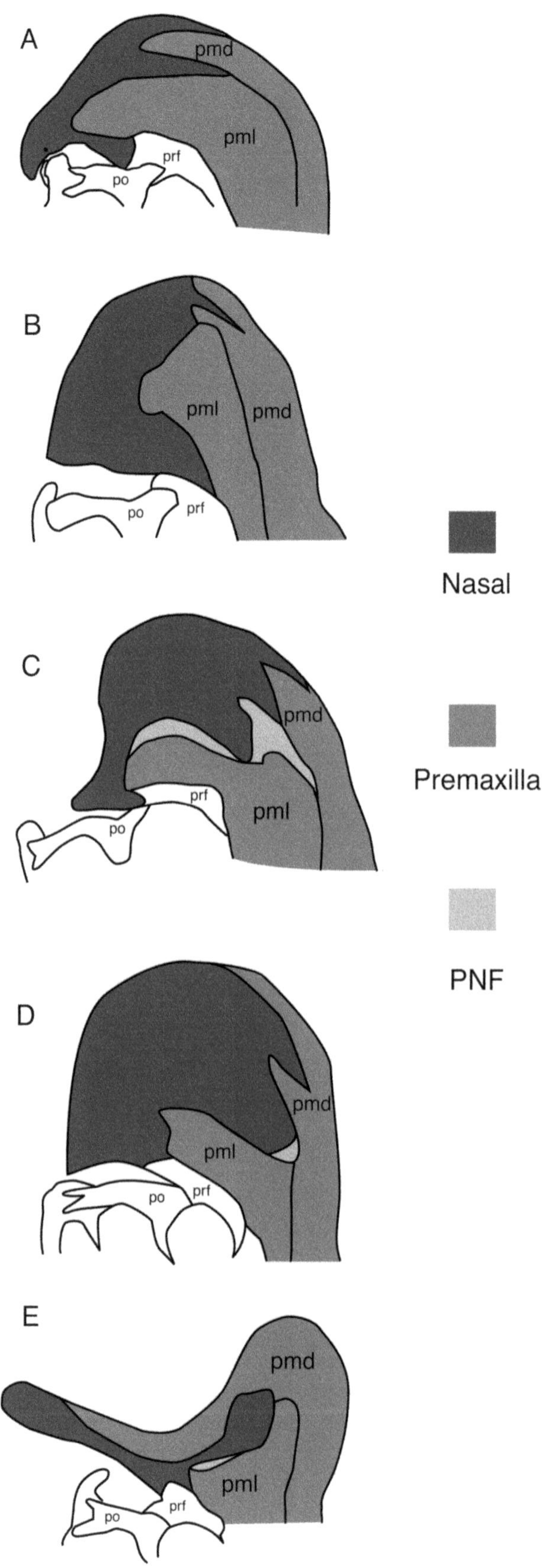

14.7. Comparison of cranial crests composition of lambeosaurine dinosaurs. (A) MOR 549, *Hypacrosaurus stebingeri;* (B) CMN 8501, *Hypacrosaurus altispinus;* (C) ROM 845, *Corythosaurus intermedius;* (D) ROM 841, *Corythosaurus casuarius;* (E) CMN 2869, *Lambeosaurus lambei.* Crests not drawn to scale.

(Gilmore, 1937; Ostrom, 1961; Evans, 2010). The caudally exposed surface bears a series of subhorizontal ridges that extend the entire width of the bone.

Basioccipital This bone contacts dorsally with the exoccipitals and prootics, and rostrally with the basisphenoid-parasphenoid complex (Figs. 14.4, 14.5). It forms the base of the foramen magnum and the occipital condyle. Syndochondroses between the overlying exoccipitals are present in caudal and lateral view. This element is similar to that of other lambeosaurines (Ostrom, 1961; Horner and Currie, 1994; Evans, 2010).

Exoccipital-Opisthotic Complex The exoccipital-opisthotic complex contacts the squamosal dorsally, supraoccipital dorsomedially, basioccipital ventrally, and the prootic rostromedially (Fig. 14.5). The paroccipital process of the exoccipital forms a lap joint with the squamosals. The dorsal surface of the paraoccipital processes is rugose for muscle attachment. The union of the exoccipitals to each other, ventral to the supraoccipital, forms the dorsal margin of the foramen magnum. The caudal margin of the skull is relatively broad and flat, and the lateral surface of the complex is perforated by many foramina (Figs. 14.4, 14.5). The fenestra ovalis is the most rostral, caudal to that is passage for the glossopharyngeal nerve (CN IX), the vagus nerve (CN X), and two small foramina for the hypoglossal nerve (CN XII) (Ostrom, 1961; Evans, 2005, 2010). The two branches of the hypoglossal are separated from each other by a small flange of bone, 0.25 mm in length. Rostral to the foramina for the hypoglossal and ventral to the foramen for the vagus is a small vascular foramen. The exoccipital and opisthotic fuse early in embryonic development, as there is no evidence of a suture between these elements (Horner and Currie, 1994).

Prootic The prootic articulates caudodorsally with the supraoccipital, dorsally with the parietal, rostrally with the laterosphenoid and caudally with the opisthotic-exoccipital complex (Fig. 14.4). It forms the caudal margin of the foramen for the trigeminal nerve (CN V), and the rostral margin of the fenestra ovalis (Ostrom, 1961; Evans, 2005, 2010). Two small foramina are present on the lateral surface of the prootic, rostral to the fenestra ovalis and caudal to the trigeminal nerve, and mark the passage of the facial nerve (CN VII, Evans, 2010). The ventral-most foramen and sulcus is for passage of the palatine branch of the facial nerve, CN VII (Ostrom, 1961; Evans, 2010).

Laterosphenoid The laterosphenoid articulates rostrally with the orbitosphenoid, rostrodorsally with the postorbital and frontal, dorsally with the parietal, caudally with the prootic, and ventrally with the basisphenoid, to form a large portion of the lateral braincase wall (Fig. 14.4). A laterally projecting process articulates into a ball-shaped depression on the medial side of the postorbital. The sutures between

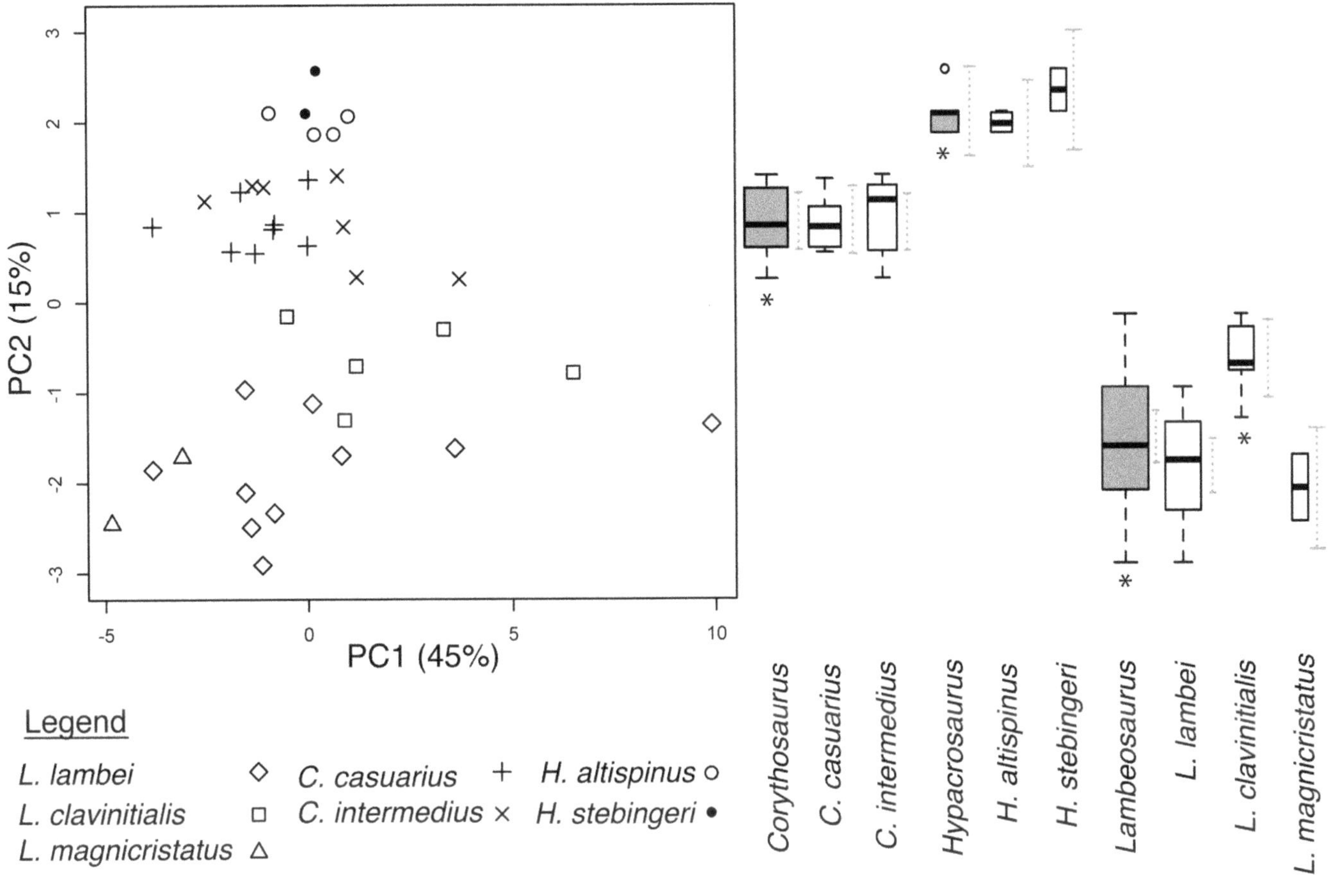

14.8. Results of the PCA. Left: graph of PC axes 1 and 2. Right: Boxplots indicating statistical significance of means of PC2 scores. Boxes are drawn with widths proportional to the square roots of the number of observations in the groups. Asterisks indicate groups that are significantly different. Gray dotted lines indicate 95% confidence intervals for each mean, calculated with a Tukey-Kramer HSD. The circle above the error bar for *Hypacrosaurus* represents an outlier.

the laterosphenoid and the prootic are fused in adult specimens. The laterosphenoid forms the rostral margin of the foramen for the trigeminal nerve (Ostrom, 1961; Evans, 2010). A large, sub-horizontal sulcus for ophthalmic branch of the trigeminal nerve (CN V_1) extends rostrally from the foramen, and a large sulcus descends ventrally for another branch, as in other hadrosaurs (Ostrom, 1961; Evans, 2005). The ophthalmic canal is open as in *H. altispinus* and *Corythosaurus*, but differs from *Lambeosaurus*, which is completely enclosed (Evans and Reisz, 2007).

Basisphenoid-Parasphenoid Complex This complex forms the rostral and basal margins of the braincase, including the rostral cultriform process and the basitubera (Figs. 14.4, 14.5). It articulates dorsally with the orbitosphenoid-presphenoid complex, laterosphenoid, and prootic, caudally with the basioccipital, and ventrally with the pterygoid. The dorsal surface of the cultriform process is concave and is thought to accommodate the median palatine artery (Ostrom, 1961). The basisphenoid-parasphenoid complex forms the ventral margin of the foramen for passage of the oculomotor nerve (CN III) and the abducens nerve (CN VI) (Ostrom, 1961). The basipterygoid processes sit loosely in a medial trough of the pterygoid. The alar processes, well preserved in MOR 455, are rounded dorsally, slightly asymmetrical, and project laterally rostral to the foramina for the internal carotid artery (Fig. 14.4). The caudal suture between the basitubera and the basioccipital is apparent in adult specimens, and the tubera are separated by a deep, rostrocaudally elongate midline fossa (Evans, 2010).

Orbitosphenoid-Presphenoid Complex The presphenoid is the most rostral bone of the braincase, and contacts the orbitosphenoid caudally and the frontal dorsolaterally (Fig. 14.4). In rostral view, the base of the presphenoid is ridged to accommodate the nerves for the olfactory system (Evans, 2006). There is no suture between the orbitosphenoid and the presphenoid, and the presphenoid is not preserved in embryonic or nestling specimens (Horner and Currie, 1994), which suggests that ossification of the presphenoid and orbitosphenoid occurs between the nestling and juvenile stage. In MOR 553s, a midline suture splits the ventral surface of the complex. A small foramen is present on the lateral side of orbitosphenoid for the passage of the trochlear nerve, CN IV (Ostrom, 1961). Rostroventral to the foramen for the trochlear nerve, a large foramen is present for passage of the optic nerve, CN II (Ostrom, 1961; Evans and Reisz, 2007).

MORPHOMETRIC ANALYSIS

The PCA demonstrates a clear distinction between *Lambeosaurus, Corythosaurus,* and *Hypacrosaurus* (Fig. 14.8; Appendix 14.2). PC1 (45% total variation) is weighted in the same direction by all variables except for the crest-snout angle and the position of the crest-snout apex and is interpreted as being driven by size, whereas PC2 (15% total variation) displays the most separation between taxonomic groups (Fig. 14.8), and is driven predominantly by crest variables: crest length at half height, total crest length, length from orbit to crest apex, crest-snout angle, and the position of crest apex. Results of the ANOVA and Tukey-Kramer HSD (Appendices 14.3, 14.4; boxplots in Fig. 14.8) on the PC2 scores suggest that *Corythosaurus, Hypacrosaurus,* and *Lambeosaurus* are significantly different from each other. At the species level, however, species of *Corythosaurus* do not differ significantly from each other, species of *Hypacrosaurus* do not differ significantly from each other, and *L. lambei* and *L. magnicristatus* do not differ significantly from each other. Among species, *Lambeosaurus clavinitialis* differs significantly from *L. lambei* and *L. magnicristatus.*

DISCUSSION

This description represents the most complete documentation of the cranial anatomy of *Hypacrosaurus stebingeri* since its original description (Horner and Currie, 1994), and the first study that specifically addresses anatomy and variation in adults. Although the embryonic-, nestling-, and subadult-sized material had been adequately described previously, the morphology of mature individuals, particularly with respect to the cranial crest, is most relevant for phylogenetic studies, as the most differentiating characters among taxa are present in the well-developed cranial crests of adult lambeosaurines (Ostrom, 1962; Dodson, 1975; Hopson, 1975).

Originally, Horner et al. (1992) and Horner and Currie (1994) noted that the crest of *H. stebingeri* closely resembled that of *Lambeosaurus* in the relative proportions of the premaxilla and the nasal in their contribution to the cranial crest. However, the overall shape of the crest in *H. stebingeri* is in fact closest to *H. altispinus* and *Corythosaurus*, particularly the *C. intermedius* morph (Evans, 2010). In the two known specimens referred to *H. stebingeri*, the crest is not identical in shape. The holotype (MOR 549) has a lower dome height when compared to the referred specimen (MOR 455), which is taller with a more symmetrical profile in lateral view. The caudolateral process of the premaxilla in the crest of MOR 549 has a rounded to nearly horizontal dorsal margin that is slightly depressed rostrally (Figs. 14.1, 14.2) and is more similar to the shape observed in *Corythosaurus*, whereas the process in MOR 455 (Fig. 14.3) is slightly elevated rostrally and is more similar to the shape observed in *Lambeosaurus.* These differences are consistent with the range of variation present in other lambeosaurine species (Fig. 14.8), including *Corythosaurus intermedius*, where the caudolateral process approaches the morphology of *H. stebingeri* in some specimens (e.g., TMP 1980.023.0004), and *C. casuarius* and *H. altispinus* in others, where the process ranges from a lancet shape to a quarter-circle shape (Evans, 2010).

The multivariate analysis on adult skulls performed here is the first multivariate analysis to include *H. stebingeri*, as Dodson's (1975) Principal Coordinates Analysis included growth series from the Dinosaur Park Formation only. The results of the PCA in this study show that the means of the PC scores for each genus are significantly different from the others (Fig. 14.8). *Lambeosaurus* is differentiated from *Corythosaurus* and *Hypacrosaurus* by the high crest-snout angle and the position of the crest apex. These variables are useful for distinguishing between the recumbent, hatchet-shaped crest of *Lambeosaurus* and the high dome-shaped crest of *Corythosaurus* and *Hypacrosaurus.* Within *Lambeosaurus, L. clavinitialis* is significantly different from the *L. magnicristatus* and *L. lambei*, and is the only species to differ significantly from conspecifics in this analysis. This could be due to the height of the crest above the orbit in these three species of *Lambeosaurus*. There is no significant overlap in morphology between *Hypacrosaurus* and *Corythosaurus*, despite similar overall crest shape (Fig. 14.8). *Hypacrosaurus stebingeri* and *H. altispinus* are differentiated from all species of *Corythosaurus* in having sloping facial profiles and shortened skull roofs, which agrees with the results of Evans (2010). It is most likely that the outlier for *H. stebingeri* in the significance test is a result of small sample size (n = 2).

The results of this morphometric analysis appear to be most consistent with recent phylogenetic hypotheses for lambeosaurines that propose a monophyletic *Hypacrosaurus*, and that both species of *Hypacrosaurus* are more closely related to *Corythosaurus* than to *Lambeosaurus* (Evans and Reisz, 2007; Gates et al., 2007; Evans, 2010). Alternate hypotheses that suggest *Hypacrosaurus* is paraphyletic (Suzuki et al., 2004; Prieto-Márquez et al., 2006; Prieto-Márquez, 2010) would infer convergence in crest shape between the two species of *Hypacrosaurus*, or that this is the primitive condition for lambeosaurines. *Hypacrosaurus stebingeri* and *H. altispinus* also share the elongation of the dorsal neural spines and the early closure of the PNF, but do not share a restricted external naris or a wide narial crest, as suggested in the original description of *H. stebingeri* (Horner and Currie, 1994; Evans, 2010). Additionally, studies of the ontogenetic series of juvenile lambeosaurines also suggest that *H. stebingeri* is more closely related to *Corythosaurus* than to *Lambeosaurus* based on the morphology of subadult skulls, especially the incipient cranial crest with a bifurcated anterior nasal (Horner and Currie, 1994; Evans et al., 2005; Brink et al., 2011).

In the original description of *Hypacrosaurus stebingeri,* Horner and Currie (1994) found no autapomorphies for the species, and suggested that this taxon was transitional between the geologically older *Lambeosaurus* and the younger *H. altispinus* due to the characters shared among these taxa (Horner et al., 1992). Previous analyses reject this hypothesis based on the phylogenetic relationships of *Lambeosaurus* and *Hypacrosaurus,* and suggested that the four-million-year interval between *H. stebingeri* and *H. altispinus* makes the argument for an ancestor-descendant relationship between these two species tenuous (Evans and Reisz, 2007; Evans, 2010). In this study, we recognize that the shape and size of the lateral process of the premaxilla is unique to *H. stebingeri.* The presence of an autapomorphy further negates the hypothesis of anagenesis within the *Hypacrosaurus* clade. However, the rate and mode of evolution within lambeosaurines is not fully understood, particularly in light of the debate as to whether "bizarre structures" are sexually selected or used in species selection, and how this relates to the evolution of crest shape within lambeosaurines (Knell and Sampson, 2011; Padian and Horner, 2011a, 2011b). It remains possible that the proportions of the caudolateral process of the premaxilla could be evolutionarily labile and change from a *Corythosaurus*- or *Lambeosaurus*-like ancestor through *H. stebingeri* to *H. altispinus* in an anagenetic lineage through time. Additional analyses of lambeosaurine crest evolution are required to test these hypotheses.

ACKNOWLEDGMENTS

For access to specimens, the authors thank M. Currie, A. McDonald, K. Shepherd (CMN); C. Ancell, B. Baziak (MOR); B. Iwama, K. Seymour (ROM); T. Courtenay, K. Leiper, R. Russell, B. Strilisky, J. Wilkie (TMP); D. Evans, and W. Ripley (TCMI). We thank J. Anderson, E. Snively, J. Theodor, N. Campione, and R. Cuthbertson for discussion, C. Brown for technical assistance, and D. Dufault for the illustrations of MOR 553s. This manuscript was greatly improved by comments from an anonymous reviewer and D. Eberth. This research was supported by a Dinosaur Research Institute Grant, the Jurassic Foundation, and the University of Calgary.

LITERATURE CITED

Brink, K. S., D. K. Zelenitsky, D. C. Evans, F. Therrien, and J. R. Horner. 2011. A sub-adult skull of *Hypacrosaurus stebingeri* (Ornithischia: Lambeosaurinae): anatomy and comparison. Historical Biology 23:63–72.

Brown, B. 1913. A new trachodont dinosaur, *Hypacrosaurus,* from the Edmonton Cretaceous of Alberta. Bulletin of the American Museum of Natural History 32:395–406.

Brown, B. 1914. *Corythosaurus casuarius,* a new crested dinosaur from the Belly River Cretaceous, with provisional classification of the family Trachodontidae. Bulletin of the American Museum of Natural History 33:559–565.

Brown, C. M., J. H. Arbour, and D. A. Jackson. 2012. Testing the effect of missing data estimation and distribution in morphometric multivariate data analyses. Systematic Biology 61:941–954.

Campione, N. E., and D. C. Evans. 2011. Cranial growth and variation in Edmontosaurs (Dinosauria: Hadrosauridae): implications for Latest Cretaceous megaherbivore diversity in North America. PLoS ONE 6:e25186.

Cope, E. D. 1869. Synopsis of the extinct Batrachia, Reptilia, and Aves of North America. Transactions of the American Philosophical Society 14:1–252.

Dodson, P. 1975. Taxonomic implications of relative growth in lambeosaurine hadrosaurs. Systematic Zoology 24:37–54.

Evans, D. C. 2005. New evidence on brain-endocranial cavity relationships in ornithischian dinosaurs. Acta Palaeontologica Polonica 50:617–622.

Evans, D. C. 2006. Nasal cavity homologies and cranial crest function in lambeosaurine dinosaurs. Paleobiology 32:109–125.

Evans, D. C. 2007. Ontogeny and evolution of lambeosaurine dinosaurs (Ornithischia: Hadrosauridae). Ph.D. dissertation, University of Toronto, Toronto, Ontario, 497 pp.

Evans, D. C. 2010. Cranial anatomy and systematics of *Hypacrosaurus altispinus,* and comparative skull growth in lambeosaurine hadrosaurids (Dinosauria: Ornithischia). Zoological Journal of the Linnean Society 159:398–434.

Evans, D. C., and R. R. Reisz. 2007. Anatomy and relationships of *Lambeosaurus magnicristatus,* a crested hadrosaurid dinosaur (Ornithischia) from the Dinosaur Park Formation, Alberta. Journal of Vertebrate Paleontology 27:373–393.

Evans, D. C., C. A. Forster, and R. R. Reisz. 2005. The type specimen of *Tetragonosaurus erectofrons* (Ornithischia: Hadrosauridae) and the identification of juvenile lambeosaurines; pp. 349–366 in P. J. Currie, and E. B. Koppelhus (eds.), Dinosaur Provincial Park: A Spectacular Ancient Ecosystem Revealed. Indiana University Press, Bloomington, Indiana.

Gates, T. A., S. D. Sampson, C. R. Delgado de Jesús, L. E. Zanno, D. Eberth, R. Hernandez-Rivera, M. C. Aguíllon Martínez, and J. I. Kirkland. 2007. *Velafrons coahuilensis,* a new lambeosaurine hadrosaurid (Dinosauria: Ornithopoda) from the Late Campanian Cerro Del Pueblo Formation, Coahuila, Mexico. Journal of Vertebrate Paleontology 27:917–930.

Gilmore, C. W. 1937. On the detailed skull structure of a crested hadrosaurian dinosaur. Proceedings of the United States National Museum 84:481–491.

Hopson, J. A. 1975. The evolution of cranial display structures in hadrosaurian dinosaurs. Paleobiology 1:21–43.

Horner, J. R., and P. J. Currie. 1994. Embryonic and neonatal morphology and ontogeny of a new species of *Hypacrosaurus* (Ornithischia, Lambeosaurinae) from Montana and Alberta; pp. 312–336 in K. Carpenter, K. F. Hirsch, and J. R. Horner (eds.), Dinosaur Eggs and Babies. Cambridge University Press, Cambridge, U.K.

Horner, J. R., D. J. Varricchio, and M. B. Goodwin. 1992. Marine transgressions and the evolution of Cretaceous dinosaurs. Nature 358:59–61.

Horner, J. R., D. B. Weishampel, and C. A. Forster. 2004. Hadrosauridae; pp. 438–463 in D. B. Weishampel, P. Dodson, and H. Osmólska (eds.), The Dinosauria, Second Edition. University of California Press, Berkeley, California.

Knell, R. J., and S. D. Sampson. 2011. Bizarre structures in dinosaurs: species recognition or sexual selection? A response to Padian and Horner. Journal of Zoology 283:18–22.

Maryańska, T., and H. Osmólska. 1979. Aspects of hadrosaurian cranial anatomy. Lethaia 12:265–273.

Oba, S., M.-A. Sato, I. Takemasa, M. Monden, K.-I. Matsubara, and S. Ishii. 2003. A Bayesian missing value estimation method for gene expression profile data. Bioinformatics 19:2088–2096.

Oksanen, J., F. G. Blanchet, R. Kindt, P. Legendre, R. G. O'Hara, G. L. Simpson, P. Solymos, M. H. H. Stevens, and H. Wagner. 2010. Vegan: Community Ecology Package. R package version 1.17-0.

Ostrom, J. H. 1961. Cranial morphology of the hadrosaurian dinosaurs of North America. Bulletin of the American Museum of Natural History 122:35–186.

Ostrom, J. H. 1962. The cranial crests of hadrosaurian dinosaurs. Postilla 62:1–29.

Owen, R. 1842. Report on British fossil reptiles, part II. Report of the British Association for the Advancement of Science 1841:60–204.

Padian, K., and J. R. Horner. 2011a. The evolution of 'bizarre structures' in dinosaurs: biomechanics, sexual selection, social selection or species recognition? Journal of Zoology 283:3–17.

Padian, K., and J. R. Horner. 2011b. The definition of sexual selection and its implications for dinosaurian biology. Journal of Zoology 283:23–27.

Parks, W. A. 1923. *Corythosaurus intermedius,* a new species of trachodont dinosaur. University of Toronto Studies, Geological Series 15:1–57.

Prieto-Márquez, A. 2010. Global phylogeny of Hadrosauridae (Dinosauria: Ornithopoda) using parsimony and Bayesian methods. Zoological Journal of the Linnean Society London 159:435–502.

Prieto-Márquez, A., L. M. Chiappe, and S. H. Joshi. 2012. The lambeosaurine dinosaur *Magnapaulia laticaudus* from the Late Cretaceous of Baja California, Northwestern Mexico. PLoS ONE 7:e38207.

Prieto-Márquez, A., R. Gaete, G. Rivas, A. Galobart, and M. Boada. 2006. Hadrosaurid dinosaurs from the Late Cretaceous of Spain: *Pararhabdodon isonensis* revisited and *Koutalisaurus kohlerorum* gen et sp. nov. Journal of Vertebrate Paleontology 26:929–943.

Rasband, W. 2008. Image J v.1.40g. National Institutes of Health, U.S.A.

Seeley, H. G. 1888. The classification of the Dinosauria. Report of the British Association for the Advancement of Science 57:31–68.

Stacklies, W., H. Redestig, M. Scholz, D. Walther, and J. Selbig. 2007. PCA methods–a bioconductor package providing PCA methods for incomplete data. Bioinformatics 23:1164–1167.

Suzuki, D., D. B. Weishampel, and N. Minoura. 2004. *Nipponosaurus sachalinensis* (Dinosauria: Ornithopoda): anatomy and systematic position within Hadrosauridae. Journal of Vertebrate Paleontology 24:145–164.

Varricchio, D. J. 1995. Taphonomy of Jack's Birthday Site, a diverse dinosaur bonebed from the Upper Cretaceous Two Medicine Formation of Montana. Palaeogeography, Palaeoclimatology, Palaeoecology 114:297–323.

Weishampel, D. B. 1981. The nasal cavity of lambeosaurine hadrosaurids (Reptilia: Ornithischia): comparative anatomy and homologies. Journal of Paleontology 55:1046–1057.

Appendix 14.1. Variables Measured for the Morphometric Analysis

Specimen	Genus	species	Skull length	Maxilla length	Quadrate height	Preorbital length	Jugal length	Postorbital length
AMNH 5338	*Corythosaurus*	*casuarius*	694	337	296	452	223	160
ROM 1933	*Corythosaurus*	*casuarius*	707	310	293	500	255	182
TMP 1980.040.0001	*Corythosaurus*	*casuarius*	719	**328**	296	463	251	178
AMNH 5240	*Corythosaurus*	*casuarius*	670	311	288	495	245	160
ROM 868	*Corythosaurus*	*casuarius*	729	340	307	529	273	**166**
ROM 871	*Corythosaurus*	*casuarius*	**761**	326	294	506	268	200
TMP 1984.121.0001	*Corythosaurus*	*casuarius*	705	**326**	285	439	243	161
UALVP 50700	*Corythosaurus*	*casuarius*	660	**324**	**289**	330	**248**	170
CMN 8676	*Corythosaurus*	*intermedius*	631	284	261	432	**243**	137
CMN 8704	*Corythosaurus*	*intermedius*	660	302	292	434	238	162
UALVP 13	*Corythosaurus*	*intermedius*	650	**320**	**287**	400	**247**	155
ROM 845	*Corythosaurus*	*intermedius*	702	**311**	297	505	262	163
TMP 1980.023.0004	*Corythosaurus*	*intermedius*	753	332	300	552	274	202
ROM 777	*Corythosaurus*	*intermedius*	764	323	296	529	259	**164**
ROM 776	*Corythosaurus*	*intermedius*	792	329	286	498	261	156
CMN 8501	*Hypacrosaurus*	*altispinus*	710	315	307	410	249	159
ROM 789	*Hypacrosaurus*	*altispinus*	732	300	312	515	242	163
ROM 702	*Hypacrosaurus*	*altispinus*	705	301	**287**	**473**	244	150
CMN 8675	*Hypacrosaurus*	*altispinus*	830	356	**289**	**480**	**249**	170
MOR 549	*Hypacrosaurus*	*stebingeri*	840	345	287	535	265	157
MOR 455	*Hypacrosaurus*	*stebingeri*	802	**321**	**288**	**474**	231	181
TMP 1982.038.0001	*Lambeosaurus*	*lambei*	**702**	**315**	288	370	231	**162**
CMN 2869	*Lambeosaurus*	*lambei*	615	287	253	450	231	144
AMNH 5373	*Lambeosaurus*	*lambei*	762	**323**	300	475	**248**	**165**
ROM 1218	*Lambeosaurus*	*lambei*	747	345	285	515	261	158
AMNH 5353	Lambeosaurus	lambei	756	348	299	**473**	**247**	177
FMNH 1479	Lambeosaurus	lambei	761	330	294	510	255	**167**
ROM 794	Lambeosaurus	lambei	786	340	307	610	251	175
CMN 8503	*Lambeosaurus*	*lambei*	495	275	227	345	193	128
CMN 351	*Lambeosaurus*	*lambei*	740	315	293	519	**246**	151
TMP 1997.012.0128	*Lambeosaurus*	*lambei*	**721**	339	**289**	505	270	**166**
ROM 869	*Lambeosaurus*	*clavinitialis*	638	290	231	380	240	**155**
TMP 1981.037.0001	*Lambeosaurus*	*clavinitialis*	674	351	298	535	246	180
YPM 3222	*Lambeosaurus*	*clavinitialis*	746	341	**286**	450	**246**	160
CMN 8703	*Lambeosaurus*	*clavinitialis*	748	342	300	501	253	185
AMNH 5382	*Lambeosaurus*	*clavinitialis*	**684**	285	290	**446**	235	**158**
CMN 8705	*Lambeosaurus*	*magnicristatus*	715	**362**	289	497	254	163
TMP 1966.004.0001	*Lambeosaurus*	*magnicristatus*	733	345	280	500	229	155

Note: Measurements are in millimeters. Details of the measurement of these variables is given in Evans (2010). Numbers in **bold** were estimated using Bayesian Principal Components Analysis.

Parietal length	Crest height above teeth	Crest height at orbit	Crest height only	Crest length at half height	Total crest length	Crest-snout angle	Position of crest apex	Orbit to crest apex	Crest area
86	551	320	282	446	506	128	475	298	1135
104	532	332	287	423	445	134	525	339	1088
84	524	261	**283**	**319**	**484**	121	477	**348**	1007
86	564	315	290	398	440	116	542	316	1183
84	**546**	**230**	**274**	**305**	**470**	125	**443**	**338**	**940**
90	643	408	378	445	466	**106**	478	410	1437
84	**546**	221	210	435	490	**117**	**443**	250	**936**
89	**535**	270	320	480	496	155	426	265	836
90	420	218	214	202	320	132	420	233	531
80	**504**	**190**	**240**	**250**	**415**	130	410	**295**	**722**
82	**516**	273	260	279	400	134	465	275	790
90	490	246	230	260	318	135	518	249	521
84	**553**	**236**	**279**	**313**	**479**	127	538	**344**	**973**
83	576	356	330	238	365	134	555	350	799
98	578	325	298	262	320	140	562	334	871
69	582	240	231	290	350	145	480	250	758
82	524	280	262	281	310	153	485	292	738
63	**521**	245	230	395	412	140	570	260	792
77	**539**	230	220	365	445	159	580	255	909
82	470	194	180	334	430	147	619	196	753
55	**524**	245	230	311	422	140	720	261	825
81	607	112	355	131	412	104	**448**	404	625
76	475	98	222	170	450	104	360	243	597
83	581	143	288	167	553	122	126	**325**	**877**
84	560	109	315	193	595	94	297	417	812
83	554	140	271	224	457	83	125	**313**	**818**
84	520	165	261	190	617	88	312	430	956
97	640	164	371	237	695	96	395	485	1243
85	370	60	155	152	202	121	**456**	**181**	**147**
82	550	138	277	191	**410**	100	400	348	683
83	510	**223**	375	193	**460**	100	392	396	**900**
78	335	91	120	180	335	116	281	155	310
80	428	85	131	179	493	92	550	345	492
82	473	95	186	207	315	100	**446**	**296**	793
80	512	140	244	218	**401**	116	400	310	651
79	371	150	145	191	370	102	**451**	290	416
93	667	360	390	615	710	62	301	595	1968
89	705	**326**	**356**	**436**	698	73	**435**	**441**	**1461**

Appendix 14.2. Results of the PCA

A

	Comp.1	Comp.2	Comp.3	Comp.4	Comp.5	Comp.6	Comp.7
Skull length	-0.277	0.121	-0.362	0.085	0.098	-0.001	-0.328
Maxilla length	-0.284	-0.095	-0.19	0.16	-0.231	0.167	-0.016
Quadrate height	-0.287	0.135	-0.209	-0.004	0.288	-0.099	0.216
Preorbital length	-0.244	-0.013	-0.329	-0.221	-0.267	-0.383	-0.329
Jugal length	-0.246	0.117	-0.299	-0.394	0.022	0.233	-0.171
Postorbital length	-0.244	0.1	-0.289	0.005	-0.033	0.282	0.724
Parietal length	-0.107	-0.18	0.282	-0.76	-0.197	0.164	0.08
Crest height above teeth	-0.316	-0.026	0.229	0.15	0.291	-0.157	-0.049
Crest height at orbit	-0.235	0.365	0.322	-0.077	0.014	0.048	-0.128
Crest height only	-0.283	-0.073	0.32	-0.082	0.425	-0.162	0.026
Crest length at half height	-0.198	0.317	0.358	0.186	-0.386	0.305	-0.05
Crest total length	-0.278	-0.242	0.046	0.283	-0.21	0.157	-0.05
Crest-snout angle	0.1	0.545	-0.056	-0.125	0.353	0.146	-0.045
Crest pos apex	0.012	0.478	0.047	-0.01	-0.396	-0.575	0.251
Crest orbit apex	-0.296	-0.276	0.087	-0.054	0.028	-0.356	0.263
Crest area	-0.346	0.038	0.162	0.13	-0.04	0.072	-0.136

B

	Comp.1	Comp.2	Comp.3	Comp.4	Comp.5	Comp.6	Comp.7
Standard deviation	2.694679	1.5376064	1.2977124	1.00524265	0.88608554	0.85699773	0.72636033
Proportion of variance	0.4538309	0.1477646	0.1052536	0.06315705	0.04907172	0.04590282	0.03297496
Cumulative proportion	0.4538309	0.6015955	0.7068491	0.77000616	0.81907789	0.86498071	0.89795566

Note: A, loadings; B, importance of components.

Appendix 14.3. Results of One-Way ANOVA for (A) Genera and (B) Species
(A)

Source of variation	Sum of squares	df	Mean square	F ratio	Probability
Among genera	77.445613	2	38.7228	109.34	<.0001
Error	12.395263	35	0.3542		
Total	89.840876	37			

(B)

Source of variation	Sum of squares	df	Mean square	F ratio	Probability
Among species	83.105616	6	13.8509	63.7509	<.0001
Error	6.735261	31	0.2173		
Total	89.840876	37			

Comp.8	Comp.9	Comp.10	Comp.11	Comp.12	Comp.13	Comp.14	Comp.15	Comp.16
0.221	0.02	0.287	0.156	0.633	0.089	-0.161	0.249	-0.027
0.689	0.049	-0.16	0.097	-0.276	-0.395	0.084	-0.125	0.013
-0.01	-0.604	-0.497	-0.108	0.012	0.089	-0.07	0.044	-0.278
-0.292	-0.259	0.335	0.035	-0.42	-0.112	0.053	0.012	-0.002
-0.177	0.54	-0.432	-0.055	-0.144	0.224	-0.101	0.009	0.067
-0.192	0.089	0.402	0.156	-0.033	0.006	0.004	-0.093	-0.033
0.282	-0.236	0.152	-0.126	0.188	0.103	-0.034	-0.047	-0.056
0.213	0.001	0.187	0.057	-0.257	0.499	-0.32	-0.381	0.26
-0.267	-0.029	-0.092	0.26	0.211	-0.514	-0.212	-0.422	0.045
0.043	0.317	0.171	-0.016	-0.197	-0.236	0.018	0.373	-0.482
-0.059	-0.169	-0.041	0.164	-0.195	0.203	-0.115	0.541	0.087
-0.194	0.035	0.067	-0.749	0.115	-0.12	-0.265	-0.069	-0.056
0.162	-0.089	0.211	-0.456	-0.142	-0.175	0.192	0.101	0.377
0.209	0.26	-0.059	-0.203	0.118	0.122	-0.037	-0.061	-0.169
-0.075	0.054	-0.188	0.028	0.179	-0.174	0.116	0.297	0.648
-0.095	0.019	-0.004	-0.04	0.153	0.233	0.815	-0.218	-0.094

Comp.8	Comp.9	Comp.10	Comp.11	Comp.12	Comp.13	Comp.14	Comp.15	Comp.16
0.65827903	0.56913011	0.50465904	0.49307304	0.39248737	0.30615089	0.22735617	0.2150734	0.17873206
0.02708321	0.02024432	0.01591755	0.01519506	0.0096279	0.00585802	0.00323068	0.00289104	0.00199657
0.92503887	0.94528319	0.96120073	0.9763958	0.98602369	0.99188172	0.99511239	0.99800343	1

Appendix 14.4. Results of Tukey-Kramer HSD Analysis for (A) Genera and (B) Species

(A)

Level	Hs	Ha	Cc	Ci	Ll	Lc	Lm
Hs	-1.468	-0.914	0.230	0.314	1.753	3.039	2.946
Ha	-0.914	-1.038	0.130	0.218	1.639	2.950	2.785
Cc	0.230	0.130	-0.785	-0.692	0.714	2.045	1.830
Ci	0.314	0.218	-0.692	-0.734	0.669	2.005	1.779
Ll	1.753	1.639	0.714	0.669	-0.929	0.391	0.205
Lc	3.039	2.950	2.045	2.005	0.391	-0.657	-0.899
Lm	2.946	2.785	1.830	1.779	0.205	-0.899	-1.468

(B)

Level	*Hypacrosaurus*	*Corythosaurus*	*Lambeosaurus*
Hypacrosaurus	-0.841	0.501	2.922
Corythosaurus	0.501	-0.532	1.893
Lambeosaurus	2.922	1.893	-0.500

Note: Alpha = 0.05, q* = 3.15 (genera) and q* = 2.447 (genera). Positive values show pairs of means that are significantly different. Abbreviations: Hs, *Hypacrosaurus stebingeri*; Ha, *Hypacrosaurus altispinus*; Cc, *Corythosaurus casuarius*; Ci, *Corythosaurus intermedius*; Ll, *Lambeosaurus lambei*; Lc, *Lambeosaurus clavinitialis*; Lm, *Lambeosaurus magnicristatus*.

Biogeography and Biostratigraphy

4

An Overview of the Latest Cretaceous Hadrosauroid Record in Europe

15

Fabio M. Dalla Vecchia

ABSTRACT

Latest Cretaceous hadrosauroids from the Upper Campanian–Maastrichtian of Ukraine, Bulgaria, Romania, Italy, Slovenia, Germany, Belgium and the Netherlands, France, and Spain are reviewed. This record includes the non-hadrosaurid *Telmatosaurus transsylvanicus* and *Tethyshadros insularis*, and the lambeosaurines *Parahabdodon isonensis*, *Arenysaurus ardevoli*, and *Blasisaurus canudoi*. *Orthomerus* is a nomen dubium and *Koutalisaurus kohlerorum* is probably a junior synonym of *Parahabdodon isonensis*. The European hadrosauroids lived in an archipelago of relatively small islands (European Archipelago) and most have a diminutive size when compared to the largest ("adult") hadrosauroids from the uppermost Cretaceous of western North America and Asia. The Romanian hadrosauroids ("Tisia-Dacia" Island) were insular "dwarfs," as were other hadrosauroids living on other European islands. However, a few relatively large individuals are reported from the relatively large "Ibero-Armorican" Island (Spain and France) and histological studies are needed to test the extent to which hadrosauroid insular dwarfism was expressed in the archipelago. The latest Cretaceous hadrosauroid assemblages from Europe more closely resemble those of eastern Asia than those of western North America.

INTRODUCTION

Uppermost Cretaceous hadrosauroid remains, as well as those of dinosaurs in general, are not as abundant in Europe as they are in western North America and eastern Asia. However, their record has notably increased during the last two decades. Regions where they were not expected to be found for paleogeographic reasons (e.g., Italy, Slovenia, Germany, and Bulgaria; Fig. 15.1) have yielded hadrosauroid specimens (e.g., Wellnhofer, 1994; Debeljak et al., 1999, 2002; Dalla Vecchia, 2009c; Godefroit and Motchurova-Dekova, 2010). Similarly, regions that had yielded hadrosauroid remains only sporadically in the preceding decades have since proved to be relatively rich in those fossils (e.g., Spain and southern France, Fig. 15.1; Laurent, 2003; Company, 2004; Riera et al., 2009; Pereda-Suberbiola, Canudo, Company, et al., 2009; Pereda-Suberbiola, Canudo, Cruzado-Caballero, et al., 2009; Bilotte et al., 2010; Cruzado-Caballero, Pereda-Suberbiola, and Ruiz-Omeñaca, 2010). This chapter builds on previous reviews by Dalla Vecchia (2006, 2009a) and provides a thorough and updated overview of the European hadrosauroid record. Also included is a comparison of element and body sizes of latest Cretaceous European hadrosauroids with those of coeval hadrosauroids from western North America and Asia.

Institutional Abbreviations AMNH, American Museum of Natural History, New York; ACKK, Ivan Rakovec Institute of Paleontology; ZRC SAZU, Ljubljana, Slovenia; BMNH, the Natural History Museum (former British Museum of Natural History), London; CGTL, Central Geological Museum Tschernyshew, Saint Petersburg; CMN, Canadian Museum of Nature, Ottawa, Ontario; FGGUB, Facultatea de Geologie şi Geofizica, Universitatea Bucureşti, Bucharest; IPS, Institut de Paleontologia Dr. M. Crusafont (currently Institut Català de Paleontologia "Miquel Crusafont"), Sabadell, Spain; IRScNB, Institut royal des Sciences naturelles de Belgique, Brussels; MAFI, Magyar Allami Földtani Intézet, Budapest; MB, Museum für Naturkunde of the Humboldt-Universität, Berlin; MCDRD, Muzeul Civilizaţiei Dacice şi Romane Deva, Deva, Romania (former Muzeul Judeţean Hunedoara); MCNA, Museo de Ciencias Naturales de Alava/Arabako Natur Zientzien Museoa, Vitoria-Gasteiz, Spain; MDE, Musée des dinosaures, Esperaza, France; MGUV, Museo de Geologia, Universidad de Valencia, Spain; MND, Museum Natura Docet, Denekamp, Netherlands; MNHN, Muséum national d'Histoire naturelle, Paris; MNMB, Magyar Nenzeti Muzeum, Budapest; MPV, Museo Paleontológico de Valencia, Valencia, Spain; MPZ, Museo Paleontólogico, Universidad de Zaragoza, Zaragoza, Spain; NHMM, Natuurhistorisch Museum Maastricht (RD, R.W. Dortangs collection), Maastricht, Netherlands; NKM, Südostbayerischen Naturkunde- und Mammut-Museum Siegdorf, Siegdorf, Germany; NMNHS, National Museum of Natural History, Sofia, Bulgaria; ROM, Royal Ontario Museum, Toronto, Ontario; SC, Italian State collection (the hadrosauroid specimens are deposited at the Museo Civico di Storia Naturale, Trieste, Italy, without a museum number); TM, Teylers Museum, Haarlem, Netherlands; TMP, Royal Tyrrell Museum of Palaeontology, Drumheller, Alberta; UBB, paleontological collections of the University Babeş-Bolyai, Cluj-Napoca, Romania; USNM, United States National Museum, Washington D.C.

TERMINOLOGY AND METHODS

Following Sereno (1998), Hadrosauroidea includes all dinosaurs more closely related to *Parasaurolophus* than *Iguanodon*. In order to ease the reference and comparison to earlier works, Hadrosauridae is considered here as including Hadrosaurinae + Lambeosaurinae (= Euhadrosauria of Weishampel et al., 1993), thus following Forster (1997) and Gates et al. (2011).

The size distribution in a sample of hadrosaurid femora and tibiae from the Dinosaur Park Formation (Late Campanian) of Alberta, Canada (Brinkman, 2011) was used as a reference for comparing the sizes of skeletal elements from European hadrosauroids (also including fibulae as a proxy for tibiae) with those from a continental North American sample. In Brinkman's analysis, the sizes of femora and tibiae are expressed as percentages with respect to the size of these same elements in reference specimen TMP 1998.058.0001, an indeterminate hadrosaurid taxon (Brinkman, 2011, this volume). In the case of incompletely preserved elements, length was estimated by comparison with complete specimens. D. Brinkman (TMP) kindly furnished the measurements of other appendicular elements from TMP 1998.058.0001 that were used for comparison with their counterparts in European hadrosauroids (humerus, ulna, scapula, ilium, ischium, pubis, and metatarsal IV). Because TMP 1998.058.0001 does not preserve the skull and lower jaws, the dentary of ROM 845 (*Corythosaurus intermedius*) was chosen to serve as a comparative reference for that element. This specimen is from the same formation and general location as TMP 1998.058.0001 (Dinosaur Park Formation at Dinosaur Provincial Park [DPFm]). Because the tibiae and femora of ROM 845 are twice the length of those of TMP 1998.058.0001, measurements of European elements are expressed as a percentage of those in ROM 845, and then transformed as a percentage of those for TMP 1998.058.0001 (50% smaller than those of ROM 845). Because the length of the maxilla of ROM 845 is not available, the maxilla of ROM 1933 (*C. casuarius*) was used as a proxy. This specimen is also from the DPFm and has a skull length very close to that of ROM 845 (Evans, 2010). The size comparisons between European and Canadian elements made here (other than tibia and femur) are approximate and must be considered with caution because of potential problems due to allometric growth during ontogeny and different proportions and morphology of the same elements between taxa (e.g., the higher skull elongation in the Hadrosaurinae with respect to the Lambeosaurinae [see below]; Guenther, this volume).

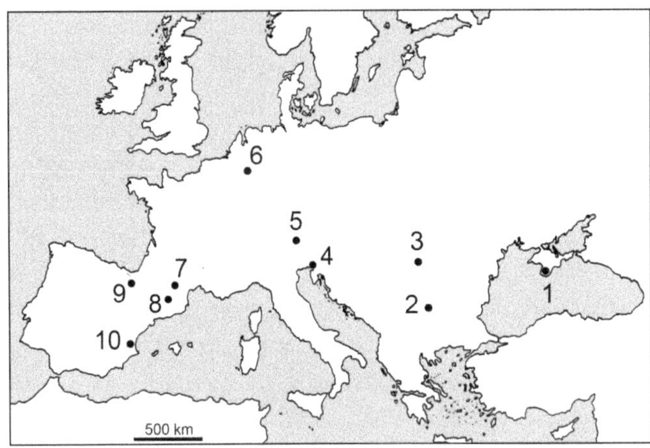

15.1. Location of the latest Cretaceous hadrosauroid-bearing localities of Europe. (1) Mt. Besh-Kosh, Crimean Peninsula (Ukraine); (2) Labirinta Cave (Bulgaria); (3) Haţeg Basin and other Transylvanian localities (Romania); (4) Villaggio del Pescatore (Italy) and Kozina (Slovenia); (5) Bad Adelholzen (Germany); (6) Limburg localities (the Netherlands and Belgium); (7) Southern France localities (Lestaillats, Jadet, Auzas, Peyrecave, Tricouté, Cassagnau, Mérigon, Larcan, and Le Bexen); (8) localities in the Tremp, Àger and Vallcebre Synclines (Spain); (9) Laño Quarry (Spain); (10) La Solana and Loma Cortada (Spain).

THE LATEST CRETACEOUS HADROSAUROID RECORD OF EUROPE

The oldest latest Cretaceous hadrosauroid evidence in Europe is possibly Late Campanian in age. In fact, hadrosauroids are unrepresented in Europe between the Early Cenomanian (Charentes site, SW France, a tooth of "*Probactrosaurus*-grade"; Vullo et al., 2007) and the Late Campanian (a single hadrosaurid tooth from the Laño quarry, Spain, see below). No hadrosauroid remains are reported in the rich Santonian site of Iharkút, Hungary (Ősi, 2004, 2005) and in the Early Campanian site of Muthmannsdorf, Austria (Weishampel et al., 2004; Sachs and Hornung, 2006), although both preserve rhabdodontid ornithopods and ankylosaurs. The ichnogenus *Apulosauripus federicianus* from the Santonian of southern Italy (Altamura, Puglia) was referred to a hadrosaurid trackmaker (Nicosia et al., 1999), but it cannot be demonstrated to be hadrosaurid nor hadrosauroidean because it has none of the apomorphies of those clades; theoretically, *Apulosauripus* could have been made by rhabdodontid iguanodontians, which were common in Europe during pre–Late Maastrichtian time (Dalla Vecchia, 2008a). Alternatively, given that several thousand ankylosaur footprints are preserved in the same site (see Dal Sasso, 2004), *Apulosauripus federicianus* might actually represent poorly preserved ankylosaur footprints (Dalla Vecchia, 2008a).

The hadrosauroids from the Italian site of the Villaggio del Pescatore (Trieste) have been wrongly reported in the literature as being Santonian or Santonian–Campanian in age (e.g., Dal Sasso, 2004; Arbulla et al., 2006; Dal Sasso and Maganuco, 2011; Weishampel and Jianu, 2011), but they are actually Late Campanian–Early Maastrichtian in age (Dalla Vecchia, 2008b, 2009c; see below). Hadrosauroids are unknown from the Lower Campanian (Le Beausset, Fuveau, and Villeveyrac; Buffetaut et al., 1996, 1997) and Upper Campanian–Lower Maastrichtian (e.g., Velaux-La Bastide Neuve, La Boucharde, Champ-Garimond, surroundings of Saint Chinian, Cruzy, Fox-Amphoux, Bellevue, Montplaisir, and the Mas-d'Azil; Buffetaut et al., 1997, 1999; Allain and Taquet, 2000; Allain and Pereda-Suberbiola, 2003; Buffetaut, 2005; Garcia et al., 2010) dinosaur sites of France, although rhabdodontids are relatively common in most of them. They are absent also from most of the supposed Upper Campanian–Lower Maastrichtian dinosaur sites of Portugal (Aveiro, Viso, and Taveiro) and Spain (Chera, Armuña, Quintanilla del Coco, and the rich site of Lo Hueco; Company, 2004 and references therein; Barroso-Barcenilla et al., 2009). As per France, rhabdodontids are found at these sites (Escaso et al., 2012). A trackway tentatively referred to a hadrosaurid (Herrero Santos, 2008) occurs in the Upper Campanian–Lower Maastrichtian site of Sierra de los Gavilanes (Yecla, Murcia Region), but this could have been made also by rhabdodontids.

In contrast, hadrosauroids are the only dinosaurs in the Upper Campanian–Lower Maastrichtian Italian site of Villaggio del Pescatore (Dalla Vecchia, 2009c) and are present but subordinate to rhabdodontids and sauropods in the Lower Maastrichtian sites of Romania (e.g., Therrien, 2005; Benton et al., 2010; Codrea et al., 2010; pers. obs. during the 1997 fieldwork of the International Paleontological Expedition along the Sibişel River valley organized by the Muzeul Civilizaţiei Dacice şi Romane of Deva and by the John Hopkins University of Baltimore; pers. obs. of the fossil collection at the Muzeul Civilizaţiei Dacice şi Romane and the Nopcsa Collection at the Natural History Museum of London). They also dominate the Upper Maastrichtian dinosaur assemblages in France and Spain, where rhabdodontids are absent (e.g., López-Martínez et al., 2001; Laurent, 2003; Company, 2004; Dalla Vecchia, 2006, 2009c; Company et al., 2009; Pereda-Suberbiola, Canudo, Company, et al., 2009; Pereda-Suberbiola, Canudo, Cruzado-Caballero, et al., 2009; Cruzado-Caballero, Pereda-Superbiola, and Ruiz-Omeñaca, 2010).

Unfortunately, the latest Cretaceous hadrosauroid record of Europe lacks complete and even partial articulated skeletons, as well as complete skulls. The only exception is the holotype of *Tethyshadros insularis* Dalla Vecchia 2009c. Five species of hadrosauroid have been named since 1900 and are considered valid to date: *Telmatosaurus transsylvanicus* (Nopcsa, 1900) from the Lower Maastrichtian of Romania; *Pararhabdodon isonensis* Casanovas-Cladellas, Santafé-Llopis and Isidro-Llorens 1993 from the Upper Maastrichtian of Spain; *Tethyshadros insularis* from the Upper Campanian–Lower Maastrichtian of Italy; *Arenysaurus ardevoli* Pereda-Suberbiola, Canudo, Cruzado-Caballero, Barco, López-Martínez, Oms, and Ruiz-Omeñaca 2009; and *Blasisaurus canudoi* Cruzado-Caballero, Pereda-Suberbiola, and Ruiz-Omeñaca 2010. The latter two are from the Upper Maastrichtian of Spain.

The European hadrosauroid record is here reviewed according to the current (2012) political geography of Europe and, from east to west, according to the probable route of hadrosauroid migration and colonization (Dalla Vecchia, 2009c). This chapter presents a complete and up-to-date summary of hadrosauroid occurrences from Europe based, in part, on previous reviews of this material by Dalla Vecchia (2006, 2009a, 2009b), updated with information that has come to light since these publications.

Ukraine

Riabinin (1945) described some hadrosauroid bones from the mount Besh-Kosh near Sebastopol, Crimea. The material includes an incomplete femur (CGTL 1/5751; estimated length 75–80 cm); a tibia (CGTL 2/5751, about 70 cm long); fragments of a fibula; three tarsal elements; metatarsals II and III (CGTL 7/5751 and 8/5751, 20 and 25 cm long, respectively); and the first phalanx of pedal digit II (CGTL 9/5751, 6.6 cm long). According to Riabinin, most of the bones are from a left hind limb, but it is unclear whether he refers to a single or multiple individuals. The specimens were found in marine limestone containing molluscs referred to "the lower part of the Danian" (1945:8); the age of the specimens is possibly latest Maastrichtian (*Belemnitella casimirovensis* Zone; Jeletzky, 1962; López-Martínez et al., 2001). Riabinin (1945) erected the new species *Orthomerus weberi* for this material, but that taxon is now considered a nomen dubium because the material has no diagnostic features (Brinkmann, 1988; Horner et al., 2004; see below).

Bulgaria

Scattered hadrosauroid bones have been reported recently from the Labirinta Cave near the town of Cherven Bryag in NW Bulgaria (Godefroit and Motchurova-Dekova, 2010). They include a distal portion of a left femur (NMNHS

F-31437; estimated length ~40 cm); a right tibia (NMNHS F-31441/31443; estimated length ~35–40 cm); the proximal part of a right fibula (NMNHS F-31438; estimated length ~35–40 cm); a left metatarsal II (NMNHS F-11899; 6.5 cm long); the phalanx 2 or 3 of a left pedal digit IV (NMNHS F-31522-2; ~1.8 cm long); the possible proximal end of a second metacarpal (NMNHS F-31517); and a caudal vertebral centrum (NMNHS F-31530; ~4.6 cm long). They were collected in the upper part of the marine Kajlâka Formation and their age is probably latest Maastrichtian based on fossils found in the same part of the Formation (Jagt et al., 2006) and Sr isotopy analysis (Mateus et al., 2010). All specimens belong to small-sized individuals as noticed by Godefroit and Motchurova-Dekova (2010).

Romania

Hadrosauroid remains are relatively common in the latest Cretaceous continental units of Transylvania (central-western Romania; see Weishampel and Jianu, 2011 and references therein; Codrea et al., 2010). They are reported from the Sânpetru Formation in the central part of the Haţeg Basin (Hunedoara County), in the Densuş-Ciula Formation in the northwestern part of the same basin, and in the "Pui beds" along the Bărbat River in the southeastern part (Therrien, 2005). Fossil-bearing sites in the Sânpetru Formation are located mainly along the Sibişel River valley; the Toteşti-Baraj and Nălaţ-Vad sites, occurring along the nearby Râul Mare, are also referred to the Sânpetru Formation (Van Itterbeeck, Marchevich, and Codrea, 2005; Bojar et al., 2005; Therrien, 2005; Therrien et al., 2009). Vălioara and Tuştea are hadrosaur-bearing localities in the Densuş-Ciula Formation. In the close Alba Iulia County, remains are reported from the basal part of the Şard Formation at the Vurpăr site and in the Sebeş Formation at the Oarda de Jos A-B, Lancrăm A-B, Sebeş-Glod and Râpa Roşie sites (Codrea et al., 2010; Vremir, 2010).

The specimens from the Haţeg Basin are reposited mainly at the BMNH (Nopcsa collection), FGGUB, MCDRD, MAFI, and MNMB; those from Alba Iulia County are mostly at the UBB.

Hadrosauroid remains from the Transylvanian sites are usually referred to *Telmatosaurus transsylvanicus* (e.g., Weishampel et al., 1991, 1993; Therrien, 2005; Grigorescu and Csiki, 2006; Benton et al., 2010; Codrea et al., 2010; Grigorescu et al., 2010; Weishampel and Jianu, 2011). However, Dalla Vecchia (2006, 2009a, 2009b) stressed the problems related to this practice, including the broad stratigraphic range of the referred specimens, and other problems associated with the identity of the type and referred materials. He suggested the possibility that *Telmatosaurus transsylvanicus* is simply a wastebasket taxon for all undiagnostic hadrosauroid elements from the uppermost Cretaceous of Transylvania. These arguments are reviewed here.

The hadrosauroid-bearing units of Transylvania are very thick. For example, the 2500 m thick Sânpetru Formation (Grigorescu, 1992) includes an 860 m thick upper portion that is exposed along the Sibişel valley south of the village of Sânpetru (Therrien et al., 2009). The lower part of that section ("lower member") yields abundant dinosaur macroremains, whereas macroremains in the uppermost 130 m ("upper member")–once supposed to be barren–are now known to be present, but rare (Weishampel et al., 1991; Grigorescu, 1992; Therrien, 2005; Therrien et al., 2009, Panaiotu and Panaiotu, 2010).

Most of the hadrosauroid remains of the Sânpetru Formation (including the holotype of *Telmatosaurus transsylvanicus*, see below) are from the lower member of the Sânpetru Formation in the Sibişel valley. Therrien et al. (2009) tentatively correlated the Toteşti-Baraj and Nalaţ-Vad sites to the upper member of the Sânpetru Formation. The Sibişel valley section was deposited mainly during an interval of reverse polarity corresponding to C31r, whereas the basal part near the Sânpetru village is Late Campanian in age (C32n.1n; Panaiotu and Panaiotu, 2010). The upper member formed during an interval of normal polarity correlable with C31n and dates to the early Late Maastrichtian. Palynomorph associations at the Toteşti-Baraj and Nalaţ-Vad sites suggest deposition at a time close in age to the Early–Late Maastrichtian boundary based on palynostratigraphic comparisons with the "lower Rogniacian" of France and the Gulpen Formation of the Netherlands (Van Itterbeeck, Marchevich, and Codrea, 2005). Thus, the purported *Telmatosaurus* material from the Sânpetru Formation spans the top of the Campanian, the entire Lower Maastrichtian, and the lower part of the Upper Maastrichtian (Panaiotu and Panaiotu, 2010:figs. 2, 7).

The Densuş-Ciula Formation is nearly 4000 m thick and is divided into lower, middle, and upper members; dinosaur remains occur mainly in the middle member, which is 2000 m thick (Bojar et al., 2005; Therrien, 2005). The middle member is usually correlated with the Sânpetru Formation because it shares similar faunal and microfloral assemblages (Grigorescu, 1992; Therrien, 2005). At least part of the middle member was deposited during an interval of reverse polarity (Panaiotu and Panaiotu, 2002), presumably C31r; it yields a palynomorph assemblage that is similar to those found at Toteşti-Baraj and Nalaţ-Vad sites (Van Itterbeeck, Marchevich, and Codrea, 2005).

The Şard Formation is over 1000 m thick and dinosaur remains are reported from its basal part, which is dated as Early Maastrichtian (Codrea and Dica, 2005). The formation is considered coeval with the Sânpetru and Densuş-Ciula Formations of the Haţeg Basin (Codrea et al., 2010).

The Sebeş Formation is 550 m thick (Csiki et al., 2010); the Sebeş-Glod site is in the lower third of the formation, and is considered as "probably late Early Maastrichtian in age, based on its position well above the Campanian-Maastrichtian boundary (dated by palynology, calcareous nanoplankton, and forams . . .)" (Csiki et al., 2010:supplementary information). The Râpa Roşie site, in the upper part of the formation, is "possibly late Maastrichtian in age" (Vremir, 2010:646), but no evidence for this age assessment is included in that study.

There is an absence of detailed stratigraphic information for the large collections from the Sânpetru and Densuş-Ciula formations (Dalla Vecchia, 2006, 2009a, 2009b). Thus, the exact stratigraphic provenance of most of the remains of *Telmatosaurus* referred to in the literature is unknown. Purported *Telmatosaurus* material comes from five different lithostratigraphic units, the fossil-bearing sections are several hundred meters thick, and it is impossible to locate the exact original stratigraphic positions of many specimens derived from them (e.g., those of the Nopcsa collection at BMNH).

Telmatosaurus transsylvanicus was erected by Nopcsa (1900) on the basis of the following material: a partial skull with mandibles (BMNH R3386, "nr.1" in Nopcsa, 1900; Fig. 15.2); a fragmentary small left maxilla (BMNH R3388) and a fragmentary small right dentary (not figured); a small braincase (BMNH R3387); the mid-part of a right maxilla (not figured); the rostral part of a right dentary (not figured, and later probably glued to the right dentary of BMNH R3386, see Fig. 15.2F2); several isolated maxillary teeth (only one crown figured, BMNH R11550), and a figured portion of a dentary tooth battery that cannot be traced back in the Nopcsa collection at BMNH (Dalla Vecchia, 2009b). All the specimens are from Szenpeterfalva (= Sânpetru), therefore, that material is from the lower part of the Sibişel Valley section of the Sânpetru Formation investigated by Panaiotu and Panaiotu (2010, see fig. 2b), which is latest Campanian–Early Maastrichtian in age. Nopcsa (1915) referred postcranial bones including femora, a tibia, metatarsals, a humerus, metacarpals, phalanges, vertebrae from all parts of the column, ribs, and apparently also chevrons to the same taxon as the skull material described in 1900. None of these latter specimens was referred to the same individual as BMNH R3386, and at least one femur was collected from the Vălioara site of the Densuş-Ciula Formation. Because the femora discovered after the publication of the 1900 paper resembled those of *Orthomerus* Seeley 1883, Nopcsa considered *Telmatosaurus* a junior synonym of *Orthomerus*. Nopsca (1915) included a brief description of the postcranium, stated that the sacrum is composed of eight vertebrae, the shaft of the femur is straight, the metatarsals are much more slender than in *Iguanodon*, and the distal phalanges are hoof shaped. Based on the presence or absence of a ventral groove in the centra of two similarly sized caudal vertebrae from the same region of the tail, Nopcsa hypothesized the presence of sexual dimorphism in the Transylvanian hadrosauroids, naming the male individuals ("grooved") *O. transsylvanicus* var. *sulcata*. Nopcsa (1928) referred several vertebrae from Szenpeterfalva and Vălioara to *Telmatosaurus* (still reported as *Orthomerus*). According to that paper, all the cervical vertebrae and some dorsal and caudal vertebrae under the collective number BMNH R3841 (including an axis and articulated cervical vertebrae 3–5, three articulated mid-distal cervicals, the articulated last cervical and the first dorsal vertebrae, four isolated and incomplete dorsal vertebrae, and four isolated and incomplete caudals) belong to the same individual as BMNH R3386 (1928:287, "R3841. Alle Halswirbel, ferner einige Rumpf- und Schwanzwirbel jenes *Orthomerus*-exemplares, dessen Schädel 1899 . . . beschrieben wurde (Exemplare A)"). Weishampel et al. (1993:362–363) redescribed *Telmatosaurus transsylvanicus*, provided a formal diagnosis, indicated a lectotype (BMNH R3386; Fig. 15.2), and listed the referred material. The diagnosis is as follows:

> A hadrosaurid dinosaur . . . of small body size (a dwarf?), having a large caudal ectopterygoidal shelf, an isosceles triangle-shaped rostral process of the jugal, a relatively long post-metotic braincase, relatively large basipterygoid processes, a relatively large scar for *m. protractor pterygoideus* on the lateral aspect of the basisphenoid, a well-developed channel for the palatine branch of the facial nerve that also accommodated the median cerebral vein, absence of a diastema between the predentary and dentary dentition, and slightly bowed femur.

Referred material is from both the Sânpetru and the Densuş-Ciula formations. Some specimens were later referred to the rhabdodontid *Zalmoxes* (some elements of BMNH R3401, BMNH R3809, R3828, R3843 [the only coracoid referred to *Telmatosaurus*], R4910, FGGUB 1000, 1018, 1040, MAFI Ob.3079, MCDRD 66 and 70; Weishampel et al., 2003).

As stressed in Dalla Vecchia (2006, 2009a) and in this chapter, comparatively small body size is common to most European hadrosauroids. The isosceles triangle–shaped rostral process of the jugal (somewhat damaged proximally in BMNH R3386, Fig. 15.2E) is not unique to *Telmatosaurus* occurring, for example, also in subadults of *Edmontosaurus* (A. Prieto-Márquez, pers. comm., 2006), in *Brachylophosaurus canadensis* and *Maiasaura peeblesorum* (see Horner et al., 2004), and in *Acristavus gagslarsoni* (see Gates et al., 2011). The absence of a diastema between the predentary and the dentary dentition is observed also in *Bactrosaurus* (Godefroit et al., 1998) and in *Tethyshadros* (Dalla Vecchia, 2009c). A bowed femur is a primitive feature common to basal iguanodontians (Norman, 2004); it is also worth stressing that

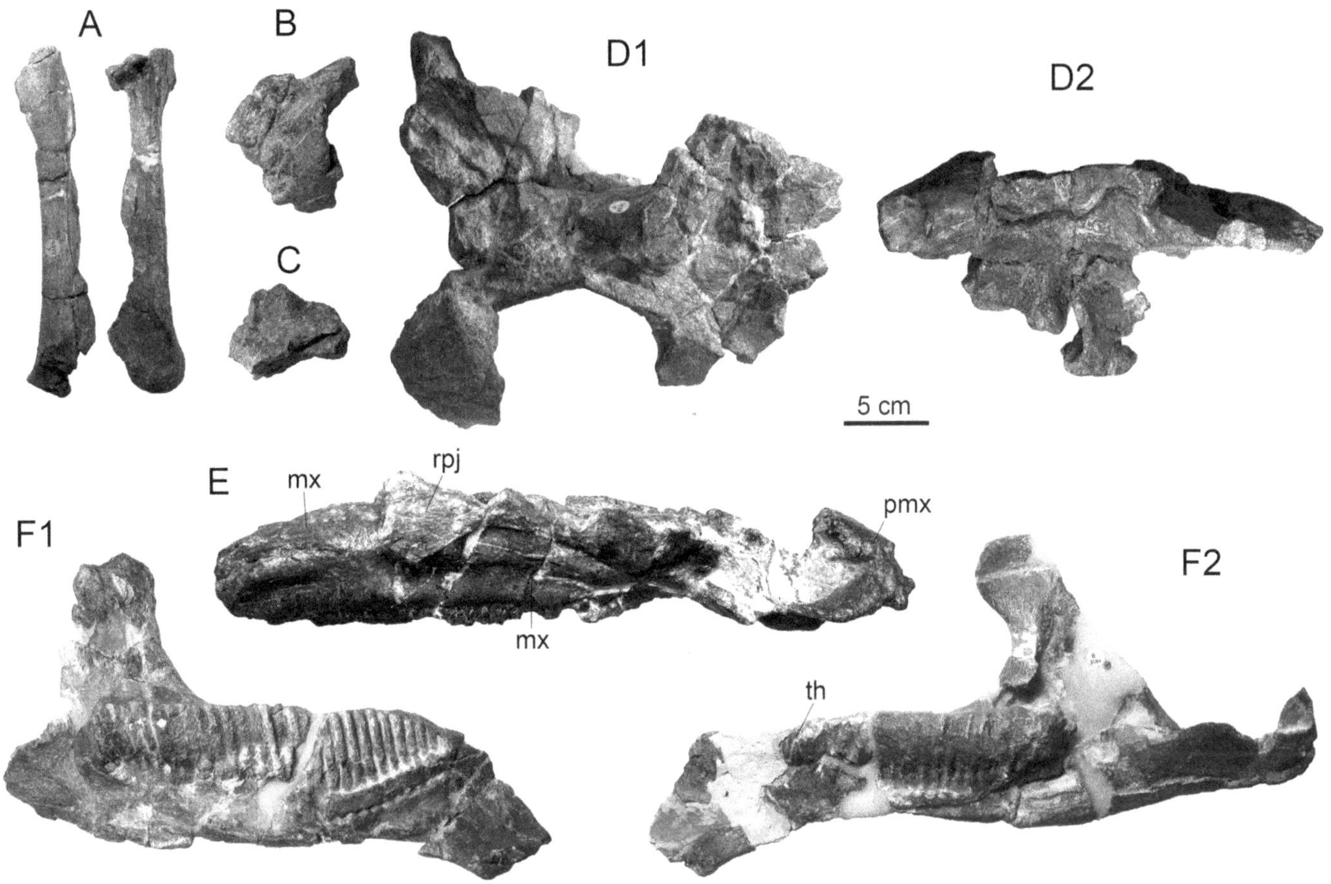

15.2. The lectotype of *Telmatosaurus transsylvanicus,* BMNH R3386, partial skull and mandibles. (A) right (left) and left (right) quadrates; (B) left squamosal with the left paroccipital process; (C), fragment possibly belonging to the rostrum; (D) skull roof and braincase in dorsal view (D1) and right lateral view (D2); (E) rostrum, right lateral view; (F) mandibular rami in lingual view, left (F1) and right (F2). Abbreviations: mx, maxilla; pmx, premaxilla; rpj, rostral process of the jugal; th, teeth. Scale bar equals 5 cm.

no femur was associated with the lectotype material (the first one was discovered in the Vălioara site after 1900; see Nopcsa, 1915). Despite the statement in Nopcsa (1928), the vertebrae from BMNH R3841 were not explicitly considered as belonging to the same individual as the lectotype.

Grigorescu and Csiki (2006) and Grigorescu et al. (2010) reported bones of hatchling hadrosauroids, and possibly newborns, too, near egg-bearing nests in the Tuştea site (Densuş-Ciula Formation) and referred the associated eggs to hadrosauroids although they are of the *Megaloolithus*-type, which is usually attributed to titanosaurian sauropods (Powell, 1992; Chiappe et al., 1998). Both bones and eggs were referred to *Telmatosaurus transsylvanicus.*

Telmatosaurus is the sister taxon of Hadrosaurinae + Lambeosaurinae in the phylogenetic analyses by Weishampel et al. (1993) and Horner et al. (2004). A similar position is found also in the analysis of Godefroit et al. (1998) and You et al. (2003). It falls in the same position in the strict consensus tree of Dalla Vecchia (2009c), but in polytomy with *Tethyshadros.* It is the sister taxon of the clade *Bactrosaurus* + Hadrosauridae in Norman (2002). In the strict consensus tree, it is in polytomy with *Tanius* and a clade formed by *Aralosaurus* + Hadrosauridae in Sues and Averianov (2009); the addition of *Arenysaurus* and *Blasisaurus* resolved the polytomy, with *Telmatosaurus* as sister taxon of the clade formed by *Aralosaurus* + Hadrosauridae (Cruzado-Caballero, Pereda-Suberbiola, and Ruiz-Omeñaca, 2010). It forms a crown polytomy with *Claosaurus agilis, Hadrosaurus foulkii, Edmontosaurus annectens,* and *Corythosaurus casuarius* in the reduced consensus tree of McDonald (2011), being more derived than *Lophorhothon atopus. Telmatosaurus* was also choosen as outgroup in phylogenetic analyses of derived hadrosauroids (e.g., Evans and Reisz, 2007; Gates et al., 2007). However, none of those analyses took into account that only the lectotype material and possibly the vertebrae of BMNH R3841 are devoid of ambiguity, while most of the other elements might belong to a different taxon. In his cladistic analysis, Prieto-Márquez (2010b) opted to use only the material described by Nopcsa (1900) plus two maxillae (FGGUB R.1010 and BMNH R4911) that are indistinguishable from those of the lectotype. In his reduced consensus tree, *Telmatosaurus* is the sister taxon of the clade *Lophorhothon* + Hadrosauridae.

Telmatosaurus was considered a "dwarf" dinosaur by Nopcsa (1914) because of its small size compared to the other hadrosauroids known at that time. This view was later supported by Weishampel et al. (1991:203) because "at ~5 m long and 500 kg, *T. transsylvanicus* was one of the smallest hadrosaurid dinosaurs. Such a body weight is ~10% of average hadrosaurids elsewhere from roughly the same time interval. Fusion of the braincase and vertebral sutures indicates that these individuals were probably fully adults at the time of their death." Using histological features of the femora, Benton et al. (2010) demonstrated that those diminutive hadrosauroids, which they refer to *T. transsylvanicus*, are small adults, not juveniles of a large-sized taxon.

Slovenia

The hadrosauroid remains from Slovenia have recently been summarized by Dalla Vecchia (2006, 2009a), and there have been no new discoveries since. Hadrosauroid teeth are reported from near Kozina, southwest Slovenia (Debeljak et al., 1999, 2002). They were found inside a paleokarst fissure formed in the Santonian–Lower Campanian Lipica Formation and are Late Campanian–Maastrichtian in age (Debeljak et al., 1999). Two specimens referred to hadrosaurids have been figured (Debeljak et al., 2002:pl. 1, fig. 4) including (1) a small maxillary tooth with a narrow crown 1.3 cm high, denticulate margins and a single median carena (ACKK-D8/20; Debeljak et al., 1999), and (2) a possible dentary tooth being apparently much wider than the other (ACKK-D8/121, crown about 2.5 cm high). Probably, the sample includes more hadrosaurid teeth because "typical hadrosaurian teeth . . . [are] the most frequent ornithopod teeth [at the site]" (Debeljak et al., 2002:194). Other possible hadrosauroid teeth are referred to the "Iguanodontidae family" and to "an unidentified ornithopod" (Debeljak et al., 2002:197, fig. 11B2).

Italy

The Villaggio del Pescatore site (Duino, Trieste Province, Friuli Venezia Giulia Region, northeast Italy) has yielded the remains of *Tethyshadros insularis* described and diagnosed by Dalla Vecchia (2009c), and there have been no new discoveries since. Those remains come from the base of the Liburnian Formation and are Late Campanian–Early Maastrichtian in age (Dalla Vecchia, 2008b). The material includes the holotype (SC 57021, a complete and articulated skeleton with tibia 55 cm long and femur 42 cm long; Fig. 15.3); a nearly complete but strongly crushed skull with mandibles exhibiting a length similar to that of the holotype skull and associated with postcranial remains from an apparently complete skeleton (SC 57026); part of an articulated and probably complete skeleton that is still unprepared (SC 57247), and that represents an individual slightly larger than the holotype (the centra of caudals 7–9 are ~6 cm long); partial, articulated forelimbs slightly more robust than the holotype, and possibly from a complete skeleton that is still in situ (SC 57022; metacarpal IV is 12.5 cm long); an isolated left pubis belonging to an individual smaller than the holotype (SC 57023; 21 cm long measured from the cranial margin of the iliac process to the distal tip of the prepubis); an isolated cervical vertebra with a fused right rib approximately the size of holotype cervicals (SC 57025; the total centrum length is 6 cm); and an isolated dorsal rib with size comparable to those of the holotype (SC 57256). *Tethyshadros insularis* was diagnosed by Dalla Vecchia (2009c:1100–1101) as follows:

> [S]kull large (skull length 1.60–1.65 times humeral length) and elongate (skull length:height ratio ~2.6); premaxillary denticles very large, slender and pointed; small paired crests in caudal part of parietals; jugal very long, slender, without ventral flange, portion caudal to dorsal process more than twice length of rostral process; infratemporal fenestra subrectangular and large, nearly twice orbit size; lateral distal condyle of quadrate flared and ventrally flat (nail-head shaped); first caudal centra longer than high; neural spines of proximal caudal vertebrae 1–6, and possibly also distal dorsals and sacrals, hatchet-shaped; ribs of caudal vertebrae 1–5 tongue-shaped, craniocaudally wide and dorsoventrally flattened; long, whip-like mid- to distal part of tail with elongated vertebral centra, 2.4 to 3.4 times longer than high at mid-centrum in caudal vertebrae 18 to 32; centra of caudal vertebrae 23 to 33 with shape of semicylinders; a long segment of proximal caudal vertebrae without haemapophyses (first chevron between caudal vertebrae 7 and 8) and consequently, there is a ventral "gap" representing 44% of the preserved holotype tail segment; distal end of haemapophyses in vertebrae 15–20 with a long posterior process; scapular blade asymmetrically expanded distally (like the primitive iguanodontians *Camptosaurus* and *Dryosaurus*); postacetabular process of ilium long, low, blade-like (no brevis shelf), triangular and tapering in lateral view; very long ischium with a sigmoid shaft and blunt, unexpanded distal end, not bent nor tapering; only three manual digits (digit V lost); flat distal articular end of metacarpals; only two phalanges in manual digit IV, distal one very reduced (lost phalanx 2 of other hadrosauroids); tibia considerably longer than femur. It also has the following unique combination of characters: low ilium with large and pendant supra-acetabular process and robust preacetabular process not markedly arched and without dorsal depression at the supra-acetabular process.

There is no evidence of immaturity in the degree of ossification and fusion of the skeletal elements; the proportions in the skull of the holotype are not those of a juvenile individual

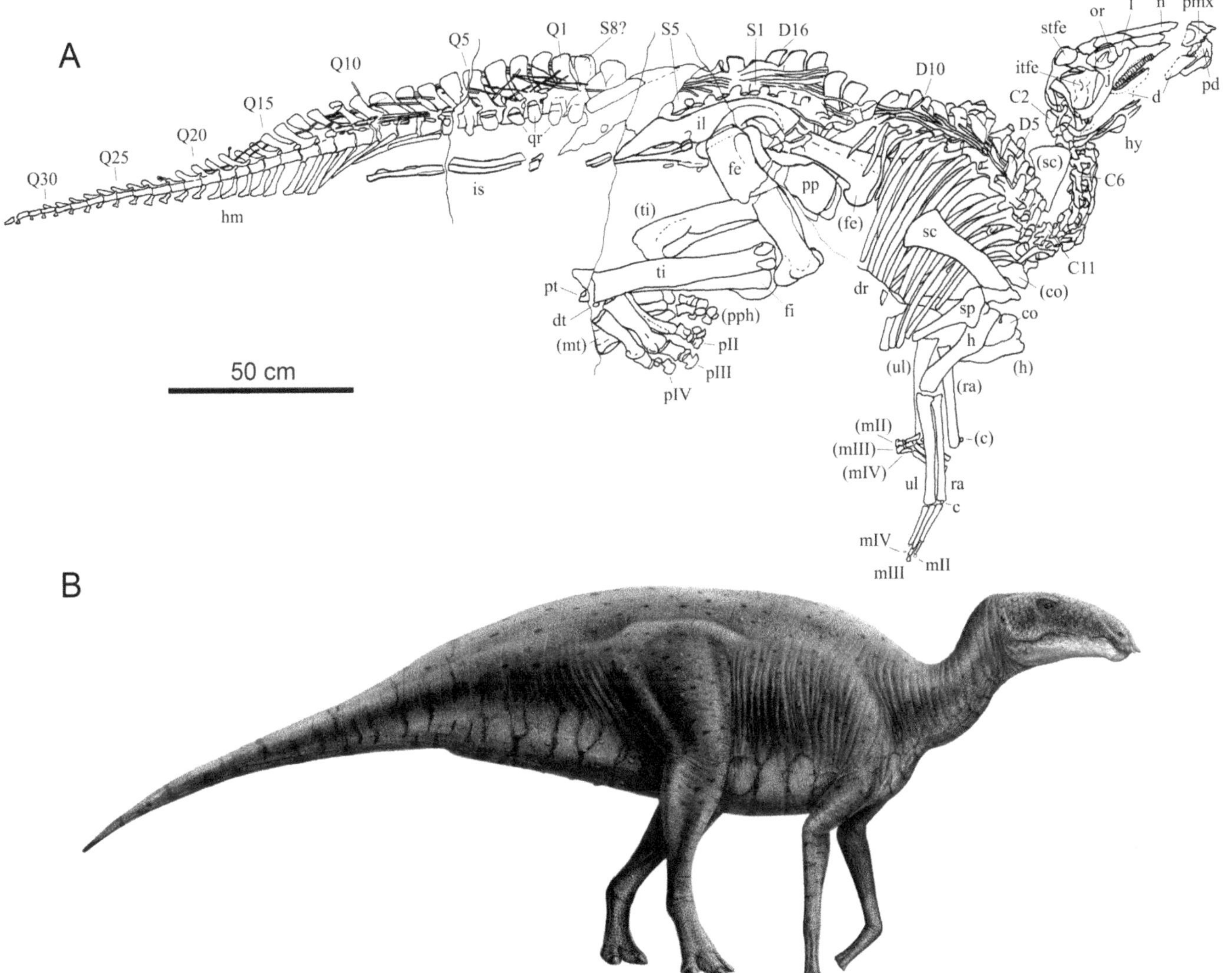

15.3. *Tethyshadros insularis* from the Villaggio del Pescatore site (Trieste, Friuli Venezia Giulia, NE Italy). (A) drawing of the holotype (SC 57021) from Dalla Vecchia (2009c); (B) reconstruction (artwork by Lukas Panzarin). Abbreviations: C, cervical vertebra; c, carpal; co, coracoid; D, dorsal vertebra; d, dentary; dr, dorsal rib; dt, distal tarsal; fe, femur; fi, fibula; h, humerus; hm, haemapophyses; hy, first ceratobranchial (hyoid apparatus); j, jugal; il, ilium; is, ischium; itfe, infratemporal fenestra; l, lacrimal; mII–IV, manal digits II–IV; mt, metatarsus; n, nasal; or, orbit; pII–IV, pedal digits II–IV; pd, predentary; pmx, premaxilla; pp, prepubic process of pubis; pph, pedal phalanges; pt, proximal tarsals; Q, caudal vertebra; qr, caudal rib; ra, radius; S, sacral vertebra; sc, scapula; sp, sternal plate; stfe, supratemporal fenestra; ti, tibia; ul, ulna. Left-side elements are in parentheses. Scale bar equals 50 cm.

(the skull is not short and deep, and orbits are not particularly large). SC 57021 is by far the most complete, articulated, latest Cretaceous hadrosauroid skeleton found to date in Europe. Its complete and well-preserved skeleton allows an accurate reconstruction of the size and body proportions of an individual of *Tethyshadros insularis* (Fig. 15.3B).

Tethyshadros insularis forms a polytomy with *Telmatosaurus transsylvanicus* and the Hadrosauridae in the strict consensus tree of the phylogenetic hypothesis of Dalla Vecchia (2009c) and is the sister taxon of the Hadrosauridae in the 50% majority rule. It forms a polytomy with *Hadrosaurus foulkii* and the Saurolophidae (i.e., it could be basal in the Hadrosauridae) in the strict consensus tree of Prieto-Márquez (2010a). It is more primitive in the reduced consensus tree of McDonald (2011), falling basal to *Bactrosaurus johnsoni* and in polytomy with the Asian *Nanyangosaurus zhugeii* and *Levnesovia transoxiana*. In the maximum agreement subtree of McDonald (2012), it falls between *Protohadros byrdi* and *Nanyangosaurus zhugeii*, and is more basal than *Bactrosaurus johnsoni*.

Germany

Hadrosauroid bones were found in the marine Gerhartsreiter Schichten of Bad Adelholzen, southern Bavaria (Wellnhofer, 1994). The horizon is dated to the Late Maastrichtian

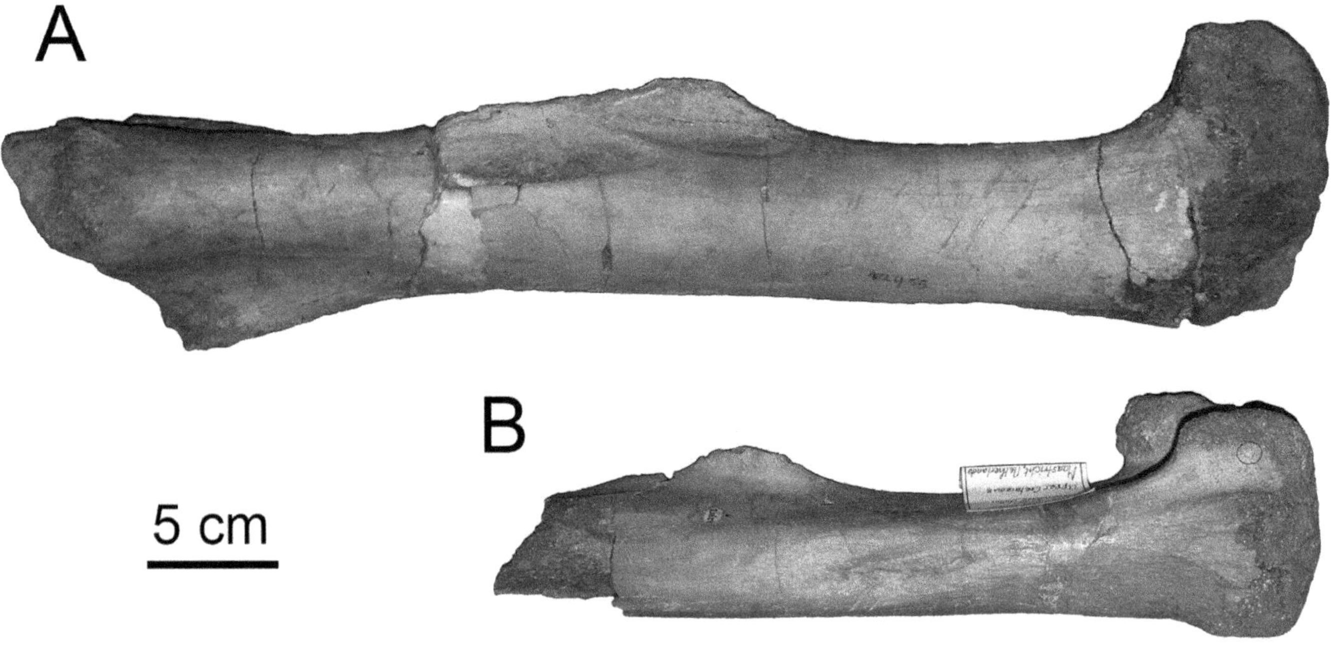

15.4. *"Orthomerus dolloi"* from the Upper Maastrichtian of Limburg. (A) BMNH 42955, right femur, medial view; (B) BMNH 42956, left femur, lateral view. Scale bar equals 5 cm.

based on its foraminifers and calcareous nanoplankton content (Wellnhofer, 1994:224); however, those microfossils have a broader range (*Gansserina gansseri* Zone; Upper Campanian–lower Upper Maastrichtian) according to López-Martínez et al. (2001:57). The specimens, which are deposited under the collective inventory number NKM 71, purportedly belong to a single individual (Wellnhofer, 1994). They consist of bones from the right hind limb (femur, metatarsal IV, and two phalanges), a fragment of a left scapula, a possible caudal vertebral centrum and a caudal neural arch. The complete femur is 34 cm long and the body length of the animal was estimated at only two meters by Wellnhofer (1994).

Belgium and the Netherlands

Isolated hadrosauroid remains are occasionally found in different levels inside the marine Maastricht Formation ("Chalk") of the Limburg, a historical province now divided into Belgian and Dutch parts. They have recently been summarized by Dalla Vecchia (2006, 2009a), and there has been only one additional caudal vertebra described since (Buffetaut, 2009). The Maastricht Formation is dated to the Late Maastrichtian (*Belemnitella junior* Zone, but with some problems; see López-Martínez et al., 2001:56). For the localities and the stratigraphic provenance of the specimens from the Maastricht Formation (which is divided into six members) see Laurent (2003) and Dalla Vecchia (2006).

The Limburg record includes the specimens on which *Orthomerus dolloi* was erected by Seeley (1883): a complete right femur (BMNH 42955, ~50 cm long, Fig. 15.4A); the distal half of a smaller left femur (BMNH 42956, Fig. 15.4B); a left tibia (BMNH 42954, ~30 cm long); and an incomplete metatarsal (BMNH 42957). No provenance data are given for the material in Seeley (1883:246) other than the vague statement, "from Maastricht," and there is no evidence of an original association of the remains. Lydekker (1888:fig. 48) referred a complete proximal caudal vertebra (centrum 5.5 cm long) from the "Maastricht beds" of Limburg to *O. dolloi*. No diagnoses for genus and species were given, and no holotype was indicated by Seeley (1883:246) who wrote, "I have referred them [the specimens] to an Iguanodont genus *Orthomerus*; and I have no doubt that the remainder of the skeleton will eventually show them to belong to a new generic type." As the material has no actual diagnostic features at the generic and specific level, *Orthomerus* and the species *O. dolloi* were considered nomina dubia by Brinkmann (1988) and Horner et al. (2004). However, that generic name was widely used during the twentieth century (see Brinkmann, 1988).

Buffetaut et al. (1985) referred to *O. dolloi* a proximal fragment of a right lower jaw (NHMM 198027), a fragmentary left ulna and a pedal phalanx (Garcet collection, no number) found in different outcrops of the Maastricht Formation. Mulder (1984) referred some fragmentary bones to *Telmatosaurus dolloi* (the distal part of a femur with an estimated total length of 60 cm, a fragment of a left tibia, and the proximal portion of a left fibula [MND K 21.04.003–5, respectively]), and considered the genus *Orthomerus* invalid, but the species "*O.*" *dolloi* valid.

Other findings consist of: three caudal vertebrae (IRScNB [Dollo, 1883] and MB.R.4450 [proximal caudal; Buffetaut, 2009]); the distal portion of a left metatarsal III (NHMM 1996001; Mulder et al., 1997); and a partial right metatarsal III and a fragmentary ?right humerus (NHMM RD 241 and TM 11253, respectively; Weishampel et al., 1999). The sample also includes five isolated maxillary and dentary teeth, probably from floating carcasses (IRScNB [no number], NHMM RD 214, NHMM 1999012, NHMM 1997274; Weishampel et al., 1999).

France

Hadrosauroid remains from several Upper Maastrichtian sites of southern France were reviewed in Laurent (2003). Lestaillats, Ausseing, Le Jadet, Tricouté 1–3, Auzas, Peyrecave, Cassagnau 1–2, and Larcan sites are in the Département de Haute-Garonne; Mérigon is in the Département de Ariège, and Le Bexen is in the Département de l'Aude (Laurent, 2003; Bilotte et al., 2010).

The two fossil-bearing outcrops of the Lestaillats site (near Mauran, Petites Pyrénées) occur in the transitional Marnes de Lestaillats (Lestaillats Marl), whose age is mid-Late Maastrichtian (Laurent, 2003; Fig. 15.5).

Lestaillats 1 yielded isolated teeth (e.g., MDE-LES1-73, dentary tooth); a fragment of a right dentary (MDE-LES1-72); a squamosal; three complete mid-posterior dorsal vertebrae (MDE-LES1-06–08); three complete proximal caudal vertebrae (MDE-LES1-4, 5, and 20); three dorsal ribs (e.g., MDE-LES1-14, 21, and 77); a left scapula (MDE-LES1-25, ~26.5 cm long); a partial left pubis (MDE-LES1-03); and three ischia with damaged extremities (a right [MDE-LES1-01], and two left [MDE-LES1-02 and 19] with estimated lengths of ~50, ~45 and ~50 cm, respectively). They were referred to indeterminate "hadrosaurids" by Laurent (2003) who also reports a rostral part of a maxilla from this outcrop. Lestaillats 2 yielded only the proximal half of a femur (MDE-LES1-10, maximum estimated length 55 cm).

The Ausseing site (near Ausseing, Petites Pyrénées) presents two close but distinct levels (Laurent, 2003:21) inside the Marnes de Lestaillats (Fig. 15.5). Hadrosauroid remains include three isolated teeth (MDE-AUS-65, 67, and 164); a fragment of a small dentary (MDE-AUS-111); a dorsal rib (MDE-AUS-160); ischia (e.g., MDE-AUS-185 [left, nearly complete, ~70 cm long]); two femora (MDE-AUS-01 [right, nearly complete, ~60 cm long] and MDE-AUS-130); three tibiae (e.g., MDE-AUS-03 [left, 60 cm long], MDE-AUS-145 and MDE-AUS-155); a right fibula (MDE-AUS-09, 49.5 cm long); a left metatarsal IV (MDE-AUS-184, 15.6 cm long); and an ungual

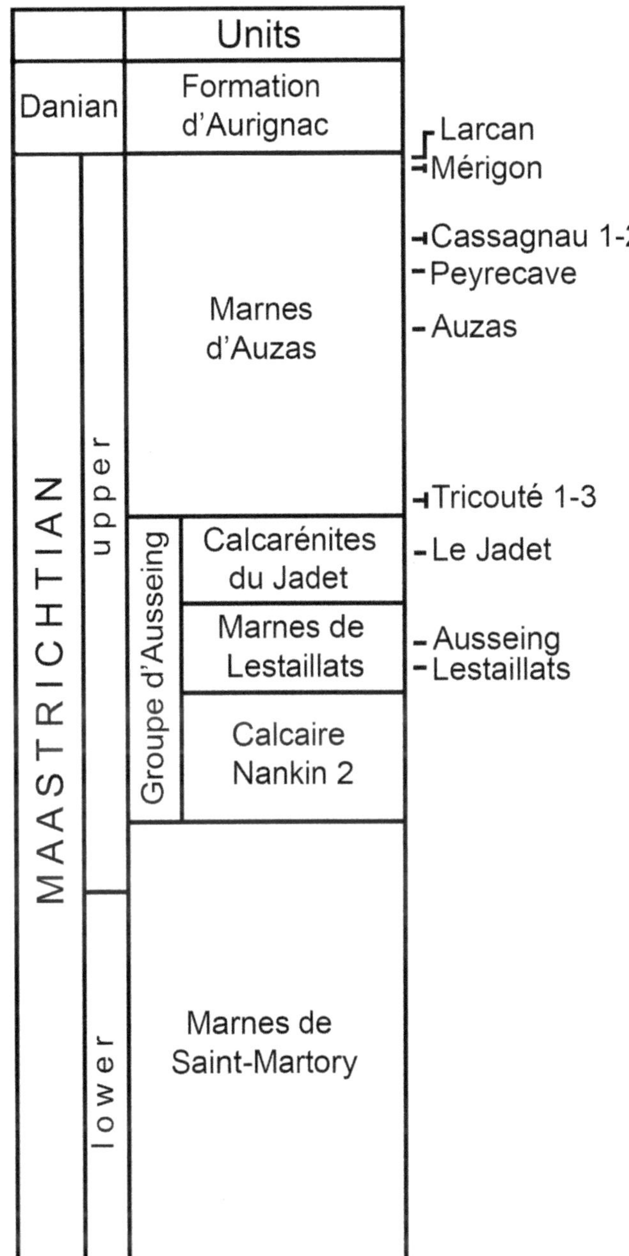

15.5. Stratigraphic position of the hadrosauroid-bearing sites in the Haute-Garonne and Ariège Départements of southern France (based on Laurent, 2003).

phalanx (MDE-AUS-47, 3.5 cm long). These were referred to indeterminate "hadrosaurids" by Laurent (2003).

Le Jadet site (near Saint-Martory, Petites Pyrénées) occurs in the marine Calcarénites du Jadet (Jadet Calcarenites; Fig. 15.5). A Late Maastrichtian age assignment is based on the micropaleontological content of the underlying units (Laurent, 2003). It yielded a fragment of a left dentary (MNHN; Paris and Taquet, 1973), which was originally referred to *Telmatosaurus transsylvanicus* because, at that time, it was the only Maastrichtian age European hadrosauroid with a preserved dentary. Laurent (2003:16) reports it as an indeterminate "hadrosaurid."

Three sites occur at distinctly different stratigraphic levels in the basal part of the Marnes d'Auzas (Auzas Marl; Fig. 15.5) near Marygnac-Lespeyres (Petites Pyrénées). From oldest to youngest these are named Tricouté 1 to 3 (Laurent, 2003:17). The Marnes d'Auzas is a transitional-continental unit that is more than 200 m thick in the type section. It is dated as late Late Maastrichtian by Laurent (2003) based on the microfossil content of its marine intercalations (e.g., the foraminifers *Hellenocyclina beotica* and *Laffiteina mengaudi*) and its rich charophyte assemblage. The upper part of the unit is mid- to late Late Maastrichtian in age according to López-Martínez et al. (2001), whereas the lower part is early Late Maastrichtian based on the charophyte biozonation of Riveline et al. (1996), which is not accepted by Laurent (2003:28).

Tricouté 1 has yielded only isolated teeth. The dentary tooth MDE-MA1-01 has a secondary mesial ridge like the dentary teeth from Tricouté 3 and Cassagnau 2 sites, and those in the dentaries from Fontllonga and La Solana sites of Spain (see below). Rostral and caudal-most dentary teeth of lambeosaurines often bear one or two secondary ridges (Horner et al., 2004), but the presence of secondary ridges is also considered a primitive feature (e.g., Casanovas-Cladellas et al., 1999). Tricouté 2 has yielded isolated teeth (including fragments of a dentary battery, MDE-MA2-01 and 02); some caudal vertebral centra; tibiae (MDE-MA1-02 [right, 47 cm long] and MDE-MA1-08 [right, ~28 cm long]); and a right metatarsal IV (MDE-MA1-03, 16 cm long).

Tricouté 3 has yielded two maxillae (MDE-MA3-15 [left] and MDE-MA3-16 [right, 17 cm long]); a left quadrate (MDE-MA3-17, 13.8 cm tall); a partial left surangular (MDE-MA3-19); a fragment of the right dentary battery (MDE-MA3-26); an isolated dentary tooth with a mesial secondary ridge (MDE-MA3-25); incomplete and fragmentary dorsal ribs (e.g., MDE-MA3-02 and 22); a left scapula (MDE-MA3-21, 23 cm long); a left sternal plate (MDE-MA3-24); a left humerus (MDE-MA3-20, 19.5 cm long); and a partial right pubis (MDE-MA3-23). Laurent (2003) considered this material as belonging to a single, small-sized individual, and referred the material provisionally to the Spanish genus *Pararhabdodon* (see below). However, referral to that taxon is unsubstantiated because the maxillae more closely resemble those found in the Larcan site (see below) and in the Asian hadrosauroid *Aralosaurus tuberiferus* (as noticed by Bilotte et al., 2010). The other bones are undiagnostic or are unknown in the Spanish taxon. The Tricouté 3 sample includes also a right prefrontal (MDE-MA3-17), a partial pterygoid (MDE-MA3-30), and a left scapular blade (MDE-MA3-12) that were not mentioned in Laurent (2003). A second sample with the same acronym (MA3) includes bones with a different, lighter, color and cannot come from the same level as the others. They include an incomplete neural arch of a cervical vertebra (MDE-MA3-01); a dorsal rib (MDE-MA3-03); a small right humerus (MDE-MA3-13, ~15 cm long); and three tibiae (MDE-MA3-11 [?left, nearly complete, ~35 cm long], MDE-MA3-09 [~23 cm long], and MDE-MA3-10 [partial]).

The Auzas site (near Auzas, Petites Pyrénées) occurs in the middle of the Marnes d'Auzas (Fig. 15.5). It yielded the proximal part of a right humerus (MDE-AUZ1-01) and the distal part of a femur referred to indeterminate "hadrosaurids" by Laurent (2003).

The Peyrecave site (near Alan, Petites Pyrénées) yielded isolated teeth that were also referred to indeterminate "hadrosaurids" by Laurent (2003). Fossils occur in two distinct levels (Peyrecave 1 and 2) inside the upper half of the Marnes d'Auzas (Fig. 15.5).

Two fossiliferous levels have been identified in the Cassagnau site (near Marygnac-Lespeyres, Petites Pyrénées), which is in the upper half of the Marnes d'Auzas (Fig. 15.5); Cassagnau 1 is less than 15 m lower stratigraphically than Cassagnau 2 (Laurent, 2003). Cassagnau 1 has yielded 20 isolated teeth (e.g., the maxillary MDE-CAS1-03, and the dentary MDE-CAS1-01, 11, and 207, with crowns sometimes having a faint secondary ridge); a small alveolar fragment (MDE-CAS1-256); the postzygapophysis of a cervical vertebra (MDE-CAS1-226); the partial neural arch of a dorsal vertebra (MDE-CAS1-07); two caudal vertebral centra (MDE-CAS1-05 and 212); the proximal portion of a relatively large left ischium (MDE-CAS1-200); three ungual phalanges (MDE-CAS1-14 and 215–216); and a phalanx of manal digit V (MDE-CAS1-166).

Hadrosauroid bones are relatively abundant in Cassagnau 2. The sample includes a small predentary (MDE-CAS2-138, 4 cm long) and a complete left dentary (MDE-CAS2-248, 16.2 cm long), supposedly from the same individual; an incomplete left dentary (MDE-CAS2-02, estimated length ~16 cm); a tooth-bearing dentary fragment (MDE-CAS2-11, crowns lack a secondary ridge); two alveolar fragments of small dentaries (MDE-CAS2-60 and 256); a larger alveolar fragment (MDE-CAS2-160); five isolated teeth (e.g., the dentary tooth MDE-CAS2-223, with a mesial secondary ridge); a dorsal neural arch with a tall neural spine (MDE-CAS2-206); six dorsal ribs (e.g., MDE-CAS2-38, 65, 260-61, and 217); a nearly complete small caudal vertebra (MDE-CAS2-62); six small caudal centra (MDE-CAS2-95, 123, 171, 251, and 252–253); an isolated caudal neural arch (MDE-CAS2-210); an isolated caudal neural spine (MDE-CAS2-161); a chevron (MDE-CAS2-239); a right humerus (MDE-CAS2-207, ~12 cm long); a left pubis (MDE-CAS2-01, 21.7 cm total length); six tibiae including two right proximal halves (MDE-CAS2-23 and MDE-CAS2-120), a nearly complete left tibia (MDE-CAS2-249, 28 cm long), a right tibia (MDE-CAS2-247,

~29 cm long), a left tibia (MDE-CAS2-24, ~36.5 cm long), a right tibia (MDE-CAS2-216, ~27.5 cm long); a right fibula (MDE-CAS2-250, 28 cm long); and a left metatarsal II (MDE-CAS2-209, 11.8 cm long). A partial articulated skeleton of a very small individual is composed of seven cervical vertebrae (MDE-CAS2-04–10); two femora, (MDE-CAS2-12 [18 cm long] and 15 [right, 19.3 cm long]); a left tibia (MDE-CAS2-13, 18 cm long according to Laurent [2003:pl. 46, fig 136], and thus longer [unusually so] than the corresponding femur); two fibulae (MDE-CAS2-14 and 16); three metatarsals (II–IV, MDE-CAS2-17–19, 5.8, 7.5 and 6 cm long, respectively); and two proximal pedal phalanges (MDE-CAS2-21–22). All the material was referred to indeterminate "hadrosaurids."

Recently, Bilotte et al. (2010) described a small right maxilla (11.5 cm long; private collection) and an associated quadrate (8.0 cm long) found in the Larcan site (near Saint-Marcet, Petites Pyrénées). They suggested a lambeosaurine affinity for the maxilla, which differs from those of *Pararhabdodon isonensis*, *Telmatosaurus transsylvanicus*, and *Tethyshadros insularis*. It might represent a distinct hadrosauroid taxon, but unfortunately the maxillae of *Arenysaurus ardevoli* and *Blasisaurus canudoi* are unknown, so comparison is impossible. It most closely resembles the maxillae from Tricouté 3 and that of *Aralosaurus tuberiferus*. The specimens come from the uppermost part of the marine Marno-calcaires jaunes de Gensac ("yellow marl-limestone of Gensac," a distal equivalent of the uppermost part of the Marnes d'Auzas; Fig. 15.5); it was collected, at most, one meter below a stratigraphic discontinuity marked by an iridium concentration that Bilotte et al. (2010) refer to the K-Pg boundary.

The Mérigon site (near Mérigon, Le Plantaurel) is located in the uppermost part of the Marnes d'Auzas (Fig. 15.5); it is mid- to late Late Maastrichtian in age according to López-Martínez et al. (2001) and latest Maastrichtian according to Laurent (2003). It yielded an isolated tooth and a fragmentary dentary that Laurent (2003) refers to indeterminate "hadrosaurids."

Le Bexen site (near Fontjoncouse, Corbierès orientales) is the easternmost and most distant from the Atlantic Gulf of all the French sites with respect to Maastrichtian paleogeography. It occurs in the Marnes Rouges de Roquelongue (Red Marl of Roquelongue), which according to Laurent et al. (1997) is Late Maastrichtian in age because it contains the charophyte *Septorella ultima*. Instead, López-Martínez et al. (2001) has proposed that this deposit may be mid-Late Maastrichtian in age. The sample includes a bone identified originally as a fragment of a right maxilla that actually may be a fragment of a dentary (MDE-FO1-11); a partial right dentary (MDE-F01-10, 30 cm maximum estimated length); two fragments of a right (MDE-FO1-01), and a left (MDE-FO1-03) dentary; three isolated teeth (MDE-FO1-12, 14, and 16); six cervical centra (MDE-FO1-4, 5, 17, 22–23, and 114); a single isolated sacral centrum (MDE-FO1-53) and two fused sacral centra (MDE-FO1-21); three proximal (MDE-FO1-64, 97, and 136) and approximately 30 mid-caudal centra; a right humerus (MDE-FO1-18, 17.5 cm long); a fragmentary left radius (MDE-FO1-45); three partial right femora (MDE-FO2-01, MDE-FO1-19 and 148, estimated lengths 21.5, 57.5, and 39.5 cm, respectively); and three partial tibiae (two left, one right; MDE-FO1-139, 143, and 115, respectively). This material was referred to *Telmatosaurus transsylvanicus* by Le Loeuff and Buffetaut (1993); later Laurent et al. (1997) and López-Martínez et al. (2001) reported them as *Pararhabdodon* sp. More recently, Laurent (2003) has referred to them as indeterminate "hadrosaurid" material. No actual diagnostic features of *Telmatosaurus* and *Pararhabdodon* are present in this material.

Spain

In Spain, hadrosauroid remains are found along the southern part of the east-central Pyrenees (Catalonia and Aragón), in the Burgos Province (NW Spain), and in the Valencia province (Fig. 15.1).

The Laño quarry site (near Laño, Condado de Treviño, Burgos) has yielded what is probably the oldest hadrosaurid specimen from the "Ibero-Armorican" Island and the only one associated with remains of rhabdodontids to date. This rich site is located in an unnamed unit that was considered Late Campanian in age (Nuñez-Betelu, 1999; Pereda-Suberbiola, Ruiz-Omeñaca, and Company, 2003), but Martin (2007) considers it to be of Late Campanian–Early Maastrichtian age and Company et al. (2009) consider it to be of Late Campanian–basal Maastrichtian age. The only hadrosauroid fossil is a single, incomplete maxillary tooth crown (MCNA 10510, only 5 mm tall) referred to an indeterminate "hadrosaur" by Pereda-Suberbiola, Ruiz-Omeñaca, and Company (2003:383).

Abundant hadrosauroid specimens were found in the "gray unit" and "lower red unit" of the Tremp Formation (Late Maastrichtian) cropping out in the Vallcebre, Tremp, and Ager Synclines of the east-central Pyrenees (see Dalla Vecchia et al., this volume).

The distal end of a small right femur was collected in the Peguera site of the Vallcebre Syncline (Berga, Barcelona Province, Catalonia; Pereda-Suberbiola, Ruiz-Omeñaca, Ullastre, and Masriera, 2003). It was found in the "Reptile sandstone" at the top of the lower red unit, which has been assigned to Chron C29r (see Oms et al., 2007) and, thus, is latest Maastrichtian in age. Also "hadrosaur" tracks are reported in the uppermost Maastrichtian sites of the Cingles

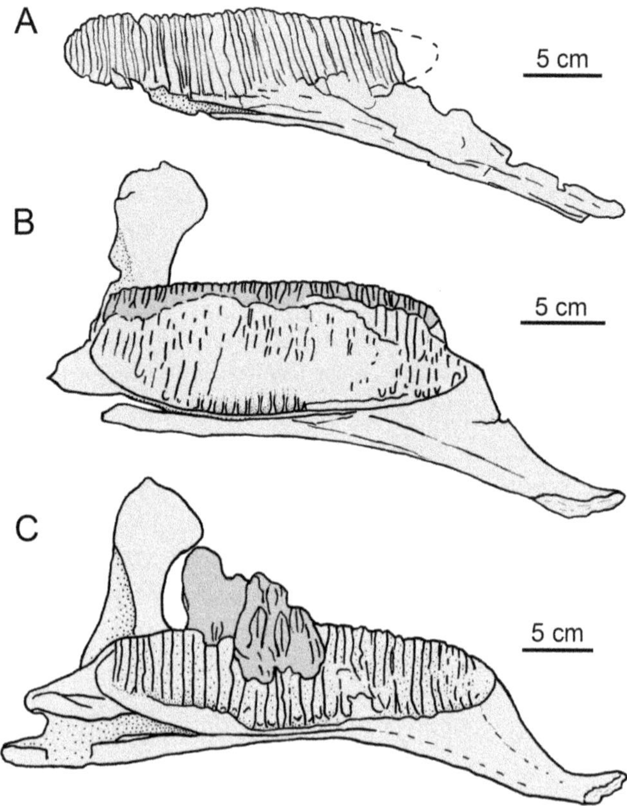

15.6. Lambeosaurine dentaries from the Upper Maastrichtian of the Tremp Syncline, Spain, in lingual view. (A) *Pararhabdodon isonensis* (IPS-29920, former IPS SRA 27), from Sant Romà d'Abella site, mirrored; (B) *Blasisaurus canudoi* (MPZ 99/665), from Blasi 1 horizon; (C) *Arenysaurus ardevoli* (MPZ2008/258), from Blasi 3 horizon. Scale bar equals 5 cm.

de Boixader, Coll de Jou, and La Pleta (Vila et al., 2005; Vila, Oms, Galobart, Marmi, and Gaete, 2006; Vila, Oms, Galobart, Gaete, et al., 2006). At least seven tracks are preserved in the Cingles de Boixader site; no information about their length is yet available, but according to Vila et al. (2005:fig. 4), they are probably not longer than 35 cm. The other sites have not been described yet.

More than 45 sites containing hadrosauroid skeletal elements and tracks occur in the eastern Tremp Syncline (Lleida Province, Catalonia) and are discussed by Dalla Vecchia et al. (this volume). Of these, the Late Maastrichtian Sant Romà d'Abella site yielded the type material of the basal lambeosaurine *Pararhabdodon isonensis* (sister taxon of the Asian *Tsintaosaurus spinorhinus* according to Prieto-Márquez and Wagner [2009]). The type material comprises: partial maxillae; cervical, dorsal and caudal vertebrae; the nearly complete sacrum; dorsal ribs; a left humerus; and parts of a left scapula and right ischium. The nearby Les Llaus site yielded the right dentary IPS-29920 (formerly IPS SRA 27; Fig. 15.6A). This was once considered the holotype of *Koutalisaurus kohlerorum* (see Prieto-Márquez et al., 2006), but was later referred to *Pararhabdodon isonensis* (see Prieto-Márquez and Wagner, 2009). At present (2012) it is under revision for the third time.

For a brief summary of the complex nomenclatural history of *Pararhabdodon* and description of referred specimens, see Prieto-Márquez et al. (2006), and Dalla Vecchia et al. (this volume). The revised diagnosis of *P. isonensis* (Prieto-Márquez and Wagner, 2009:1240) is as follows:

> Differs from all other hadrosaurids, except *Tsintaosaurus spinorhinus*, in elevation of jugal facet of maxilla such that ventralmost extent is well above level of lateral margin of ectopterygoid shelf, maxilla forms acute embayment extending ventral to jugal process between jugal facet and ectopterygoid shelf; jugal facet of maxilla anteroposteriorly foreshortened, likely with correspondingly anteroposteriorly narrow anterior jugal; anterior dentary with symphysial process projecting medially such that distance between symphysis and lateral surface of dentary is three times labiolingual thickness of alveolar chamber. Differs from *Tsintaosaurus spinorhinus* in broader, subrectangular anterodorsal region of the maxilla.

Hadrosauroid-bearing sites are less numerous in the western Tremp Syncline, being reported only from the Blasi hill near the village of Arén/Areny in the Noguera Ribagorzana valley (Huesca Province, Aragón). Five sites with six distinct fossil-bearing horizons are referred to as Blasi 1–5; an additional site, "Blasi 3,4," was recently reported by Cruzado-Caballero, Ruiz-Omeñaca, Gaete, et al. (2010). Blasi 1, 2, and 3 are very close to one another geographically (tens of meters) and occur at different horizons within a stratigraphic section that is only a few meters thick (Fig. 15.7). Blasi 1 is the topmost layer of the Arén Formation, whereas Blasi 2a–b and 3 are at the base of the overlying Tremp Formation (gray unit corresponding to La Posa Formation of Cuevas, 1992). Blasi 4 and Blasi 5 sites are located three km west of the Blasi 1–3 sites, and are 70 and 100 m stratigraphically higher, respectively, than Blasi 3 (López-Martínez et al., 2001) in the lower red unit of the Tremp Formation (corresponding to the Conquès Formation of Cuevas, 1992). "Blasi 3,4" is located in the same stratigraphic section that includes Blasi 4 and 5, but occurs in the basal unit of the Tremp Formation. All sites are Late Maastrichtian in age with Blasi 1–3 occurring in magnetochron C30n, and Blasi 5 possibly in magnetochron C29r (see Pereda-Suberbiola, Canudo, Cruzado-Caballero, et al., 2009).

About 350 hadrosauroid skeletal elements have been collected from the Blasi sites (Cruzado-Caballero, 2012). Blasi 1 yielded the type elements of the lambeosaurine *Blasisaurus canudoi* that are presumed to belong to a single individual because they were found associated (Cruzado-Caballero, Pereda-Suberbiola, and Ruiz-Omeñaca, 2010). The sample includes the left jugal (MPZ99/667, holotype); two fragments

15.7. The Blasi locality near Arèn/Areny (Huesca Province, Aragón, Spain) with the Blasi 1 to 3 hadrosaurid-bearing horizons.

of the left maxilla (MPZ99/666a, b); the right lacrimal (MPZ2009/348); the left dentary with teeth (MPZ 99/665, 34 cm long; Fig. 15.6B); and the right surangular (MPZ99/664). According to Cruzado-Caballero, Pereda-Suberbiola, and Ruiz-Omeñaca (2010:1513), the type material belongs to a "non-juvenile individual." *Blasisaurus* is characterized by a single autapomorphy in the jugal: the combination of a hook-like dorsal portion of the caudal process and a relatively narrow and D-shaped infratemporal fenestra (Cruzado-Caballero, Pereda-Suberbiola, and Ruiz-Omeñaca, 2010). Furthermore, the jugal includes a unique combination of characters: caudal edge of the rostral process well projected ventrally in a straight line (shared with *Parasaurolophus*), concave caudoventral edge beneath the infratemporal fenestra (shared with *Sahaliyania* and hadrosaurines), and length-to-height ratio lower than 1.2 (shared with *Olorotitan*, *Velafrons*, and *Tsintaosaurus*). The dentaries of *Blasisaurus* and *Pararhabdodon* have similar size and the same number of tooth positions (35), but the rostralmost portion is only modestly deflected ventrally and medially in *Blasisaurus* (Cruzado-Caballero, Pereda-Suberbiola, and Ruiz-Omeñaca, 2010). Other comparisons between the two taxa cannot be made because the jugal of *Pararhabdodon* is unknown.

The Blasi 1 horizon also yielded postcranial material that was not explicitly referred to *Blasisaurus canudoi*. This includes ribs, an ulna, ischia, femora, and metapodials of at least five individuals with "two different morphologies" (Cruzado-Caballero et al., 2007:55). The only described specimens are four more or less incomplete femora referred by Cruzado-Caballero et al. (2009) to the same morphotype (MPZ2007/930, left, estimated length ~77 cm; MPZ2007/932, right, estimated length ~69 cm; MPZ2007/702, left, estimated length ~60 cm; and MPZ2007/934, left, smaller than the others), and a nearly complete femur that is more gracile and exhibits a recurved shaft (MPZ2007/933, left, estimated length ~55 cm).

Blasi 2a yielded a dorsal and a caudal vertebra, and ribs. Blasi 2b yielded isolated teeth, a scapula, a humerus, and an ischium (Cruzado-Caballero et al., 2007).

Blasi 3 yielded the type material of the lambeosaurine *Arenysaurus ardevoli*, which is presumed to belong to a single individual because the postcranial material and the dentary are associated with the holotype and partially articulated (Pereda-Suberbiola, Canudo, Cruzado-Caballero, et al., 2009). The type material includes the caudal portion of the skull comprising the skull roof and braincase (MPZ 2008/1, holotype); fragments of the right and left maxillae (MPZ2008/256–257); the left dentary with part of the tooth battery (MPZ2008/258, 44.5 cm long, Fig. 15.6C); the right surangular (MPZ2008/259); 4 isolated teeth (MPZ2008/260–263); 7 cervical vertebrae and three cervical ribs (MPZ2007/706, 954–955, and 264–267); 1 dorsal vertebra (MPZ2008/268); 2 dorsal ribs (MPZ2008/269–270); the partial sacrum (MPZ2008/271); a pathological caudal vertebra (MPZ2004/480); 2 other isolated caudal vertebrae (MPZ2008/272 and 313); 2 isolated chevrons (MPZ2008/314 and 330); a string of 14 articulated caudal vertebrae with chevrons (MPZ2006/20); ossified tendons (MPZ2008/331, 332); a fragmentary right scapula (MPZ2008/333a, b); the right coracoid (MPZ2008/334); the incomplete right humerus (MPZ2008/336); the fragmentary right ilium (MPZ2008/335); the right pubis (MPZ2007/707); and the femora (MPZ2007/711, right; MPZ2008/337, left, 71.1

cm long). This material is considered to be from an adult individual because of the full ossification of the braincase, absence of open cranial sutures in the skull, and fusion of the elements in the axial skeleton.

Pereda-Suberbiola, Canudo, Cruzado-Caballero, et al. (2009:564) cite the apomorphies of *Arenysaurus ardevoli* as follows: "very prominent frontal dome, more developed than in other adult [lambeosaurine] specimens . . . nearly vertical prequadratic process of the squamosal and jugal process of the postorbital . . . deltopectoral crest of the humerus oriented anteriorly." They also characterized it as having the following unique combination of characters: "short frontal, with a posterior length/width ratio estimated at 0.5; midline ridge of parietal approximately at the level of the postorbital-squamosal bar; parietal not interposed between the squamosals in the occipital surface of the skull; lateral side of squamosal relatively low above the cotyloid cavity."

A comparison with *Blasisaurus canudoi*, which was found a few meters stratigraphically below *Arenysaurus ardevoli* and practically in the same place (Fig. 15.7), was limited to the dentary and teeth. According to Cruzado-Caballero, Pereda-Suberbiola and Ruiz-Omeñaca (2010), the main feature allowing distinction between the two taxa is the presence of a secondary mesial ridge in the dentary teeth of *Arenysaurus*, whereas *Blasisaurus* crowns have only the median ridge. Cruzado-Caballero et al. (2011) list further differences: the dentary of *Arenysaurus* is concave lateromedially, whereas that of *Blasisaurus* is straight in dorsal view; the distance between the first dentary tooth and the inflection point in the rostral part of the dentary is 15.5% of the dental battery length in *Arenysaurus* but only 4.5% in *Blasisaurus*; the dorsal side of the coronoid process has a caudodorsal point in *Arenysaurus*, whereas it is rounded in *Blasisaurus* (actually, the part of the process bearing that point in *Arenysaurus* seems to be broken in the only known dentary of *Blasisaurus*; Fig. 15.6B, C); the coronoid process of *Blasisaurus* has a rostral expansion that is slightly less marked than in *Arenysaurus*; the dental battery has 37 tooth positions in *Arenysaurus*, 35 in *Blasisaurus* (however, the dentary of *Arenysaurus* is 131% the length of that of *Blasisaurus*); the height-to-length ratio of the dentary teeth in the mesial positions is lower in *Arenysaurus* (3.15) than in *Blasisaurus* (3.65) (however, no mesial teeth seem to be preserved in the only reported dentary of *Arenysaurus*; Fig. 15.6C). According to Cruzado-Caballero et al. (2011:35), the two taxa also differ in the jugal morphology: the dorsocaudal side of the rostral process is straight in *Arenysaurus* and not in *Blasisaurus*; the "maxillary process" is projected less laterally and more rostrally in *Blasisaurus* than in *Arenysaurus*; the postorbital process of *Blasisaurus* is more caudodorsally oriented with regard to the long axis of the jugal than that of *Arenysaurus* (60° against 45°, respectively); the orbital fenestra is V-shaped in *Arenysaurus*, whereas it has a different outline in *Blasisaurus* (probably more U-like). However, because no jugal is reported in the type material of *Arenysaurus* (Pereda-Suberbiola, Canudo, Cruzado-Caballero, et al., 2009), Cruzado-Caballero et al. (2011) appear to be referring to new, unpublished material (see Cruzado-Caballero, 2012).

Comparison of *Arenysaurus ardevoli* with the practically coeval (see Dalla Vecchia et al., this volume) and sympatric lambeosaurine *Pararhabdodon isonensis* is limited to the humerus and the dentary (Fig. 15.6A); there is no overlap with the other taxonomically informative bones. Differences include the apomorphic orientation of the deltopectoral crest of the humerus in *Arenysaurus*, the lesser ventral deflection of the rostral end of the dentary, and its rostral edentulous portion that is shorter and less medially projecting (Pereda-Suberbiola, Canudo, Cruzado-Caballero, et al., 2009). In the phylogenetic hypothesis of Pereda-Suberbiola, Canudo, Cruzado-Caballero, et al. (2009), *Arenysaurus* is the sister taxon of *Amurosaurus* and the clade Corythosaurini + Parasaurolophini; it is relatively basal in the Lambeosaurinae, but more derived than *Tsintaosaurus* and, thus, *Pararhabdodon* (see Prieto-Márquez and Wagner, 2009).

The inclusion of *Blasisaurus canudoi* in the analysis produces a clade *Arenysaurus* + *Blasisaurus* forming a polytomy with *Amurosaurus* and the clade Corythosaurini + Parasaurolophini (Cruzado-Caballero, Pereda-Suberbiola, and Ruiz-Omeñaca, 2010). However, according to a further cladistic analysis by Cruzado-Caballero et al. (2011), *Blasisaurus* is the sister taxon of *Parasaurolophus*, whereas *Arenysaurus* is the sister taxon of *Blasisaurus* + *Parasaurolophus*.

Other material not explicitly referred to *Arenysaurus* was collected at Blasi 3 (Cruzado-Caballero et al., 2005, 2007). The sample includes sacral vertebrae and pelvic girdle bones belonging to at least two individuals; one is considered immature because the sacrum is unfused (P. Cruzado-Caballero, pers. comm.) and the other is considered a small-sized adult (Cruzado-Caballero et al., 2007). The complete right ilium (MPZ2005/90, 47 cm long) of the immature individual was attributed to an indeterminate lambeosaurine (Cruzado-Caballero et al., 2005), and although this referral is likely correct, it is not well-supported (Dalla Vecchia, 2009a). Blasi 3 yielded also another right femur (MPZ2007/704, estimated length ~75 cm; Cruzado-Caballero et al., 2009).

Blasi 3,4 yielded an incomplete left dentary (Bla 3,4-1) about 25 cm long, with 36 tooth positions and a coronoid process inclined rostrally. It was tentatively considered to be from a hadrosaurid taxon different from those already reported from the Iberian Peninsula (Cruzado-Caballero, Ruiz-Omeñaca, Gaete et al., 2010).

15.8. The track site of Arèn/Areny (Huesca Province, Aragón, Spain) in the uppermost Arèn Formation.

Cranial and postcranial elements were discovered in Blasi 4, including a fragment of a left jugal (MPZ2007/1884, the only specimen described to date) referred to Lambeosaurinae indet. (Cruzado-Caballero, Ruiz-Omeñaca, and Canudo, 2010).

Blasi 5 yielded postcranial bones (cervical and dorsal vertebrae, ribs, fragmentary pubis and femur, and a phalanx) and an incomplete right jugal (Cruzado-Caballero et al., 2007). The jugal (MPZ2007/1885, the only specimen described to date) is referred to Hadrosaurinae indet. (Cruzado-Caballero, Ruiz-Omeñaca, and Canudo, 2010).

Barco et al. (2001) reported two "hadrosaurid" trackways with footprints 50 and 41 cm long, respectively (based on the better preserved specimens; Fig. 15.8). They were found near Arén at the top of the Arén Formation (Upper Maastrichtian). The footprints probably overestimate the actual size of the pes of the trackmaker if the measurements were taken with reference to the external border of the depression (see Dalla Vecchia and Tarlao, 2000).

Hadrosauroid remains are common also in the Àger Syncline, just south of the Tremp Syncline in the Lleida Province of Catalonia (see fig. 1 in Dalla Vecchia et al., this volume). Only a left dentary lacking the symphyseal portion (IPS 36338, estimated length ~40 cm; Fig. 15.9A) from the Fontllonga site has been described to date (Casanovas-Cladellas et al., 1999). The site occurs in the upper part of the lower red unit (=Talarn Formation of Cuevas, 1992) of the Tremp Formation and is located in the highest part of Chron C30n (López-Martínez et al., 2001; Jean Le Loeuff, pers. comm.). Casanovas-Cladellas et al. (1999) referred the specimen to a basal euhadrosaurian (sensu Weishampel et al., 1993) more derived than *Telmatosaurus*, but less derived than the Hadrosaurinae + Lambeosaurinae. Pereda-Suberbiola, Ruiz-Omeñaca, and Company (2003:380) considered it as a "basal hadrosaur unlike *Telmatosaurus*" because the dentary crowns have a mesial secondary ridge and the coronoid process is perpendicular to the axis of the dentary. According to Pereda-Suberbiola, Canudo, Company, et al. (2009:940), it is a non-hadrosaurid hadrosauroid because it has several primitive features: a tooth row that terminates ventral to the apex of the coronoid process; subvertical coronoid process; tooth row that is not parallel with the lateral side of the dentary; and only 30–31 tooth positions. Furthermore, there are two functional teeth and only two replacement teeth, and the height–mesiodistal length ratio of the mid-distal crowns is relatively low (~2.7). IPS 36338 shares with the dentary of *Arenysaurus* the mesial secondary ridge in the crowns (at least the distal ones) and the presence of a caudodorsal point in the dorsal side of the coronoid process, features lacking or unknown in *Blasisaurus*. However, it is more robust, its coronoid process is comparatively stouter and slightly less inclined rostrally, and the number of tooth positions is much lower despite of being only about 10% shorter.

Tracks referred to hadrosaurs are also reported from this syncline in the Mata del Viudà site (cited also as Mas de Sauri) near the village of Millà (Llompart, 1979). The record consists of eight tridactyl tracks (positive hyporeliefs)–four of them possibly arranged in a trackway–and a dozen that are smaller and misshapen, tentatively referred to smaller individuals of the same trackmaker. The maximum length of the larger footprints ranges from 38–40 cm, the smaller ones are about 10 cm long. The larger ones were referred to cf. *Hadrosauropodus langstoni* by Vila, Oms, Galobart, Marmi, and Gaete (2006). They are preserved in the uppermost part of the lower red unit of the Tremp Formation (= Talarn Formation of Cuevas, 1992), about 25 m below the K-Pg boundary (López-Martínez et al., 1999).

La Solana site is located near the village of Carlet (Tous, Valencia Province) and occurs in an unnamed unit whose age is considered Late Maastrichtian on the basis of a charophyte assemblage (Company, 2004). Hadrosauroid remains are described in Company et al. (1998) and Company (2004). They include a fragment of a right maxilla (MPV 184); a nearly complete left dentary with 29 tooth positions and many teeth in situ (MGUV 2200, estimated length ~25 cm; Fig. 15.9B); a right dentary morphologically similar to MGUV 2200, but much larger and with 36–37 tooth positions (MPV 181, estimated length ~45 cm); a fragment of a right dentary (MPV 182); 10 isolated maxillary teeth; 4 isolated dentary teeth like those in the dentaries; 8 specimens of cervical vertebrae; 6 incomplete dorsal vertebrae; 7 caudal centra; fragmentary ribs; 2 incomplete left humeri (MGUV 2190 and 2210); 3 incomplete femora (MGUV 2204, right; MPV 186, left, estimated length ~42 cm; and one without number, but purportedly at the IPS); and 2 incomplete tibiae (MGUV 2202, left; and MGUV 2203, left, estimated length ~48 cm). Based on histology, MGUV 2203 is considered an adult by Company (2004). The dentaries were proposed as belonging to the same species because they and their teeth are alike, and given the marked size disparity of the dentaries, the different tooth counts were interpreted as resulting from ontogenetic differences (Pereda-Suberbiola, Canudo, Company, et al., 2009). This material was referred to the Hadrosauridae (sensu Weishampel et al., 1993) because the tooth row terminates posterior to the apex of the coronoid process and the number of tooth positions in the larger specimen is relatively high. However, because of the subvertical coronoid process and the presence of a secondary ridge in the crowns, the material was considered as belonging to a basal hadrosaurid (sensu Weishampel et al., 1993), more derived than *Telmatosaurus* (Pereda-Suberbiola, Canudo, Company, et al., 2009).

Lastly, the nearby Loma Cortada site yielded a distal fragment of a small left tibia (MGUV 5300; Company, 2004).

DISCUSSION

Body Size of European Hadrosauroids and Insular Dwarfism

European dinosaurs have often been considered dwarfs because of their small adult body size with respect to that of their African, Asian, and North American relatives. Nopcsa (1914) was the first to suggest that dinosaurs from the Maastrichtian of Transylvania (the hadrosauroid *Telmatosaurus*, the rhabdodontid *Zalmoxes*, and the nodosaurid *Struthiosaurus*) were insular dwarfs based on comparisons with body sizes of similar and coeval dinosaurs from elsewhere in the world, and by analogy with dwarfism of Pleistocene mammals from Mediterranean islands. Similar to Nopcsa, more recent authors have considered Transylvanian dinosaurs as insular dwarfs based on comparisons of adult body size and morphometric analyses (Weishampel et al., 1991, 1993, 2003; Jianu and Weishampel, 1999; Pereda-Suberbiola and Galton, 2001; Grigorescu, 2005). In contrast, Jianu and Boekschoten (1999) argued against their insular status, considering the Haţeg Basin as an outpost of a continental area, whereas Le Loeuff (2005) argued that the Transylvanian dinosaurs are taphonomically biased in such a way as to contain only juveniles and, consequently, small-sized individuals. In their review of this issue, Pereda-Suberbiola and Galton (2009) expressed doubts about the actual dwarf status of those dinosaurs. More recently, however, Benton et al. (2010) showed

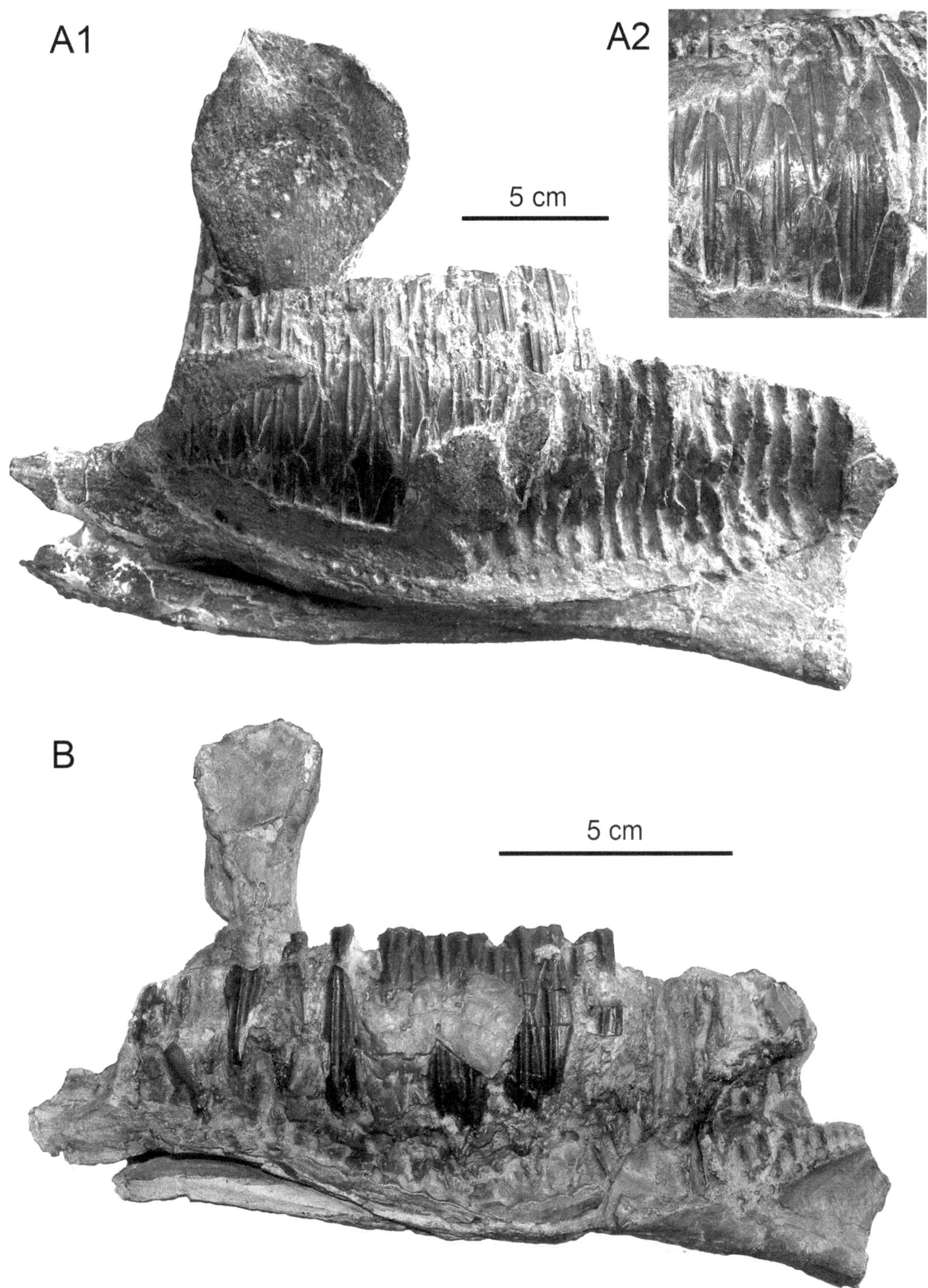

15.9. Hadrosauroid dentaries from the Upper Maastrichtian of Spain, in lingual view. (A) IPS 36338, Fontllonga site of the Àger Syncline, Lleida, Catalonia (A2, particular of the dental battery); (B) cast of MGUV 2200, La Solana site near Tous, Valencia. Scale bars equal 5 cm.

that Jianu and Boekschoten (1999) were wrong, and used histology to demonstrate that Transylvanian dinosaurs are small adults, not small-sized immature individuals (see also Stein et al., 2010). In parallel to these studies, dwarfism in the sauropod *Europasaurus holgeri* from the Upper Jurassic of Northern Germany was established by histological study (Sander et al., 2006), and Company (2011) and Klein et al. (2012) used histology to show that the comparatively small-sized sauropod *Lirainosaurus astibiae* (Late Campanian–earliest Maastrichtian, Spain) is possibly a dwarf as well. Evidence for dinosaur dwarfing has also been reported for the ornithopods of the lowermost Cretaceous fissure fillings of Cornet (Oradea, northwest Romania; Benton et al., 2006).

Dwarfism in sauropods and theropods that lived in present-day Istria (Croatia) during the Late Albian and Late Cenomanian has been hypothesized on the basis of the small size of their footprints compared to those of similar and coeval dinosaurs in continental settings, and the fact that the region was, at that time, the emergent part of an intraoceanic carbonate platform (Dalla Vecchia, 1994, 1998, 2001, 2002, 2003, 2005, 2008a; Dalla Vecchia and Tarlao, 2000; Dalla Vecchia et al., 2001). Although the dwarfism of Istrian dinosaurs and the isolation of the carbonate platform was argued by Mezga et al. (2006) – based on a single trackway of a large (non-"dwarf") titanosaurid sauropod in the Turonian-Lower Coniacian of Dalmatia (Croatia) – a single trackway does not change the overall and well established pattern as suggested by Dalla Vecchia (2008a).

Specimens of *Tethyshadros insularis* are considered insular dwarfs because of their small size, absence of evidence (osteological and morphometric) for immaturity, and because they lived on the emergent part of an intraoceanic carbonate platform (Dalla Vecchia, 2002, 2009c). Also, the other dinosaurs from the Cretaceous of Italy are considered potential dwarfs based on their small size and because they lived on the emergent part of intraoceanic carbonate platforms (Dalla Vecchia, 2003, 2008a; Dal Sasso and Maganuco, 2011 and references therein). Antunes and Sigogneau-Russell (1991, 1992) cited dwarfism to account for the small body size of dinosaurs in the Upper Campanian-Lower Maastrichtian of Portugal. Dalla Vecchia (2003, 2009c) considered dinosaur dwarfism a possible widespread phenomenon in the Albian-Maastrichtian dinosaur faunas of Europe. However, the sauropod *Ampelosaurus atacis* (Late Campanian–Early Maastrichtian, southern France) is not a "dwarf" according to Klein et al. (2012).

Dwarfism is considered as a consequence of life in insular contexts; the so-called island rule (Van Valen, 1973), according to which "large" animals evolve into smaller forms on islands than their relatives on the mainland, and otherwise "small" animals sometimes become larger. This phenomenon has recently been broadly reviewed in Benton et al. (2010), who also discuss dwarfism in dinosaurs.

According to Benton et al. (2010 and references therein), in order to support the view that the latest Cretaceous hadrosauroids found in Europe were insular dwarfs following the island rule, the following three criteria must be met: they (1) lived on islands; (2) were, on average, smaller than their nearest relatives from continents; and (3) are represented by small-sized adults, not juveniles. Here we review the evidence for dwarfism in European hadrosauroids in the context of these three criteria.

1. The European Archipelago The mid-Late Cretaceous sea level high-stand that began during the Albian and reached a maximum during the Late Cenomanian–Turonian (Haq et al., 1987) caused Europe to become an archipelago of relatively small islands in the Tethys Sea between the Afro-Arabian continent to the south and the North European landmass to the north (Fig. 15.10; Camoin et al., 1993; Yilmaz et al., 1996; Philip et al., 2000). In Figure 15.10, the sites yielding latest Cretaceous hadrosauroid remains are identified in the context of a Late Maastrichtian paleogeographic map of Europe. All Spanish and French sites are on the Ibero-Armorican Island, the largest of the archipelago. Those from Belgium and the Netherlands occur in association with the epicontinental Chalk Sea; the dinosaur remains found there probably lived on the nearby "Renish-Bohemian" Island or on the Ibero-Armorican Island and were transported seaward by marine currents. Also, the remains from southern Germany were collected in marine deposits and probably came from the nearby "Austroalpine" Island. Those from northeastern Italy and southwestern Slovenia belong to animals that dwelled on the Adriatic Island, an emergent portion of the Adriatic-Dinaric Carbonate Platform. According to Benton et al. (2010), Romanian hadrosauroids lived in the Tizia-Dacia Island that was separated from the Austroalpine Island by a deep marine trough (contra Philip et al., 2000). Bulgarian hadrosaurs were deposited in the epicontinental sea between the Tizia-Dacia, the "Ukrainian-Donetz," and the "Rhodope-Kirshehir" Islands. Finally, Ukrainian hadrosauroids, which are also found in marine deposits, possibly lived on the small "Crimean" Island, or on the much larger, but more distant, "Ukrainian-Donetz" Island. Thus, mid-Late Cretaceous European hadrosauroids were insular dwellers.

2. Hadrosauroid body size Many authors consider the latest Cretaceous hadrosauroids from the European Archipelago as "small" (e.g., Weishampel et al., 1991; Wellnhofer, 1994; Dalla Vecchia, 2006, 2009c; Godefroit and Motchurova-Dekova, 2010), but thorough quantification of size is lacking. For example, Benton et al. (2010:fig. 7) compared only a small sample of limb bones, purportedly from *Telmatosaurus transsylvanicus*, with a sample of corresponding

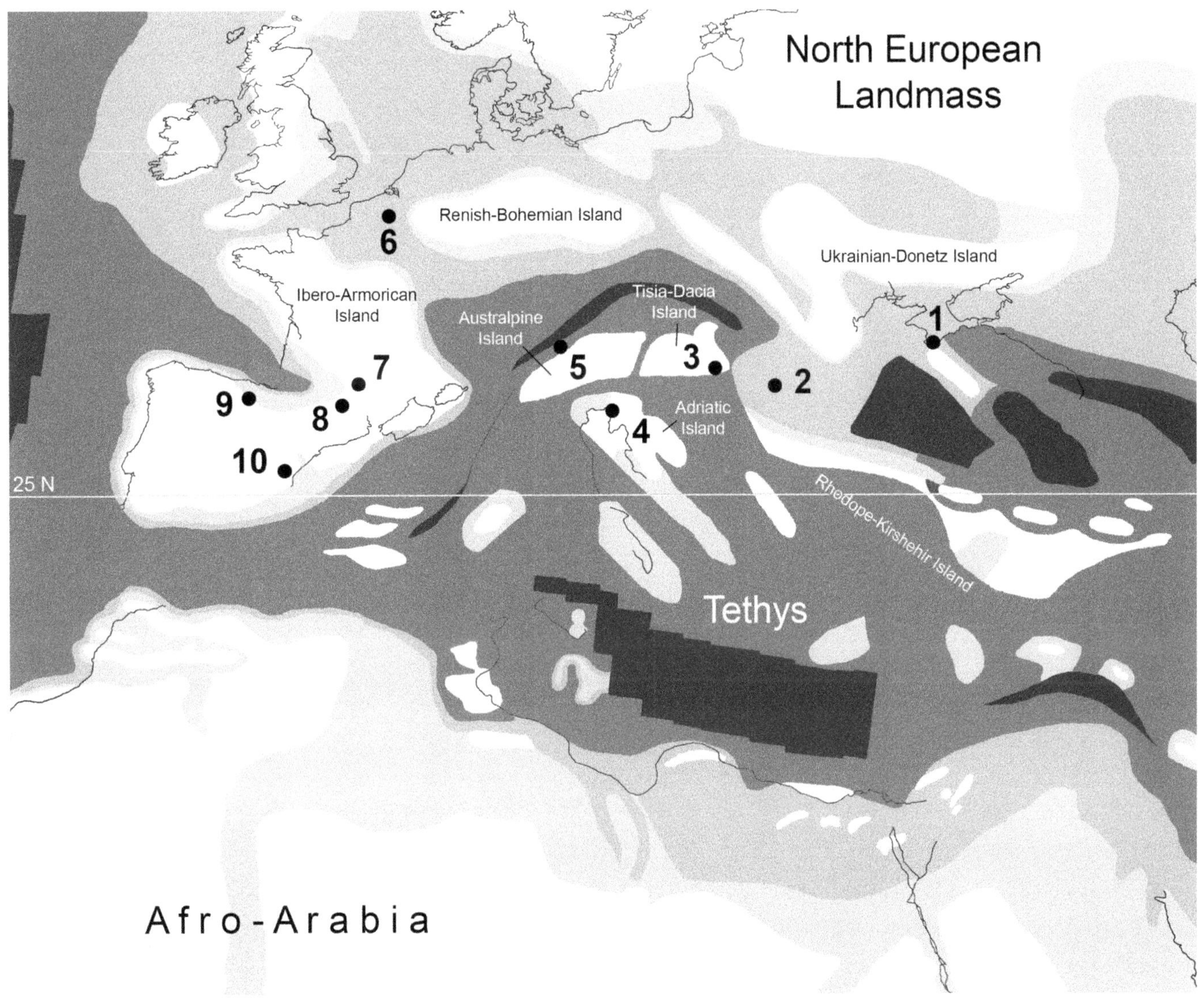

15.10. Location of the latest Cretaceous hadrosauroid localities in the Late Maastrichtian paleogeographic map of the European Archipelago (based on Philip et al., 2000, modified after Benton et al., 2010). (1) Mt. Besh-Kosh, Crimean Peninsula (Ukraine); (2) Labirinta Cave (Bulgaria); (3) Haţeg Basin and other localities of Transylvania (Romania); (4) Villaggio del Pescatore (Italy) and Kozina (Slovenia); (5) Bad Adelholzen (Germany); (6) Limburg localities (the Netherlands and Belgium); (7) Southern France localities (Lestaillats, Jadet, Auzas, Peyrecave, Tricouté, Cassagnau, Mérigon, Larcan, and Le Bexen); (8) localities in the Tremp, Àger and Vallcebre Synclines (Spain); (9) Laño Quarry (Spain); (10) La Solana and Loma Cortada (Spain). White indicates emergent land; pale gray, shallow sea; gray, Chalk Sea; dark gray, deep sea; black, oceanic basins (oceanic crust).

elements from *Maiasaura peeblesorum*. Here, I compare element lengths in a large sample from the European Archipelago (Appendices 15.1–15.3) with those from the Dinosaur Park Formation of Canada, the largest available source of hadrosauroid material from a Late Cretaceous continental area. Bones from the eastern Tremp Syncline of Spain are analyzed separately (see Dalla Vecchia et al., this volume). Here, the term "DPFm size-class" is used when referring to different size classes of Canadian hadrosauroids from the Dinosaur Park Formation that are employed to assess the relative size of the European specimens (see Terminology and Methods section for further explanation).

Lengths of hadrosaurid femora (fe) and tibiae (ti) in the sample from the Dinosaur Park Formation fall into three discrete size classes: large (fe = 78.0–114.4 cm; ti = 70.5–103.4 cm), intermediate (fe = 31.2–62.4 cm; ti = 28.2–56.4 cm), and small (fe = 5.7–20.8 cm; ti = 5.2–18.8 cm) (Fig. 15.11A). These size classes are interpreted by Brinkman (2011, this volume) as representing "adult," "juvenile" and "embryo–hatchling" growth stages, respectively. Large, presumably adult individuals of well known hadrosaurids from North America and Asia exhibit femur lengths of more than 100 cm, tibia lengths that are slightly shorter than femora (or rarely as long as femora), and dentary lengths that range from 45–90 cm

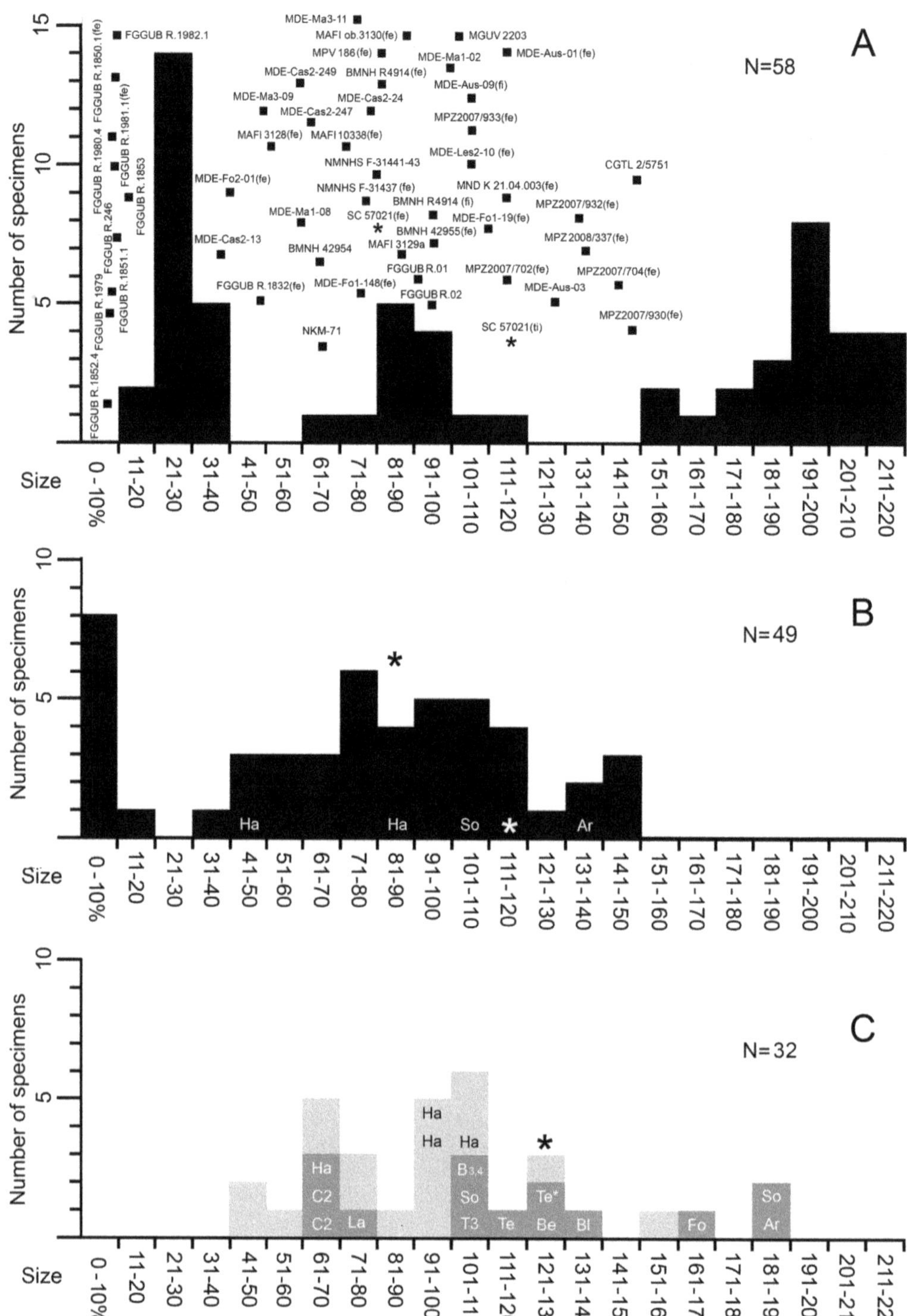

15.11. Size-frequency diagrams of some hadrosaurid skeletal elements from North America and Europe. (A) lengths of tibiae and femora from the Dinosaur Park Formation (Upper Campanian) of Alberta, Canada expressed as percentage of those of TMP 1998.058.0001 (see Brinkman, 2011); (B) lengths of tibiae, fibulae and femora from the uppermost Cretaceous of the Europe expressed as percentages of those of TMP 1998.058.0001 (each specimen is plotted in [A] and includes its museum number); (C) lengths of other appendicular elements (humerus, ulna, scapula, ilium, ischium, pubis, and metatarsal IV; gray), and dentaries plus maxillae (dark gray), expressed as percentages of the corresponding elements of TMP 1998.058.0001 (appendicular elements), of the dentary of ROM 845 (*Corythosaurus intermedius*) scaled as a percentage of that hypothetical for TMP 1998.058.0001 (dentaries), and of the maxilla of ROM 1933 (*C. casuarius*) scaled as a percentage of that hypothetical for TMP 1998.058.0001 (maxillae). See text for further explanation. Abbreviations: Ar, elements from the holotype of *Arenysaurus ardevoli* (femur in [B], dentary in [C]); B3,4, dentary from Blasi 3,4 site; Be, dentary from Le Bexen site; Bl, dentary of the holotype of *Blasisaurus canudoi;* C2, dentaries from Cassagnau 2 site; fe, femur; fi, fibula; Fo, the dentary from Fontllonga site; Ha, elements from the Haţeg Basin (smallest and largest femora in B; humerus and ulnae from Grigorescu and Csiki [2006] and dentary BMNH R3401 in [C]); La, maxilla from the Larcan site; So, elements from the La Solana site (tibia MGUV 2203 in [B] and the two dentaries in [C]); T3, maxilla from Tricouté 3 site; Te, the dentary of the lectotype of *Telmatosaurus transsylvanicus* and the maxilla FGGUB R.1010 (Te*); ti, tibia; * elements from the holotype of *Tethyshadros insularis* (SC 57021; only tibia is plotted in [B], but the asterisks indicate the position of both tibia and femur; the position of the dentary [not plotted] is shown by asterisk in [C]).

(Lull and Wright, 1942; Young, 1958; Lauters et al., 2008). Accordingly, the adult body sizes of the hadrosaurids from the Dinosaur Park Formation (Brinkman, 2011, this volume) fall within the average size range of other large/adult North American and Asian hadrosaurids.

No femora and tibiae from the European sample fall within the adult DPFm size class. Instead, most (56%) fall within the juvenile DPFm size class, 9% are distributed between the embryo–hatchling and juvenile DPFm size classes, 4% fall within the embryo–hatchling DPFm size class, and 18% are smaller than the embryo–hatchling DPFm size class (Fig. 15.11B). All of the European elements that are smaller than the embryo–hatchling DPFm size class, and half of those that fall into the embryo–hatchling DPFm size class (n = 1), belong to individuals reported as hatchlings-newborns from a single site in Transylvania (Tuştea, see above). Accordingly, that assemblage demonstrates that hatchlings-newborns from the Tizia-Dacia Island were smaller than the embryo–hatchlings from Canada. Six European specimens (13%) fall between the adult and the juvenile DPFm size classes, and represent the largest individuals in the European sample (e.g., a tibia from Crimea, and some femora from the Blasi horizons of Spain; Fig. 15.11A).

The sample of femora and tibiae from the eastern Tremp Syncline shows a similar size distribution, with the exception of the absence of individuals in the embryo–hatchling DPFm size class (see Dalla Vecchia et al., this volume). Tibiae and femora were plotted together based on the assumption that all taxa have roughly similar tibia-to-femur proportions as exhibited by the reference specimen TMP 1998.058.0001. Of course, this is not necessarily the case for European hadrosauroids, as shown by the extreme case of *Tethyshadros insularis* where the tibia is apomorphically much longer than the femur (Fig. 15.11B).

The data describing the lengths of other appendicular elements, dentaries, and maxillae in European hadrosauroids compared to the size ranges derived from the Dinosaur Park assemblage shows a slightly different picture. The appendicular element measurements fall almost completely within the juvenile DPFm size class (13 of 15 specimens), but three of the non-appendicular elements occur in the adult DPFm size class (all dentaries from Spain) and three others fall between the juvenile and the adult DPFm size classes (Fig. 15.11C). Whereas these data may suggest that the Ibero-Armorican Island hosted individuals with a body size comparable to that of those in the adult size class from Canada, this may also be an artifact of using lambeosaurine non-appendicular elements for comparisons. Lambeosaurines have a comparatively short head compared to hadrosaurines (Lull and Wright, 1942), which is further indicated by the dentary-to-femur ratio (d/fe) among a sample of hadrosaurines and lambeosaurines (Appendix 15.4; Fig. 15.12). In this context, the relative size estimate of European hadrosauroids having non-lambeosaurine skull-to-femur proportions based on dentaries and possibly maxillae was somewhat overestimated in Fig. 15.11C (certainly the case for *Tethyshadros*; see the different plot positions for the femur and dentary of that taxon in Fig. 15.11B, C, respectively).

As described above, the lambeosaurine *Arenysaurus ardevoli* includes type material supposedly from a single individual. However, the femur and the dentary plot out in two different regions of Figure 15.11B, C, with a d/fe of 0.63, which is higher than any North American lambeosaurine (Appendix 15.4). This suggests that the dentary and femur belong to distinct individuals, or the skull-to-femur proportions in the European lambeosaurine *Arenysaurus*, and potentially other European lambeosaurines (*Pararhabdodon* and *Blasisaurus*), are not like those of North American lambeosaurines. If the femur is taken as representative of *Arenysaurus ardevoli* body size, that individual is outside the adult DPFm size class (Fig. 15.11B).

The sample of other appendicular bones, dentaries and maxillae from the eastern Tremp Syncline shows a similar pattern, including large-sized individuals that are not represented by the sample of femora and tibiae (Dalla Vecchia et al., this volume:fig. 16.10C). Although the size of these individuals (estimated from maxillae and dentaries) could be somewhat overestimated, a few large-sized limb and pelvic elements from the sample also fall within the adult DPFm size class. Thus, some hadrosauroids from the Ibero-Armorican Island probably were large individuals, similar in size to some from North America. Interestingly, the European individuals that fall into the adult DPFm size class are mostly from the Ibero-Armorican Island, the largest island in the archipelago by far (see Fig. 15.10). Accordingly, this observation supports previous claims that body mass in top terrestrial herbivore species increases with increasing land area and range (Burness et al., 2001).

Lastly, the paleoichnological record, comprising abundant hadrosauroid tracks from the eastern Tremp Syncline of Spain (Dalla Vecchia et al., this volume) supports the interpretation that comparatively small-sized hadrosauroids dominated the region. Compared to adult hadrosaur tracks from western North America that range in length from ~60–100 cm, those from Spain range in length from 20–40 cm (see Dalla Vecchia et al., this volume and references therein).

3. Small-sized adults, or juveniles? The Transylvanian hadrosauroid sample includes small-sized adults (Weishampel et al., 1991; Benton et al., 2010; contra Le Loeuff, 2005). The lectotype of *Telmatosaurus transsylvanicus* is considered an adult because of the fusion of the braincase elements (Weishampel et al., 1991), but its 28 cm long dentary falls

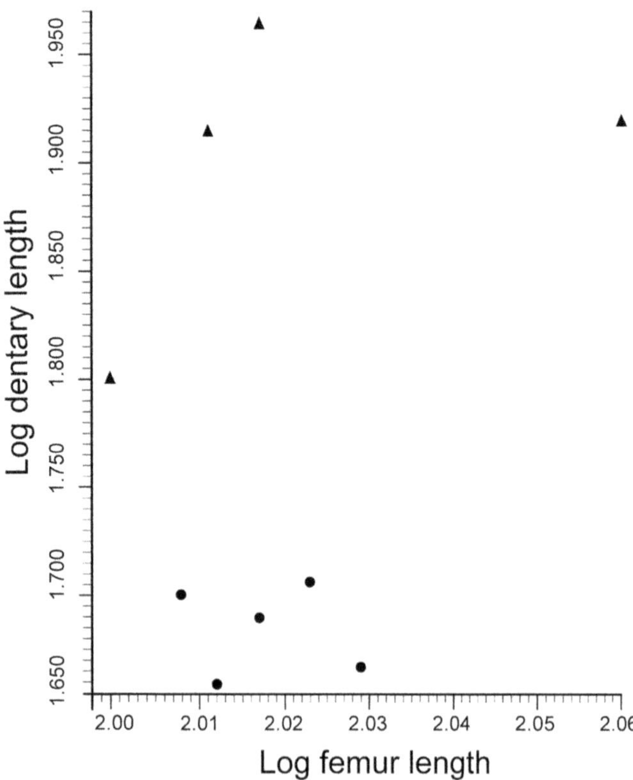

15.12. Bivariate plot of dentary length to femur length of a sample of hadrosaurines (triangle) and lambeosaurines (dots). For data see Appendix 15.4.

within the juvenile DPFm size class (Fig. 15.11C). All long bones in the sample are small. The largest femur (MAFI 3130) is 46 cm long and belongs to an adult individual (histological evidence; Benton et al., 2010). Similarly, a much smaller femur (25 cm long; FGGUB R. 1362) is considered a subadult by Benton et al. (2010). In Figure 15.11B, these two femora plot in the middle of the juvenile DPFm size class and between the juvenile and embryo–hatchling DPFm size classes, respectively.

The only known individual of *Arenysaurus ardevoli* has a fully ossified braincase and lacks open cranial sutures in the skull (Pereda-Suberbiola, Canudo, Cruzado-Caballero, et al., 2009). As seen above, with the femur as size proxy, the individual is relatively large, but it is smaller than the adults in the Canadian sample (Fig. 15.11B). Assuming that femoral length is proportional to the overall body length, and that *Arenysaurus* and *C. intermedius* had the same body shape and proportions, a comparison of their femur lengths (*Arenysaurus* femur length is 68% that of ROM 845) suggests that the body of *Arenysaurus* weighed less than one-third that of *C. intermedius*. MGUV 2203, a tibia from the La Solana site, was considered as an adult by Company (2004) based on histological evidence. However, it also falls within the juvenile DPFm size class in Fig. 15.11B. Thus, comparatively small-sized adults were present on the largest island of the archipelago, and could represent cases of insular dwarfism, although less markedly so than in Transylvanian hadrosauroids.

Other samples (mainly from the Spanish and French sides of the Pyrenees, and *Tethyshadros* material) show scarce evidence of osteological immaturity (e.g., unfused skeletal elements), but these require histological study to fully assess their relative size-age relationships, and to determine their impact in further considerations of hadrosauroid dwarfism.

Hadrosauroid Diversity in the European Archipelago During the Latest Cretaceous

During the Late Campanian–Early Maastrichtian, the islands of the central part of the archipelago (Tizia-Dacia and Adriatic Islands; Fig. 15.10) were populated by non-hadrosaurid hadrosauroids (*Tethyshadros insularis* and *Telmatosaurus transsylvanicus;* Nopcsa, 1900, 1928; Weishampel et al., 1991, 1993; Dalla Vecchia, 2009c). In the Tizia-Dacia Island, *Telmatosaurus transsylvanicus* was sympatric with rhabdodontids (*Zalmoxes;* Nopcsa, 1902, 1904, 1928; Weishampel et al., 1991, 2003; Therrien, 2005; Godefroit et al., 2009; Codrea et al., 2010), but was less abundant than them (see above). Coeval deposits in the large Ibero-Armorican Island are practically devoid of hadrosauroids (just a single tooth is known from the Laño quarry; Pereda-Suberbiola, Ruiz-Omeñaca, and Company, 2003), whereas rhabdodontids (*Rhabdodon*) are common (Matheron, 1869; Brinkmann, 1988; Buffetaut and Le Loeuff, 1991; Buffetaut et al., 1997; Garcia et al., 1999; Pereda-Suberbiola and Sanz, 1999; Barroso-Barcenilla et al., 2009; Company et al., 2009). When rhabdodontids disappeared in the archipelago during the Late Maastrichtian, hadrosauroids achieved a wider geographic distribution and became the dominant dinosaurs on the Ibero-Armorican Island (Le Loeuff et al., 1994; Buffetaut et al., 1997; Laurent et al., 1997; Company et al., 1998; Casanovas-Cladellas et al., 1999; López-Martínez et al., 2001; Allain and Pereda-Suberbiola, 2003; Laurent, 2003; Cruzado-Caballero et al., 2005, 2007, 2009, 2011; Dalla Vecchia, 2006, 2009a; Prieto-Márquez et al., 2006; Company et al., 2009; Pereda-Suberbiola, Canudo, Company, et al., 2009; Pereda-Suberbiola, Canudo, Cruzado-Caballero, et al., 2009; Riera et al., 2009; Bilotte et al., 2010; Cruzado-Caballero, Pereda-Suberbiola, and Ruiz-Omeñaca, 2010; Cruzado-Caballero, Ruiz-Omeñaca, and Canudo, 2010; Dalla Vecchia et al., this volume).

The Late Maastrichtian Ibero-Armorican hadrosauroids are a rather diverse group, apparently comprising a minimum of six taxa: two relatively primitive forms (a non-hadrosaurid hadrosauroid from Fontllonga and a possible basal hadrosaurid or a non-hadrosaurid hadrosauroid from La Solana [Casanovas-Cladellas et al., 1999; Company et al., 1998;

Pereda-Suberbiola, Canudo, Company, et al., 2009]); three lambeosaurines (*Pararhabdodon*, *Arenysaurus*, and *Blasisaurus* [Casanovas-Cladellas et al., 1993; Prieto-Márquez et al., 2006; Pereda-Suberbiola, Canudo, Cruzado-Caballero, et al., 2009; Cruzado-Caballero, Pereda-Suberbiola, and Ruiz-Omeñaca, 2010]); and a "hadrosaurine" (represented by the Blasi 5 jugal [Cruzado-Caballero, Ruiz-Omeñaca, and Canudo, 2010]). Bones referred to indeterminate lambeosaurines are reported from several sites of the eastern Tremp Syncline (Prieto-Márquez et al., 2007).

This association is very different from the coeval western North American *Triceratops* "fauna" of the "Lancian Land Vertebrate Age" (see Weishampel et al., 2004) in which ceratopsians are the dominant dinosaurs, and hadrosaurids are represented only by the hadrosaurine *Edmontosaurus* (*Anatotitan* is a junior synonym of *Edmontosaurus*; Campione and Evans, 2011). By the mid-Campanian, western North American hadrosauroids were represented only by hadrosaurids (Weishampel et al., 2004; McDonald et al., 2010; Prieto-Márquez, 2010a, 2010b; Evans et al., 2012; Campione et al., 2013). Lambeosaurines practically disappeared at the beginning of the Late Maastrichtian (Campione and Evans, 2011; Sullivan et al., 2011); *Hypacrosaurus* is reported from the 31n polarity zone of the Horseshoe Canyon Formation (Eberth and Currie, 2010; Eberth et al., 2013). The hadrosauroid assemblages of the European Archipelago are more reminiscent of those in coeval Asia. The non-hadrosaurid hadrosauroids *Bactrosaurus* and *Gilmoreosaurus* from China are possibly latest Campanian–Early Maastrichtian in age (Van Itterbeeck, Horne et al., 2005), and the non-hadrosaurid hadrosauroid *Tanius*, also from China, is Campanian in age (Buffetaut and Tong, 1995). According to Tsogtbaatar et al. (2011; this volume), a new taxon from the Djadokhta Formation of Mongolia, probably mid-Campanian in age (Van Itterbeeck, Horne et al., 2005), is the sister taxon of *Telmatosaurus* + Hadrosauridae. The Upper Maastrichtian (Van Itterbeeck, Bolotsky, et al., 2005) hadrosauroid fauna of eastern Asia (Amur Region, Russia and China) is very diverse with four lambeosaurines (*Charonosaurus*, *Sahaliyania*, *Amurosaurus*, and *Olorotitan*) and four hadrosaurines (*Wulagasaurus*, *Kerberosaurus*, and two new yet unnamed genera; Godefroit et al., 2009, 2011). Lambeosaurines comprise more than 90% of these assemblages (Godefroit et al., 2011). As in Europe, but unlike western North America, ceratopsians are absent in the Amur sites (Godefroit et al., 2009). The survival of non-hadrosaurid hadrosauroids and basal hadrosaurids during the Late Maastrichtian would suggest that the Ibero-Armorican Island was a refugium at the end of Mesozoic.

The high diversity of lambeosaurines in the Upper Maastrichtian of the Spanish part of the Ibero-Armorican Island could be an artifact due to taxonomic overevaluation of morphological differences between specimens (splitting), or reflective of an overall high diversity during a relatively short geochronological interval. Although the latter case would be unexpected in an island setting (diversity in islands is markedly lower than in nearby continental areas; MacArthur and Wilson, 1967; Benton et al., 2010), it could be explained, theoretically, by a fast species turnover rate, which is also a characteristic of insular settings (MacArthur and Wilson, 1967). The high diversity of hadrosauroids in the Upper Maastrichtian of the southern Pyrenees further suggests that caution should be utilized in acritically referring all hadrosauroid remains from the thick Transylvanian units to the single taxon, *Telmatosaurus transsylvanicus* (here regarded as a wastebasket taxon in need of further study, see above).

CONCLUSIONS

The European latest Cretaceous hadrosauroid record has increased more in the last 15 years than in the previous 100 years; hadrosauroids of that age are now reported from nine countries. The record is mainly Maastrichtian and probably mostly Late Maastrichtian in age, but a few localities could be Late Campanian. Five monospecific genera have been named and are considered valid to date, but only one (*Tethyshadros*) is represented by a complete articulated skeleton.

Insular dwarfism is strongly supported for the Romanian hadrosauroids, but comparatively small size is common in specimens from all other European localities.

The few individuals that are in the range of adults from the uppermost Cretaceous of western North America occur on the largest island of the archipelago. This pattern is in agreement with the observation that body mass in top terrestrial herbivore species increases with increasing land area. However, further histological studies are required to determine whether they (1) are the only adults in a sample dominated by juveniles, (2) represent one or more large-sized species in an association dominated by one or more small-sized species, or (3) are "giant" individuals of small-sized taxa.

ACKNOWLEDGMENTS

I thank David Eberth for assistence in preparing this chapter. I am indebted to Donald Brinkman for the measurements of TMP 1998.058.0001. I thank Julio Company for the information about the Valencia hadrosauroids, Penelope Cruzado-Caballero for that about Blasi material, Jean Le Loeuff for his assistance at the Musée des dinosaures d'Esperáza (France) and the personal communications, and

Albert Prieto-Márquez for information on North American and Asian hadrosauroids. I thank David C. Evans and an anonymous referee for the comments on the manuscript. Finally, I am grateful to Lukas Panzarin for the reconstruction of *Tethyshadros*. This research was financially supported by the Ministerio de Ciencia e Innovación of the Spanish government (CGL2008-06533-C03-01/02 and CGL2011-30069-C02-01/02) and the Departament de Cultura de la Generalitat de Catalunya. The study of the *Telmatosaurus* material at the Natural History Museum of London received support from the SYNTHESYS Project, which is financed by European Community Research Infrastructure Action under the FP6 "Structuring the European Research Area" Programme.

LITERATURE CITED

Allain, R., and X. Pereda-Suberbiola. 2003. Dinosaurs of France. Comptes Rendus Palevol 2:27–44.

Allain, R., and P. Taquet. 2000. A new genus of Dromaeosauridae (Dinosauria, Theropoda) from the Upper Cretaceous of France. Journal of Vertebrate Paleontology 20:404–407.

Antunes, M. T., and D. Sigogneau-Russell. 1991. Nouvelles données sur les dinosaurs du Crétacé Superieur du Portugal. Comptes Rendus de l'Académie des Sciences de Paris: Sciences de la Terre et des Planetès 313:113–119.

Antunes, M. T., and D. Sigogneau-Russell. 1992. La faune de petits dinosaures du Crétacé terminal portugais. Communicações dos Serviços Geológicos de Portugal 78:49–62.

Arbulla, D., F. Cotza, F. Cucchi, F. M. Dalla Vecchia, A. De Giusto, O. Flora, D. Masetti, A. Palci, P. Pittau, N. Pugliese, B. Stenni, A. Tarlao, G. Tunis, and L. Zini. 2006. Escursione nel Carso Triestino, in Slovenia e Croazia. 8 giugno. Stop 1. La successione Santoniano-Campaniana del Villaggio del Pescatore (Carso Triestino) nel quale sono stati rinvenuti i resti di dinosauro; pp. 20–27 in R. Melis, R. Romano, and G. Fonda (eds.), Guida alle escursioni/excursions guide. Società Paleontologica Italiana: Giornate di Paleontologia 2006. EUT Edizioni Università di Trieste, Trieste.

Barco, J. L., Ll. Ardevol, and J. I. Canudo. 2001. Descripción de los primeros rastros asignados a Hadrosauridae (Orrnithopoda, Dinosauria) del Maastrichtiense de la Península Ibérica (Areny, Huesca). Geogaceta 30:235–238.

Barroso-Barcenilla, F., O. Cambra-Moo, F. Escaso, F. Ortega, A. Pascual, A. Pérez-García, J. Rodríguez-Lázaro, J. L. Sanz, M. Segura, and A. Torices. 2009. New and exceptional discovery in the Upper Cretaceous of the Iberian Peninsula: the paleontological site of "Lo Hueco," Cuenca, Spain. Cretaceous Research 30:1268–1278.

Benton, M. J., N. J. Minter, and E. Posmosanu. 2006. Dwarfing in ornithopod dinosaurs from the Early Cretaceous of Romania; pp. 79–87 in Z. Csiki (ed.), Mesozoic and Cenozoic Vertebrates and Paleoenvironments: Tributes to the Career of Prof. Dan Grigorescu. Ars Docendi, Bucharest.

Benton, M. J., Z. Csiki, D. Grigorescu, R. Redelstorff, P. M. Sander, K. Stein, and D. B. Weishampel. 2010. Dinosaurs and the island rule: the dwarfed dinosaurs from Haţeg Island. Palaeogeography, Palaeoclimatology, Palaeoecology 293:438–454.

Bilotte, M., Y. Laurent, and D. Téodori. 2010. Restes d'Hadrosaure dans le Crétacé terminal marin de Larcan (Petites Pyrénées, Haute-Garonne, France). Carnets de Géologie, Letter 2010/02 (CG2010_L02):1–10.

Bojar, A.-V., D. Grigorescu, F. Ottner, and Z. Csiki. 2005. Paleoenvironmental interpretation of dinosaur- and mammal-bearing continental Maastrichtian deposits, Haţeg basin, Romania. Geological Quarterly 49:205–222.

Brinkman, D. B. 2011. The size-frequency distribution of hadrosaurs from the Dinosaur Park Formation of Alberta, Canada; pp. 16–20 in D. R. Braman, D. A. Eberth, D. C. Evans, and W. Taylor (compilers), International Hadrosaur Symposium Abstract Volume. Royal Tyrrell Museum of Palaeontology, Drumheller, Canada.

Brinkman, D. B. 2014. The size-frequency distribution of hadrosaurs from the Dinosaur Park Formation of Alberta, Canada; chapter 23 in D. A. Eberth and D. C. Evans (eds.), Hadrosaurs. Indiana University Press, Bloomington, Indiana.

Brinkmann, W. 1988. Zur Fundgeschicte und Systematik der Ornithopoden (Ornithischia, Reptilia) aus der ober-Kreide von Europe. Documenta Naturae 45:1–157.

Buffetaut, E. 2005. Late Cretaceous vertebrates from the Saint Chinian area (southern France): a review of previous research and an update on recent finds. Acta Palaeontologica Romaniae 5:39–48.

Buffetaut, E. 2009. An additional hadrosaurid specimen (Dinosauria: Ornithischia) from the marine Maastrichtian deposits of the Maastricht area. Carnets de Géologie, Letter 2009/03 (CG2009_L03):1–4.

Buffetaut, E., and J. Le Loeuff. 1991. Une nouvelle espèce de *Rhabdodon* (Dinosauria, Ornithischia) du Crétacé supérieur de l'Hérault (Sud de la France). Comptes rendus de l'Académie des Sciences de Paris: Sciences de la Terre et des Planetès 312:943–948.

Buffetaut, E., and H. Tong. 1995. The Late Cretaceous dinosaurs of Shandong, China: old finds and new interpretations; pp. 139–142 in A. Sun and Y.-Q. Wang (eds.), Sixth Symposium on Mesozoic Terrestrial Ecosystems and Biota, Short Papers. China Oceanic Press, Beijing.

Buffetaut, E., A. W. F. Meijer, P. Taquet, and G. Wouters. 1985. New remains of hadrosaurid dinosaurs (Reptilia, Ornithischia) from the Maastrichtian of Dutch and Belgian Limburg. Revue Palébiologique 4:65–70.

Buffetaut, E., G. Costa, J. Le Loeuff, M. Martin, J.-C. Rage, X. Valentin, and H. Tong. 1996. An Early Campanian vertebrate fauna from Villeveyrac Basin (Hérault, southern France). Neues Jahrbuch für Geologie und Paläontologie Monatshefte 1996:1–16.

Buffetaut, E., J. Le Loeuff, H. Tong, S. Duffaud, L. Cavin, G. Garcia, D. Ward, and Association culturelle, archéologique et paléontologique de Cruzy. 1999. Un nouveau gisement de vertébrés du Crétacé supérieur à Cruzy (Hérault, Sud de la France). Comptes rendus de l'Académie des Sciences de Paris: Sciences de la Terre et des Planetès 328:203–208.

Buffetaut, E., J. Le Loeuff, L. Cavin, S. Duffaud, E. Gheerbrant, Y. Laurent, M. Martin, J.-C. Rage, H. Tong, and D. Vasse. 1997. Late Cretaceous non-marine vertebrates from southern France: a review of recent finds. Geobios 20:101–108.

Burness, G. P., J. Diamond, and T. Flannery. 2001. Dinosaurs, dragons, and dwarfs: the evolution of maximal body size. Proceedings of the National Academy of Sciences 98:14518–14523.

Camoin, G., Y. Bellion, J. Dercourt, R. Guiraud, J. Lucas, A. Poisson, L.-E. Ricou, and B. Vrielynck. 1993. Late Maastrichtian palaeoenvironments (69.5 to 65 Ma); in J. Dercourt, L. E. Ricou, and B. Vrielynck (eds.), Atlas Tethys, Palaeoenvironmental Maps. BEICIP-FRANLAB, Rueil-Malmaison, France.

Campione, N. E., and D. C. Evans. 2011. Cranial growth and variation in Edmontosaurs (Dinosauria: Hadrosauridae): implications for Latest Cretaceous megaherbivore diversity in North America. PLoS ONE 6(9):e25186.

Campione, N. E., K. S. Brink, E. A. Freedman, C. T. McGarrity, and D. C. Evans. 2013. '*Glishades ericksoni,*' an indeterminate juvenile hadrosaurid from the Two Medicine Formation of Montana: implications for the diversity of non-hadrosauroids in the Late Campanian-Maastrichtian of North America. Palaeodiversity and Palaeoenvironments 93:65–75.

Casanovas-Cladellas, M. L., J. V. Santafé-Llopis, and A. Isidro-Llorens. 1993. *Pararhabdodon isonense* n. gen. n. sp. (Dinosauria). Estudio morfológico, radio-tomográfico y consideraciones biomecánicas. Paleontologia i Evolució 26–27:121–131.

Casanovas-Cladellas, M. L., X., Pereda-Suberbiola, J. V. Santafe, and D. B. Weishampel. 1999. A primitive euhadrosaurian dinosaur from the uppermost Cretaceous of the Ager syncline (southern Pyrenees, Catalonia). Geologie en Mijnbouw 78:345–356.

Chiappe, L. M., R. A. Coria, L. Dingus, F. Jackson, A. Chimsamy, and M. Fox. 1998. Sauropod dinosaur embryos from the Late Cretaceous of Patagonia. Nature 396:258–261.

Codrea, V., and E. P. Dica. 2005. Upper Cretaceous-lowermost Miocene lithostratigraphic units exposed in Alba Iulia–Sebeş–Vinţu de Jos area (SW Transylvanian Basin). Studia Universitatis Babeş Bolyai, Geologia 50:19–26.

Codrea, V., M. Vremir, C. Jipa, P. Godefroit, Z. Csiki, T. Smith, and C. Fărcaş. 2010. More than just Nopcsa's Transylvanian dinosaurs: a look outside the Haţeg Basin. Palaeogeography, Palaeoclimatology, Palaeoecology 293:391–405.

Company, J. 2004. Vertebrados continentales del Cretácico superior (Campaniense-Maastrichtiense) de Valencia. Ph.D. dissertation, University of Valencia, Valencia, Spain, 410 pp.

Company, J. 2011. Bone histology of the titanosaur *Lirainosaurus astibiae* (Dinosauria: Sauropoda) from the latest Cretaceous of Spain. Naturwissenschaften 98:67–78.

Company, J., À. Galobart, and R. Gaete. 1998. First data on the hadrosaurid dinosaurs (Ornithischia, Dinosauria) from the Upper Cretaceous of Valencia, Spain. Oryctos 1:121–126.

Company, J., X. Pereda-Suberbiola, and J. I. Ruiz-Omeñaca. 2009. Los dinosaurios del Cretácico Terminal del Levante ibérico en el contexto paleogeográfico europeo. Composición de las faunas e implicaciones paleobiogeográficas; pp. 17–44 in P. Huerta Hurtado and F. Torcida-Fernández Baldor (eds.), Actas de las IV Jornadas Internacionales sobre Paleontología de Dinosaurios y su Entorno. Colectivo Arqueológico-Paleontológico de Salas, Salas de los Infantes, Burgos, Spain.

Cruzado-Caballero, P. 2012. Restos directos de dinosaurios hadrosáuridos (Ornithopoda, Hadrosauridae) del Maastrichtiense superior (Cretácico superior) de Arén (Huesca). Ph.D. dissertation, University of Zaragoza, Zaragoza, Spain, 410 pp.

Cruzado-Caballero, P., J. I. Canudo, and J. I. Ruiz-Omeñaca. 2005. Nuevas evidencias de la presencias de hadrosaurios lambeosaurinos (Dinosauria) en el Maastrichtiense superior de la Península Ibérica (Arén, Huesca). Geogaceta 38:47–50.

Cruzado-Caballero, P., J. I. Canudo, and J. I. Ruiz-Omeñaca. 2007. Los dinosaurios hadrosaurios (Ornithischia: Ornithopoda) del Maastrichtiense superior (Cretácico Superior) de Arén (Huesca, España); pp. 55–56 in P. Huerta Hurtado, and F. Torcida-Fernández Baldor (eds.), IV Jornadas Internacionales sobre Paleontologia de Dinosaurios y su Entorno: Libro de resúmenes.

Cruzado-Caballero, P., J. I. Canudo, and J. I. Ruiz-Omeñaca. 2009. Los fémures de Blasi (Arén, Huesca, Spain): una contribución a los hadrosauroideos europeos del Maastrichtiense superior; pp. 197–205 in P. Huerta Hurtado and F. Torcida-Fernández Baldor (eds.), Actas de las IV Jornadas Internacionales sobre Paleontología de Dinosaurios y su Entorno. Colectivo Arqueológico-Paleontológico de Salas, Salas de los Infantes, Burgos, Spain.

Cruzado-Caballero, P., X. Pereda-Suberbiola, and J. I. Ruiz-Omeñaca. 2010. *Blasisaurus canudoi* gen. et. sp. nov., a new lambeosaurinae dinosaur (Hadrosauridae) from the latest Cretaceous of Arén (Huesca, España). Canadian Journal of Earth Sciences 47:1507–1517.

Cruzado-Caballero, P., J. I. Ruiz-Omeñaca, and J. I. Canudo. 2010. Evidencias de la coexistencia de dinosaurios hadrosaurinos y lambeosaurinos en el Maastrichtiano superior de la Península Ibérica (Arén, Huesca, España). Ameghiniana 47:153–164.

Cruzado-Caballero, P., J. I. Ruiz-Omeñaca, R. Gaete, V. Riera, O. Oms, and J. I. Canudo. 2010. Un nuevo hadrosáurido en el Maastrichtiense superior de los Pirineos (Arén, Huesca, España); pp. 51–54 in P. Huerta-Hurtado and F. Torcida-Fernández Baldor (eds.), Resúmenes de las V Jornadas Internacionales sobre Paleontología de Dinosaurios y su Entorno. Colectivo Arqueológico-Paleontológico de Salas, Salas de los Infantes, Burgos, Spain.

Cruzado-Caballero, P., J. I. Canudo, M. Moreno-Azanza, and J. I. Ruiz-Omeñaca. 2011. The complex fauna of European Maastrichtian hadrosaurids: contributions of the lambeosaurines from the Iberian Peninsula; pp. 33–37 in D. R. Braman, D. A. Eberth, D. C. Evans, and W. Taylor (compilers), International Hadrosaur Symposium Abstract Volume. Royal Tyrrell Museum of Palaeontology, Drumheller, Canada.

Csiki, Z., M. Vremir, S. L. Brusatte, and M. A. Norell. 2010. An aberrant island-dwelling theropod dinosaur from the Late Cretaceous of Romania. Proceedings of the National Academy of Sciences 107:15357–15361.

Cuevas, J. L. 1992. Estratigrafía del "Garumniense" de la Conca de Tremp, Prepirineo de Lérida. Acta Geológica Hispánica 27:95–108.

Dal Sasso, C. 2004. Dinosaurs of Italy. Indiana University Press, Bloomington, Indiana. 213 pp.

Dal Sasso, C., and S. Maganuco. 2011. Osteology, ontogenetic assessment, phylogeny, soft tissue anatomy, taphonomy and palaeobiology of *Scipionyx samniticus* (Theropoda: Compsognathidae) from the Lower Cretaceous of Italy. Memorie della Società Italiana di Scienze Naturali e del Museo Civico di Storia Naturale di Milano 37:1–281.

Dalla Vecchia, F. M. 1994. Jurassic and Cretaceous sauropod evidence in the Mesozoic carbonate platforms of southern Alps and Dinarids. Gaia 10:65–73.

Dalla Vecchia, F. M. 1998. Theropod footprints in the Cretaceous Adriatic-Dinaric carbonate platform (Italy and Croatia). Gaia 15:355–367.

Dalla Vecchia, F. M. 2001. Terrestrial ecosystems on the Mesozoic peri-Adriatic carbonate platforms: the vertebrate evidence. Proceedings VII International Symposium on Mesozoic Terrestrial Ecosystems, Asociación Paleontológica Argentina, Publicación Especial 7:77–83.

Dalla Vecchia, F. M. 2002. Cretaceous dinosaurs in the Adriatic-Dinaric carbonate platform (Italy and Croatia): paleoenvironmental implications and paleogeographical hypotheses. Memorie della Società Geologica Italiana 57:89–100.

Dalla Vecchia, F. M. 2003. I dinosauri nani dell'Arcipelago europeo. Le Scienze [Italian edition of Scientific American] 423:86–94.

Dalla Vecchia, F. M. 2005. Between Gondwana and Laurasia: Cretaceous Sauropods in an Intraoceanic Carbonate Platform; pp. 395–429 in V. Tidwell, and K. Carpenter (eds.), Thunder-Lizards: The Sauropodomorph Dinosaurs. Indiana University Press, Bloomington, Indiana.

Dalla Vecchia, F. M. 2006. *Telmatosaurus* and the other hadrosaurids of the Cretaceous European Archipelago: an overview. Natura Nascosta 32:1–55.

Dalla Vecchia, F. M. 2008a. The impact of dinosaur palaeoichnology in palaeoenvironmental and palaeogeographic reconstructions: the case of the Periadriatic carbonate platforms. Oryctos 8:89–106.

Dalla Vecchia, F. M. 2008b. I dinosauri del Villaggio del Pescatore (Trieste): qualche aggiornamento. Atti del Museo Civico di Storia Naturale di Trieste, suppl. to n. 53:111–130.

Dalla Vecchia, F. M. 2009a. European hadrosauroids; pp. 45–74 in P. Huerta Hurtado and F. Torcida-Fernández Baldor (eds.), Actas de las IV Jornadas Internacionales sobre Paleontología de Dinosaurios y su Entorno. Colectivo Arqueológico-Paleontológico de Salas, Salas de los Infantes, Burgos, Spain.

Dalla Vecchia, F. M. 2009b. *Telmatosaurus* and the other hadrosauroids of the Cretaceous European Archipelago: an update. Natura Nascosta 39:1–18.

Dalla Vecchia, F. M. 2009c. *Tethyshadros insularis,* a new hadrosauroid dinosaur (Ornithischia) from the Upper Cretaceous of Italy. Journal of Vertebrate Paleontology 29:1100–1116.

Dalla Vecchia, F. M., and A. Tarlao. 2000. New dinosaur track sites in the Albian (Early Cretaceous) of the Istrian Peninsula (Croatia), part II: paleontology. Memorie di Scienze Geologiche Padova 52/2:227–293.

Dalla Vecchia, F. M., G. Tunis, S. Venturini, and A. Tarlao. 2001. Dinosaur track sites in the upper Cenomanian (Late Cretaceous) of the Istrian Peninsula (Croatia). Bollettino della Società Paleontologica Italiana 40:25–54.

Dalla Vecchia, F. M., R. Gaete, V. Riera, O. Oms, A. Prieto-Márquez, B. Vila, A. Garcia Sellés, and À. Galobart. 2014. The hadrosauroid record in the Maastrichtian of the eastern Tremp Syncline (northern Spain); chapter 16 in D. A. Eberth and D. C. Evans (eds.), Hadrosaurs. Indiana Unversity Press, Bloomington, Indiana.

Debeljak, I., A. Kosir, and B. Otonicar. 1999. A preliminary note on dinosaurs and non-dinosaurian reptiles from the Upper Cretaceous carbonate platform succession at Kozina (SW Slovenia): razprave IV. Razreda Sazu XL(1):3–25.

Debeljak, I., A. Kosir, E. Buffetaut, and B. Otonicar. 2002. The Late Cretaceous dinosaurs and crocodiles of Kozina (SW Slovenia). Memorie Società Geologica Italiana 57:193–201.

Dollo, L. 1883. Note sur les restes de dinosauriens recontrés dans le Crétace supérieur de la Belgique. Bulletin de l'Institut royal des Sciences naturelles de Belgique 2:205–221.

Eberth, D. A., and P. J. Currie. 2010. Stratigraphy, sedimentology and taphonomy of the *Albertosaurus* bonebed (upper Horseshoe Canyon Formation; Maastrichtian), southern Alberta, Canada. Canadian Journal of Earth Sciences 47:1119–1143.

Eberth, D. A., D. C. Evans, D. B. Brinkman, F. Therrien, D. H. Tanke, and L. Russell. 2013. Dinosaur biostratigraphy of the Edmonton Group (Upper Cretaceous), Alberta, Canada: Evidence for climate influence. Canadian Journal of Earth Sciences 50:701–726.

Escaso, F., F. Ortega, J. M. Gasulla, and J. L. Sanz. 2012. New postcranial remains of *Rhabdodon* from the Upper Campanian-Lower Maastrichtian of "Lo Hueco" (Cuenxa, Spain). Fundamental 20:71–72.

Evans, D. C. 2010. Cranial anatomy and systematics of *Hypacrosaurus altispinus,* and a comparative analysis of skull growth in lambeosaurine hadrosaurids (Dinosauria: Ornithischia). Zoological Journal of the Linnean Society 159:398–434 [supporting information].

Evans, D. C., and R. R. Reisz. 2007. Anatomy and relationships of *Lambeosaurus magnicristatus,* a crested hadrosaurid dinosaur (Ornithischia) from the Dinosaur Park Formation, Alberta. Journal of Vertebrate Paleontology 27:373–393.

Evans, D. C., P. M. Barrett, and K. S. Seymour. 2012. Revised identification of a reported *Iguanodon*-grade ornithopod tooth from the Scollard Formation, Alberta, Canada. Cretaceous Research 33:11–14.

Forster, C. A. 1997. Phylogeny of the Iguanodontia and Hadrosauridae. Journal of Vertebrate Paleontology 17(3, Supplement):47A.

Garcia, G., A. Sauveur, F. Fournier, E. Thouand, and X. Valentini. 2010. A new Titanosaur genus (Dinosauria, Sauropoda) from the Late Cretaceous of southern France and its palaeobiogeographic implications. Bulletin de la Societé géologique française 181:269–277.

Garcia, G., M. Pincemaille, M. Vianey-Liaud, B. Marandat, E. Lorenz, G. Cheylan, H. Cappetta, J. Michaux, and J. Sudre. 1999. Découverte du premier squelette presque complet de *Rhabdodon priscus* (Dinosauria, Ornithopoda) du Maastrichtien inférieur de Provence. Comptes rendus de l'Académie des Sciences de Paris: Sciences de la Terre et des Planetès 328:415–421.

Gates, T. A., J. R. Horner, R. R. Hanna, and C. R. Nelson. 2011. New unadorned hadrosaurine hadrosaurid (Dinosauria, Ornithopoda) from the Campanian of North America. Journal of Vertebrate Paleontology 31:798–811.

Gates, T. A., S. D. Sampson, C. R. Delgado de Jesús, L. E. Zanno, D. Eberth, R. Hernández-Rivera, M. C. Aguillón Martínez, and J. I. Kirkland. 2007. *Velafrons coahuilensis,* a new lambeosaurine hadrosaurid (Dinosauria: Ornithopoda) from the late Campanian Cerro del Pueblo Formation, Coahuila, Mexico. Journal of Vertebrate Paleontology 27:917–930.

Godefroit, P., and N. Motchurova-Dekova. 2010. Latest Cretaceous hadrosauroid (Dinosauria: Ornithopoda) remains from Bulgaria. Comptes rendus Palevol 9:163–169.

Godefroit, P., V. Codrea, and D. B. Weishampel. 2009. Osteology of *Zalmoxes shqiperorum* (Dinosauria, Ornithopoda), based on new specimens from the Upper Cretaceous of Nălaţ-Vad (Romania). Geodiversitas 31:525–553.

Godefroit, P., Y. L. Bolotsky, L. Golovneva, W. Wenhao, and P. Lauters. 2011. New data on latest Cretaceous hadrosaurids from Russia and north-eastern China; pp. 16–20 in D. R. Braman, D. A. Eberth, D. C. Evans, and W. Taylor (compilers), International Hadrosaur Symposium Abstract Volume. Royal Tyrrell Museum of Palaeontology, Drumheller, Canada.

Godefroit, P., Z.-M. Dong, P. Bultynck, H. Li, and L. Feng. 1998. Sino-Belgian Cooperation program "Cretaceous Dinosaurs and Mammals from Inner Mongolia," 1: new *Bactrosaurus* (Dinosauria: Hadrosauroidea) material from Iren Dabasu (Inner Mongolia, P. R. China). Bulletin de l'Institut royal des Sciences naturelles de Belgique: Sciences de la Terre 68 (supplement):3–70.

Godefroit, P., J. Van Itterbeeck, P. Lauters, Y. L. Bolotsky, Z.-M. Dong, L.-Y Jin, S.-Q. Zan, S. Hai, and T. Yu. 2009. Latest Cretaceous hadrosaurid dinosaurs from Heilongjiang Province (P. R. China) and the Amur Region (Far Eastern Russia); pp. 91–120 in P. Huerta Hurtado and F. Torcida-Fernández Baldor (eds.), Actas de las IV Jornadas Internacionales sobre Paleontología de Dinosaurios y su Entorno. Colectivo Arqueológico-Paleontológico de Salas, Salas de los Infantes, Burgos, Spain.

Grigorescu, D. 1992. Nonmarine cretaceous formations of Romania; pp. 142–164 in N. Mateer, and C. Pen-Ji (eds.), Aspects of Non-marine Cretaceous Geology. China Ocean Press, Beijing.

Grigorescu, D. 2005. Rediscovery of a "forgotten land": the last three decades of research on the dinosaur-bearing deposits from the Haţeg basin. Acta Palaeontologica Romaniae 5:191–204.

Grigorescu, D., and Z. Csiki. 2006. Ontogenetic development of *Telmatosaurus transsylvanicus* (Ornithischia: Hadrosauria) from the Maastrichtian of the Haţeg Basin, Romania: evidence from the limb bones. Hantkeniana 5:20–26.

Grigorescu, R., G. Garcia, Z. Csiki, V. Codrea, and A.-V. Bojar. 2010. Uppermost Cretaceous megaloolithid eggs from the Haţeg Basin, Romania, associated with hadrosaur hatchlings: search for explanation. Palaeogeography, Palaeoclimatology, Palaeoecology 293:360–374.

Guenther, M. F. 2014. Comparative ontogenies (appendicular skeleton) for three hadrosaurids and a basal iguanodontian: divergent developmental pathways in Hadrosaurinae and Lambeosaurinae; chapter 22 in D. A. Eberth and D. C. Evans (eds.), Hadrosaurs. Indiana University Press, Bloomington, Indiana.

Haq, B. U., H. Hardenbol, and P. R. Vail 1987. Chronology of fluctuating sea-level since the Triassic. Science 235:1156–1167.

Herrero Santos, E. 2008. Nuevo yacimiento de icnitas de dinosaurio de la Sierra de los Gavilanes (Yecla, Murcia); pp. 37–38 in G. Lechuga, P. E. Collado Espejo, and M. B. Sánchez González (coordinators), XIX Jornadas de Patrimonio Cultural de la Región de Murcia, volume 1: Paleontología, Arqueología, Etnografía. Tres Fronteras Ediciones, Murcia, Spain.

Horner, J. R., D. B. Weishampel, and C. A. Forster. 2004. Hadrosauridae; pp. 438–463 in D. B. Weishampel, P. Dodson, and H. Osmólska (eds.), The Dinosauria, Second Edition. University of California Press, Berkeley, California.

Jagt, J. W. M., N. Motchurova-Dekova, P. Ivanov, H. Cappetta, and A. S. Schulp. 2006. Latest Cretaceous mosasaurs and lamniform sharks from Labirinta cave, Vratsa district (northwest Bulgaria): a preliminary note. Geološki anali Balkanskoga poluostrva 67:51–63.

Jeletzky, J. A. 1962. The allegedly Danian dinosaur-bearing rocks of the globe and the problem of the Mesozoic-Cenozoic boundary. Journal of Paleontology 36:1005–1018.

Jianu, C.-M., and G. J. Boekschoten. 1999. The Haţeg: island or outpost? Deinsea 7:195–198.

Jianu, C.-M., and D. B.Weishampel. 1999. The smallest of the largest: a new look at possible dwarfing in sauropod dinosaurs. Geologie en Mijnbouw 78:335–343.

Klein, N., P. M. Sander, K. Stein, J. Le Loeuff, J. L. Carballido, and E. Buffetaut. 2012. Modified laminar bone in *Ampelosaurus atacis* and other titanosaurs (Sauropoda): implications for life history and physiology. PLoS ONE 7(5):e36907.

Laurent, Y. 2003. Les faunes de vertébrés continentaux du Maastrichtien supérieur d'Europe: systematique et biodiversité. Strata 41:1–81.

Laurent, Y., J. Le Loeuff, and E. Buffetaut. 1997. Les Hadrosauridae (Dinosauria, Ornithopoda) du Maastrichtien supérieur des Corbières orientales (Aude, France). Revue paléobiologie 16:411–423.

Lauters, P., Y. L. Bolotsky, J. Van Itterbeeck, and P. Godefroit. 2008. Taphonomy and age profile of a latest Cretaceous dinosaur bone bed in Far Eastern Russia. Palaios 23:153–162.

Le Loeuff, J. 2005. Romanian Late Cretaceous dinosaurs: big dwarfs or small giants? Historical Biology 17:15–17.

Le Loeuff, J., and E. Buffetaut. 1993. Les derniers dinosaures du Sud de la France. Pour la Science–November 1993:50–58.

Le Loeuff, J., E. Buffetaut, and M. Martin. 1994. The last stages of dinosaur faunal history in Europe: a succession of Maastrichtian dinosaur assemblages from the Corbières (southern France). Geological Magazine 131:625–630.

Llompart, C. 1979. Yacimiento de huellas de pisadas de reptil en el Cretácico superior prepirenaico. Acta Geológica Hispánica 14:333–336.

López-Martínez, N., M. T. Fernández-Marrón, and M. F. Valle. 1999. The succession of vertebrates and plants across the Cretaceous-Tertiary boundary in the Tremp Formation, Ager valley (South-central Pyrenees, Spain). Geobios 32:617–627.

López-Martínez, N., J. I. Canudo, Ll. Ardèvol, X. Pereda-Suberbiola, X. Orue-Extebarria, G., Cuenca-Bescós, J. I. Ruiz-Omeñaca, X. Murelaga, and M. Feist. 2001. New dinosaur sites correlated with Upper Maastrichtian pelagic deposits in the Spanish Pyrenees: implications for the dinosaur extinction pattern in Europe. Cretaceous Research 22:41–61.

Lull, R. S., and N. E. Wright. 1942. Hadrosaurian Dinosaurs of North America. Geological Society of America Special Papers 40. 242 pp.

Lydekker, R. 1888. Catalogue of the fossil Reptilia and Amphibia in the British Museum, pt. I: Containing the orders Ornithosauria, Crocodilia, Dinosauria, Squamata, Rhynchocephalia, and Proterosauria. British Museum of Natural History, London, 309 pp.

MacArthur, R. H., and E. O. Wilson. 1967. The Theory of Island Biogeography. Monographs in Population Biology 1. Princeton University Press, Princeton, New Jersey, 203 pp.

Martin, J. E. 2007. New material of the Late Cretaceous globidontan *Acynodon iberoccitanus* (Crocodylia) from southern France. Journal of Vertebrate Paleontology 27:362–372.

Mateus, O., G. J. Dyke, N. Motchurova-Dekova, J. D.Kamenov, and P. Ivanov. 2010. The first record of a dinosaur from Bulgaria. Lethaia 43:88–94.

Matheron, P. P. E. 1869. Notice sur les reptiles fossiles des dépôts fluvio-lacustres crétacés du bassin à lignite de Fuveau. Mémoires de l'Académie impériale des Sciences, Belles-lettres et Arts de Marseille 1868–1869:345–379.

McDonald, A. T. 2011. The phylogeny, taxonomy and biogeography of basal iguanodonts (Dinosauria: Ornithischia); pp. 101–104 in D. R. Braman, D. A. Eberth, D. C. Evans, and W. Taylor (compilers), International Hadrosaur Symposium Abstract Volume. Royal Tyrrell Museum of Palaeontology, Drumheller, Canada.

McDonald, A. T. 2012. Phylogeny of basal Iguanodonts (Dinosauria: Ornithischia): an update. PLoS ONE 7(5):e36745.

McDonald, A. T., D. G. Wolf, and J. I. Kirkland. 2010. A new basal hadrosauroid (Dinosauria: Ornithopoda) from the Turonian of New Mexico. Journal of Vertebrate Paleontology 30:799–812.

Mezga, A., C. A. Meyer, B. Cvetko Tesović, Z. Bajraktarević, and I. Gusić. 2006. The first record of dinosaurs in the Dalmatian part (Croatia) of the Adriatic-Dinaric carbonate platform (ADCP). Cretaceous Research 27:735–742.

Mulder, E. W. A. 1984. Resten van *Telmatosaurus* (Ornithischia, Hadrosauridae) uit het Boven-Krijt van Zuid-Limburg. Grondboor en Hamer 38:108–115.

Mulder, E. W. A., M. M. M. Kuypers, J. W. M. Jagt, and H. H. G. Peeters. 1997. A new late Maastrichtian hadrosaurid dinosaur record from northeast Belgium. Neue Jahrbuch Geologische Paläontologische Monatshefte 6:339–347.

Nicosia, U., M. Marino, N. Mariotti, C. Muraro, S. Panigutti, F. M. Petti, and E. Sacchi. 1999. The Late Cretaceous dinosaur tracksite near Altamura (Bari, southern Italy). II–*Apulosauripus federicianus* new ichnogen. and new ichnosp. Geologica Romana 35:237–247.

Nopcsa, F. 1900. Dinosaurierreste aus Siebenbürgen (Schädel von *Limnosaurus transsylvanicus* nov. gen. et spec.). Denkschriften der königlichen Akademie der Wissenschaften, Wien 68:555–591.

Nopcsa, F. 1902. Dinosaurierreste aus Siebenbürgen II. (Schädelreste von *Mochlodon*). Mit einem Anhange: zur Phylogenie der Ornithopodiden. Denkschriften der königlichen Akademie der Wissenschaften. Mathematisch-Naturwissenschaftlichen Klasse, Wien 72:149–175.

Nopcsa, F. 1904. Dinosaurierreste aus Siebenbürgen III. Weitere Schädelreste von *Mochlodon*. Denkschriften der königlichen Akademie der Wissenschaften. Mathematisch-Naturwissenschaftlichen Klasse, Wien 74:229–263.

Nopcsa, F. 1914. Über das Vorkommen der Dinosaurier in Siebenbürgen. Verhandlungen der zoologische-botanischen Gesellschaft, Wien 54:12–14.

Nopcsa, F. 1915. Die Dinosaurier der siebenbürgischen Landesteile Ungarns. Mitteilungen Jahrbuch Koeniglíches Ungarische Geologische Reichsanstalt 23:1–24.

Nopcsa, F. 1928. Dinosaurierreste aus Siebenbürgen. IV. Die wirbelsäule von *Rhabdodon* und *Orthomerus*. Palaeontologia Hungarica 1(1921–1923):273–304.

Norman, D. B. 2002. On Asian ornithopods (Dinosauria: Ornithischia). 4. *Probactrosaurus* Rozhdestvensky, 1966. Zoological Journal of the Linnean Society 136:113–144.

Norman, D. B. 2004. Basal Iguanodontia; pp. 413–437 in D. B.Weishampel, P. Dodson, and H. Osmólska (eds), The Dinosauria, Second Edition. University of California Press, Berkeley, California.

Nuñez-Betelu, K. 1999. Preliminary palynological assessment of the vertebrate-rich Laño beds; age and palaeoenvironment. Estudios del Museo de Ciencias Naturales de Alava 14(número especial 1):37–42.

Oms, O., J. Dinarès-Turell, E. Vicens, R. Estrada, B. Vila, À. Galobart, and A. M. Bravo. 2007. Integrated stratigraphy from the Vallcebre Basin (southeastern Pyrenees, Spain): new insights on the continental Cretaceous-Tertiary transition in southwest Europe. Palaeogeography, Palaeoclimatology, Palaeocology 255:35–47.

Ősi, A. 2004. The first dinosaur remains from the Upper Cretaceous of Hungary (Csehbánya Formation, Bakony Mts). Geobios 37:749–753.

Ősi, A. 2005. *Hungarosaurus tormai,* a new ankylosaur (Dinosauria) from the Upper Cretaceous of Hungary. Journal of Vertebrate Paleontology 25:370–383.

Panaiotu, C. G., and C. E. Panaiotu. 2002. Palaeomagnetic studies; p. 61 in 7th European Workshop of Vertebrate Palaeontology, Abstract Volume and Excursion Field Guide. Sibiu, Romania.

Panaiotu, C. G., and C. E. Panaiotu. 2010. Palaeomagnetism of the Upper Cretaceous Sânpetru Formation (Haţeg Basin, South Carpathians). Palaeogeography, Palaeoclimatology, Palaeoecology 293:343–352.

Paris, J. P., and P. Taquet. 1973. Découverte d'un fragment de dentaire d'Hadrosaurien (Reptile Dinosaurien) dans le Crétacé supérieur des Petites Pyrénées (Haute-Garonne). Bulletin du Muséum nationale d'Histoire naturelle de Paris 130:17–27.

Pereda-Suberbiola, X., and P. M. Galton. 2001. Reappraisal of the nodosaurid ankylosaur *Struthiosaurus austriacus* from the Upper Cretaceous Gosau Beds of Austria; pp. 173–210 in K. Carpenter (ed.), The Armored Dinosaurs. Indiana University Press, Bloomington, Indiana.

Pereda Suberbiola, X., and P. M. Galton. 2009. Dwarf dinosaurs in the latest Cretaceous of Europe? pp. 263–272 in P. Huerta Hurtado, and F. Torcida-Fernández Baldor (eds.), Actas de las IV Jornadas Internacionales sobre Paleontología de Dinosaurios y su Entorno. Colectivo Arqueológico-Paleontológico de Salas, Salas de los Infantes, Burgos, Spain.

Pereda-Suberbiola, X., and J. L. Sanz. 1999. The ornithopod dinosaur *Rhabdodon* from the Upper Cretaceous of Laño (Iberian Peninsula), Estudios del Museo de Ciencias Naturales de Alava 14(número especial 1):257–272.

Pereda-Suberbiola, X., J. I. Ruiz-Omeñaca, and J. Company. 2003. Los dinosaurios hadrosaurios del registro ibérico. Descripción de nuevo material del Cretácico superior de Laño (Condado de Treviño); pp. 375–388 in F. Pérez-Lorente (ed.), Dinosaurios y otros reptiles mesozoicos en España. Ediciones Insituto de Estudios Riojanos, Logroño, Spain.

Pereda-Suberbiola, X., J. I. Ruiz-Omeñaca, J. Ullastre, and A. Masriera. 2003. Primera cita de un dinosaurio hadrosaurio en el Cretácico Superior del Pripirineo oriental (Peguera, Provincia de Barcelona). Geogaceta 34:195–198.

Pereda-Suberbiola, X., J. I. Canudo, J. Company, P. Cruzado-Caballero, and J. I. Ruiz-Omeñaca. 2009. Hadrosauroid dinosaurs from the Latest Cretaceous of the Iberian Peninsula. Journal of Vertebrate Paleontology 29:946–951.

Pereda-Suberbiola, X., J. I. Canudo, P. Cruzado-Caballero, J. L. Barco, N. López-Martínez, O. Oms, and J. I. Ruiz-Omeñaca. 2009. The last hadrosaurid dinosaurs of Europe: a new lambeosaurine from the uppermost Cretaceous of Arén (Huesca, Spain). Comptes rendus Palevol 8:559–572.

Philip, J., M. Floquet, J. P. Platel, F. Bergerat, M. Sandulescu, E. Baraboshkin, E. O. Amon, A. Poisson, R. Guiraud, D. Vaslet, Y. Le Nindre, M. Ziegler, S. Bouaziz, and J. C. Guezou. 2000. Map 16: Late Maastrichtian (69.5–65 Ma); in J. Dercourt, M. Gaetani, B. Vrielynck, E. Barrier, B. Biju-Duval, M. F. Brunet, J. P. Cadet, S. Crasquin, and M. Sandulescu (eds.), Atlas Peri-Tethys, Palaeogeographical Maps. CCGM/CGMW, Paris, France.

Powell, J. E. 1992. Hallazgo de huevos asignables a dinosaurios titanosáuridos (Saurischia, Sauropoda) de la Província de Río Negro, Argentina. Acta zoológica lilloana 41:381–389.

Prieto-Márquez, A. 2010a. *Glishades ericksoni,* a new hadrosauroid (Dinosauria: Ornithopoda) from the Late Cretaceous of North America. Zootaxa 2452:1–17.

Prieto-Márquez, A. 2010b. Global phylogeny of hadrosauridae (Dinosauria: Ornithopoda) using parsimony and Bayesian methods. Zoological Journal of the Linnean Society 159:435–502.

Prieto-Márquez, A., and J. R. Wagner. 2009. *Pararhabdodon isonensis* and *Tsintaosaurus spinorhinus:* a new clade of lambeosaurine hadrosaurids from Eurasia. Cretaceous Research 30:1238–1246.

Prieto-Márquez, A., R. Gaete, À. Galobart, V. Riera, and O. Oms. 2007. New data on European Hadrosauridae (Dinosauria: Ornithopoda) from the latest Cretaceous of Spain. Journal of Vertebrate Paleontology 27(3, Supplement):131A.

Prieto-Márquez, A., R. Gaete, G. Rivas, À. Galobart, and M. Boada. 2006. Hadrosaurid dinosaurs from the Late Cretaceous of Spain: *Pararhabdodon isonensis* revisited and *Koutalisaurus kohlerorum,* gen et sp. nov. Journal of Vertebrate Paleontology 26:929–943.

Riabinin, A. N. 1945. [Dinosaurian remains from the Upper Cretaceous of the Crimea]. Vsesoiuznyi. Nauchno-Issledovatel'skii Geologicheskij Institut, Materialy (Paleontologija i Stratigrafija) 4:4–10. [Russian]

Riera, V., O. Oms, R. Gaete, and À. Galobart. 2009. The end-Cretaceous dinosaur succession in Europe: the Tremp Basin record (Spain). Palaeogeography, Palaeoclimatology, Palaeoecology 283:160–171.

Riveline, J., J. P. Berger, M. Feist, C. Martin-Closas, M. Schudack, and I. Soulié-Märsche. 1996. European Mesozoic–Cenozoic charophyte biozonation. Bulletin de la Société géologique de France 167:453–468.

Sachs, S., and J. J. Hornung. 2006. Juvenile ornithopod (Dinosauria: Rhabdodontidae) remains from the Upper Cretaceous (Lower Campanian, Gosau Group) of Muthmannsdorf (Lower Austria). Geobios 39:415–425.

Sander, P. M., O. Mateus, T. Laven, and N. Knötsche. 2006. Bone histology indicates insular dwarfism in a new Late Jurassic sauropod dinosaur. Nature 441:739–741.

Seeley, H. G. 1883. On the dinosaurs from the Maastricht beds. Quarterly Journal of the Geological Society London 39:246–253.

Sereno, P. 1998. A rationale for phylogenetic definitions, with application to higher-level taxonomy of Dinosauria. Neue Jahrbuch Geologische Palaontologische Abhandlungen 210:41–83.

Stein, K., Z. Csiki, K. Curry Rogers, D. B.Weishampel, R. Redelstorff, J. L. Carballido, and P. M. Sander. 2010. Small body size and extreme cortical bone remodeling indicate phyletic dwarfism in *Magyarosaurus dacus* (Sauropoda:

Titanosauria). Proceedings of the National Academy of Sciences 107:9258–9263.

Sues, H.-D., and A. Averianov. 2009. A new basal hadrosauroid dinosaur from the Late Cretaceous of Uzbekistan and the early radiation of duck-billed dinosaurs. Proceedings of the Royal Society of London Biological Sciences 276:2549–2555.

Sullivan, R. M., S. E. Jasinski, M. Guenther, and S. G. Lucas. 2011. The first lambeosaurine (Dinosauria, Hadrosauridae, Lambeosaurinae) from the Upper Cretaceous Ojo Alamo Formation (Naashoibito member), San Juan basin, New Mexico. New Mexico Museum of Natural History and Science Bulletin 53:407–417.

Therrien, F. 2005. Palaeoenvironment of the latest Cretaceous (Maastrichtian) dinosaurs of Romania: insights from fluvial deposits and paleosols of the Transylvanian and Haţeg basins. Palaeogeography, Palaeoclimatology, Palaeoecology 218:15–56.

Therrien, F., D. K. Zelenitsky, and D. B. Weishampel. 2009. Palaeoenvironmental reconstruction of the Late Cretaceous Sânpetru Formation (Haţeg Basin, Romania) using paleosols and implications for the "disappearance" of dinosaurs. Palaeogeography, Palaeoclimatology, Palaeoecology 272:37–52.

Tsogtbaatar, K., D. B. Weishampel, and M. Watabe. 2011. Phylogenetic and biogeographic importance of an early Late Cretaceous (Cenomanian-Campanian) derived hadrosauroid (Ornithopoda) newly described from Mongolia; pp. 149–152 in D. R. Braman, D. A. Eberth, D. C. Evans, and W. Taylor (compilers), International Hadrosaur Symposium Abstract Volume. Royal Tyrrell Museum of Palaeontology, Drumheller, Canada.

Tsogtbaatar, K., D. B. Weishampel, M. Watabe, and D. C. Evans. 2014. A new hadrosauroid (*Plesiohadros djadokhtaensis*) from the Late Cretaceous Djadokhtan fauna of southern Mongolia; chapter 7 in D. A. Eberth and D. C. Evans (eds.), Hadrosaurs. Indiana University Press, Bloomngton, Indiana.

Van Itterbeeck, J., Y. L. Bolotsky, P. Bultynck, and P. Godefroit. 2005. Stratigraphy, sedimentology and palaeoecology of the dinosaur-bearing Kundur section (Zeya-Bureya Basin, Amur Region, Far Eastern Russia). Geological Magazine 142:735–750.

Van Itterbeeck, J., D. J. Horne, P. Bultynck, and N. Vandenberghe. 2005. Stratigraphy and palaeoenvironment of the dinosaur-bearing Upper Cretaceous Iren Dabasu Formation, Inner Mongolia, People's Republic of China. Cretaceous Research 26:699–725.

Van Itterbeeck, J., V. S. Marchevich, and V. Codrea. 2005. Palynostratigraphy of the Maastrichtian dinosaur- and mammal sites of the Râul Mare and Barbat valleys (Haţeg Basin, Romania). Geologica Carpathica 56:137–147.

Van Valen, L. M. 1973. Pattern and the balance of nature. Evolutionary Theory 1:31–49.

Vila, B., À. Galobart, and O. Oms. 2005. Nuevos yacimientos con huellas de dinosaurio en el Cretácico Superior del sinclinal de Vallcebre (Fm. Tremp, Pirineo Suroriental). Miscelánea Paleontológica 6:414–426.

Vila, B., O. Oms, À. Galobart, J. Marmi, and R. Gaete. 2006. Los últimos dinosaurios de los Pirineos y sus huellas. Enseñanza de las Ciencias de la Tierra 14:240–246.

Vila, B., O. Oms, À. Galobart, R. Gaete, J. Peralba, and J. Escuer. 2006. Nuevos hallazgos de dinosaurios y otros tetrápodos continentales en los Pirineos sur-centrales y orientales: resultados preliminares; pp. 365–378 in Colectivo Arqueológico-Paleontológico de Salas (ed.), Actas de la III Jornadas Internacionales sobre Paleontología de Dinosaurios y su Entorno. Colectivo Arqueológico-Paleontológico de Salas, Salas de los Infantes, Burgos, Spain.

Vremir, M. 2010. New faunal elements from the Late Cretaceous (Maastrichtian) continental deposits of Sebeş area (Transylvania). Terra Sebus-Acta Musei Sabesiensis 2:635–684.

Vullo, R., D. Néraudeau, and T. Lenglet. 2007. Dinosaur teeth from the Cenomanian of Charentes, western France: evidence for a mixed Laurasian-Gondwanan assemblage. Journal of Vertebrate Paleontology 27:931–943.

Weishampel D. B., and C.-M. Jianu, 2011. Transylvanian Dinosaurs. Johns Hopkins University Press, Baltimore, Maryland, 301 pp.

Weishampel, D. B., D. Grigorescu, and D. B. Norman. 1991. The dinosaurs of Transylvania: island biogeography in the Late Cretaceous. National Geographic Research and Exploration 7:68–87.

Weishampel, D. B., D. B. Norman, and D. Grigorescu. 1993. *Telmatosaurus transsylvanicus* from the Late Cretaceous of Romania: the most basal hadrosaurid dinosaur. Palaeontology 36:361–385.

Weishampel, D. B., C.-M. Jianu, Z. Csiki, and D. B. Norman. 2003. Osteology and phylogeny of *Zalmoxes* (n.g.), an unusual Euornithopod dinosaur from the latest Cretaceous. Journal of Systematic Palaeontology 1:65–123.

Weishampel, D. B., E. W. A. Mulder, R. W. Dortangs, J. W. M. Jagt, C.-M. Janu, M. M. Kuypers, H. H. G. Peeters, and A. S. Schulp. 1999. Dinosaur remains from the type Maastrichtian: an update. Geologie en Mijnbouw 78:357–365.

Weishampel, D. B., P. M. Barrett, R. A. Coria, J. Le Loeuff, X. Xing, Z. Xijin, A. Sahni, E. M. P. Gomani, and C. R. Noto. 2004. Dinosaur distribution; pp. 517–606 in D. B.Weishampel, P. Dodson, and H. Osmólska (eds.), The Dinosauria, Second Edition. University of California Press, Berkeley, California.

Wellnhofer, P. 1994. Ein Dinosaurier (Hadrosauridae) aus der Oberkreide (Maastricht, Helvetikum-Zone) des bayerischen Alpenvorlandes. Mitteilungen Bayerische Staatssammlung. Paläontologische und historisches Geologie 34:221–238.

Yilmaz, P. O., I. O. Norton, D. Leary, and R. J. Chucla. 1996. Tectonic evolution and palaeogeography of Europe; pp. 47–70 in P. A. Ziegler and T. F. Horvath (eds.), Peri-Tethys Memoir 2: Structure and Prospects of Alpine Basins and Forelands. Mémoires du Muséum national d'Histoire naturelle (Paris) 170.

You, H., Z. Luo, N. H. Shubin, L. M. Witmer, Z. Tang, and F. Tang. 2003. The earliest-known duck-billed dinosaur from deposits of late Early Cretaceous age in northwest China and hadrosaur evolution. Cretaceous Research 24:347–355.

Young, C.-C. 1958. The dinosaurian remains of Laiyang, Shantung. Palaeontologica Sinica, New Series C 16:53–138.

Appendix 15.1. Length Measurements of the Hindlimb Bones Used in the Size-Frequency Diagrams of Figure 15.11A, B

Number	Element	Site	Length
CGTL 2/5751	Tibia	Besh-Kosh (UA)	~70
NMNHS F-31437	Femur	Labirinta (BG)	40+
NMNHS F-31441/31443	Tibia	Labirinta (BG)	35–40+
MAFI 3129a	Tibia	Haţeg (RO)	41
FGGUB R.001	Tibia	Haţeg (RO)	43
FGGUB R.002	Tibia	Haţeg (RO)	45+
MAFI 10338	Femur	Haţeg (RO)	38
FGGUB R.1832(1362*)	Femur	Haţeg (RO)	25
BMNH R9414	Femur	Valioara (RO)	43
BMNH R9414	Fibula	Valioara (RO)	44+
MAFI 3130	Femur	Haţeg (RO)	46+
MAFI 3128	Femur	Haţeg (RO)	27
FGGUB R.1850.1	Femur	Haţeg (RO)	4.8
FGGUB R.1980.4	Femur	Haţeg (RO)	4.4+
FGGUB R.1981.1	Femur	Valioara (RO)	4.3
FGGUB R.1853	Tibia	Tuştea (RO)	6.3
FGGUB R.246+250	Tibia	Tuştea (RO)	4.2+
FGGUB R.1982.1	Tibia	Tuştea (RO)	4.1+
FGGUB R.1851.1	Tibia	Tuştea (RO)	3.9
FGGUB R.1979	Tibia	Tuştea (RO)	3.6
FGGUB R.1852.4	Tibia	Tuştea (RO)	3.4
SC 57021	Femur	Vill.del Pescatore (I)	42
SC 57021	Tibia	Vill.del Pescatore (I)	55
NKM 71	Femur	Bad Adelholzen (D)	34
BMNH 42955	Femur	Limburg	~50
BMNH 42954	Tibia	Limburg	~30
MND K 21.04.003	Femur	Geulhem (NL)	60+
MDE-Les2-10	Femur	Lestaillats 2 (F)	55+
MDE-Aus-01	Femur	Ausseing (F)	60+
MDE-Aus-03	Tibia	Ausseing (F)	60
MDE-Aus-09	Fibula	Ausseing (F)	49.5
MDE-Ma1-02	Tibia	Tricouté 2 (F)	47
MDE-Ma1-08	Tibia	Tricouté 2 (F)	~28
MDE-Ma3-11	Tibia	Tricouté 3 (F)	~35
MDE-Ma3-09	Tibia	Tricouté 3 (F)	~23
MDE-Cas2-249	Tibia	Cassagnau 2 (F)	~28
MDE-Cas2-247	Tibia	Cassagnau 2 (F)	~29
MDE-Cas2-24	Tibia	Cassagnau 2 (F)	~36.5
MDE-Cas2-13	Tibia	Cassagnau 2 (F)	18
MDE-Fo2-01	Femur	Le Bexen (F)	21.5
MDE-Fo1-19	Femur	Le Bexen (F)	57.5
MDE-Fo1-148	Femur	Le Bexen (F)	39.5
MPZ2007/930	Femur	Blasi 1 (SP)	77+
MPZ2007/932	Femur	Blasi 1 (SP)	69+
MPZ2007/702	Femur	Blasi 1 (SP)	60+
MPZ2007/933	Femur	Blasi 1 (SP)	55+
MPZ2008/337	Femur	Blasi 3 (SP)	71
MPZ2007/704	Femur	Blasi 3 (SP)	75+
MPV 186	Femur	La Solana (SP)	42+
MGUV 2203	Tibia	La Solana (SP)	48+

Note: Measurements are in centimeters. Data on Romanian elements are mostly taken from Grigorescu and Csiki (2006) and Benton et al. (2010); I reported only "Haţeg Basin" when the exact locality of provenance of the bone was not mentioned in literature. + estimated length; * reported as FGGUB R.1362 in Benton et al. (2010), but it is actually FGGUB R.1832.

Appendix 15.2. Length Measurements of the Postcranial Bones Used in the Size-Frequency Diagrams of Figure 15.11C

Number	Element	Site	Length
NMNHS F-11899	Metatarsal II	Labirinta (BG)	6.5
MAFI 3126	Humerus	Haţeg (RO)	22.8
MAFI 3124	Ulna	Haţeg (RO)	33
FGGUB R.128	Ulna	Haţeg (RO)	30
SC 57247	Ischium	Vill.del Pescatore (I)	~85
SC 57023	Pubis	Vill.del Pescatore (I)	21
MDE-Les1-25	Scapula	Lestaillats 1 (F)	~26.5
MDE-Les1-02	Ischium	Lestaillats 1 (F)	~45
MDE-Les1-19	Ischium	Lestaillats 1 (F)	~50
MDE-Aus-185	Ischium	Ausseing (F)	~70
MDE-Aus-184	Metatarsal IV	Ausseing (F)	15.6
MDE-Ma1-03	Metatarsal IV	Tricouté 2 (F)	16
MDE-Ma3-13	Humerus	Tricouté 3 (F)	~15
MDE-Cas2-207	Humerus	Cassagnau 2 (F)	12
MDE-Cas2-01	Pubis	Cassagnau 2 (F)	21.7
MDE-Cas2-209	Metatarsal II	Cassagnau 2 (F)	11.8
MDE-Fo1-18	Humerus	Le Bexen (F)	17.5
MPZ2005/90	Ilium	Blasi 3 (SP)	47

Note: Measurements are in centimeters.

Appendix 15.3. Length Measurements of the Dentaries and Maxillae Used in the Size-Frequency Diagrams of Figure 15.11C

Number	Element	Site	Length
BMNH R3386	Dentary	Haţeg (RO)	29+
BMNH R3401	Dentary	Haţeg (RO)	15+
FGGUB R.1010	Maxilla	Haţeg (RO)	20+
SC 57026	Dentary	V.del Pescatore (I)	30+
MDE-Ma3-16	Maxilla	Tricouté 3 (F)	17
MDE-Cas2-248	Dentary	Cassagnau 2 (F)	16.2
MDE-Cas2-02	Dentary	Cassagnau 2 (F)	16+
No number	Maxilla	Larcan (F)	11.5
MDE-Fo1-10	Dentary	Le Bexen (F)	30+
MPZ99/665	Dentary	Blasi 1 (SP)	34
MPZ2008/258	Dentary	Blasi 3 (SP)	44.5
Bla 3,4-1	Dentary	Blasi 3,4 (SP)	25+
IPS 36338	Dentary	Fontllonga (SP)	40+
MGUV 2200	Dentary	La Solana (SP)	25+
MPV 181	Dentary	La Solana (SP)	45+

Note: Measurements are in centimeters. + estimated length.

Appendix 15.4. Length Measurements of the Dentaries and Femora Used in the Bivariate Plot of Figure 15.12

Number	Taxon	Dentary	Femur	D/fe
CMN 8703	*Lambeosaurus lambei*	50	102	0.49
CMN 8501	*Hypacrosaurus altispinus*	46.5	107	0.43
ROM 845	*Corythosaurus intermedius*	49	104	0.47
ROM 776	*Corythosaurus intermedius*	51	105.5	0.48
ROM 768	*Parasaurolophus walkeri*	45.5	103	0.44
ROM 787	*Prosaurolophus maximus*	>63	100	>0.63
AMNH 5220	*Saurolophus osborni*	83	115	0.72
USNM 2414	*Edmontosaurus annectens*	82	102.5	0.80
AMNH 5730	*Edmontosaurus annectens*	92	104	0.88

Note: Measurements are in centimeters and are taken from Lull and Wright (1942).

The Hadrosauroid Record in the Maastrichtian of the Eastern Tremp Syncline (Northern Spain)

16

Fabio M. Dalla Vecchia, Rodrigo Gaete, Violeta Riera, Oriol Oms, Albert Prieto-Márquez, Bernat Vila, Albert Garcia Sellés, and Àngel Galobart

ABSTRACT

The eastern Tremp Syncline (northeastern Spain) is the richest European locality for latest Cretaceous hadrosauroids. We summarize the record of hadrosauroid skeletal remains from 33 localities and tracks from 15 sites in the Tremp Formation. Two of the bone-bearing localities occur in the Early/Late Maastrichtian "gray unit," whereas all other sites occur in the Late Maastrichtian "lower red unit." The skeletal sample represents more than 1,000 elements, mostly axial and appendicular, although several mandibular and skull bones have been recovered. The eastern Tremp Syncline includes the hitherto only known hadrosauroid bonebed from Europe, as well as the "tsintaosaur" lambeosaurine *Pararhabdodon isonensis*. Hadrosauroids are less abundant than sauropods in the gray unit, whereas they are much more abundant in the overlying lower red unit where sauropod bones are rare. This turnover is best explained as a time/event-related faunal change at the beginning of the Late Maastrichtian. The apparent absence of hadrosauroid eggshells and hatchlings suggests that the nesting grounds of these animals occurred far from the depositional areas of the Tremp Syncline, thus remaining unrecorded unlike the egg remains of coeval sauropods. The skeletal sample is mostly composed of individuals much smaller than those regarded as adults for the hadrosaurid taxa of the latest Cretaceous of western North America and Asia, although a few larger individuals in the adult size range exist. This may be explained by the prevalence of juvenile individuals compared to adults, the presence of both small and large-sized species in the sample, or by the occasional presence of unusually large individuals within populations of relatively moderate body size, and possibly dwarfs.

INTRODUCTION

The richest region in Europe for the remains of latest Cretaceous hadrosauroids is, without doubt, the Tremp Syncline of the southern Pyrenees, which is located in the Aragon and Catalonia regions of northern Spain. The western part of the syncline, in Aragon, has yielded a relatively limited sample of Late Maastrichtian hadrosaurids from three distinct sites (six fossil-bearing levels, Blasi 1 to 5; López-Martínez et al., 2001), and one footprint site (Barco et al., 2001). That assemblage has been studied in detail and exhibits a high diversity of hadrosaurids, including the lambeosaurines *Arenysaurus ardevoli* (see Pereda-Suberbiola et al., 2009) and *Blasisaurus canudoi* (see Cruzado-Caballero, Pereda-Suberbiola, and Ruiz-Omeñaca, 2010), and an indeterminate "hadrosaurine" (Cruzado-Caballero, Ruiz-Omeñaca, and Canudo, 2010).

A much larger number of sites in the eastern part of the syncline, located in Catalonia, yield abundant material. However, most of the specimens collected to date have either not been reported/described or have been mentioned only in abstracts (e.g., Martín et al., 2007; Gaete et al., 2007; Prieto-Márquez et al., 2007). Those sites and their hadrosauroid fossils are comprehensively reviewed here and some implications relating to their sizes and stratigraphic distributions are discussed.

Institutional Abbreviations PFUB, Institut für Paläontologie, Freie Universität Berlin, Germany; IPS, Institut de Paleontologia Dr. Miquel Crusafont (currently Institut Català de Paleontologia "Miquel Crusafont," ICP), Sabadell, Barcelona, Spain; MCD, Museu de la Conca Dellà, Isona, Spain; ROM, Royal Ontario Museum, Toronto, Ontario; TMP, Royal Tyrrell Museum of Palaeontology, Drumheller, Alberta.

METHODS AND TERMINOLOGY

We use the more inclusive term "hadrosauroid" (sensu Sereno, 1998) instead of "hadrosaurid" in reference to the samples from the eastern Tremp Syncline because they potentially include remains belonging to ornithopods that do not belong to the clade Hadrosauridae sensu Prieto-Márquez, 2010 (see Dalla Vecchia, this volume). We assume that all the ornithopod bones in the sample from the eastern Tremp Syncline that lack diagnostic hadrosauroid features belong to Hadrosauroidea. The only other ornithopods known in

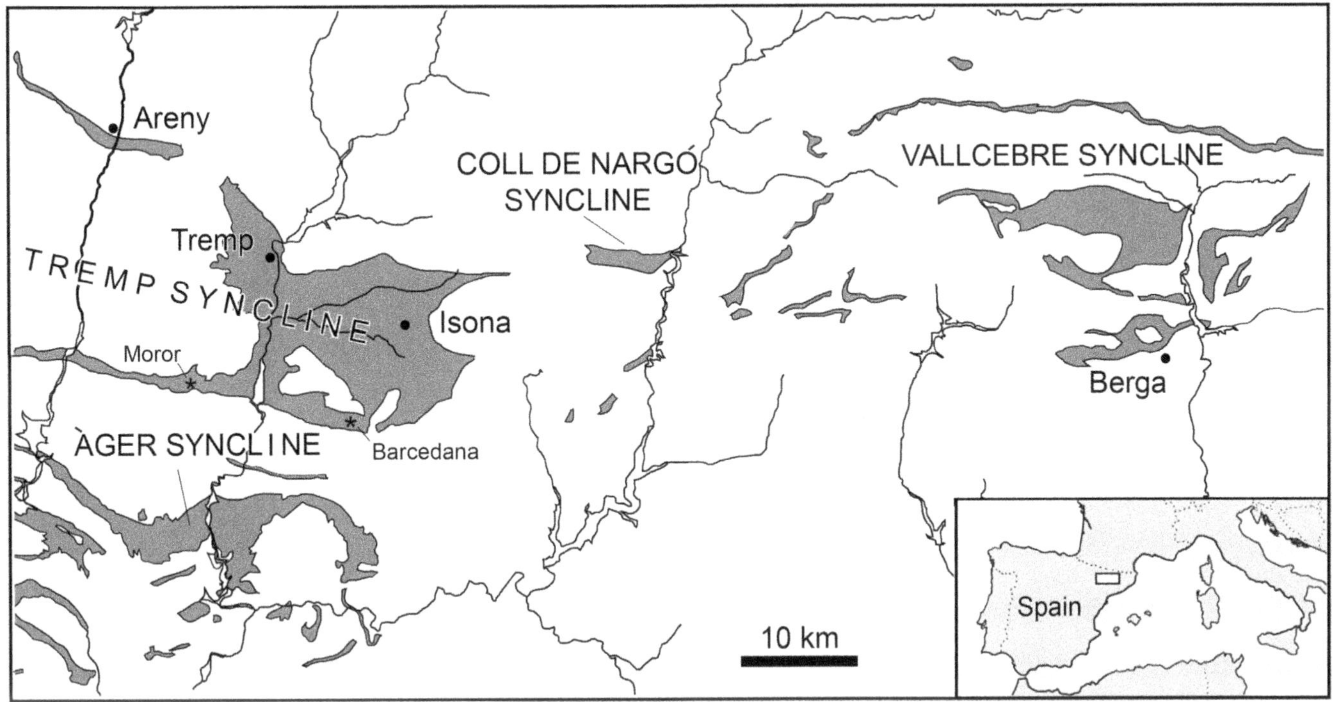

16.1. Latest Cretaceous continental deposits (gray color) in the synclines of the southern side of the Pyrenees (Spain) and location of the Tremp Syncline.

the uppermost Cretaceous of Europe are the rhabdodontids (Weishampel et al., 2004; Company et al., 2009). The possibility that the eastern Tremp Syncline deposits include the remains of rhabdodontids is excluded here because none of the several skeletal elements collected in the Tremp Syncline to date can be reliably referred to this group, and the ornithopod-bearing units in the Tremp Syncline are mostly Late Maastrichtian in age and rhabdodontids have never been reported from the Upper Maastrichtian (Company et al., 2009). Furthermore, all the appendicular and vertebral elements examined here are morphologically unlike those of rhabdodontids (cf., Pereda-Suberbiola and Sanz, 1999; Weishampel et al., 2003).

Specimens without museum repository numbers are identified here by field number if they were collected during fieldwork (e.g., LB1-01, where the acronym refers to the site [in this case Lo Bas-1]), or are referred to without number if obtained by other means. All specimens without a museum repository number, field number or museum acronym are in the collections of the MCD, unless specified otherwise.

The size distribution in a sample of hadrosaurid femora and tibiae from the Dinosaur Park Formation (Upper Campanian) of Alberta, Canada (Brinkman, 2011; Brinkman, this volume) was used as a reference for comparing the sizes of skeletal elements from Tremp hadrosauroids (also including fibulae as a proxy for tibiae) with those from a classic continental North American sample. In Brinkman's analysis, the sizes of femora and tibiae from Dinosaur Provincial Park are expressed as percentages with respect to the size of these same elements in reference specimen TMP 1998.058.0001, an indeterminate hadrosaurid taxon (Brinkman, 2011, this volume). In the case of incompletely preserved elements, the length was estimated by comparison with complete specimens. D. Brinkman (TMP) kindly furnished the measurements of other appendicular elements from TMP 1998.058.0001 that were used for comparison with their counterparts from the Tremp Syncline (humerus, ulna, ischium, metatarsal II, and pedal phalanx III-1). Because TMP 1998.058.0001 does not preserve the skull and lower jaws, the dentary of ROM 845 (*Corythosaurus intermedius*) was chosen to serve as a comparative reference for that element. This specimen is from the same formation and general location as TMP 1998.058.0001 (Dinosaur Park Formation at Dinosaur Provincial Park [DPFm]). Because the tibiae and femora of ROM 845 are twice the length of those of TMP 1998.058.0001, measurements of the Tremp elements are expressed as a percentage of those in ROM 845, and then transformed as a percentage of those for TMP 1998.058.0001 (50% smaller than those of ROM 845). Because the length of the maxilla of ROM 845 is not available, the maxilla of ROM 1933 (*C. casuarius*) was used as a proxy. This specimen is also from the DPFm and has a skull length very close to that of ROM 845 (Evans, 2010). The size comparisons between Tremp and Canadian elements made here (other than tibia

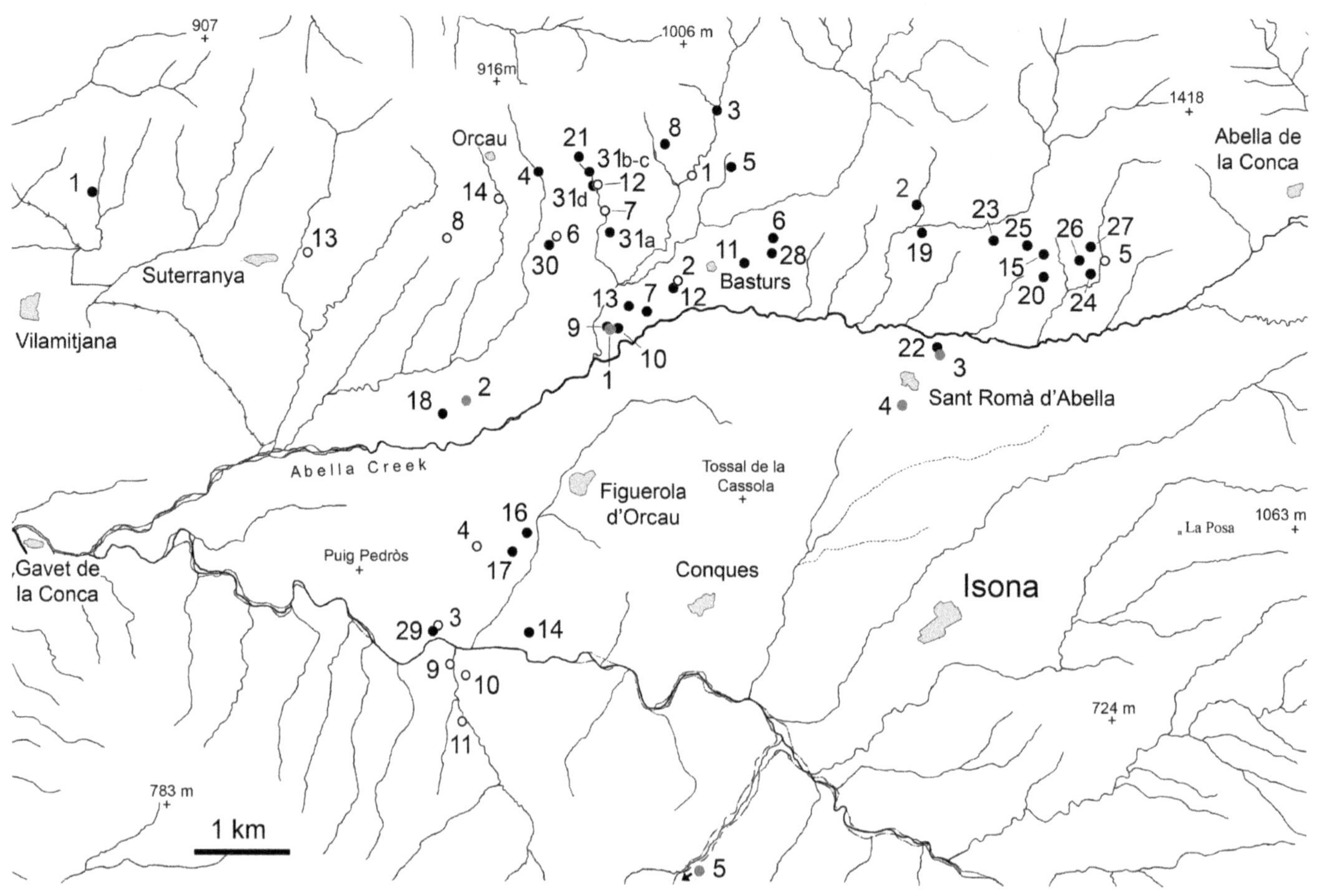

16.2. Location of the hadrosauroid-bearing sites in the easternmost Tremp Syncline. Bone sites (black dots): (1) Els Nerets; (2) Lo Bas-1, -2; (3) Barranc de la Costa Gran; (4) Casa Fabà; (5) Els Pous; (6) Les Feixes; (7) Magret; (8) Les Torres; (9) Serrat del Rostiar-1; (10) Serrat del Rostiar-3; (11) Basturs Est; (12) Basturs Poble; (13) Serrat del Rostiar-2; (14) Cabana de Gori-2; (15) Les Llaus; (16) Costa de la Serra-2; (17) Costa de la Serra-1; (18) Serrat de Pelleu; (19) Euroda Nord; (20) Les Llaus de la Doba; (21) Els Esfons; (22) Molí del Baró-1; (23) Sant Romà d'Abella; (24) Planta del Mestre; (25) Tossal de la Doba; (26) Serrat del Corb; (27) Coll de la Torre; (28) La Serra; (29) Masia de Ramon; (30) Serrat del Sanguin-1 and 2; (31a–d) Barranc de Torrebilles (TB1 to 4, respectively). Track sites (open circles): (1) Llau de la Costa; (2) Basturs Poble; (3) Masia de Ramon/Mas d'en Feliu and surroundings; (4) Costa de la Serra; (5) Tossal del Gassó; (6) Serrat del Sanguin; (7) Barranc de Torrebilles 6–7; (8) Costa Roia; (9–11) Barranc de Guixers-1 to 3; (12) Barranc de Torrebilles-8 (= Orcau 4); (13) Suterranya; (14) Torrent de Carant. Some eggshell sites of the lower red unit are marked with gray dots: (1) Serrat del Rostiar-3; (2) Costa de la Coma; (3) Molí del Baró-1; (4) Tossal de Sant Romà d'Abella; (5) Camí del Soldat. Eggshell sites Barranc de Fonguera and Coll de Faidella occur beyond the limits of the map and, thus, are not shown.

and femur) are approximate and must be considered with caution because of potential problems due to allometric growth during ontogeny and different proportions and morphology of the same elements between taxa (e.g., the higher skull elongation in the Hadrosaurinae with respect to the Lambeosaurinae).

EASTERN TREMP SYNCLINE: GEOLOGICAL SETTING

The Tremp Syncline is one of four synclines occurring along the southern Pyrenees that preserve parts of a latest Cretaceous–earliest Paleogene continental sedimentary basin (Fig. 16.1). The basin developed in the Ibero-Armorican Island, the largest island in the European Archipelago (see Dalla Vecchia, this volume).

Here, we deal only with the fossil sites occurring in the eastern Catalonian part of the syncline, east of the river Noguera Ribagorçana. Most sites are located in the easternmost part of the syncline, referred to as Conca Dellà (Isona i Conca Dellà, Abella de la Conca, Gavet de la Conca and Tremp municipalities; Fig. 16.2).

All sites are in the Mesozoic portion of the latest Cretaceous–earliest Paleogene Tremp Formation (sensu Rosell et al., 2001; Fig. 16.3). The Cretaceous part of this formation was deposited during regression, and its lower boundary with the Arén Sandstones becomes younger from east to west. The Cretaceous portion of the formation comprises two superimposed informal units, the gray unit (corresponding to the La Posa Formation of Cuevas, 1992) and the lower red unit (corresponding to the Conques Formation and the Cretaceous portion of the Talarn Formation of Cuevas, 1992).

The gray unit was deposited during the Early–Late Maastrichtian transition in the easternmost part of the Tremp Syncline (based on biostratigraphy; see references in Riera et al., 2009; Villalba-Breva and Martín-Closas, 2011) and the lower red unit is Late Maastrichtian in age based on biostratigraphic and stratigraphic data (Riera et al., 2009; Riera, 2010), magnetostratigraphy (Oms et al., 2010; Riera, 2010), and correlations with the nearby Vallcebre (Oms et al., 2007), Àger (Galbrun et al., 1993), and western Tremp synclines (Pereda-Suberbiola et al., 2009). There are no significant depositional hiatuses in the two units; the upper part of the lower red unit falls within magnetochron C29r (Oms et al., 2010; Riera, 2010) and the K-Pg boundary is located within the overlying unit that is equivalent to the Vallcebre limestone (Riera et al., 2009; Fig. 16.3).

The gray unit was deposited in a transitional environment (lagoon, tidal flat, and coastal marsh) and contains a few marine intercalations, whereas the lower red unit is composed mainly of floodplain mudstone and fluvial sandstone (point bars, channels, etc.) with sporadic intercalations of lacustrine limestone.

The stratigraphic position of the hadrosaur-bearing sites in the eastern Tremp Syncline has been detailed by Riera et al. (2009) and Riera (2010). Only two sites (Els Nerets and Moror) occur in the gray unit whereas all the others are in the lower red unit (Fig. 16.3).

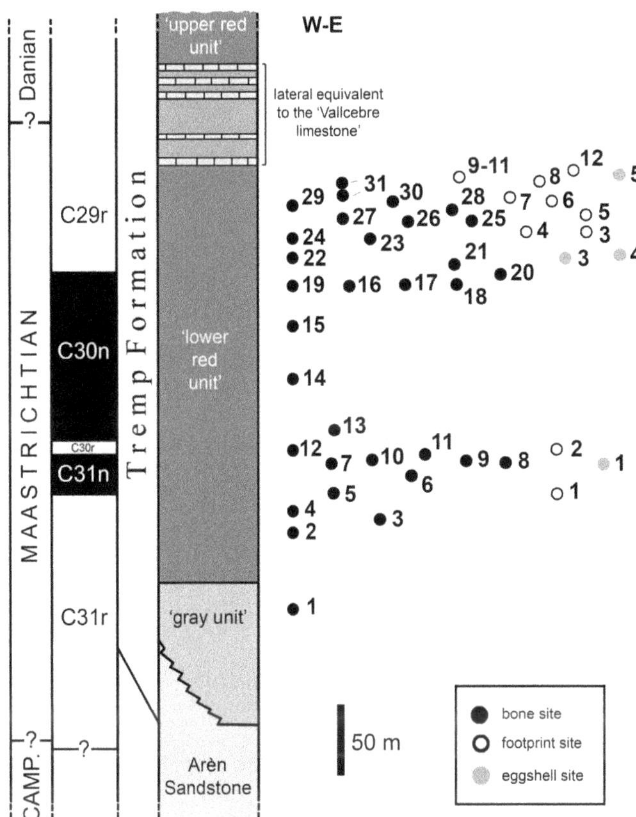

16.3. Stratigraphic occurrence of the hadrosauroid-bearing sites of the easternmost Tremp Syncline. Bone sites (black dots), track sites (open circles) and eggshell sites of the lower red unit (gray dots) have the same numbers as in Figure 16.2. The exact stratigraphic position of some sites is unknown and are not reported here. The stratigraphic extent of the lateral equivalent to the Vallcebre limestone is that reported by Riera et al. (2009). The correspondence between the magnetostratigraphic and the lithostratigraphic units is approximate. Eggshell site 2 (Costa de la Coma) is not shown; Tracksites 13 (Suterranya) and 14 (Torrent de Carant) are not shown.

HADROSAUROID SITES AND REMAINS IN THE EASTERN TREMP SYNCLINE

Thirty-three sites with hadrosauroid skeletal remains and at least 15 with tracks have been identified to date. Only one (Moror) occurs near the southern margin of the syncline (see Fig. 16.1); the others are in the northeastern part of the syncline (Fig. 16.2). To date, about one thousand skeletal elements referred to hadrosauroids and eight tracks collected from those sites are reposited at the MCD and ICP. The record includes the first hadrosaur bonebed of Europe and the type material of the first named hadrosauroid of the Iberian Peninsula, *Pararhabdodon isonensis*. Unfortunately, all the hadrosauroid skeletal record from the Tremp Formation comprises disarticulated elements, with the exclusion of a segment of caudal vertebral column from the Planta del Mestre site and a partial pelvis from Serrat del Corb site.

Hadrosauroid Bone-Bearing sites

The sites yielding hadrosauroid skeletal remains (Fig. 16.2) are reviewed in approximate stratigraphic order (see Fig. 16.3).

Moror This site yielded the proximal half of a left and a right ischium (IPFUB unnumbered specimens) probably belonging to the same individual (Brinkmann, 1984). The estimated total length is 60 cm (Brinkmann, 1984). According to Pereda-Suberbiola et al. (2003:379), a "lower jaw fragment" was also found in Moror.

Els Nerets Nine specimens from this site discovered in 1984 were referred to hadrosauroids (Casanovas-Cladellas et al., 1985), and all belong to relatively small sized individuals. They include a right femur (IPS-N-21; Fig. 16.4O), a distal portion of the right femur that is the same size as the previous one (IPS-N-3; now IPS-896, Fig. 16.4Q), a proximal caudal vertebra (IPS-N-13, now IPS-895), two proximal caudal

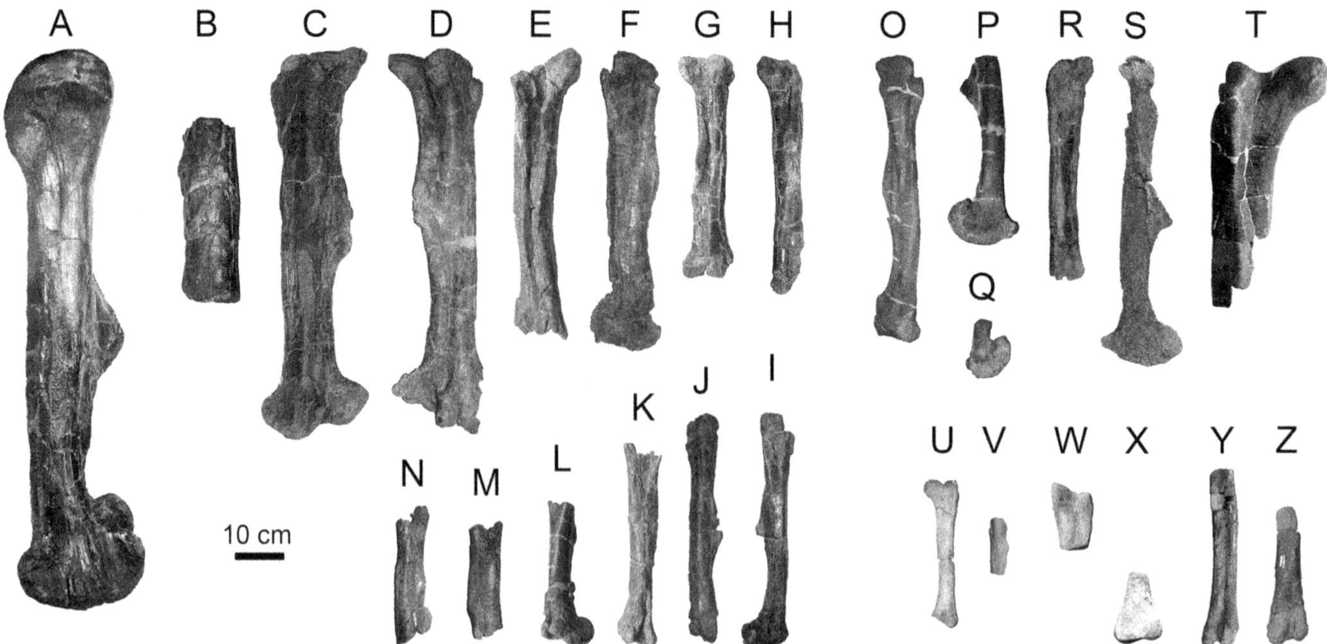

16.4. Hadrosauroid femora from the Tremp Formation of the eastern Tremp Syncline, compared for size to that of *Corythosaurus intermedius*, ROM 845 (A). Specimens (B) to (N) are from the Basturs Poble site. (B) MCD 4941; (C) MCD 5011; (D) MCD 5107; (E) MCD 4723; (F) MCD 4754; (G) MCD 4729; (H) MCD 4800; (I) MCD 4702; (J) MCD 4801; (K) MCD 4704; (L) MCD 4708; (M) MCD 4983; (N) MCD 4892; (O) IPS-N-21, Els Nerets; (P) LB-15, Lo Bas-1; (Q) IPS-896, Els Nerets; (R) MCD 4766, Casa Fabà; (S) MCD without inventory number, Magret; (T) MCD 5123, Tossal de la Doba; (U) MCD 4736, Costa de la Serra-1; (V) MCD without inventory number, Costa de la Serra-2; (W–X) MCD without inventory numbers, Serrat del Sanguin; (Y–Z) IPS-931 and IPS-932, Coll de la Torre. Scale bar equals 10 cm.

centra (IPS-N-1 [now IPS-893] and IPS-N-8 [now IPS-894]), and three mid-caudal centra (IPS-N-5 to 7). They were attributed to *Orthomerus* by Casanovas-Cladellas et al. (1985), but this is a nomen nudum based on undiagnostic elements (Brinkmann, 1988; Horner et al., 2004). During the 2003 fieldwork an isolated tooth was also found.

Lo Bas-1 and 2 Lo Bas-1 site yielded a nearly complete cervical vertebra (LB1-23), a partial left femur (LB1-15; Fig. 16.4P), a proximal portion of a left femur, and most of a straight ischial shaft. Unprepared specimens include three tibiae (LB1-7, 5, and 18) and a complete ischium with a straight (not booted) shaft (LB1-28). Lo Bas-2 site yielded a dentary fragment (LB2-2). It occurs at the same horizon as Lo Bas-1, but ~200 m to the west.

Barranc de la Costa Gran This site yielded six specimens belonging to small individuals; all are reposited at MCD without number. These include a small dentary (Fig. 16.5D), three mid-caudal centra, one proximal dorsal centrum, and some fragments of tibial shafts.

Casa Fabà This site yielded a left femur (MCD 4766; Fig. 16.4R) and a left fibula lacking the distal end (MCD 4756).

Els Pous Les Feixes, Magret, Les Torres, Serrat del Rostiar-1 and 3, Basturs Est, Basturs Poble, Serrat del Rostiar-2, and possibly Els Pous all occur within 25 m of one another, stratigraphically. The 2002 fieldwork at Els Pous yielded an incomplete proximal caudal centrum (EP-1), the diaphysis of a right femur (EP-14), a possible radius (EP-7), a metatarsal II (EP-5), and the distal part of a metatarsal III (EP-11).

Les Feixes Prospecting yielded three portions of a dentary tooth battery (the most complete crown is 2.9 cm tall apicobasally), a posterior fragment of a ?right dentary without teeth, a small fragment possibly from the latter, and a possible rostral fragment of a toothless ?left dentary. All specimens were found in a one-square-meter area on the surface, and possibly are from a single individual of modest size.

Magret At this site fossils produce from a conglomerate with clasts of lacustrine limestone. Partial long bones and vertebrae of small-sized individuals are relatively common in the fallen limestone blocks but were never collected (FMDV, pers. obs.). One block (lost when the nearby country road was enlarged) contained a small hadrosauroid femur associated with a tibia of corresponding size (FMDV, pers. obs.). A nearly complete right femur was recovered in 2011 (Fig. 16.4S).

Les Torres A tibia, the fragment of a femur, and the impression of a dentary are preserved at this site.

Serrat del Rostiar-1 This site yielded a partial left maxilla (MCD 4919), a dentary fragment, and part of a tooth battery.

Serrat del Rostiar-3 This site yielded only an incomplete caudal vertebra (SR3-1).

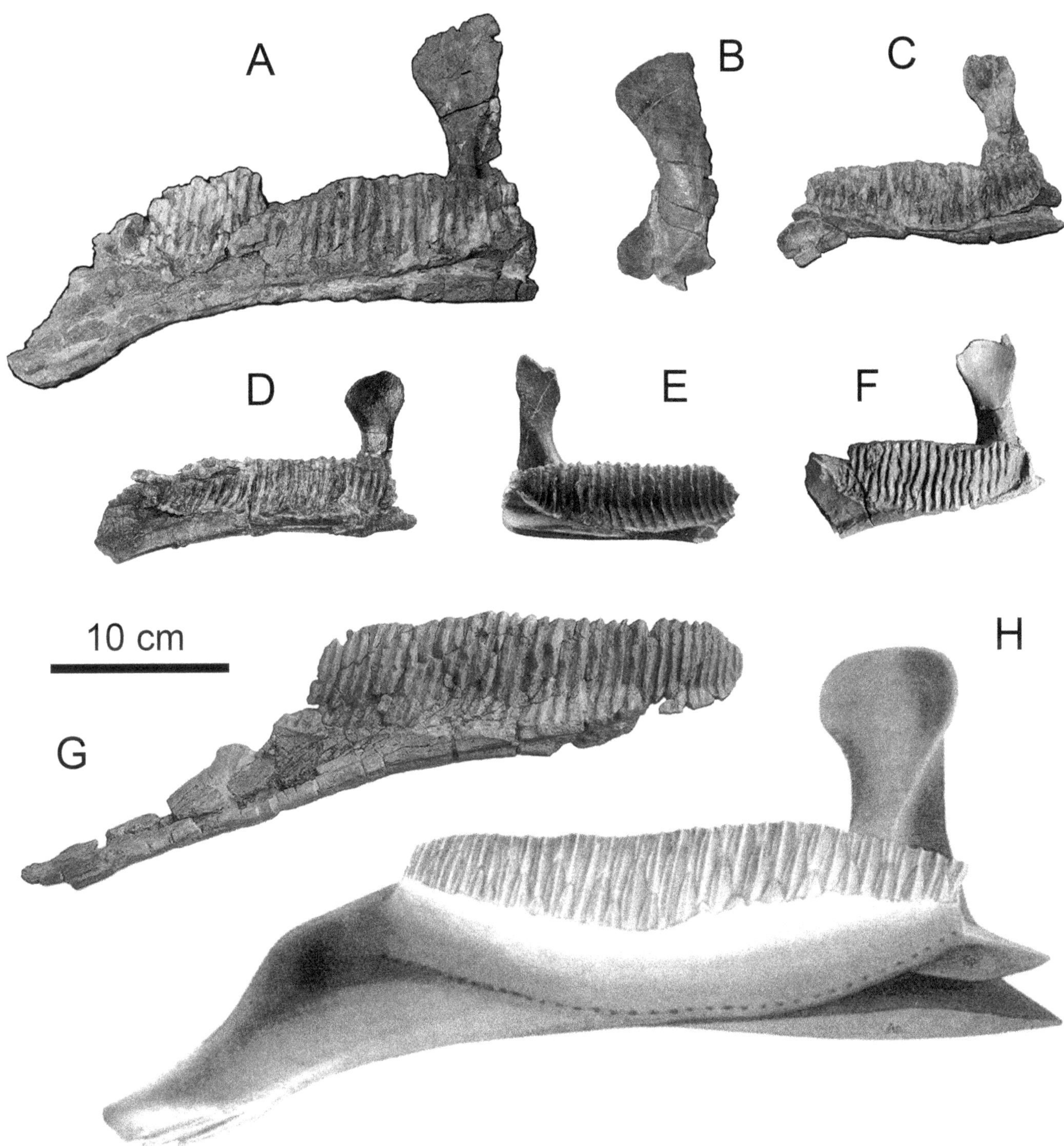

16.5. Hadrosauroid dentaries from the Tremp Formation of the eastern Tremp Syncline, compared for size to a dentary of the holotype of *Corythosaurus intermedius* (ROM 4670; from Parks, 1923) scaled to the size of that of ROM 845 (H). (A) The largest complete dentary from the Basturs Poble site, MCD 5007; (B) a coronoid process belonging to the largest dentary from the Basturs Poble site, MCD, without inventory number; (C) the smallest complete dentary from the Basturs Poble site, MCD 4945; (D) Barranc de la Costa Gran, MCD, without inventory number; (E) Molí del Baró-1, MB 33/2011; (F) Torrebilles-4, MCD 5059; (G) *Pararhabdodon isonensis,* IPS SRA 27, Les Llaus; (H) Right dentary of *Corythosaurus intermedius,* ROM 845. Scale bar equals 10 cm.

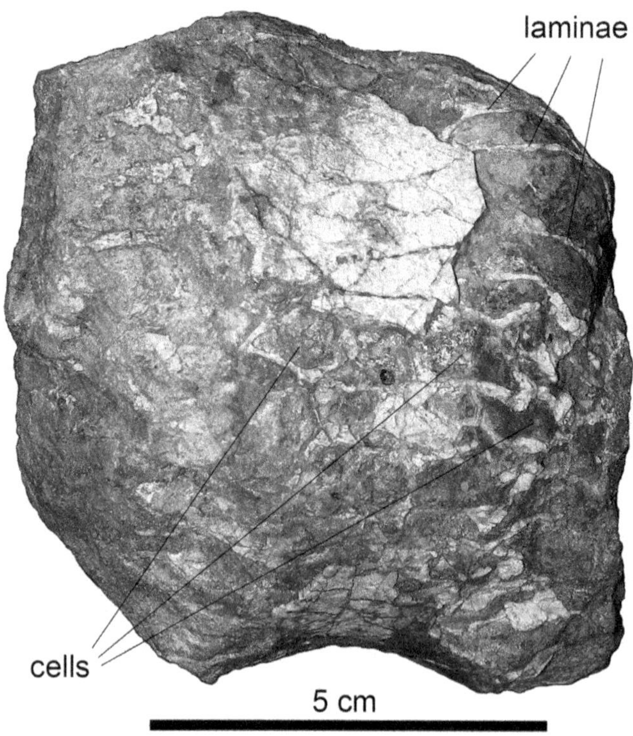

16.6. The vertebral centrum from Costa de Castelltallat site (MCD, without inventory number), lateral view. Scale bar equals 5 cm.

Basturs Est This site yielded two incomplete tibiae (BSTE-1 and 2) and a fibula (BSTE-3).

Basturs Poble This is the richest and most important site for hadrosauroid remains in the eastern Tremp Syncline. It preserves the first monodominant hadrosaurid bonebed in Europe (Martín et al., 2007). Over 950 bones have been collected during 10 years of fieldwork (2001–2011). Apart from some crocodyliform teeth and bones, and a few theropod teeth, all other identifiable dinosaurian remains pertain to hadrosauroids. To date, about 250 specimens have been prepared and can be reliably referred to this group (including 11 partial to complete dentaries). Hadrosauroid remains could number over 500 according to the field identification (updated to 2011). A few bones were attributed provisionally to Lambeosaurinae (Prieto-Márquez et al., 2007). Femora range in length from 43–72 cm (Fig. 16.4B–N). More specifically, 17 femora are relatively small (lengths <62.4 cm), 2 are of intermediate size (71 and 72 cm), and 1 is relatively large (estimated length ~80 cm). This indicates that the assemblage is dominated by small-sized individuals. A group of nearly complete dentaries range in length from 19–34 cm (Fig. 16.5A, C). Study of the material is in progress (January 2012).

Costa de Castelltallat (Not indicated in Figs. 16.2, 16.3) This locality is cited in Riera et al. (2009:table 2) as hadrosaur bearing because of a vertebral centrum attributed to "Hadrosauria indet." However, that specimen probably belongs to a somphospondylian sauropod because of its coarsely cancellous inner structure made of large cells delimited by tiny bone laminae (Fig. 16.6; see Wilson and Sereno, 1998).

Serrat del Rostiar-2 This site yielded a mid-caudal centrum (SRII-1), two dorsal centra (SRII-2 and 7), and an incomplete ischium (SRII-15).

Cabana de Gori-2 This site yielded a complete right tibia (MCD 4737) and a mid-caudal vertebral centrum (MCD 4767) mentioned in Vila et al. (2006).

Les Llaus This site yielded only an incomplete and toothless right dentary (IPS-29920, formerly IPS SRA 27; Fig. 16.5G). That dentary was first reported as belonging to *Pararhabdodon isonensis* by Casanovas-Cladellas et al. (1999). Later, it was indicated as the holotype and only specimen of *Koutalisaurus kohlerorum* Prieto-Márquez, Gaete, Rivas, Galobart, and Boada 2006 (Prieto-Márquez et al., 2006). Finally, it was referred to *Pararhabdodon isonensis* again (see Prieto-Márquez and Wagner, 2009). It has only 35 tooth positions and is characterized by an edentulous portion that is very elongate and medially projected like the Asian *Tsintaosaurus spinorhinus* (see Prieto-Márquez and Wagner, 2009).

Costa de la Serra-2 This site yielded the proximal part of a small right tibia, a small right humerus (only 17 cm long), the middle portion of the diaphysis of a small femur (Fig. 16.4V), and half of an anterior dorsal centrum (MCD 5057).

Costa de la Serra-1 This site yielded a small distal mid-caudal centrum, a small left femur (MCD 4736; Fig. 16.4U), and a right tibia (MCD 4735).

Serrat de Pelleu This site yielded a dentary fragment, a caudal centrum, a left tibia, and the fragment of a metatarsal.

Euroda Nord Only a partial right maxilla (MCD 5090) was found at this site; it was referred to the Lambeosaurinae by Gaete et al. (2007) and Prieto-Márquez et al. (2007).

Les Llaus de la Doba This site yielded a proximal caudal vertebra without neural spine (LLD-1) and a mid-caudal centrum with pedicels (LLD-2).

Els Esfons A large pedal phalanx III-1 (ES-1, 10.3 cm long; Fig. 16.7) comes from this site.

Molí del Baró-1 This is an oxbow-lake deposit (Riera et al., 2009), about 1 m thick, that is rich in plant remains and contains disarticulated scattered hadrosauroid bones. Fieldwork (undertaken in 2002, 2007, and 2010–11) yielded a maxilla fragment (MB-10); a small left dentary (MB-33/2011, Fig. 16.5E); a portion of a larger, toothless dentary (MB-8/2002); and three isolated teeth (MB-196/2010, MB-78/2011, MB-145/2011). Axial remains consist of a cervical vertebra (MB-19/2002), a dorsal vertebra (MB-62/2007), an isolated sacral vertebra (MB-18/2002), an isolated sacral centrum (MB-68/2007), a mid-caudal vertebral centrum (MB119/2011), and a complete dorsal rib (MB-50/2007). Also, 15 small segments of

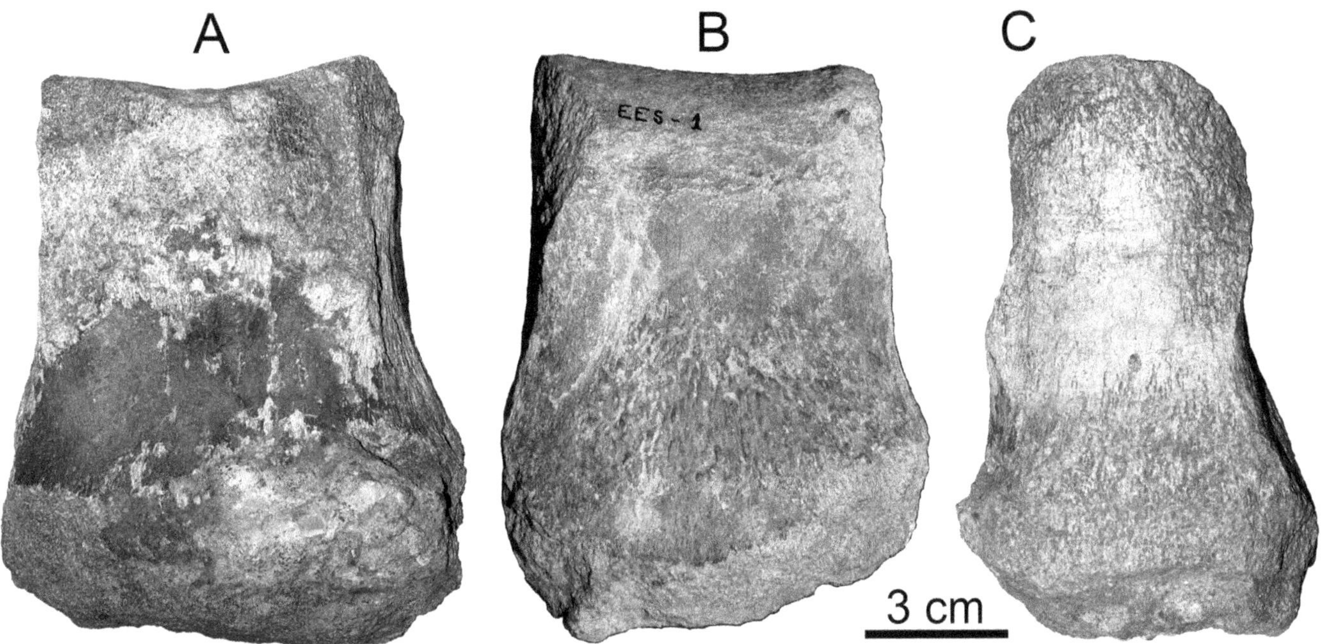

16.7. The pedal phalanx III-1 from Els Esfons site (EES-1). (A) dorsal view; (B) ventral view; (C) right lateral or medial view. Scale bar equals 3 cm.

tiny ossified tendons have been collected. The appendicular elements consist of a small sternal plate (MB-9/2002), a left ulna (MB-54/2007), the proximal part of a right ischium with portions of the shaft (MCD 5089), the fragment of an ischial shaft (MB-2/2002), a right and a left tibiae (MCD 4733 and MB-240/2010, respectively), a right fibula lacking the distal third (MCD 4734), the distal end of a right metatarsal III (MB-1/2002), and a left metatarsal II (MB-52/2007). According to Gaete et al. (2007), the ischium belongs to a lambeosaurine.

Sant Romà d'Abella The name of this site can be misleading because it is relatively far from the village of Sant Romà d'Abella (e.g., the Molí del Baró site is much closer; see Fig. 16.2). Also, the history of the specimens that are reported from this site, which were all referred to *Pararhabdodon isonensis*, is rather confused; summaries can be found in Prieto-Márquez et al. (2006) and the latest developments are presented in Prieto-Márquez and Wagner (2009). The sample consists of incomplete left and right maxillae (IPS SRA 23, now IPS-693-6, Fig. 16.8A; IPS SRA 22, now IPS-36327, the most complete and 31.5 cm long, Fig. 16.8B), a proximal cervical vertebra (MCD 4730), a proximal or middle cervical vertebra (IPS SRA 18, now IPS-693-1), a middle cervical vertebra (IPS SRA 25, now IPS-693-8), a nearly complete mid-posterior cervical vertebra (IPS SRA 1, now IPS-693, holotype), a nearly complete cervical centrum with pedicels (IPS SRA 11, now IPS-693-7), a fragmentary centrum (IPS SRA 7), a proximal dorsal vertebra (MCD 4731), a distal mid-dorsal vertebra (IPS SRA 12, now IPS-693-10), two distal-middle dorsal vertebrae (IPS SRA 19, probably corresponding to IPS-693-13, IPS SRA 20, now IPS-693-11), a dorsal centrum (IPS SRA 13 now IPS-693-4), a nearly complete sacrum made of eight fused vertebrae (IPS SRA 24), two fragmentary ribs (IPS SRA 21, now IPS-693-12), a proximal caudal vertebra (IPS SRA 17, now IPS-695-5), four proximal caudal centra (IPS-SRA-1, 3, 10, and 14), five mid-caudal centra (IPS-SRA-2, 4–6, and 9), a left humerus (IPS SRA 15, 48 cm long), the proximal part of a left scapula (IPS SRA 16, now IPS-693-3), and the massive distal end of the shaft of a large right ischium (IPS SRA 26, now IPS-693-9, Fig. 16.8C).

All specimens from Sant Romà d'Abella site "were found in a 4 × 2.5 m surface, and probably belong to a single medium-sized (c. 6-m-long) individual" (Casanovas-Cladellas et al., 1999:205). However, IPS-SRA-9 and 10 do not have the characteristic brown color of the other SRA specimens and, possibly, are not from the same outcrop or level. A proximal caudal centrum in the ICP collection (labeled IPS SRA 20) has a blackish color, so it too is probably from another outcrop or stratigraphic horizon.

Pararhabdodon isonensis was considered first as a basal lambeosaurine (Casanovas-Cladellas et al., 1999), then a non-hadrosaurid hadrosauroid (Prieto-Márquez et al., 2006), and then again as a basal lambeosaurine forming a clade with the Chinese *Tsintaosaurus spinorhinus* (Prieto-Márquez and Wagner, 2009).

Planta del Mestre This site preserves in situ the only articulated hadrosauroid specimen from the eastern Tremp

16.8. *Pararhabdodon isonensis,* type specimens from Sant Romà d'Abella site. (A) IPS-693-6, incomplete left maxilla; (B) IPS-36327, nearly complete right maxilla; (C) IPS-693-9, distal portion of a right ischium. Scale bar equals 10 cm.

Syncline. The specimen is a segment of the mid-caudal vertebral column with 12 vertebrae (Vila et al., 2006) – including neural arches and some chevrons – belonging evidently to a comparatively small sized individual. The centra are taller than long. The neural spines are inclined backward, recurved, and slightly flaring apically in lateral view, and apparently shorter than the very long, narrow, and straight chevrons.

Tossal de la Doba This site yielded only the proximal part of a left femur (47 cm long; Fig. 16.4T) with an estimated total length of at least 75 cm.

Tossal del Gassó (Not indicated in Figures 16.2, 16.3) Ardèvol et al. (1995) reported a nearly complete right dentary, the distal portion of a right femur, and a fragment of a maxilla from this site (mentioned as Abella-2 or "Tosal del Gassó"). The dentary was described in an unpublished manuscript by M. L. Casanovas-Cladellas and labeled as IPS-TG-1; it is without doubt the dentary from the Les Llaus site reported by Casanovas-Cladellas et al. (1999) and later chosen as holotype of *Koutalisaurus kohlerorum* by Prieto-Márquez et al. (2006). The distal portion of a right femur is labeled IPS-TG-2 in the manuscript; it was relabeled as IPS-931 in the inventory of the ICP collections, which identifies it as from the Sant Romà d'Abella II site (here Coll de la Torre site, see below). The maxillary fragment is listed as IPS-TG-3 in the manuscript, but it is neither figured nor described. It is now impossible to locate this specimen in the ICP collections.

Serrat del Corb A nearly complete pelvis (left ilium, MCD 4791; ischia, MCD 4787 and 4788, about 100 cm long; partial left pubis, MCD 5088) was found at this site and is associated with three or four sacral or proximal caudal vertebrae. It was referred to a lambeosaurine having a 7–8 m long body (Gaete et al., 2007; Prieto-Márquez et al., 2007). Other specimens from this site include a fragment of a tooth-bearing bone, a dorsal centrum, three mid-caudal centra, and the proximal portion of a chevron.

Coll de la Torre (= Sant Romà d'Abella II) According to Prieto-Márquez et al. (2006) the site, previously mentioned as Sant Romà d'Abella II, should be renamed Coll de la Torre. A proximal caudal centrum (IPS-SRAII-9), a mid or proximal caudal centrum (IPS-SRAII-1), and six mid-caudal centra (IPS-SRAII-3 to 8) from the site are reposited in the collection of MCD. IPS-SRAII-3–6, and 9 were described in Casanovas-Cladellas et al. (1995). Also labeled Sant Romà d'Abella II but in the collection at ICP are a small fragment of a tooth-bearing bone (IPS-SRAII-13, now IPS-37414), a proximal caudal centrum (IPS-SRAII-2, now IPS-37416), a mid-caudal centrum (IPS-SRAII-10, now IPS-923); the distal portion of a left (IPS-SRAII-15, now IPS-932; Fig. 16.4Z), and a right femur (IPS-931, former IPS-TG-2; Fig. 16.4Y). More recently, the site has yielded two fragments of a dentary and a small ungual phalanx. A site named Sant Romà-2 or "Barranco de la Llau de Doba" is mentioned by Ardèvol et al. (1995) as yielding four caudal centra.

La Serra This site yielded a pedal phalanx and the distal end of a left tibia.

Masia de Ramon This site yielded a very small right tibia and the distal end of a small femur.

Serrat del Sanguin-1 and 2 These sites are in close proximity to one another and have yielded the distal and the proximal parts of two small femora (Fig. 16.4W, X), two distal and two proximal portions of femora similar in size as those in Figure 16.4W, X, two fragments of femoral diaphysis, three mid-caudal centra, a more distal mid-caudal centrum (SS-09/9), and a manal phalanx.

Barranc de Torrebilles Four sites with hadrosauroid bones have been identified along the Torrebilles gully (TB1–4). The stratigraphic interval between the lowest (TB1; 31a in Fig. 16.2) and highest horizons (TB2; 31b in Fig. 16.2) is about 20 m thick; TB2–4 are in a stratigraphic section less than 10 m thick. Vertebrate remains occur mainly in sandstone bodies (TB2–4). Little information is available about this material.

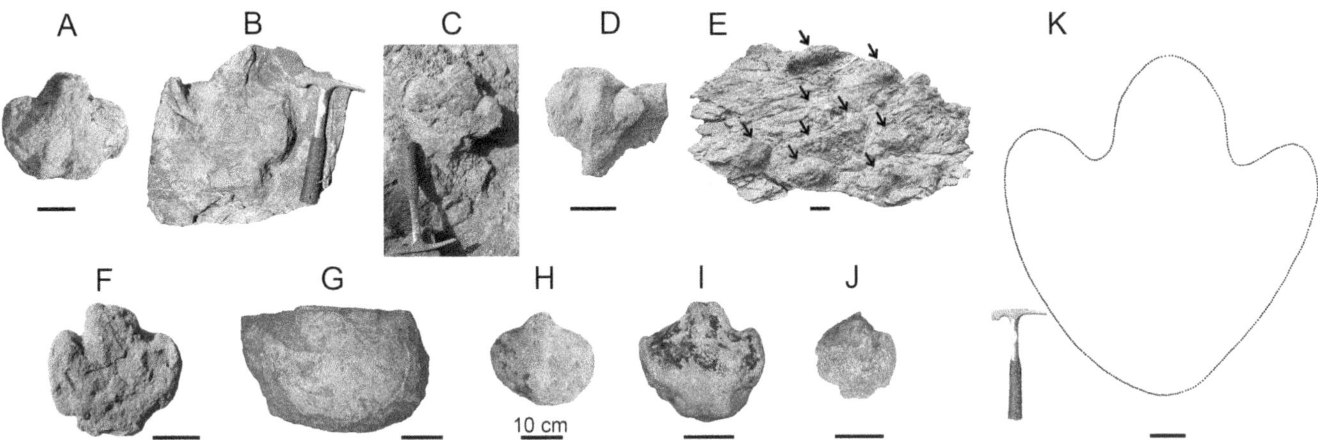

16.9. Hadrosauroid tracks from the lower red unit of the Tremp Formation, eastern Tremp Syncline. (A) from Tossal del Gassó; (B) from Serrat del Sanguin; (C–D) from Barranc de Torrebilles-6; (E) from Barranc de Torrebilles-7; (F) from Barranc de Torrebilles-8; (G) from Basturs Poble; (H) from Barranc de Guixers-3; (I–J) from Suterranya; (K) outline of an adult hadrosaur track from the uppermost Cretaceous of USA (based on Lockley and Hunt, 1995). The specimens are deposited at the MCD, without inventory number. Hammer for scale in B and C equals 11 inches (279 mm); scale bars equal 10 cm.

TB1 yielded a small manal phalanx II of digit II or III (TB1-02) and a tibia (TB1-01). TB2 yielded a very fragmented small dentary (TB2-12), a very small proximal dorsal vertebra (MCD 5063), a second but more distal and diminutive dorsal vertebra (TB2-01), two small mid-caudal centra (MCD 5058 and 5062), and a small metapodial (TB2-05). TB3 yielded a femur (TB3-20) and a caudal vertebral centrum (TB3-52). TB4 yielded a small left dentary lacking the rostral portion (MCD 5059; Fig. 16.5F), a vertebra (TB4-03), a sacrum (TB4-07), a caudal centrum (TB4-10), an ?ulna (TB4-06), the fragment of a humerus (TB4-15), two ischia (TB4-01 and 02), the distal part of an ischial shaft (TB4-11), a fragment of a femur (TB4-25), a tibia and two tibial fragments (TB4-13, 22, and 16), and part of a metatarsal (TB4-09).

Hadrosauroid Track-Bearing Sites

The complete census of the track sites is not available yet, but those fossils appear to be relatively common in the fluvial sandstones, mainly in the uppermost part of the lower red unit. Their study is in progress. To date, hadrosauroid tracks have been observed at 15 localities in the easternmost Tremp Syncline. In approximate stratigraphic order (Fig. 16.3) they are Llau de la Costa, Basturs Poble (a sandstone block with a track from a stratigraphic level close to the bonebed), Masia de Ramon/Mas d'en Feliu and surroundings, Costa de la Serra, Tossal del Gassó, Serrat del Sanguin (in the same section containing the bones), Barranc de Torrebilles 6–7, Barranc de Torrebilles 8 (= Orcau 4), Costa Roia (mentioned in Vila et al., 2005), Barranc de Guixers-1 to 3 (locality 1 is mentioned in Vila et al., 2005), Suterranya, and Torrent de Carant. Tracks are all tridactyl, usually positive hyporeliefs and are referred to hadrosauroids because: they are usually wider that long; symmetrical (the impression of digit II is not sensibly shorter than that of digit IV, and there is no indentation between digit II and the heel print) with short and broad digital marks; digit III is not much longer that the others; they lack claw marks; and have a broad heel print, often with a squared outline (Fig. 16.9A–J). The largest are about 30–35 cm long (Llau de la Costa; Serrat del Sanguin, Fig. 16.9B; and Basturs Poble, Fig. 16.9G), but those that are 20–25 cm long (Fig. 16.9A, C–F, and H–J) seem to be more abundant.

Llompart (2006) reported a hadrosaurid trackway (tracks ~38 cm long) from the middle of the gray unit in the Moror B site (southern margin of the Tremp Syncline). She referred other tridactyl tracks (17–24 cm long) to theropods. These occur in the nearby and stratigraphically overlying Moror A site (Llompart, 2006:fig. 2). These tracks probably should be referred to hadrosauroid trackmakers because of the symmetry of digital prints II and IV with respect to the heel print (see Llompart, 2006:fig. 4); the slenderness of the toe marks was plausibly caused by the collapse of the sediment.

Dinosaur Eggshell-Bearing Sites

Eggshell remains are abundant in the eastern Tremp Syncline, being reported from at least 29 sites (López-Martínez, 2000; Galobart, 2007; Garcia Sellés, pers. obs.). These are located in the uppermost part of the Arén Formation (at least 14 sites), in the gray unit of the Tremp Formation (six sites), and in the lower red unit of the Tremp Formation (seven sites; names listed in Fig. 16.2). Eggshell is mostly referred to the oogenus *Megaloolithus*, which is supposed to be titanosaurian sauropod in origin (e.g., Powell, 1992; Chiappe et al.,

1998). Eggshell referable to the oofamily Spheroolithidae, which is typical for hadrosaurids (e.g., Mikhailov, 1997), is unrepresented. This observation is unexpected, especially in the lower red unit, where hadrosauroid bones and tracks are far more common than those of sauropods.

DISCUSSION

The importance of the continental vertebrate assemblage found in the Cretaceous portion of the Tremp Formation resides in its Late Maastrichtian age and in providing additional data to improve our understanding of the still poorly known evolutionary history of European hadrosauroids.

Very few Late Maastrichtian continental vertebrate assemblages are known worldwide. Known assemblages mainly come from western North America (Lance, Hell Creek, and Scollard Formations), eastern Asia (Russia and China; Godefroit et al., 2008, 2009), and Spain and southern France (see Dalla Vecchia, this volume).

Hadrosauroid-bearing sites are abundant at two stratigraphic levels in the lower red unit: the lower- and uppermost parts (Fig. 16.3). Several hadrosauroid-bearing sites occur in levels probably corresponding to magnetochron C29r and upper C30n – i.e., the last 800,000 years (at most) of the Mesozoic (Husson et al., 2011).

The study of the hadrosaur material from the eastern Tremp Syncline is in progress, but three conclusions can be drawn from the available data: (1) hadrosauroids are, by far, the most abundantly represented dinosaurs in the younger lower red unit, whereas the remains of titanosaurian sauropods dominate in the older gray unit; (2) dinosaur egg remains are found in more than 30 sites distributed through the entire Maastrichtian section, but there are no occurrences of eggshell assignable to the oofamily Spheroolitidae, which usually refers to ornithopods; (3) with the exception of a few specimens, hadrosauroid elements (and inferred body sizes) are, on average, smaller than those of hadrosaurids from Upper Campanian–Maastrichtian sites in Asia and western North America.

Environmental Preference or Time/Event Related Faunal Change?

Hadrosauroid elements are recorded only in two sites of the gray unit (Moror and Els Nerets); they appear to be less abundant than sauropod bones at Els Nerets (Casanovas-Cladellas et al., 1987, 1995; RG, pers. obs.). Titanosaurians are the only dinosaurs reported from the Orcau-1, Prat de Sòl, and Costa de Santa Llúcia sites of the gray unit (Riera et al., 2009). Sauropod tracks are found in Orcau-2 site (Llompart et al., 1984; reinterpreted by Vila et al., 2011). Sauropods are the only dinosaurs in the Orcau-3 site, which is the lowermost vertebrate site in the lower red unit (Riera et al., 2009). In addition, the gray unit of the eastern Tremp Syncline has yielded ankylosaurian remains, although they are extremely rare (Riera et al., 2009; Escaso et al., 2010). Sauropod evidence, mainly tracks, is abundant in the gray unit of the nearby Vallcebre Syncline (Vila et al., 2008). In the lower red unit, sauropods are represented above the Orcau-3 site only by a small cervical vertebra (Lo Bas-1 site), a centrum of moderate size (Costa de Castelltallat), a small femur (Molí del Baró-2 site), some tracks (Barranc de Guixers-1 site), and probably eggshell fragments (see below). No ankylosaurian remains have been found to date in the lower red unit. Instead, the hadrosauroids are represented by the abundant sample reported above, and were clearly the dominant dinosaurs in the vertebrate fossil assemblage.

This faunal turnover has been explained as a change from a transitional environment (marine to coastal plain) during the deposition of the gray unit to a continental environment during deposition of the upper red unit (Riera et al., 2009). However, most of the hadrosaurid record occurs in the marine-transitional Arén Formation (Blasi 1 level) and in the overlying gray unit (= La Posa Formation; Blasi 2–3 levels) in the western (Aragonian) part of the Syncline, where the boundary between the units is younger due to the western migration of paleoenvironments during overall regression (Riera et al., 2009). In the western part of the syncline, the hadrosaurid record includes the holotypes of *Arenysaurus ardevoli* and *Blasisaurus canudoi*, whereas sauropods are represented only by a single tooth (López-Martínez et al., 2001). Furthermore, hadrosauroid remains are common also in the Late Maastrichtian transitional facies of southern France (Laurent, 2003). These observations support the interpretation that the change in faunal composition was on a time/event-related basis rather than just the result of local environmental changes.

The hadrosauroids seem to become dominant during the Late Maastrichtian in the nearby Àger and Vallcebre Synclines of Catalonia, in the vertebrate-bearing sites of the València Province of Spain (La Solana and Loma Cortada; Company, 2004), and in sites of that age in southern France (Dalla Vecchia, this volume). These changes in the relative abundance of different kinds of dinosaurs in the fossil assemblages of the Ibero-Armorican Island (near the Lower–Upper Maastrichtian boundary) have been noticed by several authors (e.g., Le Loeuff et al., 1994; Buffetaut and Le Loeuff, 1997; Buffetaut et al., 1997; Company et al., 2009). The record from the eastern Tremp Syncline thus further confirms these changes and the shift towards the latest Maastrichtian dominance of hadrosauroids on the Ibero-Armorican Island.

Why Are There No Hadrosauroid Eggshell Remains?

A possible explanation for the absence of spherolithid eggshell remains in the lower red unit is that some *Megaloolithus* oospecies belong to the hadrosauroids, as already suggested by Grigorescu et al. (2010). However, many other alternative explanations are possible, although some are very unlikely: hadrosauroids might have laid soft-shelled eggs or been ovoviviparous or viviparous (Varricchio et al., 2011); hadrosauroid eggshells might have a low chance of preservation because they were not buried, or were laid in mounds of vegetation (not buried and occurring in an acid environment favoring eggshell dissolution after hatching); or eggshell might have been subject to consumption or recycling. Lastly, the absence of eggshell might reflect a choice of nesting habitat, wherein hadrosauroids simply did not lay eggs in this region of the basin (represented by the eastern Tremp Syncline) and the rarer sauropods did. In this context, hadrosauroid nesting grounds did not coincide with places of sediment deposition (and subsequent exposure) in the Tremp Basin. This last hypothesis is supported by the concomitant absence of elements or tracks from embryo-hatchling size individuals at any of the sites (see below).

Are the Hadrosauroids from the Eastern Tremp Syncline Small-Sized and Insular Dwarfs?

In this study, we attempted to quantify and assess the seemingly diminutive nature of the Catalonian hadrosaur remains by comparing the lengths of some skeletal elements from the eastern Tremp Syncline with corresponding elements in a large sample of hadrosauroid material from a late Cretaceous continental area (the Late Campanian Dinosaur Park Formation of Canada [DPFm]; see Terminology and Method section for further explanation; Appendices 16.1 and 16.2). Lengths of hadrosaurid femora and tibiae in the DPFm sample fall in three size classes: a large, an intermediate, and a small size class, interpreted as representing adult, juveniles and embryo-hatchling individuals, respectively (Brinkman, 2011, this volume; Fig. 16.10A). In large individuals of hadrosaurid species (both hadrosaurine and lambeosaurine) from western North America and Asia, the femur length is typically more than 100 cm, tibiae are slightly shorter than femora or rarely as long as femora, and the corresponding dentary length ranges from 45–90 cm (Lull and Wright, 1942; Young, 1958; Lauters et al., 2008). By comparison, the presumed adult body size of the Dinosaur Park Formation hadrosaurids is similar to that of large individuals of continental hadrosaurid species in Asia and North America.

No femora and tibiae of the Tremp sample fall into the adult or embryo-hatchling size classes from DPFm. Most of them (80%) are within the juvenile size class, 7% are distributed between the embryo-hatchling and the juvenile size classes, and four specimens (13%) fall between the adult and the juvenile size classes (Fig. 16.10B). The latter four specimens are the largest femora and tibiae known from the eastern Tremp Syncline (the femur from Tossal de la Doba is the largest). Femora in the relatively large sample from Basturs Poble fall mainly into the juvenile size class, but also extend into the range between the juvenile and the adult size classes (Fig. 16.10B). One femur in the Basturs Poble

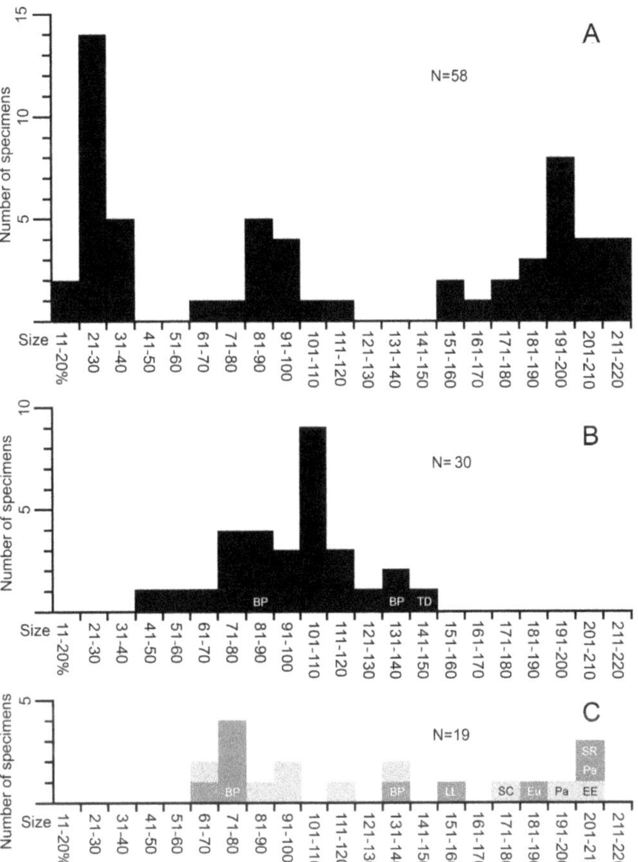

16.10. Size-frequency diagrams of hadrosauroid bones. (A) Tibiae and femora from the Dinosaur Park Formation (Upper Campanian, Alberta), with size indicated as a percentage of the length of the corresponding element in TMP 1998.058.0001; (B) tibiae, fibulae and femora from the eastern Tremp Syncline, also with size indicated as a percentage of the length of the corresponding element in TMP 1998.058.0001; (C) Lengths of other appendicular elements from the eastern Tremp Syncline (humerus, ulna, ischium, metatarsal II, and pedal phalanx III-1) expressed as percentages of the corresponding elements of TMP 1998.058.0001 (gray); Lengths of the dentaries from the eastern Tremp Syncline expressed as percentages of the dentary of ROM 845 (*Corythosaurus intermedius*) and scaled as a percentage of that hypothetical for TMP 1998.058.0001 (dark gray); Lengths of the maxillae from the eastern Tremp Syncline expressed as percentages of the maxilla of ROM 1933 (*C. casuarius*) and scaled as a percentage of that hypothetical for TMP 1998.058.0001 (dark gray). See text for further explanation. Abbreviations: BP, Basturs Poble; EE, Els Esfons; Eu, Eudora Nord; LL, Les Llaus; PA, Sant Romà d'Abella (*Pararhabdodon*); SC, Serrat del Corb; SR, Serrat del Rostiar-1; TD, Tossal de la Doba.

collection (Fig. 16.4B) might fall into the lower part of the adult size class range, but it is represented only by a portion of the diaphysis, and was not included in Figure 16.10B.

A sample of hadrosauroid femora and tibiae from the latest Cretaceous of Europe shows a similar size distribution, with the exception of the presence of individuals from a Romanian site falling into the embryo-hatchling size class (see Dalla Vecchia, this volume).

The distribution of lengths for other appendicular elements, dentaries, and maxillae, show a somewhat different picture (Fig. 16.10C). The dentaries from Basturs Poble have a size range similar to that of the femora, but a consistent part of the sample from the other sites (33%, corresponding to six individuals) falls within the DPFm adult size class. It seems that the eastern Tremp Syncline hosted a relatively low number of individuals with a body size comparable to that of those in the larger size class of the DPFm sample. This could be partly an artifact of choosing lambeosaurine dentaries and maxillae for comparisons, because lambeosaurines have a comparatively short head as shown by the dentary-to-femur ratio (d/fe), which ranges 0.43–0.47 in lambeosaurines and 0.63–0.80 in some hadrosaurines (see Dalla Vecchia, this volume). The relative size of hadrosauroids with non-lambeosaurine skull-to-femur proportions based on dentaries and possibly maxillae would be somewhat overestimated in Figure 16.10C. This is the case of *Tethyshadros insularis* (a non-hadrosaurid hadrosauroid), but also *Arenysaurus ardevoli* (a lambeosaurine) in the larger European sample (see Dalla Vecchia, this volume).

The maxillae of *Pararhabdodon isonensis* and that from Serrat del Rostiar-1 belong to individuals comparable in size with the adults of the DPFm sample (the large specimens of *Corythosaurus* and *Lambeosaurus* have maxillae 31.0–34.5 cm long; Evans, 2010). However, the humerus of *Pararhabdodon isonensis* falls into the 191–200% percent interval with respect to the humerus of TMP 1998.058.0001 (the same as the tibia of ROM 845 with respect to that of TMP 1998.058.0001), but actually is only 89% the length of ROM 845 humerus. Thus, it would fall in the lower range of the adult size class if ROM 845 were used as a reference for this element. Unfortunately, the Sant Romà d'Abella material does not include a tibia or a femur to clarify this comparison. Thus, *Pararhabdodon isonensis* most probably had a body size comparable to that of the small adults of the DPFm sample. The ischium from Serrat del Corb is approximately as long as those of large-sized hadrosaurids from western North America (see Lull and Wright, 1942), and the pedal phalanx III-1 from Els Esfons (if it actually belongs to a hadrosauroid) is also undoubtedly from a large individual.

Tracks are not subject to the same taphonomic biases as are skeletal elements (e.g., size selection and variability of mortality rates at different ontogenetic stages), thus the abundant paleoichnological record in the eastern Tremp Syncline, with hadrosauroid tracks ranging in length from 20–40 cm, provides additional evidence to support an interpretation that comparatively small-sized hadrosauroids dominated this assemblage. A 40 cm long track corresponds to an individual ~160 cm (vertical) at the hip (Henderson, 2003). In comparison, hip heights of adult hadrosaurids from western North America are more typically 2.0–2.5 m (Lucas et al., 2011). According to Thulborn (1990), North American hadrosaurid tracks are usually 70 cm to over 1 m long. Similarly, Lockley and Hunt (1995:227) document an "adult hadrosaur track" from western North America that is 90 cm long, and Lucas et al. (2011) report hadrosaurid tracks from the Upper Campanian of New Mexico that range in length from 59–94 cm.

The overall picture of hadrosauroid dinosaurs in the Late Maastrichtian of the eastern Tremp Syncline is one where small or juvenile-sized individuals (compared to adult North American hadrosaurids) dominate the assemblage, and there are sporadic occurrences of large or adult-sized individuals. This pattern can be explained in several ways:

1. Hadrosauroid populations in the eastern Tremp Syncline comprised mainly juvenile individuals with just a few adults, due to differential environmental preferences expressed at different ontogenetic stages (e.g., Horner and Woodward, 2011). In this explanation, adults were of large size (i.e., not significantly dwarfed) with respect to those living on other islands of the European Archipelago (e.g., the Tisia-Dacia Island; Benton et al., 2010). The presence of large adults on the Ibero-Armorican Island is not problematic because the island was the largest in the archipelago and body mass of the top terrestrial herbivores species seems to increase with increasing land area (Burness et al., 2001).

2. There was a dominance of one or more small-sized species with rare large-sized species. This explanation is in agreement with the presence of relatively small adults in the coeval Blasi and La Solana sites of Spain (see Dalla Vecchia, this volume).

3. Populations of hadrosauroids comprised relatively small-sized (dwarf) individuals (adults on average smaller than DPFm adults), but with a few giants due to a broad intraspecific body size variability as it is typical of animals with indeterminate growth (Andrews, 1982). This explanation also is in agreement with the presence of relatively small adults in the coeval Blasi and La Solana sites. Large individuals may have been present also if they were newly arrived from a continental area, although this seems unlikely given that the island was the westernmost in the European Archipelago (see Dalla Vecchia, this volume:fig. 15.10) and it is more likely that colonizers arrived from other islands of the archipelago.

Histological studies are now needed to establish definitively the maturity or immaturity of these elements and to test explanations of the size distributions of the hadrosauroids from the eastern Tremp Syncline.

CONCLUSIONS

The eastern Tremp Syncline has the best latest Cretaceous hadrosauroid record of Europe, with over 30 bone-bearing sites and at least 15 track sites. Sites comprise mainly Upper Maastrichtian deposits and, together with assemblages from some other Spanish and French localities, preserve a record of the last non-avian dinosaurs in Europe.

The fossil record supports an interpretation that there was a change in the local dinosaur fauna at the beginning of the Late Maastrichtian. The change resulted in an increased abundance and spread of hadrosauroids and a decrease in the abundance of sauropods, which became rarer, but did not disappear. The eggshell record (*Megaloolithus* present, Spheroolithidae absent) and the absence of hadrosauroid hatchling material suggest that hadrosauroid nesting grounds may have been separated from areas of sediment deposition and fossil preservation in the Tremp Syncline.

Individuals much smaller than those considered to be adult in hadrosaurid assemblages from the late Cretaceous of western North America and Asia are the most abundant in the sample, but a few large bones are also present that indicate the presence of individuals that may have been as large as small North American adults. This pattern can be explained in one of three ways: prevalence of juvenile individuals over adults in the local populations; sympatry of two hadrosauroid species that were large- and small-bodied; the occasional presence of giant individuals in populations of moderate body size, and possibly dwarf hadrosauroids. Histological study of the material is now the best means to test these alternative explanations.

ACKNOWLEDGMENTS

We thank Donald Brinkman for the measurements of TMP 1998.058.0001, David C. Evans and Nicolás Campione for those of ROM 845, and Enric Vicens for the help in making figure 1. Thanks also to the reviewers, Penelope Cruzado-Caballero and David C. Evans, for the useful suggestions. This research was financially supported by Ministerio de Ciencia e Innovación of the Spanish government (CGL2008-06533-C03-01/02 and CGL2011-30069-C02-01/02) and the Departament de Cultura de la Generalitat de Catalunya. We are grateful to the Municipality of Isona i Conca Dellà and the Consell Comarcal del Pallars Jussà for their support to the research project on the Tremp dinosaurs.

LITERATURE CITED

Andrews, R. M. 1982. Patterns of growth in reptiles; pp. 273–320 in C. Gans and F. H. Pough (eds.), Biology of the Reptilia, volume 13. Academic Press, London.

Ardèvol, Ll., M. L. Casanovas, and J. V. Santafé. 1995. Restos de dinosaurios del Maastrichtiense de la Conca de Tremp, Lleida (cuenca de antepaís Pirenaica Meridional); pp. 25–27 in G. López, A. Obrador, and E. Vicens (eds.), XI Jornadas de Paleontologia. Universitat Autònoma, Departament de Geologia, Unitat de Peleontologia, Barcelona, Spain.

Barco, J. L., Ll. Ardevol, and J. I. Canudo. 2001. Descripción de los primeros rastros asignados a Hadrosauridae (Ornithopoda, Dinosauria) del Maastrichtiense de la Península Ibérica (Areny, Huesca). Geogaceta 30:235–238.

Benton, M. J., Z. Csiki, D. Grigorescu, R. Redelstorff, P. M. Sander, K. Stein, and D. B. Weishampel. 2010. Dinosaurs and the island rule: the dwarfed dinosaurs from Haţeg Island. Palaeogeography, Palaeoclimatology, Palaeoecology 293:438–454.

Brinkman, D. B. 2011. The size-frequency distribution of hadrosaurs from the Dinosaur Park Formation of Alberta, Canada; pp. 16–20 in D. R. Braman, D. A. Eberth, D. C. Evans, and W. Taylor (compilers), International Hadrosaur Symposium Abstract Volume. Royal Tyrrell Museum of Palaeontology, Drumheller, Canada.

Brinkman, D. B. 2014. The size-frequency distribution of hadrosaurs from the Dinosaur Park Formation of Alberta, Canada; chapter 23 in D. A. Eberth, and D. C. Evans (eds.), Hadrosaurs. Indiana University Press, Bloomington, Indiana.

Brinkmann, W. 1984. Erster Nachweis eines Hadrosauriers (Ornithischia) aus dem unterem Garumnium (Maastrichtium) des Beckens von Tremp (Provinz Lérida, Spanien). Paläontologisches Zeitschrift 58:295–305.

Brinkmann, W. 1988. Zur Fundgeschicte un Systematik der Ornithopoden (Ornithischia, Reptilia) aus der ober-Kreide von Europe. Documenta Naturae 45:1–157.

Buffetaut, E., and J. Le Loeuff. 1997. Late Cretaceous dinosaurs from the foothills of the Pyrenees. Geology Today March–April:60–68.

Buffetaut, E., J. Le Loeuff, L. Cavin, S. Duffaud, E. Gheerbrant, Y. Laurent, M. Martin, J.-C. Rage, H. Tong, and D. Vasse. 1997. Late Cretaceous non-marine vertebrates from southern France: a review of recent finds. Geobios 20:101–108.

Burness, G. P., J. Diamond, and T. Flannery. 2001. Dinosaurs, dragons, and dwarfs: the evolution of maximal body size. Proceedings of the National Academy of Sciences 98:14518–14523.

Casanovas-Cladellas, M. L., J. V. Santafé, J. L. Sanz, and A. Delgado-Buscalioni. 1985. *Orthomerus* (Hadrosaurinae, Ornithopoda) du Crétacé supérieur du gisement de "Els Nerets" (Tremp, Lleida); pp. 99–111 in P. Taquet and C. Sudre (eds.), Les Dinosaures de la Chine à la France. Muséum d'Histoire naturelle de Toulouse, Toulouse, France.

Casanovas-Cladellas, M. L., J. V. Santafé, J. L. Sanz, and A. Delgado-Buscalioni. 1987. Arcosaurios (Crocodilia, Dinosauria) del Cretácico superior de la Conca de Tremp (Lleida, España). Estudios Geológicos, volumen extraordinario Galve-Tremp:95–110.

Casanovas-Cladellas, M. L., J. V. Santafé-Llopis, J. L. Sanz, and J. E. Powell. 1995. Nuevos restos de dinosaurios (Titanosauria y Ornithopoda) en el Cretacico Superior de las Cuenca de Tremp y Dellà (Lleida, Spain). Estudios Geológicos 51:277–283.

Casanovas-Cladellas, M. L., X. Pereda-Suberbiola, J. V. Santafé, and D. B. Weishampel. 1999. First lambeosaurine hadrosaurid from Europe: palaeobiogeographical implications. Geological Magazine 136:205–211.

Chiappe, L. M., R. A. Coria, L. Dingus, F. Jackson, A. Chimsamy, and M. Fox. 1998. Sauropod dinosaur embryos from the Late Cretaceous of Patagonia. Nature 396:258–261.

Company, J. 2004. Vertebrados continentales del Cretácico superior (Campaniense-Maastrichtiense) de Valencia. Ph.D. dissertation, University of Valencia, Valencia, Spain, 410 pp.

Company, J., X. Pereda-Suberbiola, and J. I. Ruiz-Omeñaca. 2009. Los dinosaurios del Cretácico Terminal del Levante ibérico en el contexto paleogeográfico europeo: composición de las faunas e implicaciones paleobiogeográficas; pp. 17–44 in P. Huerta-Hurtado and F. Torcida-Fernández Baldor (eds.), Actas de las IV Jornadas Internacionales sobre Paleontología de Dinosaurios y su Entorno. Colectivo Arqueológico-Paleontológico de Salas, Salas de los Infantes, Burgos, Spain.

Cruzado-Caballero, P., X. Pereda-Suberbiola, and J. I. Ruiz-Omeñaca. 2010. *Blasisaurus canudoi* gen. et. sp. nov., a new lambeosaurinae dinosaur (Hadrosauridae) from the latest Cretaceous of Arén (Huesca, España). Canadian Journal of Earth Sciences 47:1507–1517.

Cruzado-Caballero, P., J. I. Ruiz-Omeñaca, and J. I. Canudo. 2010. Evidencias de la coexistencia de dinosaurios hadrosaurinos y lambeosaurinos en el Maastrichtiano superior de la Península Ibérica (Arén, Huesca, España). Ameghiniana 47:153–164.

Cuevas, J. L. 1992. Estratigrafía del "Garumniense" de la Conca de Tremp, Prepirineo de Lérida. Acta Geológica Hispánica 27:95–108.

Dalla Vecchia, F. M. 2014. An overview of the latest Cretaceous hadrosauroid record in Europe; chapter 15 in D. A. Eberth, and D. C. Evans (eds.), Hadrosaurs. Indiana University Press, Bloomington, Indiana.

Escaso, F., A. Pérez-García, F. Ortega, and J. L. Sanz. 2010. Ankylosaurian evidence from the Upper Cretaceous of South-Central Pyrenees (Lleida, Spain): a reappraisal; pp. 33 in 8th European Association of Vertebrate Palaeontologists Meeting Abstract Volume, Aix-en-Provence, France. Available at doc.rero.ch/record/32151. Accessed January 22, 2014.

Evans, D. C. 2010. Cranial anatomy and systematics of *Hypacrosaurus altispinus,* and a comparative analysis of skull growth in lambeosaurine hadrosaurids (Dinosauria: Ornithischia). Zoological Journal of the Linnean Society 159:398–434 [also supporting information].

Gaete, R., A. Prieto-Màrquez, V. Riera, O. Oms, and À. Galobart. 2007. New discoveries of lambeosaurine hadrosaurids from the Tremp Syncline (Maastrichtian, Southern Pyrenees): description and stratigraphic setting; pp. 36–37 in J. Liston (ed.), Abstract of Presentations, 55th Symposium of Vertebrate Palaeontology and Comparative Anatomy. University of Glasgow, Glasgow, Scotland.

Galbrun, B., M. Feist, F. Colombo, R. Rocchia, and Y. Tambareau. 1993. Magnetostratigraphy and biostratigraphy of Cretaceous-Tertiary continental deposits, Ager Basin, Province of Lerida, Spain. Palaeogeography, Palaeoclimatology, Palaeoecology 102:41–52.

Galobart, À. 2007. Importancia del registro español de huevos de dinosaurio; pp. 89–177 in J. L. Sanz (ed.), Los dinosaurios en el siglo XXI. Editorial Tusquets, Barcelona.

Godefroit, P., L. Golovneva, S. Shchepetov, G. Garcia, and P. Alekseev. 2008. The last polar dinosaurs: high diversity of latest Cretaceous arctic dinosaurs in Russia. Naturwissenschaften 96:495–501.

Godefroit, P., J. Van Itterbeeck, P. Lauters, Y. L. Bolotsky, Z.-M. Dong, L.-Y Jin, S.-Q. Zan, S. Hai, and T. Yu. 2009. Latest Cretaceous hadrosaurid dinosaurs from Heilongjiang Province (P. R. China) and the Amur Region (Far Eastern Russia); pp. 91–120 in P. Huerta-Hurtado and F. Torcida-Fernández Baldor (eds.), Actas de las IV Jornadas Internacionales sobre Paleontología de Dinosaurios y su Entorno. Colectivo Arqueológico-Paleontológico de Salas, Salas de los Infantes, Burgos, Spain.

Grigorescu, D., G. Garcia, Z. Csiki, V. Codrea, and A.-V. Bojar. 2010. Uppermost Cretaceous megaloolithid eggs from the Haţeg Basin, Romania, associated with hadrosaur hatchlings: search for explanation. Palaeogeography, Palaeoclimatology, Palaeoecology 293:360–374.

Henderson, D. M. 2003. Footprints, trackways, and hip heights of bipedal dinosaurs: testing hip height predictions with computer models. Ichnos 10:99–114.

Horner, J. R., and H. N. Woodward. 2011. Riddle of the humongous hadrosaurs: what these giants reveal about dinosaur ontogeny, evolution, and ecology; pp. 66 in D. R. Braman, D. A. Eberth, D. C. Evans, and W. Taylor (compilers), International Hadrosaur Symposium Abstract Volume. Royal Tyrrell Museum of Palaeontology, Drumheller, Canada.

Horner, J. R., D. B. Weishampel, and C. A. Forster. 2004. Hadrosauridae; pp. 438–463 in D. B. Weishampel, P. Dodson, and H. Osmólska (eds.), The Dinosauria, Second Edition. University of California Press, Berkeley, California.

Husson, D., B. Galbrun, J. Laskar, L. A. Hinnov, and N. Thibault. 2011. Astronomical calibration of the Maastrichtian (Late Cretaceous). Earth and Planetary Science Letters 305:328–340.

Laurent, Y. 2003. Les faunes de vertébrés continentaux du Maastrichtien supérieur d'Europe: systematique et biodiversité. Strata 41:1–81.

Lauters, P., Y. L. Bolotsky, J. Van Itterbeeck, and P. Godefroit. 2008. Taphonomy and age profile of a latest Cretaceous dinosaur bone bed in Far Eastern Russia. Palaios 23:153–162.

Le Loeuff, J., E. Buffetaut, and M. Martin. 1994. The last stages of dinosaur faunal history in Europe: a succession of Maastrichtian dinosaur assemblages from the Corbières (southern France). Geological Magazine 131:625–630.

Llompart, C. 2006. Presencia de icnitas de dinosaurios en el Cretácico Superior del margen sur del sinclinal de Tremp (prov. de Lleida, Cataluña, España). Revista Española de Paleontología 21:1–14.

Llompart, C., M. L. Casanovas, and J. V. Santafé. 1984. Un nuevo yacimiento de icnitas de dinosaurios en las facies garumnienses de la Conca de Tremp (Lleida, España). Acta Geológica Hispánica 19:143–147.

Lockley, M., and A. P. Hunt. 1995. Dinosaur Tracks and Other Fossil Footprints of the Western United States. Columbia University Press, New York, 338 pp.

López-Martínez, N. 2000. Eggshell sites from the Cretaceous-Tertiary transition in South-Central Pyrenees (Spain); pp. 95–115 in A. M. Bravo and T. Reyes (eds.), Extended Abstracts, First International Symposium on Dinosaurs Eggs and Babies. Museu de la Conca Dellà, Isona i Conca Dellà, Isona, Spain.

López-Martínez, N., J. I. Canudo, Ll. Ardèvol, X. Pereda-Suberbiola, X. Orue-Extebarria, G., Cuenca-Bescós, J. I. Ruiz-Omeñaca, X. Murelaga, and M. Feist. 2001. New dinosaur sites correlated with Upper Maastrichtian pelagic deposits in the Spanish Pyrenees: implications for the dinosaur extinction pattern in Europe. Cretaceous Research 22:41–61.

Lucas, S. G., R. M. Sullivan, S. E. Jasinski, and T. L. Ford. 2011. Hadrosaur footprints from the Upper Cretaceous Fruitland Formation, San Juan Basin, New Mexico, and the ichnotaxonomy of large ornithopod footprints; pp. 357–361 in R. M. Sullivan, S. G. Lucas, and J. A. Spielman (eds.), Fossil Record 3. New Mexico Museum of Natural History and Science Bulletin 53.

Lull, R. S., and N. E. Wright. 1942. Hadrosaurian Dinosaurs of North America. Geological Society of America Special Papers 40. 242 pp.

Martín, M., R. Gaete, A. Galobart, V. Riera, and O. Oms. 2007. A new hadrosaurian bonebed in the Maastrichtian of the southern Pyrenees: a stratigraphic and taphonomic approach; p. 40 in J. Liston (ed.), Abstract of Presentations, 55th Symposium of Vertebrate Palaeontology and Comparative Anatomy. University of Glasgow, Glasgow, Scotland.

Mikhailov, K. E. 1997. Fossil and Recent Eggshell in Amniotic Vertebrates: Fine Structure, Comparative Morphology and Classification. Special Papers in Paleontology 56. Palaeontological Association, London, 80 pp.

Oms, O., J. Dinarès-Turell, E. Vicens, R. Estrada, B. Vila, À. Galobart, and A. M. Bravo. 2007. Integrated stratigraphy from the Vallcebre Basin (southeastern Pyrenees, Spain): new insights on the continental Cretaceous-Tertiary transition in southwest Europe. Palaeogeography, Palaeoclimatology, Palaeocology 255:35–47.

Oms, O., J. Dinarès-Turell, V. Riera, R. Estrada, E. Vicens, À. Galobart, B. Vila, R. Gaete, and F. M. Dalla Vecchia. 2010. Magnetostratigraphy of the Maastrichtian Southpyrenean succession (Spain). Resúmenes de la Reunión de la Comisión de Paleomagnetismo de la Sociedad Geológica de España, MAGIBER VI: 9–10.

Parks, W. A. 1923. *Corythosaurus intermedius,* a new species of trachodont dinosaur. University of Toronto Studies, Geological Series 15:1–57.

Pereda-Suberbiola, X., and J. L. Sanz. 1999. The ornithopod dinosaur *Rhabdodon* from the Upper Cretaceous of Laño (Iberian Peninsula). Estudios del Museo de Ciencias Naturales de Alava 14(número especial 1):257–272.

Pereda-Suberbiola, X., J. I. Ruiz-Omeñaca, and J. I. Company. 2003. Los dinosaurios hadrosaurios del registro ibérico: descripción de nuevo material del Cretácico superior de Laño (Condado de Treviño); pp. 375–388 in F. Pérez-Lorente (ed.), Dinosaurios y otros reptiles mesozoicos en España. Ediciones Instituto de Estudios Riojanos, Logroño, Spain.

Pereda-Suberbiola, X., J. I. Canudo, P. Cruzado-Caballero, J. L. Barco, N. López-Martínez, O. Oms, and J. I. Ruiz-Omeñaca. 2009. The last hadrosaurid dinosaurs of Europe: a new lambeosaurine from the uppermost Cretaceous of Arén (Huesca, Spain). Comptes rendus Palévol 8:559–572.

Powell, J. E. 1992. Hallazgo de huevos asignables a dinosaurios titanosáuridos (Saurischia, Sauropoda) de la Província de Río Negro, Argentina. Acta zoológica lilloana 41:381–389.

Prieto-Márquez, A. 2010. Global phylogeny of Hadrosauridae (Dinosauria: Ornithischia) using parsimony and Bayesian methods. Zoological Journal of the Linnean Society 159:435–502.

Prieto-Márquez, A., and J. R. Wagner. 2009. *Pararhabdodon isonensis* and *Tsintaosaurus spinorhinus:* a new clade of lambeosaurine hadrosaurids from Eurasia. Cretaceous Research 30:1238–1246.

Prieto-Márquez, A., R. Gaete, À. Galobart, V. Riera, and O. Oms. 2007. New data on European Hadrosauridae from the Latest Cretaceous of Spain. Journal of Vertebrate Paleontology 27(3, Supplement):131A.

Prieto-Márquez, A., R. Gaete, G. Rivas, À. Galobart, and M. Boada. 2006. Hadrosaurid dinosaurs from the Late Cretaceous of Spain: *Pararhabdodon isonensis* revisited and *Koutalisaurus kohlerorum,* gen et sp. nov. Journal of Vertebrate Paleontology 26:929–943.

Riera, V. 2010. Estudio integrado (Geológico y Paleontológico) de la sucesión de dinosaurios (Maastrichtiense) de la vertiente Surpirenaica. Ph.D. dissertation, Universitat Autònoma de Barcelona, Cerdanyola del Vallès, Spain, 210 pp.

Riera, V., O. Oms, R. Gaete, and À. Galobart. 2009. The end-Cretaceous dinosaur succession in Europe: the Tremp Basin record (Spain). Palaeogeography, Palaeoclimatology, Palaeoecology 283:160–171.

Rosell, J., R. Linares, and C. Llompart. 2001. El "Garumniense" Prepirenaico. Revista de la Sociedad Geológica de España 14:47–56.

Sereno, P. 1998. A rationale for phylogenetic definitions, with application to higher-level taxonomy of Dinosauria. Neue Jahrbuch Geologische Palaontologische Abhandlungen 210:41–83.

Thulborn, T. 1990. Dinosaur Tracks. Chapman and Hall, London, 410 pp.

Varricchio, D. J., D. Simon, S. Oser, D. Lawer, and F. Jackson. 2011. Dinosaur eggs in space and time. Journal of Vertebrate Paleontology, Program and Abstracts 2011:209A.

Vila B., R. Gaete, À. Galobart, and O. Oms. 2005. Icnitas de hadrosaurio de la Península Ibérica; pp. 83–84 in Libro de resúmenes del III Encuentro de Jóvenes Investigadores en Paleontología, Fumanya, St. Corneli (Cercs).

Vila, B., O. Oms, J. Marmi, and À. Galobart. 2008. Tracking Fumanya footprints (Maastrichtian, Pyrenees): historical and ichnological overview. Oryctos 8:150–130.

Vila, B., J. J. Moratalla, R. Gaete, V. Santos, and À. Galobart. 2011. New titanosaur trackways from southern Pyrenees: Orcau-2 locality (Late Cretaceous) revisited; pp. 34–35 in A. Richter and M. Reich (eds.), Dinosaur Tracks 2011: An International Symposium, Abstract Volume and Field Guide to Excursions. Obernkirchen, Germany.

Vila, B., R. Gaete, À. Galobart, O. Oms, J. Peralba, and J. Escuer. 2006. Nuevos hallazgos de dinosaurios y otro tetrápodos continentals en los Pirineos sur-centrales y orientales: resultados preliminaries; pp. 365–378 in F. Torcida-Fernández Baldor and P. Huerta Hurtado (eds.), Actas de las III Jornadas Internacionales sobre Paleontología de Dinosaurios y su Entorno. Colectivo Arqueológico-Paleontológico de Salas, Salas de los Infantes, Burgos, Spain.

Villalba-Breva, S., and C. Martín-Closas. 2011. Paleobiogeografía ecológica de las carofitas en el límite Campaniense-Maastrichtiense de las Cuencas surpirenaicas. Paleontologia i evolució: Memòria especial 5:391–394.

Weishampel, D. B., C.-M. Jianu, Z. Csiki, and D. B. Norman. 2003. Osteology and phylogeny of *Zalmoxes* (n.g.), an unusual Euornithopod dinosaur from the latest Cretaceous. Journal of Systematic Palaeontology 1:65–123.

Weishampel, D. B., P. M. Barrett, R. A. Coria, J. Le Loeuff, X. Xing, Z. Xijin, A. Sahni, E. M. P. Gomani, and C. R. Noto. 2004. Dinosaur distribution; pp. 517–606 in D. B. Weishampel, P. Dodson, and H. Osmólska (eds.), The Dinosauria, Second Edition. University of California Press, Berkeley, California.

Wilson, J. A., and P. C. Sereno. 1998. Early evolution and higher level phylogeny of sauropod dinosaurs. Society of Vertebrate Paleontology Memoir 5:1–68.

Young, C.-C. 1958. The dinosaurian remains of Laiyang, Shantung. Palaeontologia Sinica, New Series C 16:53–138.

Appendix 16.1. Length Measurements (in Centimeters) of the Hindlimb Bones Used in the Size-Frequency Diagrams of Figure 16.10B

Number	Element	Site	Length
IPS-N-21	Femur	Els Nerets	56
IPS-896	Femur	Els Nerets	55+
LB1-15	Femur	Lo Bas-1	60+
LB1-5	Tibia	Lo Bas-1	45+
LB1-7	Tibia	Lo Bas-1	~50
LB1-18	Tibia	Lo Bas-1	~50
MCD 4756	Femur	Casa Fabà	43.5
MCD 4756	Fibula	Casa Fabà	40+
In situ	Femur	Magret	~40
In situ	Tibia	Magret	~35
MCD no number	Femur	Magret	60
In situ	Tibia	Les Torres	~58
BSTE-1	Tibia	Basturs Est	50+
BSTE-2	Tibia	Basturs Est	50+
BSTE-3	Fibula	Basturs Est	35+
MCD 4802	Femur	Basturs Poble	43
MCD 5011	Femur	Basturs Poble	72
MCD 4737	Tibia	Cabana de Gori-2	42.5
MCD no number	Femur	Costa de la Serra-2	35+
MCD 4736	Femur	Costa de la Serra-1	29.5
MCD 4735	Tibia	Costa de la Serra-1	~45
MCD no number	Tibia	Serrat de Pelleu	50+
MCD 4733	Tibia	Molí del Baró -1	47
MB-240/2010	Tibia	Molí del Baró	54
MCD 4734	Fibula	Molí del Baró	37+
MCD no number	Femur	Tossal de la Doba	75+
IPS-932	Femur	Coll de la Torre	55+
IPS-931	Femur	Coll de la Torre	50+
MCD no number	Tibia	Masia de Ramon	22+
TB1-01	Tibia	Torrebilles-1	~65

Note: + estimated length.

Appendix 16.2. Length Measurements (in Centimeters) of the Bones Used in the Size-Frequency Diagrams of Figure 16.10C

Number	Element	Site	Length
IPFUB no number	Ischium	Moror	60+
LB1-28	Ischium	Lo Bas-1	~50
MCD no number	Dentary	Barranc de la Costa Gran	20+
EP-5	Metatarsal II	El Pous	21
MCD 4919	Maxilla	Serrat del Rostiar-1	30–35+
MCD 5007	Dentary	Basturs Poble	~34
MCD 4963	Dentary	Basturs Poble	~17
IPS-29920	Dentary	Les Llaus	36
MCD no number	Humerus	Costa de la Serra-2	17
MCD 5090	Maxilla	Euroda Nord	27+
EES-1	P. phalanx III-1	Els Esfons	10.3
MB 33/2011	Dentary	Molí del Baró	18+
MB54/2007	Ulna	Molí del Baró	35
MB 52/2007	Metatarsal II	Molí del Baró	15
IPS SRA 15,	Humerus	Sant Romà d'A.	48
IPS-36327	Maxilla	Sant Romà d'A.	31.5
MCD 4787	Ischium	Serrat del Corb	~100
MCD 5059	Dentary	Torrebilles-4	19+
TB2-12	Dentary	Torrebilles-2	15+

Note: + estimated length.

Hadrosaurs from the Far East: Historical Perspective and New *Amurosaurus* Material from Blagoveschensk (Amur Region, Russia)

17

Yuri L. Bolotsky, Pascal Godefroit, Ivan Y. Bolotsky, and Andrey Atuchin

ABSTRACT

In this chapter we describe the pioneering exploration of Upper Cretaceous dinosaur localities in the Amur Region by Russian scientists, ranging from the Great Siberian Expedition at the end of the 1850s, until the discovery, in 1948, of the Blagoveschensk bonebed. As a complement to the anatomical description of the lambeosaurine hadrosaur, *Amurosaurus riabinini*, several interesting bones discovered during more recent excavations of the Blagoveschensk bonebed are described. *Amurosaurus* rarely reached skeletal maturity and the largest humerus unearthed from the Blagoveschensk bonebed corresponds to an animal about 9 m long for an estimated weight of about 4000 kg. The long, caudally open external naris and the shallowly concave dorsal profile of the premaxilla in a presumed sexually mature specimen are consistent with the basal position of *Amurosaurus* in the phylogeny of lambeosaurines. Accordingly, it is possible that the nasal passage was not completely enclosed by the premaxilla in adults, as is the case for North American adult lambeosaurines, and instead was partially covered by a skin membrane. The low neural spines of the sacrals are a potential autapomorphic character of *Amurosaurus riabinini*.

INTRODUCTION

Four rich dinosaur localities have been discovered in the Amur/Heilongjiang region of eastern Asia (Fig. 17.1): Jiayin (Riabinin, 1930a; Godefroit et al., 2001) and Wulaga (Godefroit et al., 2008) in the Yuliangze Formation of northern Heilongjiang Province (China), and Blagoveschensk (Bolotsky and Kurzanov, 1991; Godefroit et al., 2004) and Kundur (Godefroit et al., 2003; Van Itterbeeck et al., 2005) in the Udurchukan Formation of southern Amur Region (Russia). All of these sites are located in the southeastern part (Lower Zeya depression) of the Zeya-Bureya sedimentary basin, near its borders with the adjacent uplifted areas: the Lesser Khingang Mountains and the Turan uplift. In the four sites, dinosaur bones form large bonebeds extending over several hundred square meters (Godefroit et al., 2000, 2003, 2004) and the dinosaur fauna is largely dominated by lambeosaurine saurolophids (sensu Prieto-Márquez, 2010), although saurolophines are also present (Bolotsky and Godefroit, 2004; Godefroit et al., 2008). The dinosaur sites from the Amur/Heilongjiang region have proven to be of fundamental importance for a better understanding of saurolophid paleobiogeography and, especially, of dinosaur extinction patterns at the end of the Cretaceous (Godefroit et al., 2003, 2004, 2008).

Although it was discovered in 1948, the Blagoveschensk dinosaur locality was neglected for a long time by the Russian scientific community because its geological context was misinterpreted by leading paleontologists from Moscow (Rozhdestvensky, 1957). It is thanks to the tenacity of local teams from the Blagoveschensk State Pedagogical University and from the Amur Integrated Scientific Research Institute of the Far Eastern Branch of the Russian Academy of Sciences that large-scale excavations were undertaken during the 1980s. Today, Blagoveschensk is regarded as one of the greatest sites in Russia in terms of abundance and preservation of dinosaur fossils.

In this chapter we present the historical context leading to the discovery of the Blagoveschensk dinosaur locality, and show that the early discovery of hadrosaur fossils in the Amur Region is intimately linked to the pioneering exploration of the Far Eastern territories by Russian scientists during the second half of the nineteenth century. We also describe some new hadrosaur fossils that complete our knowledge of the anatomy and phylogenetic relationships of the dominant dinosaur in the Blagoveschensk locality, the lambeosaurine *Amurosaurus riabinini* Bolotsky and Kurzanov, 1991.

Institutional Abbreviations AEHM, Amur Natural History Museum of the Far Eastern Institute of Geology and Nature Management, Blagoveschensk, Russia; AMNH, American Museum of Natural History, New York; CMN, Canadian Museum of Nature, Ottawa, Ontario; RAS, Russian Academy of Sciences, Moscow, Russia; ROM, Royal Ontario Museum, Toronto, Ontario; TMP, Royal Tyrrell Museum of Palaeontology, Drumheller, Alberta.

17.1. Map of the Amur region indicating the main dinosaur sites (▲). The Blagoveschensk locality is indicated by an arrow. The gray zones indicate the uplifted areas.

THE FIRST HADROSAUR DISCOVERIES IN THE AMUR REGION: A HISTORICAL PERSPECTIVE

The early discoveries of hadrosaur fossils in the Amur Region are intimately linked to the pioneering exploration of the Far Eastern territories by Russian scientists. As soon as Russian Cossacks penetrated and settled in the Amur region in the 1640s – during their eastward sweep across Siberia in quest of furs – they came into conflict with the newly established Manchu dynasty in China, who viewed the Amur River as an important part of their northern homeland (Bassin, 1983). After several armed clashes during the 1660s followed by the Treaty of Nerchinsk in 1689, the Russians relinquished all claims to the Amur and withdrew their settlements. They subsequently concentrated their activity in far northeastern Siberia, on the North Pacific coast, Kamchatka, and in Russian territories in the New World (Gibson, 1968).

In 1847, Tsar Nicholas I appointed Nikolai Nikolaevich Muraviev (1809–81) as Governor General of Eastern Siberia. As soon as he arrived in Eastern Siberia, Muraviev pleaded for reannexing the Amur region because the Amur River was an ideal supply route to Russia's fur colonies and a potentially great outlet to the rich markets of the Asiatic Far East. Moreover, large-scale incursions of European powers into China during the Opium Wars of the 1840s introduced a fear into Russia's government that Europeans might seek control over Siberia as well. According to Muraviev, it was necessary to annex the Amur region and establish a secure naval base at the mouth of the Amur River in order to fortify Russia's position in the Far East (Bassin, 1983). On December 31, 1853, Tsar Nicholas I granted Muraviev rights to negotiate with the Chinese regarding the establishment of a border along the Amur River.

As an active member of the Russian Geographical Society (founded in 1845), Muraviev had already established a Siberian branch of the Russian Geographical Society in Irkutsk, which took part in organizing the Great Siberian Expedition in the early 1850s. Starting in 1854, Muraviev negotiated with the Chinese and helped Gennady Ivanovich Nevelskoy (1813–76) to organize maritime expeditions down to the Amur estuary. Nevelskoy and Muraviev were later joined by the astronomer Ludwig Schwarz on the second expedition down the Amur as they gathered data to create the first complete map of the river.

During the expedition of 1858, Muraviev concluded the Treaty of Aigun with the Qing government of China. Accordingly, Russia received over 600,000 km^2 of land between the Stanovoy Mountains and the Amur River (Fig. 17.2), and Blagoveschensk City (the new seat of government in the Amur Region) was founded at the military outpost of Ust-Zeysky, at the mouth of the Zeya River.

The Russian Geographical Society subsequently selected Fedor Bogdanovich Schmidt (1832–1908) to lead the physical section of the Great Siberian Expedition, and from 1860–1862 Schmidt and two assistants explored the Russian Far East. The aim of the expedition was geological reconnaissance, and during three years of exploration, Schmidt discovered many fossil plant localities, including the famous Bureyinskoe Belogorje site, the type section of the Maastrichtian–Danian Tsagayan Group. Schmidt (1866:16) mentioned that he organized an excursion along the Amur River between Gomelevka and Pashkovo villages in September 1859: "close to Pashkovo village, before the Khingang Mountains, I could see a Tertiary section with white sands, big conglomerate, white clays, and a thin bed with brown coal and clay-iron concretions. . . . In the coarse conglomerate above the Tertiary layers, I saw broken hollow bones of mammalian animals. I have no further opinion." Schmidt evidently

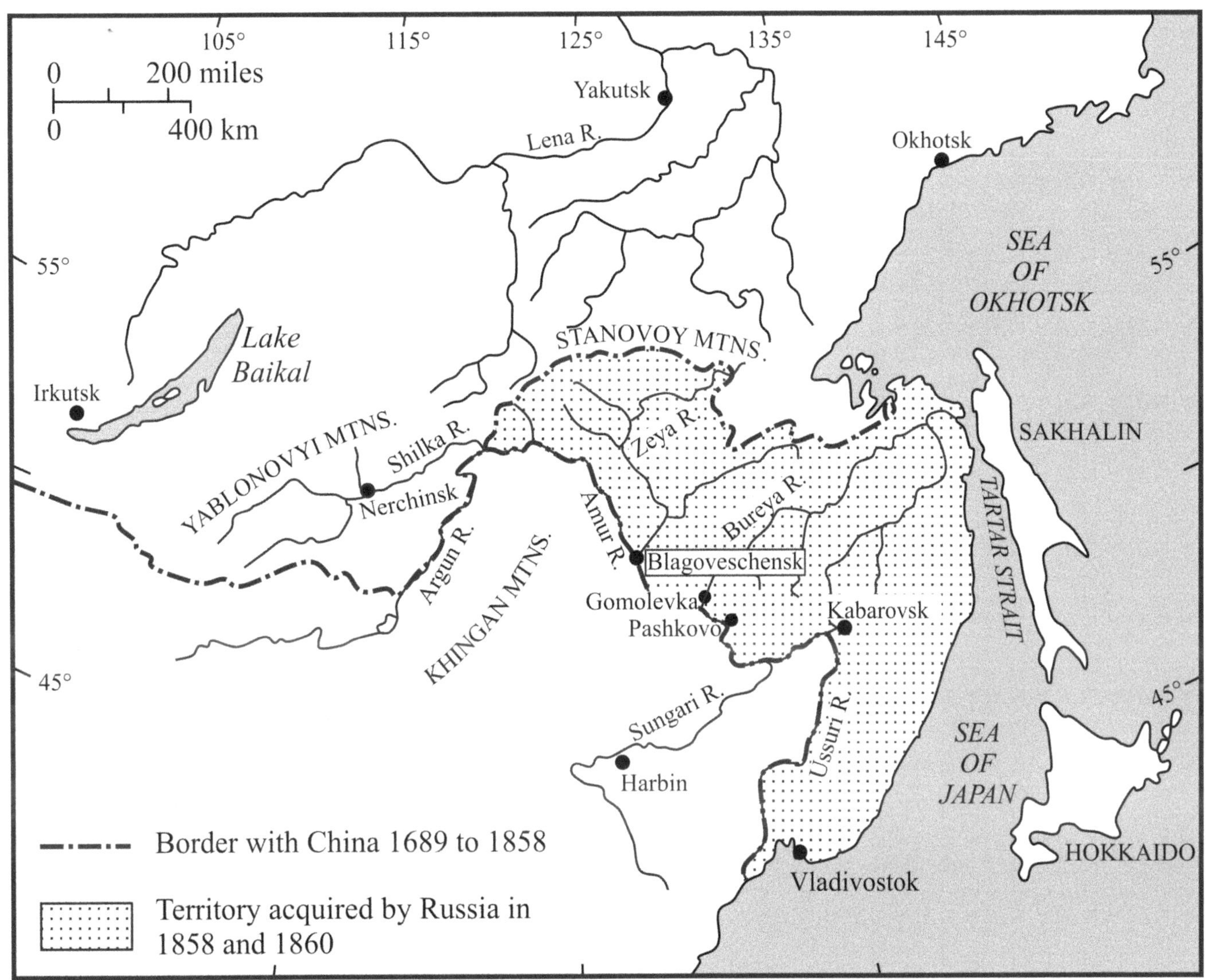

17.2. The Amur region in the mid-nineteenth century (from Bassin, 1983, modified).

misinterpreted the geological context of this section and it is likely that these bones belonged instead to dinosaurs; no fossil mammals have yet been identified in the area and hadrosaur bones are extremely abundant in the Jiayin locality, a few kilometers from Pashkovo village. It may therefore be tentatively postulated that Schmidt was the discoverer of the Jiayin locality in 1859. If this interpretation is correct, Jiayin is the first dinosaur locality reported in Asia. However, because Schmidt was a Silurian specialist, and because appropriate Mesozoic specialists were absent in Russia at that time, he sent abroad all the fossils collected during this expedition. After the return of Schmidt's expedition to St. Petersburg in January 1863, interest in the Russian Far East quickly died. Eventually, the focus of geographic exploration in the east shifted to Central Asia, reflecting Russia's military successes in that area.

In the 1890s, however, the construction of the Trans-Siberian Railway and Russia's military incursions into Manchuria and Korea led to a renewed interest in the Far Eastern territories. On August 11, 1902, the local newspaper *Amuskaya Gazeta* (Newsletter of Amur), edited at Blagoveschensk, published the following account: "Message from Colonel M. M. Manakin's headquarter: bones were found along the banks of the river, slumbering on its ancient shore, under three sazhen [a sazhen is an ancient Russian measure equal to 2.134 m] of alluvial sands. According to preliminary investigations, the skeleton is lying on its left flank, and its forelimbs are turned towards the river. The animal is about 5 sazhen long" (Fig. 17.3). It also noted, "Bones are well fossilized and therefore possibly belong to an epoch more ancient than that of the Mammoth." Apparently, Manakin did not attempt any excavation, but instead collected some isolated

АМУРСКАЯ ГАЗЕТА

ПОЛИТИЧЕСКІЙ, ОБЩЕСТВЕННЫЙ И ЛИТЕРАТУРНЫЙ ОРГАНЪ

№ 89.

ВЫХОДИТЪ ПО ВОСКРЕСЕНЬЯМЪ, СРЕДАМЪ И ПЯТНИЦАМЪ.

Годъ изданія VIII. Воскресенье, 11-го Августа 1902 года. г. Благовѣщенскъ.

Во время поѣздки по Амуру, въ минувшемъ іюнѣ мѣсяцѣ, генеральнаго штаба полковникъ Манакинъ обнаружилъ на пустынномъ правомъ берегу рѣки, между станицами Сагибовой и Касаткиной, въ 12 верстахъ ниже послѣдней, скелетъ допотопнаго животнаго. Полковникъ Манакинъ предполагаетъ, что это—кости мамонта; но, судя по тому, что кости вполнѣ окаменѣли, можно думать, что онѣ принадлежатъ къ эпохѣ болѣе древней, чѣмъ та, въ которую жилъ мамонтъ.

Кости лежатъ въ прибрежной толщѣ рѣки, покоясь на древнемъ берегу ея, состоящемъ изъ синяго конгломерата, надъ трехсаженнымъ слоемъ наноснаго песку, примѣрно на высотѣ двухъ саженъ отъ нормальнаго уровня. Насколько можно судить по наскоро произведенному изслѣдованію, скелетъ лежитъ на лѣвомъ боку, будучи обращенъ передними конечностями по направленію къ рѣкѣ, занимая въ длину до 5 саженъ.

Вокругъ костяка въ грунтѣ можно видѣть какъ бы очертанія прежде окружавшаго его мяса, изъ истлѣнія котораго грунту передалась особая желтоватая окраска.

По разсказамъ жителей Касаткиной, кости обнаружены года два тому назадъ, причемъ во время половодья понемногу размываются поднимающейся водой и постепенно падаютъ на берегъ.

Полковникомъ Манакинымъ собраны, а частью отрыты отдѣльные позвонки и части реберъ и переданы въ хабаровскій музей.

Мы слышали, что приамурскій отдѣлъ и. р. г. о. командируетъ члена совѣта отдѣла, Г. Ф. Бѣлоусова, съ порученіемъ осмотрѣть, какъ мѣсто нахожденія скелета, обнаруженнаго полковникомъ Манакинымъ, такъ

17.3. Excerpt from *Amuskaya Gazeta* (Newsletter of Amur) dated August 11, 1902, describing the discovery of fossil bones along the banks of the Amur River, and portrait of Colonel M. M. Manakin.

bones that he then deposited in the Khabarovsk Natural History Museum. Part of Manakin's collection finally arrived in the hands of Alexei Yakovlevich Gurov (1872–1932; Fig. 17.4A), a village teacher and enlightened archaeologist, who deposited several bones in the Blagoveschensk Museum of Regional Study (Fig. 17.4B, C). In 1903, Gurov drew a large archaeological map of the Amur River, precisely indicating the locality where Manakin discovered the "antediluvian animals" along the right banks of the Amur River (Fig. 17.4E). This locality is known as Belie Kruchy ("white cliffs") by Russians, and Longguchan by the Chinese. It is located close to the Chinese town of Jiayin and, today, the Jiayin Dinosaur National Geological Park is situated at the site of Manakin's original discovery.

In 1914, Gurov presented one fossil bone from Jiayin to Afrikan Nicholaïevich Kryshtofovitch (1885–1953), a geologist and paleobotanist at the Russian Geological Committee who was then working in the Amur region. On returning to St. Petersburg, Kryshtofovitch showed the bone to his colleague, Anatoly Nicholaïevich Riabinin, who identified it as the proximal end of a dinosaurian tibia (Riabinin, 1930a; Fig. 17.4D). During the winter of 1915–16, another geologist at the Russian Geological Committee, W. P. Renngarten, studied the geological section at Jiayin and discovered dinosaur bones within a greenish conglomerate. Following this preliminary research, the Russian Geological Committee, under the guidance of N. P. Stepanov, undertook two excavation campaigns during the summers of 1916 and 1917 and unearthed several bonebeds in the greenish conglomerate along the right side of the Amur River.

Before the October 1917 Revolution, all finds, including the incomplete skeleton of a hadrosaurid dinosaur, were transported to the Museum of the Geological Committee to be prepared and studied. The hadrosaurid skeleton was prepared by Stepanov between 1918 and 1923, and mounted under the supervision of Riabinin in 1924. In 1925, Riabinin produced a preliminary description of the skeleton and named it *Trachodon amurense.* He subsequently described it in detail and attributed it to a new genus, *Mandschurosaurus,* in the subfamily Hadrosaurinae (Riabinin, 1930a). Using other material collected at Jiayin, Riabinin (1930b) also named the hadrosaur *Saurolophus krychtofovici* from a fragmentary ischium, the theropod *Albertosaurus periculosus* from isolated teeth, and the turtle *Aspideretus planicostatus* from carapace fragments. The mounted skeleton of *Mandschurosaurus amurensis* reconstructed from several individuals collected during the 1916–17 campaigns is on display in the F. N. Chernishev Scientific Research Geological Institute, St. Petersburg (Fig. 17.5).

In 1925, famous writer and regionalist Vladimir Kladijevich Arseniev (1872–1930), then director of the Grodekov's Khabarovsk Museum of Regional Study, started new excavations at Jiayin and unearthed a large collection of fossils that he sent to Khabarovsk. The tibia of a small hadrosaur is still housed without any tag in the museum, but the rest of the material is apparently lost.

THE BLAGOVESCHENSK DINOSAUR LOCALITY

In 1948, a second dinosaur locality was discovered in the Amur Region, in the western part of the city of Blagoveschensk. The first bone was found by a schoolboy named Igor Bastrykin (Fig. 17.6A), who presented this fossil to the Blagoveschensk Museum of Regional Study. The director of the museum, G. S. Novikov-Daursky (Fig. 17.6A–C), and the regionalist A. G. Udod immediately started a small excavation; they were later joined by the famous Far Eastern geologist A. Z. Lazarev (Fig. 17.6B).

In 1951, a team from the Paleontological Institute of the USSR Academy of Sciences, led by A. K. Rozhdestvensky (Fig. 17.6C), visited the Amur region. They conducted a small excavation at Blagoveschensk and unearthed remains of hadrosaurs and theropods. Rozhdestvensky (1957) concluded that those bones were redeposited by the Amur River in Quaternary deposits and that the Blagoveschensk locality had little scientific significance. Blocks containing dinosaur bones were sent to Moscow, but were subsequently lost.

The Blagoveschensk dinosaur locality was rediscovered in 1972, at the occasion of large flash floods that inundated the area. Sediments containing dinosaur bones were used

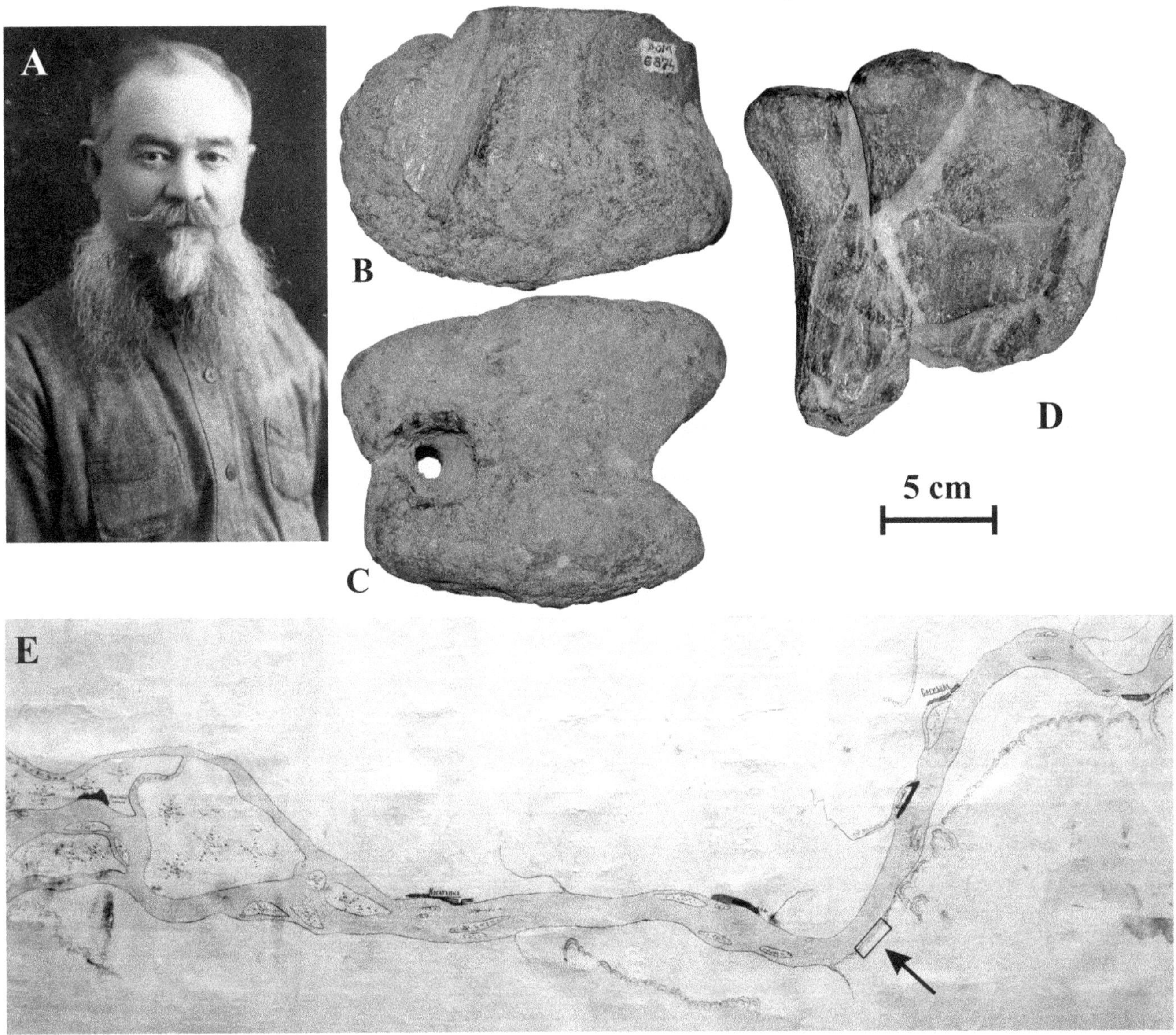

17.4. (A) portrait of A. Y. Gurov (1872–1932). (B–C) distal end of the left femur of a saurolophid dinosaur from Belie Kuchy (Jiayin) locality, deposited by Gurov in Blagoveschensk Museum of Regional Study (now, G. S. Novikov-Daursky Amur Regional Study Museum, Blagoveschensk), in lateral (B) and distal (C) views. (D) proximal part, in lateral view, of the left tibia of a saurolophid dinosaur from Belie Kuchy (Jiayin) locality, presented in 1914 by Gurov to A. N. Kryshtofovitch, and exhibited in F. N. Chernishev Scientific Research Geological Institute (St. Petersburg). (E) excerpt from the archaeological map of the Amur River drawn by A. Y. Gurov in 1903 and housed in G. S. Novikov-Daursky Amur Regional Study Museum, Blagoveschensk; the rectangle indicated by an arrow represents the discovery place of "antediluvian animals" and corresponds to the Jiayin dinosaur locality.

for the construction of embankments to protect the city. At that time, the fossil locality was guarded by military forces. Although teachers from Blagoveschensk State Pedagogical University, under the direction of N. F. Nikitenko, wanted to investigate the locality, Moscow administration would not provide the requested authorization, following Rozhdestvensky's (1957) opinion that "the locality has no scientific significance." Fortunately, the academic authorities from the Blagoveschensk State Pedagogical University saved the dinosaur site from urbanization and destruction. The locality eventually became an excursion site for students, and a small collection of bone fragments from the site was used in historical geology classes.

The systematic study of the Blagoveschensk dinosaur locality started at the beginning of the 1980s, with the foundation of the Amur Integrated Scientific Research Institute of the Far Eastern Branch of the Russian Academy of Sciences (AMUR KNII FEB RAS), in Blagoveschensk. The first excavations were undertaken in 1980–81 under the direction of A. G. Ablaev, then the head of the laboratory of stratigraphy.

17.5. Mounted skeleton of *Mandschurosaurus amurensis* in F. N. Chernishev Scientific Research Geological Institute (St. Petersburg).

Quickly it became evident that the abundant dinosaur bones were not redeposited within Quaternary sediments, but were preserved in Upper Cretaceous sediments. The palaeogeographical laboratory of the AMUR KNII FEB RAS started large-scale excavations in 1985, and many scholars from Blagoveschensk and Svobodny cities participated in the excavations.

By 1990, a surface of about 200 m² had been excavated and several hundred bones had been recovered. Most of them (more than 90%) belong to the lambeosaurine, *Amurosaurus riabinini* Bolotsky and Kurzanov, 1991 (Godefroit et al., 2004). Flat-headed saurolophines (sensu Prieto-Márquez, 2010) are represented by cranial material described as *Kerberosaurus manakini* Bolotsky and Godefroit, 2004. Isolated teeth from theropod (Alifanov and Bolotsky, 2002) and sauropod (Alifanov and Bolotsky, 2010) dinosaurs, and turtle fragments (Danilov et al., 2002) were also recovered from the bonebed.

In 2005 and 2006, the Palaeontological Laboratory of the Institute of Geology and Nature Management of the RAS and the Royal Belgian Institute of Natural Sciences organized new excavations in the Blagoveschensk dinosaur locality. A surface of about 200 m² was excavated, yielding more than 500 bones. The main purpose of these new excavations was to collect more data on the stratigraphy, sedimentology, and taphonomy of the locality. Since 2007, the Paleontological Museum of the Geological and Nature Exploration Institute RAS has organized summer excavations at the site.

Lauters et al. (2008) described in detail the geology, taphonomy, and age profile of the Blagoveschensk bonebed (base of "unit 2" in Fig. 17.7). They showed that the observed mixture of unstratified fine and coarse sediments in the bonebed is typical for sediment-gravity-flow deposits: sediment gravity flows, originating from the uplifted areas at the borders of the Zeya-Bureya Basin, reworked the dinosaur bones and teeth as a monodominant bonebed.

Although Godefroit et al. (2004) described *Amurosaurus riabinini* in detail, some parts of the skeleton, including the crest, remained completely unknown. Some of the fossils recently discovered at Blagoveschenk provide new information on the osteology and phylogenetic relationships of this Maastrichtian lambeosaurine, and are described below.

NEW *AMUROSAURUS* MATERIAL FROM BLAGOVESCHENSK AND ITS IMPORTANCE

Humerus AEHM *1/1755: What Was the Maximal Size of* Amurosaurus riabinini*?*

The size-frequency distributions of hadrosaur dentaries, humeri, femora, and tibiae in the Blagoveschensk bonebed reveal overrepresentation of late juvenile and small subadult individuals, indicative of an attritional death profile for the *Amurosaurus* fossil assemblage (Lauters et al., 2008). Until 2008, the largest humerus unearthed from the Blagoveschensk bonebed was AEHM 1/516, with a maximum length of 615 mm. It is exactly the same size as the holotype of the lambeosaurine *Olorotitan arharensis*, from the Maastrichtian of Kundur (Far Eastern Russia) and it therefore corresponds to an animal 8 m long (Godefroit et al., 2012). A much larger humerus, AEHM 1/1755, was discovered in 2009 at Blagoveschensk (Fig. 17.8). This bone is 670 mm long, the same size as the humerus of the large *Edmontosaurus annectens* (sensu Campione and Evans, 2011) AMNH 5730 (the holotype of *Anatosaurus copei* Lull and Wright, 1942; Brett-Surman, 1989:table 2), which corresponds to an animal about 9 m long (Osborn, 1909; Lull and Wright, 1942), with an estimated weight of ~4,000 kg (Anderson et al., 1985).

17.6. Discovery and first excavation of the Blagoveschensk dinosaur locality. (A) G. S. Novikov-Daursky (left) and I. Bastrykin (right) around 1948. (B) G. S. Novikov-Daursky (left) and A. Z. Lazarev (right) around 1948. (C) G. S. Novikov-Daursky (left), A. K. Rozhdestvensky (middle), and Nina S. Shevyreva (mammal specialist in the Paleontological Institute of the USSR Academy of Sciences, right) in 1951.

This humerus most likely belongs to *Amurosaurus riabinini:* although the deltopectoral crest is better developed than in typical saurolophines, it is less prominent than in other lambeosaurines described in Maastrichtian deposits from the Amur region (*Charonosaurus jiayinensis, Sahaliyania elunchunorum,* and *Olorotitan arharensis;* see Table 17.1 for a comparison with the humerus of the holotype specimen of *Olorotitan arharensis* from Kundur locality). The lateral border of the deltopectoral crest is slightly concave as in other large humeri (e.g. AEHM 1/516) referred to *Amurosaurus riabinini* (Godefroit et al., 2004). According to Horner and Woodward (2011), sexual maturity preceded skeletal maturity in hadrosaurs, and in dinosaurs in general. The relatively large number of subadults preserved at different localities where multiple individuals are preserved suggests that, in general, hadrosaurs endured high mortality rates, and that skeletal maturity was a rare event. The age-frequency profile of the humeri referred to *Amurosaurus riabinini* at the Blagoveschensk dinosaur locality (Fig. 17.9) supports this hypothesis.

Premaxilla AEHM *1/1038: The Crest of* Amurosaurus riabinini

AEHM 1/1038 is a left premaxilla (Fig. 17.10), the only element from the hollow crest of *Amurosaurus riabinini* discovered at Blagoveschensk to date. The total length of this premaxilla is about 75 cm. It corresponds to the size of the premaxilla in other lambeosaurines usually identified as adult specimens, such as AMNH 5240 (*Corythosaurus casuarius:* basal skull length = 739 mm), TMP 1980.023.0004 (*Corythosaurus intermedius:* basal skull length = 753 mm), CMN 8703 (*Lambeosaurus clavinitiatis:* basal skull length = 748 mm), and ROM 1218 (*Lambeosaurus lambei:* basal skull length = 747

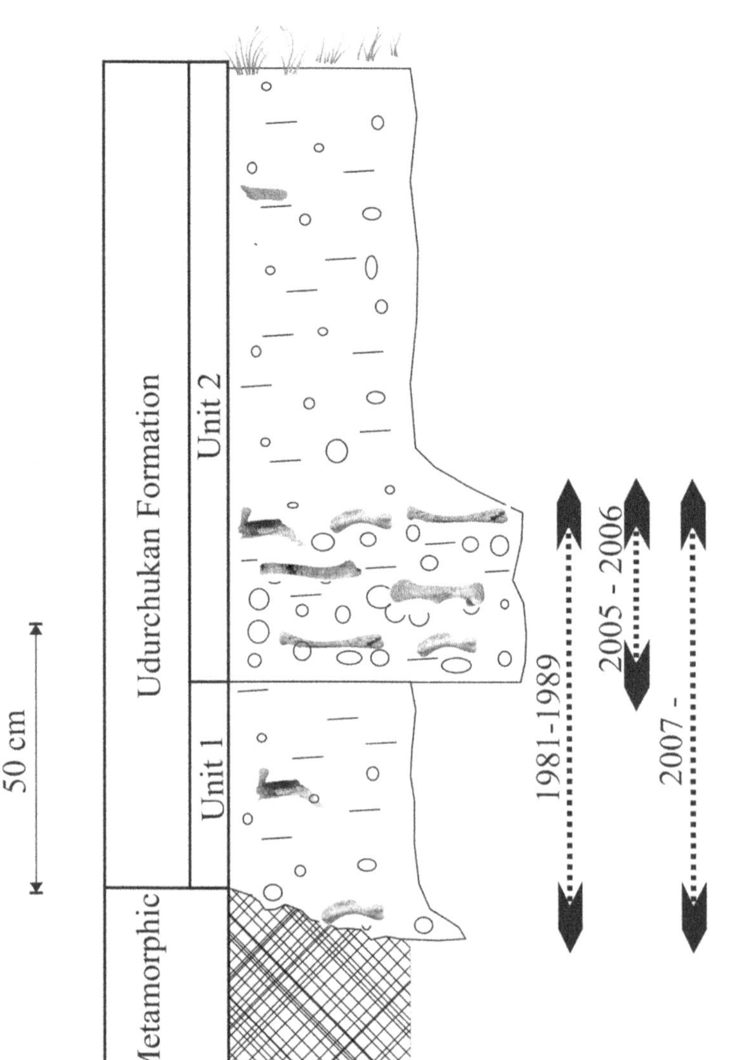

17.7. Stratigraphic column of the Blagoveschensk section. Arrows indicate levels excavated during different field seasons. Modified from Lauters et al. (2008).

mm). Rostrally, the premaxillae form a broadly arcuate oral margin in ventral view that reaches a maximum breadth of slightly less than twice the width of the postoral constriction. In ventral view, the rostral bill has a peripheral denticulate margin that is separated from a thickened palatal region by an arcuate groove that mirrors the curvature of the rostrum, as in other saurolophids (Horner et al., 2004). The lateral part of the oral margin is dorsoventrally expanded and ventrally deflected, forming a broad lip-like margin, as usually observed in lambeosaurines (Prieto-Márquez, 2010:character 62).

As is typical of lambeosaurines, the external naris is surrounded by the premaxilla only, and the left and right nasal passages are consequently completely separated in the snout region. The medial side of the premaxilla, which contacted its antimere, is perfectly flat along its entire length. A broad truncated facet along the medioventral side of the premaxilla, 30 cm from the tip of the snout, marks the contact with the rostrodorsal shelf of the maxilla. In lateral view, the external naris is roughly triangular in shape and more elongate than in any other lambeosaurine described so far. Although the caudolateral process progressively becomes higher caudally, careful preparation of AEHM 1/1038 reveals that its dorsal border does not contact the ventral border of the caudodorsal process on the completely preserved rostral portion of the specimen. The elongate and widely open external naris of *Amurosaurus riabinini* contrasts with the restricted and lacriform naris in *Hypacrosaurus altispinus* (Evans, 2010), *Parasaurolophus walkeri* (ROM 768), and *Olorotitan arharensis* (Godefroit et al., 2012). Although the external naris is also particularly elongate in *Corythosaurus casuarius* (Ostrom, 1961) and *Lambeosaurus magnicristatus* (Evans, 2010), the caudodorsal and caudolateral processes intimately contact each other in front of the angle between the snout and the rostral crest margin.

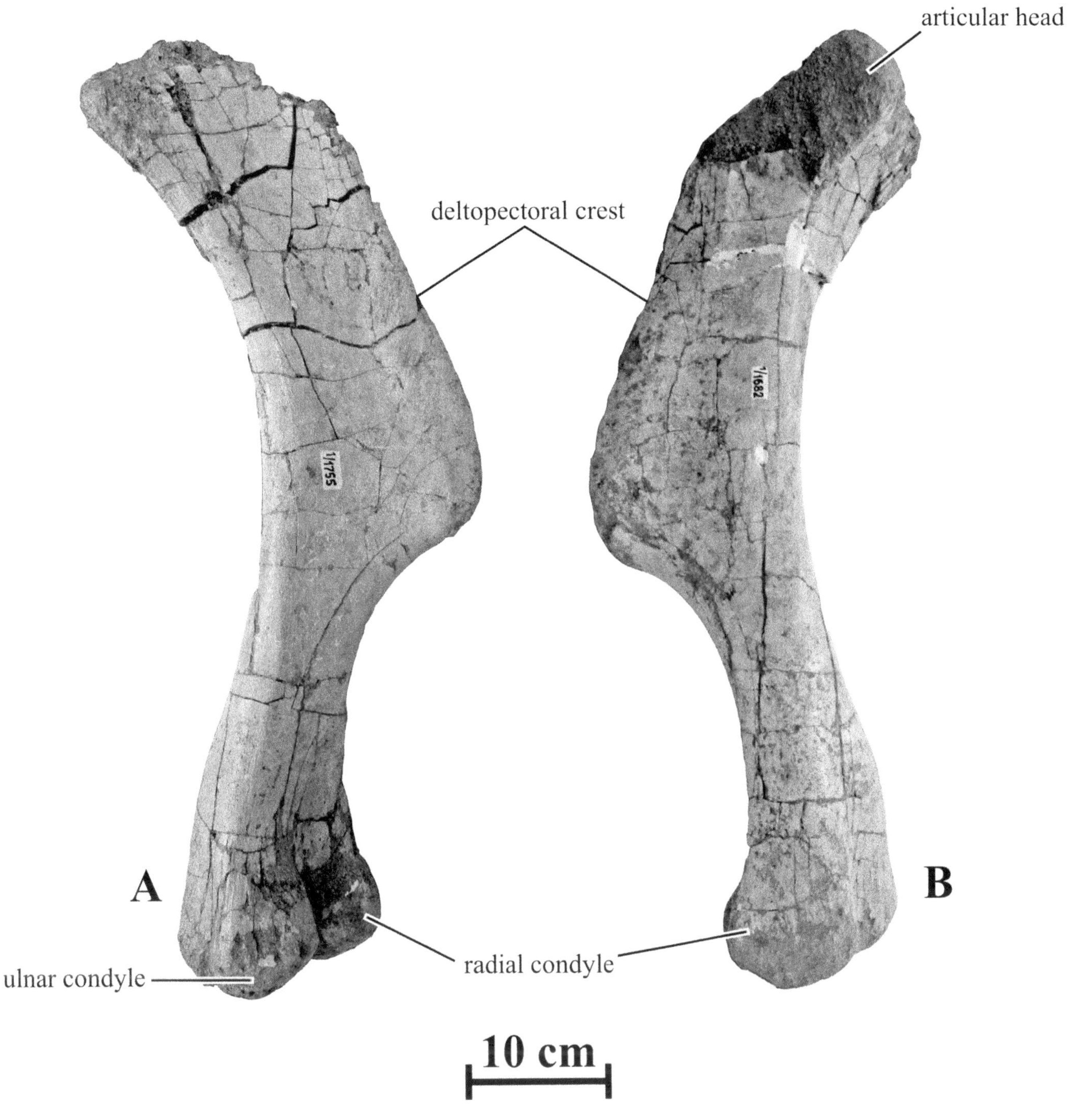

17.8. Left humerus (AEHM 1/1755) of a large *Amurosaurus riabinini* specimen from the Upper Cretaceous Blagoveschensk locality, in cranial (A) and caudal (B) views.

The degree of closure of the caudal portion of the nasal vestibule and of the premaxilla-nasal fontanelle is an ontogenetic character that displays important interspecific variations within lambeosaurines. In *Hypacrosaurus altispinus*, the premaxilla-nasal fontanelle is closed and the external naris is restricted in all known specimens, including the juvenile *Cheneosaurus tolmanensis* (CMN 2247; Evans, 2010). In *Hypacrosaurus stebingeri*, the premaxilla-nasal fontanelle is open and continuous with the narial opening in embryonic and nestling specimens, open but discontinuous in juvenile and subadult specimens, and completely closed in adult specimens (Brink, 2009). In juvenile *Corythosaurus* (*Tetragonosaurus erectofrons*; ROM 759) and *Lambeosaurus* specimens (*Tetragonosaurus praeceps*; ROM 758), the nasal aperture is open along its entire length and is confluent with the premaxilla-nasal fontanelle (Evans et al., 2005).

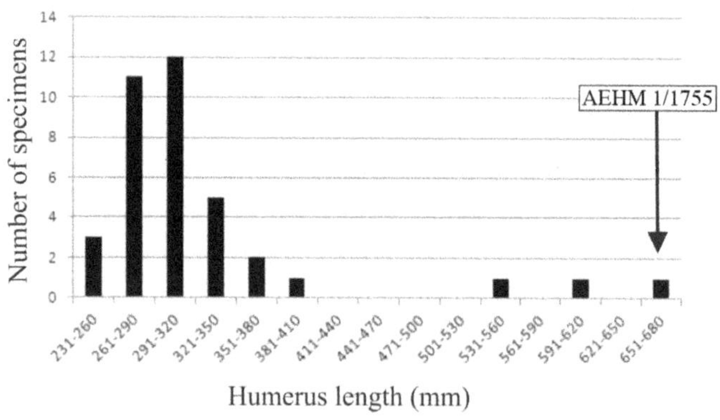

17.9. Size distribution of the humeri attributed to *Amurosaurus riabinini* from the Upper Cretaceous Blagoveschensk locality. Modified from Lauters et al. (2008).

Table 17.1. Dimensions (mm) of (1) AEHNM 1/1682, the Largest Humerus from Blagoveschensk Dinosaur Locality and Referred to *Amurosaurus riabinini*, and (2) AEHM 2/845, the Humerus of the Holotype of *Olorotitan arharensis*, from Kundur Locality

Measurement	AEHM 1/1682 *Amurosaurus riabinini*	AEHM 2/845 *Olorotitan arharensis*
Length	670	615
Length of deltopectoral crest	445	375
Width of proximal end	+/- 145	138
Width of deltopectoral crest	146 (margin eroded)	182
Minimal width of shaft	84	87
Width of distal end	154	119

The premaxilla-nasal fontanelle remains open in some *Corythosaurus* and *Lambeosaurus* adult skulls, although this closure is not consistent among all specimens and is subject to individual variation (Brink et al., 2011). Thus, the long, caudally-open external naris in AEHM 11038 might be regarded as a paedomorphic character retained in the skull of an adult *Amurosaurus* individual. However, this character is also consistent with the presumed basal position of *Amurosaurus* in the phylogeny of lambeosaurines and it might be hypothesized that the nasal passage was not completely enclosed by the caudodorsal and caudolateral processes of the premaxilla, but partially covered by a skin membrane, as in saurolophines.

Although the caudal part of the premaxilla is incompletely preserved, the dorsal profile of the premaxilla appears shallowly concave in lateral view, with a crest-snout angle probably not less than 150°. The facial profile of all juvenile lambeosaurines is broad (Brink, 2009) and it also remains shallowly concave in adults of *Hypacrosaurus altispinus*, *Hypacrosaurus stebingeri* (Evans, 2010), and *Olorotitan arharensis* (Godefroit et al., 2012). In *Corythosaurus casuarius* and *Lambeosaurus* spp., the crest-snout angle becomes progressively smaller throughout ontogeny and the crest-snout angle in adults does not exceed 135° in *Corythosaurus* and 116° in *Lambeosaurus* (Dodson, 1975; Evans, 2010). Therefore, the broad crest-snout angle of *Amurosaurus riabinini* is also compatible with its presumed basal position in the lambeosaurine phylogeny.

The caudodorsal process of the premaxilla reaches its maximal width at the level of the crest-snout angle. The caudal portion of the caudodorsal process is narrow and curves ventrally into a long and pendant hook-like ventral ramus, marking the dorsal and caudal limits of the supracranial crest. The premaxilla therefore clearly participated into the caudal border of the supracranial crest, as in *Lambeosaurus lambei*, *Lambeosaurus magnicristatus*, and *Hypacrosaurus stebingeri* (Brink et al., this volume), but unlike *Hypacrosaurus altispinus* and *Olorotitan arharensis*. The dorsal (or caudal) border of the caudal, hook-like portion of the caudodorsal process is thick, whereas its ventral (or rostral) border is very thin, so its lateral aspect is depressed. It is therefore likely that the caudal part of the caudodorsal process was covered laterally by the nasal, which in this case would have participated in a large portion of the lateral aspect of the supracranial crest as in *Corythosaurus casuarius*, *Hypacrosaurus altispinus*, *Hypacrosaurus stebingeri*, and *Olorotitan arharensis*, but unlike *Lambeosaurus lambei* and *Lambeosaurus magnicristatus*. Of course, this hypothesis remains highly speculative because no nasal has been found so far at the Blagoveschensk locality. Figure 17.11 is a new reconstruction of the skull of *Amurosaurus riabinini*, with its crest.

Predentary AEHM 1/1682

The predentary of *Amurosaurus riabinini* is gracile and shovel shaped (Fig. 17.12). The ratio between the dorsoventral depth of its rostral face and the length of its lateral process is 0.36, which, according to Prieto-Márquez (2010:character 23), is a synapomorphy for Lambeosaurinae except *Tsintaosaurus spinorhinus*. The ratio between its maximum mediolateral width and the maximum rostrocaudal length along its lateral process (Prieto-Márquez, 2010:character 22) is 1.3. Unlike *Tsintaosaurus spinorhinus* and *Olorotitan arharensis*, its rostral process is convex in dorsal view. In lateral view, it forms an angle of about 45° with the dorsal margin of the lateral processes. The oral margin of the rostral process is eroded, but appears strongly denticulate; the denticles project rostrally. A dozen nutrient foramina are distributed across the entire rostral margin. The median pair is much larger than the other foramina. The pair of median processes that extended caudally over the dentary symphysis is too incompletely preserved to be adequately described. The dorsal pair

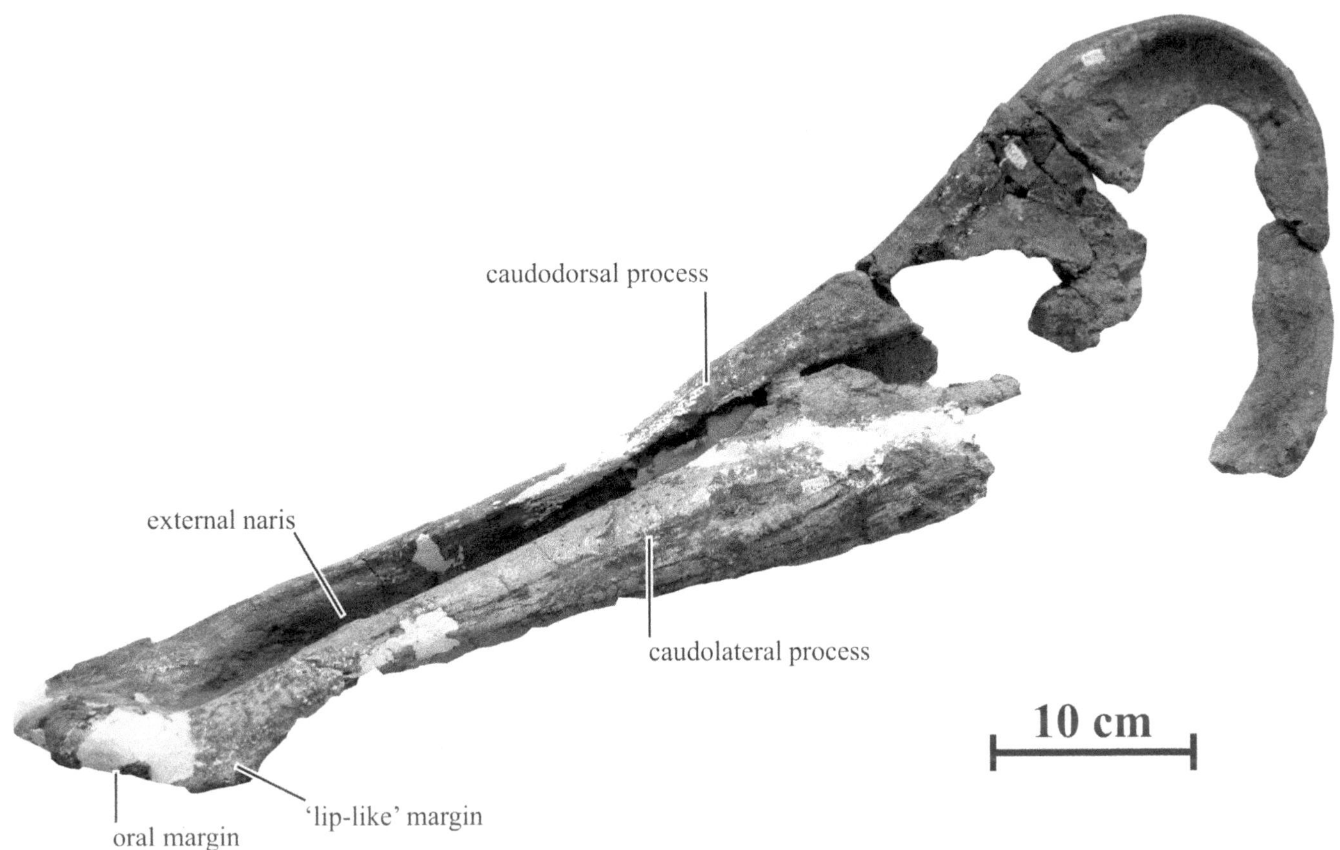

17.10. Left premaxilla (AEHM 1/1038) of *Amurosaurus riabinini* from the Cretaceous Blagoveschensk locality.

of processes is supported on a continuous ridge along the caudodorsal margin of the predentary. High dorsal ridges extend along the lateral processes. The lateral surface of the lateral processes is inclined dorsally, forming an angle of about 45° with the horizontal plane. The caudal end of the lateral process is bifid. The maximum mediolateral width of predentary AEHM 1/1682 is 134 mm; its maximum rostrocaudal length along the lateral process is 103 mm; and the depth of its rostral face is 37 mm.

Sacra AEHM 1/1715, AEHM 1/1716

Several more or less complete sacra have been unearthed recently from the *Amurosaurus* bonebed. The best-preserved specimens, on which the present description is based, are AEHM 1/1715 (Fig. 17.13A, B) and AEHM 1/1716 (Fig. 17.13C). The sacrum of *Amurosaurus riabinini* is composed of 10 fused vertebrae, including 1 dorsosacral and 2 sacrocaudals. The centra are so intimately fused together that their respective limits cannot be discerned. The dorsosacral is incompletely preserved. Its centrum seems narrower mediolaterally than those of the true sacrals and the base of each lateral side has two pairs of nutritive foramina. The centra of the succeeding sacrals gradually become wider mediolaterally. Between the articular surfaces, the centra appear less constricted than in *Charonosaurus* (Godefroit et al., 2001). The ventral side of the sacrum is longitudinally grooved from the fourth true sacral along its whole length. This groove becomes progressively wider passing through the succeeding sacrals. The pre- and postzygapophyses of the adjacent dorsosacral, sacrals, and sacrocaudals are fused together. All the transverse processes are eroded or broken off. The transverse process of the dorsosacral is directed caudodorsally. On the true sacrals and the sacrocaudals, the transverse processes are connected to the parapophyses on the lateral side of the sacrum by a prominent and sharp ridge. The ratio of the height of the highest sacral neural spine to the height of the corresponding centra is less than 2, contrary to other lambeosaurines described to date, and more closely resembling the condition observed in some hadrosaurines (Prieto-Márquez, 2011). However, because some fragmentary sacra are directly associated with typical *Amurosaurus* cranial material, it is unlikely that those sacra belong to hadrosaurines. The low sacral neural spines may therefore be regarded as an additional autapomorphic character for *Amurosaurus*, pending further evidence from articulated specimens. The

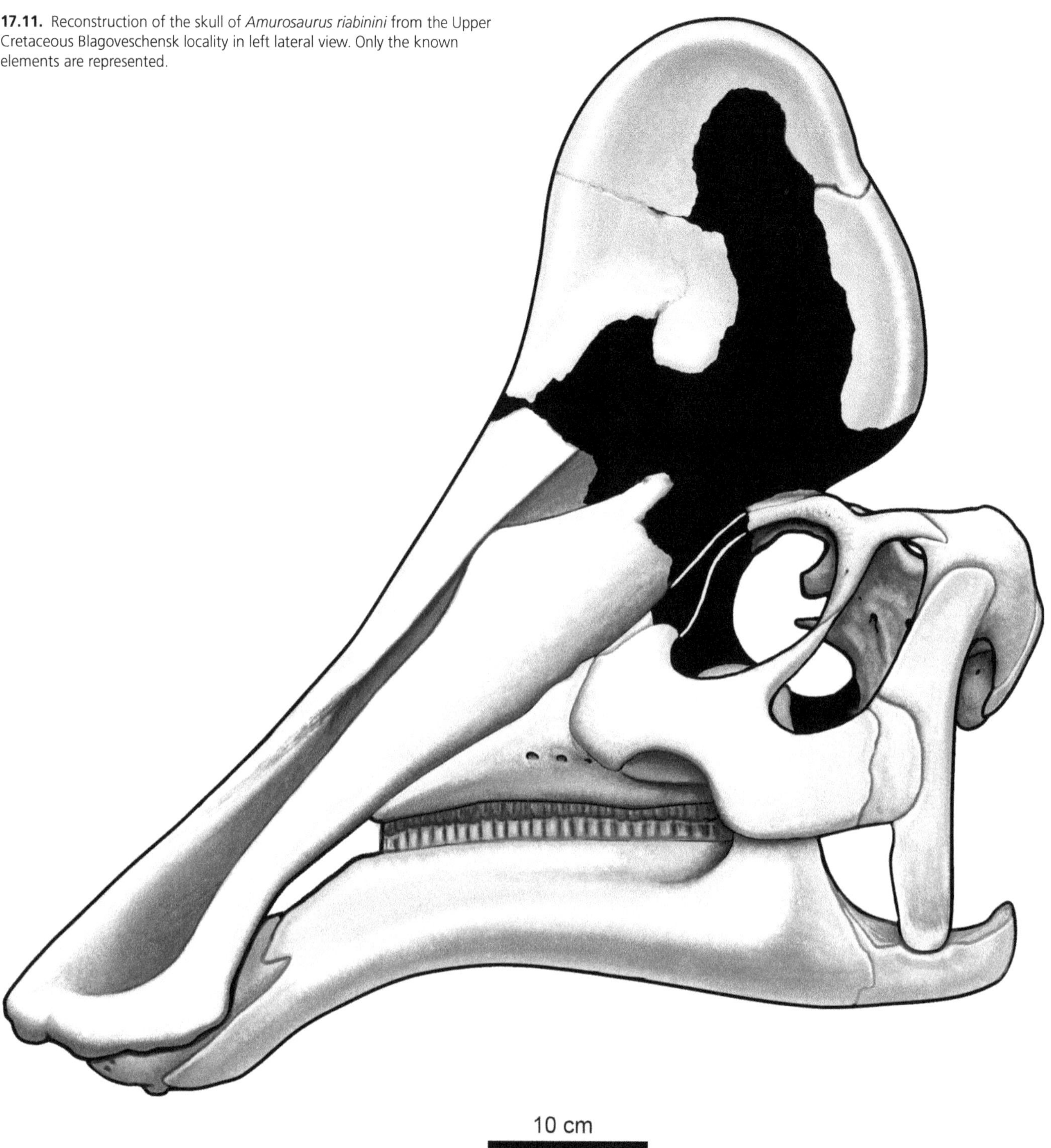

17.11. Reconstruction of the skull of *Amurosaurus riabinini* from the Upper Cretaceous Blagoveschensk locality in left lateral view. Only the known elements are represented.

neural spines of the sacral vertebrae are more inclined caudally than in *Olorotitan arharensis* (Godefroit et al., 2012). They are irregularly fused together on the different specimens preserved, probably reflecting important ontogenetic variation.

The distal ends of the seven sacral ribs are expanded to form a continuous iliac bar. Their proximal articulation is firmly fused to the ventral side of the parapophysis and to the lateral intercentral suture. The first sacral rib is fused across the articulation between the dorsosacral and the first true sacral. It is particularly massive and directed obliquely backward to participate in the iliac bar.

Seven sacral ribs are borne at the junction between adjacent co-ossified sacral centra. The base of each rib is firmly fused to the ventral surface of the transverse process; their distal ends are much expanded and fuse to form a continuous

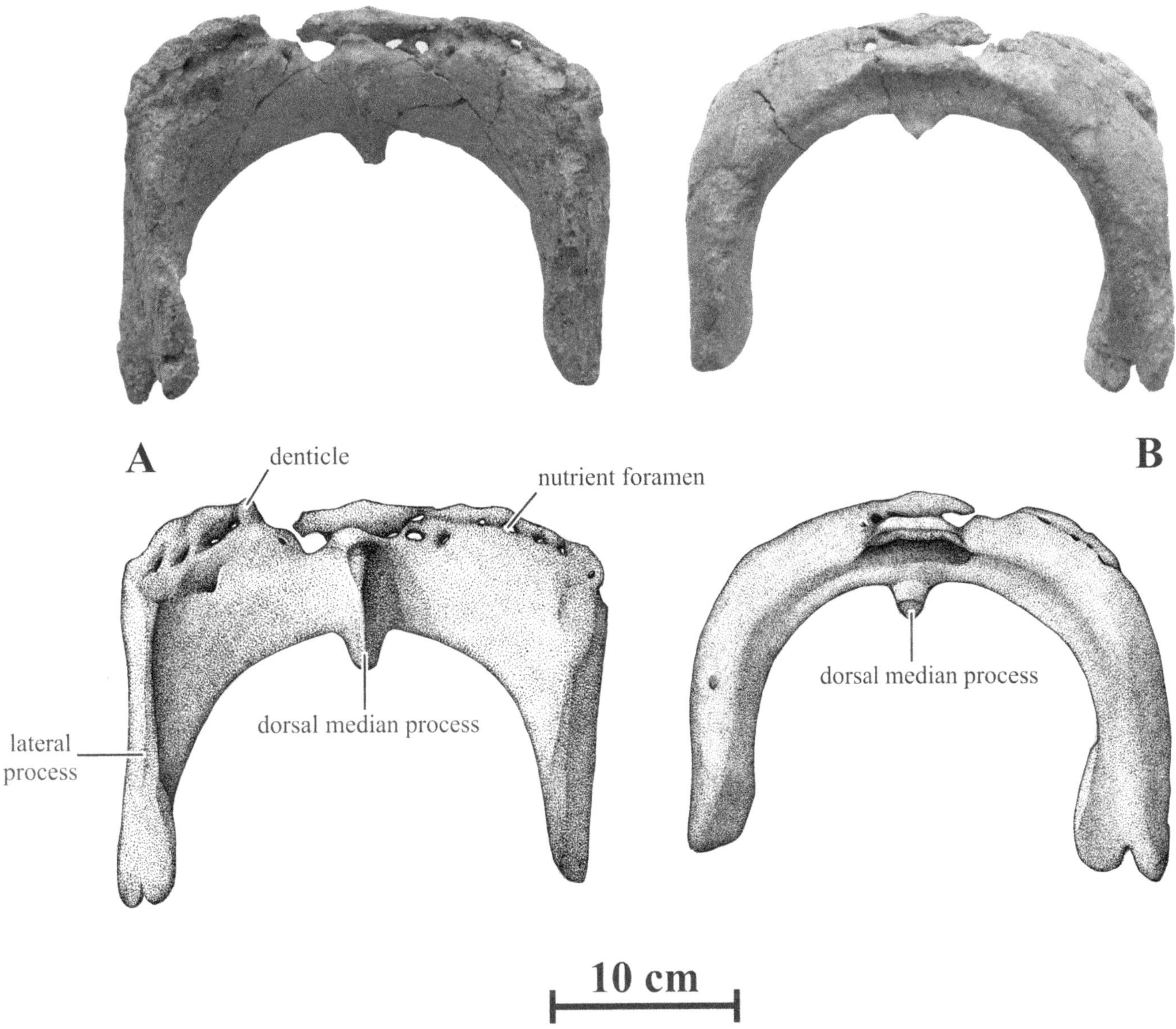

17.12. Predentary (AEHM 1/1682) tentatively referred to as *Amurosaurus riabinini*, from the Upper Cretaceous Blagoveschensk locality, in dorsal (A) and ventral (B) views.

iliac bar, beginning at the level of the junction between the first and the second vertebrae and ending at the level of the junction between the seventh and the eighth vertebrae. The iliac bar gradually diverges from the median axis of the sacrum caudally. Its lateral surface is concave and faces ventrally, following the shape of the medial surface of the ilium.

PHYLOGENETIC POSITION OF *AMUROSAURUS RIABININI*

Characters observed on the premaxilla, predentary, and sacrum of *Amurosaurus riabinini* have been added to the data matrix of the phylogenetic analysis of Lambeosaurinae recently published by Godefroit et al. (2012; emended character data for *Amurosaurus riabinini* in Appendix 17.1). Phylogenetic analysis of the modified data matrix using the same methods as those outlined in Godefroit et al. (2012) does not alter the results. The maximum parsimony analysis resulted in a single tree of 175 steps (Fig. 17.14; tree description in Appendix 17.2). The consistency index (CI) is 0.76 and the retention index (RI) is 0.83. This tree is similar with those previously published by Godefroit et al., 2003, 2004, 2008), Evans and Reisz (2007), Pereda-Suberbiola et al. (2009), and Evans (2010): a group of Asian taxa (*Aralosaurus tuberiferus*, *Tsintaosaurus spinorhinus*, *Jaxartosaurus aralensis*, and *Amurosaurus riabinini*) and European taxa (*Pararhabdodon*

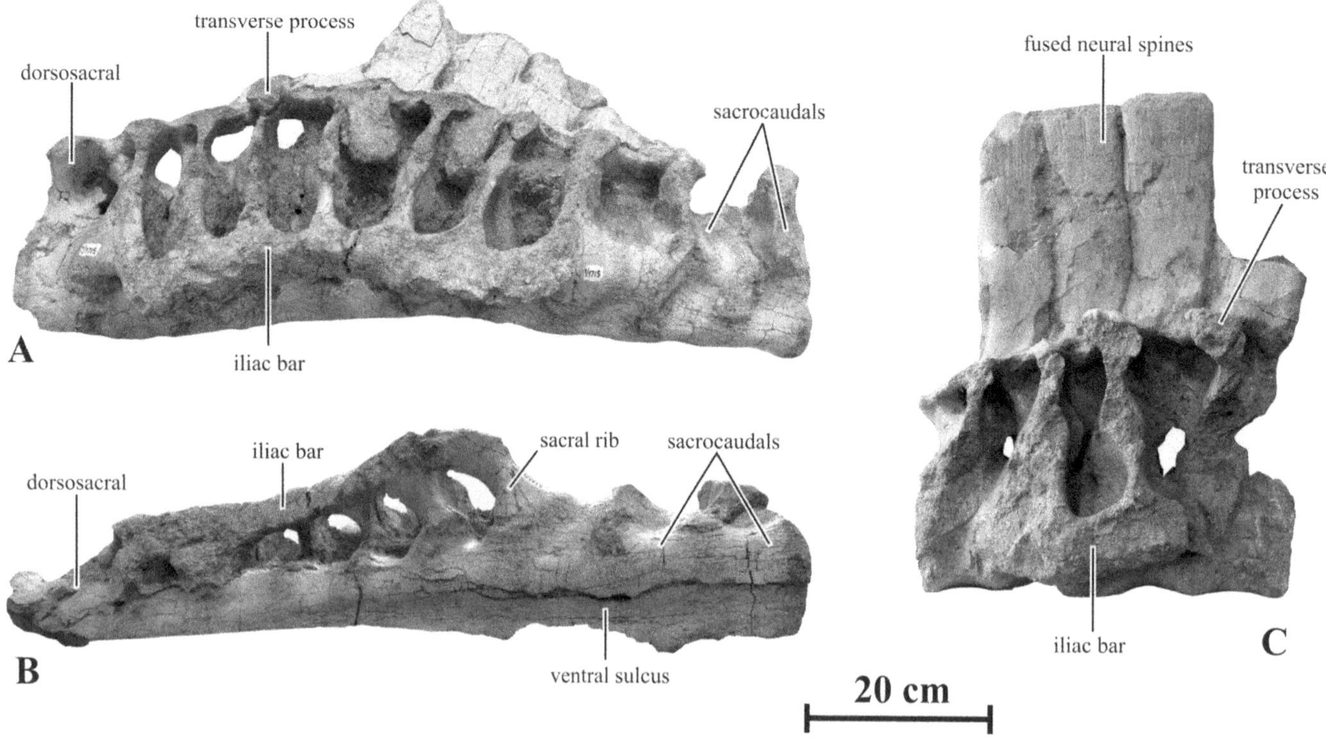

17.13. Sacra tentatively referred to as *Amurosaurus riabinini,* from the Upper Cretaceous Blagoveschensk locality. A–B, AEHM 1/1715 in left lateral (A) and ventral (B) views; C, AEHM 1/1716 in left lateral view.

isonensis and *Arenysaurus ardevoli*) form successive outgroups to a larger clade formed by Parasaurolophini and Corythosaurini. The Asian *Tsintaosaurus spinirhinus* and the European *Pararhabdodon isonensis* are regarded as sister taxa, as previously hypothesized by Prieto-Márquez and Wagner (2009) and Prieto-Márquez (2010). *Olorotitan arharensis* is placed within Corythosaurini, a quite robust clade in this analysis (bootstrap value of 96%) and not as a basal lambeosaurine, as hypothesized by Prieto-Márquez (2010).

Prieto-Márquez (2010) considered the following two synapomorphies as lacking in *O. arharensis,* but present in all other lambeosaurines except *Tsintaosaurus spinorhinus, Pararhabdodon isonensis, Jaxartosaurus aralensis,* and *Aralosaurus tuberiferus:* rostral apex of the rostral process of the jugal reduced to a blunt convexity; and dorsal margin of the infratemporal fenestra lying below the level of the dorsal margin of the orbit. However, the straight rostral margin of the rostral process of the jugal, as seen in *O. arharensis* (character 28 [2]), is a character shared with *Hypacrosaurus altispinus* (Evans, 2010). And the dorsal margin of the infratemporal fenestra is more likely above the dorsal margin of the orbit in *O. arharensis:* the impression that it is located below the dorsal margin of the orbit, on the left side of the braincase in AEHM 2/845 results from postmortem crushing of the caudal ramus of the postorbital and of the postorbital ramus of the squamosal. On the contrary, the dorsal margin of the infratemporal fenestra is located below the dorsal margin of the orbit in *Amurosaurus riabinini,* supporting its basal position in the lambeosaurine phylogeny.

CONCLUSIONS

The famous Russian explorer F. B. Schmidt was likely the first Westerner to observe and report dinosaur fossils in Asia. During the Great Siberian Expedition in 1859, he found fossil bones close to the Jiayin locality, which has subsequently yielded abundant hadrosaur material. The Blagoveschensk bonebed, the second dinosaur locality reported from the Amur region, was discovered by a schoolboy in 1948, and systematically excavated from the 1980s onward. New material recently discovered at the Blagoveschensk locality confirms that the lambeosaurine *Amurosaurus riabinini* rarely reached large sizes (9 m long; estimated weight of 4,000 kg) or attained skeletal maturity. The narial passage is not completely enclosed by the caudodorsal and caudolateral processes of the premaxilla, as in adult specimens of more advanced lambeosaurines from North America. Instead, it was likely partially covered by a skin membrane. The dorsal profile of the premaxilla is shallowly concave in lateral view. These characters were included in a phylogenetic analysis

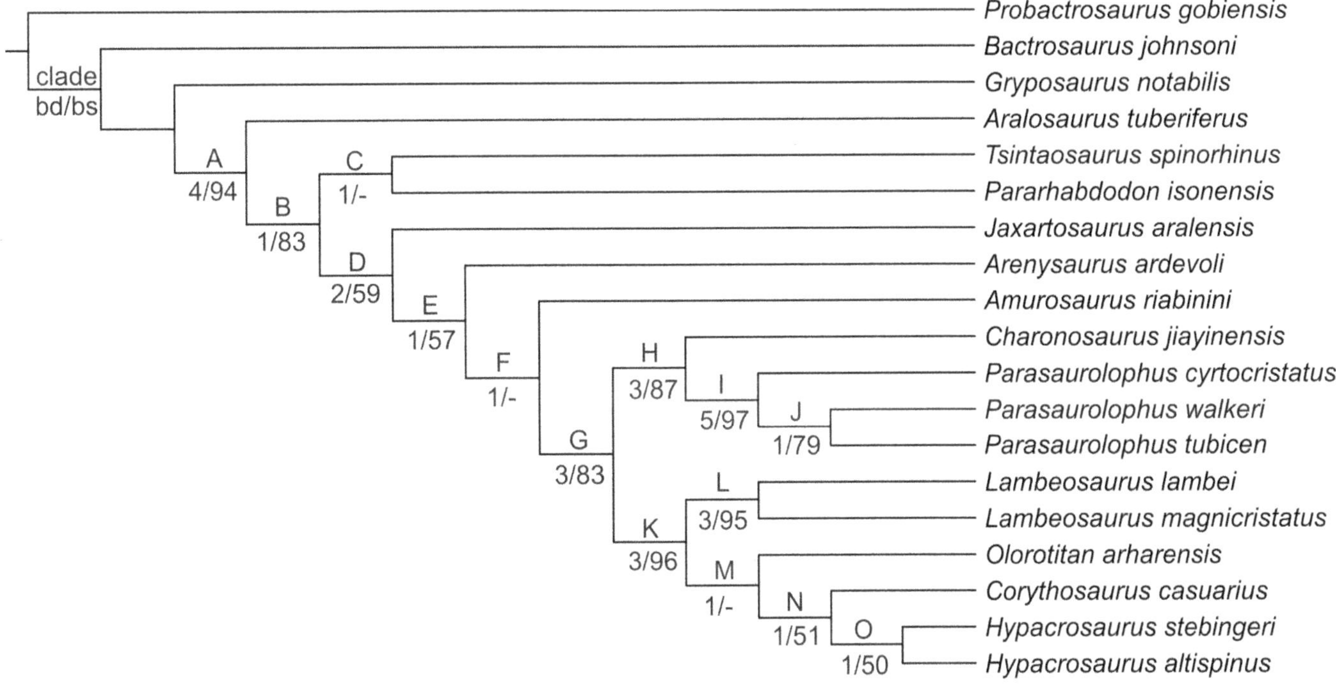

17.14. Most parsimonious tree recovered from the phylogenetic analysis of Lambeosaurinae. Tree length = 175 stps, CI = 0.76; RI = 0.83. Character list modified from Evans and Reisz (2007); see Appendix 17.1 for a list of modified and supplementary characters and Appendix 17.2 for the tree description.

recently published by Godefroit et al. (2012) and confirm that *Amurosaurus* occupies a relatively basal position within lambeosaurine phylogeny. The low neural spines of the sacral vertebrae may be regarded as an autapomorphic character of *Amurosaurus* among lambeosaurines.

ACKNOWLEDGMENTS

We warmly thank D. A. Eberth and D. C. Evans, who kindly asked us to participate in this volume; Elena I. Pastuhova (director of G. S. Novikov-Daursky Amur Regional Study Museum, Blagoveschensk), who made Gurov's map available for this study; I. Esmatsans, who found Manakin's paper; Kirstin Brink, Pascaline Lauters, and A. Prieto-Márquez for fruitful discussions; Dina Rogatnyh for drawing Figure 17.12; and H. De Potter for drawing Figure 17.2. Excavation, preparation, and study of the fossils from the Blagoveschensk locality were supported by grants from the Federal Scientific Policy of the Belgian state (S&T bilateral cooperation program BL/36/C22-R12), from the Russian Foundation of Basic Research (07-05-00168) and from the Far Eastern Branch of Russian Academy of Sciences (06-III-A-06-477). The excavations at Blagoveschensk were also supported, from a logistic point of view, by the gold-mining company Pakrovskiy Rudnik. D. A. Eberth and K. Brink made constructive and appreciated suggestions for improvement of the chapter.

LITERATURE CITED

Alifanov, V. R., and Y. L. Bolotsky. 2002. New data about the assemblages of the Upper Cretaceous carnivorous dinosaurs (Theropoda) from the Amur Region; pp. 25–26 in G. L. Kirillova (ed.), Cretaceous Continental Margin of East Asia: Stratigraphy, Sedimentation, and Tectonics. Unesco-IUGS-IGCP, Khabarovsk, Russia.

Alifanov, V. R., and Y. L. Bolotsky. 2010. *Arkharavia heterocoelica* gen. et sp. nov., a new sauropod dinosaur from the Upper Cretaceous of the Far East of Russia. Paleontological Journal 2010:76–83.

Anderson, J. F., A. Hall-Martin, and D. A. Russell. 1985. Long-bone circumference and weight in mammals, birds and dinosaurs. Journal of Zoology A 207:53–61.

Bassin, M. 1983. The Russian Geographical Society, the "Amur Epoch," and the Great Siberian Expedition 1855–1863. Annals of the Association of American Geographers 73:240–256.

Bolotsky, Y. L., and P. Godefroit. 2004. A new hadrosaurine dinosaur from the Late Cretaceous of Far Eastern Russia. Journal of Vertebrate Palaeontology 24:354–368.

Bolotsky, Y. L., and S. Kurzanov. 1991. The hadrosaurs of the Amur Region; pp. 94–103 in V. Moiseyenko (ed.), Geology of the Pacific Ocean Border. Amur KNII, Blagoveschensk, Russia.

Brett-Surman, M. K. 1989. A revision of the Hadrosauridae (Reptilia: Ornithischia) and their evolution during the Campanian and Maastrichtian. Ph.D. dissertation, George Washington University, Washington D.C., 272 pp.

Brink, K. 2009. Cranial ontogeny and evolution of the lambeosaurine dinosaur *Hypacrosaurus stebingeri* (Ornithischia: Hadrosauridae). M.Sc. thesis, University of Calgary, Calgary, Alberta, 202 pp.

Brink, K. S., D. K. Zelenitsky, D. C. Evans, J. R. Horner, and F. Therrien. 2014. Cranial morphology and variation in *Hypacrosaurus stebingeri* (Ornthischia: Hadrosauridae); chapter 14 in D.

A. Eberth, and D. C. Evans (eds.), Hadrosaurs. Indiana University Press, Bloomington, Indiana.

Brink, K. S., D. K. Zelenitsky, D. C. Evans, F. Therrien, and J. R. Horner. 2011. A sub-adult skull of *Hypacrosaurus stebingeri* (Ornithischia: Lambeosaurinae): anatomy and comparison. Historical Biology 23:63–72.

Campione, N. E., and D. C. Evans. 2011. Cranial growth in edmontosaurs (Dinosauria: Hadrosauridae): implications for latest Cretaceous megaherbivores diversity in North America. PLoS ONE 6(9):e25186.

Danilov, I. G., Y. L. Bolotsky, A. O. Averianov, and I. V. Donchenko. 2002. A new genus of lindholmemydid turtle (Testudines, Testudinoidea) from the Late Cretaceous of the Amur River Region, Russia. Russian Journal of Herpetology 9:155–168.

Dodson, P. 1975. Taxonomic implications of relative growth in lambeosaurine hadrosaurids. Systematic Zoology 24:37–54.

Evans, D. C. 2010. Cranial anatomy and systematic of *Hypacrosaurus altispinus,* and a comparative analysis of skull growth in lambeosaurine hadrosaurids (Dinosauria: Ornithischia). Zoological Journal of the Linnean Society 159:398–434.

Evans, D. C., and R. R. Reisz. 2007. Anatomy and relationships of *Lambeosaurus magnicristatus,* a crested hadrosaurid dinosaur (Ornithischia) from the Dinosaur Park Formation, Alberta. Journal of Vertebrate Paleontology 27:373–393.

Evans, D. C., C. A. Forster, and R. R. Reisz. 2005. The type specimen of *Tetragonosaurus erectofrons* (Ornithischia: Hadrosauridae) and the identification of juvenile lambeosaurines; pp. 349–366 in P. J. Currie and E. B. Koppelhus (eds.), Dinosaur Provincial Park: A Spectacular Ancient Ecosystem Revealed. Indiana University Press, Bloomington, Indiana.

Gibson, J. R. 1968. Russia on the Pacific: the role of the Amur. Canadian Geographer 12:15–27.

Godefroit P., Y. L. Bolotsky, and V. R. Alifanov. 2003. A remarkable hollow-crested hadrosaur from Russia: an Asian origin for lambeosaurines. Comptes rendus Palévol 2:143–151.

Godefroit, P., Y. L. Bolotsky, and I. Y. Bolotsky. 2012. Osteology and relationships of *Olorotitan arharensis,* a hollow-crested hadrosaurid dinosaur from the latest Cretaceous of Far Eastern Russia. Acta Palaeontologica Polonica 57:527–560.

Godefroit P., Y. L. Bolotsky, and J. Van Itterbeeck. 2004. *Amurosaurus riabinini,* a Late Cretaceous lambeosaurine dinosaur from Far Eastern Russia. Acta Palaeontologica Polonica 49:585–618.

Godefroit, P., S. Zan, and L. Jin. 2000. *Charonosaurus jiayinensis* n.g., n.sp., a lambeosaurine dinosaur from the Late Maastrichtian of northeastern China. Comptes rendus de l'Académie des Sciences de Paris: Sciences de la Terre et des Planètes 330:875–882.

Godefroit, P., S. Zan, and L. Jin. 2001. The Maastrichtian (Late Cretaceous) lambeosaurine dinosaur *Charonosaurus jiayinensis* from north-eastern China. Bulletin de l'Institut royal des Sciences naturelles de Belgique: Sciences de la Terre 71:119–168.

Godefroit, P., S. Hai, T. Yu, and P. Lauters. 2008. New hadrosaurid dinosaurs from the uppermost Cretaceous of northeastern China. Acta Palaeontologica Polonica 53:47–74.

Horner, J. R., and H. N. Woodward. 2011. Riddle of the humongous hadrosaurs: what these giants reveal about dinosaur ontogeny, evolution, and ecology; p. 66 in D. R. Braman, D. A. Eberth, D. C. Evans, and W. Taylor (compilers), International Hadrosaur Symposium Abstract Volume. Royal Tyrrell Museum of Palaeontology, Drumheller, Alberta.

Horner, J. R., D. B. Weishampel, and C. A. Forster. 2004. Hadrosauridae, pp. 438–463 in D. B. Weishampel, P. Dodson, and H. Osmólska (eds.), The Dinosauria, Second Edition. University of California Press, Berkeley, California.

Lauters, P., Y. L. Bolotsky, J. Van Itterbeeck, and P. Godefroit. 2008. Taphonomy and age profile of a latest Cretaceous dinosaur bone bed in Far Eastern Russia. Palaios 23:153–162.

Lull, R. S., and N. E. Wright. 1942. Hadrosaurian Dinosaurs of North America. Geological Society of America Special Papers 40. 242 pp.

Nixon, K. C. 2002. WinClada ver. 1.00.08. Published by the author, Ithaca, New York.

Ostrom, J. H. 1961. Cranial morphology of the hadrosaurian dinosaurs of North America. Bulletin of the American Museum of Natural History 122:33–186.

Osborn, H. F. 1909. The Upper Cretaceous iguanodont dinosaurs. Nature 81:160–162.

Pereda-Suberbiola, X., J. I. Canudo, P. Cruzado-Caballero, J. L. Barco, N. López-Martínez, O. Oms, and J. I. Ruiz-Omeñaca. 2009. The last hadrosaurid dinosaurs of Europe: a new lambeosaurine from the uppermost Cretaceous of Arén (Huesca, Spain). Comptes rendus Palévol 8:559–572.

Prieto-Márquez, A. 2010. Global phylogeny of Hadrosauridae (Dinosauria: Ornithopoda) using parsimony and Bayesian methods. Zoological Journal of the Linnean Society 159:435–502.

Prieto-Márquez, A. 2011. A reappraisal of *Barsboldia sicinskii* (Dinosauria: Hadrosauridae) from the Late Cretaceous of Mongolia. Journal of Paleontology 85:468–477.

Prieto-Márquez, A., and J. A. Wagner. 2009. *Pararhabdodon isonensis* and *Tsintaosaurus spinorhinus:* a new clade of lambeosaurine hadrosaurids from Eurasia. Cretaceous Research 30:1238–1246.

Riabinin, A. N. 1925. A mounted skeleton of the gigantic reptile *Trachodon amurense* nov. sp. Izvestija Geologicheskogo Komiteta 44:1–12.

Riabinin, A. N. 1930a. *Mandschurosaurus amurensis* nov. gen. nov. sp., a hadrosaurian dinosaur from the Upper Cretaceous of Amur River. Russkoe Paleontologicheskoe Obshchestva, Monografy 11:1–36.

Riabinin, A. N. 1930b. On the age and fauna of the dinosaur beds on the Amur River. Zapiski Rossiskogo Mineralogicheskogo Obshchestva 59:41–51.

Rozhdestvensky, A. K. 1957. On the Upper Cretaceous dinosaur localities of the Amur River. Vertebrata PalAsiatica 1:285–291.

Schmidt, F. B. 1866. Historical report on physico-geographical researches by the head of the physical section of the Siberian Expedition Professor F. B. Schmidt and his assistant P. P. Glen. Russian Geographical Society, St. Petersburg.

Van Itterbeeck, J., Y. L. Bolotsky, P. Bultynck, and P. Godefroit. 2005. Stratigraphy, sedimentology and palaeoecology of the dinosaur-bearing Kundur section (Zeya-Bureya Basin, Amur Region, Far Eastern Russia). Geological Magazine 142:735–750.

Appendix 17.1. Amended Character Data for *Amurosaurus riabinini*

Amurosaurus riabinini
11110?1001 ?01??1??11 01?11?1111 0111100111 0011010111 0101011111 11111?0011 11111110?? ??1111000? 10?1?2?011 ?00?101100 10?110?1

Note: Characters are listed in Godefroit et al. (2012).

Appendix 17.2. Tree Description

Node A (Lambeosaurinae): Unambiguous: 19, 22, 46, 99; ACCTRAN: 3, 26, 28, 61, 77, 104, 105, 109(0), 118.
Node B: Unambiguous: 32, 36, 28, 39; ACCTRAN: 20, 43; DELTRAN: 61, 71, 85, 118.
Node C: Unambiguous: 96(3), 105(2), 106; ACCTRAN: 99(2), 102, 114(0); DELTRAN: 77.
Node D: Unambiguous: 7, 40; ACCTRAN: 13, 15, 16, 52, 68, 101(2), 107, 108.
Node E: Unambiguous: 33, ACCTRAN: 42, 47; DELTRAN: 52, 105.
Node F: Unambiguous: 36(0), 44, 49, 72; DELTRAN: 3, 13, 20, 28, 43, 77, 107, 108, 114.
Node G: Unambiguous: 4, 53, 89, 98(2), 103; ACCTRAN: 110; DELTRAN: 26, 47, 68, 101(2).
Node H (Parasaurolophini): 37, 42(2), 85(0), 92, 116; ACCTRAN: 9, 13(4), 16(0), 88, 102; DELTRAN: 113.
Node I (*Parasaurolophus*): Unambiguous: 40(2), 78, 109, 111(0), 112; ACCTRAN, 36, 41, 58(0), 90, 100(2); DELTRAN: 9, 13(4), 88.
Node J: Unambiguous: 12; DELTRAN: 36, 58(0), 100(2).
Node K (Corythosaurini): Unambiguous: 6, 8, 10, 17, 23, 66, 79; ACCTRAN: 5, 87; DELTRAN: 15, 42, 110.
Node L (*Lambeosaurus*): Unambiguous: 13(3), 14, 18, 51, 97(2), 101; ACCTRAN: 104(0), 118(0); DELTRAN: 16.
Node M: Unambiguous: 11, 16(2), 95; DELTRAN: 104.
Node N: Unambiguous: 17(2), 97; DELTRAN: 5, 87.
Node O (*Hypacrosaurus*): Unambiguous: 15(2), 67, 117; ACCTRAN: 90, 114(0).

Note: Character transformations were evaluated under unambiguous, fast (ACCTRAN), and slow (DELTRAN) optimization options in Winclada (Nixon, 2002); unambiguous synapomorphies are those that diagnose a node under both fast and slow optimizations. Characters are listed in Godefroit et al. (2012). Node numbers refer to Fig. 17.14. For simple 0-1 state changes only the character number is given; for other state changes the type of change is specified between brackets.

South American Hadrosaurs: Considerations on Their Diversity

18

Rodolfo A. Coria

ABSTRACT

The record of South American hadrosaurids, which to date is exclusively from Patagonia, is reviewed from a taxonomic perspective with the goal of summarizing their diversity. *Secernosaurus koerneri* is a hadrosaurid more derived than *Telmatosaurus* and has been recently synonymized with "*Kritosaurus*" *australis*. Thus, *Secernosaurus* is diagnosed by one cranial autapomorphy and a combination of appendicular characters. However, taxonomic distinction between the holotype specimen of *Secernosaurus koerneri* and "*Kritosaurus*" *australis* should not be completely ruled out. *Willinakaqe salitralensis*, which included all hadrosaurid material known from the Allen Formation, is represented by an isolated and fragmentary premaxilla (holotype) and other cranial and postcranial elements (paratype). Its taxonomic status is controversial due to imprecise stratigraphical data and evidence of the presence of more than one taxon in the formation. *Lapampasaurus cholinoi* is another hadrosaurid from the Allen Formation distinguished by vertebral and appendicular characters. Additional hadrosaurid materials from both Allen and Coli-Toro formations are distinguished from other Patagonian hadrosaurids by a combination of several features that include robust appendicular skeletons and high and thick caudal neural spines.

INTRODUCTION

The South American hadrosaurid record is, to date, represented only by findings made in Patagonia, Argentina. These specimens are relatively abundant and represent individuals of different ontogenetic stages that were fossilized via varying preservational conditions. Within a stratigraphic framework, all hadrosaurid remains have been collected from Upper Cretaceous (Campanian–Maastrichtian) units extensively exposed in the Patagonian provinces of Chubut (Bajo Barreal or Laguna Palacio formations), Río Negro (Angostura Colorada/Coli-Toro, Los Alamitos, and Allen formations) and La Pampa (Allen Formation) (Casamiquela, 1964; Brett-Surman, 1979; Bonaparte et al., 1984; Powell, 1987a, 1987b; González Riga and Casadío, 2000; Juárez Valieri et al., 2010; Prieto-Márquez and Salinas, 2010; Fig. 18.1). All of these units seem to be correlable, and thus correspond to a related series of depositional events that took place within a short temporal interval. However, although the presence of hadrosauroid remains in these units is relatively abundant, our knowledge about their taxonomic diversity has remained sparse and controversial, mainly due to the notorious scarcity of articulated or well-preserved specimens.

At present, only *Secernosaurus koerneri* (Brett-Surman, 1979), *Willinakaqe salitralensis* (Juárez Valieri et al., 2010; Prieto-Márquez and Salinas, 2010), and *Lapampasaurus cholinoi* (Coria et al., 2012) are formally recognized as South American hadrosaurid species. In this chapter, the South American hadrosaurid record is reviewed, and considerations about a greater taxonomic diversity of this group are presented.

Institutional Abbreviations FMNH, Field Museum of Natural History, Chicago, Illinois; MACN, Museo Argentino de Ciencias Naturales "Bernardino Rivadavia," Buenos Aires, Argentina; MJG, Museo Jorge Gerhold, Ingeniero Jacobacci, Río Negro Province, Argentina; MLP, Museo de La Plata, La Plata, Buenos Aires Province, Argentina; MPCA, Museo Provincial Carlos Ameghino, Cipolletti, Río Negro Province, Argentina; MPHN, Museo Provincial de Historia Natural, Santa Rosa de La Pampa, La Pampa Province, Argentina.

SOUTH AMERICAN HADROSAURIDS

SECERNOSAURUS KOERNERI *Brett-Surman, 1979* (= "Kritosaurus" australis *Bonaparte, Franchi, Powell and Sepulveda, 1984*)

This taxon was first described and named by Brett-Surman (1979) based on fossils collected by J. B. Abbot in 1923 for the Field Museum, Chicago. The fossils were collected probably from the Upper Member of the Bajo Barreal Formation (Campanian–Maastrichtian; see Bonaparte et al., 1984) east of Colhué-Huapi Lake, southern Chubut Province. The holotype specimen (FMNH P13423) is fragmentary and represented by several elements that include a basicranium, three cervical centra, three anterior dorsal centra, one incomplete mid-dorsal neural arch, one posterior dorsal neural arch, two incomplete posterior dorsal vertebrae, several incomplete caudal neural spines, an incomplete right scapula, a right distal humeral end, a complete right ilium and partial left ilium, an incomplete right pubis, both ischiadic shafts and the distal part of a ?left femur (see Prieto-Márquez and Salinas, 2010).

Secernosaurus was originally proposed as a basal hadrosaurid closely related to *Telmatosaurus* (Brett-Surman, 1979; Weishampel et al., 1993). However, the ilium (Fig. 18.2A) possesses a sigmoidal dorsal border that is depressed over the supracetabular process and dorsally expanded at the base of the preacetabular process, a large preacetabular process longer than the postacetabular process, a squared and long postacetabular process without brevis shelf, an expanded prepubic process, and an open obturator foramen (Fig. 18.2B). These features link *Secernosaurus* with hadrosaurids more derived than *Telmatosaurus* (e.g., Horner et al., 2004; Coria, 2010; Prieto-Márquez and Salinas, 2010).

Recently, Prieto-Márquez and Salinas (2010) synonymized *Secernosaurus koerneri* and *"Kritosaurus" australis*, with the former taking taxonomic precedence. The latter taxon represents a large hadrosaurid (~8 meters long) that is based on materials from several individuals collected from the Los Alamitos Formation of southeast Río Negro Province. The material was described by Bonaparte et al. (1984) and Bonaparte and Rougier (1987) and is, by far, the best-represented and informative collection of hadrosaurid material known from South America. The material includes skull elements, lower jaws, fairly complete vertebral series, scapular and pelvic girdles, and appendicular elements from the fore- and hindlimbs (for a review see Prieto-Márquez and Salinas, 2010).

Although similarities with *Kritosaurus* from North America led Bonaparte et al. (1984) and Bonaparte and Rougier (1987) to assign the Patagonian species to that genus, that relationship is based on ambiguous characters or features that seem to be widely distributed among hadrosaurids. Horner et al. (2004) assigned the South American form to *"Kritosaurus" australis* based on the presence of a pointed and symmetrically triangular anterior process of the jugal (shared with *Saurolophus* and *Naashoibitosaurus*), and the presence of low zygapophyseal peduncles on the cervical vertebrae (shared with *Saurolophus*). Subsequently, Prieto-Márquez and Salinas (2010) recognized several apomorphies shared by the clade formed by *Kritosaurus-Gryposaurus* and the specimens from Los Alamitos, but differentiated the Patagonian taxon from the genus *Kritosaurus*. These authors also suggested that *"Kritosaurus" australis* is a junior synonym of *Secernosaurus koerneri* (Brett-Surman, 1979) on the basis of an autapomorphy and a unique combination of appendicular characters (see Prieto-Márquez and Salinas [2010] for a detailed revised diagnosis of *Secernosaurus koerneri*).

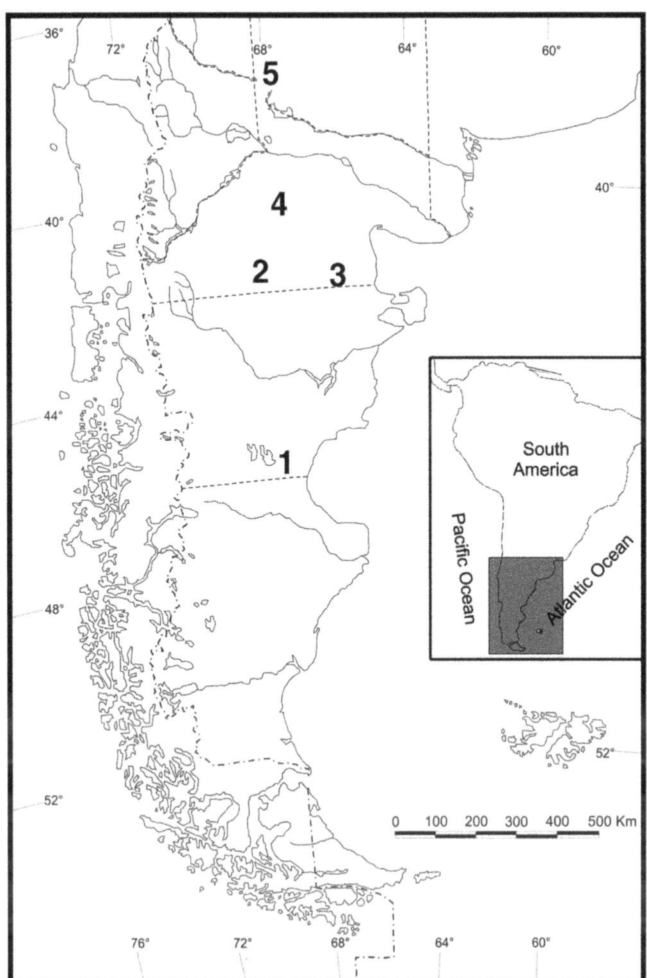

18.1. Map with the hadrosaurid bearing localities in Patagonia. (1) *Secernosaurus koerneri* (FMNH P13423); (2) Hadrosauridae indet.; (3) *Secernosaurus koerneri* (= *"Kritosaurus" australis*); (4) *Willinakaqe salitralensis;* (5) *Lapampasaurus cholinoi.*

Comments The diagnosis provided by Prieto-Márquez and Salinas (2010) for *Secernosaurus koerneri* involves one cranial character (V-shaped nasofrontal suture in which the anterior margin of each frontal forms a median triangular process that is laterally deep, anteriorly excavated, and has an anteriorly offset ventral margin) and a combination of characters present in the ilium and the pubis. Despite the skull elements known from FMNH P13423, which include an isolated basicranium and maxillary fragments (pers. obs.), only the ilium and pubis are complete enough to be compared with the specimens formerly assigned to *"Kritosaurus" australis* (Fig. 18.2).

Some morphological differences between the ilia of FMNH P13423 and those from the specimens from Los Alamitos were interpreted by Prieto-Márquez and Salinas (2010) as the

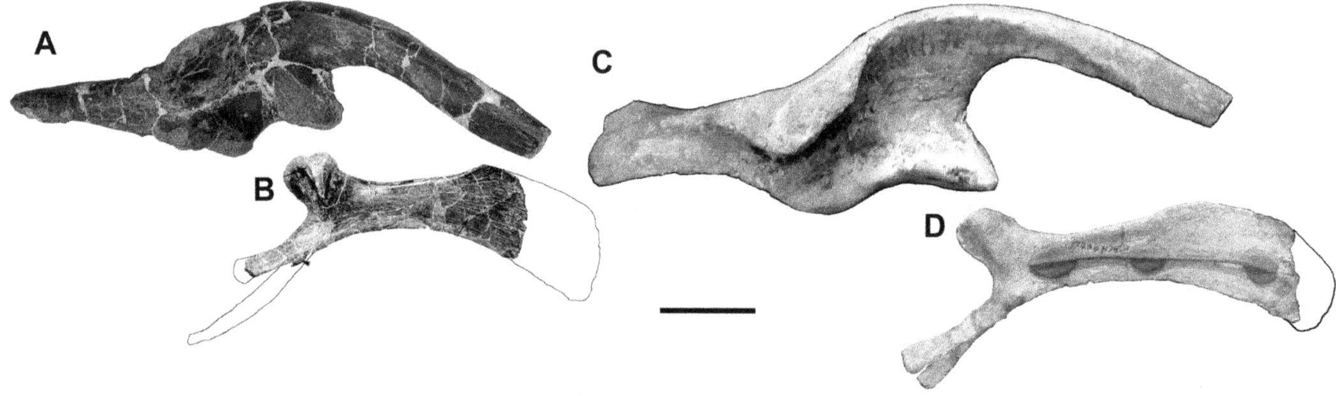

18.2. Ilium and pubis of *Secernosaurus koerneri.* (A) FMNH P13423, right ilium; (B) FMNH P13423, right pubis; (C) MACN-RN-02, right ilium; (D) MACN-RN-02, right pubis. All figures in lateral view. Scale bar equals 10 cm.

result of postdepositional deformation on the holotype specimen of *Secernosaurus koerneri,* and thus, were not regarded as significant. However, no other evidence of deformational processes is preserved or described for other bones of FMNH P13423 (pers. obs.), suggesting instead that these morphological differences may indeed be significant. For example, the left ilium of the specimen FMNH P13423, which lacks the postacetabular process, shows a narrower acetabulum with a more robust pubic peduncle (Prieto-Márquez and Salinas, 2010:fig. 15A, B), and the pubis possesses an iliac peduncle that is more robust and projects slightly more dorsally in the specimens from Los Alamitos (Prieto-Márquez and Salinas, 2010:fig. 16B, C; Fig. 18.2B–D). These same differences are present when comparing the right ilia (Fig. 18.2A–C).

Furthermore, putative ontogenetic differences between the type (proposed as a subadult) and the Los Alamitos material (adult) may be unreliable. The subadult condition of the type specimen of *Secernosaurus* was claimed based on the relatively small size of the individual (~4–5 m), and the fact that the basisphenoid-basioccipital elements are disarticulated and the neural arches are not attached to the vertebral centra (Brett-Surman, 1989; Prieto-Márquez and Salinas, 2010). However, the unfused condition of the neural arches reported by these workers is instead the result of broken neural arch bases at the level of their peduncles (Fig. 18.3). Therefore, the ontogenetic stage of the type specimen of *Secernosaurus koerneri* is likely more mature than previously thought. In this context, the type specimen may indeed represent a relatively small form. Consequently, possible morphological differences presented here between the type specimen (FMNH P13423) of *Secernosaurus koerneri* and the specimens from Los Alamitos (including both anatomical- and size-related features), provide sufficient evidence to warrant a reevaluation of the holotype specimen of *Secernosaurus koerneri* accompanied by a more thorough comparison with the Los Alamitos material. It is proposed here that such an analysis will likely result in the recognition of at least two closely related but distinct taxa of hadrosaurids.

The presence of *Secernosaurus* in the Late Cretaceous of Patagonia represents the unequivocal southernmost record of hadrosaurids in South America. The discovery of new materials of this taxon or a closely related form (Los Alamitos material) will permit a better understanding of its phylogenetic relationship to other members of Hadrosauridae, and the formulation of new hypotheses about the distribution and dispersal routes of members of this group in the high latitudes of South America.

WILLINAKAQE SALITRALENSIS *Juárez Valieri, Haro, Fiorelli, and Calvo, 2010*

This recently named taxon was proposed to include all hadrosaurid specimens collected from the Allen Formation (Campanian–Maastrichtian), a unit widely exposed in the Patagonian provinces of Río Negro, Neuquén, and La Pampa (Juárez Valieri et al., 2010). The diagnosis is based on a single and incomplete premaxilla (holotype; Fig. 18.4A, A′) and several cranial and postcranial elements (paratypes). These fossils were recovered during several fieldtrips conducted over the last decade of the twentieth century by associates of the Carlos Ameghino Provincial Museum of Cipolletti, Río Negro. They were recovered from the classic Allen Formation localities at Salitral Moreno and Salitral Ojo de Agua. In these areas, hadrosaurid elements frequently occur as isolated and transported bones, or small assemblages of bones, which are associated often with titanosaur remains or, less frequently, with theropod and ankylosaur material.

Comments The claimed taxonomic relatedness of all these fossil materials, collected from numerous localities and often separated by many kilometers, is questionable because

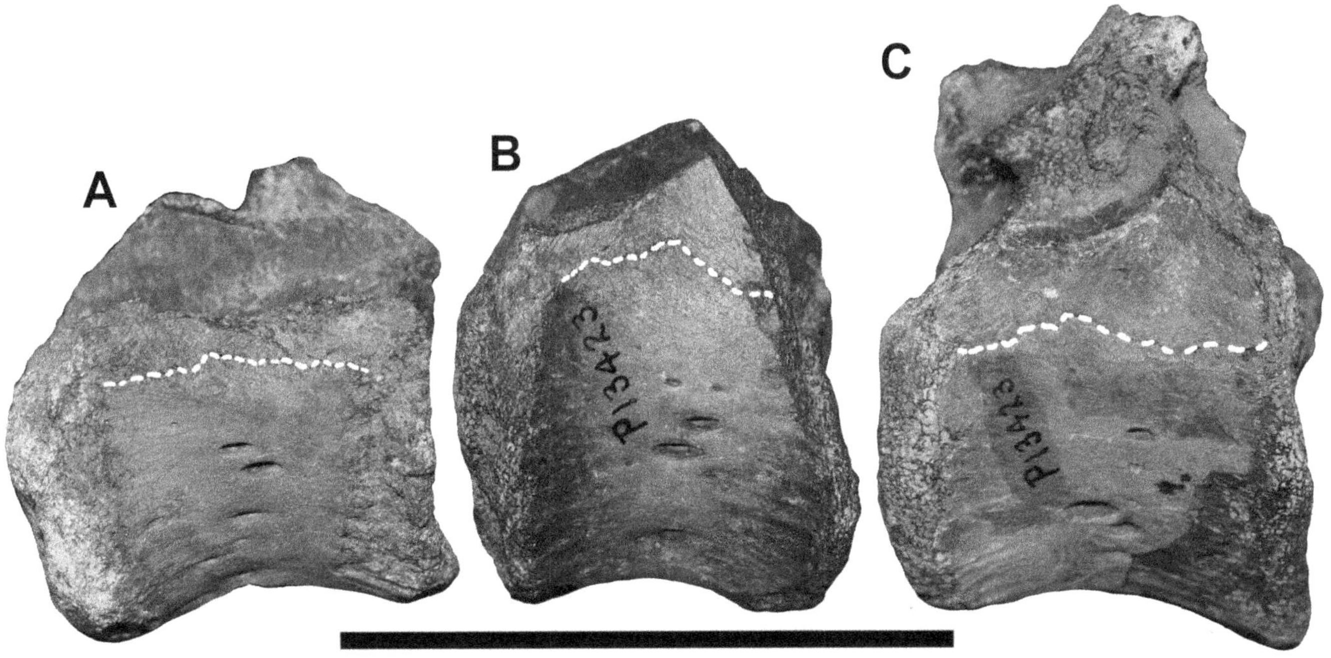

18.3. *Secernosaurus koerneri*, FMNH P13423. (A), (B) and (C) dorsal centra in right lateral view. Discontinuous white line demarcates the suture between neural arches and centra. Scale bar equals 10 cm.

they were not found in direct association. Only one specimen from Salitral Moreno, MPCA-SM-2, was found articulated and almost complete (Powell, 1987a). MPCA-SM-2 was described in brief by Powell (1987b) who assigned it to Lambeosaurinae (Fig. 18.4 D, E). However, several of the characters cited for this specimen do not allow it to be placed unambiguously in either subfamily of Hadrosauridae. For example, an asymmetrical scapular blade is found in both *Gryposaurus* and *Corythosaurus*, whereas high and robust caudal neural spines, more typical of lambeosaurines such as *Lambeosaurus* and *Corythosaurus* (Horner et al., 2004), occur also in the hadrosaurine *Barsboldia* (Prieto-Marquez, 2011).

In addition, among the collection of hadrosaurid materials housed in the Carlos Ameghino Museum, two different humeral morphologies can be identified, both collected from the Allen Formation (Fig. 18.4B, C). One type of humerus (MPCA-SM-33; Fig. 18.4B) is rather robust, with a low deltopectoral crest that typifies many hadrosaurines. In contrast, another specimen (MPCA-SM-4; Fig. 18.4C) exhibits a well-developed deltopectoral crest that is more laterally expanded distally and, thus, reminiscent of the condition exhibited by some lambeosaurines.

The putative autapomorphies present in the holotype premaxilla of *Willinakaqe* may eventually be shown to be sufficient to validate this form as a distinct genus and species. However, ascribing all other hadrosaurid material from Allen Formation to *Willinakaqe salitralensis* is problematic because the available material suggests the presence of at least two taxa. Lastly, material from the Allen Formation, collected from the northern province of La Pampa and assigned by Juárez Valieri et al. (2010) to *Willinakaqe*, have been identified recently as a different taxon (Coria et al., 2012; see below).

LAPAMPASAURUS CHOLINOI *Coria, González Riga, and Casadío, 2012*

In 2000, González Riga and Casadío reported a specimen with hadrosaurid affinities collected from strata of the Allen Formation exposed in the locality of Islas Malvinas, southwest of La Pampa Province (González Riga and Casadío, 2000). These materials, housed in the Provincial Museum of Natural History of La Pampa (MPHN-PV-01), correspond to vertebral and appendicular elements of a subadult hadrosauroid individual. Coria et al. (2012) recently identified this specimen as Hadrosauridae *incertae sedis* and named it as *Lapampasaurus cholinoi* based on the presence of anterior cervical vertebrae with lateral foramina on the dorsal side of the diapophyses, a scapula with laterally sharp deltoid crest and convex dorsal surface, a pedal ungual phalanx that is longer than wide with superficial grooves and foramina, and a shallow, longitudinal ventral ridge (Fig. 18.5).

Comments Juarez Valieri et al. (2010) had referred MPHN-PV-01 to *Willinakaqe salitralensis*. In that paper, these authors recognized the presence of cavities on the lateral side of the neural arch in a dorsal vertebra, which they included

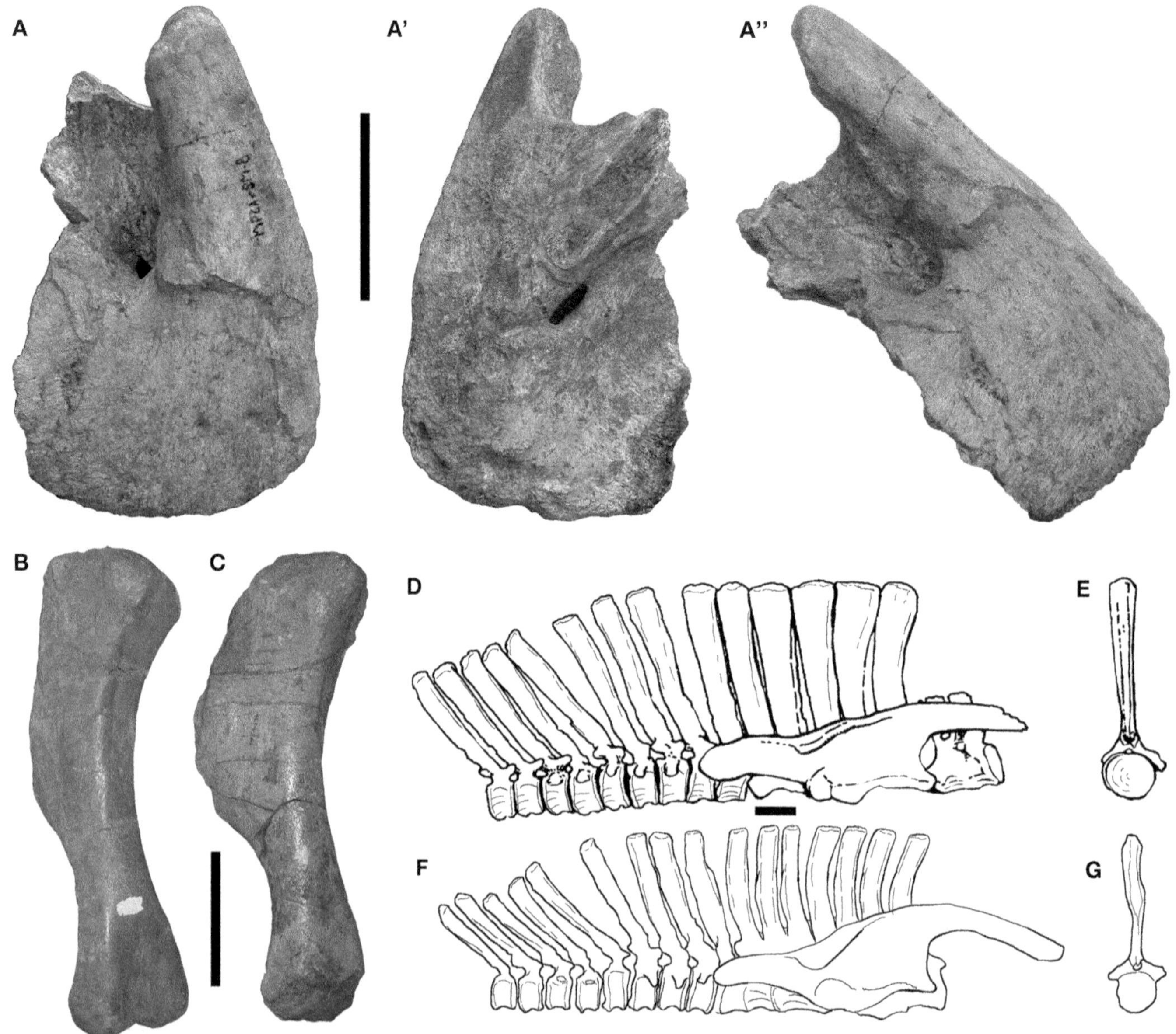

18.4. Materials assigned to *Willinakaqe salitralensis.* (A–A") MPCA Pv SM-8, holotype premaxilla in dorsolateral, ventromedial, and lateral views, respectively; (B) MPCA Pv SM-33, left humerus in posterior view; (C) MPCA Pv SM-4, left humerus in posterior view; (D) MPCA Pv SM-789, sacrum, ilium, and proximal caudals in lateral view; (E) MPCA Pv SM-789, anterior view of first caudal vertebrae; (F) *Secernosaurus koerneri* (MACN-RN-02), sacrum, ilium, and proximal caudals in lateral view; (G) *Secernosaurus koerneri* (MACN-RN-02), anterior view of first caudal vertebra in anterior view. Scale bar in (A–A") equals 5 cm; scale bar in (B–G) equals 10 cm.

as an autapomorphy of *Willinakaqe.* The illustration provided of that trait (Juárez Valieri et al., 2010:fig. 5H) does not correspond to MPHN-PV-01 nor shows such a feature. Possibly, this could be a misinterpretation of the cervical vertebrae MPHN-PV-01.30 that bears one of the autapomorphies of *Lapampasaurus* (sensu Coria et al., 2012). Regarding the cavities of that element of *Lapampasaurus*, which were described as pleurocoels by Coria et al. (2012), they probably should be referred to as "openings" because the term "pleurocoel" alludes to the existence of a postcranial skeletal pneumaticity system or apparatus, which has not yet been documented in hadrosaurids (Butler et al., 2012). Other diagnostic features of *Lapampasaurus cholinoi* (see Coria et al., 2012) support its recognition as a valid taxon.

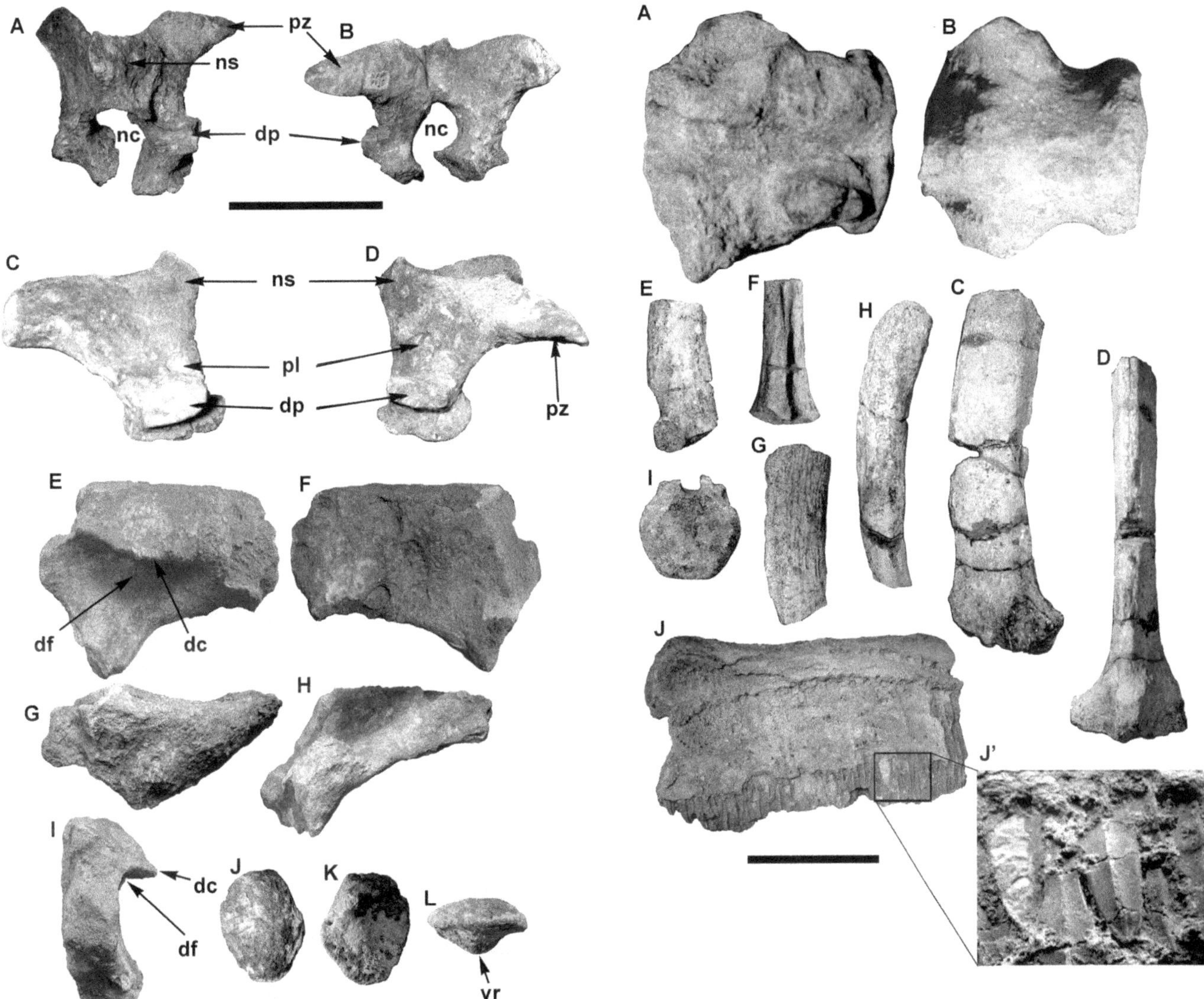

18.5. *Lapampasaurus cholinoi,* MPHN-Pv-01. Anterior cervical neural arch in dorsal (A), posterior (B), right lateral (C), and left lateral (D) views. Proximal end of left scapula in lateral (E), medial (F), dorsal (G), ventral (H), and proximal (I) views. Ungual phalanx in dorsal (J), ventral (K), and distal (L) views. Abbreviations: dc, deltoid crest; df, deltoid fossa; dp, diapophysis; nc, neural canal; ns, neural spine; pl, pleurocoel; pz, postzygapophysis; vr, ventral ridge. Scale bar equals 5 cm in (A–D); scale bar equals 10 cm in (E–L).

18.6. Hadrosauridae indet. from the Coli-Toro Formation. MLP-62-XII-13-1/2, caudal centrum in dorsal (A) and ventral (B) views; MLP-62-XII-13-1/4, caudal neural spine in lateral (C) and anterior (D) views; MLP-62-XII-13-1/5, caudal neural spine in lateral (E) and anterior (F) views; MLP-62-XII-13-1/6, caudal neural spine in lateral view (G); MLP-62-XII-13-1/8, caudal neural spine in lateral view (H); MLP-62-XII-13-1/10, caudal centrum in anterior view (I); MJG.Pa 6/6/07-3, incomplete maxilla in medial view (J) and close-up of the dentition (J'). Scale bar equals 10 cm.

HADROSAURIDAE INDET., Coli-Toro Formation (Casamiquela, 1964)

Hadrosauroid materials were collected by R. M. Casamiquela during the 1950s and 1960s. Apparently, all the specimens come from the Coli-Toro Formation that crops out at the locality of Cerro Mesa, located north of Ingeniero Jacobacci, Río Negro Province. The specimens were stored at the La Plata Museum and Jorge Gerhold Museum in Ingeniero Jacobacci, Río Negro Province. The collections from this formation include isolated and fragmentary bones, which belong to several adult and juvenile individuals (Fig. 18.6). These fossils allowed Casamiquela (1964) to report, for the first time, the presence of hadrosaurid dinosaurs in South America, and to propose that such a record was the consequence of a dispersal event of North American hadrosaurids into this continent at the end of the Cretaceous. Recently, this hypothesis has been supported by other authors (e.g., Coria, 2010; Prieto-Márquez, 2010).

Comments The morphology exhibited by all the material collected by R. M. Casamiquela from the Coli-Toro Formation is, in general terms, clearly hadrosaurid. The specimens MJG-PA 6/6/07-3 (Fig. 18.6J, J′) and four maxillary fragments show a well developed and complex dental battery composed of elongated, symmetrical teeth with a single lateral carina. Also, the tall, anterodorsally curved caudal neural spines, the proximal humeral end, the tibia, and the metatarsal distal ends of MLP-XII-13-1 are typically hadrosaurid. The same conclusions can be obtained from observation of the appendicular elements housed in the Jorge Gerhold Museum (MJG-PA 6/6/07-6 and 7, MJG-PA 7/6/07-1–6). These specimens exhibit characters present in all hadrosaurids–including a humeral head that is well separated from the lateral and medial tuberosities; and a femur with a reduced lesser trochanter that is separated from the greater trochanter, a deep anterior intercondylar groove, and an asymmetrical distal tibial end in anterior view (Horner et al., 2004). The presence of a ventral keel in the sacral vertebrae and a very robust humerus are similar to the conditions present in the hadrosaurid from Salitral Moreno (Powell, 1987b), but are also present in other hadrosaurids.

CONCLUSIONS

Based on this review, the diversity of South American hadrosaurids is represented by at least three formally named taxa: (1) *Secernosaurus koerneri*, represented by materials from the Upper Member of the Bajo Barreal Formation from Chubut Province, and Los Alamitos Formation from Río Negro Province; (2) *Willinakaqe salitralensis*, represented by an isolated and incomplete premaxilla from the Allen Formation of Río Negro Province; and (3) *Lapampasaurus cholinoi*, collected from strata of the Allen Formation of La Pampa Province.

However, a reevaluation of the specimens of *Secernosaurus koerneri* from Los Alamitos Formation (formerly known as "*Kritosaurus*" *australis*) could cause reconsideration of its current taxonomic status. Also, materials previously identified as Lambeosaurinae indet. from the Allen Formation (Powell, 1987a, 1987b) and the Coli-Toro/Angostura Colorada formations in the Río Negro province (Casamiquela, 1964; Coria, 2010) are preliminarily distinguished from the currently recognized forms from Argentina on the basis of the combination of several features, including a robust appendicular skeleton, and long and robust caudal neural spines.

The diversity of South American hadrosaurids is a relevant chapter in the evolution of terrestrial macrovertebrate faunas at the end of the Cretaceous. Currently, there is strong evidence for at least four hadrosaurid species in the Campanian–Maastrichtian deposits of Patagonia. Knowledge of the South American hadrosaurid record will be enhanced by the discovery of more complete specimens and detailed revisions of the specimens currently collected.

ACKNOWLEDGMENTS

I thank M. Becerra (MPHN, La Pampa), A. Kramarz (MACN, Buenos Aires), J. C. Muñoz (MPCA, Río Negro), P. Makovicky (FMNH) and M. Reguero (MLP, La Plata) for allowing access to specimens under their care. J. Peterson (University of Wisconsin, Oshkosh) kindly provided photographs for Figure 18.3. A. Arcucci and L. Salgado provided valuable comments on an early stage of this research. The manuscript was significantly improved by the comments from an anonymous reviewer and D. Evans. Research supported by CONICET, PIP-0233. The author also thanks the organizers of the Hadrosaur Symposium and editors of the present volume D. Evans and D. Eberth, and the Dinosaur Research Institute for the funding permitting the author's attendance at that meeting.

LITERATURE CITED

Bonaparte, J. F., and G. Rougier. 1987. The Late Cretaceous fauna of Los Alamitos, Patagonia, Argentina, part VII: the hadrosaurs. Revista del Museo Argentino de Ciencias Naturales, Paleontología 3:155–161.

Bonaparte, J. F., M. R. Franchi, J. E. Powell, and E. C. Sepulveda. 1984. La Formación Los Alamitos (Campaniano-Maastrichtiano) del sudoeste de Río Negro, con descripción de *Kritosaurus australis* nov.sp. (Hadrosauridae). Significación paleobiogeográfica de los vertebrados. Revista de la Asociación Geológica Argentina 39:284–299.

Brett-Surman, M. K. 1979. Phylogeny and paleobiogeography of hadrosaurian dinosaurs. Nature 277:560–562.

Brett-Surman, M. K. 1989. A revision of the Hadrosauridae (Dinosauria: Ornithischia) and their evolution during the Campanian and Maastrichtian. Ph.D. dissertation, George Washington University, Washington D.C., 272 pp.

Butler, R. J., P. M. Barrett, and D. J. Gower. 2012. Reassessment of the evidence for postcranial skeletal pneumaticity in Triassic archosaurs, and the early evolution of the avian respiratory system. PLoS ONE 7(3):e34094.

Casamiquela, R. M. 1964. Sobre un dinosaurio hadrosaurio de la Argentina. Ameghiniana 3:285–308.

Coria, R. A. 2010. Phylogeny and paleobiogeography of hadrosaurid dinosaurs from Argentina. Abstracts from the X Congreso Argentino de Paleontología y Bioestratigrafía VII Congreso Latinoamericano de Paleontología, La Plata, Argentina, September 2010:149–150.

Coria, R. A., B. González Riga, and S. Casadío. 2012. Un nuevo hadrosáurido (Dinosauria, Ornithopoda) de la Formación Allen, Provincia de La Pampa. Ameghiniana 49:552–572.

González Riga, B., and S. Casadío. 2000. Primer registro de Dinosauria (Ornithischia, Hadrosauridae) en la Provincia de La Pampa (Argentina) y sus implicancias paleobiogeográficas. Ameghiniana 37:341–351.

Horner, J. R., D. B. Weishampel, and C. A. Forster. 2004. Hadrosauridae; pp. 438–463 in D. B. Weishampel, P. Dodson, and H. Osmólska (eds.), The Dinosauria, Second Edition. University of California Press, Berkeley, California.

Juárez Valieri, R. D., J. A. Haro, L. E. Fiorelli, and J. O. Calvo. 2010. A new hadrosauroid (Dinosauria: Ornithopoda) from the Allen Formation (Late Cretaceous) of Patagonia, Argentina. Revista del Museo Argentino de Ciencias Naturales n.s. 12:217–231.

Powell, J. E. 1987a. Primer registro de hadrosáuridos en la Formación Allen de la Provincia de Río Negro, Argentina; pp. 10–12 in Resúmenes IV Jornadas Argentinas de Paleontología de Vertebrados. Comodoro Rivadavia, Chubut.

Powell, J. E. 1987b. Hallazgo de un dinosaurio hadrosáurido (Ornithischia, Ornithopoda) en la Formación Allen (Cretácico Superior) de Salitral Moreno, Provincia de Río Negro, Argentina. X Congreso Geológico Argentino, Actas 3:149–152.

Prieto-Márquez, A. 2010. Global historical biogeography of hadrosaurid Dinosaurs. Zoological Journal of the Linnean Society 159:503–525.

Prieto-Márquez, A. 2011. A reappraisal of *Barsboldia sicinskii* (Dinosauria: Hadrosauridae) from the Late Cretaceous of Mongolia. Journal of Paleontology 85:468–477.

Prieto-Márquez, A., and G. C. Salinas. 2010. A re-evaluation of *Secernosaurus koerneri* and "*Kritosaurus*" *australis* (Dinosauria, hadrosauridae) from the Late Cretaceous of Argentina. Journal of Vertebrate Paleontology 30:813–837.

Weishampel, D. B., D. B. Norman, and D. Grigorescu. 1993. *Telmatosaurus transsylvanicus* from the Late Cretaceous of Romania: the most basal hadrosaurid dinosaur. Palaeontology 36:361–385.

The Hadrosaurian Record from Mexico

19

Angel A. Ramírez-Velasco, René Hernández-Rivera, and Ricardo Servin-Pichardo

ABSTRACT

The record of trace and body fossils of hadrosauroid dinosaurs from Mexico is reviewed and revised, highlighting both their stratigraphic provenance and paleoenvironmental context. Studies of Mexican hadrosaurs began in 1933, and include important discoveries in Chihuahua, Michoacán, Baja California, Sonora, Coahuila and Puebla states with *Velafrons coahuilensis, Huehuecanauhtlus tiquichensis* and *Magnapaulia laticaudus* and numerous other undescribed specimens currently under study. More work is required in order to increase our knowledge of hadrosaur paleontology and paleobiology in Mexico, which is important for understanding paleobiodiversity gradients and potential dinosaur provinciality in the Late Cretaceous of North America.

INTRODUCTION

Hadrosaurs are the best represented dinosaurs in the Late Cretaceous deposits of Mexico, where they are represented by bone and ichnologic material. The first hadrosaur remains from Mexico were collected by Taliaferro (1933) in the State of Sonora (northeastern Mexico) and identified by Barnum Brown on the basis of fragmentary bones and teeth. Since then, their presence has been documented in Baja California (Langston and Oakes, 1954; Morris, 1972, 1981; Molnar, 1974; Hilton, 2003; Prieto-Márquez et al., 2012), Coahuila (Murray et al., 1960; Serrano-Brañas, 1994; Contreras-Medina, 1997; Hernández-Rivera, 1997; Aguillón-Martínez et al., 1998; Rodríguez-de la Rosa and Cevallos-Ferriz, 1998; Kirkland et al., 2000; Eberth et al., 2003; Kirkland et al., 2006; Serrano-Brañas, 2006; Serrano-Brañas et al., 2006; Gates et al., 2007; Rivera-Sylva et al., 2007; Monroy-Mújica, 2009; Rivera-Sylva, Frey, and Guzmán-Gutiérrez, 2009; Rivera-Sylva, Frey, Palomino-Sánchez, et al., 2009; Rivera-Sylva et al., 2011; Rivera-Sylva et al., 2012), Sonora (Taliferro, 1933; Lucas and González-León, 1996), Chihuahua (Montaño et al., 2009; Kappus et al., 2011), Puebla (Bravo-Cuevas and Jiménez-Hidalgo, 1996), and recently in Michoacán (Ortíz-Mendieta, 2001; Benammi et al., 2005; Mariscal-Ramos, 2006; Ramírez-Velasco et al., 2012; Fig. 19.1). Unfortunately, the vast majority of the specimens consist of isolated or fragmentary remains, complicating precise taxonomic assignments and diversity assessments. However, three new taxa have recently been described on the basis of Mexican fossils (Gates et al., 2007; Ramírez-Velasco et al., 2012; Prieto-Márquez et al., 2012), and others related to existing taxa known from more northern faunas like *Kritosaurus* sp. (Kirkland et al., 2000). These discoveries have been placed into rigorous phylogenetic context, and have contributed significantly to our knowledge of hadrosaur evolution and biogeography (Gates et al., 2007; Prieto-Márquez, 2010). In addition, there have been a large number of hadrosaur-bearing localities discovered within the last decade.

Hadrosaurs have the most complete fossil record of any dinosaur group currently known from México; they are represented by a large number of skeletal remains, footprints, skin impressions and some eggshell fragments. This large amount of evidence can help clarify the composition of dinosaur faunas in southern North America, as well as the environments in which they occur, and their patterns of relative abundance and diversity. It is important to assess how this group fits into hypotheses of dinosaur provinciality and biogeography in Late Cretaceous North America. Mexican hadrosaur fossils also contribute important data on hadrosaur anatomical variation, evolution, behavior, and pathologies.

The purpose of this chapter is to provide the first comprehensive and up-to-date overview of the hadrosauroid fossil record from Mexico, based on published work and unpublished new data, direct observation of palaeontological collections, and fieldwork carried out by the Mexican and foreign institutions working in Mexico.

Institutional Abbrevations AMNH, American Museum of Natural History, New York; BENC, Benemérica Escuela Normal de Coahuila, Saltillo, Mexico; CMN, Canadian Museum of Nature, Ottawa, Ontario; CPC, Colección Paleontológica de Coahuila, Saltillo, Mexico; DP, Colección Paleontológica del Laboratorio de Arqueozoología Subdirección del antiguo Departamento de Prehistoria, Mexico City; ERNO, Estación Regional del Noroeste de Sonora, Hermosillo, Mexico; GMH, Geological Museum of Heilongjiang, China; GMV, National Geological Museum of China, Beijing; IGLUNAM, Instituto de Geología de la Universidad Nacional Autónma

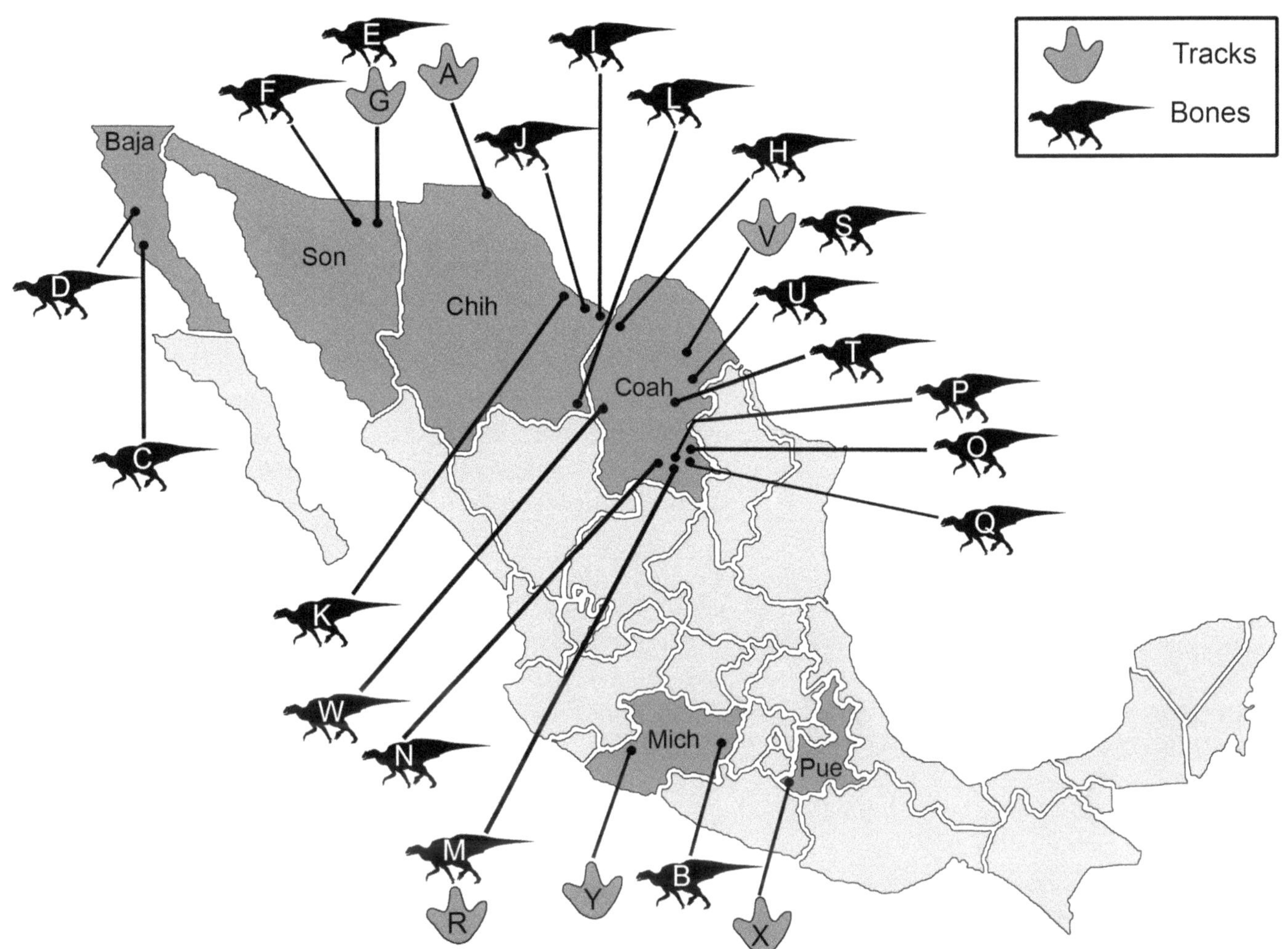

19.1. Map depicting the locations of major Mexican hadrosauroid localities. (A) Ladrillera site and *Thalassinoides* burrow site from the Cerro del Cristo Rey assemblage (Mojado Formation); (B) Barranca Los Bonetes area; (C) El Rosario area (La Bocana Roja and El Gallo formations); (D) Punta San Isidro locality (El Gallo Formation), El Destiladero locality and unknown locality, (both in El Rosario Formation); (E) "El Álamo," El Alamito, Agua de los Conejos, Puerto Viejo and Toscalar locality (Cabullona Group); (F) Naco-Cananea area (Lomas Coloradas, Areniscas Camas, Corral de En medio, and Packard formations); (G) Carro Quebrado locality (Cabullona Group); (H) Bengis Bar, Icoteas, and Altares localities (Aguja Formation); (I) Pico de Pato locality (Aguja Formation); (J) Anizul, Dueto Miserio, Álamos de Márquez, El Rebaje, La Salada, Bell Brown, Las Garzas, La Esperanza, and El Carricito localities (Aguja Formation); (K) Ojinaga area (unknown formation); (L) Arenales, Chamel, and Doctor localities; (M) Rojas I, Rojas II, Cerro de los Dinosaurios, Rincón Colorado area, El Carmen, Valle de los Tirannos, near Ceratopsian site, La Rosa, Agua de Mula, El Palmar, La Parrita, Porvenir de Jalpa, Cruce de los Caminos and Hedionda Chica localities (Cerro del Pueblo Formation); (N) B1, B2, Presa San Antonio, Las Pedreras, Tanque, and Dinosaurio Armado localities; (O) Pelillal, Fraustro, Rancho Quintanilla, Los Pinos, and El Barril localities (Cerro del Pueblo Formation); (P) Hipólito locality (Cerro del Pueblo Formation); (Q) Cañon del Oso (Olmos Formation), Hedionda and Ejido Puebla locality (both Cerro del Pueblo Formation); (R) Las Águilas and La Parrita localities (Cerro del Pueblo Formation); (S) El Mezquite, Polvorín, and Palau localities (Olmos Formation); (T) Altamira locality (Cerro Huerta Formation); (U) Phelan locality; (V) Rancho Soledad locality (Olmos Formation); (W) Rancho San Miguel; (X) Mitepec locality (Mexcala Formation); (Y) Areniscas Aguililla locality. Abbreviations: Baja, Baja California; Chih, Chihuahua; Coah, Coahuila; Mich, Michoacán; Pue, Puebla; Son, Sonora.

de México, Mexico City; IGM, Colección Nacional de Paleontología del Instituto de Geología, Mexico City; INAH, Instituto Nacional de Antropología e Historia, Teotihuacán, Mexico; INEGI, Instituto Nacional de Estadística Geografía e Informática, Aguascalientes City, Aguascalientes, Mexico; LACM, Los Angeles County Museum of Natural History, Los Angeles; MSNM, Museo Civico di Storia Naturale di Milano, Milan, Italy; MUDE, Museo del Desierto, Saltillo, Coahuila, Mexico; PASAC, Paleontological Association of Sabinas, Coahuila, Mexico; SEPC, Secretaría de Educación Pública de Coahuila, Mexico; ROM, Royal Ontario Museum, Toronto, Ontario; UABC, Universidad Autónoma de Baja California, Tijuana, Baja California, Mexico; UCMP, University of California Museum of Paleontology, Berkeley; UNAM, Universidad Nacional Autónoma de México, Mexico City; UNISON, Universidad de Sonora, Hermosillo, Mexico; YPM, Yale Peabody Museum of Paleontology, New Haven, Connecticut.

STRATIGRAPHY AND PALEOENVIRONMENTS

Fossils of Mexican hadrosaurs have been collected from outcrops of the Cretaceous, and range from the Albian to the Maastrichtian. The Albian El Cerro de Cristo Rey assemblage contains footprints from the Mojado Formation, in the state of Chihuahua (Fig. 19.1A; Kappus et al., 2011). Body fossils of Santonian age are known only from the Barranca Los Bonetes area, and originate from an unnamed formation (Fig. 19.1B; Ramírez-Velasco et al., 2012). All other hadrosauroid fossils from Mexico are derived from Campanian–Maastrichtian age sediments. For brevity, according to confirmed and published records, only the most important stratigraphic units and localities are summarized here.

Mojado Formation

The Mojado Formation is exposed in the Sierra de Juárez in the state of Chihuahua and the Sunland Park of New Mexico (Kappus et al., 2011). It is a sandstone-dominated unit, entirely siliclastic, being composed of different types of sandstone with intercalating gray and black shale. It rests atop of the Mesilla Valley Formation and the base of El Río Formation. Dinosaur footprints occur in the lower to middle portions of the Sarten Member (Lucas et al., 2010). The tracks occur in four specific types of coastal, marginal-marine facies in what appear to have been large tidal pools that were stable for long enough to accumulate hundreds of tracks. The age is Late Albian–Early Cenomanian (Kappus et al., 2011).

Barranca Los Bonetes Area

The Barranca Los Bonetes area is located in the state of Michoacán, in the lower section of an unnamed formation with continental sediments. It is predominantly composed of volcaniclastics and a few calcareous paleosols (Tovar et al., 2012). Hadrosauroid fossils occur in the lower and upper part of the stratigraphic section (Tovar et al., 2012). According to radiometric data (Benammi and Montellano-Ballesteros, 2006) and magnetostratigraphic analysis (Mariscal-Ramos, 2006), the fossiliferous strata that crop out in Barranca Los Bonetes are between 86.8 and 84 Ma in age, and correspond to the upper part of Chron C34n (Santonian: Late Cretaceous). Tovar, Fastovsky, and Bennami (2012) interpreted the palaeoenvironment as a distal volcanic zone with long periods of non-deposition and a semiarid local climate that allowed the formation of calcareous soils.

Rosario Area (La Bocana Roja Formation and El Gallo Formation)

Outcrops of the La Bocana Roja and El Gallo formations are located in the Rosario area, in Baja California. It should be noted that these names have not been formalized (Molnar, 1974; Morris, 1981), yet they are still useful for describing the general stratigraphic position of vertebrate fossils found in Baja. The La Bocana Roja Formation is described as red mudstone and tuffs (Molnar, 1974), and the El Gallo Formation consists of gray-to-yellow siltstones and sandstone with occasional tuffs. It overlies the La Bocana Roja Formation and underlies the Rosario Formation. The hadrosaur material has been recovered from the upper and the middle part of the El Gallo Formation, in the Disecado member (Morris, 1981). Using radiometric dating, Renne et al. (1991) assessed the age of the El Gallo Formation as between 74.87–73.59 Ma and, thus, Late Campanian. Fulbord and Busby (1993; Fig. 19.2) interpreted the paleoenvironmental conditions as fluvial transitioning into marine.

Cabullona Group (Corral de Enmedio, Camas, Packard and Lomas Coloradas formations)

The Cabullona Group is located to the northeast of the state of Sonora, in the area Naco-Cananea. It comprises four successive formations: the Corral de Enmedio Formation (previously defined as Snake Ridge Formation by Taliaferro [1933]) is formed by a sequence of mudstones and lesser amounts of siltstone, lenticular sandstone, and thin lenticular beds of limestone (rich in fossils); the Camas Formation is characterized dominantly by coarse to pebble conglomeratic, trough cross-bedded sandstone; the Packard Formation is dominated mainly by massive mudstone, finely laminated shale and siltstone, and planar laminated, ripple-laminated, and trough cross-bedded sandstone; and lastly, the Lomas Coloradas Formation, which gradationally overlies the Packard Formation and is characterized by abundant red siltstones (rich in nodular calcrete) alternating with trough cross-bedded sandstone and pebble conglomerate. The El Cemento Conglomerate is laterally equivalent to the Lomas Coloradas Formation and consists mainly of clast-supported conglomerate beds, sandstones, and sandy siltstones (González-Leon and Lawton, 1995; Lucas et al., 1995). The base of the Cabullona Group is not exposed, but regional stratigraphic relationships suggest that it disconformably overlies the Lower Cretaceous Cintura Formation of the Bisbee Group (Albian–Cenomanian?).

The Corral de Enmedio Formation represents a shallow, freshwater lacustrine environment, and the Camas Formation represents a fluvial depositional environment.

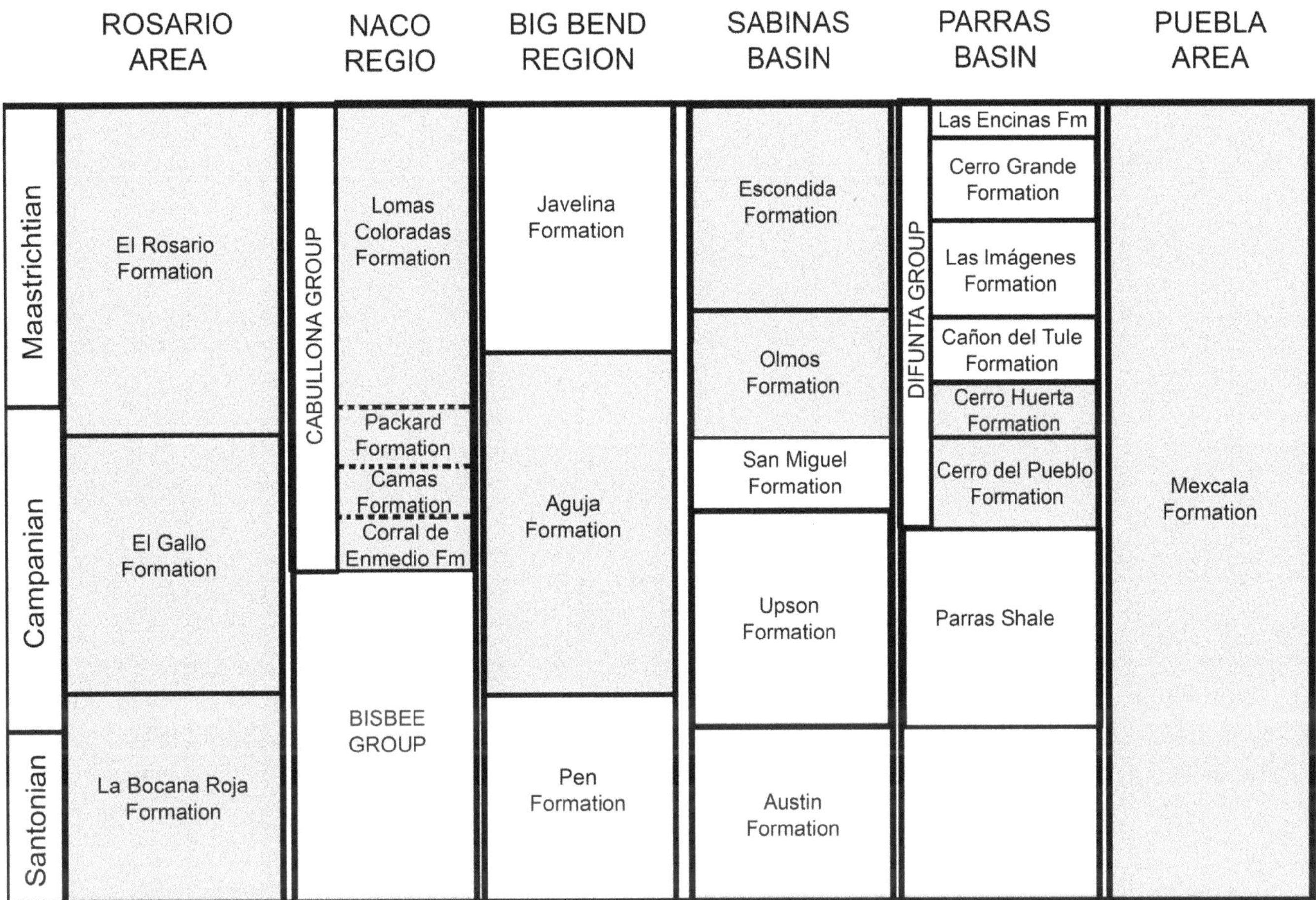

19.2. Correlation chart of the stratigraphic formations with hadrosauroid fossils from Mexico.

The Packard Formation was deposited by a complex lacustrine-deltaic system, which intertongues with distal alluvial fan deposits of the El Cemento Conglomerado. The Lomas Coloradas represents a meandering river system (González-Leon and Lawton, 1995; Lucas et al., 1995).

The age of the Cabullona Group is interpreted as Late Campanian to Late Maastrichtian based on invertebrate and vertebrate fossils (Lucas et al., 1995; Fig. 19.2).

Aguja Formation

The Aguja Formation crops out mainly in Texas, with a small part in northwest Coahuila (Ocampo municipality) and northeast of Chihuahua (Altares municipality) in Mexico. The Aguja Formation mainly comprises irregularly interbedded claystone and sandstone, and is characterized by rapid vertical and lateral facies changes (Hopkins, 1965). The Mexican outcrops correlate with the Middle Sandstone and Upper Shale, where hadrosaur fossils are found. The age has been assessed as Late Campanian, within Chron C33 1n (Benammi and Montellano-Ballesteros, 2006; Fig. 19.2). The paleoenvironment was deltaic with a narrow prodelta; marshes, oxbows, and nearshore lagoons with sandy or silty bars are documented (Hopkins, 1965).

Olmos Formation

The Olmos Formation occurs in Texas and in the northeastern part of Coahuila (Sabinas Basin). It comprises greenish-gray shales and limestone, finely stratified, with layers of coal and lignite towards the base (Serrano-Brañas, 2006). Hadrosaur fossils have been collected at the top of the Olmos Formation near the contact of the Escondido Formation (Serrano-Brañas, 2006). The Olmos Formation has been divided into two systems: a deltaic and a river system (Estrada-Ruiz et al., 2008; Estrada-Ruiz and Cevallos-Ferriz, 2009). The morphology of plant fossils suggests that the Olmos Formation represents deposition in a paratropical forest (Estrada-Ruiz et al., 2008). The age of the Olmos Formation is Late Campanian to Early Maastrichtian as determined by biochronologic analyses (Tyler and Ambrose, 1986; Pessagno, 1969; Fig. 19.2).

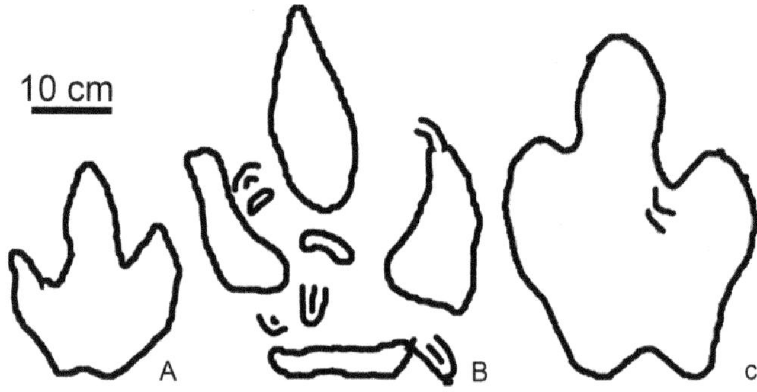

19.3. *Caririchnium* ichnites from Cerro Cristo Rey, Mojado Formation in Chihuahua, Mexico. (A) Ichnospecies A; (B) Ichnospecies A; (C) Ichnospecies B (like *Caririchnium leonardii*). Redrawn from Kappus et al. (2011).

Difunta Group: Cerro del Pueblo and Cerro Huerta Formations

The Cerro del Pueblo Formation and Cerro Huerta Formation are located in the southeast region of the state of Coahuila in the Parras Basin. They form the base of the Difunta Group. The Cerro Huerta Formation consists of a sequence of interbedded red and green mudstones (Kirkland et al., 2000). The Cerro del Pueblo Formation consists of a basal cliff-forming, marginal-marine sandstone and sequences of drab interbedded sandstones and mudstone with minor limestones (Kirkland et al., 2000). This formation preserves a large number of dinosaur and other vertebrate remains, including footprints and body fossils. In the Cerro Huerta Formation (lower Maastrichtian) the hadrosaur fossils are found only in the base, near the contact with the Cerro del Pueblo Formation (Kirkland et al., 2000).

Geological data suggests a coastal plain that included small, low-sinuosity channels, levees, plant-rich wetlands, lakes, ponds, lagoons and bays, and possibly delta front sands (Hill, 1988), with occurrences of high-order transgressive-regressive episodes and/or extreme storm events (Eberth et al., 2004). The age of the Cerro del Pueblo Formation has been established by biostratigraphy (Kirkland et al., 2000) and magnetostratigraphy (Eberth et al., 2004) as Late Campanian (32n.3r–32n.2n) close to the Campanian-Maastrichtian boundary (Fig. 19.2).

Mexcala Formation

The Mexcala Formation extends across the states of Puebla, Guerrero, and Oaxaca (southern Mexico). The unit is a deepwater–shallow marine sedimentary sequence that grades upward into shales, siltstones, sandstones, and conglomeratic lenses (Fraaije et al., 2006). The age of the formation is suggested as Turonian to late Maastrichtian (Molina-Garza et al., 2003; Fraaije et al., 2006). In the state of Puebla, hadrosaur footprints occur in fine sandstones at the top of the formation (Maastrichtian age; Fig. 19.2; Rodriguez-de la Rosa et al., 2004).

Arenisca Aguililla Unit

The Areniscas Aguililla unit is located in Michoacán State, and consists of 300 m red arkosic volcanic sandstone, with limonitic texture that represents a fluvio-lacustrine paleonvironment (Ortiz-Mendieta, 2001; Rodriguez-de la Rosa et al., 2004). Hadrosaur ichnites in the area, as discussed by Ortiz-Mendieta (2001), suggest a latest Cretaceous (Maastrichtian?) age for this unit (Fig. 19.2).

THE HADROSAUROID FOSSIL RECORD OF MEXICO

Cerro Del Cristo Rey Site, Chihuahua State

Ichnites Ichnites from the Cerro Cristo Rey site are derived from the Sarten Member, Mojado Formation, and are located in Sierra de Juárez (Ciudad Juárez; Fig. 19.1A). This unit extends north–south from the United States–Mexico international border (New Mexico, El Paso, Texas and Ciudad Juárez, Mexico). The ichnogenus *Caririchnium* has been reported (ichnosp. A and B) from the Ladrillera site and the *Thallasinoides* burrow site (Fig. 19.3). Few *Caririchnium* tracks are associated with manus prints. The smallest ornithopod ichnite from this area belongs to a juvenile individual about 3 m long (ichnosp. A, Ladrillera site). Other dinosaur ichnites from this site include theropod and ankylosaur morphotypes (Kappus et al., 2011).

It is not clear whether the track makers were hadrosaurids or non-hadrosaurid iguanodontians, as it is difficult to differentiate their tracks due to a lack of diagnostic characters. However, tracks of *Caririchnium* ichnosp. A are most often attributed to iguanodont track makers (Fig. 19.3A, B). *Caririchnium* ichnosp. B are hadrosaur-like ornithopod tracks and similar to *Caririchnium leonardii* (Fig. 19.3C). A small percentage of the ichnites represent hadrosaur-like tracks (Kappus et al., 2011). The ichnites have not been removed from the field (Kappus et al., 2011).

Barranca Los Bonetes Area, Michoacán State

Huehuecanauhtlus tiquichensis **Ramírez-Velasco et al., 2012** This taxon is diagnosed by the following unique combination of characters: two teeth exposed on the occlusal plane of the rostral third and the posterior third of the dentary and maxilla, respectively; seven sacral vertebrae; tall

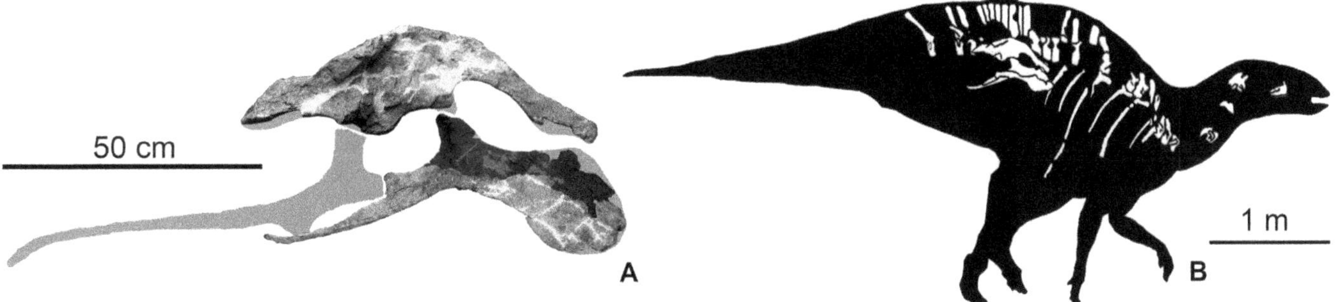

19.4. *Huehuecanauhtlus tiquichensis* holotype (IGM 6253) from Barranca Los Bonetes, Michoacán. (A) pelvic girdle elements (the gray silhouette represents the missing parts); (B) reconstruction of the holotype showing the elements found.

neural spines of posterior vertebrae, being between 3.5 and 4 times taller than their corresponding centra; supra-acetabular process as long as 75% of the length of the central iliac plate, with an apex located above the posteroventral corner of the ischiac tuberosity; and short and trapezoidal (in lateral view) postacetabular process, with a length that is less than 80% of the length of the iliac central plate (Fig. 19.4A). It differs from other hadrosauroids by possessing an extremely deflected preacetabular process of the ilium, so that the long axis of the process forms an angle less than 130° with the horizontal plane defined by the ischiac and pubic peduncles and a very deep concave profile of the dorsomedial margin of the iliac plate, adjacent to the supra-acetabular process (Ramírez-Velasco et al., 2012).

Huehuecanauhtlus tiquichensis represents the first non-hadrosaurid hadrosauroid discovered from México, and the southernmost of the basal hadrosauroids from North America (Ramírez-Velasco et al., 2012; Fig. 19.1B). The holotype (IGM 6253, Fig. 19.4B) and referred material (IGM 6254) come from Santonian strata in Michoacán. The specimens were discovered by Benammi et al. (2005) in May 2003 at the Barranca Los Bonetes locality. Subsequently, additional fossiliferous sites were found at the same general locality. These have yielded material referred to indeterminate hadrosauroid and theropod dinosaurs (Benammi et al., 2005; Mariscal-Ramos, 2006).

In a phylogenetic analysis, Ramírez-Velasco et al. (2012) posited *Huehuecanauhtlus tiquichensis* as a closely related outgroup to Hadrosauridae within Hadrosauroidea. The placement in Hadrosauroidea is supported by the following synapomorphies: preacetabular process strongly deflected ventrally, forming an angle less than 150° with the horizontal plane defined by the pubic and ischiac peduncles; and triangular and relatively short pubic peduncle of the ilium. The occurrence of *Huehuecanauhtlus tiquichensis* in Santonian strata of southern North America may represent a dispersal event from Asia to North America no later than the Albian (Ramírez-Velasco et al., 2012).

Hadrosauroidea indet. Benammi et al. (2005) collected a small tibia (IGM 8824) and caudal vertebrae in the Barranca Los Bonetes area (locality number IGM 3616; Fig. 19.1B), from the middle part of the stratigraphic column, and referred them to Hadrosauridae (Benammi et al., 2005). However, with the presence of *Huehuecanauhtlus tiquichensis* in the area, the tibia IGM 8824 may represent a non-hadrosaurid hadrosauroid, but because of the lack of long bones in the holotype of *H. tiquichensis*, it is difficult to precisely identify the specimen beyond Hadrosauroidea indet. The material is currently under study by Ramírez-Velasco.

El Rosario Area and Eréndira Area, Baja California State

***Magnapaulia laticaudus* Morris, 1981** This taxon was first diagnosed as a large hadrosaur, with adults having an estimated length of 14.0 to 15.0 m. This species differs from all other species of *Lambeosaurus* in the great depth of the tail. Caudal hemal arches are long, matching the neural spine in length (Morris, 1981). Prieto-Márquez et al. (2012) revised the diagnosis and reinterpreted the characters proposed by Morris as follows: lambeosaurine with haemal arches of proximal caudal vertebrae being at least four times longer than the depth of their respective centra, and the base of the prezygapophyses in caudal vertebrae merging to form a bowl-shaped surface, which, in the proximal-most caudals, is continuous dorsally with a deep sulcus on the cranial surface of the neural spine.

Magnapaulia laticaudus (previously known as "*Lambeosaurus*" *laticaudus*; Fig. 19.5) represents the first hadrosauroid species named from Mexican material. Its remains were collected between 1968 and 1974 by a team led by W. J. Morris from the Natural History Museum of Los Angeles County (Morris, 1981; Prieto-Márquez et al., 2012:suppl. 1). Collections were made at Arroyo El Rosario North (locality LACM 66166), Arroyo El Rosario West (locality LACM 66167), from

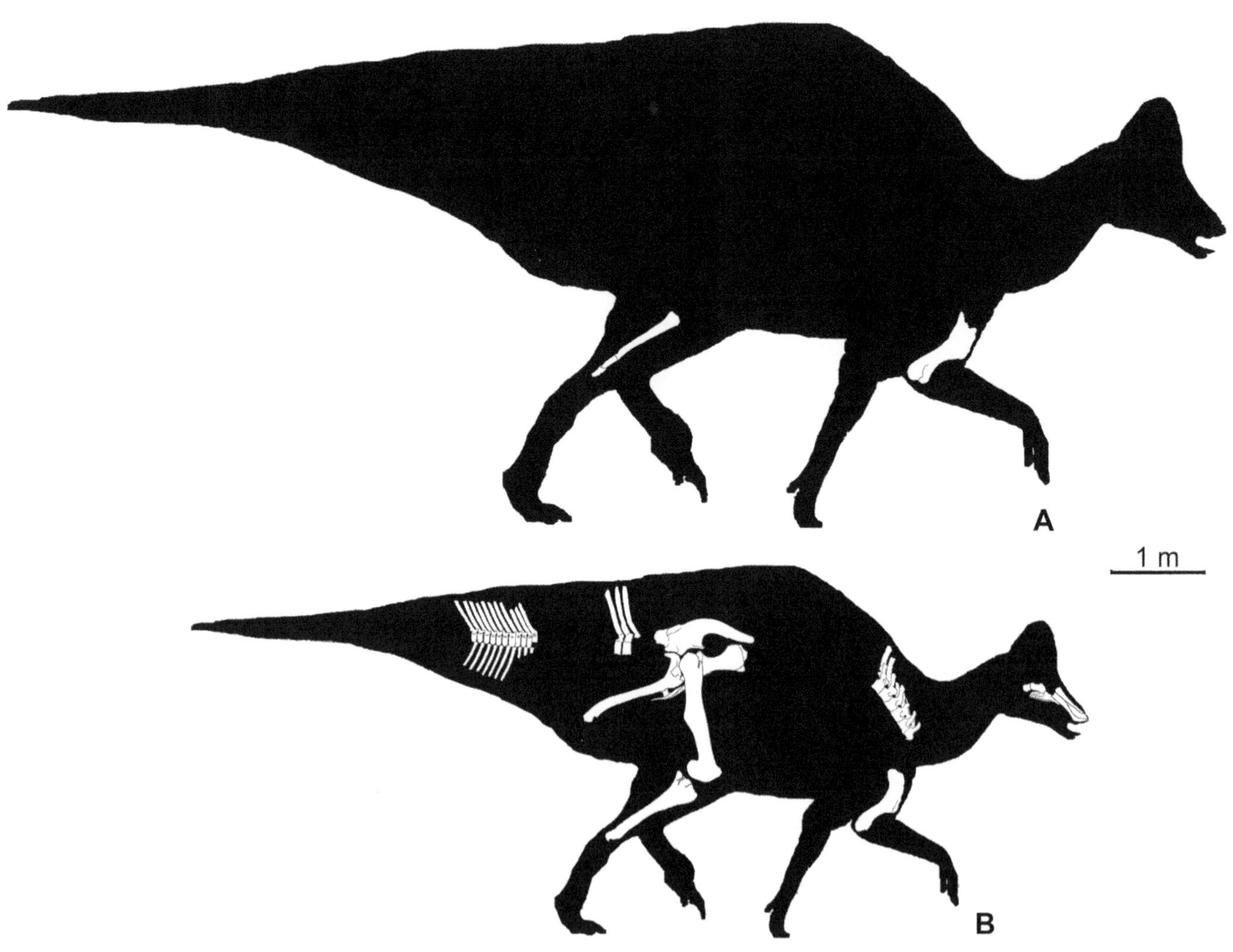

19.5. Specimens of *Magnapaulia laticaudus* from El Gallo Formation, El Rosario area, Baja California. (A) reconstruction of IGM 5845 (fibula and humerus), based on elements in the IGLUNAM; (B) composite reconstruction with elements illustrated by Morris (1981): premaxilla, maxilla, jugal, and femur (LACM 17715), tibia (LACM 17706), proximal caudal vertebrae (LACM 28990), and series of caudal vertebrae (LACM 17705); and the elements housed in the IGLUNAM: series of thoracic vertebrae (IGM 5844), humerus (IGM 5843), ilium and pubes (IGM 5845), and ischium (IGM 5846).

El Rosario (locality LACM 66168 and 6751) and El Rosario Arriba (localities LACM 6752, 6753 and 6754), and from the El Gallo Formation (Fig. 19.1C). This taxon is known from a large number of fossils, which are reposited at the LACM (LACM 17711, 28990; skin impression 17712; skull bones 17715, 17705) and the Colección Nacional de Paleontología del Instituto de Geología (UNAM). Material in the latter institution includes postcranial bones and parts of the bones from the holotype (IGM 5843–5852). In addition to the material listed above, a cast of scalation (large hexagonal and small rounded scales) with ossicles associated with a *Magnapaulia laticaudus* anterior dorsal vertebrae, mentioned by Morris (1981) and redescribed by Prieto-Márquez et al., (2012), is the first record of dinosaur integument from Mexico.

The taxonomic position and generic assignation of *Magnapaulia laticaudus* has changed several times due to the lack of complete cranial material. In the first publication in which a lambeosaurine from Baja California was mentioned, Morris (1967) noted the massively footed ischia and caudal vertebrae with high caudal spines referred to the genus *Hypacrosaurus*. Later in 1972, Morris recognized other aspects of the specimen that he felt more closely resembled the genus *Lambeosaurus*, including the open narial canal that extends two-thirds the length of the premaxilla. He therefore assigned it to a new species within the genus *Lambeosaurus* (Morris, 1981).

Gates et al. (2007) doubted the validity of the species because many other lambeosaurines, such as *Hypacrosaurus* and *Corythosaurus*, also have a similar suite of characters. Gates et al. (2007) highlighted the similarities between "*Lambeosaurus*" *laticaudus* and *Hypacrosaurus altispinus*, particularly the presence of a highly denticulated oral margin and deep tail, which are characters only possessed by *Hypacrosaurus altispinus*. Gates et al., (2007:926) therefore referred to "*Lambeosaurus*" *laticaudus* as "the Baja lambeosaurine."

Prieto-Márquez (2010) supports the inclusion of "*Lambeosaurus*" *laticaudus* in the same clade with *Velafrons coahuilensis* and *Hypacrosaurus altispinus*, a position that is supported by the following synapomorphies: angle of 113°–128° between the dentary's proximal-most edentulous slope and the horizontal; presence of brevis shelf at the base of the postacetabular process of the ilium; presence of a well-defined ridge on the medial side of the postacetabular process forming the medial margin of a medioventrally facing shelf with a postacetabular process that is progressively expanded mediolaterally toward the caudal end. Prieto-Márquez (2010) mentioned a single unambiguous synapomorphy supporting *L. laticaudus* as the sister taxon of *H. altispinus:* a very thick ischial shaft, of thickness greater than 7.5% the length of the shaft.

A more recent study by Prieto-Márquez, Chiappe, and Joshi (2012) presented a detailed osteological description of the cranial, axial, and appendicular skeletal remains available in the LACM and IGM collections, and concluded that the taxon is valid and merits a new generic name: *Magnapaulia laticaudus*. In this paper, *Magnapaulia laticaudus* is unambiguously positioned as the sister taxon to *Velafrons coahuilensis* because the ilium has a supra-acetabular crest that extends ventrally more than 25% but less than half depth of the central iliac plate, and the nasal bone has a triangular external naris gradually closing caudodorsally with a length-to-width ratio between 1.85 and 2.85. Finally the *Magnapaulia-Velafrons* clade was inferred to be the sister clade to a monophyletic group of helmet-crested lambeosaurines that includes *Hypacrosaurus, Olorotitan, Corythosaurus*, and *Lambeosaurus*.

In 1970 (Morris, 1972; Hilton, 2003) Tabrum collected a massive humerus and two cervical vertebrae, and probably one incomplete fibula (all now cataloged as IGM 5845, previously known as LACM 26757), which Morris (1972) referred to as a giant "*Lambeosaurus*" *laticaudus* (Morris, 1981). Comparisons with other humeri of North American hadrosaurs suggest that this animal was approximately 16.5 m long, thus representing one of largest hadrosaurs in the world (Morris, 1972, 1981). Prieto-Márquez et al. (2012) mentioned that they could not locate the giant humerus under LACM 26757 in the collection of the LACM or at the IGM, and therefore suggested that specimen LACM 17712 must be the same as the one that Morris illustrated in 1972, because the broken portion shows traces of the restored proximal part. Using the model proportions of *Hypacrosaurus altispinus* CMN 8501 (from Paul, 2010:313), we suggest that this taxon reached approximately 12 m in total length (Fig. 19.5). A similar estimate (12.5 m) for IGM 5845 was obtained by Prieto-Márquez et al. (2012) using a linear regression.

Hadrosauridae indet. The first hadrosaur fossils identified from Baja California were discovered in 1953 by J. Wyatt Durham and Joseph H. Peck, Jr., of the University of California, Berkeley (Hilton, 2003). They collected two metatarsals and a terminal phalanx (UCMP 43251) from two juvenile individuals in the El Gallo Formation at the Punta San Isidro locality near Eréndira (Langston and Oakes, 1954; Hilton, 2003; Johnson et al., 2006; Fig. 19.1D). In 1960, E. H. Colbert collected vertebrae, a distal femur fragment, and a fragment of dentary with teeth from the El Gallo Formation in the El Rosario area (Hilton, 2003; Fig. 19.1A). From 1966 to 1985, W. J. Morris, H. J. Garbani, M. T. Greenwald, and Tabrum collected many other many postcranial bones, fragments of dentary, and teeth of various hadrosaurs from the El Gallo and La Bocana Roja formations (Hilton, 2003; Fig. 19.1A). This material is housed in the LACM collection. The vast majority of this material has not been formally described.

In a table presented by Hilton (2003), the author notes a few phalanges (UCMP 43251) found in the Rosario Formation, but does not know the year they were recovered or the name of the collector (Fig. 19.1D). The latter are important specimens because they represent one of the few occurrences of hadrosaurs from the Maastrichtian of Baja California. Johnson et al. (2006) reported a hadrosaur tibia, broken in two parts, from the El Destiladero locality, near Eréndira, that belongs to strata from the Rosario Formation (Fig. 19.1D). This material is reposited in the Colección Paleontológica de Referencia de la Universidad Autónoma de Baja California in Ensenada under the specimen numbers UABC 2612a and UABC 2612b. Other fragmentary and indeterminate bones are included in the same catalog entry. The stratigraphic position of this material has been verified by identifying an in situ natural mold of one of the large bones (Johnson et al., 2006).

Lambeosaurinae indet. Prieto-Márquez et al. (2012) mention a jugal fragment, radius, scapula, ulna, fibula, humerus, and dentary from a juvenile individual (Fig. 19.1C) in the La Bocana Roja Formation (loc. LACM 7255, El Rosario site). Other lambeosaurine material from this site and formation (loc. IGM 7256; Fig. 19.1C) was discovered by H. L. Garbani in 1971, and includes an ischium and two thoracic vertebrae (Morris, 1981; Hilton, 2003), and a fragment of a probable preacetabular process of the ilium from the same individual (Prieto-Márquez et al., 2012). Morris (1981) first assigned the material to "*Lambeosaurus*" *laticaudus* but, later, Prieto-Márquez et al. (2012) reinterpreted it as a lambeosaurine distinct from *Magnapaulia laticaudus*. Two ischia (LACM 28990 and 17708) from the El Gallo Formation (IGM collection) probably belong to the same species of lambeosaurine,

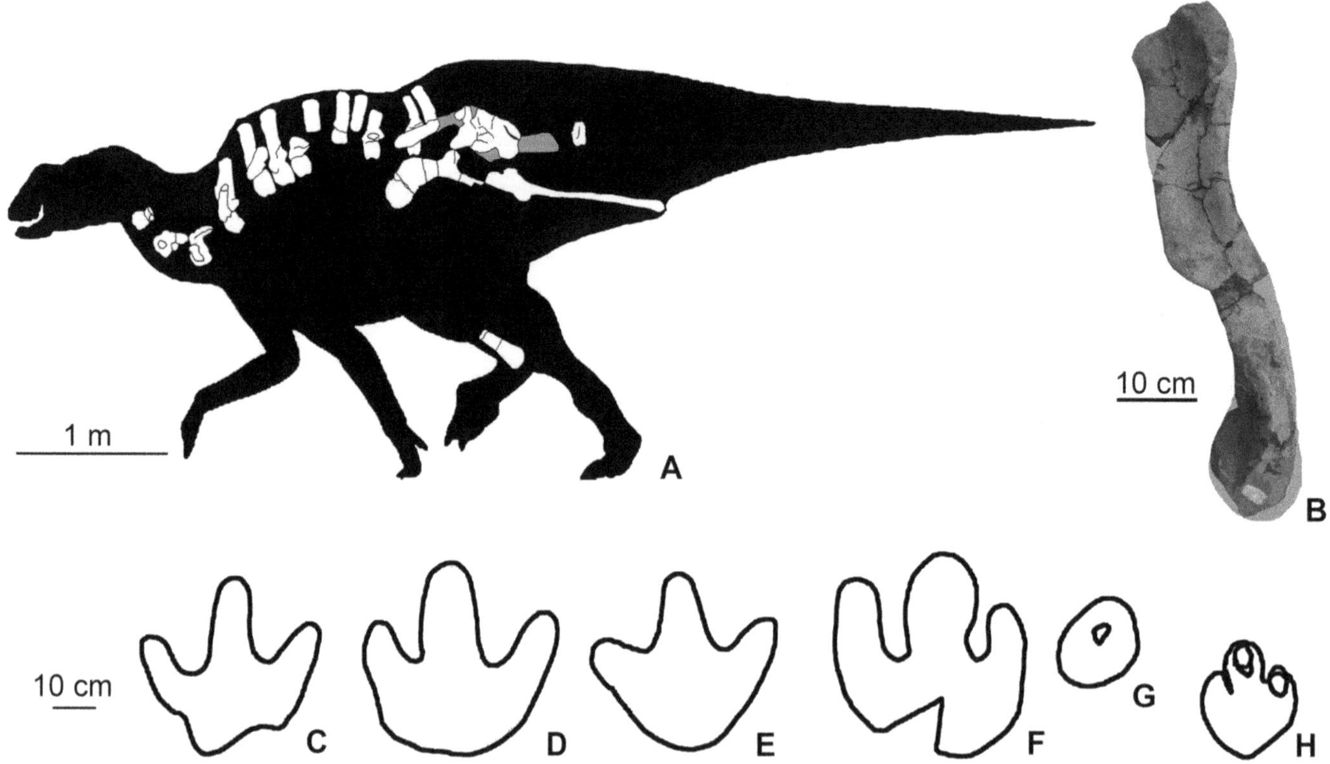

19.6. Hadrosaur fossils from the Cabullona Group in Fronteras (A, B) and Esqueda (C–H), Sonora. (A) preliminary reconstruction for the "Alamitos specimen" showing the elements found; (B) left humerus in lateral view of the hadrosaurs from Agua de los Conejos (the gray silhouette represents the missing parts); (C–E) tracks from trackway of an individual with a bipedal walking gait at the Carro Quebrado locality; (F) pes tracks from trackway of an individual with a quadruped walking gait at the Quebrado locality; (G) manus tracks from trackway of an individual with a quadruped walking gait at the Quebrado locality; (H) individual track from a juvenile.

because they have the same morphology: a marked curvature posteroventrally within the body of the ischium that results in the pubic boot projecting more ventrally (unlike the holotype ischium of *Magnapaulia*). This morphology is shared with the ischium of *Velafrons coahuilensis* CPC-59 (discussed below) and has not been formally described or discussed in other works (AAR-V, pers. obs.).

Eggshell Fragments of eggshell found in the El Gallo Formation (Fig. 19.1C) show characters (prolatospherulitic morphotype) that indicate the presence of ornithopods, likely hadrosaurs (Rodriguez de la Rosa, 1998).

Cabullona Group, Sonora State

Saurolophinae indet. Saurolophine material was collected in 2009 from the Cabullona Group at the Alamito locality in the Fronteras municipality (Fig. 19.1E). It was discovered by Guillermo Gámez Santacruz and is currently being collected by C. M. González-León, R. Hernández-Rivera, R. Pacheco Rodríguez, P. A. Sánchez Medrano, M. Gámez Santacruz, R. Duarte Bigurra, R. Servín-Pichardo and A. A. Ramírez-Velasco. Referred to as the "Alamitos specimen" (Fig. 19.6A), it represents a saurolophine of uncertain affinity. The morphology of the prepubic blade resembles *Gryposaurus*, with a short lateroventral projection of the supra-acetabular process (AAR-V, pers. obs.). The Alamitos specimen elements have not been cataloged yet, but include fragments of the ilium, pubes, ischia, ribs, and thoracic and cervical vertebrae. The Alamitos specimen represents an individual approximately 7 m long, with unusual proportions, and bones that are more robust than other Mexican hadrosaurs. This material is currently under study by R. Duarte-Bigurra from the Universidad de Sonora (UNISON).

Other material assigned to Saurolophinae indet. from the same municipality (Fig. 19.1E) includes a complete humerus (Fig. 19.6B) from the Agua de los Conejos locality collected by P. A. Sánchez Medrano and M. Gámez Santacruz. Lucas et al. (1995) illustrate another humerus (ERNO-302) and mention another left humerus from localities 16 and 17 (Naco-Cananea area; Fig. 19.1F) from the Areniscas Camas Formation (Lucas and González-León, 1996). Overall, the morphology resembles that of *Gryposaurus*, but is considered here as belonging to an indeterminate saurolophine (AAR-V, pers. obs.). Contreras-Medina (1997) mentioned that the humerus ERNO-302 could belong to a giant hadrosaur because it measures 710 mm in length, which exceeds the length of *Edmontosaurus annectens* (YPM 2185) and *Gryposaurus notabilis* (MSNM V344), although it is smaller than the humeri of *Shantungosaurus giganteus* (e.g., GMV 1780; Prieto-Márquez et al., 2012).

Hadrosauridae indet. The first hadrosaur material (teeth and bones) identified from the state of Sonora in the southern Naco-Cananea area (Fig. 19.1F) was collected in 1933 by N. L. Taliaferro of the University of California, Berkeley. Taliaferro (1933) proposed the name "Snake Ridge Formation" for the rocks that contained these hadrosaur fossils, but González-León and Lawton (1995) and Lucas et al. (1995) clarified some geological details and changed the name to Corral de Enmedio Formation (part of the Cabullona Group). In the early 1990s, C. M. González-León of the Estación Regional del Noroeste of the Instituto de Geología of UNAM (ERNO) along with S. G. Lucas studied the stratigraphy of the Cabullona Group in the Naco-Cananea area (Fig. 19.1F) and collected numerous remains of dinosaurs. Hadrosaurian elements recovered include numerous vertebrae from Localities 1, 2 and 4, and one femur (Corral de Enmedio Formation); vertebrae from Locality 32 in the Lomas Coloradas Formation; a phalanx and vertebrae from Locality 29 in the Lomas Coloradas Formation; and thoracic and caudal vertebrae from the Packard Formation (Lucas et al., 1995; Lucas and González-León, 1996).

Since 2009, prospecting in the Cabullona Group, Fronteras municipality (Fig. 19.1E), has yielded various remains of hadrosaurs from several new localities: metatarsals, fragments of a humerus, and vertebrae collected by C. M. González-León in "El Álamo"; caudal vertebrae from Puerto Viejo; and an incomplete humerus and tibia with other long bone fragments from the Agua de los Conejos (AAR-V, pers. obs.). To date, none of the material collected from the state of Sonora can be identified as lambeosaurine.

Ichnites In 2010, Angel Rubén Parra Lara and his quarry workers discovered a slab with dinosaur footprints in the Esqueda Ejido, in the Fronteras municipality, from the quarry area known as Carro Quebrado (Fig. 19.1G). These footprints are significant because they represent the first record of hadrosaurian fossils in the state of Sonora. The footprints occur within the Cabullona Group (González-León, pers. comm., May, 2010). The lithology (mainly micritic shales in horizontal beds with ripple marks in a few levels, as well as a mudstone with dry marks with intercalated ashes) and fossils (dinosaur ichnites, ostracodes, and stromatolites) all suggest a lacustrine paleoenvironment.

Two trackways are present at different stratigraphic levels in a 2.8 m thick section measured at the Carro Quebrado quarry. In the lower level, a trackway shows a hadrosaur with a bipedal gait (Fig. 19.6C–E), and in the middle part of the section, another trackway shows manual impressions (Fig. 19.6F, G). One isolated ichnite, collected from an unknown stratigraphic level within the unit by Ignacio Lara-Bustamante, a local teacher, represents a juvenile based on its small size (Fig. 19.6H). Theropod ichnites have also been discovered in beds from the same area. This unit is still under study (Servín-Pichardo et al., 2011) with the support of C. M. González-León, R. Hernández-Rivera, R. Pacheco Rodríguez, P. A. Sánchez Medrano, M. Gámez Santacruz, R. Duarte Bigurra, A. A. Ramírez-Velasco and R. Servín-Pichardo. The trackways are still in situ, and the small ichnite cited above belongs to Lara-Bustamante.

Aguja Formation, Northeastern Chihuahua State and Northwestern Coahuila State

Saurolophinae indet. Montaño et al. (2009) mentioned the first saurolophine hadrosaurid of the *Gryposaurus-Kritosaurus* clade (Prieto-Márquez, 2010) from northern Mexico in an abstract of the XI Congreso Nacional de Paleontología. The material includes cranial elements, pelvic girdle, forelimb and vertebrae of an adult individual from the Anizul locality in the state of Coahuila (Fig. 19.1H). A right maxilla and other elements of a juvenile individual have been collected from the Pico de Pato locality, Chihuahua (Fig. 19.1I). This material is under study by Montaño at Instituto de Geología UNAM.

Monroy-Mujica (2009) reports isolated teeth referred to Saurolophinae from the Bell Brown Locality and the Las Garzas site (Fig. 19.1H), also from the state of Coahuila.

In 2000, Francisco Palomino and an INEGI team discovered the Icoteas locality (Fig. 19.1J) and collected a very fragmentary hadrosauroid specimen consisting of cervical, thoracic, and caudal vertebrae, and a fragment of the proximal portion of the humerus and metatarsus belonging to an unknown saurolophine (Fig. 19.7A, B). What is notable about this saurolophine is the robust proportions of the cervical bodies of the vertebrae in a specimen with an estimated length of 7 m (Fig. 19.7B).

Another report documents a vertebra and a distal part of the scapular blade of a huge hadrosaur resembling *Kritosaurus* from the Carricito locality (Fig. 19.1H), previously known as "Las Jicoteas" (Rivera-Sylva, Frey, Palomino-Sánchez et al., 2009; Rivera-Sylva and Frey, 2011).

Lambeosaurinae indet. In 2004, M. Montellano-Ballesteros, G. Alvarez-Reyes and R. Hernández-Rivera collected an isolated humerus (Fig. 19.7C) from Las Garzas locality, state of Coahuila (Fig. 19.1H). The element was identified as lambeosaurine by A. A. Ramírez-Velasco and is reposited in the IGM collection.

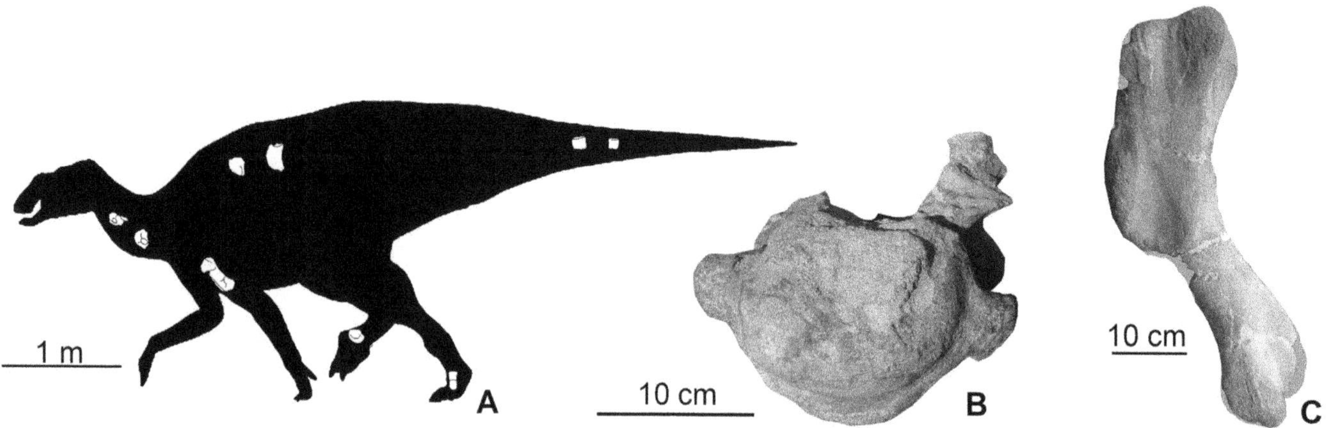

19.7. Hadrosaur fossils from the Aguja Formation in Chihuahua (A) and Coahuila (B, C). (A) preliminary reconstruction for the Icoteas saurolophine showing the elements found; (B) cervical vertebra in anterior view for the Icoteas saurolophine; (C) left humerus in lateral view of the lambeosaurine from the La Esperanza locality.

Hadrosauridae indet. The search for new localities in the Mesozoic rocks of northeast Chihuahua and northwest Coahuila began in 2000 by Montellano-Ballesteros et al. (2000), and resulted in the discovery of a large number of new localities with hadrosaur bones in the Manuel Benavides municipality of Chihuahua (Fig. 19.1J) and the Ocampo municipality of Coahuila (Fig. 19.1H). The first dinosaur remains found in northwestern Coahuila were discovered by an INEGI field team in 1996, but these were not formally reported until 2007, with MUDE paleontologists.

The Montellano-Ballesteros team collected hadrosaur vertebrae in the Altares locality in Manuel Benavides municipality, as well as sacral vertebrae, several caudal vertebrae, and other fragments of hadrosaur bones from the Bengis Bar locality in the same municipality. In the Ocampo municipality, the same team collected vertebrae and the proximal end of a tibia from the Dueto Miseria locality. The hadrosaur material recovered included a caudal vertebra and a bone fragment from the Álamos de Márquez locality (Montellano-Ballesteros, pers. comm., 2011), as well as vertebral fragments from the La Salada locality (Rivera-Sylva et al., 2007, 2011; Rivera-Sylva, Frey, and Guzmán-Gutiérrez, 2009), vertebrae, metatarsals and a dentary fragment from the Las Garzas locality, and a huge caudal vertebra from the La Esperanza locality (Montellano-Ballesteros, pers. comm., 2011).

In 2006, a team from INEGI and MUDE collected fossils from Ocampo municipality (Fig. 19.1H) and found additional localities with hadrosaur material (Rivera-Sylva et al., 2007; Rivera-Sylva, Frey, and Guzmán-Gutiérrez, 2009; Rivera-Sylva, Frey, Palomino-Sánchez, et al., 2009). From the La Salada locality, Rivera-Sylva, Frey, and Guzmán-Gutiérrez (2009) reported seven caudal vertebrae and humerus and distal metatarsal fragments (CPC-308). One caudal vertebra (CPC-309) with bite marks was interpreted as a predation mark of the giant crocodilian *Deinosuchus*. Recently Rivera-Sylva, Hone, and Dodson (2012) reported a fibula, one dorsal vertebrae, and one tibia with bite marks that they linked to a tyrannosaurine theropod based on the bite marks' size and spacing.

Ojinaga Area, Northeastern Chihuahua State

Hadrosauridae indet. The DP collection of INAH Mexico City houses some interesting postcranial material of hadrosaurids collected in the Ojinaga municipality (Fig. 19.1K), but they are of unknown locality and formation (Ana Fabiola Guzmán, pers. comm., 2012). The material is being studied by Oscar J. Polaco, as well as Ana Fabiola Guzmán and José Luis Gudiño.

Jiménez Area, Southeastern Chihuahua State

Hadrosauridae indet. In 2011 during fieldwork by the Montellano-Ballesteros team in Jiménez municipality (Arenales, Chamel, and Doctor; Fig. 19.1L), abundant but fragmentary fossil material was observed at three localities. The geology is unknown in this area, but the general lithology consists of sandstones and conglomerate beds. The fossil material includes wood (which is abundant at the Chamel locality) and fragmentary plant material, as well as abundant dinosaur bones. The lack of marine fossils suggests a terrestrial paleoenvironment. Vertebrae are the most common remains, and some small bones may represent juvenile individuals. A few bones were found in situ, corresponding mainly to reworked material. R. Hernández-Rivera, M. Montellano-Ballesteros,

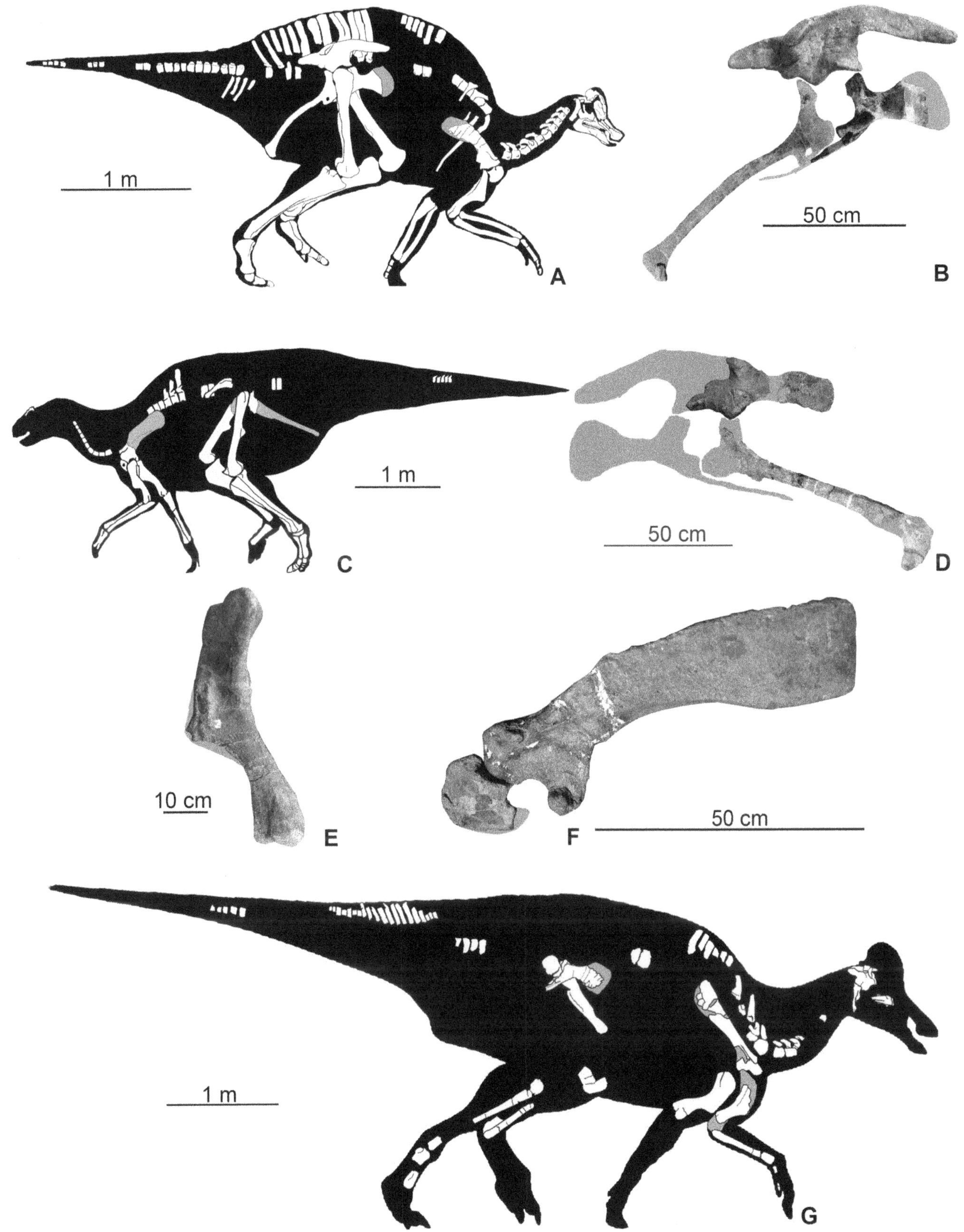

19.8. Lambeosaurinae from the Cerro del Pueblo Formation, Coahuila (the gray silhouette represents the missing parts of some bones). (A) reconstruction of the *Velafrons coahuilensis* CPC-59 showing the elements found; (B) pelvic elements of *Velafrons coahuilensis* CPC-59; (C) reconstruction of the "Isauria" specimen based on the specimen mounted in the Museo del Instituto de Geología UNAM, showing the elements found; (D) pelvic elements of the specimen Coah 14–1/3; (E) left humerus in lateral view of Isauria specimen; (F) left scapula and coracoid in lateral view of Isauria specimen; (G) reconstruction of the lambeosaurine BENC 18/1–0901.

G. Álvarez, A. Holguín, J. M. Hinojosa, R. A. Martínez-Torres, R. F. Acosta-Escobar, O. Tarango-Granados, and R. Servín-Pichardo collected vertebrae and few large bones, including a 1.3 m long femur, from the Chamel locality. These materials are under the propriety care of J. Manuel Hinojosa in the Arenales Area. A detailed paleontological and geologic study is needed for these localities.

Parras Basin: Huerta Formation and Cerro del Pueblo Formation, Coahuila State

***Velafrons coahuilensis* Gates et al., 2007** The holotype specimen (CPC-59; Fig. 19.8A–B) is characterized by the following autapomorphies: quadrate with narrow quadratojugal notch; postorbital with a dorsally positioned, high arching squamosal process; ceratobranchial with rounded anterior end; as well as the unique suite of characters, including a flat, fan-like crest created by the nasals and premaxillae, a jugal with a poorly developed posteroventral flange positioned more posteriorly than other lambeosaurine taxa and an anteriorly recurved posterior process, and a well-developed raised squamosal shelf (Gates et al., 2007).

The holotype was discovered by Hernández-Rivera and SEPC members in the Cerro de los Dinosaurios (informal name SEPC Coah 1, Quarry 7A; Fig. 19.1M). The specimen represents one of the most complete hadrosaurs known from Mexico. It is represented by much of its skull and many postcranial bones, which have yet to be described. In the phylogenetic analyses of Gates et al. (2007), the position of *Velafrons coahuilensis* was unresolved within a polytomy with *Corythosaurus*, *Olorotitan*, *Hypacrosaurus*, and *Lambeosaurus*, but the development of the crest of *Velafrons* is more similar to *Corythosaurus* and *Hypacrosaurus* than *Lambeosaurus* or *Olorotitan* (Gates et al., 2007). Subsequently, Prieto-Márquez (2010) presented two new phylogenetic hypotheses, one in which *Velafrons* and *Nipponosaurus* form a basal clade allied with *Corythosaurus*, and the other where these taxa form a clade with *Magnapaulia laticaudus* and *Hypacrosaurus altispinus*. More recently, Prieto-Márquez et al. (2012) proposed a new clade, formed by *Magnapaulia laticaudus* and *Velafrons coahuilensis*, that is inferred to be the sister clade to a monophyletic group of helmet-crested lambeosaurines (including *Hypacrosaurus*, *Olorotitan*, *Corythosaurus*, and *Lambeosaurus*).

Gates et al. (2007) proposed that CPC-59 was a juvenile (it was approximately 5 m in total body length; AAR-V, pers. obs.) and close to the developmental stage of the *Corythosaurus* specimen CMN 34825 based on the absolute size of the quadrate, nasal and dentary, as well as the presence of a frontal-prefrontal "clamp" that occurs in immature individuals of *Corythosaurus*, *Lambeosaurus*, and *Hypacrosaurus*. This observation is interesting because CMN 34825 represents an individual approximately 60% the size of the large, presumably adult *Corythosaurus casuarius* illustrated by Evans (2010). Accordingly, *Velafrons coahuilensis* probably reached a similar adult body size of 8 m or more.

Lambeosaurinae indet. During the 1980s, a team from the IGLUNAM–including Shelton P. Applegate, Hernández-Rivera, Espinosa-Arrubarrena, the SEPC team, and volunteers–excavated the remains of a dinosaur near the village of Presa de San Antonio locality in the Parras Basin (Fig. 19.1N). The specimen was found by Don Ramón López, and the subsequent site identified as quarry SPA-88-9 in field notes (field number in honor of the leader team; Espinosa-Arrubarrena et al., 1989; Contreras-Medina, 1997). This was the first dinosaur skeleton from Mexico to be mounted for public display (Fig. 19.8C). This specimen, although fairly complete, lacked a skull, but was identified as the saurolophine *Kritosaurus* sp. based on comparisons with "*Kritosaurus*" *incurvimanus*, which is now considered synonymous with *Gryposaurus notabilis* (Prieto-Márquez, 2010). Because of its importance, the skeleton was given the informal name "Isauria," and it was exhibited as a mounted cast in the Museo del Instituto de Geología, UNAM, Universum, Papalote Museum in Hidalgo State, MUDE, and several private parks. Serrano-Brañas (2006) identified one of the collected bones as a nasal, and along with the appendicular skeleton, it was referred to the genus *Gryposaurus* in her thesis. The specimen is part of a paper by Prieto-Márquez and Serrano-Brañas (2012). In the same quarry as SPA-88-9, four ischia and one fragment of an ilium (Coah 14-1/3-6; Fig. 19.8D) were collected and referred to "Corythosaurini" by Serrano-Brañas (2006).

Using the new characters proposed by Prieto-Márquez (2010) and the fact that the "Isauria" material and "Corythosaurini" material were found in the same quarry, it is presumed that all material belong to the same species, represented by several individuals of similar age. Characters including the lateral expansion of the maleolus of the fibula, the rostrodorsally curved acromion process of the scapula (Fig. 19.8E), the pronounced angulation of the deltopectoral crest of the humerus (Fig. 19.8F) and the presence of an ischial "foot" in the complete ischial elements (Fig. 19.8D) suggest that a partial herd of lambeosaurines, different from *Velafrons coahulensis* (Fig. 19.8B), was buried at this site.

Another specimen, BENC 18/1-0901 collected at the Parrita locality (Fig. 19.1M) and found by Belinda Espinosa-Chávez, Bárbara, Hernández-Rivera, and Serrano-Brañas in 2006 consists of some cranial and postcranial elements (Fig. 19.8G). The morphology of the well-developed deltopectoral crest of humerus, the acromion process of the scapula, the

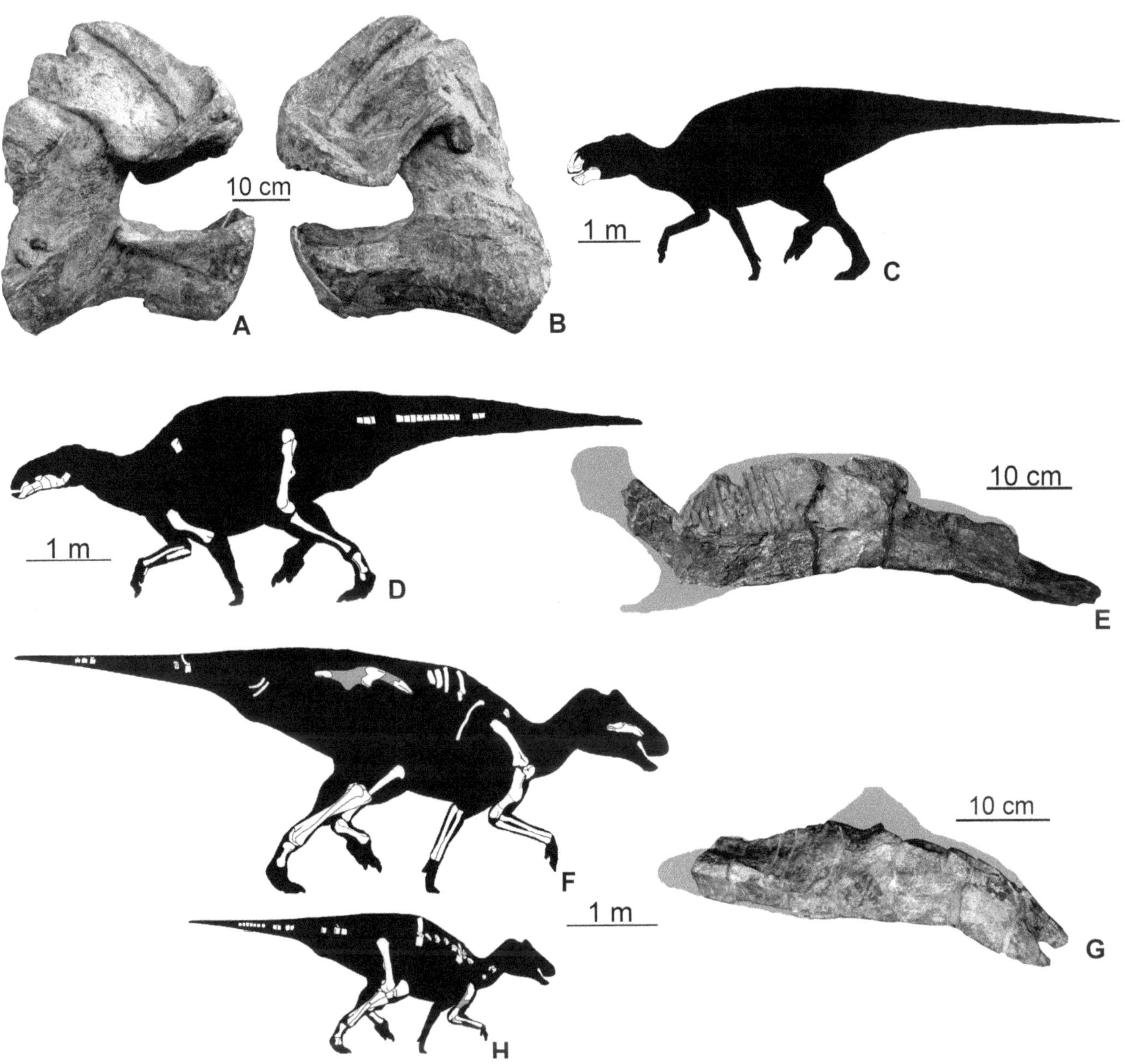

19.9. Saurolophinae from the Cerro del Pueblo Formation, Coahuila (the gray silhouette represents the missing parts of some bones). (A) anterior portion of the skull of IGM 6685 in right lateral view; (B) anterior portion of the skull of IGM 6685 in left lateral view; (C) reconstruction of the IGM 6685 showing the elements found; (D) reconstruction of the Coah 1-2/1; (E) dentary in lingual view of Coah 1-2/1; (F) reconstruction of the adult HB specimen; (G) right maxilla in lateral view of the adult HB specimen; (H) reconstruction of the juvenile HB specimen.

short pubic neck, and deep pubic blade indicate lambeosaurine affinities. BENC 18/1-0901 represents an individual approximately 8 m in length. The fossil material is still being collected (Espinosa-Chavez, pers. comm., December 2011). Eberth et al. (2003) also reported a partial lambeosaurine skull and associated postcrania in a sandstone bed hosting dinosaur tracks from Las Águilas track site (Fig. 19.1R).

Saurolophinae indet. During the expeditions of the IGLUNAM in the 1980s, José López Espinosa discovered an isolated block of sandstone containing the front portion of the skull of a hadrosaurine (IGM 6685; Fig. 19.9A–C) ~1 km north of Presa San Antonio (informal number locality Coah 14; Fig. 19.1N). Kirkland et al. (2006) referred the specimen to *Kritosaurus* sp. because the dentary is almost identical in size and shape to PASAC-1, and suggested that it could belong to the same species as the Sabinas specimen (see below).

Based on the phylogenetic analysis of Prieto-Márquez (2010), PASAC-1 was interpreted as belonging to the

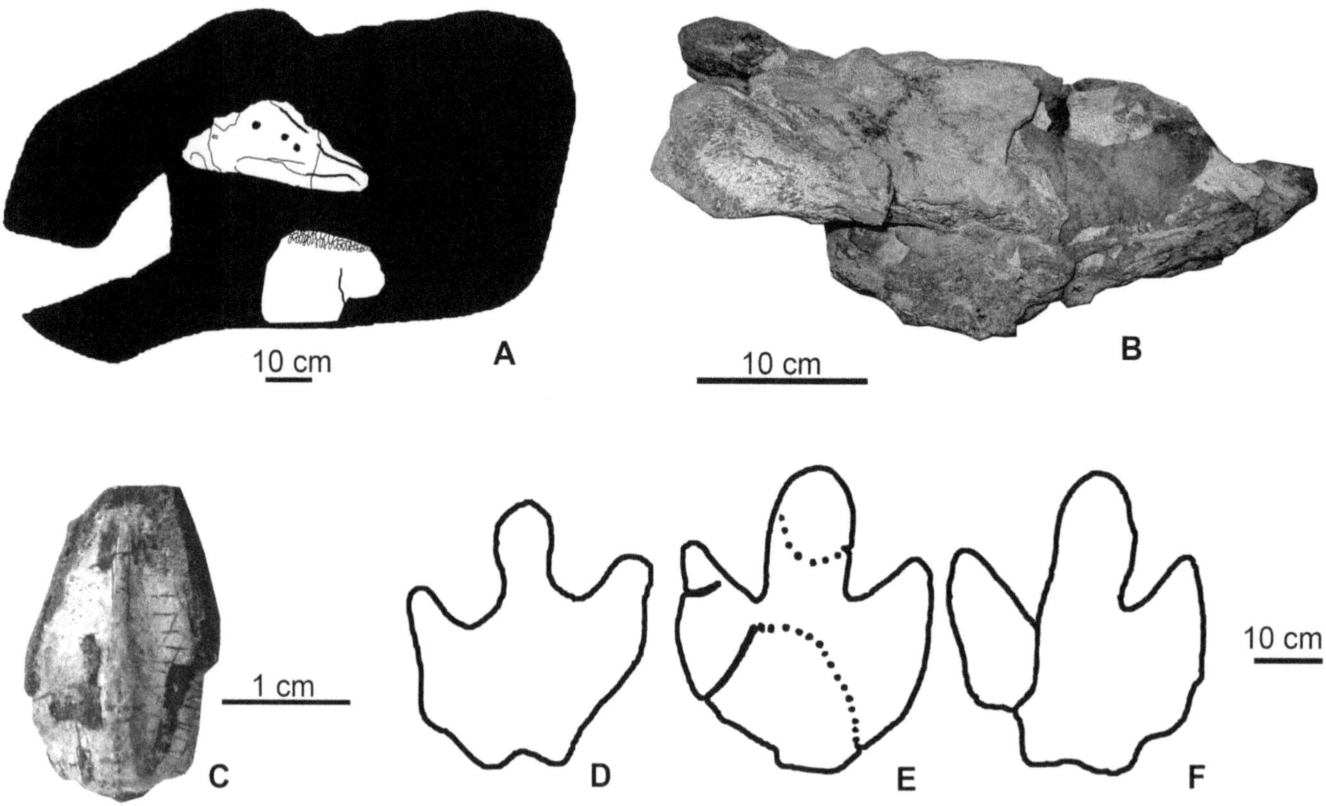

19.10. Hadrosaur body and trace fossils from the Cerro del Pueblo Formation. (A) reconstruction of the skull of BENC 1-007; (B) endocranium of BENC 4/1-0001 in right lateral view from; (C) teeth IGM 7709 in labial view; (D) silhouette from in situ track from the Las Águilas area; (E) and (F) natural casts silhouettes in MUDE exhibition from the Las Águilas area.

Saurolophus clade, and therefore IGM 6685 represents a distinct genus from *Kritosaurus*. Although the characters proposed by Prieto-Márquez (2010) – including the presence of an accessory foramen in the premaxilla and a dorsal lamina that lacks its side (Fig. 19.9A, B) – suggest affinity with the *Edmontosaurus-Brachylophosaurus* clade, it requires additional detailed study to confirm its systematic position.

Another saurolophine (Fig. 19.9D, E) collected in the Cerro de los Dinosaurios locality, in Quarry 2, was described in the thesis of Serrano-Brañas (2006). The specimen was referred to Hadrosaurinae (= Saurolophinae) mainly due to the gracile humerus proportions. One of the important features of this specimen is the elongated dentary with a coronoid process inclined caudodorsally (Serrano-Brañas, 2006; Fig. 19.9E), as in *Wulagasaurus dongi* GMH W217 (Godefroit et al., 2008), but the lack of more material and the brief description makes its taxonomic assignment difficult to assess at this time.

In the HB quarry in the Rincon Colorado locality (informal number Coah 1; Fig. 19.1M) a series of postcranial and some cranial bones belonging to several individuals (adults and juveniles) were collected that could belong to the same species (Fig. 19.9F–H). The HB saurolophine (Fig. 19.9G) likely represents a distinct taxon in comparison with the other hadrosaurs described from Mexico. The morphology of the dentary and maxilla suggest affinities with the *Saurolophus* clade (AAR-V, pers. obs.). Part of this material is housed in the Instituto de Geología, UNAM, and the rest is in the Museo del Desierto, Saltillo, Coahuila (MUDE), and is under the SEPC's custody.

At the Parrita locality (Fig. 19.1M), Espinosa Chávez collected a fragmentary saurolophine specimen (BENC 1-007, Fig. 19.10A) of robust proportions. The specimen is represented by a posterior fragment of the right dentary and a nearly complete right maxilla. When compared with other hadrosaurs, these elements exhibit some similarities to the genus *Gryposaurus* and *Kritosaurus*, but are much larger, with an estimated body length of 11 m.

Other identifiable saurolophine remains come from two localities in the municipality of Ramos Arizpe (Fig. 19.1O). Serrano-Brañas et al. (2006) reported the first braincase from the area, BENC 4/1-0001-1 (Fig. 19.10B), consisting of a natural preserved brain endocast collected from the Fraustro locality. The endocast has a similar shape to that of *Gryposaurus notabilis*, with a different morphology from that of PASAC-1. Rodriguez-de la Rosa and Cevallos-Ferriz (1998) reported only one hadrosaur tooth from the Pelillal locality (Fig. 19.10C), Ramos Arizpe municipality, found in association with theropod bones.

Hadrosauridae indet. A large number of hadrosaur remains are known from the Cerro del Pueblo Formation, but due to the poor preservation precise taxonomic assignment of the material is difficult. The hadrosaur fossils come from Ramos Arizpe, Saltillo, Parras de la Fuente, and General Cepeda municipalities in the State of Coahuila. This material has been collected by the staff of the Instituto de Geología UNAM, the Museo del Desierto, Saltillo Coahuila (MUDE), members of the Secretaría de Educación Pública de Coahuila (SEPC), the Instituto Nacional de Antropología e Historia de Saltillo (INAH), the Benemérita Escuela Normal de Coahuila (BENC), and by international collaborators – including the Utah Museum of Natural History and the Royal Tyrrell Museum of Palaeontology, among others.

The oldest collected hadrosaur remains from the Cerro del Pueblo formation come from the Hipólito locality in the municipality of Saltillo (Fig. 19.1P). These were reported by Murray et al. (1960), who collected at the base of the Difunta Group, almost in contact with the Parras Shale. According to the description of Murray et al. (1960), they may belong to the Cerro del Pueblo Formation. The fossil remains were identified as hadrosaurian by J. A. Wilson of the University of Texas and by E. H. Colbert from the American Museum of Natural History, without identification to the generic level. The Ramos Arizpe municipality (Fig. 19.1O) has yielded fragmentary remains from Rancho Quintanilla (Rivera-Sylva et al., 2011). The Los Pinos locality has produced pedal elements, an astragalus, a dentary, vertebrae, and many fragments (housed at the MUDE). A phalanx has been collected from the Barril locality (housed at MUDE), and fragmentary remains of hadrosaurs have been found at the Hedionda locality (locality number IGM 310). In addition, many caudal vertebrae from Ejido Puebla locality (housed at MUDE) have been collected in the Saltillo municipality.

Kirkland et al. (2000) reported a hadrosaur leg bone collected in 1986 by the Royal Ontario Museum (ROM) near the IGLUNAM localities B1 and B2 in the municipality of Parras de la Fuente (Fig. 19.1O). In the same area, a tooth and an incomplete maxilla from the Presa San Antonio locality (informal number Coah 14; Fig. 19.1N) were collected during the 1980s. An ilium, ulnae, femora, and metatarsals were collected from the Las Pedreras locality (locality number 13 housed in MUDE; Fig. 19.1N); fragments from Tanque (locality number IGM 2408; Fig. 19.1N); and a vertebral body and a neural arch from Dinosaurio Armado (locality number IGM 309; Fig. 19.1N).

The first report of dinosaur bones from near the town of General Cepeda was made in 1980, when the team from the IGLUNAM organized by Ismael Ferrusquia, Shelton P. Applegate, L. Espinosa Arrubarrena and Victor Torres met with José Rojas (an amateur collector from Saltillo) who showed them an almost complete skeleton with skin impressions. This material was donated to the IGLUNAM and named Rojas 1 locality and Rojas 2 locality (Hernández-Rivera, 1997; Fig. 19.1M). Material was also collected by José Rojas approximately 47 km west of Saltillo. He donated many vertebrae, long bone fragments, dentary fragments of a neonate, fibulae and a scapula from Rojas I (locality number IGM 307) and other vertebrae remains, as well as many skin impressions, ulnae, fragments of jaw, pedal elements, and fragments of a humerus from Rojas II (locality number IGM 312). The skin impressions were reported by Hernández-Rivera and Delgado-de Jesús (2000) – and it was mentioned that they were found with an associated skeleton comprised of hip, tail, and limb remains; negative relief impressions from other areas; and other isolated impressions. This material requires further detailed study because it represents one of the best preserved hadrosaurs from México.

The Cerro de los Dinosaurios locality (informal number Coah 1; Fig. 19.1M) was previously known as the Cerro de la Virgen but was renamed because of the great number of dinosaur specimens found here (Hernández-Rivera, 1997), including a tibia, vertebrae, scapulae, ulnae, ribs, phalanges and metatarsals currently housed in the Museo del Instituto de Geología UNAM. A femur, pubic bones, and fragments are housed in the SEPC collection at the MUDE.

From the Rincón Colorado area (informal number Coah 20; Fig. 19.1M), a number of limb bones from hatchlings have been collected. Vertebrae, jaws, and an astragalus were recovered from the El Carmen locality (informal number Coah 18 in MUDE; Fig. 19.1M), and vertebrae and an incomplete dentary were found at the El Valle de los Tiranos locality (in MUDE, SEPC collection; Fig. 19.1M). Other hadrosaur occurrences from this area (Fig. 19.1M) include: vertebrae and long bones from near the "Ceratopsian site" locality (informal number Coah 19); caudal vertebrae, ulna, femur, and tibia from the La Rosa locality (in the BENC collection); fragments of long bones from the Agua de Mula locality were reported by Rodriguez-de la Rosa and Ceballos-Ferriz (1998); postcranial material from the El Palmar locality; vertebrae, phalanges, and skin impressions from the Porvenir de Jalpa locality (housed in MUDE); fragments of pelvic bones from the Cruce de los Caminos (locality number IGM 318); caudal and cervical vertebrae, phalanges, and femur and tibia fragments from Hedionda Chica locality (housed in INAH, Saltillo). A large number of long bones and cranial elements was reported from the La Parrita locality by Rodriguez-de la Rosa (2007), and are housed in the BENC collections.

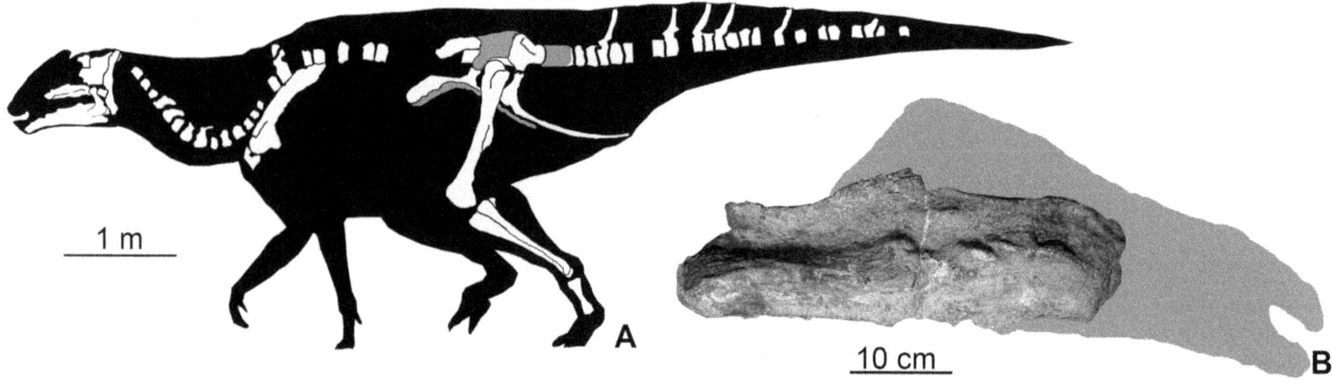

19.11. Saurolophinae specimen PASAC-1 from the Olmos Formation, Sabinas, Coahuila. (A) reconstruction of PASAC-1 showing the elements found (modified from Kirkland et al., 2006); (B) proximal part of the maxilla in lateral view (the gray silhouette represents the missing parts).

Ichnites Several localities preserving ichnites are known in the Cerro del Pueblo Formation. One of the most important sites is Las Águilas (Fig. 19.1R), near the town of Porvenir de Jalpa, Municipio General Cepeda. With an area of about 5000 m^2, it is the largest dinosaur ichnite site from Mexico (Eberth et al., 2003; Rodriguez-de la Rosa et al., 2004). This site exhibits a complex interbedded succession of coarse- and fine-grained facies, dominated by brackish-water invertebrates, and shows stratigraphic intervals with sandstones, nonmarine and marine invertebrates, as well as vertebrate fossils (bones and tracks; Eberth et al., 2003, 2004). The hadrosaur ichnites and natural casts document individuals with bipedal and quadrupedal gaits, as well as individuals of different size and age classes (ranging from 1.8 to 9 m in estimated total body lengths; Rodríguez-de la Rosa et al., 2003, 2004; Fig. 19.10D–F). In the La Parrita track site (Fig. 19.1R), footprints suggest that some of the individuals measured nearly 12 m long (Rodríguez-de la Rosa, 2007).

Despite the great amount of tracks in Las Águilas, the site has been poorly studied. Only a few abstracts and reports with little detailed information have been published to date. Further studies are necessary to understand the paleoecological and preservational context, as well as the dinosaur diversity in the area. The ichnites remain in the field, and only three natural casts are exposed in the MUDE (Fig. 19.10E, F). Casts of a hadrosaur and theropod ichnite can be found in the Museo de Sitio de Rincón Colorado.

Unknown Formation, Coahuila State

Hadrosauridae indet. Ceratopsid bones from the Soledad beds locality, Sierra Mojada municipality (Fig. 19.1W), represent the first record of dinosaur bones from Mexico (Janensch, 1926). Rivera-Sylva et al. (2007) reported a hadrosaurid vertebra from the Rancho San Miguel in the Soledad beds locality. The precise age of the rocks is unknown, but presumed by the authors to be Campanian, although detailed geological studies are lacking.

Sabinas Basin: Olmos Formation and Escondido Formation (Coahuila)

Saurolophinae indet. The second hadrosaur formally described for México, PASAC-1 (Fig. 19.11), was also the first from the Sabinas Basin within the Mezquite locality (Fig. 19.1S). This specimen was discovered in 2001 by Juan Pablo García-de la Garza and was excavated by the Association of Amateur Paleontologists of Sabinas (PASAC) in the spring of 2001. The bones were sent to the IGLUNAM in Mexico City for study under the direction of R. Hernández-Rivera (Kirkland et al., 2006). Distinctive characteristics of the PASAC-1 (Sabinas specimen) include a dorsally recurved ischium (also present in *Hadrosaurus foulkii*), a subrectangular maxilla, and a sharply down-turned anterior dentary (also present in *Kritosaurus navajovius*, AMNH 5799; Kirkland et al., 2006). Kirkland et al. (2006) identified PASAC-1 as cf. *Kritosaurus* (the rostral part of IGM 6685 was also considered as probably from the same species). Later in 2010, the systematic position of PASAC-1 was challenged by Prieto-Márquez (2010). His parsimony and Bayesian analyses suggested that the Sabinas OTU (based on PASAC-1) did not fall inside the clade composed of *Kritosaurus*, *Gryposaurus*, and *Secernosaurus*, but it appeared more closely related to the clade containing *Prosaurolophus* and *Saurolophus*. Therefore, PASAC-1 could represent a new genus and species, but the material requires more detailed study. Unfortunately, part of the material has been lost, although some bones, including the basicranium and maxilla bones (Fig. 19.11B), are housed in the Museo del Instituto de Geología UNAM.

Hadrosauridae indet. Fragmentary material of hadrosaurs from the Sabinas Basin was first collected by José Delgado in 1980, at the Polvorín locality (locality number IGM 314; Fig. 19.1S). However, it is difficult to determine the precise taxonomic level at which these specimens, as well as the material from the Palau locality (locality number IGM 313; Fig. 19.1S), can be referred. There is another area outside of the Sabinas watershed but within the Parras Basin with outcrops of the Olmos Formation, named the El Cañon del Oso locality, that belongs to the Saltillo municipality (with locality number IGM 308; Fig. 19.1Q). From here a caudal vertebra of a hadrosaur was collected in 1958 by G. P. Salas, a former director of the Instituto de Geología de la UNAM.

The collection of the IGLUNAM includes specimen IGM 322 from the Phelan locality in the Progreso municipality (Fig. 19.1U) that consists of fragmentary hadrosaur bones. The geological unit of these elements is unknown, but its geographical location within the Sabinas basin suggests they may have come from the Olmos Formation, a hypothesis that requires further verification.

Ichnites Meyer et al. (2005) reported ichnites from the Rancho Soledad locality to the southwest of Sabinas (Fig. 19.1V), which were discovered by PASAC members in 2003. This site has footprints attributed to large ornithopods. However, the report is ambiguous as to the origin of rock unit in which the ichnites were found (Olmos or Escondido formations?). Rodríguez-de la Rosa et al. (2004) report small ornithopod ichnites from Olmos Formation, but provided no illustrations or detailed description. It is our opinion that these two reports are derived from the same locality.

Mexcala Formation, Puebla State

Ichnites Bravo-Cuevas and Jiménez-Hidalgo (1996) made the first extensive report of hadrosaur ichnological remains in Mexico on the basis of footprints from Late Cretaceous Mexcala Formation near Mitepec, Puebla State (Fig. 19.1X). The material studied consists only of pes prints: 14 right (IGM-7432, IGM-7433, IGM-7434, right plaster casts and maybe IGM-7437) and 15 left (IGM-7435, IGM-7436 left plaster cast; Fig. 19.12). The tracks are from three nearby localities. Two impressions came from locality 1, and only one from locality 2. The main tracks came from locality 3.

The mean track length is 20 cm, but there are also shorter specimens. These tracks represent various small individuals that can be compared with similar samples from other parts of the world (Bravo-Cuevas and Jiménez-Hidalgo, 1996). The 10 cm long ichnites were considered by Bravo-Cuevas and Jiménez-Hidalgo (1996) as belonging to juvenile individuals, and represented a minor component of the population. They also analyzed sizes represented within the

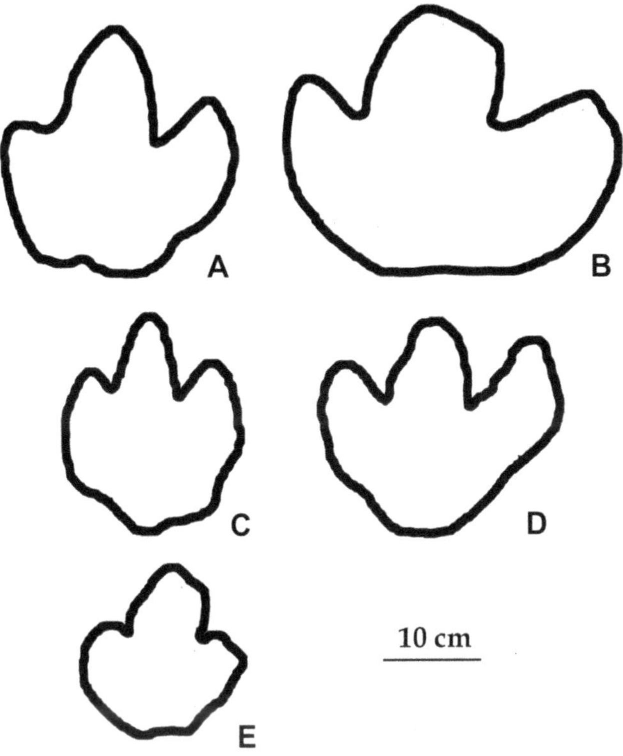

19.12. Footprint morphotypes from the Mexcala Formation, Mitepec, Puebla. (A–D) possible adult individuals; (E) possible juvenile individual. Redrawn from Bravo-Cuevas and Jiménez-Hidalgo (1996).

ichnite assemblage, and concluded that more than one age group was present. Unfortunately, neither trackways nor other defined patterns are recognizable in the distribution map published. It is important to mention that, regarding the preservation of the ichnites, most of the smallest tracks are badly preserved compared to the midsize and large footprints (Bravo-Cuevas and Jiménez-Hidalgo, 1996). It is possible that the tracks represent only juvenile individuals. In the Mitepec area, the 10–20 cm long tracks could represent individuals from approximately six months to one year in age according to the model developed by Lockley (1994). Also, a pair of isolated small sauropod ichnites have been found in the Mitepec area (right manus IGM-7439 and right pes from a single individual) and are interpreted by Bravo-Cuevas and Jiménez-Hidalgo (1996) as a juvenile sauropod. Neither hadrosaur nor sauropod tracks attain presumed adult size.

Bonebed records of the Two Medicine Formation (Varricchio and Horner, 1993; Scherzer and Varricchio, 2010) and the Cloverly Formation (Forster, 1990) of Montana suggest a tendency for juvenile hadrosaurs to travel in groups. Ichnological and osteological records suggest a period with no juvenile-adult interaction and possibly juvenile-only groups (Lockley et al., 1983; Varricchio and Horner, 1993;

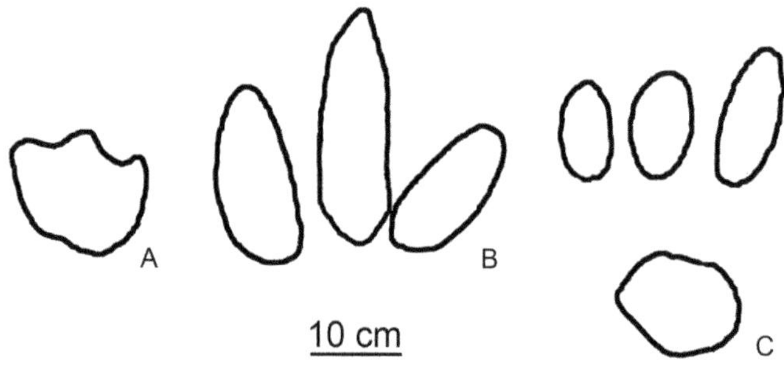

19.13. Footprint morphotypes from the Arenisca Aguililla unit, Michoacán. (A) juvenile individual; (B) and (C) adult individuals. Redrawn from Ortiz-Mendieta (2001).

Carpenter, 1999). This behavior could explain the absence of adult-sized ichnites in the Mitepec area.

As mentioned by Rivera-Sylva et al. (2006), another explanation of the presence of only small-sized individuals could be dwarfism. The authors only considered the small sauropod footprint in the area for this argument. However, this implies the possibility of an isolated environment, such as island paleogeography. There is no sufficient geologic information to affirm this scenario in the Late Cretaceous of the Mitepec area, and to demonstrate dwarfism using only tracks is very difficult (Castanera et al., 2011). With the available evidence and information, it is possible that only juvenile hadrosaur individuals are represented by these small-sized tracks.

Arenisca Aguililla Unit, Michoacán State

Ichnites The "Arenisca Aguililla" locality, Aguaje municipality, state of Michoacán (Fig. 19.1Y), shows a high paleoichnofaunal diversity which includes hadrosaur (Fig. 19.13) and theropod ichnites (Ortiz-Mendieta, 2001). However, part of this diversity could be due to variation within a single morphotype and/or extramorphologic variation (Rodríguez-de la Rosa et al., 2004). The described hadrosauroid tracks consist of nine right impressions, three left impressions, and three plaster casts (CP-XXXVI and CP-XXXVIII, catalog numbers as they appear in Ortíz-Mendieta [2001]). Due to the wide size range of the ichnites, they have been attributed to individuals of different ontogenetic ages (Ortiz-Mendieta, 2001).

CONCLUSION

Mexico is host to an extensive record of hadrosaur remains that include both body and trace fossils, including a considerable number of notable footprint localities. Hadrosaur studies based on material recovered in Mexico began around 1933 and continue to the present. These studies have been carried out by many national and international institutions, and have provided new insight into hadrosaur paleontology. Reports have increased in recent years, and new localities continue to be discovered.

Comparison of the ichnological and osteological evidence indicates that particularly large individuals of hadrosauroids are present in the following states: Baja California (*Magnapaulia laticaudus* IGM 5845); Sonora (the Alamitos specimen and the humerus ERNO-302); Chihuahua (1.30 m long femur from the Chamel locality); and from Coahuila (the Icoteas hadrosaur, the Sabinas specimen PASAC-1, IGM 6685, and the skull bones of BENC 1–007). As demonstrated by Rodriguez-de la Rosa et al. (2004) and Ortiz-Mendieta (2001), the Michoacán and Puebla ichnofaunas are the only Late Cretaceous dinosaur ichnites reported south of the Transmexican Volcanic Belt and represent the most southerly evidence of hadrosaurs in North America. The recently described *Huehuecanauhtlus tiquichensis* from Barranca Los Bonetes, Michoacán, is the southernmost of the non-hadrosaurid hadrosauroids from North America (Ramírez-Velasco et al., 2012).

Despite these advances, more work is needed on the Mexican hadrosaur assemblage. There is a wealth of undescribed fossils materials in many of the institutional and university collections in Mexico that require detailed study. The localities reported here, and others yet to be discovered, have significant potential to produce more hadrosaur fossils, and will contribute new information on hadrosaurian paleobiology, evolution, and biogeography.

ACKNOWLEDGMENTS

We thank M. Montellano-Ballesteros and G. Álvarez-Reyes (IGLUNAM), A. Holguín, J. M. Hinojosa, R. A. Martínez-Torres, R.F. Acosta-Escobar, O. Tarango-Granados, the municipal authorities in Jiménez, C.M. González-León (ERNO) and R. Pacheco Rodríguez (UNISON), J.L. Núñez Villa, E. Gámez Grijalba "Chava," P. A. Sánchez Medrano, M. Gámez Santacruz, A. R. Parra Lara and his family (R. A. Parra Ballesteros, A. E. Parra Ballesteros, J. Z. Parra Ballesteros and M. Sainz Ballesteros), and his quarry workers and municipal authorities in Fronteras and Esqueda. We thank B. Espinosa Chávez (BENC), Renata and C.A. de León, A. Amaya López, Gonzalo Ramírez, and M. R. Chavarría Martínez in Saltillo for their fieldwork support. We are grateful to M. del C. Perriliat-Montoya, V. A. Romero Mayén and J. M. Contreras Almazán (IGM), J. M. Padilla Gutiérrez (CPC), M. C. Aguillón Martínez (SEPC), Felisa Aguilar (INAH Saltillo), A. Fabiola Guzmán and J.L. Gudiño (DP), B. Espinosa Chávez (BENC), C. M. González-León (ERNO), and C.A. de León for the help on the review materials deposited in their respective paleontological collections. We thank D. Eberth, D. Evans, and O. Hinojosa Hernández for the support during the Hadrosaur Symposium. Financial support was provided by M. Montellano-Ballestero, E. Morales-Salinas, UNISON, Grupo México and PAPIIT IN-216511-2. Many thanks to L. Espinosa-Arrubarrena for careful review of the chapter. Thanks also to the anonymous reviewers for critiques that improved the chapter and to D. Evans, the editor who handled the chapter.

LITERATURE CITED

Aguillón-Martínez, M. C., I. Vallejo-González, R. Hernández-Rivera, and J. I. Kirkland. 1998. Dinosaur trackway from the Cerro del Pueblo Formation, Difunta Group (Latest Campanian, Cretaceous), Coahuila, Mexico. Journal of Vertebrate Paleontology 18(3, Supplement):23A.

Benammi, M., and M. Montellano-Ballesteros. 2006. New data from the continental Late Cretaceous faunas from northern Mexico. Journal of Vertebrate Paleontology 26(3, Supplement):41A.

Benammi, M., E. Centeno-García, E. Martínez-Hernández, M. Morales-Gámez, G. Tolson, and J. Urrutia-Fucugauchi. 2005. [Presence of dinosaurs in the Barranca Los Bonetes, southern Mexico (Tiquicheo Region, Michoacán State), and its chronostratigraphic implications]. Revista Mexicana de Ciencias Geológicas 22:429–435. [Spanish with English abstract]

Bravo-Cuevas, V., and E. Jiménez-Hidalgo. 1996. Las Dinosauricnitas de México: su significación geológico-paleontológica. Thesis, Universidad Nacional Autónoma de México, D.F., México, 147 pp.

Carpenter, K. 1999. Eggs, Nests, and Baby Dinosaurs. Indiana University Press, Bloomington, Indiana, 336 pp.

Castanera, D., J. L. Barco, I. Díaz-Martínez, J. H. Gascón, F. Pérez-Lorente, and J. I. Canudo. 2011. New evidence of a herd of titanosauriform sauropods from the lower Berriasian of the Iberian range (Spain). Palaeogeography, Palaeoclimatology, Palaeoecology 310:227–237.

Contreras-Medina, R. 1997. Origen, taxonomía y paleobiogeografía de hadrosauridae (Ornithischia: Ornithopoda) una familia de dinosaurios del Cretácico. Thesis, Universidad Nacional Autónoma de México, D.F., México, 63 pp.

Eberth, D. A., C. R. Delgado-de Jesús, J. F. Lerbekmo, D. B. Brinkman, R. A. Rodríguez-de la Rosa, and S. D. Sampson. 2004. Cerro del Pueblo Fm (Difunta Group, Upper Cretaceous), Parras Basin, southern Coahuila, Mexico: reference sections, age, and correlation. Revista Mexicana de Ciencias Geológicas 21:335–352.

Eberth, D. A., S. D. Sampson, R. A. Rodríguez-de la Rosa, M. C. Aguillón-Martínez, D. B. Brinkman, and J. López-Espinoza. 2003. Las Águilas: an unusually rich Campanian-age vertebrate locale in southern Coahuila, Mexico. Journal of Vertebrate Paleontology 23(3, Supplement):47A.

Espinosa-Arrubarrena, L., S. P. Shelton, and R. Hernández-Rivera. 1989. Crónicas de una gran expedición paleontológica. Ciencia y Desarrollo 15:23–32.

Estrada-Ruiz, E., and S. R. S. Cevallos-Ferriz. 2009. *Palmoxylon enochii* sp. nov. de la Formación Olmos (Campaniano superior-Maastrichtiano inferior), Coahuila, México. Ameghiniana 46:577–585.

Estrada-Ruiz, E., G. R. Upchurch, and S. R. S. Cevallos-Ferriz. 2008. Flora and climate of the Olmos Formation (Upper Campanian-Lower Maastrichtian), Coahuila, Mexico: a preliminary report. Gulf Coast Association of Geological Societies Transactions 58:273–283.

Evans, D. C. 2010. Cranial anatomy and systematics of *Hypacrosaurus altispinus,* and a comparative analysis of skull growth in lambeosaurine hadrosaurids (Dinosauria: Ornithischia). Zoological Journal of the Linnean Society 159:398–434.

Forster, C. A. 1990. Evidence for juvenile groups in the ornithopod dinosaur *Tenontosaurus tilletti* Ostrom. Journal of Paleontology 64:164–165.

Fraaije, R. H. B., F. J. Vega, B. W. M. van Bakel, and L. M. Garibay-Romero. 2006. Late Cretaceous dwarf decapods from Guerrero, southern Mexico and their migration patterns. Contributions to Zoology 75:121–132.

Fulbord, M. M., and C. J. Busby. 1993. Tectonic controls on non-marine sedimentation in a Cretaceous fore-arc basin, Baja California, Mexico; pp. 301–333 in L. E. Frostick, and R. J. Steel (eds.), Tectonic Controls and Signatures in Sedimentary Successions. International Association of Sedimentologists, Special Publication 20. Blackwell Scientific Publications, Oxford, U.K.

Gates, T. A., S. Sampson, C. R. Delgado de Jesús, L. E. Zanno, D. A. Eberth, R. Hernández-Rivera, M. C. Aguillón-Martínez, and J. I. Kirkland. 2007. *Velafrons coahuilensis,* a new lambeosaurine hadrosaurid (Dinosauria: Ornithopoda) from the Late Campanian Cerro del Pueblo Formation, Coahuila, Mexico. Journal of Vertebrate Paleontology 27:917–930.

Godefroit, P., H. Shulin, Y. Tingxiang, and P. Lauters. 2008. New hadrosaurid dinosaurs from the uppermost Cretaceous of northeastern China. Acta Paleontologica Polonica 53:47–74.

González-Leon, C. M., and T. Lawton. 1995. Stratigraphy, depositional environments, and origin of the Cabullona basin, northeastern Sonora; pp. 121–142 in C. Jacques-Ayala, C. M. González-León, and J. Roldán-Quintana (eds.), Studies on the Mesozoic of Sonora and Adjacent Areas. Geological Society of American Special Paper 301.

Hernández-Rivera, R. 1997. Mexican dinosaurs; pp. 433–437 in P. J. Currie, and K. Padian (eds.), Encyclopedia of Dinosaurs. Academic Press, San Diego, California.

Hernández-Rivera, R., and C. R. Delgado-de Jesús. 2000. Hadrosaur skin impressions and associated skeletal remains from Cerro del Pueblo Fm (Uppermost Campanian) southeastern Coahuila, México. Journal of Vertebrate Paleontology 20(3, Supplement):48A.

Hill, J. A. 1988. Sedimentology of delta-front sandstone, Cerro del Pueblo Formation (Upper Cretaceous), Parras Basin, Coahuila, Mexico. M.S. thesis, University of New Orleans, New Orleans, Louisiana, 163 pp.

Hilton, R. P. (ed.). 2003. Dinosaurs and other Mesozoic Reptiles of California. University of California Press, Berkeley, California, 356 pp.

Hopkins, E. M. 1965. Sedimentology of the Aguja Formation, Big Bend National Park, Brewster County, Texas. M.S. thesis, University of Texas, Austin, Texas, 165 pp.

Janensch, V. W. 1926. Dinosaurier-Reste aus Mexiko. Centralblatt für Mineralogie. Geologie und Paläontologie 1926B:192–197.

Johnson, M. E., J. Ledesma-Vázquez, and B. G. Baarli. 2006. Vertebrate remains on ancient rocky shores: a review with report on hadrosaur bones from the Upper Cretaceous of Baja California (México). Journal of Coastal Research 22:574–580.

Kappus, E. J., S. G. Lucas, and R. Langford. 2011. The Cerro de Cristo Rey Cretaceous dinosaurs tracksites, Sunland Park, New Mexico, USA, and Chihuahua, Mexico. New Mexico Museum of Natural History and Science Bulletin 53:272–288.

Kirkland, J. I., R. Hernández-Rivera, M. C. Aguillón-Martínez, C. R. Delgado-de Jesús, R. Gómez-Nuñez, and I. Vallejo-González. 2000. The Late Cretaceous Difunta Group of the Parras Basin, Coahuila, Mexico, and its vertebrate fauna; pp. 133–172 in the Society of Vertebrate Paleontology Annual Meeting Field Trip Guide Book. Universidad Autónoma del Estado de Hidalgo, Mexico.

Kirkland, J. I., R. Hernández-Rivera, T. Gates, G. S. Pau, S. Nesbitt, C. I. Serrano-Brañas, and J. P. García-de la Garza. 2006. Large hadrosaurine dinosaurs from the latest Campanian of Coahuila, Mexico. Museum of Natural History and Sciences Bulletin 35:299–315.

Langston, W., Jr., and M. H. Oakes. 1954. Hadrosaurs in Baja California. Bulletin of the Geological Society of American 65:1344. [Abstract]

Lockley, M. G. 1994. Dinosaur ontogeny and population structure: interpretations and speculations based on fossil footprints; pp. 347–365 in K. Carpenter, K. F. Hirsch, and J. R. Horner (eds.), Dinosaur Eggs and Babies. Cambridge University Press, New York.

Lockley, M. G., B. H. Young, and K. Carpenter. 1983. Hadrosaur locomotion and herding behavior: evidence from footprints in the Mesaverde Formation, Grand Mesa Coal Field, Colorado. Mountain Geologist 20:5–14.

Lucas, S. G., and C. M. González-León. 1996. Dinosaurios del Cretácico Tardío del Grupo Cabullona, Sonora. Geología del Noroeste 1:20–25.

Lucas, S. G., B. S. Kues, and C. M. González-León. 1995. Paleontology of the Upper Cretaceous Cabullona Group, northeastern Sonora; pp. 143–165 in C. Jacques-Ayala, C. M. González-León, and J. Roldán-Quintana (eds.), Studies on

the Mesozoic of Sonora and Adjacent Areas. Geological Society of America Special Paper 301.
Lucas, S. G., K. Krainer, J. A. Spielmann, and K. Durney. 2010. Cretaceous stratigraphy, paleontology, petrography, depositional environments, and cycle stratigraphy at Cerro de Cristo Rey, Doña Ana County, New Mexico. New Mexico Geology 32:103–130.
Mariscal-Ramos, C. 2006. Estudio paleontológico y magnetoestratigráfico de la localidad "Barranca los Bonetes" (Tuzantla, Michoacán). Thesis, Universidad Nacional Autónoma de México, D.F., México, 61 pp.
Meyer, C. A., E. D. Frey, B. Thüring, W. Etter, and W. Stinnesbeck. 2005. Dinosaur tracks from the Late Cretaceous Sabinas Basin (Mexico). Darmstädter Beiträge zur Naturgeschichte 14:41–45.
Molina Garza, R. S., H. N. Böhnel, and T. Hernández. 2003. Paleomagnetism of the Cretaceous Morelos and Mezcala Formations, southern Mexico. Tectonophysics 361:301–317.
Molnar, R. E. 1974. A distinctive theropod dinosaur from the Upper Cretaceous of Baja California (Mexico). Journal of Paleontology 48:1009–1017.
Monroy-Mújica, I. H. 2009. Microvertebrados fósiles cretácicos tardíos (Campaniano Tardío) de la Formación Aguja en el Noroeste de Coahuila, México. Thesis, Universidad Nacional Autónoma de México, D.F., Mexico, 111 pp.
Montaño, M. I., R. Hernández-Rivera, and M. Montellano-Ballesteros. 2009. Hadrosaurios kritosaurinos del Cretácico Tardío de Coahuila y Chihuahua, México. XI Congreso Nacional de Paleontología. Centro de Geociencias, Campus UNAM, Juriquilla, Querétaro. Abstract:47.
Montellano-Ballesteros, M., R. Hernández-Rivera, G. Álvarez-Reyes, P. Andrade-Ramos, and L. Martín-Medrano. 2000. Discovery of Late Cretaceous vertebrate local faunas in northern México. Journal of Vertebrate Paleontology 20(3, Supplement):58A–59A.
Morris, W. J. 1967. Baja California: Late Cretaceous dinosaurs. Science 155:1539–1541.
Morris, W. J. 1972. A giant hadrosaurian dinosaur from Baja California. Journal of Paleontology 46:777–779.
Morris, W. J. 1981. A new species of hadrosaurian dinosaur from the Upper Cretaceous of Baja California ?*Lambeosaurus laticaudus*. Journal of Paleontology 55:453–462.
Murray, G. E., D. R. Boyd, J. A. Wolleben, and J. A. Wilson. 1960. Late Cretaceous fossil locality, Eastern Parras Basin, Coahuila, Mexico. Journal of Paleontology 34:368–370.
Ortíz-Mendieta, J. A. 2001. Dinosauricnitas Cretácico-tardías de El Aguaje, Michoacán, región suroccidental de México y sus implicaciones geológico-paleontológicas. Thesis, Universidad Nacional Autónoma de México, D.F., México, 75 pp.
Paul, G. S. 2010. The Princeton Field Guide to Dinosaurs. Princeton University Press, Princeton, New Jersey, 320 pp.
Pessagno, E. A. 1969. Upper Cretaceous stratigraphy of the Western Gulf Coast Area of México, Texas, and Arkansas. Geological Society of America Memoir 111:1–139.
Prieto-Márquez, A. 2010. Global phylogeny of Hadrosauridae (Dinosauria: Ornithopoda) using parsimony and Bayesian methods. Zoological Journal of the Linnean Society 159:435–502.
Prieto-Márquez, A., and C. I. Serrano-Brañas. 2012. *Latirhinus uitstlani,* a "broad-nosed" saurolophine hadrosaurid (Dinosauria, Ornithopoda) from the Late Campanian (Cretaceous) of northern Mexico. Historical Biology 24:607–619.
Prieto-Márquez, A., L. M. Chiappe, and S. H. Joshi. 2012. The lambeosaurine dinosaur *Magnapaulia laticaudus* from the Late Cretaceous of Baja California, Northwestern México. PLoS ONE 7(6):e38207.
Ramírez-Velasco, A. A., M. Benammi, A. Prieto-Márquez, J. Alvarado-Ortega, and R. Hernández-Rivera. 2012. *Huehuecanauhtlus tiquichensis,* a new hadrosauroid dinosaur (Ornithischia: Ornithopoda) from the Santonian (Late Cretaceous) of Michoacán, Mexico. Canadian Journal of Earth Sciences 49:379–395.
Renne, P. R., M. M. Fulford, and C. Busby-Spera. 1991. High resolution $^{40}Ar/^{39}Ar$ chronostratigraphy of the Late Cretaceous El Gallo Formation, Baja California del Norte, Mexico. Geophysical Research Letters 18:459–462.
Rivera-Sylva, H. E., and E. Frey. 2011. The first mandible fragment of *Deinosuchus* (Eusuchia: Alligatoroidea) discovered in Coahuila, Mexico. Boletín de la Sociedad Geológica Mexicana 63:459–462.
Rivera-Sylva, H. E., E. Frey, and J. R. Guzmán-Gutiérrez. 2009. Evidence of predation on the vertebra of a hadrosaurid dinosaur from the Upper Cretaceous (Campanian) of Coahuila, Mexico. Carnets de Géologie, Letter 2:1–5.
Rivera-Sylva, H. E., D. W. E. Hone, and P. Dodson. 2012. Bite marks of a large theropod on a hadrosaur limb bone from Coahuila, Mexico. Boletín de la Sociedad Geológica Mexicana 64:155–159.
Rivera-Sylva, H. E., R. Rodríguez-de la Rosa, and J. Ortiz-Mendieta. 2006. A review of the dinosaurian record from Mexico; pp. 233–248 in F. J. Vega, T. G. Nyborg, M. d. C. Perrilliat, M. Montellano-Ballesteros, S. R. S. Cevallos-Rerriz, and S. A. Quiroz-Barroso (eds.), Studies on Mexican Paleontology. Springer, Dordrecht, Netherlands.
Rivera-Sylva, H. E, E. Frey, F. J. Palomino-Sánchez, J. R. Guzmán-Gutiérrez, and J. A. Ortiz-Mendieta. 2009. Preliminary report on a Late Cretaceous vertebrate fossil assemblage in Northwestern Coahuila, Mexico. Boletín de la Sociedad Geológica Mexicana 61:239–244.
Rivera-Sylva, H. E., J. R. Guzmán-Gutiérrez, F. J. Palomino-Sánchez, J. López-Espinosa, and I. De La Peña-Oviedo. 2007. New vertebrate fossil localities from the Late Cretaceous of Northern Coahuila, Mexico. Journal of Vertebrate Paleontology 27(3, Supplement):135A.
Rivera-Sylva, H. E., A. H. González-González, G. J. M. Padilla, O. I. De la Peña, and S. A. I. Oyervidez. 2011. Nueva localidad de la Formación Cerro del Pueblo (Cretácico Tardío: Campaniano) en Coahuila, México; pp. 118–119 in XII Congreso Nacional de Paleontología. Benemérita Universidad Autónoma de Puebla, Puebla de los Ángeles.
Rodríguez-de la Rosa, R. A. 1998. Cáscaras de huevo avianas (Neognathae) y de Ornithopoda (Dinosauria) del Cretácico Tardío de Baja California; pp. 59–60 in VI Congreso Nacional de Paleontología. Facultad de Ciencias, UNAM, México, D.F.
Rodríguez-de la Rosa, R. A. 2007. Hadrosaurian footprints from the Late Cretaceous Cerro del Pueblo Formation of Coahuila, Mexico; pp. 339–343 in E. Díaz-Martínez, and I. Rábano (eds.), 4th European Meeting on the Palaeontology and Stratigraphy of Latin America. Cuadernos del Museo Geominero 8. Instituto Geológico y Minero de España, Madrid.
Rodríguez-de la Rosa, R. A., and S. Cevallos-Ferriz. 1998. Vertebrates of the El Pelillal locality (Campanian, Cerro del Pueblo Formation), southeastern Coahuila, Mexico. Journal of Vertebrate Paleontology 18:751–764.
Rodriguez-de la Rosa, R. A., M. C. Aguillón-Martínez, J. López-Espinoza, and D. A. Eberth. 2004. The fossil record of vertebrate tracks in México. Ichnos 11:27–37.
Rodríguez-de la Rosa, R. A., D. A. Eberth, D. B. Brinkman, S. D. Sampson, and J. López-Espinoza. 2003. Dinosaur tracks from the Late Campanian Las Aguilas locality south-eastern Coahuila, México. Journal of Vertebrate Paleontology 23(3, Supplement):90A.
Scherzer, B. A., and D. J. Varricchio. 2010. Taphonomy of a juvenile lambeosaurine bonebed from the Two Medicine Formation (Campanian) of Montana, United States. Palaios 25:780–795.
Serrano-Brañas, C. I. 1994. Recolección, preparación y descripción de dinosaurios pertenecientes a la familia Hadrosauridae del Cretácico del Estado de Coahuila, México. Thesis, Universidad Simón Bolivar, México, D.F., México, 70 pp.
Serrano-Brañas, C. I. 2006. Descripción de los Dinosaurios pertenecientes a la familia Hadrosauridae del Cretácico Superior de Coahuila, México. Thesis, Universidad Nacional Autónoma de México, D.F., México, 173 pp.
Serrano-Brañas, C., R. Hernández-Rivera, R. E. Torres, and C. B. Espinosa. 2006. A natural hadrosaurid endocast from the Cerro del Pueblo Formation (upper cretaceous) of Coahuila, Mexico; pp. 317–321 in S. G. Lucas and R. M. Sullivan (eds.), Late Cretaceous Vertebrates from the Western Interior. New Mexico Museum of Natural History and Science Bulletin 35.
Servín-Pichardo, R., R. Hernández-Rivera, C. M. González-León, and R. Pacheco-Rodriguez. 2011. Primer registro de dinosauricnitas en el Grupo Cabullona (Cretácico Tardío), Esqueda, Municipio de Fronteras, Sonora; pp. 130–131 in XII Congreso Nacional de Paleontología. Benemérita Universidad Autónoma de Puebla, Puebla de los Ángeles, México.
Taliaferro, N. L. 1933. An occurrence of Upper Cretaceous sediments in northern Sonora, Mexico. Journal of Geology 41:12–37.
Tovar, R. E., D. E. Fastovsky, and M. Benammi. 2012. Calcretas pedogénicas en el Cretácico Tardío de Michoacán, México. Boletín de la Sociedad Geológica Mexicana 64:61–70.
Tyler, N., and W. A. Ambrose. 1986. Depositional systems and oil and gas plays in the Cretaceous Olmos Formation, South Texas. Bureau of Economic Geology, University of Texas Report of Investigation 152:1–42.
Varricchio, D. J., and J. Horner. 1993. Hadrosaurid and lambeosaurid bone beds from the Upper Cretaceous Two Medicine Formation of Montana: taphonomic and biologic implications. Canadian Journal of Earth Sciences 30:997–1006.

Stratigraphic Distribution of Hadrosaurids in the Upper Cretaceous Fruitland, Kirtland, and Ojo Alamo Formations, San Juan Basin, New Mexico

20

Robert M. Sullivan and Spencer G. Lucas

ABSTRACT

Hadrosaurid dinosaurs have been known from the Upper Cretaceous strata of the San Juan Basin, New Mexico, for more than 100 years. Yet their biostratigraphic distribution has been confused due to imprecise stratigraphic record keeping and taxonomic problems. Five hadrosaurid species have been named from the San Juan Basin: the hadrosaurines *Kritosaurus navajovius*, *Anasazisaurus horneri*, and *Naashoibitosaurus ostromi*; and the lambeosaurines *Parasaurolophus cyrtocristatus* and *P. tubicen*. Diagnostic specimens of these taxa are from the Kirtland Formation. No hadrosaurines have been positively confirmed in the Ojo Alamo Formation (Naashoibito Member), but a lambeosaurine, closely allied to *Corythosaurus*, is present in this unit. Fragmentary and undiagnostic hadrosaurid teeth are known from the lower part of the Fruitland Formation (Ne-nah-ne-zad Member), and large hadrosaur footprint casts (ichnogenus *Caririchnium*), presumably from a very large hadrosaurine, are known from the upper part of the Fruitland Formation (Fossil Forest Member). Whereas the two lambeosaurines, *Parasaurolophus cyrtocristatus* and *P. tubicen*, are clearly distinguished by their respective distinct narial crests, the validity of the three hadrosaurines continues to be problematic because of issues concerning inadequate type material, comparable ontogenetic stages, and intraspecific and interspecific variation. Whereas the lambeosaurines *P. cyrtocristatus* and *P. tubicen* are separated stratigraphically, the occurrences of the hadrosaurines *Kritosaurus navajovius*, *Anasazisaurus horneri*, and *Naashoibitosaurus ostromi* are clustered together stratigraphically and, in part, overlap. Of the San Juan Basin hadrosaurids, the only species-level taxon found outside of New Mexico is *P. cyrtocristatus*, which is known from Judithian strata of the Kaiparowits Formation in Utah. The genus *Parasaurolophus*, represented by the species *P. walkeri*, is known from the Judithian of Alberta, Canada. The Judithian hadrosaurid genus *Parasaurolophus* thus has its youngest record in the San Juan Basin, in strata of Kirtlandian age.

INTRODUCTION

Hadrosaurid dinosaurs have been known for more than 100 years from the Upper Cretaceous (upper Campanian–lower Maastrichtian) strata of the San Juan Basin, New Mexico, yet many of them have been the subject of much controversy regarding their taxonomic validity and their stratigraphic occurrence. Representative taxa of both subfamilies (Hadrosaurinae and Lambeosaurinae, sensu Horner et al., 2004) are known from the San Juan Basin. Although substantial numbers of isolated hadrosaurid postcranial elements have been collected over the last two decades, few nearly complete skulls have been recovered. Even though isolated dentaries, maxillae, and parts of skulls (frontals and parietals) are occasionally encountered, postcranial remains are by far the most common. Hadrosaurid specimens range throughout the entire stratigraphic section, from the lower Fruitland Formation (Ne-nah-ne-zad Member) through the lower Ojo Alamo Formation (Naashoibito Member). However, the reported formations from which they were derived have often been confused in the literature, and continue to be reported in error, and need to be clarified. Here, we reassess the stratigraphic distribution of San Juan Basin hadrosaurids by correcting erroneous stratigraphic occurrences based on recent field studies and reviewing historic documentation. This allows us to produce a more robust biostratigraphy for the San Juan Basin hadrosaurids, while noting the continued shortcomings produced by inadequate type specimens, incomplete supplemental material, and taxonomic uncertainties.

Institutional Abbreviations AMNH, American Museum of Natural History, New York; BYU, Brigham Young University, Provo, Utah; FMNH, Field Museum of Natural History, Chicago, Illinois; KUVP, University of Kansas, Lawrence, Kansas; NMMNH, New Mexico Museum of Natural History and Science, Albuquerque, New Mexico; PMU, Museum of Evolution, University of Uppsala, Uppsala, Sweden; SMP, State Museum of Pennsylvania, Harrisburg, Pennsylvania; USNM, United States National Museum of Natural History, Washington, D.C.

GEOLOGIC AND BIOSTRATIGRAPHIC OVERVIEW

Recognition and definitions of the Upper Cretaceous terrestrial strata in the San Juan Basin, New Mexico, have been in a state of flux for over a century, and disagreement as to recognition of boundaries and formal lithostratigraphic units continues to this day (Hunt and Lucas, 1992; Lucas and Sullivan, 2000; Sullivan and Lucas, 2006; Williamson and Weil, 2008; Fassett, 2010). This fact has led to some confusion with regard to the stratigraphic provenance of many important fossil vertebrates. Compounding this problem are the erroneous stratigraphic data that continue to be reported, as recently as within the last decade, for *Parasaurolophus tubicen* (Horner et al., 2004; Prieto-Márquez, 2010) and *Naashoibitosaurus ostromi* (Horner et al., 2004), despite a number of publications that have explicitly recorded the correct stratigraphic provenance of these two taxa (Sullivan and Williamson, 1999; Williamson, 1998, 2000; Williamson and Sullivan, 1998). While the formation boundaries, and recognition of subunits (members), vary among some workers, coupled with strong disagreements over usage by some (see Fassett, 2010), the stratigraphic occurrence of specimens remains firmly anchored by precise data, especially those in the form of UTMs. Most of the San Juan Basin fossils that have been collected over the last 20 years are well documented, and the implementation of GPS locators has allowed us to place some of the historic dinosaur quarries (most notably those of Sternberg from the early 1920s) in their proper stratigraphic context.

As is true in most branches of science, we recognize stratigraphic revision as a necessary outgrowth of our current knowledge and an evolving understanding of the rock record. Much as the *North American Stratigraphic Code* (North American Commission on Stratigraphic Nomenclature, 2005), so forcefully embraced by Fassett (2010), is an evolving document, so, too, do we need to recognize our evolving understanding of the rocks, their features, characterizations, and relationships–and thus to revise stratigraphic nomenclature accordingly to better reflect our current understanding of stratigraphic relationships. A synopsis of the evolution of the San Juan Basin Upper Cretaceous stratigraphic nomenclature is presented in Figure 20.1.

The recognition of distinct vertebrate faunas within the Upper Cretaceous Fruitland, Kirtland and Ojo Alamo formations was first realized by Clemens (1973), who named both the Hunter Wash and Alamo Wash local faunas for the vertebrate faunas from the upper (12 m) Fruitland and lower (17 m) Kirtland formations and the "Ojo Alamo Sandstone" (which at the time included the upper part of the Kirtland Shale; i.e., both the De-na-zin and Naashoibito members), respectively. The Hunter Wash local fauna, established mostly on isolated mammal teeth, was regarded by Clemens (1973) as having a "unique composition." Although he presented a preliminary account of fossil mammal taxa, and regarded some of them as questionable (i.e., the reported occurrence of the early Paleocene ptilodontid *Kimbetohia*), the Hunter Wash local fauna was not formally defined by Clemens (1973). Flynn (1986) reviewed some of Clemens's material and reported on additional mammalian taxa from localities in the Fruitland–Kirtland transition, which we note here are part of the Hunter Wash local fauna. A faunal list of non-mammalian vertebrates from the Hunter Wash local fauna was published by Sullivan and Lucas (2003) and revised later (Sullivan and Lucas, 2006).

The Alamo Wash local fauna was also named by Clemens (1973) for the vertebrates coming from the upper shale member (now De-na-zin Member) of the Kirtland Formation and the "Ojo Alamo Sandstone" (strata now restricted to the Naashoibito Member). Later, Lehman (1981) revised the local fauna, restricting it to fossils from the Naashoibito Member, but erroneously included some dinosaurs from the underlying De-na-zin Member. Most recently, the Alamo Wash local fauna has been revised by Jasinski et al. (2011).

Williamson and Sullivan (1998) named and defined the Willow Wash local fauna for the fossil vertebrates from the De-na-zin Member (upper shale member) of the Kirtland Formation. Faunal lists of the Willow Wash local fauna have been revised by Sullivan and Lucas (2003, 2006). Lastly, the Hunter Wash and Willow Wash local faunas collectively serve as the basis for the recognition of the Kirtlandian land-vertebrate "age," which fills the previously perceived (e.g., Russell, 1975) temporal gap between Judithian and Edmontonian time (Sullivan and Lucas, 2003, 2006).

THE SAN JUAN BASIN HADROSAURID-BEARING FORMATIONS

Fruitland Formation

Included members (in ascending order) are the Ne-nah-ne-zad Member and the Fossil Forest Member.

The Fruitland Formation was named by Bauer (1916) for an irregularly interbedded sequence of sands, shale and coal. It is a highly variable rock unit that ranges in thickness from 53 m in the northwest to 90 m in the southeast, and it was further characterized as consisting of mostly intercalated lenticular sandstones, mudstones, carbonaceous shale, and coal (Brown, 1983). Bright red clinker beds are common within the lower coaly interval of the unit. Fassett and Hinds (1971) recognized that the coal beds are the most continuous rock units within this formation and that some of these coal beds can be traced for kilometers. The lower contact of the

Fruitland Formation is conformable with the top of the Pictured Cliffs Sandstone, and following most oil geologists and consultants, Fassett and Hinds (1971) placed the contact with the overlying Kirtland Formation "at the top of the highest coal bed or carbonaceous shale bed." This Fruitland-Kirtland contact has been mapped by O'Sullivan et al. (1979), Scott et al. (1979), Brown (1983) and Mytton (1983). In contrast, Hunt and Lucas (2003b) and Lucas, Hunt, and Sullivan (2006) more closely followed Bauer's (1916) original definition (also see Bauer, 1916; Reeside, 1924) and have placed the contact between the Fruitland and Kirtland formations at the base of the Bisti bed (= base of the Kirtland Formation; see below).

Two members of the Fruitland Formation were established and recognized by Hunt and Lucas (2003b): the sparsely fossiliferous Ne-nah-ne-zad Member, characterized by a succession of thick coals and carbonaceous mudstones, with large channel sandstones and abundant sideritic concretions; and the more fossil vertebrate–rich Fossil Forest Member, consisting of sandstones, mudstones and thin coals. Lucas, Hunt, and Sullivan (2006) stated that the contact between the two members was "at the base of the first sandstone bed above the stratigraphically highest coal." Fassett (2010) criticized this subdivision on the basis of apparent ambiguity in detecting the contact in geophysical logs, not field exposures, which resulted in a thinner Ne-nah-ne-zad Member (9 m) and a thicker (75 m) Fossil Forest Member. He concluded that subdividing the Fruitland Formation into members served "no apparent useful purpose and [was] ambiguous" and rejected the member names (Fassett, 2010). Contrary to Fassett (2010), we contend that recognition of the two members not only is very useful (as specified by the North American Commission on Stratigraphic Nomenclature, 2005:1569, Article 25) in determining precise stratigraphic horizons within the formation, but in turn provides better biostratigraphic resolution for its contained fossils (plants, invertebrates, and vertebrates). Indeed, the Ne-nah-ne-zad Member is the "economic" portion of the Fruitland Formation, where its minable coal beds are located. The overlying Fossil Forest Member is the upper, more sandy interval of the Fruitland Formation that yields the stratigraphically lowest Kirtlandian vertebrate fossils. Thus, subdivision of the Fruitland Formation into the Ne-nah-ne-zad and Fossil Forest members is based on lithological differences and is of value to both economic geology and to vertebrate biostratigraphy.

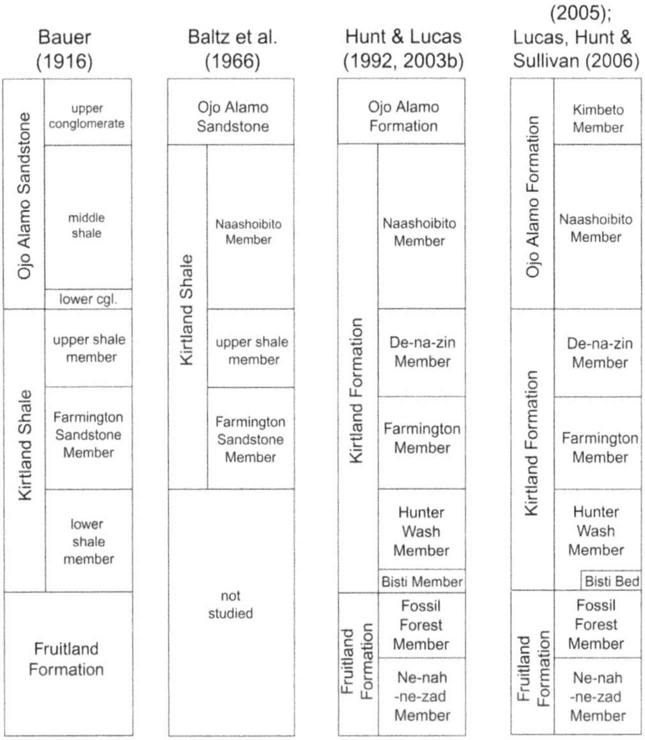

20.1. Summary of the principle changes in stratigraphic nomenclature (including groupings and ranks) for Upper Cretaceous terrestrial strata of the San Juan Basin, New Mexico.

Fruitland–Kirtland Transition

It has been recognized that the Fruitland and Kirtland formations represent a *near continuous* regressive succession of strata with the lessening accumulation of organic material, represented by coal and carbonaceous shale deposits, as the epicontinental sea regressed eastward. Therefore, placement of the boundary between the two formations has been characterized by some as being "arbitrary" (Dane, 1936; Mytton, 1983; Flynn, 1986). Two different ways of recognizing the boundary have been proposed.

Bauer (1916) recognized that the upper part of the Fruitland Formation was gradational with the lower part of the Kirtland Formation, and that in much of the west-central San Juan Basin this transition is marked by the occurrence of lenticular sandstone bodies. Reeside (1924) placed the boundary at the top of the highest sandstone, whereas Dane (1936) suggested it should be placed on top of the stratigraphically highest brown sandstone. This usage was modified by Hunt and Lucas (1992, 2003b) to identify the "stratigraphically highest brown sandstone" interval as the Bisti Member at the base of the Kirtland Formation.

Using subsurface well logs, Silver (1950) placed the Fruitland-Kirtland contact at the top of the highest coal bed. Barnes (1953) placed the contact above all major coals beds in the Fruitland Formation, whereas Barnes et al. (1954), further developing this concept, specified these coals to be ~0.4 m thick. Most subsequent geologists, working in the San Juan Basin, have placed the Fruitland-Kirtland boundary at the stratigraphically highest persistent coal bed (Fassett and

Hinds, 1971; Flynn, 1986; Fassett, 2010). This method has been adopted by such workers as Scott et al. (1979), Schneider et al. (1979), Brown (1983), Mytton (1983), and Strobell et al. (1985).

Fassett (2010) continues to advocate the usage put forth by Fassett and Hinds (1971) because "the channel-sandstone beds in the interval near the Fruitland-Kirtland contact were discontinuous, random, and erratically positioned, stratigraphically." He argued that a "relatively consistent top of the Fruitland can be located on geophysical logs (albeit with a degree of uncertainty) on the basis of the highest coal bed or carbonaceous mudstone" and went on to say that "generally speaking, there is no useful geologic purpose achieved in trying to fix this contact" (Fassett, 2010:123). However, the effectiveness of using well log analysis to identify this contact should not be the determining factor for establishing its stratigraphic utility. Rather, there is an essential need to establish an unambiguous datum from which one can stratigraphically measure sections in outcrop. This need far outweighs any subsurface correlations of coals or carbonaceous shales, which are often uncertain by Fassett's (2010) own admission.

In an attempt to clarify the stratigraphy, the boundary definition between the Fruitland and Kirtland formations was recently reviewed and revised by Lucas, Hunt, and Sullivan (2006). The Bisti bed, earlier recognized as the Bisti Member by Hunt (1984), was recognized for a laterally extensive succession of very thin sandstones at the base of the Kirtland Formation (Hunt and Lucas, 1992). It was reduced to bed status by Lucas, Hunt, and Sullivan (2006). Bauer's (1916) original usage placed the boundary between the Fruitland and Kirtland formations at the top of a series of sandstone lenses, whereas Lucas, Hunt, and Sullivan (2006), following the work of Hunt and Lucas (1992, 2003b), place the boundary below the sandstones (Bisti bed). Unfortunately, the Bisti bed is not present throughout the San Juan Basin, although the bed can be followed in outcrop from the original Fruitland type area along the San Juan River (Bauer, 1916) to the south along the arroyos that drain into the Chaco River, perhaps as far south as Escavada Wash, a distance of at least 80 km (Hunt and Lucas, 1992; Lucas, Hunt, and Sullivan, 2006). Outside of that area, it is often impossible to determine whether one is in the upper Fruitland or lower Kirtland because this bed is absent. This is especially true in parts of the south-central to southeastern regions of the San Juan Basin where exposures of the Bisti bed are intermittent or absent altogether.

Thus, the presence of the Bisti bed can be useful for delineating the Fruitland Formation from the Kirtland Formation to the north, but it does not work to the south. Importantly, the area in which the Bisti bed is well exposed is the area from which all significant vertebrate fossils have been collected from the Fruitland and Kirtland formations. The Bisti bed is a very useful lithostratigraphic construct for those developing a vertebrate biostratigraphy on outcrop in the Upper Cretaceous strata of the San Juan Basin. However, from a practical standpoint, the Bisti bed has not been mapped on any USGS topographic quadrangle, whereas the uppermost coal has (Scott et al., 1979; Schneider et al., 1979; Brown, 1983; Mytton, 1983; and Strobell et al., 1985), in order to delineate the two formations.

Kirtland Formation (Kirtland Shale)

Included members (in ascending order) are the Hunter Wash Member (= lower shale member of Bauer, 1916), the Farmington Member (= Farmington Sandstone Member of Bauer, 1916), and the De-na-zin Member (= upper shale member of Bauer, 1916).

Bauer (1916) named the Kirtland Formation and divided it into three recognizable units (parts): (1) a lower shale; (2) a medial sandy interval (which he named the "Farmington sandstone member" [*sic*]); and (3) an upper shale. Although he in part characterized the lower and upper shale units as essentially the same, and mapped them as such, he did describe their lithology as being different. Reeside (1924) refined these differences and presented a detailed measured type section for the Kirtland along the San Juan River that included the three members. He considered only one of the members (the Farmington Sandstone Member) as a formal unit.

Hunter Wash Member (Lower Shale Member) Bauer's (1916) "lower shale member" was further characterized by Reeside (1924:22) as being mainly a "gray shale with some brown and black carbonaceous layers and a minor amount of bluish, greenish, and yellow sandy shale and soft, easily weathered gray-white sandstones all irregularly bedded." For many years, the informal name "lower shale member" had been used in publications, including maps (e.g., Scott et al., 1979; Brown, 1982, 1983; Mytton, 1983). The formal name Hunter Wash Member was established for this unit by Hunt and Lucas (1992), and a type section, previously illustrated by Reeside (1924:pl. 2, section 15), was designated, that encompasses much of the lower part of the exposures in the Bisti-Hunter Wash area.

Fassett (2010) argued that the name Hunter Wash Member proposed by Hunt and Lucas (1992) violated the NACSN's (2005) *"rule of priority and preservation by well-established names."* However, this rule (NACSN, 2005:1562, Article 7[c]) applies only to formally named units, not to informally named units, such as the "lower shale member" and "upper shale member" (see below) of Bauer (1916). We agree with Fassett (2010) that the original formal naming of the Hunter Wash Member (Hunt and Lucas, 1992) lacked "clear purpose" as specified in Article 5 of the North American Stratigraphic Code (North American Commission on Stratigraphic Nomenclature, 2005:1561), but it has become well established through usage by vertebrate paleontologists in numerous scientific publications, thus meeting this criterion (see Article 5, Remark [a]). Its usage since 1992 is by far more prevalent than that of the "lower shale member" (e.g., Hunt and Lucas, 1992; Sullivan and Williamson, 1999; Lucas and Sullivan, 2000; Sullivan, 2006; Sullivan and Lucas, 2003, 2006). The same is true for the De-na-zin Member (see below).

Farmington Member (Farmington Sandstone Member) Bauer (1916) coined the formal term "Farmington sandstone member [*sic*]" for extensive, sandstone-dominated exposures along the San Juan River (138 m thick) that thin to the southeast, disappearing at the head of Coal Creek. Isolated remnant lenses of this unit were identified cropping out to the south and southeast (Reeside, 1924; Fassett and Hinds, 1971). The Farmington Sandstone Member was characterized by Bauer (1916) as consisting of variably thick, cross-bedded sandstone lenses, having its base composed of friable yellowish sandstone with clay-pebble inclusions, while the top of the Farmington Sandstone Member is distinguished by a resistant, fresh gray to weathered dark chocolate-brown colored horizon. Reeside (1924:22) recognized two parts within the Farmington Sandstone of Bauer (1916): "a lower, soft, easily eroded yellow to white sandstone containing clay pellets and locally sandstone balls of material like the matrix and reaching 4 inches in diameter, and an upper hard sandstone which is fine grained, dark gray on a fresh surface and dark ferruginous brown on weathered surfaces." Thus, this unit is not truly a homogeneous sandstone. Later, Fassett and Hinds (1971) included the overlying upper shale (i.e., the De-na-zin Member) in their definition of the Farmington Sandstone Member. The formal name of this member (Farmington Sandstone Member of Fassett and Hinds [1971]) was revised by Hunt and Lucas (1992), who designated it the Farmington Member in recognition that it also included mudstones, siltstones, and conglomerates, further illustrating that the "Farmington Sandstone" is not a homogeneous sandstone, as the original formal name implied.

Recently, Fassett (2010) rejected the term "Farmington Member" based on grounds of the long-standing usage of "Farmington Sandstone" in the literature. We continue to advocate the usage of Farmington Member following *Article 18, Remarks.* (a) *change in lithic designation* of the North American Stratigraphic Code (North American Commission on Stratigraphic Nomenclature, 2005:1565). Indeed, we find it ironic (and inconsistent) that Fassett (2010) argues for retention of this lithologic (sandstone) descriptor while he continues to include the "upper shale member" (redefined as the De-na-zin Member, see below) of Bauer (1916) in his definition and characterization of the "Farmington Sandstone" (Fassett and Hinds, 1971; Fassett, 2010).

De-na-zin Member (Upper Shale Member) The De-na-zin Member was named by Hunt and Lucas (1992) for the "upper shale member" of the Kirtland Formation first recognized by Bauer (1916). However, their characterization of the lithology ("green mudstone, siltstone and minor sandstone") and their designation of the type section/area (NW1/4, sec. 19, T24N, R11W) coincides with strata mapped by Bauer (1916:pl. LXIX), and later by Brown (1983), as the Farmington Sandstone. This area is approximately 1.2 km south to southwest of outcrops mapped as the "upper shale member." The actual De-na-zin Member is best exposed at the head of Hunter Wash between the east and west forks along the north half of section 3, T24N, R12W (see locality map and measured sections C and D in figs. 7 and 8 of Lucas and Sullivan [2000]). This area is here designated as the principle reference section for the De-na-zin Member. The lithology is consistent with that described for the "upper shale member" by Bauer (1916:275) as "shale and lenses of easily weathered grey-white sandstone, and thus very similar to the lower part. It is banded in many places with various colors, such as appear in the lower part of the Kirtland, but yellow, blue-gray, and purplish beds are more common."

Fassett and Hinds (1971) included the upper shale member (De-na-zin Member) in the Farmington Member, and Fassett (2009, 2010) continues to do so. That Fassett does not recognize the upper shale member as a stratigraphic unit distinct from the Farmington Member, coupled with his argument that formal naming of the unit and its usage is "entrenched" in the literature, contradict the evidence that it has been mapped as a lithostratigraphic unit distinct from the "lower shale member" and "Farmington Sandstone Member" by numerous geologists, and it has been referred to as the De-na-zin Member by more authors since its formalization in 1992 than those who referred to an "upper shale member" during the 76 years between Bauer (1916) and Hunt and Lucas (1992).

Ojo Alamo Formation (Ojo Alamo Sandstone)

Included members (in ascending order) are the Naashoibito Member and the Kimbeto Member.

Overview The Ojo Alamo Formation has had a complex history that has already been discussed and reviewed by numerous workers (Clemens, 1973; Fassett, 1973, 2009, 2010; Powell, 1973; Lucas and Sullivan, 2000; Fassett et al., 2002; Sullivan and Lucas, 2003; Sullivan et al., 2005). In brief, the name is derived from beds identified by Brown (1910) for dinosaur-bearing *shales* lying below a "conformable" conglomerate along Ojo Alamo Arroyo. Sinclair and Granger (1914) found that in the vicinity of Ojo Alamo and Barrel Springs Arroyo (= De-na-zin Wash of present day usage) the dinosaur-bearing shales were separated by a thin conglomerate that they called the "lower conglomerate." Bauer (1916:276) noted that Brown (1910) recognized the unit to be "essentially a sandstone including lenses of shale and conglomerate." Bauer (1916) went on to say "it seems desirable to call it Ojo Alamo sandstone and to define it as consisting on Ojo Alamo Arroyo of *two conglomeratic beds and the shale lenses which they include*" (italics ours).

Clemens (1973) presented a thorough and cogent historical review of the evolution of the lithostratigraphic construct of the Ojo Alamo Sandstone, from Bauer (1916) to 1973, and the reader is referred to his detailed account. Suffice it to say, a landmark paper by Baltz et al. (1966) changed the conceptual framework and nomenclature of the relevant strata, restricting the name Ojo Alamo Sandstone to the upper conglomerate (conglomeratic sandstones with fossil logs of Sinclair and Granger, 1914) and naming the "upper shales with dinosaurs" and the "lower conglomerate" the Naashoibito Member and placing this member in the Kirtland Formation. Although, on the face of it, the Naashoibito Member has some similarities to the upper shale member (De-na-zin Member) of the Kirtland Formation, it is predominately sandier, and thus more like the Ojo Alamo Sandstone sensu Baltz et al. (1966).

For reasons of biostratigraphic clarity, Clemens (1973) posited that it might be desirable (i.e., more logical) to return to Bauer's original lithologic definition of the Ojo Alamo, which included the lower conglomerate and upper shales with dinosaurs to avoid having an "Ojo Alamo fauna" divorced from its lithologic namesake, the Ojo Alamo Sandstone (sensu stricto) of Baltz et al. (1966). This would result in having two formal members within the Ojo Alamo Sandstone, the Naashoibito Member of Baltz et al. (1966), and the then newly named Kimbeto Member of Powell (1973) given to the Ojo Alamo Sandstone sensu stricto of Baltz et al. (1966).

Critical to our understanding of these units is not only their lithologies, but their lower and upper boundaries. Although originally characterized as being conformable, the lower and upper boundaries of the Naashoibito Member are, in fact, unconformable (e.g., Baltz et al., 1966; Sikkink, 1987; Hunt and Lucas, 1992; Lucas and Sullivan, 2000; Sullivan and Lucas, 2006). Flynn (1986:5), following Baltz et al. (1966), restricted the "'Ojo Alamo Formation' to designate the thick conglomerate capping the section at South Mesa" and argued for lumping the Naashoibito and De-na-zin members together based "on the grounds of lithological homogeneity." Fassett (2010) continues to use Bauer's (1916) concept of the Ojo Alamo Sandstone, whereas we advocate a modification of Bauer's original usage as presented by Sullivan et al. (2005), and as illustrated in Figure 20.1.

Naashoibito Member The Naashoibito Member was recognized and named by Baltz et al. (1966) and was placed in the Kirtland Formation for reasons of having similar lithology. The lower conglomerate of the Naashoibito Member has a base that is unconformable on the underlying De-na-zin Member of the Kirtland Formation. At this unconformity, a hiatus of approximately 4 Ma separates the Naashoibito Member (Ojo Alamo Formation) from the underlying De-na-zin Member (Kirtland Formation sensu stricto) based on the highest Ash J date of 73.04 Ma (Fassett and Steiner, 1997) for the top of the De-na-zin Member (Kirtland Formation) and correlation of *Alamosaurus sanjuanensis* in the Big Bend region of Texas, which has been dated at approximately 69 Ma (Sullivan and Lucas 2003, 2006; Lehman et al., 2006; Jasinski et al., 2011). The top of the Naashoibito Member is also unconformable with the overlying Kimbeto Member.

Lucas and Sullivan (2000) placed the Naashoibito Member in the Ojo Alamo Formation based on similar lithology and recognized that the Ojo Alamo Formation was thus not exclusively Paleocene in age. The Naashoibito Member was rejected as a valid member by Fassett (2010) based on his equivocal assertion that the unit, as part of the Ojo Alamo Sandstone, is of Paleocene age – an age which has been rejected by numerous workers, most recently Lucas et al. (2009), Koenig et al. (2012), Ludwig (2012), and Renne and Goodwin (2012). Moreover, Fassett (2010:115) advocated that the name "'Naashoibito' be abandoned because there is no distinct, mappable, lithologically consistent group of rocks to which the name can be consistently applied, as required by the North American Stratigraphic Code." First, being "mappable" is not a requirement of formal member status (NACSN, 2005:Article 25, Remarks [a]:1569), despite the fact that, contrary to Fassett's assertion, the Naashoibito Member is mappable (see, for example, Lehman, 1981:text-fig. 9.2; Jasinski et al., 2011:fig.1) and has a type locality and type section (Baltz et al., 1966:D20, locality 5). Moreover, the Naashoibito Member is lithologically distinct from the overlying Kimbeto and underlying De-na-zin members, and its upper and lower contacts are marked by unconformities (Sullivan et al., 2005). Indeed, it may be defensible to raise its rank to that of formation, but for now we follow Sullivan et al. (2005) and retain it as a member of the bipartite Ojo Alamo Formation. Lastly, the Naashoibito Member is characterized by a distinctive early Maastrichtian (*not* Paleocene) vertebrate fauna, the Alamo Wash local fauna (see below).

Kimbeto Member This unit was named by Powell (1973) for the upper part (= "upper conglomerate" of Bauer, 1916) of the Ojo Alamo "Sandstone" (believed to be Paleocene age) that lies unconformable above the Upper Cretaceous Naashoibito Member of Baltz et al. (1966). Concomitantly, Powell (1973) proposed that the name Naashoibito Member be retained for the middle "shales" and lower conglomerate of the Ojo Alamo "Sandstone." Lucas and Sullivan (2000), based largely on the presence of dinosaur fossils from a locally sandstone dominated Naashoibito Member in the region of Betonnie Tsosie Wash, recognized the need to include the Naashoibito Member and the Kimbeto Member in the Ojo Alamo Formation as originally advocated by Powell (1973). Fassett (2010) rejected this formal member status on the same spurious grounds already discussed above for the Naashoibito Member.

LOCAL FAUNAS

Three local faunas have been recognized in the Upper Cretaceous rocks of the San Juan Basin (oldest to youngest): the Hunter Wash local fauna, the Willow Wash local fauna and the Alamo Wash local fauna.

Hunter Wash Local Fauna

The Hunter Wash local fauna was named for the fossil vertebrate assemblage from the upper Fruitland Formation (Fossil Forest Member) and from the lower Kirtland Formation (Hunter Wash Member). The original definition was based largely on isolated mammal teeth and the recognition that the fauna was distinct from the younger mammalian faunas of the Lance and Hell Creek formations and Bug Creek Ant Hills (Clemens, 1973). Flynn (1986) regarded Clemens's "local fauna" as an assemblage and added other stratigraphically younger assemblages, from laterally distant localities, to be included in the concept of the Hunter Wash local fauna. In recognition that many lower vertebrates have biostratigraphic utility, Sullivan and Lucas (2003, 2006) produced a list of lower vertebrate taxa for the Hunter Wash local fauna and for the Willow Wash local fauna (see below). Since 2006, additional taxa have been recovered that further serve to differentiate the Hunter Wash local fauna (Burns and Sullivan, 2011; Jasinski and Sullivan, 2011; Sullivan and Fowler, 2011) from the younger Willow Wash local fauna (Williamson and Sullivan, 1998). The Hunter Wash local fauna and the younger Willow Wash local fauna, in combination, include the vertebrate fossil assemblages that characterize the Kirtlandian land vertebrate age of Sullivan and Lucas (2003, 2006).

Willow Wash Local Fauna

The Willow Wash local fauna was established by Williamson and Sullivan (1998) for the vertebrate fossil assemblage from the De-na-zin Member of the Kirtland Formation. Key taxa include *Parasaurolophus tubicen* and *Naashoibitosaurus ostromi*. If valid, *Kritosaurus navajovius* is also a member of this local fauna. Some taxa have been removed from this local fauna (notably *Pentaceratops sternbergi*, which has not been unequivocally documented for this fauna) since its inception, although others have been added. The Willow Wash local fauna and the older Hunter Wash local fauna in combination include the vertebrates that characterize the Kirtlandian land vertebrate age (Sullivan and Lucas, 2003, 2006).

Alamo Wash Local Fauna

Clemens (1973) named the Alamo Wash local fauna for the fossil vertebrate assemblage from the upper part (i.e., "upper shale," now De-na-zin Member) of the Kirtland Formation and from the Naashoibito Member in the vicinity of the old Ojo Alamo trading post. Unfortunately, the fossil vertebrates from this fauna were not listed or defined by Clemens (1973). Clemens (1973) recognized early on that there was an "incomplete understanding of terrestrial faunal evolution within the San Juan Basin area during the Late Cretaceous." In fact, relatively little collecting had occurred up to Clemens's time, and the little collecting that had transpired was sporadic at best and far from systematic.

With the reconnaissance collecting that was brought about by the 1977 Bureau of Land Management Paleontological Survey of the Bisti–Star Lake area (Kues et al., 1977) and subsequent collecting, new and more specimens from the Kirtland and Ojo Alamo formations emerged. Lehman (1981) was the first to publish on the vertebrates from the Naashoibito Member in any comprehensive way. In his important paper, Lehman (1981) restricted the term "Alamo Wash local fauna" to the vertebrates from the Naashoibito Member, but he unwittingly included some vertebrate taxa (dinosaurs), notably *Parasaurolophus tubicen* (see Sullivan and Williamson, 1999), from the underlying De-na-zin Member (upper part of the Kirtland Shale of Bauer [1916]). Unfortunately, at the time of Lehman's study the vertebrate fauna from the Naashoibito Member had not been well sampled and thus was not very taxonomically diverse. The only dinosaur that stood out in this assemblage was the titanosaurian *Alamosaurus sanjuanensis*.

In an attempt to clarify the biostratigraphic distribution of fossil vertebrates in the Naashoibito Member, as well as the underlying De-na-zin Member, intensive collecting by the State Museum of Pennsylvania and the New Mexico Museum of Natural History and Science commenced in the late 1990s. During the last two decades, a number of papers have been published on new fossil vertebrates from both the Naashoibito and De-na-zin members as a direct result of these combined field efforts, as well as a major synthetic work published recently by Jasinski et al. (2011) redefining the taxa of the Alamo Wash local fauna (Naashoibito Member).

HADROSAURIDS FROM THE SAN JUAN BASIN: AN OVERVIEW

In 1904, Brown collected an incomplete skull of the first hadrosaurid from New Mexico's Cretaceous strata ("Ojo Alamo beds") and named it *Kritosaurus navajovius* (Brown, 1910). Ten years later, Sinclair and Granger (1914:302) reported on "a trachodont maxilla (AMNH 5797; see Williamson, 2000:205), that was collected a few feet above the conglomerate separating the two horizons at which the dinosaurs were found." In a subsequent paper, Gilmore (1916:283) noted that Brown identified the maxilla "as pertaining to *Kritosaurus navajovius*." However no justification was given for this identification. Gilmore (1916) also listed a number of indeterminate hadrosaurid specimens without reference to institutional catalog numbers (Gilmore, 1919, 1935). Gilmore (1935) reported on the occurrence of a partial hadrosaurid skull (USNM 8629) which he referred to *Kritosaurus navajovius*, and a nearly complete tail (USNM 13492; Fig. 20.2), which he referred to *Parasaurolophus tubicen*. Both these specimens are discussed below.

Charles H. Sternberg, venerable dinosaur collector of the late nineteenth and early twentieth centuries, collected dinosaurs and other fossil vertebrates from the Upper Cretaceous rocks of the San Juan Basin from 1921–23, in what are now known as the Ah-shi-sle-pah Wilderness Study Area, Fossil Forest Research Natural Area, and the Bisti/De-na-zin Wilderness (Hunt, 1991; Hunt and Lucas, 2003b; Mateer, 1981; Sternberg, 1933; Sullivan, 2006; Sullivan and Lucas, 2011). Two notable Sternberg discoveries were the holotypes of the lambeosaurine *Parasaurolophus tubicen*, described and named by Wiman (1931), and *Parasaurolophus cyrtocristatus*, described and named many years after its discovery by Ostrom (1961, 1963). These specimens most certainly came from the present day Bisti/De-na-zin Wilderness and Fossil Forest Research Natural Area, respectively (Sullivan and Williamson, 1999; Hunt and Lucas, 2003b).

These two regions, together with others, were the focus of a paleontological survey in the mid-1970s for the Bureau of Land Management that identified numerous fossil vertebrate localities in an unpublished report (Kues et al., 1977) that resulted in the establishment of the protected lands that contain these sites. Scientific papers published as a result of this survey include those by Lehman (1981, 1985) on the Alamo Wash local fauna from the Naashoibito Member and that erroneously included the type of *Parasaurolophus tubicen* in the Naashoibito Member (see Sullivan and Williamson, 1999; and below), while Mateer (1981) reviewed some of the fossil vertebrates, most notably the holotype of *Kritosaurus navajovius* from the Kirtland Formation. Lucas (1981) published a review of the paleoecology and fossil vertebrate assemblages of the Upper Cretaceous Fruitland and Kirtland formations. Lucas et al. (1987) summarized the previously reported occurrences of hadrosaurids and other dinosaurs from the San Juan Basin Cretaceous. Later, Hunt and Lucas (1991) reported on the first lambeosaurine from the Naashoibito Member based on an incomplete and disarticulated skeleton. Hall (1993) provided the first account of a juvenile hadrosaurid (KUVP-96980) from the Fruitland/Kirtland Formation of the Fossil Forest Research Natural Area.

Williamson (2000) provided the first detailed comprehensive review of the San Juan Basin hadrosaurids and provided a list of specimens in the NMMNH collections in an appendix. He argued for the synonymy of *P. cyrtocristatus* with *P. tubicen*, proposing that the former is a subadult of the latter; i.e., that the short-crested forms are representative of an earlier (younger) ontogenetic stage. He also suggested that *Naashoibitosaurus ostromi* and *Anasazisaurus horneri* were synonyms of *Kritosaurus navajovius* based on unproven ontogenetic arguments and similar stratigraphic provenance. The taxon *Anasazisaurus horneri* was recently reassessed–and its taxonomic validity substantiated–by Lucas, Spielmann, et al. (2006). In another study, Prieto-Márquez (2010) also considered *Kritosaurus navajovius* to be valid, with *Naashoibitosaurus ostromi* and *Anasazisaurus horneri* as synonyms. Sullivan, Lucas, and Jasinski (2011) reported on a humerus (SMP VP-2202) pertaining to a hatchling *Parasaurolophus tubicen* from the De-na-zin Member, and Sullivan, Jasinski, et al. (2011) reported on additional lambeosaurine material (a nearly complete left humerus SMP VP-2263 and a badly weathered right jugal SMP VP-1534) from the Naashoibito Member.

The principal repositories for hadrosaurid specimens from the San Juan Basin are those listed in the institutional abbreviations (see above). In addition, we provide a list of

20.2. USNM 13492, articulated tail attributed to *Parasaurolophus* (Gilmore, 1935) in the field in 1929. The exact locality is unknown and presumably the specimen is from the Hunter Wash Member (Kirtland Formation) near Brimhall Wash (see text). Courtesy of Smithsonian Institution, National Museum of Natural History.

hadrosaur specimens from the collections of the State Museum of Pennsylvania for the first time, and update the list of hadrosaurid specimens from the Fruitland, Kirtland, and Ojo Alamo formations acquired by the NMMNH since 2000 (Appendix 20.1).

STRATIGRAPHIC DISTRIBUTION OF THE SAN JUAN BASIN HADROSAURIDS

HADROSAURIDAE Cope, 1869
HADROSAURINAE Lambe, 1918
KRITOSAURUS NAVAJOVIUS Brown, 1910
Holotype: AMNH 5799
Formation: Kirtland Formation
Member: De-na-zin Member
Age: Late Campanian (late Kirtlandian)
Fauna: Willow Wash local fauna

Kritosaurus navajovius is based on an incomplete skull that is missing the greater anterior part in front of the orbits. The holotype skull (AMNH 5799) was described by Brown (1910) as being that of an "old aged individual" with most of the sutures of the skull "obliterated." The skull was originally restored without the nasals, and its muzzle was modeled after "*Trachodon,*" with a subsequent restoration supposedly incorporating one of the nasals to show the nasal arch. Lull and Wright (1942:166) showed that the skull is missing the greater anterior part in front of the orbits and noted that the "form of the nasal hump [is] quite conjectural." Williamson (2000) figured and described an "incomplete left nasal" that was left out of the snout reconstruction on the holotype skull of *Kritosaurus navajovius,* contrary to the claim by Lull and Wright (1942) that they had been incorporated, placed in their "normal position" (Williamson, 2000:figs. 6d, e). Williamson (2000) concluded that the preserved "nasal" has the same morphology as that seen in *Gryposaurus, Anasazisaurus* and *Naashoibitosaurus.* However, Kirkland et al. (2006) identified the "nasal" element of Williamson (2000) as the lower portion of the right premaxilla (Kirkland et al., 2006:figs. 17c–f). A separate fragment (Kirkland et al., 2006:figs. 17a, b) was identified as pertaining to the region behind the narial opening, thus shedding doubt on Williamson's (2000) interpretation and synonymies. Horner et al. (2004) considered *Kritosaurus navajovius* a nomen dubium and it was excluded from their analysis. But, in the same paper, they contradicted this assessment by stating that *Kritosaurus navajovius* (and *Anasazisaurus horneri*) were "diagnostic at the species level" (Horner et al., 2004:460).

Gilmore (1916) revised the stratigraphic provenance of the holotype skull of *Kritosaurus navajovius* and determined that it came from the "Kirtland Shale" (i.e., De-na-zin Member), as its stratigraphic occurrence was not mentioned by Brown (1910).

Two key problems have plagued this dinosaur taxon: its precise stratigraphic occurrence, and disagreement on the validity of the taxon. Compounding its taxonomic problems, it has been considered by some to be synonymous with *Gryposaurus* Lambe, 1914 (in part) from the chronostratigraphically older, and geographically separate, Dinosaur Park Formation, Alberta, Canada. Lull and Wright (1942:164) synonymized *Gryposaurus* with *Kritosaurus* based on the fact that "no feature of generic value found in the one which is not present or possible in the other" and that all other differences were specific ones.

Mateer (1981:55) described a few hadrosaurid specimens from the University of Uppsala, Museum of Evolution collections that he referred to "*Kritosaurus* cf. *navajovius*" although he regarded this taxon to be a "dustbin species." The specimens were collected by C. H. Sternberg in 1921 from the Meyers Creek region, now known as Ah-shi-sle-pah Wash (Sullivan, 2006; Sullivan and Lucas, 2011). The principal dinosaur-bearing exposures along the south branch and main section of the wash drainage are all in the lower Kirtland Formation (Hunter Wash Member). The majority of the fossil specimens in the Uppsala collections are from this region. The elements Mateer (1981) briefly described (left humerus, left ilium, left pubis, left ischium, and sternum) are just some of the many hadrosaurid specimens (many of which were incorrectly numbered by Mateer [1981]; see Sullivan and Williamson [1997]) collected by Sternberg from the exposures of Ah-shi-sle-pah Wash. Numerous incomplete vertebrae (dorsal, caudal), incomplete ribs, haemal spines, phalanges, and maxillae fragments were also recovered.

Gilmore (1935) referred a partial skull (USNM 8629) to *Kritosaurus navajovius* without discussion and without illustration. According to Gilmore (1935:161), the specimen was collected by J. B. Reeside, Jr., in 1916, "4 miles southwest of Kimbetoh [*sic*]," which places the locality in Ah-shi-sle-pah Wash, in the lower part of the Hunter Wash Member of the Kirtland Formation. This is well below the stratigraphic interval (De-na-zin Member) from which the holotype of *Kritosaurus navajovius* originated. USNM 8629 consists of a number of cranial elements, including the greater posterior portion of the skull, together with both quadrates, the left dentary/coronoid, and the first three cervical vertebrae. The dentary of this specimen is identical to another (SMP VP-2705) from the upper Fruitland Formation (Fossil Forest Member) (RMS, pers. obs.).

USNM 8629 was listed by Lull and Wright (1942:25) but not included under the taxonomic account as a referred specimen. Later, the specimen was referred to the genus *Kritosaurus* by Lucas et al. (1987) and referred to *K. navajovius* by Horner (1992), in both papers without discussion. These two taxonomic assignments were presumably based solely on Gilmore's original taxonomic assignment. Williamson (2000) was the first to illustrate this specimen, but like those who preceded him, he did not provide a detailed description of it.

ANASAZISAURUS HORNERI Hunt and Lucas, 1993
Holotype: BYU 12950
Formation: Kirtland Formation
Member: upper part of the Hunter Wash Member
Age: Late Campanian (middle Kirtlandian)
Fauna: Hunter Wash local fauna

Anasazisaurus horneri was named based on an incomplete crushed skull (BYU 12950) by Hunt and Lucas (1993), purportedly from the Farmington Member of the Kirtland Formation of Kimbeto Wash. Horner (1992) had previously identified the specimen as *Kritosaurus navajovius* and presented a reconstruction of the skull. He considered BYU 12950 and NMMNH P-16106 to be *Kritosaurus navajovius*, and commented that the differences in the shape of the nasals could be attributed to ontogeny. Later, Horner et al. (2004) considered *Anasazisaurus horneri* to be valid at the species level, but did not include it their analysis, nor did they suggest what genus would be considered valid for the reception of this species. They also cited it as coming from the "Lower Kirtland Formation" without specific reference to member.

Williamson (2000) correctly identified the geographic site of the holotype skull as Betonnie Tsosie Wash, not Kimbeto Wash, and later, Lucas, Spielmann, et al. (2006) reiterated its Farmington Member provenance. However, the geology here has been mapped as the lower Kirtland by Brown (1983), which is now the Hunter Wash Member. The site from which the skull was obtained is located in Section 8, T22N, R9W, and more precisely is the historic site of "KU-16," known by the University of Kansas Natural History Museum as the Black Ridge locality. One of us (RMS) has collected there a number of times and rejects the notion that this is the Farmington Member, whereas the other (SGL) maintains that the locality is in the Farmington Member (and see Hunt and Lucas, 1992; Lucas and Sullivan, 2000; Lucas, Spielmann, et al., 2006). The Farmington Member, previously referred to as the Farmington Sandstone, has a rather limited exposure, confined to areas near the heads of Hunter and Willow and along the east part near the heads of De-na-zin and Alamo washes. It pinches out to the southeast and it is absent in regions of Ah-shi-sle-pah and Kimbeto washes (Brown, 1983). Moreover, this unit is rather unfossiliferous – compared to the lower and upper parts of the Kirtland Formation, Hunter Wash, and De-na-zin members, respectively – and few fossils have been collected from this unit.

The Farmington Member is made up largely of dusky yellow, yellowish gray to yellowish brown sandstones, interbedded with siltstones and mudstones. The exposures at KU-16 are consistent with the lithology of the upper part of the Hunter Wash Member – light tan mudstones and sandstones known elsewhere in the San Juan Basin, notably those exposed in the eastern part of Ah-shi-sle-pah WSA and exposures southeast of Alamo Mesa (Bisti/De-na-zin Wilderness). Regardless of which nomenclature one follows, the stratigraphic position is the same relative to the fossiliferous lower part of the Kirtland Formation and the overlying upper part (De-na-zin Member). Thus, the holotype of *Anasazisaurus horneri* comes from the upper part of the Hunter Wash Member, probably only a few meters below the occurrences of the holotypes of *Kritosaurus navajovius* and *Naashoibitosaurus ostromi*.

NAASHOIBITOSAURUS OSTROMI Hunt and Lucas, 1993
Holotype: NMMNH P-16106
Formation: Kirtland Formation
Member: De-na-zin Member
Age: Late Campanian (late Kirtlandian)
Fauna: Willow Wash local fauna

Hunt and Lucas (1993) erected the taxon *Naashoibitosaurus ostromi* based on a skull (NMMNH P-16106) that Horner (1992) had previously assigned to *Kritosaurus navajovius*. The specimen was first figured by Hunt and Lucas (1992) with the assigned number NMMNH P-1041, and identified as "*Edmontosaurus saskatchewanensis*" based on the jugal and concomitant shape of the temporal fenestra, elongation of the postorbital process, and the right angle formed by the ventral margin of the lateral temporal fenestra. Horner et al. (2004) suggested that *Naashoibitosaurus* belonged to a clade including *Saurolophus* and "*Kritosaurus*" *australis* by virtue of "having a jugal whose pointed rostral process is symmetrically triangular in shape" known also in a few lambeosaurines, thus arguably a primitive feature with little or no phylogenetic weight.

The holotype of *Naashoibitosaurus ostromi* was mistakenly cited by Horner et al. (2004) as coming from the "Lower Kirtland Formation," but, in fact, it was originally published as coming from the Naashoibito Member (Hunt and Lucas, 1992), which was previously considered the uppermost member of the Kirtland Formation but is now recognized as part of the Ojo Alamo Formation (Sullivan et al., 2005; Jasinski et al., 2011). However, it has been determined that the holotype of *Naashoibitosaurus ostromi* is from the lower part of the De-na-zin Member (= "upper shale member" of Bauer, 1916) of the Kirtland Formation (Williamson, 1998, 2000; Williamson and Sullivan, 1998).

LAMBEOSAURINAE Parks, 1923
PARASAUROLOPHINI Evans and Reisz, 2007
PARASAUROLOPHUS Parks, 1922

Two species of *Parasaurolophus* occur in the Upper Cretaceous deposits of the San Juan Basin, *P. cyrtocristatus*, from the lower Kirtland (Hunter Wash Member), and *P. tubicen*, from the upper Kirtland Formation (De-na-zin Member; Sullivan and Williamson, 1999). There is at least 1.5 million years separating them based on ash dates published by Fassett and Steiner (1997), and there is no evidence for co-occurrence of these two species in either member. Regardless, Williamson (2000) considered *Parasaurolophus cyrtocristatus* to be a synonym of *P. tubicen* based on the thesis that the shorter crest in the former species represents a younger individual. However, given the small sample size of these dinosaurs, plus the fact that all known short-crested forms (including three specimens from the Kaiparowits Formation, Utah) occur in older (lower) strata and that the two long-crested specimens of *P. tubicen* occur in the higher (younger) strata, dated around 73.4–73.0 Ma (Sullivan and Williamson, 1999), and in the absence of any histological evidence to support synonymy, we regard these two forms as distinct species. Evans et al. (2009) reported that the long-crested species *P. walkeri* from Alberta, Canada, is from the base of the Dinosaur Park Formation, which has been dated at 76.5–75.3 Ma. This date is similar to, but slightly older than, the dates for *P. cyrtocristatus* specimens from the Kaiparowits and upper Fruitland/lower Kirtland formations of 75.5–74.5 Ma.

PARASAUROLOPHUS CYRTOCRISTATUS
Ostrom, 1961
Holotype: FMNH P27393
Formation: Kirtland Formation
Member: lower Hunter Wash Member
Age: Late Campanian (early Kirtlandian)
Fauna: Hunter Wash local fauna

Parasaurolophus cyrtocristatus was first named and diagnosed in a brief note by Ostrom (1961), and the holotype specimen (FMNH P27393), an incomplete skull consisting mostly of a short narial crest and nearly complete postcranial skeleton, was cited by him as coming from the "Maastrichtian? Fruitland Formation." Ostrom's short paper was followed two years later by a more thorough contribution describing both the skull and postcranial skeleton (Ostrom, 1963). The stratigraphic position of *P. cyrtocristatus* has been reviewed by Wolberg, Hall, Bellis et al. (1988), and more recently by Sullivan and Williamson (1999), who confirmed that it originated from the area of Coal Creek, which is approximately 9 miles (14.4 km) northeast of Tsaya. This area is today known as the Fossil Forest Research Natural Area and is one of the parcels of federally protected lands in the San Juan Basin under the jurisdiction of the Bureau of Land Management. The area was mapped by Brown (1983) and Strobell et al. (1985), and both maps show the Fruitland and Kirtland contact (i.e., the highest coal bed) along the western part of the study area.

Hunt (1991) reported a series of abandoned quarries in the Fossil Forest Natural Research area. Of the five prominent quarries, two may have been made by Charles H. Sternberg in 1922–23 (Hunt, 1991; Hunt and Lucas, 2003b). One may be the type locality of *Parasaurolophus cyrtocristatus* and the other the type locality of *Pentaceratops sternbergi*. It is also possible that one of these two quarries may be the site of Stovall's *Pentaceratops* skeleton, OMNH (Oklahoma Museum of Natural History) 10165, collected in 1940 (Hunt and Lucas, 2003b). Regardless, these quarries, which include "Quarry 1" of Rigby and Wolberg (1987), are located on the eastern side of the Fruitland-Kirtland boundary (as mapped by Brown, 1983 and Strobell et al., 1985). Thus, they are in the lower part of the Kirtland Formation (Hunter Wash Member), not the Fruitland as interpreted by Hunt (1991) and Hunt and Lucas (2003b). Consequently, all of the vertebrate fossils reported by Rigby and Wolberg (1987), Wolberg, Hall, Bellis et al. (1988) and Hall (1991, 1993) are also from the lower part of the Kirtland Formation and not from the Fruitland Formation. The fossil stump field, from which the Fossil Forest Member derives its name, and which is the subject of a recent study by Davies-Vollum et al. (2011), is also in the lower part of the Kirtland Formation (Hunter Wash Member) based on the mapping by Brown (1983) and Strobell et al. (1985). This stump field is similar to another extensive stump field in the Hunter Wash Member located within the Ah-shi-sle-pah WSA and may be correlative to it (RMS, pers. obs.). The Ah-shi-sle-pah stump field is located on a flat, adjacent to the famous Kirtland bluffs on the south side of "Meyers Creek," where C. H. Sternberg collected many fossil vertebrates that are now housed in the collections of the Museum of Evolution, Uppsala University (Sullivan, 2006; Sullivan and Lucas, 2011). If this correlation is correct, then the exposures of the Fossil Forest Member are restricted to the western side of the Fruitland-Kirtland boundary in the Fossil Forest Natural Research area.

PARASAUROLOPHUS TUBICEN Wiman, 1931
Holotype: PMU R.1250
Formation: Kirtland Formation
Member: De-na-zin Member

Age: Late Campanian (late Kirtlandian)
Fauna: Willow Wash local fauna

Wiman (1931) named *Parasaurolophus tubicen* for an incomplete skull and complete narial crest (PMU R.1250). The specimen was found in the upper ("Ojo Alamo") beds (sensu Sinclair and Granger, 1914) of the Kirtland Formation and presumably at the same level as the holotype of *Kritosaurus navajovius* (i.e., below the lower conglomerate, in the De-na-zin Member). No specific locality was given by Charles H. Sternberg, who collected the specimen in 1921. This runs contrary to Lehman (1981) and Mateer (1981), who assumed that the specimen originated from the Naashoibito Member. Sullivan and Williamson (1999) reported on a new, nearly complete skull (NMMNH P-25100) from the De-na-zin Member, strongly supporting the likely stratigraphic position of the holotype as coming from below the lower conglomerate (i.e., the De-na-zin Member). Thus, *Parasaurolophus tubicen* was removed as a component of the Alamo Wash local fauna, and included as a member of the Willow Wash local fauna (Williamson and Sullivan, 1998). Unfortunately, to this day, there still continue to be erroneous citations of its provenance, such as coming from the "lower Kirtland" (i.e., below the De-na-zin Member, formerly the "upper shales"; Horner et al., 2004:442), which only confuses matters further.

PARASAUROLOPHUS sp.
Referred material: USNM 13492, SMP VP-1090
Formation: Kirtland Formation
Member: lower Hunter Wash Member
Age: Late Campanian (early Kirtlandian)
Fauna: Hunter Wash local fauna

Two important specimens, referable to *Parasaurolophus* sp. from the Kirtland Formation, are worth mentioning. The first, USNM 13492, was collected in 1929 by N. H. Boss and described by Gilmore (1935), consisting of an articulated tail (Fig. 20.2), posterior end of the left ilium, left femur and posterior half of the right maxilla. The right maxilla has since been determined not to be from the same specimen and is now cataloged under a different number (USNM 358621) and is from an indeterminate locality. USNM 13492 "was collected in T25N, R13W, about 6 miles [~10 km] north of Hunter's Store (Bisti P.O.)" (Gilmore, 1935). Unfortunately this is a rather imprecise geographic description covering a vast 93 km^2 area, so the precise stratigraphic horizon from which the specimen originated is uncertain. It is said to have come from the Kirtland Formation, but from exactly where in the formation is not known. This region is now part of the Navajo Nation. The specimen is likely from the Hunter Wash Member because the outcrop area of this unit widens towards the northwest, and it is likely referable to *P. cyrtocristatus* based on biostratigraphic grounds, although this is somewhat conjectural.

The second specimen is an incomplete juvenile skull (SMP VP-1090) lacking the crest (Sullivan and Bennett, 2000). It is from the Kirtland Formation, and not the Fruitland Formation as initially reported (Sullivan, 2006). The specimen comes from the lower part of the Hunter Wash Member and is also probably referable to *Parasaurolophus cyrtocristatus*, although this cannot be demonstrated with any degree of certainty. It comes from about the same stratigraphic level (low in the Hunter Wash Member) as the holotype (FMNH P27393) of *P. cyrtocristatus*, assuming that the holotype came from one of the quarries of the Fossil Forest NRA.

LAMBEOSAURINI Sullivan, Jasinski, Guenther, and Lucas, 2011
LAMBEOSAURINI INDET.
Referred material: NMMNH P-19147, SMP VP-1534, 2263
Formation: Ojo Alamo Formation
Member: Naashoibito Member
Age: early Maastrichtian (late Edmontonian)
Fauna: Alamo Wash local fauna

In 1991, Hunt and Lucas described a partial hadrosaurid skeleton (NMMNH P-19147) from the Naashoibito Member (Ojo Alamo Formation) that they interpreted as belonging to the subfamily Lambeosaurinae, either *Lambeosaurus* or *Corythosaurus*, based on the morphology of the pubis. Williamson (2000) reinterpreted it as being a hadrosaurine using different criteria. Sullivan, Jasinski, et al. (2011) also reassessed the identity of NMMNH P-19147, using characters of the pubis and scapula identified by Prieto-Márquez (2010), and confirmed the specimen to be a lambeosaurine, contrary to the conclusion reached by Williamson (2000).

Sullivan, Jasinski, et al. (2011) also reported on two additional hadrosaurid specimens from the Naashoibito Member, a badly weathered right jugal (SMP VP-1534) and a nearly complete left humerus (SMP VP-2263). Based on characters used by Prieto-Márquez (2010), Sullivan, Jasinski, et al. (2011) indicated that these specimens are close to the Tribe Lambeosaurini–defined by them as the clade (*Lambeosaurus* [*Corythosaurus* + *Hypacrosaurus*]).

The presence of lambeosaurines in the Naashoibito Member is consistent with the interpretation that the Naashoibito Member is late early or early late Maastrichtian age (~69 Ma), though we note that there is no agreement on boundary criteria for the substages within the Maastrichtian Stage (Ogg et al., 2004). The presence of lambeosaurines would seem to exclude the possibility of the Naashoibito Member being late Maastrichian, as lambeosaurines are not known from late Maastrichtian strata, represented by the Lancian land vertebrate age in North America (Campione and Evans, 2011; Jasinski et al., 2011; Sullivan, Jasinski, et al., 2011). Moreover, the Naashoibito Member cannot be construed to be of "Lancian age," which has a maximum duration of 65.58–67.61 Ma, with the base of the *Triceratops* zone at 66.8 ± 1.1 Ma (Cifelli et al., 2004), and therefore does not correspond to the entire late Maastrichtian substage. Curiously, for the present, no hadrosaurines have been positively identified from the Naashoibito Member, only lambeosaurines. Based on their presence elsewhere in younger strata in North America (see Campione and Evans, 2011), we predict the presence of hadrosaurines in the San Juan Basin during this time interval.

ICHNOTAXON
CARIRICHNIUM Leonardi, 1984
CARIRICHNIUM isp.
Referred material: NMMNH P-41045, 63031–63035
Formation: Fruitland Formation
Member: Fossil Forest Member (lower part)
Age: Late Campanian (early Kirtlandian)
Fauna: Hunter Wash local fauna

Wolberg, Hall, and Bellis (1988) briefly reported on one complete and two incomplete tridactyl footprint casts from the Fruitland Formation near, but outside the southwestern part of the present day Bisti/De-na-zin Wilderness. The footprint casts were located in a succession of gray-weathering mudstones, coals and clinker deposits of the Ne-nah-ne-zad Member and were not referenced by specimen numbers. Presumably the specimens were left in the field.

Later, Hunt and Lucas (2003a) described an additional footprint cast (NMMNH P-7145) from the lower part of the Fossil Forest Member (Fruitland Formation) in the Fossil Forest Research Natural Area. This occurrence was later followed by the report of a number of large sandstone footprint casts, also in the lower part of the Fossil Forest Member (Lucas et al., 2011). Lucas et al. (2011) characterized these footprint casts (NMMNH P-nos. 63031–63035) as being tridactyl, with a relative large central digit, flanked on both sides by smaller and shorter digits, with a rounded and broad heel. These tracks were referred to the ichnotaxon *Caririchnium* (Lucas et al., 2011). In all three reports it was suggested that the trackmaker was probably a very large hadrosaur, more specifically, a hadrosaurine.

TAXONOMIC DIVERSITY OF THE SAN JUAN BASIN HADROSAURIDS

Understanding the taxonomic diversity of hadrosaurids from the San Juan Basin has been thwarted by the fact that many taxa are known from incomplete or inadequate material. To complicate matters, generic and specific identity has been largely based on skull material, and even where more than half the skull is preserved, there is often doubt as to which taxon one is dealing with. This is certainly true concerning the holotypes of *Kritosaurus navajovius, Anasazisaurus horneri* and *Naashoibitosaurus ostromi.* While the stratigraphic positions of the holotype specimens are now largely agreed upon, the taxonomic validity of these hadrosaurids, and thus hadrosaurid taxonomic diversity, is not.

Here is what we know based on our present understanding of San Juan Basin hadrosaurid taxa and their respective stratigraphic distribution (Fig. 20.3). The hadrosaurines *Kritosaurus navajovius, Anasazisaurus horneri* and *Naashoibitosaurus ostromi* are clustered together and, in part, overlap, coming from the upper part of the Hunter Wash/Farmington and De-na-zin members. Moreover, these three hadrosaurine taxa are also of questionable taxonomic status. All may represent a single taxon ("*Kritosaurus*"), or there may be as many as three distinct taxa, but this is not agreed on. It is hoped that additional studies will resolve some of these uncertainties. Additional material (SMP VP-2705 and USNM 8926) from the Fossil Forest Member (Fruitland Formation) and the Hunter Wash Member (Kirtland Formation) may be referable to yet another hadrosaurine taxon. Thus, the biostratigraphic ranges of one, two, three, or four of these hadrosaurines rests on their respective taxonomic validity. Of the San Juan Basin hadrosaurids, the only species-level taxon found outside of New Mexico is *P. cyrtocristatus,* which is known from Judithian strata of the Kaiparowits Formation in Utah. The genus *Parasaurolophus* is also known from the Dinosaur Park Formation (Judithian) of Alberta, Canada. The hadrosaur genus *Parasaurolophus* thus has its youngest record in the San Juan Basin, in strata of Kirtlandian age.

CONCLUSIONS

Hadrosaurids occur throughout the Upper Cretaceous Fruitland, Kirtland, and Ojo Alamo formations of the San Juan Basin, New Mexico. Confirmation of the generic and specific identities of the San Juan Basin hadrosaurines–*Kritosaurus, Anasazisaurus,* and *Naashoibitosaurus*–still remain to be fully resolved, and more complete specimens, especially skulls, need to be discovered to better understand the taxonomic diversity and biostratigraphic distribution of the San Juan Basin hadrosaurids. The hadrosaurid genus

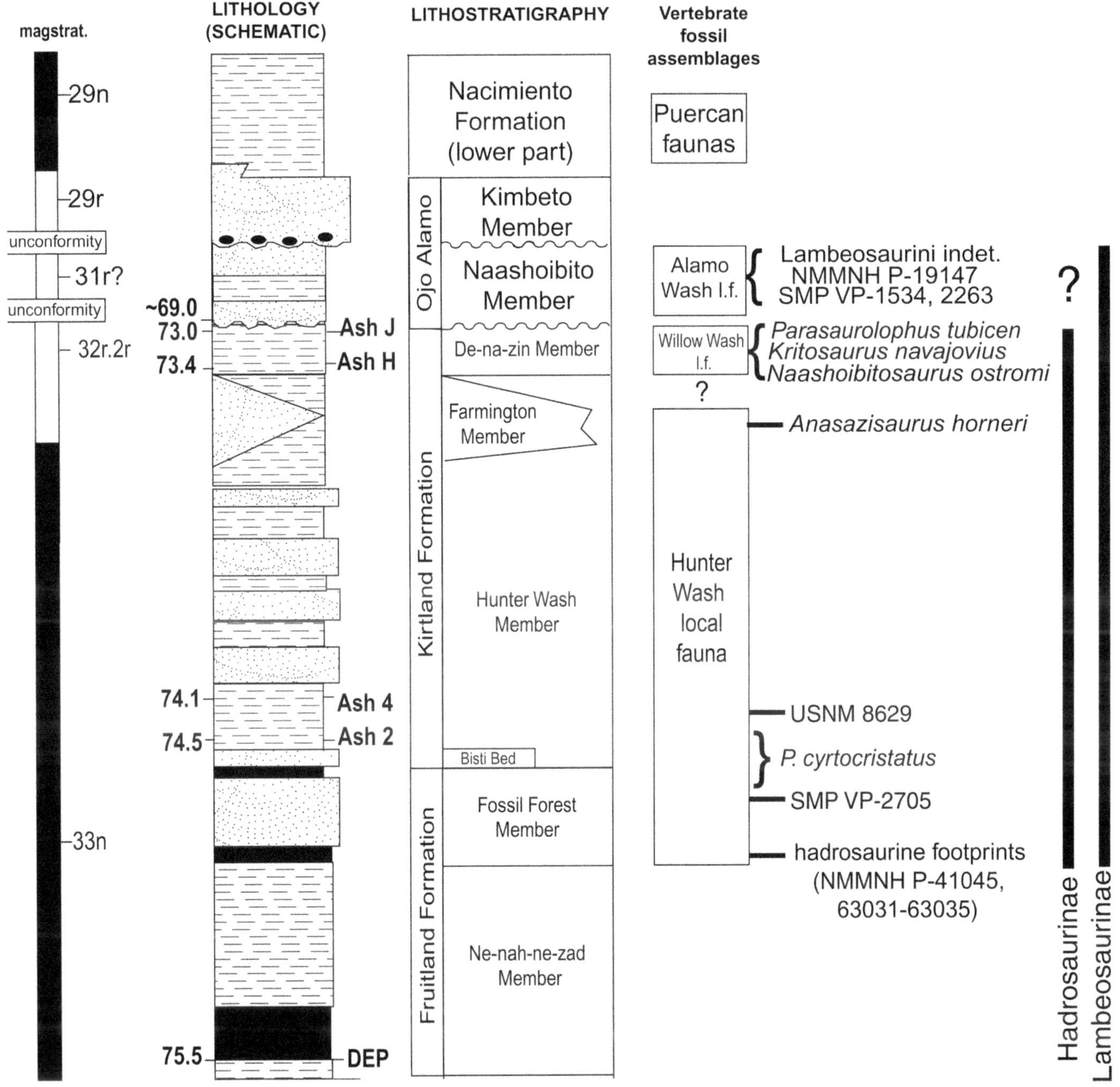

20.3. Stratigraphic distribution of hadrosaurid taxa and some key specimens in the San Juan Basin, New Mexico. Holotype specimens (indicated by their binomial) and other key specimens (identified by institutional number) indicate their relative stratigraphic positions.

Parasaurolophus has its youngest record in the San Juan Basin, in strata of Kirtlandian age, whereas the Naashoibito lambeosaurines, exemplified by the *Corythosaurus*-like jugal (Sullivan, Jasinski, et al., 2011), are the youngest known occurrence of this clade, and thus support a "pre-Lancian age" for the Naashoibito Member and all of its vertebrates (Alamo Wash local fauna).

ACKNOWLEDGMENTS

We thank the late Larry Martin and Miao Desui (University of Kansas) for access to the University of Kansas field data; the late Solweig Stuenes and Jan Ove Ebbestad (Museum of Evolution, Uppsala University) for access to the collections and supporting collections data over the course of two separate visits by RMS; and William Simpson (Field Museum of Natural History) for data concerning *Parasaurolophus cyrtocristatus.* Special thanks are extended to Michael Brett-Surman and Charyl Ito (United State National Museum) for access to specimens and field notebooks; Thomas Jorstad (USNM) for the field photo of USNM 1349; and Mark Norell and Carl Mehling (American Museum of Natural History) for access to the holotype of *Kritosaurus navajovius.* Thanks are extended to Justin A. Spielmann (formerly of the New Mexico Museum of Natural History and Science) for compiling and providing us with the updated list of hadrosaurids in the NMMNH collections. We thank Patty Ralrick (Royal Tyrrell Museum of Palaeontology) for her help in editing the manuscript. David C. Evans (Royal Ontario Museum) and Denver Fowler (Museum of the Rockies, Montana State University) provided thoughtful comments and suggestions that helped to improve the manuscript, and we thank them for their input. RMS thanks David Eberth (Royal Tyrrell Museum of Palaeontology) for many discussions concerning stratigraphic and correlation issues over the past few years. Lastly, we thank him, and D. Evans, for inviting us to participate in this special volume.

LITERATURE CITED

Baltz, E. H., S. R. Ash, and R.Y. Anderson. 1966. History of nomenclature and stratigraphy of rocks adjacent to the Cretaceous-Tertiary boundary western San Juan Basin, New Mexico. U. S. Geological Survey Professional Paper 524-D:D1–D23.

Barnes, H. 1953. Geology of the Ignacio area, Ignacio and Pagosa Springs quadrangles, La Plata and Archuleta counties, Colorado. 1:63,360. U.S. Geological Survey Oil and Gas Investigations Map OM-138. U.S. Geological Survey, Reston, Virginia.

Barnes H., E. H. Baltz, Jr., and P. T. Hayes. 1954. Geology and fuel resources of the Red Mesa area. La Plata and Montezuma counties, Colorado. 1:62,500. U.S. Geological Survey Oil and Gas Investigations Map OM 139. U.S. Geological Survey, Reston, Virginia.

Bauer, C. M. 1916. Contributions to the geology and paleontology of San Juan County, New Mexico. 1. Stratigraphy of a part of the Chaco River Valley. United States Geological Survey Professional Paper 98:271–278.

Brown, B. 1910. The Cretaceous Ojo Alamo beds of New Mexico with a description of the new dinosaur genus *Kritosaurus.* Bulletin of the American Museum of Natural History 28:267–274.

Brown, J. 1982. Geologic map of the Alamo East Quadrangle, San Juan County, New Mexico. 1:24,000. U.S. Geological Survey Miscellaneous Field Studies Map MF-1497. U.S. Geological Survey, Reston, Virginia.

Brown, J. 1983. Geologic and isopach maps of the Bisti, De-na-zin and Ah-she-sle-pah [*sic*] Wilderness Study areas, New Mexico. 1:50,000. U. S. Geological Survey Miscellaneous Field Studies Map MF-1508-A. U.S. Geological Survey, Reston, Virginia.

Burns, M. E., and R. M. Sullivan. 2011. A new ankylosaurid from the Upper Cretaceous Kirtland Formation, San Juan Basin, with comments on the diversity of ankylosaurids in New Mexico; pp. 169–178 in R. M. Sullivan, S. G. Lucas, and J. A. Spielmann (eds.), Fossil Record 3. New Mexico Museum of Natural History and Science Bulletin 53.

Campione, N. E., and D. C. Evans. 2011. Cranial growth and variation in edmontosaurs (Dinosauria: Hadrosauridae): implications for latest Cretaceous megaherbivore diversity in North America. PloS ONE 6(9):e25186.

Cifelli, R. L., J. J. Eberle, D. L. Lofgren, J. A. Lillegraven, and W. A. Clemens. 2004. Mammalian biochronology of the latest Cretaceous; pp. 21–42 in M. O. Woodburne (ed.), Late Cretaceous and Cenozoic Mammals of North America. Columbia University Press, New York.

Clemens, W. A. 1973. The roles of fossil vertebrates in interpretation of Late Cretaceous stratigraphy of the San Juan Basin, New Mexico; pp. 154–167 in J. E. Fassett (ed.), Cretaceous and Tertiary Rocks of the Southern Colorado Plateau. Four Corners Geological Society, Durango, Colorado.

Cope, E. D. 1869. Remarks on *Eschrichtius polyporus, Hysibema crassicauda, Hadrosaurus tripos,* and *Polydectes biturgidus.* Proceedings of the Academy of Natural Sciences of Philadelphia 21:192.

Dane, C. H. 1936. The La Ventana–Chacra Mesa coal field, part 3 of Geology and fuel resources of the southern part of the San Juan Basin, New Mexico. United States Geological Survey Bulletin 860-C:81–161.

Davies-Vollum, K. S., L. D. Boucher, P. Hudson, and A. Y. Proskurowski. 2011. A Late Cretaceous coniferous woodland from the San Juan Basin, New Mexico. Palaios 26:89–98.

Evans, D. C., and R. R. Reisz. 2007. Anatomy and relationships of *Lambeosaurus magnicristatus,* a crested hadrosaurid dinosaur (Ornithischia) from the Dinosaur Park Formation, Alberta. Journal of Vertebrate Paleontology 27:373–393.

Evans, D. C., R. Bavington, and N. E. Campione. 2009. An unusual hadrosaurid braincase from the Dinosaur Park Formation and the biostratigraphy of *Parasaurolophus* (Ornithischia: Lambeosaurinae) from southern Alberta. Canadian Journal of Earth Sciences 46:791–800.

Fassett, J. E. 1973. The saga of the Ojo Alamo Sandstone; or the rock-stratigrapher and the paleontologist should be friends; pp. 123–130 in J. E. Fassett (ed.), Cretaceous and Tertiary Rocks of the Southern Colorado Plateau. Four Corners Geological Society, Durango, Colorado.

Fassett, J. E. 2009. New geochronologic and stratigraphic evidence confirms the Paleocene age of the dinosaur-bearing Ojo Alamo Sandstone and Animas Formation in the San Juan Basin, New Mexico and Colorado. Palaeontologia Electronica 12.1.3A:1–150.

Fassett, J. E. 2010. Stratigraphic nomenclature of rock strata adjacent to the Cretaceous-Tertiary interface in the San Juan Basin; pp. 113–124 in J. E. Fassett, K. E. Zeigler, and V. W. Lueth (eds.), Geology of the Four Corners Country: New Mexico Geological Society 61st Annual Field Conference, September 22–25, 2010. New Mexico Geological Society Guidebook 61.

Fassett, J. E., and J. S. Hinds. 1971. Geology and fuel resources of the Fruitland Formation and Kirtland Shale of the San Juan Basin, New Mexico. United States Geological Survey Professional Paper 676:1–76.

Fassett, J. E., and M. B. Steiner. 1997. Precise age of C33N-C32R magnetic-polarity reversal, San Juan Basin, New Mexico and Colorado, pp. 239–247 in O. J. Anderson, B. S. Kues, and S. G. Lucas (eds.), Mesozoic Geology and Paleontology of the Four Corners Region: New Mexico Geological Society, Forty-eighth Annual Field Conference, October 1–4, 1997. New Mexico Geological Society Guidebook 48.

Fassett, J. E., R. A. Zielinski, and J. R. Budahn. 2002. Dinosaurs that did not die: evidence for Paleocene dinosaurs in the Ojo Alamo Sandstone, San Juan Basin, New Mexico. Geological Society of America Special Paper 356:307–336.

Flynn, L. J. 1986. Late Cretaceous mammal horizons from the San Juan Basin, New Mexico. American Museum Novitates 2845:1–30.

Gilmore, C. W. 1916. Contributions to the geology and paleontology of San Juan County, New Mexico, 2: vertebrate faunas of the Ojo Alamo, Kirtland, and Fruitland formations. United States Geological Survey Professional Paper 98:279–302.

Gilmore, C.W. 1919. Reptilian faunas of the Torrejon, Puerco, and underlying Upper Cretaceous formations of the San Juan Basin, New Mexico. United States Geological Survey Professional Paper 19:1–68.

Gilmore, C. W. 1935. On the Reptilia of the Kirtland Formation of New Mexico, with descriptions of new species of fossil turtles. Proceedings of the United States National Museum 83:159–188.

Hall, J. P. 1991. Lower vertebrate paleontology of the upper Fruitland Formation, Fossil Forest area, New Mexico, and implications for Late Cretaceous terrestrial biostratigraphy. M.S. thesis, University of Kansas, Lawrence, Kansas, 126 pp.

Hall, J. P. 1993. A juvenile hadrosaurid from New Mexico. Journal of Vertebrate Paleontology 13:367–369.

Horner, J. R. 1992. Cranial morphology of *Prosaurolophus* (Ornithischia: Hadrosauridae) with descriptions of two new hadrosaur species and an evaluation of hadrosaurid phylogenetic relationships. Museum of the Rockies Occasional Paper 2:1–119.

Horner, J. R., D. B. Weishampel, and C. A. Forster. 2004. Hadrosauridae; pp. 438–463 in D. B. Weishampel, P. Dodson, and H. Osmólska (eds.), The Dinosauria, Second Edition. University of California Press, Berkeley, California.

Hunt, A. P. 1984. Stratigraphy, sedimentology, taphonomy and magnetostratigraphy of the Fossil Forest area, San Juan County, New Mexico. M.S. thesis, New Mexico Institute of Mining and Technology, Socorro, New Mexico, 338 pp.

Hunt, A. P. 1991. Integrated vertebrate, invertebrate and plant taphonomy of the Fossil Forest area (Fruitland and Kirtland formations: Late Cretaceous), San Juan County, New Mexico, U.S.A. Palaeogeography, Palaeoclimatology, Palaeoecology 88:85–107.

Hunt, A. P., and S. G. Lucas. 1991. An associated Maastrichtian hadrosaur and a Turonian ammonite from the Naashoibito Member, Kirtland Formation (Late Cretaceous: Maastrichtian), northwestern New Mexico. New Mexico Journal of Science 31:27–35.

Hunt, A. P., and S. G. Lucas. 1992. Stratigraphy, paleontology and the age of the Fruitland and Kirtland formations (Upper Cretaceous), San Juan Basin, New Mexico. pp. 217–239 in S. G. Lucas (ed.), San Juan Basin IV: New Mexico Geological Society, Forty-third Annual Field Conference, September 30–October 3, 1992. New Mexico Geological Society Guidebook 43.

Hunt, A. P., and S. G. Lucas. 1993. Cretaceous vertebrates of New Mexico; pp. 77–91 in S. G. Lucas and J. Zidek (eds.), Vertebrate Paleontology in New Mexico. New Mexico Museum of Natural History and Science Bulletin 2.

Hunt, A. P., and S. G. Lucas. 2003a. A new hadrosaur track from the Upper Cretaceous Fruitland Formation of northwestern New Mexico; pp. 379–381 in S. G. Lucas (ed.), Geology of the Zuni Plateau: New Mexico Geological Society, Fifty-fourth Annual Field Conference, September 24–27, 2003. New Mexico Geological Society Guidebook 54.

Hunt, A. P., and S. G. Lucas. 2003b. Origin and stratigraphy of historic dinosaur quarries in the Upper Cretaceous Fruitland Formation of the Fossil Forest Research Natural Area, northwestern New Mexico; pp. 383–388 in S. G. Lucas (ed.), Geology of the Zuni Plateau: New Mexico Geological Society, Fifty-fourth Annual Field Conference, September 24–27, 2003. New Mexico Geological Society Guidebook 54.

Jasinski, S. E., and R. M. Sullivan. 2011. Re-evaluation of pachycephalosaurids from the Fruitland-Kirtland transition (Kirtlandian, late Campanian), San Juan Basin, New Mexico, with a description of a new species of *Stegoceras* and a reassessment of *Texacephale langstoni;* pp. 202–215 in R. M. Sullivan, S. G. Lucas, and J. A. Spielmann (eds.), Fossil Record 3. New Mexico Museum of Natural History and Science Bulletin 53.

Jasinski, S. E., R. M. Sullivan, and S. G. Lucas. 2011. Taxonomic composition of the Alamo Wash local fauna from the Upper Cretaceous Ojo Alamo Formation (Naashoibito Member) San Juan Basin, New Mexico; pp. 216–271 in R. M. Sullivan, S. G. Lucas, and J. A. Spielmann (eds.), Fossil Record 3. New Mexico Museum of Natural History and Science Bulletin 53.

Kirkland, J. I., R. Hernández-Rivera, T. Gates, G. S. Paul, S. Nesbitt, C. I. Serrano-Brañas, and J. P. Garcia-De La Garza. 2006. Large hadrosaurine dinosaurs from the latest Campanian of Coahuila, Mexico; pp. 299–315 in S. G. Lucas and R. M. Sullivan (eds.), Late Cretaceous Vertebrates from the Western Interior. New Mexico Museum of Natural History and Science Bulletin 35.

Koenig, A. E., S. G. Lucas, L. A. Neymark, A. B. Heckert, R. M. Sullivan, S. E. Jasinski, and D. W. Fowler. 2012. Direct U-Pb dating of Cretaceous and Paleocene dinosaur bones, San Juan Basin, New Mexico. Geology Forum, April 2012, e262 doi:10.1130/G32154C.1.

Kues, B. S., J. W. Froehlich, J. A. Schiebout, and S. G. Lucas. 1977. Paleontological survey, resource assessment, and mitigation plan for the Bisti-Star Lake area, northwestern New Mexico. Report to the Bureau of Land Management, Albuquerque. [Unpublished]

Lambe, L. M. 1914. On *Gryposaurus notabilis,* a new genus and species of trachodont dinosaur from the Belly River of Alberta, with a description of the skull of *Chasmosaurus belli.* Ottawa Naturalist 27:145–154.

Lambe, L. M. 1918. On the genus *Trachodon* of Leidy. Ottawa Naturalist 31:135–139.

Lehman, T. M. 1981. The Alamo Wash Local Fauna: a new look at the Ojo Alamo Fauna; pp. 189–221 in S. Lucas, J. K. Rigby, Jr., and B. Kues (eds.), Advances in San Juan Basin Paleontology. University of New Mexico Press, Albuquerque, New Mexico.

Lehman, T. M. 1985. Depositional environments of the Naashoibito Member of the Kirtland Shale, Upper Cretaceous, San Juan Basin, New Mexico. New Mexico Bureau of Mines and Mineral Resources Circular 195:55–79.

Lehman, T. M., F. W. McDowell, and J. N. McConnelly. 2006. First isotope for the Late Cretaceous *Alamosaurus* vertebrate fauna of West Texas, and its significance as a link between two faunal provinces. Journal of Vertebrate Paleontology 26:922–928.

Leonardi, G. 1984. Le impronte fossili di dinosaur; pp. 165–186 in J. F. Bonaparte (ed.), Sulle orme de dinosauri. Errizzo, Venice.

Lucas, S. G. 1981. Dinosaur communities of the San Juan Basin: a case for lateral variations in the composition of Late Cretaceous dinosaur communities; pp. 337–393 in S. G. Lucas, J. K. Rigby, Jr., and B. S. Kues (eds.), Advances in San Juan Basin Paleontology. University of New Mexico Press, Albuquerque, New Mexico.

Lucas, S. G., and R. M. Sullivan. 2000. Stratigraphy and vertebrate biostratigraphy across the Cretaceous-Tertiary boundary, Betonnie Tsosie Wash, San Juan Basin, New Mexico; pp. 95–104 in S. G. Lucas and A. B. Heckert (eds.), Dinosaurs of New Mexico. New Mexico Museum of Natural History and Science Bulletin 17.

Lucas, S. G., A. P. Hunt, and R. M. Sullivan. 2006. Stratigraphy and age of the Upper Cretaceous Fruitland Formation, west-central San Juan Basin, New Mexico; pp. 1–6 in S. G. Lucas and R. M. Sullivan (eds.), Late Cretaceous Vertebrates from the Western Interior. New Mexico Museum of Natural History and Science Bulletin 35.

Lucas, S. G., N. J. Mateer, A. P. Hunt, and F. M. O'Neill. 1987. Dinosaurs, the age of the Fruitland and Kirtland formations, and the Cretaceous-Tertiary boundary in the San Juan Basin, New Mexico. Geological Society of America Special Papers 209:33–50.

Lucas, S. G., R. M. Sullivan, S. E. Jasinski, and T. L. Ford. 2011. Hadrosaur footprints from the Upper Cretaceous Fruitland Formation, San Juan Basin, New Mexico and the ichnotaxonomy of large ornithopod footprints; pp. 357–362 in R. M. Sullivan, S. G. Lucas, and J. A. Spielmann (eds.), Fossil Record 3. New Mexico Museum of Natural History and Science Bulletin 53.

Lucas, S. G., J. A. Spielmann, R. M. Sullivan, A. P. Hunt, and T. A. Gates. 2006. *Anasazisaurus,* a hadrosaurian dinosaur from the Upper Cretaceous of New Mexico; pp. 293–297 in S. G. Lucas and R. M. Sullivan (eds.), Late Cretaceous Vertebrates from the Western Interior. New Mexico Museum of Natural History and Science Bulletin 35.

Lucas, S. G., R. M. Sullivan, S. M. Cather, S. E. Jasinski, D. W. Fowler, A. B. Heckert, J. A. Spielmann, and A. P. Hunt. 2009. No definitive evidence of Paleocene dinosaurs in the San Juan Basin. Palaeontologia Electronica 12.2.9A:1–10.

Ludwig, K. R. 2012. Direct U-Pb dating of Cretaceous and Paleocene dinosaur bones, San Juan Basin, New Mexico. Geology Forum, April 2012, e258 doi:10.1130/G32503C.1.

Lull, R. S., and N. E. Wright. 1942. Hadrosaurian Dinosaurs of North America. Geological Society of America Special Papers 40. 242 pp.

Mateer, N. J. 1981. The reptilian megafauna from the Kirtland Shale (Late Cretaceous) of the San Juan Basin, New Mexico; pp. 49–75 in S. G. Lucas, J. K. Rigby, Jr., and B. S. Kues (eds.), Advances in San Juan Basin Paleontology. University of New Mexico Press, Albuquerque, New Mexico.

Mytton, J. W. 1983. Geologic map of the Chaco Canyon 30′ × 60′ quadrangle, showing coal zones of the Fruitland Formation, San Juan, Rio Arriba and Sandoval counties, New Mexico. 1:100,000. U.S. Geological Survey Investigation Map C92-A. U.S. Geological Survey, Reston, Virginia.

North American Commission on Stratigraphic Nomenclature. 2005. North American Stratigraphic Code. American Association of Petroleum Geologists Bulletin 89:1547–1591.

Ogg, J. G., F. P. Agterberg, and F. M. Gradstein. 2004. The Cretaceous Period; pp. 344–383 in F. Gradstein, J. Ogg, and A. Smith (eds.), A Geologic Time Scale 2004. Cambridge University Press, Cambridge, U.K.

Ostrom, J. H. 1961. A new species of hadrosaurian dinosaur from the Cretaceous of New Mexico. Journal of Paleontology 35:575–577.

Ostrom, J. H. 1963. *Parasaurolophus cyrtocristatus,* a crested hadrosaurian dinosaur from New Mexico. Fieldiana: Geology 14:143–168.

O'Sullivan, R. B., G. R. Scott, and J. H. Heller. 1979. Reconnaissance geologic map of the Bisti Trading Post Quadrangle, San Juan County, New Mexico. 1:24,000. U.S. Geological Survey Miscellaneous Field Studies Map MF-1075. U.S. Geological Survey, Reston, Virginia.

Parks, W. A. 1922. *Parasaurolophus walkeri,* a new genus and species of crested trachodont dinosaur. University of Toronto Studies 13:5–32.

Parks, W. A. 1923. *Corythosaurus intermedius,* a new species of trachodont dinosaur. University of Toronto Studies 15:5–57.

Powell, J. S. 1973. Paleontology and sedimentation models of the Kimbeto Member of the Ojo Alamo Sandstone; pp. 111–122 in J. E. Fassett (ed.), Cretaceous and Tertiary Rocks of the Southern Colorado Plateau. Four Corners Geological Society, Durango, Colorado.

Prieto-Márquez, A. 2010. Global phylogeny of Hadrosauridae (Dinosauria: Ornithopoda) using parsimony and Bayesian methods. Zoological Journal of the Linnean Society 159:435–502.

Reeside, J. B., Jr. 1924. Upper Cretaceous and Tertiary formations of the western part of the San Juan Basin Colorado and New Mexico. United States Geological Survey Professional Paper 134:1–70.

Renne, P. R., and M. B. Goodwin. 2012. Direct U-Pb dating of Cretaceous and Paleocene dinosaur bones, San Juan Basin, New Mexico. Geology Forum, April 2012, e259, doi:10.1130/G32521C.1.

Rigby, J. K., Jr., and D. L. Wolberg. 1987. The therian mammalian fauna (Campanian) of Quarry 1, Fossil Forest study area, San Juan Basin, New Mexico. Geological Society of America Special Paper 209:51–79.

Russell, L. S. 1975. Mammalian faunal succession in the Cretaceous System of Western North America. Geological Association of Canada Special Paper 13:137–161.

Scott, G. R., R. O'Sullivan, and J. W. Mytton. 1979. Reconnaissance geologic map of the Alamo Mesa West quadrangle, San Juan County, New Mexico. 1:24,000. U.S. Geological Survey Miscellaneous Field Studies Map MF-1074. U.S. Geological Survey, Reston, Virginia.

Schneider, G. B., D. L. Weide, J. W. Mytton, and G. R. Scott. 1979. Geologic Map of the Pueblo Bonito NW Quadrangle, San Juan County, New Mexico. 1:24,000. U.S. Geological Survey Miscellaneous Field Studies Map MF-1117. U.S. Geological Survey, Reston, Virginia.

Sikkink, P. G. L. 1987. Lithofacies relationships and depositional environment of the Tertiary Ojo Alamo Sandstone and related strata, San Juan Basin, New Mexico and Colorado. Geological Society of America Special Paper 209:81–104.

Silver, C. 1950. The occurrence of gas in Cretaceous rocks of the San Juan Basin, New Mexico and Colorado; pp. 109–123 in San Juan Basin, New Mexico and Colorado: New Mexico Geological Society 1st Field Conference Guidebook. New Mexico Geological Society, Socorro, New Mexico.

Sinclair, W. J., and W. Granger. 1914. Paleocene deposits of the San Juan Basin, New Mexico. Bulletin of the American Museum of Natural History 33:297–316.

Sternberg, C. H. 1933. Hunting Dinosaurs in the Badlands of the Red Deer River Alberta, Canada. Published by the author, San Diego, California, 261 pp.

Strobell, J. D., Jr., R. B. O'Sullivan, J. W. Mytton, and M. F. Erpenbeck. 1985. Preliminary geologic map of the Pretty Rock Quadrangle. 1:24,000. U.S. Geological Survey, Miscellaneous Field Studies Map MF-1788. U.S. Geological Survey, Reston, Virginia.

Sullivan, R. M. 2006. Ah-shi-sle-pah Wilderness Study Area (San Juan Basin, New Mexico): a paleontological (and historical) treasure and resource; pp. 169–174 in S. G. Lucas (ed.), America's Antiquities: 100 Years of Managing Fossils on Federal Lands. New Mexico Museum of Natural History and Science Bulletin 34.

Sullivan, R. M., and G. E. Bennett, III. 2000. A juvenile *Parasaurolophus* (Ornithischia: Hadrosauridae) from the Upper Cretaceous Fruitland Formation of New Mexico; pp. 215–220 in S. G. Lucas and A. B. Heckert (eds.), Dinosaurs of New Mexico. New Mexico Museum of Natural History and Science Bulletin 17.

Sullivan, R. M., and D. W. Fowler. 2011. *Navajodactylus boerei,* n. gen., n. sp. (Pterosauria, ?Azhdarchidae) from the Upper Cretaceous Kirtland Formation (upper Campanian) of New Mexico; pp. 393–404 in R. M. Sullivan, S. G. Lucas, and J. A. Spielmann (eds.), Fossil Record 3. New Mexico Museum of Natural History and Science Bulletin 53.

Sullivan, R. M., and S. G. Lucas. 2003. The Kirtlandian, a new land-vertebrate "age" for the Late Cretaceous of western North America; pp. 369–377 in S. G. Lucas (ed.), Geology of the Zuni Plateau: New Mexico Geological Society, Fifty-fourth Annual Field Conference, September 24–27, 2003. New Mexico Geological Society Guidebook 54.

Sullivan, R. M., and S. G. Lucas. 2006. The Kirtlandian land-vertebrate "age": faunal composition, temporal position and biostratigraphic correlation in the nonmarine Upper Cretaceous of western North America; pp. 7–29 in S. G. Lucas and R. M. Sullivan (eds.), Late Cretaceous Vertebrates from the Western Interior. New Mexico Museum of Natural History and Science Bulletin 35.

Sullivan, R. M., and S. G. Lucas. 2011. Charles Hazelius Sternberg and his San Juan Basin Cretaceous dinosaur collections: correspondence and photographs (1920–1925); pp. 429–471 in R. M. Sullivan, S. G. Lucas, and J. A. Spielmann (eds.), Fossil Record 3. New Mexico Museum of Natural History and Science Bulletin 53.

Sullivan, R. M., and T. E. Williamson. 1997. Additions and corrections to Sternberg's San Juan Basin collection, Paleontological Museum, University of Uppsala, Sweden; pp. 255–257 in O. J. Anderson, B. S. Kues, and S. G. Lucas (eds.), Mesozoic Geology and Paleontology of the Four Corners Region: New Mexico Geological Society, Forty-eighth Annual Field Conference, October 1–4, 1997. New Mexico Geological Society Guidebook 48.

Sullivan, R. M., and T. E. Williamson. 1999. A New Skull of *Parasaurolophus* (Dinosauria: Hadrosauridae) from the Kirtland Formation of New Mexico and a Revision of the Genus. New Mexico Museum of Natural History and Science Bulletin 15. 52 pp.

Sullivan, R. M., S. E. Jasinski, M. Guenther, and S. G. Lucas. 2011. The first lambeosaurine (Dinosauria, Hadrosauridae, Lambeosaurinae) from the Upper Cretaceous Ojo Alamo Formation (Naashoibito Member), San Juan Basin, New Mexico; pp. 405–417 in R. M. Sullivan, S. G. Lucas, and J. A. Spielmann (eds.), Fossil Record 3. New Mexico Museum of Natural History and Science Bulletin 53.

Sullivan, R. M., S. G. Lucas, and D. R. Braman. 2005. Dinosaurs, pollen, and the Cretaceous-Tertiary boundary in the San Juan Basin, New Mexico; pp. 395–407 in S. G. Lucas (ed.), Geology of the Chama Basin: Fifty-sixth Annual Field Conference September 21–24, 2005 New Mexico Geological Society Guidebook 56.

Sullivan, R. M., S. G. Lucas, and S. E. Jasinski. 2011. The humerus of a hatchling lambeosaurine (Dinosauria: Hadrosauridae) referable to cf. *Parasaurolophus tubicen* from the Upper Cretaceous Kirtland Formation (De-na-zin Member), San Juan Basin, New Mexico; pp. 472–474 in R. M. Sullivan, S. G. Lucas, and J. A. Spielmann (eds.), Fossil Record 3. New Mexico Museum of Natural History and Science Bulletin 53.

Williamson, T. E. 1998. Review of dinosaurs of the Alamo Wash local fauna, Naashoibito Member, San Juan Basin, New Mexico. New Mexico Geology 20:54.

Williamson, T. E. 2000. Review of Hadrosauridae (Dinosauria, Ornithischia) from the San Juan Basin, New Mexico; pp. 191–213 in S. G. Lucas and A. B. Heckert (eds.), Dinosaurs of New Mexico. New Mexico Museum of Natural History and Science Bulletin 17.

Williamson, T. E., and R. M. Sullivan. 1998. A new local fauna, the Willow Wash local fauna, from the Upper Cretaceous (Campanian) Kirtland Formation, San Juan Basin, New Mexico. Journal of Vertebrate Paleontology 18(3, Supplement):86A.

Williamson, T. E., and A. Weil. 2008. Stratigraphic distribution of sauropods in the Upper Cretaceous of the San Juan Basin, New Mexico, with comments on North America's Cretaceous "sauropod hiatus." Journal of Vertebrate Paleontology 28:1218–1223.

Wiman, C. 1931. *Parasaurolophus tubicen* n. sp. aus der Kreide in New Mexico. Nova Acta Regiae Societatis Scientiarum Upsaliensis 7:1–11.

Wolberg, D. L., J. P. Hall, and D. Bellis. 1988. First record of dinosaur footprints from the Fruitland Formation, San Juan County, New Mexico. New Mexico Bureau of Mines and Mineral Resources Bulletin 122:33–34.

Wolberg, D. L., J. P. Hall, D. Bellis, W. X. Chavez, O. Anderson, R. Moro, and A. Gil. 1988. Regional historic, stratigraphic, and paleontologic framework of the Late Cretaceous (Campanian-Maastrichtian) Fossil Forest locality near Split Lip Flats, San Juan Basin, New Mexico. New Mexico Bureau of Mines and Mineral Resources Bulletin 122:7–21.

Appendix 20.1. Hadrosaurid Specimens from the Fruitland, Kirtland, and Ojo Alamo Formations in the Collections of the NMMNH and SMP

Collection	No.	Genus	Species	Loc. no.	Formation	Member	Description
NMMNH P-	32578	Hadrosauridae indet.		4506	Kirtland	De-na-zin	Partial tooth
NMMNH P-	32605	Hadrosauridae indet.		3228	Kirtland	De-na-zin	Tooth fragments
NMMNH P-	32642	Hadrosauridae indet.		4015	Fruitland		Radius, partial humerus, etc.
NMMNH P-	32756	Hadrosauridae indet.		3228	Kirtland	De-na-zin	Tooth fragments
NMMNH P-	32780	Hadrosauridae indet.		3228	Kirtland	De-na-zin	Tooth
NMMNH P-	32798	Hadrosauridae indet.		4534	Kirtland	Naashoibito	Jaw and postcranial fragments
NMMNH P-	32803	Hadrosauridae indet.		4010	Fruitland		Tooth fragment
NMMNH P-	32810	Hadrosauridae indet.		4010	Fruitland		Tooth
NMMNH P-	32840	Hadrosauridae indet.		4010	Fruitland		Tooth
NMMNH P-	32843	Hadrosauridae indet.		4010	Fruitland		Coronoid process
NMMNH P-	32856	Hadrosauridae indet.		4541	Kirtland	De-na-zin	2 tooth fragments
NMMNH P-	32859	Hadrosauridae indet.		4535	Kirtland	Naashoibito	2 teeth
NMMNH P-	32874	Hadrosauridae indet.		4537	Kirtland	Naashoibito	Tooth
NMMNH P-	33030	Hadrosauridae indet.		4275	Kirtland	Hunter Wash	2 teeth
NMMNH P-	37624	Hadrosauridae indet.		5142	Fruitland/Kirtland?		Incomplete juvenile femur
NMMNH P-	37797	Hadrosauridae indet.		5220	Kirtland	Naashoibito	Tooth
NMMNH P-	37944	Hadrosauridae indet.		5221	Kirtland	Naashoibito	Tooth fragment
NMMNH P-	38363	Hadrosauridae indet.		5241	Kirtland	Naashoibito	Incomplete tooth
NMMNH P-	38364	Hadrosauridae indet.		5241	Kirtland	Naashoibito	Tooth
NMMNH P-	38372	Hadrosauridae indet.		5238	Kirtland	De-na-zin	Incomplete, concreted, fragmentary jaw
NMMNH P-	38375	*Parasaurolophus*		5242	Kirtland	Hunter Wash	Left maxilla and skull fragment
NMMNH P-	38766	Hadrosauridae indet.		4500	Kirtland	Naashoibito	19 rib fragments
NMMNH P-	39629	Hadrosauridae indet.		5362	Kirtland		Phalanx
NMMNH P-	40513	Hadrosauridae indet.		4268	Fruitland	Fossil Forest	Right dentary
NMMNH P-	42463	Hadrosauridae indet.		5614	Kirtland	Farmington?	Right metatarsal IV (juvenile)
NMMNH P-	42843	Hadrosauridae indet.		1708	Kirtland		2 costals(?), fragmentary vertebra
NMMNH P-	44303	Hadrosauridae indet.		3517	Kirtland	Hunter Wash	Partial dentary and associated fragments in matrix
NMMNH P-	44518	Hadrosauridae indet.		4000	Kirtland?	Hunter Wash?	Proximal end of tibia
NMMNH P-	44519	Hadrosauridae indet.		4000	Kirtland?	Hunter Wash?	Phalanx
NMMNH P-	44520	Hadrosauridae indet.		4000	Kirtland?	Hunter Wash?	Numerous teeth and tooth fragments
NMMNH P-	44528	Hadrosauridae indet.		4000	Kirtland?	Hunter Wash?	Distal end of metatarsal
NMMNH P-	44529	Hadrosauridae indet.		4000	Kirtland?	Hunter Wash?	Articulated caudal centra
NMMNH P-	46415	Hadrosauridae indet.		6134	Kirtland		2 tibia fragments
NMMNH P-	46416	Hadrosauridae indet.		6133	Kirtland		Caudal vertebrae
NMMNH P-	46417	Hadrosauridae indet.		6133	Kirtland		Phalanx
NMMNH P-	46418	Hadrosauridae indet.		6134	Kirtland		Rib fragment
NMMNH P-	46421	Hadrosauridae indet.		6133	Kirtland		Coracoid? fragment
NMMNH P-	46422	Hadrosauridae indet.		6134	Kirtland		100+ femur fragments
NMMNH P-	49947	Hadrosauridae indet.		6765	no data	no data	7 ischia fragments
NMMNH P-	49950	Hadrosauridae indet.		6203	Fruitland	Fossil Forest	Rib fragment
NMMNH P-	49951	Hadrosauridae indet.		6203	Fruitland	Fossil Forest	Rib
NMMNH P-	49963	Hadrosauridae indet.		6765	no data	no data	Dentary
NMMNH P-	49964	Hadrosauridae indet.		6765	no data	no data	Right quadrate
NMMNH P-	49966	Hadrosauridae indet.		6765	no data	no data	Right dentary
NMMNH P-	49879	Hadrosauridae indet.		6369	Fruitland	Hunter Wash	Dentary fragment
NMMNH P-	49901	Hadrosauridae indet.		6544	Kirtland?	Hunter Wash?	Tooth
NMMNH P-	49909	Hadrosauridae indet.		6378	Fruitland	Hunter Wash	Dentary fragment and pieces
NMMNH P-	49919	Hadrosauridae indet.		6367	Kirtland		2 dentary fragments
NMMNH P-	50903	Hadrosauridae indet.		6791	Kirtland	Hunter Wash	Partial tooth
NMMNH P-	51635	Hadrosauridae indet.		6854	Kirtland	Hunter Wash	Numerous partial maxilla fragments
NMMNH P-	58076	Hadrosauridae indet.		7720	Fruitland	Fossil Forest?	Large track (sandstone cast)

Appendix 20.1. (continued)

Collection	No.	Genus	Species	Loc. no.	Formation	Member	Description
NMMNH P-	64422	Hadrosauridae indet.		8633	Kirtland		Limb bone
NMMNH P-	64423	Hadrosauridae indet.		8502	Ojo Alamo Fm	Naashoibito	Ilium and fragments
NMMNH P-	63031	Hadrosauridae indet.		7779	Fruitland	Fossil Forest	Track (sandstone cast)
NMMNH P-	63032	Hadrosauridae indet.		7779	Fruitland	Fossil Forest	Track (sandstone cast)
NMMNH P-	63033	Hadrosauridae indet.		7779	Fruitland	Fossil Forest	Track (sandstone cast)
NMMNH P-	63034	Hadrosauridae indet.		7779	Fruitland	Fossil Forest	Track (sandstone cast)
NMMNH P-	63035	Hadrosauridae indet.		7779	Fruitland	Fossil Forest	Track (sandstone cast)
NMMNH P-	64442	Hadrosauridae indet.		8577	Kirtland	De-na-zin	Partial juvenile mandible
NMMNH P-	64443	Hadrosauridae indet.		8519	Kirtland	De-na-zin	Caudal vertebra
NMMNH P-	64444	Hadrosauridae indet.		8512	Kirtland	De-na-zin	Partial jaw with teeth
NMMNH P-	64445	Hadrosauridae indet.		8588	Ojo Alamo Fm	Naashoibito	Postorbital
NMMNH P-	64447	Hadrosauridae indet.		8583	Kirtland	De-na-zin	Limb bone end
NMMNH P-	64460	Hadrosauridae indet.		8507	Kirtland	De-na-zin	Vertebra
NMMNH P-	64475	Hadrosauridae indet.		8578	Kirtland	De-na-zin	Jaw fragment with teeth
SMP VP-	815	Hadrosauridae indet.		360a	Kirtland	De-na-zin	?Proximal metatarsal
SMP VP-	827	Hadrosauridae indet.		210	Kirtland	De-na-zin	Incomplete scapula
SMP VP-	834	*Kritosaurus*	*navajovius*	360a	Kirtland	De-na-zin	Incomplete skull roof; parietal, frontals, postorbitals, basicranium and associated fragments
SMP VP-	968	Hadrosauridae indet.		311	Kirtland	De-na-zin	Pes phalanx
SMP VP-	974	cf. Hadrosauridae indet.		210	Kirtland	De-na-zin	?Paroccipital process
SMP VP-	977	Hadrosauridae indet.		228	Kirtland	Hunter Wash	Pes phalanx
SMP VP-	980	Hadrosauridae indet.		349	Kirtland	Hunter Wash	Anterior proximal caudal vertebra
SMP VP-	989	Hadrosauridae indet.		313a	Kirtland	De-na-zin	Tooth
SMP VP-	1090	*Parasaurolophus*	sp.	365	Kirtland	Hunter Wash	Incomplete r. dentary, both surangulars and jugals, left quadratojugal, left quadrate, partial ?postorbital, and fragments
SMP VP-	1101	Hadrosauridae indet.		361	Kirtland	De-na-zin	Radius
SMP VP-	1108	Hadrosauridae indet.		368	Kirtland	Hunter Wash	Caudal vertebra
SMP VP-	1110	Hadrosauridae indet.		309a	Kirtland	De-na-zin	Left surangular
SMP VP-	1112	Hadrosauridae indet.		309a	Kirtland	De-na-zin	Distal end of metatarsal 3
SMP VP-	1137	Hadrosauridae indet.		373	Kirtland	Hunter Wash	Incomplete radius
SMP VP-	1144	Lambeosaurinae indet.		365	Kirtland	Hunter Wash	Incomplete basicranium and associated bone fragments
SMP VP-	1161	Hadrosauridae indet.		363a	Kirtland	De-na-zin	4 teeth
SMP VP-	1169	Hadrosauridae indet.		313a	Kirtland	De-na-zin	2 trunk vertebrae, neural spines and associated fragments
SMP VP-	1247	Hadrosauridae indet.		374	Ojo Alamo	Naashoibito	Incomplete caudal vertebra
SMP VP-	1259	cf. *Parasaurolophus*	*tubicen*	364a	Kirtland	De-na-zin	Fragmentary left mandible
SMP VP-	1263	Hadrosauridae indet.		363a	Kirtland	De-na-zin	Incomplete right ilium
SMP VP-	1267	*Kritosaurus*	*navajovius*	363a	Kirtland	De-na-zin	Right humerus
SMP VP-	1268	cf. *Kritosaurus*	*navajovius*	363a	Kirtland	De-na-zin	Incomplete trunk vertebra
SMP VP-	1269	cf. *Kritosaurus*	sp.	360a	Kirtland	De-na-zin	Incomplete right maxilla with teeth
SMP VP-	1288	cf. *Parasaurolophus*	sp.	362	Kirtland	De-na-zin	Distal end of left ischium (boot)
SMP VP-	1291	Hadrosauridae indet.		309a	Kirtland	De-na-zin	Carpal
SMP VP-	1298	Hadrosauridae indet.		360a	Kirtland	De-na-zin	Distal end of femur
SMP VP-	1316	Hadrosauridae indet.		382	Kirtland	De-na-zin	Incomplete ungual
SMP VP-	1326	Hadrosauridae indet.		313a	Kirtland	De-na-zin	Coronoid process? -associated with dentary fragment
SMP VP-	1333	Hadrosauridae indet.		385	Kirtland	Hunter Wash	Incomplete caudal vertebra
SMP VP-	1339	Hadrosauridae indet.		358	Kirtland	De-na-zin	Badly weathered braincase

Appendix 20.1. (continued)

Collection	No.	Genus	Species	Loc. no.	Formation	Member	Description
SMP VP-	1344	cf. *Parasaurolophus*	*tubicen*	309a	Kirtland	De-na-zin	Complete right and left ischia, left ilium, right pubis; 2 dorsal vert. with ossified tendons; 4 additional vertebrae
SMP VP-	1351	*Parasaurolophus*	sp.	309a	Kirtland	De-na-zin	Proximal end of rib
SMP VP-	1361	Hadrosauridae indet.		388a	Kirtland	De-na-zin	Tibia
SMP VP-	1365	Hadrosauridae indet.		389a	Kirtland	De-na-zin	?Radius
SMP VP-	1367	Hadrosauridae indet.		389a	Kirtland	De-na-zin	Caudal vertebra
SMP VP-	1383	Hadrosauridae indet.		370a	Kirtland	De-na-zin	Medial part of right maxilla
SMP VP-	1388	Hadrosauridae indet.		228	Kirtland	Hunter Wash	Tooth
SMP VP-	1396	Hadrosauridae indet.		309a	Kirtland	De-na-zin	?Proximal end of radius
SMP VP-	1400	Hadrosauridae indet.		388a	Kirtland	De-na-zin	Proximal end of metatarsal?
SMP VP-	1410	Hadrosauridae indet.		319a	Kirtland	De-na-zin	Distal end of fibula?
SMP VP-	1412	cf. *Parasaurolophus*	*tubicen*	309a	Kirtland	De-na-zin	4 incomplete vertebrae, femur and associated bone fragments
SMP VP-	1440	Hadrosauridae indet.		365	Kirtland	Hunter Wash	Distal end of right femur
SMP VP-	1441	Hadrosauridae indet.		281	Kirtland	Hunter Wash	Incomplete cervical prong of vertebra
SMP VP-	1448	Hadrosauridae indet.		392a	Kirtland	De-na-zin	Proximal end of scapula
SMP VP-	1468	Hadrosauridae indet.		363a	Kirtland	De-na-zin	Nearly complete left femur
SMP VP-	1472	Hadrosauridae indet.		281	Kirtland	Hunter Wash	2 incomplete caudal vertebra
SMP VP-	1487	Hadrosauridae indet.		281	Kirtland	Hunter Wash	Caudal vertebra
SMP VP-	1492	Hadrosauridae indet.		319a	Kirtland	De-na-zin	Nearly complete tibia
SMP VP-	1495	Hadrosauridae indet.		319a	Kirtland	De-na-zin	Complete ulna
SMP VP-	1508	Hadrosauridae indet.		228	Kirtland	Hunter Wash	3 incomplete vertebrae, including an anterior dorsal vertebra and dorsal vertebra, and associated fragments
SMP VP-	1511	Hadrosauridae indet.		365	Kirtland	Hunter Wash	Caudal vertebra
SMP VP-	1534	Lambeosaurinae indet.		392b	Ojo Alamo	Naashoibito	Jugal, right
SMP VP-	1537	Hadrosauridae indet.		319a	Kirtland	De-na-zin	Nearly complete left femur
SMP VP-	1542	Hadrosauridae indet.		372	Kirtland	Hunter Wash	Caudal vertebra
SMP VP-	1546	Hadrosauridae indet.		394	Kirtland	Hunter Wash	Incomplete right humerus
SMP VP-	1591	Hadrosauridae indet.		396	Kirtland	Hunter Wash	Fragments of maxilla
SMP VP-	1592	cf. *Kritosaurus*	*navajovius*	396	Kirtland	Hunter Wash	Nearly complete right maxilla with teeth
SMP VP-	1600	Hadrosauridae indet.		398	Kirtland	Hunter Wash	Dentary fragments
SMP VP-	1607	Hadrosauridae indet.		398	Kirtland	Hunter Wash	?Carpal
SMP VP-	1608	Hadrosauridae indet.		398	Kirtland	Hunter Wash	Caudal vertebra
SMP VP-	1609	Hadrosauridae indet.		398	Kirtland	Hunter Wash	Incomplete caudal vertebra
SMP VP-	1617	Hadrosauridae indet.		402	Kirtland	Hunter Wash	Incomplete right pubis
SMP VP-	1619	cf. *Parasaurolophus*	*cyrtocristatus*	401	Kirtland	Hunter Wash	Incomplete right humerus
SMP VP-	1623	Hadrosauridae indet.		396	Kirtland	Hunter Wash	Nearly complete tibia
SMP VP-	1624	Hadrosauridae indet.		384b	Ojo Alamo	Naashoibito	Vertebrae and associated fragments
SMP VP-	1629	Hadrosauridae indet.		319a	Kirtland	De-na-zin	Right scapula
SMP VP-	1630	Hadrosauridae indet.		319a	Kirtland	De-na-zin	Coronoid and dentary fragments
SMP VP-	1639	Hadrosauridae indet.		309a	Kirtland	De-na-zin	Tooth
SMP VP-	1656	Hadrosauridae indet.		396	Kirtland	Hunter Wash	Incomplete left ilium
SMP VP-	1662	cf. *Kritosaurus*	*navajovius*	398	Kirtland	Hunter Wash	Upper margin of left maxilla and associated fragments
SMP VP-	1677	Hadrosauridae indet.		400	Kirtland	Hunter Wash	Caudal vertebra with associated fragments
SMP VP-	1680	Hadrosauridae indet.		400	Kirtland	Hunter Wash	Incomplete radius
SMP VP-	1684	Hadrosauridae indet.		400	Kirtland	Hunter Wash	Caudal vertebra

Appendix 20.1. (continued)

Collection	No.	Genus	Species	Loc. no.	Formation	Member	Description
SMP VP-	1703	Hadrosauridae indet.		409	Kirtland	Hunter Wash	?Skin impression on bone fragment
SMP VP-	1723	Hadrosauridae indet.		309a	Kirtland	De-na-zin	Caudal vertebra
SMP VP-	1738	Hadrosauridae indet.		361	Kirtland	De-na-zin	3 large vertebrae and associated fragments
SMP VP-	1740	cf. Hadrosauridae indet.		382	Kirtland	De-na-zin	Vertebra and fragments
SMP VP-	1742	Hadrosauridae indet.		361	Kirtland	De-na-zin	Tooth
SMP VP-	1794	Hadrosauridae indet.		228	Kirtland	Hunter Wash	Tooth
SMP VP-	1816	Hadrosauridae indet.		365	Kirtland	Hunter Wash	Incomplete caudal vertebra
SMP VP-	1818	Hadrosauridae indet.		365	Kirtland	Hunter Wash	Incomplete tooth
SMP VP-	1855	Hadrosauridae indet.		281	Kirtland	Hunter Wash	Maxillary fragment without teeth
SMP VP-	1858	Hadrosauridae indet.		365	Kirtland	Hunter Wash	Incomplete caudal
SMP VP-	1867	Hadrosauridae indet.		410b	Ojo Alamo	Naashoibito	Rib and vertebra fragments
SMP VP-	1897	Hadrosauridae indet.		350	Kirtland	De-na-zin	Tooth
SMP VP-	1913	Hadrosauridae indet.		350	Kirtland	De-na-zin	Numerous (shed) teeth
SMP VP-	1933	Hadrosauridae indet.		417	Kirtland	Hunter Wash	Tooth
SMP VP-	1941	Hadrosauridae indet.		281	Kirtland	Hunter Wash	Carpal?
SMP VP-	1960	Hadrosauridae indet.		210	Kirtland	De-na-zin	Coracoid
SMP VP-	1971	Hadrosauridae indet.		418	Kirtland	Hunter Wash	3 vertebrae and associated fragments
SMP VP-	1972	Hadrosauridae indet.		419	Kirtland	Hunter Wash	Incomplete carpal
SMP VP-	1973	Hadrosauridae indet.		419	Kirtland	Hunter Wash	Carpal?
SMP VP-	1975	Hadrosauridae indet.		419	Kirtland	Hunter Wash	Carpal
SMP VP-	1987	Hadrosauridae indet.		419	Kirtland	Hunter Wash	Vertebra
SMP VP-	1993	Hadrosaurine indet.		412	Kirtland	Hunter Wash	Incomplete parietal
SMP VP-	2007	Hadrosauridae indet.		350	Kirtland	De-na-zin	Numerous teeth and tooth fragments
SMP VP-	2012	Hadrosauridae indet.		419	Kirtland	Hunter Wash	Weathered and incomplete fibula
SMP VP-	2020	Hadrosauridae indet.		420a	Kirtland	De-na-zin	2 fragmentary teeth
SMP VP-	2028	Hadrosauridae indet.		376a	Kirtland	De-na-zin	Neural spine
SMP VP-	2041	Hadrosauridae indet.		421	Kirtland	Hunter Wash	2 tooth fragments
SMP VP-	2046	Lambeosaurinae indet.		388a	Kirtland	De-na-zin	Right humerus and ?carpal
SMP VP-	2050	Hadrosauridae indet.		420b	Ojo Alamo	Naashoibito	Weathered fibula
SMP VP-	2053	Hadrosauridae indet.		421	Kirtland	Hunter Wash	Tooth fragments
SMP VP-	2060	Hadrosauridae indet.		421	Kirtland	Hunter Wash	Ischium
SMP VP-	2061	Hadrosauridae indet.		421	Kirtland	Hunter Wash	Chevron
SMP VP-	2062	Hadrosauridae indet.		421	Kirtland	Hunter Wash	Neural spine
SMP VP-	2086	Hadrosauridae indet.		421	Kirtland	Hunter Wash	Scapula
SMP VP-	2087	Hadrosauridae indet.		383	Ojo Alamo	Naashoibito	Skull fragments
SMP VP-	2091	Hadrosauridae indet.		388a	Kirtland	De-na-zin	Incomplete scapula
SMP VP-	2094	Hadrosauridae indet.		420a	Kirtland	De-na-zin	Caudal vertebra
SMP VP-	2135	Hadrosauridae indet.		228	Kirtland	Hunter Wash	Metacarpal?
SMP VP-	2145	Hadrosauridae indet.		228	Kirtland	Hunter Wash	Fragmentary maxilla
SMP VP-	2146	Hadrosauridae indet.		228	Kirtland	Hunter Wash	Incomplete metacarpal and two fragments
SMP VP-	2148	Hadrosauridae indet.		228	Kirtland	Hunter Wash	2 tooth fragments
SMP VP-	2152	Hadrosauridae indet.		281	Kirtland	Hunter Wash	Fragmentary tooth
SMP VP-	2169	cf. *Kritosaurus*	*navajovius*	421	Kirtland	Hunter Wash	Incomplete left dentary
SMP VP-	2184	Hadrosauridae indet.		421	Kirtland	Hunter Wash	Incomplete caudal vertebra
SMP VP-	2195	Hadrosauridae indet.		433	Kirtland	Hunter Wash	Incomplete cervical vertebrae, at least two, with zygopophyses
SMP VP-	2196	Hadrosauridae indet.		370a	Kirtland	De-na-zin	Vertebra
SMP VP-	2199	Hadrosauridae indet.		421	Kirtland	Hunter Wash	Vertebra centrum
SMP VP-	2202	*Parasaurolophus*	*tubicen*	319a	Kirtland	De-na-zin	Right humerus
SMP VP-	2206	Hadrosauridae indet.		421	Kirtland	Hunter Wash	Incomplete dorsal vertebra

Appendix 20.1. (continued)

Collection	No.	Genus	Species	Loc. no.	Formation	Member	Description
SMP VP-	2215	Hadrosaurine indet.		433	Kirtland	Hunter Wash	Incomplete ischium and other elements
SMP VP-	2219	Hadrosauridae indet.		433	Kirtland	Hunter Wash	Vertebra
SMP VP-	2221	Hadrosauridae indet.		433	Kirtland	Hunter Wash	Tooth
SMP VP-	2226	Hadrosauridae indet.		424a	Kirtland	De-na-zin	Axis vertebra
SMP VP-	2228	Hadrosauridae indet.		421	Kirtland	Hunter Wash	Dorsal vertebra with neural spine attached
SMP VP-	2239	Hadrosauridae indet.		421	Kirtland	Hunter Wash	Vertebrae and phalanges
SMP VP-	2240	Hadrosauridae indet.		421	Kirtland	Hunter Wash	Ungual
SMP VP-	2241	Hadrosauridae indet.		435	Kirtland	Hunter Wash	?Fibula
SMP VP-	2248	Hadrosauridae indet.		435	Kirtland	Hunter Wash	Tooth
SMP VP-	2257	Hadrosauridae indet.		435	Kirtland	Hunter Wash	2 teeth
SMP VP-	2263	Lambeosaurinae indet.		403b	Ojo Alamo	Naashoibito	Nearly complete left humerus
SMP VP-	2354	Hadrosauridae indet.		313a	Kirtland	De-na-zin	Caudal vertebrae and associated bone fragments
SMP VP-	2378	Hadrosauridae indet.		389a	Kirtland	De-na-zin	Left femur, nearly complete
SMP VP-	2380	Hadrosauridae indet.		382	Kirtland	De-na-zin	Ossified tendon
SMP VP-	2386	Hadrosauridae indet.		376a	Kirtland	De-na-zin	Tooth
SMP VP-	2409	Hadrosauridae indet.		450	Fruitland	Fossil Forest	Incomplete caudal vertebra
SMP VP-	2422	Hadrosauridae indet.		350	Kirtland	De-na-zin	Numerous teeth
SMP VP-	2437	Hadrosauridae indet.		451	Kirtland	Hunter Wash	Caudal vertebra
SMP VP-	2451	Hadrosauridae indet.		228	Kirtland	Hunter Wash	Tooth
SMP VP-	2470	Hadrosauridae indet.		360a	Kirtland	De-na-zin	Tooth
SMP VP-	2471	Hadrosauridae indet.		360a	Kirtland	De-na-zin	3 teeth
SMP VP-	2479	Hadrosauridae indet.		319a	Kirtland	De-na-zin	Tooth
SMP VP-	2496	Hadrosauridae indet.		389a	Kirtland	De-na-zin	Right metatarsal IV
SMP VP-	2497	Hadrosauridae indet.		382	Kirtland	De-na-zin	Ulna
SMP VP-	2508	Hadrosauridae indet.		370b	Ojo Alamo	Naashoibito	28 teeth and tooth fragments
SMP VP-	2510	Hadrosauridae indet.		398	Kirtland	Hunter Wash	Incomplete squamosal and pedal phalanx
SMP VP-	2552	Hadrosauridae indet.		450	Fruitland	Fossil Forest	2 teeth
SMP VP-	2560	Hadrosauridae indet.		450	Fruitland	Fossil Forest	2 teeth and unidentified element
SMP VP-	2563	Hadrosauridae indet.		450	Fruitland	Fossil Forest	Incomplete tooth
SMP VP-	2587	Hadrosauridae indet.		450	Fruitland	Fossil Forest	2 teeth
SMP VP-	2597	Hadrosauridae indet.		319a	Kirtland	De-na-zin	Surangular and nearby bone fragments
SMP VP-	2599	Lambeosaurinae indet.		319a	Kirtland	De-na-zin	Anterior portion of right jugal
SMP VP-	2604	Hadrosauridae indet.		319a	Kirtland	De-na-zin	Tooth
SMP VP-	2613	Hadrosauridae indet.		319a	Kirtland	De-na-zin	2 teeth
SMP VP-	2622	Hadrosauridae indet.		313a	Kirtland	De-na-zin	3 teeth
SMP VP-	2629	Hadrosauridae indet.		313a	Kirtland	De-na-zin	3 teeth
SMP VP-	2636	Hadrosauridae indet.		450	Fruitland	Fossil Forest	Tooth
SMP VP-	2641	Hadrosauridae indet.		450	Fruitland	Fossil Forest	Tooth
SMP VP-	2645	Hadrosauridae indet.		319a	Kirtland	De-na-zin	Tooth
SMP VP-	2676	Hadrosauridae indet.		228	Kirtland	Hunter Wash	3 incomplete teeth
SMP VP-	2679	Hadrosauridae indet.		228	Kirtland	Hunter Wash	2 teeth
SMP VP-	2684	Hadrosauridae indet.		450	Fruitland	Fossil Forest	Tooth
SMP VP-	2688	Hadrosauridae indet.		450	Fruitland	Fossil Forest	2 caudal vertebrae and vertebra fragments (with associated indet. bone material)
SMP VP-	2690	Hadrosauridae indet.		450	Fruitland	Fossil Forest	Metatarsal 4?
SMP VP-	2692	Hadrosauridae indet.		370b	Ojo Alamo	Naashoibito	18 teeth
SMP VP-	2705	Hadrosauridae indet.		450	Fruitland	Fossil Forest	Left dentary
SMP VP-	2708	Hadrosauridae indet.		450	Fruitland	Fossil Forest	Caudal vertebrae
SMP VP-	2710	Hadrosauridae indet.		228	Kirtland	Hunter Wash	Terminal phalanx (ungual)

Appendix 20.1. (continued)

Collection	No.	Genus	Species	Loc. no.	Formation	Member	Description
SMP VP-	2711	Hadrosauridae indet.		228	Kirtland	Hunter Wash	Dentary fragment
SMP VP-	2717	Hadrosauridae indet.		228	Kirtland	Hunter Wash	Right frontal
SMP VP-	2730	Hadrosauridae indet.		450	Fruitland	Fossil Forest	Tooth
SMP VP-	2734	Hadrosauridae indet.		349	Kirtland	Hunter Wash	2 caudal vertebrae, neural spine, and vertebrae fragments
SMP VP-	2737	cf. *Parasaurolophus*	*cyrtocristatus*	408	Kirtland	Hunter Wash	Incomplete left pubis
SMP VP-	2738	Hadrosauridae indet.		372	Kirtland	Hunter Wash	Ulna, metacarpal 2, ?metatarsal 2 and associated fragments
SMP VP-	2750	Hadrosauridae indet.		228	Kirtland	Hunter Wash	Incomplete jugal
SMP VP-	2763	Hadrosauridae indet.		450	Fruitland	Fossil Forest	12 teeth
SMP VP-	2767	Hadrosauridae indet.		450	Fruitland	Fossil Forest	Nearly complete right ilium, left frontal, nueral spine, other indeterminate bone fragments
SMP VP-	2776	Hadrosauridae indet.		450	Fruitland	Fossil Forest	8 teeth
SMP VP-	2793	Hadrosauridae indet.		461	Kirtland	Hunter Wash	Metacarpal (with associated indet. bone material)
SMP VP-	2800	Hadrosauridae indet.		450	Fruitland	Fossil Forest	8 teeth
SMP VP-	2833	Hadrosauridae indet.		450	Fruitland	Fossil Forest	3 teeth
SMP VP-	2838	Hadrosauridae indet.		450	Fruitland	Fossil Forest	9 teeth
SMP VP-	2862	Hadrosauridae indet.		349	Kirtland	Hunter Wash	Tooth
SMP VP-	2863	Hadrosauridae indet.		450	Fruitland	Fossil Forest	2 teeth
SMP VP-	3333	Hadrosauridae indet.		392a	Kirtland	De-na-zin	Nearly complete phalanx
SMP VP-	3342	Hadrosauridae indet.		281	Kirtland	Hunter Wash	Incomplete tooth and 2 additional fragments (probably tooth fragments)
SMP VP-	3347	Hadrosauridae indet.		409	Kirtland	Hunter Wash	Incomplete tooth
SMP VP-	3367	Hadrosauridae indet.		450	Fruitland	Fossil Forest	Tooth
SMP VP-	3379	Hadrosauridae indet.		450	Fruitland	Fossil Forest	4 teeth (one is questionable)

Note: The NMMNH collection includes only specimens acquired by the NMMNH since the publication of Williamson (2000).

NOTE ADDED IN PROOF

While in press, a crucial paper on *Kritosaurus navajovius* and related hadrosaurine taxa was published by Prieto-Márquez (2014) and has major bearing on the hadrosaurines discussed in our paper. Owing to our continued studies regarding the validity of *Kritosaurus* and *Anasazisaurus*, we regard the latter as a distinct genus, separate from *Kritosaurus*. We also remain unconvinced of Prieto-Márquez's (2014) assignment of USNM 8629 to *Kritosaurus navajovius*.

REFERENCE

Prieto-Márquez, A. 2014. Skeletal morphology of *Kritosaurus navajovius* (Dinosauria: Hadrosauridae) from the Late Cretaceous of the North American south-west, with an evaluation of the phylogenetic systematics and biogreography of Kritosaurini. Journal of Systematic Palaeontology 12: 133–175

Relocating the Lost *Gryposaurus incurvimanus* Holotype Quarry, Dinosaur Provincial Park, Alberta, Canada

21

Darren H. Tanke and David C. Evans

ABSTRACT

During the past century, hundreds of major dinosaur specimens were collected from today's Dinosaur Provincial Park in southern Alberta, Canada. Precise locality information for many of these skeletons, including holotypes, was recorded inadequately or not at all, seriously diminishing the scientific value of these specimens. One specimen that was not properly documented was the holotype of the hadrosaurid *Gryposaurus incurvimanus* (ROM 764), excavated in 1918. This long-lost site was relocated in 2006 using the combination of fortuitous discovery of on-site quarry trash and archival field photographs. The methodology used to relocate the site is outlined in this chapter. Relocating dinosaur quarries not only increases the scientific value of the fossils, but bolsters the history of Albertan paleontology, multidisciplinary research on Dinosaur Provincial Park biostratigraphy, faunal turnover, and paleoecology of this rich assemblage.

INTRODUCTION

The approximately 80 km^2 badlands of Campanian terrestrial deposits in Dinosaur Provincial Park (DPP), southern Alberta, Canada, have long been recognized as a valuable source for the fossilized remains of dinosaurs and other biota (Russell, 1966; Currie, 2005). Many thousands of skeletons and countless fossils remain under the surface awaiting discovery, and rapid erosion ensures fossils will be found there for many years to come (Henderson and Tanke, 2010). Fieldwork conducted by paleontological expeditions began in the area of what is now DPP in 1897, leading to the Great Canadian Dinosaur Rush in the early part of the twentieth century (Russell, 1966; Spalding, 1999). The University of Toronto (UT) conducted small-scale fieldwork in DPP in 1908 and in the Drumheller area in 1912. The UT and the Royal Ontario Museum (ROM) have had a long and close association (Fritz, 1939; Spalding, 1999; Anonymous, 2012). Because of this association and a larger crew, ROM's first expedition to collect major dinosaur specimens is widely considered to have occurred in 1918. This marked the beginning of a series of fossil collecting expeditions to the province that took place periodically over the next half century. This fieldwork produced a diverse array of fossil specimens and many new taxa. The expedition in 1918 resulted in the collection of what is largely recognized as ROM's first complete dinosaur skeleton, the beautiful holotype of the hadrosaurid now called *Gryposaurus incurvimanus* (Parks, 1919).

The ROM crew in the summer of 1918 was led by Dr. William A. Parks (1868–1936; Fig. 21.1), and included assistant Robert Wilson; Dorothy, Alberta resident Wilfred G. Hodgson (1885–1971; Hodgson and Hodgson, 1971); and several other unidentified local men. Despite the crew's relative lack of experience at collecting large dinosaur specimens (only Hodgson had prior experience; helping excavate isolated dinosaur material on the UT 1912 fieldtrip), they collected the nearly complete, articulated skeleton of *Gryposaurus incurvimanus* (Figs. 21.1–21.4, 21.6A) and successfully shipped it to Toronto. The skeleton, ROM 764, was prepared and made into a panel-mount (Fig. 21.4), which went on public display at ROM in May 1920. However, the premiere showing of Toronto's first dinosaur skeleton evidently did not rouse much media interest, with only a few small articles resulting (Anonymous, 1920a, b, c, d), and it was reportedly almost replaced by Greek and Roman statuary soon after being exhibited (Anonymous, 1920e, f). However, dinosaurs ultimately prevailed at ROM, and despite numerous spectacular discoveries and several major paleontology gallery renovations, ROM 764 has remained a fixture in the dinosaur gallery for nearly a century. By the late 1970s and early 1980s, casts of a number of ROM's dinosaurs, including ROM 764, were sold to other museums (Grant, 1982) and are on display today (e.g., Royal Tyrrell Museum *Gryposaurus incurvimanus* holotype, TMP 1980.051.0002).

A major modern tenet of professional fossil collecting is to record locality data and collection methods in as much detail as possible (e.g., Brunton, 1995), so that the stratigraphic, paleoenvironmental, and taphonomic context are available in the future. Some early fossil collectors in DPP were usually good at recording discovery, collection, and rudimentary taphonomic data, but often failed to precisely record locality information. This was particularly unfortunate because this area of southern Alberta had already been mapped using the

21.1. Dr. William A. Parks at the ROM 764 *Gryposaurus incurvimanus* holotype quarry. Photograph ROM2007_9356_1.

21.2. Unidentified ROM field assistants with a small plaster block from the *Gryposaurus* get ready to remove it by horse-drawn stoneboat. Uncataloged ROM photograph.

Dominion Land Survey (legal land description) method and the system had been in use for several decades for the development and settlement of the land by immigrant farmers (Drummond, 1882; Lester, 1963; MacGregor, 1981; Barnett, 2001; Larmour, 2005). Basic, but still useful, topographic maps of western Canada, including badland regions, existed for over a quarter century before the dinosaur collectors arrived, and were produced at scales that would have been useful when trying to mark the location of a quarry or relocate an old quarry today. However, these maps were only rarely used to mark dinosaur localities; the first known use was by George F. Sternberg in 1915 and 1916 on his expeditions upriver from Drumheller. In some of those rare cases in which a land description was actually given (i.e., Langston, 1956), it was found to be incorrect, leading searchers well away from the actual site (Tanke and Ralrick, 2009, 2010). Whether such errors are related to collector or museum secrecy, unfamiliarity of the legal land description system, disinterest, user error, or simply reflect what was acceptable for the times is uncertain. In the course of all this busy multi-institutional collecting and poor record keeping, mistakes and confusion were bound to happen.

Knowledge of who collected what, when, and where was forgotten or became confused, especially in the complex and quickly eroding badland terrains. Recognizing this problem in the mid-1930s, ROM technician Levi Sternberg and his brother Charles M. Sternberg of the Geological Survey of Canada began marking important dinosaur skeleton and other fossil localities with an iron pipe post topped with a data-bearing bronze head. The post was set using concrete poured into a hole drilled into the floor of the excavated quarry (Danis, 1988; Tanke, 1994). These information bearing quarry stakes served much like a graveyard headstone, marking the exact spot where a dinosaur or other important find had been collected, thus preserving important geological, spatial, and stratigraphic information for contemporary and future researchers. The quarry staking work was eventually tied into a survey-grade GPS mapping project in DPP (MacDonald et al., 2005), and has directly supported dinosaur biostratigraphic studies (Béland and Russell, 1979; Currie and Russell, 2005; Ryan and Evans, 2005).

Unfortunately, many quarries excavated prior to 1930 (especially those that lacked the involvement of the Sternbergs, such as the quarries of the AMNH) were not marked in this way, and their exact locations have seemingly been lost forever. The locality for the holotype of *Gryposaurus incurvimanus*, collected by UT a year before they hired Levi Sternberg, is one of the more significant dinosaur sites that did not get a quarry stake and its whereabouts remained unknown until 2006. In 1997, Tanke began a project to relocate these lost quarries in DPP and identify "mystery quarries": sites in which excavation took place but a museum specimen could not be tied to the apparent activity (Tanke, 2001, 2005a, b). Thus far, nearly 35 major fossil quarries have been relocated in and near DPP by using a variety of investigative techniques including field notes, field photographs, and datable quarry garbage left behind at a site by the original collectors. These quarries represent more than one-third of the pre-1954 DPP quarries identified to date. Fifteen of these relocated quarries yielded hadrosaurs, two of which were holotype specimens (Table 21.1) including ROM 764.

Based on prior successes at relocating lost quarries in Drumheller (Russell, 1986; Tanke, 2005b) and DPP (Tanke, 2001, 2005a, b, 2006; Tanke and Ralrick, 2009, 2010), the

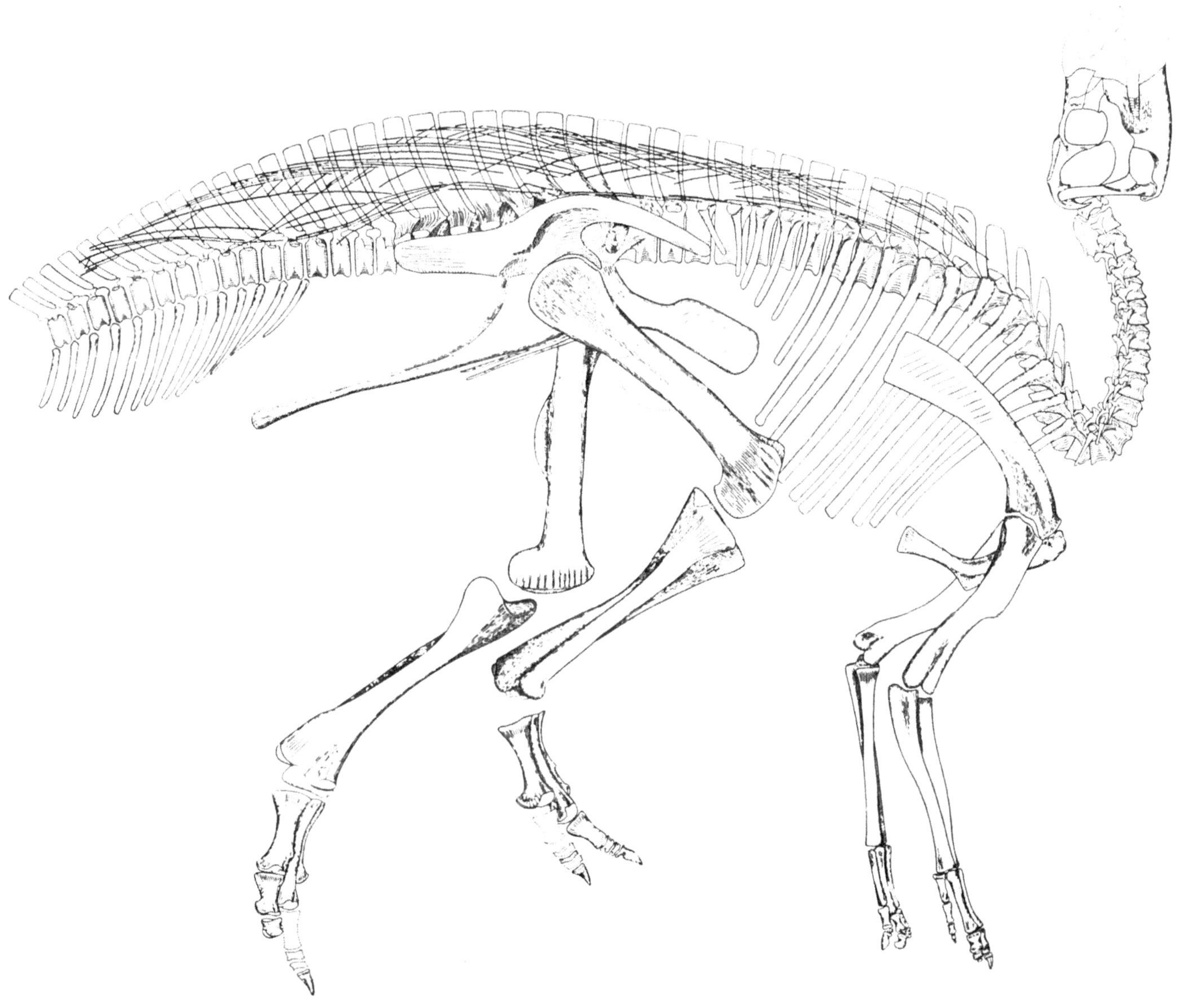

21.3. ROM 764 *Gryposaurus incurvimanus* skeleton (after Parks, 1920) as found in the field.

authors discussed the feasibility of finding the ROM *Gryposaurus incurvimanus* holotype quarry in fall 2005. No one knew the quarry's location except that it was within the limits of DPP, and anyone who had been involved in the dig had passed away. However, relocating the site was recognized as important for several reasons. First of all, the excavated specimen is a holotype and a valuable research specimen that had been well described and illustrated, and used in a number of comparative studies (e.g., Lull and Wright, 1942; Horner, 1992; Gates and Sampson, 2007; Prieto-Márquez, 2010). Second, the specimen would be important to further document high-resolution patterns of dinosaur biostratigraphy in DPP and to test hypotheses about the nature of variation in *Gryposaurus* from that biostratigraphic perspective (Currie and Russell, 2005; Ryan and Evans, 2005; Evans and Reisz, 2007; Ryan et al., 2007; Mallon et al., 2012; Prieto-Marquez, 2010). Relocating the *Gryposaurus* quarry site (and others) would also add to the preliminary biostratigraphic observations made by C. M. Sternberg over half a century earlier (Sternberg, 1950). Finally, the skeleton is of historical importance as the first full dinosaur skeleton excavated and exhibited by ROM.

In this chapter, quarry photographs (e.g., Q99_b001012 [4].psd) can be found in Humber et al. (2005) on the CD ROM supplement to *Dinosaur Provincial Park* (Currie and Koppelhus, 2005).

Institutional Abbreviations AMNH, American Museum of Natural History, New York; BMNH, British Museum

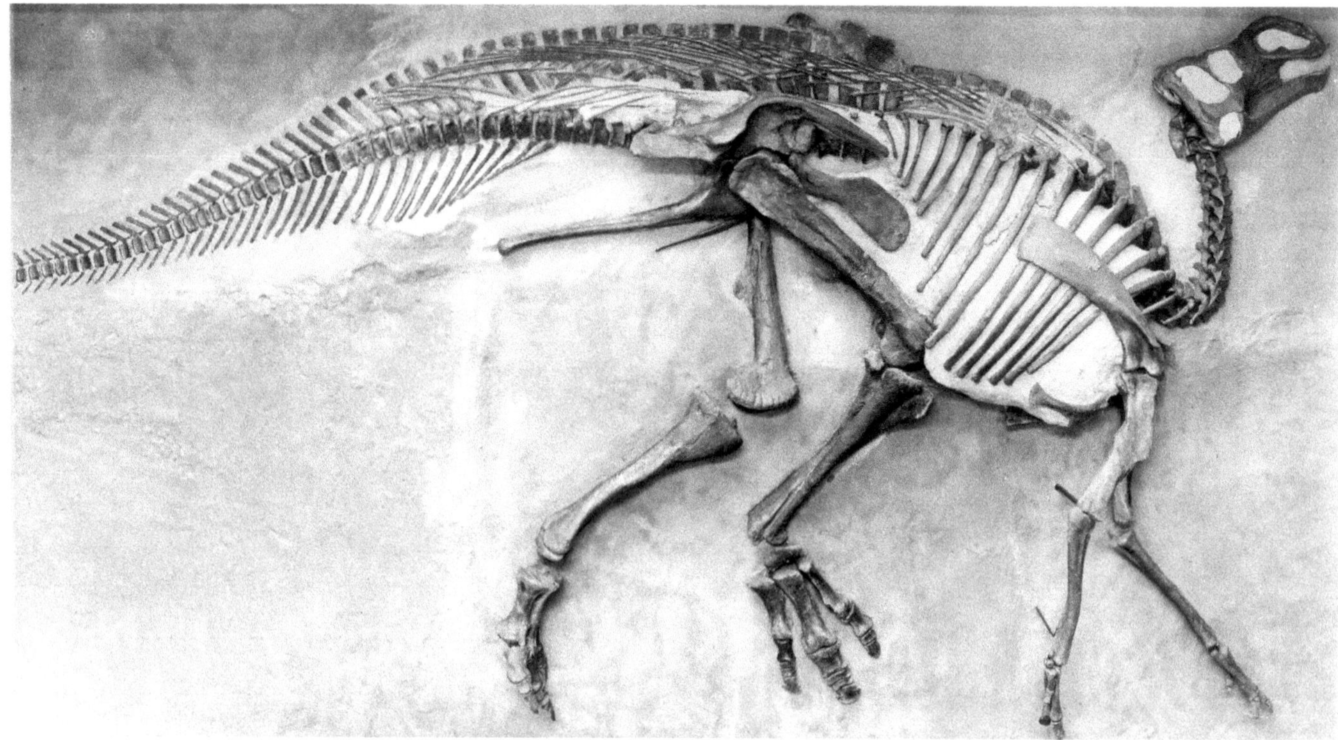

21.4. Historical photograph showing panel mount of ROM 764 *Gryposaurus incurvimanus* soon after being placed on public exhibit.

(Natural History) [now Natural History Museum], London, U.K.; CMN, Canadian Museum of Nature, Ottawa, Ontario; FMNH, Chicago Field Museum, Chicago, Illinois; MSNM V, Museo di Storia Naturale di Milano, Milan, Italy; ROM, Royal Ontario Museum, Toronto, Ontario; TMP, Royal Tyrrell Museum of Palaeontology, Drumheller, Alberta; UA, University of Alberta, Edmonton, Alberta; UAD, Uncollected Articulated Dinosaur (TMP field number designation); UALVP, University of Alberta Laboratory of Vertebrate Palaeontology, Edmonton, Alberta; UT, University of Toronto, Toronto, Ontario.

RELOCATION EFFORT

The impetus to undertake a major effort to relocate the quarry from which ROM 764 was collected was the realization that approximately one dozen archival photographs of the 1918 quarry existed at ROM, and were properly printed and labeled. The photographs showed the progress of the excavation from mapping to plaster jacketing, and, finally, horse-drawn wagon removal. Although rates of erosion are high in the badlands (Bryan and Campbell, 1986; Campbell, 1987), the general or regional topography still looks basically the same a hundred years later. Although fine geographic details and contours are likely to have been modified by erosion, the larger-scale shapes of hills and coulees, along with their preserved stratigraphy, remain essentially the same. Therefore, the photographs were a useful reference for identifying the general area within DPP where the excavation took place. Unfortunately, despite having worked in the DPP badlands since 1979, an examination of the photographs by Tanke revealed that none of the scenery was familiar.

In addition to photographs, W. A. Parks's 1918 field notes (on file at ROM and TMP) were consulted for any possible quarry location information as well as his initial description of the *Gryposaurus* skeleton (Parks, 1919). The locality information was vague and indicated only that the specimen was found "at a height of about 100 feet above the Red Deer River, in the bad lands of the Belly River formation on Sand Hill creek" (Parks, 1919:51). This description matches countless places within DPP and is thus of little use.

The monographic description of the skeleton reported better, but still poor, locality information: "about 2.5 miles above Happy Jack Ferry and 1 mile south of the [Red Deer] river" (Parks, 1920:5). It was uncertain whether this description was based on straight-line distance from the ferry or by following the curves of the meandering river. The Happy Jack Ferry ceased to operate in 1924 (Haestie, 1986; Tanke, 2010), but its former location is well known. The distances given in Parks (1920) were assumed to be correct and two reference points calculated: one by following the curves of the river (using a piece of string cut to the correct scaled length)

Table 21.1. Fifteen Hadrosaur Quarry Sites and Mystery Quarries at DPP and in the Area That Have Been Relocated, or Whose Contents Have Been Identified (Mystery Quarries)

Year of dig	Institution	Taxonomic identification	Specimen type	Specimen number	Quarry number	Notes
1913	AMNH (for Calgary Public Museum)	Juvenile *Gryposaurus*	Skull and partial skeleton	now CMN 8784	252	
1914	CMN	Hadrosaur	Skeleton, uncovered and abandoned	——	UAD 156	
1915	CMN	*Corythosaurus*	Skull and partial skeleton	1915–9 (field number)	243	now TMP 1984.121.0001
1915	?AMNH	Hadrosaur indet.	?	AMNH ?	73	
1916	CH and L Sternberg for BMNH	Hadrosaur indet.	?	——	17	Sunk on SS *Mount Temple*, 1916
1916	CH and L Sternberg for BMNH	Hadrosaur indet.	?	——	18	Sunk on SS *Mount Temple*, 1916
1916	CH and L Sternberg for BMNH	cf. *Parasaurolophus*	Partial skeleton, no skull	——	196	Sunk on SS *Mount Temple*, 1916
1918	ROM	*Gryposaurus incurvimanus* (holotype)	Skull and most of skeleton	ROM 764	99	Incorrectly plotted in Sternberg (1950); Tanke and Evans, this chapter
1921	George F. Sternberg for UA	*Corythosaurus excavatus* (holotype)	Partial skull, rest of skeleton abandoned	UALVP 13	259	Tanke and Russell, 2011, 2012
1922	FMNH	Hadrosaur indet.	Section of tail with skin impressions	FMNH PR 1811	95	
1922	FMNH	Headless lambeosaurine	Most of skeleton	FMNH PR 380	69	
1922	FMNH	*Gryposaurus*	Most of skeleton	Now MSNM V 345	137	In Milan, Italy
1922	FMNH	Hadrosaur indet.	Partial skeleton	FMNH PR ?	232	
1930	ROM	Juvenile *Lambeosaurus*	Partial skeleton	ROM 758	45	Incorrectly marked in Sternberg (1950)
1954	ROM or CMN	Hadrosaur, indet.	Partial skeleton	——	UAD 154	

Note: Arranged by year of collection. Updated from Tanke (2005a).

and the other by utilizing straight-line distances plotted on a topographic map of dinosaur quarries in the Park (Sternberg, 1950). A search area around each reference point was created by expanding out in all directions by 500 m. This exercise resulted in two adjoining circular search areas totaling 0.785 km^2 apiece (Fig. 21.5).

Examination of the 1918 field photographs revealed some commonality of subject matter in the separate pictures. For example, the same outcrop, with distinctive bands of alternating clay and siltstone and a small eroded cave-like structure, appeared in two different pictures (compare Figs. 21.1 and 21.2) allowing these images to be roughly overlapped. High quality photocopies were made of the photographs and each photocopied image cut out. Images were then overlapped until the best match could be made and then taped together. By doing this, we developed a bigger and better idea of the surrounding terrain. More image manipulation informed us that the *Gryposaurus* quarry was nestled within a small, shallow box canyon–like area. The photographs revealed both sides and one end of the canyon's stratigraphy and topography. While none of the images were to the same scale (having been taken at different distances), they were quite useful nevertheless. Shadows cast by people or equipment in the photographs helped orientate the view in each picture (north, west, and east) and revealed the box canyon was aligned roughly in a north–south direction. Among the 1918 photographs were some scenery pictures showing several large isolated badlands hills which were thought to be located near the *Gryposaurus* site. One of the proposed search areas on the topographic map included several large isolated hills, giving much promise.

Fieldwork dedicated to relocating the site was conducted on July 18, 2006. The authors and several assistants explored the two search areas. It quickly became apparent that the 1918 scenery photographs with the large isolated hills were a dead lead and likely had nothing to do with the quarry because none of the hills in the search area matched those in the photographs. Although the general terrain, stratigraphic position, and geology of the first search area looked similar to the 1918 quarry pictures, no matches could be made. The second search area was ruled out for two reasons: it is dominated by geologically older Oldman Formation outcrops with

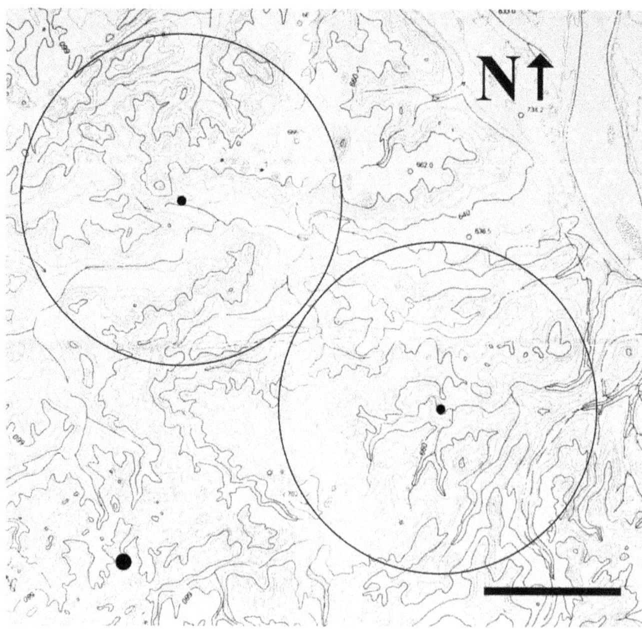

21.5. The two search areas (large circles) within DPP for the ROM 764 *Gryposaurus* location as based on poor data provided in Parks (1919, 1920) with actual location of site (black circle, lower left) indicated. Scale bar equals 500 meters.

differing low topography largely overgrown with shrubby vegetation; and the related hadrosaurine *Brachylophosaurus*, not *Gryposaurus*, is known from that horizon in DPP (Ryan and Evans, 2005).

About a week later, another TMP field crew was prospecting for fossils in badlands roughly 1 km south of the center point of the first area previously explored (Fig. 21.5). One of the crew found rusty nails and pieces of old milled lumber, items that when found in the badlands are indicative of an old quarry or work associated with one. Such accumulations of garbage deep in the badlands can easily be differentiated from that created by local farmers and ranchers or oilfield activities–such garbage is simply tipped into the badlands directly off the edge of the prairie. The fieldworkers recorded the GPS coordinates of the site and gave them to Tanke. The next day he led a small group to the site. The nails and lumber pieces were the scrap remnants of wooden crates that would be built on site to receive large plaster of Paris field jackets, a common field practice many years ago. Crated jackets were then loaded onto a horse-drawn wagon with a large wooden tripod and chain block hoist (Fig. 21.6A).

At first glance the garbage appeared to be associated with a 1954 ROM *Corythosaurus* skull quarry (Quarry 117) excavated by Ralph R. Hornell and staked about 30 m away to the northeast. Also, another quarry (Quarry 99; excavator, date, and specimen collection unknown) which had been relocated and later staked by Tanke in 2003 was located about 30 m to the north-northwest; it, too, contained old quarry garbage and the highly decayed nature of the wood there (Fig. 21.7) also suggested greater antiquity than the 1954 ROM site, a point first realized by Currie (1999). Pieces of a glazed earthenware "moonshine jug" were found in Quarry 99 (Humber et al., 2005:Q99_b001012.psd). These jugs were used to hold white shellac for gluing and stabilizing fossil bone. The presence of such jug pieces suggested the age of the quarry dated to c. 1917–21, the timeframe during which such jugs were used in the area (DHT, pers. obs.). All three sites (Q117, Q99, "new site") were roughly equally spaced in a triangular configuration and in a flat bottomed canyon, with each site visible from the others (Fig. 21.8A).

Quarry trash discarded by fieldworkers long ago can subsequently be moved about by wind, water, animals, and probably curious fieldworkers who pick items up while hiking and drop them later. It is possible to find isolated pieces of old quarry trash up to 50 m from where it originated (DHT, pers. obs.). Also, in some cases, food and drink items were stored away from the quarries in flat, shaded, or breezy areas where the workers rested or ate. Years ago, TMP staff routinely returned such strayed trash back to any nearby staked quarries. Therefore, rarely, quarry trash from two different adjacent quarries may be found at the wrong quarry site. Our investigation of these three sites indicated that, indeed, some cross-contamination had happened in this area. Specifically, rare pieces of highly corroded wood were found in the 1954 Q117, where otherwise better preserved wood dominates (Humber et al., 2005:Q117_001012.psd). In contrast, all the wood and nails at the new site were quite old, as indicated by lichen-encrusted wood and deeply pitted and corroded nails (Tanke, 2006), which clearly indicate that this site is older than Q117. Also, based on the large number of discarded nails and the thicknesses and lengths of the discarded plank pieces at the new site, the crate(s) constructed at the new site would have been large and thus used to remove a large specimen.

Quarry 117 nearby was worked by ROM in 1954 and was the final expedition to use horses as draft animals in DPP, but archival photographs of the excavation show that horses were used to pull stone boats loaded with plaster field jackets and that no wagons were used to haul crates out of the badlands that year.

In a small alcove (small arrow in Fig. 21.8A) near the new rusty nails site, one of the investigating group found a hand-sized wad of old eroded newspaper stuck together and to the outcrop with dried mud and silt. The private and concealed nature of the alcove suggested this might be an early dinosaur collector's latrine, and the newspaper unused toilet paper. There was no indication of a quarry there. The discovery of the old newspaper was especially critical as it has important ramifications to date historical sites. In all but one case of DPP dinosaur quarries at which old newspaper

21.6. (A) Uncataloged ROM photograph from 1918 field season showing a crate of ROM 764 being lifted. A small, white tent for storing field supplies appears in the background. Later in the 1918 field season this tent's location was also used to pile garbage from the ROM 764 (*Gryposaurus incurvimanus*) excavation. This garbage pile (see Fig. 21.7) was later misidentified as Quarry 99 in Sternberg (1950); misdiagnosed as Quarry 99 (relocated) in Humber et al., 2005:Q99_b001012 (1).psd and Q99_001012 (4).psd; and then erroneously marked as such by TMP with a quarry stake in 2003. (B) Best match photograph of the same area taken in summer 2011.

21.7. Rotted wood plank "cut off" pieces seen in 2011 at the site marked as Quarry 99 in Sternberg (1950) and listed in Humber et al. (2005). These wood pieces, scraps from crate construction, have sat at or on the surface and been exposed to the elements for 93 years.

has been found and used for dating purposes, the date of the newspaper is the same as the year of excavation (Tanke, 2005a, b, 2006). The newspaper typeface appeared to be of an old style, but the age was not immediately apparent. The newspaper wad was collected for further examination. The individual pages on the newspaper were stuck together by dried clay, mud, and sand that had infiltrated the spaces between the pages during decades of weathering. The paper was soaked in warm water and the wet pages were then carefully teased apart one by one. Although no actual publication date was found, various datable clues were revealed. News story datelines of May 16 showed the paper was issued on or shortly after that date. Publication data showed it was a *Calgary Daily Herald* newspaper. A variety of news stories spoke of American ground forces having joined the war effort in France. Other terms in these articles referred to the enemy as the "Hun" and British aviators as being part of the "Royal Flying Corps." These passages indicated the paper was of World War I vintage. The American ground troops arriving in France narrowed the years down to 1917 or 1918. Used car ads listing various 1916, 1917, and one 1918 model for sale supported this hypothesis. Another important clue was an advertisement announcing an upcoming sale on Saturday, May 18. A perpetual calendar indicated that May 18 was indeed a Saturday in 1918. Final confirmation was made by comparing the newspaper pieces to a microfilm copy of 1918 *Calgary Daily Herald* newspapers, which showed the paper to be issued on Friday, May 17, 1918. Identification of the newspaper pieces took less than one hour of combined time.

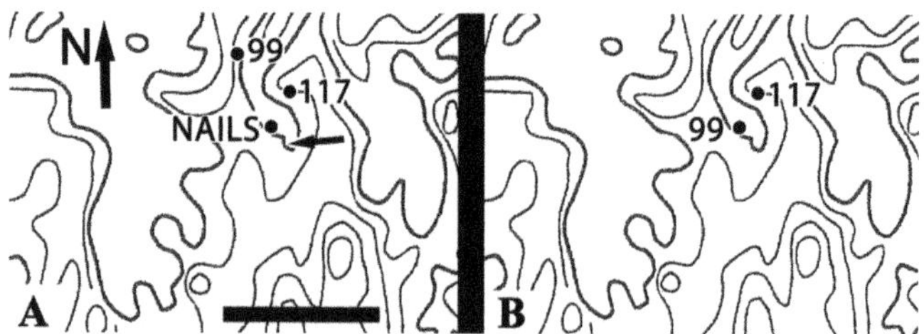

21.8. (A) Topographic map of part of DPP, showing the distribution of quarries 99 (excavation date unknown), 117 (excavated 1954), and rusty nails "new site" discovered in 2006 (excavation date unknown) as understood in late July 2006. Small arrow points to location of 1918 newspaper. Scale bar equals 50 m. (B) Topographic map showing the same area reinterpreted, showing the distribution of quarries 99 (excavated 1918) and 117 (excavated 1954) as now resolved through the results of this project. Note that Quarry 99 has been moved to its proper location; the previous interpretation for it was represented by a pile of 1918 quarry trash.

This speed demonstrates again the value of newspaper in old paleontological sites; if enough of the paper is preserved, various clues contained therein can be combined to quickly establish its publication date (Tanke, 2005a, 2006).

The identification of the 1918 newspaper enhanced the significance of the specific location where it was found. ROM was the only museum in DPP in the summer of 1918 and the *Gryposaurus* was the only major dinosaur skeleton they collected. Could the *Gryposaurus* quarry be in the vicinity of the 1918 newspaper, heavily corroded nails, and wood of the new site? No old quarry was visible there. Armed with copies of the 1918 photographs, the local badlands terrain was immediately matched to the photographs. Remarkably, the actual quarry site had been completely obliterated by 88 years of erosion, yet the crate building new site only a few meters away and the nearby newspaper had survived. The quarry was excavated in a flat-bottomed area and did not create the typical notch in the side of an outcrop, but rather, was simply a pit dug down about 2 m into a flat pediment flanking the base of a nearby outcrop. After the field crew removed the specimen and abandoned the site, the excavation pit quickly filled with sediment. The infilling and erosion of extracted rock must have been rapid as only 18 years later, in 1936, Charles M. Sternberg did not accurately relocate the site but did document something else related to it close by (see below).

Additional examination of the relocated quarry site revealed about a dozen more nails, both unused straight nails and discarded ones bent during on-site crate construction. Some of these crates can be seen in the old photographs (Fig. 21.6A). While matching the photographs to surrounding terrain it was observed in one archival photograph that the 1918 team had erected a small tent-like structure about 30 m north-northwest from the quarry (Fig. 21.6A). When that spot was visited, garbage was discovered that consisted of rotted lumber scraps (Fig. 21.7), bits of plaster of Paris, lumber, wire, tin cans, broken pieces of a glazed crockery moonshine jug, a partial metal pail, and an in situ quarry stake identified as Quarry 99 (Sternberg, 1950). The 1918 photographs also showed what is believed to be the same jug (which originally contained shellac to stabilize and preserve the fragile bones) and the same metal pail.

We now know that Quarry 99, mapped and plotted by C. M. Sternberg in 1936 as an unidentified quarry (Sternberg, 1950), is simply the garbage pile left by the 1918 crew, and the supposed quarry wall is nothing more than an erosional artifact; a result of rock slumping that occurred post-1918 and pre-1936 and that is still visible today (Humber et al., 2005:Q99_001012.psd). A revised understanding of the two old ROM quarries in the field area is presented in Figure 21.8B. The ROM 764 quarry is stratigraphically low in the Dinosaur Park Formation (16 m above the Dinosaur Park–Oldman Formation contact) and corroborates other stratigraphically low occurrences of *Gryposaurus* in DPP.

So how close were Parks's site descriptions from the actual site? The center point in both search areas was about 1.1 km away. Therefore, the closest our first search parties ever got to the actual ROM 764 site was some 500 m away, a long distance in the tortuous badlands landscape.

SCIENTIFIC SIGNIFICANCE

The Sternbergs' effort to relocate and permanently mark the precise quarry locations of major specimens from the area now know as Dinosaur Provincial Park has left an unparalleled scientific legacy manifest as the largest dataset of precise biostratigraphic, paleoecologic, and taphonomic information for any dinosaur assemblage. This has resulted in pioneering studies of detailed taphonomy (Eberth and Getty, 2005) and biostratigraphy (Béland and Russell, 1979; Currie and Russell, 2005; Ryan and Evans, 2005; Mallon et al., 2012) that have greatly increased knowledge of dinosaur paleoecology and evolution, and contributed to the current status of DPP as the datum by which all other dinosaur localities are compared.

Quarry relocation efforts have demonstrated their utility in dinosaur biostratigraphy and other studies in DPP (e.g., Evans, 2001; Ryan and Evans, 2005; Tanke, 2010) and no doubt will be useful for current and future studies of similar nature in the Drumheller valley corridor.

Ongoing relocation of old quarries increases the scientific value of each specimen and enhances the growing database of dinosaur biostratigraphic and evolutionary studies. As the ROM 764 *Gryposaurus* example shows, such relocation efforts can be done cheaply and relatively quickly (some sites have been resolved in as little as 10 minutes) if the prerequisite original site photographs and/or quarry trash are available for reference and dating analysis. Since 1997, 15 lost DPP hadrosaur quarries have been relocated by this project, which enhances the hadrosaur biostratigraphic dataset currently available. The three *Gryposaurus* quarries that have been relocated will be important for testing hypotheses on the nature of variation in this taxon. Some authors recognize two species, *G. notabilis* and *G. incurvimanus*, from DPP (Gates and Sampson, 2007), while others recognize only one variable taxon (Prieto-Márquez, 2010). Currie and Russell (2005) suggested that the two species may represent transitional taxa in a phyletic series. The precise biostratigraphic data now available will allow these hypotheses to be tested more rigorously.

CONCLUSIONS

1. Quarry 99, as plotted on the Sternberg (1950) map sheet, listed and figured (Humber et al., 2005:Q99_b001012 [1].psd; Q99_001012 [4].psd) on the CD ROM supplement to Currie and Koppelhus (2005) is not a quarry. It is a trash pile related to the 1918 ROM 764 *Gryposaurus incurvimanus* quarry which was excavated only ~25 m away (Fig. 21.8b). The UTM for this trash pile is 12U; 466,368mE; 5,622,358mN (WGS 84).

2. The correct location for the ROM 764 *Gryposaurus* quarry is 12U; 466,384mE; 5,622,311mN (WGS 84), deep within the preserve area of DPP. The Quarry 99 stake placed in 2003 has been removed and placed in its correct location.

3. The ROM 764 quarry is stratigraphically low in the Dinosaur Park Formation (16 m above the Dinosaur Park–Oldman Formation contact), which corroborates the low stratigraphic position occurrences of other *Gryposaurus* finds in DPP.

4. Some dispersed quarry trash from the incorrect (or nearby true location) of Quarry 99 (1918) was likely picked up by TMP field staff over the past decades and unknowingly put "back" into Quarry 117 (1954). Such items, easily recognized by their extra corrosion or rotted appearance (on wood items) have been returned to the corrected quarry site or trash pile close by.

5. Quarry 117, a *Corythosaurus* skull (ROM 1933), is presumed to be accurately quarry staked as that work was done soon after excavation was completed and is positioned at 12U; 466,398mE; 5,622,330mN (WGS 84).

6. Relocation of lost early twentieth century quarries in DPP can be achieved via archival photographs and analysis of trash left behind. Analysis of old newspaper, particularly its source and vintage, recovered from these sites continues to play a critical interpretive role.

7. In addition to recording accurate locality coordinates and possibly quarry staking their sites, current collectors are urged to take numerous site photographs during fieldwork. Such photographs should be taken from different angles, from all four cardinal points, and include the nearby and distant horizons to serve as reference points for future researchers. Copies of these photographs should be fully labeled and filed with the catalog collection records and used in technical publications as a relocation aid for future workers. GPS and other land description techniques are useful methods to record (and relocate) a site, but supplemental photographs will stand the test of time and are the best for site relocation purposes.

FUTURE PROJECT PERSPECTIVES

Owing to incomplete or lacking data, precise locality information for well over 100 DPP dinosaur quarries, 150+ excavated further upstream of Drumheller, and smaller numbers elsewhere across southern Alberta are still missing, so the relocation work for these sites continues (Tanke and Russell, 2011, 2012). There have been recent quarry relocation successes upstream from Drumheller of mostly hadrosaur bonebeds or associated skeletons. As in DPP, these specimens have only general locality data now associated with them, such as distance below a current or former geographic location, a now long extinct ranch house or ferry crossing, which side of a river the quarry was worked, or rough elevation above the water (typically rounded to the nearest 50 ft). Such locality information is not precise enough for the high-resolution needs of modern paleontological studies, and has very limited value for actual relocation purposes (e.g., Sato and Wu, 2006). This problem affects researchers globally; there must be many thousands of dinosaur and other important fossil localities whose original whereabouts remain unknown though others are trying to relocate sites in their field areas and there have been some recent successes (e.g., Everhart, 2011; Puckett, 2012). However daunting site relocating seems, in DPP once a site is relocated, it makes all the later discoveries a little easier to solve. Resolution of each of the previously unidentified Albertan sites occurs at a rate of roughly two per year, with numerous quarries that yielded major specimens remaining lost. With the longevity of the project in mind, a new generation of lost-quarry hunters and mystery-quarry researchers is now being introduced to the objectives and methodologies of the project and are being educated both in lab and field settings on how to solve a mystery quarry and search for and find lost ones. The scientific value of artifacts recovered from these sites, critical for dating purposes, are archaeological artifacts documenting paleontological history in their own right. Efforts are underway to ensure these artifacts are curated in an Albertan museum with proper facilities for storing and conserving such items.

ACKNOWLEDGMENTS

The authors thank Don Brinkman (TMP) and his field crew for finding the old nails which helped resolve this case. We also thank Randy Lyons (West Chester, Pennsylvania) for discovering the old newspaper near the 1918 ROM *Gryposaurus* quarry and bringing it to our attention. Jennifer Bancescu (TMP) helped with early renditions of Figures 21.5 and 21.8A–B. We also thank Kevin Seymour (ROM) for assistance with the historic photographs. Patricia E. Ralrick assisted with reviewing, editing, and formatting. We also thank David A. E. Spalding and Dave Eberth (TMP) for reviewing the manuscript. Assistance by Jose Ruben Guzman and Alessandro Chiarenza is also acknowledged.

LITERATURE CITED

Anonymous. 1920a. Museum gets the dinosaur: gigantic fossil skeleton presented by Professor W. A. Parks. Toronto Globe and Mail May 11:10.

Anonymous. 1920b. University news. Toronto Star May 11:25.

Anonymous. 1920c. Meet Mr. Fossil–3,000,000 years old. Edmonton Bulletin May 12:14.

Anonymous. 1920d. Alberta dinosaur is given to Ontario. Calgary Daily-Herald May 11:1.

Anonymous. 1920e. Fight for a dinosaur. Toronto Globe and Mail June 14:6.

Anonymous. 1920f. Grecian art preferred to the dinosaur–Antediluvian monsters from Red Deer ousted from National Geological Museum. Edmonton Journal November 13:11.

Anonymous. 2012. Our history. www.rom.on.ca /en/about-us/rom/our-history. Accessed March 6, 2013.

Barnett, D. 2001. Early surveys and settlements in central Alberta. Provincial Archives and City of Edmonton Archives, Edmonton, Alberta, Canada.

Béland, P., and D. A. Russell. 1979. Ectothermy in dinosaurs: paleoecological evidence from Dinosaur Provincial Park, Alberta. Canadian Journal of Earth Sciences 26:250–255.

Brunton, H. 1995. Documentation of palaeontological material; pp. 5–13 in C. Collins (ed.), The Care and Conservation of Palaeontological Material. Butterfield-Heinemann, London.

Bryan, R. B., and I. A. Campbell. 1986. Runoff and sediment discharge in a semiarid ephemeral drainage basin. Zeitschrift für Geomorphologie 58:121–143.

Campbell, I. A. 1987. Badlands of Dinosaur Provincial Park, Alberta. Canadian Geographic 31:82–87.

Currie, P. J. 1999. July 27, 1999 field notes, copy on file in library of the Royal Tyrrell Museum, Drumheller, Alberta, Canada.

Currie, P. J. 2005. History of research; pp. 3–33 in P. J. Currie and E. B. Koppelhus (eds.), Dinosaur Provincial Park: A Spectacular Ancient Ecosystem Revealed. Indiana University Press, Bloomington, Indiana.

Currie, P. J., and E. B. Koppelhus. 2005. Dinosaur Provincial Park: A Spectacular Ancient Ecosystem Revealed. Indiana University Press, Bloomington, Indiana, 648 pp.

Currie, P. J., and D. A. Russell. 2005. The geographic and stratigraphic distribution of articulated and associated remains; pp. 537–569 in P. J. Currie and E. B. Koppelhus (eds.), Dinosaur Provincial Park: A Spectacular Ancient Ecosystem Revealed. Indiana University Press, Bloomington, Indiana.

Danis, J. 1988. Quarry staking at Dinosaur Provincial Park, Alberta: a unique type of locality data conservation; p. 16 in Society for the Preservation of Natural History Collections, Program and Abstracts. Carnegie Museum of Natural History, Pittsburgh, Pennsylvania.

Drummond, T. 1882. Annual report of the Department of the Interior for the year 1882, Canada.

Eberth, D. A., and M. A. Getty. 2005. Ceratopsian bonebeds: occurrence, origins, and significance; pp. 501–536 in P. J. Currie and E. B. Koppelhus (eds.), Dinosaur Provincial Park: A Spectacular Ancient Ecosystem Revealed. Indiana University Press, Bloomington, Indiana.

Evans, D. C. 2001. The hadrosaurs of Dinosaur Provincial Park. Canadian Palaeobiology 6:5–9.

Evans, D. C., and R. R. Reisz. 2007. Anatomy and relationships of *Lambeosaurus magnicristatus,* a crested hadrosaurid dinosaur (Ornithischia) from the Dinosaur Park Formation, Alberta. Journal of Vertebrate Paleontology 27:373–393.

Everhart, M. J. 2011. Rediscovery of the *Hesperornis regalis* Marsh 1871 holotype locality indicates earlier stratigraphic occurrence. Kansas Academy of Science Transactions 114:59–68.

Fritz, M. A. 1939. Outline of the history and development of the Royal Ontario Museum of Palaeontology. Contributions of the Royal Ontario Museum of Palaeontology 1:1–19.

Gates, T. A., and S. D. Sampson 2007. A new species of *Gryposaurus* (Dinosauria: Hadrosauridae) from the late Campanian Kaiparowits Formation, southern Utah, USA. Zoological Journal of the Linnean Society 151:351–376.

Grant, D. 1982. Replicas made of dinosaurs to replace crumbling bones. Globe and Mail (Toronto) July 12:10.

Haestie, E. 1986. Ferries and Ferrymen in Alberta. Glenbow-Alberta Institute, Calgary, Alberta, 187 pp.

Henderson, D. M., and D. H. Tanke. 2010. Estimating past and future dinosaur skeletal abundances in Dinosaur Provincial Park, Alberta, Canada. Canadian Journal of Earth Sciences 47:1291–1304.

Hodgson, M., and B. Hodgson. 1971. Wilfred Garstang Hodgson; pp. 268–269 in H. B. Roen (ed.), The Grass Roots of Dorothy 1895–1970. Northwest Printing, Calgary, Alberta.

Horner, J. R. 1992. Cranial morphology of *Prosaurolophus* (Ornithischia: Hadrosauridae) with descriptions of two new hadrosaurid species and an evaluation of hadrosaurid phylogenetic relationships. Museum of the Rockies Occasional Paper 2:1–119.

Humber, J., B. Spencer, P. J. Currie, and E. B. Koppelhus. 2005. Dinosaur Provincial Park: GPS survey, GIS mapping, and site pictures. CD-ROM supplement in P. J. Currie and E. B. Koppelhus (eds.), Dinosaur Provincial Park: A Spectacular Ancient Ecosystem Revealed. Indiana University Press, Bloomington, Indiana.

Langston, W., Jr. 1956. The shell of *Basilemys varialosa* (Cope). National Museum of Canada Bulletin 142:155–165.

Larmour, J. 2005. Laying Down the Lines: A History of Land Surveying in Alberta. Brindle and Glass, Victoria, British Columbia, 325 pp.

Lester, C. 1963. Dominion land surveys. Alberta Historical Review 11:20–28.

Lull, R. S., and N. E. Wright. 1942. Hadrosaurian dinosaurs of North America. Geological Society of America Special Papers 40:1–242.

MacDonald, M., P. J. Currie, and W. A. Spencer. 2005. Precise mapping of fossil sites in the Park using survey grade GPS technology; pp. 478–485 in P. J. Currie and E. B. Koppelhus (eds.), Dinosaur Provincial Park: A Spectacular Ancient Ecosystem Revealed. Indiana University Press, Bloomington, Indiana.

MacGregor, J. G. 1981. Vision of an Ordered Land: The Story of the Dominion Land Survey. Western Producer Prairie Books, Saskatoon, Saskatchewan, 202 pp.

Mallon, J. C., D. C. Evans, M. J. Ryan, and J. S. Anderson. 2012. Megaherbivorous dinosaur turnover in the Dinosaur Park Formation (upper Campanian) of Alberta, Canada. Palaeogeography, Palaeoclimatolology, Palaeoecology 350–352:124–138.

Parks, W. A. 1919. Preliminary description of a new species of trachodont dinosaur of the genus *Kritosaurus, Kritosaurus incurvimanus.* Transactions of the Royal Society of Canada, 3rd Series, 13(Section 4):51–59.

Parks, W. A. 1920. The osteology of the trachodont dinosaur *Kritosaurus incurvimanus.* University of Toronto Studies, Geological Series 11:1–76.

Prieto-Márquez, A. 2010. The braincase and skull roof of *Gryposaurus notabilis* (Dinosauria, Hadrosauridae), with a taxonomic revision of the genus. Journal of Vertebrate Paleontology 30:838–854.

Puckett, K. 2012. Couple pinpoints likely site of historic Montana dino find. www.greatfallstribune .com/article/20121026/NEWS01/310260015. Accessed October 27, 2012.

Russell, L. S. 1966. Dinosaur hunting in western Canada. Royal Ontario Museum, Life Sciences Contribution 70:1–37.

Russell, L. S. 1986. Exploring a great dinosaur graveyard. Rotunda 19:20–29.

Ryan, M. J., and D. C. Evans. 2005. Ornithischian dinosaurs; pp. 312–348 in P. J. Currie and E. B. Koppelhus (eds.), Dinosaur Provincial Park: A Spectacular Ancient Ecosystem Revealed. Indiana University Press, Bloomington, Indiana.

Ryan, M. J., R. Holmes, and A. P. Russell. 2007. A revision of the late Campanian centrosaurine ceratopsid genus *Styracosaurus* from the western interior of North America. Journal of Vertebrate Paleontology 27:944–962.

Sato, T., and X.-C. Wu. 2006. Review of plesiosaurians (Reptilia: Sauropterygia) from the Upper Cretaceous Horseshoe Canyon Formation, in Alberta, Canada. Paludicola 5:150–169.

Spalding, D. A. E. 1999. Into the Dinosaur's Graveyard: Canadian Digs and Discoveries Revealed. Doubleday, Toronto, 305 pp.

Sternberg, C. M. 1950. Steveville west of the 4th meridian, with notes on fossil localities. Geological Survey of Canada, Map 969A.

Tanke, D. H. 1994. Macrovertebrate collecting: Site documentation; pp. 81–92 in P. Leiggi and P. May (eds.), Vertebrate Paleontological Techniques. Cambridge University Press, New York.

Tanke, D. H. 2001. Historical archaeology: solving the mystery quarries of Drumheller and Dinosaur Provincial Park, Alberta; pp. 70–76 in H. Allen (ed.), Alberta Palaeontological Society, 5th Annual Meeting, Abstracts volume. Mount Royal College, Calgary, Alberta.

Tanke, D. H. 2005a. Identifying lost quarries; pp. 34–53 in P. J. Currie and E. B. Koppelhus (eds.), Dinosaur Provincial Park: A Spectacular Ancient Ecosystem Revealed. Indiana University Press, Bloomington, Indiana.

Tanke, D. H. 2005b. Buried history: identification of early quarries, paleontological camp sites, and work sites in Dinosaur Provincial Park and the Drumheller valley, Alberta; pp. 126–133 in D. R. Braman, F. Therrien, E. B. Koppelhus, and W. Taylor (compilers), Dinosaur Park Symposium: Short Papers, Abstracts, and Program. Royal Tyrrell Museum of Palaeontology, Drumheller, Alberta.

Tanke, D. H. 2006. Read all about it! The interpretive value of newspaper from old unidentified dinosaur quarries and paleontological field camps in Alberta, Canada. Paludicola 6:22–30.

Tanke, D. H. 2010. Ceratopsian discoveries and work in Alberta, Canada: historical review and census; CD-ROM supplement in M. J. Ryan, B. J. Chinnery-Allgeier, and D. A. Eberth (eds.), New Perspectives on Horned Dinosaurs. Indiana University Press, Bloomington, Indiana.

Tanke, D. H., and P. E. Ralrick. 2009. The case of the missing *Basilemys* turtle quarry (Dinosaur Provincial Park, Alberta); p. 177 in D. Braman (compiler), Gaffney Turtle Symposium, Abstracts and Program. Royal Tyrrell Museum of Palaeontology, Drumheller, Alberta.

Tanke, D. H., and P. E. Ralrick. 2010. Successful relocation of a lost 1914 *Basilemys* (Testudines: Nanhsiungchelyidae) quarry in Dinosaur Provincial Park, Alberta; pp. 44–48 in H. Allen (ed.), Alberta Palaeontological Society, 14th Annual Meeting, Abstracts. Mount Royal University, Calgary, Alberta.

Tanke, D. H., and R. C. Russell. 2011. Headless wonder: ongoing analysis of a head-hunted hadrosaur quarry in Dinosaur Provincial Park, Alberta. tyrrellmuseum.academia.edu/DarrenTanke/Talks/55900. Accessed August 10, 2013.

Tanke, D. H., and R. C. Russell. 2012. Headless wonder: Possible evidence of a head-hunted dinosaur skeleton in Dinosaur Provincial Park, Alberta; pp. 14–17 in H. Allen (ed.), Alberta Palaeontological Society, 16th Annual Symposium, Mount Royal University, Calgary, Alberta.

5 Function and Growth

Comparative Ontogenies (Appendicular Skeleton) for Three Hadrosaurids and a Basal Iguanodontian: Divergent Developmental Pathways in Hadrosaurinae and Lambeosaurinae

22

Merrilee F. Guenther

ABSTRACT

The hadrosaurid taxa *Maiasaura peeblesorum*, a hadrosaurine, and *Hypacrosaurus stebingeri*, a lambeosaurine, are represented by abundant postcranial elements and two of the best growth series known among dinosaurs. Here, I examine appendicular elements and their growth series in both taxa, and use those observations to compare their appendicular ontogenies. Ultimately, the data are used to assess the developmental origins and pathways of the two different adult morphologies. In addition, observations from the less well represented hadrosaurine hadrosaurid *Brachylophosaurus canadensis* and basal iguanodontian *Tenontosaurus tilletti* are included in order to provide a broader context for comparing appendicular ontogenies within Hadrosauridae and Iguanodontia. Analysis of the appendicular skeleton shows that hadrosaurids, in early growth stages, are quite similar, regardless of whether they are hadrosaurine or lambeosaurine. It is possible that the more gracile adult hadrosaurine appendicular elements of *Maiasaura peeblesorum* and *Brachylophosaurus canadensis* reflect pedomorphosis resulting from the postdisplacement of the onset of the development of a number of morphological characteristics. Conversely, the robust appendicular skeleton of the lambeosaurine, *Hypacrosaurus stebingeri*, may have resulted from peramorphosis characterized by acceleration and/or predisplacement of the development of appendicular characters, particularly in the pectoral and pelvic girdles.

INTRODUCTION

In order to have a better understanding of the evolutionary relationships within a group, it is important to study the group from an ontogenetic perspective (Alberch et al., 1979; Smith, 1996, 1997, 2001; Rice, 1997; Velhagen, 1997; Mabee et al., 2000; Dilkes, 2001; Jeffery et al., 2002; Carr, 2003; McNulty et al., 2006; Guenther, 2009; Kilbourne and Makovicky, 2010). For many dinosaurian groups, the lack of embryonic and juvenile specimens makes that challenging. Remains of eggs and embryonic bones have been reported for *Troodon formosus* (Horner and Weishampel, 1988); an oviraptor (Norell et al., 1994); a therizinosaur (Kundrát et al., 2008); an unidentified theropod (Mateus et al., 1997); *Massospondylus* (Reisz et al., 2012); and a sauropod (Chiappe et al., 1998); as well as the hadrosaurs, *Maiasaura peeblesorum* (Horner, 1999), *Hypacrosaurus stebingeri* (Horner and Currie, 1994), and an unidentified lambeosaurine (Sullivan et al., 2011). Within Dinosauria, iguanodontians – specifically the hadrosaurids – offer some of the best opportunities to study ontogeny because of the preservation not only of embryonic material, but also of a variety of intermediate growth stages as well. In this chapter I focus on the hadrosaurids *M. peeblesorum* and *H. stebingeri.*

The Two Medicine Formation of Montana, United States, and the Oldman Formation of Alberta, Canada, have yielded many juvenile and embryonic dinosaur fossils. These remains have primarily been found on nesting horizons (Horner and Currie, 1994; Tanke and Brett-Surman, 2001). One of the largest collections of baby dinosaur material represents the lambeosaurine hadrosaurid *Hypacrosaurus stebingeri*, which has been found both at Devil's Coulee, Alberta, and in adjacent nesting horizons across the international border in Montana. Another hadrosaurid represented by a large collection of nesting and embryonic dinosaur material is the Two Medicine Formation hadrosaurine, *Maiasaura peeblesorum.* The large number of embryonic and juvenile individuals allows not only for a study of ontogeny in the species, but also illustrates the variation within individuals of the same growth stage.

In addition to these very well represented hadrosaurid taxa, the taxa *Brachylophosaurus canadensis* and *Tenontosaurus tilletti* are represented by fewer growth stages, but still provide useful insights about the appendicular ontogenies of *H. stebingeri* and *M. peeblesorum.* The hadrosaurid *B. canadensis* is represented by bonebed material from Montana. Although this species lacks embryonic and hatchling material, it adds data to the ontogenetic story of the Hadrosauridae (Prieto-Márquez, 2005). Likewise, the more basal iguanodontian *T. tilletti* is also represented by a growth series, ranging from hatchling-sized to adult individuals. And although this species does not have as complete a growth series as *H. stebingeri* and *M. peeblesorum*, inclusion of these data allows the comparative ontogenies of all of the taxa to be assessed within a broader iguanodontian perspective.

A focus on the appendicular skeleton was chosen due to abundant available skeletal data and my interest in establishing a baseline of postcranial development that could be used to analyze the biomechanical development of the girdles and limbs.

Each growth series is considered with the intention of identifying a series of developmental events and determining the relative developmental rates for different morphological features of an element. The growth series is examined on both an element-by-element and species-by-species basis with the assumption that within a bonebed or within a grouping of specimens, those elements that are of a similar size range belong to individuals of approximately the same growth stage. None of the taxa is represented by a continuous size series of specimens, making the recognition of clusters of similarly sized individuals easier to accomplish and interpret in the context of growth stages.

When studying the ontogenies of basal and derived (hadrosaurid) iguanodontians, specific characters of the appendicular skeleton can be used to track differential morphological development within limbs and girdles of each species. The appearance and relative growth of these appendicular characters can be used to establish a timeline of developmental events during ontogeny. Identifying differences in these ontogenetic sequences can lead to an understanding of the developmental origins of variation in the adult appendicular morphology of different species. Such variation is present in several appendicular characters in the mature postcranial skeletons of hadrosaurines and lambeosaurines (Brett-Surman, 1975; Horner et al., 2004; Prieto-Márquez, 2010). The hadrosaurine humerus retains a relatively short deltopectoral crest that does not project from the shaft as in lambeosaurines, where it projects more than twice the diameter of the humeral shaft (Horner et al., 2004; Prieto-Márquez, 2010). Differences also exist throughout the pelvic girdle. Hadrosaurines also possess a dorsoventrally shallow ilium, whereas lambeosaurines have a deeper ilium with a more massive supra-acetabular process and a more strongly arched cranial process (Lull and Wright, 1942; Brett-Surman, 1975; Horner et al., 2004; Prieto-Márquez, 2010). The pubis has a prepubic blade whose shape can be diagnostic to the generic level, with hadrosaurines having a long, slender proximal shaft with a short expanded blade and lambeosaurines having proximal blades that expand in the cranial direction. The ischium of hadrosaurines is more delicately constructed and lacks the distally expanded boot of lambeosaurines (Horner et al., 2004; Prieto-Márquez, 2010).

Determination of variation in developmental sequences and the relative developmental rates for iguanodontian taxa allows for an analysis of the impact of heterochrony on the evolution and development of the postcranial skeleton of this group as well. Heterochrony refers to a phyletic change in the onset or timing of development so that the appearance or rate of development of a feature in a descendant ontogeny is either accelerated or retarded relative to the appearance or rate of development of the same feature in the ontogeny of an ancestor (de Beer, 1954:140–158). Heterochrony has been considered in terms of changing developmental rates and appearances (Gould, 1977:70–99) and in morphological terms, with the separation of organismal shape and size from developmental time (Alberch et al., 1979). Although rate and timing modification in the strict sense cannot be determined for fossil taxa, relative rates and timing modifications can be, as can modifications in number and location of a morphological feature (Webster and Zelditch, 2005).

The nomenclature that will be utilized for the description of heterochrony in this study is based on Gould (1977) and Alberch et al. (1979) and includes pedomorphosis (the retention of ancestral juvenile characters by later ontogenetic stages of descendants) and peramorphosis (a descendant developing beyond the adult stage of the ancestor and thus resulting in previously unseen traits or morphologies). These changes can result from an increase (acceleration) or a decrease (neotony) in the rate at which the trajectory of shape change is followed. They can also result from an early (predisplacement) or late (postdisplacement) onset of the trajectory of shape change, or from the early (progenesis) or late (hypermorphosis) termination of the trajectory of shape change.

Institutional Abbreviations CMN, Canadian Museum of Nature, Ottawa, Ontario; MOR, Museum of the Rockies, Bozeman, Montana; OMNH, Sam Noble Museum, University of Oklahoma, Norman, Oklahoma; TMP, Royal Tyrrell Museum of Palaeontology, Drumheller, Alberta; YPM, Yale Peabody Museum of Natural History, New Haven, Connecticut.

MATERIALS AND SPECIMENS

In analyzing the ontogenetic changes within a species or a group of taxa, it is advantageous to have as much developmental data as possible, beginning as early in ontogeny as possible. However, embryological data, other than from late-stage embryos, are not available for iguanodontians, or for most fossil groups. This results, in part, from a lack of

Table 22.1. Specimens Used in This Study and the Elements Available for Each Taxon

Taxon	Specimen numbers	Growth stages represented	Elements available
Maiasaura	MOR 547; MOR 005; YPM 22405; YPM 22432; YPM 22400; YPM 22472	5	Scapula, coracoid, humerus, radius, ulna, pubis, ilium, ischium, femur, tibia, fibula
Hypacrosaurus	MOR 355; MOR 548; MOR 49; MOR 553; MOR 559; MOR 600; TMP 1994.385.0001	5	Scapula, coracoid, humerus, radius, ulna, pubis, ilium, ischium, femur, tibia, fibula
Brachylophosaurus	CMN 8893; MOR 1071	2	Scapula, coracoid, humerus, radius, ulna
Tenontosaurus	OMNH 58340; OMNH 34191; OMNH 10144; OMNH 34785	3	Scapula, pubis, ischium, ilium, femur, tibia, fibula

preservation of the cartilaginous precursors of bones and of soft tissue morphology. As a result, in analyzing the ontogeny of extinct taxa, ontogenetic studies are necessarily restricted to osteological characters of the ossified skeleton of taxa that are represented by specimens from at least two growth stages.

Despite these limitations, there are three iguanodontian species for which well-preserved embryonic and juvenile specimens exist (*Tenontosaurus tilletti*, *Maiasaura peeblesorum*, and *Hypacrosaurus stebingeri*). Another species, *Brachylophosaurus canadensis*, lacks embryonic material, but is represented by juvenile and later growth stages. Specimens of these four taxa were therefore included in this study (Tables 22.1, 22.2). The included specimens provide a reasonable representation of hadrosaurid diversity and provide an opportunity for comparison with a basal iguanodontian species. This analysis focuses on macro-morphological changes rather than histological changes and utilizes a qualitative approach to the description of ontogenetic changes. For each species, morphological changes during growth are identified for each element, with special attention to changes in proportion and appearance of new morphological structures.

DESCRIPTIVE APPENDICULAR ONTOGENIES

Maiasaura peeblesorum

Maiasaura peeblesorum is represented by one of the most diverse and abundant growth series known for dinosaurs (Fig. 22.1). Certain appendicular elements of *M. peeblesorum* show pronounced changes in morphology, whereas others are more conservative during growth, exhibiting more isometric development. Specimens used in the study include those from the Yale Peabody Museum of Natural History and the Museum of the Rockies (Tables 22.1, 22.2).

The scapula of *M. peeblesorum* undergoes moderate ontogenetic changes (Fig. 22.1A). In approaching full size, the scapula becomes relatively longer than the humerus. The scapular blade remains rather narrow throughout development. The blade maintains a uniform width from the scapular neck to the distal margin of the element throughout ontogeny, with the distal tip becoming slightly expanded in older individuals compared to hatchlings and juveniles. The primary change in the scapular blade morphology relates to the curvature of the blade; in hatchlings and juveniles it is slightly curved dorsoventrally and mediolaterally. By adulthood, the scapular blade becomes straighter, and no longer conforms to the rib cage. The proximal portion of the element does not change significantly. The relative sizes of the glenoid and coracoid articular surfaces remain constant during growth, as does the relative size of the acromion process. Hatchlings and juveniles have a scapular morphology that is typically hadrosaurine, and accordingly, the elements retain the more subtle hadrosaurine morphology to adulthood.

The coracoid of *M. peeblesorum* changes minimally during ontogeny (Fig. 22.1B). Although some of the coracoids examined here are in poor condition, some basic morphological information is preserved. The overall profile of the coracoid becomes more crescentic with the cranial half of the element becoming more ventrally curved during growth. The majority of change occurs in the cranial region of the element. The cranial process becomes proportionally longer and more ventrally hooked during growth. Across the rest of the element, proportions do not change drastically, except for the width of the coracoid, which increases relative to length in the adults. The coracoid foramen becomes proportionally smaller in diameter compared to the overall length of the element and is situated more caudally in adults.

The forelimb elements show few ontogenetic changes, except for the humerus. The most significant change in the humerus during development relates to the change in the size of the humerus relative to the overall size of the pectoral girdle and other forelimb elements (Fig. 22.1C). The element becomes relatively shorter than both the scapula and the ulna. In hatchlings, the humerus is approximately 80% the length of the scapula; by adulthood, the humerus is 50–67% the length of the scapula. Similarly, the humerus is approximately 80% the length of the ulna, but ultimately in mature individuals the humerus is only 67% the length of the ulna.

In general, the humerus retains the slender hadrosaurine condition throughout growth, with the deltopectoral crest region of the element only half again as wide as the humeral shaft. The relative proportions of the element remain consistent during growth; the proximal deltopectoral crest region of the element is approximately 50% of the proximodistal length of the element. The humeral head does not change in proportion significantly compared to the rest of the element. It projects from the proximal margin slightly more in adults than in hatchlings and juveniles, but does not become more mediolaterally extensive. The distal condyles do not change significantly during ontogeny, and do not become more extensive during growth.

The shaft of the radius (not shown) maintains a semicircular shape in cross section with growth and retains the same proportion to the ulna throughout ontogeny. The only change recognized in the ulna (not shown) is subtle and involves the increasing prominence of the olecranon process, which becomes wider and more robust compared to the length of the element.

The ilium of *M. peeblesorum* becomes more robust and taller as the animal grows (Fig. 22.1D). The cranial process of the ilium is arched to a greater degree in mature individuals than in juveniles. The supra-acetabular process becomes more robust with age, projecting farther laterally and ventrally in more mature individuals. The process does not significantly overlap the lateral side of the ilium as it does in many lambeosaurines. The morphology and proportions of the caudal process of the ilium are consistent throughout ontogeny. The caudal process retains its general morphology, with a squared-off caudal margin, and does not change in size relative to the overall size of the element. The acetabular margin becomes slightly more concave as the animal grows, but otherwise does not change. The pubic and ischiac peduncles retain the same morphology throughout growth, but become slightly more robust in adult individuals.

The morphology and proportions of the pubis of *M. peeblesorum* do not change significantly during ontogeny (Fig. 22.1E). The pubic blade is thin throughout development and retains its rounded cranial margin into maturity. The iliac and ischiac peduncles remain relatively short compared to the overall length of the element. Although the iliac peduncle appears more robust in mature individuals, the ischiac peduncle has a consistent morphology throughout development.

The ischium of *M. peeblesorum* develops isometrically (Fig. 22.1F). Though the distal tips of many of the embryonic and hatchling specimens are missing, those that are present suggest that the length and morphology of the ischial shaft relative to other pelvic elements is consistent throughout growth. Throughout growth, there is a long, slender ischial shaft that tapers slightly at the distal tip. As expected in hadrosaurines (*M. peeblesorum*) there is no enlargement at the distal tip of the shaft at any growth stage. The pubic and iliac peduncles have a consistent relative size compared to

Table 22.2. Table of Elements Represented for Each Taxon, Including the Number of Growth Stages Represented and the Size Range (of the Length) of the Elements

Taxon	Element	Growth stages represented	Size range of elements (mm)
Maiasaura	Scapula	5	14–547
Maiasaura	Coracoid	3	9–78
Maiasaura	Humerus	5	9–276
Maiasaura	Radius	3	8–256
Maiasaura	Ulna	3	8–249
Maiasaura	Ilium	4	12–409
Maiasaura	Pubis	3	17–308
Maiasaura	Ischium	4	21–402
Maiasaura	Femur	5	11–578
Maiasaura	Tibia	4	12–564
Maiasaura	Fibula	4	10–575
Hypacrosaurus	Scapula	5	14–270
Hypacrosaurus	Coracoid	3	8–67
Hypacrosaurus	Humerus	5	12–234
Hypacrosaurus	Radius	3	10–209
Hypacrosaurus	Ulna	3	15–214
Hypacrosaurus	Ilium	3	20–247
Hypacrosaurus	Pubis	3	14–234
Hypacrosaurus	Ischium	3	15–289
Hypacrosaurus	Femur	5	11–409
Hypacrosaurus	Tibia	5	9–425
Hypacrosaurus	Fibula	3	14–401
Brachylophosaurus	Scapula	3	201–247
Brachylophosaurus	Coracoid	2	45–76
Brachylophosaurus	Humerus	3	94–161
Brachylophosaurus	Radius	2	87–154
Brachylophosaurus	Ulna	2	83–168
Brachylophosaurus	Ilium	2	194–412
Brachylophosaurus	Pubis	2	182–390
Brachylophosaurus	Ischium	1	280
Brachylophosaurus	Femur	1	420
Brachylophosaurus	Tibia	1	428
Brachylophosaurus	Fibula	1	413
Tenontosaurus	Scapula	2	11–161
Tenontosaurus	Coracoid	1	84
Tenontosaurus	Humerus	1	167
Tenontosaurus	Radius	2	12–154
Tenontosaurus	Ulna	2	9–160
Tenontosaurus	Ilium	1	209
Tenontosaurus	Pubis	2	178
Tenontosaurus	Ischium	2	130–176
Tenontosaurus	Femur	3	15–256
Tenontosaurus	Tibia	3	12–245
Tenontosaurus	Fibula	3	10–247

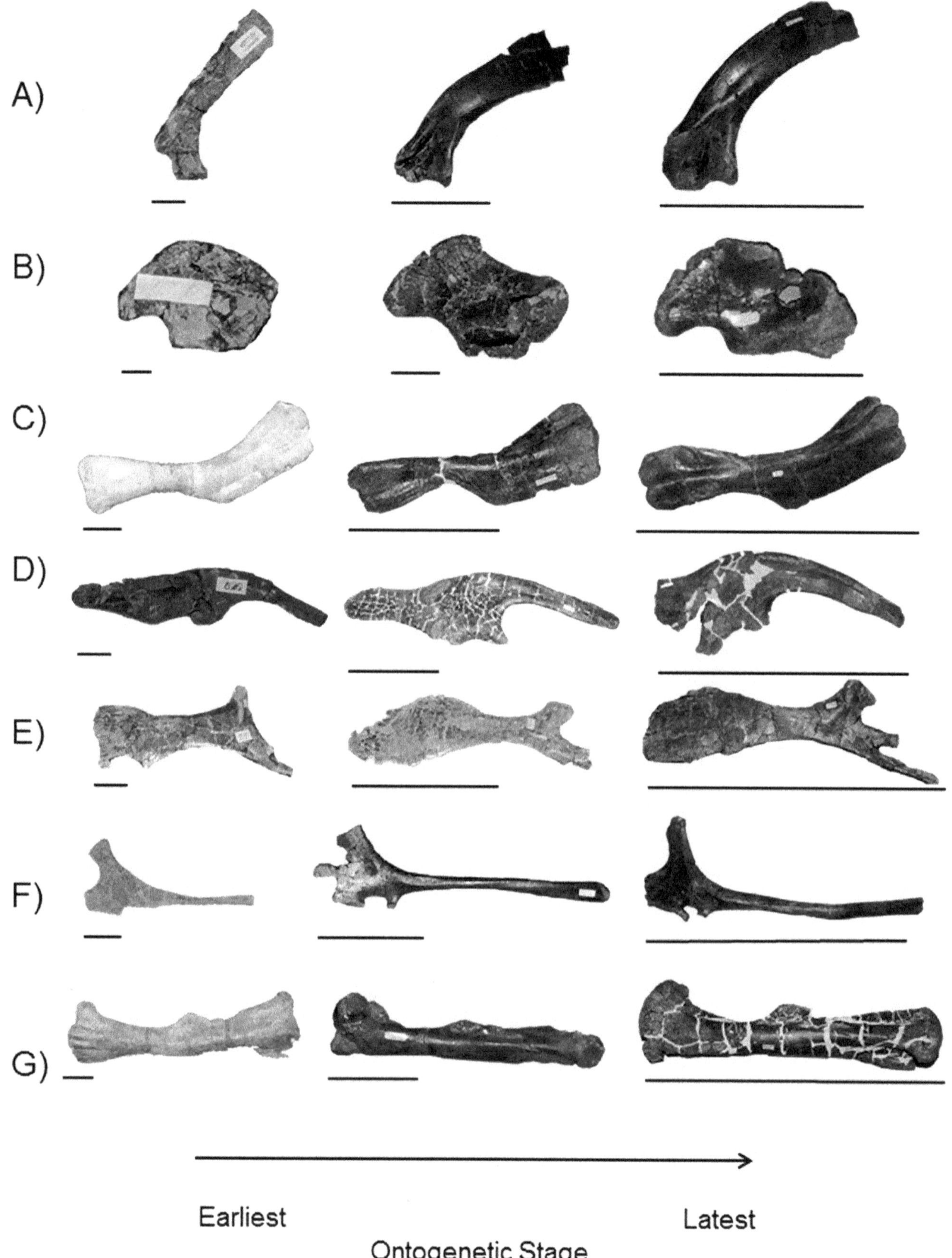

22.1. Postcranial elements of *Maiasaura peeblesorum* (MOR 547, 005; YPM 22400, 22432, 22472) in lateral view and elements from some of the represented growth stages. Elements that show morphological changes during ontogeny include (A) Scapula; (B) Coracoid; (C) Humerus; (D) Ilium; (E) Pubis; (F) Ischium; and (G) Femur. Note: elements are not to scale; scale bars below each specimen indicate the size of each element relative to the adult elements shown at far right.

the size of the element as a whole. In certain specimens the peduncles seem to be relatively more robust and mediolaterally broader in the adults. However, this does not seem to be strictly due to growth, since mature individuals of similar size show variation in this characteristic.

In general, the femur of *M. peeblesorum* becomes more robust as the animal ages (Fig. 22.1G). One especially notable change is in the size of the fourth trochanter. In hatchlings, the fourth trochanter extends for approximately one-quarter the length of the element and is located mid-shaft. In adults, the fourth trochanter extends for roughly one-third the length of the element. The craniocaudal projection of the fourth trochanter also increases relative to the diameter of the shaft as the animal ages. The fourth trochanter generally becomes broader and wing-like in the adult, increasing proportionally compared to the diameter of the shaft. The shaft becomes more robust as the animal ages, with the shaft becoming proportionally wider compared to the length of the femur. The prominence of the femoral head remains consistent in all but the earliest hatchling stage of ontogeny. The greater and lesser trochanters are also relatively well developed, even in earlier juvenile stages of development. However, distally, the length of the distal condyles appears to increase relative to the width of the shaft as the animal grows. This is particularly true of the caudal extension of the distal condyles.

The tibia (not shown) of *M. peeblesorum* changes slightly during ontogeny. The primary change is in the width of the proximal and distal ends of the element compared to the mid-shaft width. Both proximal and distal ends become relatively wider compared to the shaft as the animal ages. In older individuals, the proximal and distal ends begin to expand at the midpoint of the shaft of the element moreso than they do in hatchlings and juvenile individuals. In other words, the central shaft is narrow and of uniform diameter for approximately half the length of the element in younger individuals, whereas it makes up only one-third of the element's length in mature individuals.

The morphology of the fibula (not shown) does not change significantly during ontogeny. The simple construction of the element–straight shaft and slightly enlarged proximal and distal ends–remains constant during ontogeny. The only notable change in morphology is the slight enlargement of the crescentic distal surface of the fibula. As the animal grows, the distal articulation with the tibia becomes better developed.

Hypacrosaurus stebingeri

Like *M. peeblesorum*, *Hypacrosaurus stebingeri* is represented by one of the more complete growth series known among dinosaurs (Fig. 22.2) and provides a unique opportunity to study ontogenetic changes in the postcranial skeleton of a lambeosaurine.

The scapula of *H. stebingeri* undergoes significant changes throughout ontogeny (Fig. 22.2A); in the earliest stages, the scapulae appear more hadrosaurine, but as the animal grows, the scapulae look progressively more lambeosaurine. Specifically, the scapular blade broadens dorsoventrally, becoming much wider than the scapular neck–morphology characteristic of lambeosaurines. In hatchlings, the scapula is short relative to the length of the humerus, but during growth, the blade becomes relatively longer than the humerus. In hatchlings, the entire element has uniform width and the scapular blade has a uniform dorsoventral width, but in juveniles and adults, the neck is tapered compared to the blade, which widens by expanding dorsoventrally starting at the neck. The blade is comparatively straight in juveniles and adults; however, in hatchlings the blade is curved. In adults and juveniles, the dorsal margin of the element is straight with the ventral margin curving caudoventrally distal to the scapular neck. The acromion process projects progressively more laterally as the animal ages, resulting in a more concave lateral face of the scapula. By maturity, the acromion process also migrates to become more centrally located on the lateral face of the scapula. In the hatchlings, the acromion process is located more cranially, but with growth, the process shifts caudally. The size of the articular surfaces relative to one other remains constant throughout ontogeny. The surface for articulation with the coracoid is larger than the scapular contribution to the glenoid. Both surfaces are concave and become deeper with age.

The coracoid of *H. stebingeri* becomes more robust during growth (Fig. 22.2B). The thickness of the element increases relative to its length, as in *M. peeblesorum*. The relative length and width of the element also change during growth, with the element becoming longer. The coracoid foramen also changes in several ways during ontogeny. The coracoid foramen is positioned more proximally and caudally in older individuals. In hatchlings, the coracoid foramen is semicircular, dorsoventrally narrow, and craniocaudally wide. With growth, the coracoid foramen becomes more circular and wider compared to the rest of the element. The visible suture in hatchlings that marks where the coracoid foramen becomes enclosed, disappears in older individuals. The articular surfaces of the coracoid are concave and oval in outline, and thus similar to those of the adults. The concave surface becomes slightly deeper as the animal ages.

The relative size of the articular surfaces of the pectoral girdle varies even among individuals of the same approximate age. In many of the hatchlings, the scapular and glenoid surfaces are subequal in length, although some hatchlings appear to have larger glenoid surfaces. In adults, however,

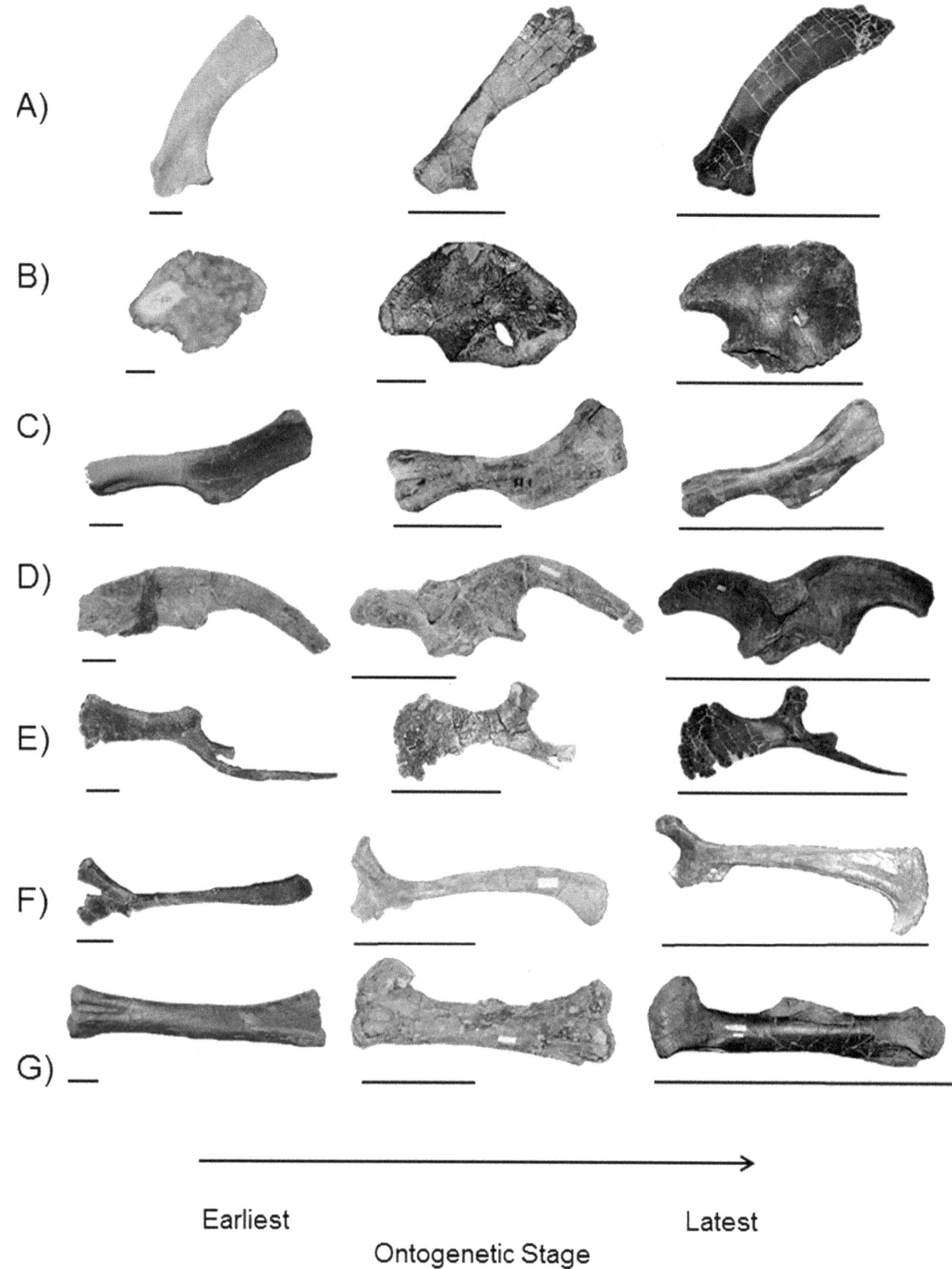

22.2. Postcranial elements of *Hypacrosaurus stebingeri* (MOR 49, 355, 548, 553, 559; TMP 1994.385.0001) in lateral view and elements from some of the represented growth stages. Elements that show morphological changes during ontogeny include (A) Scapula; (B) Coracoid; (C) Humerus; (D) Ilium; (E) Pubis; (F) Ischium; and (G) Femur. Scale bars as in Figure 22.1.

the articular surface for the scapula is larger than the glenoid surface. The cranial process of the coracoid also becomes more arched during growth. In hatchlings and juveniles, the cranial process projects in a straight manner. However, during growth, the cranial process becomes more hooked and extends to the level of the articular surfaces. The shape of the caudal process remains the same, with the tip tapering to a slightly rounded end.

Like the scapula, the humerus of *H. stebingeri* achieves a progressively more lambeosaurine morphology during ontogeny (Fig. 22.2C). Adult lambeosaurine humerus morphology is characterized by a robust deltopectoral crest that is more than twice the width of the humeral shaft. In hatchlings, the proximal portion of the element, including the deltopectoral crest, makes up approximately one-half the length of the element. As the animal grows, this proximal segment becomes proportionally longer than the distal shaft and condyles (Fig. 22.3). In adults, the deltopectoral crest portion of the element makes up two-thirds the length of the element. The craniocaudal proportions of the element also change through ontogeny. In hatchlings, the width of the deltopectoral crests ranges from slightly less than twice the width of the humeral shaft to twice the width of the shaft (Fig. 22.3). In juveniles, this proportion increases and ultimately, in the adults, the deltopectoral crest can be almost three times the width of the humeral shaft. This development is in line with the observation that lambeosaurines have more robust humeri and wider deltopectoral crests than do hadrosaurines. Accordingly, it appears that more robust humeri result from the disproportionally faster growth of the proximal portion of the element, including the deltopectoral crest. The deltopectoral crest also changes its orientation during growth. In hatchlings and juveniles, the crest is directed cranially and medially, resulting in a relatively flat proximal half of the element. However, with growth, the deltopectoral crest becomes increasingly directed cranially. The medial and lateral condyles are approximately the same relative size and constitute a slight enlargement of the humeral shaft, being approximately half again the width of the shaft. This arrangement remains the same throughout ontogeny. The placement and size of the humeral head in relation to the rest of the element changes with growth. The humeral head is centrally placed in the hatchling specimens and has a width that is approximately one-third of the total width of the proximal margin of the element. By adulthood, the humeral head constitutes only one-fourth of the width of the proximal margin of the humerus. However, despite its relatively diminished size, the humeral head projects outward from the element more that it does in the younger individuals, and the ridge that extends distally from the apex of the humeral head extends farther distally from the humeral head.

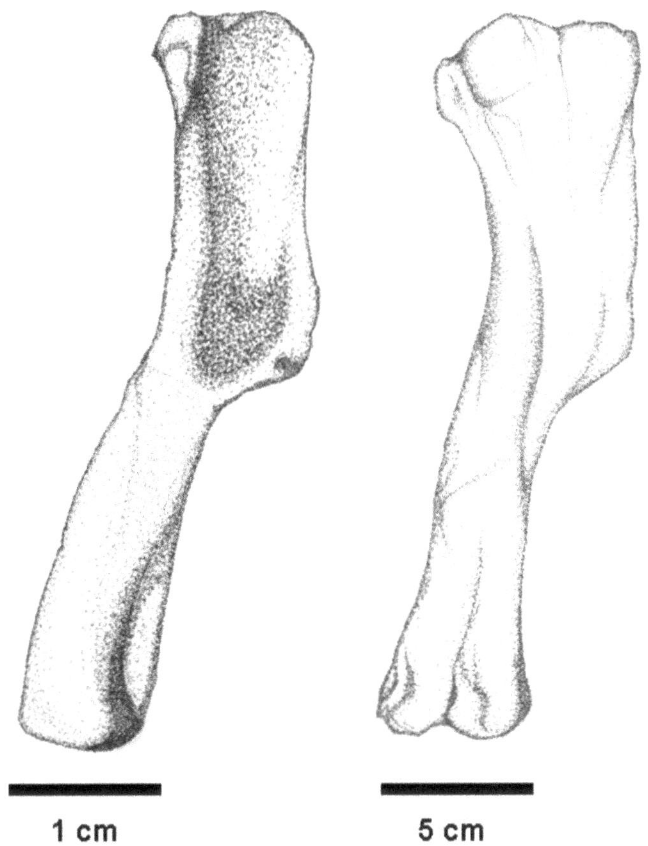

22.3. Illustration of hatchling (left) and subadult humeri of *Hypacrosaurus stebingeri* in lateral view. Illustration by Arielle Wydra.

As is true of *M. peeblesorum*, there are no noticeable morphological changes to the radius (not shown) or ulna (not shown) during growth. The proportions of the radial shaft and the proportions of the proximal and distal expansions of the element remain consistent throughout the different stages of growth. The ulna also does not show significant morphological change throughout ontogeny. The olecranon process is prominent at the proximal end, even in hatchlings.

The ilium shows subtle changes during ontogeny (Fig. 22.2D). At all ontogenetic stages available for study, the element has a long, slender profile that is extensive craniocaudally, thin mediolaterally, and slender dorsoventrally. The ilium has a rather straight dorsal margin that becomes slightly more curved in adults; specifically, the cranial process – just dorsal to the pubic peduncle – becomes more dorsally extensive as the animal grows. Ventrally, the acetabular margin becomes more concave and better defined in more mature individuals. Accompanying this, the pubic and ischiac peduncles become more robust, becoming more ventrally extensive and mediolaterally wide. The most noticeable change in the acetabular region of the element is the relative growth of the supra-acetabular process. In hatchlings, the supra-acetabular

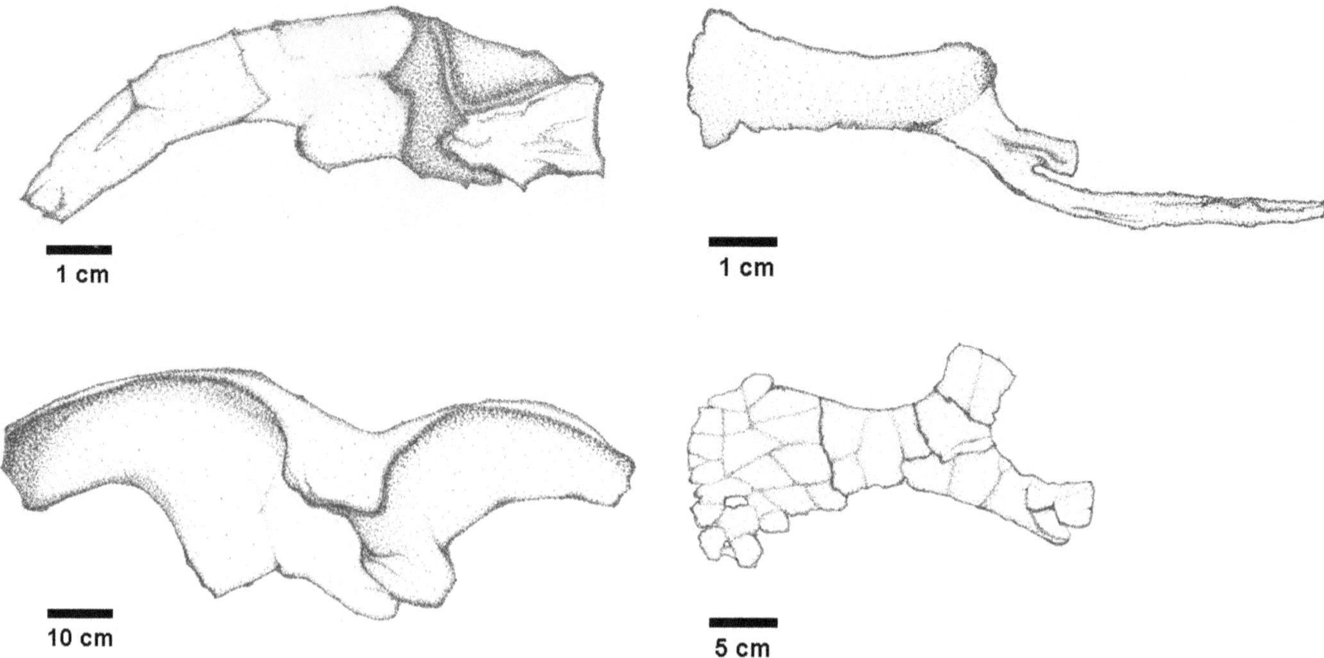

22.4. Illustration of hatchling (top) and subadult ilia of *Hypacrosaurus stebingeri* in lateral view. Illustration by Arielle Wydra.

22.5. Illustration of hatchling (top) and subadult pubes of *Hypacrosaurus stebingeri* in lateral view. Illustration by Arielle Wydra.

process projects only laterally (Fig. 22.4). In juveniles the supra-acetabular process begins to extend both cranially and ventrally, and by adulthood the supra-acetabular process extends laterally and folds over the lateral side of the element in an almost pendant manner. Caudally, the element becomes rectangular with growth, the caudal margin becoming blunt in adults. The cranial process remains relatively consistent throughout ontogeny. The process is widest at the neck and then tapers slightly cranially and remains uniformly thick for the remainder of its length. At all growth stages, the cranial process trends slightly ventrally, toward the cranial tip of the element. In general, both the cranial and caudal processes of the ilium become taller relative to the overall length of the element.

During ontogeny, the pubis of *H. stebingeri* undergoes subtle changes. The neck of the pubic blade becomes wider dorsoventrally (Fig. 22.2E) and the pubic blade becomes longer relative to the overall length of the element. Throughout ontogeny, the pubic blade maintains a blunt hatchet-shaped morphology. In all developmental stages the pubic blade has a flat cranial margin (Fig. 22.5). The iliac peduncle becomes more robust during ontogeny, and although its relative length does not increase, its relative width increases relative to the rest of the element. From a dorsal view, the iliac peduncle takes on a triangular profile as the animal increases in size. Though many of the juvenile pubes are missing the distal tip of the pubic shaft, it appears that the relative length and cross sectional diameter of the pubic shaft remain consistent throughout ontogeny. The ischiac peduncle becomes slightly more robust as the animal grows. The short shaft of the peduncle retains similar proportions throughout growth, but the distal articular end of the peduncle becomes more robust and expands to a wide, circular articular surface.

The ischium of *H. stebingeri* changes markedly during ontogeny (Fig. 22.2F). At the hatchling stage, the ischium looks hadrosaurine, with a shaft that does not widen distally. However, as the animal grows, the element rapidly develops the distally expanded ischial boot of lambeosaurines. By the juvenile stage, the ischium already exhibits this more robust feature. In hatchlings, the ischial shaft is of uniform thickness for the proximal half, widening distally starting at the midpoint of the shaft. In juveniles the booted structure has begun to develop and is proximodistally wide and rounded, but does not yet expand as far cranially as it will in the adult. In adults the boot is craniocaudally long, but accounts for a very small portion of the distal length of the shaft. It is a narrow structure that tapers to a point cranially. The distal margin of the boot is flattened compared to the rounded distal margin in the juveniles. In hatchlings through to adults, the proximal peduncles remain unchanged. The pubic peduncle is wider craniocaudally, but transversely thinner than the iliac peduncle, which is almost cylindrical in some specimens. The margin of the acetabulum also changes little during growth; the acetabular margin is concave, but remains shallow in individuals of all age ranges. Proximally, there are changes in the obturator process. In the youngest individuals, the obturator process is barely visible. In juvenile specimens, the obturator process projects almost the width of the shaft and in adult specimens it projects approximately the width of the shaft and in some cases can coalesce with the peduncle to form a foramen.

The femur of *H. stebingeri* undergoes subtle changes during growth, becoming more robust as the animal develops (Fig. 22.2G). The most noticeable change is in the relative size of the fourth trochanter. In hatchlings, the fourth trochanter accounts for approximately 20–25% of the length of the element. In adults, the fourth trochanter makes up one-third of the length of the femur. In comparison with the diameter of the shaft, the height of the fourth trochanter also increases during ontogeny. The muscle scars on the medial side of the fourth trochanter become more prominent as the animal ages. The distal condyles also become well developed as the animal grows. In hatchlings and juveniles the distal condyles project caudally for approximately one-half the diameter of the shaft, whereas in subadults and adults the condyles project caudally for a distance equal to the diameter of the shaft. Both the greater and lesser trochanters are well developed throughout ontogeny. Finally, the femoral head becomes relatively larger compared to the overall size of the femur and becomes more spherical during growth.

The tibia of *H. stebingeri* (not shown) undergoes several changes in proportion and orientation in the element. The proximal and distal ends of the tibia are both expanded, with the axis of expansion for the proximal end being craniocaudal and the axis of expansion on the distal end mediolateral. The cnemial crest becomes better developed with growth. In hatchlings and juveniles, the proximal end of the tibia expands almost equally medially and laterally, but as the animal grows, the lateral side of the cnemial crest develops relatively faster than the medial side. In adults, the cnemial crest projects laterally and also expands caudally creating a concave space for the reception of the fibula. The shaft remains cylindrical throughout the growth, however, the narrow part of the shaft becomes relatively longer compared to its length in hatchlings. As the animal grows, the proximal and distal regions of the shaft widen. The narrowest part of the shaft initially makes up one-half the length of the element but decreases to one-third the element's length in adults.

In the fibula (not shown) of *H. stebingeri*, only the proportions of the element change during growth. Throughout ontogeny, the element consists of a shaft with an oval cross section, and expanded proximal and distal ends. In hatchlings and juveniles, the middle one-third of the element has a uniform diameter; beyond that, the element increases in width both proximally and distally. The proximal and distal tips are both somewhat flattened mediolaterally. In adults the shaft is relatively longer compared with the diameter of the element and the middle three-quarters of the shaft is thin and of uniform diameter, with only the distal- and proximal-most tips of the element expanding. The shape of the shaft in cross section remains consistent throughout growth, but the shapes of the proximal and distal ends change. The proximal and distal ends of the element are not flattened in adults. In cross section, the proximal end of the fibula forms a half circle with the flattened side receiving the tibia. Distally, the fibula is also flattened medially, but is otherwise oval in cross section.

Brachylophosaurus canadensis

Although there is not a very complete growth series for *Brachylophosaurus canadensis*–there are no hatchling or embryonic specimens of the species–the ontogenetic changes observable in *B. canadensis* (Fig. 22.6) can be used to further assess the significance of the developmental differences observed between *M. peeblesorum* and *H. stebingeri*. The available juvenile specimens (MOR 1071) are from a bonebed in the Judith River Formation of Montana and are approximately one-half to two-thirds the average size of adult individuals of *B. canadensis*.

With few growth stages to compare, the scapula of *B. canadensis* appears to undergo subtle changes in morphology (Fig. 22.6A). The scapula retains the broad, distally blunt, squared scapular blade, which remains of uniform width for most of its length, tapering only at the neck and expanding only at the distal tip. At all stages, the blade is mediolaterally thin, although it appears to become relatively thinner, compared to the general width of the blade, as the animal ages. The scapula has a prominent acromion process, even at the juvenile stage. Muscle scars that are obvious on the face of the acromion process are present at all available stages, but become more distinct and proportionally larger as the animal ages. With growth, the distal articular surfaces become more concave and the glenoid becomes deeper and more concave.

The coracoid of *B. canadensis* becomes more robust as the animal ages (Fig. 22.6B). The thickness of the element increases proportional to the overall length of the element. The enclosure of the coracoid foramen does not appear to be contingent on growth; the coracoid foramen is already fully enclosed in the juvenile individuals. The foramen does, however, become proportionally smaller as the animal grows; at the juvenile stage, the foramen is already the same absolute size as it is in the adult specimens. The anatomical position of the foramen in relation to the rest of the element also

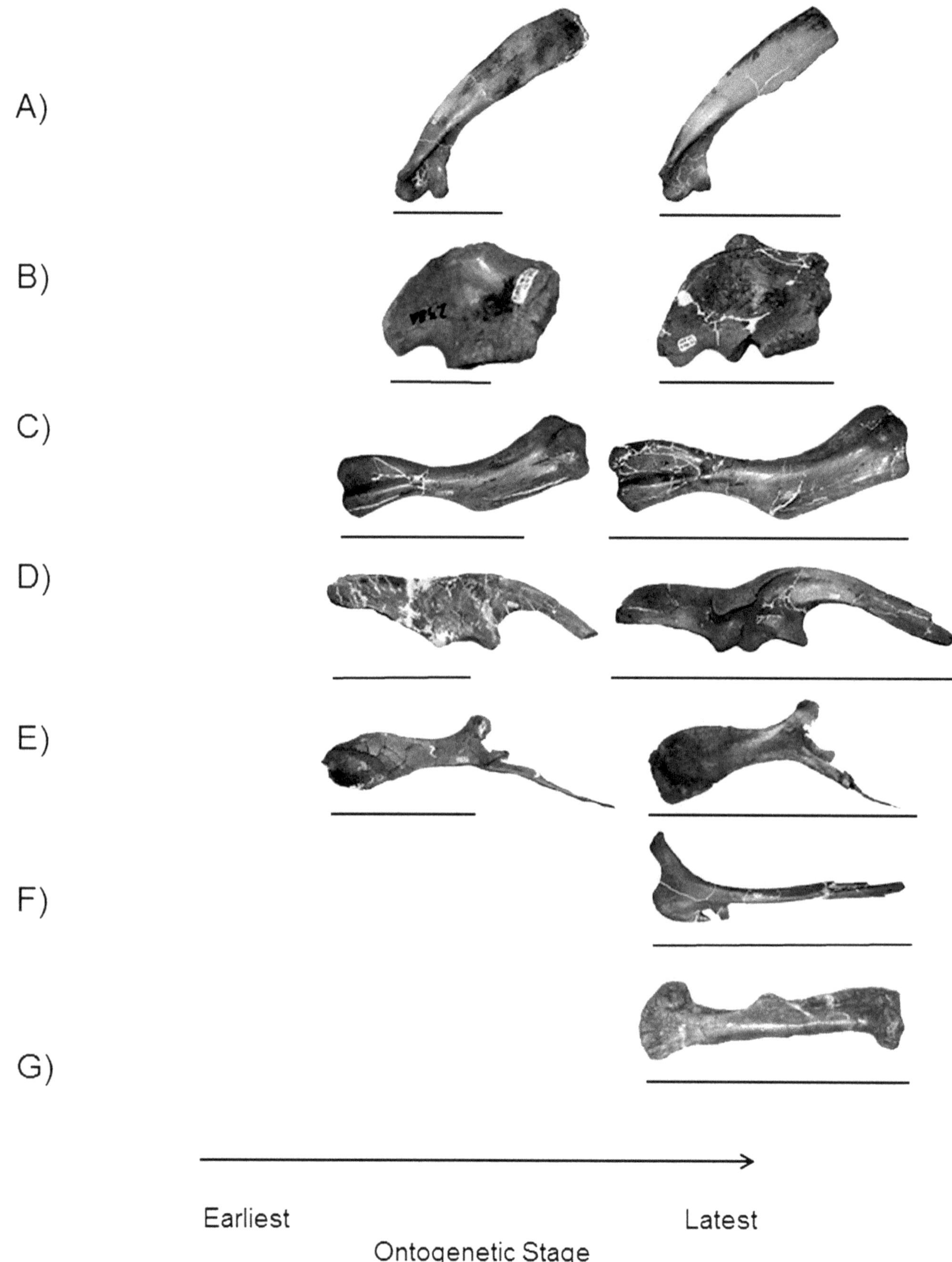

22.6. Postcranial elements of *Brachylophosaurus canadensis* (CMN 8893; MOR 1071) in lateral view and elements from the represented growth stages. Elements that show morphological changes during ontogeny include (A) Scapula; (B) Coracoid; (C) Humerus; (D) Ilium; (E) Pubis; (F) Ischium; and (G) Femur. Scale bars as in Figure 22.1.

changes. There appears to be a relative increase in growth in the cranial part of the element, particularly in the hook like extension of the cranial process, compared with the caudal portion of the element. Therefore, the coracoid foramen appears more centrally placed in juveniles. Related to this is the relative decrease in length of the articular surfaces compared to the overall length of the element. Conversely, the widths of the articular surfaces increase with ontogeny. This is expected because the overall width of the element increases during ontogeny. During growth, the cranial process extends farther relative to the rest of the element and becomes more arched. The tip of the cranial process is in line with the glenoid surfaces in adults; it is not in juveniles. The tip of the cranial process also becomes more pointed in adults, whereas it is more rounded in juveniles. In adults, the dorsal ridge of the element becomes more rugose during growth, and there is a prominent protuberance on the surface that similarly becomes more robust.

Many of the available humeri for *B. canadensis* are from mature individuals; however, there are several juvenile and subadult specimens that shed light on the transitions in this element during growth (Fig. 22.6C). The humerus is hadrosaurine with its slender construction and deltopectoral crest that is less than twice the width of the humeral shaft, and a shape, straight profile, and proportions that remain consistent through ontogeny. Distinct ontogenetic changes relate primarily to muscle scars, which are better preserved in *B. canadensis* than in other taxa. The muscle scar on the caudolateral side of the ridge that leads distally from the humeral head continues for one-quarter of the length of the deltopectoral crest. This muscle scar is proportionally smaller in the subadult specimens and is concave rather than convex, as it is in the adult specimens. The muscle scars located on the cranial margin of the humerus are also more prominent in the adult specimens. In general, the adult humeri are somewhat more robust than those of the subadults. In particular, the humeral head is one-quarter of the width of the proximal margin of the humerus in the subadult specimens, whereas it becomes one-third of the width of the margin in the adult individuals. Distally, there are few changes in the element during ontogeny. In adults, the medial and lateral condyles become somewhat wider in relation to the width of the shaft.

A range of growth stages are not represented for the radius (not shown), and therefore ontogenetic changes cannot be determined for that element. However, the radius has a conservative morphology and, based on *M. peeblesorum*, it would be reasonable to expect little change in the element during growth.

The specimens for the ulna (not shown) of *B. canadensis* are limited in number and range in size from subadult to adult. The morphology of the ulna remains conservative throughout the portion of growth known for this species. The ulnae have prominent olecranon processes that are well developed in the juveniles as well as in the adults. However, the olecranon process does appear to project more distally, relative to the width of the shaft of the element, as the animal reaches maturity. Distally, the element flares at its tip, with the expansion remaining a consistent size, relative to the width of the shaft, at all available growth stages. The only ontogenetic change distally is the change of the distal articular surface from flat in the juveniles to convex in the mature individuals.

There are several ilia exhibiting a range of sizes for *B. canadensis*. They do not represent the full range of the growth series, however, but do shed light on the transitions between larger juveniles, subadults, and mature individuals (Fig. 22.6D). The smallest specimens (juveniles) are compressed dorsoventrally, relative to those of the adults. The juvenile ilia also have a very flat dorsal margin and little overall curvature to the element compared to adults. In the adults, the cranial process of the ilium curves ventrally so that its cranial tip is level with the distal tip of the pubic peduncle. Throughout ontogeny, the acetabular margin remains shallow and the ischiac peduncle becomes more robust and wider mediolaterally relative to the rest of the element. The supra-acetabular process has a unique orientation in *B. canadensis*; it projects laterally, perpendicular to the rest of the element. Typically, in hadrosaurids the process folds over the lateral side of the element as seen in its most exaggerated form in lambeosaurines, particularly *Parasaurolophus*. In adults, it extends even farther relative to the mediolateral width of the element.

There is a range of sizes, from juvenile to adult, of well-preserved pubes for *B. canadensis* (Fig. 22.6E). The juveniles have a dorsal and ventral expansion of the blade not observed in the adults, which have relatively flat dorsal and ventral margins. The articular surface of the iliac peduncle of the pubis is flat in the younger individuals and becomes concave in some of the mature individuals. The iliac peduncle is prominent at all stages, with a distinct articular surface that is triangular in dorsal view. The lateral point of this triangular surface leads ventrally into a ridge on the lateral face of the element. As the animal ages, this dorsoventrally oriented ridge becomes positioned more caudally relative to the rest of the element. The neck of the pubis becomes gradually shorter relative to the length of the pubic blade during growth.

The ischia in this collection of *B. canadensis* elements are all large and approximately the same size, making ontogenetic changes impossible to determine. Nonetheless, in all of these specimens, there is a prominent obturator process and in some, the obturator process almost coalesces with the base of the pubic peduncle to form a foramen (Fig. 22.6F). This coalescence is caused by the proximal extension of the obturator process in some ischia and the distal expansion of the pubic process in others, indicating variation within growth stages.

Although the femora and tibiae included in MOR 1071 are very well preserved, they do not represent a range of growth stages and are limited to those of adults, thus preventing a better understanding of ontogenic change in those elements within *B. canadensis* (Fig. 22.6G).

Several fibulae (not shown) occur in a range of sizes for *B. canadensis*. The smallest juvenile fibulae are essentially straight shafts with slightly expanded proximal and distal ends. The proximal end of all fibulae, of all size ranges, is flat and crescentic. In the largest specimen (an adult), the proximal end is even more enlarged, relative to the width of the shaft, than in the juveniles. The lateral side of the fibula is convex, but the medial side is flattened for articulation with the tibia. This flattened medial side creates a ridge along the face of the shaft that begins in the distal medial corner of the element and leads to the proximal lateral corner, essentially crossing the shaft. This ridge becomes more and more prominent as the animal becomes larger.

Tenontosaurus tilletti

Juvenile and hatchling specimens of *Tenontosaurus tilletti* have been analyzed previously, but the studies have had a histological focus (Werning, 2005). Appendicular elements from hatchlings and juveniles exist; however, because the number is small, so are the number of growth stages represented by specimens (Fig. 22.7). Accordingly, the overall ontogenetic changes in the appendicular skeleton are less clear than for the hadrosaurid species. Partial ontogenetic series are available only for the scapula, radius, ischium, femur, tibia, and fibula.

There are several ontogenetic changes of note in the scapula of *T. tilletti*. The primary changes (Fig. 22.7A) include the proximal articular end of the scapula becoming broader in adults than it is in hatchlings and juveniles. The articular surfaces of the proximal end of the scapula also become deeper and more concave during growth. The scapular neck is wider compared to the scapular blade in juveniles. In adults of *T. tilletti*, the scapular neck is more tapered, with the scapular blade expanding distally, particularly in the distal third. The width of the scapular blade retains a more uniform width in individuals of earlier ontogenetic stages. The mediolateral thickness of the scapula changes slightly through ontogeny as well, with the distal tip of the scapular blade becoming relatively thinner compared to the rest of the blade in adults. It is hard to know what changes in the coracoid accompany these changes to the scapula since no growth series is preserved for the coracoid (Fig. 22.7B).

Although ontogenetic changes to the humerus of *T. tilletti* would serve as an important comparison to the hadrosaurid growth series, no growth series is preserved for the humerus. Adults have a robust humeral morphology (Fig. 22.7C), similar to that of lambeosaurines, but it remains unknown whether this species follows a similar developmental pathway as in *H. stebingeri*.

As in hadrosaurid taxa, there is little ontogenetic change to the radius (not shown) or (ulna) for *T. tilletti*. The radius retains its simple construction and exhibits a narrow, straight shaft that is enlarged at both its proximal and distal ends. There is a slight change in the proportions of different parts of the element, with the width of the shaft becoming relatively wider compared to the overall length of the element.

Like the pectoral girdle, only some elements of the pelvic girdle are represented by a growth series. The ilium lacks specimens from multiple stages (Fig. 22.7D). Due to the poor preservational quality of the hatchling specimens, it is difficult to determine the transitions that take place between the earliest growth stage and the juvenile stage for the pubis.

The primary change in the pubis occurs in the caudal half of the element where the ischiac peduncle becomes more robust as the animal ages and develops a more convex articular surface (Fig. 22.7E). The caudal blade of the pubis is wider in mature individuals than in juveniles. The postacetabular region makes up a larger proportion of the length of the element as the animal reaches maturity. The acetabular margin of the pubis becomes slightly more convex as the animal grows due to the relative increase in the height of the peduncles compared to the overall height of the element.

The fragmentary nature of the smallest ischia makes it difficult to recognize the earliest transitions taking place in the element. However, the changes seen during the shift from juvenile to adult are dramatic (Fig. 22.7F). In general, the element becomes more robust. The pubic and iliac peduncles become more robust and relatively larger compared to the overall length of the element. Both peduncles develop broader surfaces for articulation and both have a craniocaudal width that becomes proportionally larger compared to the overall length of the element. The obturator process becomes more pronounced during ontogeny, ultimately projecting the full width of the ischial shaft. The obturator process remains narrow, becoming thinner proportional to the shaft of the element. The shaft becomes relatively shorter and wider as the animal grows, and the shaft becomes more and more flattened in cross section during ontogeny. The acetabular margin remains shallow and flattened throughout ontogeny.

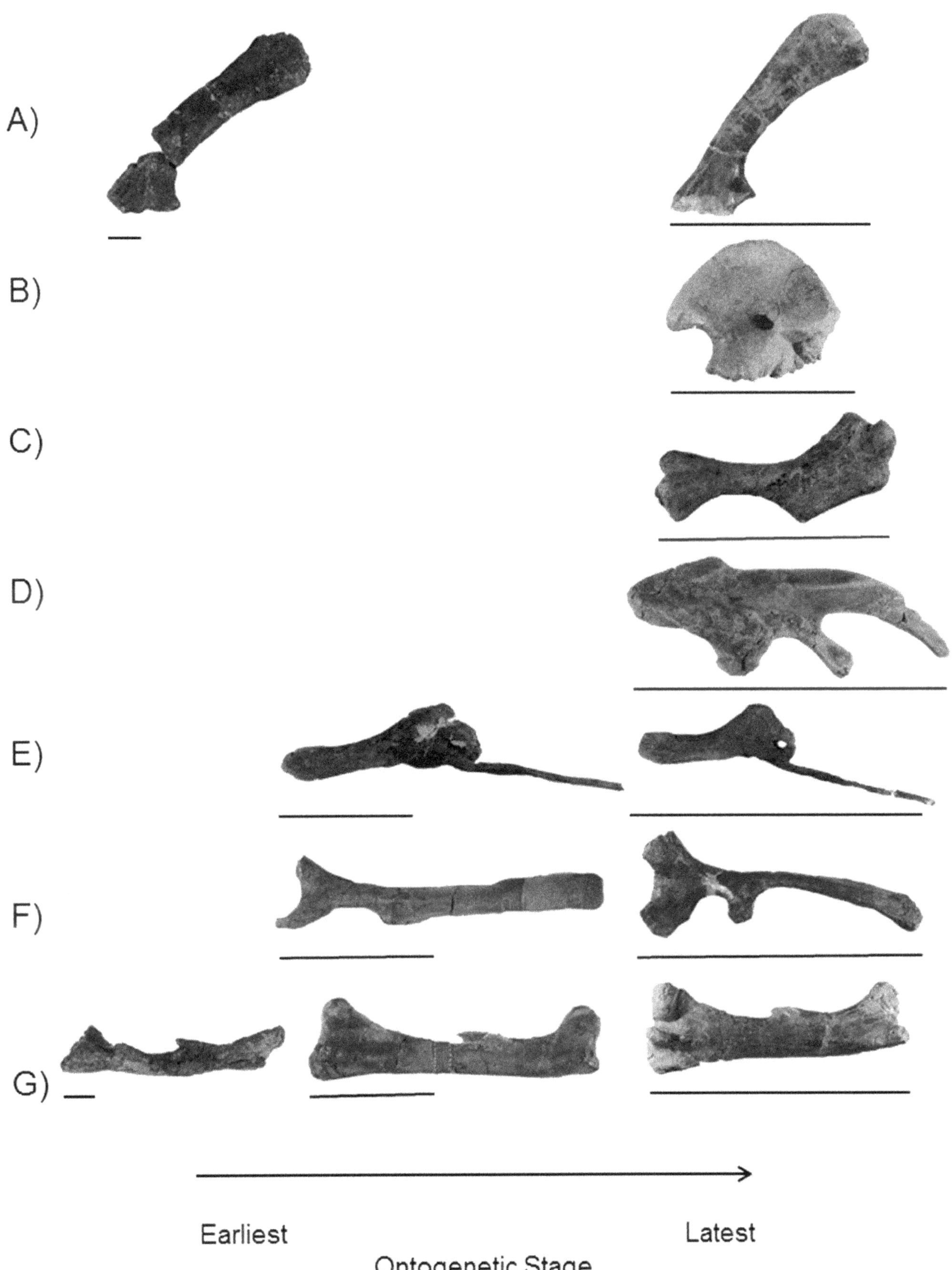

22.7. Postcranial elements of *Tenontosaurus tilletti* (OMNH 58340, 34191, 10144, 34785) in lateral view presented in the represented growth stages. Elements that show morphological changes during ontogeny include (A) Scapula; (B) Coracoid; (C) Humerus; (D) Ilium; (E) Pubis; (F) Ischium; and (G) Femur. Scale bars as in Figure 22.1.

The femora of hatchlings and juveniles in *T. tilletti* are similar in morphology to those of adults (Fig. 22.7G). The fourth trochanter is prominently pendant, even in these early growth stages. The fourth trochanter becomes less sharply pendant during growth and does not project as far laterally (compared to the width of the femoral shaft) as it does in the earliest stages of growth. Conversely, the fourth trochanter becomes longer compared to the length of the shaft, changing from one-fifth the length of the femur in hatchlings to one-fourth the length of the femur in adults. The fourth trochanter also becomes proportionally wider during growth. The femoral head and the greater and lesser trochanters are already well formed in hatchlings. In juveniles and adults, the femoral head is oriented medially at a 90° angle to the shaft, whereas in hatchlings the femoral head is oriented slightly mediodorsally rather than strictly medially. Distally, the medial and lateral condyles of femora in hatchlings and juveniles are well developed and resemble those of mature individuals. With growth, the femur increases the width of the shaft faster than it increases the length of the element. In cross section, the shaft changes from oval in hatchlings to more circular in mature individuals.

In contrast to the femur, the tibia and fibula (neither shown) of *T. tilletti* change little during ontogeny. The expanded ends retain a relatively constant maximum width proportional to the maximum diameter of the tibial shaft. In the smallest hatchlings, the shaft is approximately one-third of the proximodistal length of the element, but in slightly larger juvenile individuals, the shaft makes up one-half the overall length of the element, a ratio that remains consistent during the remainder of growth. During ontogeny, the fibula becomes more robust as the width of the shaft becomes proportionally greater relative to its length.

DISCUSSION

The appendicular anatomy of hadrosaurids and iguanodontians is conservative, with morphological variations typically restricted to changes in proportions. The youngest hadrosaurines and lambeosaurines have very similar appendicular morphologies and proportions, with their developmental pathways diverging later in development, beyond the hatchling stage. Thus, there is support for the hypothesis that changes in morphology–specifically changes in proportion–stem from ontogenetic changes coupled with varied relative growth rates.

The sister-taxon relationship of *B. canadensis* and *M. peeblesorum* is well established (Weishampel and Horner, 1990; Horner et al., 2004) and the ontogenetic analysis presented here further supports that interpretation. Specifically, the growth trajectories of these two hadrosaurines exhibit consistent differences with the lambeosaurine *H. stebingeri*.

A consideration of the basal iguanodontian *T. tilletti* provides a broader evolutionary context for considering these differing ontogenies. The ontogenetic development of the appendicular skeleton in *T. tilletti* suggests that the plesiomorphic developmental pathway involves beginning with simple, gracile elements that then develop into robust elements with exaggerated characteristics. In this context, the occurrence of robust appendicular elements in adult lambeosaurines, such as *H. stebingeri*, appears to be plesiomorphic rather than derived. This contrasts sharply with the derived state of many cranial characteristics in lambeosaurines. Conversely, the more gracile appendicular skeletons of hadrosaurines would appear to be a derived state. If this interpretation is correct then the more gracile condition of hadrosaurine appendicular elements would suggest the occurrence of pedomorphosis in the developmental history of hadrosaurines resulting from the postdisplacement of the onset of the development of some appendicular morphological characteristics. Alternatively, the more robust condition of lambeosaurine appendicular elements suggests the occurrence of peramorphosis in the developmental history of that group, resulting from the acceleration and/or predisplacement of the growth trajectories of appendicular characters.

In either case, using the adult appendicular morphologies of basal iguanodontians as a baseline to identify plesiomorphic versus derived morphological conditions in hadrosaurids points to heterochrony as an important factor in the evolutionary divergence of hadrosaurines and lambeosaurines. In the pectoral girdle of *M. peeblesorum* and *B. canadensis* there is evidence of pedomorphosis with the retention of a juvenile morphology in the adult. The length of the scapula appears to increase in size at approximately the same rate as in *H. stebingeri*, meaning that the length of the scapula is similar in proportion to overall body size to that of *H. stebingeri* (Fig. 22.8). However, the width of the element does not increase at the same rate for the hadrosaurines and the lambeosaurines. It would seem then that the element has the same duration of growth as in *H. stebingeri*. In turn, this suggests that it is the rate of growth along the dorsoventral axis of the scapula that decelerates in hadrosaurines.

In contrast, it appears that peramorphosis occurs in the humerus of *H. stebingeri*. In *H. stebingeri* and other lambeosaurines, the width of the deltopectoral crest is proportionally larger than in other hadrosaurids or iguanodontians. In order to achieve this robust humerus, a change in the rate of growth must occur in *H. stebingeri* in comparison to its hadrosaurine relatives (Fig. 22.9). As with the scapula of the hadrosaurines, it appears that the humerus, and specifically the deltopectoral crest, has the same duration of growth as in the hadrosaurines. The presence of a similar deltopectoral crest in both the hadrosaurine and lambeosaurine embryonic and hatchling specimens suggests that the development of the crest begins at the same time. Therefore, it seems reasonable to conclude that the rate of development in the humerus accelerates in *H. stebingeri* and is responsible for the robust lambeosaurine condition. The impact of this change in growth rate can be observed early in ontogeny. Despite the slender nature of the early stage lambeosaurine humerus, the lambeosaurine humerus is already more robust than its hadrosaurine counterpart by the young juvenile stage.

In the pelvic girdle the two elements affected by heterochronic influences are the ilium and the ischium. The adult ilium of *H. stebingeri* and many lambeosaurines exhibits a well-developed, ventrally folded supra-acetabular process. The development of this feature exceeds the condition present in hadrosaurine taxa, and is unique among iguanodontians. The process does not appear to develop for a longer duration across growth stages than in hadrosaurines, so, an accelerated rate of development best explains iliac adult morphology in lambeosaurines and, as is true of the deltopectoral crest of the humerus, it appears that peramorphosis is responsible for the exaggerated morphology of the supra-acetabular process in *H. stebingeri*.

In the ischium, the primary characteristic to be considered is the distal boot that develops on the ischia of lambeosaurines but not on the tapered distal tip of the hadrosaurine ischium. In this case, there is the possibility of heterochronic influences on development of the ischia of all three taxa (Fig. 22.10). For the ischia of *M. peeblesorum* and *B. canadensis* it would seem that the tapered distal tip is the result of pedomorphosis and, specifically, postdisplacement. The distal tip of hadrosaurine ischia not only does not fully develop a distal boot, but does not develop one at all, suggesting that the starting point for the development of that feature is permanently delayed. The booted ischium of *H. stebingeri*, in contrast, could have a slight peramorphic acceleration of the rate of development of the distal boot. The distal expansion of the ischium is seen throughout Iguanodontia; the lambeosaurines have furthered that expansion.

CONCLUSIONS

The recognition that variation exists within the growth strategies of iguanodontian dinosaurs allows for the possibility that heterochrony was a crucial influence in the evolution of the group. Whereas there are previously recorded instances of sequence heterochrony within the Iguanodontia (Guenther, 2009), this study indicates that changes in the relative growth rates of different parts of the appendicular elements can result in distinctly different adult morphologies.

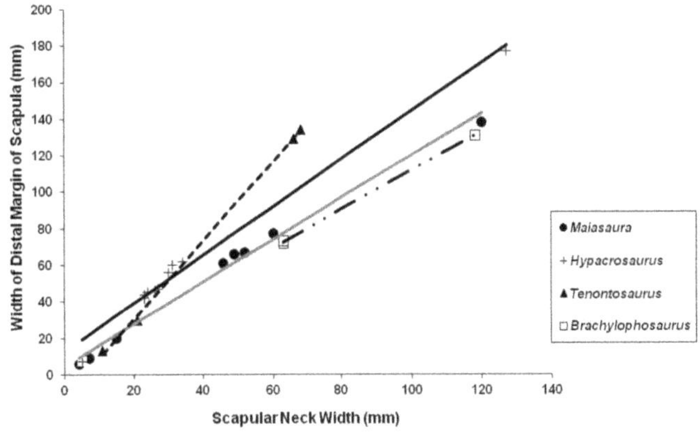

22.8. Graph of the development of the scapula showing the growth trajectories for the four taxa examined herein, and the relationship between the growth of the distal portion of the element and the scapular neck.

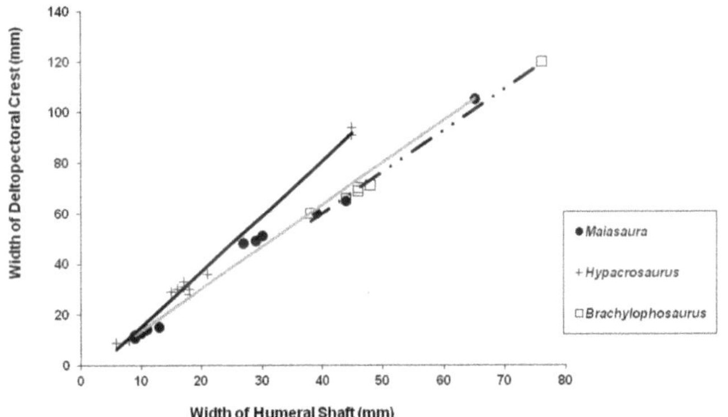

22.9. Graph of the development of the humerus showing the growth trajectories for the taxa examined herein (except *Tenontosaurus tilletti*) and the relationship between the growth of the humeral shaft and the development of the deltopectoral crest.

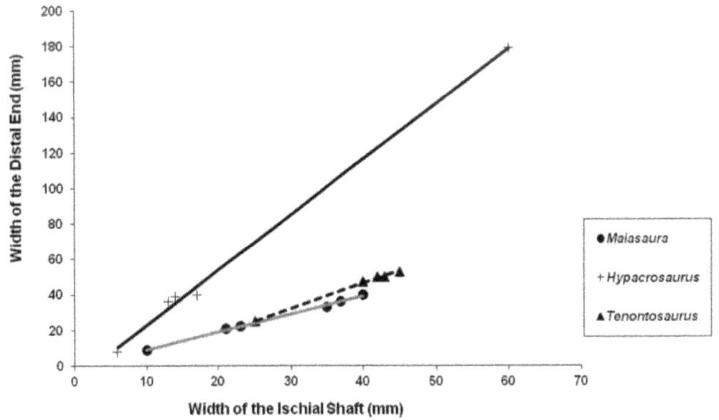

22.10. Graph of the development of the ischium showing the growth trajectories for the taxa examined herein (except *Brachylophosaurus canadensis*) and the relationship between the growth of the main shaft and development of the distal shaft.

There are biomechanical implications for the differing adult appendicular skeletons, but there are also biomechanical implications for individuals at each growth stage. Variation in the weight-bearing and locomotive capabilities of the girdles and limbs could therefore vary between hadrosaurines and lambeosaurines not just in adulthood, but also during the juvenile and subadult stages. Individuals are subject to strong selective pressures at each stage. Rather than serving as a means to achieve an adult morphology, the different growth strategies could also represent different approaches among iguanodontians for optimizing functional capabilities at each growth stage.

The study of the ontogeny of iguanodontians is limited by the availability of specimens of different growth stages. The discovery of more growth series within Hadrosauridae will help to illuminate the nuances of differing growth series within the Lambeosaurinae and the Hadrosaurinae. More importantly, discoveries of growth series of additional basal iguanodontians and ornithopods will allow more robust testing of ontogenetic hypotheses, specifically with regard to optimization of biomechanics at different growth stages.

ACKNOWLEDGMENTS

I thank P. Dodson, H. Pfefferkorn, and F. Scatena. I also thank C. Ancell, P. Currie, M. Feuerstack, J. Gardner, J. Horner, J. Person, K. Seymour, and K. Shepherd for help with and access to specimens; and A. Wydra for the illustrations. Thanks to the reviewers and the editors, D. A. Eberth and D. C. Evans. This project was funded by the Delaware Valley Paleontological Society and the Binns-Williams Research Fund, and supported also by resources from Elmhurst College.

LITERATURE CITED

Alberch P., S. J. Gould, G. F. Oster, and D. B. Wake. 1979. Size and shape in ontogeny and phylogeny. Paleobiology 5:296–317.

Brett-Surman, M. K. 1975. The Appendicular Anatomy of Hadrosaurian Dinosaurs. M.S. thesis, University of California, Berkeley, California, 122 pp.

Carr, T. 2003. New insight into the evolution of tyrannosauroid theropods: "event pair cracking" and the integration of ontogenetic and phylogenetic data in paleontology. Journal of Vertebrate Paleontology 23(3, Supplement):38A.

Chiappe, L. M., R. A. Coria, D. Lingus, J. Jackson, A. Chinsamy, and M. Fox. 1998. Sauropod dinosaur embryos from the Late Cretaceous of Patagonia. Nature 396:258–261.

de Beer, G. R. 1954. Embryos and Ancestors. Revised edition. Clarendon Press, Oxford, U.K., 159 pp.

Dilkes, D. W. 2001. An ontogenetic perspective on locomotion in the Late Cretaceous dinosaur *Maiasaura peeblesorum* (Ornithischia, Hadrosauridae). Canadian Journal of Earth Sciences 38:1205–1227.

Gould, S. J. 1977. Ontogeny and Phylogeny. Harvard University Press, Cambridge, Massachusetts, 501 pp.

Guenther, M. F. 2009. Influence of sequence heterochrony on hadrosaurid dinosaur postcranial development. Anatomical Record 292:1427–1441.

Horner, J. R. 1999. Egg clutches and embryos of two Hadrosaurian Dinosaurs. Journal of Vertebrate Paleontology 19:607–611.

Horner, J. R., and P. J. Currie. 1994. Embryonic and neonatal morphology and ontogeny of a new species of *Hypacrosaurus* (Ornithischia, Lambeosauridae) from Montana and Alberta; pp. 312–336 in K. Carpenter, K. F. Hirsch, and J. R. Horner (eds.), Dinosaur Eggs and Babies. Cambridge University Press, Cambridge, U.K.

Horner, J. R., and D. B. Weishampel. 1988. A comparative embryological study of two ornithischian dinosaurs. Nature 332:256–257.

Horner, J. R., D. B. Weishampel, and C. A. Forster. 2004. Hadrosauridae; pp. 438–463 in D. B. Weishampel, P. Dodson, and H. Osmólska (eds.), The Dinosauria, Second Edition. University of California Press, Berkeley, California.

Jeffery, J. E., M. K. Richardson, M. I. Coates, and O. R. P. Bininda-Emonds. 2002. Analyzing developmental sequences within a phylogenetic framework. Systematic Biology 51:478–491.

Kilbourne, B. M., and P. J. Makovicky. 2010. Limb bone allometry during postnatal ontogeny in non-avian dinosaurs. Journal of Anatomy 217:135–152.

Kundrát, M., A. R. Cruickshank, T. W. Manning, and J. Nudds. 2008. Embryos of therizinosaurid theropods from the Upper Cretaceous of China: diagnosis and analysis of ossification patterns. Acta Zoologica 89:231–251.

Lull, R. S., and N. E. Wright. 1942. Hadrosaurian Dinosaurs of North America. Geological Society of America Special Papers 40. 242 pp.

Mabee, P. M., K. L. Olmstead, and C. C. Cubbage. 2000. An experimental study of intraspecific variation, developmental timing, and heterochrony in fishes. Evolution 54:2091–2106.

Mateus, I., H. Mateus, A. M. Telles, O. Mateus, P. Taquet, V. Ribeiro, and G. Manuppella. 1997. Couvée, oeufs et embryons d'un Dinosaure Théropode du Jurassique supérieur de Lourinhã (Portugal). Comptes rendus de l'Académie des Sciences, Paris 325:71–78. [French with English abstract]

McNulty, K. P., S. R. Frost, and D. S. Strait. 2006. Examining affinities of the Taung child by developmental simulation. Journal of Human Evolution 51:274–296.

Norell, M. A., J. M. Clark, D. Dashzveg, R. Barsbold, L. M. Chiappe, A. McKenna, A. Perle, and M. J. Novacek. 1994. A theropod dinosaur embryo and the affinities of the Flaming Cliffs dinosaur eggs. Science 266:779–782.

Prieto-Márquez, A. 2005. New information on the cranium of *Brachylophosaurus canadensis*, with a revision of its phylogenetic position. Journal of Vertebrate Paleontology 25:144–156.

Prieto-Márquez, A. 2010. Global phylogeny of Hadrosauridae (Dinosauria: Ornithopoda) using parsimony and Bayesian methods. Zoological Journal of the Linnaean Society 159:435–502.

Reisz, R. R., D. C. Evans, E. M. Roberts, H-D. Sues, and A. M. Yates. 2012. Oldest known dinosaurian nesting site and reproductive biology of the Early Jurassic sauropodomorph *Massospondylus*. Proceedings of the National Academy of Sciences 109:2428–2433.

Rice, S. H. 1997. The analysis of ontogenetic trajectories: when a change in size or shape is not heterochrony. Proceedings of the National Academy of Sciences 94:907–912.

Smith, K. K. 1996. Integration of craniofacial structures during development in mammals. American Zoologist 36:70–79.

Smith, K. K. 1997. Comparative patterns of craniofacial development in eutherian and metatherian mammals. Evolution 51:1663–1678.

Smith, K. K. 2001. Heterochrony revisited: the evolution of developmental sequences. Biological Journal of the Linnean Society 73:169–186.

Sullivan, R. M., S. E. Jasinski, M. Guenther, and S. G. Lucas. 2011. The first lambeosaurine (Dinosauria, Hadrosauridae, Lambeosaurinae) from the Upper Cretaceous Ojo Alamo Formation (Naashoibito Member), San Juan Basin, New Mexico. New Mexico Museum of Natural History and Science Bulletin 53:405–417.

Tanke, D. H., and M. K. Brett-Surman. 2001. Evidence of hatchling and nestling-size hadrosaurs (Reptilia: Ornithischia) from Dinosaur Provincial Park (Dinosaur Park Formation: Campanian), Alberta, Canada; pp. 206–218 in D. H. Tanke and K. Carpenter (eds.), Mesozoic Vertebrate Life. Indiana University Press, Bloomington, Indiana.

Velhagen, W. A. 1997. Analyzing developmental sequences using sequence units. Systematic Biology 46:204–210.

Webster, M., and M. L. Zelditch. 2005. Evolutionary modifications of ontogeny: heterochrony and beyond. Paleobiology 31:354–372.

Weishampel, D. B., and J. R. Horner. 1990. Hadrosauridae; pp. 534–561 in D. B. Weishampel, P. Dodson, and H. Osmólska (eds.), The Dinosauria. University of California Press, Berkeley, California.

Werning, S. 2005. Ontogeny and osteohistology of the ornithopod dinosaur *Tenontosaurus tilletti*. Journal of Vertebrate Paleontology 25(3, Supplement):128A–129A.

The Size-Frequency Distribution of Hadrosaurs from the Dinosaur Park Formation of Alberta, Canada

23

Donald B. Brinkman

ABSTRACT

The size-frequency distribution of hadrosaurs from the Dinosaur Park Formation at Dinosaur Provincial Park, Alberta, is quantified using both articulated skeletons and isolated elements. Two well-defined subadult peaks and a broad cluster of adult individuals are present. The presence of subadult peaks is interpreted as a result of annual cycles of high birth rate, high mortality rates, and a rapid growth rate. This pattern differs from the size-frequency distribution of hadrosaurs from Montana in that a peak representing young adults is absent, and suggests that the hadrosaurs from Dinosaur Provincial Park retained a high growth rate after their first year of growth, and reached an adult size after two years of growth. The absence of distinct adult peaks indicates that growth had slowed so that the size of age classes overlapped. Thus, size alone is a poor predictor of relative age in adult individuals.

INTRODUCTION

Size-frequency distributions can be an important source of data for estimating biological attributes of vertebrates (e.g., Brinkman et al., 2007). In extant populations of reptiles and amphibians, age classes may be represented by distinct size classes. When this is the case, size classes can be used to establish the age structure of a population and to infer growth rates (e.g., Andrews, 1982; Halliday and Verrell, 1988). The paleobiological significance of size-frequency distributions of fossil vertebrates was first considered by Olsen (1957). He recognized that rates of growth and mortality are fundamental properties that affect the size-frequency distributions of extinct organisms. Later, in one of the first uses of computers in paleontology, Craig and Oertel (1966) examined the effect of different patterns of recruitment, growth rates, and mortality rates on size-frequency distributions of death assemblages. They concluded that size classes representing age classes should be present in death assemblages derived from organisms that had seasonally distinct times of recruitment, seasonally high death rates, and high rates of growth. Examples of the use of size-frequency distributions of fossil vertebrates to establish growth rates and mortality patterns include studies of fish (Wilson, 1984) and pterosaurs (Bennett, 1993).

In dinosaurs, size-frequency distribution patterns have been discussed primarily in studies of catastrophic death assemblages and have been used as a basis for arguing that some bonebeds are the results of mass kills (Currie and Dodson, 1984; Weishampel and Westphal, 1986). Size-frequency distributions were used to infer growth rates in hadrosaurs by Varricchio and Horner (1993). In a study of hadrosaur remains from six hadrosaur-dominated bonebeds in Montana, they observed that a distinct cluster of individuals about 3.5 m long was present. They argued that since size was related to age, as had been documented by histological studies (Horner and Weishampel, 1988; Chinsamy, 1991; Varricchio, 1993), then the size-frequency distributions in their samples should approximate age profiles. Thus, they concluded that the hadrosaurs represented in their bonebeds either had very rapid juvenile growth rates, reaching a length of 3.4 m in 1 year, or (less likely) the juveniles had a delayed association with mixed-age herds, possibly living in strictly juvenile groups. Further, in *Maiasaura peeblesorum*, they concluded that reproduction was synchronous and seasonal among herd members. In a subsequent study of size distribution of bones from a *Maiasaura* monospecific bonebed in the Willow Creek Anticline of Montana, Schmitt et al. (1998) recognized distinct size classes indicative of juveniles, subadults, and adults.

The *Maiasaura* growth dynamics proposed by Varricchio and Horner (1993) were assessed by Horner et al. (2000) in a study of the changes in long bone histology during growth. In this study, the relationship between histological structure, size classes, growth rates, and age classes was considered. Six growth stages were recognized on the basis of relative size and patterns of histological differences: small nestlings (body length of approximately 45 cm), large nestlings (body length of approximately 90 cm), early juveniles (body length of approximately 120 cm), late juveniles (body length of approximately 3.5 m), sub-adults (body length of approximately 4.7 m), and adults (body length of approximately 7 m). The first three stages are represented by non-overlapping size classes. The latter three were those recognized by Schmitt et al. (1998) on the basis of discrete size classes in the population of *Maiasaura* preserved in the catastrophic bonebed that they studied.

Horner et al. (2000) first estimated the length of time it took for *Maiasaura* to reach an adult stage by using the size of individuals in successive growth stages, together with an estimate of the daily rate of deposition of bone. Using a conservative estimate of 15 µm/day for radial deposition of laminar bone tissue, this approach suggested that the adult stage would be reached in seven years. However, this approach also resulted in the unrealistic conclusion that the one-year-old age cohort was missing in the sample (using a bone growth rate of 15 µm/day, one-year-olds should be represented by individuals with a size intermediate between "nestlings" and "late juveniles").

An alternative hypothesis was then considered in which individuals 3.5 m long and smaller represent the small and large nestlings, as well as members of the one-year cohort. In this hypothesis late juveniles are missing. This hypothesis was tested by assessing the occurrence of lines of arrested growth (LAGs) in these individuals. LAGs were not used to age individuals directly because it was noted that there are differences in the number of lines of growth in different elements of a single skeleton (Horner et al., 1999). However, it was argued that LAGs can be used to assess age indirectly because, within each element, there are consistent trends in absolute spacing and decreasing distances between LAGs over time. These consistent trends indicate that LAGs record growth cycles and that growth slows with age (Chinsamy, 1993). The hypothesis that yearlings are present in the sample was supported by observation that LAGs are rare in bones from individuals up to 3.5 m in length (single LAGs were observed in only two of these specimens). Thus, in this alternative hypothesis, long bone growth was estimated to have been very rapid during the first year (40 µm/day) and then slower (15 µm/day) afterward. Using this differential growth rate approach, it was estimated that an adult size would be reached in just under 5 years. Ultimately, Horner et al. (2000) combined the two hypotheses and estimated that *Maisaura* took six years to reach the adult stage (body length of about 7 m).

In these studies, size-frequency distributions have played a key role in developing hypotheses of rates of growth independent of histological studies. However, the data on size-frequency distributions in hadrosaurs have been geographically and taxonomically limited. The recognition that different taxa of hadrosaurs have a shared size-frequency distribution pattern by Varricchio and Horner (1993) suggests that the growth patterns are typical for hadrosaurs as a group. One way of testing this is by examining the size-frequency distributions of hadrosaurs from other localities and environments of deposition.

Table 23.1. Measurements of Limb Elements of Articulated Hadrosaur Specimens from the Dinosaur Park Formation of Alberta, Canada

Specimen	Identification	Femur (cm)	Tibia (cm)
TMP 1998.058.0001	gen. indet.	52	47
ROM 768	*Parasaurolophus*	102	—
TMP 1983.064.0003	*Prosaurolophus*	59	51
AMNH 5461	*Corythosaurus*	52	51
AMNH 5340	*Lambeosaurus*	59	55
TMP 1980.022.0001	*Gryposaurus*	85	79
ROM 787	*Prosaurolophus*	99	82
ROM 764	*Gryposaurus*	104	90
ROM 1218	*Lambeosaurus*	105	91
AMNH 5338	*Corythosaurus*	99	92
ROM 845	*Corythosaurus*	104	94
TMP 1982.038.0001	*Lambeosaurus*	105	98
CMN 8532	*Corythosaurus*	104	98
AMNH 5240	*Corythosaurus*	108	100
CMN 8703	*Lambeosaurus*	102	100
TMP 1966.004.0001	*Lambeosaurus*	110	102
ROM 873	*Gryposaurus*	114	103

In this chapter, the size-frequency distribution of hadrosaurs from the Dinosaur Park Formation of Alberta, Canada, is documented and its significance to interpretations of growth dynamics of hadrosaurs based on studies of material from Montana are considered. The Dinosaur Park Formation sample differs from that used by Varricchio and Horner (1993) and Horner et al. (2000) in that it is an aggregate of articulated skeletons and isolated elements from different locations and stratigraphic levels. Also, the environmental conditions under which the Dinosaur Park Formation hadrosaurs lived differs from the Montana sample. The hadrosaurs from Montana used in studies by Varricchio and Horner (1993) and Horner et al. (2000) lived in a comparatively dry environment subject to drought (Rogers, 1990), whereas the Dinosaur Park Formation was deposited in a very wet environment, possibly with seasonally intense tropical storms (Eberth and Getty, 2005; Eberth et al., 2009). Furthermore, the taxonomic composition of the hadrosaurs of the Dinosaur Park Formation differs since a diverse assemblage of five genera of hadrosaurs is present (Table 23.1). Given these differences, the presence of a size-frequency distribution matching that documented by Varricchio and Horner (1993) would provide strong support for the hypothesis that to the growth dynamics of hadrosaurs from Montana are common to hadrosaurs, rather than being taxon or environment specific.

Institutional Abbreviations AMNH, American Museum of Natural History, New York; CMN, Canadian Museum of Nature, Ottawa, Ontario; ROM, Royal Ontario Museum, Toronto, Ontario; TMP, Royal Tyrrell Museum of Palaeontology, Drumheller, Alberta.

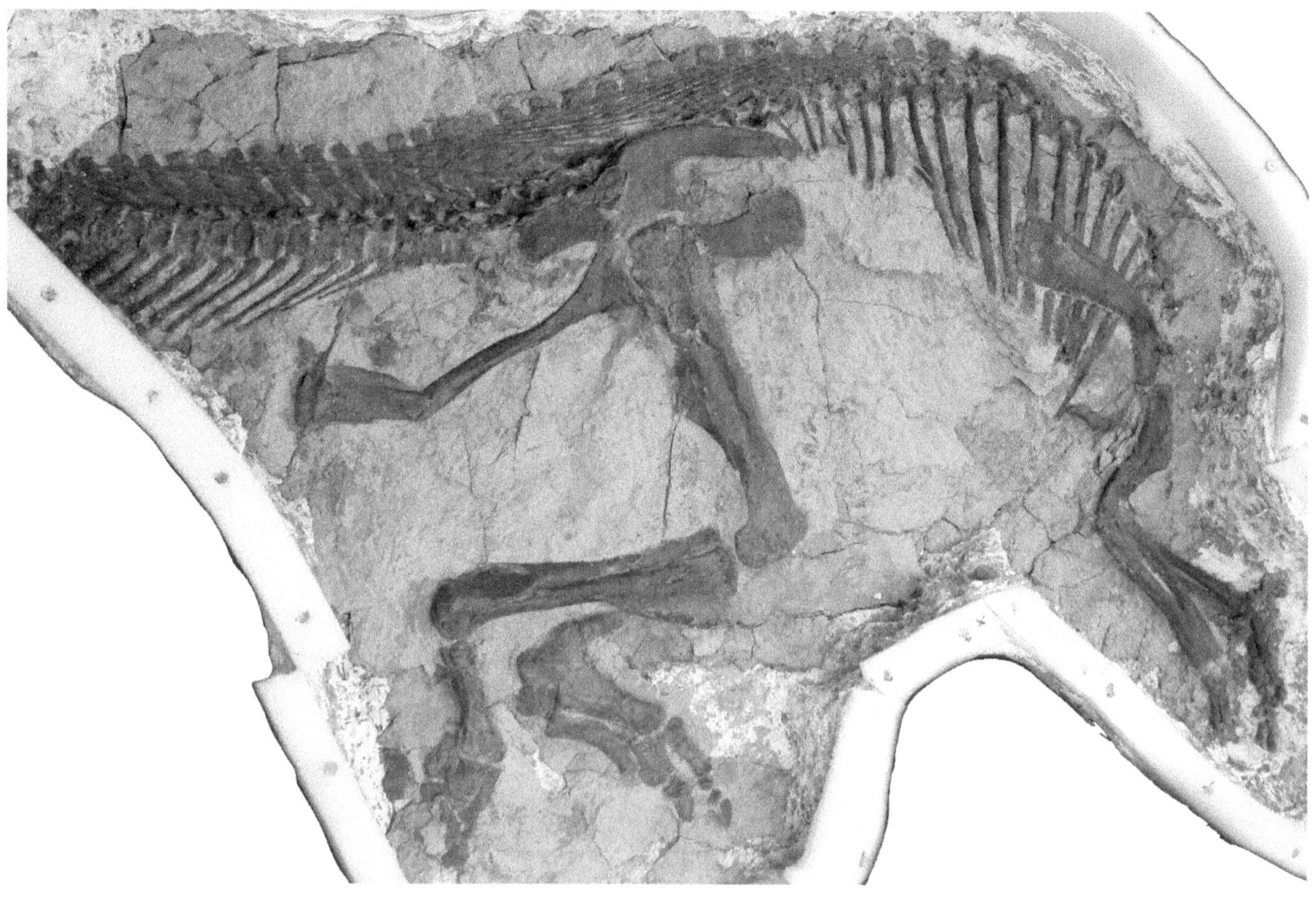

23.1. Juvenile hadrosaur skeleton, specimen TMP 1998.058.0001, used as a standard of comparison for quantifying the relative size of hadrosaur remains from Dinosaur Provincial Park.

METHODS

Linear measurements of tibiae and femora are used to evaluate the size of hadrosaurs from the Dinosaur Park Formation of Alberta. In order to incorporate data from both elements in a single histogram, measurements are expressed as a percentage of the length of the corresponding element in a complete, articulated, reference skeleton. Specimen TMP 1998.058.0001 was chosen as a reference specimen because it includes a complete tibia, fibula, and femur (Fig. 23.1). Measurements were taken from specimens in the collections of the Royal Tyrrell Museum and the Royal Ontario Museum. Measurements from specimens housed in other institutions published by Lull and Wright (1942) were also included in the data set. In general, the isolated elements included in this dataset were found as isolated specimens or from bonebeds that show evidence of wear, alteration, and reworking. Thus, each element was assumed to be from a different individual. In the case of articulated specimens, the length of the tibia was taken as the measure of the size of that specimen. If tibiae were absent, the femur was used. The resulting data base includes 56 specimens (Tables 23.1, 23.2), including 17 articulated specimens (Table 23.1), 20 isolated tibiae, and 19 isolated femora (Table 23.2).

RESULTS

The elements in this data set range from 19% to 219% the length of comparable elements in specimen TMP 1998.058.0001. A histogram showing the size-frequency distribution of these remains (Fig. 23.2) shows two well-defined subadult peaks, and a broad cluster of adult individuals with two ill-defined peaks.

The smallest size class includes 21 individuals that range from 19% to 38% of the linear dimensions of TMP 1998.058.0001, with a peak in the 21–30% size range. Thus, individuals in this size class would have had a body length ranging from about 70 cm to about 105 cm. This is the size class that Horner et al. (2000) referred to as "large nestlings." Thus, following Varricchio and Horner (1993), this size class is interpreted as consisting of hatchling individuals and individuals that had recently left the nest, rather

Table 23.2. Measurements of Isolated Hadrosaur Femora and Tibiae from the Dinosaur Park Formation of Alberta, Canada

Isolated tibiae	Length (cm)	Isolated femora	Length (cm)
TMP 1989.039.0113	10	TMP 1992.036.0920	10
TMP 1994.019.0822	11	TMP 1992.036.0921	10
TMP 1992.036.0536	11	TMP 1989.036.0415	11
TMP 1984.067.0060	11	TMP 1992.036.0600	12
TMP 1994.045.0008	12	TMP 1990.036.0412	12
TMP 1985.036.0138	12	TMP 1989.036.0173	13
TMP 1967.020.0239	18	ROM 706	13
TMP 1967.008.0063	38	TMP 1994.012.0465	19
TMP 1981.019.0128	39	TMP 1994.666.0082	33
TMP 1967.010.0082	39	TMP 1990.036.0073	38
TMP 1983.180.0010	42	TMP 1987.078.0005	51
TMP 1979.014.0308	43	TMP 1984.001.0001	90
TMP 1965.016.0004	71	TMP 1990.002.0028	95
TMP 1966.042.0023	74	TMP 1987.062.0029	100
TMP 1965.010.0018	92	TMP 1996.012.0172	16
TMP 1981.019.0126	93	TMP 1996.012.0175	12
ROM 26032	97	TMP 1997.012.0173	13
ROM 655	98	TMP 1999.055.0350	11
TMP 1994.012.0840	40	TMP 2001.012.0089	20
TMP 1997.012.0216	14		

than a first-year age class. Not surprisingly, the elements in this size class all show clear morphological evidence that they are at a very immature stage of development. This includes the presence of a punctate surface texture and the presence of poorly defined articular surfaces. It is striking that the peak including the hatchling elements is particularly high. This is consistent with the assumption that when individuals left the nest it was a time of particularly high mortality.

The second peak includes 13 individuals that range from 63% to 117% of the linear dimensions of TMP 1998.058.0001, with a peak in the 81–90% size range. Thus, individuals in this size class would have had a body length ranging from approximately 2–4 m; the size class that Varricchio and Horner (1993) and Horner et al. (2000) hypothesize includes the one-year-old size class. The bones in this peak generally have a striated texture and their articular ends are not strongly rugose.

The third size class contains 22 individuals that range from 181% to 219% of the linear dimensions of TMP 1998.058.0001, with a peak in the 191–200% size range. All the articulated specimens that are included in this size class have been considered previously to be adult, including one of the smallest individuals in this group, TMP 1980.022.0001, a *Gryposaurus* specimen, with a tibia that is 181% the length of the tibia in TMP 1998.058.0001. Isolated elements are also considered to be from adults because they have a smooth, finished external texture and rugose articular surfaces.

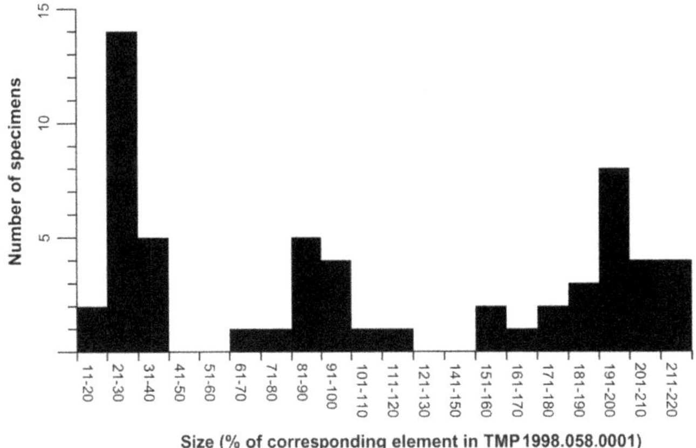

23.2. Size-frequency diagram of hadrosaurs from the Dinosaur Park Formation of Alberta, Canada.

DISCUSSION

The occurrence of discrete size/age clusters in the sample of hadrosaurs from Dinosaur Park Formation is initially surprising because this sample constitutes an attritional assemblage accumulated over an approximately 1.5-million-year period (Eberth, 2005; Eberth and Currie, 2005). Numerous potential sources of variation in size of animals in such an assemblage could have swamped the age-related size clusters. For example, given that six genera of hadrosaurids are known from Dinosaur Provincial Park, it is reasonable to expect that adult size differences between these taxa would be represented in this assemblage. However, given that distinct age/size classes are present in the sample, the taxa included in it must have had reasonably similar growth rates and mortality patterns. Variation in growth rates over time is another potential source of variation that could have swamped the age-based size clusters. Again, the data presented here suggest that growth rates and mortality patterns must have been consistent among these different taxa during the approximately 1.5 million years represented by the sample. Lastly, the presence of size and age clusters in an attritional assemblage of dinosaurs is unexpected, because it is generally assumed that these would be obscured by random mortality events. However, as shown by Craig and Oertel (1966), size classes should be preserved even in an attritional assemblage provided that the same seasonality in the time of hatching and high mortality was present across these taxa during the overall time duration represented by the sample. Dinosaur Provincial Park was located at approximately 58° N paleolatitude so major changes in the length of the photoperiod over the course of the year would have been present. At 58° N the day length during midsummer is about 18 hours, and in midwinter it is about 6 hours. This difference in the number of daylight hours would have resulted in seasonality in primary productivity. A seasonal period of hatching, as has been proposed for *Maiasaura*, is likely because it would have taken advantage of the time of high productivity. A seasonal period of high mortality also is likely since sedimentological evidence has suggested the presence of seasonal periods of intense tropical storms and that these had a major negative impact on the biota (Eberth and Currie, 2005; Eberth and Getty, 2005; Eberth et al., 2009).

The size-frequency distributions of hadrosaurs from the Dinosaur Park Formation are similar to those documented by Varricchio and Horner (1993) in that a distinct size class containing individuals of about 3.5 m in length is present. The presence of a similar size class in the Dinosaur Park assemblage is consistent with the interpretation presented by Varricchio and Horner (1993) and Horner et al. (2000) that this is a one-year-old size class, and suggests that the rapid growth rate required to reach this size was typical for hadrosaurs.

The size-frequency distribution of the Dinosaur Park Formation hadrosaurs differs from that of *Maiasaura* as described by Schmitt et al. (1998) in that animals at the subadult stage, the stage characterized by individuals of about 4.7 m in length, are rare to absent. Although Horner et al. (2000) considered the number of lines of arrested growth to be an unreliable indicator of age, they noted that some individuals at this size class in *Maiasaura* have two lines of arrested growth, suggesting that this cluster includes animals in the two-year cohort. As suggested in the studies of Varricchio and Horner (1993), the absence of a size class could have been because the individuals of that age were living somewhere else. However, given the geographic and stratigraphic area sampled, this is unlikely. A more probable explanation for the absence of this size class in the Dinosaur Park Formation hadrosaurs is that the rapid rate of growth of the juvenile resulted in adult size being attained after two years' growth.

The lack of distinct size classes in the adult stage of development is consistent with the expectation that once the adult stage was reached, growth had slowed. With a sufficiently slow growth rate, successive age classes would pile up in a single broad-size class. Taxonomic differences in adult size also may have smeared out the adult size peaks. Thus, within the adult size class, size is not a reliable indicator of age. Histological studies are required to resolve patterns of size, age, and growth rate in adults.

ACKNOWLEDGMENTS

This study was made possible by the many collectors who recognized the importance of isolated limb elements of hadrosaurs and spent considerable effort in collecting and preserving these. Thanks also to the collection managers who have ensured that this material is available for study. In particular, I thank Brandon Strilisky for his work on the collections of the Royal Tyrrell Museum and Kevin Seymour for his work on the collections of the Royal Ontario Museum. Finally I thank David Eberth for extensive discussions that aided in the development of ideas presented herein, and David Eberth and David Evans for encouragement in including this presentation at the Hadrosaur Symposium and in this volume.

LITERATURE CITED

Andrews, R. M. 1982. Patterns of growth in reptiles; pp. 273–320 in C. Gans and F. H. Pough (eds.), Biology of the Reptilia, volume 13. Academic Press, London.

Bennett, S. C. 1993. The ontogeny of *Pteranodon* and other pterosaurs. Paleobiology 15:92–106.

Brinkman, D. B., D. A. Eberth, and P. J. Currie. 2007. From bonebeds to paleobiology: applications of bonebed data; pp. 221–236 in R. R. Rogers, D. A. Eberth, and A. R. Fiorillo (eds.), Bonebeds: Genesis, Analysis, and Paleobiological Significance. University of Chicago Press, Chicago, Illinois.

Chinsamy, A. 1991. Physiological implications of the bone histology of *Syntarsus rhodesiensis* (Saurischia: Theropoda). Palaeontologia Africana 27:77–82.

Chinsamy, A. 1993. Bone histology and growth trajectory of the prosauropod dinosaur *Massospondylus carinatus* Owen. Modern Geology 18:319–329.

Craig, G. Y., and G. Oertel. 1966. Deterministic models of living and fossil populations of animals. Quarterly Journal of the Geological Society of London 122:315–355.

Currie, P. J., and P. Dodson. 1984. Mass death of a herd of ceratopsian dinosaurs; pp. 52–60 in W. E. Reif and F. Westphal (eds.), Third Symposium of Mesozoic Terrestrial Ecosystems. Attempto Verlag, Tübingen, Germany.

Eberth, D. A. 2005. The geology; pp. 54–82 in P. J. Currie and E. B. Koppelhus (eds.), Dinosaur Provincial Park: A Spectacular Ancient Ecosystem Revealed. Indiana University Press, Bloomington, Indiana.

Eberth, D. A., and P. J. Currie. 2005. Vertebrate taphonomy and taphonomic modes; pp. 453–477 in P. J. Currie and E. B. Koppelhus (eds.), Dinosaur Provincial Park: A Spectacular Ancient Ecosystem Revealed. Indiana University Press, Bloomington, Indiana.

Eberth, D. A., and M. A. Getty. 2005. Ceratopsian bonebeds: occurrence, origins, and significance; pp. 501–536 in P. J. Currie and E. B. Koppelhus (eds.), Dinosaur Provincial Park: A Spectacular Ancient Ecosystem Revealed. Indiana University Press, Bloomington, Indiana.

Eberth, D. A., D. B. Brinkman, and V. Barkas. 2009. A centrosaurine mega-bonebed from the Upper Cretaceous of Southern Alberta: implications for behaviour and death events; pp. 495–508 in M. J. Ryan, B. J. Chinnery-Allgeier, and D. A. Eberth (eds.), New Perspectives on Horned Dinosaurs. Indiana University Press, Bloomington, Indiana.

Halliday, T. R., and P. A. Verrell. 1988. Body size and age in amphibians and reptiles. Journal of Herpetology 22:253–265.

Horner, J. R., and D. B. Weishampel. 1988. A comparative embryological study of two ornithischian dinosaurs. Nature 332:256–257.

Horner, J. R., A. de Ricqlès, and K. Padian. 1999. Variation in dinosaur skeletochronology indicators: implications for age assessment and physiology. Paleobiology 25:295–304.

Horner, J. R., A. de Ricqlès, and K. Padian. 2000. Long bone histology of the hadrosaurid dinosaur *Maiasaura peeblesorum:* growth dynamics and physiology based on an ontogenetic series of skeletal elements. Journal of Vertebrate Paleontology 20:115–129.

Lull, R. S., and N. E. Wright. 1942. Hadrosaurian Dinosaurs of North America. Geological Society of America Special Papers 40. 242 pp.

Olsen, E. C. 1957. Size-frequency distributions in samples of extinct organisms. Journal of Geology 65:309–333.

Rogers, R. R. 1990. Taphonomy of three dinosaur bone beds in the Upper Cretaceous Two Medicine Formation of northwestern Montana: evidence for drought-related mortality. Palaios 5:394–413.

Schmitt, J. G., J. R. Horner, R. R. Laws, and F. Jackson. 1998. Debris flow deposition of a hadrosaur-bearing bone bed, Upper Cretaceous Two Medicine Formation, northwest Montana. Journal of Vertebrate Paleontology 18(3, Supplement):76A.

Varricchio, D. L. 1993. Bone microstructure of the Upper Cretaceous theropod dinosaur *Troodon formosus.* Journal of Vertebrate Paleontology 13:99–104.

Varricchio, D. J., and J. R. Horner. 1993. Hadrosaurid and lambeosaurid bone beds from the Upper Cretaceous Two Medicine Formation of Montana. Canadian Journal of Earth Sciences 30:997–1006.

Weishampel, D. B., and F. Westphal. 1986. Die Plateosaurier von Trossingen im Geologischen Institut der Eberhard-Karls-Universität Tübingen. Ausstellungskatalog der Universität Tübingen, 19.

Wilson, M. V. H. 1984. Year classes and sexual dimorphism in the Eocene catostomid fish *Amyzon aggregatum.* Journal of Vertebrate Paleontology 3:137–142.

Osteohistology and Occlusal Morphology of *Hypacrosaurus stebingeri* Teeth throughout Ontogeny with Comments on Wear-Induced Form and Function

24

Gregory M. Erickson and Darla K. Zelenitsky

ABSTRACT

Like mammalian herbivores, the chewing surfaces with which adult hadrosaurid dinosaurs orally processed plant matter wore with usage to create their final functional morphology. The distributions and differential wear resistance of dental tissues across the chewing surfaces were the primary determinants of the morphology of such surfaces. The nature of topographical changes to hadrosaurid occlusal surfaces throughout ontogeny are poorly documented, and whether morphological modifications were mediated by changes in tissue biomechanics is not known. Herein we describe the morphology and osteohistology of *Hypacrosaurus stebingeri* teeth throughout development. We use the results to make inferences regarding changes in dietary ecology during ontogeny, and to test whether tissue modifications facilitated such changes. The results show that embryos and neonates possessed cup-like teeth, subadults possessed shearing plane dentitions, and adults had teeth with dual slicing and crushing surfaces. These results point to substantial changes in diet during *H. stebingeri* development, and to the possibility that concurrent changes in chewing biomechanics occurred. The tissue ensemble for *H. stebingeri* shows the same complexity seen in adults of other taxa. However, during ontogeny, substantial changes in the presence and distributions of these tissues occurred. These changes had a major influence on tooth wear throughout development, and contributed to making developmental stage-specific morphologies and diets possible.

INTRODUCTION

Hadrosaurid dinosaurs were the dominant megaherbivores (1000 kg+, sensu Owen-Smith, 1988) of Late Cretaceous North America, Europe, and Asia (Evans, 2007; Williams et al., 2009; Prieto-Márquez, 2010). These animals also gained niches in Gondwanan South America and Antarctica (Case et al., 2000; Prieto-Márquez and Salinas, 2010; Prieto-Márquez, 2010). The key innovation that is thought to have facilitated their extensive radiation is a feeding system that allowed for the grinding (Lull and Wright, 1942; Ostrom, 1961) and/or slicing (Erickson et al., 2012) of tough and often abrasive plants such as horsetail, fern, and angiosperm groundcover, as well as conifers (Kräusel, 1922; Currie et al., 1995; Chin, 1997, 2007; Murphy et al., 2007; Tweet et al., 2008). The interlocking teeth with which these plants were orally processed are known as dental batteries. Like most mammalian herbivores (Peyer, 1968; Hillson, 1986; Janis and Fortelius, 1988; Janis, 1990; Lucas, 2004), wear-induced topographical changes in dental occlusal surfaces allowed the teeth to morph into their taxon-specific functional morphologies (Erickson et al., 2012)

The dental batteries with which hadrosaurids orally processed plant matter consist of interlocked columns of developing and functional teeth (Fig. 24.1A). In the batteries of adults the functional teeth formed occlusal surface pavements composed of dozens of teeth, of which two to four were functional per tooth row (= family sensu Edmund, 1969) (Prieto-Márquez, 2008; Fig. 24.1B). The teeth were constantly in motion as they developed, erupted, and moved across the chewing surface. The dentary batteries of some taxa wore to have steep-angled leading edge slicing faces (= Lingual Worn Zone sensu Weishampel, 1984) followed by more horizontal, and typically coarser, grinding and/or crushing pavements (= Buccal Worn Zone sensu Weishampel, 1984). The batteries of others wore exclusively to steep-angled shearing faces or to just grinding pavements (Erickson et al., 2012). Hadrosaurid tooth wear rates of up to 0.5 mm/day rivaled the highest values known among living animals (Erickson, 1996). Remnants of tooth roots were shed from each tooth row at rates of 50–80 days, and such rates were maintained throughout life (Erickson, 1996). Differential wear to the dental tissues exposed on the chewing surfaces, at least in adults, led to the strategic formation of crests and basins in some taxa that are reminiscent of extant mammalian grinding dentitions such as those seen in the cheek teeth of horses (*Equus caballus*) and bison (*Bison bison*) (Carrano et al., 1999, Erickson et al., 2012; Fig. 24.1C).

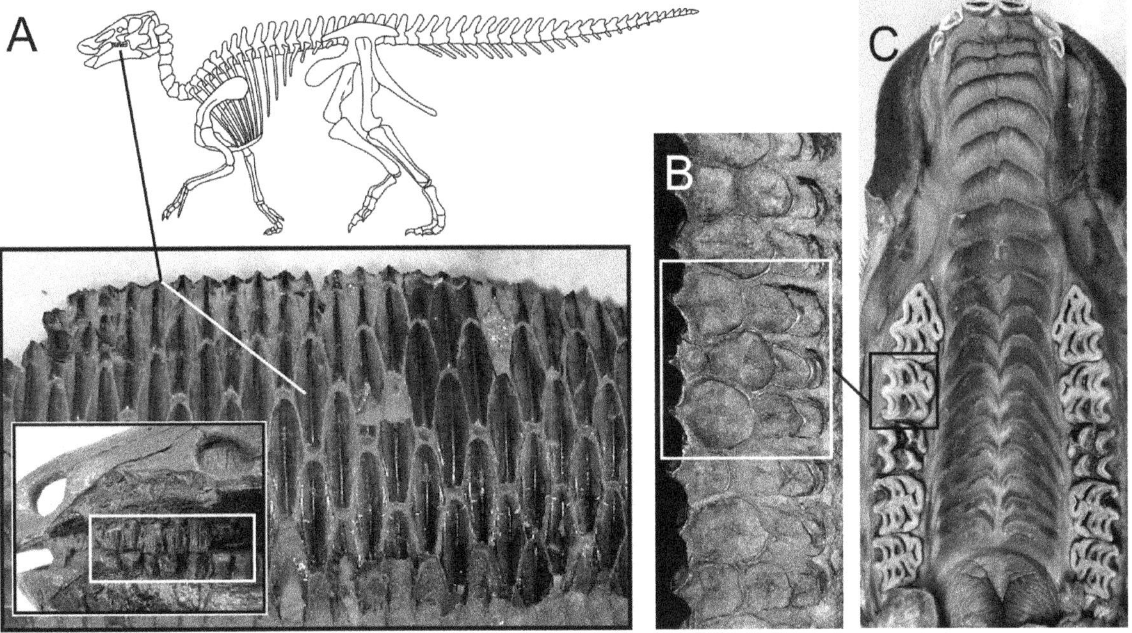

24.1. Comparison between dentitions. (A) Hadrosaurid skeleton (*Edmontosaurus regalis*) and lower dental battery (lingual view) showing developing teeth. *Camptosaurus dispar* (inset) possessing individual, shearing teeth at each tooth position, an outgroup condition to that seen in hadrosaurids. (B) Hadrosaurid (*Edmontosaurus annectens*) multi-crested chewing surface. Note that the rows of miniaturized teeth are in various stages of wear moving from left to right. (C) Analogous grinding dentition of a horse (*Equus caballus*). Photo: Uwe Gille, Creative Commons.

Cuthbertson et al. (2012) questioned whether the teeth composing Buccal Worn Zones were functional at all because their computerized renderings and analysis of possible kinematics for specimens of *Brachylophosaurus canadensis* and *Edmontosaurus regalis* could not realistically allow for the maxillary teeth to be drawn completely across the dentary chewing surfaces. However, the analysis in Mallon et al. (2011) of hadrosaurid tooth wear shows that the Buccal Worn Zones have wear striae with different trajectories than those in the Lingual Worn Zones and thus renders this hypothesis highly unlikely. Furthermore, Erickson et al. (2012) showed that (1) different histological and topographical features were exposed at each tooth position as the teeth progressed across the chewing surfaces, (2) the area and shape of the wear surfaces of the individual teeth continuously changed as the teeth progressed across the chewing surfaces, and (3) the labial-most dentary teeth characteristically show the lowest relief relative to other teeth in the Buccal Worn Zones (see Fig. 24.1C). All of these features attest that the maxillary teeth typically caused substantial wear to all of the teeth present on dentary battery chewing surfaces during chewing strokes, including those composing the Buccal Worn Zones. Just the same, there is evidence that some hadrosaurid individuals did not draw their upper dentition entirely across the dentary chewing surfaces. Weishampel (1984) pointed out that specimens are occasionally found with a longitudinal groove traversing the buccal edge of the exposed dentary teeth, and Cuthbertson et al. (2012:fig. 10) illustrated a specimen showing the labial-most dentary teeth elevated above the middle-most teeth. We agree that these likely represent cases in which the upper teeth were not fully drawn across the lowers, but our own examinations of hadrosaurid dentitions suggest that this was atypical. We found no evidence for such grooves or elevated teeth in the *H. stebingeri* batteries studied here.

The histological complexity of adult hadrosaurid teeth has only recently been identified, and reveals a sophisticated architecture rivaling that of herbivorous mammals (Erickson et al., 2012; Fig. 24.2A). Major constituents include the primitive reptilian tissues, enamel and orthodentine, as well as independently derived secondary dentine, coronal cementum, giant tubules, and a thick mantle dentine. Correlations with relief seen in naturally worn batteries (Fig. 24.2B, C), and tribological (= wear) testing and simulations, show that highly wear resistant enamel and mantle dentine promoted the formation of crests (Erickson et al., 2012). The latter sometimes acted alone in this regard since enamel is not present in most teeth in the lower batteries. Inclined slicing faces are largely composed of giant tubule curtains with intermediate wear resistance that often span between the wear-resistant mantle dentine crests and high-wear orthodentine basins. Large individual and branched giant tubules formed intermediate-height ridges partitioning the basins and influenced basin depth at each tooth position (greater sliding distance = greater scour; Archard, 1953; Archard and Hirst, 1956). They also provided

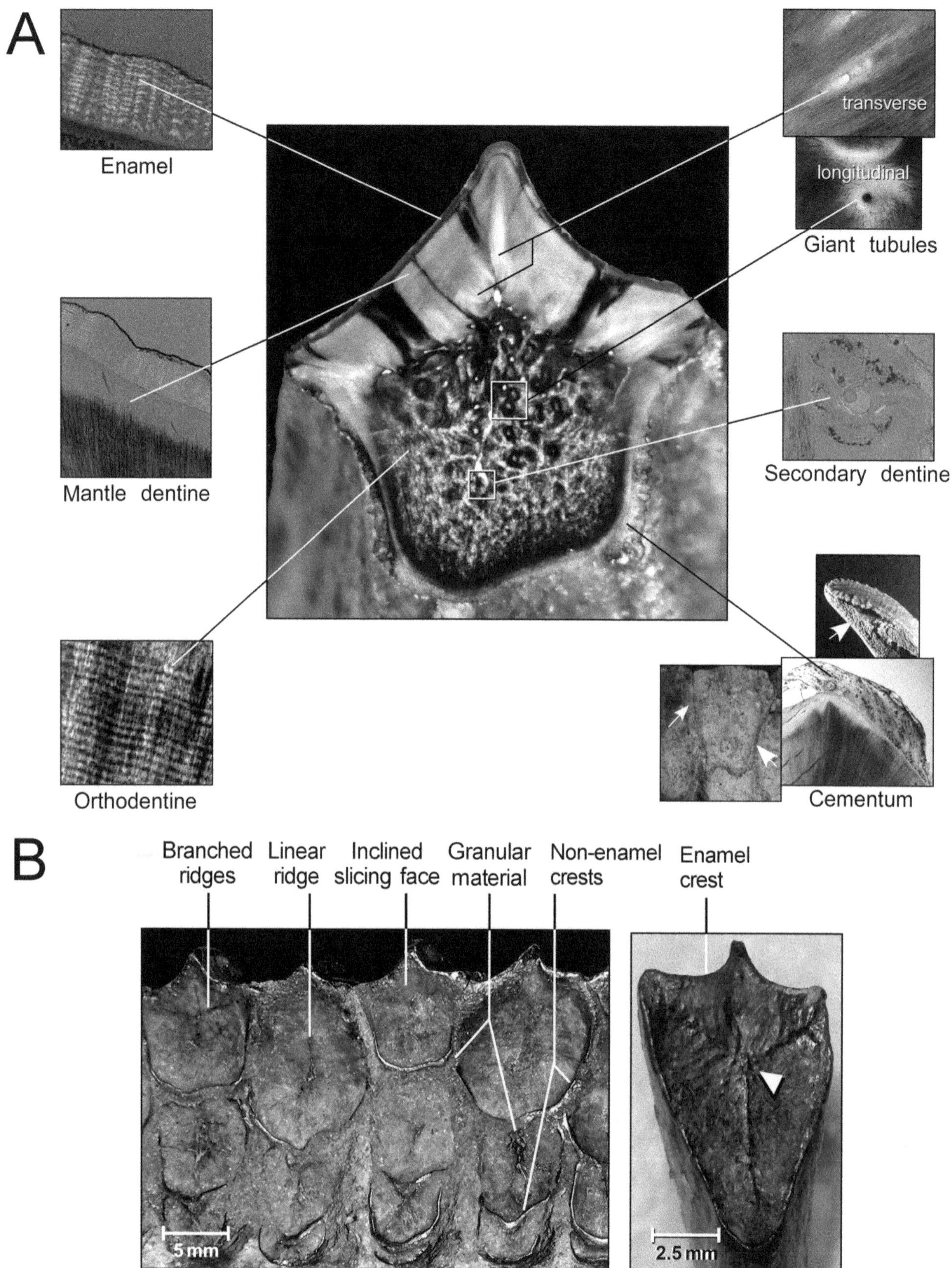

24.2. Hadrosaurid dental organization. (A) Major histological types. (B) *Edmontosaurus annectens* dentary chewing surface (left) showing major wear-induced topographic features. Branched and linear ridges are composed of transversely oriented giant tubules. Inclined slicing faces are composed of transversely oriented giant tubule curtains. The granular material between teeth is composed of coronal cementum. The intra-basin granular material is composed of secondary dentine. *Hadrosaurus foulkii* tooth (right) with leading edge crest composed of enamel and a branched ridge (arrow) composed of transversely oriented giant tubule dentine (arrow).

for finer grinding of plants than major crests. Besides binding the teeth together, the former root attachment tissue, coronal cementum, served as a stress bridge, minimizing singularities on the brittle crests. Finally, abscess-preventing secondary dentine is present where the pulp cavity was locally breached, but unlike mammalian grinding teeth, this tissue did not substantially contribute to basin formation through wear.

Erickson et al. (2012) recently showed that evolutionary changes in dental tissue compositions facilitated changes in the wear-induced functional morphology of hadrosaurid dental batteries and allowed them to diversify into specialized ecological niches. From the primitive slicing and grinding condition that was presumably suited for processing of moderately tough fodder, some taxa (e.g., *Corythosaurus casuarius*, *Lambeosaurus lambei*) evolved coarse semi-planar grinding pavements for processing extremely tough plant matter. In other lineages (e.g., *Saurolophus osborni*, *Prosaurolophus blackfeetensis*), grinding capacity was largely lost, and the adult teeth were specialized for high-angle slicing and shearing of presumably bulky, somewhat tough, folivorous plant matter (Lucas, 2004).

Although the detailed dental histology of adult hadrosaurids is known, these aspects remain undocumented for young hadrosaurids, including embryos. Notably, embryonic and neonate hadrosaurids possessed only single functional teeth at each tooth position (Horner and Currie, 1994; Prieto-Márquez, 2008), and subaults typically had one or two less functional teeth than somatically mature adults (Prieto-Márquez, 2008). Thus, the capacity for forming elaborate dental pavements like those of adults was infeasible. Furthermore, wear-relevant changes in tissue architecture, similar to those that occurred interspecifically among adults, are undocumented. Here we describe and compare the histology of embryonic, subadult, and adult dentary teeth from the lambeosaurine *Hypacrosaurus stebingeri*, the only hadrosaurid for which specimens spanning development (i.e. embryonic to adult stages) are known (Horner and Currie, 1994; Brink et al., 2010; Brink et al., this volume). How the architectural tissue suites may have contributed to the wear-induced morphology of the chewing surfaces and dietary change is inferred using the hadrosaurid tissue wear attributes (Erickson et al., 2012).

Institutional Abbreviations MOR, Museum of the Rockies, Bozeman, Montana; TMP, Royal Tyrrell Museum of Palaeontology, Drumheller, Alberta.

MATERIALS AND METHODS

Specimens used for both morphological and histological characterization in our study were discovered in the Upper Cretaceous Oldman Formation of Devil's Coulee, southern Alberta, a locality that has produced abundant remains of dinosaur eggs and embryos (Horner and Currie, 1994; Brink et al., 2010). Jaws with teeth and isolated teeth of embryonic, subadult, and adult hadrosaurids were examined from three sites within Devil's Coulee, which have yielded nests with dense concentrations of embryonic remains as well other bonebed materials. The specific fossils we studied are housed under the specimen number TMP 1987.077.0099 at the Royal Tyrrell Museum of Paleontology (TMP), Drumheller, Alberta. The hadrosaurid embryos from Devil's Coulee along with hatchling and adult lambeosaur specimens from areas just south of this locality in northern Montana, United States, have been ascribed to a single taxon, *Hypacrosaurus stebingeri* (Horner and Currie, 1994). We assume that the isolated teeth studied from these nests and concentrations all belong to this taxon (also see Horner and Currie, 1994; Brink et al., 2010). Specimens of hatchling and adult *H. stebingeri* discovered in the Two Medicine Formation in northern Montana that are housed in the Museum of the Rockies (MOR), Bozeman, Montana, were used to supplement our morphological characterizations of the dental battery occlusal surfaces for this taxon.

To minimize the extent of destructive sampling, we restricted our analysis to isolated dentary teeth, presumably dislodged from intact batteries, that were available for sampling. Dentary teeth also show greater topographical relief in naturally worn batteries making it easier to correlate histological distributions with wear patterns and the mechanical attributes of their fodder.

Embryonic (TMP 1987.079.0266), hatchling (MOR 548), subadult (TMP 1988.151.0075), and adult (MOR 549) dental battery occlusal surfaces and their morphologies were examined, characterized, and photographed using dissecting microscopy.

Isolated, newly erupted dentary teeth (identified from the location of the replacement tooth facet relative to the enamel face) from an embryo (5.5 mm crown height), subadult (24 mm unworn crown height), and adult (34 mm crown height) *H. stebingeri* were embedded in clear epoxy resin, and serially sectioned transversely in 1.2 mm increments using a slow-speed bone saw with a diamond-tipped blade. The sections were affixed to petrographic microscope slides and the histological structuring photographed through the glass. The histology was described using the terminology of Peyer (1968), Schmidt and Keil (1971), Hillson (1986), Bhaskar (1991), and Erickson et al. (2012). The specimens were then wet sanded with water as an irrigant to ~100 micron thicknesses using silicon-carbide sandpaper (120–1,200 grit) and fine polished using wet aluminum oxide on a felt pad. These were viewed using polarizing and dissecting microscopy, and the histology further described.

RESULTS

Embryonic and Hatchling Teeth

Our results show the occlusal surfaces of embryonic and hatchling *H. stebingeri* dentary teeth are angled at approximately 18–20 degrees (where horizontal = 0° and vertical = 90°) and the worn tooth surfaces are more elongate mesiodistally than are those of adult teeth (Fig 24.3A, B). Topographically the in ovo and neonate dinosaur tooth wear surfaces are cupped, showing an elevated rim surrounding a roughened shallow basin (Fig. 24.3B). This is similar to the condition seen in hatchling *Maiasaura peeblesorum* and *Edmontosaurus regalis* (GME, pers. obs., and illustrated for *H. stebingeri* by Horner and Currie, 1994:fig. 21.23). This non-grinding morphology is reminiscent of that for smaller taxa from outgroup clades, *Heterodontosaurus tucki* and *Dryosaurus altus* in particular (GME, pers. obs.). Presumably these teeth functioned in the hatchlings to slice and crush succulent plant matter, and perhaps some insects and small vertebrate prey as well (neonates of extant herbivorous reptiles are often initially omnivores, or even predators; Pough et al., 2004).

The histological composition of the embryonic, and presumably the hatchling teeth, is simpler than is typical of most adult hadrosaurid teeth (see above) and conspecifics from later developmental stages (see below). As in adult hadrosaurids, the enamel is restricted to the lingual faces of the dentary teeth (Fig. 24.3C), and polarized microscopy reveals that the wavy microstructure described by Sander (1999) and Hwang (2005) in adult hadrosaurids is prevalent in these younger individuals (Fig. 24.3D). Cellular coronal cementum surrounds the remainder of each tooth (Fig. 24.3C, E). Mantle dentine, which lies adjacent to the enamel and cementum, forms a ring around the orthodentine, the latter of which composes the bulk of each tooth (Fig. 24.3C). The mantle dentine is considerably thicker adjacent to the enamel and shows extensive peritubules, whereas such processes are nonexistent elsewhere. Longitudinally oriented giant tubules are sparse but appear in a row along the mesiodistally oriented pulp cavity (Fig. 24.3C). Secondary dentine locally lines the pulp cavity (Fig. 24.3C).

With regard to tribological attributes, the distribution of these tissues with respect to the chewing surfaces in the embryos provides a biomechanical explanation for their having worn to simple basins. Enamel, the hardest, most wear-resistant material ensured that the leading edge of newly erupting teeth was higher than the remainder of the chewing surface. The adjacent, thick, mantle dentine, which is slightly less wear resistant, contributed to this effect, but also, by surrounding the remainder of each tooth, created a circumferential rim. Coronal cementum likely helped to mechanically support these brittle prominences. The centers of the teeth are composed of soft orthodentine whose poor wear resistance led to the formation of intra-tooth basins. The longitudinally oriented giant tubules had negligible exposure on the chewing surface and so contributed little to the topography. Nevertheless they helped partition the pulp cavity so that major expanses were not exposed as the crown was breached with usage. Abscess-preventing secondary dentine lines the remainder of the pulp cavity where giant tubules do not exist. Unlike the grinding teeth of mammals, where secondary dentine is a major contributor to basin formation (Peyer, 1968; Schmidt and Keil, 1971; Hillson, 1986), the small footprint of this tissue in embryonic *H. stebingeri* provided little contribution to the resultant wear topography.

Subadult Teeth

Our results show the occlusal surfaces of subadult *H. stebingeri* dentary batteries are angled at approximately 25 degrees (Fig. 24.4A). Mesiodistally and labiolingually, the worn surfaces of individual teeth are more evenly proportioned than the embryonic and hatchling teeth (Fig. 24.4A, B). Topographically the chewing surfaces, consisting of two (and occasionally three; Brink et al., 2010) functional teeth, show a planar blade with only slight bowling out to the labial third of most teeth (Fig. 24.4B). This morphology is reminiscent of that seen in outgroup *Probactrosaurus gobiensis* (GME, pers. obs.). The absence of substantially elevated inter-tooth crests in these batteries suggests that they functioned primarily to shear bulky, tough plant matter.

The dental histology of the subadult teeth (Fig. 24.4C) is more complex than those of the embryos (see above). Enamel is restricted to the lingual faces of the teeth (Fig. 24.4C), and polarized microscopy reveals that the wavy microstructure described by Sander (1999) and Hwang (2005) in adult hadrosaurid teeth is prevalent (Fig. 24.4D). Cellular coronal cementum surrounds the remainder of each tooth (Fig. 24.4C, F, G). Mantle dentine, which lies adjacent to the enamel and cementum, forms a ring around the orthodentine and giant tubules (Fig. 24.4C). The mantle dentine is thicker adjacent to the enamel and shows less extensive peritubules than embryonic teeth. The mantle dentine adjacent to the coronal cementum lacks these processes. Transversely oriented giant tubules are prevalent (Fig. 24.4C, E). The largest unite to form a trident shape extending from the center of the pulp cavity and radiating towards the median and lateral carinae (Fig. 24.4C). It extends only a short distance labially (Fig. 24.4C). Extending from the major branches of the trident are dense curtains of transversely oriented giant tubules (Fig. 24.4E). The transversely oriented giant tubules span (moving labially) across approximately 50% of

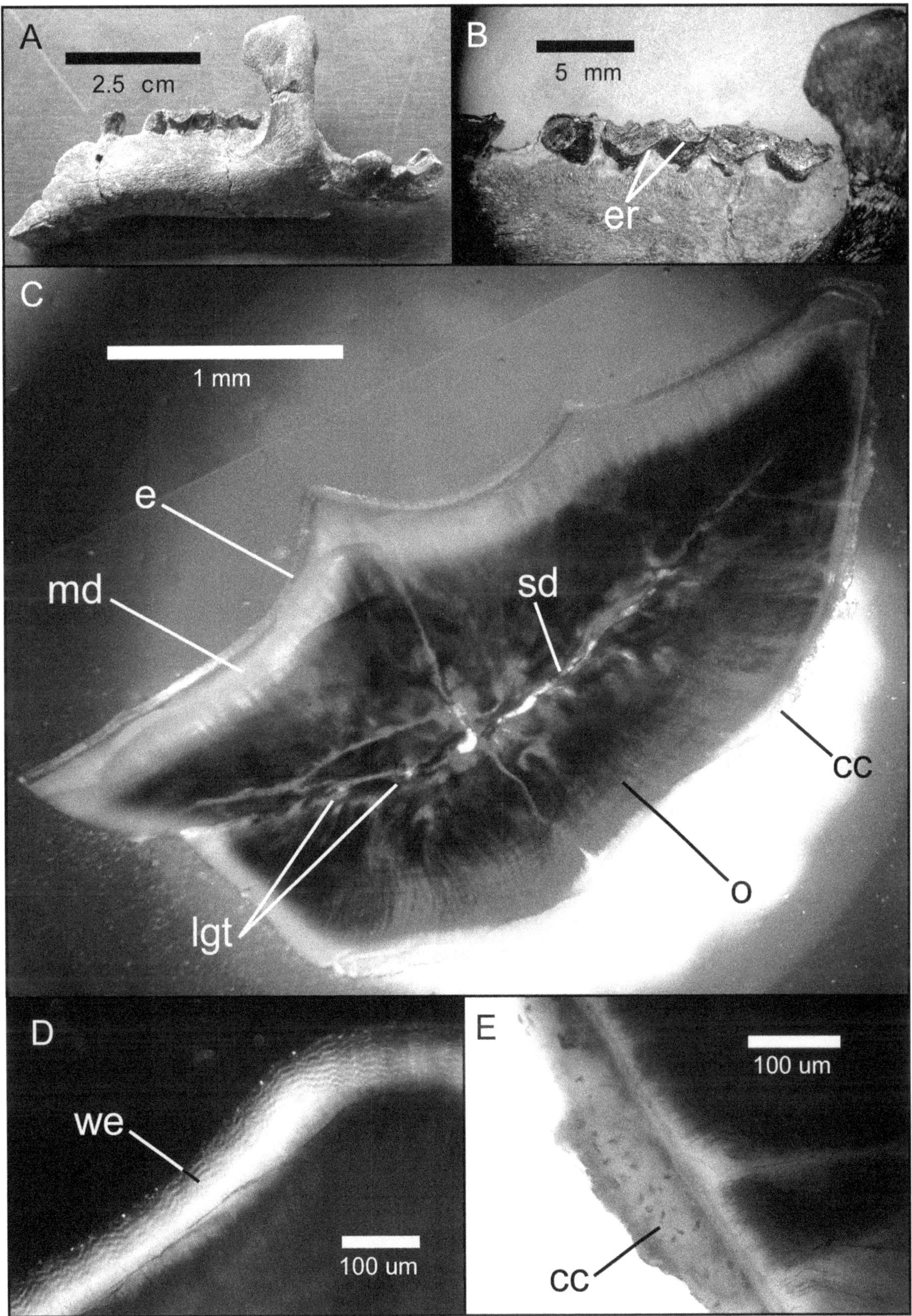

24.3. *Hypacrosaurus stebingeri* embryonic dental morphology and histology. (A) Dentary of TMP 1987.079.0266. (B) Occlusal surface of teeth in TMP 1987.079.0266 showing elevated rim (er). (C) Transverse histological section through a dentary tooth of TMP 1987.077.0099 showing enamel (e), mantle dentine (md), orthodentine (o), longitudinally oriented giant tubules (lgt), coronal cementum (cc) and secondary dentine (sd). (D) Wavy enamel in TMP 1987.077.0099 (we). (E) Coronal cementum in TMP 1987.077.0099 (cc).

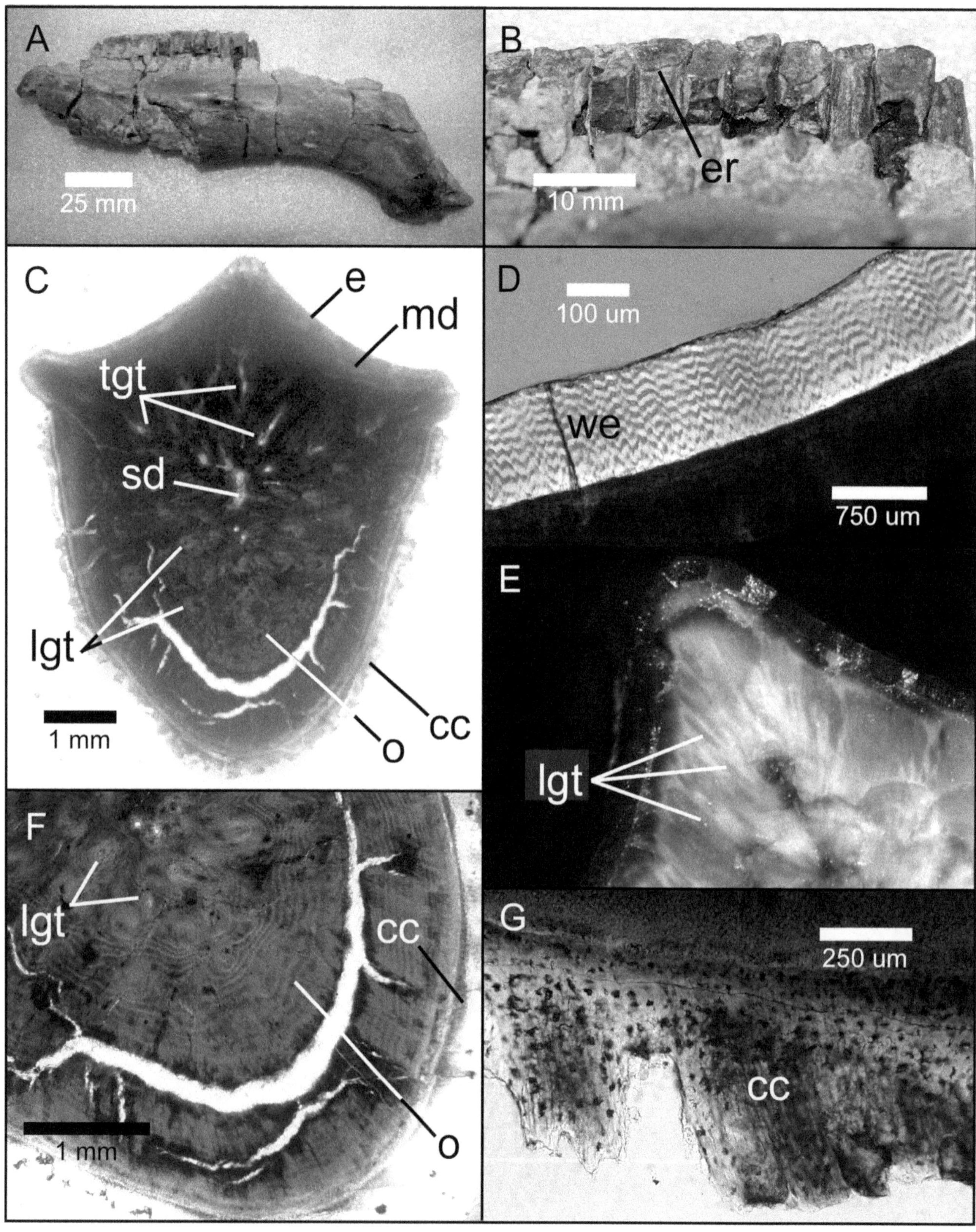

24.4. *Hypacrosaurus stebingeri* sub-adult dental morphology and histology. (A) Dentary of TMP 1988.151.0075. (B) Occlusal surface of dentary teeth in TMP 1988.151.0075 showing slicing faces, and slightly elevated rim (er) composed of mantle dentine. (C) Transverse histological section through the dentary tooth of TMP 1987.077.0099 showing enamel (e), mantle dentine (md), orthodentine (o), transversely oriented giant tubules (tgt), longitudinally oriented giant tubules (lgt), coronal cementum (cc) and secondary dentine (sd). (D) Wavy enamel in TMP 1987.077.0099 (we). (E) Radiating longitudinally oriented giant tubules (lgt) in the leading (lingual) edge of a dentary tooth from TMP 1987.077.0099. (F) Distal (labial) half of a dentary tooth from TMP 1987.077.0099 showing the transition from longitudinally oriented giant tubules (lgt) to orthodentine (o) composition. (G) Coronal cementum in TMP 1987.077.0099 (cc).

the exposed tooth crowns, at which point longitudinal giant tubules become prevalent (Fig. 24.4C). The labial-most portions of the worn teeth are composed of orthodentine (Fig. 24.4F). Secondary dentine lines small breaches in the pulp cavity (Fig. 24.4C).

With regard to tribological attributes, the distribution of these tissues with respect to the occlusal surface topography provides a biomechanical explanation for their wearing to shearing planes. Enamel, the hardest, most wear-resistant material ensured that the leading edge of the teeth was higher than the remainder of the chewing surface. The adjacent, thick mantle dentine, which is slightly less wear-resistant, contributed to the slicing plane, as did the extensive radiating curtains of moderately wear resistant, transversely oriented giant tubules and prevalent longitudinally oriented giant tubules. The footprint of low wear resistance material is small, being composed of soft orthodentine that wore to a very shallow basin. This occurred since recession magnitude is dependent not only on the wear resistance of the material in abrasive wear systems, but also on sliding distance (Archard, 1953; Archard and Hirst, 1956). Little travel over soft orthodentine occurred before the hard mantle dentine was contacted (Fig. 24.4B). Coronal cementum served to bind the teeth together but probably had a negligible role in mechanically supporting the mantle dentine crests, owing to their low relief. Secondary dentine is present in the tiny sections of pulp cavity that is exposed where giant tubules do not exist. Unlike the case in mammalian grinding dentitions (Peyer, 1968; Schmidt and Keil, 1971; Hillson, 1986), the negligible exposure of this tissue provided little contribution to the resultant wear topography.

Adult Teeth

Our results show the occlusal surfaces of adult *H. stebingeri* lower dental battery teeth are dual-planar (also see Weishampel, 1983, 1984; Fig. 24.5A, B). The exposed teeth are more elongate along the labiolingual axis than embryonic and subadult *H. stebingeri* teeth (Fig. 24.5A), and the lingual half of newly erupted functional teeth have a steep slicing edge that is angled at approximately 56–60° (Fig. 24.5B). The labial portions of such teeth, and the entirety of the subsequent more heavily worn teeth compose a more horizontal surface angled at approximately 30° (Fig. 24.5B). The basins of the leading and second functional teeth possess a labiolingually oriented central ridge, with a deep trough on each side (Fig. 24.5B). The tiny labial-most functional teeth are simply remnants of the roots. Their wear surfaces are cupped and have no intra-basin ridging. Presumably these batteries had a dual function, allowing for both coarse slicing and crushing of tough plant matter.

The dental histology of adult *H. stebingeri* teeth is more complex than in the embryonic and subadults (see above; Fig. 24.5C). Enamel is restricted to the lingual faces of the teeth (Fig. 24.5C), and polarized microscopy reveals characteristic adult hadrosaurid wavy microstructure (Fig. 24.5D). Cellular coronal cementum surrounds the remainder of each tooth (Fig. 24.5C, E). Mantle dentine, which lies adjacent to the enamel and cementum, surrounds a thin expanse of orthodentine (Fig. 24.5C). The mantle dentine is thicker adjacent to the enamel and is like subadult *H. stebingeri* teeth in showing less extensive peritubules than embryonic teeth. The mantle dentine adjacent to the coronal cementum lacks these processes. Rows of longitudinally oriented giant tubules and a few transversely oriented giant tubules form a trident shape extending from the center of the pulp cavity and radiating toward the median and lateral carinae and extensively down the midline of the tooth in the labial direction (Fig. 24.5C, F). The labial branch terminates approximately 80% of the way across the tooth. Emanating from the branches of the trident are dense curtains of transversely oriented giant tubules, few of which extend all of the way to the mantle dentine (Fig. 24.5C, F). These structures have the largest exposure on the chewing surface. Orthodentine lies between these giant tubule curtains and the mantle dentine (Fig. 24.5C). Abscess-preventing secondary dentine (not shown) locally lines in the remainder of the pulp cavity where giant tubules are absent.

With regard to the wear attributes, the positioning of these tissues with respect to the chewing surfaces provide a biomechanical explanation for their wearing to both high-angled blades and a flattened pavement with a central ridge. Enamel, the hardest, most wear resistant material ensured that the leading edge of the tooth was higher than the remainder of the chewing surface. The adjacent mantle dentine is exceptionally thick, and, being slightly less wear resistant, contributed to the high-angled slicing plane. The dense curtains of intermediate wear-resistant, transversely oriented giant tubules facilitated slicing plane formation. The remainder of the tooth (and the mid-battery teeth) formed a lower crushing plane whose central ridge was composed of intermediate wear resistant giant tubules surrounded by orthodentine with poor wear resistance. This naturally wore to create the ridge with labiolingually oriented basins on each side. Coronal cementum served to bind the teeth together but probably had a negligible role in mechanically supporting the mantle dentine crests, owing to their low relief. Secondary dentine shows negligible exposure on the chewing surface lining sections of pulp cavity that are exposed where giant tubules do not exist. It was not a major contributor to the topography. The tiny labial-most functional teeth are composed solely of crescent-shaped

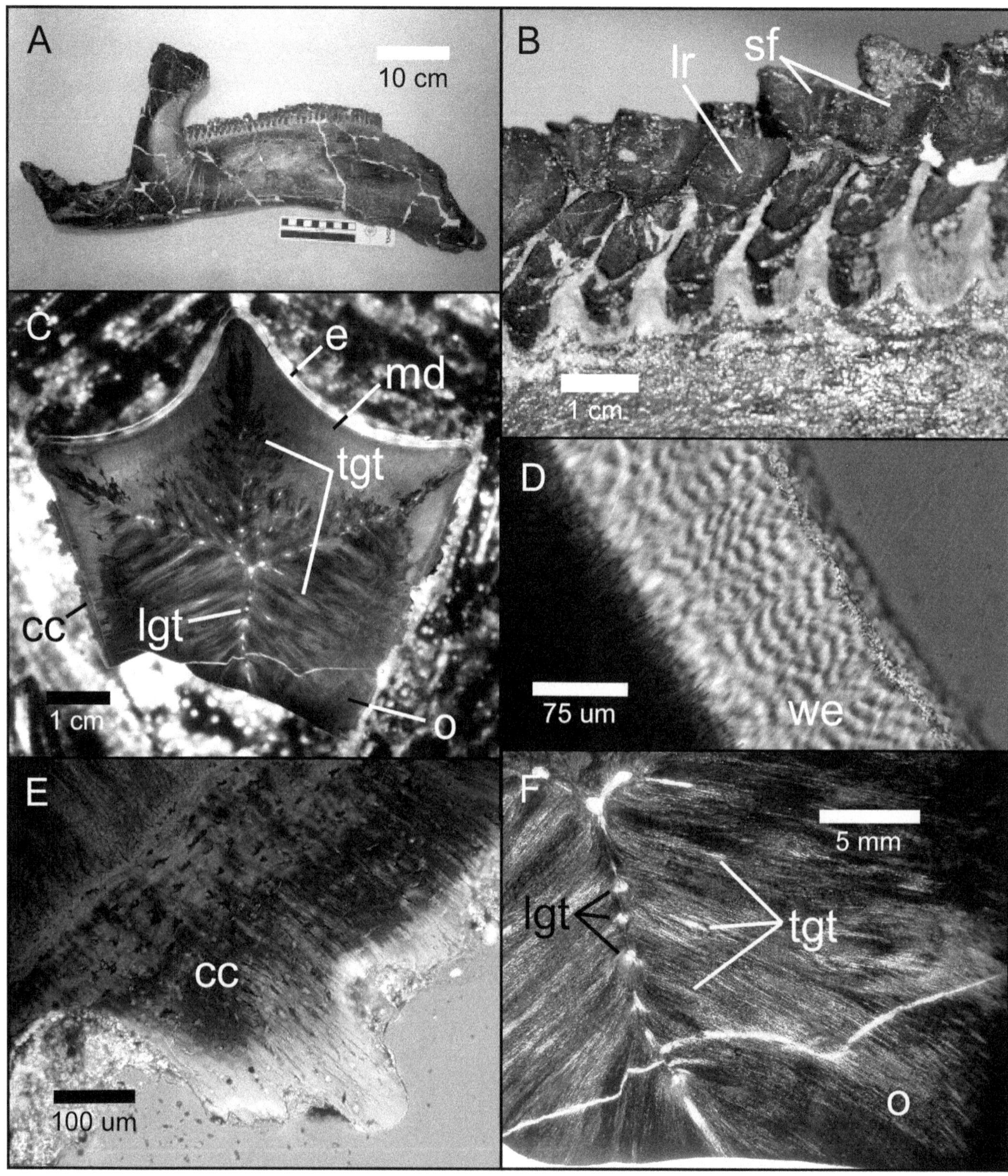

24.5. *Hypacrosaurus stebingeri* adult dental morphology and histology. (A) Dentary of MOR 548. (B) Occlusal surface of dentary teeth in MOR 548 showing slicing faces (sf) and a longitudinal ridge (lr). (C) Transverse histological section through the dentary tooth of TMP 1987.077.0099 showing enamel (e), mantle dentine (md), orthodentine (o), transversely-oriented giant tubules (tgt), longitudinally oriented giant tubules (lgt), coronal cementum (cc). (D) Wavy enamel in TMP 1987.077.0099 (we). (E) Coronal cementum in TMP 1987.077.0099 (cc). (F) Extensive radiating transversely oriented giant tubules (lgt) extending down the midline in the distal (labial) half of a dentary tooth from TMP 1987.077.0099 bounded laterally by curtains of transversely oriented giant tubules (tgt) followed by a short expanse of orthodentine (o).

rings of hard mantle dentine surrounding expanses of soft orthodentine (not shown). These simply wore to shallow basins. Coronal cementum adheres to the mantle dentine, but again, probably provided little in the way of mechanical support to the mantle dentine crests, owing to their low relief.

DISCUSSION

During the hadrosaurid radiation, evolutionary changes in the types and distributions of osseous tissues within the teeth facilitated modifications in the topography of the chewing surfaces through wear (Erickson et al., 2012). Such changes made it possible for these animals to specialize and/or generalize on different plant types as adults. The present research highlights that in addition to developmental changes in the number of functional teeth within each tooth row (from one to three or four) in *H. stebingeri* (Prieto-Márquez, 2008), wear-mediated topographic changes also occurred during ontogeny. These changes presumably reflect major changes in dietary ecology concurrent with the substantial changes in body size that these animals underwent during their lives. What those specific plants (or even prey) were is not known. Regardless of their diet, *H stebingeri* clearly progressed from having dentary teeth with inclined cup-like structures in embryos and neonates, to primarily shearing teeth as subadults, to combined slicing and crushing surfaces as adults. At no stage do they appear to have possessed what would be considered a grinding dentition by the standards used to categorize modern ungulate, marsupial, and rodent dentitions (Peyer, 1968; Schmidt and Keil, 1971; Hillson, 1986; Janis and Fortelius, 1988). In other words they did not possess coarse, semi-planar chewing surfaces involving two or more prominent crests over which plant cell contents were liberated and cell walls comminuted through pulverization. As revealed by Erickson et al. (2012), shearing is a derived, and perhaps not uncommon condition within hadrosaurids. Thus, generalizations that hadrosaurids exclusively possessed grinding dentitions are inaccurate, in our opinion.

Besides changes in dietary ecology during *H. stebingeri* ontogeny, it is notable that the stages of morphological changes to the numbers, shapes, and occlusal surface topography of the teeth are similar to several of the stages that appeared in the transformation series leading to the evolution of the hadrosaurid dental battery (Erickson et al., 2012). This is interesting in that it suggests young *H. stebingeri* may have been exploiting the same or mechanically similar forage as their ancestors, and their dietary preferences as adults may have came about through peramorphosis (Gould, 1977).

The feeding biomechanics of adult ornithopods have been studied extensively (Weishampel, 1983; Norman and Weishampel, 1985, 1991; Rybczynski et al., 2008; Bell et al., 2009; Williams et al., 2009). These point to major differences between hadrosaurids and outgroup taxa in tooth form, translation of the jaws and teeth past one another, and possibly cranial kinesis. Because of the aforementioned ontogenetic changes in the feeding anatomy of *H. stebingeri* (and presumably other hadrosaurids), it is plausible that young *H. stebingeri* had substantially different feeding biomechanics than the adults. This will make an interesting avenue for future investigation. Also, since there are major differences in the chewing pavements among hadrosaurid adults (e.g., slicing and grinding, pure grinding, pure slicing), the possibility that interspecific differences in jaw movements and cranial kinesis also existed is a distinct possibly that is worthy of investigation.

How were the morphological changes to the chewing surfaces of *H. stebingeri* during development brought about? Our results suggest that they were achieved through abrasive wear via developmental changes in the histological makeup of the teeth, tooth size, form, and the number of teeth encompassing the chewing surfaces. Notably no major tissue types were found that are unique to any one particular developmental stage. The same ensemble of tissues seen in adult *H. stebingeri*, and found variably among outgroup and adults of other hadrosaurids (Erickson et al., 2012), were utilized during ontogeny. The most notable tissue suite difference that we found occurs among embryonic (and presumably neonate) teeth, that lack transversely oriented giant tubules. Functionally these tissues wear to form slicing planes and intra-basin ridges in subadult and adult *H. stebingeri*. Thus, the absence of broad slicing faces and intra-basin ridges is not unexpected in the very young *H. stebingeri*.

Several growth series spanning from neonate through adult stages are already known for hadrosaurids (e.g., *Maiasaura peeblesorum* and *Edmontosaurus regalis*), and more will certainly be discovered in the future. It will be interesting to compare and contrast the morphology of the chewing surfaces in these animals and the histological influences on their formation with those data we present here. The results will contribute to a more comprehensive understanding of how dietary shifts were made throughout development in these fascinating animals.

ACKNOWLEDGMENTS

We thank John Horner, David Eberth, Don Brinkman and the curatorial staff of the Museum of the Rockies and Royal Tyrrell Museum of Palaeontology for their assistance in gaining access to the *H. stebingeri* specimens in their respective museums. In addition we thank David Evans and David Eberth for inviting us to present our findings at the International Hadrosaurid Symposium held at the TMP, and Ken Womble for producing our graphics. Finally a special thanks goes out to David Weishampel, whose lifetime contributions toward our understanding of hadrosaurids has helped inspire the next generation of researchers to study the paleobiology and evolution of these remarkable animals. This research was funded by NSF grants (EAR 0959029 and EAR 027149) presented to GME.

LITERATURE CITED

Archard, J. F. 1953. Contact and rubbing of flat surfaces. Journal of Applied Physics 24:981–988.

Archard, J. F., and W. Hirst. 1956. The wear of materials under unlubricated conditions. Proceedings of the Royal Society of London A 236:397–410.

Bell, P. R., E. Snively, and L. Shychoski. 2009. A comparison of the jaw mechanics in hadrosaurid and ceratopsid dinosaurs using finite element analysis. Anatomical Record 292:1338–1351.

Bhaskar, S. N. (ed.). 1991. Orban's Oral Histology and Embryology. Mosby Year Book, 11th edition, London, 489 pp.

Brink, K. S., D. K. Zelenitsky, D. C. Evans, F. Therrien, and J. R. Horner. 2010. A sub-adult skull of *Hypacrosaurus stebingeri* (Ornithischia: Lambeosaurinae): anatomy and comparison. Historical Biology 23:63–72.

Brink, K. S., D. Zelenitsky, D. C. Evans, J. R. Horner, and F. Therrien. 2014. Cranial morphology and variation in *Hypacrosaurus stebingeri* (Ornithischia: Hadrosauridae); chapter 13 in D. A. Eberth and D. C. Evans (eds.). Hadrosaurs. Indiana University Press, Bloomington, Indiana.

Carrano, M. T., C. M. Janis, and J. J. Sepkoski, Jr. 1999. Hadrosaurs as ungulate parallels: lost lifestyles and deficient data. Acta Palaeontologica Polonica 44:237–261.

Case, J. A., J. E. Martin, D. C. Chaney, M. Reguero, S. A. Marenssi, S. M. Santillana, and M. O. Woodburne. 2000. The first duck-billed dinosaur (Family Hadrosauridae) from Antarctica. Journal of Vertebrate Paleontology 20:612–614.

Chin, K. 1997. What did dinosaurs eat? Coprolites and other direct evidence of dinosaur diets; pp. 371–382 in J. O. Farlow and M. K. Brett-Surman (eds.), The Complete Dinosaur. Indiana University Press, Bloomington, Indiana.

Chin, K. 2007. The paleobiological implications of herbivorous dinosaur coprolites from the Upper Cretaceous Two Medicine Formation of Montana: why eat wood? Palaios 22:554–566.

Currie, P. J., E. B. Koppelhus, and A. F. Mohammad. 1995. Stomach contents of a hadrosaur from the Dinosaur Park Formation (Campanian, Upper Cretaceous) of Alberta, Canada; pp. 111–114 in A. Sun and Y. Wang (eds.), Sixth Symposium on Mesozoic Terrestrial Ecosystems and Biota: Short Papers. China Ocean Press, Beijing, China.

Cuthbertson, R. S., A. Tirabasso, N. Rybczynski, and R. B. Holmes. 2012. Kinetic limitations of intracranial joints in *Brachylophosaurus canadensis* and *Edmontosaurus regalis* (Dinosauria: Hadrosauridae), and their implications for the chewing mechanics of hadrosaurids. Anatomical Record 295:968–979.

Edmund, A. G. 1969. Dentition; pp. 115–200 in A. C. Gans (ed.), Biology of the Reptilia, Volume 1: Morphology. Academic Press, London, U.K.

Erickson, G. M. 1996. Incremental lines of von Ebner in dinosaurs and the assessment of tooth replacement rates using growth line counts. Proceedings of the National Academy of Sciences 93:14623–14627.

Erickson, G. M., B. A. Krick, M. Hamilton, G. R. Bourne, M. A. Norell, E. Lilleodden, and W. G. Sawyer. 2012. Complex dental structure and wear biomechanics in hadrosaurid dinosaurs. Science 338:98–101.

Evans, D. C. 2007. Ontogeny and Evolution of Lambeosaurine Dinosaurs (Ornithischia: Hadrosauridae). Ph.D. dissertation, University of Toronto, Toronto, Ontario, 497 pp.

Gould, S. J. 1977. Ontogeny and Phylogeny. Belknap Press, Cambridge, U.K., 501 pp.

Hillson, S. 1986. Teeth, Second Edition. Cambridge University Press, Cambridge, U.K., 375 pp.

Horner, J. R., and P. J. Currie. 1994. Embryonic and neonatal morphology and ontogeny of a new species of *Hypacrosaurus* (Ornithischia, Lambeosauridae) from Montana and Alberta; pp. 312–336 in K. Carpenter, K. F. Hirsch, and J. R. Horner (eds.), Dinosaur Eggs and Babies. Cambridge University Press, Cambridge, U.K.

Hwang, S. H. 2005. Phylogenetic patterns of enamel microstructure in dinosaur teeth. Journal of Morphology 266:208–240.

Janis, C. M. 1990. The correlation between diet and dental wear in herbivorous mammals, and its relationship to the determination of diets of extinct species; pp. 241–259 in A. J. Boucot (ed.), Evolutionary Paleobiology of Behavior and Coevolution. Elsevier, New York.

Janis, C., and M. Fortelius. 1988. On the means whereby mammals achieve increased functional durability of their dentitions, with special reference to limiting factors. Biological Reviews 63:197–230.

Kräusel, R. 1922. Die Nahrung von *Trachodon*. Paläontologische Zeitschrift 4:80.

Lucas, P. W. 2004. Dental Functional Morphology: How Teeth Work. Cambridge University Press, Cambridge, U.K., 355 pp.

Lull, R. S., and N. E. Wright. 1942. Hadrosaurian Dinosaurs of North America. Geological Society of America Special Papers 40. 242 pp.

Mallon J. C., R. S. Cuthbertson, and A. Tirabasso. 2011. Hadrosaurid jaw mechanics as revealed by cranial joint limitations and dental microwear analysis; pp. 87–90 in D. R. Braman, D. A. Eberth, D. C. Evans, and W. Taylor (compilers), Hadrosaur Symposium Abstract Volume. Royal Tyrrell Museum of Palaeontology, Drumheller, Alberta.

Murphy, N. L., D. Trexler, and M. Thompson. 2007. "Leonardo," a mummified *Brachylophosaurus* (Ornithischia: Hadrosauridae) from the Judith River Formation of Montana; pp. 117–133 in K. Carpenter (ed.), Horns and Beaks: Ceratopsian and Ornithopod Dinosaurs. Indiana University Press, Bloomington, Indiana.

Norman, D. B., and D. B. Weishampel. 1985. Ornithopod feeding mechanisms: their bearing on the evolution of herbivory. American Naturalist 126:151–164.

Norman, D. B., and D. B. Weishampel. 1991. Feeding mechanisms in some small herbivorous dinosaurs: processes and patterns; pp. 161–181 in J. M. V. Rayner and R. J. Wootton (eds.), Biomechanics in Evolution. Cambridge University Press, Cambridge, U.K.

Ostrom, J. H. 1961. Cranial morphology of the hadrosaurian dinosaurs of North America. Bulletin of the American Museum of Natural History 122:33–186.

Owen-Smith, R. N. 1988. Megaherbivores: The Influence of Very Large Body Size on Ecology. Cambridge University Press, Cambridge, U.K., 369 pp.

Peyer, B. 1968. Comparative Odontology. University of Chicago Press, Chicago, Illinois, 347 pp.

Pough, F. H., R. M. Andrews, J. E. Cadle, M. L. Crump, A. H. Savitsky, and K. D. Wells. 2004. Herpetology, Third Edition. Pearson Prentice Hall, Upper Saddle River, New Jersey, 726 pp.

Prieto-Márquez, A. 2008. Phylogeny and Historical Biogeography of Hadrosaurid Dinosaurs. Ph.D. dissertation, Florida State University, Tallahassee, Florida, 859 pp.

Prieto-Márquez, A. 2010. Global phylogeny of Hadrosauridae (Dinosauria: Ornithopoda) using parsimony and Bayesian methods. Zoological Journal of the Linnean Society 159:435–502.

Prieto-Márquez, A., and G. Salinas. 2010. A re-evaluation of *Secernosaurus koerneri* and *Kritosaurus australis* (Dinosauria, Hadrosauridae) from the Late Cretaceous of Argentina. Journal of Vertebrate Paleontology 30:813–837.

Rybczynski, N., A. Tirabasso, P. Bloskie, R. Cuthbertson, and C. Holliday. 2008. A three-dimensional animation model of *Edmontosaurus* (Hadrosauridae) for testing chewing hypotheses. Palaeontologia Electronica 11.2.9A.

Sander, P. M. 1999. The microstructure of reptilian tooth enamel: terminology, function, and phylogeny. Münchner Geowissenschaftliche Abhandlungen Reihe A: Geologie und Paläontologie 38:1–102.

Schmidt, W. J., and A. Keil. 1971. Polarizing Microscopy of Dental Tissues. Pergamon Press, Oxford, U.K., 584 pp.

Tweet, J. S., K. Chin, D. R. Braman, and N. L. Murphy. 2008. Probable gut contents within a specimen of *Brachylophosaurus canadensis* (Dinosauria: Hadrosauridae) from the Upper Cretaceous Judith River Formation of Montana. Palaios 23:624–635.

Weishampel, D. B. 1983. Hadrosaurid jaw mechanics. Acta Palaeontologica Polonica 28:271–280.

Weishampel, D. B. 1984. Evolution of jaw mechanisms in ornithopod dinosaurs. Advances in Anatomy, Embryology and Cell Biology 87:1–110.

Williams, V. S., P. M. Barrett, and M. A. Purnell. 2009. Quantitative analysis of dental microwear in hadrosaurid dinosaurs, and the implications for hypotheses of jaw mechanics and feeding. Proceedings of the National Academy of Science 106:11194–11199.

Three-Dimensional Computational Modeling of Pelvic Locomotor Muscle Moment Arms in *Edmontosaurus* (Dinosauria, Hadrosauridae) and Comparisons with Other Archosaurs

25

Susannah C. R. Maidment, Karl T. Bates, and Paul M. Barrett

ABSTRACT

A computational model is used to estimate three-dimensional pelvic muscle moment arms for the saurolophine hadrosaur *Edmontosaurus*. The results are compared to models of the ornithischians *Lesothosaurus* and *Kentrosaurus*, the crocodilian *Alligator*, and the theropod *Allosaurus* in order to examine changes related to the evolution of quadrupedalism and to set the results in an evolutionary context. *Edmontosaurus* and *Kentrosaurus* generally had higher flexion moment arms than the other taxa, suggesting a correlation with ornithischian quadrupedalism. *Edmontosaurus* had relatively high flexor and extensor moment arms compared to other dinosaurs, which could indicate increased locomotor performance in conjunction with decreased femoral excursion relating to a flexed hind limb stance. *Alligator* had the lowest flexion-extension and abduction moment arms and the highest adduction moment arms of all taxa modeled. Differences between *Alligator* and the dinosaurs probably reflect the different gaits utilized by these taxa. Dinosaurs use an upright gait emphasizing flexion and extension, while in the sprawling gait of *Alligator*, long-axis rotation and abduction-adduction joint excursions are also important. High adduction moments in *Alligator* can be attributed to the control of abductor moments generated around the hip by the ground reaction force acting on the horizontal femur; in contrast, adductor moments are generated around the hip in dinosaurs, and are controlled using abductor musculature.

INTRODUCTION

Three-dimensional (3-D) modeling of locomotor muscle moment arms provides a repeatable and quantitative method for examining the similarities and differences in muscle leverage between taxa. The method has recently been used in several studies to examine specific hypotheses relating to locomotor evolution in archosaurs (Hutchinson et al., 2005, 2008; Bates and Schachner, 2012; Bates, Benson, and Falkingham, 2012; Bates, Maidment, et al., 2012), and is particularly powerful because sensitivity analyses can be carried out that allow the definition and quantification of uncertainty within these models. Comparisons with previously published models of other taxa are relatively easy to make because the method is quantitative and repeatable. Although muscle function is dictated by a large number of factors, including muscle mass, architecture, and contractile properties (Alexander, 2003), quantification of muscle moment arms does provide a basic understanding of individual muscle force orientations. Furthermore, the facility to manipulate 3-D anatomy to produce a spectrum of muscle reconstructions means that the level of uncertainty with which muscle insertions and paths are reconstructed in fossil animals can be accounted for explicitly. This is the crucial first step in producing more complex, dynamic locomotory models (e.g., Sellers et al., 2009).

Debate has surrounded interpretations of the stance and gait of hadrosaurid dinosaurs since their discovery over 150 years ago (e.g., Leidy, 1858; Cope, 1883; Lambe, 1914). Early authors (e.g., Leidy, 1858; Cope, 1883; Colbert, 1951) regarded them as aquatic animals, using their broad hind feet to walk on soft mud, their dorsoventrally deep tails to scull through water, and the mitten-like morphology of the manus to paddle. Ostrom (1964) demonstrated that the evidence for an aquatic mode of life was unconvincing, and suggested that they were terrestrial bipeds, citing evidence such as a poorly ossified carpus to indicate that the forelimbs could not have been used to bear weight: other authors of the time generally agreed (e.g., Galton, 1970; Maryańska and Osmólska, 1984). However, a general consensus has now been reached that hadrosaurids employed their forelimbs in locomotion and were at least facultatively quadrupedal (Carrano, 2001; Horner et al., 2004; Sellers et al., 2009). Indeed, evidence from limb-bone scaling (Dilkes, 2001) and trackways (Lockley and Wright, 2001) suggests that their main form of progression may have been quadrupedal, although Godefroit et al. (2012) have suggested that the lambeosaurine *Olorotitan* was predominantly bipedal.

The aim of this work is to use three-dimensional computational modeling to estimate the moment arms of the pelvic locomotor muscles in *Edmontosaurus*. We compare the results to those of bipedal (*Lesothosaurus* [Bates, Maidment, et al., 2012]) and quadrupedal (*Kentrosaurus*) ornithischians in order to investigate changes in muscle moment arms that may be related to functional changes correlated with the evolution of quadrupedality. We set the model in an evolutionary context by comparing estimated moment arms with the bipedal theropod *Allosaurus* (Bates, Benson, and Falkingham, 2012) and the extant crocodilian *Alligator* (Bates and Schachner, 2012).

Quadrupedalism evolved in three ornithischian lineages (Thyreophora, Ceratopsidae, and Ornithopoda) during the group's 170 million year evolutionary history. The transition from bipedality to quadrupedality is exceedingly rare in tetrapod evolution, having occurred on only one occasion outside of Dinosauria (in the silesaurid dinosauriformes: Nesbitt et al., 2010). The driving forces for the repeated evolution of quadrupedality, and the mechanisms by which it occurred in ornithischians, are currently poorly understood (Maidment and Barrett, 2011). The results of this work will allow us to formulate hypotheses about functional changes that occurred as quadrupedalism evolved, and will form a framework for further study that will test these hypotheses using a more comprehensive sample of quadrupedal and bipedal ornithischians.

Institutional Abbreviations CMN, Canadian Museum of Nature, Ottawa, Ontario; NHMUK, Natural History Museum, London, U.K.; ROM, Royal Ontario Museum, Toronto, Ontario; TMP, Royal Tyrrell Museum of Palaeontology, Drumheller, Alberta; USNM, National Museum of Natural History, Smithsonian Institution, Washington D.C.

MATERIALS AND METHODS

Myological Reconstruction

The postcranial skeletons of 17 hadrosaurids were examined for gross morphology and muscle scars (Table 25.1). Myological reconstruction used the extant phylogenetic bracket, whereby the reconstruction of unpreserved soft anatomical features in non-avian dinosaurs can be inferred by their presence/absence in crocodiles and birds (Bryant and Russell, 1992; Witmer, 1995). We dissected numerous extant archosaurs and drew additional information from the literature (e.g., Romer, 1923; George and Berger, 1966; Norman, 1986; Dilkes, 2000; Hutchinson, 2001a, b; Smith et al., 2006; Maidment and Barrett, 2011) to inform our myological reconstruction (Fig. 25.1), which was set in an evolutionary context by examination of extinct archosaurs and over 150 ornithischian specimens (SCRM and PMB, unpubl. data). Abbreviations of the names of muscles used throughout are given in Table 25.2.

Table 25.1. Hadrosaurid Specimens Examined

Taxon	Specimen number
Brachylophosaurus canadensis	CMN 8893
Brachylophosaurus canadensis	TMP 1990.104.0001
Corythosaurus casuarius	ROM 1947
Corythosaurus casuarius	TMP 1980.040.0001
Corythosaurus casuarius	ROM 845
Edmontosaurus regalis	CMN 8399
Edmontosaurus regalis	ROM 801
Edmontosaurus regalis	CMN 2289
Edmontosaurus sp.	NHMUK R6862
Edmontosaurus sp.	NHMUK R5915
Edmontosaurus sp.	NHMUK R14369
Lambeosaurinae indet.	NHMUK R14368
Gryposaurus notabilis	ROM 764
Gryposaurus notabilis	TMP 1980.022.0001
Hadrosauridae indet.	TMP 1998.058.0001
Hypacrosaurus altispinus	CMN 8501
Hypacrosaurus sp.	TMP 2007.010.0002
Lambeosaurus lambei	ROM 1218
Lambeosaurus magnicristatus	TMP 1966.004.0001
Parasaurolophus walkeri	ROM 768

Model Construction

Hadrosaur material from the Sternberg Collections of the Natural History Museum, London (NHMUK) was used to build a three-dimensional model of the pelvis of *Edmontosaurus* (Fig. 25.2). An ilium (NHMUK R4862) from the Maastrichtian Lance Formation of Wyoming (collected by C. H. Sternberg, purchased by the NHMUK in 1910, and referable to *Edmontosaurus* sp.) was chosen as the basis for the model due to its completeness and good-quality preservation. No complete *Edmontosaurus* pubis or ischium could be identified in the collections of the NHMUK, and so a complete ischium (NHMUK R14368) of an indeterminate lambeosaurine, likely from the Campanian Dinosaur Park Formation of Alberta, was substituted. This specimen was unregistered prior to this study and no locality or acquisition information could be found; however, the style of preservation and conservation are very similar to those of the pubis described above, suggesting the ischium was acquired by the NHMUK at the same time as the pubis and was therefore also part of the 1915 Sternberg Collection from, what is now called, the Dinosaur Park Formation. The morphology of the ischium in hadrosaurids is rather conservative, and the only major difference between saurolophines and lambeosaurines is the absence of the ischiadic boot in the former (Prieto-Márquez,

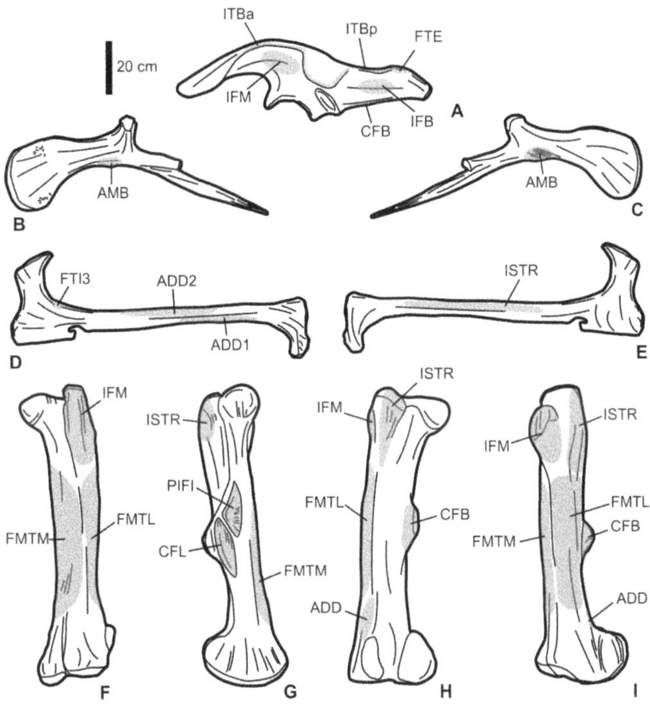

25.1. Myological reconstruction of the pelvis of *Edmontosaurus.* (A) Left ilium in lateral view, based on *Edmontosaurus regalis* (ROM 801) and *E. regalis* (CMN 8399); left pubis, based on *Edmontosaurus regalis* (ROM 801) in (B) lateral and (C) medial view; left ischium, based on *Lambeosaurus lambei* (ROM 1218) in (D) lateral and (E) medial view; left femur, based on *Edmontosaurus regalis* (CMN 2289; ROM 801) in (F) anterior, (G) medial, (H) posterior and (I) lateral view. Abbreviations as in Table 25.2. Scale bar equals 20 cm.

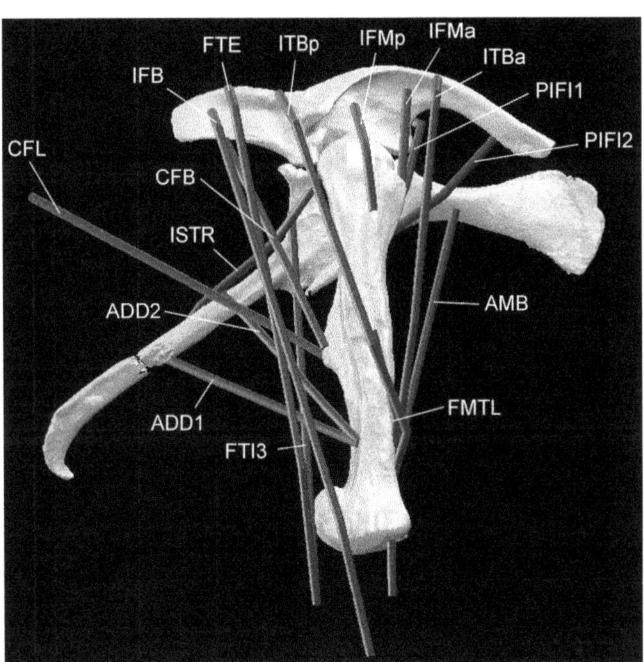

25.2. Three-dimensional model of the pelvis of *Edmontosaurus.* Abbreviations as in Table 25.2.

2010). Since no pelvic locomotor muscles attach to the distal end of the ischium, the substitution of a lambeosaurine ischium into our three-dimensional model will have no effect on the results. Similarly, a complete, well-preserved pubis for *Edmontosaurus* was not available for this study. A pubis of an indeterminate lambeosaurine (NHMUK R5915) from the Dinosaur Park Formation of Alberta (collected by C. H. Sternberg in 1915) was used here. The single muscle attachment site on the neck of the pubis relevant to this study does not differ significantly in its size or position between saurolophines and lambeosaurines, and substitution of the lambeosaurine pubis is also unlikely to affect the model significantly. The femur of an indeterminate hadrosaur (NHMUK R14369) was chosen because it was complete, well preserved, and undistorted. The specimen was unregistered prior to this study and, as with the ischium, locality and acquisition information are unknown. However, the surface of the bone bears paint very similar to that used by Sternberg to label other specimens from the 1915 collection, so it seems very likely that the femur is also from the Dinosaur Park Formation of Alberta. Hadrosaur femora cannot be distinguished at genus level (Prieto-Márquez, 2010), so our use of an indeterminate femur in the model will have no effect on the results.

The pelvis and femur were surface-scanned using a Viuscan handheld laser scanner and the resulting files were imported into the computer-aided design software Maya (www.autodesk.com). The elements were rescaled according to measurements taken directly from more complete *Edmontosaurus* specimens (ROM 801, CMN 8399) and supplemented with data from Lull and Wright (1942; USNM 2414). The pelvis and femur were manually articulated in a "neutral" posture (e.g., Hutchinson et al., 2005) with the femur extending vertically from the acetabulum. The hip joint was modeled as a ball-and-socket joint.

The articulated 3-D model was imported into the forward dynamics modeling software GaitSym (www.animalsimulation.org) and the origin and insertion points of muscles specified. Intermediate or "via" points and wrapping cylinders (Sellers et al., 2003) were used to guide 3-D muscle paths from origin to insertion. The effective moment arm of each muscle for flexion-extension, abduction-adduction, and femoral long-axis rotation (Fig. 25.3) was recorded, and sensitivity analyses were conducted to account for uncertainty in certain muscle attachment sites (Fig. 25.4, see below).

The moment arm is generally defined as the shortest perpendicular distance from the joint center of rotation to the muscle's line of action, and measures the capacity of a muscle to generate rotational force (torque) at a joint from the linear force of its contraction. Here we calculate the moment arm for each muscle by dividing the change in angle of the hip by the change in length of the muscle over a wide spectrum of

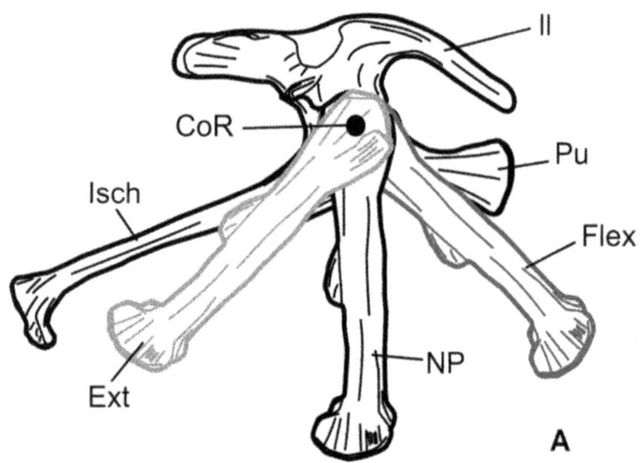

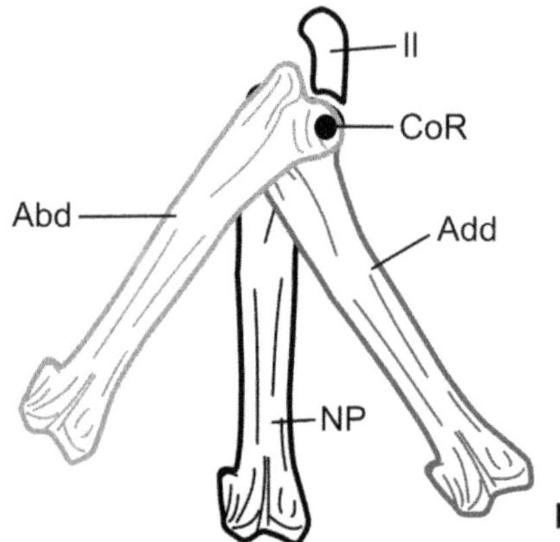

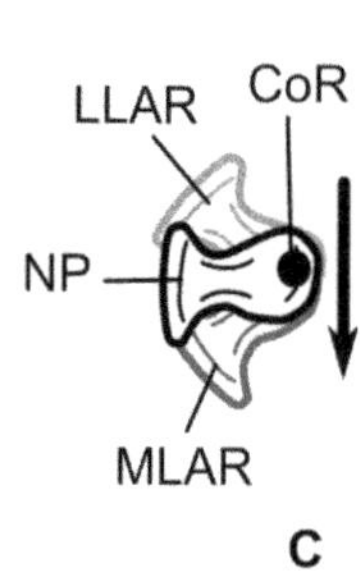

25.3. (A) Lateral view of the hip of *Edmontosaurus* showing hip flexion (movement of the femur anteriorly [dark gray]), and hip extension (movement of the femur posteriorly [light gray]). (B) Anterior view of the hip of *Edmontosaurus* showing a cross-section through the ilium and the femur in anterior view. Hip abduction is movement of the femur away from the sagittal plane (light gray), whereas hip adduction is movement of the femur towards the sagittal plane (dark gray). (C) Dorsal view of the femoral head of *Edmontosaurus;* medial is to the right, whereas the arrow points anteriorly. Medial long-axis rotation causes the lateral side of the femur to move anteriorly, whereas lateral long-axis rotation causes the lateral side of the femur to move posteriorly. Abbreviations: Abd, abduction; Add, adduction; CoR, center of rotation; Ext, extension; Flex, flexion; Il, ilium; Isch, ischium; LLAR, lateral long-axis rotation; MLAR, medial long-axis rotation; NP, neutral posture; Pu, pubis.

3-D hip-joint angles. During locomotion in an animal with a parasagittal gait, flexion-extension movements predominate over abduction-adduction and long-axis rotation at the hip. Moment arm data in Figures 25.5–25.10 are therefore plotted against hip flexion-extension angle with the femur held in neutral posture for abduction-adduction (around 10 degrees abduction) and long-axis rotation (around 0 degrees rotated), as this is most informative about how moment arms change during the step cycle. However, data from other postures are available from SCRM.

Sensitivity Analysis

The myological reconstruction used herein is founded on the extant phylogenetic bracket and set in an evolutionary context (see Myological Reconstruction, above), limiting speculation and defining levels of certainty with which the soft tissues are reconstructed. However, during reconstruction of any unpreserved feature, some subjectivity is inevitable. For example, muscles that have fleshy insertions on bones often have extensive areas of attachment; defining their centroid can therefore be subjective. In order to assess the effects of this subjectivity on our results, we identified three muscles (ADD1, ADD2, PIFI2) that lacked osteological correlates for their origin or insertion, and for which we were able to produce alternative reconstructions that we considered to be equally as likely as our original reconstruction.

The ischiadic shaft in hadrosaurids is elongate and slender, and lacked osteological correlates for the origin of the adductor musculature (ADD1&2) in any of the specimens examined. ADD1&2 could therefore be reconstructed originating from any location on the ischiadic shaft. In fact, these muscles may well have originated from an extended area along the whole of the shaft (Fig. 25.1). In order to assess the effects of this uncertainty, we produced two alternative reconstructions. In the first, ADD1&2 have a more proximal origin than in the original reconstruction, whereas in the second, ADD1&2 originated more distally (Fig. 25.4B, C).

PIFI2 originates on the sides of the dorsal vertebrae in extant crocodilians (Romer, 1923), and it is usually reconstructed in the same location in dinosaurs (e.g., Galton, 1969; Norman, 1986; Dilkes, 2000). However, there is an osteological correlate to suggest that this muscle moved laterally onto the medial surface of the preacetabular process in ornithischians (Maidment and Barrett, 2011) and in our original reconstruction, PIFI2 originates from this derived ornithischian location. In order to assess the effect of this movement on the moment arm of PIFI2, and to provide a comparison with other reconstructions of dinosaurs, an alternative origin on the sides of the dorsal vertebrae was also modeled (Fig. 25.4A).

The insertion of the PIFI musculature on the femur has a robust osteological correlate in *Edmontosaurus* in the form of a deep pit medial to the fourth trochanter and dorsal to a similar pit for CFL (Fig. 25.1). However, in crocodiles, PIFI2 also inserts more anteriorly and dorsally on the femur (Romer, 1923). Although no osteological correlates for this insertion were visible in *Edmontosaurus*, we examined the effect of PIFI2 inserting more anteriorly by producing an alternative reconstruction of its insertion (Fig. 25.4A). A comparison of estimated moment arm magnitudes between our original reconstruction and these six alternative reconstructions is presented below.

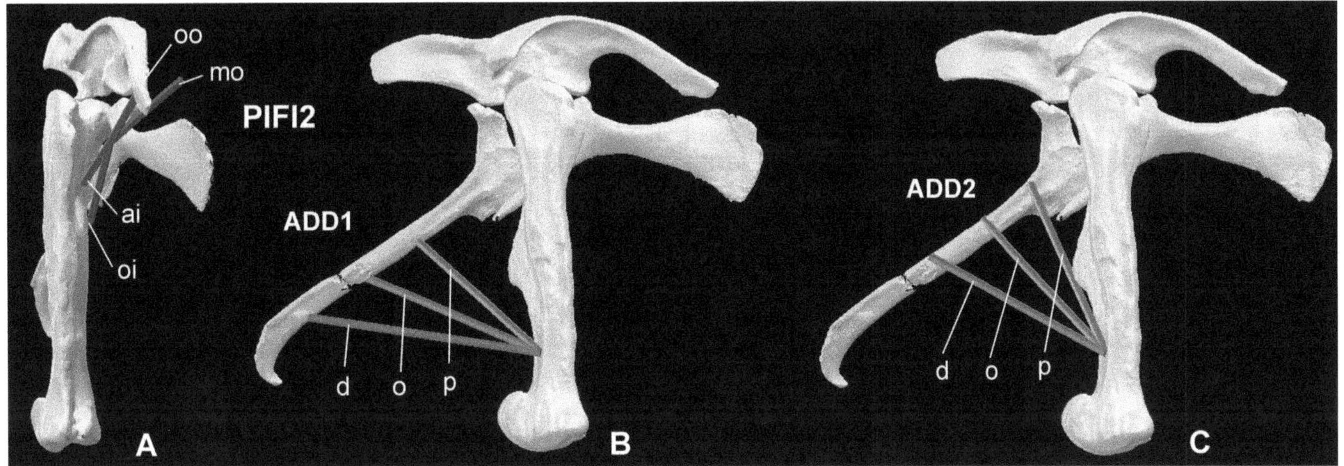

25.4. Sensitivity analysis. (A) Pelvis of *Edmontosaurus* in oblique anterior view showing alternative origins and insertions for PIFI2; (B, C) pelvis of *Edmontosaurus* in lateral view showing (B) alternative origins for ADD1; and (C) alternative origins for ADD2. Abbreviations: ai, anterior insertion; d, distal origin; mo, medial origin; o, original origin (ADD1&2); oi, original insertion; oo, original origin (PIFI2); p, proximal origin.

Comparative Analysis

In order to set the results in an evolutionary context and to examine trends related to changing locomotor strategy in ornithischians, estimated moment arm magnitudes for *Edmontosaurus* were compared to those derived from similar models of the extant quadrupedal crocodilian *Alligator* (Bates and Schachner, 2012), the bipedal theropod dinosaur *Allosaurus* (Bates, Benson, and Falkingham, 2012), the bipedal basal ornithischian *Lesothosaurus* (Bates, Maidment, et al., 2012), and an unpublished model of the quadrupedal stegosaurian dinosaur *Kentrosaurus*, built using a myological reconstruction by SCRM. In order to remove the influence of body size on our comparisons, we normalized moment arms by femoral length. Rather than compare each individual muscle in each taxon, which would require a total of 225 individual comparisons, we focus on elucidating the evolutionary and functional implications of general trends and clear differences between taxa.

RESULTS

Edmontosaurus *Moment Arms*

Flexion-Extension Moment Arms Of the 15 muscles modeled, eight had moment arms for extension (ADD1, ADD2, CFB, CFL, FTE, FTI3, IFB, ITBp; Fig. 25.5), while five had moment arms for flexion (AMB, IFMa, ITBa, PIFI1, PIFI2; Fig. 25.5). ISTR and IFMp had extremely small moment arms for flexion and extension, but changed their moment arms from very weak flexion at extended hip-joint angles to very weak extension at flexed hip-joint angles (Fig. 25.5; weak extension occurred at hip flexion angles beyond those shown). ADD1 had the highest moment arm of any of the extensors (0.53m), whereas ITBa had the highest moment arm of any of the flexors (-0.29m). The moment arms of all extensors and flexors varied considerably with hip-joint angle. Extensors with origins posterior to the ilium (ADD1, ADD2, CFL) had peak moment arms when the femur was within 10 degrees of being vertical, and moment arms significantly decreased at high hip flexion and extension angles. Extensors originating on the posterior iliac process (CFB, FTE, IFB, ITBp) had peak moment arms at extended hip-joint angles, and moment arms decreased significantly at flexed hip-joint angles. FTI3, originating on the proximal ischium, was the only muscle to experience its peak moment for extension at flexed hip-joint angles, with a decrease in moment arms for extension as the femur was retracted. Flexors originating on or near the preacetabular process (ITBa, PIFI1, PIFI2) experienced peak moments for flexion at flexed hip-joint angles, with moment arms significantly decreasing as the hip extended. IFMa and AMB both experienced peak moments for flexion when the femur was within 10 degrees of vertical.

Adduction-Abduction Moment Arms The majority of muscles modeled (CFB, AMB, FTE, FTI3, IFMa, IFMp, IFB, ITBa, ITBp, PIFI1, PIFI2) have moment arms for abduction, while CFL and ISTR, originating medially, are the only muscles to consistently adduct the limb at all hip flexion-extension (Fig. 25.5) and abduction-adduction (SCRM, unpubl. data) angles.

With hip flexion and extension (Fig. 25.5), the majority of muscles maintained a relatively constant moment arm for abduction (AMB, CFB, FTE, FTI3, IFMp, ITBa) or adduction (CFL, ISTR). Several muscles originating posteriorly and

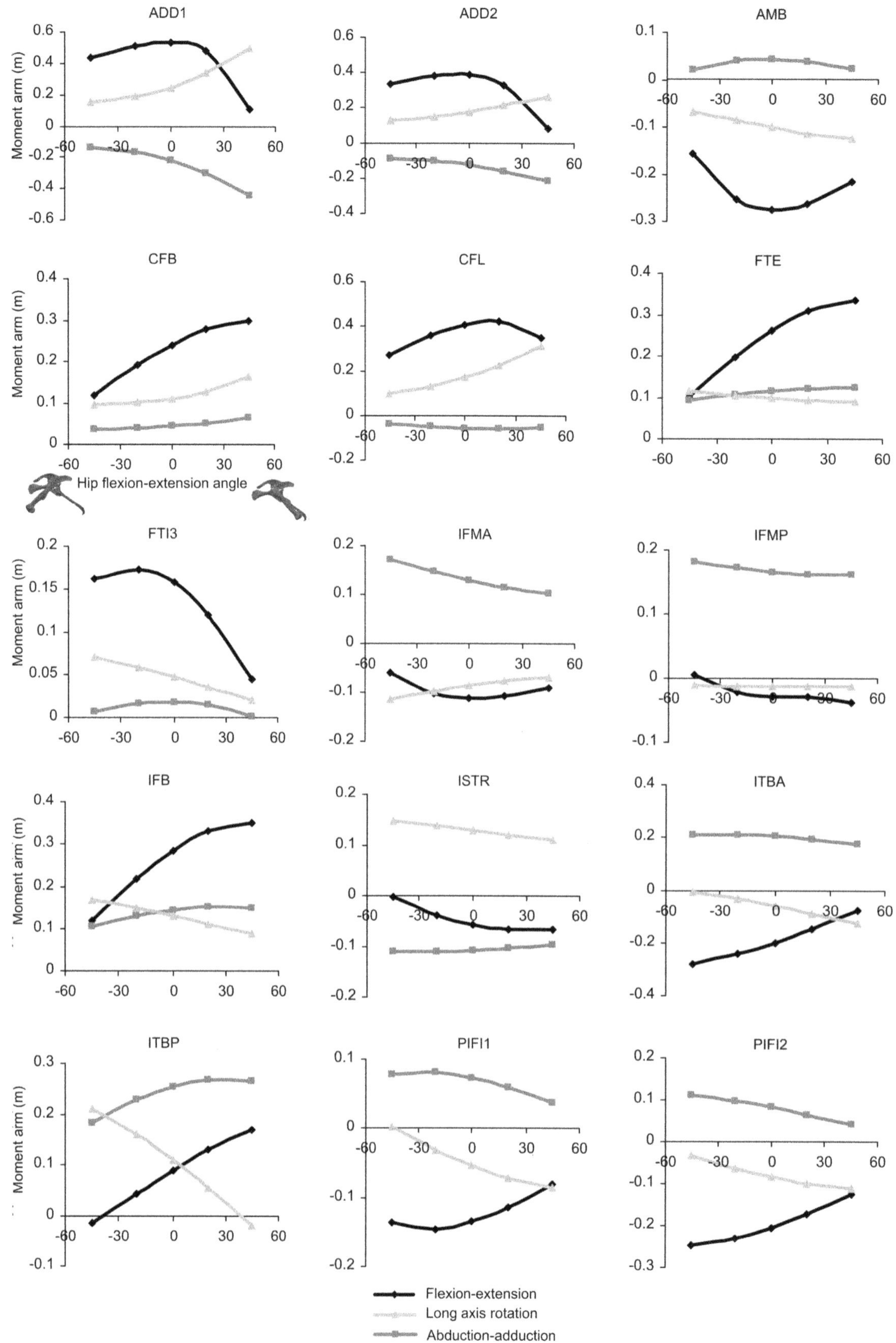

25.5. Pelvic muscle moment arms for flexion-extension, abduction-adduction, and long-axis rotation in *Edmontosaurus*. The *y* axis of each graph is the moment arm. Negative moment arms indicate flexion, adduction and medial long-axis rotation, whereas positive moment arms indicate extension, abduction, and lateral long-axis rotation. The *x* axis is the flexion-extension hip-joint angle in all graphs (as labeled for CFB). Negative values indicate hip flexion; positive values indicate hip extension, as illustrated in the graph for CFB. Abbreviations as in Table 25.2.

inserting distolaterally experienced an increase in leverage for adduction (ADD1, ADD2) or abduction (IFB, ITBp) at extended joint angles, while PIFI1, PIFI2, and IFMa, originating anteriorly and inserting proximomedially, experienced an increase in abduction leverage at flexed hip-joint angles.

When plotted against abduction-adduction angles, ADD1 and ADD2 are adductors when the limb is abducted, but switch their leverage to abduction when the limb is adducted beyond the ischium, where they originate. ITBa had the highest moment arm of any abductor (0.21 m), whereas ADD1 had the highest moment arm of any adductor (-0.44 m). Abductor muscles originating on the iliac blade (CFB, FTE, IFMa, IFMp, IFB, ITBa, ITBp) generally experienced peak moments for abduction at highly abducted hip-joint angles, and moments for abduction decreased as the limb was adducted. AMB and FTI3, which originate ventrally on the pubis and ischium respectively, experienced peak moments for abduction when the limb was adducted, and their leverage for abduction decreased sharply as the limb was abducted. PIFI1&2 maintained relatively constant leverage for abduction at all abduction-adduction hip-joint angles. ADD1&2, CFL, and ISTR were less affected by abduction-adduction of the hip than the abductors, maintaining relatively constant moments for adduction throughout the range of abduction-adduction hip-joint angles modeled.

Long-Axis Rotation Moment Arms Nine of the fifteen modeled muscles, all of which originated posterior to the hip joint, laterally rotated the femur about its long axis at all long-axis rotation and flexion-extension hip-joint angles (ADD1, ADD2, CFB, CFL, FTE, FTI3, IFB, ISTR, ITBp). The remaining six, originating anteriorly, were medial rotators of the femur (AMB, IFMa, IFMp, ITBa, PIFI1, PIFI2; Fig. 25.5). ADD1 had the highest moment arm for lateral rotation (0.50 m), while AMB had the highest moment arm for medial rotation (-0.19 m). Almost all muscles experienced peak moment arms at medially rotated femoral angles, regardless of whether they were lateral rotators (FTI3, FTE, IFB, ISTR, CFB, CFL, ITBp) or medial rotators (AMB, IFMa, PIFI1, PIFI2, ITBa). ADD1 and ADD2, however, experienced peak moments for lateral rotation at laterally rotated femoral angles. ADD1, ADD2, CFB, and CFL experienced higher moments for lateral rotation when the hip was extended, while FTI3, FTE, IFB, ISTR, and ITBp experienced higher moments for lateral rotation when the hip was flexed. All medial rotators (AMB, ITBa, PIFI1, PIFI2) except IFMa experienced peak moments at extended joint angles. The leverage for medial rotation of IFMa increased with increasing hip flexion. IFMp maintained approximately the same moment for medial rotation throughout the range of long-axis rotation and flexion-extension angles modeled.

Sensitivity Analysis

Alternative reconstructions of ADD1, ADD2, and PIFI2 were generated in order to address specific areas of uncertainty in our model (see Materials and Methods). Modeling a more proximal insertion for both ADD1 and ADD2 resulted in lower moment arms for flexion-extension, abduction-adduction, and long-axis rotation, whereas more distal insertions resulted all moments being higher (Fig. 25.6A–E). Changes to the abduction-adduction and long-axis rotation moments were more profound at extended joint angles than at flexed joint angles (maximum difference in adduction moment: ADD2 original vs. ADD2 proximal = 62.3%; maximum difference in lateral rotation moment: ADD2 original vs. ADD2 proximal = 65.8%; Fig. 25.6B, C, E, F), but peaks in moment arms were experienced at the same hip-joint angles.

An alternative origin for PIFI2 on the lateral sides of the dorsal vertebrae (i.e., shifting the origin medially) resulted in no change to flexion-extension, abduction-adduction, or long-axis rotation moment arms (Fig. 25.6G–I). In contrast, an alternative insertion of PIFI2 on the anteromedial surface of the femur resulted in significant changes to all moment arms. The moment arm for flexion is similar at extended and vertical hip-joint angles, but decreases rapidly and the leverage switches to extension at highly flexed joint angles (Fig. 25.6G). This is due to differences in the location of via points that prevent the muscle extending through the femur at highly flexed joint angles. The difference in moment arms is therefore likely to be an artifact, and in any case it occurs at hip-joint angles far more extreme than a living animal would be expected to achieve. A rapid decrease in flexion moment arms was obtained for a similar sensitivity analysis carried out on *Lesothosaurus* (Bates, Maidment, et al., 2012), and Hutchinson et al. (2008) obtained a similar result for *Velociraptor* and *Tyrannosaurus*. The alternative insertion of PIFI2 results in an overall slightly lower abduction moment arm (Fig. 25.6H). Although the peak in abduction leverage with abduction-adduction hip-joint angle occurs at the same hip-joint angle and has a similar value in both models (0.102 m for the original insertion, and 0.095 m for the alternative insertion: a difference of 6.8%), the moment arm decreases more rapidly as the limb is abducted, so that when the limb is abducted by 30 degrees from vertical the difference in moment arms is 38.6%. Furthermore, the alternative insertion leads to a slight decrease in abductor moment arms as the hip is flexed, resulting in a peak moment arm when the femur is held vertically, rather than when the hip is flexed, as was the case in the original model (Fig. 25.6H). The alternative insertion also leads to an overall increase in medial rotator moment arms, which is more profound at laterally rotated joint angles (an increase of 11% at 30 degrees of

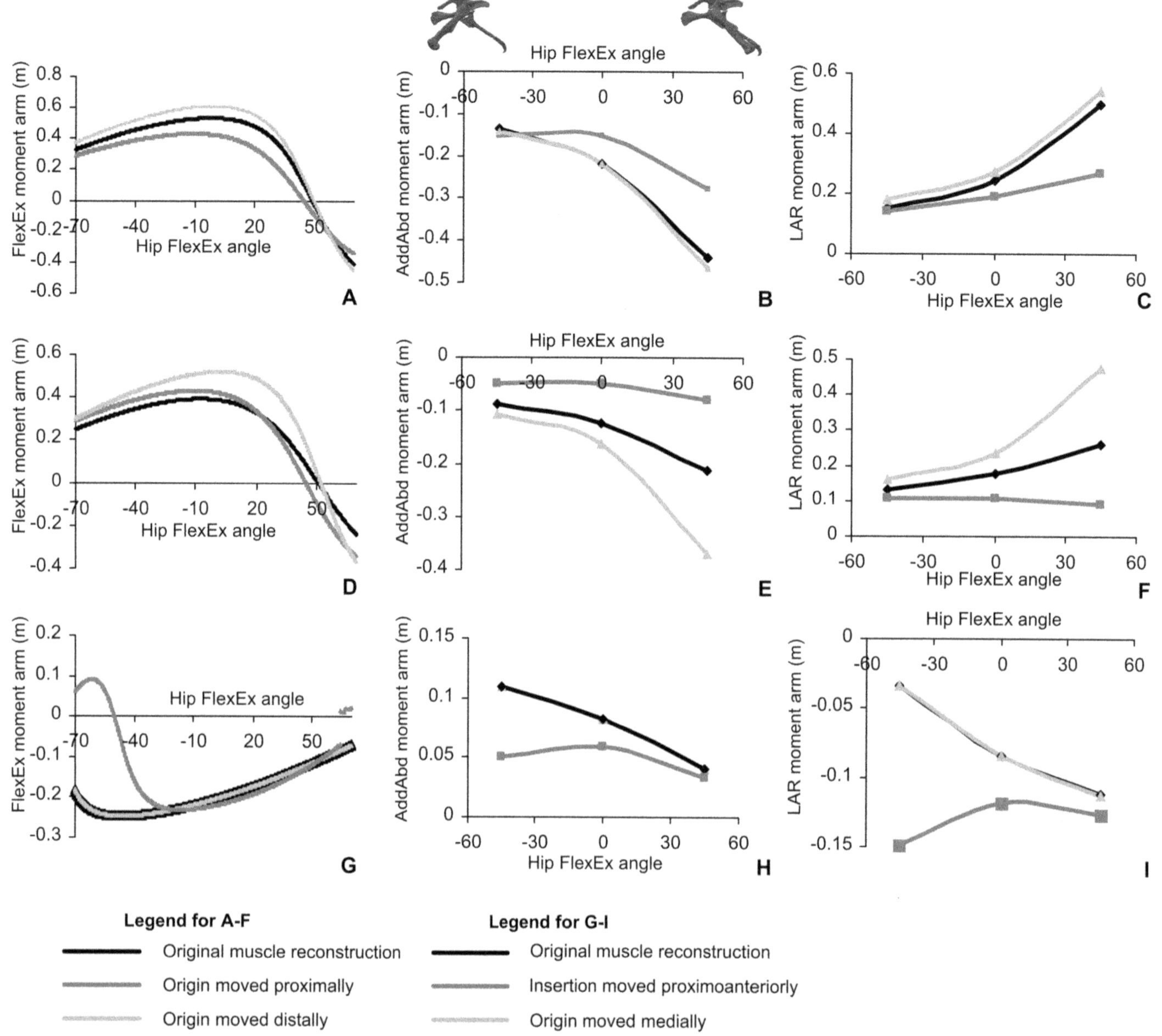

25.6. Results of the sensitivity analysis. (A–C) ADD1; (D–F) ADD2; (G–I) PIFI2. (A, D, G) Flexion-extension moment arms; (B, E, H) adduction-abduction moment arms; (C, F, I) long-axis rotation moment arms. The *y* axis of each graph is the moment arm. Negative moment arms indicate flexion, adduction and medial long-axis rotation, whereas positive moment arms indicate extension, abduction, and lateral long-axis rotation. The *x* axis is the flexion-extension hip-joint angle. Negative values indicate hip flexion; positive values indicate hip extension, as illustrated in (B). Abbreviations: muscles as in Table 25.2; AddAbd, adduction-abduction; FlexEx, flexion extension; LAR, long-axis rotation.

medial rotation; an increase of 30.1% at 30 degrees of lateral rotation). A change in peak medial rotation leverage with hip flexion and extension is also observed: the moment arm for medial rotation increases significantly at flexed joint angles, resulting in a peak moment arm when the hip is flexed rather than when the hip is extended (Fig. 25.6I).

Comparative Moment Arms

As would be expected for terrestrial tetrapods that share a close common ancestor, the majority of muscles have similar leverage in *Alligator*, *Allosaurus*, *Lesothosaurus*, *Kentrosaurus*, and *Edmontosaurus*. Small variations in the hip flexion-extension angle at which peak moments are experienced may be due to variations in modeling parameters such as the exact location of via points and the use of wrapping surfaces. Minor subjective differences in muscle reconstruction

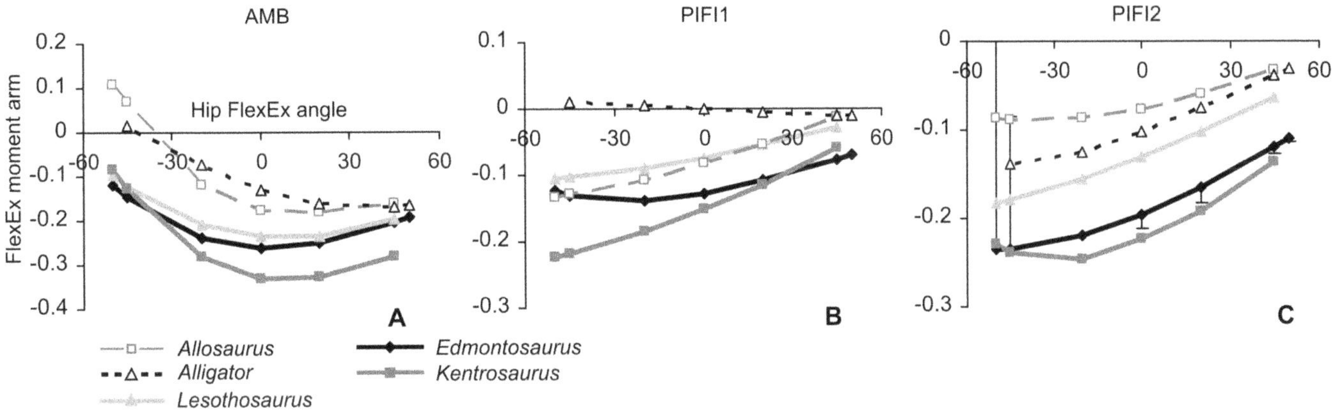

25.7. Examples of comparative flexion moment arms. In each graph, the *y* axis is the moment arm. Negative values indicate flexion, whereas positive values indicate extension. The *x* axis is hip flexion-extension angle (as labeled in [A]), with negative values indicating hip flexion and positive values indicating hip extension. (A) AMB; (B) PIFI1; (C) PIFI2. Error bars for *Edmontosaurus* in (C) represent the results of the sensitivity analysis (see Fig. 25.4G). Abbreviations: muscles as in Table 25.2; FlexEx, flexion-extension.

between workers may also lead to slight differences in the moment arms obtained (KTB produced muscle reconstructions for *Alligator* and *Allosaurus*, whereas reconstructions of *Lesothosaurus*, *Kentrosaurus*, and *Edmontosaurus* were produced by SCRM), although our previous study suggests that investigator bias is likely to have a minimal effect on the overall results obtained (Bates, Maidment, et al., 2012). We therefore discuss what we consider to be significant differences in muscle moment arms and explore the evolutionary and functional implications of these differences in the Discussion section.

Flexion-Extension Moment Arms Five muscles consistently flex the hip in all taxa: AMB, IFMa, ITBa, PIFI1, and PIFI2. *Edmontosaurus* and *Kentrosaurus* generally have higher moment arms for flexion than the other taxa (Fig. 25.7), although for ITBa *Allosaurus* has the highest moment arm when the femur is flexed, but shows strong angular dependency (change in moment arm with change in hip angle). High flexion moment arms in *Edmontosaurus* and *Kentrosaurus* are due to the elongation of the preacetabular process (the origin for PIFI1 in both taxa and PIFI2 in *Kentrosaurus*) and the prepubis (the origin for AMB), resulting in movements of the origins of these muscles anteriorly relative to the other taxa. IFMa also originates more anteriorly in *Edmontosaurus* and *Kentrosaurus* as a result of development of the supra-acetabular crest dorsal to the acetabulum. In general, *Alligator* had the lowest moment arms for flexion.

Eight muscles consistently extend the hip in all taxa: ADD1, ADD2, CFL, CFB, FTE, FTI3, IFB, and ITBp (Fig. 25.8). Extensor moment arms for ADD1&2, originating on the ischial shaft, were highest in the ornithischian taxa because the ischium of ornithischians is elongated and extended posteriorly in comparison with *Alligator* and *Allosaurus* (e.g., Fig. 25.8C). In contrast, FTI3, originating on the proximal ischium, had similar moment arms in all taxa modeled. *Edmontosaurus* had the highest moment arms for CFB (Fig. 25.8B) and IFB, as a consequence of elongation of the postacetabular process, on which they originate. *Allosaurus* had the highest moment arms for FTE and ITBp, which were reconstructed more posteriorly on the ilium than in *Edmontosaurus*. Once again, *Alligator* had generally low moment arms for extension in comparison with the other taxa (Fig. 25.8A–C).

ISTR was an extensor in *Lesothosaurus*, *Kentrosaurus*, and *Allosaurus*, but a flexor in *Edmontosaurus*. In *Alligator* the muscle leverage was dependent on hip flexion-extension angle, being an extensor when the hip was flexed but switching to become a flexor as the hip extended (Fig. 25.9A). Flexion leverage in *Edmontosaurus* is achieved at all hip flexion-extension angles because of a more dorsal insertion of ISTR on the femur than in the other taxa. IFMp had relatively weak leverage for flexion-extension in all taxa, but functioned as an extensor in *Alligator*, *Allosaurus*, and *Lesothosaurus*. In *Edmontosaurus* and *Kentrosaurus*, this muscle was a flexor, however (Fig. 25.9B). This is due to anterior movement of the origin as a consequence of development of the supra-acetabular crest dorsal to the acetabulum.

Abduction-Adduction Moment Arms ADD1, ADD2, CFL, and ISTR consistently adducted the femur in all taxa modeled. *Alligator* had the highest moment arms for adduction in all of these muscles, while the dinosaur taxa had generally similar magnitude moment arms. Slightly higher moment arms in *Alligator* for ADD1&2 are a consequence of a slightly more ventral origin, while in ISTR slightly higher moment arms result from a more ventral insertion on the femur. CFL is shorter in *Alligator* than it is in the other taxa

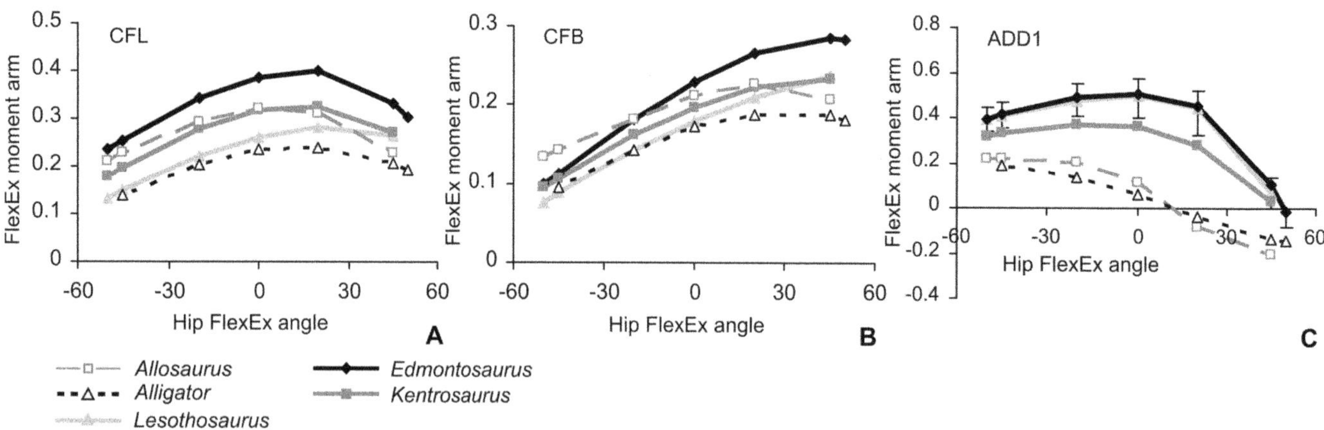

25.8. Examples of comparative extension moment arms. The *y* axis is the moment arm. Negative values indicate flexion, whereas positive values indicate extension. The *x* axis is hip flexion-extension angle, with negative values indicating hip flexion and positive values indicating hip extension. (A) CFL; (B) CFB; (C) ADD1. Error bars for *Edmontosaurus* in C represent the results of the sensitivity analysis (See Fig. 25.4A). Abbreviations: muscles as in Table 25.2; FlexEx, flexion-extension.

and is modeled originating closer to the pelvis. However, the origins of this fan-shaped muscle extend along the sides of the caudal vertebrae, and its true extent has not been fully reproduced in any of our models, partly due to the absence of caudal vertebrae from our models and/or the specimens on which they are based. In addition, the passage of the CFL from the proximal tail is highly dependent on both its size and those of the surrounding muscles. The results for CFL may therefore be somewhat subjective, despite the presence of robust osteological correlates for this muscle in most archosaurs.

AMB, FTE, IFB, IFMa, IFMp, ITBa, ITBp, and PIFI2 consistently abducted the femur in all modeled taxa. *Kentrosaurus* had consistently the highest moment arms for abduction of any of the taxa modeled (e.g., Fig. 25.10), and this results from the lateral expansion of the ilium, which moves the origins of the abductor muscles farther from a plane extending through the center of rotation for abduction and adduction. In general, *Alligator* and *Lesothosaurus* had the lowest abduction moment arms, although some variation was observed across individual muscles (Fig. 25.10).

PIFI1 was an abductor in all dinosaurian taxa, but it was an adductor in *Alligator* (Fig. 25.10A). Although this muscle originates in a similar location in all taxa, the path of the muscle extends ventral to the center of rotation for abduction and adduction in *Alligator*, but dorsal to it in all dinosaurs. CFB was an adductor in *Alligator* and *Allosaurus*, but an abductor in the ornithischians (Fig. 25.10C). This muscle originates on the ventral margin of the postacetabular process (the brevis shelf and its homologous surface), which extends farther ventrally in *Alligator* and *Allosaurus* than it does in ornithischians, meaning that the muscle extends under the center of rotation for abduction-adduction in *Allosaurus* and *Alligator*, but dorsal to it in ornithischians (Bates, Maidment, et al., 2012).

Long-Axis Rotation Moment Arms ADD2, CFB, CFL, and ISTR, consistently laterally rotated the femur in all taxa. The moment arms are quite similar, but *Alligator* and *Lesothosaurus* generally had lower moment arms for lateral rotation than the other taxa. PIFI2 is the only muscle that consistently medially rotated the femur in all taxa. In the ornithischian taxa, the peak moment arm occurred when the hip was extended; however, in *Alligator* and *Allosaurus*, the peak moments occurred during hip flexion. The sensitivity analysis in *Edmontosaurus* showed that changing the insertion of PIFI2 to a location more similar to that modeled for *Alligator* and *Allosaurus* would result in peak moments occurring during hip flexion, so it appears that changes in peak moments in different taxa result from uncertainties in muscle reconstruction. Since the insertion for PIFI1 is the same as that for PIFI2, differences between ornithischian taxa and *Allosaurus* and *Alligator* can be explained in the same way.

The remaining muscles show a wide variation in moment arms between taxa, with few consistent patterns observable. Muscles inserting on the lower limb (AMB, FTE, IFB, FTI3, ITBa, ITBp) showed a very high degree of angular dependency in *Kentrosaurus*. All of these muscles had a high moment arm for lateral rotation when the hip was flexed, but switched to a high moment arm for medial rotation when the hip was extended, because of the lateral expansion of the pelvis.

ADD1 had a similar moment arm to ADD2 and was a lateral rotator in all dinosaurs. However, at flexed hip-joint angles,

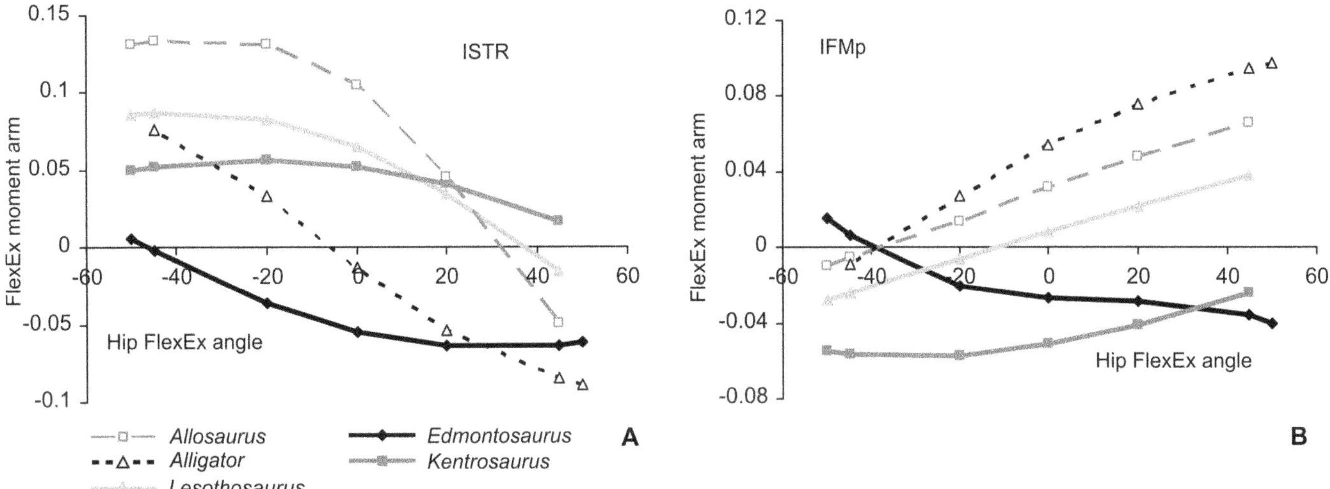

25.9. Examples of muscles with different leverage in different taxa. In each graph, the *y* axis is the moment arm. Negative values indicate flexion, whereas positive values indicate extension. The *x* axis is hip flexion-extension angle, with negative values indicating hip flexion and positive values indicating hip extension. (A) ISTR; (B) IFMp. Abbreviations: muscles as in Table 25.2; FlexEx, flexion-extension.

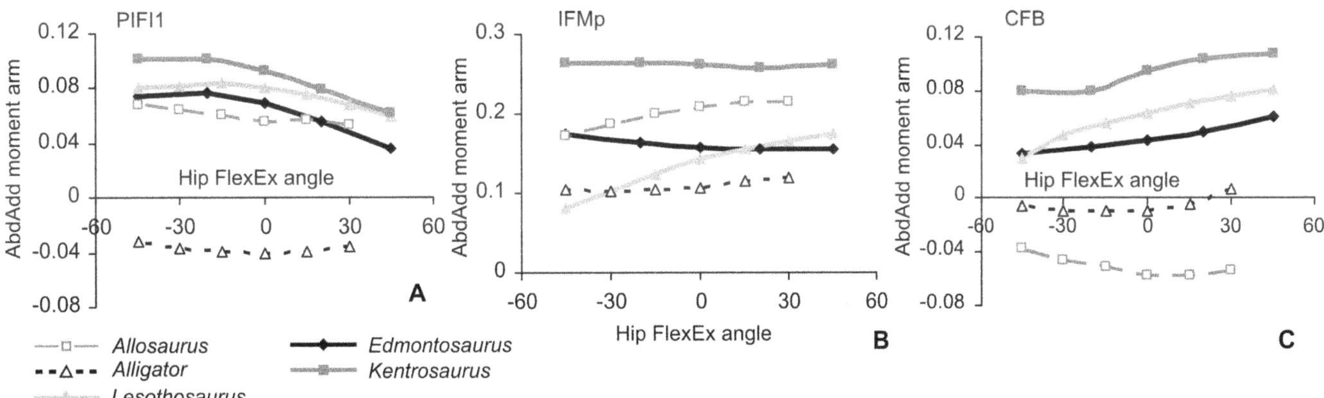

25.10. Examples of comparative abduction-adduction moment arms. In each graph, the *y* axis is the moment arm. Negative values indicate adduction, whereas positive values indicate abduction. The *x* axis is hip flexion-extension angle, with negative values indicating hip flexion and positive values indicating hip extension. (A) PIFI1: in *Alligator* this muscle is an adductor, whereas in the dinosaurs examined here it is an abductor. (B) IFMp: *Kentrosaurus* has the highest moment arm for abduction, whereas that of *Alligator* is the lowest. (C) CFB: an abductor in the ornithischians but an adductor in *Allosaurus* and *Alligator*. Abbreviations: muscles as in Table 25.2; AbdAdd, abduction-adduction; FlexEx, flexion-extension.

ADD2 medially rotated the femur in *Alligator*. This is because the muscle path extends medial to the center of long-axis rotation when the femur is flexed in *Alligator*, in contrast to the condition in dinosaurs, where the muscle path remains lateral to the center of long-axis rotation throughout all modeled hip angles. IFMa was a weak medial rotator in all taxa except *Alligator*, in which it laterally rotated the femur. This is due to its insertion, which is more distal and posterolateral than it is in the dinosaurian taxa. Finally, IFMp was a weak lateral rotator in all taxa except *Edmontosaurus*, in which it had no moment arm for long-axis rotation. This appears to be because it inserts on the center of long-axis rotation in *Edmontosaurus*, but inserts slightly anterior to it in the other taxa. Our previous study has suggested ambiguity in the function of muscles inserting very close to the center of rotation (Bates, Maidment, et al., 2012), and that result is reinforced here.

DISCUSSION

Muscle Leverage in Edmontosaurus

The potential contribution of an individual muscle to an animal's gait is not only dependent on its mechanical leverage, but also its mass, architecture, and contractile and elastic properties (e.g., contraction speed). The role of a muscle during locomotion is complicated further by dynamic variables such as the timing of contraction, length at contraction, and its interaction with other soft tissues (Bates, Maidment, et al., 2012). Our models, which elucidate the moment arm of specific muscles at specific joint angles, are unconstrained by soft tissues, and the modeled muscles do not interact with each other. Therefore these models are unable to fully explicate pelvic muscle function in *Edmontosaurus*. However, muscle leverage can be informative about what was possible or not possible for the muscles, and can be used to test hypotheses of muscle function.

Locomotor muscle function in *Edmontosaurus* has not been previously addressed. Indeed, Dilkes (2000) is the only author to have produced a muscle reconstruction for a hadrosaur, and he did not comment explicitly about muscle function. However, Romer (1927), Galton (1969), and Norman (1986) investigated muscle function in other ornithopods (*Thescelosaurus*, *Hypsilophodon*, and *Mantellisaurus*, respectively).

By comparing the moment arms for each function (flexion-extension, abduction-adduction, and long-axis rotation) in a specific muscle, we are able to determine for which of the functions that specific muscle had the greatest leverage (Fig. 25.5). ADD1&2 in *Edmontosaurus* had the greatest leverage for hip extension, as has been previously suggested (Romer, 1927; Galton, 1969). Their leverage for lateral rotation was slightly greater than their leverage for adduction. Previous authors (Romer, 1927; Galton, 1969; Norman, 1986) have suggested that the caudofemorales (CFL and CFB) were the important limb retractors, and our results support this hypothesis. The hamstrings (FTI3 and FTE) were cited as retractors and adductors by Romer (1927) and we find their highest moment arm is for retraction (hip extension), although they have relatively weak moments for adduction, and in the case of FTI3, a greater leverage for lateral rotation than adduction. IFB also has highest leverage for extension, as previously hypothesized (Romer, 1927; Galton, 1969).

The muscles of the triceps femoris (AMB, ITBa, ITBp), particularly ITBa, have previously been cited as protractors (hip flexors). Our results show that while AMB has the highest moment for flexion, ITBa has equally high leverage for flexion and abduction, and ITBp has higher leverage for abduction than it does for the other functions. IFMa and IFMp (or their avian homologues, IFE and ITC) are usually considered to be femoral protractors (hip flexors; Romer, 1927; Galton, 1969; Norman, 1986). In contrast, both IFMa and IFMp have greater leverage for abduction than for flexion in *Edmontosaurus*. Indeed, IFMp has little to no moment arm for flexion. Galton (1969) considered ISTR to be a hip extensor, whereas Romer (1927) suggested a primary role in hip stabilization. We find that ISTR has a greater moment arm for lateral rotation than for the other functions, and because of its insertion dorsal to the center of rotation for flexion-extension, it has a weak moment for flexion rather than extension.

The functions of PIFI1&2 have been controversial in the literature. Romer (1927) suggested that both muscles were protractors and adductors, but Galton (1969) and Norman (1986) suggested that PIFI1 protracted the femur while PIFI2 retracted it. PIFI2 was hypothesized to retract the femur by both Norman (1986) and Galton (1969) despite the fact that they reconstructed its insertion in different places. It was reconstructed as inserting at the top of the greater trochanter by Norman (1986) and at the top of the lesser trochanter by Galton (1969). However, both of these insertions actuated hip extension because the tops of the lesser and greater trochanters both project dorsal to the center of rotation for flexion-extension in ornithopods, resulting in femoral retraction as the muscle contracted. Recent work by Hutchinson (2001b) has demonstrated that PIFE inserted on the greater trochanter, whereas parts of the iliofemoralis musculature inserted on the lesser trochanter, and the muscle reconstruction used herein (Fig. 25.1) follows that interpretation. In our models, PIFI1&2 insert farther ventrally on the femur, and both have greater leverage for hip flexion than they do for other functions.

Functional and Evolutionary Trends and Implications

Edmontosaurus and *Kentrosaurus* had higher flexion moment arms than observed in the other taxa as a result of the elongation of the anterior pelvis (Fig. 25.7). The possession of a preacetabular process that extends farther anteriorly than the pubic peduncle is a synapomorphy of Ornithischia (e.g., Butler et al., 2008), and the trend for preacetabular process elongation continued independently in all of the quadrupedal ornithischian lineages (thyreophorans, hadrosaurids, and ceratopsids). It is possible that the selective pressures favoring elongation of the preacetabular process were related to benefits associated with increasing flexion moment arms as quadrupedalism evolved convergently in these lineages, but it is difficult to envisage why flexion would be more important for quadrupedal locomotion without additional information on muscle properties. High extensor moment arms in *Edmontosaurus* in muscles originating on the posterior pelvis (Fig. 25.8) are due to the elongation of the postacetabular process, which is a synapomorphy of Neornithischia (Butler et al., 2008).

The confounding effects of constraints on hip motion by soft tissues, unknowns regarding posture during stance (Gatesy et al., 2009), and muscle properties (Bates et al.,

2010) render it difficult to speculate on the functional implications of high extensor and flexor leverage in *Edmontosaurus*, and a number of interpretations are possible. For example, in comparison with the other taxa modeled, high flexor and extensor moment arms in *Edmontosaurus* would generate higher torques at the hip, potentially allowing *Edmontosaurus* to counter larger ground reaction forces that would be generated during absolutely higher performance activities (e.g., faster running). However, given a constant muscle fiber length, the increase in overall muscle length associated with the higher moment arm means that the potential increase in torque production occurs at the expense of joint angular velocity (Lieber and Friden, 2001; Payne et al., 2006). This form-function correlation is therefore far from straightforward. The force-length characteristics of vertebrate skeletal muscle also mean that relatively longer muscles (with higher moment arms) are typically associated with relatively reduced joint excursions (Zajac, 1992; Payne et al., 2006; Smith et al., 2007). It is therefore tempting to suggest that larger flexion and extension moment arms in *Edmontosaurus* are indicative of relatively smaller arcs of femoral excursion during habitual locomotion, but such an interpretation inherently assumes no absolute change in muscle fiber length across the taxa we modeled.

Hadrosaurids are usually reconstructed and mounted with more flexed hind limbs (i.e., a more crouched posture) than other quadrupedal ornithischians. It is possible that high flexion-extension moment arms in *Edmontosaurus* relate to increased locomotor performance in combination with a decrease in femoral excursion relating to a more flexed hindlimb and greater derivation of power and acceleration from the lower limb than seen in the other taxa modeled. These hypotheses warrant further examination.

Another feature characteristic of Neornithischia is the movement of the insertion of ISTR from a location ventral to the lesser trochanter (Maidment and Barrett, 2011) to a position on the posterior side of the greater trochanter (Fig. 25.1). This results in a change in the function of ISTR from extension in non-neornithischians and birds (Jacobson and Hollyday, 1982) to flexion in *Edmontosaurus* (Fig. 25.9A), because ISTR inserts dorsal to the center of rotation for flexion-extension, thereby protracting the femur as it contracts. We predict that ISTR would have a flexor moment arm in other neornithischians, including ceratopsians.

IFMp also changes function from extension in *Alligator*, *Allosaurus*, and *Lesothosaurus* to flexion in *Edmontosaurus* and *Kentrosaurus* (Fig. 25.9B). This is because the dorsal margin of the ilium dorsal to the acetabulum is laterally folded and thickened in both *Kentrosaurus* and *Edmontosaurus* to generate a supra-acetabular crest. The presence of this crest means that the origin of IFMp must be farther anterior on the iliac blade than it is in the other taxa modeled. The supra-acetabular crest or an equivalent lateral eversion of the dorsal margin of the ilium evolved convergently in all of the quadrupedal ornithischian lineages (Romer, 1927; Coombs, 1978), but its function is unclear. Its absence in the facultatively quadrupedal *Iguanodon bernissartensis*, which attained the size of hadrosaurids such as *Edmontosaurus* (Norman, 1980), indicates that this feature is not a correlate of large body size (Romer, 1927), and it may be associated with obligate quadrupedality. *Iguanodon bernissartensis* is considered to be facultatively quadrupedal rather than an obligate quadruped herein. While *I. bernissartensis* possesses several features of the manus indicative of weight bearing, such as hoof-like ungual phalanges (Norman, 1980), it lacks some morphological features of the forelimb that are present in hadrosaurids that we consider indicative of obligate quadrupedalism in the clade. The manal phalanges of *I. bernissartensis* have concave facets proximally and convex facets distally, indicating that flexion of the digits was possible (Norman, 1980:fig. 80), unlike in hadrosaurids, where the phalanges are flat proximally and distally making digital flexion impossible. Trackway evidence suggests the manus was placed lateral to the pes during quadrupedal locomotion and supinated, allowing it to be used for grasping (Wright, 1999), whereas the manus of hadrosaurids was placed anterior to the pes during quadrupedal locomotion and pronated (Lockley and Wright, 2001), preventing grasping of objects between the hands. Supination of the manus in *I. bernissartensis* suggests a grasping function of the manus was still possible. Finally, the possession of the spike-like pollex in *I. bernissartensis*, which is absent in hadrosaurids, suggests a function of the manus other than weight bearing (Norman, 1980). The position of the supra-acetabular crest dorsal to the acetabulum suggests that it may have provided a buttress to the femoral head, and in concert with connective tissues, may have constrained the range of motion of the femur. It is worth noting that a similar anterior movement of the iliofemoralis group musculature also occurs convergently in birds, associated with the development of the elongate preacetabular process. In birds, ITC has a high medial rotator moment arm (Bates and Schachner, 2012; Bates, Maidment, et al., 2012), which is thought to contribute to the rotation-based mode of lateral limb support of extant avians (Hutchinson and Gatesy, 2000), but an increase in medial rotation is not observed in *Edmontosaurus* because the muscle inserts on the center of long-axis rotation.

In general, *Alligator* has the lowest flexion-extension moment arms (Figs. 25.7, 25.8). Dinosaurs have a parasagittal gait that emphasizes flexion and extension over abduction, adduction, and long-axis rotation of the femur. In contrast, *Alligator* has a sprawling gait, in which transverse and rotational movements of the limbs are utilized far more during locomotion. Given these different habitual strategies it is not surprising that the dinosaur taxa in our study have larger flexion-extension moment arms than *Alligator*.

Kentrosaurus had the highest abduction moment arms as a result of lateral enlargement of the ilium (Fig. 25.10), but in general, the dinosaur taxa had higher abduction moment arms and lower adduction moment arms than *Alligator*. Hutchinson and Gatesy (2000) reasoned that dinosaur taxa with upright, parasagittal gaits utilized abduction to counter adductor moments generated about the hip by the ground reaction force during locomotion, while sprawling taxa required adductors to counter abductive moments at the hip generated because the femur is held horizontally. Similarly to Bates and Schachner (2012), our results demonstrate a quantitative difference in lateral limb support between sprawling and upright taxa and strongly support the hypothesis of Hutchinson and Gatesy (2000).

PIFI1 acted as an abductor in dinosaurs, but as an adductor in *Alligator* (Fig. 25.10A), and this change in function is due to a fundamental difference in the anatomy of the hip in *Alligator* and dinosaurs. All dinosaurs possess an in-turned, ball-shaped femoral head; and in our models, the center of rotation is placed towards the medial-most margin of the ball. In *Alligator*, the femoral head is not in-turned, so the center of rotation lies farther lateral with respect to the acetabulum. Because the center of rotation for abduction and adduction is more medially placed in the dinosaurs, PIFI1 extends dorsal to it, resulting in a moment for abduction. In contrast, the more laterally placed center of rotation in *Alligator* results in the muscle passing ventral to it, and an adduction moment being generated. The more laterally placed center of rotation is also the cause for the switch in function of ADD1 from lateral to medial long-axis rotation at flexed hip-joint angles in *Alligator*. This switch in function is not observed in the dinosaurs because of their more medially placed center of rotation, which allows the path ADD1 to extend lateral to the center of long-axis rotation throughout all modeled hip angles.

CONCLUSIONS

The development of an upright, parasagittal gait, and the morphological changes associated with it–such as an in-turned, ball-shaped femoral head that fitted snugly in the acetabulum–are characteristic of dinosaurs (Hutchinson, 2001b). These changes had clear effects on the moment arms of locomotor muscles and are exemplified by several trends in our study. The increased emphasis on hip flexion-extension over abduction and adduction related to the parasagittal gait is exemplified by the much lower moment arms for most flexors and extensors experienced by *Alligator* in comparison with the dinosaurs in the study. *Alligator* generally had higher moment arms for adduction, while the dinosaurs experienced higher muscular leverage for abduction. PIFI1 was an adductor in *Alligator*, while the same muscle was an abductor in the dinosaurs. This difference in emphasis of abduction in dinosaurs compared with adduction in *Alligator* is reflective of the need to control different moments generated about the hip by the ground reaction force. In *Alligator*, the femur is held horizontally and abduction moments are generated by the ground reaction force, whereas in dinosaurs, which hold their femora vertically, the ground reaction force generates adductor moments about the hip (Hutchinson and Gatesy, 2000).

The change from bipedality to quadrupedality in ornithischian dinosaurs might be expected to have produced further pronounced changes to pelvic locomotor musculature, and the results show some changes in moment arms unique to *Kentrosaurus* and *Edmontosaurus* that might be correlated with their quadrupedal habits. Flexor moment arms are higher in *Kentrosaurus* and *Edmontosaurus* than they are in the other taxa modeled, and this is related to the elongate preacetabular process of the ilium and the enlarged prepubis in these taxa. The functional implications of increased flexion moment arms for quadrupeds are not immediately clear. While various hypotheses can be proposed based on very general mechanical relationships (see Discussion, above), it is clear that the functional implications of these findings will be best illuminated by a better quantitative understanding of moment arms and their variation in extant bipedal and quadrupedal animals. However, there are reasons to think that modern quadrupeds are not good analogs for quadrupedal ornithischians. Quadrupedal reptiles (e.g., crocodilians, lizards) tend to have a sprawling gait, rather than the parasagittal gait of ornithischians, and quadrupedal mammals lack a muscular tail and possess a mobile scapula, enabling them to increase forelimb stride length (Eaton, 1944).

Edmontosaurus and *Kentrosaurus* both possess a supra-acetabular crest, which has resulted in the anterior movement of the iliofemoralis musculature and increased flexion moments generated by this muscle in comparison with other dinosaurs. Again, the functional implications of the supra-acetabular crest are not immediately obvious; it does not seem to be correlated with large body size, because other large-bodied ornithischians lack this feature (see Discussion). A larger sample size of bipedal and quadrupedal ornithischians and more detailed modeling of musculature arising from the dorsolateral margin of the ilium may help to elucidate the function of the supra-acetabular crest and its relationships with quadrupedalism in ornithischians.

Finally, the results indicate some features unique to *Edmontosaurus* among the taxa modeled. High extensor moments are generated by the development of an elongate postacetabular process, which may indicate that hadrosaurids had increased their locomotor performance but decreased their femoral excursion relative to other taxa modeled. Interpretations of this nature are complicated, however, by considerations of the unknown muscle architecture, soft tissue constraints and stance phase limb posture in dinosaurs (see Discussion). The preacetabular process, although lengthened relative to basal ornithischians, is not as elongate in hadrosaurids as it is in ceratopsids and thyreophorans (SCRM and PMB, unpubl. data) and it may be that *Edmontosaurus* increased flexion moments by the addition of an extra flexor, rather than extreme elongation of the preacetabular process. Again we stress that these hypotheses should be investigated further by increasing the sample size and diversity of taxa modeled, by comparison with living quadrupeds, and by developing a clearer understanding of how moment arms relate to function in living taxa.

ACKNOWLEDGMENTS

The following people allowed access to specimens in their care and provided hospitality during data collection: D. Evans, K. Seymour, and B. Iwama (ROM), M. Currie, K. Shepherd, A. Macdonald, and C. Kennedy (CMN), and D. Henderson (TMP). D. Evans (ROM) identified *Edmontosaurus* and other hadrosaurian material in the Sternberg Collection of the NHMUK, and we thank him for useful discussion. D. Evans (ROM) and D. Dilkes (University of Wisconsin Oshkosh) provided insightful reviews that improved an earlier version of this manuscript. We also thank W. Sellers (University of Manchester) for help and discussion using GaitSym, which is freely available from his website, www.animalsimulation.org. SCRM is funded by a Natural Environment Research Council Standard Grant number NE/G001898/1 awarded to PMB.

LITERATURE CITED

Alexander, R. M. 2003. Principles of Animal Locomotion. Princeton University Press, Princeton, New Jersey, 371 pp.

Bates, K. T., and E. R. Schachner. 2012. Disparity and convergence in bipedal archosaur locomotion. Journal of the Royal Society Interface 9:1339–1353.

Bates, K. T., R. B. J. Benson, and P. L. Falkingham. 2012. A computational analysis of locomotor anatomy and body mass evolution in Allosauroidea (Dinosauria: Theropoda). Paleobiology 38:486–507.

Bates, K. T., S. C. R. Maidment, V. Allen, and P. M. Barrett. 2012. Computational modeling of locomotor muscle moment arms in the basal dinosaur *Lesothosaurus diagnosticus:* assessing convergence between birds and basal dinosaurs. Journal of Anatomy 220:212–232.

Bates, K. T., P. L. Manning, L. Margetts, and W. I. Sellers. 2010. Sensitivity analysis in evolutionary robotic simulations of bipedal dinosaur running. Journal of Vertebrate Paleontology 30:458–466.

Bryant, H. N., and A. P. Russell. 1992. The role of phylogenetic analysis in the inference of unpreserved attributes of extinct taxa. Philosophical Transactions of the Royal Society B 337:405–418.

Butler, R. J., P. Upchurch, and D. B. Norman. 2008. The phylogeny of the ornithischian dinosaurs. Journal of Systematic Palaeontology 6:1–40.

Carrano, M. T. 2001. Implications of limb bone scaling, curvature and eccentricity in mammals and non-avian dinosaurs. Journal of Zoology 254:41–55.

Colbert, E. H. 1951. Environment and adaptations of certain dinosaurs. Biological Reviews 26:265–284.

Coombs, W. P., Jr. 1978. The families of the ornithischian dinosaur order Ankylosauria. Palaeontology 21:143–170.

Cope, E. D. 1883. On the characters of the skull in the Hadrosauridae. Proceedings of the Academy of Natural Sciences of Philadelphia 35:97–107.

Dilkes, D. W. 2000. Appendicular myology of the hadrosaurian dinosaur *Maiasaura peeblesorum* from the Late Cretaceous (Campanian) of Montana. Transactions of the Royal Society of Edinburgh, Earth Sciences 90:87–125.

Dilkes, D. W. 2001. An ontogenetic perspective on locomotion in the Late Cretaceous dinosaur *Maiasaura peeblesorum* (Ornithischia: Hadrosauridae). Canadian Journal of Earth Sciences 38:1205–1227.

Eaton, T. H. 1944. Modifications of the shoulder girdle related to reach and stride in mammals. Journal of Morphology 75:167–171.

Galton, P. M. 1969. The pelvic musculature of the dinosaur *Hypsilophodon* (Reptilia: Ornithischia). Postilla 131:1–64.

Galton, P. M. 1970. The posture of hadrosaurian dinosaurs. Journal of Paleontology 44:464–473.

Gatesy, S. M., M. Baker, and J. R. Hutchinson. 2009. Constraint-based exclusion of limb poses for reconstructing dinosaur locomotion. Journal of Vertebrate Paleontology 29:535–544.

George, J. C., and A. J. Berger. 1966. Avian Myology. Academic Press, New York, 500 pp.

Godefroit, P., Y. L. Bolotsky, and I. Y. Bolotsky. 2012. Osteology and relationships of *Olorotitan arharensis,* a hollow-crested hadrosaurid dinosaur from the latest Cretaceous of far eastern Russia. Acta Palaeontologica Polonica 57:527–560.

Horner, J. R., D. B. Weishampel, and C. A. Forster. 2004. Hadrosauridae; pp. 438–463 in D. B. Weishampel, P. Dodson, and H. Osmólska (eds.), The Dinosauria, Second Edition. University of California Press, Berkeley, California.

Hutchinson, J. R. 2001a. The evolution of pelvic osteology and soft tissues on the line to extant birds (Neornithes). Zoological Journal of the Linnean Society 131:123–168.

Hutchinson, J. R. 2001b. The evolution of femoral osteology and soft tissues on the line to extant birds (Neornithes). Zoological Journal of the Linnean Society 131:169–197.

Hutchinson, J. R., and S. M. Gatesy. 2000. Adductors, abductors, and the evolution of archosaur locomotion. Paleobiology 26:734–751.

Hutchinson, J. R., F. C. Anderson, S. S. Blemker, and S. L. Delp. 2005. Analysis of hindlimb muscle moment arms in *Tyrannosaurus rex* using a three-dimensional musculoskeletal computer model: implications for stance, gait and speed. Paleobiology 31:676–701.

Hutchinson, J. R., C. E. Miller, G. Fritsch, and T. Hildebrand. 2008. The anatomical foundation for multidisciplinary studies of animal limb function: examples from dinosaur and elephant limb imaging studies; pp. 23–38 in R. Frey and H. Endo (eds.), Anatomical Imaging Techniques: Towards a New Morphology. Springer-Verlag, Berlin.

Jacobson, R. D., and M. Hollyday. 1982. A behavioral and electromyographic study of walking in the chick. Journal of Neurophysiology 48:238–256.

Lambe, L. M. 1914. On *Gryposaurus notabilis,* a new genus and species of trachodont dinosaur from the Belly River Formation of Alberta, with a description of the skull of *Chasmosaurus belli.* Ottawa Naturalist 27:145–154.

Leidy, J. 1858. *Hadrosaurus foulkii,* a new saurian from the Cretaceous of New Jersey. Proceedings of the Academy of Natural Sciences of Philadelphia 1858:215–218.

Lieber, R. L., and J. Friden. 2001. Clinical significance of skeletal muscle architecture. Clinical Orthopedics and Related Research 383:40–151.

Lockley, M. G., and J. L. Wright. 2001. Trackways of large quadrupedal ornithopods from the Cretaceous: a review; pp. 428–442 in D. H. Tanke and K. Carpenter (eds.), Mesozoic Vertebrate Life. Indiana University Press, Bloomington, Indiana.

Lull, R. S., and N. E. Wright. 1942. Hadrosaurian Dinosaurs of North America. Geological Society of America Special Papers 40. 242 pp.

Maidment, S. C. R., and P. M. Barrett. 2011. The locomotor musculature of basal ornithischian dinosaurs. Journal of Vertebrate Paleontology 31:1265–1291.

Maryańska, T., and H. Osmólska. 1984. Postcranial anatomy of *Saurolophus angustirostris* with comments on other hadrosaurs. Palaeontologia Polonica 46:119–141.

Nesbitt, S. J., C. A. Sidor, R. B. Irmis, K. D. Angelczyk, R. M. H. Smith, and L. A. Tsuiji. 2010. Ecologically distinct dinosaurian sister group shows early diversification of Ornithodira. Nature 464:95–98.

Norman, D. B. 1980. On the Ornithischian Dinosaur *Iguanodon bernissartensis* from the Lower Cretaceous of Bernissart (Belgium). Mémoire de l'Institut royal des Sciences naturelles de Belgique 178.103 pp.

Norman, D. B. 1986. On the anatomy of *Iguanodon atherfieldensis* (Ornithischia: Ornithopoda). Bulletin de l'Institut royal des Sciences naturelles de Belgique: Sciences de la Terre 56:281–372.

Ostrom, J. H. 1964. A reconsideration of the paleoecology of hadrosaurian dinosaurs. American Journal of Science 262:975–997.

Payne, R. C., R. H. Crompton, K. Isler, R. Savage, E. E. Vereecke, M. M. Günther, S. K. S. Thorpe, and K. D'Août. 2006. Morphological analysis of the hindlimb in apes and humans, II: Moment arms. Journal of Anatomy 208:725–742.

Prieto-Márquez, A. 2010. Global phylogeny of Hadrosauridae (Dinosauria: Ornithopoda) using parsimony and Bayesian methods. Zoological Journal of the Linnean Society 159:435–503.

Romer, A. S. 1923. Crocodilian pelvic muscles and their avian and reptilian homologues. Bulletin of the American Museum of Natural History 48:533–552.

Romer, A. S. 1927. The pelvic musculature of ornithischian dinosaurs. Acta Zoologica 8:225–275.

Sellers, W. I., L. Dennis, and R. H. Crompton. 2003. Predicting the metabolic energy costs of bipedalism using evolutionary robotics. Journal of Experimental Biology 206:1437–1448.

Sellers, W. I., P. L. Manning, T. Lyson, K. A. Stevens, and L. Margetts. 2009. Virtual palaeontology: gait reconstruction of extinct vertebrates using high performance computing. Palaeontologia Electronica 12.3.11A.

Smith, N. C., A. M. Wilson, K. J. Jespers, and R. C. Payne. 2006. Muscle architecture and functional anatomy of the pelvic limb of the ostrich (*Struthio camelus*). Journal of Anatomy 209:765–779.

Smith, N. C., R. C Payne, K. J. Jespers, and A. M. Wilson. 2007. Muscle moment arms of pelvic limb muscles of the ostrich (*Struthio camelus*). Journal of Anatomy 211:313–324.

Witmer, L. M. 1995. The Extant Phylogenetic Bracket and the importance of reconstructing soft tissues in fossils; pp. 19–33 in J. J. Thomason (ed.), Functional Morphology in Vertebrate Palaeontology. Cambridge University Press, Cambridge, U.K.

Wright, J. L. 1999. Ichnological evidence for the use of the forelimb in iguanodontid locomotion. Special Papers in Palaeontology 60:209–219.

Zajac, F. E. 1992. How muscle-tendon architecture and joint geometry affect the capacity of muscles to move and exert force on objects: a review with application to arm and forearm tendon transfer design. Journal of Hand Surgery 17:799–804.

Duckbills on the Run: The Cursorial Abilities of Hadrosaurs and Implications for Tyrannosaur-Avoidance Strategies

26

W. Scott Persons IV and Philip J. Currie

ABSTRACT

Hadrosaurs, or duckbilled dinosaurs, were a speciose and highly successful group of herbivorous dinosaurs. Unlike contemporary herbivorous dinosaurs, hadrosaurs have long appeared puzzlingly devoid of obvious predator defenses. An assessment of the primary locomotive muscle masses and arrangements in hadrosaurs and tyrannosaurs is based on new qualitative considerations of skeletal anatomy and quantitative digital restoration techniques. It suggests that, under the right circumstances, adult hadrosaurs could have been successful at outrunning potential predators. The locomotive musculature of tyrannosaurs was well adapted for sprinting but poorly adapted for endurance. In contrast, hadrosaurs had femoral retractor muscles that, while comparable in relative mass to those of tyrannosaurs, were well adapted for slower and sustained locomotion.

THE MYSTERY OF SUCCESSFUL DUCKBILLS

Hadrosaurs are a group of large herbivorous ornithopod dinosaurs with characteristic broad premaxilla/predentary beaks that give the group its colloquial title: "duckbilled dinosaurs" or simply "duckbills." During the Late Cretaceous, hadrosaurs flourished throughout the northern hemisphere despite competitive coexistence with many other dinosaurian herbivores, including horned ceratopsians, armored ankylosaurs, enormous titanosaurs, and a variety of smaller and, presumably, more agile ornithopod herbivores (Brett-Surman, 1979; Forster, 1997). In North America, hadrosaur fossils easily rank among the most abundant of all Late Cretaceous dinosaurs (Horner et al., 2004). This ubiquity has earned duckbills the additional nickname, "cows of the Cretaceous."

Dinosaur paleontology has recognized many of the adaptations that were likely key to hadrosaur success. Foremost among these were the densely packed rows of teeth, or dental batteries, which formed large and complex grinding surfaces that rivaled the sophisticated chewing apparatuses of modern artiodactyl grazers and gave hadrosaurs an advantage in processing vegetation (Galton, 1973; Bakker, 1978). The discoveries of mass-death bonebed assemblages (Varricchio and Horner, 1993; Eberth et al., this volume), unidirectional trackways (Lockley et al., 1983; Currie, 1983), and fossil nests (Horner, 2000) suggest that hadrosaurs were socially gregarious and efficient reproducers with large clutch sizes. Insights into cerebral anatomy confirm that hadrosaurs had well-developed sight and hearing (Hopson, 1975), and large encephalization quotients compared to those of other herbivorous dinosaurs (Evans, 2005). However, one major aspect of hadrosaur success has remained unexplained. Roughly concurrent with the Late Cretaceous rise of hadrosaurids was that of tyrannosaurids, which came to dominate the niche of large terrestrial carnivores across the Laurasian landmass. How, in a world filled with some of the largest and most advanced of all carnivorous dinosaurs (Holtz, 2004), did duckbills overcome the challenge of predation?

Ecological theory predicts that, as terrestrial animals increase in size, the topography of their environments becomes increasingly irrelevant and their world becomes functionally more two dimensional (Thompson, 1961; Bakker, 1980; Levin, 1992; Kaspari and Weiser, 1999). Larger animals face fewer insurmountable barriers, but can squeeze into fewer crevices and shelters, and have a reduced potential to climb and burrow. As a result, predator defenses that are based on refuge, concealment, and crypsis lose practicality. It is, therefore, an ecological requirement that large herbivores evolve adaptive strategies for dealing with direct predator confrontations, although these strategies need not necessarily be violent (Estes, 1974; Lovegrove, 2001).

At no point in the history of life was there a richer ecology of large terrestrial herbivores than during the age of dinosaurs. Modern megafaunas offer potential insight into the anti-predator strategies that may have been employed by large herbivorous dinosaurs. Focusing on the most megafauna-rich environment of the modern world, the great savannas of Africa, a number of distinct strategies can be observed (although few are mutually exclusive and there are gradients between each). For example, modern elephants employ gigantism, a strategy that takes advantage of the greater abundance of fodder that is available to herbivores and allows them to grow to a size that dauntingly exceeds that of potential predators (Owen-Smith, 1988; Feldhamer et al., 2007:372).

Rhinoceros and Cape buffalo can be classified as weapon strategists. They are armed with horns and wield them with formidable force and reflexes, such that would-be assailants must risk death or mutual injury (Goddard, 1966; Berger and Cunningham, 1996; Prins, 1996). Ground pangolins and the crested porcupine are fortresses, having evolved protective coverings of armored scales or sharp quills (Corsini et al., 1995; Feldhamer et al., 2007:310). Antelope exemplify a cursorial strategy. They are long legged and agile sprinters, capable of fleeing even Africa's swift feline predators (Estes, 1991).

It is easy to recognize potential anatomical evidence for all these strategies in various dinosaur groups. However, none of these strategies match well with the morphology of hadrosaurs. The osteology of duckbills provides no evidence of osteological armor or weapons, such as horns or spikes, and hadrosaur fossil skin impressions and so-called "mummy" specimens (cf. Manning et al., this volume) confirm that hadrosaurs had no prominent integumentary defenses (Osborn, 1911). While some forms may have been protected in adulthood by their enormous bulk (Hu, 1973; Morris, 1981), most adult hadrosaurs were smaller than–or equal in size to–the tyrannosaurs that shared their environments. Previous examinations of the relative cursoriality of hadrosaurs and tyrannosaurs have also failed to offer an explanation (Lockley et al., 1983; Gatesy, 1991; Holtz, 1995; Hutchinson, 2004; Paul, 2008). Tyrannosaurs appear to have been faster runners, with relatively longer tibiae and metatarsals and, therefore, longer strides with biomechanical advantages for speed. Additionally, hadrosaurs lacked the advanced arctometatarsus of tyrannosaurs, which has been interpreted as a critical cursorial adaptation (Holtz, 1995). Indeed, duckbilled dinosaurs seem so devoid of traditional predator defenses that some paleontologists writing for a popular audience have felt compelled to propose novel defenses, for which fossil evidence could not be expected (such as the use of infrasound trumpeting to disorient and deter predators [Bakker, 2010]).

This chapter considers the issue of hadrosaur defense in the context of the relative locomotory abilities of predator (large tyrannosaurids) and prey (hadrosaurids). Specifically, we examine hadrosaurid running capabilities as a defense strategy. We move beyond skeletal anatomy and consider the properties of the primary hindlimb retractor muscle set–the M. caudofemoralis–in both hadrosaurids and tyrannosaurids. In modern crocodilians and many other non-avian sauropsids, the M. caudofemoralis is literally the driving force behind terrestrial locomotion. It is the largest femoral retractor muscle and has been shown to be the major contributor, in all gaits, to the power stroke of the hindlimbs (Snyder, 1949, 1962, 1967; Gatesy, 1990, 1997). Based on the prominence of the femoral fourth trochanter, a caudofemoral-dominated locomotive system has been inferred for most non-avian dinosaurs, including ornithopods (Dollo, 1883; Romer, 1927; Gatesy, 1990). The size of the M. caudofemoralis is thought to be the functional reason that most dinosaurs have large, robust tails (Persons and Currie, 2011a). Recent biomechanical studies have established the critical importance of the M. caudofemoralis in assessing the running abilities of dinosaurs and how its properties can be inferred from fossil material (Allen et al., 2009; Arbour, 2009; Mallison, 2011; Persons and Currie, 2011a, 2011b).

Hadrosaur caudofemoral and other tail musculature are here assessed using two exceptional fossil specimens: an adult specimen of *Lambeosaurus lambei* (TMP 1982.038.0001); and a subadult specimen (TMP 1998.058.0001) with a fully articulated caudal series. Musculature assessments can be made both qualitatively, through consideration of gross anatomy, and quantitatively, using osteological muscle insertion correlates and new innovative digital muscle modeling techniques (Persons and Currie, 2011a, 2011b). These methods have been shown to be accurate muscle-mass predictors in a variety of modern reptiles (Persons and Currie, 2011a). Previous work on tyrannosaurs (Persons and Currie, 2011a) provides a basis of comparison between the locomotive potential of hadrosaurs and that of their presumed primary predators.

Institutional Abbreviations BHI, Black Hills Institute of Geological Research, Hill City, South Dakota; TMP, Royal Tyrrell Museum of Palaeontology, Drumheller, Alberta.

MATERIALS, METHODS, AND QUANTITATIVE RESULTS

Detailed measurements were made on the caudal and pelvic skeletons of two hadrosaur specimens: TMP 1982.038.0001–an adult specimen of *Lambeosaurus lambei*, and TMP 1998.058.0001–a subadult specimen of an unknown genus. Both specimens are from the Dinosaur Park Formation (76.5–75.0 Ma) of southern Alberta. The digital caudal muscle reconstruction method outlined in Persons and Currie (2011a) and subsequently refined in Persons and Currie (2011b) was applied to TMP 1982.038.0001 and TMP 1998.058.0001. All digital models were created using the software Rhinoceros® (McNeel and Associates, 2007). First, digital skeletons were created based on the skeletal measurements. Then, musculature was digitally sculpted over the digital skeletons according to the methods and muscular insertion patterns observed in dissections of modern sauropsids (Persons and Currie, 2011a).

Dinosaur caudal musculature can be subdivided into four major sets: M. spinalis, M. longissimus, M. ilio-ischiocaudalis, and M. caudofemoralis. Throughout the caudal series, the M. spinalis inserts onto the tips and lateral surfaces of the

Table 26.1. Hadrosaur and Tyrannosaur Tail Muscle Mass Estimations

Hadrosaurs	M. spinalis	M. longissimus	M. ilio-ischiocaudalis	M. caudofemoralis
Lambeosaurus lambei **TMP 1982.038.0001**	32000 g	53000 g	91000 g	114000 g
Total tail muscle mass: 290000 g	11.0%	18.3%	31.4%	39.3%
Total body mass: 2500000 g	1.3%	2.1%	3.6%	4.6%
Hadrosaur TMP 1998.058.0001 (subadult)	2200 g	5300 g	8700 g	10400 g
Total tail muscle mass: 26600 g	8.3%	9.4%	32.7%	39.1%
Total body mass: 183000 g	1.2%	2.9%	4.8%	5.6%

Tyrannosaurs	M. spinalis	M. longissimus	M. ilio-ischiocaudalis	M. caudofemoralis
Tyrannosaurus rex **BHI 3033**	65200 g	154200 g	159400 g	522200 g
Total tail muscle mass: 901000 g	7.2%	17.1%	17.7%	58.0%
Total body mass: 5622000 g	1.2%	2.7%	2.8%	9.3%
Gorgosaurus libratus **TMP 1991.036.0500 (subadult)**	3900 g	6900 g	10300 g	17300 g
Total tail muscle mass: 38300 g	10.2%	18.0%	26.9%	45.2%
Total body mass: 400000 g	1.0%	1.7%	2.6%	4.3%

Note: Tyrannosaur data taken from Persons and Currie (2011a). Total body mass estimations taken from Paul (2010) with mass estimation of similar specimens used and scaled according to femur size when necessary. Results indicate a tyrannosaur caudofemoral/body mass range of 4.3–9.3% for a gracile *Gorgosaurus* subadult and robust *Tyrannosaurus* adult. Both hadrosaur caudofemoral/body mass estimations fall within this range.

neural spines. Prior to the termination of the caudal ribs at what is termed the "caudal transition point," the M. longissimus inserts onto the lateral surfaces of the neural arches and the dorsal surfaces of the caudal ribs. After the termination of the caudal ribs, the M. longissimus inserts only onto the lateral surfaces of the neural arches. The M. spinalis and M. longissimus were reconstructed together by assuming both were bound medially by each vertebral neural arch and neural spine, ventrally by each caudal rib or (when vertebrae lacked caudal ribs) by the base of the neural arch, and dorsolaterally by an arc swung from the dorsal tip of each neural spine to a point level with the dorsal surface of each caudal rib or each neural arch base. Later, the M. spinalis and M. longissimus were divided by assuming the septa that separated the two originated and extended in line with the dorsolateral edges of the prezygapophyses.

The M. caudofemoralis is subdivided into the m. caudofemoralis brevis and the m. caudofemoralis longus. The m. caudofemoralis brevis fills the brevis fossa of the ilium, whereas the m. caudofemoralis longus extends posteriorly across a series of anterior chevrons and vertebral centra, before typically tapering and ultimately terminating at the caudal transition point. Anterior to the caudal transition point, the m. caudofemoralis longus was reconstructed by extending an arc in dorsoventral cross-section from the ventrolateral surface of the caudal ribs to the ventral tip of the chevrons. At the caudal transition point, the m. caudofemoralis longus was reconstructed by assuming the muscle tapered smoothly and did not extend beyond the final caudal rib set. The mass and form of the m. caudofemoralis brevis could be estimated directly from measurements and volume calculations of the brevis fossa and was not digitally reconstructed.

The M. ilio-ischiocaudalis extends from the ilium and ischium, and persists to the tip of the tail. The M. ilio-ischiocaudalis attaches to the ventral surfaces of the caudal ribs and envelopes the m. caudofemoralis longus, until the m. caudofemoralis longus terminates and the M. ilio-ischiocaudalis covers the full lateral surfaces of the chevrons and centra. The M. ilio-ischiocaudalis was reconstructed by assuming it was bound dorsally by the ventral surface of each caudal rib and formed a lateral ring of uniform thickness around the M. caudofemoralis and, posterior to the caudal transition point, extended from the base of each neural arch to below each chevron.

Volume measurements for each reconstructed muscle were generated directly by the Rhinoceros® software. These values were then multiplied by the known standard muscle density of 1.06 g·cm^3 (Mendez and Keys, 1960) to calculate muscle mass. Comparing these muscle mass estimations to previous total body mass estimations offers one means of examining the relative size of the M. caudofemoralis across taxa, and it requires the fewest assumptions (Table 26.1). Dinosaur body mass estimation is a contentious field of study (see Paul, 1997; Hutchinson et al., 2011 [and comments therein]). However, for the purpose of comparisons, what matters most is that body mass estimation techniques are consistent for all taxa. All body mass estimations utilized here are taken, or derived, from the work of Paul (2010).

Simple ratios of caudofemoralis mass to body mass are not ideal for assessing the potential locomotive contribution of the M. caudofemoralis, because muscle force generation does not scale directly with muscle mass. A better assessment would be the ratio of M. caudofemoralis anatomical cross-sectional area (ACSA) to body mass. ACSA is the

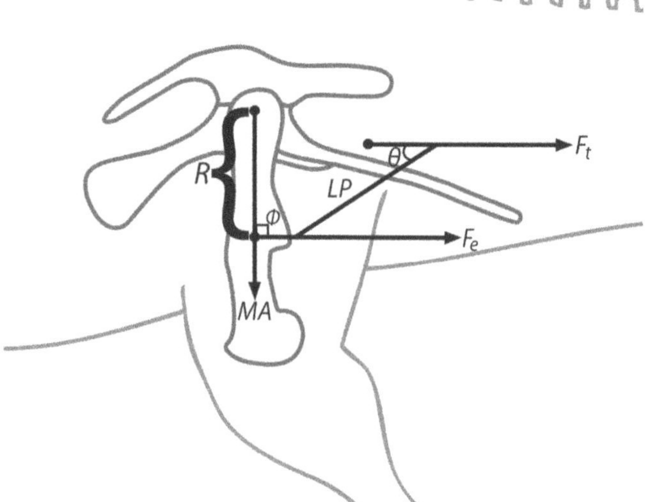

26.1. Vector relationships used in the biomechanical equations, illustrated in the hind limb of a generic hadrosaur. Abbreviations: F_e, effective force vector; F_t, total force vector; LP, line of pull; MA, moment arm vector; R, moment arm length.

cross-sectional area through a muscle at its greatest width (Snively and Russell, 2007) and was measured directly from the digital models.

The total contractile force (F_t) of the M. caudofemoralis can be estimated by multiplying the calculated ACSA by an assumed specific tension (ST), following Equation 1:

Equation 1: $F_t = ACSA{\cdot}ST$

Specific tension, sometimes referred to as specific force (Brooks and Faulkner, 1994), is a ratio of muscle strength to muscle area. ST varies among different muscles and among different taxa. Based on limb muscles of modern animals, Persons and Currie (2011a) considered values within the range of 25–35 N·cm^2 to be reasonable ST estimations for the M. caudofemoralis of dinosaurs. Here, a conservative ST estimation of 25 N·cm^2 is assumed for all taxa.

Another meaningful measure of the locomotive contribution of the M. caudofemoralis is its capacity to generate torque. A muscle's torque-generating capacity is proportional to the force the muscle exerts in a direction orthogonal to its moment arm, where the moment arm is measured from the joint's center of rotation to the muscle's insertion (Snively and Russell, 2007). This orthogonal force, or effective force (F_e), is calculated following Equation 2, where θ is the angle between the vector of line of pull of the M. caudofemoralis and a vector orthogonal to the moment arm (the dorsal-ventral distance from the midpoint of the femoral head to the midpoint of the femoral fourth trochanter; Fig. 26.1). The vector of line of pull of the M. caudofemoralis is equivalent to the orientation of the caudofemoral tendon, which is inferred from the orientation of muscle scars on the fourth trochanter. All F_e calculations were made for a femur positioned perpendicular to the ground.

Equation 2: $F_e = F_t{\cdot}\cos\theta$

The potential torque generation of the M. caudofemoralis can then be calculated with Equation 3, where Φ is the angle between the vectors of the moment arm and the effective force (this angle is assumed to be 90° and the sinΦ is, therefore, equal to 1), and R is the length of the moment arm (Fig. 26.1). Again, a femur positioned perpendicular to the ground is assumed.

Equation 3: $\tau_m = R{\cdot}F_e{\cdot}\sin\Phi$

The results of all these biomechanical calculations are presented in Table 26.2 and confirm the qualitative assessments. The hadrosaur caudofemoral complex is estimated to overlap that of tyrannosaurs with regard to most locomotive contributions. In terms of torque generation, the hadrosaur caudofemoral complex is estimated to be substantially superior.

TAIL ANATOMY

The chevrons and caudal neural spines of adult hadrosaurs are greatly extended in comparison to those of most other similarly sized dinosaurs, including more primitive ornithopods, such as iguanodontids and camptosaurids. We propose that these extensions suggest an enlarged tail musculature. The results of digital modeling (Fig. 26.2) indicate large hypaxial musculature – resulting from the proportions of hadrosaur chevrons – but, despite the extended neural spines, only moderately sized epaxial muscles. This result stems from the lack of a corresponding elongation of hadrosaur caudal ribs, which are important in determining the mass of the epaxial musculature, but not the hypaxial musculature (Persons and Currie, 2011a, b). Unexpanded epaxial muscles conflict with the previous suggestion that hadrosaur tails were specifically adapted for swimming (Ostrom, 1964), because large epaxial muscles (particularly the M. longissimus) are important in powering aquatic sculling in modern crocodilians (Frey, 1982; Bakker, 1986).

In the anterior caudal region, the M. caudofemoralis is positioned lateral to the chevrons and ventral to the caudal ribs, and the dorsoventral distance from the tips of the chevrons to the caudal ribs is proportional to caudofemoral girth (Wilhite, 2003; Persons and Currie, 2011a, b). As noted by Persons and Currie (2011a), the relative mass and cross-sectional area of the M. caudofemoralis of tyrannosaurs (and most other non-avian theropods) were substantially greater than that of

Table 26.2. Summary of Biomechanical Calculations

	M. caudofemoralis mass	M. caudofemoralis ACSA	M. caudofemoralis F_t	M. caudofemoralis F_e	M. caudofemoralis torque
***Lambeosaurus lambei* TMP 1982.038.0001**	61500 g	900 cm^2	22500 N	19100 N	8900 Nm
Ratio to estimated body mass (2500000 g)	0.025 g/g	3.6×10^{-4} cm^2/g	9×10^{-3} N/g	7.63×10^{-3} N/g	3.56×10^{-3} Nm/g
Hadrosaur TMP 1998.058.0001 (subadult)	5200 g	200 cm^2	5000 N	4240 N	770 Nm
Ratio to estimated body mass (183000 g)	0.028 g/g	1.09 x 10-3 cm^2/g	0.0273 N/g	0.0232 N/g	4.2×10^{-3} Nm/g
***Tyrannosaurus rex* BHI 3033**	261100 g	1790 cm^2	44750 N	44750 N	22580 Nm
Ratio to estimated body mass (56220000 g)	0.046 g/g	3.18×10^{-5} cm2/g	7.96×10^{-4} N/g	7.96×10^{-4} N/g	4.02×10^{-4} Nm/g
***Gorgosaurus libratus* TMP 1991.036.0500** (subadult)	8650 g	140 cm2	3500 N	3500 N	510 Nm
Ratio to estimated body mass (400000 g)	0.022 g/g	3.5×10^{-4} cm^2/g	8.75×10^{-3} N/g	8.75×10^{-3} N/g	1.28×10^{-3} Nm/g

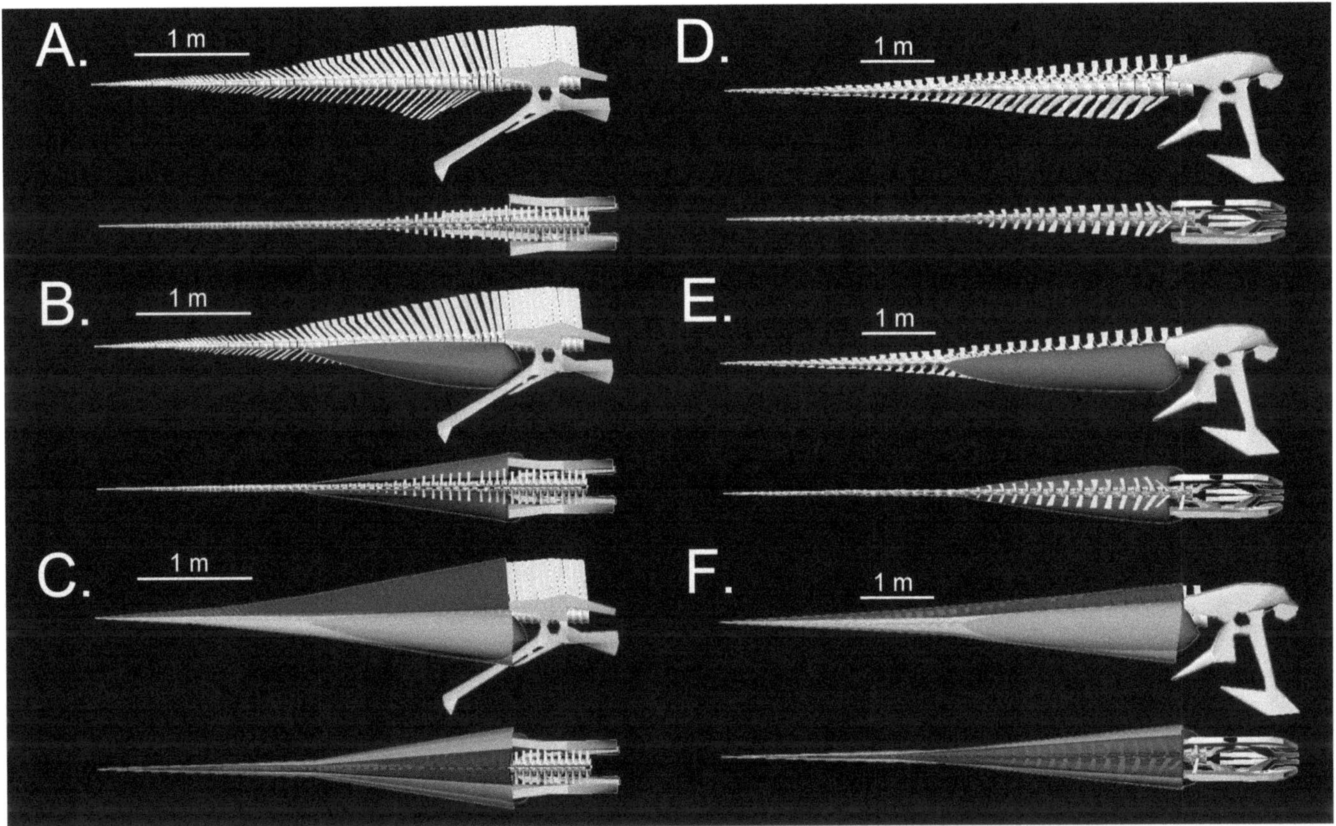

26.2. Digital models of the caudal osteology and musculature of the hadrosaur *Lambeosaurus lambei* (A, B, C) and the tyrannosaur *Tyrannosaurus rex* (D, E, F). Three stages of reconstruction are shown in right-lateral and dorsal views: the caudal skeleton modeled based on specimen measurements (A, D); the M. caudofemoralis longus modeled over the digital skeleton (B, E); and the full muscle reconstruction with M. spinalis/M. longissimus and M. ilio-ischiocaudalis (C, F).

crocodilians and other modern reptiles, because the anterior caudal ribs of tyrannosaurs are positioned high on the neural arches–well above the dorsal edges of the caudal centra. Elevation of the caudal ribs extended the lateral attachment surfaces and permitted the expansion of the M. caudofemoralis. The caudal ribs of hadrosaurs are not elevated and are positioned below, or level with, the dorsal edges of the caudal centra. Nonetheless, there is reason to suspect that hadrosaurs also had relatively large caudofemoral muscles, and the digital modeling results confirm this suspicion. The increased size of the hadrosaur M. caudofemoralis was accomplished by extending the lateral attachment surfaces ventrally, rather than dorsally, through the deepening of the anterior chevrons (Fig. 26.3).

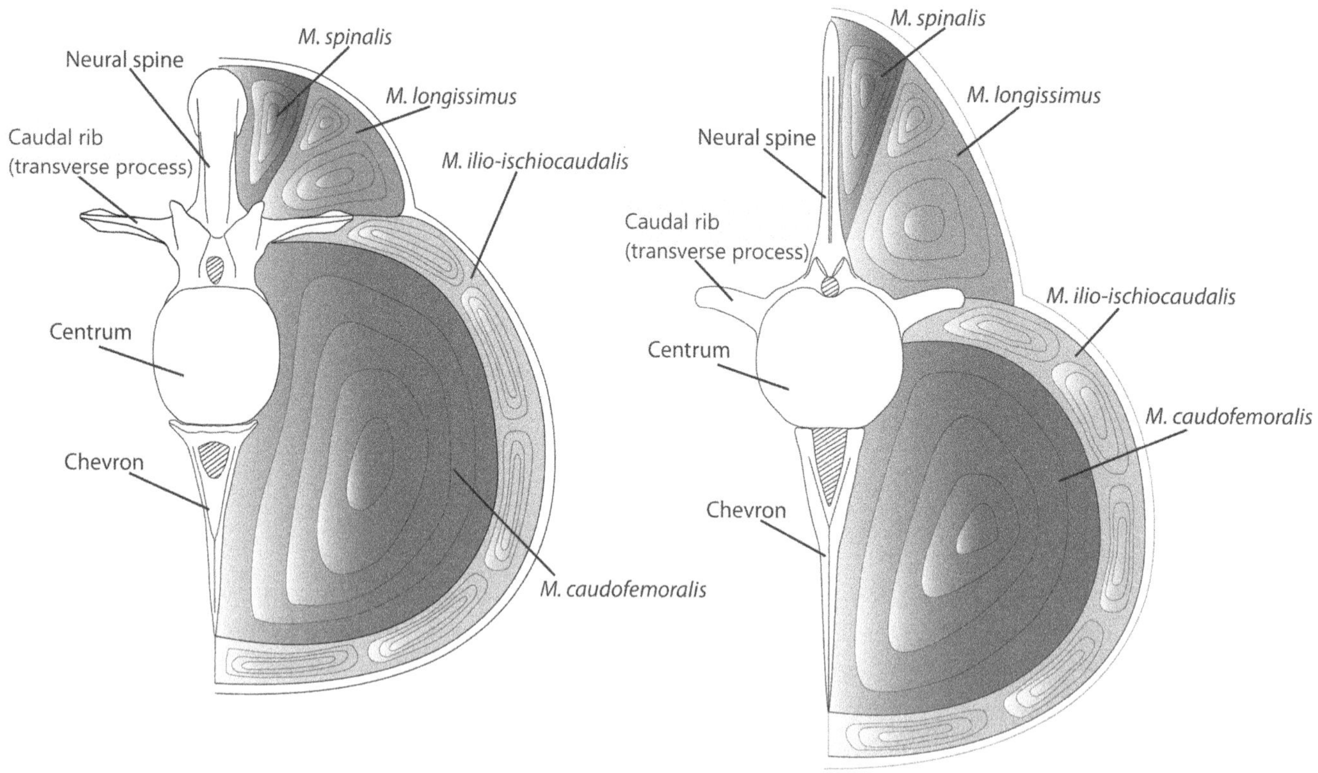

26.3. Hadrosaur (*Lambeosaurus lambei;* left) and tyrannosaur (*Tyrannosaurus rex;* right) tail vertebrae and chevrons in anterior view, with one half of caudal musculature illustrated in each case. Note the elevated caudal ribs and dorsally expanded M. caudofemoralis in the tyrannosaur and the extended chevron and ventrally expanded M. caudofemoralis in the hadrosaur. Both vertebrae and accompanying chevrons are scaled to equal centrum height.

DISCUSSION

The Biomechanical Thought Experiment of Racing Dinosaurs

Dorsoventrally, the M. caudofemoralis is roughly symmetrical, and ventral or dorsal expansion of the M. caudofemoralis should, therefore, result in a corresponding shift of the centerline of the muscle and the averaged force vector of its contractions. Ideally, to maximize muscle efficiency, the average force vector of the contractions of M. caudofemoralis should be in line with the vector of pull on the femur. In the caudofemoral complex, the vector of pull on the femur is equivalent to the orientation of the caudofemoral tendon, which runs from the caudofemoralis proper to the femoral fourth trochanter. In tyrannosaurs and all other theropods, the fourth trochanter is positioned proximally on the femur shaft–roughly one-fifth of the way down the femur in the genus *Tyrannosaurus*. In hadrosaurs, the fourth trochanter is positioned at roughly the proximodistal center of the femur (Fig. 26.4). The position of the femoral fourth trochanter in both groups is consistent with, and confirms, the inference of directionally differing M. caudofemoralis expansions. In tyrannosaurs, the more dorsal M. caudofemoralis results in a contraction force vector more in line with the high position of its femoral fourth trochanter, whereas in hadrosaurs, the more ventral M. caudofemoralis results in a contraction force vector more in line with the low position of its femoral fourth trochanter.

Unlike other contemporaneous large herbivorous dinosaurs, which also have ventrally positioned caudal ribs but not extended chevrons, hadrosaurs had caudofemoral proportions that overlapped with those of tyrannosaurs. However, while the ventral versus dorsal enlargement of the M. caudofemoralis in hadrosaurs and tyrannosaurs both constitute investments in locomotive power, the two are not simply different anatomical means of achieving the same functional result. Rather, it is likely that the two muscle arrangements imparted fundamentally different athletic abilities.

To understand the importance of these two limb-muscle arrangements, it is helpful to imagine a tyrannosaur and a duckbill running on two enormous treadmills, and to consider the arcs swung by their femora during the backward strokes of the step cycle (Fig. 26.4). For the sake of comparison, let us temporarily assume that all aspects not related to

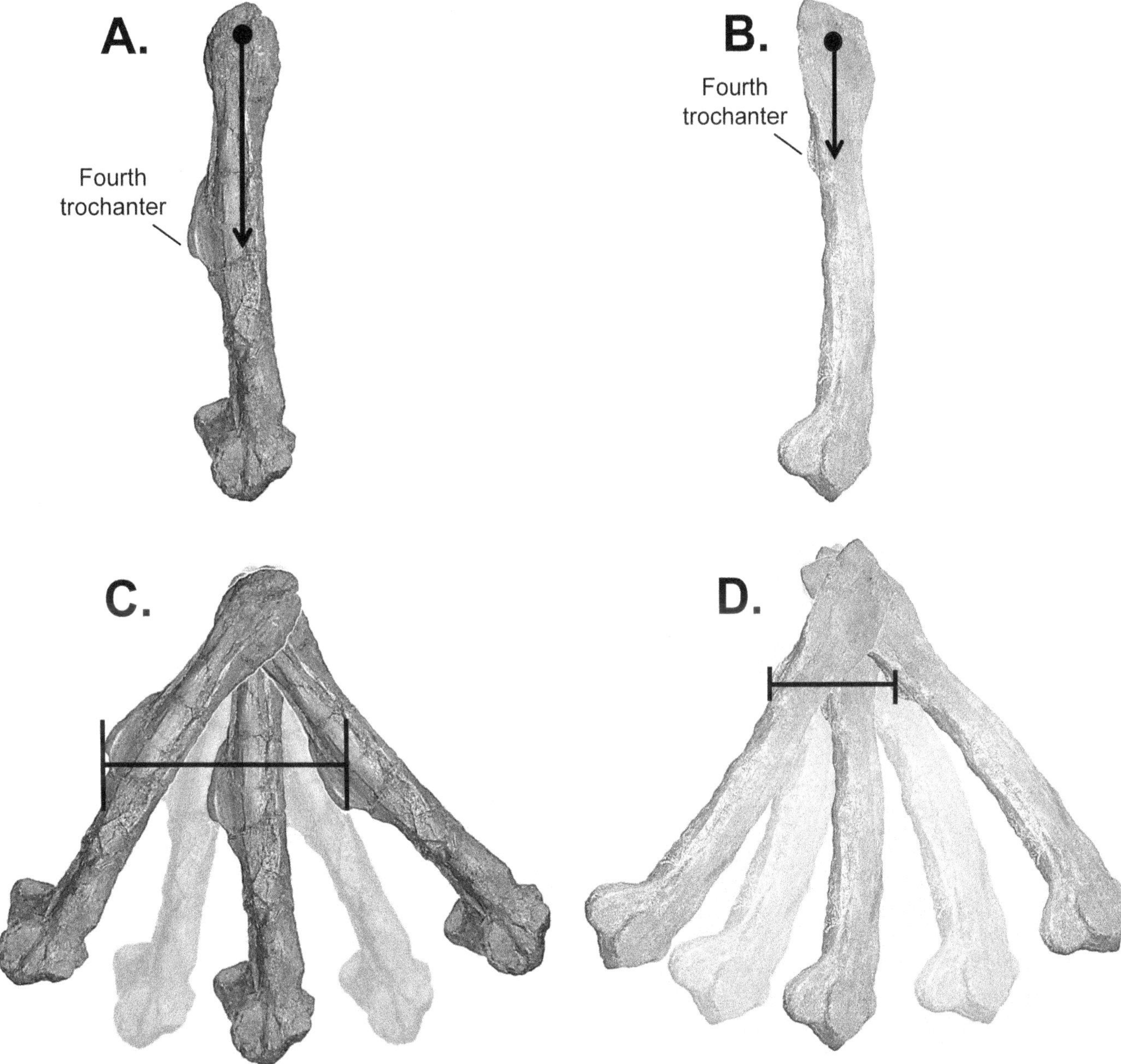

26.4. Right femora of a hadrosaur (*Lambeosaurus lambei* [A, C]) and a tyrannosaur (*Albertosaurus sarcophagus* [B, D]), each in lateral view scaled to equal proximodistal height. Arrows illustrate the distance from the center of the femoral head to the peak of the fourth trochanter (A, B). Brackets illustrate the distance traveled by the fourth trochanter during retraction, assuming strides of equal length (C, D).

the workings of the caudofemoral complex are made equal between the two dinosaurs and that both swing their femora through arcs of equal size.

With a more dorsally positioned caudofemoral complex, the tyrannosaurs should be capable of achieving a more rapid rate of femur retraction, because the arc that the proximal fourth trochanters must travel in each step is relatively shorter; therefore, the muscle contraction required to bring each retraction to fruition is shorter and quicker. In contrast, with a more ventrally positioned caudofemoral complex, the hadrosaur would have slower femoral retraction, because the arc that the distal fourth trochanter must travel in each step is relatively longer and, therefore, longer muscle contractions are required. The positive impact that proximal limb-muscle insertions have on running speed has been well established within the comparative and experimental literature (see Smith and Savage, 1955; Kardong, 1995; Christiansen, 2002).

In reality, it is unlikely that the caudofemoral contraction lengths of hadrosaurs and tyrannosaurs greatly differed. Muscles have a limited maximum excursion distance

(Oatis, 2009), and the force generated by muscles decreases when shortened over relatively long distances (due to reduced muscle tension) (Kardong, 1995). Hadrosaurs were, therefore, probably not capable of such long contractions. Instead, to maintain a reasonable degree of muscle excursion, hadrosaurs would have been forced to reduce the length of the femoral retraction arcs. This, in turn, would reduce the length of each stride (without an increase in stride frequency), and ultimately produce the same relative effect: slower maximum attainable running speed. The prediction that hadrosaurs took relatively shorter strides than theropods is consistent with evidence from fossil trackways (Currie, 1983).

Although the caudofemoral complex of hadrosaurs would have produced a slower maximum running speed, it was also capable of conferring a major locomotive advantage. The low femoral attachment would have given the hadrosaur arrangement longer moment arms (roughly twice the length of those in the tyrannosaur arrangement) and, therefore, imparted superior leverage. Muscle leverage is important to locomotive energetics (McMahon et al., 1987; Roberts et al., 1998), and, because limb muscles serve the additional function of supporting body weight, limb muscle moment arms are particularly important in considerations of locomotion in large terrestrial animals (Biewener, 1989; Hutchinson et al., 2005, Ren et al., 2010). Relative to the tyrannosaur, each retraction of the hadrosaur leg should have required less muscle exertion, ultimately reducing muscular fatigue and the metabolic cost of locomotion.

The significance of these differences becomes clear if we shift the two dinosaurs from treadmills to an actual racetrack. The question of whether the hadrosaur or tyrannosaur would win a race entirely depends on the length of the track. In a brief sprint, the tyrannosaurs would be predicted to have a clear anatomical advantage, having both longer legs and the ability to take longer strides. In a prolonged run, the duckbill, with an ideal arrangement for efficient strides (and, therefore, greater endurance) would be expected to have the athletic edge.

Endurance Running as a Predation Defense Strategy

That hadrosaurs were well adapted for sustained locomotion and that tyrannosaurs, along with other carnivorous dinosaurs with similar caudofemoral arrangements, were poorly adapted for sustained locomotion has several broad implications. First, this result is consistent with previous evidence that hadrosaurs were capable of large-scale migrations (Bell and Snively, 2008) and that many carnivorous dinosaurs were not (Bakker and Bir, 2002). It also lends considerable support to the conclusion that tyrannosaurs hunted through short swift chases or ambushes, more analogous to the big-game hunting tactics of modern felines (Elliott et al., 1977; Hunter, 2006) than to those of canids or orca whales–which hunt through prolonged chases and serial skirmishes (Estes and Goddard, 1967; Silber et al., 1990). Lastly, an explanation for how hadrosaurs met the challenge of predation is finally revealed.

Endurance running is a common anti-predator strategy among members of modern megafaunas. To return to the African savanna analog, zebras may offer the best comparison. Like duckbills, zebras lack defensive weapons but are nonetheless one of the most common large herbivores in their environment. At full gallop a zebra can reach a maximum speed of 55 kph (Estes, 1991), far slower than the maximum sprint of common feline predators, including lions and cheetahs. However, zebras have superior endurance, whereas large cats tire quickly. So, zebras can win predator-initiated chases, even against cheetahs, provided the zebras are sufficiently vigilant to avoid breaching the short striking range of a large cat (Elliott et al., 1977; Kingdon, 1982; Estes, 1991). As in zebras, the keen senses and herding behavior of hadrosaurs would have been especially helpful in spotting and keeping predators at a safe distance. The endurance-runner strategy was potentially even more viable for hadrosaurs than modern mammalian herbivores, because the size of tyrannosaurs would have made it easier to spot stealthy ambushes and avoid close encounters.

ACKNOWLEDGMENTS

This work was made possible by the financial support of the Dinosaur Research Institute. We thank G. Housego (TMP), I. Macdonald (TMP), and B. Strilisky (TMP) for their help in photographing specimens. We also thank R. Bakker (Houston Museum of Natural Science), M. Habib (Chatham University), J. Hall (University of Southern California), and R. Holmes (University of Alberta) for their advice and counsel during this manuscript's development. We thank the two reviewers and the editors of this volume.

LITERATURE CITED

Allen, V., H. Paxton, and J. R. Hutchinson. 2009. Variation in center of mass estimates for extant sauropsids and its importance for reconstructing inertial properties of extinct archosaurs. Anatomical Record 292:1442–1461.

Arbour, V. M. 2009. Estimating impact forces of tail club strikes by ankylosaurid dinosaurs. PLoS ONE 4(8):e6738.

Bakker, R. T. 1978. Dinosaur feeding behavior and the origin of flowering plants. Nature 274:661–663.

Bakker, R. T. 1980. Dinosaur heresy–dinosaur renaissance, why we need endothermic archosaurs for a comprehensive theory of bioenergetic evolution; pp. 351–462 in R. D. K. Thomas, and F. C. Olson (eds.), A Cold Look at the Warm-Blooded Dinosaurs. Westview Press, Boulder, Colorado.

Bakker, R. T. 1986. The Dinosaur Heresies. Kensington, New York, 165 pp.

Bakker, R. T. 2010. Dino Babies! Random House, New York, 24 pp.

Bakker, R. T., and G. Bir. 2002. Dinosaur crime scene investigations: theropod behavior at Como Bluff, Wyoming, and the evolution of birdness; pp. 301–342 in P. J. Currie, E. B. Koppelhus, M. A. Shugar, and J. L. Wright (eds.), Feathered Dragons. Indiana University Press, Bloomington, Indiana.

Bell, P. R., and E. Snively. 2008. Polar dinosaurs on parade: a review of dinosaur migration. Alcheringa 32:271–284.

Berger, J., and C. Cunningham. 1996. Is rhino dehorning scientifically prudent? Pachyderm 21:60–68.

Biewener, A. A. 1989. Scaling body support in mammals: limb posture and muscle mechanics. Science 245:45–48.

Brett-Surman, M. K. 1979. Phylogeny and palaeobiogeography of hadrosaurian dinosaurs. Nature 277:560–562.

Brooks, S. V., and J. A. Faulkner. 1994. Skeletal muscle weakness in old age: underlying mechanisms. Medicine and Science in Sports Exercise 26:432–439.

Christiansen, P. 2002. Locomotion in terrestrial mammals: the influence of body mass, limb length and bone proportions on speed. Zoological Journal of the Linnean Society 136:658–714.

Corsini, M. T., S. Lovari, and S. Sonnino. 1995. Temporal activity patterns of crested porcupines *Hystrix cristata*. Journal of Zoology 236:43–54.

Currie, P. J. 1983. Hadrosaur trackways from the Lower Cretaceous of Canada. Acta Palaeontologica Polonica 28:63–73.

Dollo, L. 1883. Note sur la présence, sur les Oiseaux, du "troisième trochanter" des Dinosauriens et sur la fonction de celui-ci. Musée royal d'Histoire naturelle de Belgique, Bulletin Tome II, 13.

Eberth, D. A., D. C. Evans, and D. H. Lloyd. 2014. Occurrence and taphonomy of the first documented hadrosaurid bonebed from the Dinosaur Park Formation (Belly River Group, Campanian) at Dinosaur Provincial Park, Alberta, Canada; chapter 30 in D. A. Eberth and D. C. Evans (eds.), Hadrosaurs. Indiana University Press, Bloomington, Indiana.

Elliott, J. P., I. McTaggart Cowan, and C. S. Holling. 1977. Prey capture by the African lion. Canadian Journal of Zoology 55:1811–1828.

Estes, R. D. 1974. Social organization of the African Bovidae; pp. 166–205 in V. Geist and F. Walther (eds.), The Behavior of Ungulates and its Relation to Management. Morges, Switzerland.

Estes, R. D. 1991. The Behavior Guide to African Mammals. University of California Press, Los Angeles, California, 611 pp.

Estes, R. D., and J. Goddard. 1967. Prey selection and hunting behavior of the African wild dog. Journal of Wildlife Management 31:52–70.

Evans, D. C. 2005. New evidence on brain-endocranial cavity relationships in ornithischian dinosaurs. Acta Palaeontologica Polonica 50:617–622.

Feldhamer, G. A., L. C. Drickamer, S. H. Vessey, J. F. Merritt, and C. Krajewski. 2007. Mammalogy. The Johns Hopkins University Press, Baltimore, Maryland, 643 pp.

Forster, C. A. 1997. Hadrosauridae; pp. 293–299 in P. J. Currie and K. Padian (eds.), Encyclopedia of Dinosaurs. Academic Press, New York.

Frey, E. 1982. Ecology, locomotion and tail muscle anatomy of crocodiles. Neues Jahrbuch für Geologie und Paläontologie, Abhandlungen 164:194–199.

Galton, P. M. 1973. The cheeks of ornithischian dinosaurs. Lethaia 6:67–89.

Gatesy, S. M. 1990. Caudofemoral musculature and the evolution of theropod locomotion. Paleobiology 16:170–186.

Gatesy, S. M. 1991. Hind limb scaling in birds and other theropods: implications for terrestrial locomotion. Journal of Morphology 209:83–96.

Gatesy, S. M. 1997. An electromyographic analysis of hindlimb function in *Alligator* during terrestrial locomotion. Journal of Morphology 234:197–212.

Goddard, J. 1966. Mating and courtship of the black rhinoceros in East Tsavo Park, Kenya. East African Wildlife Journal 3:118–119.

Holtz, T. R., Jr. 1995. The arctometatarsalian pes, an unusual structure of the metatarsus of Cretaceous Theropoda (Dinosauria: Saurischia). Journal of Vertebrate Paleontology 14:480–519.

Holtz, T. R., Jr. 2004. Tyrannosauroidea; pp. 111–136 in D. B. Weishampel, P. Dodson, and H. Osmólska (eds.), The Dinosauria, Second Edition. University of California Press, Berkeley, California.

Hopson, J. A. 1975. The evolution of cranial display structures in hadrosaurian dinosaurs. Paleobiology 1:21–43.

Horner, J. R. 2000. Dinosaur reproduction and parenting. Annual Review 28:19–45.

Horner, J. R., D. B. Weishampel, and C. A. Forster. 2004. Hadrosauridea; pp. 438–463 in D. B. Weishampel, P. Dodson, and H. Osmólska (eds.), The Dinosauria, Second Edition. University of California Press, Berkeley, California.

Hu, C. 1973. A new hadrosaur from the Cretaceous of Chucheng, Shantung. Acta Geologica Sinica 2:179–206.

Hunter, L. 2006. Cats of Africa: Behavior, Ecology, and Conservation. Johns Hopkins University Press, Baltimore, Maryland, 63 pp.

Hutchinson, J. R. 2004. Biomechanical modeling and sensitivity analysis of bipedal running ability, 2: Extinct taxa. Journal of Morphology 262:441–461.

Hutchinson, J. R., F. C. Anderson, S. S. Blemker, and S. L. Delp. 2005. Analysis of hindlimb muscle moment arms in *Tyrannosaurus rex* using a three-dimensional musculoskeletal computer model: implications for stance, gait, and speed. Paleobiology 31:676–701.

Hutchinson, J. R., K. T. Bates, J. Molnar, V. Allen, and P. J. Makovicky. 2011. A computational analysis of limb and body dimensions in *Tyrannosaurus rex* with implications for locomotion, ontogeny, and growth. PLoS ONE 6(10):e26037.

Kardong, V. K. 1995. Vertebrates: Comparative Anatomy, Function, Evolution. Wm. C. Brown, Dubuque, Iowa, 777 pp.

Kaspari, M., and M. Weiser. 1999. The size-grain hypothesis and interspecific scaling in ants. Functional Ecology 13:530–538.

Kingdon, J. S. 1982. East African Mammals: An Atlas of Evolution in Africa, Volume 3, Part B. Academic Press, New York, 450 pp.

Levin, S. 1992. The problem of pattern and scale in ecology. Ecology 73:1943–1967.

Lockley, M. G., B. H. Young, and K. Carpenter. 1983. Hadrosaur locomotion and herding behavior: evidence from footprints in the Mesaverde Formation, Grand Mesa Coal Field, Colorado. Mountain Geologist 20:5–14.

Lovegrove, B. 2001. The evolution of body armor in mammals: plantigrade constraints of large body size. Evolution 55:1464–1473.

Mallison, H. 2011. Defense capabilities of *Kentrosaurus aethiopicus* Hennig, 1915. Palaeontologia Electronica 14.2.10A.

Manning, P. L., R. A. Wogelius. B. van Dongen, T. R. Lyson. U. Bergmann, S. Webb, M. Buckley, V. M. Egerton, and W. I. Sellers. 2014. The role and biochemistry of melanin pigment in the exceptional preservation of hadrosaur skin; chapter 36 in D. A. Eberth and D. C. Evans (eds.), Hadrosaurs. Indiana University Press, Bloomington, Indiana.

McMahon, T. A., G. Valiant, and E. C. Frederick. 1987. Journal of Applied Physiology 62:2326–2337.

McNeel, R., and Associates. 2007. Rhinoceros NURBS modeling 4.0 (Windows). Seattle, Washington.

Mendez, J., and A. Keys. 1960. Density and composition of mammalian muscle. Metabolism 9:184–188.

Morris, W. J. 1981. A new hadrosaurian dinosaur from the Upper Cretaceous of Baja California–?*Lambeosaurus laticaudus*. Journal of Paleontology 55:453–462.

Oatis, C. A. 2009. Kinesiology: The Mechanics and Pathomechanics of Human Movement. Lippincott Williams and Wilkins, Baltimore, Maryland, 899 pp.

Osborn, H. E. 1911. A dinosaur mummy. American Museum Journal 2:7–11.

Ostrom, J. H. 1964. A reconsideration of the paleoecology of hadrosaurian dinosaurs. American Journal of Science 262:975–997.

Owen-Smith, N. 1988. Megaherbivores: The Influence of Very Large Body Size on Ecology.

Cambridge University Press, Cambridge, U.K., 382 pp.

Paul, G. S. 1997. Dinosaur models: the good, the bad, and using them to estimate the mass of dinosaurs; pp. 129–154 in D. L. Wolberg, E. Stump, and G. D. Rosenberg (eds.), Dinofest International: Proceedings of a Symposium held at Arizona State University. Academy of Natural Sciences, Philadelphia, Pennsylvania.

Paul, G. S. 2008. The extreme lifestyles and habits of the gigantic tyrannosaurid superpredators of the Late Cretaceous of North America and Asia; pp. 307–349 in P. L. Larson and K. Carpenter (eds.), *Tyrannosaurus rex,* the Tyrant King. Indiana University Press, Bloomington, Indiana.

Paul, G. S. 2010. Dinosaur Mass Estimate Table, Official Website of Gregory S. Paul–Paleoartist, Author, and Scientist. Available at gspauldino.com/data.html. Accessed January 10, 2012.

Persons, W. S., IV, and P. J. Currie. 2011a. The tail of *Tyrannosaurus:* reassessing the size and locomotive importance of the *M. caudofemoralis* in non-avian theropods. Anatomical Record 294:119–131.

Persons, W. S., IV, and P. J. Currie. 2011b. Dinosaur speed demon: the caudal musculature of *Carnotaurus sastrei* and implications for the evolution of South American abelisaurids. PLoS ONE 6(10):e25763.

Prins, H. H. T. 1996. Ecology and Behavior of the African Buffalo. Chapman and Hall, London, U.K., 293 pp.

Ren, L., C. E. Miller, R. Lair, and J. R. Hutchinson. 2010. Integration of biomechanical compliance, leverage, and power in elephant limbs. Proceedings of the National Academy of Sciences 107:7078–7082.

Roberts, T. J., M. S. Chen, and C. R. Taylor. 1998. Energetics of bipedal running II. Limb design and running mechanics. Journal of Experimental Biology 201:2753–2762.

Romer, A. S. 1927. The pelvic musculature of the ornithischian dinosaurs. Acta Zoologica 8:225–275.

Silber, G. K., M. W. Newcomer, and H. Pérez-Cortés. 1990. Killer whales (*Orcinus orca*) attack and kill a Bryde's whale (*Balaenoptera edeni*). Canadian Journal of Zoology 68:1603–1606.

Smith, J. M., and R. J. G. Savage. 1955. Some locomotory adaptations in mammals. Zoological Journal of the Linnean Society 42:603–622.

Snively, E., and A. P. Russell. 2007. Craniocervical feeding dynamics of *Tyrannosaurus rex.* Paleobiology 33:610–638.

Snyder, R. C. 1949. Morphological evidence for bipedal locomotion of the lizard *Basiliscus basiliscus.* Copeia 2:129–137.

Snyder, R. C. 1962. Adaptations for bipedal locomotion of lizards. American Zoologist 2:191–203.

Snyder, R. C. 1967. Adaptive values of bipedalism. American Journal of Physical Anthropology 26:131–134.

Thompson, D. 1961. On Growth and Form. Cambridge University Press, Cambridge, U.K., 345 pp.

Varricchio, D. J., and J. R. Horner. 1993. Hadrosaurid and lambeosaurid bone beds from the Upper Cretaceous Two Medicine Formation of Montana: taphonomic and biologic implications. Canadian Journal of Earth Sciences 30:997–1006.

Wilhite, R. 2003. Biomechanical reconstruction of the appendicular skeleton in three North American Jurassic sauropods. Ph.D. dissertation, Louisiana State University, Baton Rouge, Louisiana, 225 pp.

Duck Soup: The Floating Fates of Hadrosaurs and Ceratopsians at Dinosaur Provincial Park

27

Donald M. Henderson

ABSTRACT

To investigate the hypothesis that hadrosaurs were more resistant to drowning compared to ceratopsians, three-dimensional, digital models of representatives of both groups found in Dinosaur Provincial Park, Alberta, Canada (DPP) were created and tested for their degrees of body immersion while floating in water, and their ability be keep their head above the water surface. Examples of the crested and non-crested hadrosaurs *Lambeosaurus lambei* and *Gryposaurus notabilis*, respectively; and the short-frilled and long-frilled ceratopsians *Centrosaurus apertus* and *Chasmosaurus belli*, respectively, were examined. Axial body densities were set to 1000 gm/l for the post-cervical region, while the limbs were set to 1100 gm/l. Head densities of the ceratopsians were increased by 80–100% to account for the additional masses of the frills and horns. The models include air-filled lung cavities equal to approximately 10% of axial body volume. It was found that both the hadrosaurs and the ceratopsians were able to float passively at the surface with the body sub-horizontal, but the hadrosaurs had their heads clear of the water surface, thus enabling the animals to breathe. In contrast, *Centrosaurus*, with its head fully immersed, would have struggled to elevate its head to breathe. *Chasmosaurus* had its head deeply immersed, and would have quickly drowned. Breathing difficulties experienced by immersed *Centrosaurus* may explain their mass mortalities documented in the many bonebeds found in the Park. The exceptional abundance of hadrosaur remains in the Park may be a reflection of their better adaptive fit to the fluvial environment recorded in the rocks of DPP.

INTRODUCTION

Ceratopsian-dominated bonebeds are relatively common at Dinosaur Provincial Park (DPP), and are attributed to mass mortality events during widespread coastal-plain flooding events (Eberth and Getty, 2005; Eberth et al., 2010). In contrast, hadrosaur-dominated bonebeds are rare within the Park (Eberth and Currie, 2005; Eberth et al., this volume). The rarity of hadrosaur bonebeds is surprising given that hadrosaurs are by far the most abundant form of dinosaur fossil at DPP (Dodson, 1971; Brinkman et al., 2005), with 179 skulls and partial and complete hadrosaur skeletons compared to 45 of those of ceratopsians (Currie and Russell, 2005; Henderson and Tanke, 2010). The rocks exposed at DPP are mainly those of the Dinosaur Park Formation, comprising a record of alluvial, estuarine, and paralic paleoenvironments (Eberth, 2005). Additionally, abundant remains of non-dinosaurian taxa from the Park–such as freshwater turtles, crocodilians, choristoderes, amphibians and fishes (Currie and Koppelhus, 2005, and chapters therein)–support the interpretation of DPP as having numerous bodies of standing and flowing water. Fossil vertebrates at DPP are found in palaeochannels, crevasse splay, and overbank deposits (Eberth, 1990; Eberth and Currie, 2005; Eberth and Getty, 2005), all of which are indicative of high water levels and, at times, fast-flowing water.

This difference in the fossil abundances of the two types of ornithischians, combined with the common preservation of *Centrosaurus* in mass mortality bonebeds, suggest that hadrosaurs and ceratopsians may have fared differently during periods of high water. The body shapes and proportions of these animals must have influenced how they would have experienced and survived flooding events. There are two fundamental questions for a normally terrestrial tetrapod that finds itself immersed in deep water: Does the animal float or sink? And if it does float, can the animal breathe easily and avoid drowning? Large extant ungulates such as rhinos, hippos, or water buffalo are possible modern analogs with which to test ideas about buoyancy and resistance to drowning in ceratopsians and hadrosaurs. However, none of these mammals have large bony frills projecting from the back of the head, nor a substantial and muscled tail. Hadrosaurs, with their tall and narrow bodies, hindlimbs that are much larger than their forelimbs, and a long muscled tail, have body shapes and mass distributions unlike those of any living animal. Having no truly appropriate living analogs with body shapes and sizes similar to those of hadrosaurs or ceratopsians, an alternative way to test ideas about the functions of extinct animals is to use digital models that can be analyzed when subject to basic physical laws such as the forces of gravity and buoyancy.

METHODS

Body Model Generation

Digital models of four types of ornithischian dinosaur commonly found in DPP were based on the isometric restorations prepared by Gregory S. Paul and published in various articles over the years (see figure captions for sources). The hadrosaurs were represented by two examples from the crested and non-crested hadrosaurs–*Lambeosaurus lambei* and *Gryposaurus notabilis*, respectively (Fig. 27.1). The short-frilled and long-frilled ceratopsians were represented by *Centrosaurus apertus* and *Chasmosaurus belli*, respectively (Fig. 27.2). The lateral and dorsal views of the axial body and limbs provide sagittal and frontal outlines. These outlines are then used to provide the semi-major and semi-minor radii for ellipses used to define three-dimensional slices through the axial body and limbs. The perimeters of the slices are used to delimit the body surface. The surfaces of the bodies and limbs are treated as meshes of polygons. Full details of the slicing technique and mass estimation method can be found in Henderson (1999), and the mass estimation and flotation properties of digital models of extant reptiles and mammals have been tested extensively (Henderson, 2003, 2004, 2006; Henderson and Naish, 2010).

Assignment of Body Densities

Not only do the slices define the body shape, but in combination with the appropriate tissue density values, they also facilitate calculation of the mass of the body model. Two contiguous slices will bound a thick slab of the body or a limb, and a density can be assigned to the slab. An added benefit of the slicing technique is that densities can vary along the length of the body to more accurately reflect the variable distribution of the air spaces within the body by assigning different densities to different slabs. The basic tissue density for the axial bodies was assumed to be 1000 gm/l, the same as that of water. The head and neck densities were reduced to 900 gm/l to reflect the presence of oral and nasal cavities in the head and the trachea and esophagus in the neck. This latter value is an estimate, but a similar value has been successfully used with digital models of alligators (Henderson, 2003), sea turtles (Henderson, 2006), elephants (Henderson, 2004), and horses and giraffes (Henderson and Naish, 2010). The modeling technique allows for the inclusion of separately defined volumes to be included in the model, and a separate lung space was generated for each model. It has been observed that the lungs of extant reptiles represent about 8% of body volume (Milsom, 1975). In the case of the dinosaurs employed here, the lungs were made slightly larger, on the order of 9–10%, on account of their presumed more active mode of life. The presence of the lung will lower the density in the thoracic region to approximately 700 gm/l, depending on the animal and the shape of its thoracic region. The limbs, with their higher proportion of dense muscle and large bones, had densities set to 1100 gm/l.

The heads of the ceratopsians have significant additions in the form of the frill and nasal horns when compared to those of the hadrosaurs. To incorporate the effects of this added mass on total mass and center of mass of a model, the masses of the frills and horns were computed separately. The mean thicknesses of the frills of *Centrosaurus* and *Chasmosaurus* were set to 2.0 cm and 1.5 cm, respectively. These low thickness estimates also account for the large gaps in the frills formed by the paired frontal fontanelles (not explicitly included in the frill models, and assumed to be covered with skin in life). The external area of the frill was computed and this value was multiplied by the mean frill thickness to get a volume estimate for the bone comprising the frill. The density of the frill and horn material was set to 1850 g/l, similar to that of compact adult human bone, and this value was used to estimate the mass of the frill. Holding the volume of the unadorned head constant, a new density value for the head was computed for each of the two ceratopsians incorporating the combined masses of the ornaments and the head. This new, higher, density value was assigned to the slices defining the head region, and used in computing the overall mass and center of mass of each model. The new head densities for *Centrosaurus* and *Chasmosaurus* are 1850 g/l and 2000 g/l, respectively.

Determination of Flotation Equilibrium

See Henderson (2003) for full details of the calculations of the weight force, the center of mass (CM), the buoyant force, the center of buoyancy (CB), and the process for bringing a model to translational and rotational equilibrium. What follows is a brief summary.

Vertical Position in the Water Column The first part of the process of determining the final equilibrium state of the immersed model is to calculate its final vertical position. This position is a result of the competition between the downward weight force due to gravity, and the upward buoyant force resulting from the volume of water displaced by the immersed body. Both the immersed portions of the axial body and the limbs are used in determining the magnitude of the buoyant force. The weight force does not change, but the buoyant force will change, depending on the degree of immersion of the model at any instant. The method of attaining the final vertical position of the model consists of comparing the upward and downward forces, and making

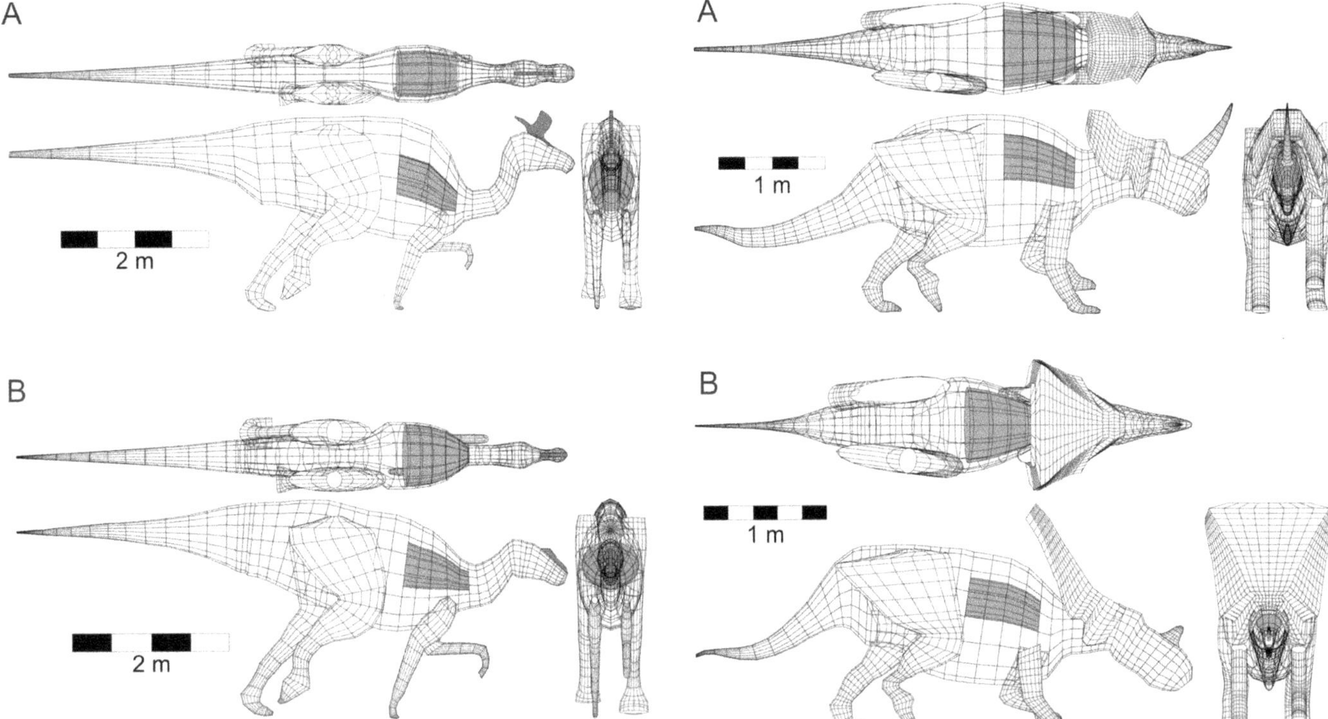

27.1. Isometric views (lateral, dorsal and anterior) of the two hadrosaur models tested for buoyancy and final equilibrium positions. (A) *Lambeosaurus lambei,* with body form adapted from the illustration of *Corythosaurus* in Paul (1987) and the crest outline taken from Lull and Wright (1942); (B) *Gryposaurus annectens,* adapted from the illustration by Paul (1997) in (Dodson, 1990). Medium gray shape in chest region represents the estimated lung volume and position for the models. See Table 27.1 for details of each model.

27.2. Isometric views (lateral, dorsal, and anterior) of the two ceratopsian models tested for buoyancy and final equilibrium positions. (A) *Centrosaurus apertus;* (B) *Chasmosaurus belli.* Both models were adapted from Paul (1997), but the *Chasmosaurus* frill was based on that illustrated in Godfrey and Holmes (1995). Light gray shape in chest region represents the estimated lung volume and position for the models. See Table 27.1 for details of each model.

small up or down adjustments to the model in the water column, depending on whether the upward buoyant force is greater or whether the downward weight force is greater. After the application of the up or down displacement, the buoyant force is then recalculated to reflect the new degree of immersion of the model. This calculate-compare-adjust cycle is continued until the absolute value of the difference between the competing upwards and downwards forces is less than 0.5% of the weight force.

Angular Orientation in Water The second part of finding the final equilibrium state involves determining the equilibrium inclination of the model as measured in the sagittal plane when the models are immersed. Initially, the CM of a model, derived during the mass calculation, will not be perfectly aligned with its CB, derived during the buoyant force calculation. The buoyant force will act upward through the CB, whereas the weight force will act downward through the CM. Even the slightest degree of misalignment of these forces will act to rotate the body about its CM. If the CB lies anterior to the CM, the head end will tip up, lifting the front part of the body out of the water, while lowering the rear portion and increasing its degree of immersion. If the CB lies posterior to the CM, the rear portion will be elevated while the front portion is lowered. An added effect is that the greater the horizontal separation of the CB and CM the greater the turning force due to the long lever arm represented by the separation. The rotational displacements will change both the magnitude of the buoyant force and the position of the CB. Bringing the model to equilibrium within the sagittal plane follows a similar process as for the attainment of vertical equilibrium. A calculation of the magnitude of the buoyant force and the CB is made, followed by a test of the turning moment (buoyant force multiplied by the horizontal distance between the CM and CB). If the moment is significant enough to have a noticeable effect, a rotational displacement is applied to the model based on the magnitude and direction of the turning moment, followed by a recalculation of the CB and buoyant force. A final resting state occurs when the magnitude of the turning moment is less than 0.5% of a predefined reference torque.

Table 27.1. Body Masses and Component Masses for the Hadrosaur and Ceratopsian Models of Figures 27.1 and 27.2

	Total mass	Body length (m)	Axial mass (%)	Single leg mass (%)	Single arm mass (%)	Lung volume (% axial volume)
Lambeosaurus lambei	2370	8.21	1790 (75.5)	271 (11.4)	21.2 (0.894)	168 (8.51)
Gryposaurus notabilis	2260	7.24	1570 (69.4)	315 (13.9)	31.3 (1.38)	174 (9.15)
Centrosaurus apertus	1480	5.07	1200 (81.0)	109 (7.36)	31.9 (2.15)	114 (9.40)
Chasmosaurus belli	926	4.39	699 (75.4)	94.6 (10.2)	19.2 (2.07)	70.9 (10.5)

Note: All masses are in kilograms with component relative fractions expressed as percentages of the total mass in parentheses. Body length is the "path length" along the central axis of the axial body. Axial mass accounts for presence of lung cavity. Lung volume measured in liters. All results reported to three significant figures.

Table 27.2. Model Results at Buoyant Equilibrium

	Mean body density	Center of mass (x,y)	Center of buoyancy (x,y)	Relative CM position (%)	Relative immersion depth (%)	Fineness ratio
Gryposaurus notablis	943	4.15, -0.560	4.15, -0.603	13.4	8.33	8.73
G. notabilis 50% lung deflation	980	4.17, -0.599	4.18, -0.622		8.59	
Lambeosaurus lambei	933	4.59, -0.524	4.59, -0.583	17.2	7.10	10.9
L. lambei 50% lung deflation	973	4.64, -0.588	4.61, -0.621		7.56	
Centrosaurus apertus	913	2.73, -0.368	2.73, -0.442	44.1	8.34	5.78
Chasmosaurus belli	989	2.06, -0.471	2.05, -0.481	41.0	10.9	5.22

Note: Body density measured in kilograms per cubic meter. Center of mass and buoyancy coordinates are in meters, with the X-coordinate measured from the tip of the tail and the Y-coordinate measured from the water surface. Relative CM Position is the horizontal distance of CM from the acetabulum divided by the horizontal gleno-acetabular distance expressed as a percentage. Relative immersion depth is the center of buoyancy Y-coordinate divided by body length expressed as a percentage.

RESULTS

Figures 27.1 and 27.2 show the basic models in their terrestrial walking mode and demonstrate the full surfaces and internal structure (lungs) of the models via their transparent axial body meshes. The total estimated body masses, the absolute and relative limb masses, along with the absolute and relative lung volumes, are summarized in Table 27.1. The reported body lengths are measured along the central axis of the body, and can be thought of as a "path length." Hence, the lengths will be longer than the straight line horizontal separation between tail and snout tips.

Figures 27.3A, 27.3B, 27.4A and 27.4B demonstrate the passive floating states attained by the hadrosaur models with fully inflated lungs. Details of the floating states are summarized in Table 27.2. The first thing to note is that both models, with no additional dorsi-flexion of the head and neck relative to the basic terrestrial models of Figure 27.1, have their heads clear of the water surface. Most importantly, the rostrodorsally positioned external nares are well clear of the water surface, indicating that the animals would be able to breathe. The elevation of the anterior part of the body as a whole is a result of a large, heavily muscled tail that acts to displace the CM rearward, causing the posterior portion of the body to sit deeper in the water. As a test of just how resistant the hadrosaurs would be to drowning, a second flotation test was done with the lungs of both models deflated by 50%, and these results are shown graphically in Figures 27.3C and 27.4C, and also summarized in Table 27.2. The choice of 50% deflation was an arbitrary one, as complete deflation would result in a mean body density greater than that of water, and the subsequent sinking of the model. Both of these lung-deflated models show increases in mean body density of approximately 4%, and as a result they sit slightly lower in the water as indicated by the slight increase in the relative depth of immersion. Even with the deeper immersion, *Gryposaurus* could still easily breathe, and with a modest elevation of the head, *Lambeosaurus* could have its external nares fully

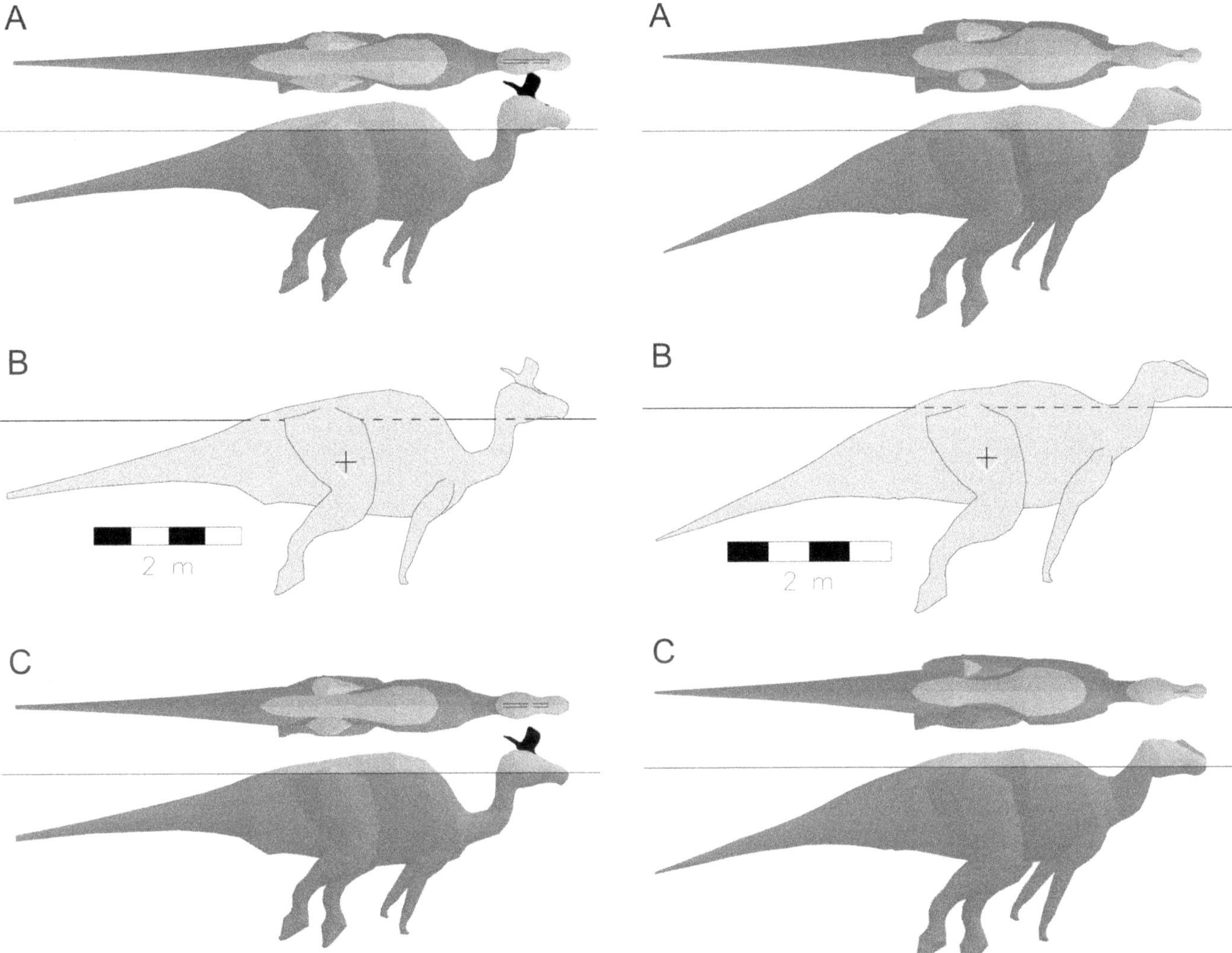

27.3. Immersed *Lambeosaurus* model in buoyant equilibrium. The animal is upright and the head and nostrils are clear of the water surface enabling the animal to see and breathe successfully. The animal is passively floating with the slightly flexed limbs hanging freely from the body. The center of mass is located in the posterior quarter of the trunk, resulting in a positive inclination of the axial body. Horizontal line indicates the water surface. (A) Dorsal and lateral views of the buoyant model with fully inflated lungs. The light body shading highlights the dry, exposed parts of the body, while the dark shading indicates the wet, immersed portions. (B) Outline lateral view of the model with fully inflated lungs and the centers of mass and buoyancy indicated by the plus sign and the diamond shape, respectively. (C) Repeat of the flotation test, but with the lungs deflated by 50% to test the sensitivity of the floating state. See Table 27.2 for details of both floating states.

27.4. Demonstration of the *Gryposaurus* model in buoyant equilibrium. As with the model of *Lambeosaurus,* this posteriorly positioned center of mass tilts the anterior end of the animal upward, where the animal is able to breathe and see while floating. Details as per Figure 27.3. See Table 27.2 for details of the floating states of the models.

clear of the water surface. Should the animals be actively paddling with their forelimbs, this would raise the head and anterior portion of the body, further distancing the mouth and external nares from the water surface.

Figures 27.5 and 27.6 show the equilibrium floating states of the two ceratopsian models. The key feature of these models is that they float with their heads, and in particular, their mouths and external nares, fully immersed. Attempts to raise the head might lift the nares clear of the water, but there are two complications: the fusion of the first few cervical vertebrae of ceratopsians suggest limited capability for neck dorsiflexion; and all living birds and reptiles have the external nares antero-ventrally positioned (Witmer, 2001), suggesting that extinct dinosaurs would have had the same configuration for their nares. This implies that the ceratopsian head and neck would have to be maximally elevated to enable any chance of breathing. The unfortunate combination of the fully immersed, ventrally located external nares, and the difficulty of maintaining an elevated head suggest that ceratopsians would be at risk of drowning in water deep enough for them to float freely.

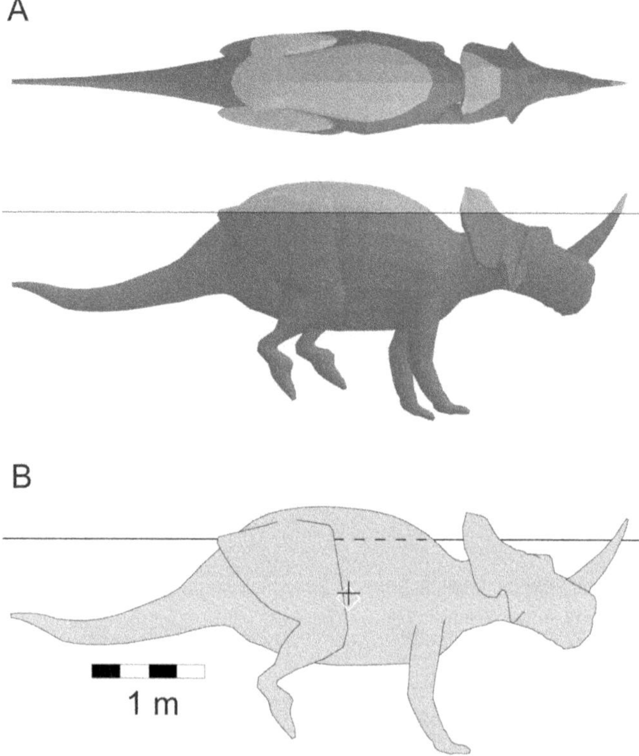

27.5. Demonstration of the *Centrosaurus* model in dorsal and lateral views in buoyant equilibrium with the lungs fully inflated. The center of mass is centrally located in the trunk, resulting in a horizontal body inclination. The head is fully immersed and the rostrally positioned nostrils are deeply immersed making it impossible to breathe. There seems to be little potential to sufficiently elevate the head to enable breathing. Details as per Figure 27.3. See Table 27.2 for details of the floating state of the model.

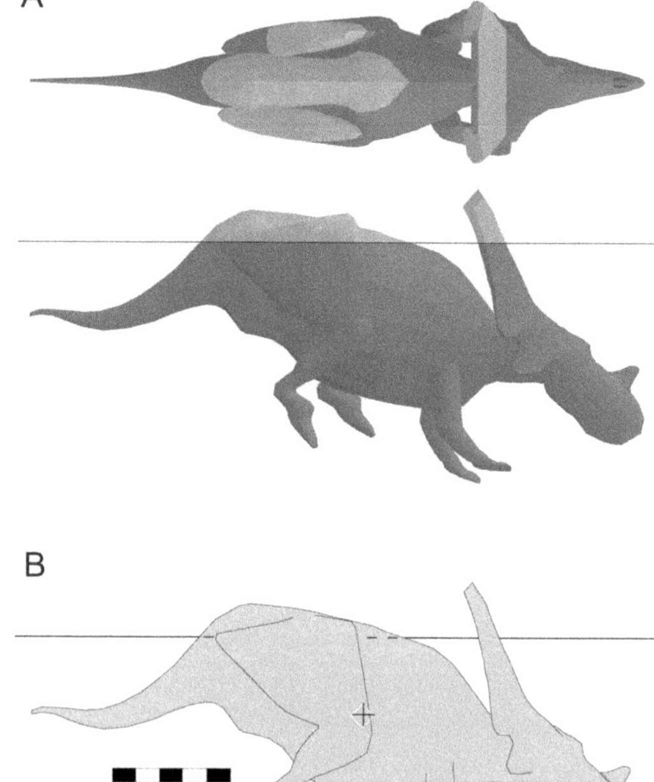

27.6. Demonstration of the *Chasmosaurus* model in buoyant equilibrium with fully inflated lungs. The large and heavy head with its massive frill is fully underwater and the nostrils are deeply immersed making it impossible to breathe. The depth of immersion of the head would rule out any possibility of active dorsi-flexion of the head and neck to enable breathing. Details as per Figure 27.3. See Table 27.2 for details of the floating state of the model.

DISCUSSION

The mean body density computed for the model of *Chasmosaurus* is higher than that in the other models. This is interpreted to be a result of the large expanse of dense bone forming the frill of *Chasmosaurus*. The sequencing of the models in Tables 27.1 and 27.2 is in order of both longest and heaviest, *Gryposaurus*, to shortest and lightest, *Chasmosaurus*. Increasing body density will make a model sit lower in the water, and this is highlighted by calculation of the relative depth of immersion (RDI) where the depth (Y-coordinate) of the center of buoyancy is divided by body length and expressed as a percentage. The roughly 7–8% RDI of the hadrosaurs is less than the 8–10% RDI of the ceratopsians. Unfortunately, the small sample size of the current study means that there is no statistical significance to these differences in body density, body length, or RDI.

The fineness ratio is computed by dividing the length of an object by its maximum transverse dimension and gives a measure of how streamlined an object is and how easily it could move through a fluid (McGowan, 1999). For immersed objects, a fineness ratio of between 3 and 7 has been found to be close to the optimum value as a compromise between pressure drag related to cross-sectional area and friction drag related to total body surface area (Weihs and Webb, 1983). Table 27.2 shows the fineness ratios for the models, and surprisingly, the fineness ratio for the ceratopsians lies within this ideal range, whereas that of the hadrosaurs lies beyond the high end of the preferred range. However, the fineness ratio for a complicated shape such as that of a ceratopsian with a large head frill is an extremely crude measure of the degree of streamlining, and a more thorough analysis – possibly using scale models and flume tanks – would be needed. Intuitively, the large frill, squat body, and relatively short limbs of ceratopsians, when compared to hadrosaurs, would appear to put them at a disadvantage when immersed in water.

Hadrosaurs were initially interpreted to be semiaquatic animals (e.g., Leidy, 1858), but a thorough reexamination of their skeletal remains and the various fossils that provide evidence of soft tissue structures demonstrated that they are better interpreted as fully terrestrial animals (Ostrom, 1964). However, despite the widely accepted modern view that they were principally terrestrial, the remains of hadrosaurs are often found associated with well-watered environments, such as the Dinosaur Park Formation (cf. Horner et al., 2004), and may have even foraged semiaquatically (Dodson, 1971). Most terrestrial vertebrates can at least float, and many can swim if they have to, although some are better than others (reviewed in Henderson and Naish, 2010). The results of the present study suggest that hadrosaurs were not liable to drown when immersed in water, and that they would have been able to cope with the high-water conditions that frequently occurred at Dinosaur Provincial Park during the Late Cretaceous. The combination of floating upright, with the head clear of the water surface to enable breathing, and large body size to slow the loss of body heat into cooler water, all suggests that immersed hadrosaurs would be able to survive for hours, possibly days, in a flooded landscape.

Assuming that many hadrosaurs were gregarious, the large abundance of isolated and disarticulated hadrosaur remains in the Park may indicate that they actually preferred this type of habitat (Dodson, 1971), whereas the fact that they are only rarely found in bonebeds (Eberth and Currie, 2005; Eberth et al., this volume), suggest that they were not as susceptible to mass drownings as appears to have been the case for ceratopsians.

CONCLUSION

This study has tested the hypothesis that the hadrosaurs of Dinosaur Provincial Park would have been better able to survive flooding events than the sympatric ceratopsians. It was found that hadrosaurs had a more posteriorly positioned center of mass, which resulted in an elevation of the anterior portion of the body, and that their more elongate necks allowed the head and external nares to be clear of the water surface to facilitate breathing. In contrast, ceratopsians floated with the anterior portions of their bodies deeper than the posterior, and with their heads fully immersed. The combination of limited potential for dorsi-flexion of the ceratopsian neck and the large frill projecting from the back of the head would have impaired the ability of ceratopsians to raise their heads in an attempt to breathe, rendering them more susceptible to drowning during the large-scale coastal plain flooding events that characterized this region of Western Canada during the Late Cretaceous. Hadrosaurs are the most common type of dinosaur fossil at Dinosaur Provincial Park, suggesting that they may have had a better adaptive fit to the environment, and were better able to survive the frequent catastrophic flooding events that otherwise decimated ceratopsians such as *Centrosaurus*.

LITERATURE CITED

Brinkman, D., A. P. Russell, and J.-H. Peng. 2005. Vertebrate microfossil sites and their contribution to studies of paleoecology; pp. 88–98 in P. J. Currie and E. B. Koppelhus (eds.), Dinosaur Provincial Park: A Spectacular Ancient Ecosystem Revealed. Indiana University Press, Bloomington, Indiana.

Currie, P. J., and E. B. Koppelhus (eds.). 2005. Dinosaur Provincial Park: A Spectacular Ancient Ecosystem Revealed. Indiana University Press, Bloomington, Indiana, 648 pp.

Currie, P. J., and D. A. Russell. 2005. The geographic and stratigraphic distribution of articulated and associated dinosaur remains; pp. 537–569 in P. J. Currie and E. B. Koppelhus (eds.), Dinosaur Provincial Park: A Spectacular Ancient Ecosystem Revealed. Indiana University Press, Bloomington, Indiana.

Dodson, P. 1971. Sedimentology and taphonomy of the Oldman Formation (Campanian), Dinosaur Provincial Park, Alberta (Canada). Palaeogeography, Palaeoclimatology, Palaeoecology 10:21–74.

Dodson, P. (ed.). 1990. Encyclopedia of Dinosaurs. Publications International, Lincolnswood, Illinois, 256 pp.

Eberth, D. A. 1990. Stratigraphy and sedimentology of vertebrate microfossil sites in the uppermost Judith River Formation (Campanian), Dinosaur Provincial Park, Alberta, Canada. Palaeogeography, Palaeoclimatology, Palaeoecology 78:1–36.

Eberth, D. A. 2005. The geology; pp. 54–82 in P. J. Currie and E. B. Koppelhus (eds.), Dinosaur Provincial Park: A Spectacular Ancient Ecosystem Revealed. Indiana University Press, Bloomington, Indiana.

Eberth, D. A., and P. J. Currie. 2005. Vertebrate taphonomy and taphonomic modes; pp. 453–477 in P. J. Currie and E. B. Koppelhus (eds.), Dinosaur Provincial Park: A Spectacular Ancient Ecosystem Revealed. Indiana University Press, Bloomington, Indiana.

Eberth, D. A., and M. A. Getty. 2005. Ceratopsian bonebeds: occurrence, origins, and significance; pp. 501–536 in P. J. Currie and E. B. Koppelhus (eds.), Dinosaur Provincial Park: A Spectacular Ancient Ecosystem Revealed. Indiana University Press, Bloomington, Indiana.

Eberth, D. A., D. B. Brinkman, and V. Barkas. 2010. A centrosaurine mega-bonebed from the Upper Cretaceous of southern Alberta: implications for behavior and death events; pp. 495–508 in M. J. Ryan, B. Chinnery-Allgeier, and D. A. Eberth (eds.), New Perspectives on Horned Dinosaurs. Indiana University Press, Bloomington, Indiana.

Eberth, D. A., D. C. Evans, and D. H. Lloyd. 2014. Occurrence and taphonomy of the first documented hadrosaurid bonebed from the Dinosaur Park Formation (Belly River Group, Campanian) at Dinosaur Provincial Park, Alberta, Canada; chapter 30 in D. A. Eberth and D. C. Evans (eds.), Hadrosaurs. Indiana University Press, Bloomington, Indiana.

Godfrey, S. J., and R. Holmes. 1995. Cranial morphology and systematics of *Chasmosaurus* (Dinosauria: Ceratopsidae) from the Upper Cretaceous of western Canada. Journal of Vertebrate Paleontology 15:726–742.

Henderson, D. M. 1999. Estimating the masses and centers of mass of extinct animals by 3-D mathematical slicing. Paleobiology 25:88–106.

Henderson, D. M. 2003. Effects of stomach stones on the buoyancy and equilibrium of a floating crocodilian: a computational analysis. Canadian Journal of Zoology 81:1346–1357.

Henderson, D. M. 2004. Tipsy punters: sauropod dinosaur pneumaticity, buoyancy and aquatic habits. Proceedings of the Royal Society of London B 271:S180–S183. [Supplementary material]

Henderson, D. M. 2006. Floating point: a computational study of buoyancy, equilibrium and gastroliths in plesiosaurs. Lethaia 39:227–244.

Henderson, D. M., and D. Naish. 2010. Predicting the buoyancy, equilibrium and potential swimming ability of giraffes by computational analysis. Journal of Theoretical Biology 265:151–159.

Henderson, D. M., and D. H. Tanke. 2010. Estimating past and future dinosaur skeletal abundances in Dinosaur Provincial Park, Alberta,

Canada. Canadian Journal of Earth Sciences 47:1291–1304.

Horner, J. R., D. B. Weishampel, and C. A. Forster. 2004. Hadrosauridae; pp. 438–463 in D. B. Weishampel, P. Dodson, and H. Osmólska (eds.), The Dinosauria, Second Edition. University of California Press, Berkeley, California.

Lull, R. S., and N. E. Wright. 1942. Hadrosaurian Dinosaurs of North America. Geological Society of America Special Papers 40. 242 pp.

Leidy, J. 1858. *Hadrosaurus foulkii,* a new saurian from the Cretaceous of New Jersey. Proceedings of the Academy of Natural Sciences of Philadelphia 1858:215–218.

McGowan, C. 1999. A Practical Guide to Vertebrate Mechanics. Cambridge University Press, Cambridge, U.K., 301 pp.

Milsom, W. K. 1975. Development of buoyancy control in juvenile Atlantic loggerhead turtles, *Caretta c. caretta.* Copeia 1974:758–762.

Ostrom, J. H. 1964. A reconsideration of the paleoecology of hadrosaurian dinosaurs. American Journal of Science 262:975–997.

Paul, G. S. 1987. The science and art of restoring the life appearance of dinosaurs and their relatives: a rigorous how-to guide; pp. 5–49 in S. J. Czerkas and E. C. Olson (eds.), Dinosaurs Past and Present, Volume 2. University of Washington Press, Seattle, Washington.

Paul, G. S. 1997. Dinosaur models: the good, the bad, and using them to estimate the mass of dinosaurs; pp. 129–154 in D. L. Wolberg, E. Stump, and G. D. Rosenburg (eds.), Dinofest International: Proceedings of a Symposium held at Arizona State University. Academy of Natural Sciences, Philadelphia, Pennsylvania.

Weihs, D., and P. W. Webb. 1983. Optimization of locomotion; pp. 339–371 in P. W. Webb and D. Weihs (eds.), Fish Biomechanics. Praeger Press, New York.

Witmer, L. 2001. Nostril position in dinosaurs and other vertebrates and its significance for nasal function. Science 293:850–853.

Hadrosauroid Jaw Mechanics and the Functional Significance of the Predentary Bone

28

Ali Nabavizadeh

ABSTRACT

The predentary is a single bone found in the lower jaw complex of all ornithischian dinosaurs. Located rostral to the paired dentary bones of the mandible, it occludes with the premaxilla (or apomorphically with the rostral bone in ceratopsians). Although it is universally accepted that the predentary was used in nipping vegetation, its absence in all other fossil and extant vertebrate herbivores (including sauropodomorphs) indicates that the functional significance of this element is not well understood. The overall morphology of the lower jaw elements, the articular surface between the predentary and dentary, other associations of the mandible with the cranium, and dental wear orientation of hadrosauroids are described here with emphasis on function. Widened expansions and bifurcated processes on the predentary as well as an absence of a firm, clasping junction with the dentary (or between the dentaries themselves) suggest mobility at this junction. A distinct medial curvature of the rostral portion of the dentary, medially recurved coronoid processes, and ball-and-socket articulation between the quadrate and articular of the mandible suggest a rotating surface and range of movement at the predentary-dentary junction, which in turn allows medial rotation of both dentary bones independent of the predentary. These morphologies, and dental micro- and mesowear, suggest palinal jaw movement to initially shear vegetation and medial rotation of the dentaries against the maxillae, maneuvering vegetation into the oral cavity independently on both sides.

INTRODUCTION

The evolution of ornithopod jaw mechanisms has been subject to investigation for over a century, with in-depth analyses of craniomandibular structure, intracranial joint morphology, and dental microwear (Marsh, 1893; Nopcsa, 1900; Versluys, 1910, 1912, 1923; Lambe, 1920; Kripp, 1933; Lull and Wright, 1942; Ostrom, 1961; Thulborn, 1971; Galton, 1974; Hopson, 1980; Sues, 1980; Weishampel, 1984; Norman, 1984; Norman and Weishampel, 1985; Rybczynski et al., 2008; Bell et al., 2009; Williams et al., 2009; Cuthbertson et al., 2012). Ornithopod skulls, mandibles, and dentition are of such unique morphology that there is no modern analog with which to compare them sufficiently. With the herbivorous nature of dental morphology in Ornithopoda, some of which include flat occlusal surfaces used for oral processing, they have been likened to modern, large ungulates. However, there are numerous differences in their jaw morphologies that complicate this analogy. These differences include placement of the craniomandibular joint, jaw curvature, muscular attachment sites, dental morphology, and morphology of the mandibular symphysis. Consequently, researchers proposed novel mechanisms for oral food processing in ornithopods that are unrecognized in modern herbivores. These are based on morphological properties of distinctive traits.

For over two decades, pleurokinesis has been the most widely accepted feeding mechanism proposed for most ornithopods (Weishampel, 1984; Norman, 1984; Norman and Weishampel, 1985). Pleurokinesis is a type of chewing facilitated by a unique form of cranial kinesis that involved movement of the quadrate against the squamosal, creating a domino effect of kinesis in cranial elements that ultimately causes the maxillae to be pushed and rotated laterally as the teeth occlude (Fig. 28.1). This accounts for their unusual transverse dental microwear. This mechanism has recently been challenged with reexamination of hadrosaur skulls and detailed computer modeling of pleurokinesis (Holliday and Witmer, 2008; Rybczynski et al., 2008; Bell et al., 2009; Cuthbertson et al., 2012). These studies indicated the necessity of unlikely secondary movements at various intracranial joints in order for the primary movements of pleurokinesis to take place. Also, they suggested that there is substantial separation of cranial elements in response to these movements. Despite these challenges to the pleurokinetic hypothesis, the transverse dental microwear and the unusual wear pattern in which the occlusal surface is angled laterally on each side of the mandible, still needs to be accurately accounted for in a hadrosaurid chewing model. Since neither solely propalinal nor orthal chewing can explain this orientation of tooth wear, and the angle at which teeth occlude would not allow the entire jaw to chew transversely as one unit, there is

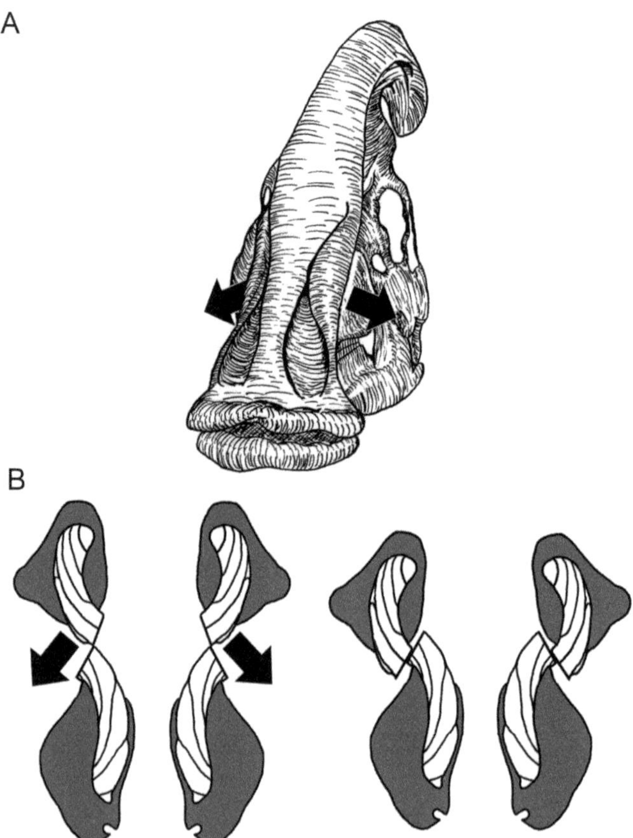

28.1. Summary of pleurokinesis in hadrosaurids. (A) Oblique view of *Parasaurolophus* with arrows showing lateral movement of maxillae; (B) coronal cross-section (illustration based on Lambe, 1920) of hadrosaur maxillae (on top) and dentaries (on bottom) before pleurokinetic power stroke with arrows showing movement (on left) and after power stroke (on right).

likely another aspect of the jaw mechanism that is not well understood. For this reason, it is crucial to understand jaw mechanisms in large modern herbivores such as ungulates, while being aware that the actual mechanisms of ornithopods themselves are likely quite different.

Although mammals are not closely related to ornithischian dinosaurs, the biomechanical implications of mammalian mandibular symphyses described in previous studies still apply in the fundamental biomechanics of ornithischian mandibular symphyses, aside from the predentary. Previous studies of unfused reptilian mandibular symphyses are phylogenetically more relevant (such as studies in crocodilians [Porro et al., 2011], lizards [Herrel et al., 2000; Holliday et al., 2010], and *Sphenodon* [Jones et al., 2012]); however, they lack biomechanical comparisons to species with a completely fused mandibular symphysis. In modern herbivorous mammals, fusion at the mandibular symphysis is variable. In a fused mandibular symphysis, seen in such forms as perissodactyls, elephants, camels, and hippopotamuses, antagonistic muscle forces are transmitted through the symphysis, restricting bilateral rotation of the mandibular corpora around their long axes. In species with an unfused mandibular symphysis, including all ruminant artiodactyls, the forces do not transmit as well across this area due to the joint between the corpora. Consequently, there is medial torsion or rotation of the dentaries accommodating their chewing (Greaves, 1978; Lieberman and Crompton, 2000; Hogue and Ravosa, 2001). Because ornithopod taxa (and ornithischians as a whole) do not fuse both dentaries together rostrally, the median, rostrally positioned predentary may have assisted in a jaw mechanism similar to an unfused symphysis. The plausibility of mobility at the predentary-dentary junction described in hadrosaurs by Holliday and Witmer (2008), Bell et al. (2009), and Cuthbertson et al. (2012), calls for a chewing mechanism that involves rotation of each individual mandibular corpus around its long axis against the predentary. This style of mandibular kinesis was previously proposed by Nopcsa (1900), Versluys (1923), and Kripp (1933), but was later rejected by Ostrom (1961). Cuthbertson et al. (2012) further tested the long-axis rotation hypothesis with computer modeling software, examining ranges of movement at each joint and verifying that it is in fact a plausible mechanism.

The hypothesis of mobility of each individual mandibular corpus against the predentary has been proposed for other ornithischian taxa as well. Weishampel (1984) investigated the skull of *Heterodontosaurus* and proposed a jaw mechanism involving long-axis rotation of the dentaries against the predentary. Crompton and Attridge (1986), Porro (2007), and Norman et al. (2011) rejected this assertion based on restrictions in jaw morphology; however, they proposed the possibility of a separate lateral "wish-boning" motion at the predentary-dentary junction. This mechanism is almost unknown in other vertebrates. Sereno (1991) and Rybczynski and Vickaryous (2001) proposed similar movements at the predentary-dentary junction for *Lesothosaurus* and *Euoplocephalus*, respectively, with the observation that the dentaries were separate entities that did not firmly articulate with the predentary, suggesting mobility. As *Heterodontosaurus* and *Lesothosaurus* are considered among the most basal ornithischians and *Euoplocephalus*, an ankylosaurid, represents a much more derived ornithischian (Butler et al., 2007), the evolutionary and functional significance of the ornithischian predentary comes into question.

The predentary is a single bony mandibular element that is found in all Ornithischia (Fig. 28.1; Weishampel, 2004). Located rostrally at the midline of the mandibular symphysis, it occludes with the premaxilla (or rostral bone in ceratopsians) and was likely covered by a rhamphotheca, a keratinous bill similar to that of a modern bird or turtle (Morris, 1970). In many cases, there are numerous foramina on the rostral surface, indicating neurovasculature in the keratinous sheath. It has been universally accepted that the predentary in hadrosauroids was part of a plant-gathering beak used in nipping vegetation (Morris, 1970). However, due in part to an absence of modern analogs, the functional significance of this element in jaw mechanisms of hadrosauroids (and ornithischians in general) has yet to be fully understood. In this chapter, the hadrosaurid predentary is described in detail with particular attention to its association with the rest of the lower jaw and its functional morphology.

Institutional Abbreviations AMNH, American Museum of Natural History, New York; CMNH, Carnegie Museum of Natural History, Pittsburgh, Pennsylvania; KUVP, University of Kansas Museum of Natural History, Lawrence, Kansas; TMP, Royal Tyrrell Museum of Palaeontology, Drumheller, Alberta; USNM, National Museum of Natural History (Smithsonian), Washington, D.C.

Anatomical Abbreviations **as**, articular surface; **clp**, caudolateral process; **cp**, coronoid process; **d**, denticle; **db**, dental battery; **de**, dentary; **di**, diastema of dentary; **dll**, dorsal lateral lip of surangular; **dp**, dorsal process; **em**, emargination; **f**, foramen; **ju**, jugal; **mc**, mandibular canal; **mx**, maxilla; **p**, pit (see Discussion section); **pr**, predentary; **qu**, quadrate; **rap**, retroarticular process; **sa**, surangular; **sy**, tongue-and-groove symphysis; **vp**, ventral process; **vq**, spherical ventral portion of quadrate.

MATERIALS AND METHODS

In this study, various hadrosauroid predentaries, dentaries, and postdentary elements as well as their associations with the cranium were examined. Because the purpose of this study is to outline an alternative jaw mechanism hypothesis for hadrosauroids using mandibular elements rather than to challenge pleurokinesis as in Holliday and Witmer (2008), Rybczynski et al. (2008), Bell et al. (2009), and Cuthbertson et al. (2012), intracranial joint morphologies were not pertinent for investigation, although further investigation of the pleurokinetic model remains necessary. Due to the fragmentary nature of many fossil specimens (and casts of specimens) and the overall interspecific similarity in jaw morphology, several hadrosauroid genera were examined to increase sample size. These include *Bactrosaurus, Corythosaurus, Edmontosaurus, Gryposaurus, Hypacrosaurus, Kritosaurus, Lambeosaurus, Parasaurolophus, Prosaurolophus, Saurolophus, Telmatosaurus* (cast), and other unidentified hadrosauroid material (see Appendix 28.1 for specimens examined in this study). These specimens were collectively examined at the following institutions: AMNH, CMNH, KUVP, TMP, and USNM. Some specimens used here have also been included in previous studies of hadrosauroid jaw mechanics for consistency; however, many other specimens have been included as well. Specimens were photographed with a D-SLR camera as well as surface scanned three-dimensionally using a NextEngine surface scanner, which allowed further observation in three-dimensional shape of many of the elements.

The predentary-dentary, dentary-dentary, and craniomandibular articular contact morphologies (i.e., qualitative curvature and broadness) were used to assess the type of mobility that could have occurred at these junctions. The physical shape of the mandibular elements themselves, such as rostral curvature of the dentary, shape of the coronoid process, and morphology of predentary processes, as well as associations with the premaxilla, maxilla, and quadrate, was also examined with a focus on functional morphology and interpretation. Consideration was given to any distortion that might have affected the form of the specimens. Dental micro- and macrowear morphology and orientation of three hadrosaurid dentaries from the San Juan Basin of New Mexico (KUVP 17400; KUVP 96884; KUVP 96887) were also examined under light microscopy to obtain general information about directionality of mastication (i.e., motion at the maxilla-dentary occlusal surfaces) in more advanced hadrosaurids and to verify findings of previous studies.

DESCRIPTION

Hadrosauroid predentary and dentary morphology is discussed in detail with an emphasis on functional interpretation. Because some of the postdentary elements do not appear to have had functional significance, only a brief description of each is provided. These descriptions illustrate the general shape of the postdentary region and help to emphasize its larger functional significance. For more detailed descriptions of postdentary elements, refer to Lambe (1920), Ostrom (1961), and Horner (1992). An overview of dental micro- and mesowear is also provided here.

Predentary

The predentary (Figs. 28.2–28.5) is a single, unpaired, median element that rests on the rostral end of the articulated paired dentaries. It is edentulous and forms the lower bill that comes in contact with the premaxilla when the oral cavity is closed. It is the only part of the mandible that contacts the premaxilla, occluding rostrally. A keratinous rhamphotheca likely surrounded the rostral and lateral surfaces of the predentary (Morris, 1970). Depending on the taxon of hadrosauroid, the predentary ranges from a broadly rounded, spoon-shaped rostral edge, as seen in *Kritosaurus* (AMNH 5799; see Fig. 28.5), to a flat, expanded, shovel-like rostral

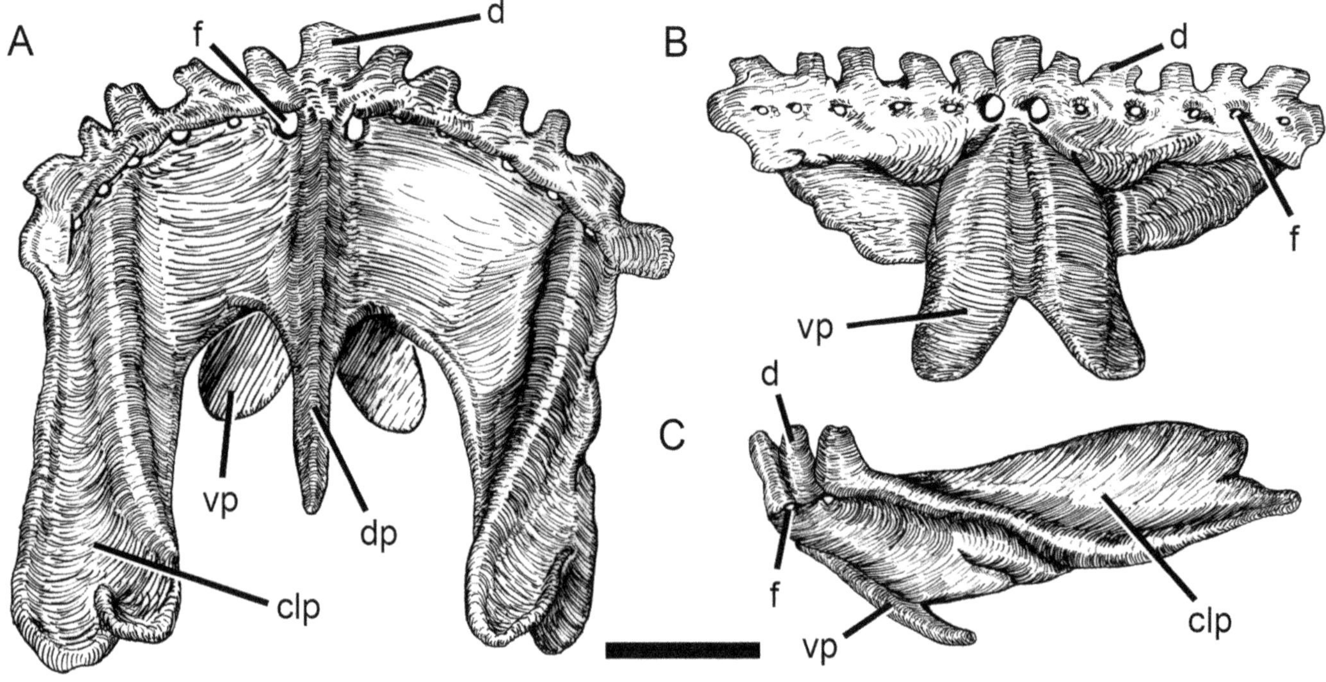

28.2. Generalized hadrosauroid predentary (based mainly off of *Lambeosaurus*, USNM 10309). Note that there is considerable variation between species. (A) Dorsal view; (B) rostroventral view; (C) lateral view. Scale bar equals 5 cm.

edge, as seen in *Lambeosaurus* (USNM 10309; see Fig. 28.2D–F) and *Corythosaurus* (USNM 11893; USNM 16600; see Fig. 28.4). Along the rostrodorsal border, there are between four to eight blunt, sub-rectangular (and sometimes conical) projections that protrude dorsally on either side of the median sagittal plane (with one midline projection in some cases) forming a denticulate dorsal edge, much like the ventrally oriented denticulate edge of the premaxilla. Just ventral to these projections are a series of foramina piercing rostrocaudally through the rostral ridge, the largest of which being the first foramen just lateral to either side of the median sagittal plane. These foramina undoubtedly housed branches of the mental artery and nerve (c.n. V) branching from the inferior alveolar artery that provided blood and nerve supply to the keratinous bill (Ostrom, 1961).

A total of four processes contact the paired dentaries. Two extend caudally along the sagittal plane, both on the dorsal and ventral aspect of the predentary (Fig. 28.2). The dorsal process is elongate, straight, and rod-like with a subtriangular dorsal ridge and a conical caudal tip (Fig. 28.3). It rests on the dorsal aspect of the dentary symphysis. The ventral process is much broader mediolaterally and flat. As it extends caudally, ventral to the dentary symphysis, it becomes bilobed, with each lobe supporting its corresponding dentary at its rostroventral aspect. In lambeosaurines, the dorsal process seems longer than the ventral process, whereas in hadrosaurines it is the opposite, although due to the significant amount of broken specimens, it is difficult to make this distinction with certainty.

There are also two larger processes that project caudolaterally to rest upon the rostrodorsal edge of each dentary (Fig. 28.5). In dorsal view, the processes of the predentary provide its characteristic U shape with both caudolateral processes forming a broad, triangular plane on each side as they continue rostrally, forming a butterfly-like shape (e.g., AMNH 5285, AMNH 6372, USNM 11893, USNM 10309). In caudal view, the caudal end of each process is broadened, laterally compressed, bilobed, and sometimes bowed slightly medially. Each process is angled, forming a flat, dorsomedial to lateroventral plane. The flattened plane on the medial side of each process continues rostrally and becomes less broadened and shallowly concave on the medial side midway to the rostral border (Figs. 28.3C, 28.4). This makes for an osteologically indeterminate articulation with the sharp, rostral ridge of the dentary on both sides; there is no tight, clasping junction, and it provides a surface on which the dentary was able to freely move (Figs. 28.4, 28.5). When the dentary is articulated with the ventral portion of the medial aspect of the large caudolateral process of the predentary, a gap is formed on the dorsal portion of that articular surface. Conversely, when the dentary is articulated with the dorsal portion of that articular surface, there is another gap on the

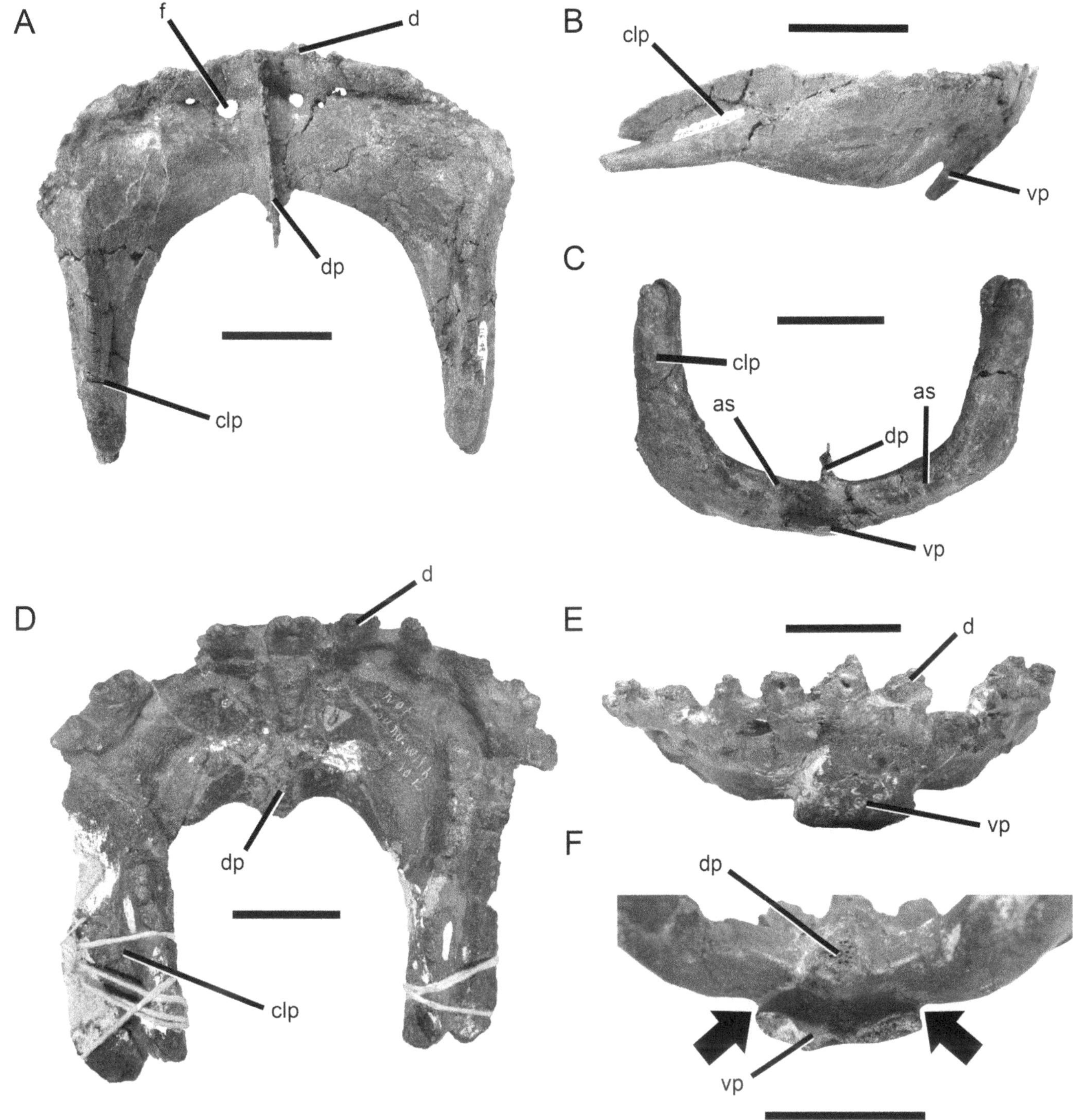

28.3. Hadrosauroid predentaries. (A–C) Hadrosaurid predentary (TMP 1991.036.0311): (A) dorsal view; (B) lateral view; (C) caudoventral view depicting articular surface. (D–F) *Lambeosaurus* predentary (USNM 10309): (D) dorsal view; (E) rostral view; (F) close-up of caudoventral view with arrows depicting caudomedial space of predentary between the dorsal and ventral processes where both dentaries articulated with the predentary and each other. Scale bars equal 5 cm.

ventral region between the smooth surface of the predentary and dentaries. This indicates a broad, flat surface area against which the dentaries were potentially able to articulate more freely. Together, these spaces show that membranous or ligamentous tissue, at least in Hadrosauroidea, likely held the two elements together.

Dentary

The dentary (Fig. 28.6) is the largest element in the hadrosauroid mandible. It articulates with the predentary and the opposite dentary rostrally, the splenial laterally, and the surangular and angular caudally. In lateral view, it is elongate

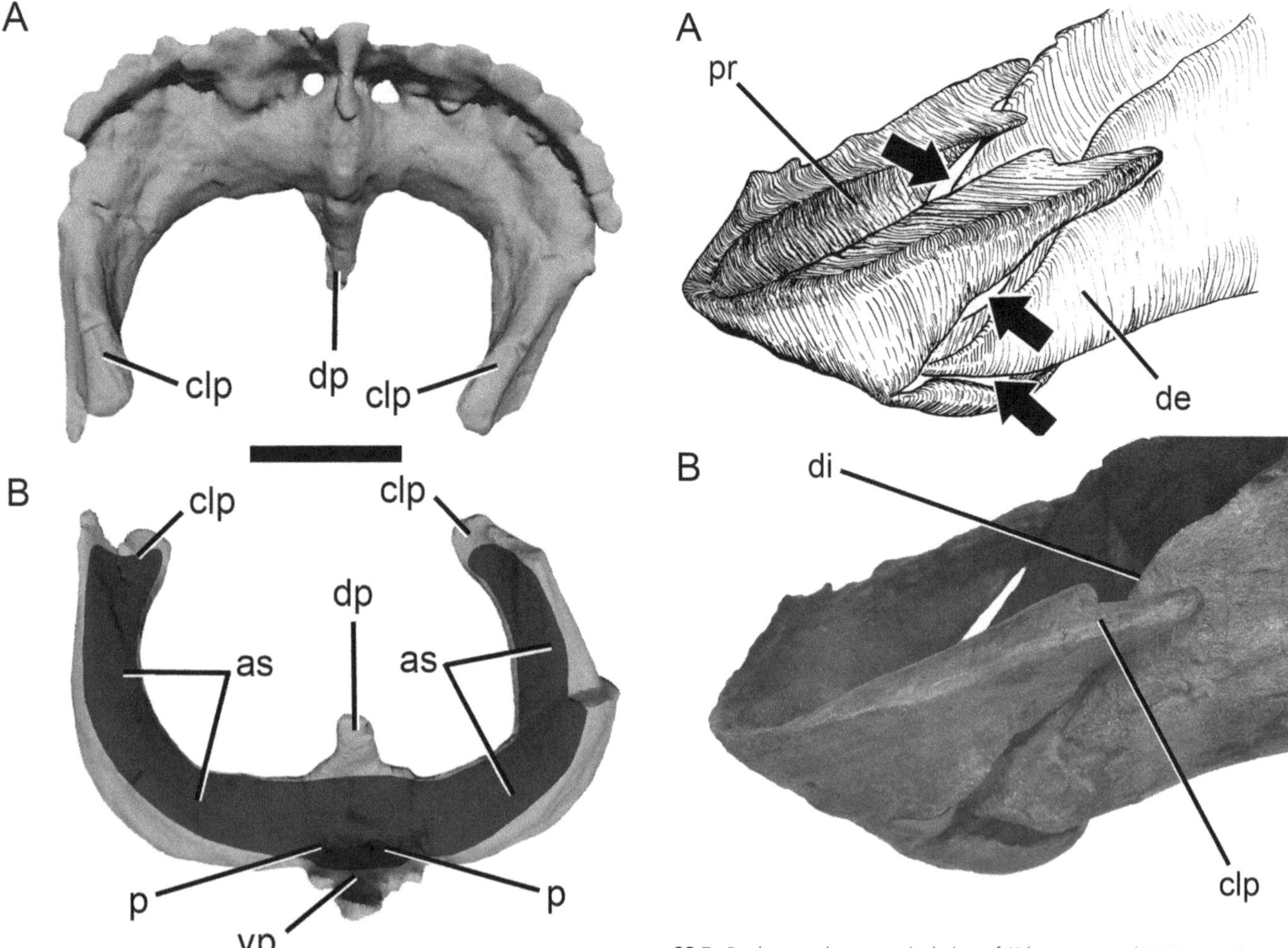

28.4. Surface scan of *Corythosaurus* (USNM 11893) predentary showing the articular surface. (A) Dorsal view showing predentary shape; (B) caudoventral view with darker shaded area depicting wide range on which the sharp, dorsal ridge of the rostral end of the dentary variably articulates. Scale bar equals 3 cm.

28.5. Predentary-dentary articulation of *Kritosaurus* sp. (AMNH 5799). (A) Restored illustration with arrows depicting gaps between the elements; (B) photograph of articulated specimen indicating mediolaterally expanded caudolateral process of predentary resting perpendicularly upon thin, grooved ridge of dentary diastema.

and tapers ventrally at its rostral end. The lateral surface is emarginated lateral to the dental battery, forming a low ridge. There are several foramina on the lateral surface, some of which are lined up ventral to the dental battery along the emargination, which likely supplied neurovasculature to the skin on the lateral surface of the mandible. One large foramen at the rostral tip of the dentaries supplied neurovasculature (mental artery and nerve) rostrally to the predentary. Three or more foramina are also distinctive and placed more caudally but still in the rostral end of the lateral surface; these likely acted as the mental foramina supplying both the predentary and the skin of the more rostral surface of the dentaries themselves. The rostral tip of the dentary is recurved medially and ends in a rostrodorsally expanded straight edge, which comes in contact with its counterpart at the symphysis. It is very blunt with numerous unorganized, tongue-and-groove ridges suggesting bone remodeling induced by forces from movement against the opposite dentary in life. These ridges are seen in nearly all hadrosauroid dentary specimens and are especially exaggerated in more adult forms. Articulation with the predentary at the rostral end is discussed above.

As the dentary extends caudally, it bows out laterally and then straightens. Within the area between the symphysis and the dental battery, there is a distinct diastema that is recurved, flat, and concave medially forming a cupped medial surface (Fig. 28.7A). A distinct thin, recurved ridge with a slight groove upon which the predentary rested, likely supported by membranous or ligamentous tissue, characterizes the dorsal border of this region. This dorsal ridge recurves to eventually become parallel with, and close to, the dorsal ridge of its counterpart before reaching the dental battery. The lateral edge then starts to diverge more laterally mid-length in an arch until reaching the

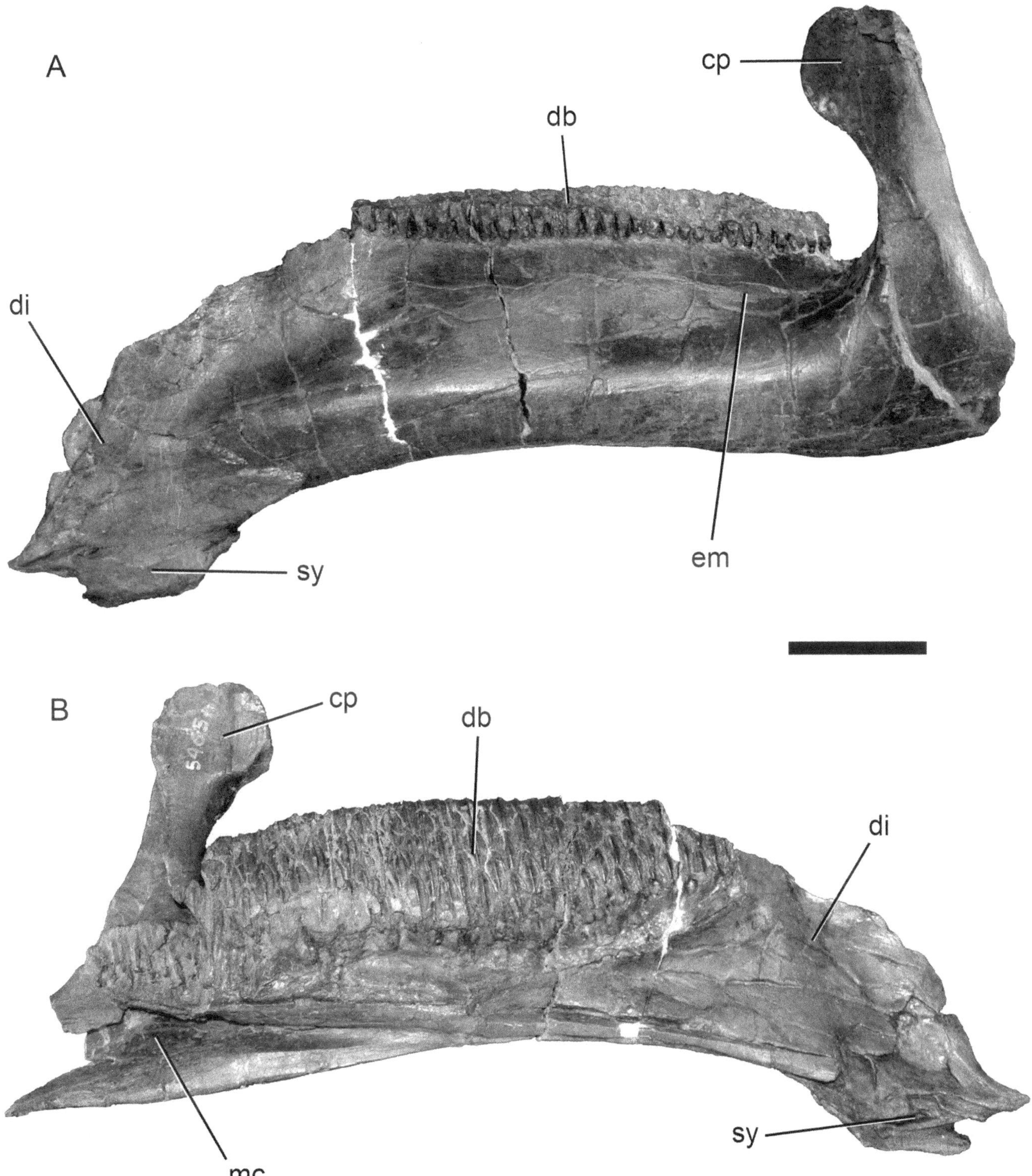

28.6. *Gryposaurus* dentary (AMNH 5465). (A) Lateral view; (B) medial view. Scale bar equals 10 cm.

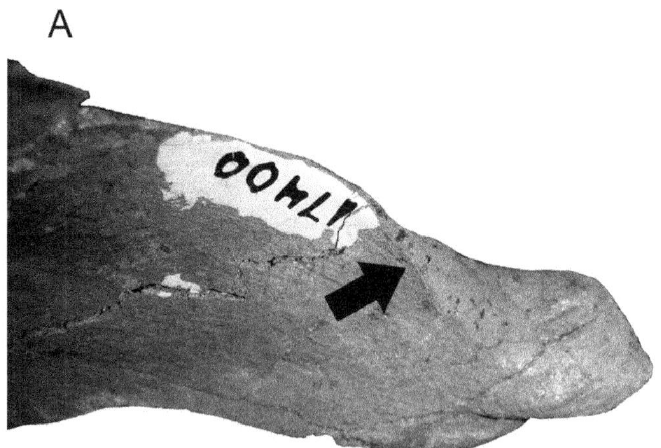

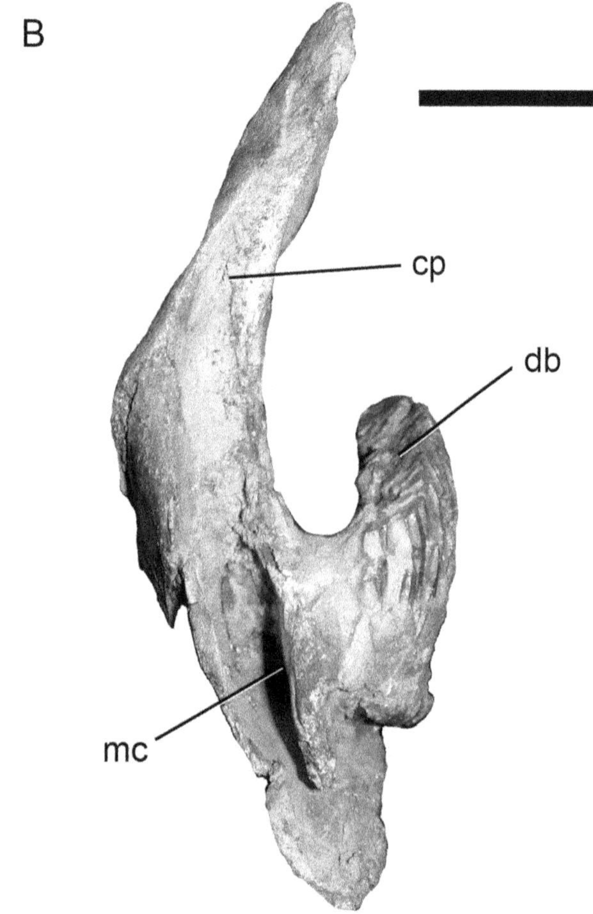

28.7. Hadrosaurid subadult dentary (KUVP 17400). (A) Medial view of anterior diastema with arrow depicting curved articular surface for predentary to rest upon; (B) caudal view of dentary depicting medial curvature of coronoid process and mandibular canal. Scale bar equals 2 cm.

dorsally oriented coronoid process, which is the most lateral portion of the dentary.

The coronoid process (Fig. 28.7B) is usually large and extends dorsally; it is convex on the rostral and lateral sides and concave with thin outer edges on the caudal side of its body. As it ascends dorsally, it becomes oriented rostrally toward the orbit and narial region rather than caudally as in most vertebrates (Ostrom, 1961). Ventrally, it bows laterally from the lateral surface of the dentary body, and forms a convex surface flush with the rest of the dentary. It extends dorsally from the dentary body, medial to the jugal. The dorsal aspect of the dental battery lies just medial to the coronoid process. The medial aspect is much less convex, and sometimes ventrodorsally straight, in hadrosaurids. The most dorsal aspect of the coronoid process is slightly expanded rostrocaudally.

In caudal view, the mandibular canal is large and runs from immediately ventral to the concavity of the caudal surface of the coronoid process to the ventromedial aspect of the dentary body ventral to the dental battery. Two elongate, flat, triangular processes make up the dorsal and ventral parts of the mandibular groove on the medial side and articulate with the surangular. Ventral to the mandibular groove is a flat surface for articulation with the angular. At the caudal end of the dental battery there is also a bony caudal process supporting the splenial on its medial and dorsal surface.

The most distinctive feature of the hadrosauroid dentary is the dental battery, which covers more than half the length of the dentary beginning just caudal to the diastema. It typically contains 40 or more tightly packed columns of teeth, variable with ontogeny and species, that extend dorsally from separate alveoli. The toothrow reaches the caudal-most extremity of the dentary, extending even farther caudally than the coronoid process, which lies lateral to it. All tooth columns jointly form a medially convex configuration that extends dorsally and form a flat, smooth occlusal surface that faces dorsolaterally and occludes with the ventrolateral orientation of the dentition of the maxilla dorsally. In medial view, the alveoli are V-shaped, with both rostral and caudal ends closest to the dorsal surface with the least amount of teeth in the column. A thin, rugose layer of bone conceals the ventral half of the medial surface of the dental battery.

The individual teeth are diamond shaped and apicobasally elongate. There is a thin midline ridge on the lingual side running the height of the tooth. This ridge, when reaching the occlusal surface, creates a serrated edge along the mesiodistal length of the lingual side of the tooth row when combined with the rest of the tooth columns. This serrated edge forms a saw-blade configuration of the tooth row that is bowed dorsally in the center. A description of dental micro- and mesowear is given below.

Splenial

The splenial is a thin, sheet-like bone that covers the ventromedial aspect of the dentary and dental battery, blanketing the dental foramina on that side, but is not firmly adhered to it. It is triangular rostrally where it meets the splenial process of the dentary by means of a shallow groove. It also articulates with the surangular, angular, and articular caudally with a jagged articular surface.

Angular

The angular is an elongate, narrow, laterally compressed bone positioned ventral to the mandibular groove on the medial surface of the dentary rostrally and surangular caudally. It forms the ventromedial edge of the caudal portion of the mandibular corpus, and articulates with the splenial rostrodorsally and the articular caudodorsally. Along with the surangular, it helps brace the articular for sturdy articulation with the quadrate.

Surangular, Articular, and Their Association with the Quadrate

The surangular makes up the largest portion of the mandibular body caudal to the dentary. It forms the walls of the caudal end of the mandibular canal and also possesses a thin, ascending sheet of bone that forms a thin part of the caudal portion of coronoid process. It also articulates with the splenial medially and the angular ventromedially. Caudally, it articulates with the articular, forming the retroarticular process. A dorsally extended lateral lip, or buttress, forms the lateral portion of the articular surface on which the quadrate of the cranium rests, which comprises the synovial craniomandibular joint.

The articular is small, flattened, and subrectangular. It articulates with the surangular rostrally and rostrolaterally, the angular ventrally, and splenial rostromedially. It has a very small contribution to the most medial aspect of the articular surface with the quadrate (mostly made up by the surangular) at the craniomandibular joint. Along with the surangular, it forms the retroarticular process caudally.

As the dorsal aspect of the surangular-articular complex extends caudally, it is elongate and becomes shallowly concave (both rostrocaudally and mediolaterally), sloping upwardly in each direction (Fig. 28.8A). When articulated with the ventral portion of the quadrate, the quadrate covers roughly one-half to two-thirds of the rostrocaudal length of the articular surface, showing that there was a range of movement for the quadrate to translate in that direction against the articular, as the mandible moves. The shallow concavity on the mediolateral surface of the articular also shows that the mandible was not locked in place across this plane. It also allows mediolateral rotation of the mandibular corpus against the quadrate, although there is more room for medial rotation as the lateral side of the articular surface is

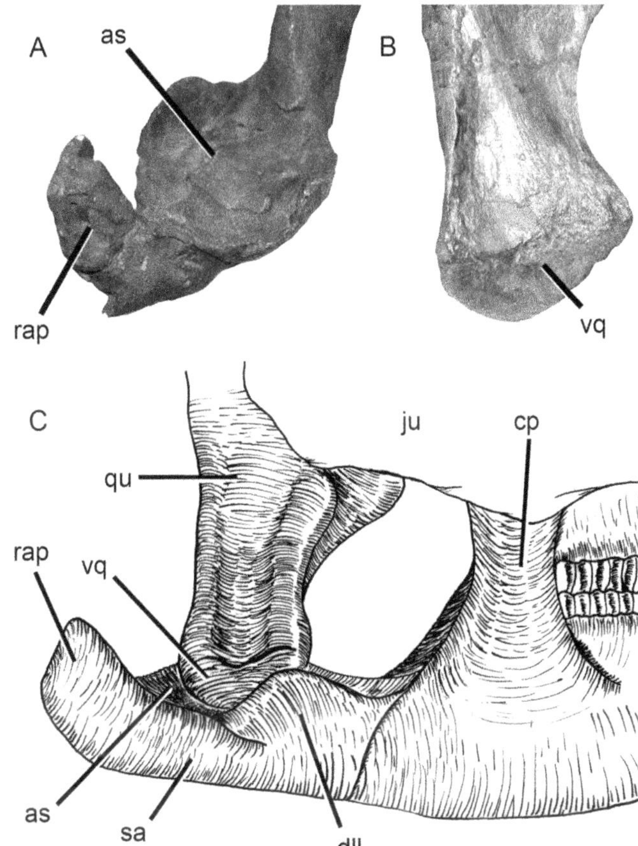

28.8. Hadrosauroid jaw joint quadrate-articular articulation. (A) Ventral portion of hadrosaurid quadrate (AMNH 5220) depicting spherical end; (B) surangular-articular basin (*Telmatosaurus* cast) depicting where quadrate sits; (C) depiction of craniomandibular articulation caudal to dentary.

obstructed by the dorsally extended lateral lip of the surangular. The quadrate is notably spherical at its ventral end (Fig. 28.8B) to the point where it is almost completely ball-shaped in more derived hadrosauroids. In more basal hadrosauroids and ornithopods there are two condyles situated mediolaterally. This effectively creates a ball-and-socket synovial joint at which the mandibular corpus could potentially rotate mediolaterally as well as rostrocaudally within the synovial capsule (Fig. 28.8C).

Dental Micro- and Mesowear

Hadrosauroid dental microwear is well understood (Ostrom, 1961; Weishampel, 1984; Williams et al., 2009), so for the purposes of this study it is briefly discussed based on general observations of three hadrosaurid dentaries from the San Juan Basin of New Mexico (KUVP 96884; KUVP 17400; KUVP 96887), all of which show equivalent wear patterns to each other, and with the morphology described in previous studies.

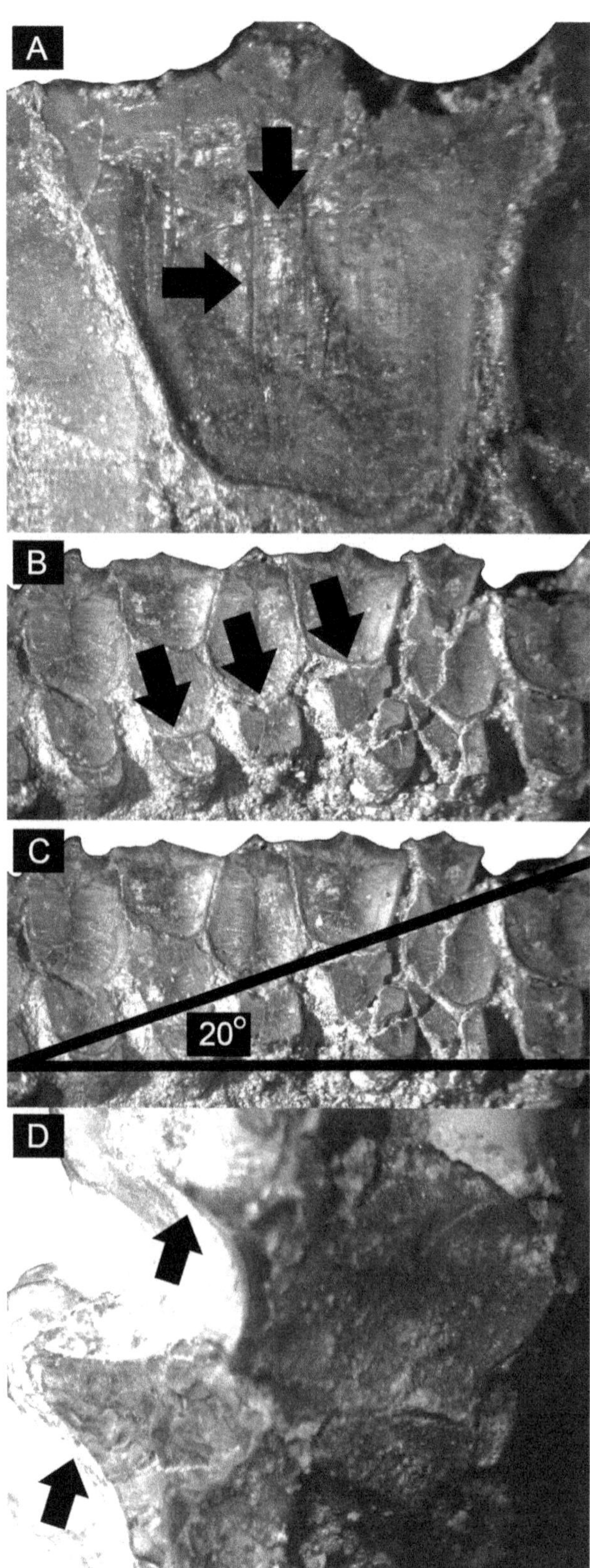

28.9. Hadrosauroid dental micro- and mesowear. (A) Dentary tooth (KUVP 96884) occlusal surface (lingual edge at top) with arrows depicting an example of both mesiodistal and buccolingual wear; (B) occlusal surface of multiple dentary teeth (KUVP 17400) with denticulate serrated edge of dental battery seen on lingual edge (at top). Arrows depict step-ladder mesowear where the edge of the maxilla occludes with dentary; (C) same dentary teeth with lines showing 20 degree angle of maxilla-dentary occlusion; (D) maxillary teeth (KUVP 96887) with arrows depicting curling of denticulate edge from dentary rotation.

When observed head-on with the midline denticle (on the lingual side) on top of the field of view, the occlusal surface of a single dentary tooth is U-shaped capped with a three-point crown, the middle point being represented by the midline ridge that forms a single denticle of the serrated tooth row. Microwear observations show wear in both the mesiodistal and buccolingual directions (Fig. 28.9A). The mesiodistal microwear is largely present along the occlusal surface of the lingual denticulate edge. United they form a shelf of wear along that edge, displaying a rostrocaudal aspect of its jaw mechanism. The buccolingual microwear is very prominent along the majority of the occlusal surface and many of the scratches stretch nearly the entire height of the occlusal surface. This microwear displays a strong transverse aspect of the mandibular corpus as well on both sides of the jaw. Given the dorsolateral orientation of the occlusal surfaces of both dentary tooth rows, a transverse mechanism of the entire lower jaw acting as one unit is unlikely (Weishampel, 1984).

Mesowear patterns occur at two levels on the dentary tooth row: a higher step on the labial edge and a lower step on the lingual edge that ends at the lingual edge of the occlusal surface of the tooth (Fig. 28.9B). The sections at which these two steps meet represent the point at which the maxillary teeth began or ended occlusion with the dentary teeth. Between the dentaries, these wear edges are produced on opposite sides of each other, meaning that on both dentaries the higher steps are both positioned on the labial side of the dentary and the lower steps are both positioned on the lingual side of the dentary. Also, there is an acute angle (~20°) at which the higher mesowear level meets the lower level along the tooth row, showing an offset occlusion (Fig. 28.9C).

On the maxillary occlusal surface, there is more of a concave pit present. This represents the region where the dentary teeth initially occluded with the maxillary teeth. On the lingual edge, instead of a prominent sharp corner, there is a more rounded margin permitting rotation at the end of occlusion of the dentary against the maxilla (Fig. 28.9D). This was most plausibly accomplished by medial rotation of the dentary teeth such that the maxillary teeth wore in a rounded manner. Roughly half of the specimens of hadrosauroid maxillae examined show that this edge is chipped away throughout the tooth row, thus indicating extensive wear. This, along with the offset occlusal patterns and microwear of the dentary teeth, indicate that the dentaries are somehow moving in opposite directions to maneuver the vegetation into the oral cavity, therefore allowing both dentaries to independently rotate mediolaterally simultaneously.

DISCUSSION

Feeding Mechanism

The predentary bone is unusual because it is the only unpaired element in the ornithischian mandible (Weishampel, 2004). This bone remains unfused from the rest of the jaw throughout life, strongly suggesting a kinetic function at this junction (Cuthbertson et al., 2012). Many aspects of the predentary, dentary, and postdentary elements in hadrosauroids, as well as dental micro- and mesowear patterns support an interpretation of independent mobility of both dentaries with respect to the predentary bone attached at their rostral ends, as hypothesized by Bell et al. (2009) and Cuthbertson et al. (2012). Due to its flexible articulation, the predentary likely acted as an axial point allowing both dentaries to independently rotate mediolaterally in relation to it (Fig. 28.10). Its range of movement was restricted by the mediolateral width of the caudolateral processes of the predentary with its ventromedially slanted articular surface. Also, the volume of the assumed flexible membranous or ligamentous tissue that would have suspended the predentary from the dentaries also would have played a role in restricting the range of motion. The rotational surfaces observed on the groove located on the rostrally placed diastema of the dentaries further support this. The roughened, tongue-and-groove edge of the rostralmost aspect of the dentary where it meets its counterpart suggests extensive rotational motion at this junction. This long-axis rotation of the dentaries is analogous to the restricted long-axis rotation seen in modern reptiles such as crocodilians (Porro et al., 2011), lizards (Herrel et al., 2000; Holliday et al., 2010), and *Sphenodon* (Jones et al., 2012) as well as mammals with an unfused symphysis (such as ruminants; Greaves, 1978; Hogue and Ravosa, 2001) and herbivorous marsupials (Crompton et al., 2010), although it is much more exaggerated. It has also been suggested that early extinct ornithurine birds possessing predentaries used independent kinesis of the dentaries against the predentary, although via a different mechanism (Zhou and Martin, 2011).

It is also necessary to demonstrate other possible sources of mobility at the symphysis, other than the likely occurence of membranous or ligamentous tissue. In the space between the dorsal and ventral processes of the predentary of some specimens (i.e., USNM 11893; see Fig. 28.4B), caudal to the rostral border, a variable number of small pits occur bilaterally adjacent to the dentary symphysis. These pits, although possibly suggestive of ligamentous attachment, might also be indicative of an attachment site for M. genioglossus, the main tongue muscle. If this tongue muscle did extend to attach to the predentary, then it might have acted in the control of or, possibly, independent movement of the predentary relative to the dentaries when they were in motion. This is a similar to the manner in which frogs use their mentomeckelian bone, which is similarly located rostral to the dentaries at the symphysis, in tongue flipping (Regal and Gans, 1976). Also, the potential of an autapomorphic synovial cavity at the predentary-dentary joint cannot be ruled out, although such a joint is unknown in vertebrate mandibular symphyses. If a synovial cavity existed at this joint, it would provide even greater mobility at the symphysis than previously postulated. In order to test these speculations, further analysis in hadrosauroids is required to examine the articular surfaces (for signs of calcified cartilage that would indicate a synovial cavity) and histology of the symphyseal region.

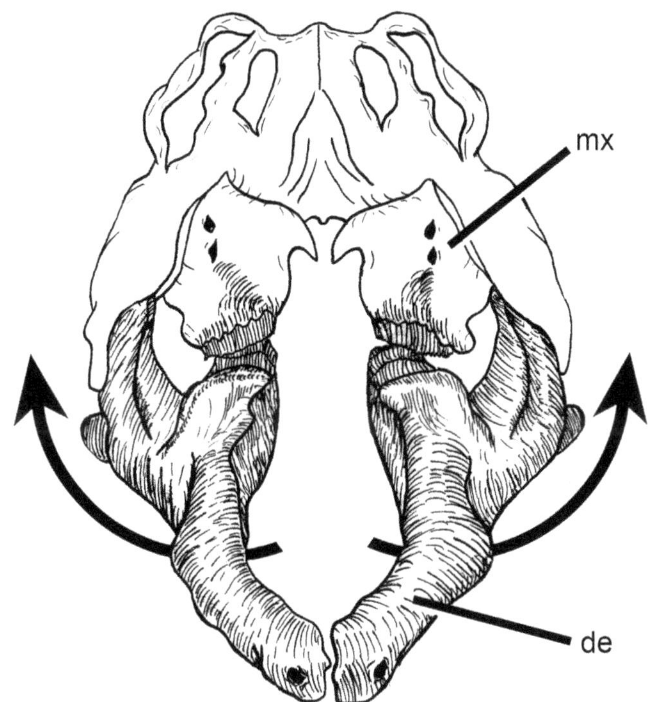

28.10. Rostral view of *Edmontosaurus* skull with premaxilla and predentary removed. Arrows show rotation of dentaries against maxillae. Line drawing based on screen capture of surface scan in video from Rybczynski et al. (2008).

The spherical ventral end of the quadrate and its articulation with the broadened, shallowly bowl shaped basin of the dorsal surangular and articular bone surfaces suggest a great deal of mediolateral rotation of the mandibular corpus at the synovial craniomandibular joint. Also, the fact that the ventral end of the quadrate does not occupy the entire rostrocaudal length of the articular surface provides potential for propalinal motion as well. The overall cylindrical, mediolaterally curved long axis of the dentary suggests mandibular long-axis rotation, where the postdentary elements would not be affected because they act as a continuation of the overall shape of the mandibular corpus. Furthermore, the flexible

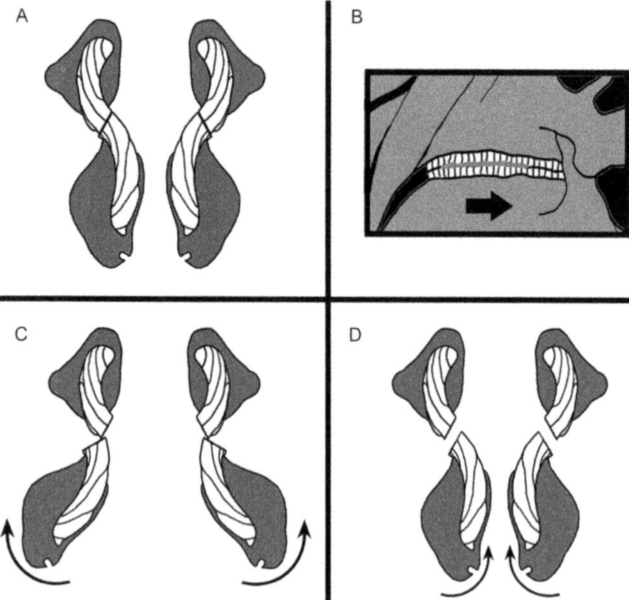

28.11. Proposed four-step feeding mechanism (with coronal jaw cross-section illustrations based on Lambe (1920) with maxillae on top and dentaries on bottom). (A) Occlusion of dental batteries with simultaneous palinal motion in closing stroke; (B) lateral view of hadrosauroid jaw stripping bark with palinal motion of the jaw; (C) medial side of dentaries rotating medially in power stroke with maxillary teeth pressing against the dentary teeth maneuvering vegetation into the oral cavity; (D) returning to resting position in opening stroke before repeating the process.

articulation of the splenial on the medial side of the dentary would allow more room for potential bending to occur in the corpus.

The unusual, medially recurved shape of the coronoid process also suggests medial dentary rotation, as it would allow the dentary to sweep ventromedial to the ventral edge of the jugal during rotation. M. adductor mandibulae externus musculature, originating at the squamosal and temporal region and inserting onto the coronoid process (Holliday, 2009) would have acted as one large fan of muscle that, when contracted, could have acted in both raising and medially rotating the lower jaw, as the muscles would be directed in a ventrolateral orientation. Both the unusual rostral orientation of the coronoid process and its medially recurved shape would have increased lever-arm length in both the rostrocaudal and mediolateral directions respectively, thereby providing a greater mechanical advantage in this respect. When the muscles contracted, they would have pulled the coronoid process caudodorsally as well as medially, rotating them. Future lever-arm studies would be beneficial in quantifying this inference.

The mesiodistal microwear pattern suggests a palinal aspect to the mechanism, and the more prominent labiolingual microwear pattern suggests mediolateral motion of the mandibular corpus. The "step-ladder" dental mesowear pattern described above also documents a second medial rotation. Following occlusion, the two dentaries would have rocked caudally. The ventral ends swinging laterally would have caused the dorsal ends (i.e., the dental battery) to move medially and cut along the labial edge of the maxillary teeth, wearing the teeth down in a rounded fashion, as described above. This imitates a bolt-cutter motion in which branches and conifer needles could be snapped in half and maneuvered into the oral cavity instead of falling out of the mouth. This accounts for the occlusal surface of the dentary teeth facing labially rather than lingually, making it difficult to push food already caught between the dentition into the oral cavity without an additional movement of the jaw in the opposite direction. After bringing the neck forward to obtain vegetation, the hadrosauroid would have to pull it backward to strip the bark and leaves off of the branches. Correspondingly, it may have been possible for the hadrosauroid to drag the serrated edges of its dental battery along the branch as it sheared off bark into its mouth. The denticles along the edge of the premaxilla and predentary could have acted as additional cutting edges, assuming that the keratinous sheath covering them also possessed denticles. This motion corresponds to the tooth wear along the edges of the serrations of the denticles on the top ridge of the dental batteries.

Given this anatomical evidence, a four-step feeding mechanism is proposed for hadrosauroid mandibular complexes (Fig. 28.11): (1) occlusion of the dental batteries with simultaneous palinal motion after picking off bark with its keratinous bill in the closing stroke; (2) stripping bark with palinal motion of the jaw; (3) dentaries rotating medially in the power stroke with the maxillary teeth pressing against the dentary teeth as the jaw is pulled caudally, maneuvering vegetation into the oral cavity; (4) returning to resting position in opening stroke before repeating the process. This mechanism would have allowed hadrosauroids to use a bilateral chewing mechanism with independent, simultaneous motion of each side of the lower jaw, rather than unilateral chewing seen in most modern ungulates. The voluminous nature of predentary-dentary articulation, as described in the morphological analysis above, would have potentially accommodated this bilateral displacement of the mandibular corpora against each other. This would have allowed these animals to process more vegetation in a single bite. The assertion of bilateral occlusion contradicts Cuthbertson et al. (2012), who stated that the feeding mechanism likely used more unilateral occlusion than bilateral occlusion based on computer modeling and animation of an *Edmontosaurus* skull. This skull unfortunately lacked a predentary. Further analyses are required, however, to test the validity of this assertion. Studies of *Heterodontosaurus* (Weishampel, 1984; Crompton and Attridge, 1986; Porro, 2007; Norman et al., 2011), *Lesothosaurus* (Sereno, 1991), and *Euoplocephalus* (Rybczynski and Vickaryous, 2001) also suggest mobility at the predentary-dentary junction and serve as analogs for movements in hadrosaur jaws and, in turn, show the plausibility of this kinesis in the mandible of many other ornithischian dinosaurs.

Comparisons with Other Ornithopods Basal ornithopods such as *Hypsilophodon* have a triangular-shaped predentary (Galton, 1974). There is a single median ventral process cradling the inferior border of the dentaries as well as a groove on the caudal edge of two caudally bifurcating processes on which the rostral edge of the dentary rested. The dentary itself possesses leaf-shaped teeth and is curved medially at the rostral tip, although to a minimal extent. Like hadrosauroids, as well as *Heterodontosaurus, Lesothosaurus,* and *Euoplocephalus,* there is potential for mobility at the predentary-dentary junction. The same holds true for *Thescelosaurus,* although its predentary is much more rostrocaudally elongate, and the rostral tip is more pointed (Boyd et al., 2009).

Zalmoxes shows the first signs of a dorsoventral heightening of the predentary, as well as increased curvature of its rostral end (Weishampel et al., 2003). This curvature begins to form two more distinct bifurcating processes directed caudally over the dentaries, along with the same single ventral process that evolved to become more broadened. In ornithopod evolution, iguanodontian predentaries evolved more distinct denticles on the dorsal edge and the caudal bifurcating processes are expanded to overlap the rostral edge of the dentaries on either side. These widened expansions do not clasp onto the dentaries, but merely allow the dentaries to rest up against them (Norman, 1986). As in hadrosauroids, the paired dentaries do not form a fused symphysis but in fact more loosely press against each other. There are tongue-and-groove ridges along the symphysis of the dentaries, which are interpreted here as due to bone remodeling caused by forces acting at this junction, plausibly from the rubbing of the two dentaries against each other during mastication. Most non-hadrosauroid iguanodontians such as *Dryosaurus* and *Tenontosaurus* also have a bifurcated ventral process of the predentary, which likely provided more stability at this location (Galton, 1983; Winkler et al., 1997). *Ouranosaurus* is a special case in which the caudolateral processes of the predentary are not present, and its articulation with the dentary is by means of its flat caudal surface. Extreme mobility is likely for *Ouranosaurus* as the dentary tooth row also exhibits strong curvature from a ventrolateral orientation rostrally to a dorsomedial orientation caudally (Taquet, 1976).

Derived hadrosauroids exhibit the most extreme cases of widening and expansion of the caudolaterally bifurcating processes, to the extent that they curve dorsally at an angle while still maintaining a flat articular surface. The median ventral bifurcating process is still present. There is no indication of a firm, clasping junction at the predentary-dentary articulation (or between the dentaries themselves). There are gaps between the predentary and dentary regardless of what orientation they are articulated.

Flexible predentary articulations are not restricted to Ornithopoda. Thyreophorans also possess a seemingly mobile predentary bone. Stegosaur predentaries have similar bifurcating processes to more basal ornithopods and *Lesothosaurus* (Galton and Upchurch, 2004). Ankylosaur predentaries possess the least developed articulation of all, with their predentaries shaped roughly like one cylindrical body with a flattened surface on the caudoventral side that merely rests upon the dentaries (Rybczynski and Vickaryous, 2001; Vickaryous et al., 2004). More derived ceratopsians, with parrot-like beaks, seem to be the only case where the predentary locks into the dentaries by means of a slit in the predentary (Dodson et al., 2004); however, this is a derived feature within Ornithischia. Due to the anatomy of the predentary in a large variety of ornithischian dinosaurs, it can be hypthesized that the predentary evolved as a key element in feeding mechanisms of ornithischians, and future work could provide insight into its paleoecological importance throughout ornithischian evolution.

CONCLUSION

Many morphological aspects of the predentary-dentary junction, the postdentary elements, and dental microwear support the presence of independent kinesis of the paired dentary bones relative to the predentary. The predentary would have served as an axial point for the paired dentaries to simultaneously rotate mediolaterally with accommodation of the adductor musculature of the jaw (Ostrom, 1961). Various other mandibular features also suggest a rotating surface and range of movement likely with membranous or ligamentous tissue at the predentary-dentary junction, thus allowing medial rotation of both dentary bones, although muscular or synovial mobility of the predentary-dentary joint are also possible alternatives that require further investigation. Two orientations of tooth wear on the serrated edges and occlusal surfaces suggest both palinal jaw movement to shear vegetation and medial rotation of the dentary teeth against the maxillary teeth. This would maneuver the vegetation into the oral cavity independently on both sides (i.e., work both sides of the jaw simultaneously rather than one side at a time as in most herbivores). More in-depth, and possibly quantitative, analysis of the predentary bone and associated jaw elements in ornithopods is essential to understand its functional implications with respect to pleurokinesis. Analysis of predentary and mandibular morphologies of other ornithischian clades within an evolutionary context is also important, as each one has its own unique form and joint morphologies. This will help elucidate the significance of the predentary bone in ornithischian evolution.

ACKNOWLEDGMENTS

I thank my graduate mentor, D. Weishampel, for his knowledgeable input and helpful discussion, encouragement, and support. I thank the following people for access to specimens: M. Carrano and M. Brett-Surman (USNM); A. Henrici (CMNH); C. Mehling (AMNH); F. Therrien (TMP); L. Martin, D. Burnham, and D. Miao (KUVP). I thank L. Martin for advising me as an undergraduate and sparking the question of the origins of the predentary bone in ornithischians that has continued on as my Ph.D. dissertation project. I also thank P. Dodson and P. Manning for helpful and stimulating discussion as well as the faculty and students of the Functional Anatomy and Evolution department at the Johns Hopkins University School of Medicine for their encouragement and support. I thank the anonymous reviewer, as well as D. Evans and D. Eberth for their helpful suggestions, which helped improve this manuscript. Funding from the NSF Graduate Fellowship Program as well as a University of Kansas Undergraduate Research Grant helped support my research.

LITERATURE CITED

Bell, P. B., E. Snively, and L. Shychoski. 2009. A comparison of the jaw mechanics in hadrosaurid and ceratopsid dinosaurs using finite element analysis. Anatomical Record 292:1338–1351.

Boyd, C. A., C. M. Brown, R. D. Scheetz, and J. A. Clarke. 2009. Taxonomic revision of the basal neornithischian taxa *Thescelosaurus* and *Bugenasaura*. Journal of Vertebrate Paleontology 29:758–770.

Butler, R. J., P. Upchurch, and D. B. Norman. 2007. The phylogeny of the ornithischian dinosaurs. Journal of Systematic Palaeontology 6:1–40.

Crompton, A. W., and J. Attridge. 1986. Masticatory apparatus of the larger herbivores during Late Triassic and Early Jurassic times; pp. 223–236 in K. Padian (ed.), The Beginning of the Age of Dinosaurs: Faunal Change Across the Triassic-Jurassic Boundary. Cambridge University Press, London, U.K.

Crompton, A. W., T. Owerkowicz, and J. Skinner. 2010. Masticatory motor pattern in koala (*Phascolarctos cinereus*): a comparison of jaw movements in marsupial and placental herbivores. Journal of Experimental Zoology 313A:564–578.

Cuthbertson, R. S., A. Tirabasso, N. Rybczynski, and R. B. Holmes. 2012. Kinetic limitations of intracranial joints in *Brachylophosaurus canadensis* and *Edmontosaurus regalis* (Dinosauria: Hadrosauridae), and their implications for the chewing mechanics of hadrosaurids. Anatomical Record 295:968–979.

Dodson, P., C. A. Forster, and S. D. Sampson. 2004. Ceratopsidae; pp. 494–513 in D. B. Weishampel, P. Dodson, and H. Osmólska (eds.), The Dinosauria, Second Edition. University of California Press, Berkeley, California.

Galton, P. M. 1974. The ornithischian dinosaur *Hypsilophodon* from the Wealdon of the Isle of Wight. Bulletin of the British Museum (Natural History) Geology 25:1–152.

Galton, P. M. 1983. The cranial anatomy of *Dryosaurus,* a hypsilophodontid dinosaur from the upper Jurassic of North America and East Africa, with a review of hypsilophodontids from the Upper Jurassic of North America. Geologica et Palaeontologica 17:207–243.

Galton, P. M., and P. Upchurch. 2004. Stegosauria; pp. 343–362 in D. B. Weishampel, P. Dodson, and H. Osmólska (eds.), The Dinosauria, Second Edition. University of California Press, Berkeley, California.

Greaves, W. S. 1978. The jaw lever system in ungulates: a new model. Journal of Zoology in London 184:271–285.

Herrel, A., P. Aerts, and F. De Vree. 2000. Cranial kinesis in geckoes: functional implications. Journal of Experimental Biology 203:1415–1423.

Hogue, A. S., and M. J. Ravosa. 2001. Transverse masticatory movements, occlusal orientation, and symphyseal fusion in selenodont artiodactyls. Journal of Morphology 249:221–241.

Holliday, C. M. 2009. New insights into dinosaur jaw muscle anatomy. Anatomical Record 292:1246–1265.

Holliday, C. M., and L. M. Witmer. 2008. Cranial kinesis in dinosaurs: intracranial joints, protractor muscles, and their significance for cranial evolution and function in diapsids. Journal of Vertebrate Paleontology 28:1073–1088.

Holliday, C. M., N. M. Gardner, S. M. Paesani, M. Douthitt, and J. L. Ratliff. 2010. Microanatomy of the mandibular symphysis in lizards: patterns in fiber orientation and Meckel's Cartilage and their significance in cranial evolution. Anatomical Record 293:1350–1359.

Hopson, J. A. 1980. Tooth function and replacement in early Mesozoic ornithischian dinosaurs: implications for aestivation. Lethaia 13:93–105.

Horner, J. R. 1992. Cranial Morphology of *Prosaurolophus* (Ornithischia: Hadrosauridae) with Descriptions of Two New Hadrosaurid Species and an Evaluation of Hadrosaurid Phylogenetic Relationships. Museum of the Rockies Occasional Paper 2. 119 pp.

Jones, M. E. H., P. O'Higgins, M. J. Fagan, S. E. Evans, and N. Curtis. 2012. Shearing mechanics and the influence of a flexible symphysis during oral food processing in *Sphenodon* (Lepidosauria: Rhynchocephalia). Anatomical Record 295:1075–1091.

Kripp, D., von. 1933. Die Kaubewegung und Lebensweise von *Edmontosaurus* spec. auf Grund der mechanisch-konstruktiven Analyse. Palaeobiologica 5:409–422.

Lambe, L. M. 1920. The Hadrosaur *Edmontosaurus* from the Upper Cretaceous of Alberta. Canadian Geological Survey Memoir 120. 79 pp.

Lieberman, D. E., and A. W. Crompton. 2000. Why fuse the mandibular symphysis? A comparative analysis. American Journal of Physical Anthropology 112:517–540.

Lull, R. S., and N. E. Wright. 1942. Hadrosaurian Dinosaurs of North America. Geological Society of America Special Papers 40. 242 pp.

Marsh, O. C. 1893. The skull and brain of *Claosaurus*. American Journal of Science, Series 3 55:83–86.

Morris, W. J. 1970. Hadrosaurian dinosaur bills: morphology and function. Contributions in Science, Los Angeles County Museum of Natural History 193:1–14.

Nopsca, F. B. 1900. Dinosaurierreste aus Siebenbergen. I. Schadel von *Limnosaurus transsyvanicus* nov. gen. et spec. Denkschr Akad Wiss Wien 68:555–591.

Norman, D. B. 1984. On the cranial morphology and evolution of ornithopod dinosaurs. Symposium of the Zoological Society in London 52:521–547.

Norman, D. B. 1986. On the anatomy of *Iguanodon atherfieldensis* (Ornithischia: Ornithopoda). Bulletin de l'Institut royal des Sciences naturelles de Belgique, Sciences de la Terre 56:281–372.

Norman, D. B., and D. B. Weishampel. 1985. Ornithopod feeding mechanisms: their bearing on the evolution of herbivory. American Naturalist 126:151–164.

Norman, D. B., A. W. Crompton, R. J. Butler, L. B. Porro, and A. J. Charig. 2011. The lower Jurassic ornithischian dinosaur *Heterodontosaurus tucki* Crompton & Charig, 1962: cranial anatomy, functional morphology, taxonomy, and relationships. Zoological Journal of the Linnean Society 163:182–276.

Ostrom, J. H. 1961. Cranial morphology of the hadrosaurian dinosaurs of North America. Bulletin of the American Museum of Natural History 122:39–186.

Porro, L. B. 2007. Feeding and jaw mechanisms in *Heterodontosaurus tucki* using finite element analysis. Journal of Vertebrate Paleontology 27(3, Supplement):131A.

Porro, L. B., C. M. Holliday, F. Anapol, L. C. Ontiveros, L. T. Ontiveros, and C. F. Ross. 2011. Free body analysis, beam mechanics, and finite element modeling of the mandible of

Alligator mississippiensis. Journal of Morphology 272:910–937.

Regal, P. J., and C. Gans. 1976. Functional aspects of the evolution of frog tongues. Evolution 30:718–734.

Rybczynski, N., and M. K. Vickaryous. 2001. Evidence of complex jaw movement in the Late Cretaceous ankylosaurid *Euoplocephalus tutus* (Dinosauria: Thyreophora); pp. 299–317 in K. Carpenter (ed.), The Armored Dinosaurs. Indiana University Press, Bloomington, Indiana.

Rybczynski, N., A. Tirabasso, P. Bloskie, R. Cuthbertson, and C. Holliday. 2008. A three-dimensional animation model of *Edmontosaurus* (Hadrosauridae) for testing chewing hypotheses. Palaeontologica Electronica 11.2.9A.

Sereno, P. C. 1991. *Lesothosaurus,* "fabrosaurids," and the early evolution of Ornithischia. Journal of Vertebrate Paleontology 11:168–197.

Sues, H. D. 1980. Anatomy and relationships of a new hypsilophodontid dinosaur from the Lower Cretaceous of North America. Palaeontographica A 169:51–72.

Taquet, P. 1976. Ostéologie d'*Ouranosaurus nigeriensis,* Iguanodontide du Crétacé Inférieur du Niger; pp. 57–168 in Géologie et Paléontologie du Gisement de Gadoufaoua (Aptien du Niger). Éditions du Centre national de la Recherche scientifique, Paris.

Thulborn, R. A. 1971. Tooth wear and jaw action in the Triassic ornithischian dinosaur *Fabrosaurus.* Journal of Zoology in London 164:165–179.

Versluys, J. 1910. Streptostylie bei Dinosauriern nebst Bemerkungen uber die Verwandtschaft der Vogel und Dinosaurier. Zoologische Jahrbucher Abteilung fur Anatomie und Ontogenie der Tiere 30:175–260.

Versluys, J. 1912. Das Streptostylie-Problem und die Bewegung im Schadel bei Sauropsida. Zoologische Jahrbucher Abteilung fur Systemik (Okologie) Geographie und Biologie der Tiere (Supplement) 15:545–716.

Versluys, J. 1923. Der Schadel des Skeletts von *Trachodon annectus* im Senckenberg-Museum. Abhandlungen Herausgegeben von der Senckenbergischen Naturforschenden Gesellschaft 38:1–19.

Vickaryous, M. K., T. Maryańska, and D. B. Weishampel. 2004. Ankylosauria; pp. 363–392 in D. B. Weishampel, P. Dodson, and H. Osmólska (eds.), The Dinosauria, Second Edition. University of California Press, Berkeley, California.

Weishampel, D. B. 1984. Evolution of jaw mechanisms in ornithopod dinosaurs. Advances in Anatomy Embryology and Cell Biology 87:1–109.

Weishampel, D. B. 2004. Ornithischia; pp. 323–324 in D. B. Weishampel, P. Dodson, and H. Osmólska (eds.), The Dinosauria, Second Edition. University of California Press, Berkeley, California.

Weishampel, D. B., C.-M. Jianu, Z. Csiki, and D. B. Norman. 2003. Osteology and phylogeny of *Zalmoxes* (n.g.), an unusual euornithopod dinosaur from the latest Cretaceous of Romania. Journal of Systematic Palaeontology 1:1–56.

Williams, V. S., P. M. Barrett, and M. A. Purnell. 2009. Quantitative analysis of dental microwear in hadrosaurid dinosaurs, and the implications for hypotheses of jaw mechanics and feeding. Proceedings of the National Academy of Sciences 106:11194–11199.

Winkler, D. A., P. A. Murry, and L. L. Jacobs. 1997. A new species of *Tenontosaurus* (Dinosauria: Ornithopoda) from the Early Cretaceous of Texas. Journal of Vertebrate Paleontology 17:330–348.

Zhou, Z., and L. D. Martin. 2011. Distribution of the predentary bone in Mesozoic ornithurine birds. Journal of Systematic Palaeontology 9:25–31.

Appendix 28.1. Hadrosauroid Specimens Examined in This Study

Taxon	Specimen	Elements
Bactrosaurus	AMNH 6553	Paired dentaries, left maxilla
	AMNH 6372	Predentary
	AMNH 6365	Partial skull
	AMNH 6366	Skull roof
Corythosaurus	AMNH 3971	Left dentary
	AMNH 5240	Complete jaw and skull
	AMNH 5338	Complete jaw and skull
	AMNH 5461	Complete jaw and skull
	CMNH 1074	Right dentary, partial right maxilla
	CMNH 11376	Paired dentaries
	USNM 11893	Predentary, paired dentaries
	USNM 16600	Predentary, right dentary, complete skull
Edmontosaurus	AMNH 5046	Right dentary, skull
	AMNH 5879	Paired maxillae
	CMNH 9970	Right dentary
	USNM 2155	Paired jaw (no predentary), partial skull
	USNM 3814	Complete jaw and skull
	USNM (uncataloged)	Left dentary
Gryposaurus	TMP 1980.022.0001	Complete jaw and skull
Hypacrosaurus	AMNH 5278	Paired jaws (no predentary), complete skull
	AMNH 5357	Left dentary, left maxilla
Kritosaurus	AMNH 5799	Complete jaws and skull
	USNM 8629	Left jaw ramus
Lambeosaurus	AMNH 5353	Complete jaw and skull
	AMNH 5373	Complete jaw and skull
	AMNH 5382	Partial jaw (no predentary), partial skull
	USNM 10309	Predentary; right dentary, quadrate
	TMP 1966.004.0001	Complete jaw and skull
Parasaurolophus	KUVP (uncataloged–cast of holotype)	Complete jaw and skull
Prosaurolophus	USNM 12712	Complete jaw and skull
Saurolophus	AMNH 5220	Complete jaw and skull
	AMNH 5221	Paired dentaries, paired premaxillae, paired maxillae
Telmatosaurus	AMNH 5145 (cast of holotype)	Paired jaws (no predentary), partial skull
Unidentified hadrosaurids	AMNH 5285	Predentary
	AMNH 21523	Left dentary
	AMNH 21524	Left dentary
	AMNH 21529	Predentary, left dentary, maxilla
	AMNH 5899	Paired dentaries, quadrate
	KUVP 17400	Left dentary
	KUVP 96884	Dentary
	KUVP 96887	Dentary
	TMP 1975.011.0045	Predentary
	TMP 1991.036.0311	Predentary

Preservation, Tracks, and Traces

Debris Flow Origin of an Unusual Late Cretaceous Hadrosaur Bonebed in the Two Medicine Formation of Western Montana

29

James G. Schmitt, Frankie D. Jackson, and Rebecca R. Hanna

ABSTRACT

Sedimentologic and taphonomic analyses of a paucitaxic bonebed (0.4–1.5 m thick) containing thousands of hadrosaur skeletal elements in the Two Medicine Formation (Upper Cretaceous) of western Montana (Willow Creek anticline) indicate transport and deposition by an unconfined fluid debris flow to hyperconcentrated flow. Deposition occurred as a single sedimentation event across a low relief alluvial plain characterized by small, low-gradient sandy fluvial channels and adjacent interchannel areas composed of crevasse splay and vertical accretion deposits. The normally graded, clast- to matrix-supported conglomerate consists of predominantly pebble- to boulder-size intraformational clasts of micrite, mudstone, siltstone, and sandstone mixed with disarticulated hadrosaur bones in a sandy mudstone matrix. Subrounded lithic clasts, abraded bones, and bones broken across cobbles indicate significant turbulence and clast interaction during transport. Bones are minimally weathered, unburned, and show only rare evidence of gnawing or boring, suggesting minimal residence time (10^{-1} to 10^{1} years) at the surface before transport. The sediment transport event collected disarticulated bones from the land surface and ripped up underlying floodplain deposits, likely incorporating some previously buried bones in the process, and transported them together in a mud-rich flow. Skeletal disarticulation suggests the debris flow–producing event was not responsible for death of the hadrosaurs. Our analysis fails to support previous interpretations of this bonebed as a catastrophic mass-death assemblage of a large hadrosaur herd that resulted from a volcanic eruption.

INTRODUCTION

The majority of bone-bearing horizons preserved in the fossil record of fluvial environments are paucitaxic to polytaxic time-averaged assemblages recording either sequential accumulation of skeletal elements through time and/or their reworking and concentration by hydrodynamic processes (e.g., Behrensmeyer, 1982; Rogers and Kidwell, 2007). Conversely, monotaxic or monodominant bonebeds that preserve the size and age structure of a biological population are less common (e.g., Eberth et al., 2007). These mass-mortality assemblages require a sedimentation event coinciding both temporally and spatially with the catastrophic death of large numbers of a single population, and burial by the deposits of the same event (Behrensmeyer and Hook, 1992). For example, death by suffocation during volcanic ashfall may produce both catastrophic mass mortality and simultaneous burial of animals (Voorhies, 1985).

In other cases, sedimentation events, causally and temporally unrelated to the animals' death, may facilitate transport and deposition of bones or carcasses (Siebe et al., 1999). For example, disease or starvation may result in death of large numbers of individuals of diverse species, followed by rainfall on destabilized ash-covered hill slopes that transport and bury the remains. These sedimentary processes may obscure age-related size classes in these attritional deposits. Recognition of remnant class distributions becomes increasingly difficult with longer durations of reworking and higher rates of bone modification between death and deposition. Other bonebeds result from mass mortality by a sediment transport event, but sample only a random portion of a population, providing a potentially biased and inaccurate snapshot of population structure (Turnbull and Martill, 1988).

The Upper Cretaceous (Campanian) Two Medicine Formation of western Montana contains numerous paucitaxic dinosaur bonebeds interpreted as drought-induced mass-death assemblages (Rogers, 1990; Varricchio and Horner, 1993; Varricchio, 1995). In contrast to these parautochthonous assemblages, our study focuses on a hadrosaur bonebed (MOR-TM-003) located in Two Medicine Formation exposures at the Willow Creek anticline (Fig. 29.1) west of Choteau, Montana.

The original taphonomic study of the Willow Creek anticline bonebed resulted from research conducted as part of an unfinished master's thesis. In an abstract of this study, Hooker (1987) hypothesized that gas and ash from a volcanic

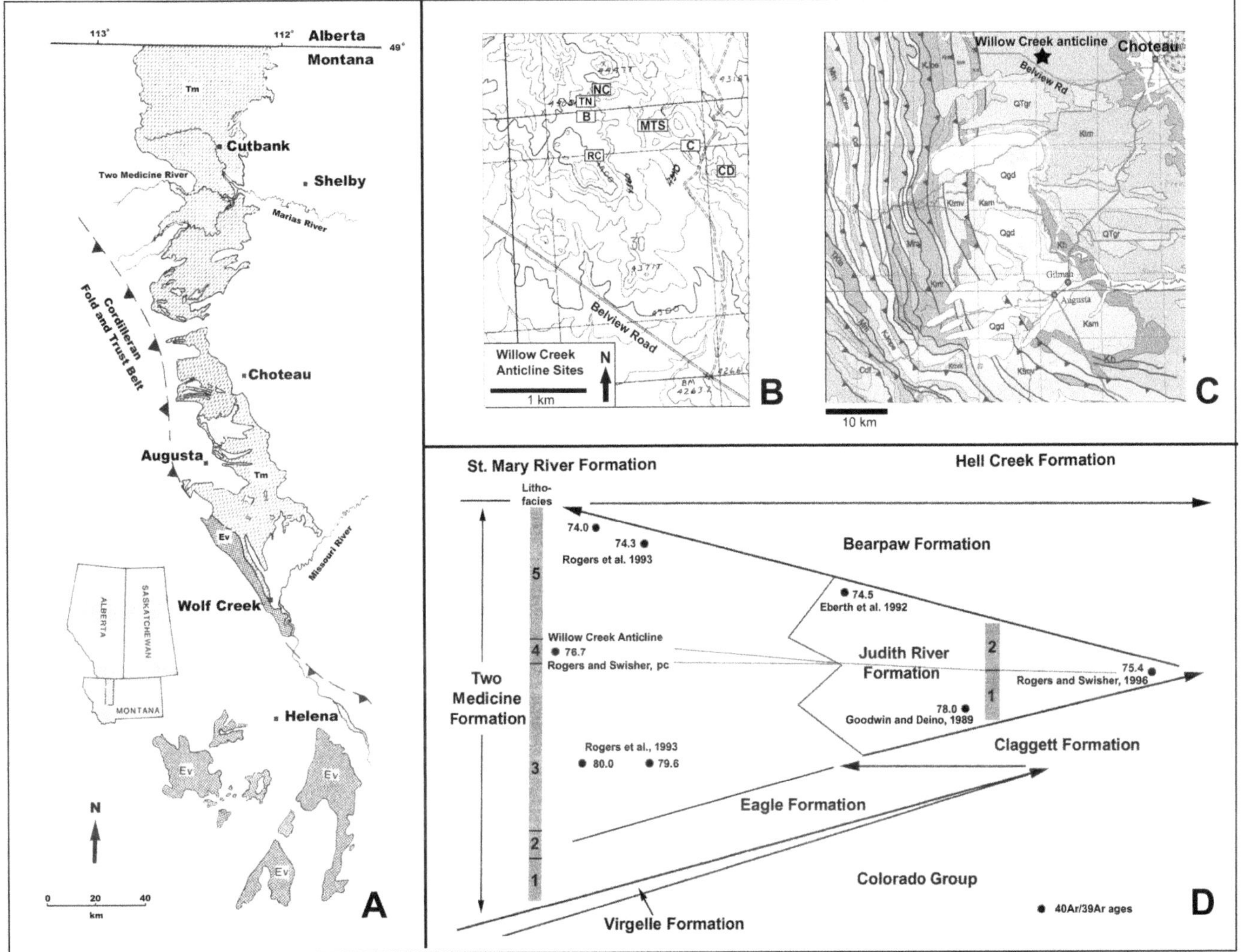

29.1. Geologic, geographic, and stratigraphic context of the study area. (A) Outcrop extent of the Upper Cretaceous (Campanian) Two Medicine Formation and volcanic rocks of the Elkhorn Mountains (Ev) and Adel Mountain (Av) volcanic fields in western Montana; (B) the Willow Creek anticline study area and location of measured sections including Children's Dig (CD), Camposaur (C), *Maiasaura* Type Section (MTS), Nose Cone (NC), That's Nice (TN), Brandvold (B), and Rainbow Cliff (RC) sections (scale bar equals 1 km); (C) geologic setting of the Willow Creek anticline (star); and (D) stratigraphic diagram showing the position of the five Two Medicine Formation lithostratigraphic units recognized at the Willow Creek anticline and distribution of radiometric dates from ash beds in the Two Medicine–Judith River clastic wedge. Modified from Rogers (1998) and Horner et al. (2001).

eruption killed and partially buried members of a hadrosaur herd. Following death, disarticulation, decay, bone dissolution, and partial permineralization occurred at the land surface, followed by a flood that remobilized and transported the ash and bones in a sediment slurry. Other interpretations include inferences of gregarious behavior and estimated hadrosaur herd size exceeding 10,000 individuals (Horner and Gorman, 1988).

Here we present the results of the first detailed taphonomic study of the bonebed. We evaluate previous interpretations of the origin and preservation of the fossil assemblage through detailed taphonomic and sedimentologic analyses. Compared to other Two Medicine bonebeds, the Willow Creek anticline bonebed represents an unusual deposit because of its relative abundance of lithic clasts mixed with skeletal elements. This study, therefore, constitutes an important contribution to understanding both the origin and paleoebiological constraints of this fossil assemblage.

GEOLOGIC SETTING

The Upper Cretaceous Two Medicine Formation was deposited in the foredeep of the Cordilleran foreland basin system during the Campanian (Fig. 29.1A; Lorenz, 1981; Rogers, 1998). Comprising 600 to 1500 m of nonmarine strata, the Two Medicine records deposition in a variety of fluvial distributary channel systems and their accompanying floodplain environments (Lorenz, 1981; Lorenz and Gavin, 1984;

Rogers, 1994; 1998; Roberts, 1999; Shelton, 2007). Rogers (1998) infers that Two Medicine deposition approximates nonmarine coastal plain deposition synchronous primarily with the R8 regression and subsequent T9 transgression of the western shoreline of the Western Interior seaway (Two Medicine Judith River clastic wedge; Fig. 29.1D). Changes in alluvial architecture (channel belt morphology) recorded in Two Medicine strata are likely related primarily to both eustatic (R8 regression) and local to regional tectonic (T9 transgression) drivers (Gill and Cobban, 1973; Rogers, 1998; Roberts, 1999).

At the Willow Creek anticline (Fig. 29.1B, C), the Upper Cretaceous (Campanian) Two Medicine Formation consists primarily of anastomosed and ephemeral fluvial channel sandstone and related floodplain mudrock interbedded with lacustrine micrite-dominated intervals (Lorenz, 1981; Lorenz and Gavin, 1984; Shelton, 2007). These strata range in age from 80.0 ± 0.1 Ma to 74.1 ± 0.7 Ma (Rogers et al., 1993) and record terrestrial foreland basin sedimentation along the western margin of the Cretaceous seaway coeval with eruptive activity in the explosive Elkhorn Mountains (81 to 74 Ma) and effusive Adel Mountain volcanic fields to the south (76.0 to 73.6 ± 0.7 Ma) (Skipp and McGrew, 1977; Sheriff and Gunderson, 1990; Gunderson and Sheriff, 1991; Harlan et al., 2005; Fig. 29.1D). Varricchio et al. (2009) dated a single detrital zircon crystal in the bonebed at 75.65 ± 0.45 Ma and a tuff located 4.3 m above the bonebed at 75.53 ± 0.32 Ma. Using the informal lithostratigraphic subdivision of Two Medicine strata at the Willow Creek anticline proposed by Lorenz (1981) and Lorenz and Gavin (1984) and augmented with vertebrate fossil data by Horner et al. (2001), the bonebed is located at or near the base of Lithofacies 4 (see Fig. 29.1D) that records maximum lowstand separating the underlying regressive fluvial systems tract of Lithofacies 2 and 3 from the overlying transgressive alluvial systems tract of Lithofacies 5 (Horner et al., 2001). Placed in a paleogeographic context, bonebed deposition occurred when the Cretaceous seaway shoreline was located at or near its maximum distance (several hundred kilometers to the east) from the area of the Willow Creek anticline during Campanian time.

Two Medicine fluvial systems were supplied with lithic detritus derived from erosion of thrust allochthons in the adjacent Cordilleran fold and thrust belt to the west and volcaniclastic sediment from the explosive Elkhorn Mountain volcanic field and its satellites (e.g., Wolf Creek volcanic center) to the south-southwest (Schmidt, 1978; Lorenz, 1981; Bibler and Schmitt, 1986; Smith, 1998). Explosive volcanism especially affected sedimentation in Two Medicine fluvial systems in the area to the south between Wolf Creek and Augusta, Montana, where 1500 m of Two Medicine strata contain numerous volcaniclastic channel conglomerates (lahar deposits), ash-flow tuffs, and both rhyolite and basalt lava flows (Smith, 1998; LaBranche, 1999). Although direct effects of explosive activity on channels and floodplains as far north as the Willow Creek anticline was less, ash-fall tuffs are common (Rogers et al., 1993; Rogers, 1998; Roberts, 1999; Shelton, 2007). In addition, Roberts and Hendrix (2000) document ignition and burial of a Two Medicine floodplain forest by a pyroclastic flow (80.002 ± 0.114 Ma) several km south of the Willow Creek anticline.

METHODS

Fossil Assemblage

Excavations of several sites in the bonebed at the Willow Creek anticline by the Museum of the Rockies over the last several decades have produced several thousand bones. Unfortunately, most specimens were assigned the bonebed number MOR-TM-003 and therefore the specific localities are unknown. This study includes only those previously excavated bones (n = 189) cataloged in the Museum of the Rockies collections with unique specimen numbers referable to a quarry map. Additional data were recovered from a 2 m^2 test pit near the Camposaur site, which yielded 58 bones, for a combined data set of 247 elements. The test pit was sequentially excavated in 10 cm intervals and all bones and lithic clasts were mapped and numbered at each level to determine the distribution, spatial relationships, and quantity of bone versus lithic clasts.

Parameters recorded for each bone or bone fragment include quantity, vertical and horizontal positions, orientation (strike and dip), taxon, anatomical position, and other taphonomic features (see list below). Following Klein and Cruz-Uribe (1984), quantity was calculated so that a complete element equals one and a partial element equals some fraction less than one. This allows more accurate calculation of Minimum Number of Individuals (MNI), and avoids artificial inflation of the MNI in cases where partial elements potentially belong to one individual (e.g., one-half of a proximal right humerus and one-third of a distal right humerus).

Provenance data include quarry coordinates and (when available) depth below surface or excavation level. Taxonomic identification reflects the most specific level possible for confident assignment. Anatomical position includes element, portion of element, side, and size measurements (e.g., length, width, diameter). Other taphonomic features documented include: (1) degree of abrasion; (2) degree of weathering; (3) presence or absence of root traces; (4) presence or absence of borings; (5) presence or absence of burning; (6) presence or absence of gnaw and bite marks (i.e., depressed fractures, punctures, and/or tooth scrape marks); (7) characterization of breakage pattern (i.e., angular, crushed, irregular longitudinal, irregular perpendicular, smooth longitudinal, smooth perpendicular, spiral, unbroken, and unspecified); (8) presence or absence of plastic deformation; and (9) presence or absence of diagenetic crushing. Characterization of breakage pattern follows Shipman (1981), modified to accommodate a wider range of breakage types and to provide more easily understood nomenclature. Abrasion categories follow Fiorillo (1988) and Cook (1995): Stage 0, very angular; Stage 1, subangular; Stage 2, subrounded; Stage 3, rounded; and Stage 4, extremely rounded. Intermediate categories were also recorded. Assessment of degree of weathering follows Fiorillo's (1988) scheme, which follows Behrensmeyer's (1978) stages for modern bones and modified for application to fossilized material. Intermediate weathering stages were recorded for elements between discrete stages. Weathering stages include Stage 0, bone surface that lacks cracks or flakes; Stage 1, cracks confined to outermost cortical bone, which are usually parallel to the bone grain; Stage 2, flaking and cracking of outer bone surface, with cracks starting to penetrate the cortex; and Stage 3, superficial cortical layers absent and most cracks penetrating into bone cavities (Fiorillo, 1988).

Sedimentology and Stratigraphy

Detailed stratigraphic sections were measured at seven locations along the outcrop belt of the hadrosaur bonebed in the Two Medicine Formation at the Willow Creek anticline to establish the stratigraphic framework of the bonebed (Fig. 29.1B). Strata up to approximately 20 m below and above the bonebed, as well as the bonebed itself, were measured. Locations were selected on the basis of exposure quality, particularly of the bonebed unit, whereas geographic separation of the stratigraphic sections ensured recognition of any lateral variations in strata and bonebed characteristics. Lithofacies were delineated following Miall's (1996) methods for alluvial deposits. Detailed lithofacies profiles were constructed during measurement and description of each stratigraphic section. In addition, lateral tracing of sandstone bodies and mudrock intervals permitted recognition of lateral lithofacies variation and unit geometry. Photomosaics of several individual sandstone body outcrops were used to map lithofacies to ascertain the size and nature of the alluvial channels.

The bonebed was trenched and examined at all seven measured sections (Fig. 29.1B). The detailed lithologic and sedimentologic characteristics of the bonebed were derived primarily from exposures created by quarrying at the Nose Cone site. In addition, all rocks and bones in the aforementioned Camposaur test pit were mapped and numbered to evaluate the distribution, relationship, and quantity of bones versus lithic clasts in the deposit. Textural and sedimentary structure data were collected during the course of this excavation, whereas shovel trenching allowed examination of the bonebed at the remaining five stratigraphic sections.

Lithic clasts were identified and counted during excavation of the Camposaur test pit. Thin sections of several micrite clasts were prepared and examined to aid in determination of carbonate lithology. Eight matrix samples were analyzed for mineral composition using X-ray diffraction (XRD). Samples were run using a Scintag X1 Powder Diffraction Spectrometer in the Imaging and Chemical Analysis Laboratory (ICAL) at Montana State University. Glycolated and air-dried oriented mounts were prepared for each sample to assess expandable layer clay mineral content.

RESULTS

Assemblage Data

Of the 247 skeletal elements examined in this study, 171 (69.2%) and 15 (6.1%) bones are referable to the Hadrosauridae and Maiasaurinae, respectively, and 7 elements (2.8%) are assignable at the species level to *Maiasaura peeblesorum*. The remaining sample includes 5 theropod (2.0%) and 49 (19.8%) unidentified skeletal elements. The Number of Individual Specimens (NISP) for hadrosaurs is 193, whereas the Minimum Number of Individuals (MNI) is calculated as 7 (using right tibiae), with an NISP/MNI ratio of 27. Hadrosaur material includes cranial, axial, and appendicular elements, with greater representation of elements that are more abundant within the skeleton (e.g., caudal and dorsal vertebrae). Theropod elements are rare and include two caudal vertebrae, four shed tyrannosaurid teeth, and a crushed theropod limb bone.

Results for weathering, abrasion, breakage pattern, gnawing, plastic deformation, and diagenetic crushing are summarized in Figure 29.2. Histograms for borings and root traces are not included because of the insignificant representation of the former and our lack of confidence that the latter represent fossil rather than modern attributes. Most bones exhibit minimal to no weathering, with 60% judged as Stage 1 and 37% as unweathered (Fig. 29.2A). Bones exhibit a wide range of abrasion stages, although most are ranked as subangular to subrounded (Fig. 29.2B).The overwhelming majority (84%) of bones are broken and most of these consist of angular (32.3%) and irregular perpendicular (32.3%) style breaks (Fig. 29.2C). The only other well-represented breakage pattern is smooth perpendicular (6%). Only 3.9% of the specimens examined show evidence of gnawing in the form of tooth scrape marks or punctures (Fig. 29.2D). Invertebrate borings occur on 1.6% of the specimens, and

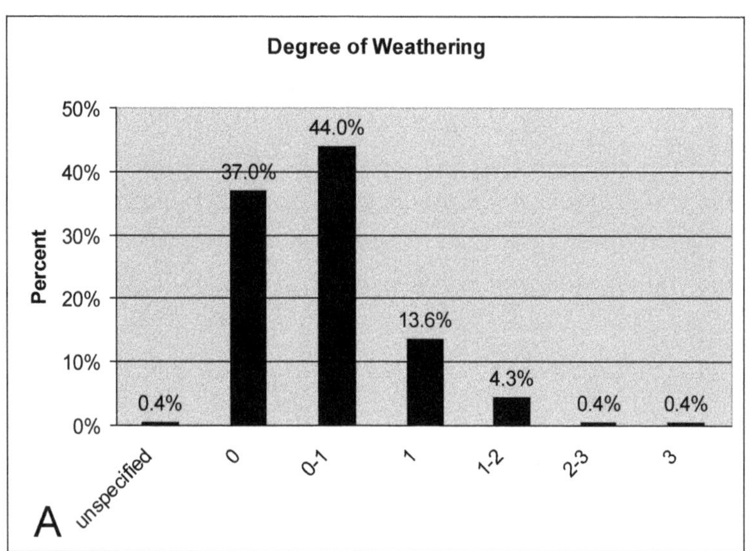

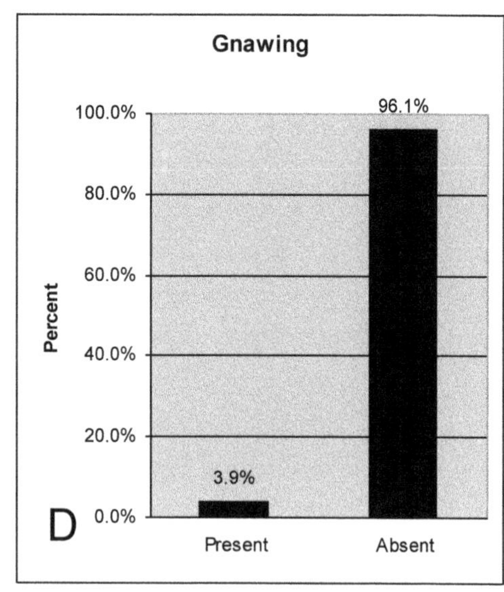

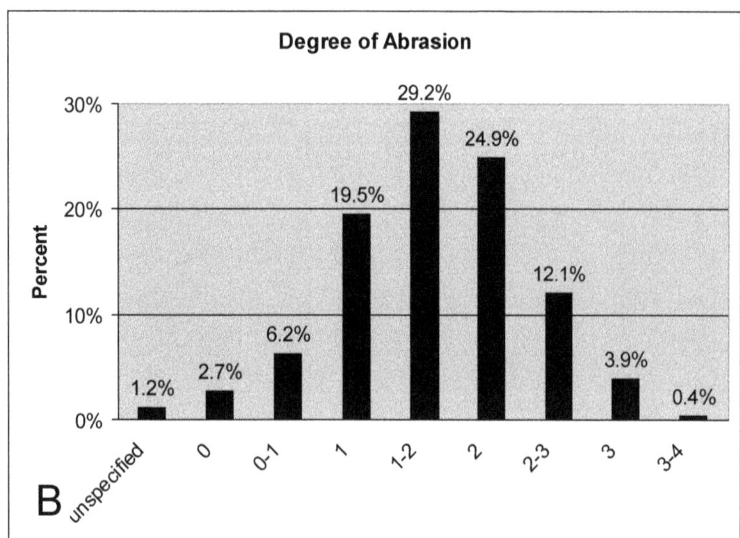

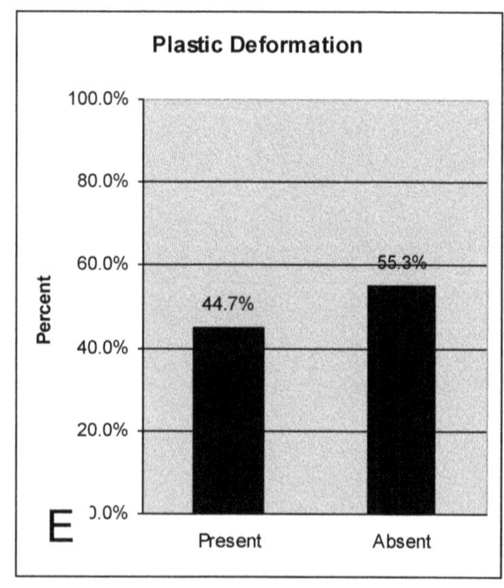

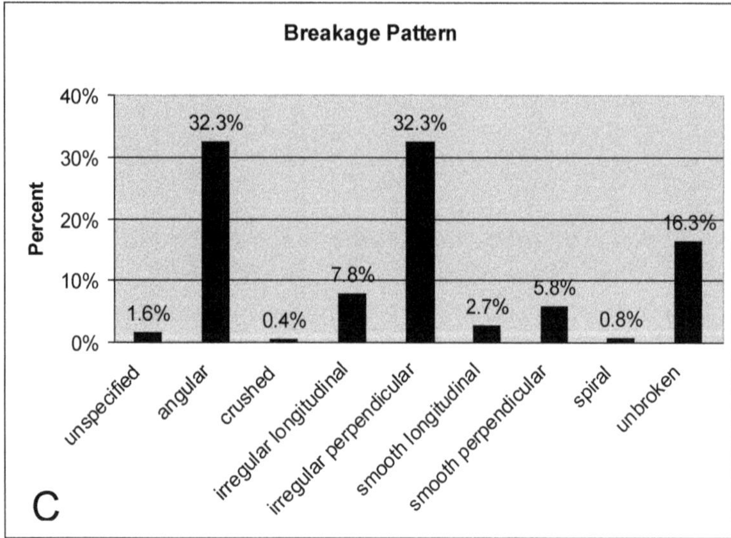

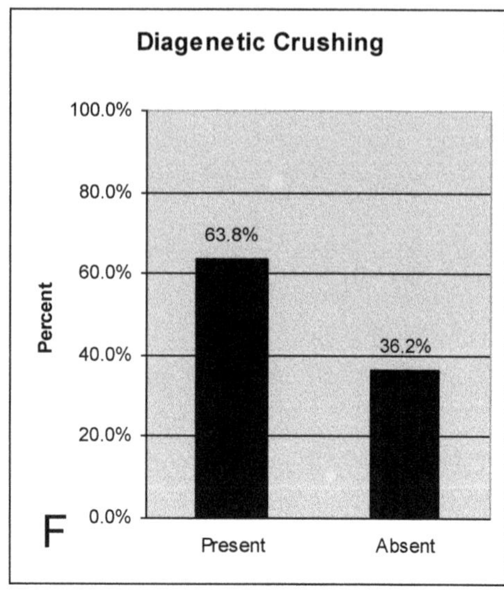

29.2. Histograms of taphonomic features from 247 skeletal elements. (A) Degree of weathering; (B) degree of abrasion; (C) breakage pattern; (D) gnawing; (E) plastic deformation; (F) diagenetic crushing.

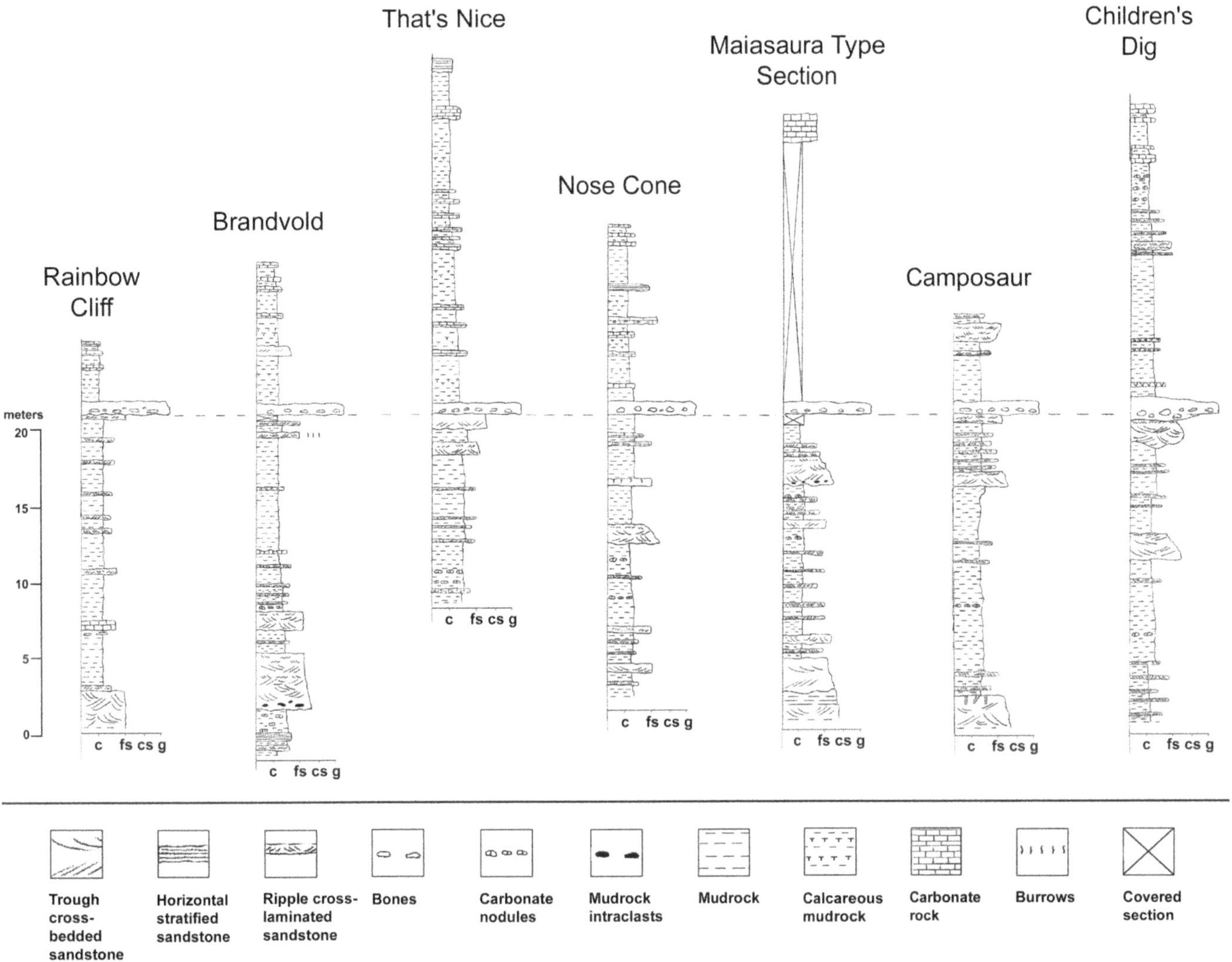

29.3. Willow Creek anticline stratigraphic sections. Seven sections were measured in the study area, with dashed line representing datum.

2.3% have features caused by either boring or acid etching. No specimens in this study show evidence of burning. Root traces are present on 78.6% of the bones, and likely represent a recent, near-surface phenomenon. Other post-fossilization processes produced distortion, with plastic deformation present on 44.7% (Fig. 29.2E) and diagenetic crushing on 63.8% of the specimens (Fig. 29.2F).

Sedimentologic Attributes of the Bonebed

The 0.4–1.5 m thick bonebed stratum is a structureless clast- to matrix-supported conglomerate laterally traceable for a distance of approximately 2 km (Fig. 29.3). The gravel-size framework grains include lithic fragments (80%; n = 224) and fossilized bones (20%; n = 58). Lithic fragments include pebbles and cobbles (subrounded to subangular) and boulders (subrounded to rounded). Overall, clast lithologies include micrite (51%; n = 116), tuffaceous mudstone (33%; n = 73), siltstone (14%; n = 31), and sandstone (2%; n = 4). All cobble- and boulder-size clasts are composed of micrite; more labile clast types (mudstone, siltstone, sandstone) are exclusively granule- to pebble-size, although some pebble-size micrite clasts are also present. Petrographic observation of micrite clasts reveals the presence of small (1–3 mm long), sinuous spar-filled tubes and clotted texture, with scattered silt- to very fine sand–size siliciclastic grains. In contrast to micrites at the nearby Egg Mountain locality, no glass shards are present (Varricchio et al., 1999). In addition, no invertebrate skeletal fragments or other allochems were observed in either hand sample or thin-section.

The bonebed matrix consists of a poorly sorted mixture of medium- to fine-grained sand, silt, and clay. Sand and silt consist of siliclastic material, including quartz, feldspar, and biotite. The clay-size matrix fraction consists primarily of

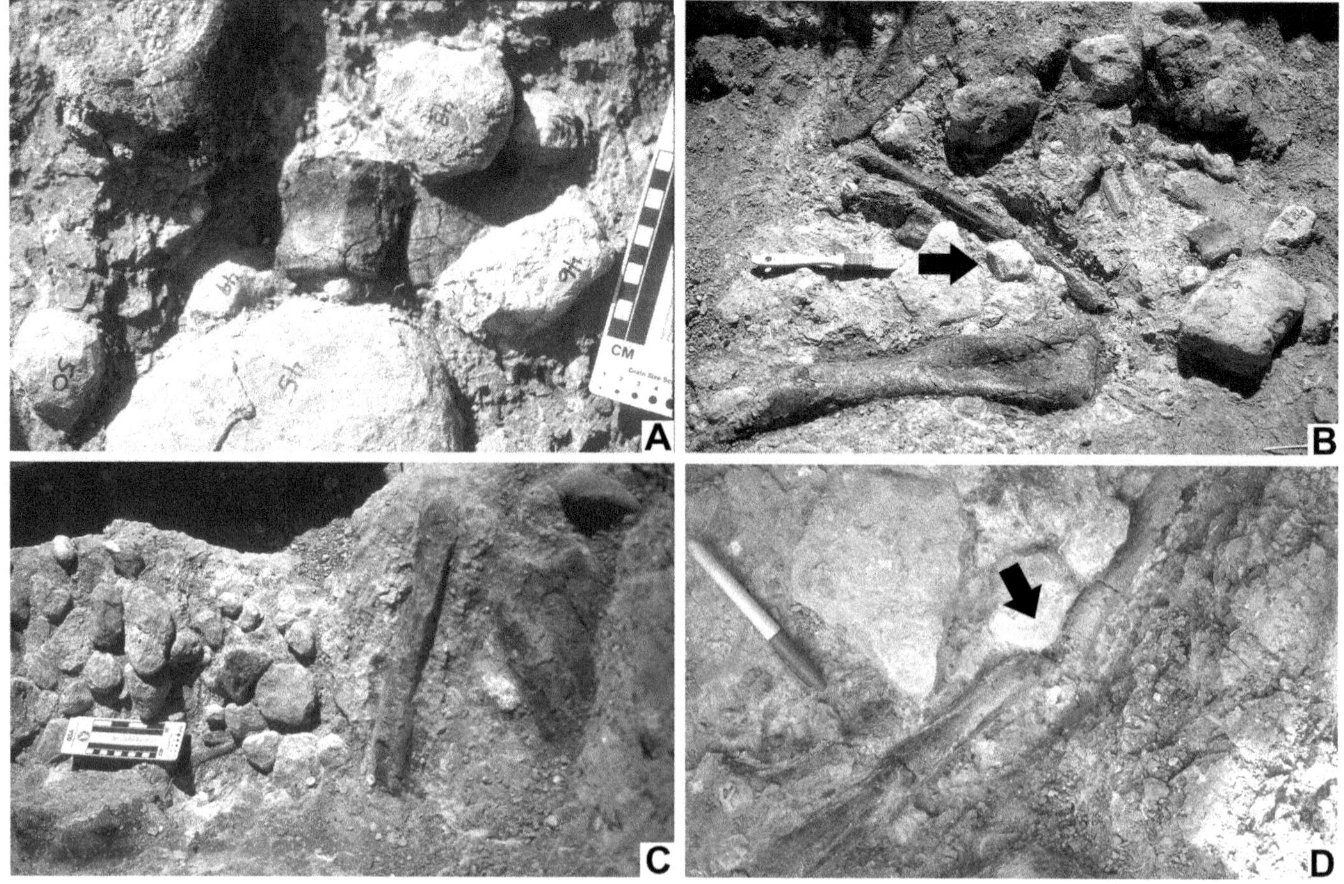

29.4. Plan views of lowermost portion of bonebed in test pit. (A) Abraded vertebrae surrounded by numbered micrite pebbles and cobbles (scale bar equals 10 cm); (B) long bone fractured at contact with micrite cobble indicated by arrow (brush [22 cm long] for scale); (C) cluster of micrite pebbles and cobbles adjacent to long bone (scale bar equals 10 cm); (D) micrite pebbles and cobbles in contact with long bone (pen [14.5 cm long] for scale).

montmorillonite and illite, with minor amounts of chlorite. The matrix also contains abundant carbonized plant debris and lesser amounts of charcoal.

The lithic clasts in the bonebed occur in close association with disarticulated skeletal elements up to 1 m in length (Fig. 29.4). In addition, a large (~65 cm) mudstone cast of a plant was noted at the Nose Cone quarry. The skeletal elements are predominantly flat-lying to slightly tilted (<29 degrees), and decrease in size upward from the base of the unit (Fig. 29.5). Long bone orientation (azimuth and dip) shows no preferred arrangement (Fig. 29.6). Maps of the test pit excavation show that pebble- to boulder-size lithic clasts are more abundant than bones in the lower portion of the deposit (Figs. 29.5, 29.7), an attribute observed at other excavations and in outcrops. In the lower part of the bonebed, clasts commonly occur in direct contact with bones (Figs. 29.5, 29.7), and hollow long bones often contain mudstone and small to medium pebble-size clasts. The ends of skeletal elements show strong abrasion (Fig. 29.8) and in some cases bones are broken across micrite cobbles. For example, in the excavated pit an 8 cm cobble was observed at the apex of a break, approximately one-third the length along the 54 cm long hadrosaur radius (Fig. 29.4B).

The largest skeletal elements and lithic fragments (micrite cobbles and boulders up to 30 cm in diameter) are concentrated in the clast-supported, lowermost 20–40 cm of the deposit (Fig. 29.5). Granule- to mostly pebble-size lithic fragments (primarily mudstone, siltstone, and sandstone) and smaller (<15 cm) bone fragments occur randomly distributed through the deposit. A gradual transition characterizes the change to the matrix-supported upper 60–80 cm of the deposit (Fig. 29.5A). Soft, low-density tuffaceous mudrock clasts are most common in the uppermost 40 cm. These relations within the conglomerate depict a well-developed coarse-tail (gravel size portion of grain size distribution curve) normal grading in both the lithic fragment and skeletal element components.

The bone-bearing deposit overlies primarily fine-grained, ripple cross-laminated sandstone and massive mudrock (Figs. 29.3, 29.9A, B) and is overlain by interbedded micrite and

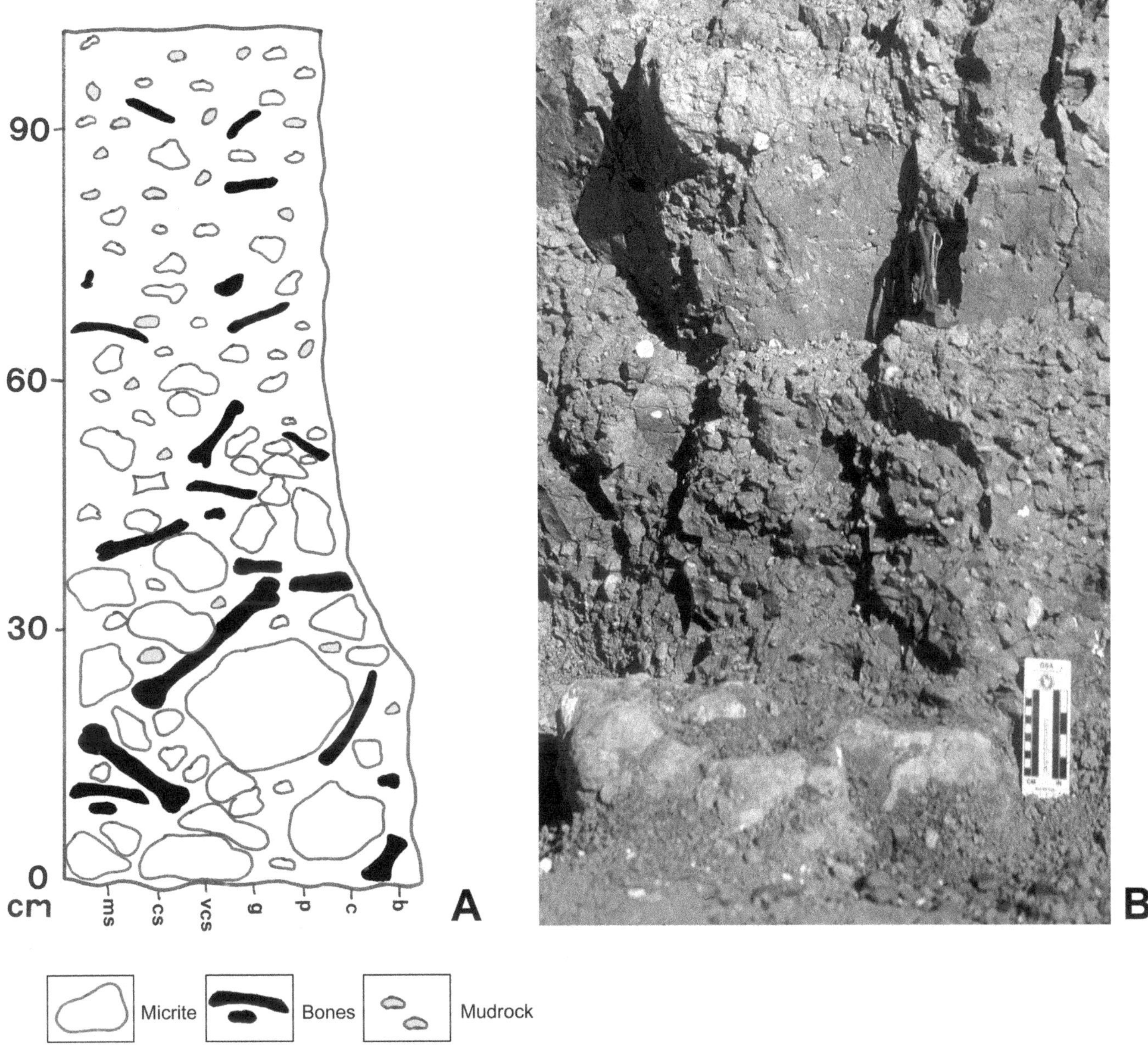

29.5. Cross-section of bonebed. (A) Schematic drawing based on exposure at Nose Cone; (B) photograph depicting relative location of bone elements, micrite and mudrock clasts, and maximum particle size profile. Scale bar equals 10 cm.

sandy, dark gray to green mudrock (Fig. 29.9C). Underlying mudrock units contain carbonate (micrite) nodules; some nodules cross cut the underlying cross-stratified sandstone bodies, suggesting a post-depositional pedogenic origin as caliche nodules or concretions. Interbedded trough cross-bedded, fining-upward sandstone lenses and sheets typically range from 1.2 to 2.5 m in thickness (Fig. 29.9B). Less commonly, the bonebed fills underlying depressions in small-scale (<2 m thick) trough cross-bedded lenticular sandstone bodies, up to 1.5 m thick. These strata are laterally traceable onto overbank facies beyond the margins of the lenticular sandstone bodies (Fig. 29.10). The lowermost contact with underlying units is abrupt and undulatory, indicating some erosion. Where the bonebed base rests directly on underlying sandstone, large bones and impressions of bones are apparent, suggesting that the sandy substrate was unlithified.

Overlying units comprising interbedded micrite and dark gray to green, calcareous mudrock represent lacustrine deposits (Lorenz, 1981; Shelton, 2007). As best exposed at the Children's Dig section, the bonebed is directly overlain by

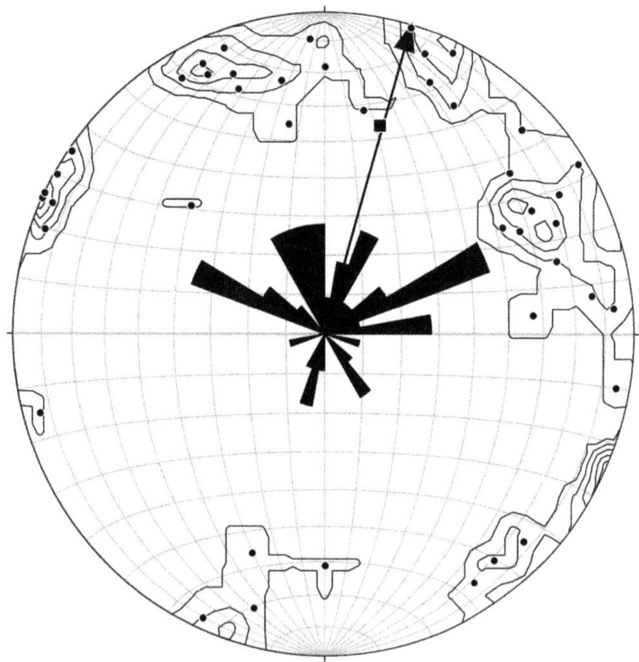

29.6. Rose diagram of lower hemisphere projection showing trend and plunge of long axis of elongate skeletal elements (n = 50). Small dots represent intersection of bone long axes with lower hemisphere. Contours are of distribution of lower hemisphere intersection data. Large arrow is mean azimuth; square is mean azimuth and plunge.

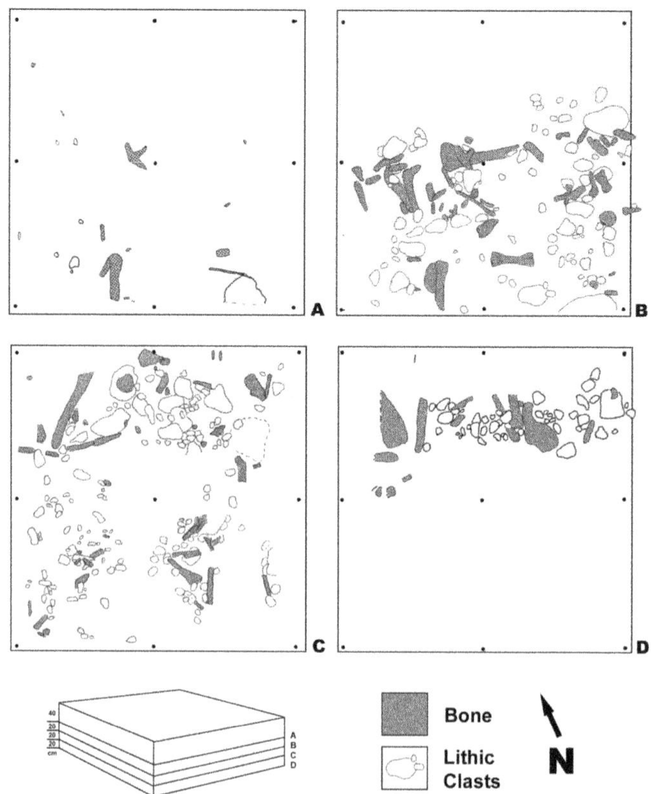

29.7. Spatial (lateral and vertical) distribution of bone elements and lithic clasts. (A–D) correspond to levels A–D, shown in descending order of excavation. Vertical interval between maps is 20 cm.

approximately 1.0 m of mudrock indicating a change from dominantly fluvial to lacustrine conditions after bonebed deposition.

DISCUSSION

Bonebed Sedimentology

We interpret the bonebed deposit as that of a fluid, non-cohesive debris flow (pseudoplastic debris flow of Schultz [1984]) to hyperconcentrated flow. This flow likely varied temporally and spatially in its internal characteristics with introduction of sediment from the underlying substrate (bulking) or water from channels or as precipitation (dilution) (Postma, 1986; Costa, 1988). Such lateral and temporal flow transformations are common in sediment gravity flows (e.g., Pierson and Scott, 1985; Iverson, 1997; Sohn et al., 1999). The coarse-tail normal grading indicates that clasts could readily move downward due to gravitational settling in the flow during transport, suggesting that the flow lacked significant cohesion and was fluid (Lowe, 1982; Benvenutti and Martini, 2002). Evidence of bone and lithic clast abrasion and rounding strongly support our interpretation that the sediment flow was likely turbulent with low cohesion and matrix strength and at times dominated by clast interactions in its lower portion (Postma, 1986). Increased dilution would spatially and temporally produce areas of hyperconcentrated flow characterized by greater turbulence in which particles and water behave as separate phases (Coussot and Meunier, 1996). Consistent with debris flow/hyperconcentrated flow transport, the long bones lack orientation (Fig. 29.6; Coussot and Meunier, 1996).

The lithic clasts (micrite, tuffaceous mudrock, siltstone, and sandstone) are exclusively intraformational in origin and reflect the immediate floodplain substrate sedimentary rock types, rather than Paleozoic and Mesozoic carbonate and siliciclastic units from the adjacent fold and thrust belt. Clotted micrite texture and presence of spar-filled tubules interpreted as rootlet structures indicate that some of these clasts represent caliche reworked from the substrate. The dominance of hard, resistant micrite in the lithic clast population suggests that softer clast types (mudstone, siltstone, sandstone) experienced high rates of clast attrition during flow. We infer that turbulence at the base of the flow drove local erosion of the substrate and facilitated clast interactions throughout much of the flow during transport. This erosion likely occurred more commonly along channel margins, where bank collapse provided sediment to the flow where

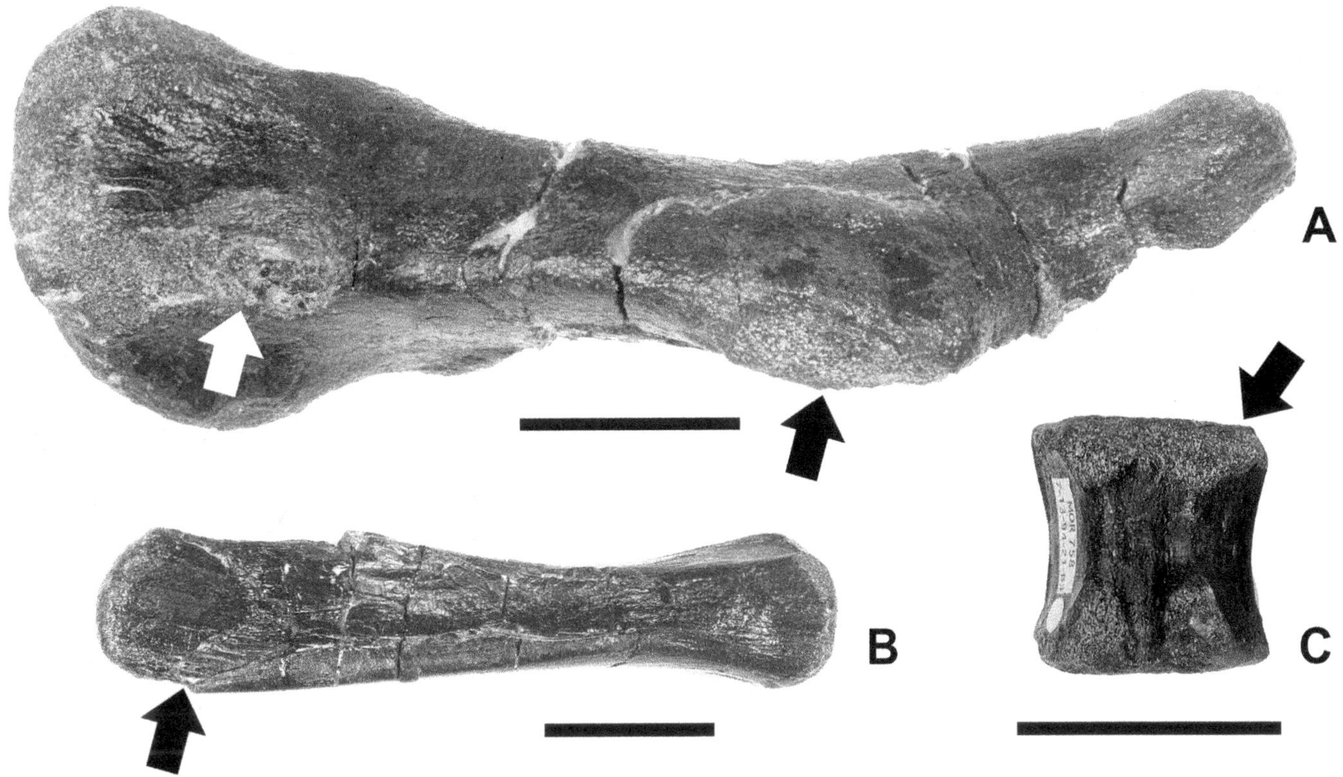

29.8. Abraded hadrosaur elements. (A) Metatarsal; (B) metacarpal; (C) vertebrae. Arrow indicates areas of abrasion. Scale bars equal 5 cm.

confined. Across floodplain surfaces, the eroding, turbulent basal portion of the flow picked up surficial matter such as plant debris, granule- to boulder-sized clasts, and bones as accidental clasts from the underlying, near-surface soil complex (e.g., Mohrig et al., 1999).

The lack of substantial erosional relief at the bonebed base suggests that scouring of the substrate, where active, was likely limited to the underlying 10 to 30 cm of material. This depth, however, is adequate for eroding bones and pedogenic caliche nodules and possible early cemented lacustrine/paludal micrite from near-surface soil complex (e.g., Mohrig et al., 1999). Incorporation of substrate sediment and clasts (i.e., bone and rock) implies that a numerically indeterminate component of the bonebed skeletal element assemblage may include bones previously buried by floodplain vertical accretion deposits. These likely included theropod elements (i.e., shed teeth, vertebrae, limb bones), some of the small, more highly rounded pebble- and cobble-size bones, and larger hypsilophodont (Varricchio and Horner, 1993) and hadrosaur bones. Given the low weathering stages of all bones and lack of extensive scouring at the bonebed base, exhumed skeletal elements probably represent a relatively small component of the assemblage from a period of time spanning the age range of the scoured floodplain sediment. Nonetheless, using modern rates of floodplain sediment accumulation, 30 cm of scour suggests 10^2 years as the minimum time averaging in the bonebed skeletal assemblage due to erosional exhumation of bones (Aalto et al., 2008).

Skeletal Assemblage Modification

Hadrosaur bones are the dominant skeletal elements in the bonebed, comprising at least 97.5% (n = 205) of the identifiable skeletal components. The bones are disarticulated and most lack significant weathering features. Weathering stages (Stage 0–1) indicate a time-averaging estimation of 10^{-1} to 10^1 years (Rogers, 1993; time-averaging patterns in Kidwell and Behrensmeyer (1993). Thus, it is possible that the hadrosaurs died over a period of time lasting at least several years and possibly decades, thus representing one or more mass-death episodes. This appears plausible because drought and disease played major roles in the mass deaths of hadrosaurs preserved in other Two Medicine bonebeds (e.g., Rogers, 1990; Varricchio and Horner, 1993; Varricchio, 1995).

Pre-transport scavenging and consumption of the decaying animals was minimal as indicated by the low percentage of bones with gnawing or boring traces. Breakage patterns show that bone fracturing preceded fossilization and likely

29.9. Exposures of the bonebed and related units. (A) Children's Dig bonebed showing underlying fluvial and overlying lacustrine sequences; (B) fluvial/channel sandstone at Children's Dig (map board [0.75 m wide] for scale); (C) lacustrine units at Nose Cone. Arrow indicates level of the bonebed; second quarry located to the far right out of the picture. Tarp-covered supplies to right of arrow are about 0.75 m in height.

resulted from rock-bone interactions during transport, thus casting doubt on interpretations of bone permineralization prior to transport (Hooker, 1987; Horner and Gorman, 1988). In addition, a hadrosaur radius broken across an 8 cm diameter cobble further supports this interpretation (Fig. 29.4B), and filling of broken bones by mudstone matrix and clasts (bedload) suggest rapid infill (Martin, 1999:231) and that most bones were brittle and hollow during transport.

Abrasion represents a pervasive feature of this assemblage, resulting in extensive modification of the broken edges of bones and/or removal of cortical bone on the ends of unbroken elements. Figure 29.8 shows the high degree of abrasion affecting the ends of long bones. The bone surfaces lack deep scratches, a feature produced by bone trampling (Lyman, 1994; Britt et al., 2009). In addition, trampling typically sorts by size and surface area, with small bones becoming more deeply buried than larger bones or those with larger surface areas (Lyman, 1994:187). We observed no evidence for this inverse size relation in the bonebed. In some places, excavation of the bonebed exposed subrounded bone fragments of indeterminate type ranging in size from large pebble to small cobble. Although some trampling of bones may have taken place prior to entrainment, the pervasive modification of original bone morphology suggests that the majority of bone abrasion happened during transport. The mixture of similar size, subrounded indurated micrite and bone clasts and fluid nature of the flow suggest a high frequency of collisions (e.g., bone-bone, bone–lithic clast, and lithic clast–lithic clast) during transport.

Debris Flows in Alluvial Settings

Deposits of debris flows and hyperconcentrated flows are relatively uncommon in alluvial plain settings not directly influenced by explosive volcanic activity. Martin and Turner (1998) interprets sheet-like sandstone bodies in the braided alluvial plain deposits of the Triassic "Bunter" Pebble Beds, England, as flood-generated hyperconcentrated flow deposits in both channel and floodplain settings. Benvenuti and Martini (2002) report similar deposits from Late Carboniferous–Triassic strata in England, Australia, and the eastern United States. Jorgensen and Fielding (1999) attribute deposition of debris flow conglomerates in Upper Triassic coal deposits of Queensland, Australia, to result from intrabasinal faulting, and seismic destabilization of saturated sediment. Similarly, termination of coal mire deposition by influx of debris flows caused by channel overbank flooding (crevasse-splay events) are documented in Eocene strata of the Bighorn basin, Wyoming (Roberts et al., 1994). Zaleha and Wiesemann (2005) interpret deposition of muddy gastrolith-bearing diamictites and wackes in the Lower Cretaceous Cloverly Formation, central Wyoming, as the byproduct of ashfall produced hyperconcentrated flows emanating from the adjacent Cordilleran thrust-belt highlands.

From the perspective of the vertebrate fossil record, bone-bearing horizons interpreted as deposits of debris flow or hyperconcentrated flow are far less common than those deposited by channelized streamflow in alluvial successions (see review by Britt et al. [2009] and dataset in Eberth et al. [2007]). Despite a large range in their physical and taphonomic attributes (e.g., Fastovsky et al., 1995; Eberth et al., 2000, 2006; Rogers, 2005; Van Itterbeeck et al., 2005; Lauters et al., 2008; Britt et al., 2009; Scherzer and Varricchio, 2010; Liu et al., 2011), several general similarities exist, including (1) broadly lenticular to tabular shape; (2) thickness ranging from decimeters to one meter; (3) composed of poorly sorted pebbly to conglomeratic, massive or ungraded to normally graded deposits; (4) presence of some floating skeletal elements and intra- or extra-basinal clasts; (5) lack of imbrication and preferred orientation of elongate elements; and (6) frequent occurrence above a paleosol or sequence boundary.

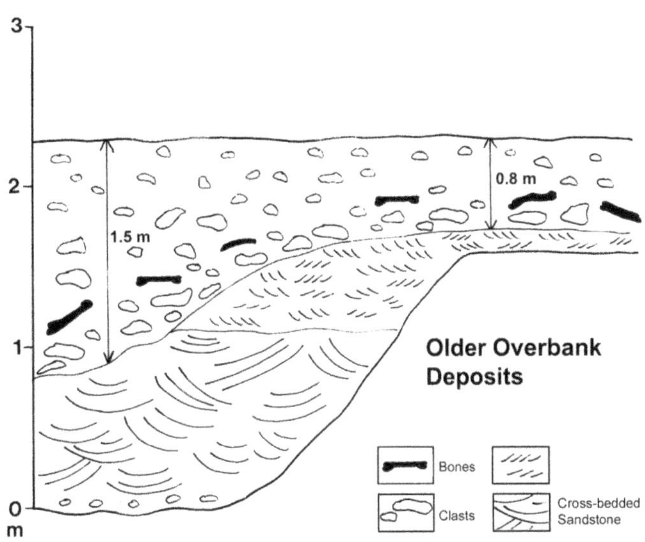

29.10. Schematic cross-section of bonebed at Children's Dig locality showing lateral change in thickness from channel to floodplain overlap above older fluvial deposits.

Origin of the Assemblage

The original hypothesis explaining the formation of this hadrosaur bonebed involved the death of large numbers of individuals by volcanic gas and ash, with partial burial of carcasses in related ash fall deposits (Hooker, 1987). Later, disarticulation and defleshing of carcasses presumably occurred at the land surface, accompanied by partial silica permineralization (Hooker, 1987). Finally, generation of a mudflow by breaching of a distant mountain lake transported the bones cross the land surface (Lorenz, 1981; Hooker, 1987; Horner and Gorman, 1988). Here, we examine these previous hypotheses, namely the number of individuals represented in the deposits and their purported catastrophic death in a volcanic eruption.

Number of Individuals The bonebed fossil assemblage consists primarily of hadrosaur elements, suggesting that this taxon lived in large numbers in the surrounding area and represented a dominant component of the Two Medicine vertebrate ecosystem. However, the time averaging indicated by weathering stages (10^{-1} to 10^{1} years) suggests that this assemblage could represent one or possibly more mass-death episodes, as reported for drought- and disease-related hadrosaur bonebeds from the Two Medicine (Rogers, 1990; Varricchio and Horner, 1993; Varricchio, 1995). Erosional scouring of 10 to 30 cm of the substrate extends this time averaging of skeletal elements to a minimum 10^{-2} years (Aalto et al., 2008). Although a single population of individuals has been suggested for the Willow Creek anticline bonebed using discrete class sizes (Varricchio and Horner, 1993), Western and Behrensmeyer (2009) report that time averaging of bone assemblages over four decades accurately record the living community structure in the Amboseli ecosystem of southern Kenya. Similarly, we suggest that the bonebed assemblage records community dominance by hadrosaurs over an extended period of time. Furthermore, we cannot eliminate the possibility that the skeletal material in the bonebed records the deaths of small groups of or individual animals over time and across the landscape, producing an attritional assemblage. However, we would expect this mode of origin to result in a higher percent of non-hadrosaur skeletal material in the fossil assemblage and greater variation in weathering stages.

Finally, the relatively low numbers of individuals (MNI = 7; NISP/MNI = 27) represented in the bones sampled from the test pit and identifiable quarries in our study are comparable to MNI (20) and NISP/MNI (23) values calculated by Varricchio and Horner (1993) for the Willow Creek anticline bonebed. Comparable MNI (4 to 18) and NISP/MNI (21 to 81) are reported from other hadrosaur bonebeds in the Two Medicine Formation (Varricchio and Horner, 1993; Scherzer and Varricchio, 2010). Based on these MNI/NISP values we suggest that 101 to 102 individuals represents a more conservative and reasonable estimate for the Willow Creek anticline bonebed assemblage.

Volcanic Eruption The original interpretation of death by a volcanic eruption relied primarily on a volcanic ash (bentonite) layer located "18 inches" above the top of the bonebed (Hooker, 1987; Horner and Gorman, 1988). However, the lack of abundant volcaniclastic detritus in the bonebed precludes a volcanic eruption mechanism for animal death. Both epiclastic and syn-eruptive (lahar) volcanic debris flow deposits confined to channels are present in the Two Medicine Formation to the south near Augusta, Montana. At these sites, epiclasitic debris flow deposits contain abundant large (cobble to boulder) volcanic rock fragments, whereas syn-eruptive (lahar) deposits show evidence of thermal alteration at bed lower contacts (Smith, 1998). Both provide direct evidence for their volcanic origin. No volcanic rock fragments are present in the gravel size component suggesting an absence of an extrabasinal source of volcanic detritus. No evidence exists for thermal alteration at bed contacts, or blackening of bones due to thermal effects; in contrast, Roberts and Hendrix (2000) report such features in a pyroclastic flow and ash deposit at a lower stratigraphic level in the Two Medicine Formation at the Sevenmile Hill locality south of Choteau, Montana. In addition, the bonebed matrix consists primarily of mud-size sediment, with no clear evidence of a primary ash component. Although montmorillonite in the bonebed matrix may have formed by alteration of volcanic ash, the abundance of tuffaceous mudstone clasts in the deposit suggests that any ash components likely represent material reworked from older volcanic layers.

The stratigraphic position of the ash also rules out its involvement in a catastrophic mass death of hadrosaurs represented by the bonebed. In our measured sections (see Fig. 29.3), approximately 1 m of sandy, dark gray to green mudstone that accumulated by vertical accretion in a ponded water environment overlies the bonebed. Given modern rates of floodplain sediment accumulation (Aalto et al., 2008), we interpret the stratigraphically higher ash layer to postdate deposition of the bonebed by as much as 10^2 to 10^3 years. This stratigraphic relationship and the lack of a dominant ash component to the bonebed matrix lead us to conclude that the bonebed sedimentation event was not a response to remobilization of volcanic ash from a recently ash-mantled landscape.

Alternative Hypotheses

As an alternative to a volcanic origin for the bone assemblage, Varricchio et al. (2009) tested a hypothesis relating formation of the Willow Creek anticline bonebed to the Manson impact event. Based on differences in age, they rejected this hypothesis. However, they did use the presence of soot and the palynomorph *Ulmoideipites* sp., a postfire succession taxon, to suggest that a major wildfire may have preceded the bonebed sedimentation event. Indeed, volcanic activity in the region at times led to wildfire development, as evidenced by the pyroclastic flow and preservation of the in situ burned forest (Roberts and Hendrix, 2000). Wildfire often induces debris flow events after rainfall (e.g., Cannon et al., 2001); for example, Lovelace (2006) proposed entombment of an isolated sauropod skeleton in the Upper Jurassic Morrison Formation near Douglas, Wyoming, as resulting from such an event. However, the aforementioned absence of a volcaniclastic component in the Willow Creek anticline bonebed suggests that wildfire–if it indeed scarified the landscape, burning vegetation and generating hydrophobic soils as precursor to the bonebed sedimentation event–was likely of a more mundane lightning than volcanic origin. Nonetheless, Brown et al. (2012) suggest that because of elevated atmospheric oxygen levels the Cretaceous Earth was a "high-fire" world, and infer that bonebeds in the Campanian of Dinosaur Provincial Park, Alberta, may be due to postfire erosion and sedimentation events.

Cyclonic weather systems (hurricanes) provide another possible mechanism for generating sediment transport events. Cyclonic storm systems were common in the Western Interior seaway, especially along the western margin (Erickson and Slingerland, 1990). A storm system moving inland along the western margin of the Cretaceous seaway would produce significant amounts of precipitation on the alluvial plain, and possibly trigger sediment transport events. For example, Eberth et al. (2010) interpreted an *Albertosaurus* bonebed in the Upper Cretaceous (Maastrichtian) Horseshoe Canyon Formation, Alberta, Canada as generated by a large-scale rainfall event, perhaps associated with hurricane conditions (Eberth and Currie, 2010). Indeed, the impacts of wildfire across the Two Medicine landscape could have operated in consort with episodic cyclonic rainfall events to facilitate generation of the hadrosaur bonebed debris flow.

Finally, it is tempting to assign a causative factor to the stratigraphic location of the hadrosaur bonebed at a significant upward transition from underlying alluvial plain to overlying lacustrine deposits at or in close proximity to the sequence boundary at the base of Lithofacies 4 at the Willow Creek anticline. Eberth et al. (2006) attributed similar stratigraphic occurrences of bonebeds at sequence boundaries in the Lower Cretaceous Cedar Mountain Formation of Utah to changes in accommodation space related to forebulge migration. Conversely, Eberth and Currie (2010) relate the occurrence of an *Albertosaurus* bonebed in Alberta at a stratigraphic boundary representing an up-section change from non-coaly to coal deposition, to establishment of a regionally higher water table in response to climatic change. We lack sufficient data to determine the origin of the stratigraphic boundary at which the hadrosaur bonebed occurs. However, we hypothesize that the noted presence (e.g., Rogers and Kidwell, 2000; Britt et al., 2009) of debris flow–deposited bonebeds at stratigraphic boundaries may be a function of increased preservation potential through inundation (as opposed to increased frequency of development) due to an increase in base level, regardless of the driver (climatic or tectonic).

CONCLUSION

The paucitaxic hadrosaur bonebed in the Two Medicine Formation (Upper Cretaceous) of western Montana (Willow Creek anticline) contains thousands of hadrosaur skeletal elements transported and deposited by an unconfined fluid debris flow to hyperconcentrated flow. The 0.4–1.5 m thick bonebed fills small (<1 m deep) alluvial channels and extends up to 2 km beyond channel margins. This suggests that deposition occurred as a single sedimentation event across a low relief alluvial plain characterized by small, low-gradient sandy fluvial channels and adjacent interchannel areas comprising crevasse splay and vertical accretion deposits. The normally-graded, clast- to matrix-supported bonebed conglomerate contains pebble–boulder size clasts primarily of disarticulated hadrosaur bones and reworked intraformational clasts including micrite (caliche), tuffaceous mudstone, siltstone, and sandstone in a sandy mudstone matrix. Subrounded micrite clasts, abraded bones, and bones broken across boulders indicate significant turbulence and clast interaction during transport. The relatively unaltered nature of the bones (low degree of weathering, unburned, and rare evidence of gnawing or boring) suggests minimal residence time (10^{-1} to 10^{1} years) at the surface prior to transport. The sediment transport event collected disarticulated bones from the land surface, ripped up underlying floodplain deposits, and likely incorporated some previously buried bones in the process, and transported them together in a mud-rich slurry. Lack of skeletal articulation and evidence for transport of bones without soft tissue suggest the debris flow-producing event was not responsible for death of the hadrosaurs. Further, the lack of abundant volcaniclastic detritus in the bonebed precludes a volcanic eruption mechanism for animal death. The thorough mixing of skeletal fragments with lithic clasts, combined with evidence of bone abrasion due to clast interactions, and derivation of bones from the land surface and substrate, suggest that assessing hadrosaur population structure and herding behavior from this deposit should not be attempted. Indeed, caution should be exercised in making population structure and behavioral (e.g., herding) inferences from vertebrate bonebeds showing evidence of transport of skeletal elements and lacking clear evidence of a lethal sedimentation event (obrution) (e.g., Rogers and Kidwell, 2007).

ACKNOWLEDGMENTS

We thank Dave Hanna, former manager of the Nature Conservancy Pine Butte Preserve for granting permission to conduct fieldwork, the Museum of the Rockies for logistical support, and the Department of Earth Sciences, Montana State University for financial support. Discussions with J. Horner, D. Eberth, R. Rogers, D. Varricchio and B. Jackson, and manuscript reviews by B. Britt and D. Eberth proved extremely beneficial.

LITERATURE CITED

Aalto, R., J. W. Lauer, and W. E. Dietrich. 2008. Spatial and temporal dynamics of sediment accumulation and exchange along Strickland River floodplains (Papua New Guinea) over decadal-to-centennial timescales. Journal of Geophysical Research 113:1–22.

Behrensmeyer, A. K. 1978. Taphonomic and ecologic information from bone weathering. Paleobiology 4:150–162.

Behrensmeyer, A. K. 1982. Time resolution in fluvial vertebrate assemblages. Paleobiology 8:211–227.

Behrensmeyer, A. K., and R. W. Hook. 1992. Paleoenvironmental contexts and taphonomic modes; pp. 15–138 in A. K. Behrensmeyer, J. D. Damuth, W. A. DiMichele, R. Potts, H-D. Sues, and S. L. Wing (eds.), Terrestrial Ecosystems through Time: Evolutionary Paleoecology of Plants and Animals. University of Chicago Press, Chicago, Illinois.

Benvenutti, M., and I. P. Martini. 2002. Analysis of terrestrial hyperconcentrated flows and their deposits; pp. 167–193 in I. P. Martini, V. R. Baker, and G. Garzon (eds.), Flood and Megaflood Processes and Deposits: Recent and Ancient Examples. International Association of Sedimentologists Special Publication no. 32. Blackwell Science, London.

Bibler, C. J., and J. G. Schmitt. 1986. Barrier-coastline deposition and paleogeographic implications of the Upper Cretaceous Horsethief Formation, northern Disturbed Belt, Montana. Mountain Geologist 23:113–127.

Britt, B. B., D. A. Eberth, R. D. Scheetz, B. W. Greenhalgh, and K. L. Stadtman. 2009. Taphonomy of debris-flow hosted dinosaur bonebeds at Dalton Wells, Utah (Lower Cretaceous, Cedar Mountain Formation, USA). Palaeogeography, Palaeoclimatology, Palaeoecology 280:1–22.

Brown, S. A. E., A. C. Scott, I. J. Glasspool, and M. E. Collinson. 2012. Cretaceous wildfires and their impact on the Earth system. Cretaceous Research 36:162–190.

Cannon, S. H., R. M. Kirkham, and M. Parise. 2001. Wildfire-related debris-flow initiation processes, Storm King Mountain, Colorado. Geomorphology 39:171–188.

Cook, E. 1995. Taphonomy of two non-marine Lower Cretaceous bone accumulations from southeastern England. Palaeogeography, Palaeoclimatology, Palaeoecology 116:263–270.

Costa, J. E. 1988. Rheologic, geomorphic, and sedimentologic differentiation of water floods, hyperconcentrated flows and debris flows; pp. 113–122 in V. R. Baker, R. C. Kochel, and P. C. Patton (eds.), Flood Geomorphology. John Wiley and Sons, New York.

Coussot, P., and M. Meunier. 1996. Recognition, classification, and mechanical description of debris flows. Earth Science Reviews 40:209–227.

Eberth, D. A., and P. J. Currie. 2010. Stratigraphy, sedimentology, and taphonomy of the *Albertosaurus* bonebed (upper Horseshoe Canyon Formation, Maastrichtian), southern Alberta, Canada. Canadian Journal of Earth Sciences 47:1119–1143.

Eberth, D. A., D. B. Brinkman, and V. Barkas. 2010. A centrosaurine mega-bonebed from the Upper Cretaceous of southern Alberta: implications for behavior and death events; pp. 495–508 in M. J. Ryan, B. Chinnery-Allgeier, and D. A. Eberth (eds.), New Perspectives on Horned Dinosaurs. Indiana University Press, Bloomington, Indiana.

Eberth, D. A., M. Shannon, and B. G. Noland. 2007. A bonebeds database: classification, biases, and patterns of occurrence; pp. 103–221 in R. R. Rogers, D. A. Eberth, and A. R. Fiorillo (eds.), Bonebeds: Genesis, Analysis, and Paleobiological Significance. University of Chicago Press, Chicago, Illinois.

Eberth, D. A., D. S. Berman, S. S. Sumida, and H. Hopf. 2000. Lower Permian terrestrial paleoenvironments and vertebrate paleoecology of

the Tambach Basin (Thuringia, central Germany). Palaios 15:293–313.
Eberth, D. A., B. B. Britt, R. Scheetz, K. L. Stadtman, and D. B. Brinkman. 2006. Dalton Wells: geology and significance of debris-flow-hosted dinosaur bonebeds in the Cedar Mountain Formation (Lower Cretaceous) of eastern Utah, USA. Palaeogeography, Palaeoclimatology, Palaeoecology 236:217–245.
Erickson, M. C., and R. Slingerland. 1990. Numerical simulations of tidal and wind-driven circulation in the Cretaceous Interior Seaway of North America. Geological Society of America Bulletin 102:1499–1516.
Fastovsky, D. E., J. M. Clark, N. H. Strater, M. R. Montellano, R. Hernandez, and J. A. Hopson. 1995. Depositional environments of a Middle Jurassic terrestrial vertebrate assemblage, Huizachal Canyon, Mexico. Journal of Vertebrate Paleontology 15:561–575.
Fiorillo, A. R. 1988. Taphonomy of Hazard Homestead Quarry (Ogallala Group), Hitchcock County, Nebraska. Contributions to Geology, University of Wyoming 26:57–97.
Gill, J. R., and W. A. Cobban. 1973. Stratigraphy and geologic history of the Montana Group and equivalent rocks, Montana, Wyoming, and North and South Dakota. United States Geological Survey Professional Paper 776. 37 pp.
Gunderson, J. A., and S. D. Sheriff. 1991. A new Late Cretaceous paleomagnetic pole from the Adel Mountain Volcanics, west-central Montana. Journal of Geophysical Research 96:317–326.
Harlan, S. S., L. W. Snee, M. W. Reynolds, H. H. Mehnert, R. G. Schmidt., S. D. Sheriff, and A. D. Irving. 2005. $^{40}Ar/^{39}Ar$ and K-Ar Geochronology and Tectonic Significance of the Upper Cretaceous Adel Mountain Volcanics and Spatially Associated Tertiary Igneous Rocks, Northwestern Montana. United States Geological Survey Professional Paper 1696. 29 pp.
Hooker, J. S. 1987. Late Cretaceous ashfall and the demise of a hadrosaurian "herd." Geological Society of America Abstracts with Programs 19:284.
Horner, J. R., and J. Gorman. 1988. Digging Dinosaurs. Workman, New York, 210 pp.
Horner, J. R., J. G. Schmitt, F. Jackson, and R. Hanna. 2001. Bones and rocks of the Upper Cretaceous Two Medicine-Judith River clastic wedge complex, Montana; pp. 3–13 in C. L. Hill (ed.), Guidebook for the Field Trips: Mesozoic and Cenozoic Paleontology in the Western Plains and Rocky Mountains. Society of Vertebrate Paleontology, 61st Annual Meeting. Museum of the Rockies Occasional Paper 3.
Iverson, R. M. 1997. The physics of debris flows. Reviews of Geophysics 35:245–296.
Jorgensen, P. J., and C. R. Fielding. 1999. Debris-flow deposits in an alluvial-plain succession: the Upper Triassic Callide Coal Measures of Queensland, Australia. Journal of Sedimentary Research 69:1027–1040.
Klein, R. G., and K. Cruz-Uribe. 1984. The Analysis of Animal Bones from Archaeological Sites. The University of Chicago Press, Chicago, Illinois, 249 pp.
Kidwell, S. M., and A. K. Behrensmeyer. 1993. Summary: estimates of time-averaging; pp. 301–302 in S. M. Kidwell, and A. K. Behrensmeyer (eds.), Taphonomic Approaches to Time Resolution in Fossil Assemblages. Short Courses in Paleontology 6.
LaBranche, J. V. 1999. Sedimentary deposits and processes of the Late Cretaceous Adel Mountain volcanclastic apron, west-central Montana. M.S. thesis, Montana State University, Bozeman, Montana, 114 pp.
Lauters, P., Y. L. Bolotsky, J. Van Itterbeeck, and P. Godefroit. 2008. Taphonomy and age profile of a latest Cretaceous dinosaur bonebed in far eastern Russia. Palaios 23:153–162.
Liu, Y., H. Kunag, N. Peng, H. Xu, and Y. Liu. 2011. Sedimentary facies of dinosaur trackways and bonebeds in the Cretaceous Jiaolai Basin, eastern Shandong, China, and their paleogeographical implications. Earth Science Frontiers 18:9–24.
Lorenz, J. C. 1981. Sedimentary and tectonic history of the Two Medicine Formation, Late Cretaceous (Campanian), northwestern Montana. Ph.D. dissertation, Princeton University, Princeton, New Jersey, 215 pp.
Lorenz, J. C., and W. Gavin. 1984. Geology of the Two Medicine Formation and the sedimentology of a dinosaur nesting ground; pp. 172–186 in J. D. McBane and P. B. Garrison (eds.), Northwestern Montana Guidebook, 1984 Field Conference. Montana Geological Society, Billings, Montana.
Lovelace, D. M. 2006. An Upper Jurassic Morrison Formation fire-induced debris flow: taphonomy and paleoenvironment of a sauropod (Sauropoda: *Supersaurus vivianae*) locality, east-central Wyoming; pp. 47–56 in J. R. Foster and S. G. Lucas (eds.), Paleontology and Geology of the Upper Morrison Formation. New Mexico Museum of Natural History and Science Bulletin 46.
Lowe, D. R. 1982. Sediment gravity flows: II, Depositional models with special reference to the deposits of high-density turbidity currents. Journal of Sedimentary Petrology 52:279–297.
Lyman, L. R. 1994. Vertebrate Taphonomy. Cambridge University Press, Cambridge, U.K., 524 pp.
Martin, C. A. L., and B. R. Turner. 1998. Origins of massive-type sandstones in braided river systems. Earth Science Reviews 44:15–38.
Martin, R. E. 1999. Taphonomy: A Process Approach. Cambridge University Press, Cambridge, U.K., 526 pp.
Miall, A. D. 1996. The Geology of Fluvial Deposits: Sedimentary Facies, Basin Analysis, and Petroleum Geology. Springer, New York, 598 pp.
Mohrig, D., A. Elverhoi, and G. Parker. 1999. Experiments on the relative mobility of muddy subaqueous and subaerial debris flows, and their capacity to remobilize antecedent deposits. Marine Geology 154:117–129.
Pierson, T. C., and K. M. Scott. 1985. Downstream dilution of a lahar: transition from debris flow to hyperconcentrated streamflow. Water Resources Research 21:1511–1524.
Postma, G. 1986. Classification for sediment gravity-flow deposits based on flow conditions during sedimentation. Geology 14:291–294.
Roberts, E. M. 1999. A stratigraphic and sedimentologic analysis of the southern Two Medicine Formation, central Montana. M.S. thesis, University of Montana, Missoula, Montana, 80 pp.
Roberts, E. M., and M. S. Hendrix. 2000. Taphonomy of a petrified forest in the Two Medicine Formation (Campanian), northwest Montana: implications for palinspastic restoration of the Boulder batholith and Elkhorn Mountains volcanics. Palaios 15:476–482.
Roberts, S. B., R. W. Stanton, and R. M. Flores. 1994. A debris flow deposit in alluvial, coal-bearing facies, Bighorn Basin, Wyoming, USA: evidence for catastrophic termination of a mire. International Journal of Coal Geology 25:213–241.
Rogers, R. 1990. Taphonomy of three dinosaur bonebeds in the Upper Cretaceous Two Medicine Formation of northwestern Montana: evidence for drought-related mortality. Palaios 5:394–413.
Rogers, R. R. 1993. Systematic patterns of time-averaging in the terrestrial vertebrate record; pp. 228–249 in S. M. Kidwell and A. K. Behrensmeyer (eds.), Taphonomic Approaches to Time Resolution in Fossil Assemblages. Short Courses in Paleontology 6. The Paleontological Society, Knoxville, Tennessee.
Rogers, R. R. 1994. Nature and origin of through-going discontinuities in nonmarine foreland basin strata, upper Cretaceous, Montana: implications for sequence analysis. Geology 22:1119–1122.
Rogers, R. R. 1998. Sequence analysis of the Upper Cretaceous Two Medicine and Judith River formations, Montana: nonmarine response to the Claggett and Bearpaw marine cycles. Journal of Sedimentary Research 68:615–631.
Rogers, R. R. 2005. Fine-grained debris flows and extraordinary vertebrate burials in the Late Cretaceous of Madagascar. Geology 33:297–300.
Rogers, R. R., and S. M. Kidwell. 2000. Associations of vertebrate skeletal concentrations and discontinuity surfaces in terrestrial and shallow marine records: a test in the Cretaceous of Montana. Journal of Geology 108:131–154.
Rogers, R. R., and S. M. Kidwell. 2007. A conceptual framework for the genesis and analysis of vertebrate skeletal concentrations; pp. 1–63 in R. R. Rogers, D. A. Eberth, and A. R. Fiorillo (eds.), Bonebeds: Genesis, Analysis, and Paleobiological Significance. University of Chicago Press, Chicago, Illinois.
Rogers, R. R., C. C. Swisher, and J. R. Horner. 1993. $^{40}Ar/^{39}Ar$ age and correlation of the nonmarine Two Medicine Formation (Upper Cretaceous), northwestern Montana, U.S.A. Canadian Journal of Earth Sciences 30:1066–1075.
Scherzer, B. A., and D. J. Varricchio. 2010. Taphonomy of a juvenile Lambeosaurine bonebed from the Two Medicine Formation (Campanian) of Montana, United States. Palaios 25:780–795.
Schmidt, R. G. 1978. Rocks and Mineral Resources of the Wolf Creek Area, Lewis and Clark and Cascade Counties, Montana. United States Geological Survey Bulletin 1441. 91 pp.
Schultz, A. W. 1984. Subaerial debris-flow deposition in the Upper Paleozoic Cutler formation, western Colorado. Journal of Sedimentary Petrology 54:759–772.
Shelton, J. A. 2007. Application of sequence stratigraphy to the nonmarine Upper Cretaceous Two Medicine Formation, Willow Creek Anticline, northwestern Montana. M.S. thesis, Montana State University, Bozeman, Montana, 238 pp.
Sheriff, S. D., and J. A. Gunderson. 1990. Age of the Adel Mountain volcanic field, west-central Montana. Isochron/West 56:21–23.

Shipman, P. 1981. Life History of a Fossil: An Introduction to Taphonomy and Paleoecology. Harvard University Press, Cambridge, Massachusetts, 222 pp.

Siebe, C., P. Schaaf, and J. Urrutia-Fucugauchi. 1999. Mammoth bones embedded in a late Pleistocene lahar from Popocatepetl volcano, near Tocuila, central Mexico. Geological Society of America Bulletin 111:1550–1562.

Skipp, B., and L. W. McGrew. 1977. The Maudlow and Sedan Formations of the Upper Cretaceous Livingston Group on the west edge of the Crazy Mountains Basin, Montana. United States Geological Survey Bulletin 1422-B. 68 pp.

Smith, A. E. 1998. Volcanism and associated sedimentation in a retro-arc foreland basin: the Upper Cretaceous Two Medicine Formation of west-central Montana. M.S. thesis, Montana State University, Bozeman, Montana, 170 pp.

Sohn, Y. K., C. W. Rhee, and B. C. Kim. 1999. Debris flow and hyperconcentrated flood-flow deposits in an alluvial fan, northwestern part of the Cretaceous Youngdong Basin, central Korea. Journal of Geology 107:111–132.

Turnbull, W. D., and D. M. Martill. 1988. Taphonomy and preservation of a monospecific titanothere assemblage from the Washakie Formation (Late Eocene), southern Wyoming: an ecological accident in the fossil record. Palaeogeography, Palaeoclimatology, Palaeoecology 63:91–108.

Van Itterbeeck, J., Y. Boltsky, P. Bulytynck, and P. Godefroit. 2005. Stratigraphy, sedimentology, and paleoecology of the dinosaur-bearing Kundur section (Zeya-Bureya Basin, Amur Region, far eastern Russia). Geological Magazine 142:735–750.

Varricchio, D. J. 1995. Taphonomy of Jack's Birthday Site: a diverse dinosaur bone bed from the Upper Cretaceous Two Medicine Formation of Montana. Palaeogeography, Palaeoclimatology, Palaeoecology 114:297–323.

Varricchio, D. J., and J. R. Horner. 1993. Hadrosaurid and lambeosaurid bone beds from the Upper Cretaceous Two Medicine Formation of Montana: taphonomic and biologic implications. Canadian Journal of Earth Sciences 30:997–1006.

Varricchio, D. J., F. Jackson, and C. N. Trueman. 1999. A nesting trace with eggs for the Cretaceous theropod dinosaur *Troodon formosus.* Journal of Vertebrate Paleontology 49:379–417.

Varricchio, D. J., C. Koeberl, R. R. Raven, W. S. Wolbach, W. C. Elsick, and D. P. Miggins. 2009. Tracing the Manson impact across the Western Interior Cretaceous seaway; pp. 269–299 in R. L. Gibson and W. U. Reimold (eds.), Large Meteorite Impacts and Planetary Evolution. Geological Society of America Special Paper 465.

Voorhies, M. R. 1985. A Miocene rhinoceros herd buried in volcanic ash. National Geographic Research Reports, Projects 1978:671–688.

Western, D., and A. K. Behrensmeyer. 2009. Bone assemblages track animal community structure over 40 years in an African savannah ecosystem. Science 324:1061–1064.

Zaleha, M., and S. A. Wiesemann. 2005. Hyperconcentrated flows and gastroliths: sedimentology of diamictites and wackes of the upper Cloverly Formation, Lower Cretaceous, Wyoming, U.S.A. Journal of Sedimentary Research 71:880–894.

Occurrence and Taphonomy of the First Documented Hadrosaurid Bonebed from the Dinosaur Park Formation (Belly River Group, Campanian) at Dinosaur Provincial Park, Alberta, Canada

30

David A. Eberth, David C. Evans, and David W. H. Lloyd

ABSTRACT

The Princess Bonebed (PBB) at Dinosaur Provincial Park (the Park) is a monodominant hadrosaurid bonebed containing large numbers of hadrosaurid macrofossils and a small number of taxonomically diverse microfossils. The PBB occurs in the middle of the Dinosaur Park Formation (~76 Ma) and is hosted by a lag deposit at the base of a paleochannel succession. Although the bonebed covers >15,000 m^2, so far only 36 m^2 have been excavated from four discrete locations, resulting in a collection of 176 specimens. Skeletal recovery is low (7%) and large elements (long bones, girdle elements) are over represented, both of which indicate that the assemblage is a hydraulically sorted lag deposit. Taphonomically, the assemblage is characterized by ubiquitously broken elements, with some exhibiting bone corrosion/rot and abrasion. We infer that at least one mass-kill event (coastal plain flooding?) decimated a group of tens to hundreds of hadrosaurids (most likely lambeosaurines). Carcasses then experienced a multi-event biostratinomic history involving thorough disarticulation, trampling and minor abrasion, scavenging, and bone corrosion or rot. Later, during a separate flooding event, the assemblage was reworked into the base of a fluvial channel where it was sorted, oriented, and interred. Although data are scant, there is the possibility that this bonebed assemblage was sourced from an age-segregated biocoenose of subadults and adult lambeosaurines containing no juveniles. Because hadrosaurid remains are abundant throughout the Park's strata, systematic and quantitative approaches of assessment are required in the future to positively identify monodominant hadrosaurid bonebeds.

INTRODUCTION

The Belly River Group at Dinosaur Provincial Park (southern Alberta, Canada) is famous for preserving a broad diversity and unusually large numbers of fossil vertebrates, most notably dinosaurs (Eberth et al., 2001; Currie and Koppelhus, 2005). Among the variety of preservational states for the Park's fossils, geographically extensive macrofossil bonebeds are exceedingly common, numbering comfortably in the hundreds (Currie and Koppelhus, 2005; Eberth and Currie, 2005; Eberth and Evans, 2011). Although multitaxic bonebeds represent more than 90% of bonebed occurrences at the Park (Eberth and Currie, 2005), it is the monotaxic/monodominant bonebeds (Eberth, Shannon, and Noland, 2007) that have attracted considerable scientific and popular attention, most likely because of the opportunities they provide to address paleobiological questions relating to growth and behavior (e.g., Currie and Dodson, 1984; Ryan et al., 2001; Fiorillo and Eberth, 2004; Eberth and Currie, 2005; Eberth and Getty, 2005; Brinkman et al., 2007; Eberth et al., 2010).

Until now, each of the Park's monotaxic/monodominant dinosaur bonebeds (~20 known occurrences) has been shown to be dominated by one of three centrosaurine ceratopsians–*Centrosaurus apertus, Centrosaurus brinkmani,* and *Styracosaurus albertensis* (Eberth and Currie, 2005; Eberth and Getty, 2005). This pattern has been proposed as evidence that each of these ceratopsian taxa was gregarious to some degree, possibly congregating in "herds" during seasonal migrations (Currie, 1981; Currie and Dodson, 1984; Brinkman et al., 1998; Eberth and Getty, 2005; Eberth et al., 2010).

In contrast, the absence of reports of monotaxic/monodominant bonebeds dominated by hadrosaurids at the Park is peculiar given the large abundance of hadrosaurid fossils there (Dodson, 1971; Brinkman, 1990; Brinkman et al., 2005; Brinkman, 2011), and the numerous monodominant hadrosaurid bonebeds that occur elsewhere in Upper Cretaceous strata throughout Canada, the Western Interior, and globally (e.g., Rogers, 1990; Varricchio and Horner, 1993; Schmitt et al., 1998; Hu et al., 2001; Lauters et al., 2008; Gangloff and Fiorillo, 2010; Scherzer and Varricchio, 2010; Ryan et al., 2011; Schmitt et al., this volume). Here we partly

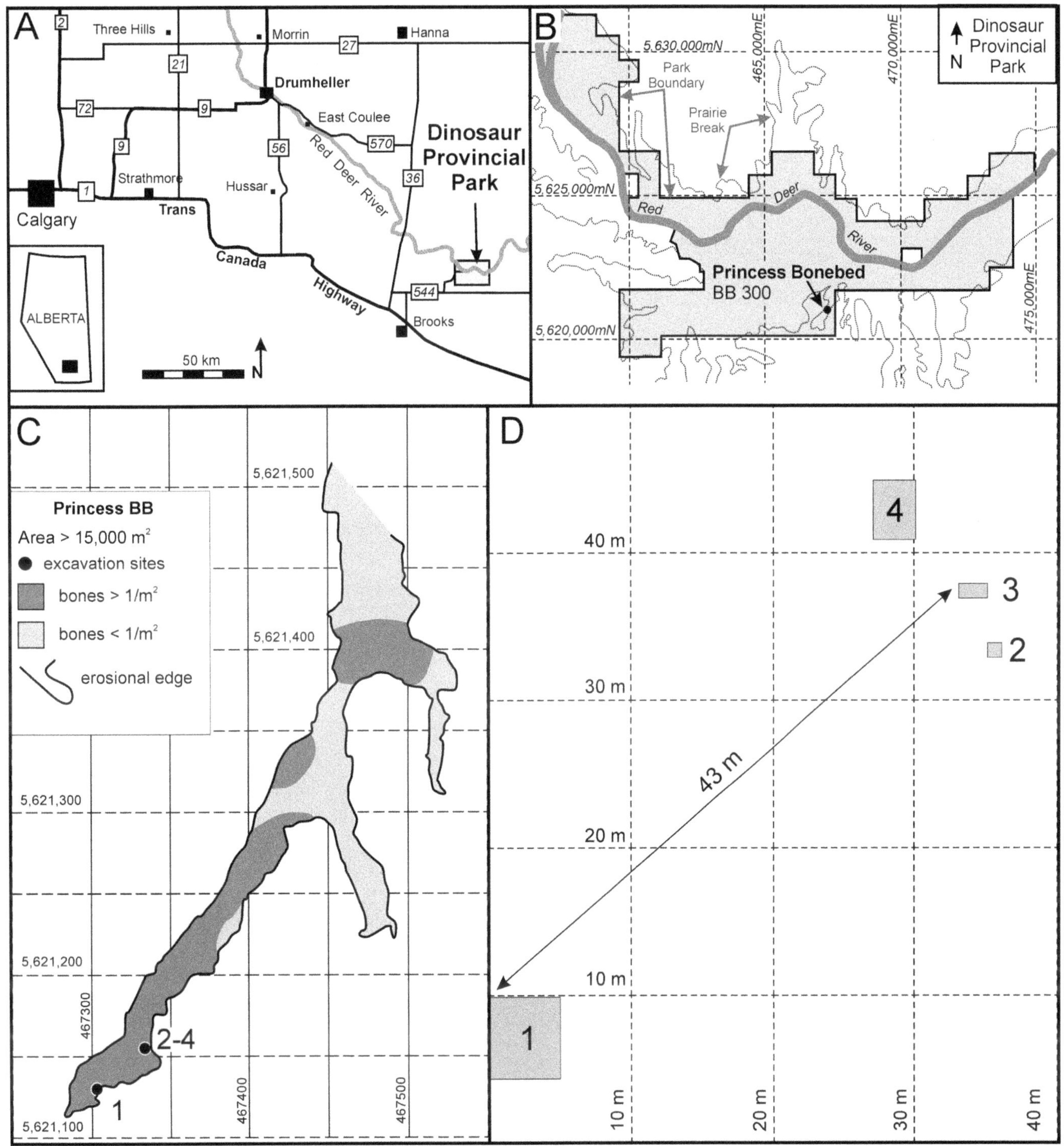

30.1. Locality maps for the Princess Bonebed. (A) Location of Dinosaur Provincial Park in southern Alberta; (B) location of Princess Bonebed at Dinosaur Provincial Park (shaded area indicates extent of the Park; grid spacings are 5 km); (C) extent of the exposed Princess Bonebed (shaded areas indicate different concentrations of fossils; black circles indicate excavation sites 1 and 2–4; grid spacings are 50 m); (D) locations of excavation sites 1–4 (grid spacings are 10 m).

resolve this paradox by documenting the first monodominant hadrosaurid bonebed from the Park – the Princess Bonebed (Royal Tyrrell Museum [TMP] locality L2369; TMP bonebed JR300) – and discuss its significance. We propose that the absence of hadrosaurid-dominated bonebeds is an artifact of non-quantitative survey methods used to document the relative abundances of the Park's fossil resources. In short, because dinosaur fossils are unusually abundant throughout the Park's strata (Currie and Koppelhus, 2005) and hadrosaurids are the most common macrovertebrates (Dodson, 1971,

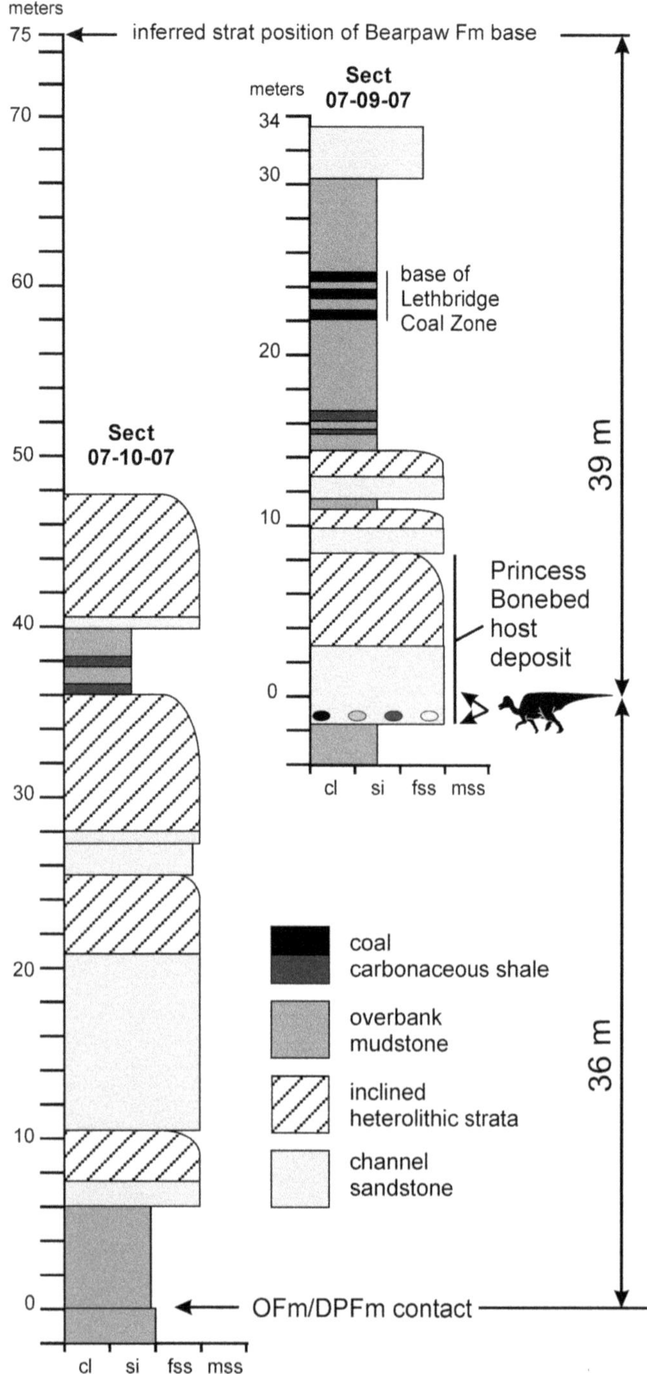

30.2. Measured sections in the area of the Princess Bonebed. Stratigraphic position of bonebed in section 07-09-07 indicated by *Corythosaurus* silhouette.

1983, 1987; Brinkman, 1990; Brinkman et al., 2005; Ryan and Evans, 2005), it has been difficult to recognize hadrosaurid bonebeds as distinct from the "background" richness of hadrosaurid remains using only casual assessments of surface material. We propose that the Park's uniquely rich fossil abundance has made it difficult to see the forest for the trees, especially in the case of hadrosaurids.

METHODS

A pre-excavation survey of exposed portions of the PBB identified areas of highly concentrated fossils, and helped determine the overall content, preservational conditions, and geological context of the fossils (Eberth, Rogers, and Fiorillo, 2007). The survey revealed that, as a whole, the bonebed is a macrofossil site that contains a small proportion of vertebrate microfossils. Accordingly, the site can also be interpreted as a mixed macrofossil/microfossil bonebed (cf. Eberth, Shannon, and Noland, 2007). Although the site was initially targeted for excavation and study because it was thought to be a multitaxic bonebed, the pre-excavation survey revealed that it was dominated by hadrosaurs. This observation intensified our interest in the site because no hadrosaur bonebeds had been documented previously from the Park.

Within the bonebed, four discrete sites (Sites 1–4; Fig. 30.1C, D) totaling 36 m^2 were selected for excavation. At each site, surface bone "scrap" was collected, a baseline was installed, and a conventional alphanumeric meter grid system was established in order to expedite mapping and data collection (cf. Eberth, Rogers, and Fiorillo, 2007). During the summer of 2007 specimens were uncovered, mapped and then removed for preparation and storage at the Royal Tyrrell Museum of Palaeontology (TMP). Three-dimensional orientations were recorded as azimuth and plunge measurements for elongate elements (length >5 cm). Notable non-vertebrate objects, such as coalified tree parts, were also mapped, but not collected. Sediment samples were collected with many of the elements for further lithologic, petrographic and geochemical analyses (not part of this study).

A total of 176 specimens were collected during 2007. These were prepared and curated during the period 2007–9. Specimens were reexamined in the lab to identify taxon and skeletal element, and to identify and quantify taphonomic features such as breaks and surface modifications (Appendix 30.1; cf. Eberth, Rogers, and Fiorillo, 2007).

LOCATION AND GEOLOGY

The PBB occurs in a narrow, steep-walled, northeastward draining coulee along the south-central border of the Park (Fig. 30.1B, C). Although the fossil producing extent of the bonebed is quite large (Fig. 30.1C), the primary collecting area (Site 1) is located at 12U; 467312mE; 5621129mN (Universal Transverse Mercator coordinates obtained from a handheld Garmin global positioning system calibrated to World Geodesic Survey [WGS] 1984/North American Datum [NAD] 1983).

The bonebed was placed into stratigraphic context by assessing its stratigraphic position relative to the

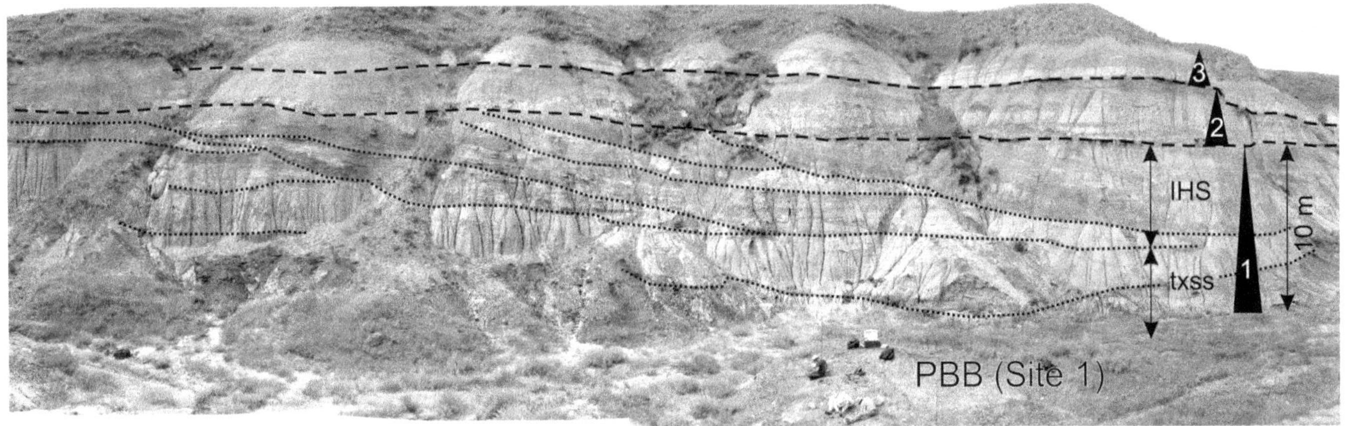

30.3. The Princess Bonebed (PBB) occurs in an extensive, sandy lag deposit at the base of a ~10 m thick paleochannel rhythm (black triangle "1"). The lower half of the deposit is dominated by trough cross-bedded, fine-to-medium-grained sandstones (txss), intraclast conglomerates, and fossils. The upper half of the deposit comprises inclined heterolithic stratification (IHS). IHS beds dip to the northeast and define point bar surfaces in a channel that was ~60 m wide and 5 m deep at bankfull, and oriented east-west.

Oldman–Dinosaur Park formational contact (OFm-DPFm contact), and the base of the Lethbridge coal zone. Although the OFm-DPFm contact is time transgressive along a north-south trend across southern Alberta (Eberth and Hamblin, 1993), it serves as a valuable lithostratigraphic surface and chronostratigraphic datum within the confines of the Park (Eberth, 2005). Similarly, the base of the Lethbridge coal zone is a useful datum across the Park (Eberth and Hamblin, 1993; Eberth, 2005).

Stratigraphically, the PBB lies in the middle of the Dinosaur Park Formation (DPFm), ~36 m above the contact with the Oldman Formation (OFm) and ~39 m below the base of the Bearpaw Formation (BPFm)(Fig. 30.2). Neither of these contacts is present at the bonebed site, but the OFm-DPFm contact is preserved 1.2 km north of the site, and the base of the Lethbridge coal zone, which consistently occurs 17 m below the base of the BPFm at the Park, is preserved 750 m south of the PBB. Figure 30.2 displays two measured sections that, together, show a typical vertical section through most of the DPFm (e.g., note the lower sandier and upper muddier nature of the composite section [cf. Eberth, 2005]). These two sections were correlated by tracing the strata between the sections.

Following Kuiper et al. (2008) we have revised the ^{40}Ar/^{39}Ar radiometric ages for the Park reported by Eberth (2005) as 0.6% older. Accordingly, we estimate the age of the site at ~76.0 Ma.

Overall depositional context was determined by sedimentological study of the host lithesome–a 10 m thick, laterally extensive multistoried sandstone body that dominates exposures in the coulee (Fig. 30.3). The PBB occurs at or near the base of the multistoried sandstone deposit. The sandstone comprises an assemblage of facies, and sedimentological and architectural features that indicate a paleochannel origin for the deposit (cf. Wood, 1985, 1989; Eberth, 2005). These include

1. An overall upward-fining grain size pattern and thinning-upward set thicknesses. Grain sizes range from pebbly lag deposits at the base to very fine grained sandstone and mudstones from the middle of the deposit upward. Cross-bed set thicknesses range from ~1 m at the base (trough cross beds) up to cm-scale ripple cross lamination near the top.
2. A basal lag of vertebrate fossils (the bonebed) mixed with ironstone intraclasts, angular to rounded mudstone clasts, and coalified plant remains. The lag varies in thickness, but has been observed as being up to 50 cm thick; it is laterally discontinuous or patchy over its 15,000 m^2 distribution (Fig. 30.1), and is often associated with shallow centimeter-scale scour fills. Other lag deposits mark erosional boundaries or surfaces between stacked paleochannel deposits ("rhythms" of Wood [1989]). All erosional surfaces are broadly concave up.
3. Inclined heterolithic strata (IHS) dominate the upper one-half of the sandstone body and comprise alternating, decimeter-thick beds of organic-rich mudstone and fine grained sandstone that dip consistently toward the northeast (Fig. 30.3).

Table 30.1. Assemblage Data by Taxon from the Princess Bonebed

Taxon	Macrofossil (>5 cm)	Microfossil (<5 cm)	Totals	%	Hadrosaur/non-hadrosaur
Ornithischia/hadrosaur	37	25	62	35%	
Hadrosauridae (lambeosaurine)	8		8	5%	84%
Hadrosauridae (subfamily uncertain)	60	18	78	44%	
Testudines (Trionychidae)	3	4	7	4%	
Tyrannosauridae		4	4	2%	
Amiidae		3	3	2%	
Lepisosteidae		3	3	2%	
Dromaeosauridae		2	2	1%	
Champsosauridae		2	2	1%	
Testudines (Chelydridae)		1	1	1%	16%
Testudines (*Basilemys*)	1		1	1%	
Testudines (Macrobaeinidae?)		1	1	1%	
Ornithomimidae		1	1	1%	
Ankylosauridae		1	1	1%	
Polyodontidae		1	1	1%	
Myledaphus		1	1	1%	
TOTALS	109	67	176	100%	100%
%	62%	38%			

Interpretation

These features indicate that the channel was emplaced at the PBB location during an avulsion event that resulted in widespread erosion and reworking of underlying sediments and vertebrate elements (cf. Rogers and Brady, 2010). The mixture of intraclasts, disarticulated fossils, and coalified wood debris in the basal lag suggests that large amounts of skeletal debris and wood were probably scattered across the landscape prior to flooding and were transported, reworked, and concentrated at the base of the channel during erosion and initial infilling.

The up-section transition from cross-bedded sandstones into IHS is a common feature of paleochannel sandstones at the Park (Wood, 1989; Eberth, 2005) and indicates either an up-point-bar-slope transition from sandy thalweg deposits into more muddy, deposits, or the transition of an initially straight sandy channel (emplaced during avulsion) into a mixed bedload/suspended load meandering channel through time (cf. Wood, 1989; Eberth, 2005).

The IHS beds reflect the preservation of point bar deposits, and the orientation and size of the channel can be approximated by assessing the direction of maximum dip and by using the two-thirds rule applied to a cross-section of the IHS (Williams, 1988), as well as a direct measure of the vertical relief exhibited by IHS (point bar) surfaces. In the case of the PBB host paleochannel, the meandering channel portion at bankfull flow was ~60 m wide, and ~5 m deep. We estimate that paleochannel trend at this location was northwest-southeast, perpendicular to the trend of the exposed outcrop. These dimensions and orientation are typical for paleochannels in Dinosaur Park Formation elsewhere in the Park (Wood, 1989; Eberth, 2005).

ASSEMBLAGE DATA

Assemblage data from the 176 elements collected from the PBB are presented in Table 30.1 and Appendix 30.1. The bonebed is a "mixed site" (Eberth, Shannon, and Noland, 2007) comprising 62% macrofossils (>5 cm) and 38% microfossils (<5 cm). However, a significant number of the microfossils (n = 43; 65%) are fragmentary pieces from large bones of ornithischians. This indicates that the overwhelming majority of fossils that were eventually reworked into this site began their taphonomic history as macrofossils.

Where identifiable to family level, macrofossil remains are ubiquitously those of hadrosaurids. In contrast, microfossils are taxonomically diverse. During the pre-excavation survey and examination of the entire bonebed area we also identified mostly hadrosaurid remains and a few fragments of turtle and rare microfossils; no large ceratopsian or theropod elements were identified.

One hundred and fourteen specimens can be identified to family or to a finer taxonomic level (Table 30.1). Among these, 86 (75%) are hadrosaurid. Among these hadrosaurid elements, 8 can be further diagnosed as Lambeosaurinae (Fig. 30.4).

We regard the unidentified ornithischian specimens (n = 62) as most likely hadrosaurids because (1) the specimens are derived from large macrofossils; (2) the specimens exhibit no bone texture characteristic of ankylosaurs; and (3) there are no identifiable ceratopsian remains in this assemblage,

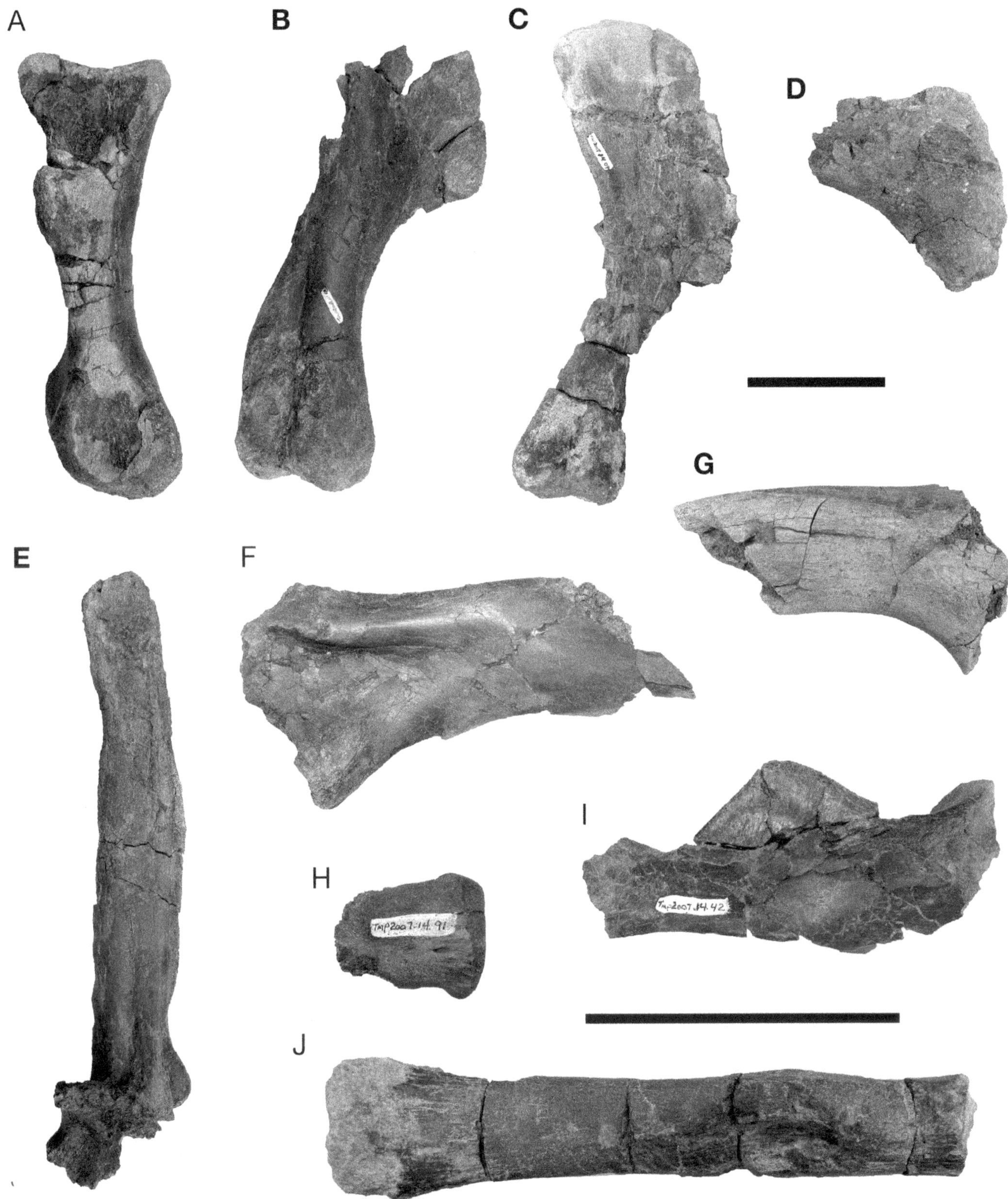

30.4. Hadrosaurid bones from the Princess Bonebed. (B–E) and (G) (in boldface) are from a lambeosaurine hadrosaurid. (A) Metatarsal IV in medial view; (B) adult humerus in posterior view; (C) subadult humerus in posterior view; (D) incomplete distal terminus of ischium in medial view; (E) dorsal neural spine in lateral view; (F) incomplete scapula in lateral view; (G) distal segment of ischium in lateral view; (H) ungual phalanx from manus in plantar view; (I) surangular in lateral view; (J) metacarpal III in posterior view. Scale bars equal 10 cm.

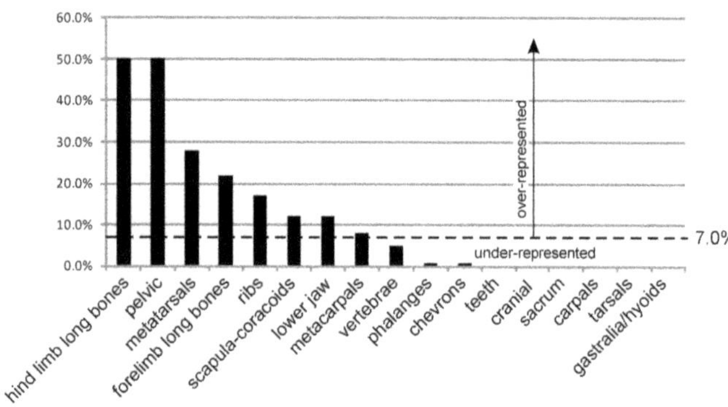

30.5. Skeletal-element representation in the Princess Bonebed collection. Overall recovery is ~7% (dashed line). All elements with a relative abundance greater than 7% (black columns) are over-represented; those with relative abundances less than 7% (white) are under-represented. Vertebrae are variably represented by complete elements, centra, neural arches, and spines.

nor were any identified from surface materials examined during the pre-excavation survey. Including this unidentified ornithischian material with the other hadrosaurid specimens in the bonebed is similar to the practice used in our ceratopsian bonebed analyses at the Park and other locations in southern Alberta (Currie and Dodson, 1984; Ryan et al., 2001; Eberth and Getty, 2005; Eberth et al., 2010). In those cases, material that is not diagnostic at a level finer than "Ceratopsia" is grouped with the most finely diagnosed specimens for purposes of assemblage analysis (e.g., *Centrosaurus apertus* at BB 30; Eberth and Getty, 2005) unless there is evidence that there are two or more ceratopsian taxa at the site. Accordingly, we regard the collection of hadrosaurid material from the PBB as comprising a total NISP of 148 (84%), and likely representing a taxon of lambeosaurine.

The 28 identifiable elements that are non-hadrosaurid (16% of the total assemblage) are microfossil remains from 13 different taxa of aquatic and terrestrial vertebrates, including fish, turtles, champsosaur, small and large theropods, and an ankylosaur. This material is similar in composition to assemblages that can be recovered from most channel lag deposits in the DPFm at the Park and, accordingly, is regarded as background material. Because of this status, we regard it as unlikely that this material experienced the same taphonomic history as the hadrosaurid elements, and we do not consider the material further in our assessment of the taphonomic history of the site.

The hadrosaurid material exhibits size and ontogenetic variation from subadult to fully grown individuals. One complete humerus (TMP 2007.014.0009) is 340 mm long and typical of subadult size, whereas one almost complete femur (TMP 2007.014.0005) is 1038 mm in length, and is the size of typical adult lambeosaurines at DPP (Brinkman, 2011:table 1; Brinkman, this volume) and elsewhere (Lull and Wright, 1942). No specimens from the bonebed are juvenile; however, juvenile hadrosaurid specimens are common elsewhere at the Park (Brinkman, 2011:fig. 1; Brinkman, this volume).

The minimum number of individuals (MNI) of hadrosaurids represented in the collection is three, based on a size comparison of femora and humeri. Four femora (TMP 2007.014.0001, 0002, 0004, and 0005) and three humeri (TMP 2007.014.0008, 0009, and 0036) in the PBB collection are from at least three significantly different size ranges, as indicated by differences in preserved or inferred element lengths of more than 10%. Further MNI analysis was hampered by the fragmentary nature of the specimens.

Given that a lambeosaurine skeleton consists of approximately 389 elements (not counting tendons and teeth), complete recovery of all skeletal parts for three individuals should include approximately 1167 elements. Our collection of 86 hadrosaurid elements was adjusted to 76 by removing nine tendons and one tooth. Accordingly, this collection of 76 identifiable elements from a minimum of three individuals represents a recovery of ~7% (76/1167). Skeletal-element representation analysis (Table 30.2; Fig. 30.5) indicates that the PBB collection contains an over-representation of large, massive elements (limb long bones, hip/shoulder elements, ribs, metatarsals and metacarpals) and is under-represented in small, lighter elements (phalanges, chevrons, teeth, cranial elements, carpals, tarsals, gastralia). A few element groups do not fit this pattern: sacra are absent, and there is a slight over-abundance of lower jaw elements versus a slight under-abundance of vertebrae.

Interpretation

In general, hadrosaurids comprise 40%–50% of identifiable dinosaurian remains in the Park's fossil assemblages, regardless of whether the data are sourced from macrofossil or microfossil assemblages (Dodson, 1971, 1987; Brinkman, 1990; Brinkman et al., 1998; Brinkman et al., 2005:table 5.1; Eberth and Currie, 2005). Relative to all vertebrate taxa, hadrosaurid remains at the Park typically represent 10%–20% of any assemblage (only assessed in microfossil studies; e.g., Brinkman, 1990:table 2).

In the PBB assemblage, hadrosaurids comprise 91% of all dinosaurs identified to family and 39% of the microfossil assemblage identified to family, indicating that the bonebed is hadrosaurid dominated. This interpretation is more robust if we accept the assignation of the 62 unidentified ornithischian specimens to Hadrosauridae. In that case 95% of all dinosaurs and 61% of all microfossils are hadrosaurids. Thus, by any measure, the PBB is a hadrosaurid-dominated bonebed. However, due to our limited dataset that lacks

Table 30.2. Hadrosaur Skeletal Element/Part Representation in the Princess Bonebed

Element Class	Number of elements per skeleton	× MNI 3	PBB Collection	NISP (%) of MNI 3	Over-underrepresented
Hindlimb long bones	6	18	9	50.0%	Over
Pelvic	6	18	9	50.0%	Over
Metatarsals	6	18	5	27.8%	Over
Forelimb bones	6	18	4	22.2%	Over
Scapula-coracoid	4	12	2	16.7%	Over
Lower jaw	11	33	4	12.1%	Over
Ribs	64	192	22	11.5%	Over
Metacarpals	8	24	2	8.3%	Over
Vertebrae	112	336	18	5.4%	Under
Phalanges	48	144	1	0.7%	Under
Chevrons	70	210	1	0.5%	Under
Teeth	100s	1000s	1	0.0%	Under
Cranial	41	123	0	0.0%	Under
Sacrum	1	3	0	0.0%	Under
Carpals	2	6	0	0.0%	Under
Tarsals	2	6	0	0.0%	Under
Gastralia, etc. (hyoids)	2	6	0	0.0%	Under
Totals	389	1167	78	6.7%	

Note: Teeth and tendons were excluded from the analysis.

bones diagnostic to species, we may only speculate that this monodominant hadrosaurid bonebed is dominated by a single species of lambeosaurine. Given the stratigraphic position of the bonebed in the Dinosaur Park Formation and the known distribution of lambeosaurines within that unit (Ryan and Evans, 2005), the taxon most likely represented in the bonebed is *Lambeosaurus lambei.*

This disarticulated assemblage is preserved in a channel lag deposit, is over-represented in large skeletal elements, and is under-represented in smaller elements. This supports an interpretation that the assemblage was hydraulically sorted prior to final burial. Given the absence of skeletal element associations and the ubiquity of breaks it seems likely that the assemblage was reworked from a thanatocoenose in which disarticulation had already occurred and many taphonomic modifications were already present. This interpretation is compatible with the geological history of the site that indicates elements were reworked and transported during channel incision and infilling.

ELEMENT DISTRIBUTIONS AND ORIENTATIONS

All elements are flat lying with respect to the underlying erosional surfaces. At Site 1, where 141 (80%) of the elements were collected, the erosional surface dips toward the south across an area of ~27 m^2 (Fig. 30.6). All of the elements at this site lie at or slightly above this surface with their long axes parallel to it. In contrast, the erosional surfaces that underlie the bonebed at each of the other three sites (2–4) are more or less horizontal, and the bones lie parallel to them. The concentration of elements is variable and patchy. In the areas of greatest bone concentration there are as many as 10 elements/m^2.

Azimuths of 42 elongate elements at Site 1 were analyzed to assess whether or not the elements are distributed uniformly around 360°. Figure 30.7 shows the distribution of these data in a bimodal (mirror-image) rose diagram and also includes the results of statistical tests for uniformity (Rayleigh, Rao's Spacing, and Kuiper's tests). All tests fail to reject the hypothesis of uniform distribution. Also, the length of the mean vector (129°–309°) is short (0.16) and standard deviation is large (55.2°); thus, there is no statistical evidence that this group of elements was preferentially oriented along this trend by fluvial currents or any other influences. However, further inspection of the rose diagram suggests the possible presence of two trends oriented perpendicular to one another (northwest-southeast and northeast-southwest). Although there are no reliable statistical analyses that can be used to test the significance of an apparent quadrumodal pattern of azimuths, we observed (1) seven vertebrate microfossils concentrated along the northwest-facing side of a northeast-southwest oriented femur at Site 4 (Fig. 30.8), (2) a coalified log at Site 1 trending northwest-southeast (Fig. 30.6), (3) a few large-scale cross beds with eastward dipping foresets (measured parallel to the central axis) a few tens of meters east of Site 1, and (4) point bar surfaces dipping uniformly to the northeast and indicative of a northwest-southeast paleochannel trend (see above). In combination, these observations suggest that flow was from the northwest,

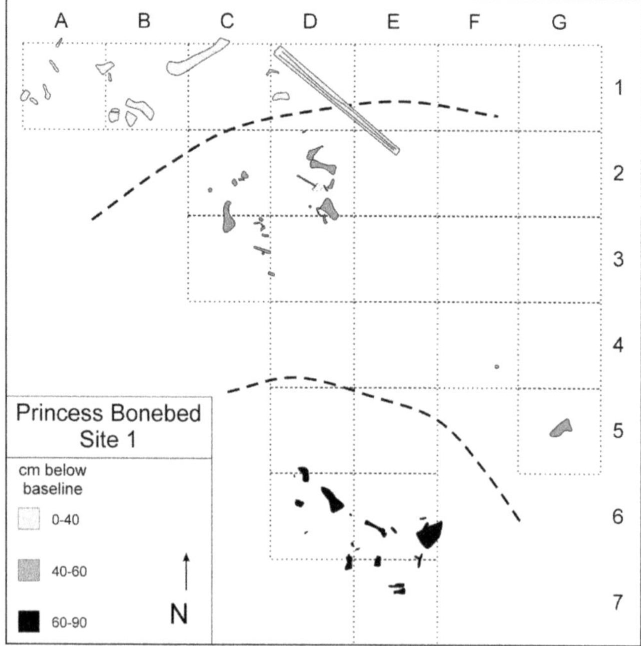

30.6. In situ distribution of hadrosaurid elements and coalified tree (grids D1–E2) at Site 1. Bedding surface slopes to the south as indicated by specimen depth below baseline (dashed lines are simple surface contours bounding 0–40, 40–60, and 60–90 cm below baseline). Note the extreme patchiness of fossils at this site. Patchiness typifies the bonebed throughout its extent and probably reflects variations in fluvial flow and depositional conditions and events. Letters, numbers, and dotted lines indicate the alphanumeric grid system.

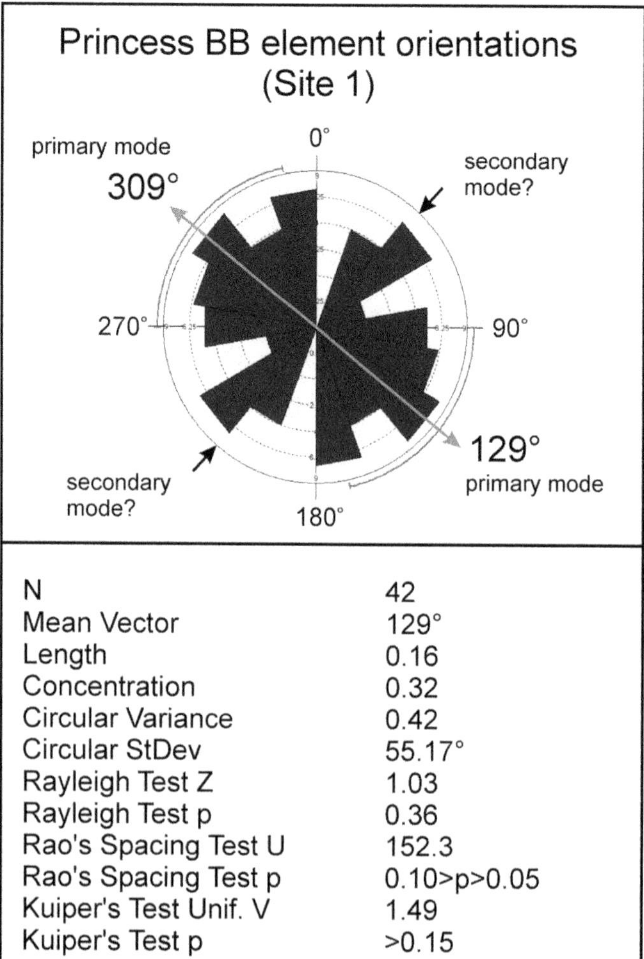

N	42
Mean Vector	129°
Length	0.16
Concentration	0.32
Circular Variance	0.42
Circular StDev	55.17°
Rayleigh Test Z	1.03
Rayleigh Test p	0.36
Rao's Spacing Test U	152.3
Rao's Spacing Test p	0.10>p>0.05
Kuiper's Test Unif. V	1.49
Kuiper's Test p	>0.15

30.7. Bimodal rose diagram and statistical analysis of azimuths of 42 elongate macrofossil elements at Site 1. Statistical analyses indicate the absence of any significant trends and thus support an interpretation of uniform circular distribution. Visual inspection, however, suggests the presence of a northwest-southeast primary trend (129°–309°) and a southwest-northeast secondary trend.

compatible with previous sedimentological interpretations of overall flow directions in DPFm paleochannels at the Park (e.g., Eberth, 2005). Accordingly, we propose that many elongate elements were deposited in or became oriented parallel to southeastward flowing waters (primary trend; Fig. 30.7), but that some elements became oriented perpendicular to these primary southeastward currents, thus forming a secondary, southwest-northeast trend. Elements may align perpendicular to the current in shallow water or under the influence of weaker currents (Behrensmeyer, 1990), or as a result of being only partially submerged (Voorhies, 1969).

ELEMENT MODIFICATIONS

Almost every element in the collection shows evidence for post-fossilization fracturing and crushing due to differential burial-compaction. Such diagenetic fracturing and crushing is typical of most of the Park's fossils and is ignored in the following biostratinomic (perthotaxic) taphonomic assessment. Bone surface preservation is generally very good (ignoring diagenetic fracturing and crushing), and ranges from mostly unaltered to slightly "weathered," as exhibited by two occurrences of checking and cracking (Appendix 30.1).

Element modifications due to biostratinomic processes fall into two groups: those assessed as common and exhibited by more than one-half of the specimens, and those that are assessed as uncommon and present in less than one-half of the specimens.

Common Element Modifications

- 163 (93%) of the elements are incomplete (Fig. 30.4; Fig. 30.9);
- 143 (81%) of the elements are <50% complete to fragmentary; conversely, 33 (19%) of the elements are >50% complete, with 13 deemed complete. Among the complete elements, 7 are very small microfossils;
- Bone pebbles and fragments consisting solely of trabecular bone are common;

- 95 elements preserve evidence of laminar-bone breakage (does not include trabecular bone pebbles);
- Three varieties of breaks are present (Fig. 30.9A–C, E–G). In descending abundance, these are transverse/stepped (n = 78); parallel/longitudinal (n = 40); and oblique or spiral ("green," n = 9). Different kinds of breaks may occur on single elements.

Uncommon Element Modifications

- Corrosion/bone rot and collagen degradation are indicated by trabecular bone loss at the broken ends of ~20 elements (Fig. 30.9F–G);
- Evidence for slight abrasion and "polishing" occurs along the otherwise sharply broken edges of 8 elements (Fig. 30.9C–D);
- Highly focused, circular to oval areas of crushed laminar bone on 4 elements (Fig. 30.9E) interpreted as evidence for biting;
- Light-colored "plant or fungal-hyphae discoloration" marks are present on 4 elements (Fig. 30.9G);
- Slight cracks and flaking, tentatively interpreted as evidence for weathering, are preserved on 2 elements;
- A possible tooth mark is preserved on 1 element.

Interpretation

The combination of ubiquitously broken elements, many of which exhibit trabecular bone loss (interpreted as corrosion/rot) and minor abrasion, is the characteristic element-modification signature of the PBB assemblage. This signature allows us to infer that most elements were chemically altered, trampled, and scavenged while residing in a wet, muddy substrate–a significant event in their biostratinomic history. Fresh, collagen-rich bones are easy to abrade but difficult to break (Behrensmeyer, 1982, 1991; Lyman, 1994). Accordingly, we regard it as unrealistic that element transport in a sandy river channel during flood could have resulted in such thorough breakage without resulting in significant amounts of abrasion. In contrast, highly localized and focused forces–such as chewing and trampling–often result in thoroughly broken elements, but only trampling results in large amounts of breakage with minor to modest amounts of abrasion (resulting from elements being trampled in or on mildly to highly abrasive substrates [Brain, 1967; Andrews and Cook, 1985; Behrensmeyer et al., 1986]). Our interpretation that breakage in the PBB is due to trampling is also supported by the fact that the same interpretation explains the breakage and minor abrasion of elements in many of the ceratopsian monodominant bonebeds at the Park (Eberth and Getty, 2005; Eberth et al., 2010).

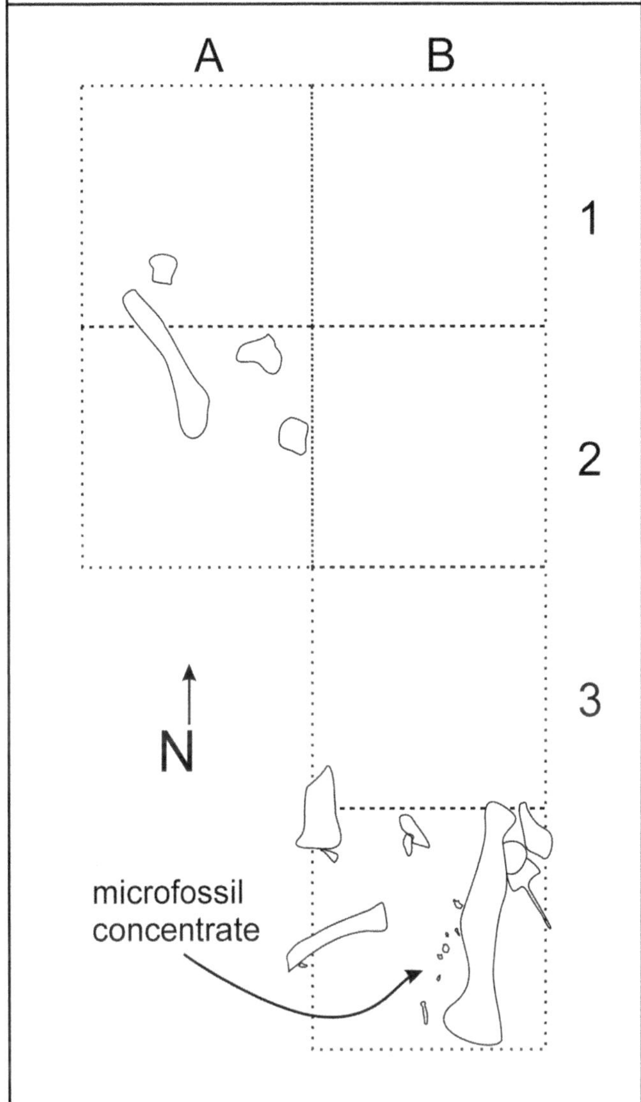

30.8. In situ distribution of hadrosaurid elements at Site 4. Note the concentration of vertebrate microfossils west of the large femur in grid square B4.

The large percentage of transverse, stepped, and longitudinal breaks compared to spiral (green fracture) breaks suggests that many of the bones may have lost some elasticity through the breakdown of collagen, the first step toward recrystallization before the trampling event(s) (cf. Rogers et al., 2010). In turn, this suggests that the PBB elements may have been altered to varying degrees even prior to trampling. The lack of any in situ skeletal or bone-piece associations indicate that trampling and breakage preceded the hydraulic

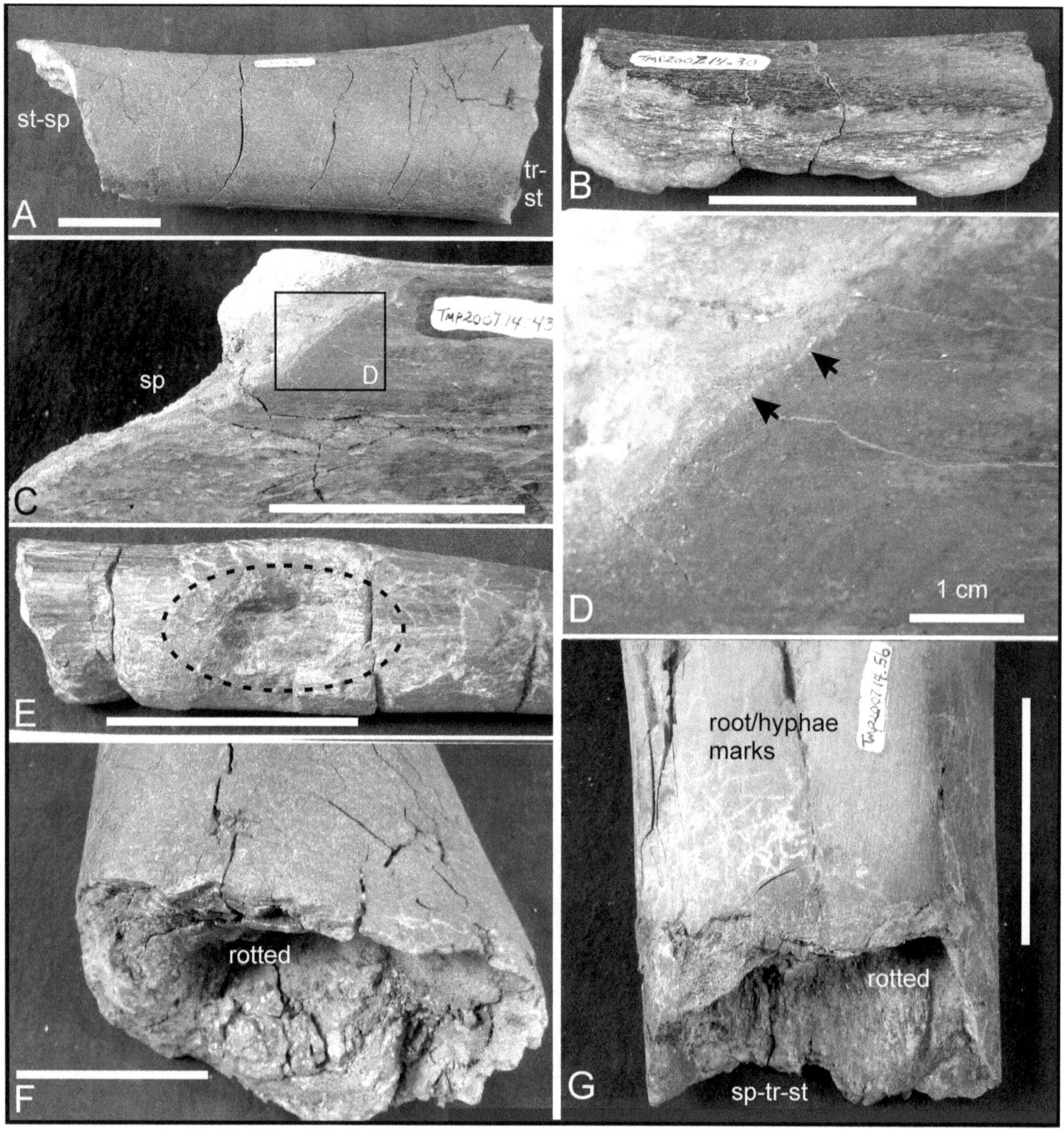

30.9. Taphonomic features of the Princess Bonebed. (A) Stepped, spiral, and transverse breaks on a tibial shaft (TMP 2007.014.0011); (B) longitudinal break along the long axis of a rib (TMP 2007.014.0030); (C–D) spiral break on an ischium (TMP 2007.014.0043) and low-grade abrasion (arrows) along its broken edge; (E) oval-shaped load/impact feature (dashed oval) on a metacarpal (TMP 2007.014.0022); (F) broken end of tibia (TMP 2007.014.0011) with excavated trabecular bone indicative of bone corrosion/rot; (G) broken end of ischium (TMP 2007.014.0056) showing rotted interior and root/hyphae marks on bone surface. All scale bars equal 5 cm, except in (D), where it is 1 cm. Abbreviations: sp, spiral; st, stepped; tr, transverse.

reworking event(s) that deposited and buried elements in the PBB; they were not trampled at the PBB sites.

A multistep biostratinomic history is supported by abundant evidence for corrosion/rot, root-like patterns of discoloration on eight elements, the presence of cracked and flaked bone in two specimens, a single, possible tooth mark, and four specimens that exhibit highly localized, circular-to-oval areas of crushed bone (Fig. 30.9E), interpreted here as crushing due to biting by large meat-eating dinosaurs. These features indicate residency for some time on or in a saturated, muddy substrate in a landscape where trabecular bone was chemically altered or destroyed, plants were present, and scavengers had access to decomposing carcasses, and other dinosaurs could trample the elements (e.g., channel bar, levee, or floodplain).

DISCUSSION AND CONCLUSIONS

Our geologic and taphonomic data indicate that the PBB assemblage records a multistep biostratinomic history as follows: (1) death of several hadrosaurids (most likely a group of lambeosaurines); (2) skeletal disarticulation, trampling and breakage of elements accompanied by bone rot and breakdown of collagen, and variable surface modification of elements including abrasion, bioerosion, and biomarking; (3) reworking and sorting of the modified elements into a paleochannel lag deposit. We regard this succession of three 'steps' as the minimum required to explain all the observations presented here.

Although we are only able to calculate an MNI of three, the vast size of the bonebed (>15,000 m^2) and the fact that we excavated only a total of 36 m^2 suggests to us that the number of individual hadrosaurids that contributed skeletal elements to this site was much larger than three, and likely ranged from tens to hundreds of individuals when scaled across the exposed area of the bonebed. Such numbers are not unrealistic for other monodominant hadrosaurid bonebeds. Gangloff and Fiorillo (2010) described an *Edmontosaurus* bonebed from the north slope of Alaska that comprises the remains of at least 36 individuals (based on a collection of 2769 identifiable elements). Similarly, the *Shantungosaurus* bonebed of Zhucheng comprises the remains of many hundreds of individuals (Hu et al., 2001; DAE, pers. obs., 2010) that were entrained in a succession of mud-flows over time (DAE, unpubl.).

Although the apparent absence of juveniles in the PBB is only very weakly supported by our data (MNI = 3), our pre-excavation surveys of the PBB (qualitative examinations of other exposed portions of the bonebed) failed to reveal any juvenile elements. However, because remains of juvenile hadrosaur are common at other fossil localities in the Park (e.g., Brinkman, 2011, this volume), we speculate that there may have been an age group bias against juveniles in the original biocoenose (living assemblage). If so, the absence of juveniles in the PBB would be compatible with paleobiologic hypotheses that juveniles remained ecologically segregated from groups of sub-adults and adults (cf. Forster, 1990; Lauters et al., 2008; Scherzer and Varricchio, 2010). Although we regard this interpretation as highly speculative in the case of the small PBB dataset, the hypothesis can be tested further at this and other fossil localities in the Park.

Our data offer no clear answers as to what killed these animals, the length of time over which they died, or whether they died together or individually over time. However, given the overwhelming relative abundance of hadrosaurids in this assemblage and the consistent taphonomic signature exhibited by the assemblage, it is a reasonable interpretation that many of these elements are derived from individuals that died at approximately the same time and were then exposed to similar taphonomic influences and events. In this context, we infer that these individuals were part of a biocoenose of hadrosaurids that experienced at least one mass kill.

The better known ceratopsian monodominant bonebeds at the Park are interpreted as having formed in response to major coastal plain flooding events that drowned groups of ceratopsians numbering in the tens to more than a thousand (Eberth and Currie, 2005; Eberth and Getty, 2005; Eberth et al., 2010). Furthermore, Eberth and Currie (2005) reviewed taphonomic modes and patterns across the Park's entire fossil assemblage and proposed that coastal plain flooding was likely a major cause of mass mortalities among the Park's macrovertebrates (dinosaurs), and the most consistent influence on patterns of vertebrate preservation at the Park.

In the context of the foregoing, we speculate that the PBB assemblage comprises the remains of a group of subadult to adult lambeosaurines that died during a mass kill event (coastal-plain flooding?). Carcasses were deposited across a limited portion of the coastal plain and continued to disarticulate under the influence of soft-tissue decomposition, trampling and abrasion, scavenging and bone corrosion/rot. Ultimately, the modified assemblage was reworked and concentrated into the channel of an eastwardly flowing river during a significant channel-belt avulsion event.

It has been difficult to recognize hadrosaurid bonebeds at the Park as distinct from the background richness of macrofossil remains because dinosaur fossil resources are unusually abundant throughout the Park's strata, and hadrosaurids are the most common dinosaurs preserved there (Dodson, 1971; Brinkman, 1990; Brinkman et al., 2005; Ryan and Evans, 2005). Hadrosaurid bonebeds are probably quite common at the Park, and quantitative methods should be used to identify them in the future. Otherwise, these assemblages run the risk of being misidentified as multitaxic bonebeds.

SUMMARY

- The PBB is the first documented hadrosaurid-dominated (monodominant) bonebed at the Park and in the Belly River Group of southern Alberta.
- The PBB is a mixed macrofossil/microfossil assemblage that contains a large number of hadrosaurid macrofossil elements and a small number of taxonomically diverse microfossils. Similar kinds of microfossils are present in all channel lag deposits at the Park, and thus are considered background noise in this study.
- The PBB occurs in the middle of the Dinosaur Park Formation, has an age of ~76 Ma, and is hosted by channel lag deposits that are at least 15,000 m^2 in extent. Fossil elements are concentrated in small patches and, so far, only 36 m^2 of the bonebed have been excavated, yielding a total of 176 elements.
- An MNI of three hadrosaurids (likely lambeosaurines) is documented in the PBB, and skeletal element recovery is estimated at ~7%. Overall, large elements are over-represented (limb long bones, girdle elements) and small elements are under-represented (phalanges, teeth, cranials), compatible with an interpretation that elements were hydraulically reworked and sorted.
- The characteristic element-modification signature of the assemblage is ubiquitously broken elements and significant, but less frequent, occurrences of bone corrosion/rot and abrasion.
- The biostratinomic history of the site is interpreted as follows: (1) at least one mass mortality event (possibly a coastal-plain flooding event) killed a group of tens to hundreds of hadrosaurids; (2) carcasses experienced a multistep biostratinomic history involving thorough disarticulation and trampling, but lesser amounts of bone corrosion/rot, abrasion, and scavenging; and (3) the assemblage was eventually reworked into the base of a channel where it was sorted, oriented, and interred.
- Based on very weak data, there is a possibility that this bonebed assemblage was sourced from an age-segregated biocoenose comprising subadults and adult lambeosaurines, but no juveniles. This hypothesis requires further testing.
- Monodominant hadrosaurid bonebeds are probably common at the Park, but require systematic and quantitative assessments to identify them in the future.

ACKNOWLEDGMENTS

We thank D. Brinkman for pointing out this site to DAE in 2007. We thank the staff at Dinosaur Provincial Park for their kind assistance in providing access to the site. Thanks to Y. Sato for preparation assistance. We thank collections staff at the Royal Tyrrell Museum for curatorial assistance and access to prepared specimens. DAE thanks M. LaFramboise, R. Cooke, R. Sissons, S. Klipper, D. McLeod, S. Sabrowski, K. Brink, and D. Larson for assistance with the bonebed excavation during the summer of 2007. DCE thanks the National Sciences and Engineering Research Council of Canada for Discovery Grant funding.

LITERATURE CITED

Andrews, P., and J. Cook. 1985. Natural modification to bones in a temperate setting. Man 20:675–691.

Behrensmeyer, A. K. 1982. Time resolution in fluvial vertebrate assemblages. Paleobiology 8:211–228.

Behrensmeyer, A. K. 1990. Transport-hydrodynamics: bones; pp. 232–235 in D. E. G. Briggs and P. R. Crowther (eds.), Palaeobiology: A Synthesis. Blackwell Scientific Publications, Oxford, U.K.

Behrensmeyer, A. K. 1991. Terrestrial vertebrate accumulations; pp. 291–335 in P. A. Allison and D. E. G. Briggs (eds.), Taphonomy: Releasing the Data Locked in the Fossil Record. Plenum, New York.

Behrensmeyer, A. K., K. D. Gordon, and G. T. Yanagi. 1986. Trampling as a cause of bone surface damage and pseudo-cutmarks. Nature 319:768–771.

Brain, C. K. 1967. Bone weathering and the problem of bone pseudo-tools. South African Journal of Science 63:97–99.

Brinkman, D. B. 1990. Paleoecology of the Judith River Formation (Campanian) of Dinosaur Provincial Park, Alberta, Canada: evidence from microfossil localities. Palaeogeography, Palaeoclimatology, Palaeoecology 78:37–54.

Brinkman, D. B. 2011. The size-frequency distribution of hadrosaurs from the Dinosaur Park Formation of Alberta, Canada; pp. 16–20 in D. R. Braman, D. A. Eberth, D. C. Evans, and W. Taylor (compilers), The International Hadrosaur Symposium, Program and Abstracts. Royal Tyrrell Museum, Drumheller, Alberta.

Brinkman, D. B. 2014. The size-frequency distribution of hadrosaurs from the Dinosaur Park Formation of Alberta, Canada; chapter 23 in D. A. Eberth and D. C. Evans (eds.), Hadrosaurs. Indiana University Press, Bloomington, Indiana.

Brinkman, D. B., D. A. Eberth, and P. J. Currie. 2007. From bonebeds to paleobiology: applications of bonebed data; pp. 221–263 in R. R. Rogers, D. A. Eberth, and A. R. Fiorillo (eds.), Bonebeds: Genesis, Analysis, and Paleobiological Significance. University of Chicago Press, Chicago, Illinois.

Brinkman, D. B., A. P. Russell, and J.-H. Peng. 2005. Vertebrate microfossil sites and their contribution to studies of paleoecology; pp. 88–98 in P. J. Currie and E. B. Koppelhus (eds.), Dinosaur Provincial Park: A Spectacular Ancient Ecosystem Revealed. Indiana University Press, Bloomington, Indiana.

Brinkman, D. B., M. J. Ryan, and D. A. Eberth. 1998. The paleogeographic and stratigraphic distribution of ceratopsians (Ornithischia) in the upper Judith River Group of Western Canada. Palaios 13:160–169.

Currie, P. J. 1981. Hunting dinosaurs in Alberta's great bonebed. Canadian Geographic 10:34–39.

Currie, P. J., and P. Dodson. 1984. Mass death of a herd of ceratopsian dinosaurs; pp. 52–60 in W. E. Reif and F. Westphal (eds.), Third Symposium of Mesozoic Terrestrial Ecosystems. Attempto Verlag, Tübingen.

Currie, P. J., and E. B. Koppelhus (eds.). 2005. Dinosaur Provincial Park: A Spectacular Ancient Ecosystem Revealed. Indiana University Press, Bloomington, Indiana, 648 pp.

Dodson, P. 1971. Sedimentology and taphonomy of the Oldman Formation (Campanian), Dinosaur Provincial Park, Alberta (Canada). Palaeogeography, Palaeoclimatology, Palaeoecology 10:21–74.

Dodson, P. 1983. A faunal review of the Judith River (Oldman) Formation, Dinosaur Provincial Park, Alberta. The Mosasaur 1:89–118.

Dodson, P. 1987. Microfaunal studies of dinosaur paleoecology, Judith River Formation of southern Alberta; pp. 70–75 in P. J. Currie and E. H. Koster (eds.), Fourth Symposium on Mesozoic Terrestrial Ecosystems: Short Papers. Occasional Paper of the Tyrrell Museum of Palaeontology 3. Tyrrell Museum of Palaeontology, Drumheller, Alberta.

Eberth, D. A. 2005. The geology; pp. 54–82 in P. J. Currie and E. B. Koppelhus (eds.), Dinosaur Provincial Park: A Spectacular Ancient Ecosystem Revealed. Indiana University Press, Bloomington, Indiana.

Eberth, D. A., and P. J. Currie. 2005. Vertebrate taphonomy and taphonomic modes; pp. 453–477 in P. J. Currie and E. B. Koppelhus (eds.), Dinosaur Provincial Park: A Spectacular Ancient Ecosystem Revealed. Indiana University Press, Bloomington, Indiana.

Eberth, D. A., and D. C. Evans. 2011. Geology and Palaeontology of Dinosaur Provincial Park, Alberta: Post-Symposium Field Trip Guidebook, International Hadrosaur Symposium, 2011. Special Publication of the Royal Tyrrell Museum. Tyrrell Museum of Palaeontology, Drumheller, Alberta, 52 pp.

Eberth, D. A., and M. A. Getty. 2005. Ceratopsian bonebeds: occurrence, origins, and significance; pp. 501–536 in P. J. Currie and E. B. Koppelhus (eds.), Dinosaur Provincial Park: A Spectacular Ancient Ecosystem Revealed. Indiana University Press, Bloomington, Indiana.

Eberth, D. A., and A. P. Hamblin. 1993. Tectonic, stratigraphic, and sedimentologic significance of a regional discontinuity in the upper Judith River Group (Belly River wedge) of southern Alberta, Saskatchewan, and northern Montana. Canadian Journal of Earth Sciences 30:174–200.

Eberth, D. A., P. J. Currie, D. B. Brinkman, M. J. Ryan, D. R. Braman, J. D. Gardner, V. D. Lam, D. N. Spivak, and A. G. Neuman. 2001. Alberta's dinosaurs and other fossil vertebrates: Judith River and Edmonton Groups (Campanian-Maastrichtian); pp. 47–75 in C. L. Hill (ed.), Guidebook for the Field Trips: Mesozoic and Cenozoic Paleontology in the Western Plains and Rocky Mountains. Society of Vertebrate Paleontology, 61st Annual Meeting Museum of the Rockies Occasional Paper 3.

Eberth, D. A., R. R. Rogers, and A. R. Fiorillo. 2007. A practical approach to the study of bonebeds; pp. 265–331 in R. R. Rogers, D. A. Eberth, and A. R. Fiorillo (eds.), Bonebeds: Genesis, Analysis, and Paleobiological Significance. University of Chicago Press, Chicago, Illinois.

Eberth, D. A., M. Shannon, and B. G. Noland. 2007. A bonebeds database: classification, biases, and patterns of occurrence; pp. 103–219 in R. R. Rogers, D. A. Eberth, and A. R. Fiorillo (eds.), Bonebeds: Genesis, Analysis, and Paleobiological Significance. University of Chicago Press, Chicago, Illinois.

Eberth, D. A., D. B. Brinkman, and V. Barkas. 2010. A centrosaurine mega-bonebed from the Upper Cretaceous of southern Alberta: implications for behavior and death events; pp. 495–508 in M. J. Ryan, B. Chinnery-Allgeier, and D. A. Eberth (eds.), New Perspectives on Horned Dinosaurs. Indiana University Press, Bloomington, Indiana.

Fiorillo, A. R., and D. A. Eberth. 2004. Dinosaur taphonomy; pp. 607–613 in D. B. Weishampel, P. Dodson, and H. Osmólska (eds.), The Dinosauria, Second Edition. University of California Press, Berkeley, California.

Forster, C. A. 1990. Evidence for juvenile groups in the ornithopod dinosaur *Tenontosaurus tilletti.* Journal of Paleontology 64:164–165.

Gangloff, R. A., and A. R. Fiorillo. 2010. Taphonomy and paleoecology of a bonebed from the Prince Creek Formation, North Slope Alaska. Palaios 25:299–317.

Hu, C., Z. Cheng, Q. Pang, and X. Fang. 2001. *Shantungosaurus giganteus.* Geological Publishing House, Beijing, 139 pp.

Kuiper, K. F., A. Deino, F. J. Hilgen, W. Krijgsman, P. R. Renne, and J. R. Wijbrans. 2008. Synchronizing rock clocks of Earth history. Science 320:500–504.

Lauters, P., Y. L. Bolotsky, J. Van Itterbeeck, and P. Godefroit. 2008. Taphonomy and age profile of a Latest Cretaceous dinosaur bone bed in far eastern Russia. Palaios 23:153–162.

Lull, R. S., and N. E. Wright. 1942. Hadrosaurian Dinosaurs of North America. Geological Society of America Special Papers 40. 242 pp.

Lyman, R. L. 1994. Vertebrate Taphonomy. Cambridge manuals in archaeology. Cambridge University Press, Cambridge, U.K., 524 pp.

Rogers, R. 1990. Taphonomy of three dinosaur bone beds in the Upper Cretaceous Two Medicine Formation of northwestern Montana: evidence for drought-related mortality. Palaios 5:394–413.

Rogers, R. R., and M. E. Brady. 2010. Origins of microfossil bone beds: insights from the Upper Cretaceous Judith River Formation of north-central Montana. Paleobiology 36:80–112.

Rogers, R. R., H. C. Fricke, V. Addona, R. R. Canava, C. N. Dwyer, C. L. Harwood, A. E. Koenig, R. Murray, J. T. Thole, and J. Williams. 2010. Using laser ablation-inductively coupled plasma-mass spectrometry (LA-ICP-MS) to explore geochemical taphonomy of vertebrate fossils in the Upper Cretaceous Two Medicine and Judith River formations of Montana. Palaios 25:183–195.

Ryan, M. J., and D. C. Evans. 2005. Ornithischian dinosaurs; pp. 312–348 in P. J. Currie and E. B. Koppelhus (eds.), Dinosaur Provincial Park: A Spectacular Ancient Ecosystem Revealed. Indiana University Press, Bloomington, Indiana.

Ryan, M. J., A. P. Russell, D. A. Eberth, and P. J. Currie. 2001. The taphonomy of a *Centrosaurus* (Ornithischia: Ceratopsidae) bonebed from the Dinosaur Park Formation (Upper Campanian), Alberta, Canada, with comments on cranial ontogeny. Palaios 16:482–506.

Ryan, M. J., D. A. Eberth, D. C. Evans, D. B. Brinkman, and M. Clemens. 2011. Taphonomy and behavioral implications of *Edmontosaurus* (Ornithischia: Hadrosauridae) bonebeds from the Horseshoe Canyon Formation (Upper Campanian), Alberta Canada; pp. 16–20 in D. R. Braman, D. A. Eberth, D. C. Evans, and W. Taylor (compilers), The International Hadrosaur Symposium, Program and Abstracts. Royal Tyrrell Museum of Palaeontology, Drumheller, Alberta.

Scherzer, B. A., and D. J. Varricchio. 2010. Taphonomy of a juvenile lambeosaurine bonebed from the Two Medicine Formation (Campanian) of Montana, United States. Palaios 25:780–795.

Schmitt, J. G., J. R. Horner, R. R. Laws, and F. Jackson. 1998. Debris-flow deposition of a hadrosaur-bearing bone bed, Upper Cretaceous Two Medicine Formation, Northwest Montana. Journal of Vertebrate Paleontology 18(3, Supplement):76A.

Schmitt, J. G., F. D. Jackson, and R. R. Hanna. 2014. Debris flow origin of an unusual Late Cretaceous hadrosaur bone bed in the Two Medicine Formation of western Montana; chapter 29 in D. A. Eberth and D. C. Evans (eds.), Hadrosaurs. Indiana University Press, Bloomington, Indiana.

Varricchio, D. J., and J. R. Horner. 1993. Hadrosaurid and lambeosaurid bone beds from the Upper Cretaceous Two Medicine Formation of Montana: taphonomic and biologic implications. Canadian Journal of Earth Sciences 30:997–1006.

Voorhies, M. 1969. Taphonomy and Population Dynamics of an Early Pliocene Vertebrate Fauna, Knox County, Nebraska. University of Wyoming Contributions to Geology Special Paper 1. 69 pp.

Williams, G. P. 1988. Paleofluvial estimates from dimensions of former channels and meanders; pp. 321–334 in V. R. Baker, R. C. Kochel, and P. C. Patton (eds.), Flood Geomorphology. Wiley, New York.

Wood, J. M. 1985. Sedimentology of the Late Cretaceous Judith River Formation, "Cathedral" area, Dinosaur Provincial Park, Alberta. M.Sc. Thesis, University of Calgary, Calgary, Alberta, 215 pp.

Wood, J. M. 1989. Alluvial architecture of the Upper Cretaceous Judith River Formation, Dinosaur Provincial Park, Alberta Canada. Bulletin of Canadian Petroleum Geology 37:169–181.

Appendix 30.1. Assemblage, Orientation, and Taphonomic Modification Data

TMP 2007.014.00	Order	Family	Subfamily	Genus	Element type	Complete
01	Ornithischia	Hadrosauridae			Femur	>50%
02	Ornithischia	Hadrosauridae			Femur	>50%
03	Ornithischia	Hadrosauridae			Tibia	<50%
04	Ornithischia	Hadrosauridae			Femur	<50%
05	Ornithischia	Hadrosauridae			Femur	Complete
06	Ornithischia	Hadrosauridae			Metatarsal	Complete
07	Ornithischia	Hadrosauridae			Vertebra	>50%
08	Ornithischia	Hadrosauridae	Lambeosaurinae		Humerus	>50%
09	Ornithischia	Hadrosauridae	Lambeosaurinae		Humerus	Complete
10	Ornithischia	Hadrosauridae	Lambeosaurinae		Ischium	Complete
11	Ornithischia	Hadrosauridae			Tibia	>50%
12	Ornithischia	Hadrosauridae			Metatarsal	>50%
13	Ornithischia	Hadrosauridae	Lambeosaurinae		Vertebra	>50%
14	Ornithischia	Hadrosauridae			Ilium	<50%
15	Ornithischia	Hadrosauridae			Vertebra	>50%
16	Ornithischia	Hadrosauridae			Ulna	<50%
17	Ornithischia	Hadrosauridae			Vertebra	>50%
18	Ornithischia				Bone—incomplete	Frag
19	Ornithischia	Hadrosauridae			Ilium	<50%
20	Ornithischia	Hadrosauridae			Vertebra	>50%
21	Ornithischia	Hadrosauridae			Metatarsal	<50%
22	Ornithischia	Hadrosauridae			Metacarpal	>50%
23	Ornithischia	Hadrosauridae			Fibula	<50%
24	Ornithischia	Hadrosauridae			Scapula	<50%
25	Saurischia	Tyrannosauridae			Tooth	<50%
26	Saurischia	Tyrannosauridae			Tooth	>50%
27	Testudines	Nanhsiungchelyidae		*Basilemys*	Shell plate	<50%
28	Ornithischia	Hadrosauridae			Rib	<50%
29	Ornithischia	Hadrosauridae			Rib	<50%
30	Ornithischia	Hadrosauridae			Metatarsal	<50%
31	Testudines	Trionychidae			Plate	<50%
32	Ornithischia				Bone—incomplete	Frag
33	Ornithischia				Unidentified	Frag
34	Ornithischia	Hadrosauridae			Dentary	<50%
35	Ornithischia	Hadrosauridae			Rib	<50%
36	Ornithischia	Hadrosauridae			Humerus	<50%
37	Ornithischia	Hadrosauridae			Rib	<50%
38	Ornithischia	Hadrosauridae			Neural spine	<50%
39	Ornithischia	Hadrosauridae			Scapula	<50%
40	Ornithischia	Hadrosauridae			Rib	<50%
41	Ornithischia	Hadrosauridae			Tibia	<50%
42	Ornithischia	Hadrosauridae			Surangular	>50%
43	Ornithischia	Hadrosauridae	Lambeosaurinae		Ischium	<50%
44	Ornithischia	Hadrosauridae			Rib	<50%
45	Eosuchia	Champsosauridae			Vertebra	Complete
46	Ornithischia	Hadrosauridae			Tendon	<50%
47	Testudines	Trionychidae			Shell plate	<50%
48	Ornithischia	Hadrosauridae			Vertebra	<50%
49	Ornithischia	Hadrosauridae			Tooth	<50%
50	Ornithischia	Hadrosauridae			Vertebra	<50%
51	Saurischia	Ornithomimidae			Ungual	>50%
52	Ornithischia				Bone—incomplete	Frag
53	Ornithischia	Hadrosauridae			Vertebra	<50%
54	Ornithischia				Bone—incomplete	Frag
55	Ornithischia				Bone—incomplete	Frag

Size (5 cm)	Azimuth	Polarity	Baseline	Abrasion	Breaks	Marks	Exposure
>	55		29.5		Spiral		
>	115				Spiral		Rotted bone
>	50		88.5		Spiral	Chatter	
>							Rotted bone
>	4						
>	100		52			Load feature	
>							
>	178		52.5		Step	Groove—single	
>	110		33				
>	59					Load feature	Flaking
>	65			0–1			Etching
>	165		74				
>	115	295	61.5				
>	143			0–1			Rotted bone
>	137	137			Longitudinal		Rotted bone
>	79			0–1	Longitudinal		
>	159	339	39		Transverse		Rotted bone
>					Longitudinal and transverse		Rotted bone
>	88				Transverse		Rotted bone
>							
>	50		32		Longitudinal and transverse		
>	120		68		Transverse	Load feature	
>	139		61		Transverse	Grooves—paired	
>	100		40		Spiral?	Groove—single	
<	65		53		Transverse		
<					Transverse		
>			54	0–1	Transverse		
>	119		68		Longitudinal and transverse		
>	42		64		Spiral		
>	25		60.5		Longitudinal and transverse		
>	165		59		Load feature?		
<			52		Transverse		
<				0–1			
>			65		Transverse		Rotted bone
>	160		65		Transverse?		
>					Longitudinal and transverse		
>	130		69		Transverse	Groove—single	
>						Root/hyphae	
>	140		71.5		Transverse		Rotted bone
>	85		92.75		Transverse		
>	87		45		Longitudinal and transverse		Rotted bone
>	85		88.7				
>	40		62	0–1	Transverse		
>					Longitudinal	Root/hyphae	
<						Root/hyphae	
>					Spiral?		
<					Transverse		
<					Spiral?		
<					Longitudinal and transverse		
>			72				
<					Transverse		
<							
>			62		Transverse		
>	20		89.5		Longitudinal and transverse		
>	166				Longitudinal and transverse		

Appendix 30.1. (continued)

TMP 2007.014.00	Order	Family	Subfamily	Genus	Element type	Complete
56	Ornithischia	Hadrosauridae	Lambeosaurinae		Ischium	<50%
57	Ornithischia	Hadrosauridae	Lambeosaurinae		Ischium	<50%
58	Ornithischia	Hadrosauridae			Rib	<50%
59	Ornithischia	Hadrosauridae			Rib	<50%
60	Ornithischia	Hadrosauridae			Dentary	<50%
61	Ornithischia	Hadrosauridae			Tibia	<50%
62	Testudines	Trionychidae			Hypoplastron	Complete
63	Ornithischia	Hadrosauridae			Rib	<50%
64	Saurischia	Dromaeosauridae			Phalanx	Complete
65	Ornithischia	Hadrosauridae			Neural spine	>50%
66	Ornithischia				Bone—incomplete	Frag
67	Ornithischia				Bone—incomplete	Frag
68	Ornithischia	Hadrosauridae			Rib	<50%
69	Ornithischia				Bone—incomplete	Frag
70	Ornithischia	Hadrosauridae			Vertebra	<50%
71	Ornithischia				Bone—incomplete	Frag
72	Ornithischia	Hadrosauridae			Tendon	<50%
73	Ornithischia	Hadrosauridae			Tendon	<50%
74	Ornithischia	Hadrosauridae			Vertebra-caudal	>50%
75	Ornithischia				Bone—incomplete	Frag
76	Ornithischia				Bone—incomplete	Frag
77	Ornithischia	Hadrosauridae			Rib	<50%
78	Ornithischia	Hadrosauridae			Rib	<50%
79	Ornithischia	Hadrosauridae			Pelvic girdle bone	<50%
80	Testudines	Trionychidae			Plate	<50%
81	Ornithischia				Bone—incomplete	Frag
82	Ornithischia				Bone—incomplete	Frag
83	Ornithischia	Hadrosauridae			Rib	<50%
84	Ornithischia	Hadrosauridae	Lambeosaurinae		Ischium	<50%
85	Ornithischia	Hadrosauridae			Neural arch	<50%
86	Ornithischia				Bone—incomplete	Frag
87	Ornithischia				Limb bone large	<50%
88	Ornithischia	Hadrosauridae			Rib	<50%
89	Ornithischia				Bone—incomplete	Frag
90	Saurischia	Tyrannosauridae			Tooth	>50%
91	Ornithischia	Hadrosauridae			Ungual	Complete
92	Ornithischia	Hadrosauridae			Rib	<50%
93	Ornithischia	Hadrosauridae			Tendon	<50%
94	Ornithischia				Bone—incomplete	Frag
95	Ornithischia	Hadrosauridae			Neural arch	<50%
96	Ornithischia				Bone—incomplete	Frag
97	Ornithischia				Bone—incomplete	Frag
98	Ornithischia				Bone—incomplete	Frag
99	Ornithischia				Bone—incomplete	Frag
100	Ornithischia				Bone—incomplete	Frag
101	Ornithischia				Bone—incomplete	Frag
102	Ornithischia				Bone—incomplete	Frag
103	Ornithischia				Bone—incomplete	Frag
104	Ornithischia				Bone—incomplete	Frag
105	Ornithischia				Bone—incomplete	Frag
106	Ornithischia				Bone—incomplete	Frag
107	Ornithischia				Bone—incomplete	Frag
108	Ornithischia				Bone—incomplete	Frag
109	Ornithischia				Bone—incomplete	Frag
110	Ornithischia				Bone—incomplete	Frag

Size (5 cm)	Azimuth	Polarity	Baseline	Abrasion	Breaks	Marks	Exposure
>	146		80.5		Transverse	Root/hyphae	
>	19				Transverse		Rotted bone
<	135		37		Longitudinal and transverse		
>	135		90		Longitudinal		
>	127				Step?	Groove—single	
>	84				Longitudinal and transverse		Rotted bone
>					Tips		
<	85				Longitudinal and transverse		
<	31						
>							
>	124				Longitudinal and transverse		Rotted bone
<					Longitudinal and transverse		
>	129				Longitudinal and transverse		
>	116				Longitudinal? and transverse		
>					Transverse		
>							Rotted bone
>	65				Transverse		
>					Transverse		
<					Spiral?		
<							
>	40				Longitudinal and transverse		
>	30		87.5		Step	Load feature	
>	169		65		Longitudinal		
>	50	50	59.5		Longitudinal	Grooves—paired	
<			65		Transverse		Flaking
<	55		52		Transverse		
>	160		65		Longitudinal and transverse		
>	85		88		Longitudinal and transverse		
>	165		76		Transverse		Rotted bone
>	166				Transverse		
<	178				Longitudinal and transverse		
>	106				Longitudinal and transverse		
<	15				Transverse		
>	63				Transverse		
<	91				Transverse		
<	155*		75				
<			81		Longitudinal and transverse		
<					Transverse		
<			82		Longitudinal and transverse		
<	47				Transverse		
<	145				Longitudinal and transverse		
>	129		44		Longitudinal and transverse		Rotted bone
>							
>							
>							
>							
>							
>							
>							
>							
>							
>						Grooves—paired	
<							
<							
>							

Appendix 30.1. (continued)

TMP 2007.014.00	Order	Family	Subfamily	Genus	Element type	Complete
111	Ornithischia				Bone—incomplete	Frag
112	Ornithischia				Bone—incomplete	Frag
113	Ornithischia				Bone—incomplete	Frag
114	Ornithischia				Bone—incomplete	Frag
115	Ornithischia				Bone—incomplete	Frag
116	Ornithischia				Bone—incomplete	Frag
117	Ornithischia				Bone—incomplete	Frag
118	Ornithischia	Hadrosauridae			Rib	<50%
119	Ornithischia				Bone—incomplete	Frag
120	Ornithischia				Bone—incomplete	Frag
121	Ornithischia				Bone—incomplete	Frag
122	Ornithischia				Bone—incomplete	Frag
123	Ornithischia	Hadrosauridae			Vertebra	<50%
124	Ornithischia				Bone—incomplete	Frag
125	Ornithischia				Bone—incomplete	Frag
126	Ornithischia				Bone—incomplete	Frag
127	Ornithischia				Bone—incomplete	Frag
128	Ornithischia				Bone—incomplete	Frag
129	Ornithischia				Bone—incomplete	Frag
130	Ornithischia				Bone—incomplete	Frag
131	Ornithischia				Limb bone	<50%
132	Ornithischia				Bone—incomplete	Frag
133	Ornithischia				Bone—incomplete	Frag
134	Lepisosteiformes	Lepisosteidae			Gar scale	Complete
135	Ornithischia	Hadrosauridae			Vertebra	>50%
136	Ornithischia				Bone—incomplete	Frag
137	Ornithischia				Bone—incomplete	Frag
138	Ornithischia				Bone—incomplete	Frag
139	Ornithischia				Bone—incomplete	Frag
140	Ornithischia				Bone—incomplete	Frag
141	Ornithischia	Hadrosauridae			Tendon	<50%
142	Ornithischia				Bone—incomplete	Frag
143	Testudines	Macrobaenidae?			Shell	<50%
144	Ornithischia	Hadrosauridae			Tendon	<50%
145	Testudines	Trionychidae			Shell	<50%
146	Lepisosteiformes	Lepisosteidae		*Lepisosteus*	Scale	>50%
147	Ornithischia	Hadrosauridae			Tendon	<50%
148	Ornithischia	Hadrosauridae			Tendon	<50%
149	Ornithischia	Hadrosauridae			Rib	<50%
150	Ornithischia				Bone—incomplete	Frag
151	Ornithischia	Hadrosauridae			Rib	<50%
152	Ornithischia				Vertebra	<50%
153	Ornithischia	Hadrosauridae			Rib	<50%
154	Ornithischia	Hadrosauridae			Tendon	<50%
155	Ornithischia	Hadrosauridae			Rib	<50%
156	Testudines	Trionychidae			Dermal plate	<50%
157	Eosuchia	Champsosauridae			Ulna	<50%
158	Ornithischia	Hadrosauridae			Rib	<50%
159	Ornithischia	Hadrosauridae			Metatarsal	<50%
160	Ornithischia	Hadrosauridae			Rib	<50%
161	Ornithischia	Hadrosauridae			Neural arch	<50%
162	Ornithischia	Hadrosauridae			Metacarpal	Complete
163	Ornithischia	Hadrosauridae			Pubis	<50%
164	Ornithischia	Hadrosauridae			Neural arch	<50%
165	Saurischia	Dromaeosauridae			Phalanx	<50%

Size (5 cm)	Azimuth	Polarity	Baseline	Abrasion	Breaks	Marks	Exposure
>							
>							
<							
>							
>							
<							
<							
>							
<							
<							
<							
>							
<							
<							
>							
>							
>							
<							
<							
<							
>					Longitudinal and transverse		
<							
>							
<							
>							
<							
<							
>							
>							
>					Longitudinal and transverse		
<					Transverse		
<					Longitudinal and transverse		
<					Transverse		
<							
<							
<							
>	75						
<			75		Transverse		
>	76				Transverse		
>	99						
>	32		34.5		Longitudinal and transverse		
>	107				Longitudinal and transverse		
>	106		49		Transverse		
<					Transverse		
<	137		83		Longitudinal and transverse		
<					Longitudinal and transverse		
<					Longitudinal and transverse		
<					Step?		
>					Spiral?		
>				0–1	Transverse		
>			65.5		Transverse		Rotted bone
>							
>	137		64		Transverse		Rotted bone
>							
<							

Appendix 30.1. (continued)

TMP 2007.014.00	Order	Family	Subfamily	Genus	Element type	Complete
166	Amiiformes	Amiidae			Operculum	<50%
167	Acipenseriformes	Polyodontidae			Skull-incomplete	<50%
168	Amiiformes	Amiidae			Skull-incomplete	<50%
169	Lepisosteiformes	Lepisosteidae			Vertebra	<50%
170	Testudines	Chelydridae			Costal	<50%
171	Ornithischia	Ankylosauridae			Dermal plate	Complete
172	Ornithischia	Hadrosauridae			Dentary	<50%
173	Saurischia	Tyrannosauridae			Tooth	<50%
174	Amiiformes	Amiidae		*Cyclurus*	Vertebra	Complete
175	Testudines	Trionychidae			Hypoplastron	>50%
176	Rajiformes			*Myledaphus*	Tooth	Complete

Size (5 cm)	Azimuth	Polarity	Baseline	Abrasion	Breaks	Marks	Exposure
<							
<							
<							
<							
<							
<							
<					Longitudinal and transverse		Rotted bone
<							
<							
>				1–2			
<							

Body Size Distribution in a Death Assemblage of a Colossal Hadrosaurid from the Upper Cretaceous of Zhucheng, Shandong Province, China

31

David W. E. Hone, Corwin Sullivan, Qi Zhao, Kebai Wang, and Xing Xu

ABSTRACT

Here we examine the distribution of body sizes in a monospecific assemblage of hadrosaurines from the Upper Cretaceous of Zhucheng, Shandong Province, China. The material is referred to cf. *Shantungosaurus*, a taxon widely regarded as the largest ornithischian dinosaur. Hadrosaur bonebeds are quite common and typically comprise assemblages that consist of mixed age classes of adults and juveniles, or juveniles alone. In contrast, the cf. *Shantungosaurus* assemblage examined here is composed almost exclusively of adult individuals. At least 55 individuals are represented, and almost every one is estimated to have had a mass in excess of 6 metric tons. This assemblage provides tentative support for the hypothesis that adult and juvenile hadrosaurs were ecologically partitioned and likely formed separate herds, with maturing subadults eventually joining adult herds when they reached large size.

INTRODUCTION

Dinosaur remains were first discovered in Zhucheng, Shandong Province, eastern China in 1964 (Ji et al., 2011). Recovery of a colossal hadrosaurine excavated from the Longgujian Quarry eventually led to the erection of the taxon *Shantungosaurus giganteus* (Hu, 1973). Although significant quantities of material were quarried from Longgujian in the 1960s, and from both Longgujian and the nearby Kugou Quarry in the 1980s (Ji et al., 2011), a flurry of more recent excavations has led to the recovery of thousands of elements (primarily hadrosaurian) from Longgujian, Kugou, and the major new quarry of Zangjiazhuang, all located within the borders of the county-level city of Zhucheng (Fig. 31.1). Some specimens from these sites have been identified as representatives of new taxa – including leptoceratopsid (Xu, Wang, Zhao, Sullivan, and Chen, 2010) and ceratopsid (Xu, Wang, Zhao, and Li, 2010) ceratopsians, a tyrannosaurine theropod (Hone et al., 2011), and the additional nominal hadrosaurines *Zhuchengosaurus maximus* (Zhao et al., 2007) and *Huaxiaosaurus aigahtens* (Zhao et al., 2011). Additional material representing a number of non-avian dinosaur clades, and likely representing new taxa, awaits description. The Zhucheng quarries are therefore now established as a major source of Upper Cretaceous dinosaur fossils.

Shantungosaurus was regarded by Hu (1973) as a member of the non-crested hadrosaurid lineage Hadrosaurinae (Saurolophinae of Prieto-Márquez, 2010, and some other authors), and recent phylogenetic analyses (e.g., Prieto-Márquez, 2010; Gates et al., 2011) support this conclusion. *Shantungosaurus* is also claimed to have been the largest ornithischian dinosaur in terms of both total length and mass, with estimated values in excess of 15 m (total length, e.g., Ji et al. [2011]) and 20 t (metric ton; Seebacher, 1999). *Zhuchengosaurus* (Zhao et al., 2007) and *Huaxiaosaurus* (Zhao and Li, 2009) were both described as being even larger, with body lengths of 16.6 m and 18.7 m, respectively.

Here we examine the size distribution of femora referable to *Shantungosaurus* that remain in situ in a dense bonebed exposed in the Kugou Quarry at Zhucheng. Because equations for estimating dinosaur body masses from femoral dimensions are available in the literature (e.g., O'Gorman and Hone, 2012), measurements of the femora make it possible to reevaluate the probable body mass of *Shantungosaurus* in addition to exploring the size distribution of the hadrosaur sample preserved in the Kugou Quarry.

LOCALITY INFORMATION

The Longgujian Quarry in Kugou Village, Lübiao Town, Zhucheng is the classic *Shantungosaurus* locality, discovered in 1964. Kugou Village lies approximately 11 km southwest of the urban center of Zhucheng. The site called the Kugou Quarry has been known since at least 1988 and represents the second major exposure of fossil-bearing deposits in Kugou Village. The Longgujian and Kugou quarries are on opposite sides of a long hill, and are separated by a distance of only ~100 m. The third, more recently discovered site, Zangjiazhuang Quarry, occurs a few kilometers north-northeast of the Kugou Village sites within Zangjiazhuang Village, Lübiao Town.

All three quarries expose sediments from the middle to upper part of the Upper Cretaceous Wangshi Group, near the transition between the Xingezhuang Formation and the overlying Hongtuya Formation (Liu et al., 2010, 2011). An age of 73.5 Ma was obtained by $^{40}Ar/^{39}Ar$ dating of a basalt layer in a part of the Hongtuya Formation stratigraphically above the beds exposed in the quarries (Liu et al., 2010), and suggests a Campanian age for the Zhucheng fauna. Liu et al. (2010) described the quarry sediments as dominated by debris flow, braided channel and floodplain deposits. Each quarry contains two major bonebeds, which appear to be separated by a stratigraphic interval of about 32 m in the Kugou/Longgujian section, and about 4 m in the Zanjiazhuang section (Liu et al., 2010, 2011). It is not clear whether the upper and lower bonebeds in the Kugou and Longgujian quarries correspond temporally to the more narrowly separated upper and lower bonebeds at Zangjiazhuang. In the Kugou and Longgujian quarries the lower bonebed is the main exposed fossiliferous unit, and both the lower and upper beds are debris flow deposits composed of pebble-bearing fine sandstone. In the Zangjiazhuang Quarry, by contrast, the main fossiliferous unit is the upper bed, and both beds are floodplain deposits composed of pebble-bearing siltstone (Liu et al., 2010).

Within the trench-like Kugou Quarry, strata dip strongly to the north and the surface of the fossil-bearing strata is exposed along the south bank of the quarry. Exposures of the bonebed are terminated at the east and west ends by sheer quarry walls. The exposed part of the surface is sub-rectangular in shape and approximately 300 m long and 20 m wide, although at some locations the width approaches 30 m (Fig. 31.2).

31.1. Map showing the location of the Kugou Quarry near Zhucheng, Shandong Province, China (femur symbol). Modified from Hone et al. (2011).

TAXONOMIC IDENTITY

Referral of the Kugou Quarry specimens to *Shantungosaurus giganteus* is complicated by the putative presence of two additional large hadrosaurids, *Zhuchengosaurus maximus* (Zhao et al., 2007) and *Huaxiaosaurus aigahtens* (Zhao and Li, 2009; Zhao et al., 2011), both of which are based on material from the other Zhucheng quarries. *Z. maximus* is based on material from the Longgujian Quarry, where *S. giganteus* also occurs. As proposed by Ji et al. (2011) it is likely that *Z. maximus* represents a junior synonym of the latter species. Other than suggested differences in presacral and caudal vertebral counts, and total body length (difficult to confirm given that almost all the material assigned to each taxon is disarticulated), the only clear distinction between the two taxa adduced by Zhao et al. (2007) is that *Z. maximus* purportedly has 9 rather than 10 sacral vertebrae and lacks the ventral groove on the sacrum seen in *S. giganteus*. However, as pointed out by Ji et al. (2011), the difference in sacral count might reflect either incompleteness or immaturity of the sacrum, and the ventral groove is a minor feature that could easily be the result of individual or ontogenetic variation (our interpretation). Thus, until a detailed comparison between the material referred to *Z. maximus* and a large representative sample of *S. giganteus* is undertaken, it appears most likely that the Longgujian Quarry contains a single hadrosaurine species, *S. giganteus*.

H. aigahtens was erected by Zhao and Li (2009) to encompass the hadrosaurine material from the Zangjiazhuang Quarry. Again, setting aside body length and vertebral counts, several skeletal details–including the shape of the upper temporal fenestra and the proportions of the anterior articular faces of the cervical vertebrae–were postulated to distinguish *H. aigahtens* and *S. giganteus* (Zhao et al., 2011). It is possible that both taxa are valid and that hadrosaurine from the Zangjiazhuang Quarry should be referred to the former rather than the latter taxon. Alternatively, examination of large samples from the Longgujian and Zangjiazhuang quarries might indicate that the purported differences fall within the range of anatomical and preservational variation present in *S. giganteus*.

Although Xu, Wang, Zhao, Sullivan, and Chen (2010) considered the abundant hadrosaurine material from the Kugou Quarry as representing *Shantungosaurus*, the taxonomic identity of this material has never been extensively discussed in the literature. Given the uncertainty regarding the number of colossal hadrosaurid taxa present in the Zhucheng

31.2. The Kugou Quarry, Zhucheng, China. Hadrosaurid elements are exposed on the left-hand side. The quarry is ~300 m long and up to 30 m across. To provide scale, the supporting poles holding up the temporary roof are spaced 3–4 m apart.

quarries, and because there is currently no evidence that there is more than one hadrosaurid taxon present at the Kugou site, we take a conservative position and identify all the Kugou site's hadrosaur material as cf. *Shantungosaurus*.

HADROSAUR MATERIAL AT KUGOU QUARRY

The hadrosaurid bones examined in this study are exposed either near the surface of the main exposed fossil-bearing layer at the Kugou Quarry, which represents a single fossil-producing horizon with a stratigraphic thickness of ~1 m (Liu et al., 2010), or at slightly greater depth in the same horizon. The bonebed almost certainly contains additional elements that are not exposed because it appears to continue beneath the overburden that makes up the floor of the trench-like quarry as well as beneath the east and west walls. However, only specimens that were already exposed in situ were examined and measured. All of the specimens in the bonebed were considered to represent components of a single depositional event.

Our survey of more than 3000 cranial, axial and appendicular bones (Fig. 31.3) included all those elements that were visible across approximately 60% of the exposed area of the bonebed. Thus, we estimate that ~5000 bones, in total, are exposed within the currently exposed portions of the Kugou Quarry.

Only a very limited number of non-hadrosaurid elements were identified during this survey: one tooth, one small femur, and one small third metatarsal from a tyrannosaurid; one crocodilian osteoderm; and parts of a turtle plastron. The leptoceratopsid *Zhuchengceratops* (Xu, Wang, Zhao, Sullivan, and Chen, 2010) was also recovered prior to our survey from the Kugou Quarry. Given these observations and assuming that only one hadrosaurid species is present, we regard the bonebed assemblage as essentially monospecific.

No articulated hadrosaurid bones were observed in the quarry (see Fig. 31.3) and the elements did not appear to be size sorted. In all cases, long axes of limb elements were oriented parallel to depositional surfaces, exhibiting little to no plunge. Overall, the material was in good condition and the majority of bones were observed to be largely complete and unbroken. Although elements exhibit some minor surface damage, we observed no obvious surface modifications on the bones due to tooth marks.

The generally low numbers of broken bones and the absence of shed theropod teeth and tooth marks all suggest an

31.3. Bones of colossal hadrosaurs in the Kugou Quarry at Zhucheng, China. Visible are a number of disarticulated pieces, including vertebrae, femur, tibia, chevron, metatarsal, and indeterminate elements. Specimen labels are 10 cm in length.

absence of trampling and scavenging, and thus relatively rapid burial after death. The overall good condition of the material, lack of size sorting, and, in particular, lack of widespread fracturing of elements suggest limited transport or reworking (Liu et al., 2010; Scherzer and Varricchio, 2010), although the robustness of many of the bones presumably made them somewhat resistant to damage.

Liu et al. (2010) suggested that the main bonebeds in both the Kugou and Longgujian quarries were deposited by a debris flow that killed and immediately buried a herd of hadrosaurids, and Ji et al. (2011) proposed a similar scenario. In contrast, Liu et al. (2010) interpreted the main bonebed at Zangjiazhuang as containing bones that were washed onto a floodplain by alluvial fans. The interpretation for the Kugou and Longgujian sites implies that these hadrosaurid taphocoenoses are the result of mass-mortality events, and that their compositions should closely reflect those of the corresponding biocoenoses. Although a debris-flow burial event appears to be a reasonable interpretation based on the evidence presented by Ji et al. (2011), we question whether such an event involving such relatively fine sediments could have also dismembered the hadrosaurids sufficiently to account for the thorough disarticulation observed in the Kugou Quarry assemblage. It seems equally or more plausible to us that death and burial (both by mass sediment flow) were separate events that occurred close enough in time to preclude widespread trampling, scavenging, and other forms of bone modification, but were also separated in time sufficiently to allow some degree of soft-tissue decomposition or skeletal disarticulation prior to final burial (cf. Britt et al. 2009). Thorough comparative taphonomic studies of these sites are now required to answer these and other related questions.

Table 31.1. Identities and Measurements of in situ Femora, and Individual Mass Estimates for cf. *Shantungosaurus* from the Kugou Quarry

Left(L)/Right(R) femora	Maximum length (cm)	Minimum width (cm)	Estimated mass (t)
L	107	15	3.0
L	108	17	3.1
L	111	15	3.4
R	127	20	5.1
R	134	20	6.0
?	134	26	6.0
R	135	20	6.1
L	135		6.1
R	137	19	6.4
R	139	22	6.7
R	142	26	7.1
R	142	24	7.1
?	142	21	7.1
R	142	22	7.1
R	142		7.1
L	142		7.1
R	142		7.1
?	143	21	7.3
R	143	23	7.3
?	143		7.3
R	145	21	7.6
R	145	25	7.6
L	145	23	7.6
?	145	22	7.6
L	145	18	7.6
?	145*	20	7.6
L	146	25	7.8
L	146		7.8
L	146	18	7.8
R	147	22	7.9
L	147	23	7.9
R	147	22*	7.9
L	147		7.9
R	148	23	8.1
L	148	22	8.1
R	148	24	8.1
R	149	23	8.2
?	149	22	8.2
L	149	18	8.2
R	149	20	8.2
?	150	24	8.4
?	150*	24	8.4
L	150	20	8.4
R	150	20	8.4
?	151	25	8.6
?	151*		8.6
L	152	22	8.8
L	153	22	8.9
L	153	25	8.9
?	153	21	8.9
R	153	23	8.9
L	153		8.9
?	153*	21	8.9
L	153	20	8.9
L	154	24	9.1
L	154	25	9.1
R	154	22	9.1
?	154	23	9.1
R	154	20	9.1
R	155	26	9.3
R	155	21	9.3
R	155	30*	9.3
?	157	23*	9.7
L	157	22	9.7
R	157		9.7
L	157	19	9.7
R	157*	29*	9.7
?	159	23	10.0
?	159	24	10.0
L	159	22	10.0
L	159	22	10.0
L	160	23	10.2
R	160	22	10.2
R	160	23	10.2
L	160	23	10.2
?	163*	25	10.8
R	163	23	10.8
L	164		11.0
?	164*	22	11.0
R	166	22	11.4
R	166	19	11.4
L	167		11.7
R	167	19	11.7
L	169	23	12.1
R	171	23	12.5
R	172	24	12.7
?		26	
?		25	
?		24	
R		23	
?		23	
L		22	
?		21*	
?		20	
?		20*	
L		20	
L		20	
?		19*	
R		18	
?		16*	
R			
R			
R			
L			
L			
L			
L			

Note: Data ordered by increasing femoral length. Nine additional femora could not be measured or identified (left/right). Asterisks indicate measurements that are slightly uncertain owing to damage. Mass estimates are based on femoral length and derived using methods described in text.

METHODS AND RESULTS

Given that the elements we examined and measured are part of an in situ exhibit and are not cataloged in the conventional sense, there is a strong possibility that the sample described here will eventually be unavailable for data confirmation or further testing. Nonetheless, given that large collections of disarticulated skeletal elements continue to be amassed from the Kugou site and placed under the care of the Zhucheng Dinosaur Culture Research Center, we believe that the basic observations and size-distribution patterns described here will be testable in the future.

In total, 110 femora (all in excess of 50% complete) were observed in the quarry and indicate that a minimum of 55 individuals are represented in the assemblage that we examined. Of these, 100 femora were sufficiently well preserved and accessible to be measured (2 were inaccessible and 8 were unsuitable for measurement due to incompleteness or inadequate exposure).

We attempted to measure maximum proximodistal length and minimum mediolateral width for every femur (Table 31.1). Seventy-four elements were measured for both length and width; 12 for length but not width, and 14 for width but not length. We identified 38 left femora and 41 right femora. Thirty-one femora could not be identified as left or right.

The shortest femur is 107 cm long and 15 cm wide, and the longest is 172 cm long and 24 cm wide. The average (mean) length and width of the femora in our sample are 149.8 cm and 22.0 cm, respectively. Until recently, the longest femur reported in the literature for any ornithischian was 1650 mm (*Shantungosaurus*; Brett-Surman [1997]) but Zhao et al. (2007) reported a length of 1700 mm for *Zhuchengosaurus*, and Zhao and Li (2009) reported a length of 172 cm for *Huaxiaosaurus*. Thus, the 172 cm long femur reported here matches the longest femur on record for an ornithischian. The measured lengths of both bones are, of course, susceptible to error, but the generally good condition of the limb bones preserved in the Zhucheng quarries implies that the amount of inaccuracy introduced by distortion and damage should be small.

Two models were fitted using maximum likelihood estimation using the packages mle and sn in R (R core team, 2012). The first was a standard normal distribution and the second was a skewed normal. The skewed normal was a significantly better fit than the standard normal distribution (normal distribution: mean = 150, standard deviation = 11.8, skew = -0.56, AIC = 558.1457; skewed normal: mean = 149, sd = 11.5, skew = -0.5642768, AIC = 325.4838). Thus, the main block of data (histogram in Figure 31.4) is normally distributed with the outliers being responsible for the skew.

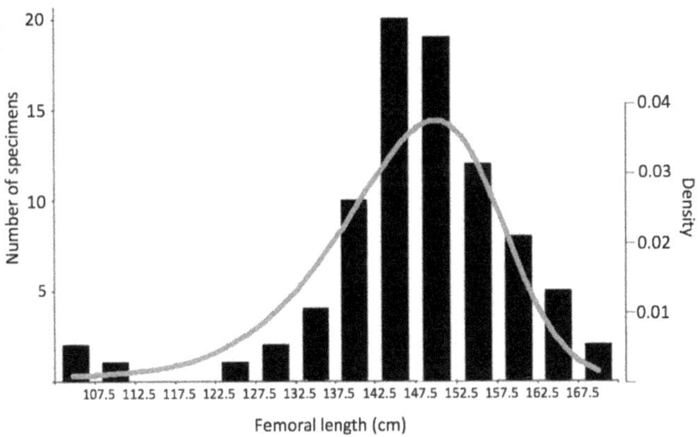

31.4. Histogram showing length distribution (5 cm divisions, from 105.0–109.9 cm through 170.0–174.9 cm) of 86 cf. *Shantungosaurus* femora from the Kugou Quarry, Zhucheng, China. The gray line represents the fit of the skewed normal maximum likelihood model.

DISCUSSION

Although the status of *Shantungosaurus* as an extremely large hadrosaurid is uncontroversial, a variety of numerical estimates of its size have appeared in the literature. A mounted composite skeleton on which the original description of the genus was partly based has a length of 14.72 m (Hu, 1973). Brett-Surman (1997) reported the length of the largest *Shantungosaurus* femur known at the time as 1650 mm, and estimated that this bone came from an individual 17 m in length. Using these data, Seebacher (1999) calculated a mass of up to 22 t. Horner et al. (2004) suggested the largest individuals had a mass of 16 t, but that these were rare. More recently, Ji et al. (2011) suggested a total length of 15 m for *Shantungosaurus*, but a mass of just 7 t.

Given the disarticulated nature of the material in Kugou and the other Zhucheng quarries, length estimates for Zhucheng hadrosaurid specimens must be considered profoundly uncertain. Furthermore, Hone (2012) demonstrated that it is difficult to estimate the length of incomplete dinosaur skeletons because tail length is highly variable and correlates poorly with the snout-to-sacrum length (i.e., head + neck + body). Body mass however, can be estimated with fewer complications. Although this is most rigorously done via a detailed method such as 3-D slicing (Henderson, 1999), rough estimates can be generated from simple equations derived from the correlation between mass and femur length in extant taxa (Christiansen, 1999) or other estimates (O'Gorman and Hone, 2012), and can be used to suggest a rank order by mass for a given sample of animals (Carrano, 2005).

The longest femur recorded here measures 172 cm. Apart from other Zhucheng hadrosaur femora, this is longer than any femur previously reported for an ornithischian (cf.

O'Gorman and Hone, 2012); indeed, few adult ornithischians have femora that are over 1 m long (comparable to the shortest femora in our dataset). The available data suggest that the Zhucheng hadrosaurids have maximal femoral lengths that are comparable to some large sauropods such as *Diplodocus* (165 cm), *Haplocanthosaurus* (175 cm), *Antarctosaurus* (177 cm), *Apatosaurus* (179 cm) and *Giraffititan* (191 cm) (O'Gorman and Hone, 2012). Application of the femur-mass regression of O'Gorman and Hone (2012:fig. 1) suggests a mass of ~12.7 t for the individual represented by the longest femur in our sample (length = 172 cm), and a mass of ~3.0 t for the individual represented by the shortest femur (length = 107 cm). Using this method, the average mass of the animals at Kugou is 8.7 t (all femora transformed to mass, and a mean calculated; Table 31.1). Although these values must be treated with caution, it is clear that the hadrosaurids of Zhucheng must rank as extraordinarily large by ornithischian standards.

The lengths of the femora in our sample (Fig. 31.4) are mostly distributed within a narrow range. Some 70 of 84 (approximately 83%) of the femoral length measurements in our dataset fall between 135 and 175 cm, with the three smallest lengths representing discrete outliers at the lower end of the distribution. Because this assemblage (1) comprises mostly similarly sized, large individuals; (2) is preserved in a monospecific bonebed; and (3) formed partly as the result of a mass-mortality event, we conclude that the individuals described here were part of a gregarious population of, presumably, adult individuals. We did not observe any material in the quarry that could unambiguously be referred to a small individual (i.e., a humerus or sacrum from an animal considerably smaller than the smallest individual identified by femoral length) and whereas these may have been overlooked given the large volume of material at the site, if they are present at all, they are a very minor component of the assemblage.

Some other studies of hadrosaur bonebeds have presented size distributions for multiple skeletal elements (e.g., Lauters et al., 2008; Scherzer and Varricchio, 2010; Brinkman, this volume). Brinkman (this volume) carried out an analysis similar to ours based on numerous isolated elements within the Dinosaur Park Formation in Alberta. His size distribution of various hadrosaurid skeletal elements included two narrow peaks that he interpreted as discrete subadult age classes, and a single broad peak of relatively large elements that he interpreted as a group of slow-growing adults of varying ages. His results contrast sharply with the single, narrow peak of large individuals detected in our study, which suggests that nearly all the specimens recovered from the Kugou Quarry represent adults. Lauters et al. (2008) studied an *Amurosaurus* bonebed at Blagoveschensk and obtained results that were rather different from ours. In that study, all but one of the femora examined were from late-stage juveniles or subadults, with a lone example belonging to a rather large, apparently adult individual.

In the case of the Kugou site, we are confident that most of the hadrosaurids in the quarry were at an adult ontogenetic stage when they died. This is supported by the fact that sacra noted during our survey of the quarry were fully fused and thus from osteologically mature individuals. Although absolute size is typically a poor indicator of exact age in large dinosaurs, the large size of the vast majority of the femora we measured–as large or larger than those of any other known adult hadrosaurids (also see below)–suggests that they belonged to mature individuals. Although the few individuals with femora only 1 m or so in length may not have been adults, it is unlikely that they were young juveniles; a 1 m long femur would make them larger than the adults of most other hadrosaurid genera. Histological analyses are now required to fully test these interpretations.

The adult-dominated nature of the Kugou Quarry assemblage contrasts with at least some previously described hadrosaurid body fossil and trackway assemblages. For example, a number of known assemblages contain only juvenile hadrosaurids (Varricchio, 2010) or subadult animals (Scherzer and Varricchio, 2010), whereas others represent mixtures of juveniles (of a variety of sizes) and adults (Varricchio and Horner, 1993; Horner et al., 2004). However, we are not aware of any bonebeds that are so biased toward adults as those exposed at Zhucheng (but see Eberth et al. (this volume) for a possible example from southern Alberta).

A number of authors (e.g., Matthew, 1915; Currie, 1983; Lauters et al., 2008) have suggested that juvenile and adult hadrosaurids formed separate herds, with juveniles eventually joining adult herds as they matured. This hypothesis is consistent with the pattern seen here, in that the shortest femora in our dataset represent relatively small but not tiny individuals that may have been older juveniles attached to an aggregation of presumably mature and full-sized animals.

SUMMARY

Following Liu et al. (2010) and Ji et al. (2011) we interpret the taphocoenoses preserved in the Kugou Quarry as resulting from a mass-sediment flow that impacted a large group of *Shantungosaurus*, although here we suggest the animals may have already been dead from a mass-mortality event prior to final burial. The absence of bedding, sedimentary structures, and the presence of poorly sorted and sizable hadrosaurid elements (bioclasts) and other large clasts at different depths in the 1 m thick deposit are all consistent with such an interpretation.

Carcasses were disarticulated and buried either by the mass-sediment flow that killed the individuals (cf. Liu et al., 2010), or by a subsequent flow following a mass death. In either case, the lack of evidence for scavenging or widespread element modification suggests relatively rapid burial, and the absence of evidence for size-based sorting of elements suggests only a limited amount of transport and reworking. The monospecific makeup of the bonebed and the large number of individuals represented in the preserved assemblage suggests that at least some *Shantungosaurus* adults formed large herds. The size distribution of the femora observed within the bonebed and the co-occurrence of skeletally mature sacra at the site indicate that the death assemblage was dominated by large, adult individuals, representing the largest ornithischians known. This assemblage of nearly exclusively adult individuals provides additional evidence for ecological partitioning in hadrosaurids, with juveniles and adults distributed in separate herds, and maturing subadults eventually joining adult herds when they reached large size.

ACKNOWLEDGMENTS

We extend thanks to Dave Eberth and David Evans for inviting us to contribute to this volume and editing our manuscript with the efficiency needed, and to Patty Ralrick for her help in proofing the work. We thank David Eberth and two anonymous referees for their constructive comments that helped improve the manuscript. We are grateful to Chen Yanming, Zou Qingzhong, Zhang Yanxia, and Chen Ruxia for access to material and assistance in Zhucheng, and to Mark Stevenson for assistance with the data analysis. This work was supported by grants from the Zhucheng City Government, the National Natural Science Foundation of China, and the Chinese Academy of Sciences.

LITERATURE CITED

Brett-Surman, M. K. 1997. Ornithopods; pp. 330–346 in J. O. Farlow and M. K. Brett-Surman (eds.), The Complete Dinosaur. Indiana University Press, Bloomington, Indiana.

Brinkman, D. B. 2014. The size-frequency distribution of hadrosaurs from the Dinosaur Park Formation of Alberta, Canada; chapter 23 in D. A. Eberth and D. C. Evans (eds.), Hadrosaurs. Indiana University Press, Bloomington, Indiana.

Britt, B. B., D. A. Eberth, R. D. Scheetz, B. W. Greenhalgh, and K. L. Stadtman. 2009. Taphonomy of debris-flow hosted dinosaur bonebeds at Dalton Wells, Utah (Lower Cretaceous, Cedar Mountain Formation, USA). Palaeogeography, Palaeoclimatology, Palaeoecology 280:1–22.

Carrano, M. T. 2005. Body-size evolution in the Dinosauria; pp. 225–268 in M. T. Carrano, R. W. Blob, T. J. Gaudin, and J. R. Wible (eds.), Amniote Paleobiology: Perspectives on the Evolution of Mammals, Birds, and Reptiles. University of Chicago Press, Chicago, Illinois.

Christiansen, P. 1999. Long bone scaling and limb posture in non-avian theropods: evidence for differential allometry. Journal of Vertebrate Paleontology 19:666–680.

Currie, P. J. 1983. Hadrosaur trackways from the Lower Cretaceous of Canada. Acta Palaeontologica Polonica 28:63–73.

Eberth, D. A., D. C. Evans, and D. W. H. Lloyd. 2014. Occurrence and taphonomy of the first documented hadrosaurid bonebed from the Dinosaur Park Formation (Belly River Group, Campanian) at Dinosaur Provincial Park, Alberta, Canada; chapter 30 in D. A. Eberth and D. C. Evans (eds.), Hadrosaurs. Indiana University Press, Bloomington, Indiana.

Gates, T. A., J. R. Horner, R. R. Hanna, and C. R. Nelson. 2011. New unadorned hadrosaurine hadrosaurid (Dinosauria, Ornithopoda) from the Campanian of North America. Journal of Vertebrate Paleontology 31:798–811.

Henderson, D. M. 1999. Estimating the masses and centers of mass of extinct animals by 3-D mathematical slicing. Paleobiology 25:88–106.

Hone, D. W. E. 2012. Variation in the tail length of non-avian dinosaurs. Journal of Vertebrate Paleontology 32:1082–1089.

Hone, D. W. E., K. Wang, C. Sullivan, X. Zhao, S. Chen, D. Li, S. Ji, Q. Ji, and X. Xu. 2011. A new, large tyrannosaurine theropod from the Upper Cretaceous of China. Cretaceous Research 32:495–503.

Horner, J. R., D. B. Weishampel, and C. A. Forster. 2004. Hadrosauridae; pp. 438–463 in D. B. Weishampel, P. Dodson, and H. Osmólska (eds.), The Dinosauria, Second Edition. University of California Press, Berkeley, California.

Hu, C. 1973. A new hadrosaur from the Cretaceous of Zhucheng, Shantung. Acta Geologica Sinica 2:179–202.

Ji, Y., X. Wang, Y. Liu, and Q. Ji. 2011. Systematics, behavior and living environment of *Shantungosaurus giganteus* (Dinosauria: Hadrosauridae). Acta Geologica Sinica (English Edition) 85:58–65.

Lauters, P., Y. L. Bolotsky, J. Van Itterbeeck, and P. Godefroit. 2008. Taphonomy and age profile of a Latest Cretaceous dinosaur bone bed in Far Eastern Russia. Palaios 23:153–162.

Liu, Y., H. Kuang, N. Peng, H. Xu, and Y. Liu. 2011. Sedimentary facies of dinosaur trackways and bonebeds in the Cretaceous Jiaolai Basin, eastern Shandong, China, and their paleogeographical implications. Earth Science Frontiers 18:9–24.

Liu, Y., H. Kuang, N. Peng, S. Ji, X. Wang, S. Chen, Y. Zhang, and H. Xu. 2010. Sedimentary facies and taphonomy of Late Cretaceous deaths of dinosaur, Zhucheng, eastern Shandong. Geological Review 56:457–468. [Chinese with English abstract]

Matthew, W. D. 1915. Climate and evolution. Annals of the New York Academy of Science 24:171–318.

O'Gorman, E., and D. W. E. Hone. 2012. Body size distribution of the dinosaurs. PLoS ONE 7(12):e51925.

Prieto-Márquez, A. 2010. Global phylogeny of Hadrosauridae (Dinosauria: Ornithopoda) using parsimony and Bayesian methods. Zoological Journal of the Linnean Society 159:435–502.

Scherzer, B. A., and D. J. Varricchio. 2010. Taphonomy of a juvenile lambeosaurine bonebed from the Two Medicine Formation (Campanian) of Montana, United States. Palaios 25:780–795.

Seebacher, F. 1999. A new method to calculate allometric length-mass relationships of dinosaurs. Journal of Vertebrate Paleontology 21:51–60.

Varricchio, D. J. 2010. A distinct dinosaur life history? Historical Biology 23:91–107.

Varricchio, D. J., and J. R. Horner. 1993. Hadrosaurid and lambeosaurid bone beds from the Upper Cretaceous Two Medicine Formation of Montana: taphonomic and biologic implications. Canadian Journal of Earth Sciences 30:997–1006.

Xu, X., K. Wang, X. Zhao, and D. Li. 2010. First ceratopsid dinosaur from China and its biogeographical implications. Chinese Science Bulletin 16:1631–1635.

Xu, X., K. Wang, X. Zhao, C. Sullivan, and S. Chen. 2010. A new leptoceratopsid (Ornithischia: Ceratopsia) from the Upper Cretaceous of Shandong, China and its implications for neoceratopsian evolution. PLoS ONE 5(11):e13835.

Zhao, X., and D. Li. 2009. *Huaxiaosaurus aigahtens* zhao, gen, et sp, nov. Dinosaur Research 8:1–36. [Chinese with English abstract]

Zhao, X.-J., K.-B. Wang, and D.-J. Li. 2011. *Huaxiaosaurus aigahtens*. Geological Bulletin of China 30:1671–1688. [Chinese with English abstract]

Zhao, X., D. Li, G. Han, H. Zhao, F. Liu, L. Li, and X. Fang. 2007. *Zhuchengosaurus maximus* from Shandong Province. Acta Geoscientica Sinica 2:111–1222.

First Hadrosaur Trackway from the Upper Cretaceous (Late Campanian) Oldman Formation, Southeastern Alberta

32

François Therrien, Darla K. Zelenitsky, Kohei Tanaka, and Wendy J. Sloboda

ABSTRACT

A hadrosaur track site discovered in the Milk River Natural Area (MRNA) of southeastern Alberta represents the first dinosaur tracks reported from the Upper Cretaceous (late Campanian) Oldman Formation. The MRNA track site consists of a series of concretions exposed on top of fine-grained deposits, some recognizable as tridactly pedal tracks. Seven consecutive concretions aligned into a narrow 12 m long trackway and one isolated track are preserved. The tracks are robust with blunt digit termination, slightly longer than wide (length ~58cm, width ~54cm), and up to 17.5 cm thick. Divarication angle of digits II–IV is estimated at 63°, where the interdigital angle of digits II–III (22°) is smaller than that of digits III–IV (41°). Morphometric variables indicate the tracks were formed by a large hadrosaur approaching 3.4 m at the hip. The occurrence of a hadrosaur trackway in the upper Oldman Formation expands the stratigraphic and geographic ranges of dinosaur tracks in Canada.

INTRODUCTION

Although dinosaur tracks have been reported from a number of Upper Cretaceous rock formations in southern Alberta (Currie, 1989; McCrea et al., 2005), these ichnofossils are rare in comparison to fossil bones. One notable exception is the St. Mary River Formation, where theropod and hadrosaur tracks are the most common types of vertebrate fossils (Currie et al., 1991; Nadon, 1993). Dinosaur tracks found in Alberta are predominantly natural molds at the surface of sandstone beds or natural casts on the undersurface of sandstone beds. Recently, however, dinosaur tracks preserved as siderite concretions were identified in the Dinosaur Park Formation in Dinosaur Provincial Park (DPP; McCrea et al., 2005).

Here we report on the discovery of the first dinosaur tracks from the Upper Cretaceous (late Campanian) Oldman Formation. The track site, inferred to have been produced by a large hadrosaur, consists of a series of brown concretions exposed on a small plateau in the badlands of the Milk River Natural Area (MRNA). This new trackway extends the stratigraphic and geographic range of known dinosaur tracks in Canada.

MATERIALS AND METHODS

The MRNA track site was documented and mapped using a Brunton compass, a measuring tape, and a digital camera during summer 2007. Documentation of the tracks included their geographic and stratigraphic location, morphology, and the distance between tracks. A variety of morphometric variables of the tracks and trackway, including length and divarication and interdigital angles, useful for taxonomic identification of the track maker (Moratalla et al., 1988; Romilio and Salisbury, 2011), were measured on the original tracks or on digital photos using the software ImageJ version 1.41o. These variables were subsequently plotted in a chart, following the approach of Romilio and Salisbury (2011), to determine the taxonomic identity of the track maker. A stratigraphic section was measured in the immediate vicinity of the trackway to document the stratigraphic setting of the MRNA track site. Sedimentary structures, grain size, pedological features, and color of the lithological units were documented.

GEOLOGIC SETTING

The MRNA track site occurs in the Belly River Group (formerly known as the Judith River Group; Eberth and Hamblin, 1993; Eberth, 2005) of southern Alberta. The Belly River Group is an eastward-thinning clastic wedge that was deposited in the Western Canada Basin during the Campanian in response to tectonic uplift of the Canadian Cordillera. Based on lithological and architectural differences, the Belly River Group is subdivided into three formations, the Foremost, Oldman, and Dinosaur Park formations.

The Milk River Natural Area is located along the Milk River, immediately north of the Canada–United States border, in southeastern Alberta (Fig. 32.1). The area preserves exposures of the Foremost and Oldman formations; exposures of the Dinosaur Park Formation are limited in southernmost

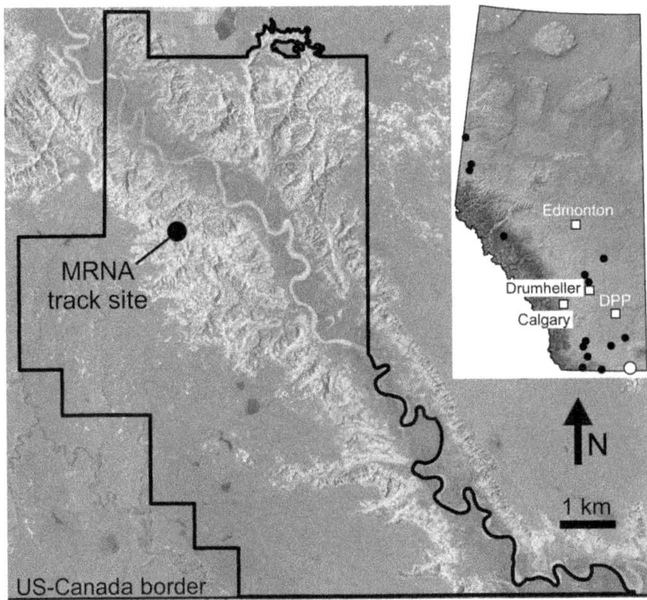

32.1. Location of hadrosaur track site in the Milk River Natural Area (MRNA, black outline). Inset illustrates the location of the MRNA (white circle) relative to Dinosaur Provincial Park (DPP) and other landmarks in Alberta. The location of previously known Alberta dinosaur tracks is indicated by black circles (after Currie, 1989). Tracks are also known from DPP and Drumheller (not indicated).

Alberta due to the wedge-shaped stratigraphic architecture of the formation in this region (Eberth and Hamblin, 1993). Because of this stratigraphic architecture and the time-transgressive contact between the Oldman and Dinosaur Park formations, the upper part of the Oldman Formation in southernmost Alberta is time equivalent to the exposures of the Dinosaur Park Formation in DPP (Eberth and Hamblin, 1993; Eberth, 2005).

The Oldman Formation of southernmost Alberta is over 160 m thick and consists of interbedded sandstones, siltstones, and mudstones that were deposited in an ephemeral, low-sinuosity fluvial setting (Eberth and Hamblin, 1993). The presence of carbonate nodules and slickensides has been interpreted as evidence of deposition in a semi-arid to arid climate, characterized by seasonal precipitation (Eberth and Hamblin, 1993). Based on stratigraphic architecture and sedimentological features, the Oldman Formation has been divided into three informal units: a lower, a middle (= Comrey Sandstone zone), and an upper member. These three members were deposited during high-order regressive, low-stand, and transgressive cycles of the Bearpaw Sea, respectively (Eberth and Hamblin, 1993; Eberth, 2005). Based on radiometric dating of volcanic ash deposits, the upper Oldman Formation of southernmost Alberta is estimated to have been deposited between 76.2 Ma and 74.9 Ma (Eberth and Hamblin, 1993).

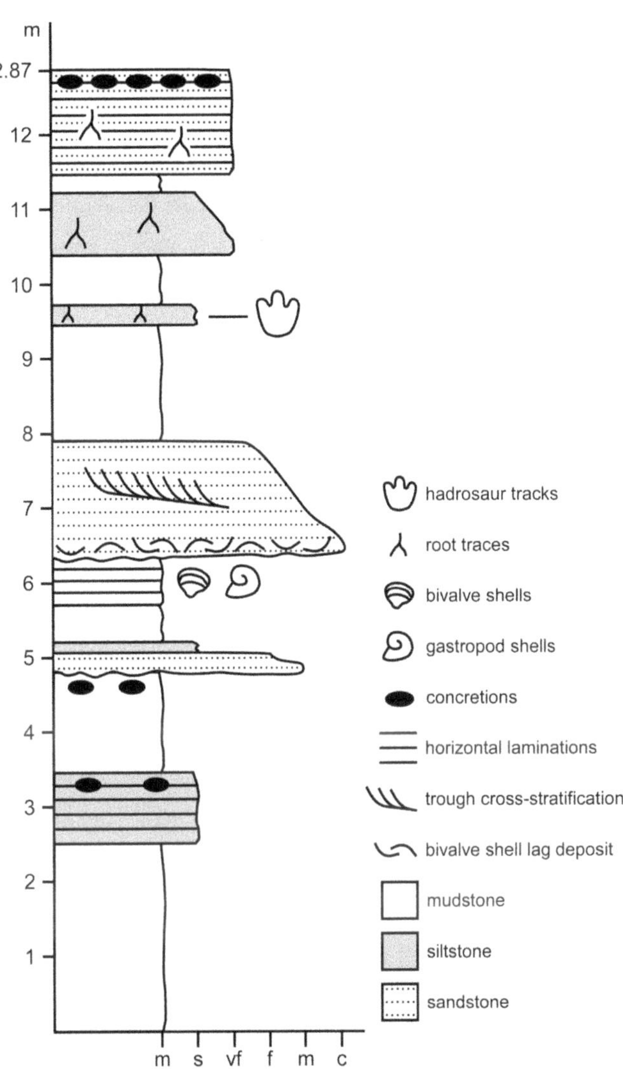

32.2. Stratigraphic section measured in study area demonstrating succession of lithologic units consistent with deposition in an alluvial setting. The hadrosaur tracks occur at the surface of a thick, drab mudstone and are hosted within a light gray siltstone.

THE MILK RIVER NATURAL AREA TRACK SITE

The MRNA track site is located in the upper member of the Oldman Formation, approximately 25 m above the top of the Comrey Sandstone, in badlands exposed on the south side of the Milk River (Fig. 32.1). Strata exposed in the study area consist of interbedded mudstones, siltstones, and cross-stratified to laminated sandstones (Fig. 32.2), typical of sediments deposited in alluvial settings. The tracks are exposed atop a drab green, structureless mudstone unit (Fig. 32.3A, B); however, their host sediment is a thin, light gray, sandy siltstone layer that immediately overlies this mudstone (Fig. 32.2). Field observations reveal the siltstone to be structureless other than for the occurrence of root traces.

32.3. Photographs of track site area. (A) Trackway consists of a series of concretions (arrows) exposed at the surface of a plateau. A few concretions preserve three-toed track morphology (Tracks A and B) whereas the others are incomplete and indistinguishable from concretions nearby (seen on far left). Arrows pointing from the left represent right tracks, arrows pointing from the right represent left tracks. (B) View of the strata exposed in the study area. The surface on which the tracks occur is indicated by arrow. Plateau on which tracks occur is approximately 9.5 m high. (C) Microstructure of track host siltstone. Soft-sediment deformation (ss), root traces (rt), and redoximorphic features (rd) are present. Scale bar equals 5 mm.

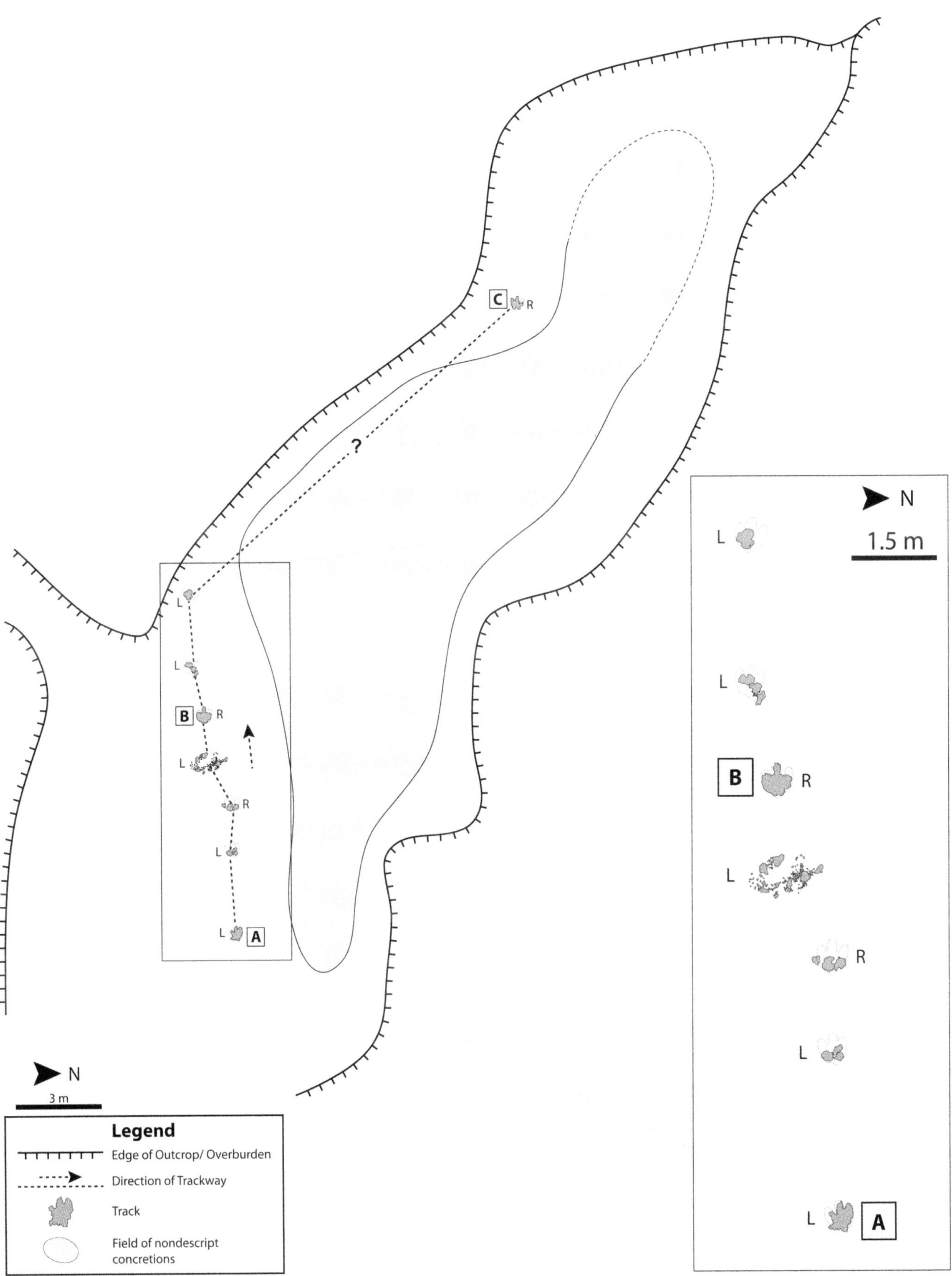

32.4. Map of MRNA hadrosaur tracksite. Field of non-descript concretions is located north of trackway. Best preserved tracks are labeled A, B, and C. Relationship of Track C to trackway is unclear. Inset shows details and spacing of seven consecutive tracks.

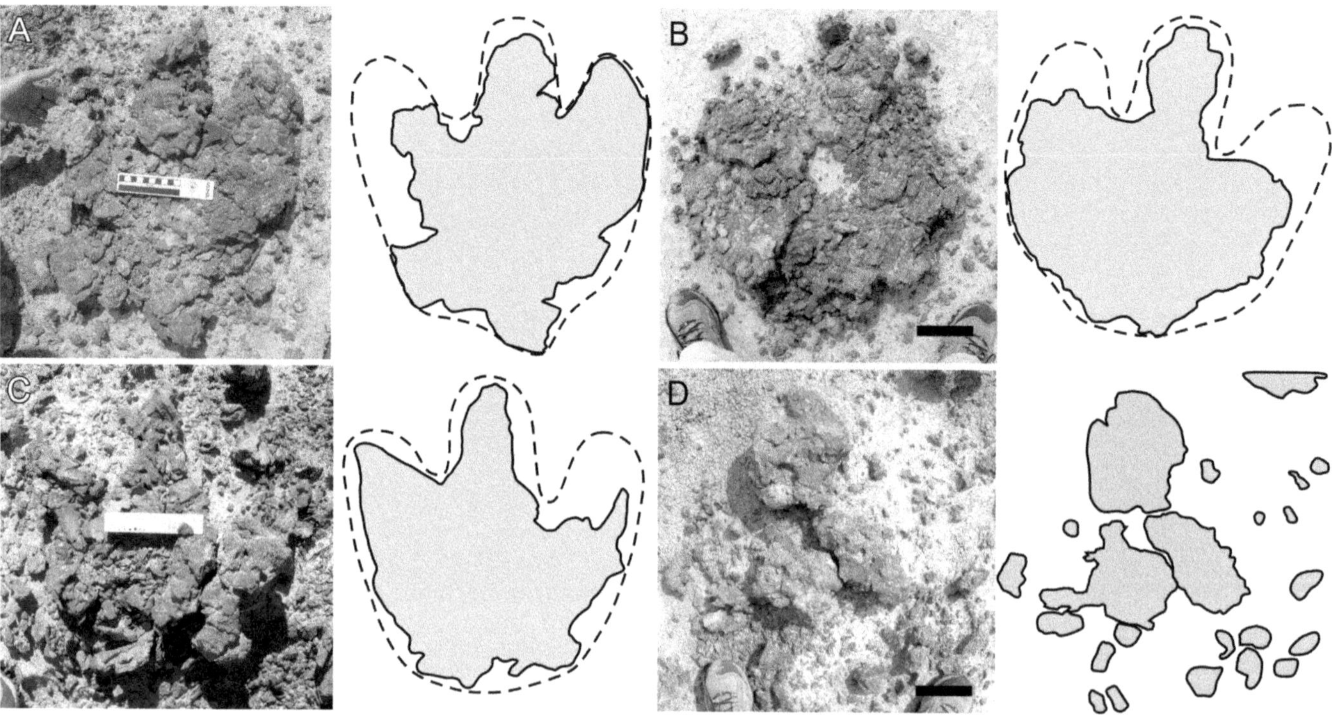

32.5. MRNA tracks. (A) Photograph and schematic representation (gray) of Track A with hypothesized outline of complete track (dash line); (B) photograph and schematic representation (gray) of Track B with hypothesized outline of complete track (scale bar equals 10 cm); (C) photograph and schematic representation (gray) of Track C with hypothesized outline of complete track; (D) photograph and schematic representation (gray) of heavily weathered and broken track. Scale bar equals 10 cm. See Figure 32.3 for location of individual tracks. Track shown in (D) is second track in trackway.

Petrographic analysis of the siltstone provides information about the nature of the substrate at the time the tracks were formed. Thin-sections show the siltstone is characterized by an abundance of distorted and convoluted laminations, indicative of soft-sediment deformation, and localized accumulation of iron oxides (Fig. 32.3C). Iron oxide accumulation occurs as small zones of moderately to strongly impregnated groundmass and occasionally forms iron coatings on grains. These characteristics suggest that those accumulations represent redoximorphic features. Redoximorphic features are pedogenic ferruginous accumulations that formed in response to fluctuations in the level of the water table, precipitating when soil conditions fluctuate between reducing and oxidizing (PiPujol and Buurman, 1994; Vepraskas, 1992; Vepraskas and Caldwell, 2008). These lines of evidence indicate that the substrate was soft and saturated with water at the time the tracks were formed.

The track site consists of a series of seven consecutive and one isolated brown concretions exposed at the surface of a small plateau (Figs. 32.3, 32.4). The concretions are heavily fractured and weathered, attesting to a relatively long period of exposure. Three of the tracks are sufficiently well preserved to allow for their recognition as tridactyl, pedal tracks (Fig. 32.5A–C). The remaining tracks are so weathered (Fig. 32.5D) that they are not recognizable as tracks and are indistinguishable from other concretions found nearby (Fig. 32.4).

The spatial arrangement of seven consecutive concretionary tracks clearly establishes them as forming a trackway (Fig. 32.4); the relationship of one isolated track (labeled C) to the trackway is unclear. The trackway is narrow, oriented nearly due west, and can be traced over a distance of 12 m (Fig. 32.4). Based on the angles separating digits and the relative position of the concretions, five left tracks and three right tracks are recognized. The absence of concretionary material for the two missing right tracks indicates that they were not preserved. The distance between tracks varies slightly along the trackway: an initial stride length of 272 cm is followed by a pace length of 160 cm at mid-trackway, and ends with a stride length of 264 cm.

The three best preserved tracks (Tracks A, B, and C in Figs. 32.4 and 32.5) permit a more detailed documentation of their size and shape. The tracks are robust with blunt digit termination, slightly longer than wide (length ~58cm, width ~54cm, l/w = 1.07), and up to 17.5 cm thick. The total divarication angle of digits II–IV is estimated to be ~ 63°, where the interdigital angle of digits II–III (22°) is smaller than that of digits III–IV (41°). Morphometric variables consistently fall within the range expected for an ornithopod track maker (Fig. 32.6; Table 32.1).

The most complete track also displays two protuberances along its posterior margin (Fig. 32.5A). These protuberances may indicate the track maker had an asymmetrically bilobed heel, although it cannot be ruled out that the protuberances are posterior drag marks or weathering artifacts. Bilobed heels are typical of large ornithopods from the Cretaceous of North America (Currie et al., 1991, 2003; Lockley and Hunt, 1995; Lockley et al., 2004).

The track size and spacing provide insight into the body size and speed of the track maker. Based on the relationship between track size and leg length in large ornithopods, where hip height = 5.9·foot length (Thulborn, 1990), the track maker is estimated to have measured approximately 3.4 m at the hip. Based on the relationship between the distance separating tracks and locomotion speed, where speed = $0.25 \cdot (\text{gravity})^{0.5} \cdot (\text{stride length})^{1.67} \cdot (\text{hip height})^{-1.17}$ (Alexander, 1976), the track maker is inferred to have been moving slowly. From an estimated 3.7 km/h (SL/h ~0.82) at the onset of the trackway, the animal subsequently increased its speed to 4.9 km/h (SL/h ~0.97) before decelerating to 3.4 km/h (SL/h ~ 0.78) at the end of the trackway. Because the SL/h ratios are lower than 2.0, the animal can be inferred to have been walking rather than trotting or running (Alexander, 1976; Thulborn, 1990). These relatively slow speeds are similar to those reported for the Early Cretaceous *Amblydactylus* of northwestern Alberta and northeastern British Columbia (Currie, 1983). Based on sedimentological evidence for a water-saturated substrate in the area of the track site (see above), one can hypothesize that the animal walked slowly as it traversed an area characterized by a wet and soft substrate.

DISCUSSION AND CONCLUSION

The size and shape of the MRNA tracks, quantified by morphometric variables, suggest that the track maker was a large ornithopod (Fig. 32.6; Table 32.1). The tracks are morphologically similar to ornithopod tracks assigned to the ichnotaxon *Amblydactylus* from Early Cretaceous deposits of northwestern Alberta and northeastern British Columbia (Sternberg, 1932; Currie, 1983). However, given the incomplete nature of the MRNA tracks, we refrain from assigning them to an ichnotaxon.

The MRNA tracks were most likely produced by a hadrosaur, as these animals are the only large-bodied ornithopods known from the late Campanian of North America. The estimated height of the track maker (3.4 m at the hip; see above) is consistent with large hadrosaurs exceeding 10 m in length (Brown, 1913, 1916; Lull and Wright, 1942). Although hadrosaurs are known from the upper Oldman Formation exposed in the MRNA, their precise taxonomic identity is unknown because fossil remains are too fragmentary. Other

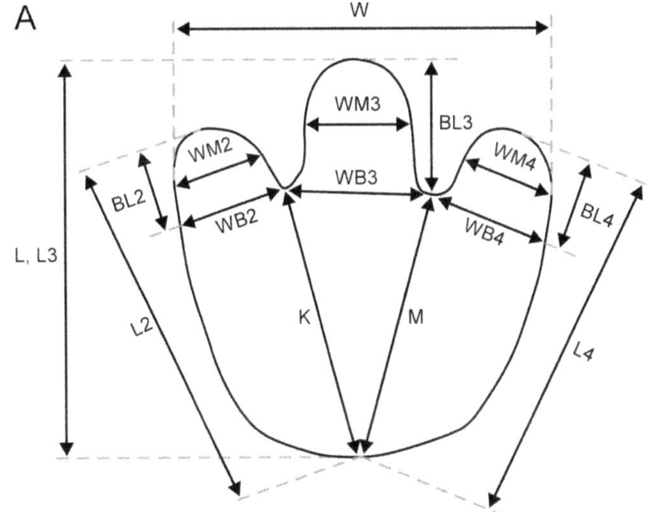

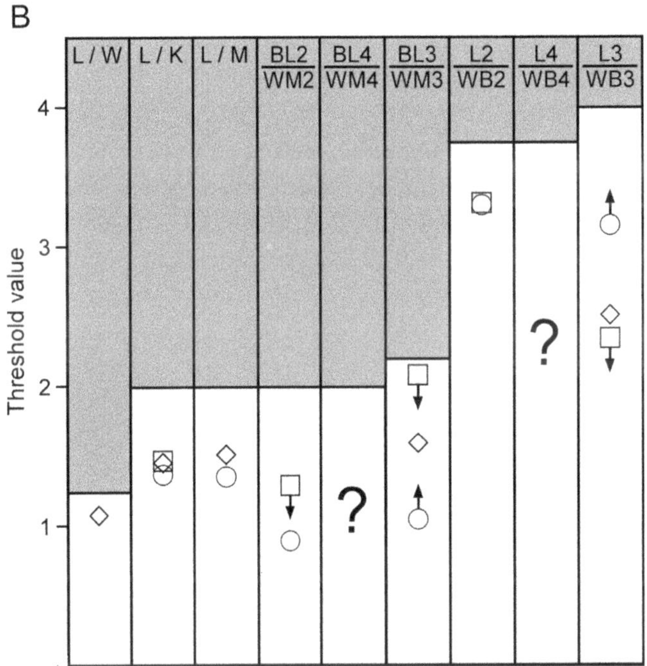

32.6. Morphometric analysis of MRNA tracks. (A) Morphometric variables measured to determine the taxonomic identity of the MRNA tracks (after Moratalla et al., 1988). Abbreviations: L, track length; W, track width; L2-L4, whole digit lengths; BL2–4, basal digit lengths; WB2–4, basal digit widths; WM2–4, middle digit widths; K and M, heel-to-interdigit lengths. (B) Multivariate analysis of MRNA track morphology. Circles indicate track A, diamonds indicate track B, and squares indicate track C. White intervals indicate ornithopod track morphology, gray intervals indicate theropod track morphology. Due to the incomplete nature of the tracks, the results represent the best approximation of the morphometric variables for each track. Arrows indicate instances where values are expected to be slightly higher or lower if tracks were more complete. Occasionally a given track may have been too incomplete to evaluate a specific variable.

Table 32.1. Morphometric Variables Used to Differentiate Ornithopod and Theropod Tracks

Morphometric variable	Threshold value and associated probability that track belongs to theropod or ornithopod	Track A	Track B	Track C
L / W	80% theropod >**1.25** >88.2% ornithopod	?	1.07	?
L / K	70.5% theropod >**2.00** >88% ornithopod	1.36	1.45	1.46
L / M	65% theropod >**2.00** >90.7% ornithopod	1.35	1.51	?
BL2 / WM2	76.1% theropod >**2.00** >97.7% ornithopod	0.89	?	1.29*
BL4 / WM4	76.1% theropod >**2.00** >97.7% ornithopod	?	?	?
BL3 / WM3	72.7% theropod >**2.20** >97.7% ornithopod	1.04**	1.59	2.08*
L2 / WB2	84.6% theropod >**3.75** >90.2% ornithopod	3.30	?	3.31
L4 / WB4	73.7% theropod >**3.75** >93.4% ornithopod	?	?	?
L3 / WB3	70.6% theropod >**4.00** >91.5% ornithopod	3.15**	2.51	2.34*

Note: Variables, threshold values, and probability values of ornithopod/theropod affinity are from Moratalla et al. (1988). See Figure 32.6A for explanation of morphometric variables. Single asterisks indicate maximum value due to incompleteness of track. Double asterisks indicate minimum value due to incompleteness of track.

hadrosaur taxa (e.g., *Corythosaurus*, *Lambeosaurus*, *Parasaurolophus*, *Gryposaurus*, *Prosaurolophus*) are known however from the temporally equivalent Dinosaur Park Formation in DPP and Judith River Formation of Montana (Eberth et al., 2001; Weishampel et al., 2004; Ryan and Evans, 2005).

Dinosaur tracks in Canada have been reported previously from the Late Triassic and Early Jurassic of Nova Scotia and the Late Jurassic through Late Cretaceous of Alberta and eastern British Columbia (Currie, 1989; Weishampel and Young, 1996; Weishampel et al., 2004). The MRNA track site is the first documented occurrence of dinosaur tracks in the Oldman Formation of southern Alberta and expands the stratigraphic and geographic ranges of dinosaur tracks in Canada (see Fig. 32.1). Dinosaur tracks are a rare occurrence in the Belly River Group of Canada, having been previously reported exclusively from Redcliff (Foremost Formation) and DPP (Dinosaur Park Formation) (Currie, 1989; McCrea et al., 2005), and have yet to be reported from the correlative Judith River and Two Medicine formations of Montana (Weishampel et al., 2004). The unusual preservation of tracks as concretions could have potentially hindered their recognition in the field.

ACKNOWLEDGMENTS

The authors wish to thank Cam Lockerbie for granting access to the MRNA, Raymond Strom for producing high-quality thin sections, and Jennifer Bancescu for assistance with Figure 32.1. Rich McCrea and David Evans provided constructive comments that improved the manuscript. This research was partly funded by the Royal Tyrrell Museum, a Killam Postdoctoral Fellowship (DKZ), and a Natural Sciences and Engineering Research Council Discovery grant (DKZ).

LITERATURE CITED

Alexander, R. M. 1976. Estimates of speeds in dinosaurs. Nature 261:129–130.

Brown, B. 1913. The skeleton of *Saurolophus*, a crested duck-billed dinosaur from the Edmonton Cretaceous. Bulletin of the American Museum of Natural History 32:387–393.

Brown, B. 1916. *Corythosaurus casuarius:* skeleton, musculature and epidermis. Bulletin of the American Museum of Natural History 35:709–716.

Currie, P. J. 1983. Hadrosaur trackways from the Lower Cretaceous of Canada. Acta Palaeontologica Polonica 28:63–73.

Currie, P. J. 1989. Dinosaur footprints of Western Canada; pp. 293–300 in D. D. Gillette and M. G. Lockley (eds.), Dinosaur Tracks and Traces. Cambridge University Press, Cambridge, U.K.

Currie, P. J., D. Badamgarav, and E. B. Koppelhus. 2003. The first Late Cretaceous footprints from the Nemegt Locality in the Gobi of Mongolia. Ichnos 10:1–13.

Currie, P. J., G. C. Nadon, and M. G. Lockley. 1991. Dinosaur footprints with skin impressions from the Cretaceous of Alberta and Colorado. Canadian Journal of Earth Sciences 28:102–115.

Eberth, D. A. 2005. The geology; pp. 54–82 in P. J. Currie and E. B. Koppelhus (eds.), Dinosaur Provincial Park: A Spectacular Ancient Ecosystem Revealed. Indiana University Press, Bloomington, Indiana.

Eberth, D. A., and A. P. Hamblin. 1993. Tectonic, stratigraphic, and sedimentologic significance of a regional discontinuity in the upper Judith River Group (Belly River wedge) of southern Alberta, Saskatchewan, and northern Montana. Canadian Journal of Earth Sciences 30:174–200.

Eberth, D. A., P. J. Currie, D. B. Brinkman, M. J. Ryan, D. R. Braman, J. D. Gardner, V. D. Lam, D. N. Spivak, and A. G. Neuman. 2001. Alberta's dinosaurs and other fossil vertebrates: Judith River and Edmonton groups (Campanian-Maastrichtian); pp. 49–75 in C. L. Hills (ed.), Guidebook for the Field Trips: Mesozoic and Cenozoic Paleontology in the Western Plains and Rocky Mountains. Society of Vertebrate Paleontology, 61st Annual Meeting. Museum of the Rockies Occasional Paper 3.

Lockley, M. G., and A. P. Hunt. 1995. Dinosaur Tracks and Other Fossil Footprints of the Western United States. Columbia University Press, New York, 338 pp.

Lockley, M. G., G. C. Nadon, and P. J. Currie. 2004. A diverse dinosaur-bird footprint assemblage from the Lance Formation, Upper Cretaceous, eastern Wyoming: implications for ichnotaxonomy. Ichnos 11:229–249.

Lull, R. S., and N. E. Wright. 1942. Hadrosaurian Dinosaurs of North America. Geological Society of America Special Papers 40. 242 pp.

McCrea, R. T., P. J. Currie, and S. G. Pemberton. 2005. Vertebrate ichnology; pp. 405–416 in P. J. Currie and E. B. Koppelhus (eds.), Dinosaur Provincial Park: A Spectacular Ancient Ecosystem Revealed. Indiana University Press, Bloomington, Indiana.

Moratalla, J. J., J. L. Sanz, and S. Jimenez. 1988. Multivariate analysis on Lower Cretaceous dinosaur footprints: discrimination between ornithopods and theropods. Geobios 21:395–408.

Nadon, G. C. 1993. The association of anastomosed fluvial deposits and dinosaur tracks, eggs, and nests: implication for the interpretation of floodplain environments and a possible survival strategy for ornithopods. Palaios 8:31–44.

PiPujol, M. D., and P. Buurman. 1994. The distinction between ground-water gley and surface-water gley phenomena in Tertiary paleosols of the Ebro Basin, NE Spain. Palaeogeography, Palaeoclimatology, Palaeoecology 110:103–113.

Romilio, A., and S. W. Salisbury. 2011. A reassessment of large theropod dinosaur tracks from the mid-Cretaceous (late Albian-Cenomanian) Winton Formation of Lark Quarry, central-western Queensland, Australia: a case for mistaken identity. Cretaceous Research 32:135–142.

Ryan, M. J., and D. C. Evans. 2005. Ornithischian dinosaurs; pp. 312–348 in P. J. Currie and E. B. Koppelhus (eds.), Dinosaur Provincial Park: A Spectacular Ancient Ecosystem Revealed. Indiana University Press, Bloomington, Indiana.

Sternberg, C. M. 1932. Dinosaur tracks from Peace River, British Columbia. National Museum of Canada, Annual Report 1930:59–85.

Thulborn, T. 1990. Dinosaur Tracks. Chapman and Hall, London, 410 pp.

Vepraskas, M. J. 1992. Redoximorphic Features for Identifying Aquic Conditions. North Carolina Agricultural Research Services Technical Bulletin 301. 33 pp.

Vepraskas, M. J., and P. V. Caldwell. 2008. Interpreting morphological features in wetland soils with a hydrologic model. Catena 73:153–165.

Weishampel, D. B., and L. Young. 1996. Dinosaurs of the East Coast. Johns Hopkins University Press, Baltimore, Maryland, 275 pp.

Weishampel, D. B., P. M. Barrett, R. A. Coria, J. Le Loeuff, X. Xu, X. Zhao, A. Sahni, E. M. P. Gomani, and C. R. Noto. 2004. Dinosaur distribution; pp. 517–606 in D. B. Weishampel, P. Dodson, and H. Osmólska (eds.), The Dinosauria, Second Edition. University of California Press, Berkeley, California.

Paleopathology in Late Cretaceous Hadrosauridae from Alberta, Canada

33

Darren H. Tanke and Bruce M. Rothschild

ABSTRACT

Late Cretaceous hadrosaurs from Alberta provide the best preserved and most numerous examples of dinosaurian osteopathy and are ideal for invasive or noninvasive studies, including histology and scanning (X-ray, CT, or MRI). Enough specimens have been observed in the field and museum collections to demonstrate predictable patterns of osteopathy, not only in representative types, but in some cases body position and relative ontogenetic stage. Hadrosaurs account for about 60% of all dinosaurian paleopathology seen in Royal Tyrrell Museum collections. Whereas osteopathy is observed in animals of all ages (including neonate), most affected elements are adult pedal phalanges, ribs, and especially the caudal vertebrae. Examples often reflect active healing of fractures, although there are relatively few cases of complete bone repair. Caution is urged in the interpretation of caudal injuries in hadrosaurs as failed predation attempts by tyrannosaurids. Many tail injuries have crush etiologies and appear related to accidental intraspecific encounters such as trampling.

INTRODUCTION

Paleopathology in its strictest definition is the study of diseases or pathologic conditions in ancient organisms. Analyzing paleopathologies in fossil assemblages can provide unique and important insights into paleobiology and paleoecology (Rothschild and Martin, 1987, 1993, 2006; Rothschild, 1997; Tanke and Currie, 2000; Rothschild et al., 2001, 2005, 2012; Happ, 2008; Rothschild and Molnar, 2008; Peterson et al., 2009; Tanke and Rothschild, 2010).

Paleopathologies are particularly common in hadrosaur fossils from the Upper Cretaceous (Campanian) Oldman and Dinosaur Park formations (OFm, DPFm) of the Belly River Group at Dinosaur Provincial Park (DPP, the Park; Fig. 33.1), and their large abundance suggests that pathologies were simply a fact of life for these animals. In this chapter we document the range of ancient osteopathies in DPP's hadrosaurs (and some found elsewhere both within Alberta and worldwide) and interpret their origins and significance. Osteopathies are so abundant in Alberta's Upper Cretaceous hadrosaurs that a thorough documentation and quantification of their occurrences must await future studies. Here we describe and figure only selected representatives. Thus, like the recently published Albertan ceratopsian osteopathy review (Tanke and Rothschild, 2010), this chapter presents a broad overview of the types of known hadrosaur osteopathies with special reference to Alberta.

Institutional and Formation Abbreviations AEHM, Amur Natural History Museum of the Far Eastern Institute of Mineral Resources, Blagoveschensk, Russia; AMNH, American Museum of Natural History, New York; BLASI, Museo Palaeontológico de la Universidad de Zaragoza, Zaragoza, Spain; CMN, Canadian Museum of Nature, Ottawa, Ontario; DMNH, Denver Museum of Natural History, Denver, Colorado; DPFm, Dinosaur Park Formation; GNSM, Gwacheon National Science Museum, Seoul; HCFm, Horseshoe Canyon Formation; JRF, Judith River Foundation/Judith River Dinosaur Institute, Malta, Montana; MOR, Museum of the Rockies, Bozeman, Montana; NHMUK, Natural History Museum, London; OFm, Oldman Formation; PBMNH, Palm Beach Museum of Natural History, Palm Beach, Florida; RAM, Raymond M. Alf Museum of Paleontology, Claremont, California; ROM, Royal Ontario Museum, Toronto, Ontario; TMP, Royal Tyrrell Museum of Palaeontology, Drumheller, Alberta; UAD, uncollected articulated dinosaur, field site designation by TMP; UNO, University of New Orleans, New Orleans, Louisiana; USNM, United States National Museum, Washington, D.C.; ZPAL MgD, Institute of Paleobiology, Polish Academy of Sciences, Warsaw.

MATERIALS AND METHODS

Unless otherwise noted, the specimens described or discussed here are isolated elements from adult-size hadrosaurids of indeterminate genus and species from the DPFm at DPP. We examined specimens in the field at DPP and collected a substantial number of specimens from there (curated at the TMP with specimen numbers typically reported). We also include a significant number of observations from specimens collected from the Upper Cretaceous Horseshoe Canyon

Formation (HCFm) exposures at Drumheller (Campanian–Maastrichtian) and curated at the TMP. Other museums also contain pathological specimens collected from DPP and elsewhere in Alberta during the past 100 years. Many were examined during this study. We repeatedly refer readers to Figure 33.1 to show the general locations of major collecting areas in the Belly River Group and Horseshoe Canyon Formation of southern Alberta (e.g., Drumheller, Rumsey, Dinosaur Park, and Sandy Point).

Presently there are about 1200 pathological dinosaur specimens catalogued in the TMP. Most of these are from DPP and nearly 60% are hadrosaurian. This number does not include pedal phalanges with osteochondrosis (Rothschild and Tanke, 2007), which would substantially increase this percentage.

Specimens were prepared for gross examination and photography using standard small fossil preparation techniques and hand tools (e.g., water, brushes, and dental picks), and mechanical air scribes, as required. Many fossils were found fully exposed on the surface and required only a light scrub with brush and water to remove adhering lichens, silt, and sand. Some were further cleaned with an air-abrasive unit charged with sodium bicarbonate (baking soda) or aluminum oxide powders.

Some specimens underwent additional noninvasive study. Fluoroscopy work was done with a portable Xi Tec, Model 1000 fluoroscope at 40–65 KVp on all cataloged hadrosaur caudal vertebrae (Rothschild et al., 2003). CT scans of some specimens were made at the Drumheller Health Centre with a Toshiba Aquilion 16 scanner with settings of 120 kV and 450 mAs, images taken at 0.5 mm slices. A few small specimens were sliced and made into histological thin sections to determine the quality of the end product and were found to preserve excellent cellular details.

33.1. Map of Alberta, Canada, with localities mentioned in text designated by stars and numbers. (1) Edmonton (HCFm); (2) Rumsey (HCFm); (3) Drumheller (HCFm); (4) Dinosaur Provincial Park (OFm, DPFm); (5) Sandy Point (OFm, DPFm); (6) Grassy Lake (Foremost Formation); (7) Devil's Coulee (OFm); and (8) Milk River badlands (OFm).

PREVIOUS WORK

Work on Albertan hadrosaur osteopathy got off to a slow start. Despite hadrosaurs being known from the province for over 120 years, only a handful of observations concerning their associated osteopathies have appeared in the literature. Most of these are briefly included as part of anatomical descriptions (e.g., Parks, 1922; Moodie, 1923; Sawyer and Erickson, 1985), or within popular books, technical reports, or soft science news stories (e.g., Langston, 1961; Swinton, 1970; Monastersky, 1990; also see literature review in Tanke and Rothschild, 2002).

Roy Lee Moodie (1880–1934), the father of paleopathology studies in the Western Hemisphere, wrote one short paper dedicated to the topic (Moodie, 1930), and had written a general overview manuscript on Albertan hadrosaur and other dinosaur osteopathy, but died before it was published. Dedicated work on hadrosaur paleopathology has been conducted only recently (Rothschild and Tanke, 2007; Straight et al., 2009). With this chapter, the authors hope to establish a foundation on which future studies of hadrosaur and large ornithopod osteopathy may be conducted and compared.

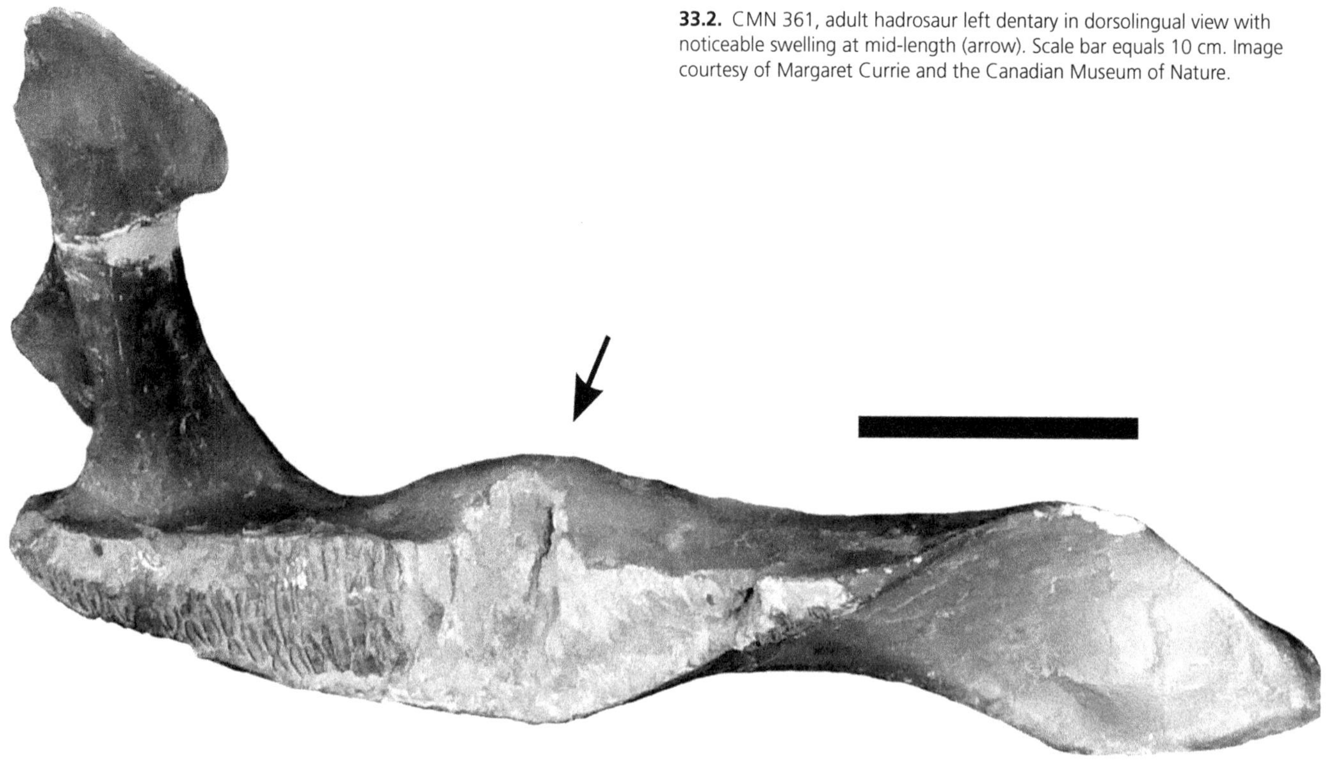

33.2. CMN 361, adult hadrosaur left dentary in dorsolingual view with noticeable swelling at mid-length (arrow). Scale bar equals 10 cm. Image courtesy of Margaret Currie and the Canadian Museum of Nature.

DESCRIPTION

Skull and Teeth

Among the first dinosaur specimens collected and described from the province were an unassociated hadrosaur maxilla (CMN 362) and dentary (CMN 361; Fig. 33.2) with deeply eroded or expanded pockets of bone and premortem dentary tooth loss. Although Lambe (1902) figured but did not describe their pathological conditions, Moodie (1930) later considered these as dental abscesses, a diagnosis with which we concur. It is curious that Lambe should find two examples in one field season and that none have been reported subsequently, despite intensive fieldwork and hundreds of hadrosaur jaws seen or collected.

The skull of *Prosaurolophus* (TMP 1993.081.0001) preserves what appears to represent a massive fracture with extensively thickened but evidently noninfectious callus (120 mm × 95 mm) on the external left dentary, centered about 180 mm posterior to the hypothetically restored symphysis. There is slight ventral shifting and angulation of the distal portion of the dentary. A shallow groove on the new bone growth, orientated roughly dorsoventrally, may represent the original fracture line. The lingual portion of this dentary is still in the matrix, as is the other dentary, so it is unknown if that element was affected. The skull of *Brachylophosaurus* (TMP 1990.104.0001), from the Milk River badlands (Fig. 33.1), exhibits a deep circular pit of unknown etiology on the external right postorbital. Additionally, there is a minor disruption and depression of the bone grain texture on the external right dentary, about 10 cm anterior to the coronoid process and just above the ventral edge. Directly below the anterior tooth rows of the left dentary, close to the ventral aspect of the lingual edge, there is an area 9.4 cm in length with disrupted bone texture and swelling anteriorly. The anteriormost left dentary teeth (about 6 rows) appear smaller than the same teeth in the opposite dentary. These combined features are suggestive of a well-healed bilateral jaw fracture, although a CT scan is needed for confirmation. Another anomalous skull bone includes an isolated subadult asymmetrical occipital from DPP (TMP 1993.066.0011).

Hadrosaur teeth are common at DPP. Despite large museum and field samples, dental developmental conditions appear to be extremely rare. Dental caries and premortem tooth fractures are unknown. However, the absence of caries and fractures is not surprising, given how quickly the teeth were replaced and the way in which the tightly packed teeth are deeply embedded in the jaws. TMP 1985.036.0161 (Fig. 33.3A) has a dorsoventrally compressed appearance. The etiology of this condition is unknown, but it appears developmental, suggesting tooth bud damage that resulted in the deformity. Dental anomalies in hadrosaurs beyond Alberta are also extremely rare; Prieto-Márquez (2012:522, fig. 9C) briefly notes and figures the splitting of a single median

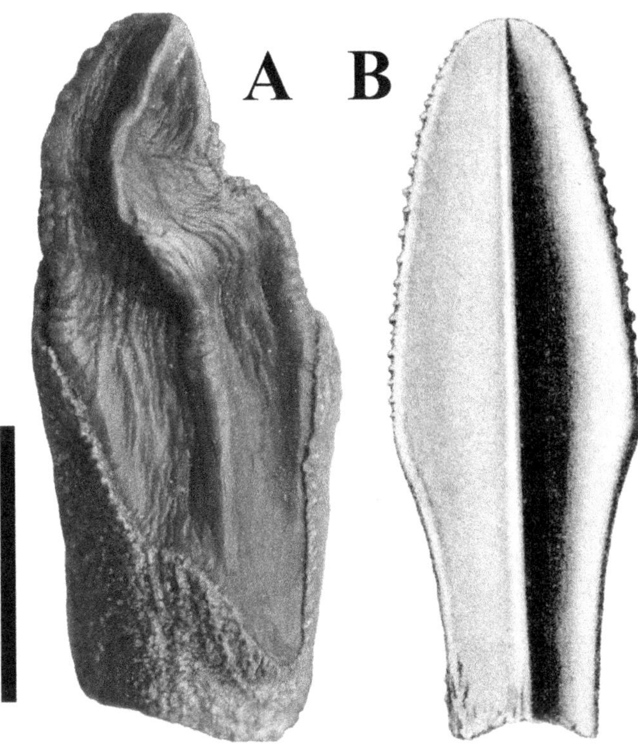

33.3. (A) TMP 1985.036.0161, pathologic hadrosaur tooth compared to (B) a tooth of normal morphology (after Lambe, 1902). Scale bar equals 1 cm.

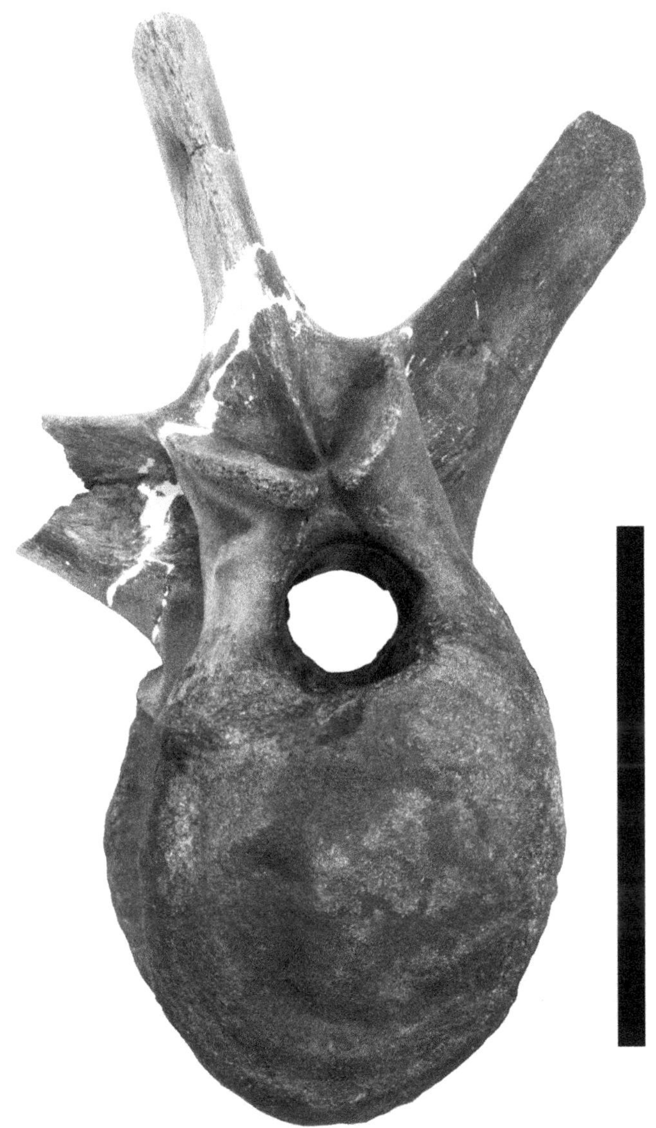

33.4. TMP 1991.036.0289, dorsal vertebra in anterior view showing strong bilateral asymmetry. Scale bar equals 10 cm.

ridge into two smaller and finer ridges in a *Gryposaurus* from Montana.

Vertebrae

Vertebrae, especially those from the tail, are the most frequently injured element in the hadrosaur skeleton. Rothschild et al. (2003) reported occurrences of benign neoplasia (hemangioma, desmoplastic fibroma, osteoblastoma) and malignant neoplasia (metastatic carcinoma) in Asian and American hadrosaurs from many locations. Frequency of occurrence was similar to that observed in modern animals (Natarajan et al., 2007). As no cases have so far been identified in hadrosaurs from Alberta, this pathology will not be discussed further.

Cervical TMP 1986.012.0022, a fragment of a cf. *Edmontosaurus* cervical vertebra from the HCFm at Drumheller (Fig. 33.1), exhibits minor pitting and erosions of the diapophysis articulating face. Osteopathy from the neck is unreported in the literature or rarely observed. This is expected, as any source of bone injury (trauma) in this body region would likely prove fatal.

Dorsal A scoliotic congenital deformity of an embryonic or neonate *Hypacrosaurus* (TMP 1987.079.0012) was recovered from the OFm at Devil's Coulee (Fig. 33.1). One dorsal centrum, 8.5 mm wide and 6.5 mm tall, has a deep, rounded pit of small diameter on the exposed articular face. The opposing face is slightly beveled, and articulates to a wedge-shaped centrum with one neural arch base instead of two. Both vertebrae are fused ventrally over a distance of 1 mm. A third incomplete articulated centrum also appears to have been wedge shaped. While scoliosis related to wedge-shaped centra (hemi-vertebrae) have been cited as congenital defects in mammals (including *Homo*; Brothwell, 1967; Baker and Brothwell, 1980), this appears to be its first recognition in an embryonic dinosaur. Recently a hemi-vertebra was reported in *Dysalotosaurus* from the Late Jurassic of Tanzania (Witzmann et al., 2008). Another possible scoliotic example, TMP 1991.036.0289, is unusual in that there is asymmetrical growth, with one side of the neural arch base being about 16 mm taller than the other (Fig. 33.4), resulting in strong lateral tilting. Subadult centra TMP 1980.016.0016 and TMP 1992.036.1165 each exhibit a circular pit or an elongate curved depression on one endplate that possibly represents osteochondrosis.

Parasaurolophus (ROM 786) has the tips of three neural spines angled together, with the posterior pair joined at the tip by a roughened bony pad. Parks (1922) suggested that this pad served as a ligamentous attachment point for a skin flap between the tip of the posteriorly elongated crest and the back. However the anomaly (see Naish [2008] for more discussion) is no doubt related to nearby healing rib fractures (see below). There is no complementary roughened bone on the distal end of the crest to which such a flap would attach. Despite this, many *Parasaurolophus* life reconstructions figure the skin flap and thus have been influenced by this minor pathology and flawed interpretation. An indeterminate subadult dorsal neural arch (TMP 1993.036.0273) preserves two well-healed longitudinal fractures of the neural spine without angulation, one on the right side 50 mm below tip and one of the left side 22 mm lower.

Several cf. *Edmontosaurus* specimens from the HCFm in the Drumheller area express osteopathy. Dorsal centrum TMP 1984.164.0001 bears three deep circular holes (one with an overhanging margin) on one endplate. TMP 1993.036.0148, a fragmentary neural arch, bears prominent erosions of the articulating facets of the pre- and postzygapophyses, especially the former. Although the causative agent is unknown, infection or a form of arthritis referred to as spondyloarthropathy are the most likely candidates. TMP 1993.031.0183, recovered from an *Edmontosaurus*-dominated bonebed near Drumheller appears to be a congenital defect in which two dorsal centra share one neural arch.

Sacral Osteopathy has rarely been observed from this region. TMP 1974.010.0007 is a complete adult indeterminate hadrosaur sacrum consisting of eight vertebrae in which neural spines IV–VI (counted from the cranial end) preserve well-healed longitudinal fractures near mid-length with little angulation. All are in the same state of repair, indicating these fractures were received simultaneously. Several large, broad, and flat neural spine fragments with fully remodeled fractures and angulation in TMP collections (e.g., TMP 1990.036.0409) may derive from the sacrum, but are too incomplete for verification.

Caudal (General Comments) Caudal injuries are frequently seen and are the most common type of dinosaur osteopathy found in the specimens we examined. This is not surprising given the large number of vertebrae in hadrosaur tails. An adult *Corythosaurus* with a complete 4.55 m long tail preserves 75 vertebrae and nearly as many chevrons (Gilmore, 1946). The large abundance of caudal vertebrae in hadrosaur skeletons combined with the large number of hadrosaur remains in Upper Cretaceous nonmarine deposits throughout North America, and their robust compact shapes that resist wear help explain why isolated hadrosaur caudal centra and complete caudal vertebrae are so abundant at DPP. Because of this, a variety of osteopathy types and frequencies of occurrence can be discerned, although these patterns still require quantification. First, there appears to be a positive correlation between body size and osteopathy. In this study, hadrosaur caudal vertebrae were divided into several ontogenetic classes based on a combination of relative size (compared to mounted skeletons), degree of neural arch fusion, and cortical surface bone texture.

Neonate to juvenile material (animals <2.5 m long) is present but rare at DPP (Tanke and Brett-Surman, 2001), and we have not observed osteopathy in the caudals of these individuals. We propose that these individuals either did not sustain caudal injuries or, alternatively, did not survive major wounds to their tails. Subadult material (animals 2.5–4.5 m long) is abundant, although their neural arches–the area most frequently injured in hadrosaur vertebrae–are unfused to the centrum, and thus poorly represented in museum collections. Accordingly, the relatively few subadult neural arches observed provide limited data regarding osteopathy frequency. However, comparisons of subadult versus adult centra indicate that centrum osteopathy is rare in subadults. One possible reason for this reduced occurrence of osteopathy in subadults may be related to their life history. Segregation of subadults ("bachelor herds") and adults in hadrosaurs has been proposed by a variety of workers (Horner, 1984; Rogers, 1985; Forster, 1990; Varricchio and Horner, 1993). If true, subadults may have engaged in different behaviors or lived under different circumstances that simply did not result in the kinds of tail injuries seen in adults. Hadrosaur caudal vertebrae from adults (individuals >4.5 m long) are extremely abundant in museum collections (numbering in the thousands) and are frequently encountered in the field. Osteopathy is regularly observed, and the widest range of osteopathic conditions are preserved in adult specimens.

In the descriptions and discussions below, caudal vertebrae are assigned to one of three positions along the tail: proximal, medial, and distal. Proximal vertebrae are from the base of the tail. We estimate that there are approximately 16 proximal caudals (Gilmore, 1946). They bear caudal ribs of varying lengths (elongate to low, rounded bumps), and have elongate, erect, and laterally flattened neural spines, angled slightly posteriorly. Some neural spines may develop parallel vertically oriented rows of smooth grooves and ridges near the tip, forming a fluted surface. Neural spines are longer in lambeosaurines than in hadrosaurines. The zygapophyses

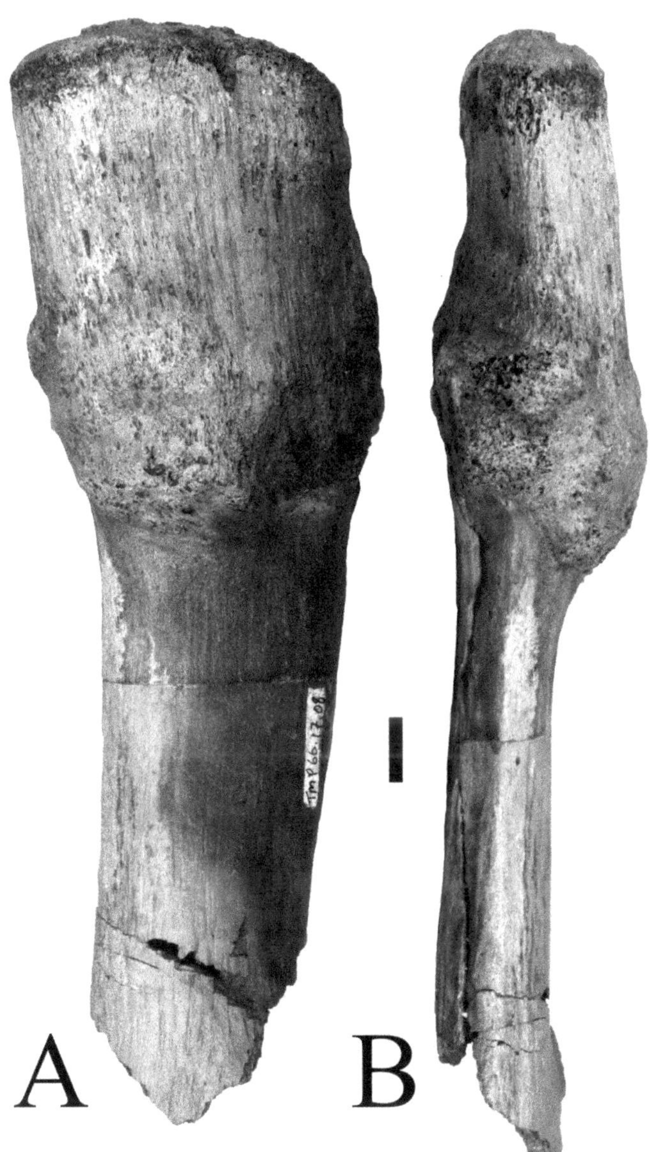

33.5. (A) TMP 1966.017.0008, proximal hadrosaur caudal neural spine in left lateral view; and (B) anterior view. Swelling marks the site of a fracture callus. Scale bar equals 1 cm.

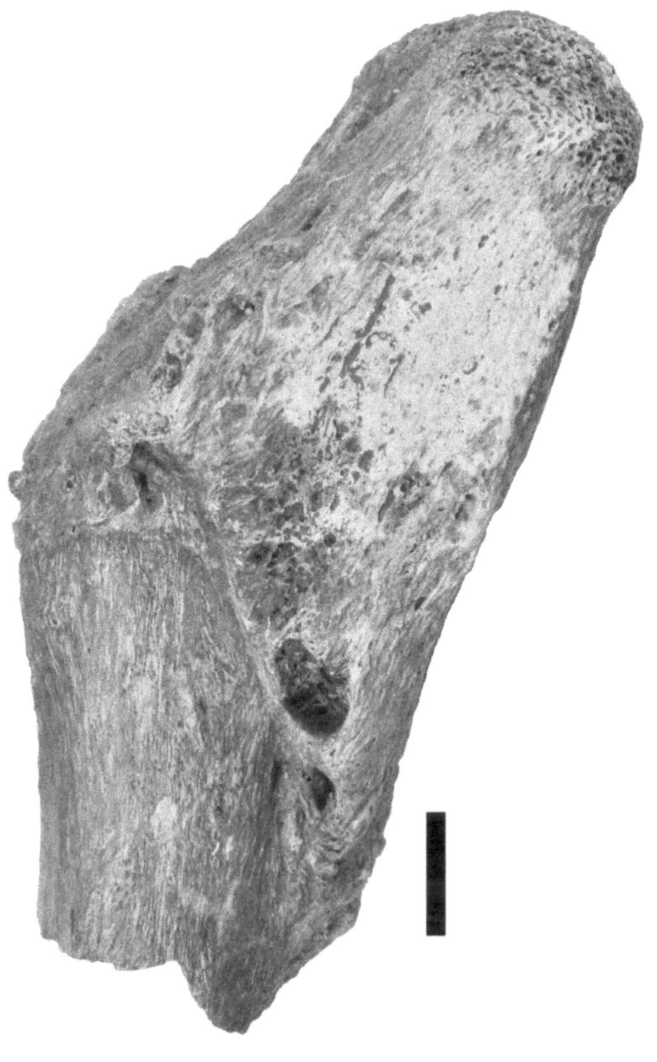

33.6. TMP 1987.036.0401, proximal caudal neural spine with fracture, healing, and angulation of tip. Semicircular depressions are vascular openings, not tooth marks. Scale bar equals 1 cm.

are well developed. Ventrally, there are well-developed chevron facets (except those closest to the sacrum, which lack chevrons). Most proximal centra have a tall trapezoidal shape when viewed from the front or back, but those closest to the sacrum are round.

Medial caudal vertebrae are those from the center of the tail. These are characterized by the absence of caudal ribs, relatively short neural spines (compared to proximal caudals) that tend to be more rounded in cross section, and more inclined posteriorly than vertically. They possess well-developed but relatively smaller chevron facets and zygapophyses compared to the proximal vertebrae, and centra have a more hexagonal shape when viewed from front or back.

Distal caudal vertebrae are those located near and at the end of the tail. These are readily identified by their small size in adults (<~5 cm overall length), more noticeable taper distally, and rounded-hexagonal to round endplates. They have a simple, spool-like morphology (more so distally), with greatly reduced to absent neural spines, relatively small zygapophyses (without facets), and reduced, or absent chevron facets. Anteroposterior orientations of medial and distal centra that lack complete neural arches was based on the presence or absence of prezygapophyses (or their bases), anteropostero tapering of the centrum, and location of chevron facets.

Proximal Caudal Vertebrae and Chevrons A proximal caudal vertebra with a "tumor" is figured in Norman and Milner (1989) and identified as coming from Alberta (Capasso, 2004). However, this specimen (NHMUK R4862)

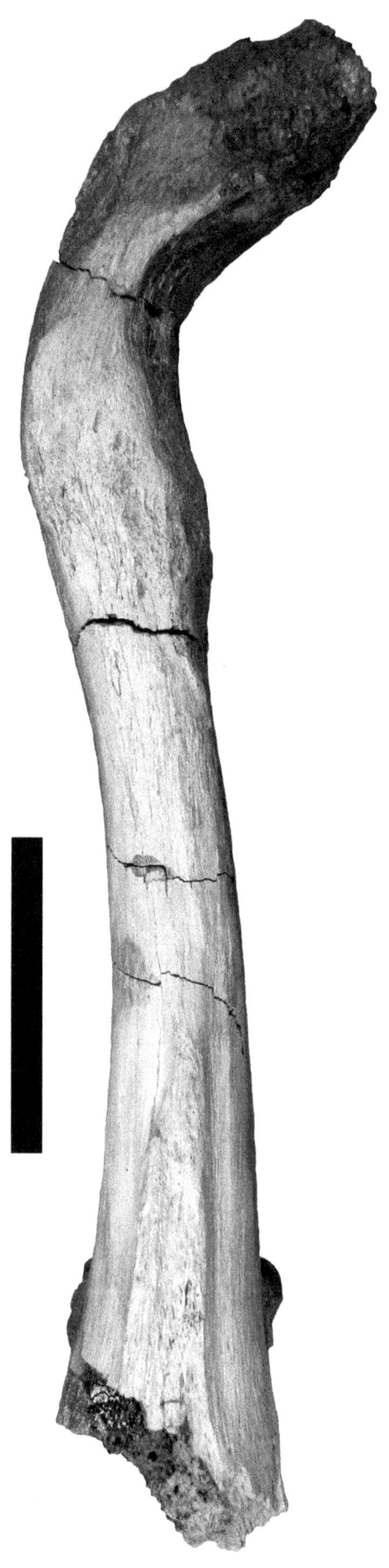

33.7. TMP 1985.047.0009, proximal neural spine in anterior or posterior view. The large elongate depression is vascular in nature and not a tooth mark. Note how the spine is angled strongly laterally. Scale bar equals 1 cm.

33.8. (right) TMP 1981.027.0051, proximal caudal neural spine in anterior view showing strong lateral angulation at tip. Scale bar equals 5 cm.

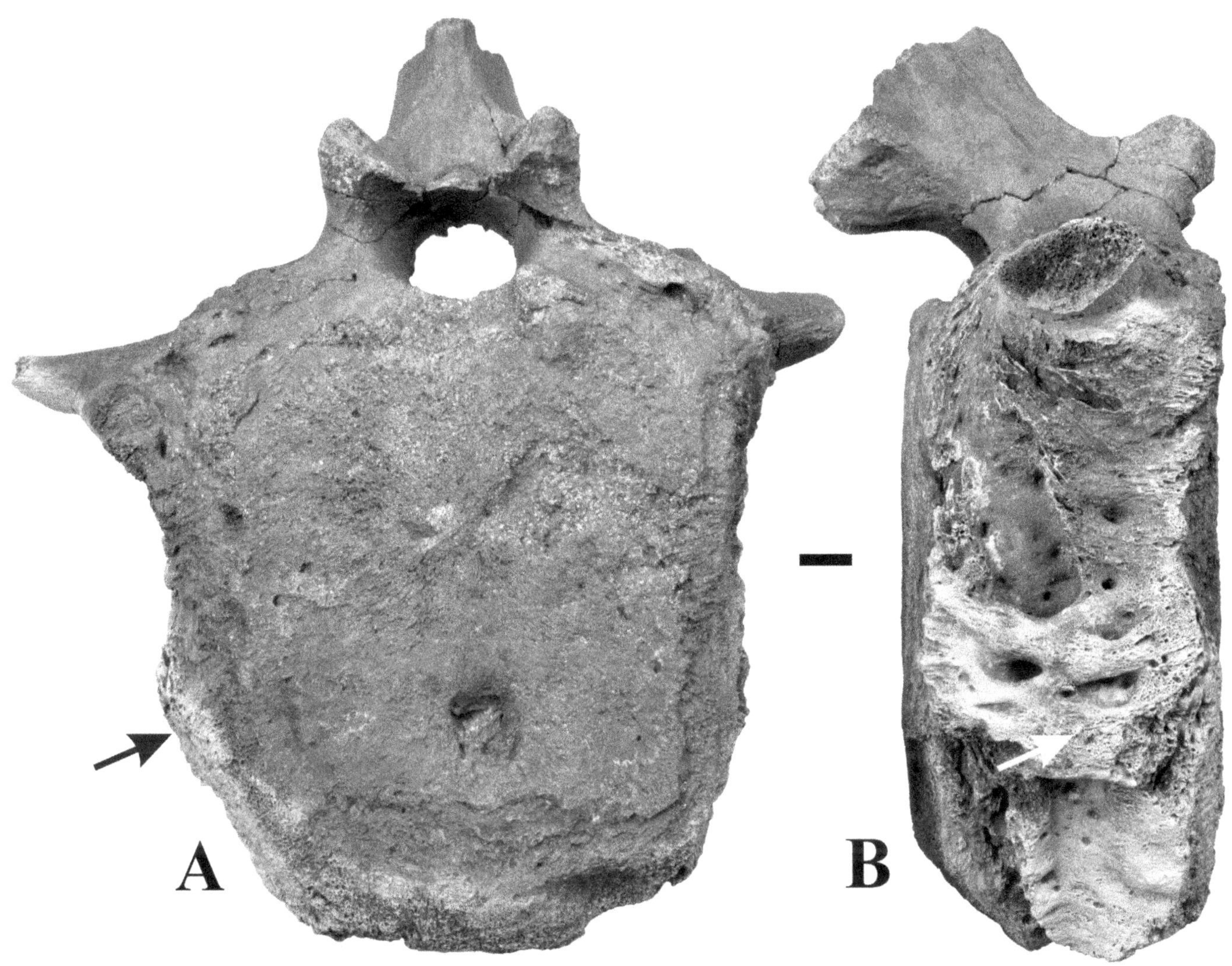

33.9. (A) TMP 1992.036.0714, partial proximal caudal vertebra in anterior view; and (B) right lateral view, showing endplate destruction and new bone growth (arrows). Scale bar equals 1 cm.

is actually from Wyoming (Chapman, pers. comm., 2012), and thus is not considered further here. One of the most common pathologies observed in hadrosaurs from DPP are healed (or healing) proximal neural spines (Figs. 33.5–33.8). Typically, the neural spine is broken once and leans laterally or forms (when viewed from front or back) sigmoidal curves (TMP 1979.014.0746). Occasionally, two healing fractures are present on one spine (TMP 1981.019.0154, 1986.036.0520, 1991.036.0394, 1993.666.0043, 1994.012.0901, 1998.059.0001; Straight et al., 2009). In specimens with two fractures, those lesions are both in the same stage of repair, indicating the fractures were incurred at the same time. Numerous examples are from isolated vertebrae or spines, but articulated specimens preserving the pathology are also known in Alberta and elsewhere (Appendix 33.1). In TMP 1987.036.0401, the distal end of the neural spine was snapped off and had slid or been pulled (via muscle contractions?) partway down the side of the injured spine, where it became fused onto the lateral surface of the spine, but at an angle (Fig. 33.6). Injured neural spines exhibit a variety of morphologies at their tips. They can appear normal, or exhibit roughened, swollen, saddle-like (TMP 1985.047.0009; Fig. 33.7), or club-like shapes (e.g., TMP 1967.008.0053, 1985.036.0024, 1993.666.0043), occasionally with deep, hollowed-out pockets. Some (e.g., TMP 1981.027.0051) exhibit extreme angulation at the tip (Fig. 33.8). The broken segment often is dislocated and fuses in an abnormal position, typically leaning to the side and usually up to 45° off the vertical plane. Fracture callus textures range from rough and fibrous (similar to undersides of ankylosaur scutes) to smooth where healing is advanced or completed. A few specimens preserving a fracture with advanced healing then a more recent fracture also undergoing repair have been seen, demonstrating that the bone was injured on separate occasions. A number of

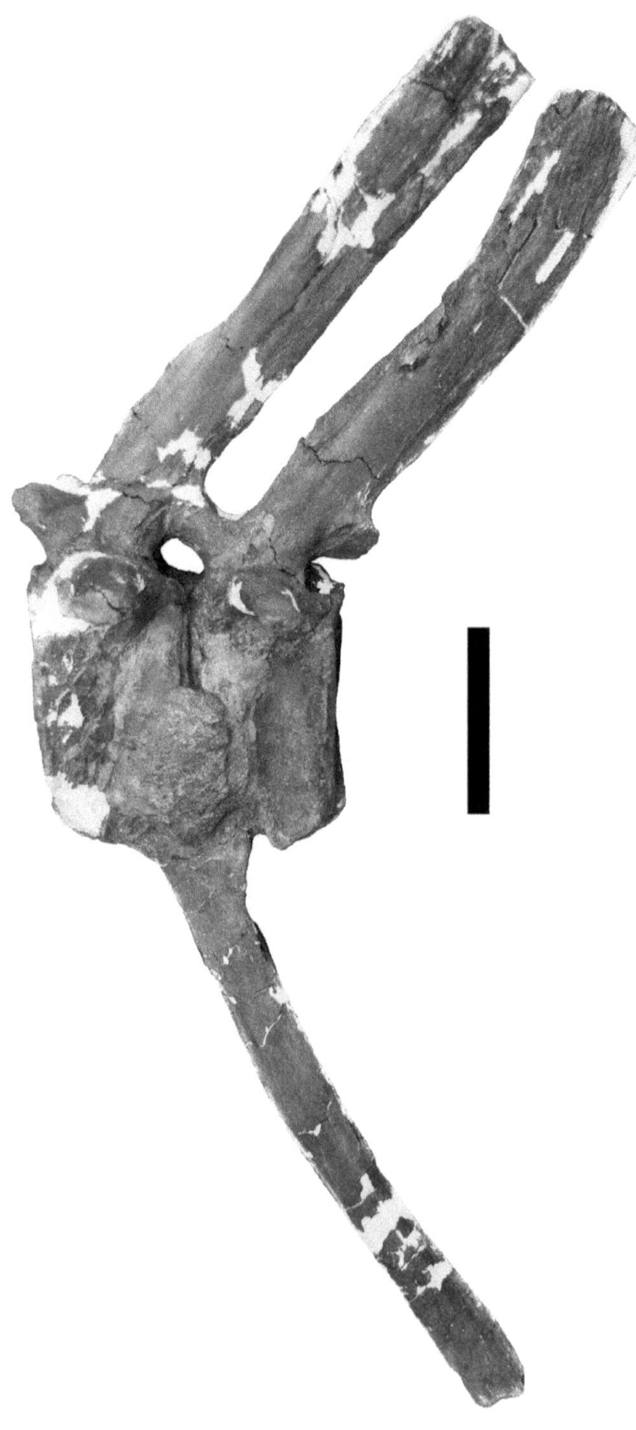

33.10. TMP 1978.004.0001, articulated and fused proximal caudal vertebrae and chevron, left lateral view. Scale bar equals 10 cm.

pseudoarthroses (non-unions of fracture components) positioned above the zygapophysis complex were observed (e.g., TMP 1991.036.0589, 1992.036.0390, and 1993.066.0030). Proximal caudal ribs protrude laterally from the vertebrae. Only one (TMP 1990.002.0033) from Sandy Point (Fig. 33.1) was injured, and was represented by an irregular rugose mass.

Chevrons are rare in the sample and rarely seen in the field, except in articulated skeletons. TMP 1994.012.0768 bears a well-healed longitudinal fracture distally. TMP 2005.012.0071, a distally positioned chevron, bears a slight swelling distally that may represent an old healed fracture.

Diffuse idiopathic skeletal hyperostosis (DISH) is known in several dinosaur families (Rothschild, 1985; Rothschild and Berman, 1991; Rothschild and Martin, 1993, 2006; Tanke and Rothschild, 2010), where two or more vertebrae fuse together through the ossification of lateral spinal ligaments, but centrum endplates and zygapophyseal articulations are uninvolved. Only one Albertan specimen, a single proximal caudal (TMP 1992.036.0714; Fig. 33.9), appears to express this condition. As it has not fused to the next vertebra, diagnosis cannot be confirmed. In dinosaurs, particularly sauropods (Rothschild and Berman, 1991), DISH is believed to help support the weight of the tail. If so, one might expect to find more examples in hadrosaurs than is the case, although these animals do have a complex latticework of tendons (which sauropods lacked) to support the tail. One of the authors (BMR), has previously hypothesized that the tendon lattice served a similar function to DISH and would render the latter redundant, perhaps explaining its rarity in those groups (Rothschild and Martin, 2006).

A curious case involves two proximal caudal vertebrae and their associated chevron from an indeterminate lambeosaurine (TMP 1978.004.0001; Fig. 33.10) collected from Sandy Point. The bones are fused together by the presence of a large, ovoid mass on the left side, 110 mm long and 70 mm wide. The lesion appears to be a syndesmophyte, diagnostic of spondyloarthropathy. More of the tail in this individual expresses healing injuries (Appendix 33.1). A similar, but highly eroded and broken specimen (TMP 1989.036.0347) has massive thickening of the endosteum. Although a tumor is possible, infection is the most likely explanation for these pathologic findings. X-ray examination should clarify the question.

Medial Caudal Vertebrae Mid-caudal osteopathy generally resembles that in proximal caudals, with occurrences of healing fractures at or close to the tip of the neural spine (although some can be more proximally located; Fig. 33.11). Pseudoarthrosis of the basal neural spine occurs regularly (TMP 1989.036.0321), with the fracture occurring between the neural arch and zygapophyses. Healing fractures (swellings) can also be found at the base of the neural arch where it surrounds the neural canal. Caudal postzygapophyseal destruction occasionally complicates infectious processes. In TMP 1973.008.0365 the neural spine base probably suffered a greenstick fracture, which subsequently healed in association with deep holes eroding into the postzygapophyses. The rough and pitted bone surface texture on the anterior edge

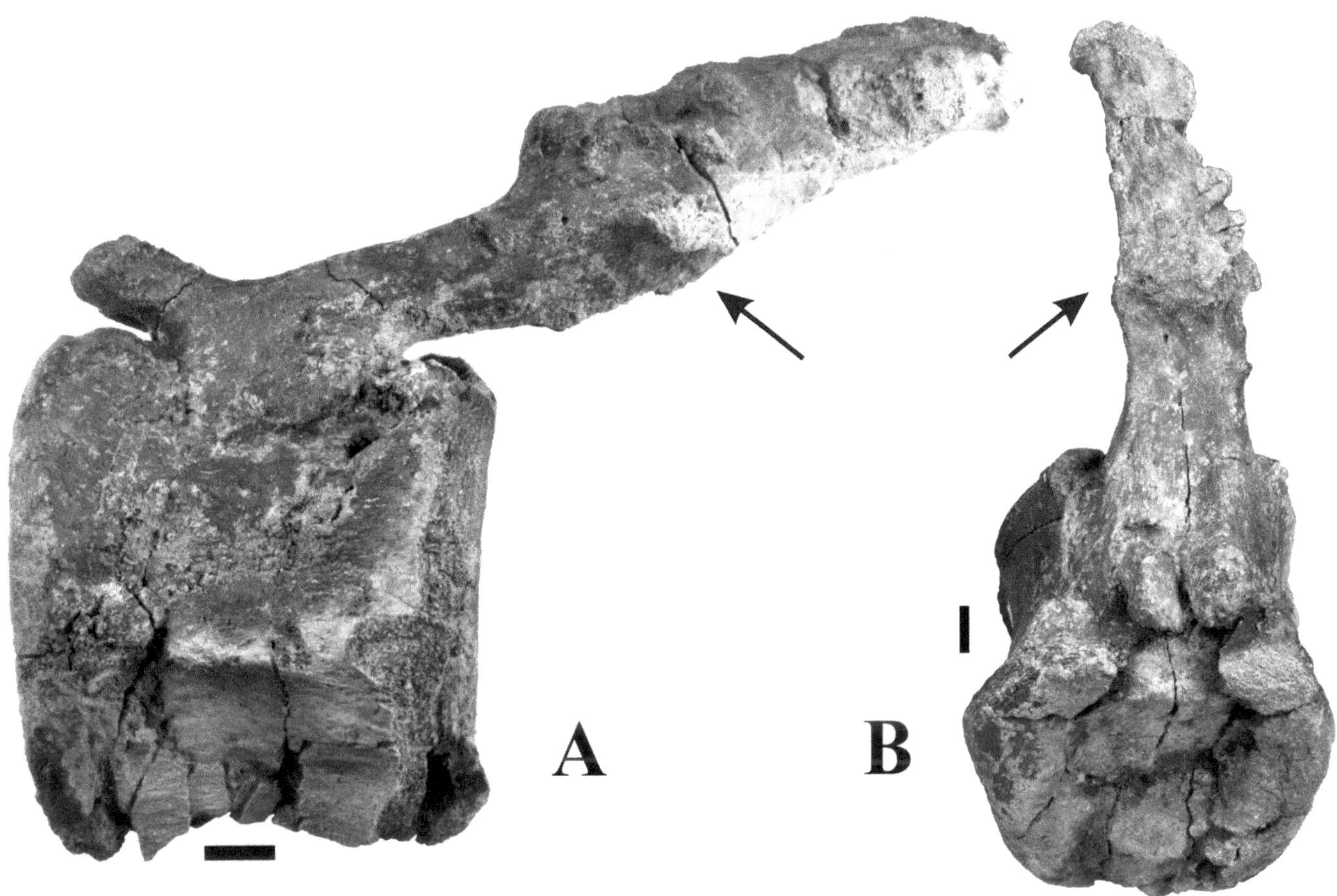

33.11. (A) TMP 2011.012.0134, mid-caudal vertebra in left lateral view; (B) anterodorsal view with poorly healed fracture (arrows) of mid-neural spine. Scale bars equals 1 cm.

and tip of a mid-caudal neural spine (TMP 1989.017.0007) was associated with infectious damage to the postzygapophyses. Curiously, the prezygapophyses do not seem affected. TMP 2011.012.0132 is unusual in that the neural spine curves laterally over much of its length for no apparent reason, although it does bear a healing fracture near the very tip.

TMP 1978.009.0064 and TMP 1980.016.0973 each consist of two completely fused caudal centra from Sandy Point. Cross-sectional examination of the former revealed normal trabeculae. There was no line of separation between the centra, although remnants of the disc space were faintly visible on the surface in a few places. The latter specimen also exhibits prominent, deep meandering cracks running dorsoventrally on both endplates. An injury early in development is suggested. These cases, while visually impressive, are not as massively swollen as in an American cf. *Edmontosaurus* specimen described by Partridge (2007).

A form of osteopathy not seen in proximal caudals is the centrum endplate with development of narrow (usually), and relatively flat bottomed cracks with rounded edges indicating repair. These cracks are associated with healing bone, suggesting, remarkably, that the centrum was split fully in two or more pieces by a traumatic event. They resemble healing near-vertical cracks in cervical vertebrae of extant soay sheep *Ovis aries*, where head-butting behavior was implicated (Clutton-Brock et al., 1990). A few hadrosaur specimens exhibit multiple cracks radiating in varying directions, indicating the centrum was more seriously injured and broken into multiple pieces; three pieces are evident in TMP 1991.050.0140. Endplate cracks occur vertically (e.g., TMP 1982.014.0003; Fig. 33.12), diagonally (Tanke and Rothschild, 2010:fig. 2), or transversely, and usually are linear. Centra injured this way were broken into two pieces, either roughly in half or into subequal portions, with one piece slightly larger than the other. In some cases only a corner of the endplate and adjoining bone were broken away. The direction of the crack is often mirrored on the opposing endplate, although single endplate cracks and curving cracks (TMP 1967.019.0031; Fig. 33.13) are known. The authors have observed medical cases in *Homo* resulting in similar centrum splitting (Fig. 33.14), involving falls resulting in spinal compression, or derived from motorcycle accidents.

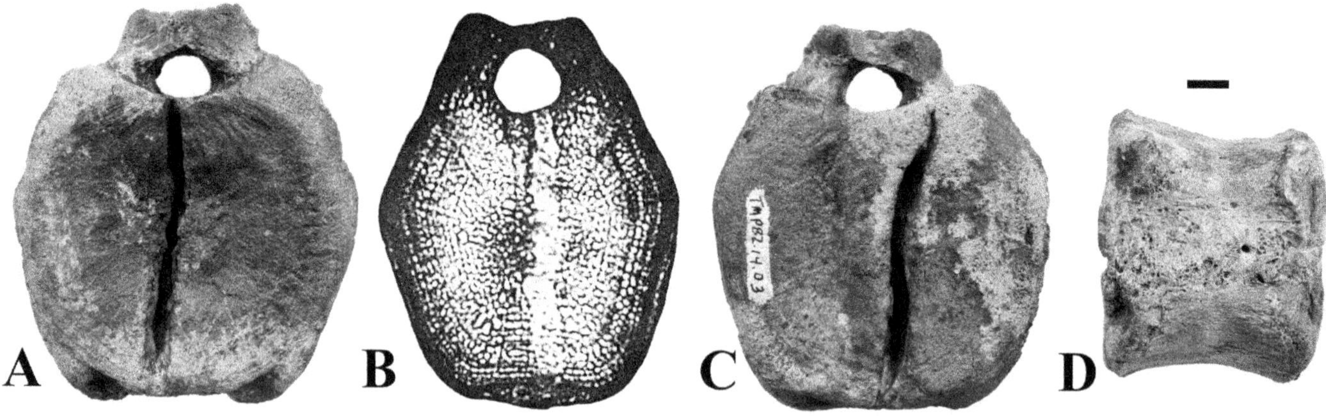

33.12. (A) TMP 1982.014.0003, mid-caudal vertebral centrum with dorsoventral healing crack in posterior view; (B) transverse CT scan view; (C) anterior view; and (D) ventral views. Scale bar equals 1 cm.

Deep, wide, and tapering cracks with healing but not infilled with reparative bone have also been observed (Fig. 33.15). CT scans, taken in transverse (Fig. 33.12B) and coronal planes through afflicted centra show the internal continuation of disrupted bone beyond the endplate cracking, confirming a traumatic injury broke the centrum into two or more segments, which were in the process of healing when the individual perished. Most specimens healed back together in proper alignment, but in TMP 1991.036.0256, one half of the centrum shifted longitudinally several millimeters before healing was achieved.

Gangloff and Fiorillo (2010:fig. 14A) figured (upside down) a subadult Alaskan *Edmontosaurus* mid-caudal centrum with two short and curved opposing shallow endplate depressions. They attribute these depressions to osteochondrosis. Endplate depressions of this morphology are common on DPP hadrosaur medial and distal caudal vertebrae, and some bear a marked resemblance to centrum lesions in *Homo* attributed to burst fractures, resulting from vertical compression that ruptures the intervertebral disc through the vertebral endplate, and forces disc tissue into the vertebral body (Lovell, 1997:143, fig. 3; Lovell, 2008:355, fig. 11.8). There are so many Albertan examples of this condition that it raises questions of whether or not they are related to ontogeny, normal anatomy, or represent true osteopathy. Confusing the issue are vertebrae with these depressions and endplate cracking (TMP 1990.117.0009; Fig. 33.16), suggesting multiple origins (developmental anomalies and/or trauma) of these unusual lesions. In TMP 1990.117.0009 the lateral quarter of the centrum appears to have spalled off and fused back on, the crack following the same lines as the osteochondrosis lesion. Possibly, these lesions or developmental artifacts created plane(s) of weakness that allowed the vertebra to break apart along these lines when traumatized (see below). Some rare specimens (e.g., subadult mid-caudal TMP 1980.008.0321 [Fig. 33.17A], 2011.012.0418 [Fig. 33.17B]) have depressed defects that penetrate so deeply they fully pierce the centrum from one endplate to the other. The former appears possibly to be developmental in origin, whereas the latter somewhat resembles insect boring damage to dinosaur bones reported from a number of western North American and overseas localities (e.g., Rogers, 1992; Roberts et al., 2007; Kirkland and Bader, 2010; Saneyoshi et al., 2011). However, Albertan dinosaur bones with possible insect boring traces are extremely rare, and we are aware of only a few possible examples: a distal hadrosaur humerus (TMP 1993.049.0001),

33.13. TMP 1967.019.0031, distal caudal vertebral centrum preserving broadly curving inverted crack (arrows) affecting endplate. Scale bar equals 1 cm.

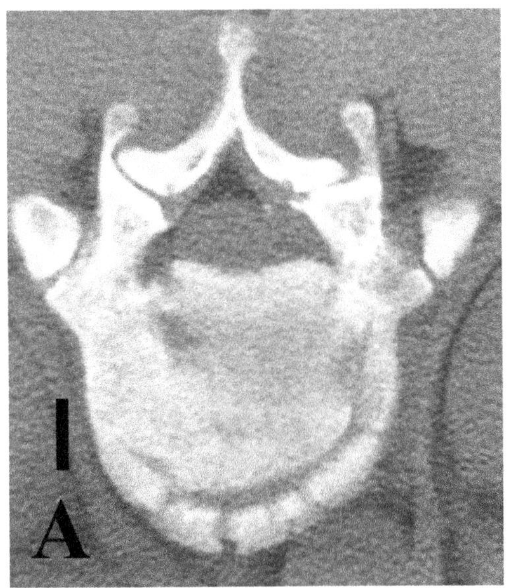

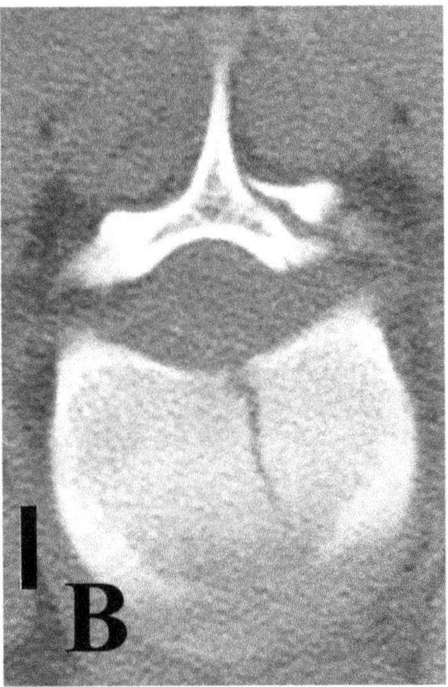

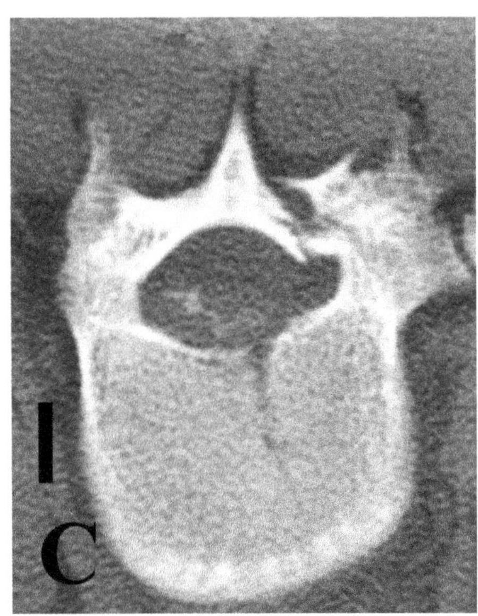

33.14. (A–C) Young adult male *Homo* thoracic vertebrae with fracture trauma related to a fall resulting in compression fractures of the centra and unilateral fracture of the neural arch (B–C). Scale bars equal 1 cm.

a small distal caudal centrum (TMP 1967.009.0145), and a hadrosaur ilium fragment (TMP 1966.017.0005). Complicating the matter of open-ended centrum cracks is TMP 1993.036.0059 (Fig. 33.17C), which appears trauma related. The deep crack through the bilaterally distorted centrum presents finished bone with rounded edges indicating that some split centra did not fully fuse back together, leaving the crack open. Splitting of centra through the neural canal must have resulted in spinal cord damage and possibly paralysis, but was not detrimental as the specimens demonstrate at least some individuals lived an extended period of time after being injured.

Some non-pathologic mid-caudal centra have a normally smooth and low longitudinal ridge at or near mid-height on the side of the centrum. In pathological specimens of this morphotype, this ridge may become more prominent, forming a keel with the development of associated osteophytes occurring uni- or bilaterally on, or dorsal to, the keel; osteophytes ventral to the keel have only rarely been observed.

Other endplate anomalies include circular, symmetrically placed, low, rounded lesions (TMP 1979.008.0096), and a strong unilateral keel (TMP 1981.041.0139). In the latter, articulation with the preceding vertebra must have been seriously compromised, yet no eburnation was present.

Distal Caudal Vertebrae Distal caudal vertebrae exhibit the widest variety of injuries and examples are common. These vertebrae often exhibit massive comminuted crush fracture trauma. Some of the more serious cases suggest spinal cord damage, and extreme distal tail infection with amputations or sloughing of the distal segment (see below).

33.15. (A) TMP 1992.036.0300, distal caudal centrum in left lateral; (B) anterior; and (C) right lateral views showing a deep penetrating crack across centrum face. The posterior endplate appears normal. Scale bar equals 1 cm.

33.16. TMP 1990.117.0009, distal caudal centrum with deep curving crack; trauma predisposed to related osteochondrosis lesion? Scale bar equals 1 cm.

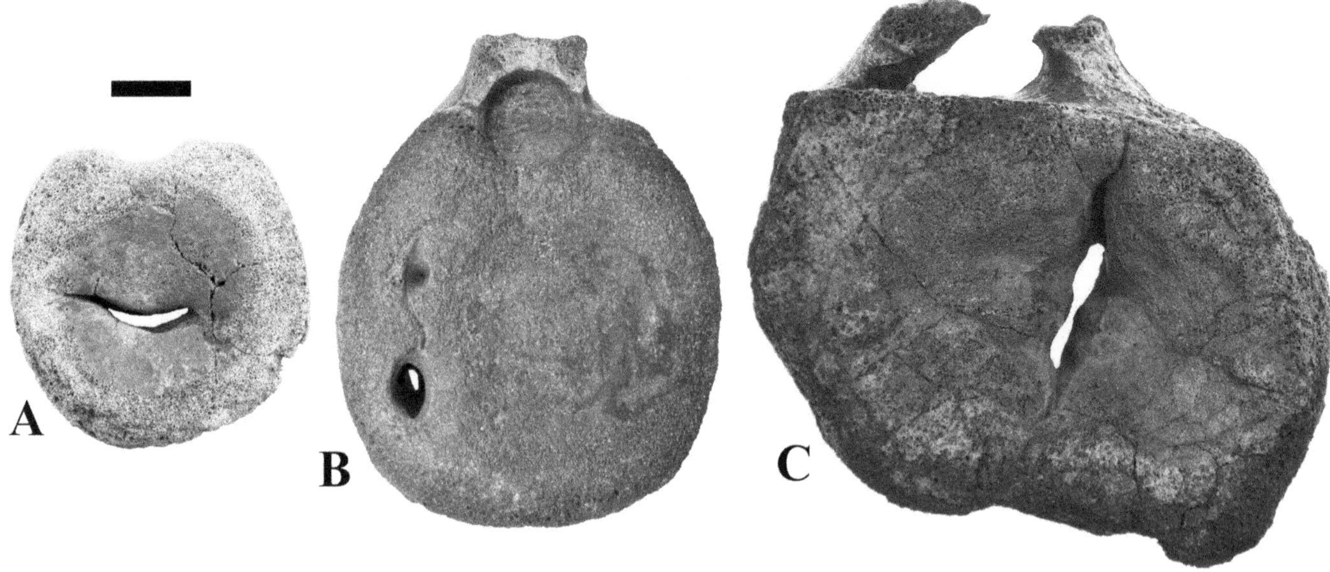

33.17. (A) TMP 1980.008.0321; (B) TMP 2011.012.0418; and (C) TMP 1993.036.0059, caudal centra with fully penetrating crack defects. The bilateral asymmetry in (C) is related to trauma. Scale bar equals 1 cm.

Centra can express some or nearly all of the conditions summarized as follows:

1. The crack (cleft) phenomenon seen in medial caudals with a visible healing longitudinal crack on the neural canal floor in some specimens (TMP 1966.030.0005, 1993.036.0285, 1993.150.0005), and often associated with a longitudinal smooth and rounded ridge on ventral midline representing a remodeled fracture callus. Some cracks run longitudinally through a neural arch base (TMP 1980.016.0225). Lesions on endplates, ventral centrum, neural arch, and canal floor, while of different appearance/texture, are connected when examined closely. Endplate cracks may be vertical, transverse, or diagonal. The angle or direction of the defect is often mirrored on the opposing endplate (complete fracture), although single endplate cracks (incomplete fracture) are known. Centra are split roughly in half bilaterally, but in some specimens, only an edge is affected.
2. An overall swollen or lumpy condition, with the cortical bone perforated by numerous small openings giving it a spongy appearance (osteomyelitis). Some specimens are so deformed as to no longer resemble a vertebra, although the endplates are still present, often with centrally positioned notochordal projections (Figs. 33.18, 33.23A, C). Some centra have thin and finely textured patches of periosteal bone, which flakes off easily and reveals normal bone underneath.
3. When viewed dorsally, the normally straight centrum may be slightly curved with both endplates curving laterally in the same direction.
4. Fusions of adjoining vertebrae, sometimes at angles, demonstrate that the tail healed with a noticeable upward or lateral kink similar to that described in *Majungasaurus* from Madagascar (Farke and O'Connor, 2007). Tanke and Rothschild (2011:146) noted fusions of up to four vertebrae

33.18. TMP 1980.016.0162, two fused distal caudal vertebrae in left lateral view, showing slight angulation, osteomyelitis, and pointed notochordal projection on anterior centrum (arrow). Scale bar equals 1 cm.

33.19. (A) TMP 2011.012.0374, five fused distal caudal vertebrae with osteomyelitis in left lateral view; and (B) ventral view. Scale bar equals 1 cm.

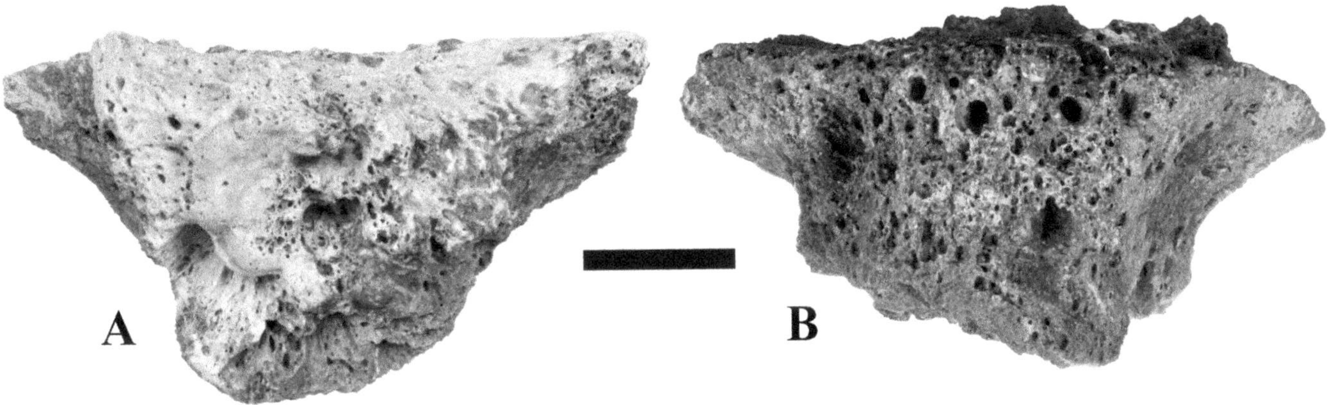

33.20. (A) TMP 2011.012.0417, badly deformed and osteomyelitic distal caudal vertebra in ?anterior view; and (B) ventral view. Scale bar equals 1 cm.

(but typically two). TMP 2011.012.0374 (Fig. 33.19), from the extreme tip of the tail, exhibits five fused vertebrae with the fusions occurring along the vertical midline of each centrum. The specimen superficially resembles a dinosaur pygostyle (Barsbold et al., 2000) but is truly pathological. Other vertebrae fuse completely, although lines of demarcation between the centra are still defined in some specimens. Other osteopathic fusions of tail tips are present in TMP 1992.036.0136 (two vertebrae) and TMP 1997.012.0160 (three vertebrae). Fused or badly infected specimens, such as TMP 2011.012.0417 (Fig. 33.20), may show evidence of reworking, and may be so abraded and waterworn prior to burial that they resemble simple amorphous lumps of bone with little symmetry. In such cases, CT scanning reveals greatly thickened trabecular bone but no centrum.

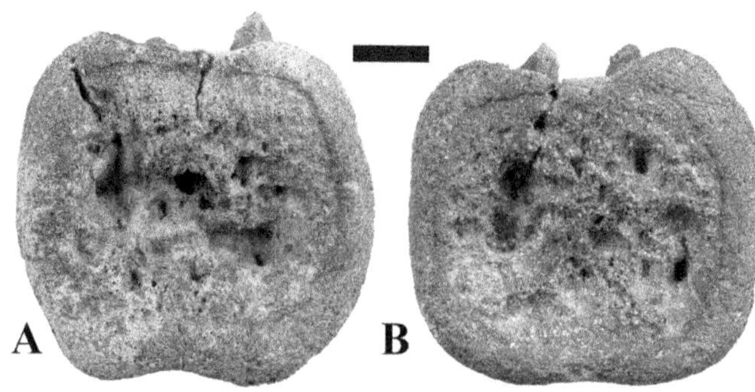

33.21. TMP 1993.036.0062, distal caudal centrum with endplate erosions. Scale bar equals 1 cm.

5. Moderately to deeply concave endplate(s) with numerous small pitted subcircular erosions (Fig. 33.21), and/or centrally positioned pointed notochordal projections (Figs. 33.18 and 33.23A, C). Some large endplate erosions can penetrate deeply inside the centrum and expand internally (TMP 2011.012.0420; Fig. 33.22). Endplates may also exhibit asymmetrical bulges.
6. Osteophytes that occur singly, in small groups, or sometimes in spectacularly crowded and radiating spicular growths that entirely encircle the vertebra (endplates excepted) and greatly increase its diameter (TMP 1991.036.0409, 2006.012.0048, 2008.012.0066; Fig. 33.23). These suggest rampant osteomyelitis. Small singular osteophytes sometimes have an accessory blood vessel opening onto the bone surface.
7. Anteriorly positioned neural arch base (TMP 1965.030.0007, 1992.030.0181, 1994.012.0791, 1994.012.0697; Fig. 33.24), visible in broken specimens suggesting posterior elongation of the centrum during healing. Some specimens suggest posterior and anterior centrum elongation after trauma.
8. Bilateral or coronal asymmetry of endplate(s). Asymmetry in fossils can also result from burial compaction and deformation (Ross, 1978; White, 2003; Forrest, 2004; Eberth and Currie, 2005; Boyd and Motani, 2008). However, deformed vertebrae and phalanges of hadrosaurs from the Park always demonstrate true pathological conditions upon close inspection, and therefore we interpret them as being asymmetrical due to biological responses to injury.
9. Attenuation of endplate margins with a low and sharp-edged encircling ridge (TMP 1966.030.0005). This specimen and others (TMP 1966.011.0050) preserve a marked color difference between endplates (gray) and centrum body (off-white), the significance of which is unknown. Tanke and Currie (2000) also noted the different colors of face bite lesions versus normal bone in tyrannosaurids. Sawyer and Erickson (1998) noted similar color differences in pathologies affecting Paleocene *Leidyosuchus* crocodiles. DPP baenid turtle shells with possible fungal infections (e.g., Hutchinson and Frye, 2001) also have color differences with light gray to white-colored lesions on tan to dark gray normal-colored bone.
10. Hollowed out bone. This feature suggests pyogenic osteomyelitis. An isolated, extreme distal centrum (TMP 2011.012.0417; Fig. 33.25), with a hollowed interior indicative of osteolysis, is missing half its length, with the "broken" edge exhibiting signs of rounding and repair, suggesting the very tip of the tail was lost premortem. TMP 2011.012.0374 (Fig. 33.19) may also be missing part of the posterior-most centrum.
11. Hypertrophic development of normal anatomical features, such as muscle attachment points, dorsoventrally arranged ridging on endplates, and chevron facet margins.
12. Strong lateral compression of vertebrae. When viewed end on, the normal round centrum outline is flat sided and more rectangular (TMP 1992.036.1168, 1994.012.0900).
13. Asymmetrical anteroposterior compression of centra. Viewed dorsally, one side of the centrum is longer than the other. In TMP 2011.012.0291 opposite sides are 51 mm and 42 mm long; in TMP 1996.666.0009 (Fig. 33.26) opposite sides are 61 and 39 mm long. Although these specimens could be misinterpreted as resulting from scoliosis-caused spinal curvature, the short side of TMP 2011.012.0291 exhibits bilateral anomalies and a ventral ridge suggestive of trauma. Similarly, TMP 1996.666.0009 exhibits endplate cracking indicative of trauma.
14. Some rare specimens may exhibit dense smooth-surfaced cortical bone with a high degree of polish (TMP 1973.023.0011), similar to that seen in contemporaneous saurischian bone. The significance of this change is currently not understood.
15. Some centra exhibit a small hairline cleft or slit on the midline that occupies the upper one-quarter to one-half of the endplate; the cleft widens slightly as it nears the neural canal (Fig. 33.27A, B). These can occur on one or both endplates. They seem to be developmental in nature and resemble "Type

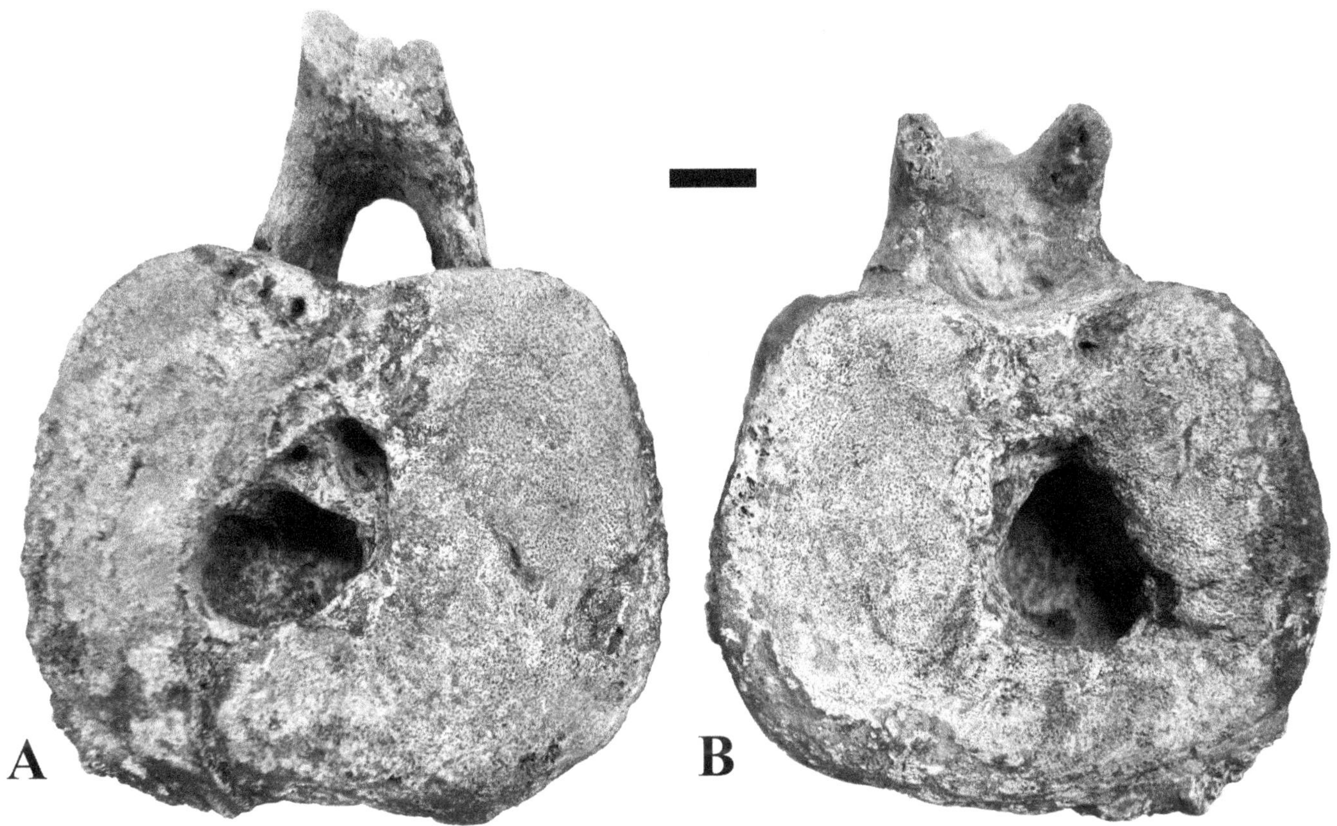

33.22. (A) TMP 2011.012.0420, associated distal caudal vertebrae in posterior view; and (B) anterior view, with deep endplate erosions. Found ex situ ~1 m apart at the same site, these two vertebrae may have formed an articulated pair with the pathological endplates face to face. Scale bar equals 1 cm.

33.23. (A) TMP 2008.012.0066, distal caudal centrum wholly encased in osteomyelitis forming crowded radiating spicular growths in ?anterior; (B) lateral; and (C) ?posterior views. The bases of notochordal projections are seen on both endplates. Scale bar equals 1 cm.

2" articular surface lesions in cattle phalanges from European archaeological sites where their presence was suggested as being a "benign developmental condition" (Thomas and Johannsen, 2011:53). Endplate cracking due to trauma (Fig. 33.27C) suggests several etiologies for the cleft phenomenon. Complicating the interpretation are other specimens where these clefts are absent and, instead, a low and very narrow (~0.25 mm) osseous ridge occurs (TMP 1988.036.0185, 1990.036.0193).

33.24. TMP 1965.030.0007, distal caudal centrum, right lateral view, with anteriorly placed (to the right) neural arch. Scale bar equals 1 cm.

16. Broken specimens may exhibit trabecular bone with thicker walls compared to specimens that are normal.
17. Elongate centrum; extends anteriorly and/or posteriorly (more common).
18. Extra blood vessel openings at/near basal walls of neural canal, and/or on the floor of the neural canal.
19. Presence of elongate, low and sharply defined ridges on sides of centrum, ~1 mm wide and 1 mm high. These ridges resemble the lines of fracture callus seen on neural arch bases and are here interpreted as fracture lines as well. Such ridges can run longitudinally along the sides of the centrum (TMP 1992.036.1167), at angles (TMP 1993.036.0060), or even with the ridges intersecting, forming an X shape (TMP 1987.036.0221; Fig. 33.28).
20. TMP 1971.027.0002, 1987.036.0162 (Fig. 33.29), and 2005.012.0046 are examples of non-traumatized vertebrae with a thickened ring of new bone that encircles the centrum. The endplates are normal on one side but pitted and grooved on the other. These specimens have some attributes of early stage DISH but have failed to fuse to the next vertebra. Perhaps these are examples of infectious fusion that can mimic DISH.

Neural arches of distal caudal vertebrae may exhibit pathological conditions as follows:

1. Mangled appearance, not resembling normal morphology.
2. Positioned more anteriorly than normal, suggestive of post-traumatic posterior elongation of the centrum (Fig. 33.24).
3. A single raised fracture callus line or ridge may occur along the neural arch base, just dorsal to where it fuses with the centrum.
4. The neural arch base may be offset, with one side more anteriorly placed than the other. The thickness of the bone forming the basal walls of the neural canal may be thicker on one side than the other, sometimes markedly so.
5. Noticeable thinning of the neural arch bone immediately surrounding the neural canal. Alternatively, in cases of serious osteomyelitis, the neural arch walls surrounding the neural canal can be greatly thickened.
6. Non-union (pseudoarthrosis) of bone at or near base of neural spine (TMP 1989.036.0321, 1991.036.0301). In TMP 1993.036.0601, which consists of 3.5 distalmost caudal vertebrae; non-union is exhibited through the most proximal centrum in a dorsoventral plane.
7. Tall neural arches due to hypertrophic dorsally oriented growth of the walls of bone lateral to the spinal cord (Figs. 33.27, 33.29B; Godefroit et al., 2012:fig. 17E).

Osteopathy in Articulated or Associated Tails The caudal vertebrae examples described above are interesting cases, but tell us nothing about how other vertebrae within the same series also may have been affected. Examples of multiple caudal vertebrae osteopathy are summarized in Appendix 33.1. Here we describe two specimens from the Park that each exhibit serious caudal trauma and disease response across a series of vertebrae.

The first, TMP 1980.023.0002, is an unidentified in situ adult hadrosaur skeleton on public display at the Park. It includes a series of fused vertebrae at the end of the tail that Langston (1961:9) suggested formed a "solid bar of bone." Nine vertebral units (single or paired fused vertebrae) were removed and later prepared; eight of them preserve progressively more serious osteopathy distally, suggestive of premortem loss of the tail tip. The first two vertebrae recovered were articulated to the rest of the tail and in situ, but the rest of the distal tail is disarticulated in situ (partially prepared) or loose on the surface. Some caudal neural arches were lost after being uncovered. For purposes of description, the vertebral units in this series are labelled A–I in the caudal direction. Although we reassociated the vertebrae using the

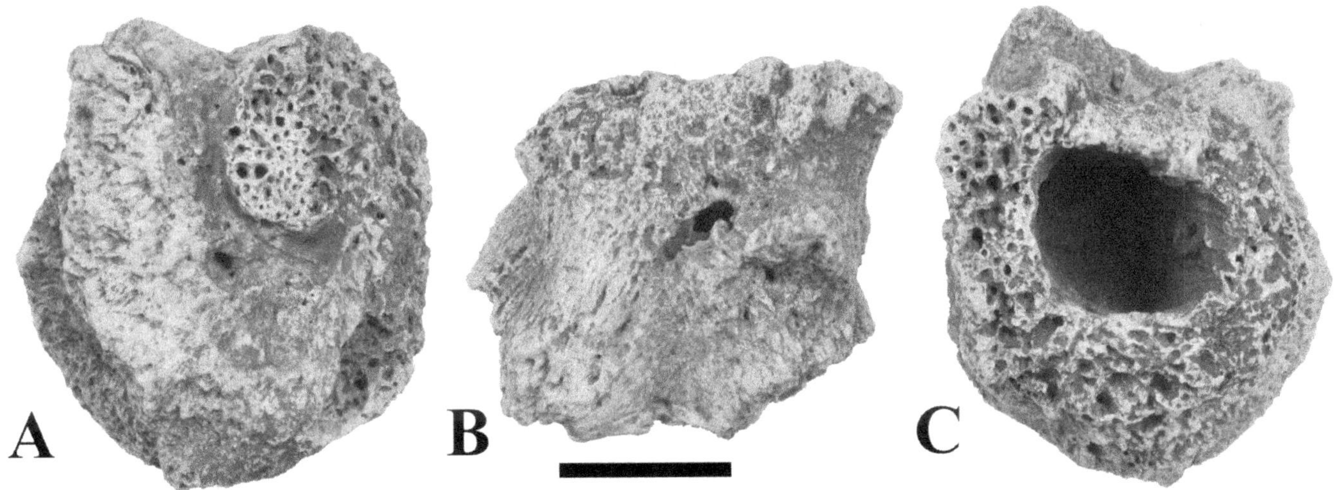

33.25. (A) TMP 2011.012.0417, distal caudal centrum in presumed proximal view; (B) dorsal view; and (C) distal view. The centrum is hollowed out internally and with rounded distal end suggestive of pyogenic infection and distal tail loss. The small foramen visible in the center of (B) connects to the cavity inside the centrum. Scale bar equals 1 cm.

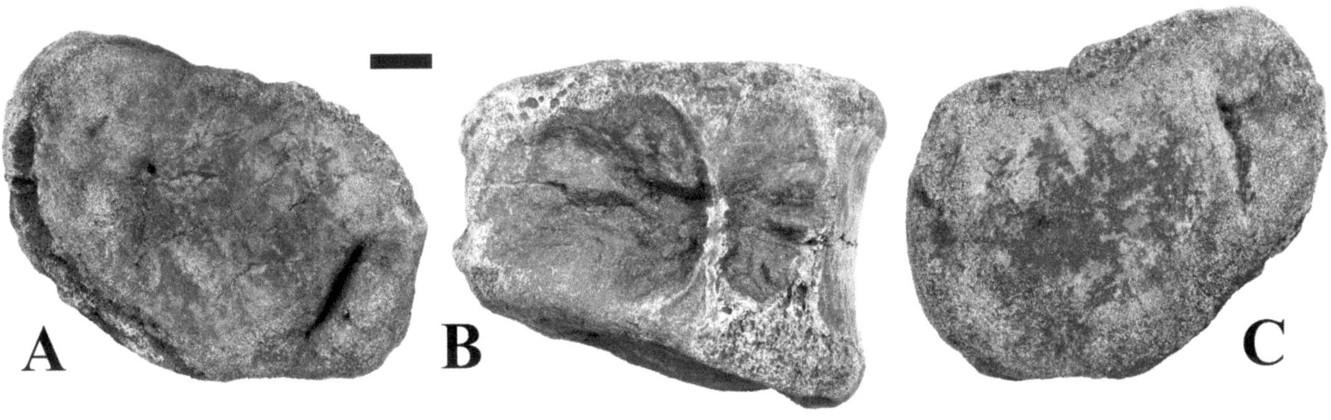

33.26. (A) TMP 1996.666.0009, compressed distal caudal centrum in ?anterior; (B) dorsal view; and (C) ?posterior view. Scale bar equals 1 cm.

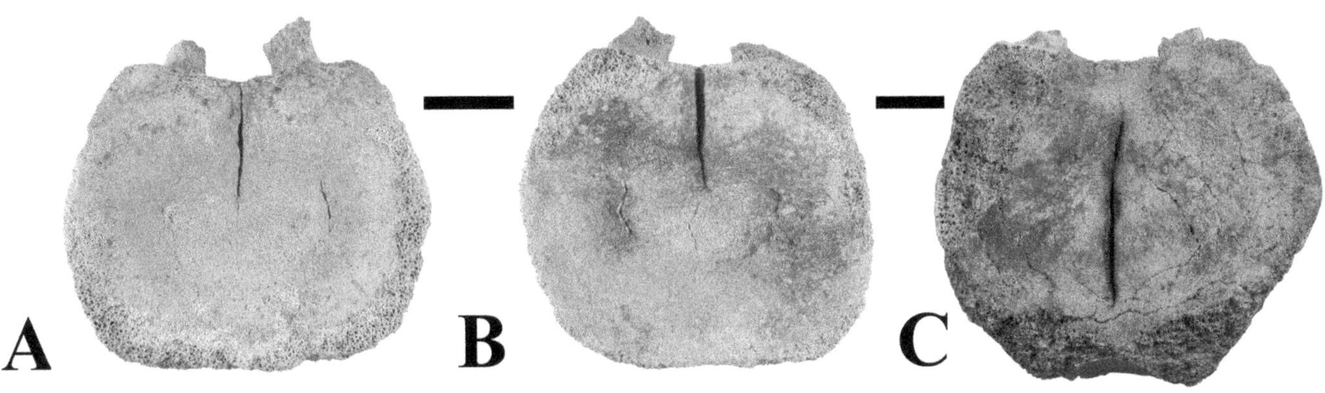

33.27. Distal caudal centra exhibiting vertical midline cracks on endplates; varying etiologies. (A) TMP 1993.036.0305 in ?anterior view; (B) ?posterior view (developmental); (C) TMP 1994.012.0509 (trauma). Scale bar equals 1 cm.

33.28. TMP 1987.036.0221, two fused distal caudal vertebrae in right lateral view exhibiting raised ridges (arrows) on sides of centra bodies. Ridges on the right centrum intersect, forming an X shape. Scale bar equals 1 cm.

relative sizes and lengths of the centra (Fig. 33.30), there is some uncertainty in the reconstruction as indicated by the possibility that there may be a vertebra missing between C and D. Vertebra A is normal and non-pathologic. Vertebrae B and C both preserve longitudinal fracture calluses on both sides of and near the base of the neural arch; the calluses are more prominent in C. Vertebra D has low and mildly rugose posterodorsal ridges on both sides of the neural arch. The ridges start at the base of the arch and extend up along the sides of the neural spine near its base. Vertebra E is missing the neural arch, but preserves offset neural arch bases, with the right side more anteriorly placed than the left. Centrum E bears a wrinkled bone texture on the floor of the neural canal and a low longitudinal ridge along the ventral midline. The center region of the posterior endplate is finely pitted. Vertebra F bears the same posterodorsal ridges seen in D, as well as a low ventral midline ridge, and a dorsally elevated neural arch (cf. Fig. 33.28), which is positioned anteriorly. Vertebrae G–I are all badly malformed, some are fused, and all exhibit cortical bone textures indicative of osteomyelitis. G consists of two fused vertebrae, although atypically, they appear fused at the neural arches, not the centra, whose degree of fusion (if any) cannot be determined because of the presence of matrix. G bears a deep circular depression on the anterior endplate. The neural arch is shifted slightly anteriorly. The anterior centrum is markedly swollen and twisted slightly to the right, and has a short, low longitudinal ridge along the ventral midline. The base of the left neural arch in the posteriormost of the vertebral pair appears to have broken away from its base at the time of trauma and is angled anteriorly where it is fused to the malformed neural arch of the preceding vertebra. The posterior centrum is twisted to the left; so despite both vertebrae in the fused pair being asymmetrically twisted in opposing directions, the pair are in a relatively straight line. The distal endplate is anteroposteriorly beveled, malformed, and concave. Vertebral unit H is difficult to interpret and may represent one or two vertebrae. The centrum is swollen and exhibits osteomyelitis. A strongly developed dorsoventral ridge is present on the left side. The neural arch is hyperdeveloped dorsally with oversized prezygapophyses, especially on the right side. The beveled distal endplate of G results in the articulation surface of G and H curving to the left at a strong angle. The distal articular end of unit H is anteroposteriorly beveled. Vertebral unit I appears to consist of two fused centra, the neural arches of which are now lost. The centra are swollen, exhibit osteomyelitis, and – like vertebral unit H – bear a dorsoventral keel on the left side, but with a shallow, smooth-bottomed groove along the keel's edge, probably marking the line of separation between the two centra. There is a healing crack on the proximal endplate that is oriented roughly dorsoventrally. As noted, the neural arches are missing, but sutural evidence indicates that the two arch bases on the right side were co-joined; but only one is present on the left side. Units H and I articulate at an angle, adding more to the slight curvature at the tip of the tail of this individual. The posterior endplate is malformed, with a deep pit bisected by a low, dorsoventrally oriented ridge. The last vertebra appears truncated, suggesting that possibly 30 cm of the tail tip was lost premortem.

Our examination of this preserved tail indicates that the pathologies appear to be the result of serious post-trauma infections related to crush fractures, most likely due to trampling. Crushing seriously damaged the surrounding skin and muscles, exposing the vertebrae to infection. Although it is possible that premortem tail wounding and/or amputation may have resulted from predation (e.g., Plotz and Linklater, 2011; Butler et al., 2013), there is no evidence of a cartilaginous regenerated section as seen in a fossil crocodilian with a premortem tail tip amputation (Buffetaut, 1985). Premortem tail amputations in dinosaurs have been reported elsewhere (Farke and O'Connor, 2007; Butler et al., 2013) so occurrences in hadrosaurs is perhaps not unexpected. TMP 1980.023.0002 is interesting in that it exhibits a variety of paleopathologies and reparative bone responses over a relatively short distance of tail.

The second specimen that exhibits a series of pathological caudal vertebrae (TMP 2011.012.0420) is from an

33.29. (A) TMP 1987.036.0162, distal caudal centrum in distal; (B) right lateral; and (C) proximal views. Scale bar equals 1 cm.

33.30. (A) TMP 1980.023.0002, reconstructed pathologic distal tail in dorsal view; and (B) left lateral view. Anterior is to the left. Scale bar equals 10 cm.

uncollected hadrosaur skeleton, possibly lambeosaurine, at DPP (UAD 91). Eleven disarticulated but associated distal caudal vertebrae were collected on the surface. Of these, 8 express pathological conditions including fracture calluses at the base of a neural arch and a bulging endplate, but 6 preserve large cavernous erosions that penetrate and expand deeply into the centra (Fig. 33.22). These unusual lesions have not been recognized previously among the thousands of pathological hadrosaur centra available to us, and appear to represent a type of osteopathy new to hadrosaurs at DPP; and recently observed in a specimen from Montana (Freedman et al., 2012).

Ribs

Rib fractures with well-aligned healing or with signs of infection are known from various Albertan sites and occur in all non-neonate individuals, especially adults. Single fractures are typical, although two affecting the same rib are known (TMP 1986.036.0121, 1989.017.0004 [cf. *Edmontosaurus*]). *Prosaurolophus* (ROM 787) has a series of six well-healed lower rib fractures on the right side. Parks (1935:plate III) illustrated a healed mid-dorsal rib fracture in an *Edmontosaurus* (ROM 807) collected near Drumheller. *Parasaurolophus* (ROM 786) has three healed rib fractures just superior to the left

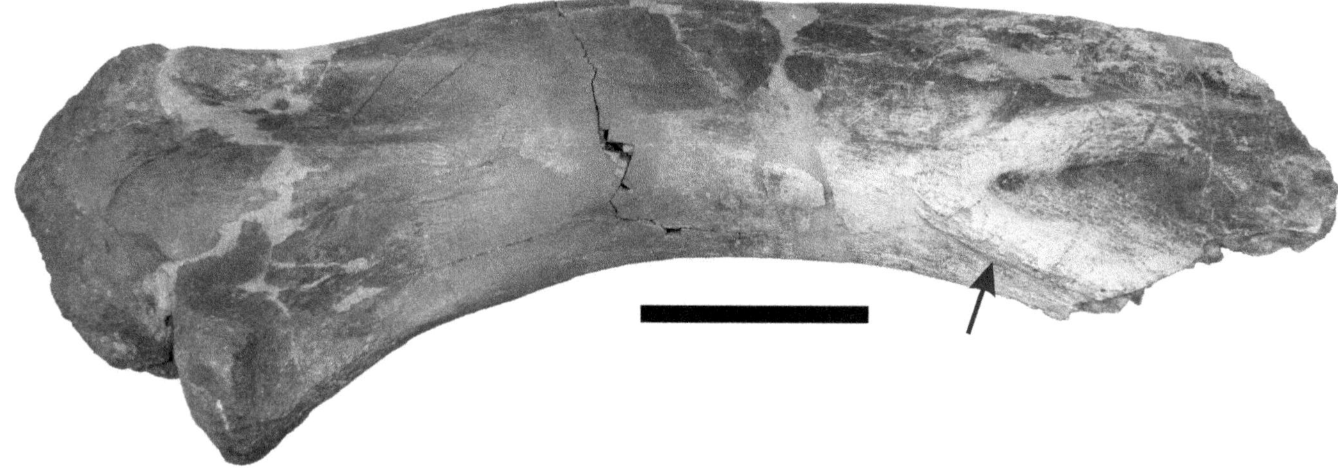

33.31. TMP 1966.031.0084; left scapula, external view, preserving a raised C-shaped lesion (arrow) on posterior blade. Scale bar equals 10 cm.

scapula (Parks, 1922). The injury appears to have affected the associated vertebrae, as the neural spines tips are angled together and fused. A disarticulated cf. *Parasaurolophus* (TMP 1992.053.0021), includes 19 dorsal ribs, 3 of which preserve well-healed breaks. Two are from the right side of the body (mid-torso and farther back), preserving well-healed and aligned fractures at mid-length and just below the rib head, respectively. A left rib from the posterior portion of the dorsal series was injured just below the rib head. All are in the same state of repair, suggesting that they were broken at the same time. A juvenile *Hypacrosaurus* dorsal rib (TMP 1988.151.0031) from Devil's Coulee exhibits a swollen area on the posterior surface of the rib neck. Two healed fractures (TMP 1989.017.0004) are present in an adult rib from an *Edmontosaurus* bonebed in Edmonton (Fig. 33.1). One is mid-shaft and the other is located below the rib head. Again, both breaks probably occurred at the same time, given their similar states of repair. Some unfractured specimens (TMP 1987.048.0003, 1993.666.0003) exhibit swelling and signs of osteomyelitis where the head of the rib articulates with the vertebra.

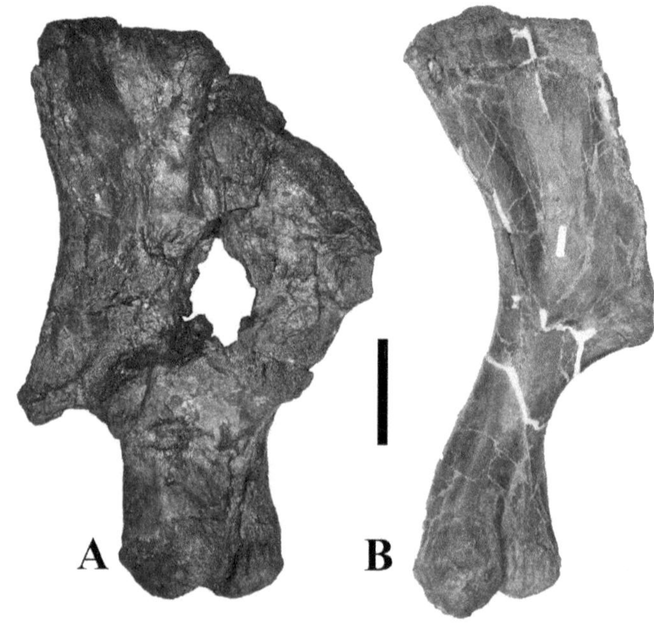

33.32. (A) TMP 1998.011.0001 (cast of pathologic humerus AMNH 5207); (B) normal right humerus (reversed, ~same length) of cf. *Parasaurolophus* (TMP 1992.053.0021) for comparison. Scale bar equals 10 cm.

Shoulder Girdle and Forelimb

Shoulder girdle osteopathy is rare and appears of minor significance. A scapula fragment (TMP 1994.009.0001 [cf. *Edmontosaurus*]) bears a low rounded lesion with a drainage sinus with no indication of fracture or osteomyelitis. The external surface of a left distal scapula (TMP 1966.031.0084; Fig. 33.31) from DPP bears a low and smooth C-shaped lesion of undetermined etiology, likely an ossified subperiosteal hematoma or other residual damage from trauma.

Few traumatic forelimb injuries were observed and some were serious in nature. The sole example of humeral pathology occurs in an isolated specimen collected from the upper HCFm, north of Drumheller (AMNH 5207; TMP 1998.011.0001 [cast]). This specimen was originally incorrectly identified as that of a ceratopsian (Moodie, 1923), but more recently has been correctly identified as a heavily modified hadrosaur humerus, likely belonging to *Hypacrosaurus* (Swinton, 1970). The element was originally fractured transversely or

33.33. TMP 1996.012.0434, radius and ulna, completely fused at mid-length. Scale bar equals 10 cm.

obliquely in a dorsoventral plane. The two pieces then separated, rotated slightly into misalignment, and then were poorly healed with the subsequent development of osteomyelitis (Fig. 33.32A). The bone is so heavily modified that only the distal articulating end confirms that it is, in fact, a humerus.

A well-healed radius fracture with mild arching in a DPP specimen is figured in Sawyer and Erickson (1985). A more serious example (TMP 1996.012.0434; Fig. 33.33) involves the complete fracture of both radius and ulna at mid-length. The periosteum of both bones was ruptured and reparative bone cells could freely proliferate into a large fracture callus, which fused the bones together in life position. Much of the fracture callus had been remodeled, indicating the injury was long-standing. Lastly, TMP 1966.017.0009 appears to represent a section of radius shaft completely encased in an infectious fracture callus.

Osteopathy was not observed in the few wrist elements we examined. TMP 1996.012.0182 is a metacarpal with extensive osteomyelitis. TMP 1989.036.0317, another metacarpal, preserves a thin irregular layer of osteomyelitic new bone that easily flakes off of the normal bone. Despite a good sample size from DPP (n = 56), occurrences of metacarpal osteopathy in the TMP collections and in the field are rare, and restricted to a few cases of mild osteomyelitis (TMP 1989.036.0319) or osteophyte development (TMP 1991.036.0224). Metacarpal pathology in hadrosaurs is better known elsewhere. Spectacular healing fracture trauma of metacarpals III and IV has been reported from Montana in *Edmontosaurus* (USNM V3814; Moodie, 1926; Rothschild and Martin, 2006:fig. 17-8 [lower]; TMP 1999.032.0001 [cast]). Prieto-Márquez and Brañas (2012) reported a Mexican specimen of *Latirhinus* with fused shafts of metacarpals III and IV. The rarity of metacarpal injuries suggests that they were either rare events or incompatible with survival. Some large metacarpals express low ridging and mild new bone development, but these appear restricted to very large animals and may be related to size and late ontogeny. No pathologic manal phalanges were observed despite the large sample we examined.

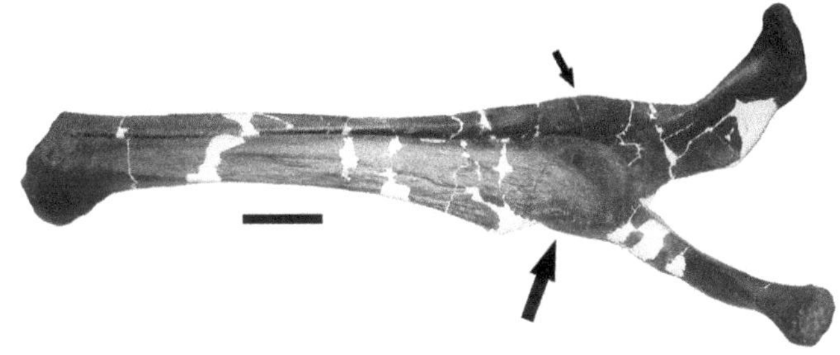

33.34. TMP 1979.014.0381, paired adult ischia in dorsolateral view with well-healed fractures proximally (arrows). Scale bar equals 10 cm.

Pelvic Girdle and Hindlimb

An articulated pair of lambeosaurine ischia (TMP 1979.014.0381) are both fractured proximally and healed in a twisted position with massive callus formation (Fig. 33.34). The shafts appear to have fused much of their length, including the expanded distal boot. This specimen exhibits the most serious deep-body skeletal fracture so far recorded in a dinosaur, and it is indeed remarkable that the individual survived. Other fractured ischia with healing are reported by Charig (1979) and Blows (1989) for *Iguanodon*. Blows (1989) suggested such injuries may have been caused by unsuccessful mating attempts, but could also conceivably have resulted from bad falls. The only other examples of pelvic osteopathy in hadrosaurs are Waldman's (1969:574) note of "a possible pathological condition" on the prepubis of a juvenile *Kritosaurus* (CMN 8784), and an enlarged external face on an iliac

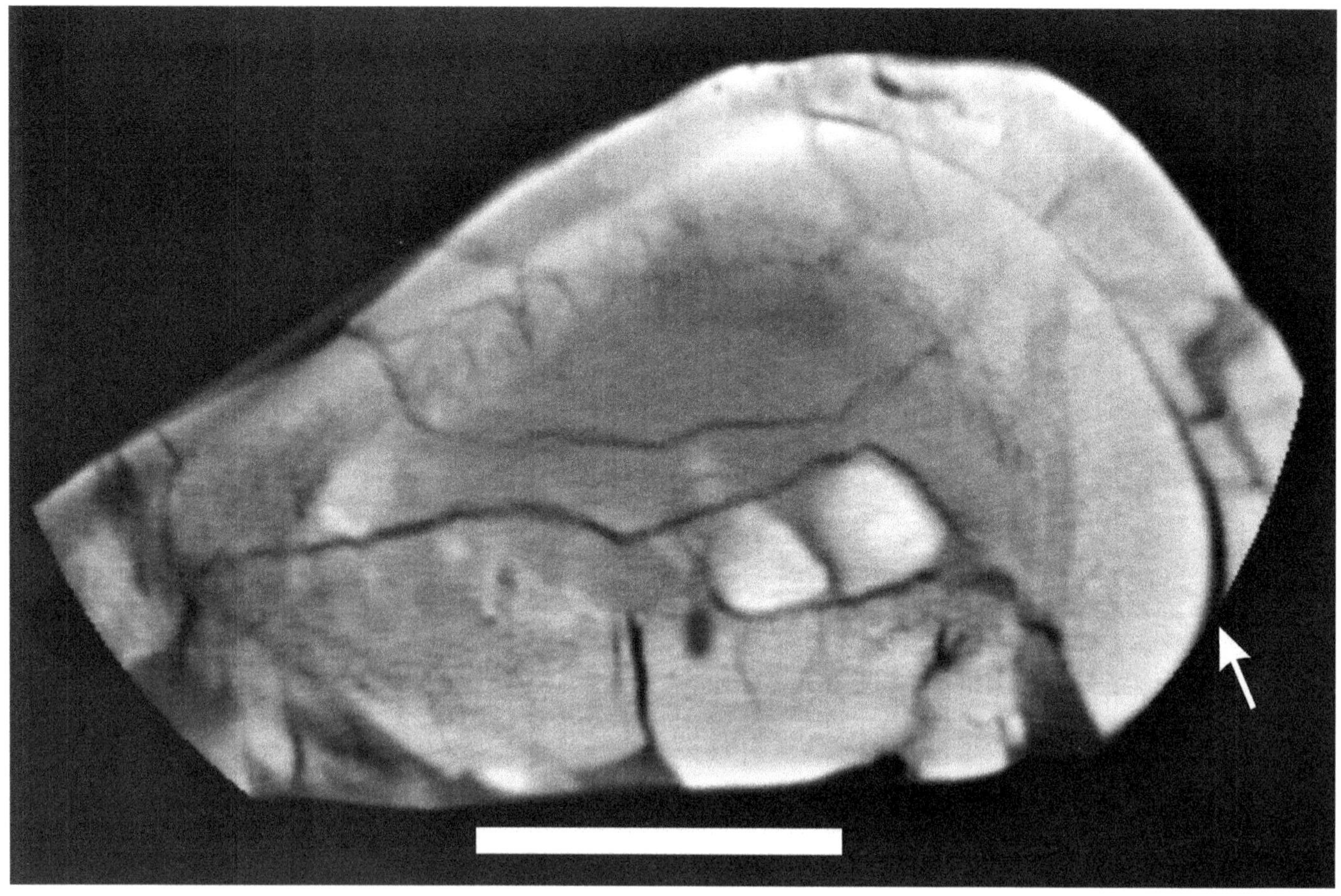

33.35. Transverse CT scan of distal of region CMN 41201, an adult tibia with severe osteomyelitis. On the far right, new bone is separating (arrow) from the normal bone. Scale bar equals approximately 10 cm.

pubic peduncle in *Parasaurolophus* (ROM 786) that partly overlaps the pubis (Lull and Wright, 1942).

Hindlimb injuries are rare. A thickened lesion, measuring 232 mm by 149 mm on a tibia shaft (CMN 41201; Edmund, 1954; Lindblad, 1954; Fig. 33.35) appears osteomyelitic and was apparently active at the time of death. Two large pieces of tibia (TMP 1991.036.0306) also appear to have been badly infected. A complete fracture with massive pseudoarthrosis development at mid-length was observed in an incomplete, putative fibula preserved in two sections (TMP 1995.405.0072; Fig. 33.36). A broken cross-section shows the original bone to be about 14.3 cm in diameter. Yet just 16 cm away, the bone is sharply swollen, forming a massive pseudoarthrosis, 48 cm in diameter–a 335.7% increase over the normal bone. This lesion must have been endured over a long time, and likely had a crippling effect on the individual as it enlarged.

The paucity of major fracture trauma in large, weight-bearing hindlimb elements is expected. Femoral or tibia fractures would be anticipated to impair mobility, resulting in death from trauma, dehydration, starvation, or predation. Osteopathy of bones in the ankle region was not observed, in contrast to what was recognized in *Iguanodon* (Rothschild, 1990). Whereas no evidence of trauma was observed in the 72 metatarsals examined in TMP collections, a metatarsal III fragment (TMP 2008.012.0062) preserves evidence of infection to such a degree that it suggests distal sloughing of the entire digit. Missing toes are documented in theropods (citations in Tanke and Rothschild, 2002:85–86, 89–90; see also Ishigaki and Lockley, 2010), and one possible example of an infected and swollen toe in a hadrosaur ichnite was reported by Currie et al. (2003:fig. 7E) from Mongolia. In a *Hypacrosaurus* (CMN 8501) from Rumsey (Fig. 33.1), right metatarsal IV is also extensively covered by callus formation of an undetermined nature.

Pedal injuries and infections appear confined to fully mature individuals. The deformed and barely recognizable pedal phalanx (TMP 1975.008.0007) has several large openings present on the infected bone surface and is extensively hollowed out internally, suggesting massive pyogenic infection and perhaps sloughing off of the distal segment. Osteochondrosis regularly affects pedal phalanges, especially the proximal articular faces of the fourth digit. The lesions

33.36. End view of fibula shaft TMP 1995.405.0072, showing original bone (ob) in cross section and exuberant new bone development (nb). Scale bar equals 1 cm.

consist of circular to elongate-shaped shallow pits with a smooth or finely pitted floor, and occur singly or as multiple depressions. These common lesions are considered more fully in Rothschild and Tanke (2007). Stress fractures of pedal phalanges, well documented in contemporaneous ceratopsians (Rothschild, 1988; Rothschild and Martin, 2006; Tanke and Rothschild, 2010:fig. 25.22A), are unreported in hadrosaurs.

Ossified Tendons

Ossified tendons in hadrosaurs and other dinosaurs have been reported as pathological or protective (Moodie, 1927, 1928; Campbell, 1966; Reid, 1984; Rothschild, 1987; Organ and Adams, 2005; Organ, 2006). Two isolated ossified tendon sections, believed to be hadrosaurian (TMP 1966.017.0013, 1985.047.0005), have mushroom-shaped, tumor-like growths of an as yet undetermined nature. The tendons may have been injured at the same time as the associated caudal vertebrae, although Organ (2006) noted caudal vertebrae fracture trauma in Montanan specimens, where nearby tendons were unaffected. TMP 1966.017.0013 might represent an ossified hematoma, radiologic and histologic studies are required for confirmation. A healing ossified tendon fracture may also be present in TMP 1998.093.0100. An unusual anomaly involving a tendon is noted by Maier (1982). During the excavation of a posterior skeleton of an unidentified hadrosaur (TMP 1982.017.0001), Maier (1982:n.p.) made the following observation: "An ossified tendon had actually looped from one side of a neural spine to the opposite side of the following spine."

Eggs and Skin

Zelenitsky and Hills (1997) reported multilayered eggshell pathology in the hadrosaurian oogenus *Spheroolithus albertensis* from the OFm at Devil's Coulee. Fossilized dinosaur skin impressions also preserve pathology as indicated by a specimen with a healing laceration (PBMNH.P.06.016.T) from

33.37. Speculative scenario showing accidental intraspecific tail trampling in a resting *Edmontosaurus*. The abundance of tail injuries in DPP hadrosaurs (not *Edmontosaurus*) suggests that this scenario was common among hadrosaurs. Drawing by Tracy L. Ford.

South Dakota (Rothschild and DePalma, 2013). Lucas et al. (2000:85) suggest that the numerous tubercles in hadrosaur skin "increased the resistance of the skin to tearing and puncturing." Martill (1991) noted possible parasite damage to skin preserved in a Lower Jurassic *Scelidosaurus* from England. In reptiles, visible skin pathologies may be observed due to disease (Buenviaje et al., 1998; Alibardi and Toni, 2005) and can relate to physical trauma like cuts, tears, and nicks received under varying circumstances such as falls, scrapes against protruding objects, or physical interactions with conspecifics or predators (Webb and Manolis, 1989:101). Hadrosaur skin impressions are common in Alberta and are associated both with single bones and skeletons, but no examples of healing, scarring, or other damage has been observed yet.

DISCUSSION

Hadrosaurs account for about 50% of the dinosaur fauna at DPP (Currie and Russell, 2005). Specimens occur as skeletons (Henderson and Tanke, 2010), small elements in vertebrate microfossil localities (Dodson, 1983), and overwhelmingly occur as disarticulated elements in multitaxic bonebeds, and isolated elements in paleochannel deposits (Eberth and Currie, 2005). The types of dental and skeletal osteopathy described and categorized here predominantly reflect the bumps and knocks incurred during the life of a hadrosaur, and are likely related to simple events such as falls, congenital defects, teratisms, and intraspecific encounters, be they accidental or driven by interactive behaviors. The abundance of pathological elements among hadrosaurs and their common occurrence in caudal vertebrae has resulted in an historical oddity. Because they are so frequently encountered as isolated elements in the field but yield little taxonomic or phylogenetic information, caudal vertebrae are often ignored by prospectors, who, typically, are in search of fossils that are otherwise poorly represented, can be identified to genus and species, or represent display-quality specimens. This collecting bias is unfortunate because it is clear from our study that abundant but previously uninteresting fossils, such as caudal vertebrae, actually can provide a wealth of information about the paleobiology of hadrosaurs. Specifically, based on the brief survey presented here, one cannot help but be impressed by the large numbers of tail injuries in hadrosaurs that exhibit advanced but incomplete healing. The large numbers of these fossil injuries beg an explanation of their etiologies.

Earlier interpretations of proximal caudal neural spine injuries (Tanke, 1989; Rothschild and Tanke, 1992; Rothschild, 1994) considered mating trauma, with the heavy weight of the mounted male resting on and crushing the female's proximal tail. If this interpretation is true, then lambeosaurines, with their more elongate neural spines may have been more prone to injury. With the exception of Isles (2009:176) this suggestion was not well received by the scientific community, even though dinosaur mating trauma had been previously suggested (Gilmore, 1909, 1912; Blows, 1989; Vance,

1989; more citations in Tanke and Rothschild, 2002:91), and courtship/mating trauma and death are well documented in the modern world (e.g., Baker and Brothwell, 1980; Droscher, 1976). An adult hadrosaur tail was a large appendage, extending up to 4.5 m from the torso (Gilmore, 1946). In gregarious animals such as hadrosaurs, a long tail must have gotten in the way on occasion. An animal making a sudden turn could swing its tail into unyielding objects nearby, be they conspecifics or inanimate objects such as trees. Furthermore, given that modern herding animals tend to exhibit complex behaviors related to, in some cases, agonistic behaviors and defense from predators, it is also reasonable to speculate that hadrosaurs may have used their tails in aggressive interactions with conspecifics, or to thwart predators, thus sustaining caudal injuries. However, the numerous examples of crushed centra from the distal tail region and those injured vertebrae located more proximally are best explained by highly focused loading on the vertebrae. In the context of well-documented gregarious behaviors in many hadrosaur taxa (Horner et al., 2004), the most likely cause of highly focused loading would be trampling by nearby and large conspecifics (Fig. 33.37). Although limited, there is some trackway evidence showing bipedal dinosaurs did on occasion drag or place their tails on the ground (e.g., Irby and Albright, 2002; Pérez-Lorente and Herrero Gascon, 2007; see also Mallon et al., 2013:fig. 2) leaving them potentially exposed and vulnerable to trampling.

Alternatively, there have been suggestions of predation-related injuries on hadrosaur tails (Canudo-Sanagustín, 2004, Canudo, Cruzado-Caballero, et al., 2005; Canudo, Barco, et al., 2005; Carpenter, 1988, 2000; DePalma et al., 2013; Murphy et al., 2013; Rega, 2013; see also Farlow and Holtz, 2002; Holtz, 2002, 2003, 2008; Hone and Watabe, 2010) and several lines of evidence demonstrate conclusively that tyrannosaurids fed on hadrosaurs. However, for a scenario that infers that most of the tail damage was due to failed predation attempts, the ubiquity of damaged hadrosaur caudals requires that each adult hadrosaur would have experienced multiple failed predation attempts. Such a conclusion seems unreasonable to us given the inferred relatively large numbers of hadrosaurs versus tyrannosaurids in these ecosystems.

SUMMARY AND CONCLUSIONS

Osteopathy is well known from hadrosaur fossils collected from the Upper Cretaceous of Dinosaur Provincial Park and elsewhere in Alberta where bone preservation is good and specimens are plentiful.

Osteopathy in juveniles and subadults is not well documented. Among juveniles this is possibly related to either the small sample size of material available to us or the relative fragility of juvenile skeletons and the reasonable expectation that a serious injury would result in death. Among subadults this pattern is potentially related to segregation of subadults from adults, as suggested for contemporaneous hadrosaur faunas in nearby Montana. In this scenario, segregated subadults may have avoided bone-modifying contact with larger individuals. This interpretation is supported by the observation that most of the fracture injuries seen in adults were actively healing, thus demonstrating that they were received as adults, and not accumulated throughout earlier growth stages.

Among all forms of osteopathy, fractures predominate, with adults most frequently affected.

Hadrosaurs often survived debilitating fractures and injuries to many parts of the skeleton (skull, vertebrae, forelimbs, and pelvis). However, no major (weight-bearing) hindlimb fractures with healing are known, indicating that such injuries were invariably lethal.

Fractures often show advanced, but rarely total, healing, indicating that the afflicted animal survived its trauma for extended periods of time (months to years). Although total healing of fractures and resorption of the fracture callus can make them difficult to see, angulation of the affected parts can still reveal the presence of old fractures. Why so few animals survived long enough to achieve total bone healing remains unexplained.

The tail is the most frequently injured skeletal region in hadrosaurs. Distally, injuries increase in frequency, diversity, and severity. All caudal neural spine factures are longitudinal in orientation. Fractured (single or paired) and healing caudal neural spine tips with one (typically) or two fractures occur proximally along the tail, and often exhibit some degree of lateral angulation. Caudal vertebral fusions are rare, typically involving only two vertebrae. Medial caudals have single or paired fractures of the neural spine tip and fracture non-union (pseudoarthroses) above and below the zygapophyses. Centra can have deep premortem fracture-related cracks with variable orientations on one or both endplates. Intervertebral fusions are uncommon; examples seen involve two vertebrae. Distal caudals typically exhibit massive crush trauma with fusions of seriously malformed vertebrae (up to five in number, but usually only two). Affected centra can fuse back together at an angle causing a kink in the tail. Osteomyelitis is known and some specimens suggest distal tail amputation or sloughing.

Given the widespread evidence for gregariousness in many hadrosaur taxa, we propose distal tail injuries were most likely related to intraspecific trampling that involved stepping onto the mid-to-distal tail of adjacent individuals causing fractures and crushing trauma, or clipped and fractured caudal neural spines while stepping over prone individuals.

The ubiquity of caudal injuries shows that, although serious in nature, they were not immediately fatal.

There is no evidence of healing bites or healing tooth mark trauma (by predaceous tyrannosaurs) among the caudal vertebrae surveyed here.

Although modern large vertebrates with serious injuries are usually subject to heavy predation pressure, we have documented that numerous individual hadrosaurs with serious, debilitating injuries managed to survive long after the accident (months to years). Perhaps these injured hadrosaurs were overlooked by predators when immersed in a herd, or adopted some other strategy to avoid detection until their injuries were sufficiently healed.

Given their large adult sizes, similar overall morphologies, and broadly expressed (taxonomically) gregarious tendencies, we hypothesize that wherever adult hadrosaurs (and possibly iguanodonts) are found and well represented, the types of osteopathies described here will also be found, especially those related to the tail. Conversely, the absence of these osteopathic features in other kinds of large, long-tailed dinosaurs may provide evidence of limited physical interactions, or the inaccessibility of tails for trampling.

Finally, we encourage artists who render hadrosaur reconstructions to include some with caudal injuries. This should include animals with distal ends of tails missing, swollen, or with the tail tip deflected dorsally or laterally at angles of about 10–25°.

FUTURE DIRECTIONS

The numbers and diverse types of osteopathy described herein are impressive and provide paleobiological insights about hadrosaurs. Clearly these animals (especially adults) were injured regularly, and sometimes quite severely. Many of the wounded survived and show advanced bone healing indicative of multi-month and multi-year survivability, yet perished before repairs were completed. We find this point interesting as it suggests hadrosaurs were not only accident prone, but that their injuries were likely fatal over the long term. However, until more inclusive and quantitative work is undertaken, we remain uncertain whether the Albertan fossil record accurately records the true story of hadrosaur osteopathy and survivability. The data and discussions presented here represent the current state of knowledge. However, much remains to be done. Although Alberta's hadrosaurs likely will continue to be in the forefront of dinosaur paleopathology, we suggest some future research directions:

1. Establish a library of osteopathic conditions and identify similar pathologies in the modern animal record, human or otherwise, to allow their characterization and evidence-based pursuit of mechanisms.
2. Compare the character and frequency of Albertan hadrosaur osteopathy to those in other settings. For example, Ramírez-Velasco (pers. comm., 2011) noted that healing fibular fractures are regularly seen in Mexican hadrosaurs he examined, yet they are rare in Albertan material. Is such a disparity related to behavior, differing terrain, or other conditions?
3. Increased sample sizes of pathological elements among hadrosaurs will allow quantifiable comparisons among neonates, juveniles, subadults, and adults, perhaps resulting in insights into possible differences in niche or habitat selection among these forms.
4. Explanations of caudal vertebrae endplate depressions, healing cracks, and other defects require better comparisons with human cases. Biomechanical testing of replica specimens, cast in a material with the same properties as bone, could be done to determine etiology and mechanism of injuries in the crushed and split centra described above. Similar testing on centra in large extant vertebrates to determine mechanisms of injury may also be insightful.
5. More in-depth examinations of fracture callus formation in caudal neural spines in order to search for embedded teeth of predators. If such injuries are predation-related as widely believed, then some teeth are likely to be present.

ACKNOWLEDGMENTS

We thank Patricia E. Ralrick for transcription of an earlier version of this chapter, editing, and discussions. Howard Allen drafted Figure 33.1. Dave Eberth and Andy Farke reviewed the manuscript and offered valuable editing opinions. Tracy Ford did some citation searches and fact checking, and drafted Figure 33.37. The authors benefited from discussions and cooperation with Andrey Atuchin, Danny Barta, Laurie Bryant, Richard Butler, Canadian Museum of Nature, Sandra Chapman, Allesandro Chiarenza, Margaret Currie, Philip J. Currie, Dinosaur Provincial Park staff, Fabio Dalla Vecchia, Andy Farke, Denver Fowler, Liz Freedman Fowler, Ruben Guzman-Gutierrez, Graeme Housego, Jordan Mallon, Anthony Maltese, Andrew Milner, David Norman, Clayton Pilbro, Angel Ramírez-Velasco, Mark Van Tomme, Angelica Torices, Allison Tumarkin-Deratzian, and others over the many years this work has slowly come together. Jay Guidos provided photo-editing software. We are also grateful to Brandon Strilisky and the

TMP Collections staff for their decades of assistance and support. Chris Capobianco and Sue Sabrowski provided photographic assistance. Sandra Mills, CT Technologist at the Drumheller Health Centre, did some CT scan work. Thanks to our colleagues on the Facebook social group WIKIPALEO, who provided digital copies of obscure or hard-to-find papers.

We dedicate this chapter to the memory of our friend and colleague of paleopathological studies Larry D. Martin (1943–2013).

LITERATURE CITED

Alibardi, L., and M. Toni. 2005. Wound keratins in the regenerating dermis of lizard suggest that the wound reaction is similar in the tail and limb. Journal of Experimental Zoology 303A:845–860.

Anonymous. 2012. Hadrosaur tail video footage on YouTube. http://www.youtube.com/watch?v=f0WFsfkRu0I. Accessed February 26, 2012.

Baker, J., and D. Brothwell. 1980. Animal Diseases in Archaeology. Academic Press, London, U.K., 235 pp.

Barsbold, R., P. J. Currie, N. P. Myhrvold, H. Osmólska, K. Tsogtbaatar, and M. Watabe. 2000. A pygostyle from a non-avian theropod. Nature 403:155.

Blows, W. T. 1989. A pelvic fracture in *Iguanodon*. Archosaurian Articulations 1:49–50.

Boyd, A. A., and R. Motani. 2008. Three-dimensional re-evaluation of the deformation removal technique based on "jigsaw puzzling." Palaeontologia Electronica 11.2.7A:1–7.

Brothwell, D. 1967. Major congenital anomalies of the skeleton: evidence from earlier populations; pp. 423–443 in D. Brothwell, and A. T. Sandison (eds.), Diseases in Antiquity. Charles C. Thomas, Springfield, Illinois.

Buenviaje, G. N., P. W. Ladds, and Y. Martin. 1998. Pathology of skin diseases in crocodiles. Australian Veterinary Journal 76:357–363.

Buffetaut, E. 1985. Ein *Steneosaurus* (Crocodilia, Mesosuchia) mit regeneriertem Schwanzende aus dem Lias Epsilon (Toarcium) von Schwaben. Stuttgarter Beiträge zur Naturkunde, Series B (Geology und Paläeontology) 113:1–7.

Butler, R. J., A. M. Yates, O. W. M. Rauhut, and C. Foth. 2013. A pathological tail in a basal sauropodomorph dinosaur from South Africa: evidence of traumatic amputation? Journal of Vertebrate Paleontology 33:224–228.

Campbell, J. G. 1966. A dinosaur bone lesion resembling avian osteopetrosis with some remarks on the mode of development of the lesions. Journal of the Royal Microscopical Society 85:163–174.

Canudo-Sanagustín, J. I. 2004. El caso del hadrosaurio de Arén (Huesca): ¿muerte accidental o asesinato? Naturaleza Aragonesa 13:4–14.

Canudo, J. I., J. L. Barco, P. Cruzado-Caballero, G. Cuenca-Bescós, J. I. Ruiz-Omeñaca, and R. Royo-Torres. 2005. Evidencias de predación des dinosaurios terópodos en el Maastichtiense Superior, Cretácico Superior de Arén (Huesca). Lucas Mallada 12:29–58.

Canudo, J. I., P. Cruzado-Caballero, and M. Moreno-Azanza. 2005. Possible theropod predation evidence in hadrosaurid dinosaurs from the Upper Maastrichtian (Upper Cretaceous) of Arén (Huesca, Spain); pp. 9–13 in G. Gruber (ed.), Current Research in Vertebrate Palaeontology. Third Annual Meeting European Association of Vertebrate Paleontology, Därmstadter Kaupia, Beiträge zur Naturgeschichte, 14, Därmstadter, 2005.

Capasso, L. L. 2004. Antiquity of cancer. International Journal of Cancer 113:2–13.

Carpenter, K. 1988. Evidence of predatory behavior by *Tyrannosaurus;* in J. R. Horner (ed.), International Symposium on Animal Behavior as Derived from the Fossil Record. Museum of the Rockies, Bozeman, Montana. [No pagination]

Carpenter, K. 2000. Evidence of predatory behavior by carnivorous dinosaurs. Gaia 15:135–144.

Charig, A. J. 1979. A New Look at the Dinosaurs. Heinemann, London, in association with the British Museum (Natural History), 160 pp.

Clutton-Brock, J., K. Dennis-Bryan, P. L. Armitage, and P. A. Jewell. 1990. Osteology of the soay sheep. Bulletin of the British Museum (Natural History), Zoology 56:1–56.

Currie, P. J., and D. A. Russell. 2005. The geographic and stratigraphic distribution of articulated and associated remains; pp. 537–569 in P. J. Currie, and E. B. Koppelhus (eds.), Dinosaur Provincial Park: A Spectacular Ancient Ecosystem Revealed. Indiana University Press, Bloomington, Indiana.

Currie, P. J., D. Badamgarav, and E. B. Koppelhus. 2003. The first Late Cretaceous footprints from the Nemegt locality in the Gobi of Mongolia. Ichnos 10:1–13.

DePalma, R. A. II, D. A. Burnham, L. D. Martin, B. M. Rothschild, and P. L. Larson. 2013. Physical evidence of predatory behavior in *Tyrannosaurus rex.* Proceedings of the National Academy of Science www.pnas.org/cgi/doi/10.1073/pnas.1216534110

Dodson, P. 1983. A faunal review of the Judith River (Oldman) Formation, Dinosaur Provincial Park, Alberta. Mosasaur 1:89–118.

Droscher, V. B. 1976. They Love and Kill: Sex, Sympathy and Aggression in Courtship and Mating. E. P. Dutton, New York, 363 pp.

Eberth, D. A., and P. J. Currie. 2005. Vertebrate taphonomy and taphonomic modes; pp. 453–477 in P. J. Currie, and E. B. Koppelhus (eds.), Dinosaur Provincial Park: A Spectacular Ancient Ecosystem Revealed. Indiana University Press, Bloomington, Indiana.

Edmund, G. 1954. 1954 Expedition to Oldman Formation, Little Sandhill Creek badlands, Alberta. Fieldnotes on file at Royal Tyrrell Museum of Palaeontology library, Drumheller, Alberta.

Farke, A. A., and P. M. O'Connor. 2007. Pathology in *Majungasaurus crenatissimus* (Theropoda: Abelisauridae) from the Late Cretaceous of Madagascar. Society of Vertebrate Paleontology Memoir 8:180–184.

Farlow, J. O., and T. R. Holtz. 2002. The fossil record of predation in dinosaurs. Paleontological Society Papers 8:251–265.

Forrest, R. 2004. Taphonomic distortion of cervical vertebrae of a specimen of cf. *Pleisosaurus dolichodeirus* (Reptilia; Plesiosauria) from the Lower Lias of Charmouth. Proceedings of the Dorset Natural History and Archaeological Society 125:105–108.

Forster, C. A. 1990. Evidence for juvenile groups in the ornithopod dinosaur *Tenontosaurus tilletti* Ostrom. Journal of Paleontology 64:164–165.

Freedman, E. A., D. H. Tanke, and E. D. S. Wolff. 2012. Osteopathology in hadrosaurines (Dinosauria: Ornithischia) of the Judith River Formation (Campanian) of north-central Montana. Journal of Vertebrate Paleontology, Program and Abstracts 2012:97A.

Gangloff, R. A., and A. R. Fiorillo. 2010. Taphonomy and paleoecology of a bonebed from the Prince Creek Formation, North Slope, Alaska. Palaios 25:299–317.

Gilmore, C.W. 1909. Osteology of the Jurassic reptile *Camptosaurus,* with a revision of the species of the genus, and descriptions of two new species. Proceedings of the U.S. National Museum 36:197–332.

Gilmore, C. W. 1912. The mounted skeletons of *Camptosaurus* in the United States National Museum. Proceedings of the U.S. National Museum 41:687–696.

Gilmore, C. W. 1946. Notes on recently mounted reptile fossil skeletons in the United States National Museum. Proceedings of the U.S. National Museum 96:195–203.

Godefroit, P., Y. Bolotsky, and V. Alifanov. 2003. A remarkable hollow-crested hadrosaur from Asia: an Asian origin for lambeosaurines. Comptes Rendus Palevol 2:143–151.

Godefroit, P., Y. L. Bolotsky, and I. Y. Bolotsky. 2012. Osteology and relationships of *Olorotitan arharensis,* a hollow-crested hadrosaurid dinosaur from the latest Cretaceous of far eastern Russia. Acta Palaeontologica Polonica 57:527–560.

Happ, J. 2008. An analysis of predator-prey behavior in a head-to-head encounter between *Tyrannosaurus rex* and *Triceratops;* pp. 354–368 in P. Larson, and K. Carpenter (eds.), *Tyrannosaurus rex:* The Tyrant King. Indiana University Press, Bloomington, Indiana.

Henderson, D. M., and D. H. Tanke. 2010. Estimating past and future dinosaur skeleton abundances in Dinosaur Provincial Park, Alberta, Canada. Canadian Journal of Earth Sciences 47:1291–1304.

Holtz, T. R. 2002. The fossil record of predation in dinosaurs. Paleontological Society Papers 8:251–265.

Holtz, T. R. 2003. Dinosaur predation: evidence and ecomorphology; pp. 325–340 in P. H. Kelley, M. Kowaleski, and T. A. Hanse (eds.), Predator-prey Interactions in the Fossil Record. Kluwer Academic/Plenum Press, New York.

Holtz, T. R. 2008. A critical reappraisal of the obligate scavenging hypothesis for *Tyrannosaurus rex* and other tyrant lizards; pp. 371–396 in P. Larson, and K. Carpenter (eds.), *Tyrannosaurus rex:* The Tyrant King. Indiana University Press, Bloomington, Indiana.

Hone, D. W. E., and M. Watabe. 2010. New information on scavenging and selective feeding behavior of tyrannosaurids. Acta Palaeontologica Polonica 55:627–634.

Horner, J. R. 1984. The nesting behavior of dinosaurs. Scientific American 250:130–137.

Horner, J. R., D. B. Weishampel, and C. A. Forster. 2004. Hadrosauridae; pp. 438–463 in D. B. Weishampel, P. Dodson, and H. Osmólska (eds.), The Dinosauria, Second Edition. University of California Press, Berkeley, California.

Hutchinson, J. H., and F. L. Frye. 2001. Evidence of pathology in early Cenozoic turtles. PaleoBios 21:12–19.

Irby, G. V., and L. B. Albright III. 2002. Tail-drag marks and dinosaur footprints from the Upper Cretaceous Toreva Formation, northeastern Arizona. Palaios 17:516–521.

Ishigaki, S., and M. G. Lockley. 2010. Didactyl, tridactyl, and tetradactyl theropod trackways from the Lower Jurassic of Morocco: evidence of limping, labouring and other irregular gaits. Historical Biology 22:100–108.

Isles, T. E. 2009. The socio-sexual behavior of extant archosaurs: implications for understanding dinosaur behavior. Historical Biology 21:139–214.

Kirkland, J. I., and K. Bader. 2010. Insect trace fossils associated with *Protoceratops* carcasses the Djadokhta Formation (Upper Cretaceous), Mongolia; pp. 509–519 in M. J. Ryan, B. Chinnery-Allgeier, and D. A. Eberth (eds.), New Perspectives on Horned Dinosaurs. Indiana University Press, Bloomington, Indiana.

Lambe, L. M. 1902. On Vertebrata of the Mid-Cretaceous of the Northwest Territory 2. New genera and species from the Belly River Series (Mid-Cretaceous). Contributions to Canadian Paleontology 3:25–81.

Langston, W. 1961. New from members. Society of Vertebrate Paleontology News Bulletin 63:7–9.

Lindblad, G. E. 1954. Steveville, Alberta. Fieldnotes on file at the Royal Tyrrell Museum of Palaeontology library.

Lovell, N. C. 1997. Trauma analysis in paleopathology. Yearbook of Physical Anthropology 40:139–170.

Lovell, N. C. 2008. Analysis and interpretation of skeletal trauma; pp. 341–386 in M. A. Katzenberg, and S. R. Saunders (eds.), Biological Anthropology of the Human Skeleton. Second Edition. John Wiley & Sons, Hoboken, New Jersey.

Lucas, S. G., A. B. Heckert, and R. M. Sullivan. 2000. Cretaceous dinosaurs in New Mexico. New Mexico Museum of Natural History Science Bulletin 17:83–90.

Lull, R. S., and N. E. Wright. 1942. Hadrosaurian Dinosaurs of North America. Geological Society of America Special Papers 40. 242 pp.

Maier, G. 1982. Field notes. On file at the Royal Tyrrell Museum of Palaeontology library, Drumheller, Alberta.

Mallon, J. C., D. C. Evans, M. J. Ryan, and J. S. Anderson. 2013. Feeding height stratification among the herbivorous dinosaurs from the Dinosaur Park Formation (upper Campanian) of Alberta, Canada. BMC Ecology 13:14 doi:10.1186/1472-6785-13-14.

Martill, D. M. 1991. Organically preserved dinosaur skin: taphonomic and biological implications. Modern Geology 16:61–68.

Maryańska, T., and H. Osmólska. 1981. First lambeosaurine dinosaur from the Nemegt Formation, Upper Cretaceous, Mongolia. Acta Palaeontologica Polonica 26:243–255.

Monastersky, R. 1990. Reopening old wounds: physicians and paleontologists learn new lessons from ancient ailments. Science News 137:40–42.

Moodie, R. L. 1923. Paleopathology: An Introduction to the Study of Ancient Evidences of Disease. University of Illinois Press, Urbana, Illinois, 567 pp.

Moodie, R. L. 1926. Excess callus following fracture of the fore foot in a Cretaceous dinosaur. Annals of Medical History 8:73–77.

Moodie, R. L. 1927. Studies in paleopathology XX. Vertebral lesions in the sabre-tooth, Pleistocene of California, resembling the so-called myositis ossificans progressive, compared with certain ossifications in the dinosaurs. Annals of Medical History 9:91–102.

Moodie, R. L. 1928. The histological nature of ossified tendons. American Museum Novitates 311:1–15.

Moodie, R. L. 1930. Studies in paleodontology: dental abscesses in a dinosaur millions of years old, and the oldest yet known. Pacific Dental Gazette 38:435–440.

Murphy, N. L., K. Carpenter, and D. Trexler. 2013. New evidence for predation in a large tyrannosaurid; pp. 279–285 in J. M. Parrish, R. E. Molnar, P. J. Currie, and E. B. Koppelhus (eds.), Tyrannosaurid Paleobiology. Indiana University Press, Bloomington, Indiana.

Naish, D. 2008. Pads and notches in the necks of tube-crested hadrosaurs . . . SAY WHAT? scienceblogs.com/tetrapodzoology/2008/09/11/tube-crested-hadrosaur-neck.

Natarajan, L. C., A. L. Melott, B. M. Rothschild, and L. D. Martin. 2007. Bone cancer rates in dinosaurs compared to modern vertebrates. Transactions of the Kansas Academy of Science 110:155–158.

Norman, D. B., and A. Milner. 1989. Dinosaur. Stoddardt, Toronto, Ontario, 63 pp.

Organ, C. L. 2006. Biomechanics of ossified tendons in ornithopod dinosaurs. Paleobiology 32:652–665.

Organ, C. L., and J. Adams. 2005. The histology of ossified tendons in dinosaurs. Journal of Vertebrate Paleontology 25:602–613.

Parks, W. A. 1922. *Parasaurolophus walkeri:* a new genus and species of crested trachodont dinosaur. University of Toronto Studies, Geological Series 13:5–32.

Parks, W. A. 1935. New species of trachodont dinosaurs from the Cretaceous formations of Alberta: with notes on other species. University of Toronto Studies, Geological Series 37. 45 pp.

Partridge, D. 2007. Spinal lesions in a hadrosaur, *Edmontosaurus annectens.* Prototype X Magazine 2:118–121.

Patchus, R. J., and K. Derstler. 1998. A new *Edmontosaurus* mummy from the Lance Formation of Wyoming; pp. 103–108 in R. A. Hunter (ed.), Tate '98 Life in the Late Cretaceous, Field Conference, June 5–7, 1998. Tate Museum Guidebook no. 3. Tate Geological Museum, Casper, Wyoming.

Patchus, R. J., and K. Derstler. 2002. A hadrosaurine dinosaur mummy from the Lance Formation (Upper Maastrichtian), Niobrara County, Wyoming. Journal of Vertebrate Paleontology 22(3, Supplement):95A.

Pérez-Lorente, F., and J. Herrero Gascon. 2007. [A dinosaur movement deduced from a theropod trackway with mud immersion footprints and tail drag structures (Villar del Arzobispo Formation. Galve, Teruel, Spain)]. Revista Española de Paleontología 22:157–174. [Spanish]

Peterson, J. E., M. D. Henderson, R. P. Scherer, and C. P. Vittore. 2009. Face biting on a juvenile tyrannosaurid and behavioral implications. Palaios 24:780–784.

Plotz, R. D., and W. L. Linklater. 2011. Black Rhinoceros (*Diceros bicornis*) calf succumbs after lion predation attempt: implications for conservation management. African Zoology 44:283–287.

Prieto-Márquez, A. 2011. A reappraisal of *Barsboldia sicinskii* (Dinosauria: Hadrosauridae) from the Late Cretaceous of Mongolia. Journal of Paleontology 85:468–477.

Prieto-Márquez, A. 2012. The skull and appendicular skeleton of *Gryposaurus latidens,* a saurolophine hadrosaurid (Dinosauria: Ornithopoda) from the early Campanian (Cretaceous) of Montana, USA. Canadian Journal of Earth Sciences 49:510–532.

Prieto-Márquez, A. and I. S. Brañas. 2012. *Latirhinus uitstlani,* a 'broad-nosed' saurolophine hadrosaurid (Dinosauria, Ornithopoda) from the late Campanian (Cretaceous) of northern Mexico. Historical Biology 24:607–619.

Rega, E. 2012. Disease in dinosaurs; pp. 666–711 in M. K. Brett-Surman, T. R. Holtz, Jr., and J. O. Farlow (eds.), The Complete Dinosaur, Second Edition. Indiana University Press, Bloomington, Indiana.

Reid, R. E. H. 1984. The histology of dinosaur bone, and its possible bearing on dinosaurian physiology. Symposium of the Zoological Society of London 52:629–663.

Roberts, E. M., R. R. Rogers, and B. Z. Foreman. 2007. Continental insect borings in dinosaur bone: examples from the Late Cretaceous of Madagascar and Utah. Journal of Paleontology 81:201–208.

Rogers, K. L. 1985. Possible physiological and behavioral adaptations of herbivorous dinosaurs. Journal of Vertebrate Paleontology 5:371–372.

Rogers, R. R. 1992. Non-marine borings in dinosaur bones from the Upper Cretaceous Two Medicine Formation, northwestern Montana. Journal of Vertebrate Paleontology 12:528–531.

Ross, C. A. 1978. Distortion of fossils in shales. Journal of Paleontology 52:943–945.

Rothschild, B. M. 1985. Diffuse idiopathic skeletal hyperostosis: misconceptions and reality. Clinical Rheumatology 4:207–212.

Rothschild, B. M. 1987. Diffuse idiopathic skeletal hyperostosis as reflected in the paleontologic record: dinosaurs and early mammals. Seminars in Arthritis and Rheumatism 17:119–125.

Rothschild, B. M. 1988. Stress fracture in a ceratopsian phalanx. Journal of Paleontology 62:302–303.

Rothschild, B. M. 1990. Radiologic assessment of osteoarthritis in dinosaurs. Annals of the Carnegie Museum 59:295–301.

Rothschild, B. M. 1994. Paleopathology and sexual habits of dinosaurs, as determined from study of their fossil remains. Paleontological Society Special Publication 7:275–283.

Rothschild, B. M. 1997. Dinosaurian paleopathology; pp. 426–448 in J. O. Farlow and M. K. Brett-Surman (eds.), The Complete Dinosaur. Indiana University Press, Bloomington, Indiana.

Rothschild, B. M., and D. Berman. 1991. Fusion of caudal vertebrae in Late Jurassic sauropods. Journal of Vertebrate Paleontology 11:29–36.

Rothschild, B. M., and R. Depalma. 2013. Skin pathology in the Cretaceous: evidence for probable failed predation in a dinosaur. Cretaceous Research 42:44–47.

Rothschild, B. M., and L. D. Martin. 1987. Avascular necrosis: occurrence in diving Cretaceous mosasaurs. Science 236:75–77.

Rothschild, B. M., and L. D. Martin. 1993. Paleopathology: Disease in the Fossil Record. CRC Press, Boca Raton, Florida, 386 pp.

Rothschild, B. M., and L. D. Martin. 2006. Skeletal Impact of Disease. New Mexico Museum of Natural History and Science Bulletin 33. 226 pp.

Rothschild, B. M., and R. E. Molnar. 2008. Tyrannosaurid pathologies as clues to nature and nurture in the Cretaceous; pp. 287–304 in P. Larson and K. Carpenter (eds.), *Tyrannosaurus rex:* The Tyrant King. Indiana University Press, Bloomington, Indiana.

Rothschild, B. M., and D. H. Tanke. 1992. Paleopathology of vertebrates: insights to lifestyle and health in the fossil record. Geoscience Canada 19:73–82.

Rothschild, B. M., and D. H. Tanke. 2007. Osteochondrosis in Late Cretaceous Hadrosauria: a manifestation of ontologic failure; pp. 171–183 in K. Carpenter (ed.), Horns and Beaks: Ceratopsian and Ornithopod Dinosaurs. Indiana University Press, Bloomington, Indiana.

Rothschild, B. M., L. D. Martin, and A. S. Schulp. 2005. Sharks eating mosasaurs, dead or alive? Netherlands Journal of Geosciences 84:335–340.

Rothschild, B. M., H.-P. Schultze, and R. Peligrini. 2012. Herpetological Osteopathology: Annotated Bibliography of Amphibians and Reptiles. Springer-Verlag, Heidelberg, Germany, 450 pp.

Rothschild, B. M., D. H. Tanke, and T. L. Ford. 2001. Theropod stress fractures and tendon avulsions as a clue to activity; pp. 331–336 in D. H. Tanke and K. Carpenter (eds.), Mesozoic Vertebrate Life. Indiana University Press, Bloomington, Indiana.

Rothschild, B. M., D. H. Tanke, M. Helbling, and L. D. Martin. 2003. Epidemiologic study of tumors in dinosaurs. Naturwissenschaften 90:495–500.

Saneyoshi, M., M. Watabe, S. Suzuki, and K. Tsogtbaatar. 2011. Trace fossils on dinosaur bones from Upper Cretaceous eolian deposits in Mongolia: taphonomic interpretation of paleoecosystems in ancient desert environments. Palaeogeography, Palaeoclimatology, Palaeoecology 311:38–47.

Sawyer, G. T., and B. R. Erickson. 1985. Injury and diseases in fossil animals: the intriguing world of paleopathology. Encounters 8:25–28.

Sawyer, G. T., and B. R. Erickson. 1998. Paleopathology of the Paleocene Crocodile *Leidyosuchus* (*Borealosuchus*) *formidabilis.* Science Museum of Minnesota Monograph 4. 38 pp.

Straight, W. H., G. L. Davis, C. W. Skinner, A. Haims, B. L. McClennan, and D. H. Tanke. 2009. Bone lesions in hadrosaurs: Computed tomographic imaging as a guide for paleohistologic and stable-isotopic analysis. Journal of Vertebrate Paleontology 29:315–325.

Swinton, W. E. 1970. The Dinosaurs. George Allen and Unwin, London, 331 pp. [Originally published in 1934]

Tanke, D. H. 1989. Paleopathologies in Late Cretaceous hadrosaurs (Reptilia: Ornithischia) from Alberta, Canada. Journal of Vertebrate Paleontology 9(3, Supplement):41A.

Tanke, D. H., and M. K. Brett-Surman. 2001. Evidence of hatchling and nestling-size hadrosaurs (Reptilia: Ornithischia) from Dinosaur Provincial Park (Dinosaur Park Formation: Campanian), Alberta, Canada; pp. 206–218 in D. H. Tanke and K. Carpenter (eds.), Mesozoic Vertebrate Life. University of Indiana Press, Bloomington, Indiana.

Tanke, D. H., and P. J. Currie. 2000. Head-biting in theropods: paleopathological evidence. Gaia 15:167–184.

Tanke, D. H., and B. M. Rothschild. 2002. DINOSORES: An Annotated Bibliography of Dinosaur Paleopathology and Related Topics, 1838–2001. New Mexico Museum of Natural History and Science Bulletin 20. 96 pp.

Tanke, D. H., and B. M. Rothschild. 2010. Paleopathologies in Albertan Ceratopsids and their behavioural significance; pp. 355–384 in M. J. Ryan, B. J. Chinnery-Allgeier, and D. A. Eberth (eds.), New Perspectives on Horned Dinosaurs. Indiana University Press, Bloomington, Indiana.

Tanke, D. H., and B. M. Rothschild. 2011. Osteopathy in Hadrosauridae from Alberta, Canada; pp. 145–147 in D. R. Braman, D. A. Eberth, D. C. Evans, and W. Taylor (compilers), International Hadrosaur Symposium, Abstract Volume. Royal Tyrrell Museum of Palaeontology, Drumheller, Alberta.

Thomas, R., and N. Johannsen. 2011. Articular depressions in domestic cattle phalanges and their archaeological relevance. International Journal of Paleopathology 1:43–54.

Vance, T. 1989. Probable use of the vestigial forelimbs of the tyrannosaurid dinosaurs. Bulletin of the Chicago Herpetological Society 24:41–47.

Varricchio, D. J., and J. R. Horner. 1993. Hadrosaurid and lambeosaurid bone beds from the Upper Cretaceous Two Medicine Formation of Montana: taphonomic and biologic implications. Canadian Journal of Earth Sciences 30:997–1006.

Waldman, M. 1969. On an immature specimen of *Kritosaurus notabilis* (Lambe), (Ornithischia: Hadrosauridae) from the Upper Cretaceous of Alberta, Canada. Canadian Journal of Earth Sciences 6:569–576.

Webb, G., and C. Manolis. 1989. Australian Crocodiles: A Natural History. Reed Books, Sydney, Australia, 160 pp.

White, T. 2003. Early hominids: diversity or distortion? Science 299:1994–1997.

Witzmann, F., P. Asbach, K. Remes, O. Hampe, A. Hilger, and A. Paulke. 2008. Vertebral pathology in an ornithopod dinosaur: a hemivertebra in *Dysalotosaurus lettowvorbecki* from the Jurassic of Tanzania. Anatomical Record 291:1149–1155.

Zelenitsky, D., and L. V. Hills. 1997. Normal and pathological eggshells of *Spheroolithus albertensis* oosp. nov. from the Oldman Formation (Judith River Group; Late Campanian), southern Alberta. Journal of Vertebrate Paleontology 17:167–171.

Appendix 33.1. Articulated/Associated Hadrosaur Skeletons from Alberta, Montana, South Dakota, Wyoming, Mongolia, Russia, and Spain that Exhibit Multiple Osteopathy of the Tail

Identification	Catalog #	Place collected	Total number of caudal neural spines injured versus tail position (vertebra number)	Information source
cf. *Edmontosaurus*	CMN 9872	Drumheller, Alberta	10; near tips of nos. 6, 12–15, 22, 25–26, 28–29	Undescribed; pers. obs., DHT
cf. *Edmontosaurus*	TMP 1970.018.0001	Drumheller, Alberta	22; nos. 1–14, 16–17, 20, 25, 28, 31, 35–36	Undescribed; pers. obs., DHT
cf. *Edmontosaurus*	MOR uncataloged	Montana; "Warwick's Duck" locality	22+; ? (unprepared, but possibly entire tail with pathologic neural spines, anterior caudals most seriously affected)	Undescribed; D. Fowler, pers. comm., 2011
Edmontosaurus	RAM 7150	McCone County, Montana	44 caudals present (1st 3–4 missing); of these nos. 14–16 and 21–29 (counting from the 1st vertebra preserved) show injured neural spines	Rega, 2012; Fig. 33.11
cf. *Edmontosaurus*	UNO, number not provided	Wyoming	~13; total number and anatomical position uncertain (specimen only partly prepared; mid- to distal caudals affected so far); "punctured fractured and/or infected caudal neural spines, fused caudal vertebrae"	Patchus and Derstler, 1998, 2002:95A
Edmontosaurus "Big Ed"	GNSM PV 320201	20 miles NW of Buffalo, South Dakota	Fused centra: 9–10, 36–37, 40–41; fused chevrons: 9, 20–21, 23–25; damaged chevrons: 41, 49; damaged neural spines: 7–11, 17–26	A. Maltese, pers. comm., 2012
Edmontosaurus	DMNH 1943	Dry Creek, Dawson County, Montana	5; nos. 13–16	Carpenter, 2000
Unidentified	TMP 1981.035.0001; quarry 162	DPP, Alberta	5+; extreme end of tail; specimen not fully prepared	Undescribed; DHT, pers. obs.
Unidentified	TMP 1980.023.0002; quarry 126	DPP, Alberta	8 (+?); extreme end of tail	Tanke and Rothschild, this chapter; Fig. 33.30A, B
Unidentified	TMP 1998.059.0001	DPP, Alberta	3 +?; 4 associated proximal caudal vertebrae	Undescribed; DHT, pers. obs.
Olorotitan	AEHM 2/845 (type)	Amur region, eastern Russia	"proximal third of the tail" (Godefroit et al., 2003); wounding to proximal, medial and extreme distal caudals evident (Godefroit et al., 2012)	Godefroit et al., 2003:fig 1c, d; Godefroit, et al., 2012:fig. 17b–e; Atuchin, pers. comm., 2012

Appendix 33.1. (continued)

Identification	Catalog #	Place collected	Total number of caudal neural spines injured versus tail position (vertebra number)	Information source
Blasisaurus	Blasi3/140	Arén, Spain	3; of 30+; not articulated, but affected vertebrae are mid-caudals	Canudo-Sanagustín, 2004; Canudo, Barco et al., 2005; Canudo, Cruzado-Caballero et al., 2005.
Unidentified	TMP 2011.012.0420; UAD 91 (= quarry 107?)	DPP, AB	11 distal caudal vertebrae found on surface, 8 have abnormal centrum shapes and/or deep internal vacuities; some have fractured and healing neural arch bases	Tanke and Rothschild, this chapter; Fig. 33.22A, B
Brachylophosaurus	MOR 2919	Rudyard, Hill County, Montana	4 mid-caudals with healing fractures of neural spine tips	Freedman, pers. comm., 2012
Barsboldia	ZPAL MgD-1/110	Ömnogöv, Mongolia	Club-shaped distal caudal neural spines; are these normal morphology (Prieto-Márquez, 2011) or osteopathy?	Maryańska and Osmólska, 1981
Corythosaurus	USNM 15493; quarry 26	DPP, Alberta	Mid-caudals ~27–31 appear pathologic in figures.	Gilmore, 1946:pls. 17–1.
Unidentified lambeosaurine	TMP 1978.004.0001	Sandy Point, Alberta	Disarticulated proximal and mid-caudals; 12 of 16 show simple, non-infectious neural spine fractures near tips; centra normal	Tanke and Rothschild, this chapter
Lambeosaurus	TMP 1982.038.0001; quarry 175	DPP, Alberta	Caudals 12–13 exhibit fractures near tips of neural spines, 12 especially so	Undescribed; DHT, pers. obs.
Subadult *Brachylophosaurus;* complete articulated tail	JRF 1002	Phillips County, Malta, Montana	Neural spines of sacrals 8–9 and caudals 1–3 exhibit calluses and missing tip (caudal 1)	Anonymous, 2012; Murphy et al., 2013
Unidentified; includes 9.5 articulated mid-caudal vertebrae	MOR 2580	Locality JR-429, Hill County, Montana	Two articulating centra expanded and fused ventrally; appears similar to condition in Fig. 33.22	Freedman et al., 2012
Unidentified and uncollected; includes 11 articulated exposed proximal and mid-caudal vertebrae	UAD 136	DPP, Alberta	4 of the posteriormost proximal caudals show healing fractures near neural spine tips	

Note: Barsboldia and *Corythosaurus* USNM 15493 neural spines need to be examined for confirmation. Curiously, *Edmontosaurus* or cf. *Edmontosaurus* predominates; the significance of this (if any) is unknown.

A Review of Hadrosaurid Skin Impressions

34

Phil R. Bell

ABSTRACT

Among dinosaurs, the fossil record of integument is most diversely represented in Hadrosauridae. A review of the literature, supplemented by additional undescribed specimens, identifies skin impressions from virtually all parts of the body, including rare examples from the face. In most cases, individual taxa are incompletely represented by only one or two specimens, and additional material is required to characterize the integumentary covering across the entire body of most taxa. Nevertheless, skin impressions on commonly represented areas, such as the tail, illustrate the diversity of scale architecture within Hadrosauridae. Where multiple individuals are available for a given species, consistency in scale characteristics and configurations supports the hypothesis that scale morphology and patterning is indeed species specific and does not vary significantly between individuals of the same species. Conversely, scale configuration is highly variable at both the inter- and intrageneric levels. Although these results are preliminary and more specimens are required before they can be fully incorporated into species diagnoses, closer attention should be paid to the standardized description of scale morphology, which should accompany osteological descriptions where possible.

INTRODUCTION

Dinosaur integument is known from most major groups including theropods, sauropods, thyreophorans, and cerapodans (Osborn, 1911, 1912; Lambe, 1914a, 1914b; Brown, 1917; Horner, 1984; Briggs et al., 1997; Czerkas, 1992, 1997; Dodson et al., 1998; Ji and Bo, 1998; Ji, 1999; Xu, Tang, and Wang, 1999; Xu, Wang, and Wu, 1999; Xu et al., 2001; Coria and Chiappe, 2007; Siber and Möckli, 2009). Skin traces are preserved most frequently in Hadrosauridae (sensu Horner et al., 2004). The first hadrosaurid skin impressions were discovered in 1884 by J. L. Wortman on the holotype specimen of "*Trachodon mirabilis*" (AMNH 5730; now *Edmontosaurus annectens* [Campione and Evans, 2011]). Although skin impressions were said to have largely covered the specimen, they were destroyed during excavation and only small patches from the tail were saved (Osborn, 1912). Since these unfortunate beginnings, hadrosaurid integument has become well known due in part to several remarkable "mummies" from North America, including famous specimens of *Corythosaurus* (AMNH 5240), *Edmontosaurus* (AMNH 5060, MRF-03), and *Brachylophosaurus* (JRF 115). Such specimens preserve not only skin impressions, but also epidermal microstructure, putative muscle bulk, and the keratinous coverings of the "beak" and terminal pedal phalanges (Osborn, 1911, 1912; Brown, 1916; Versluys, 1923; Murphy et al., 2007; Manning, 2009; Manning et al., 2009).

Hadrosaurid scales, like those of many dinosaurs, differ from most modern reptiles in that they are tubercle-like and non-imbricating, similar to extant varanid and helodermatid lizards (Osborn, 1912). Unlike many small theropods, there is no evidence to suggest a filamentous covering in adult hadrosaurs; however, skin has not yet been described for hatchlings or nestlings. Within Ornithischia, only the heterodontosaurid *Tianyulong* and the ceratopsian *Psittacosaurus* are known to have had filamentous or bristle-like integument structures (Zheng et al., 2009). Soft-tissue anatomy is relatively well understood for the Hadrosauridae: the integument consisted of an intricate and often mosaic arrangement of epidermal scales along the length of the body; the cervical vertebrae were held within a thick and muscular neck quite unlike the slender and sinuous morphology reflected by the bones themselves. Keratinous ramphothecae lined the predentary and premaxillae providing an ever-growing and resistant surface for food procurement. Enlarged, tabular scales ("segments" of Horner [1984]) along the dorsal midline in some taxa may have been used for visual display (Horner, 1984). The manus was enclosed within a fleshy "mitten" that restricted movement of the metacarpals and phalanges, which may have been useful for swimming and/or afforded the forelimb greater weight-bearing capacity (Osborn, 1912; Currie, 1983; Czerkas, 1997).

Despite the relatively rich fossil record of hadrosaurid integument, few attempts have been made to synthesize what is currently known of these structures (Lull and Wright, 1942). Nevertheless, preliminary studies suggest the taxonomic utility of hadrosaurid scale morphology (Brown, 1916; Lull and Wright, 1942; Negro, 2001; Bell, 2012). The purpose of this chapter is to summarize what is known of hadrosaurid scale architecture and distribution, using the standardized terminology of Bell (2012), based on the published literature and additional hitherto undescribed specimens.

Institutional Abbreviations AMNH, American Museum of Natural History, New York; CMN, Canadian Museum of Nature, Ottawa, Ontario; JRF, Judith River Foundation/Great Plains Dinosaur Museum, Malta, Montana; MOR, Museum of the Rockies, Bozeman, Montana; MRF, Marmarth Research Foundation, Marmarth, North Dakota; NMMNH, New Mexico Museum of Natural History and Science, Albuquerque, New Mexico; NMSG, Naturmuseum St. Gallen, St. Gallen, Switzerland; OTM, Old Trail Museum, Choteau, Montana; PIN, Palaeontologiceski Institut, Academii Nauk, Moscow, Russia; ROM, Royal Ontario Museum, Toronto, Ontario; SM, Senckenberg Museum, Frankfurt, Germany; TMP, Royal Tyrrell Museum of Palaeontology, Drumheller, Alberta; UALVP, University of Alberta, Edmonton, Alberta; UMNH VP, Natural History Museum of Utah, Salt Lake City, Utah; UW, University of Wyoming Geological Museum, Wyoming; ZPAL, Institute of Palaeobiology of the Polish Academy of Sciences, Warsaw.

METHODS

In an effort to simplify descriptions and facilitate the comparison of skin impressions, the standardized terminology of Bell (2012) is used (Table 34.1). Herein, the terms "basement" or "basement-scales" are used to describe the scales that form the major part of the integumentary surface. Collections of similar-sized basement-scales in a mosaic-like arrangement were described as "cluster areas" by Osborn (1912:42), and this terminology is retained here. Basement-scales form the background pattern onto which larger and sporadically arranged scales are often imposed. These larger scales are referred to as "feature-scales." Feature-scales present along the dorsal midline above the neural spines (referred to as segmented frill or non-osseous dermal scutes by various authors) are termed "midline feature-scales" (MFSs). "Interstitial tissue" refers to the integument between the scales and presumably afforded the skin its ability to flex and fold.

Anatomical directions such as dorsal or ventral reference the scales' relationship relative to the axial midline of the animal as they appear in most two-dimensional fossils. For example, in the case of a three-dimensional animal, the ventral edge of a scale underhanging the belly where the outer (superficial) surface faced the ground refers to the edge that may have been facing somewhat medially in life.

In most cases, specimens were examined by the author or (as in the case of *Brachylophosaurus*) photos were supplied to validate the observations of the various authors cited in the descriptions. However, other specimens could not be accessed, were not available for study, or were destroyed after the initial publication and so descriptions here are reliant on information presented in the relevant literature.

Table 34.1. Scale Morphology and Descriptive Terminology from Bell (2012)

	Size	Type	Ornamentation
Basement scales	1–10 mm	Polygonal	Smooth, striated
		Pebbles	Smooth
		Irregular	Smooth, corrugated
		Shells	Corrugated
Feature scales	>7 mm	Polygonal	Smooth
		Irregular	Smooth, corrugated
		Shield	Smooth, corrugated, striated
		Multi-pointed shield	Smooth, corrugated

DESCRIPTION OF INTEGUMENT IN HADROSAURIDAE

Saurolophinae (sensu Prieto-Márquez, 2010)

Brachylophosaurus canadensis **Sternberg, 1953** Skin impressions in *Brachylophosaurus* were made famous by the discovery of a well-preserved mummy (JRF 115) from the Judith River Formation near Malta, Montana (Murphy et al., 2007). The skeleton is complete except for portions of the proximal and distal tail, and integumentary traces are present in all areas except the face (the keratinous ramphothecae are preserved, however) and an area on the lateral surface of the right forelimb. The integument over the neck, shoulders, and flanks is composed of imperfectly preserved basement-scales 2–4 mm in diameter. The shapes of the individual scales were not described by Murphy et al. (2007). A small patch preserved on the mid-caudal vertebrae is similar to that described for the aforementioned regions.

The lateral and dorsal surfaces of the forearms were covered in a regular basement of polygonal (often pentagonal) scales considerably larger (~10 mm in diameter) than the basement-scales from elsewhere on the body (Murphy et al., 2007). These grade into smaller and less distinct basement-scales across the metacarpals and into an indistinct trace that forms a mitten-like covering over the manus. Digit I was apparently not included in the manal sheath (Murphy et al., 2007).

Large (~10 mm in diameter) polygonal basement-scales also persisted on the dorsal and lateral surfaces of the hindlimbs, similar to those described for the forelimbs. Scales from the medial part of the legs and hind feet were not described.

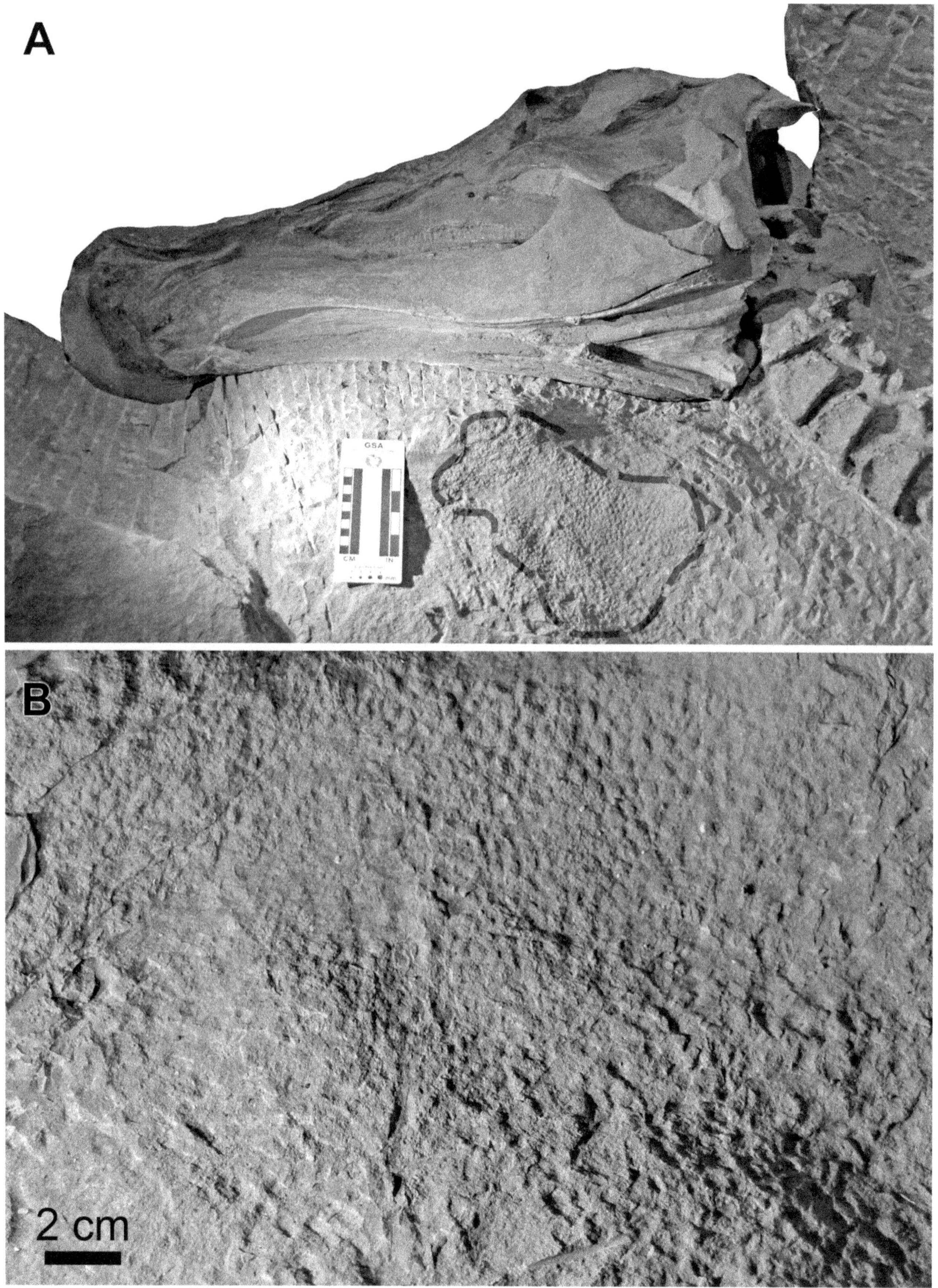

34.1. *Edmontosaurus annectens* (NMSG P 0001). (A) Ventrolateral view of skull and anterior cervical vertebrae showing skin impressions (outlined) within the throat region. Left scale bar equals 10 cm. (B) Close-up of throat scales showing undifferentiated polygonal basement scales.

The most striking integumentary feature is the series of triangular midline feature-scales (MFSs) that extend along the dorsal midline from the base of the neck to the base of the tail. It is possible that MFSs persisted along the tail; however, most of the tail was lost to erosion prior to discovery. MFSs occur dorsal to the tips of the neural spines, although details regarding the size and height above the neural spines were not reported (Murphy et al., 2007). Feature-scales were not identified from any other part of the body.

***Edmontosaurus annectens* (Marsh, 1892)** The integument of *Edmontosaurus* is perhaps the best known of any hadrosaurid and is preserved in association with several exceptional skeletons with skin impressions. The first hadrosaurid mummies were discovered in the Hell Creek Formation of Montana by G. H. Sternberg and C. H. Sternberg in 1908 (AMNH 5060; Osborn, 1911, 1912) and 1910 (SM 4036; Versluys, 1923). AMNH 5060 is still one of the best-preserved mummies in existence and preserves integumentary impressions on all parts of the body, excepting the hindlimbs and tail (the latter of which had eroded away prior to discovery). Soft-tissue structures within the circumnarial depression on the cranium are reported by Prieto-Márquez and Wagner (2013). An undescribed specimen (MRF-03) from the Hell Creek Formation of North Dakota may rival this specimen in terms of preservation quality and extent of integumentary covering (Manning, 2009; Manning et al., 2009).

In virtually all areas of the body, the integument of AMNH 5060 is characterized by cluster areas that include two types of basement-scales: smooth, polygonal scales forming clusters ("pavement tubercles" of Osborn [1912]) within a more continuous basement of smaller, pebbly scales. Polygonal basement-scales typically measure <5 mm in diameter and decrease in size towards the periphery of each cluster where they grade into pebbly basement-scales. Pebbles range from 1–3 mm in diameter. Individual clusters are small (2.0–4.5 cm in diameter) ventrally (i.e., throat, chest, and abdomen) where they are arranged in roughly longitudinal lines. More dorsally, the cluster areas increase in size (5–10 cm), expanding into more or less confluent but irregular patches of polygonal basement-scales over the pelvis. Osborn (1912) speculated that this pattern persisted along the dorsalmost parts of the body; however, integument in these areas is not preserved in AMNH 5060.

The forearms differ from this general plan in that cluster areas are absent. Instead, the dorsal surface of the forearm is covered in large polygonal scales (~10 mm; see also Manning et al. [2009]) and the ventral surface of the forearm is covered by pebbly basement-scales.

Horner (1984) first described MFSs on the tail of MOR V 007, an incomplete specimen he tentatively attributed to *Edmontosaurus*. The remainder of that skeleton (NMSG P 0001; Fig. 34.1) now at the Naturmuseum St. Gallen in Switzerland, includes a well-preserved skull that confirms its assignment to *E. annectens*. MFSs have now been confirmed on additional *Edmontosaurus* specimens (e.g., MRF-03; Manning, 2009). The MFSs in MOR V 007 are rectangular (50 mm long; 45 mm tall) and correspond to the dorsal tip of each neural spine. Each scale is separated by about 10 mm from the adjacent scale and the lateral surfaces are ornamented with vertical striations (Horner, 1984). Manning (2009) also commented on the considerable depth of the tail in the North Dakota specimen. In both MOR V 007 and MRF-03, the integumentary traces extend the tail's dorsoventral depth up to 50% more than is suggested by the caudal skeleton alone.

***Edmontosaurus regalis* Lambe, 1917** Skin impressions from an unknown part of the body (ROM 801) show cluster areas of irregular basement-scales that differ from those described by Osborn (1912) for *E. annectens:* cluster areas comprise irregular scales with deeply inscribed radial corrugations around the periphery (Fig. 34.2). The center of these scales forms an elevated and smooth plateau. Such corrugations have not been identified in *E. annectens*. Scales at the center of each cluster measure 5–7 mm in greatest dimension, and diminish to 4 or 5 mm in diameter at the edge of the cluster. The clusters measure approximately 3 cm in diameter, suggesting they may come from the ventral part of the body (Osborn, 1912). The intervening basement-scales are roughly polygonal and otherwise featureless – measuring, on average, 3 mm in diameter.

***Gryposaurus notabilis* Lambe, 1914b** A small area of skin impressions is preserved on the abdomen near the ischium of the holotype of *Gryposaurus notabilis* (CMN 2278). This area is characterized by a uniform covering of polygonal basement-scales averaging 5 mm in diameter and devoid of patterns or other features (Lull and Wright, 1942). Several other patches are preserved on a second specimen (ROM 764; holotype of *G. "incurvimanus"*); however, skin impressions that covered the ventral surface of the neck in that specimen were destroyed during preparation (Parks, 1920). Parks (1920) described a patch of skin (45 cm in diameter) in ROM 764 that apparently spanned most of the cervical vertebrae and consisted of a uniform covering of small (~3 mm in diameter) polygonal basement-scales lacking any other pattern or feature-scales. Other skin impressions are retained on the distal part of the right scapula and a small patch spanning thoracic ribs seven and eight at about mid-shaft. In both of these areas, the basement-scales are small (~2.5 mm and 5.0 mm in diameter on the scapula and ribs, respectively) and polygonal. Feature-scales are preserved on small patches of skin from along the right flank and a patch spanning the neural spines of dorsal vertebrae 7–9 (Fig. 34.3A). These are sub-conical, 12–15 mm

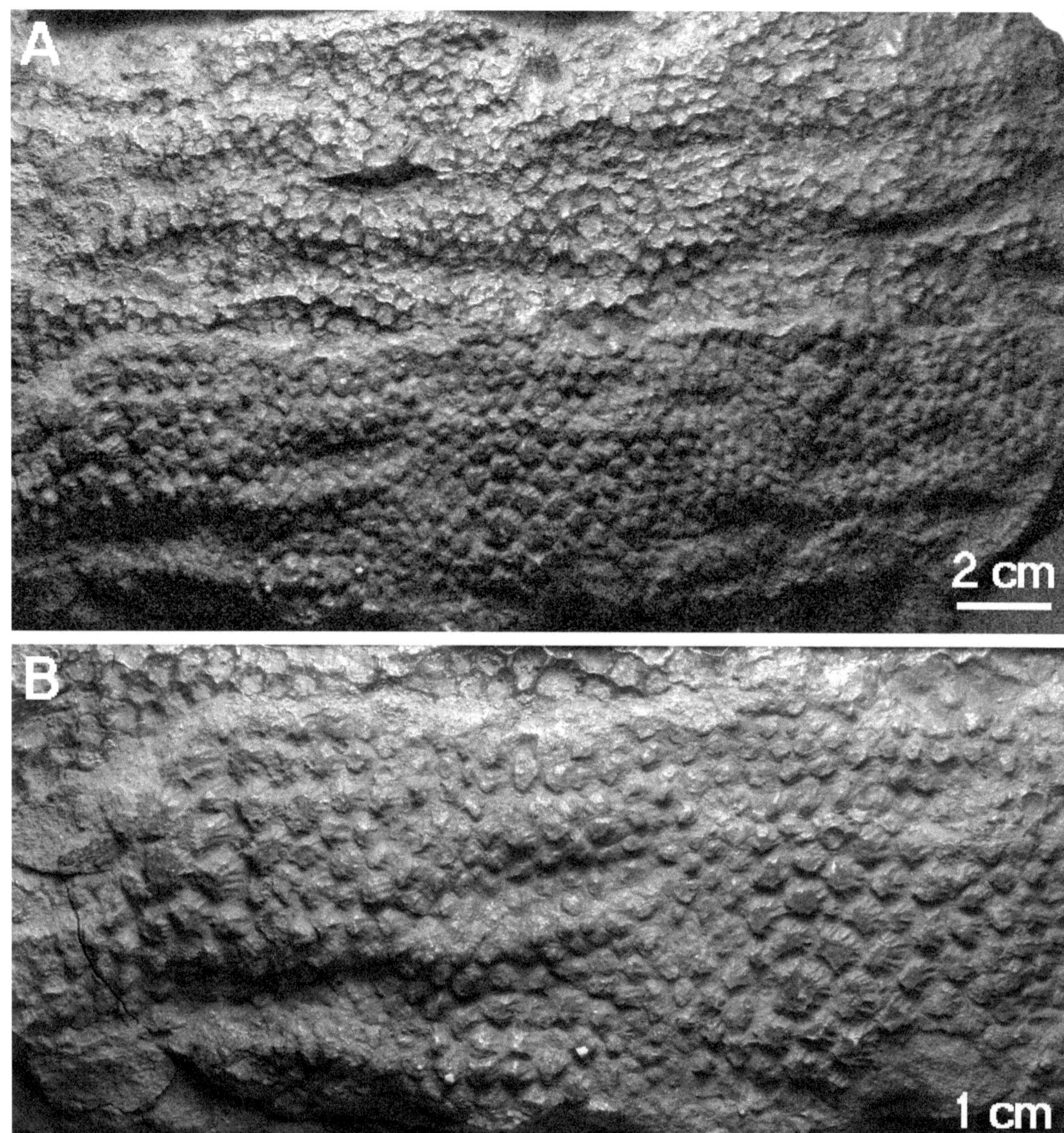

34.2. *Edmontosaurus regalis* (ROM 801). Cluster areas of irregular, corrugated basement-scales from an unknown part of the body. (A) Overview of specimen showing the underlying "negative" impressions with raised interstitial areas (top half of image) and layer of "positive" scales (bottom half). (B) Close-up of cluster areas showing radial ornamentation of larger scales.

in diameter, and radially corrugated. Although the patches of skin impressions are too small to discern a pattern of arrangement in the feature-scales, where two or more feature-scales occur, they are separated by a gap of 40–60 mm. Parks (1920) described these scales as composed of 12 radial sectors; however, this claim could not be confirmed. The surrounding basement-scales are polygonal, ranging from ~5 mm in diameter along the flank to 3 mm closer to the dorsal vertebrae.

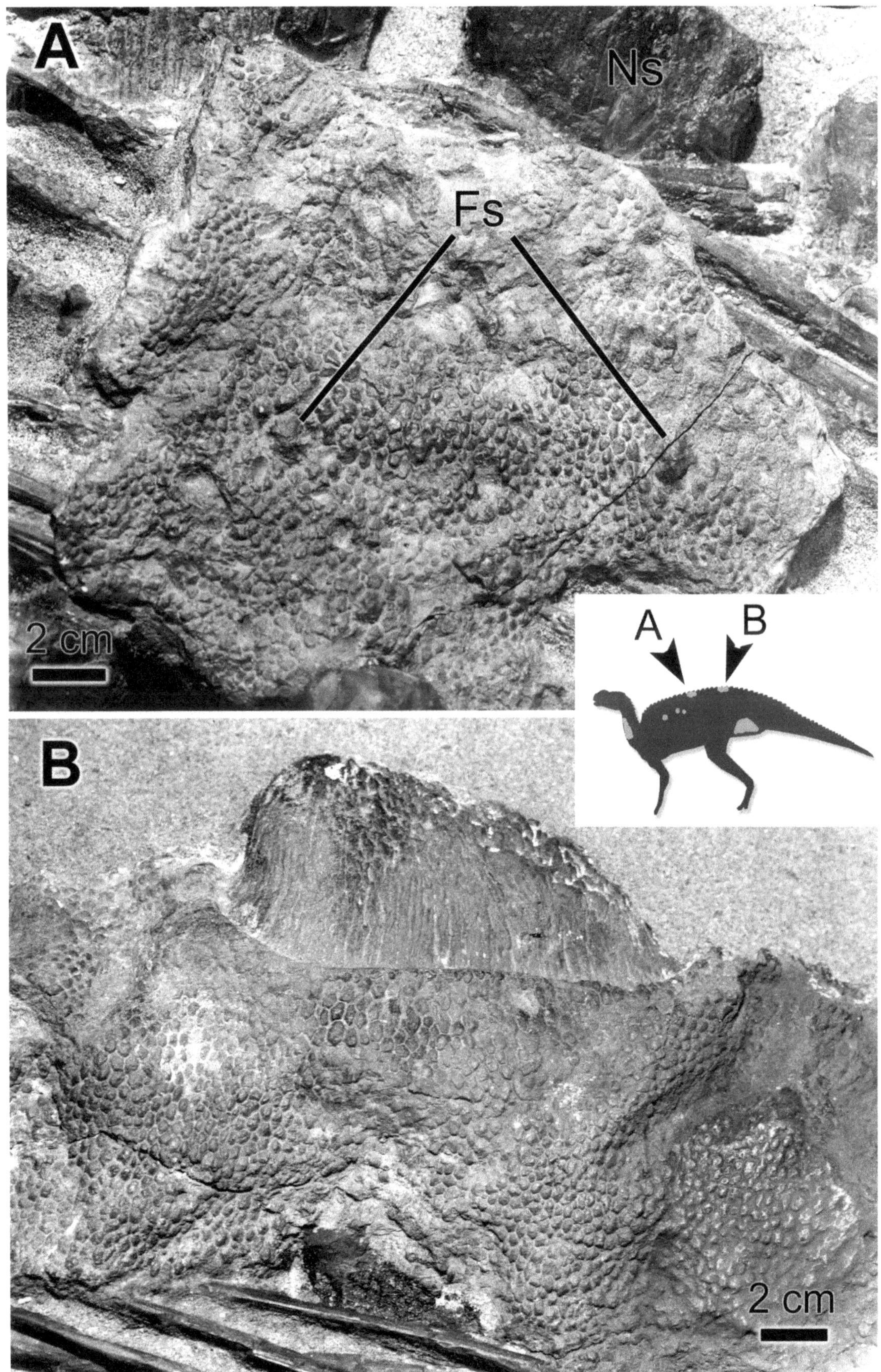

34.3. *Gryposaurus notabilis* (ROM 764). (A) Skin impressions from the mid-dorsals showing feature-scales (Fs); (B) midline feature-scale. Inset showing location of skin impressions figured in (A) and (B). Ns, neural spine. Photos courtesy of ROM archives.

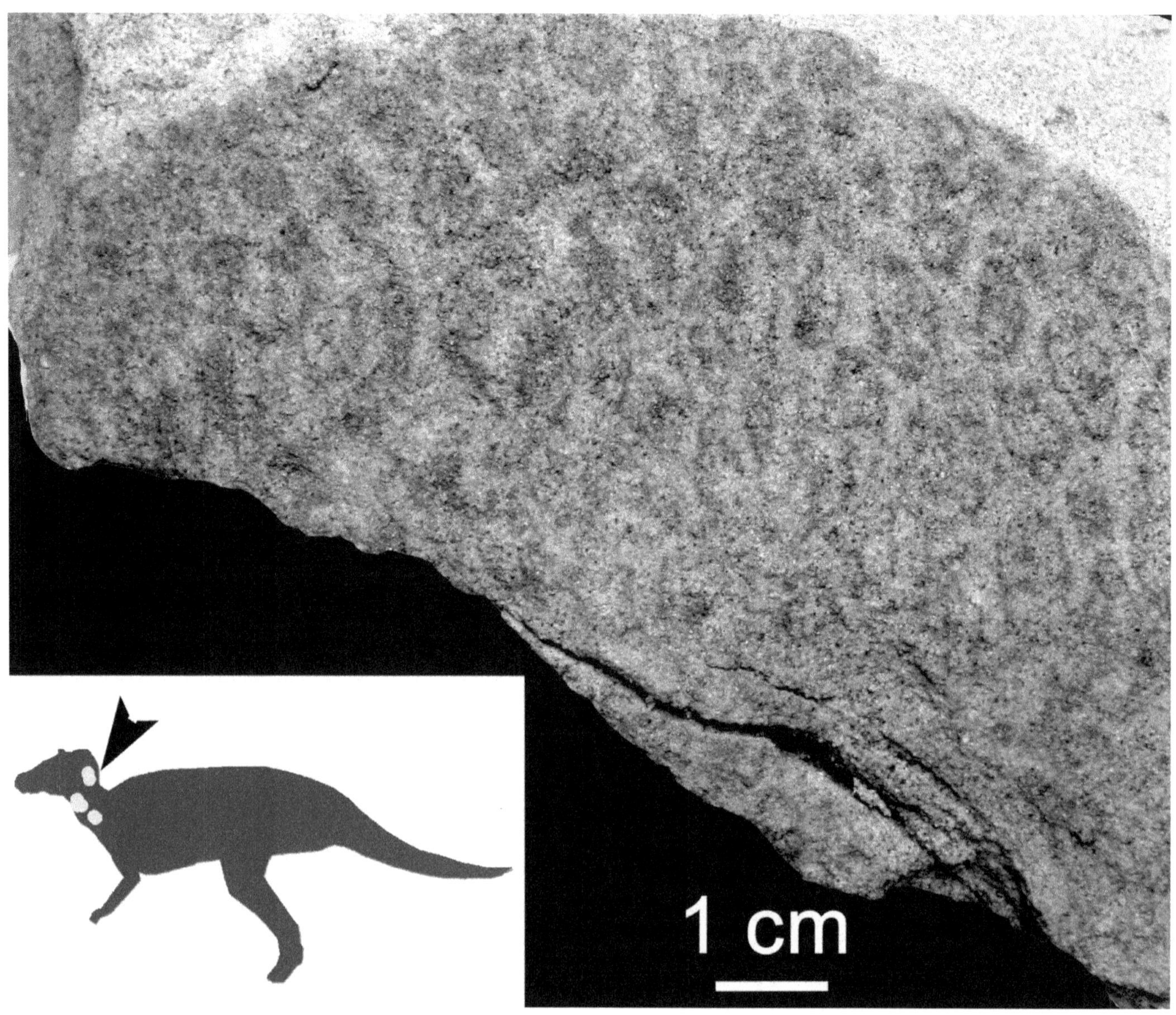

34.4. *Maiasaura peeblesorum* (ROM 44770). Polygonal basement-scales from the neck preserved as a faint mineral rind. Inset shows the position of figured skin impressions.

Midline feature-scales are preserved on a small patch of skin over the neural spines of the posterior dorsals and anterior sacral vertebrae (Fig. 34.3B). Although four vertebrae are enclosed by integumentary covering, only two MFSs are preserved, suggesting they occurred only on every second vertebra in this region of the body. MFSs are roughly trapezoidal, measuring 125 mm long and 60 mm tall. The overall structure of these scales is evocative of rhinoceros horns, which are composed of an arrangement of hair-like keratinized tubules (Heironymus et al., 2006). In ROM 764, the "hairs" are 2 mm wide and up to 35 mm long, extending from the base towards the dorsal midline of the MFS. The dorsal half of each MFS is composed of irregular "scales" that Parks (1920) asserted were similar to the basement-scales elsewhere on the body. However, on close examination, it appears that these scales are the hair-like scales viewed in section.

***Maiasaura peeblesorum* Horner and Makela, 1979** Skin impressions in *Maiasaura* (ROM 44770) are known only from the neck immediately behind the skull. Reports of a possible throat wattle or dewlap by Czerkas (1997) are unsubstantiated (pers. obs.). Rather, these impressions indicate a deep muscular neck as has been observed on other hadrosaurid specimens (e.g., *Corythosaurus* Brown, 1916). ROM 44770 shows a uniform covering of polygonal (usually five- or six-sided) basement-scales that are longer than they are dorsoventrally tall (Fig. 34.4). Scales measure on average 8–10 mm long and 4–6 mm tall. The interstitial spaces are relatively wide, typically measuring 1–2 mm and – in some places – up to 3 mm.

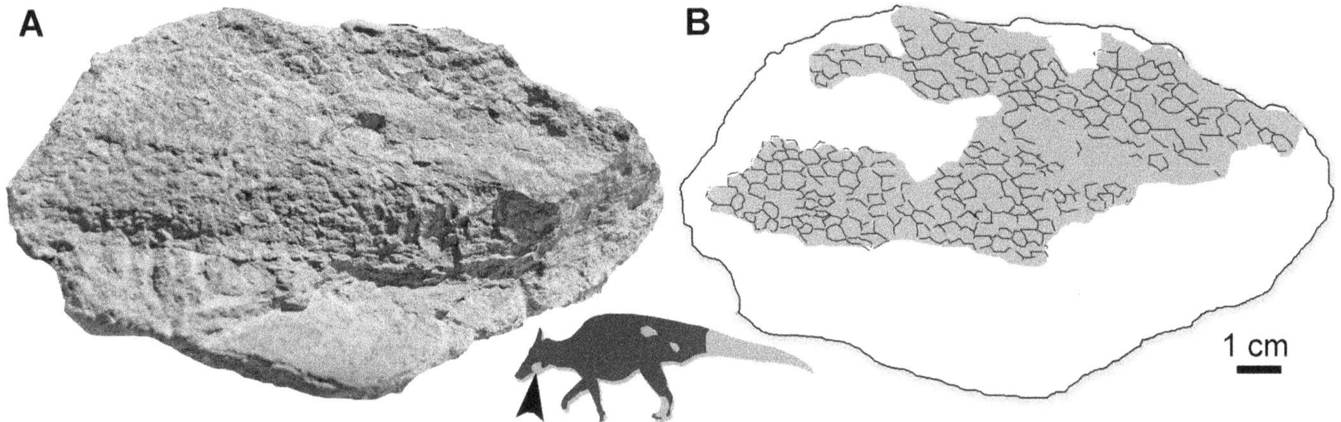

34.5. *Saurolophus osborni.* (A, B) Skin impressions from the right dentary of AMNH 5220. Skin impressions are shown in gray. Photograph by E. Snively.

Trexler (1995) also noted possible integumentary vestiges on the premaxilla and nasal crest of a specimen (OTM F138) from the Two Medicine Formation (Campanian) of Montana; however, they were not preserved well enough to exhibit individual scales, and thus cannot be conclusively identified as skin impressions at this time.

Saurolophus osborni **Brown, 1912** Patches of skin associated with the mandible, pelvis, and pes were found with the holotype specimen of *Saurolophus osborni* (AMNH 5220). The mandible had a uniform covering of polygonal basement-scales that are anteroposteriorly longer than tall (5–6 mm long, 3–4 mm tall; Fig. 34.5; Bell, 2012). Isolated skin impressions recovered from the lateral body of the ilium and the pes are composed of polygonal basement-scales (4 mm in diameter). Scales on the anterior surface of the metatarsus are considerably smaller (2 mm in diameter) and pebbly, becoming larger (up to 7 mm in diameter) over the digits. A number of additional skin impressions from unknown regions of the body of AMNH 5220 and AMNH 5221 (paratype) demonstrate further diversity in the scale architecture of *S. osborni*. Several feature-scales are interspersed within a basement-of polygonal scales. Feature-scales are up to 45 mm in diameter, the edges of which occupy the angles formed by the adjacent polygonal basement-scales, thus forming a multi-pointed star (Bell, 2012).

A third specimen (AMNH 5271) preserves discontinuous skin impressions along much of the tail. Proximally, scales form cluster areas of small (1–2 mm in diameter) and large (4–5 mm in diameter) polygonal basement-scales on the lateral surfaces of the centra (Bell, 2012). Clusters grade into one another. More distally, clusters of small scales are absent, with the area consisting entirely of large (4–6 mm in diameter) polygonal (sometime irregular) basement-scales. Basement-scales decrease to 3 mm in diameter toward the end of the tail. Rare, dorsoventrally elongate feature-scales are present close to the centrum-chevron articulations. These feature-scales have irregular outlines and range from 6–8 mm in anteroposterior length (Bell, 2012).

Saurolophus angustirostris **Rozhdestvensky, 1952** Skin impressions of *Saurolophus angustirostris* are represented by numerous individuals spanning a wide range of sizes and ages. Consequently, it is one of the best known taxa in terms of scale architecture (Bell, 2012). Patches of skin over the infratemporal fenestra and jugal-quadratojugal-quadrate of a juvenile specimen (ZPAL MgD I-159) show small (1 mm) pebbly scales in this region of the face. Pebbly scales are also present along the length of the entire forelimb and are only slightly larger (2–3 mm) over the ribs.

In the same specimen, irregular basement-scales (3 mm in diameter) occupy the proximal hindlimb and knee joints, which also include domical shield feature-scales. Feature-scales are ~7 mm across and are aligned in rows 20–25 mm apart in a grid-like arrangement. Feature-scales are ornamented with radial striations around their margins.

Scale patterns along the length of the tail are remarkably diverse, as exemplified by several large individuals (PIN 3747, PIN 3738, UALVP 52748). Proximally, the integument is divided into alternating vertical bands of morphologically distinct scales, which Bell (2012) categorized as zones "A" and "B" (Fig. 34.6). Zone A is typified by polygonal basement-scales interspersed with shield feature-scales (15–25 mm in diameter). Basement-scales are largest (10 mm in diameter) at the center of each band, and decrease in diameter towards the edges where they grade into zone B. Zone B bands are comparatively narrow (20–30 mm wide, compared with 50 mm in zone A) and comprise only corrugated shell basement-scales. More distally, the tail apparently lacks vertical zoning and feature-scales, and instead is composed of

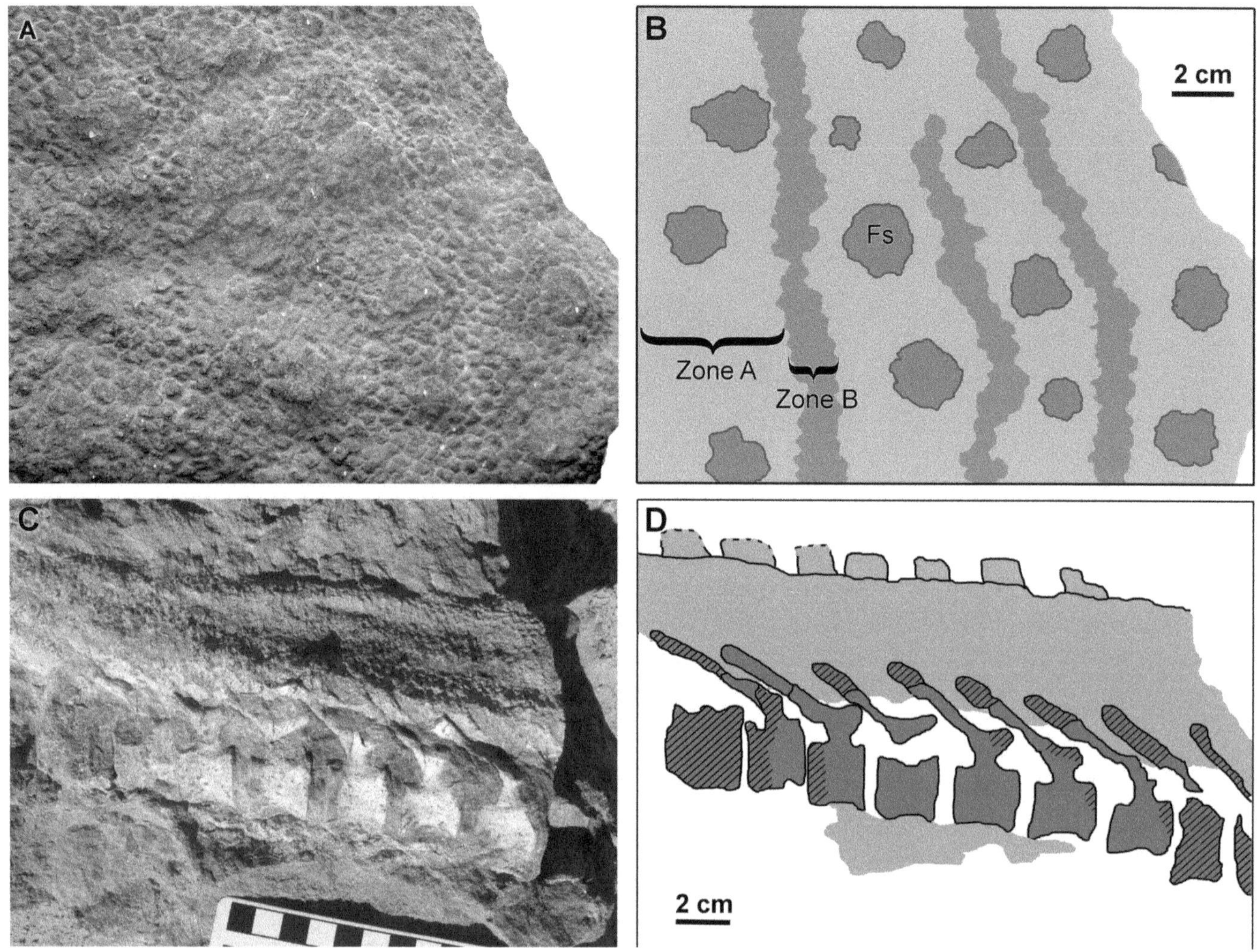

34.6. *Saurolophus angustirostris* tail skin. (A, B) Proximal tail skin (PIN 3747-2) showing vertical banding and large shield feature-scales. (C, D) Field photograph of a juvenile specimen (UALVP 52824) showing tabular midline feature-scales on the distal tail. Note that vertical banding and regular feature-scales are absent. Fs, feature-scale. Cross-hatching denotes broken bone surface. Skin impressions in (D) are highlighted in light gray.

regular polygonal scales, 5 or 6 mm in diameter. MFSs are present along the entire length of the tail (Bell, 2012). In general, midline feature-scales are longer than they are tall, but increase in size with the increased absolute size of the individual. The lateral sides of each MFS are ornamented by several posterodorsally oriented corrugations that contrast with the fine, hair-like ornamentation on the MFSs of *G. notabilis* (ROM 764). Midline feature-scales were present on both juveniles (UALVP 52824) and adults (UALVP 52748).

Lambeosaurinae (sensu Prieto-Márquez, 2010)

***Corythosaurus casuarius* Brown, 1914** The holotype of *Corythosaurus casuarius* (AMNH 5240) is one of the most complete hadrosaur mummies in existence. Brown (1916) described the integument of this specimen as being preserved in virtually all parts of the body except the cranium and appendicular skeleton. As in *Brachylophosaurus* (JRF 115; Murphy et al., 2007), preservation of the skin envelope around the cervical skeleton indicates that the neck was considerably deeper and more muscular that the skeleton alone suggests. The greater part of the body is typified by a uniform covering of polygonal basement-scales. Although Brown (1916) did not comment on the actual size of the basement-scales, he did note that the scales are smaller along the inner thigh and the anterodorsal region of the trunk than in other areas. Scales on the flank and belly of a second specimen (TMP 1980.040.0001) are polygonal, and many are triangular, 3–4 mm in maximum length. Basement-scales on the distal tail and dorsal to the proximal caudal skeleton are somewhat larger than elsewhere on the body (Brown, 1916). Although Brown (1916) noted that the dorsal margin of the integument above the axial skeleton is not preserved, midline feature-scales were not observed elsewhere.

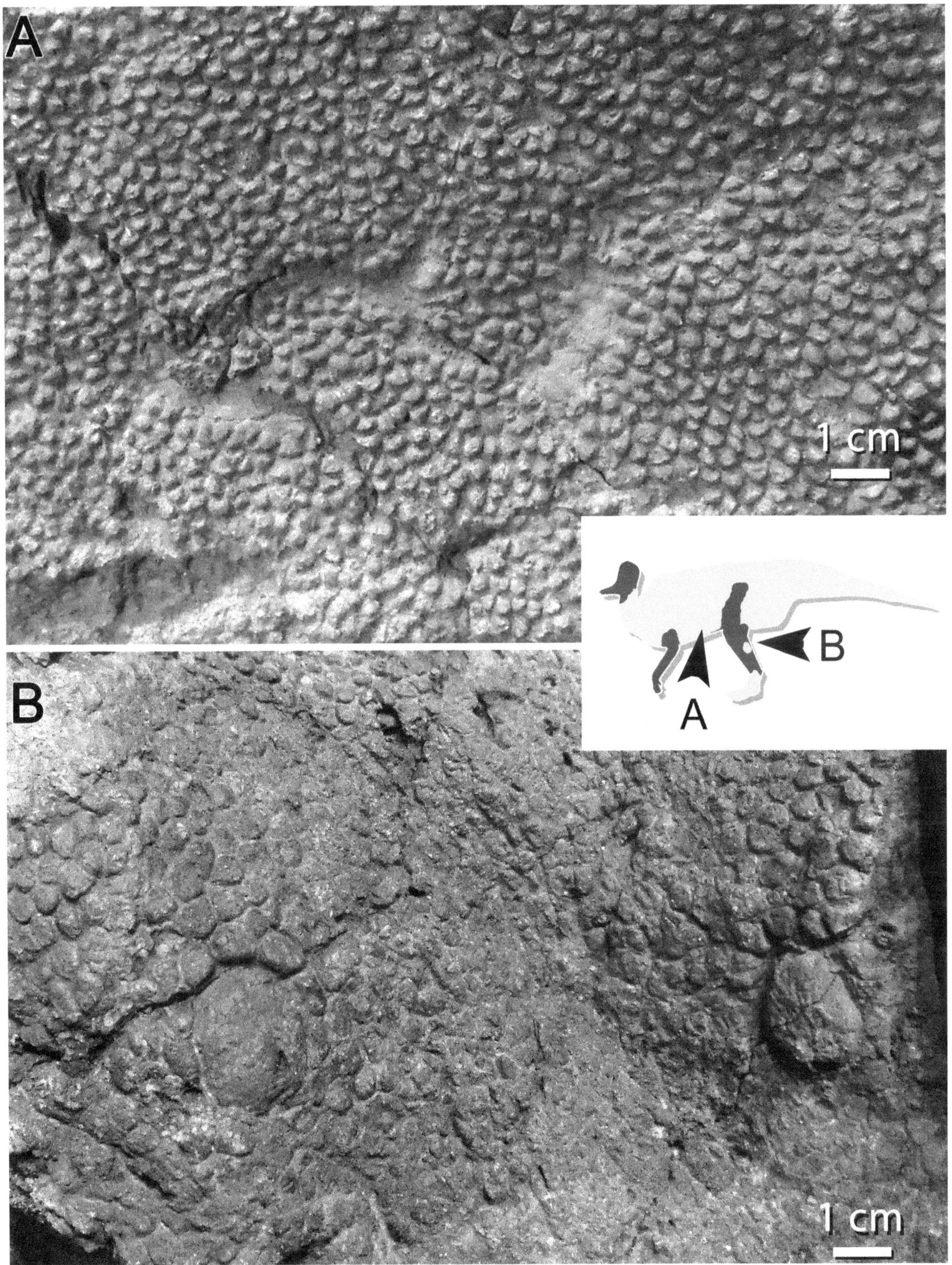

34.7. *Corythosaurus casuarius* (TMP 1980.040.0001). (A) Undifferentiated pebbly basement-scales on the soft part of the abdomen; (B) skin impressions on the distal hindlimb showing domical feature-scales (arrowheads). Inset shows the positions of figured skin impressions.

34.8. *Lambeosaurus lambei* (ROM 1218). (A) regular polygonal basement-scales from an unknown part of the body; (B, C) isolated segmented feature-scale.

The most notable feature of the integument is the presence of domed shield feature-scales within the region of the crus, ventral to the pelvis of AMNH 5240. Feature-scales measure 45 mm in length and 30 mm in dorsoventral height. At least two horizontal rows of such feature-scales are visible in this area and are spaced ~5 cm apart. Within each row, individual feature-scales are spaced approximately 2–3 cm apart. Surficial ornamentation on the scales varies from smooth to irregular undulations to coarse radial corrugations; however, it is possible that both undulations and corrugations may be a taphonomic feature caused by the wrinkling and/or desiccation of the original soft tissue. Basement-scales in this region are polygonal and comparable in size to those observed on the tail. A patch of skin adjacent to the left fibula of TMP 1980.040.0001 conforms to this general pattern (Fig. 34.7). Basement-scales are smooth and polygonal (3–4 mm in diameter) and interspersed with shield feature-scales 16 mm in diameter. The disposition of feature-scales on the lower leg in addition to the crus suggests the entire hindlimb, with exception of the pes, was covered in such a pattern. The pedes of AMNH 5240 are covered in a uniform arrangement of polygonal basement-scales typical of that on the rest of the body.

An immature specimen (CMN 8676, = *C. "excavatus"*) described by Sternberg (1935) preserves the integument from the dorsal surface of the forearm. The skin is dominated by large (18 mm in diameter) polygonal basement-scales, which grade into an area of smaller (4 mm in diameter) polygonal scales along the periphery of the specimen. These smaller scales are likely displaced from the ventral surface of the forearm similar to the configuration in *Edmontosaurus annectens* (AMNH 5060).

***Lambeosaurus lambei* Parks, 1923** In CMN 8703 (= *L. "clavinitialis"*), skin impressions are preserved over the posterior ribs and hindlimb anterior to the femur. The entire region is covered by a uniform arrangement of "small

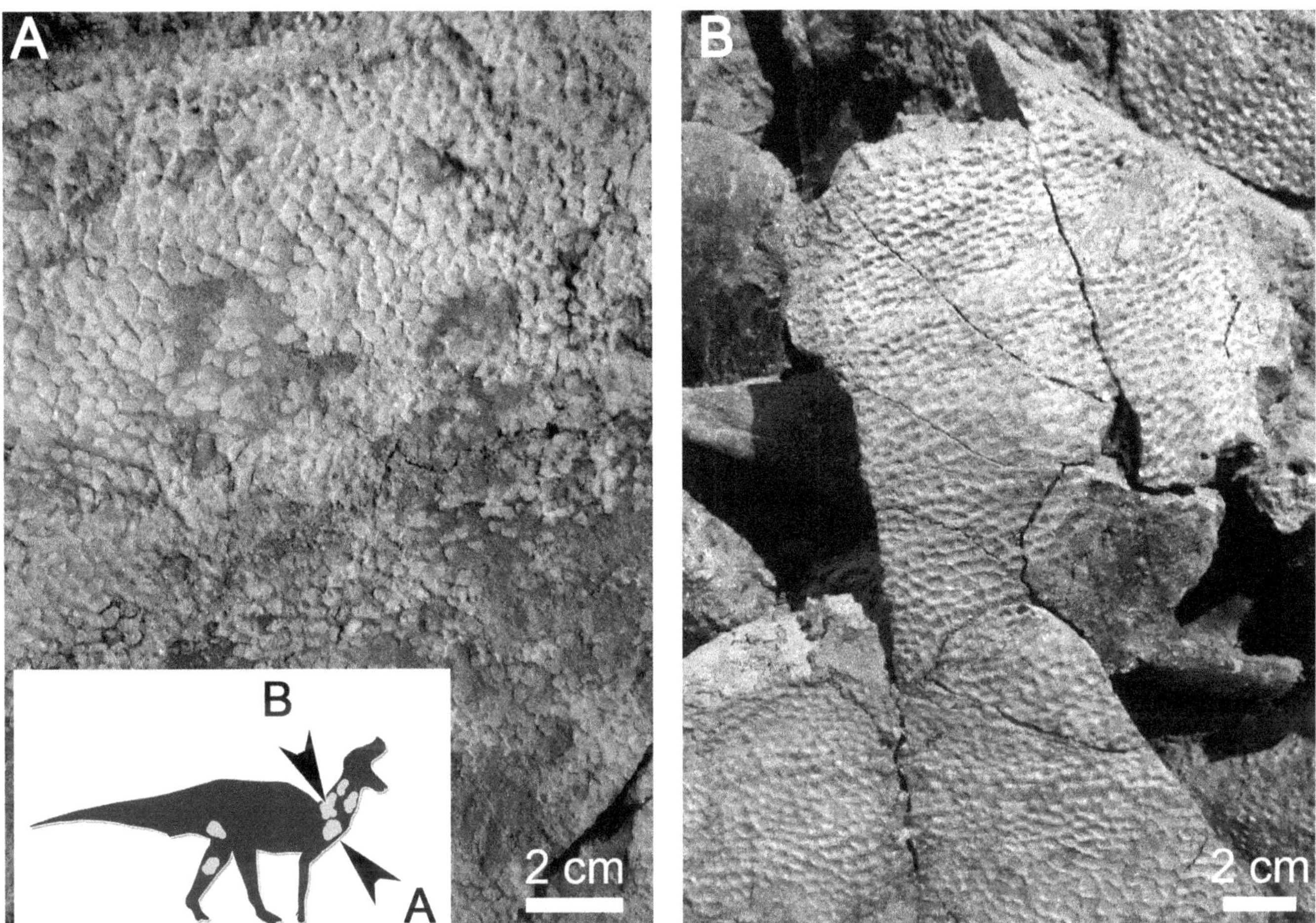

34.9. *Lambeosaurus magnicristatus* (TMP 1966.004.0001). Undifferentiated, polygonal basement-scales from the neck. (A) Close-up of throat area; (B) overview of neck scales. Inset shows the position of figured skin impressions.

tubercles without differentiation of size or pattern" (Sternberg, 1935:19) and feature-scales are absent (Fig. 34.8A). A large section of skin impression across the proximal 1.2 m of the tail preserves a similar integumentary covering devoid of feature-scales; however, the basement-scales in this area are, on average, larger than those on the flank and hindlimb. Sternberg (1935) also observed a lack of folding in the skin that extends from the torso to the femur, which he suggested was evidence for the proximal hindlimb being largely enclosed within the flank as in modern equids or bovids.

A second specimen of skin impression (ROM 1218) from an unknown part of the body preserves relatively large (7–9 mm in diameter) polygonal basement-scales. The most unusual feature in this specimen is a subcircular arrangement of scales 12 mm in diameter (Fig. 34.8B, C). The circle comprises eight wedge-shaped sectors that converge on the central point in an arrangement that is similar to the feature-scales described from the flank of *G. notabilis* by Parks (1920).

***Lambeosaurus magnicristatus* Sternberg, 1935** Evans and Reisz (2007) described skin impressions in TMP 1966.004.0001 from the dorsal and ventral regions of the neck, the proximal humerus, and crus. In all areas, the basement-scales are polygonal and average 5 mm in diameter (Fig. 34.9). They form a uniform covering devoid of feature-scales or cluster areas (Evans and Reisz, 2007). The authors also noted an unusual relationship between raised polygonal scales and scales demarcated by raised interstitial tissue, similar to what would be expected in a negative impression. In the area of the neck, the raised scales lie immediately external (superficial) to the grid of raised interstitial tissue. This suggests the preservation of both epidermal and deeper (dermal?) tissues (cf. Manning et al., this volume). Regardless of their relationship, the size and morphology of the scales and deeper integument are consistent with one another. A similar relationship between raised scales and interstitial tissue is present also on a specimen of *Edmontosaurus regalis* (ROM 801) from an unknown region of the body (Fig. 34.2).

Parasaurolophus walkeri Parks, 1922 Skin impressions were found with the holotype of *Parasaurolophus walkeri* (ROM 768) and described briefly by Lull and Wright (1942). Although Parks (1922) also noted their presence on the holotype, in neither of these publications were skin impressions figured, nor was any specific information given regarding their location on the body or the size of the individual scales. The scales are preserved in patches on the flank and comprise a uniform arrangement of polygonal basement-scales devoid of feature-scales (Parks, 1922; Lull and Wright, 1942).

Hadrosauridae indet.

A considerable number of publications exist that deal with the scale architecture of incomplete and/or undiagnostic hadrosaurid material. Such studies serve to document the increasing diversity of scale morphology within Hadrosauridae. This section is not exhaustive but includes a number of the most significant integument traces that cannot be assigned to a particular genus.

With the exception of *Edmontosaurus* and both species of *Saurolophus*, only a single other case of skin impressions has been reported from the face of a hadrosaur. Wegweiser et al. (2006) described polygonal basement-scales from the area immediately ventral to the dentary and posterior to the mandibular symphysis of an indeterminate hadrosaurid (UW-39449) from the upper Lance Formation (Maastrichtian) of Wyoming. The basement-scales average 5 mm in diameter and preserve faint, but clear, ridges that fan out from one side of the scale rather than radiating from a central point.

In hadrosaurids, skin impressions are best represented on the tails and, in that region, they show the greatest variation in scale architecture. A partial hadrosaur skeleton from the Nelsen Formation (upper Campanian), Utah, preserves large areas of integumentary impressions that extend from the mid-torso region to the middle of the caudal series (Anderson et al., 1999). The specimen apparently preserves a gradation of polygonal basement-scales that increase in size from along the flanks and tail (average scale diameter 7 mm) to nearly twice that size (average 13 mm) in the area dorsal to the dorsal vertebrae. Although Anderson et al. (1999) described the latter scales as polygonal, deep radial corrugations give these scales a more irregular outline. The authors also note radially corrugated shield feature-scales on the tail, dorsal surface, and flanks. However, there is no indication of a regular arrangement (e.g., horizontal lines) in these feature-scales.

Similarly, corrugated scales were identified on a hadrosaurid (NMMNH P-2611) from the Ringbone Formation (Campanian–Maastrichtian) of New Mexico (Anderson et al., 1998). Skin impressions from the proximal to mid-caudal series comprise rather conical, subcircular basement-scales that range in size from 3 × 3 mm to 13 × 16 mm. No clear arrangement between large and small scales is apparent. Virtually all scales are ornamented with deep radial corrugations that converge (or nearly so) at the central apex.

Clayton et al. (2011) provided a preliminary description of a partial saurolophine with exquisite skin impressions from the Kaiparowits Formation of Utah. Those authors tentatively assigned the specimen (UMNH VP 12265) to *Gryposaurus*, the only positively identified saurolophine from that formation (Gates and Sampson, 2007). Skin impressions extend approximately two-thirds the length of the tail and comprise irregular, corrugated basement-scales ranging in diameter from 5–12 mm. Interestingly, the average diameter of the basement-scales increases distally, a feature also observed in polygonal basement-scales on the distal tail of an unidentified hadrosaur from the Wapiti Formation, Alberta (UALVP 50688). Clayton et al. (2011) also identified heart-shaped feature-scales (7–11 cm in length) associated apparently with the tips of the caudal neural spines. These scales represent midline feature-scales that are dorsoventrally shorter than previously described MFSs (Parks, 1920; Horner, 1984; Bell, 2012). They are keeled with fine, radiating, but nonparallel grooves that converge along the midline, similar to the midline feature-scales described for *G. notabilis* (Parks, 1920).

Lambe's (1902) "*Kritosaurus marginatus*" (AMNH 5894) from the Dinosaur Park Formation of southern Alberta preserves skin impressions at the base of the tail. Based on the dimensions of the associated humerus (Lambe, 1902:pl. 6), this specimen may represent a saurolophine. Skin impressions at the tail base exhibit an arrangement of conical feature-scales (~13 mm in diameter) dispersed within a basement of smaller polygonal scales (Lull and Wright, 1942) that are reminiscent of the abdominal scales of *Corythosaurus casuarius* (AMNH 5240). Feature-scales are separated from one another by a distance of 5–7 cm.

Gates and Farke (2009) described isolated skin impressions from a partial skeleton (AMNH 3651) from the Maastrichtian Almond Formation (Wyoming), which those authors identified as *Saurolophus*. This assignment however is tenuous (Bell, 2010); here the specimen is referred to Saurolophinae indet. Skin impressions from this specimen were removed from the skeleton and an understanding of their anatomical positions has been lost. Impressions consist of polygonal to circular basement-scales (up to 4.5 mm in diameter) with no evidence of feature-scales (Gates and Farke, 2009). Large (>2 mm) circular basement-scales have not been previously recognized, although circular feature-scales and pebbly basement-scales have been observed in other hadrosaurs such as *Saurolophus* (Bell, 2012).

Footprints A small number of unequivocal hadrosaurid footprints from North America reveal impressions of scales on

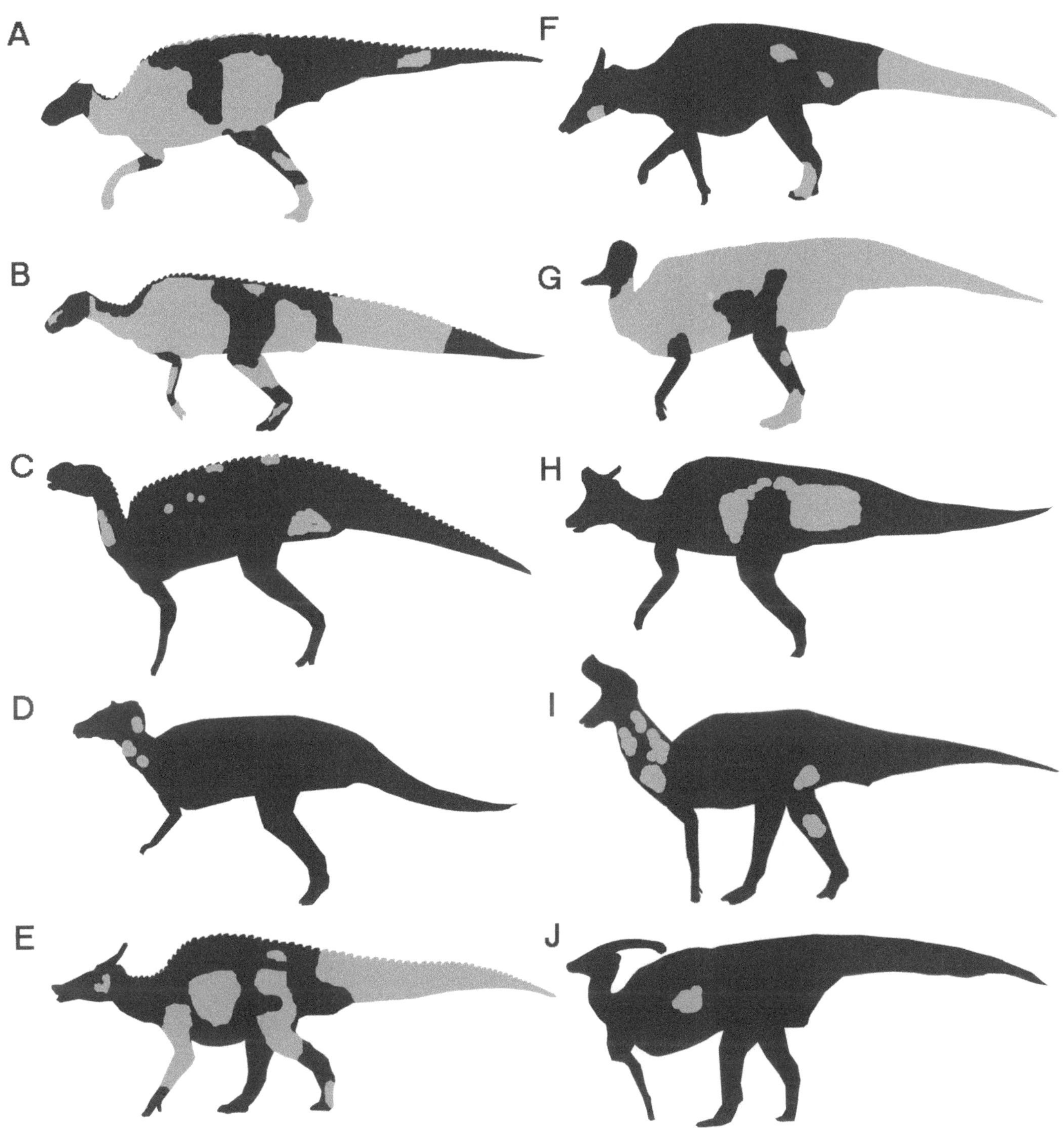

34.10. Preserved areas of skin (gray) known for Hadrosauridae. (A) *Brachylophosaurus canadensis;* (B) *Edmontosaurus annectens;* (C) *Gryposaurus notabilis;* (D) *Maiasaura peeblesorum;* (E) *Saurolophus angustirostris;* (F) *Saurolophus osborni;* (G) *Corythosaurus casuarius;* (H) *Lambeosaurus lambei;* (I) *Lambeosaurus magnicristatus;* (J) *Parasaurolophus walkeri.* Not to scale.

the sole of the pes (Currie et al., 1991; Lockley et al., 2003). These tracks, attributed to the ichnotaxon *Hadrosauropodus langstoni*, are restricted to the St. Mary River Formation (Alberta) and the Lance Formation (Wyoming) and preserve discontinuous patches of integumentary impressions on the digital and metatarsophalangeal (heel) pads. Skin impressions on the digital pads consist of relatively uniform, and small sub-ovatepebbles (1.5–3.5 mm in diameter) but are larger (3.5–5.0 mm in diameter) on the heel (Currie et al., 1991). Footprints from the Lance Formation conform to this general pattern (Lockley et al., 2003); however, the larger scales on the heel are distinctly polygonal (pentagonal and hexagonal).

DISCUSSION

Characterization of Hadrosaurid Skin Impressions

Fossilized integument is preserved to some degree in approximately one-third (11 of the 34) of valid hadrosaurid taxa (based on Prieto-Márquez, 2010; Fig. 34.10). More specifically, among saurolophines (n = 15 valid taxa) integument is known in 7 taxa (47%) and among lambeosaurines (n = 19 valid taxa) integument is known in 4 taxa (21%). Curiously, however, skin impressions have only been reported for a single taxon outside of North America (*Saurolophus angustirostris*).

In general, skin impressions from the cranium and mandibles are exceptionally rare, with scant remains described in only four hadrosaurid specimens. Specifically, cluster areas of polygonal scales interspersed within a larger area of pebbly basement-scales are present on the quadrate and the region immediately posterior to it in *E. annectens* (Osborn, 1912). Pebbly basement-scales are present on the posterolateral part of the face in a juvenile *S. angustirostris* and undifferentiated polygonal basement-scales cover the mandible in *S. osborni*. An indeterminate hadrosaur from the Lance Formation preserves polygonal basement-scales on the anterior dentary similar to *S. osborni*, but these are ornamented by fine corrugations that fan out from the ventral margin of each scale (Wegweiser et al., 2006).

The cervical region is invariably surrounded by a mass of muscle that gave the neck a considerably deeper profile than is suggested by the cervical skeleton alone. The cervical skin of *Brachylophosaurus*, *Corythosaurus*, *Lambeosaurus magnicristatus*, and *Maiasaura* all consist of polygonal basement-scales, although those of *Maiasaura* are up to twice as large as those in other taxa. Only *E. annectens* deviates from this pattern in the presence of cluster areas of polygonal basement-scales dispersed within a larger area of pebbles.

In *E. annectens*, pebbly basement-scales are present on the ventral surface of the forearm and medial surface of the humerus; however, the humerus also bears small cluster areas of polygonal basement-scales (Osborn, 1912). The dorsal surface of the entire arm in *E. annectens* and the ventral surface of the forearm in *B. canadensis* are covered in a uniform basement of polygonal scales up to 10 mm in diameter (Osborn, 1912; Murphy et al., 2007). *Corythosaurus* matches the condition in *Edmontosaurus* in the disposition of polygonal basement-scales on the dorsal surface of the forearm and pebbly basement-scales on the ventral surface. In contrast, the entire shoulder and arm of *S. angustirostris* is covered in an undifferentiated basement of pebbly scales. Undifferentiated polygonal basement-scales are also present on the anterior surface of the humerus of *L. magnicristatus* (Evans and Reisz, 2007).

Lull and Wright (1942) recognized two categories of scale patterns on the flanks of hadrosaurs (Table 34.2). The first comprises an undifferentiated covering of polygonal basement-scales (*B. canadensis*, *C. casuarius*, *G. notabilis*, *L. lambei*, *P. walkeri*, and *S. angustirostris*). The second category consists of a mosaic of polygonal basement-scales arranged into cluster areas restricted to *E. annectens* (Osborn, 1912).

The scale architecture of hadrosaurs is best represented along the tail and demonstrates the morphological diversity of integument in this group. Midline feature-scales are known in *B. canadensis* (Murphy et al., 2007), *E. annectens* (Horner, 1984; Manning, 2009), *G. notabilis* (Parks, 1920), *S. angustirostris*, and cf. *Gryposaurus* from Utah (Clayton et al., 2011). In all except *Brachylophosaurus*, where they are not fully described (Murphy et al., 2007), the midline feature-scales are ornamented with parallel striae that converge along the midline of each scale. The striae in *Gryposaurus* appear to be smaller and finer than in other taxa and are nonparallel in the Utah specimen (Clayton et al., 2011). It is unclear whether the triangular profile of the midline feature-scales of *Brachylophosaurus* and *Edmontosaurus* (Manning et al., 2011:fig. 1) is in fact different from the more tabular profile in *G. notabilis* or whether this is a preservational (deformation) artifact (Clayton et al., 2011; pers. obs.). Tail integument can be classified into several groups (Table 34.2): (1) undifferentiated and smooth polygonal basement-scales (*C. casuarius*, *L. lambei*, distal tail of *S. angustirostris*); (2) basement-scales showing fine radial striae (Gillette et al., 2002) or radial corrugations (Anderson et al., 1999; Clayton et al., 2011) (several indeterminate hadrosaurs); (3) polygonal basement-scales punctuated by shield feature-scales, which may bear radial ornamentation on both basement- and feature-scales (Anderson et al., 1998) or feature-scales only ("*T. marginatus*"; Lambe, 1914a). Furthermore, the tail integument in species of *Saurolophus* is the most complex for any hadrosaurid and includes two additional categories: (4) cluster areas of polygonal basement-scales occur proximally and graduate to polygonal basement-scales with rare, irregular feature-scales more distally (*S. osborni*); (5) alternating vertical bands of shell and polygonal basement-scales accompanied by shield feature-scales present on the proximal tail (*S. angustirostris*).

Skin surrounding the pelvic and abdominal regions of *G. notabilis*, *L. magnicristatus*, and species of *Saurolophus* is characterized by an undifferentiated basement of polygonal scales (Lambe, 1914b; Evans and Reisz, 2007) whereas *E. annectens* retains the mosaic of cluster areas that is present elsewhere on the body (Osborn, 1912). A third category is present in *Corythosaurus*, where a grid-like arrangement of shield feature-scales punctuates a basement of polygonal scales.

Table 34.2. Distribution and Variation of Scale Morphology in Hadrosauridae Exclusive of Unidentified Specimens

Body region	Scale configuration	Taxa
Face	Polygonal basement scales only	*Saurolophus angustirostris, Saurolophus osborni*
Neck	Polygonal basement scales only	*Brachylophosaurus canadensis, Corythosaurus casuarius, Gryposaurus notabilis, Lambeosaurus magnicristatus, Maiasaura peeblesorum*
	Pebbly basement scales with cluster areas of polygonal basement scales	*Edmontosaurus annectens*
Flank	Polygonal basement scales only	*Brachylophosaurus canadensis, Corythosaurus casuarius, Gryposaurus notabilis, Lambeosaurus lambei, Parasaurolophus walkeri, Saurolophus angustirostris*
	Pebbly basement scales with cluster areas of polygonal basement scales	*Edmontosaurus annectens*
Abdomen/Pelvis	Polygonal basement scales only	*Gryposaurus notabilis, Lambeosaurus magnicristatus, Saurolophus angustirostris, Saurolophus osborni*
	Polygonal basement scales with shield feature scales	*Corythosaurus casuarius*
Hindlimb	Polygonal basement scales only	*Edmontosaurus annectens, Lambeosaurus lambei, Lambeosaurus magnicristatus*
	Polygonal basement scales with shield feature scales	*Corythosaurus casuarius, Saurolophus angustirostris*
Tail	Polygonal basement scales only	*Corythosaurus casuarius, Lambeosaurus lambei, Saurolophus angustirostris* (distal tail only)
	cluster areas of polygonal basement scales	*Saurolophus osborni*
	Alternating vertical bands of shell and polygonal basement scales; shield feature scales present on bands of polygonal scales only	*Saurolophus angustirostris*
Midline feature scales	N/A	*Brachylophosaurus canadensis, Edmontosaurus annectens, Gryposaurus notabilis, Saurolophus angustirostris*

Skin impressions are poorly represented on the legs of most hadrosaurids, and two categories are recognized: polygonal basement-scales devoid of feature-scales (*E. annectens*, species of *Lambeosaurus*, and the pes of *S. osborni*); domical shield feature-scales distributed across a basement of polygonal scales (*S. angustirostris* and *C. casuarius*).

Comments on Variation and Conservatism in Scale Architecture

With relatively few exceptions, skin impressions are represented in only a single specimen for any given taxon. Where multiple specimens of a taxon are present with skin impressions, multiple occurrences of body parts with integument is rare. Because of these insufficiencies, the wholesale characterization of scale architecture for virtually all hadrosaurids is not possible. Similarly, individual and sexual variations of integumentary morphologies and distributions are not yet understood. Accordingly, an outstanding question is this: To what degree do the documented differences in scale architecture between taxa constitute actual taxonomic differences? Although a thorough analysis is beyond the scope of this chapter, a comparison with modern squamates and archosaurs provides some insight into this problem. Scale morphology (especially scale counts) in modern squamates and crocodilians is typically conserved intraspecifically and is highly important in species identification (Spearman, 1973; Brazaitis, 1987; Hall, 1989; Cox et al., 1993; Charette, 1995; Branch, 1998). However, scale size (inferred by the inverse relation to number of scales) has been shown to vary positively with body size in some lizard species (e.g. *Sceloporus*, Oufiero et al., 2011). Similarly, body size (and hence, scale count) is variable both intra- and interspecifically among some squamates in relationship (both positive and negative) to Bergmann's Rule (i.e., body size decreases at lower latitudes or in warmer environments), which is correlated with minimum annual temperature and aridity (Sears and Angilletta, 2004; Oufiero et al., 2011). Despite these variations, scale morphology is one of the most reliable in the identification of extant squamates and crocodilians, especially in closely related species (Brazaitis, 1987; Charette, 1995). It is notable also that tarsal- and toe-scale patterns have been used successfully to identify individual species of extant avians, particularly raptors (Clark, 1972; Stauber, 1984, 1985; Palma, 1996). Considering these findings in relationship to the extant phylogenetic bracket (EPB; Witmer, 1995), it is highly probable that phylogeny and taxonomy reflect interspecific scale variation in Hadrosauridae. However, the potential for individual and sexual variation cannot be discounted at this time.

CONCLUSIONS

Among Hadrosauridae, the distribution of various scale morphologies is highly variable across the body. Similarly, interspecific variation is strikingly diverse even between closely related taxa (such as species of *Saurolophus*). However, scale architecture appears to have been consistent within certain species where multiple specimens are available (e.g., *Edmontosaurus annectens, Corythosaurus casuarius, Saurolophus angustirostris*). These findings, supported by comparisons to living archosaurs, lend credence to the hypothesis that scale architecture is species specific as suggested by earlier studies (Brown, 1916; Lull and Wright, 1942) and provides the rationale for attempting to characterize hadrosaurid skin impressions. The historic (and present) reliance on osteological data in taxonomy may benefit from supplementary information gained from skin impressions; however, apparent differences in scale architecture have yet to be formally tested in a phylogenetic context (Negro, 2001). Nevertheless, the current results present a promising direction for future studies of hadrosaurid taxonomy.

Skin impressions are widely represented across Hadrosauridae from virtually all parts of the body. Morphological diversity between taxa is most apparent in the tail, although considerable variation also exists in the flank and hindlimb. Although the integument of some taxa, such as *Corythosaurus* and *Saurolophus angustirostris*, is relatively well documented, more specimens are needed to fully characterize the skin of all taxa. Other aspects, such as ontogenentic and intraspecific variation, are virtually unknown.

Appropriate documentation of skin impressions is essential during and after preparation and final curation of such specimens. Museum specimens often lack information regarding the position and orientation of hand samples after they have been removed from larger, more complete specimens. Such missing information limits the utility of skin impressions, especially when attempting to characterize scale architecture within particular parts of the body. Photographs and, ideally, maps of the skin impressions should be made prior to their removal from a skeleton in order to correctly match their position on the body at a later date.

ACKNOWLEDGMENTS

I am indebted to P. Currie, D. Evans, and M. Ryan for stimulating and encouraging my research on hadrosaur skin impressions. E. Snively, K. Brink, and T. Bürgin kindly provided photos of AMNH 5220, ROM 764, and NMSG P 0001, respectively. C. Mehling (AMNH), B. Strilisky (RTMP), D. Evans, K. Seymour, I. Morrison (ROM), M. Borsuk-Biatynickia (ZPAL), V. Alifanov, and T. Tumanova (PIN) are gratefully acknowledged for access to specimens. N. Campione (ROM) and T. Bürgin (NMSG) provided valuable information on several *Edmontosaurus* specimens. Detailed reviews and editing by D. Evans and D. Eberth, K. Clayton, and P. Ralrick greatly improved the final version. Financial assistance was received from the Dinosaur Research Institute, Hwaseong City, Gyeonggi Province and the Basic Research Project (No. 12-3613-1) of Korea Institute of Geoscience and Mineral Resources, South Korea.

LITERATURE CITED

Anderson, B. G., R. E. Barrick, M. L. Droser, and K. L. Stadtman. 1999. Hadrosaur skin impressions from the Upper Cretaceous Nelsen Formation, Book Cliffs, Utah: morphology and paleoenvironmental context; pp. 295–301 in D. D. Gillette (ed.), Vertebrate Fossils of Utah. Miscellaneous Publication 99-1. Utah Geological Survey, Salt Lake City, Utah.

Anderson, B. G., S. G. Lucas, R. E. Barrick, A. B. Heckert, and G. T. Basabilvaso. 1998. Dinosaur skin impressions and associated skeletal remains from the Upper Cretaceous of southwestern New Mexico: new data on the integument morphology of hadrosaurs. Journal of Vertebrate Paleontology 18:739–745.

Bell, P. R. 2010. Redescription of the skull of *Saurolophus osborni* Brown 1912 (Ornithischia: Hadrosauridae). Cretaceous Research 32:30–44.

Bell, P. R. 2012. Standardized terminology and potential taxonomic utility of hadrosaurid skin impressions: a case study for *Saurolophus* from Canada and Mongolia. PLoS ONE 7(2):e31295.

Branch, B. 1998. Field Guide to Snakes and other Reptiles of Southern Africa. New Holland, Cape Town, South Africa, 131 pp.

Brazaitis, P. 1987. The identification of crocodilian skins and products; pp. 373–386 in G. W. Webb, S. C. Maniolis, and P. J. Whitehead (eds.), Wildlife Management: Crocodiles and Alligators. Surrey Beatty and Sons, Sydney, Australia.

Briggs, D. E. G., P. R. Wilby, B. P. Pérez-Moreno, J. L. Sanz, and M. Fregenal-Martínez. 1997. The mineralisation of dinosaur soft tissue in the Lower Cretaceous of Las Hoyas, Spain. Journal of the Geological Society, London 154:587–588.

Brown, B. 1912. A crested dinosaur from the Edmonton Cretaceous. Bulletin of the American Museum of Natural History 31:131–136.

Brown, B. 1914. *Corythosaurus casuarius,* a new crested dinosaur from the Belly River Cretaceous, with provisional classification of the family Trachodontidae. Bulletin of the American Museum of Natural History 33:559–565.

Brown, B. 1916. *Corythosaurus casuarius:* skeleton, musculature and epidermis. Bulletin of the American Museum of Natural History 35:709–723.

Brown, B. 1917. A complete skeleton of the horned dinosaur *Monoclonius,* and description of a second skeleton showing skin impressions. Bulletin of the American Museum of Natural History 37:281–306.

Campione, N. E., and D. C. Evans. 2011. Cranial growth and variation in edmontosaurs (Dinosauria: Hadrosauridae): implications for latest Cretaceous megaherbivore diversity in North America. PLoS ONE 6(9):e25186.

Charette, R. 1995. CITES Identification Guide–Crocodilians: Guide to the Identification of Crocodilian Species Controlled under the Convention of International Trade in Endangered Species of Wild Fauna and Flora. Environment Canada, Ottawa, Canada, 148 pp.

Clark, G. A., Jr. 1972. Passerine foot scutes. Auk 89:594–558.

Clayton, C. E., R. B. Irmis, M. A. Getty, E. K. Lund, W. J. Nicholls, and M. A. Loewen. 2011. Non-osseous dermal scutes and integument impressions from an exceptionally preserved hadrosaurid dinosaur skeleton, Upper Cretaceous Kaiparowits Formation of Utah; pp. 23–27 in D. R. Braman, D. A. Eberth, D. C. Evans, and W. Taylor (compilers), Hadrosaur Symposium Abstract Volume. Royal Tyrrell Museum of Palaeontology, Drumheller, Alberta.

Coria, R. A., and L. M. Chiappe. 2007. Embryonic skin from Late Cretaceous sauropods (Dinosauria) of Auca Mahuevo, Patagonia, Argentina. Journal of Paleontology 81:1528–1532.

Cox, J. H., R. S. Frazier, and R. A. Maturbongs. 1993. Freshwater crocodiles of Kalimantan (Indonesian Borneo). Copeia 1993:564–566.

Currie, P. J. 1983. Hadrosaur trackways from the Lower Cretaceous of Canada. Acta Palaeontologica Polonica 28:63–73.

Currie, P. J., G. Nadon, and M. G. Lockley. 1991. Dinosaur footprints with skin impressions from the Cretaceous of Alberta and Colorado. Canadian Journal of Earth Sciences 28:102–115.

Czerkas, S. A. 1992. Discovery of dermal spines reveals a new look for sauropod dinosaurs. Geology (Boulder) 20:1068–1070.

Czerkas, S. A. 1997. Skin; pp. 669–675 in P. J. Currie and K. Padian (eds.), The Encyclopedia of Dinosaurs. Academic Press, San Diego, California.

Dodson, P., D. W. Krause, C. A. Forster, S. D. Sampson, and F. Ravoavy. 1998. Titanosaurid (Sauropoda) osteoderms from the Late Cretaceous of Madagascar. Journal of Vertebrate Paleontology 18:563–568.

Evans, D. C., and R. R. Reisz. 2007. Anatomy and relationships of *Lambeosaurus magnicristatus,* a crested hadrosaurid dinosaur (Ornithischia) from the Dinosaur Park Formation, Alberta. Journal of Vertebrate Paleontology 27:373–393.

Gates, T. A., and A. A. Farke. 2009. Biostratigraphic and biogeographic implications of a hadrosaurid (Ornithopoda: Dinosauria) from the Upper Cretaceous Almond Formation of Wyoming, USA. Cretaceous Research 30:1157–1163.

Gates, T. A., and S. Sampson. 2007. A new species of *Gryposaurus* (Dinosauria: Hadrosauridae) from the late Campanian Kaiparowits Formation, southern Utah, USA. Zoological Journal of the Linnean Society 151:351–376.

Gillette, D. D., L. B. Albright III, A. L. Titus, and M. H. Graffam. 2002. Skin impressions from the tail of a hadrosaurian dinosaur in the Kaiparowits Formation (Upper Cretaceous), Grande Staircase-Escalante National Monument. Geological Society of America Abstracts with Programs 34:6.

Hall, P. M. 1989. Variation in geographic isolates of the New Guinea Crocodile (*Crocodylus novaeguineae* Schmidt) compared with the similar, allopatric Philippine Crocodile (*C. mindorensis* Schmidt). Copeia 1989:71–80.

Heironymus, T. L., L. M. Witmer, and R. C. Ridgely. 2006. Structure of white rhinoceros (*Ceratotherium simum*) horn investigated by x-ray computed tomography and histology with implications for growth and external form. Journal of Morphology 267:1172–1176.

Horner, J. R. 1984. A "segmented" tail frill in a species of hadrosaurian dinosaur. Journal of Paleontology 58:270–271.

Horner, J. R., and R. Makela. 1979. Nest of juveniles provides evidence of family structure among dinosaurs. Nature 282:296–298.

Horner, J. R., D. Weishampel, and C. Forster. 2004. Hadrosauridae; pp. 438–463 in D. Weishampel, P. Dodson, and H. Osmòlska (eds.), The Dinosauria, Second Edition. University of California Press, Berkley, California.

Ji, S.-A. 1999. Initial report of fossil psittacosaurid skin impressions from the uppermost Jurassic of Sihetun, northeastern China. Earth Science 53:314–316.

Ji, S.-A., and H. C. Bo. 1998. Discovery of the psittacosaurid skin impressions and its significance. Geological Review 44:603–606.

Lambe, L. 1902. New genera and species from the Belly River Series (mid-Cretaceous). Geological Survey of Canada 3:23–81.

Lambe, L. 1914a. On the fore-limb of a carnivorous dinosaur from the Belly River Formation of Alberta, and a new genus of Ceratopsia from the same horizon, with remarks on the integument of some Cretaceous herbivorous dinosaurs. Ottawa Naturalist 27:129–135.

Lambe, L. 1914b. On *Kritosaurus notabilis,* a new genus and species of trachodont dinosaur from the Belly River Formation of Alberta, with a description of the skull of *Chasmosaurus belli.* Ottawa Naturalist 27:145–155.

Lambe, L. 1917. A new genus and species of crestless hadrosaur from the Edmonton Formation of Alberta. Ottawa Naturalist 31:65–73.

Lockley, M. G., G. Nadon, and P. J. Currie. 2003. A diverse dinosaur-bird footprint assemblage from the Lance Formation, Upper Cretaceous, Eastern Wyoming: implications for ichnotaxonomy. Ichnos 11:229–249.

Lull, R. S., and N. E. Wright. 1942. Hadrosaurian Dinosaurs of North America. Geological Society of America Special Papers 40. 242 pp.

Manning, P. L. 2009. Grave Secrets of Dinosaurs: Soft Tissues and Hard Science. National Geographic Society, Washington, D.C., 316 pp.

Manning, P. L., R. A. Wogelius, M. Buckley, B. E. van Dongen, T. Lyson, U. Bergmann, S. Webb, and W. I. S. Sellers. 2011. A multidisciplinary approach to the analysis of fossil hadrosaur integument and the taphonomic role of skin pigment in exceptional preservation; pp. 23–27 in D. R. Braman, D. A. Eberth, D. C. Evans, and W. Taylor (compilers), Hadrosaur Symposium Abstract Volume. Royal Tyrrell Museum of Palaeontology, Drumheller, Alberta.

Manning, P. L., R. A. Wogelius, B. van Dongen, T. R. Lyson, U. Bergmann, S. Webb, M. Buckley, V. M. Egerton, and W. I. Sellers. 2014. The role and biochemistry of melanin pigment in the exceptional preservation of hadrosaur skin; chapter 36 in D. A. Eberth and D. C. Evans (eds.), Hadrosaurs. Indiana University Press, Bloomington, Indiana.

Manning, P. L., P. M. Morris, A. McMahon, E. Jones, A. Gize, J. H. S. Macquaker, G. Wolff, A. Thompson, J. Marshall, K. G. Taylor, T. Lyson, S. Gaskell, O. Reamtong, W. I. Sellers, B. E. van Dongen, M. Buckley, and R. A. Wogelius. 2009. Mineralized soft-tissue structure and chemistry in a mummified hadrosaur from the Hell Creek Formation, North Dakota (USA). Proceedings of the Royal Society B 276:3429–3437.

Marsh, O. C. 1892. Notice of new reptiles from the Laramie Formation. American Journal of Science, Series 3 43:449–453.

Murphy, N. L., D. Trexler, and M. Thompson. 2007. "Leonardo," a mummified *Brachylophosaurus* from the Judith River Formation; pp. 117–133 in K. Carpenter (ed.), Horns and Beaks: Ceratopsian and Ornithopod Dinosaurs. Indiana University Press, Bloomington, Indiana.

Negro, G. 2001. Phylogenetic interpretations of hadrosaurian skin. Journal of Vertebrate Paleontology 21(3, Supplement):83A.

Osborn, H. F. 1911. A dinosaur mummy. American Museum Journal 2:7–11.

Osborn, H. F. 1912. Integument of the iguanodont dinosaur *Trachodon.* American Museum of Natural History Memoir, Part II:33–54.

Oufiero, C. E., G. E. A. Gartner, S. C. Adolph, and T. Garland. 2011. Latitudinal and climatic variation in body size and dorsal scale counts in *Scleroporus* lizards: a phylogenetic perspective. Evolution 65:3590–3607.

Palma, C. R. 1996. The use of tarsal scale patterns to identify individual birds of prey. M.S. thesis, McGill University, Montreal, Quebec, 125 pp.

Parks, W. A. 1920. The osteology of the trachodont dinosaur *Kritosaurus incurvimanus.* Transactions of the Royal Society of Canada 13:5–76.

Parks, W. A. 1922. *Parasaurolophus walkeri,* a New Genus and Species of Crested Trachodont Dinosaur. University of Toronto Studies, Geological Series 13. 32 pp.

Parks, W. A. 1923. *Corythosaurus intermedius,* a new species of trachodont dinosaur. University of Toronto Studies, Geological Series 15. 57 pp.

Prieto-Márquez, A. 2010. Global phylogeny of Hadrosauridae (Dinosauria: Ornithopoda) using parsimony and Bayesian methods. Zoological Journal of the Linnean Society 159:435–502.

Prieto-Márquez, A., and J. R. Wagner. This volume. Soft-tissue structures of the nasal vestibular region of saurolophine hadrosaurids (Dinosauria, Ornithopoda) revealed in a "mummified" specimen of *Edmontosaurus annectens;* chapter 35 in D. A. Eberth and D. C. Evans (eds.), Hadrosaurs. Indiana University Press, Bloomington, Indiana.

Rozhdestvensky, A. K. 1952. [A new representative of duck-billed dinosaurs from the Upper Cretaceous of Mongolia]. Doklady Akademii SSSR 86:405–408. [Russian]

Sears, M. W., and M. J. Angilletta. 2004. Body size clines in *Scleroporus* lizards: proximate mechanisms and demographic constraints. Integrated Comparative Biology 44:433–442.

Siber, H. J., and U. Möckli. 2009. The Stegosaurs of the Sauriermuseum Aathal. Sauriermuseum Aathal, Switzerland, 56 pp.

Spearman, R. I. C. 1973. The Integument: A Textbook of Skin Biology. Cambridge University Press, Cambridge, U.K., 211 pp.

Stauber, E. H. 1984. Footprinting raptors for identification. Raptor Research 18:67–71.

Stauber, E. H. 1985. Raptor identification by photographic analysis of toe scale patterns. Hawk Chalk 24:40–42.

Sternberg, C. M. 1935. Hooded hadrosaurs of the Belly River Series of the Upper Cretaceous. National Museum of Canada Bulletin 77:1–38.

Sternberg, C. M. 1953. A new hadrosaur from the Oldman Formation of Alberta: discussion of nomenclature. Bulletin of the Department of Natural Resources of Canada 128:1–12.

Trexler, D. 1995. Detailed description of newly discovered remains of *Maiasaura peeblesorum* (Reptilia: Ornithischia) and a revised diagnosis of the genus. M.S. thesis, University of Calgary, Calgary, Alberta, 217 pp.

Versluys, J. 1923. [The skull of *Trachodon annectens* in the Senckenberg Museum]. Abhandlungen der Senckenbergischen Naturforschenden Gesellschaft 38:1–19. [German]

Wegweiser, M. D., S. A. Hartman, and D. M. Lovelace. 2006. Duckbill dinosaur chin scales: ups, downs and arounds of surficial morphology of Upper Cretaceous Lance Formation dinosaur skin; pp.119–125 in S.G. Lucas and R. M. Sullivan (eds.), Late Cretaceous Vertebrates from the Western Interior. New Mexico Museum of Natural History and Science Bulletin 35.

Witmer, L. M. 1995. The Extant Phylogenetic Bracket and the importance of reconstructing soft tissues in fossils; pp. 19–33 in J. J. Thomason (ed.), Functional Morphology in Vertebrate Paleontology. Cambridge University Press, New York.

Xu, X., Z.-L.Tang, and X.-L. Wang. 1999. A therizinosaurid with integumentary structures from China. Nature 399:350–354.

Xu, X., X.-L. Wang, and X.-C. Wu. 1999. A dromaeosaurid dinosaur with a filamentous integument from the Yixian Formation of China. Nature 401:262–266.

Xu, X., Z.-H. Zhou, and R. O. Prum. 2001. Branched integumental structures in *Sinornithosaurus* and the origin of feathers. Nature 410:200–204.

Zheng, X.-T., H.-L. You, X. Xu, and Z.-M. Dong. 2009. An Early Cretaceous heterodontosaurid dinosaur with filamentous integumentary structures. Nature 458:333–336.

Soft-Tissue Structures of the Nasal Vestibular Region of Saurolophine Hadrosaurids (Dinosauria, Ornithopoda) Revealed in a "Mummified" Specimen of *Edmontosaurus annectens*

35

Albert Prieto-Márquez and Jonathan R. Wagner

ABSTRACT

Reexamination of the nasal vestibular region of AMNH FARB 5060, the renowned *Edmontosaurus annectens* "mummy," reveals hitherto undescribed details of the circumnarial fossae of the specimen. The caudodorsal margin of the apertura ossis nasi (bony nostril) is extended rostrally by a flange of soft tissue. This flange may represent a cartilaginous extension of the caudodorsal margin of the foramen, which may have been partially occluded in life. This suggests that the passage between the skull and the circumnarial structure may have been in the caudal portion of the apertura ossis nasi, rather than in the rostral end of the opening. The skull also preserves a thin septum that appears to have bounded an extracranial extension of the nasal vestibulum across the circumnarial space and out to the fleshy nostril. Finally, a patch of skin impressions lies within the rostral circumnarial fossa. Observation of these nasal vestibular soft-tissue structures provides new information on the anatomy of the circumnarial space of saurolophine hadrosaurids, adding to the data previously derived only from skeletal morphology.

INTRODUCTION

Along with ceratopsians (Dodson et al., 2004) and ankylosaurs (Witmer and Ridgely, 2008), hadrosaurid dinosaurs are distinctive among ornithischians in possessing greatly hypertrophied nasal vestibular structures (Hopson, 1975; Weishampel, 1981). In lambeosaurine hadrosaurids the nasal passage is enclosed by thin sheets of bone, extending caudodorsally and arching in a broad loop to enter the skull dorsally rostral to the orbits (Ostrom, 1962); this enclosure of the nasal passage permits comparative study of its anatomy in hadrosaurids and other fossil reptiles (Weishampel, 1981; Evans, 2006; Evans et al., 2009). In the other major clade of hadrosaurids, Saurolophinae (sensu Prieto-Márquez, 2010a), the nasal passage remains rostral to the orbits in the ancestral iguanodontian location (Wagner, 2001, 2004; Horner et al., 2004).

The saurolophine apertura ossis nasi (bony nasal aperture or, more simply, bony nostril) and nasal vestibulum lie within an extensive excavated region on the lateral surface of the nasal and premaxilla. This excavation, the circumnarial depression (Horner, 1992; Horner et al., 2004; Bell, 2011a), and the nasal passage are not enshrouded by bone as in lambeosaurines (Hopson, 1975; Wagner, 2001, 2004). Variation in the depth of the circumnarial depression and length, proportions, and the number and placement of various foramina, ridges, and subordinate fossae have been documented within the clade (Lambe, 1920; Lull and Wright, 1942; Sternberg, 1953; Hopson, 1975; Horner, 1983, 1992; Wagner, 2001, 2004; Prieto-Márquez, 2008; Gates and Sampson, 2007; Bell, 2011a, 2011b; Campione and Evans, 2011; Gates et al., 2011). Thus, for example, the caudal region of the circumnarial depression is only gently incised in *Gryposaurus* (Prieto-Márquez, 2010b) and *Brachylophosaurus* (Cuthbertson and Holmes, 2010), whereas it is deeply excavated in *Edmontosaurus* (Lambe, 1917). Likewise, the caudodorsal extension of the caudal margin of the depression ranges in adults from lying rostral to the orbit in *Edmontosaurus* to probably overlying the supratemporal fenestra in *Brachylophosaurus* (Prieto-Márquez, 2005), and even extending beyond the occiput in *Saurolophus* (Bell, 2011b). Despite speculation (e.g., Hopson, 1975; Wagner, 2004), to date there has been little direct evidence of the soft-tissue structures that may have been bounded within the circumnarial depression of saurolophine hadrosaurids.

Here, we report on soft-tissue structures preserved within the circumnarial depression of an exquisitely preserved specimen of *Edmontosaurus annectens*, AMNH FARB 5060. This nearly complete articulated skeleton is the classic dinosaur "mummy" collected by Charles H. Sternberg in 1908 from late Maastrichtian strata of the Lance Formation in Converse County, Wyoming, and subsequently described by Osborn (1911, 1912). The specimen preserves multiple impressions and natural molds of the integument throughout the caudal region of the skull, neck, shoulder, forelimb, rib cage, pelvic girdle, and hindlimb regions of the body (Figs. 35.1, 35.2).

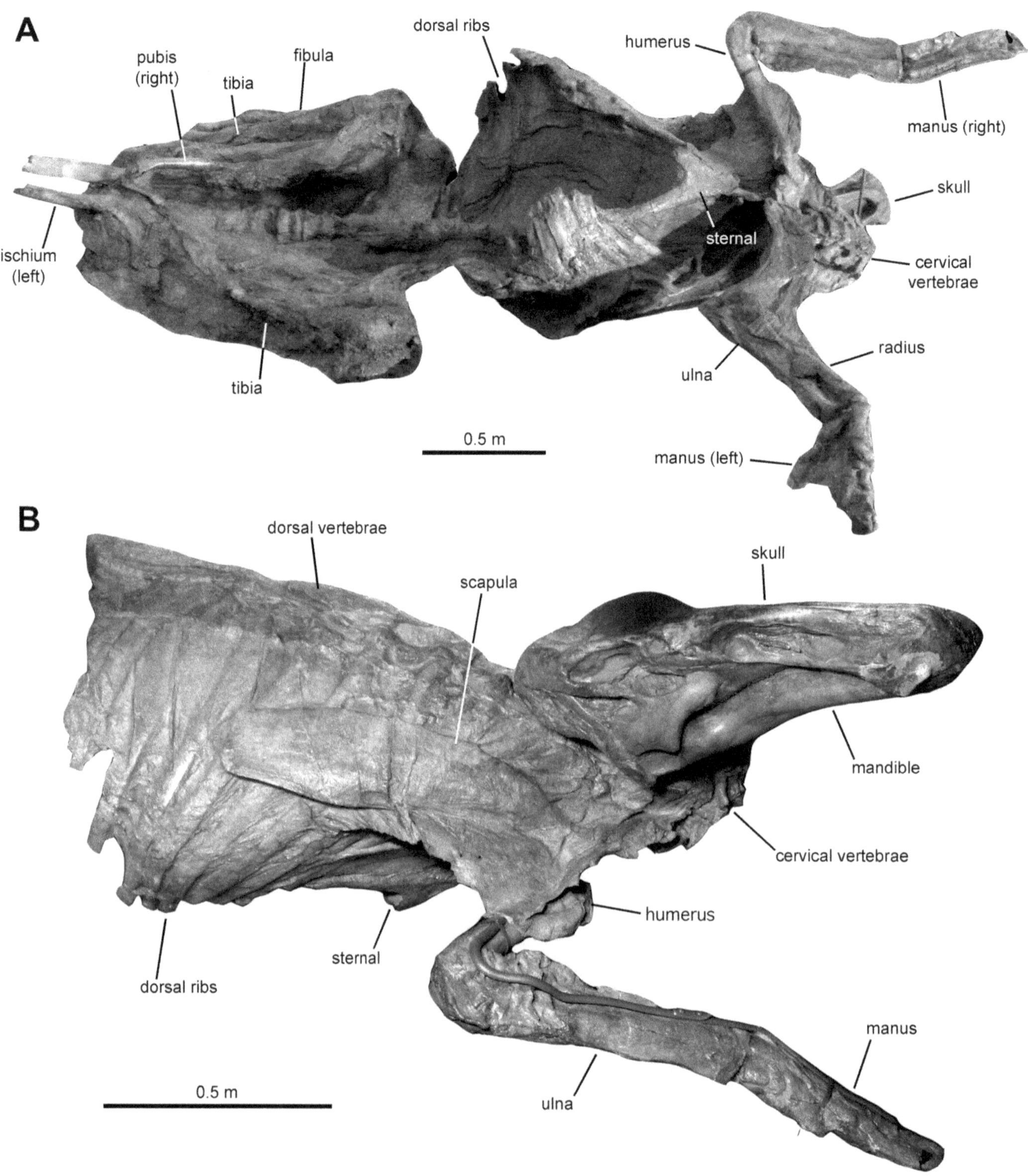

35.1. Articulated skeleton of *Edmontosaurus annectens* (AMNH FARB 5060) bearing multiple impressions and natural casts of the integument. (A) Ventral view of forelimbs, and pectoral, thoracic, and pelvic regions; (B) right lateral view of the cranial half of the skeleton.

Lull and Wright (1942:113) already noted the presence of the nasal vestibular structures that constitute the subject of the present contribution: "In addition to the beak, the only integumentary structure preserved in the skull is that of the soft tissue around the narial opening of the mummied *Anatosaurus* (No. 5060 A.M.N.H.)." According to these authors, "there is no indication of a tubercular skin like that which covers the rest of the body, but the entire bony aperture is filled with

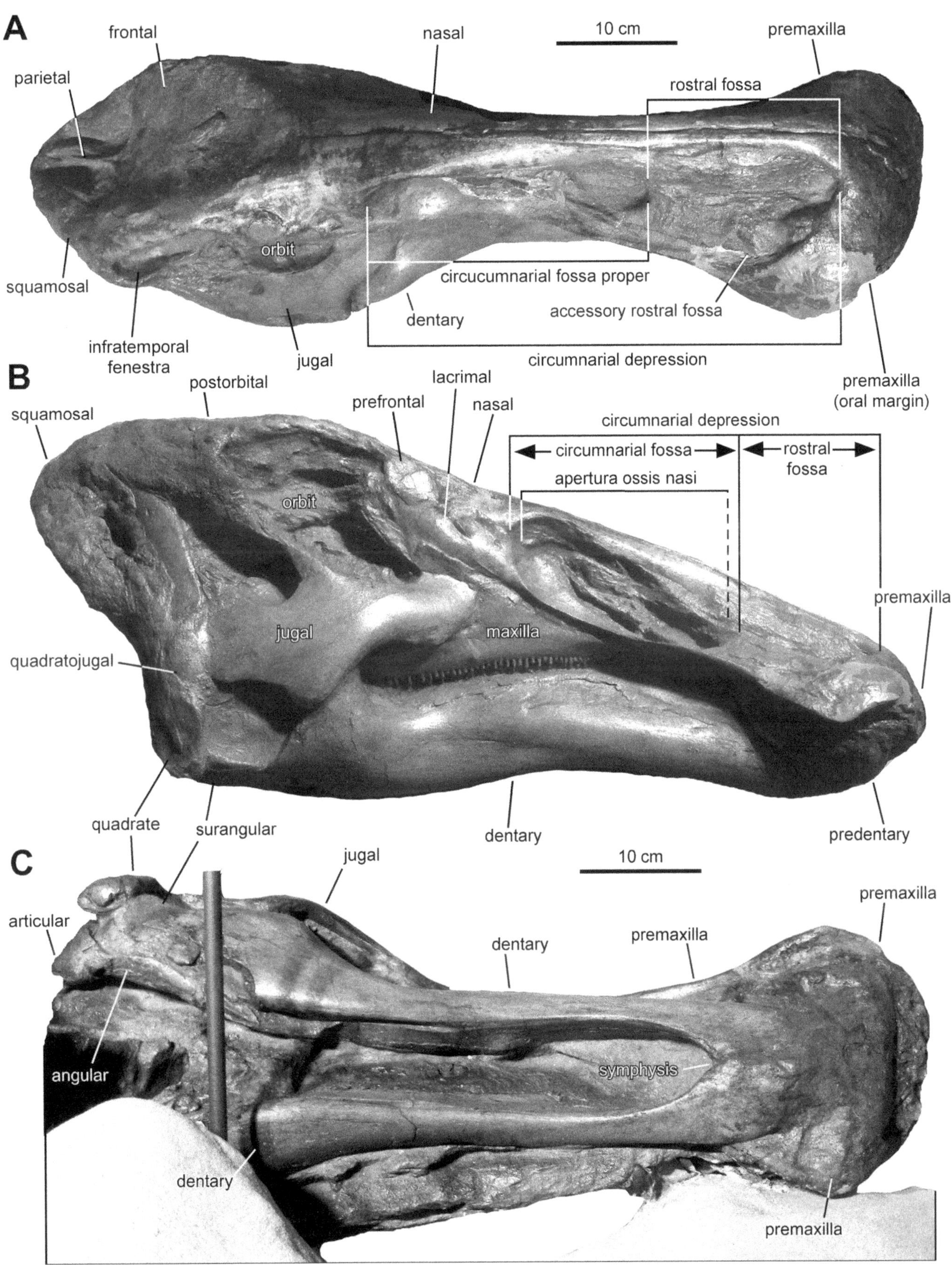

35.2. Complete articulated skull of the mummified *Edmontosaurus annectens* specimen AMNH FARB 5060. (A) Right laterodorsal; (B) right lateral; and (C) ventral views.

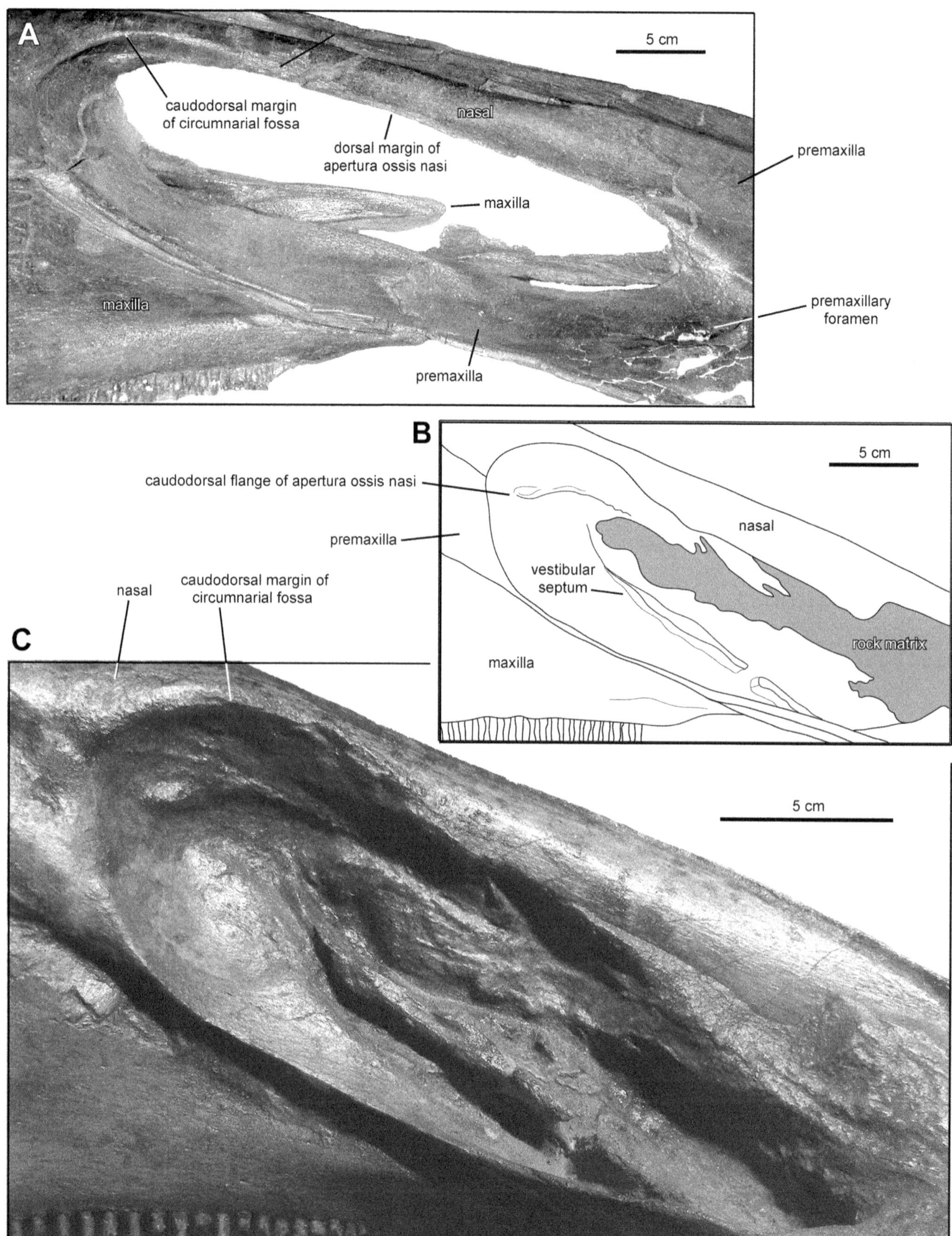

35.3. Comparison of the circumnarial fossa of *Edmontosaurus annectens* specimens with and without fossilized soft-tissue structures. (A) MOR 003 in left lateral view (reversed); (B) line drawing of the nasal vestibular structures observed in AMNH FARB 5060; (C) circumnarial fossa of AMNH FARB 5060 in right lateral view.

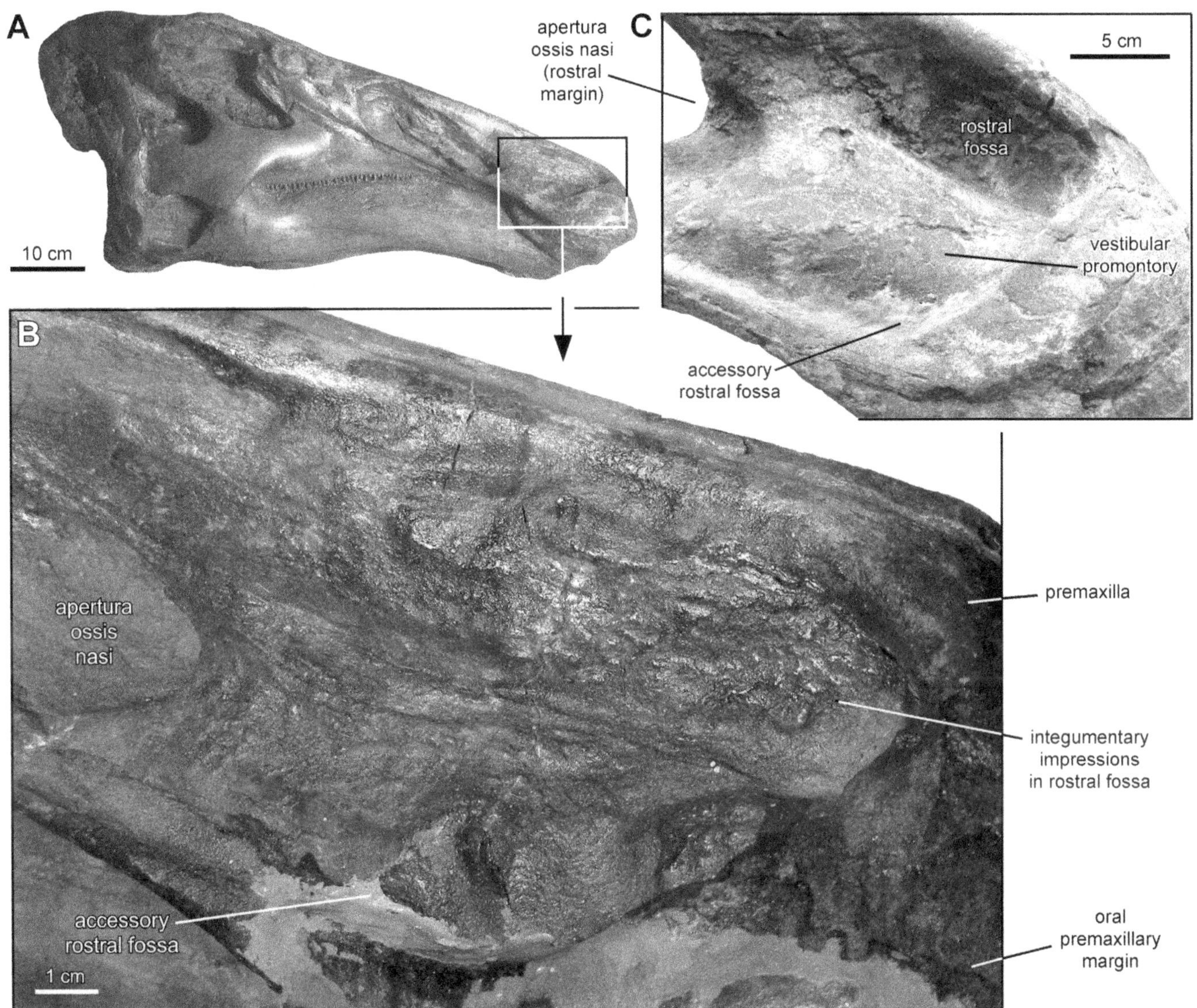

35.4. Comparison of the rostral fossa of the circumnarial depression of *Edmontosaurus annectens* specimens missing and preserving fossilized integument. (A) Right lateral view of the AMNH FARB 5060 skull, photographed under different lighting conditions than in Fig. 35.2B; (B) rostral fossa of AMNH FARB 5060 in lateral view, showing integumentary impressions; (C) rostral fossa of LACM 23502, showing bony structures commonly present in *Edmontosaurus*.

what appears to be soft tissue thrown into folds" (Lull and Wright, 1942:113). To date, AMNH FARB 5060 remains one of the best preserved and most complete dinosaur specimens ever collected; among hadrosaurids, it is rivaled only by a handful of pristinely preserved exemplars of *Brachylophosaurus canadensis* from the Judith River Formation of Montana (Murphy et al., 2007).

Institutional Abbreviations AMNH FARB, American Museum of Natural History (Fossil Amphibians, Reptiles, and Birds collection), New York; CMN, Canadian Museum of Nature, Ottawa; LACM, Natural History Museum of Los Angeles County, Los Angeles; MOR, Museum of the Rockies, Bozeman, Montana; ROM, Royal Ontario Museum, Toronto; SM, Senckenberg Museum, Frankfurt am Main, Germany.

DESCRIPTION

The saurolophine circumnarial depression contains two distinct fossae (Fig. 35.2A, B). An extensive and rostrocaudally elongate circumnarial fossa proper surrounds the apertura ossis nasi (Fig. 35.3A). Rostral to this lies the smaller rostral fossa (the "outer narial fossa" of Horner et al. [2004]), set against the rostrodorsal margin of the depression, bounded rostrally by the oral margin of the premaxilla and caudally by low and relatively smooth transverse ridge set entirely within the circumnarial margin (e.g., *Gryposaurus notabilis* CMN 2278 in Lambe [1914a:pl. 18]; *G. latidens* AMNH FARB 5465 in Prieto-Márquez [2012:fig. 1C]). In species of *Edmontosaurus* there is an accessory rostral fossa ventral to this, adjacent to

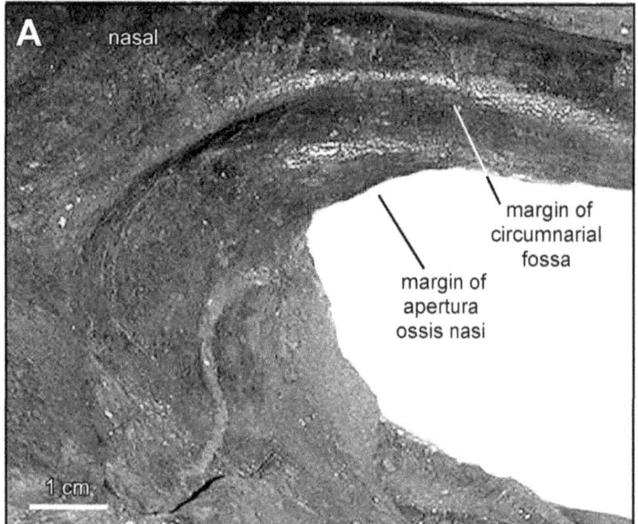

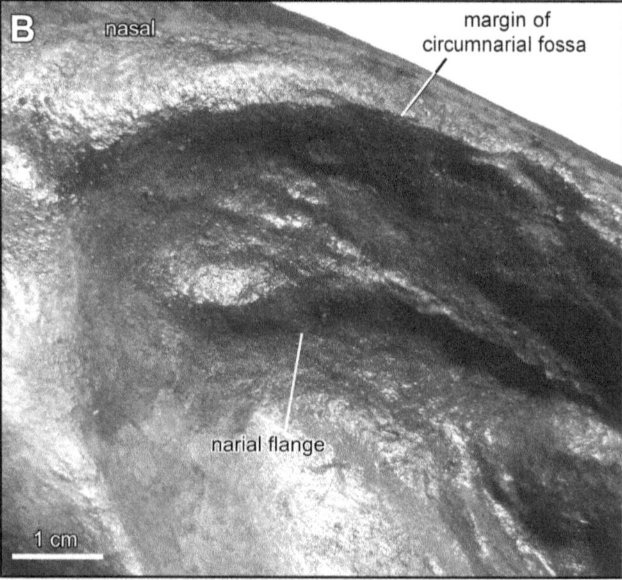

35.5. Comparison of the caudodorsal region of the apertura ossis nasi and circumnarial fossa of *Edmontosaurus annectens* specimens missing and preserving the caudodorsal narial flange. (A) MOR 003, left lateral view (reversed); (B) AMNH FARB 5060, right lateral view.

the rostroventral oral margin of the premaxilla (Fig. 35.4C). This accessory fossa is separated from the remainder of the rostral fossa by a vestibular promontory of roughly triradiate morphology (e.g., SM R4036 in Versluys, 1923:pl. 1) (Fig. 35.4C). This promontory may be homologous to the relatively undepressed portion of the floor of the circumnarial fossa ventral to the rostral fossa in other saurolophines (e.g., *Prosaurolophus maximus* ROM 787 or *Saurolophus osborni* AMNH 5220), differing primarily due to the absence of the accessory rostral fossa in those taxa.

In the numerous known skulls of *Edmontosaurus annectens* (e.g., AMNH 5730, AMNH 5886, MOR 003, SM R4036, LACM 23502), the caudodorsal region of the circumnarial fossa is

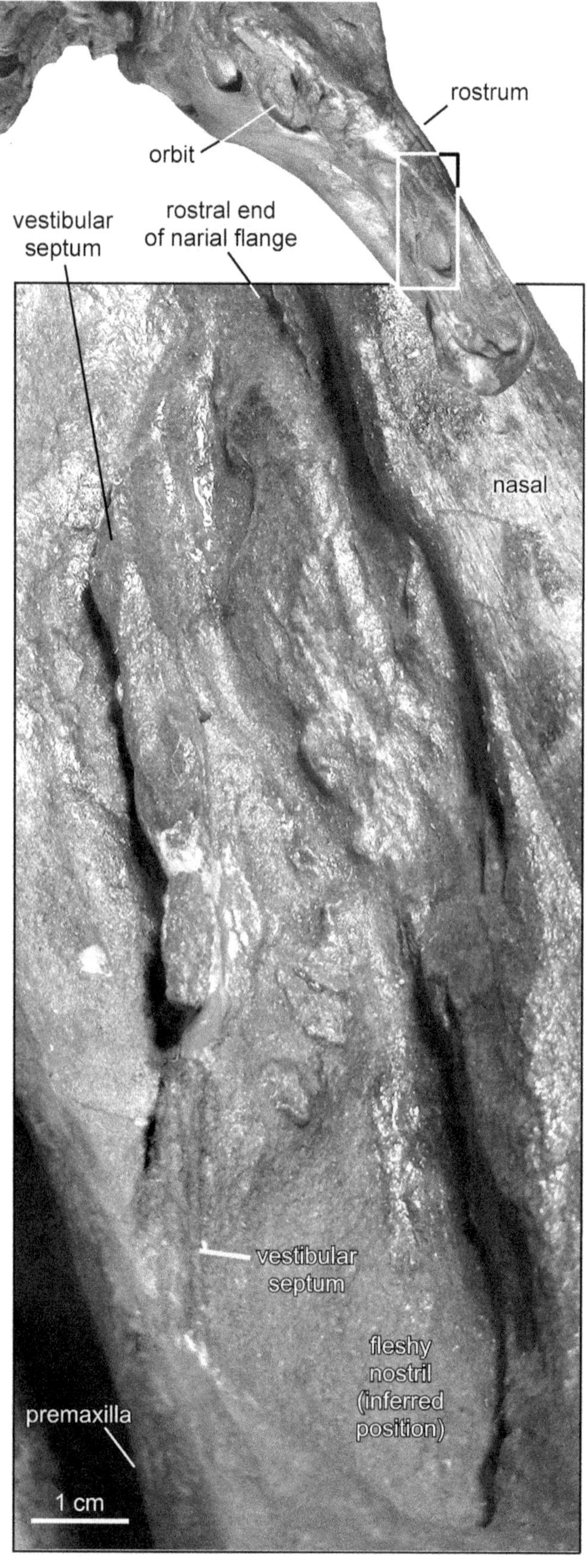

35.6. Laterodorsal view of the vestibular septum preserved within the circumnarial fossa of the *Edmontosaurus annectens* specimen AMNH FARB 5060. The rostral end of the caudodorsal narial flange is also observable near the upper margin of the image.

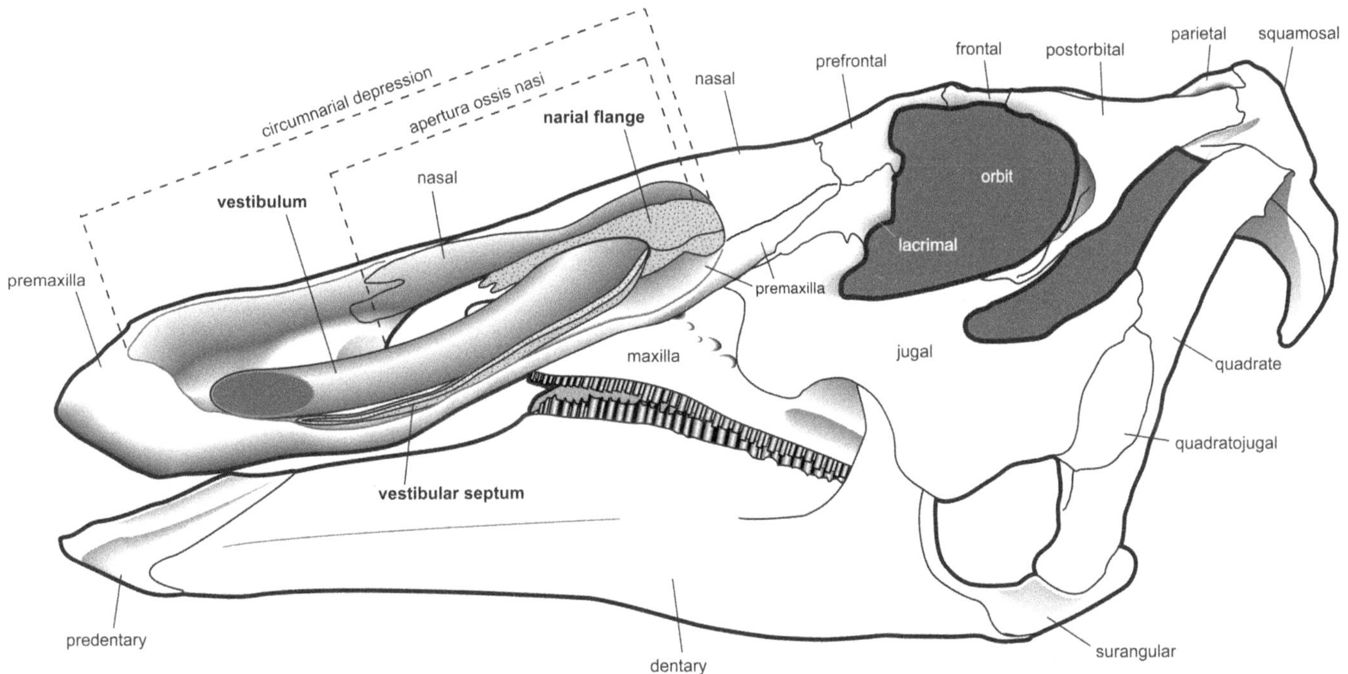

35.7. Tentative reconstruction of part of the soft-tissue anatomy of the nasal vestibular region of *Edmontosaurus annectens* (osteology based on the left lateral side of the skull of SM R4036). The dotted gray pattern represents cartilage. This reconstruction hypothesizes the presence of a cartilaginous septum supporting ventrally the vestibulum and separating it from the rest of circumnarial structures. We speculate that the nasal flange that underlies the dorsal margin of the circumnarial depression, along the nasal and above the vestibulum, extended a greater distance rostrally than preserved in AMNH 5060. The flange might have also extended farther ventrally to completely surround the posterior margin of the apertura ossis nasi.

deeply excavated and bounded by a prominent arcuate margin of the nasal (Fig. 35.3A; see also Campione and Evans [2011:fig. 2]). The caudodorsal margin of the apertura ossis nasi, set within the circumnarial fossa, extends rostrally a short distance from the caudodorsal border of the circumnarial depression (Fig. 35.5A). In AMNH FARB 5060, however, the caudodorsal margin of the apertura ossis nasi is further extended rostroventrally by an arcuate flange of soft tissue (Figs. 35.3B, C, 35.5B); the bony margin of the apertura ossis nasi is obscured by rock matrix. Caudally, this flange displays a convexity that extends ventrally from the narial margin (Fig 35.5B). Rostrally the flange curves, following the contour of the caudodorsal margin of the apertura ossis nasi; and projects rostroventrally, forming an extremely thin and wavy ridge. This ridge is parallel to the medial dorsal border of the circumnarial depression of the nasal dorsal process (Figs. 35.5B, 35.6).

Rostral and ventral to the caudodorsal narial flange, AMNH FARB 5060 preserves a long septum that extends obliquely (rostroventrally) throughout three-quarters of the length of the circumnarial fossa of the nasal vestibular region, passing across and apparently resting upon the dorsal surface of the ventrolateral process of the premaxilla (Figs. 35.3B, C, 35.6). The lateral edge of this septum becomes substantially thicker than the lateral margin of the premaxillary lateral process along its middle third. Rostroventrally, the septum converges with the rostroventral border of the circumnarial fossa, caudoventral to the premaxillary foramen (which is covered in matrix in AMNH FARB 5060).

Finally, AMNH FARB 5060 also preserves a small mosaic of poorly demarcated polygonal impressions covering the dorsal region of the rostral circumnarial fossa, adjacent to the rostral segment of the medial process of the premaxilla (Fig. 35.4B). The tight arrangement and geometry of these polygonal tubercles is consistent with that of the numerous findings of hadrosaurid integumentary casts and impressions (Cope, 1883; Lambe, 1914b; Brown, 1916; Lull and Wright, 1942; Horner, 1984; Anderson et al., 1998; Negro and Prieto-Márquez, 2001; Wideman and Lofgren, 2001; Manning et al., 2009; Herrero and Farke, 2010) and that of the various postcranial regions of AMNH FARB 5060 (Osborn, 1912), such as the skin impressions on the medial side of the deltopectoral crest of the right humerus or some of the natural molds found on the coracoids. This leads to the interpretation of the mosaic of structures in the rostral fossa of AMNH FARB 5060 as a patch of skin scalation. Each polygonal impression is 3–5 mm in diameter. Near the dorsal margin of the rostral fossa, the impressions of the scalation are rostrocaudally elongated relative to the more equidimensional polygons located near the ventral border of the imprinted skin patch.

DISCUSSION AND CONCLUSION

We consider the sedimentary ridges within the nasal vestibular region of AMNH FARB 5060 as representatives of soft-tissue structures rather than preservation or preparation artifacts because their morphology and position are consistent with previous reconstructions and hypotheses (Versluys, 1936; Lull and Wright, 1942; Hopson, 1975) of the functional and soft-tissue anatomy of the nasal vestibular region in hadrosaurids. The structures of the circumnarial fossa described above are consistent with the presence of a soft-tissue circumnarial structure within the circumnarial depression – perhaps composed of cartilaginous nasal capsules and air-filled accessory spaces (Versluys, 1936; Hopson, 1975), and a diverticulum of the nasal vestibulum covered by flexible and inflatable skin (Hopson, 1975; Wagner, 2004). Specifically, the soft-tissue flange may represent a cartilaginous extension of the apertura ossis nasi and may have occluded some of the latter in life (Fig. 35.7). However, the apparently complete edge paralleling the margin of the osseous structure suggests that only the caudal-most portion of the narial foramen was covered. This suggests that the opening through which the nasal vestibule entered the intracranial cavity was housed in the caudal part of the narial foramen.

The septum described above appears to follow a direct path from the rostroventral extremity of the circumnarial fossa, the most logical position for the fleshy nostril (Witmer, 2001), and the putative aperture for transmission of the vestibulum into the cranium. It follows that this septum may have underlain an extracranial extension of the nasal vestibulum across the circumnarial space, through the circumnarial structure (Fig. 35.7). Extracranial analogs to this structure among extant vertebrates include the nasal cartilages of mammals (Salazar et al., 1995) and the atrial or rostral vestibular concha present in some birds (Jin et al., 2008). Likewise, osteological evidence of the influence of structures associated with the nasal vestibulum is not at all uncommon within the cranium (e.g., the slot for the cartilaginous nasal septum passing along the dorsal surface of the intermaxillary suture in crocodylians [Witmer, 1995]).

It is uncertain whether this putative vestibular septum is analogous or homologous to the only other described osseous structure within the circumnarial fossa of a saurolophine, the short longitudinal ridges described by Maryańska and Osmólska (1981) in a juvenile *Saurolophus angustirostris*. The latter structures are indeed located on the circumnarial fossa that excavates the dorsal surface of the elongate supracranial crest of this dinosaur. Two low ridges are present on each side of the crest: one located medial to the rostrodorsal margin of the orbit, and the other near the caudal margin of the circumnarial fossa. Both are located on the distal end of the crest, caudal to the apertura ossis nasi (Maryańska and Osmólska, 1981:fig. 3). If they are homologous, then the vestibular passage of *Saurolophus angustirostris* may have formed a loop along the crest, much as it does in lambeosaurines. A more parsimonious interpretation is that they represent other parts of the circumnarial structure, possibly related to the vestibulum but not directly related to the direct path of airflow into the cranium.

Finally, the skin impressions described here support the interpretation that the fleshy nostril did not fill the entire circumnarial fossa (Witmer, 2001), and they provide clear evidence of the nature of the integument surrounding the circumnarial structure. The latter has been represented in some artistic reconstructions as a scaleless, inflatable bladder. If it was so, the bladder did not extend to the rostral terminus of the fossa. Hopson's (1975) suggestion of a scale-covered bladder, imagined as having brightly colored skin between the scales that were exposed when inflated, seems more likely.

Regardless, these details reveal more complexity within the saurolophine circumnarial depression than previously assumed based on the skeletal record alone, yet hypothesized by several authors (Hopson, 1975; Wagner, 2001, 2004). Comparison with the narial structures of lambeosaurines may support previously proposed hard-tissue homologies within the crests of saurolophids (e.g., Wagner, 2004), and may reciprocally reinforce our interpretation of such structures as presented here.

ACKNOWLEDGMENTS

We thank M. A. Norell for allowing investigation of AMNH FARB 5060. D. C. Evans and L. M. Witmer provided insightful comments that improved the quality of the manuscript. Thanks also to J. R. Horner and L. M. Chiappe for allowing the reproduction of images of MOR 003 and LACM 23502, respectively. This study was supported by the American Museum of Natural History through a Kalbfelisch Postdoctoral Fellowship and an Alexander von Humboldt Fellowship for Postdoctoral Researchers.

LITERATURE CITED

Anderson, B. G., S. G. Lucas, R. E. Barrick, A. B. Heckert, and G. T. Basabilvazo. 1998. Dinosaur skin impressions and associated skeletal remains from the Upper Campanian of southwestern New Mexico: new data on the integument morphology of hadrosaurs. Journal of Vertebrate Paleontology 18:739–745.

Bell, P. R. 2011a. Redescription of the skull of *Saurolophus osborni* Brown 1912 (Ornithischia: Hadrosauridae). Cretaceous Research 32:30–44.

Bell, P. R. 2011b. Cranial osteology and ontogeny of *Saurolophus angustirostris* from the Late Cretaceous of Mongolia with comments on *Saurolophus osborni* from Canada. Cretaceous Research 32:30–44.

Brown, B. 1916. *Corythosaurus casuarius:* skeleton, musculature and epidermis. Bulletin of the American Museum of Natural History 35:709–716.

Campione, N. E., and D. C. Evans. 2011. Cranial growth and variation in *Edmontosaurus* (Dinosauria: Hadrosauridae): implications for latest Cretaceous megaherbivore diversity in North America. PLoS ONE 6(9):e25186.

Cope, E. D. 1885. The ankle and skin of the dinosaur, *Diclonius mirabilis.* American Naturalist 19:1208.

Cuthbertson, R. S., and R. B. Holmes. 2010. The first complete description of the holotype of *Brachylophosaurus canadensis* Sternberg, 1953 (Dinosauria: Hadrosauridae) with comments on intraspecific variation. Zoological Journal of the Linnean Society 159:373–397.

Dodson, P., C. A. Forster, and S. D. Sampson. 2004. Ceratopsidae; pp 494–513 in D. B. Weishampel, P. Dodson, and H. Osmólska (eds.), The Dinosauria, Second Edition. University of California Press, Berkeley, California.

Evans, D. C. 2006. Nasal cavity homologies and cranial crest function in lambeosaurine dinosaurs. Paleobiology 32:109–125.

Evans, D. C., R. Ridgely, and L. M. Witmer. 2009. Endocranial anatomy of lambeosaurine hadrosaurids (Dinosauria: Ornithischia): a sensorineural perspective on cranial crest function. Anatomical Record 292:1315–1337.

Gates, T. A., and S. D. Sampson. 2007. A new species of *Gryposaurus* (Dinosauria: Hadrosauridae) from the late Campanian Kaiparowits Formation, southern Utah, USA. Zoological Journal of the Linnean Society 151:351–376.

Gates, T. A., J. R. Horner, R. R. Hanna, and C. R. Nelson. 2011. New unadorned hadrosaurine hadrosaurid (Dinosauria, Ornithopoda) from the Campanian of North America. Journal of Vertebrate Paleontology 31:798–811.

Herrero, L., and A. A. Farke. 2010. Hadrosaurid dinosaur skin impressions from the Upper Cretaceous Kaiparowits Formation of southern Utah, USA. PalArch's Journal of Vertebrate Palaeontology 7:1–7.

Hopson, J. 1975. The evolution of cranial display structures in hadrosaurian dinosaurs. Paleobiology 1:21–43.

Horner, J. R. 1983. Cranial osteology and morphology of the type specimen of *Maiasaura peeblesorum* (Ornithischia: Hadrosauridae), with discussion of its phylogenetic position. Journal of Vertebrate Paleontology 3:29–38.

Horner, J. R. 1984. A "segmented" epidermal tail frill in a species of hadrosaurian dinosaur. Journal of Paleontology 58:270–271.

Horner, J. R. 1992. Cranial Morphology of *Prosaurolophus* (Ornithischia: Hadrosauridae) with Descriptions of Two New Hadrosaurid Species and an Evaluation of Hadrosaurid Phylogenetic Relationships. Museum of the Rockies Occasional Paper 2. 119 pp.

Horner, J. R., D. B. Weishampel, and C. A. Forster. 2004. Hadrosauridae; pp. 438–463 in D. B. Weishampel, P. Dodson, and H. Osmólska (eds.), The Dinosauria, Second Edition. University of California Press, Berkeley, California.

Jin, E. H., K. M. Peng, J. X. Wang, A. N. Du, L. Tang, L. Wei, Y. Wang, S. H. Li, and H. Song. 2008. Study of the morphology of the olfactory organ of African ostrich chick. Anatomia Histologia Embryologia 37:161–165.

Lambe, L. M. 1914a. On *Gryposaurus notabilis,* a new genus and species of trachodont dinosaur from the Belly River Formation of Alberta, with description of the skull of *Chasmosaurus belli.* Ottawa Naturalist 27:145–155.

Lambe, L. M. 1914b. On the fore-limb of a carnivorous dinosaur from the Belly River Formation of Alberta, and a new genus of Ceratopsia from the same horizon, with remarks on the integument of some Cretaceous herbivorous dinosaurs. Ottawa Naturalist 27:129–135.

Lambe, L. M. 1917. A new genus and species of crestless hadrosaur from the Edmonton Formation of Alberta. Ottawa Naturalist 31:65–73.

Lambe, L. M. 1920. The hadrosaur *Edmontosaurus* from the Upper Cretaceous of Alberta. Memoirs of the Department of Mines, Geological Survey 120. 79 pp.

Lull, R. S., and N. E. Wright. 1942. Hadrosaurian Dinosaurs of North America. Geological Society of America Special Papers 40. 242 pp.

Manning, P. L., P. M. Morris, A. McMahon, E. Jones, A. Gize, J. H. S. McQuaker, G. Wolff, A. Thompson, J. Marshall, K. G. Taylor, T. Lyson, S. Gaskell, O. Peamtong, W. I. Sellers, B. E. Van Dongen, M. Buckley, and R. A. Wogelius. 2009. Mineralized soft-tissue structure and chemistry in a mummified hadrosaur from the Hell Creek Formation, North Dakota (USA). Proceedings of the Royal Society B, Biological Sciences 276:3429–3437.

Maryańska, T., and H. Osmólska. 1981. Cranial anatomy of *Saurolophus angustirostris* with comments on the Asian Hadrosauridae (Dinosauria). Palaeontologia Polonica 42:5–24.

Murphy, N. L., D. Trexler, and M. Thompson. 2007. "Leonardo," a mummified *Brachylophosaurus* (Ornithischia: Hadrosauridae) from the Judith River Formation of Montana; pp. 117–133 in K. Carpenter (ed.), Horns and Beaks: Ceratopsian and Ornithopod Dinosaurs. Indiana University Press, Bloomington, Indiana.

Negro, G., and A. Prieto-Márquez. 2001. Hadrosaurian skin impressions from the Judith River Formation (Lower Campanian) of Montana. PaleoBios 21(Supplement, 2):96.

Osborn, H. F. 1911. A dinosaur mummy. American Museum Journal 11:7–11.

Osborn, H. F. 1912. Integument of the iguanodont dinosaur *Trachodon.* Memoirs of the American Museum of Natural History 1:33–54.

Ostrom, J. H. 1962. The cranial crests of hadrosaurian dinosaurs. Postilla 62:1–29.

Prieto-Márquez, A. 2005. New information on the cranium of *Brachylophosaurus canadensis* (Dinosauria, Hadrosauridae), with a revision of its phylogenetic position. Journal of Vertebrate Paleontology 25:144–156.

Prieto-Márquez, A. 2008. Phylogeny and historical biogeography of hadrosaurid dinosaurs. Ph.D. dissertation, Florida State University, Tallahassee, Florida, 934 pp.

Prieto-Márquez, A. 2010a. Global phylogeny of Hadrosauridae (Dinosauria: Ornithopoda) using parsimony and Bayesian methods. Zoological Journal of the Linnean Society 159:435–502.

Prieto-Márquez, A. 2010b. The braincase and skull roof of *Gryposaurus notabilis* (Dinosauria, Hadrosauridae), with a taxonomic revision of the genus. Journal of Vertebrate Paleontology 30:838–854.

Prieto-Márquez, A. 2012. The skull and appendicular skeleton of *Gryposaurus latidens,* a saurolophine hadrosaurid (Dinosauria: Ornithopoda) from the early Campanian (Cretaceous) of Montana, USA. Canadian Journal of Earth Sciences 49:510–532.

Salazar, I., P. Sánchez Quintero, and J. M. Cifuentes. 1995. Comparative anatomy of the vomeronasal cartilage in mammals: mink, cat, dog, pig, cow and horse. Annals of Anatomy 177:475–481.

Sternberg, C. M. 1953. A new hadrosaur from the Oldman Formation of Alberta: discussion of nomenclature. Bulletin of the National Museum of Canada 128:275–286.

Versluys, J. 1923. Der Schadel des Skelettes von *Trachodon annectens* im Senckenberg-Museum. Abhandlungen der Senckenbergischen Naturforschenden Gesellschaft 38:1–19.

Versluys, J. 1936. Kranium and Visceralskelett der Sauropsiden. 1. Reptilien; pp. 699–808 in L. Bolk, E. Göppert, E. Kallius, and W. Lubosch (eds.), Handbuch der Vergleichenden Anatomie der Wirbeltiere, Vol. 4. Urban & Schwarzenberg, Berlin.

Wagner, J. R. 2001. The hadrosaurian dinosaurs (Ornithischia: Hadrosauria) of Big Bend National Park, Brewster County, Texas, with implications for Late Cretaceous Paleozoogeography. M.S. thesis, Texas Tech University, Austin, Texas, 417 pp.

Wagner, J. R. 2004. Hard-tissue homologies and their consequences for interpretation of the cranial crests of lambeosaurine dinosaurs (Dinosauria: Hadrosauria). Journal of Vertebrate Paleontology 24(Supplement, 3):125A–126A.

Weishampel, D. B. 1981. The nasal cavity of lambeosaurine hadrosaurids (Reptilia: Ornithischia): comparative anatomy and homologies. Journal of Paleontology 55:1046–1057.

Wideman, N. K., and D. L. Lofgren. 2001. Partial hadrosaur skeleton with skin impression, Hell Creek Formation, McCone County, Montana. PaleoBios 21(Supplement, 2):134.

Witmer, L. M. 1995. Homology of facial structures in extant archosaurs (birds and crocodilians), with special reference to paranasal pneumacity and nasal conchae. Journal of Morphology 225:269–327.

Witmer, L. M. 2001. Nostril position in dinosaurs and other vertebrates and its significance for nasal function. Science 293:850–853.

Witmer, L. M., and R. C. Ridgely. 2008. The paranasal air sinuses of predatory and armored dinosaurs (Archosauria: Theropoda and Ankylosauria) and their contribution to cephalic architecture. Anatomical Record 291:1362–1388.

The Role and Biochemistry of Melanin Pigment in the Exceptional Preservation of Hadrosaur Skin

36

Phillip L. Manning, Roy A. Wogelius, Bart Van Dongen, Tyler R. Lyson, Uwe Bergmann, Sam Webb, Michael Buckley, Victoria M. Egerton, and William I. Sellers

ABSTRACT

The recognition of both structure and composition in dinosaur soft tissues has been revolutionized by the application of sophisticated new analytical techniques to fossil material. The identification of mineralized soft tissue has not been restricted to that contained within bones and claws. The structure and macromolecular composition of dinosaur skin has also been identified. The suite of techniques available for the study of soft tissue in the fossil record is now extending to infrared spectroscopy and synchrotron-based X-ray fluorescence, yielding impressive results. Here we review the application of multiple analytical techniques applied to the exceptionally preserved remains of the hadrosaurine dinosaur MRF-03 (cf. *Edmontosaurus* sp.) from the Hell Creek Formation (Upper Cretaceous) of North Dakota. In addition to these techniques, synchrotron-based spectroscopic techniques are applied to skin samples from MRF-03 for the very first time. When combined with the results from earlier work, the synchrotron-based results in this study provide evidence for the presence of Cu-O/Cu-N complexation, which is indicative of endogenous pigments. The trace metals might be associated with keratin, but this would have sulfur in the first coordination shell, which is not present. The presence of trace metals in the skin potentially played a taphonomic role in the spectacular soft tissue preservation shown by MRF-03.

INTRODUCTION

In the past 100 years there have been only a handful of "mummified" dinosaurs discovered, with the first being that of a hadrosaur by Charles Sternberg in 1908 (Osborn, 1909). While the term mummy is more commonly associated with exceptionally preserved or embalmed human remains, it is now liberally applied to well-preserved remains of animals (including dinosaurs) that often show skin and other soft tissues (Osborn, 1911; Schweitzer, 2011 for review). The term has its roots in the Arabic word for "bitumen," *mumiyyah*, which was applied to embalmed Egyptian bodies thought to be wrapped in tar or bitumen-soaked bandages. The medieval world transformed Arabic to the Latin, giving us *mumia*, which has become the word we use today when applied to the exceptional preservation of a body (Manning, 2008).

The recent remains of human mummies provide useful analogs to review the taphonomic processes that might lead to the exceptional preservation of soft-tissue structures. However, we are quick to point out that it is only the natural aspects of mummification that are relevant to soft tissue preservation in some dinosaurs. Such unique preservation is intimately linked to the fixing and recycling of organic compounds and their breakdown products in the carbon cycle (Tissot and Welte, 1984). Once an organism dies and its remains enter a sedimentary system, burial environment and microbial activity immediately affect the organic inventory comprising the organism. The altered biomolecules and decay products derived from the endogenous organic compounds can vary greatly, depending upon the timing of specific fixing processes that impact an organism before and after burial (Carpenter, 2007; Manning, 2008). Preservation of hard structural tissues, such as bones, teeth, and shells are well known, but rare occurrences of soft tissues, such as skin, keratin, and tendons challenge our understanding of this facet of the carbon cycle. Nonetheless, given that the hydrocarbon industry is founded upon the endurance of organic molecules in the geologic record, it is our conclusion that the paleontological sciences should be more open to seeking and identifying original organic molecules in fossil animals as well as products associated with their preservation and decomposition.

The carbon cycle is remarkably efficient at recycling organic material, but under certain preservational circumstances some of the chemical building blocks of an organism avoid recycling. In exceptionally preserved fossils, it is possible that remnants of structural proteins and associated organic molecules survive and can be mapped to help resolve original biological structures. The potential for new techniques to compositionally or spatially resolve such "chemical fossils" (Bergmann et al., 2010) is being realized with the recognition of elemental and organic residues that once

36.1. Lateral view of skin envelope from the lateral craniodorsal portion of tail of MRF-03. A distinct midline "frill" along the tail (uppermost portion of image) consisted of a series of keratinous plates running along the dorsal ridge. Scale bar equals 10 cm.

comprised living tissue (Morton and Witherspoon, 1993; Edwards et al., 2011; Wogelius et al., 2011; Glass et al., 2012; Lindgren et al., 2012). Until the advent of techniques sensitive enough to resolve trace amounts of organic compounds and organically bound elements (Wogelius et al., 2011; Glass et al., 2012; Lindgren et al., 2012), it was difficult to untangle potential material transfer from microbes, geochemical fluids, and the contamination from sampling and conservation techniques applied to a sample. However, the advent of synchrotron-based imaging and infrared spectroscopy has revolutionized sample analysis (Bergmann et al., 2010; Edwards et al., 2011; Wogelius et al., 2011; Glass et al., 2012; Lindgren et al., 2012), enabling high-resolution scans that spatially resolve reaction aureoles, precipitates, and so on.

The first dinosaur mummy to be discovered (AMNH 5060) was collected from the Lance Formation by the Sternberg field team in 1908 (Osborn, 1909, 1911). This specimen remains one of the best examples of a mummified dinosaur and is currently on display at the American Museum of Natural History (New York). Osborn (1912) attributed the exceptional preservation of AMNH 5060 to dehydration of the body prior to its final rapid burial by a flood. Based upon evidence from the observation of vertebrate carcasses on modern floodplains, Carpenter (2007) reiterated the importance of tissue dehydration in the mummification process. Carpenter (2007) went on to suggest that the term "mummy," when applied to dinosaurs, refers to impressions of the skin and not the preservation of biomaterials. The Sternberg mummy (AMNH 5060) shows clear evidence of consolidants having been applied to its surface that would compromise any chemical analyses of this famous example. However, the fossil remains of another hadrosaur mummy were discovered by Tyler Lyson and subsequently excavated by a team led by the Marmarth Research Foundation (MRF) and the University of Manchester in 2006 (Manning, 2008). The specimen (MRF-03) displays large areas of an uncollapsed

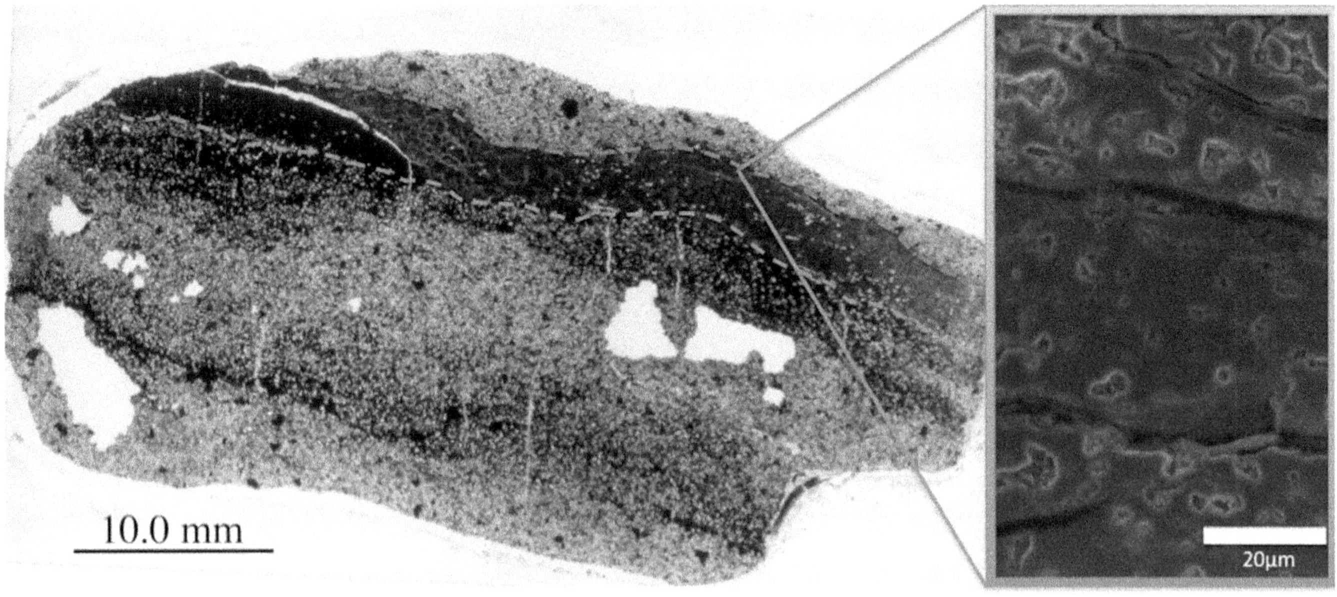

36.2. Visible light image of polished thin-section (left) and expanded BSE image of MRF-03 skin envelope (right). The dashed line on the thin-section delineates the skin envelope. Accelerating voltage 15.0 kV, working distance 9.8 mm at 0.5 Torr.

skin "envelope" around the tail, and parts of the torso, legs, and arm. The uniquely inflated nature of much of this fossil has already yielded important information that has aided muscle mass estimates for locomotion studies (Sellers et al., 2009). The preservation of the skin is three-dimensional, with clear raised scales and interconnecting hinge areas (Fig. 36.1), similar to that observed by Evans and Reisz (2007) in a small region of the neck of the lambeosaur hadrosaur, TMP 1966.004.0001. The skin of TMP 1966.004.0001 predominately displays skin traces with raised ridges that resemble a negative impression of the integument (Evans and Reisz, 2007). Thus, the impression of skin observed in TMP 1966.004.0001 is likely to be a function of bacterially mediated cements forming a skin mold as the integument was consumed by bacteria, leaving the resistant connected ridges observed by Evans and Reisz (2007). In contrast, MRF-03 exhibits both depth and structure (Manning et al., 2009), and thus indicates that more than just an impression or mold of the skin has been preserved (previously suggested for these kinds of specimens by Carpenter [2007]). In this chapter, we review previous work on a skin sample from MRF-03, and employ complementary synchrotron-based imaging techniques to more fully characterize the exceptional preservation of this unique hadrosaurid mummy.

Institutional Abbreviations AMNH, American Museum of Natural History, New York; MRF, Marmarth Research Foundation, Marmarth, North Dakota; TMP, Royal Tyrrell Museum of Palaeontology, Drumheller, Alberta.

MULTIDISCIPLINARY APPROACHES TO STUDYING EXCEPTIONAL PRESERVATION

The preservation of MRF-03 raises the question as to how such large areas of soft tissue avoided being recycled by bacteria that had swamped the carcass. The fossilized skin is clearly not an impression (Fig. 36.2) as indicated by thin-sections showing it to be several millimeters thick. Fossilized skin raises additional questions as to whether any structural information might still be retained in the fossil epidermis, or indeed whether any endogenous biomolecules (or their decomposition products) are present in the now-mineralized tissue. A multidisciplinary team was assembled to help answer these questions (Manning, 2008).

A small sample of fossil skin taken from the base of the tail of MRF-03 was used to make polished thin-sections (Fig. 36.2). These provide suitable samples for analysis within an environmental scanning electron microscope (ESEM) used in backscattered electron (BSE) mode (Manning et al., 2009). The same thin-sections were also used in the most recent synchrotron rapid scanning X-ray fluorescence (SRS-XRF) analyses. The BSE microscopy provided information on both the structure and composition of the skin envelope. Given that each element has a different number of electrons, there is a direct correlation between atomic number and electron backscattering efficiency; a more dense electronic cloud yields more backscattering of an incident electron beam. Thus, the BSE mode on an ESEM yields images that reflect the average surficial element composition of a sample (Goldstein et al., 1981).

The samples from MRF-03 (taken from the ventral surface at the base of the tail) were ~2.5–3.5 mm thick and preserved internal structure (Fig. 36.2) that was visible in BSE mode (Manning et al., 2009). Compositional data were also recovered using energy dispersive X-ray (EDX) analysis, but the results were not sensitive enough to provide information on the spatial distribution of the elemental inventory of the skin. Similarly, the BSE data indicated the presence of organic compounds (that appear black in BSE mode) restricted to within the upper and lower boundaries of the sectioned skin envelope, but EDX did not permit their identification. A thin-section was then analyzed by both infrared and mass spectroscopic techniques to further resolve the presence and identity of organic compounds within the skin.

Fourier transform infrared spectroscopy (FTIR) was used as a screening tool by Manning et al. (2009) to spatially map the presence of infrared active organic functional groups in skin, hoof and tendon samples from MRF-03. FTIR exploits the fact that different molecules absorb specific infrared frequencies reflecting their structure (Griffiths and Haseth, 2007). The frequency of the absorbed infrared spectra matches that of the bond or group that vibrates. FTIR of biomaterials recovered from the skin and terminal ungual phalanx of MRF-03 indicated the presence of compounds containing amide groups. These amide groups absorb infrared light at two positions called amide I and amide II, and dominate the absorption spectrum of keratin. Amides are the most stable of all carbonyl functional groups (Killops and Killops, 2005). The position and appearance of the FTIR bands of the amide I and II groups present within the skin regions of MRF-03 were similar to what was measured in extant beta-keratin samples taken for bird and reptile tissues, constraining analogous biomaterials used by the Extant Phylogenetic Bracket (Witmer, 1995; Manning et al., 2009). The presence of amide groups was interpreted by the earlier study as a sign that intact proteins or their breakdown products might possibly have survived in MRF-03.

Proteins form one of the major building blocks of life, carrying out a range of different functions–from structural biomaterials to cell maintenance and the regulation of biosynthetic pathways (Nelson and Cox, 2008). They are made up of chemical units called amino acids, of which there are 20 encoded for by the genetic code. Amino acid–composition analysis is a method used to hydrolyze (break up) proteins present in a given sample into its constituent amino acid units in order to provide a general impression of the dominant proteins within a sample (or to screen fossils to see if any proteins could be present) (Buckley et al., 2008). Due to the asymmetric structure of most amino acids (all except glycine), they exist in two different mirror-image configurations often called D (dextro) and L (levo) amino acids, but only the L forms are typically incorporated into proteins during synthesis. However, once protein synthesis stops (i.e., with the death of an organism), L-amino acids start to spontaneously convert to D-amino acids via a process called racemization (Wehmiller and Miller, 2000). Although various environmental factors need to be taken into account, this amino acid racemization process can be used to assess the state of preservation of proteins in fossils. In both of the above methods the amino acids (both L and D forms) are usually separated by some form of chromatography (in this case, liquid chromatography) and detected by a variety of methods (in this case, fluorescence). The amino acid composition and racemization analyses of a skin envelope sample from MRF-03 exhibited a distinct composition clearly different from the surrounding matrix (Manning et al., 2009). High glycine-to-alanine ratios were observed, potentially indicative of fibrous structural proteins such as collagens and keratins. After using amino acid composition and racemization analyses to confirm the presence of protein, there are a variety of protein isolation techniques that can be used to study how intact the proteins are within the fossil.

The most common protein isolation technique is called polyacrylamide gel electrophoresis (PAGE). This involves taking a sample of proteins (i.e., not hydrolyzed as with amino acid analyses) and applying them to a polyacrylamide gel with a tracking dye (Gordon, 1975). During electrophoresis the samples are induced to separate according to size and charge (electrophoretic mobility) using an electric field over a given amount of time. Following electrophoresis the gel can be stained, allowing for visualisation of how far each band of separated proteins has traveled and, if used with molecular weight markers, can help estimate the size of the protein. The PAGE protocol successfully removed beta-keratin from extant samples in the prior study, but was not able to repeat the same for skin and tendon samples from MRF-03 (Manning et al., 2009). However, when size exclusion was removed from the PAGE protocol, low–molecular weight fractions (less than 12 kDa) were observable on the electrophoresis gel for MRF-03 samples, indicating the organic material was present. Additional analyses using matrix-assisted laser desorption/ionization-mass spectrometry (MALDI-MS) and liquid chromatography–electrospray ionization (LC-ESI) were consistent with the results produced by the PAGE protocol in that low–molecular weight peaks at m/z 1100–2200 were observed (Manning et al., 2009). Therefore, the presence of intact proteins could not be confirmed using protein mass spectrometry, even after such promising amino acid analysis, but breakdown products of proteins were likely. To further unravel the macromolecular composition of the skin, some

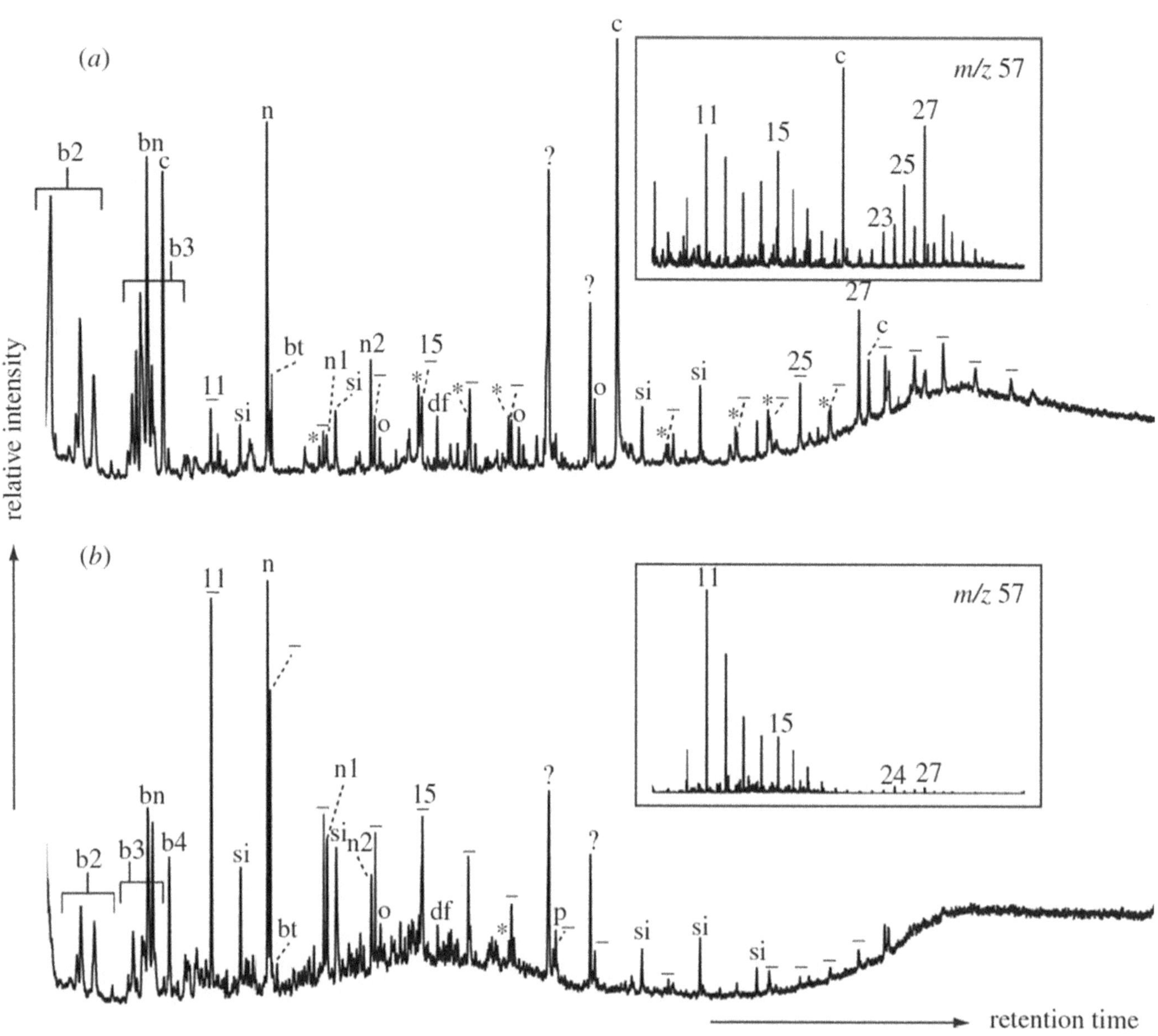

36.3. Partial Py-GCMS total ion current chromatograms of (A) the skin envelope and (B) the surrounding sediment associated with the skin envelop of MRF-03. The insets show the m/z 57 mass chromatograms revealing the distribution of *n*-alkanes moieties with numbers indicating the carbon chain length (from Manning et al., 2009).

destructive analysis was required – in this case, pyrolysis gas chromatography mass spectrometry (Py-GCMS).

Py-GCMS works on the principle that when a sample is rapidly heated to decomposition it produces smaller molecules that are then separated by gas chromatography and subsequently may be detected using mass spectrometry (Halket and Zaikin, 2006). The application of Py-GCMS was more successful and revealed a substantial difference in the aliphatic polymer between the skin and associated sediment of MRF-03 (Manning et al., 2009).

The Py-GCMS of the skin generated *n*-alkanes/*n*-alken-1-ene homologues ranging in carbon number from C9 to C36, with a trimodal distribution of *n*-alkanes (Fig. 36.3A; maxima C11, C15, and C27). In comparison, the *n*-alkane/*n*-alken-1-ene homologues distribution pattern in the enclosing sediment differed considerably, ranging from C9 to C30 but dominated by the C10 to C18 *n*-alkanes with a maximum at C11 (Fig. 36.3B). The observed differences were inconsistent with an origin solely via migration from enclosing sediment and must, thus, have been derived endogenously. This suggested that the organics present in the skin consisted of a macromolecule that was in part aliphatic. Comparable to earlier studies on plant and insect fossils, these aliphatic components are presumably the result of in situ polymerization (Gupta, Briggs, et al., 2007; Gupta, Michels, et al., 2007).

The evidence from the analyses undertaken by Manning et al. (2009) suggests that the preservation of the organic material was probably caused by a rapid burial in combination with intensely reducing pore waters causing oxidized iron species to be reduced and feldspar and rock fragments to partially dissolve. The reduced iron in solution rapidly replaced the soft tissue with carbonate minerals, outpacing microbial decay processes. Besides preservation of soft tissue structures this ensured that some breakdown products of organic molecules were preserved within the mineral matrix. However, the paradox of skin surviving, but more structurally robust biomaterials (such as ligaments and tendons) being rare (only epaxial tendons evident), suggest something else was aiding the preservation of the skin.

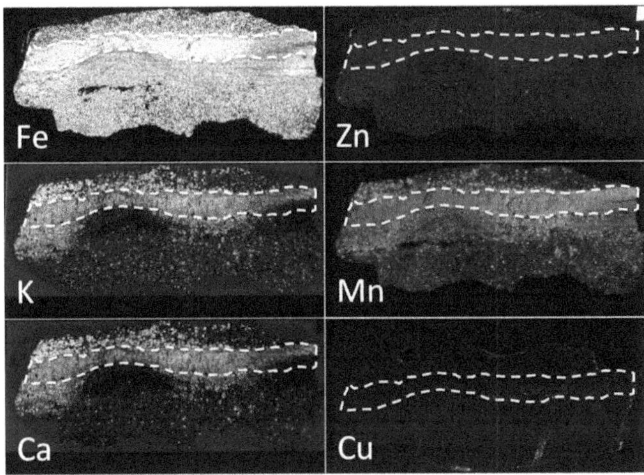

36.4. Synchrotron rapid scanning X-ray fluorescence maps of polished thin-section (see Fig. 36.2 [right] for scale) of skin from MRF-03 undertaken at the Stanford Synchrotron Radiation Lightsource beam line 2–3. The dashed white line delineates the boundary of the skin envelope and indicates the concentration of mapped elements in the sample.

SYNCHROTRON RAPID SCANNING X-RAY FLUORESCENCE

The recent application of synchrotron rapid scanning X-ray fluorescence (SRS-XRF) to specimens of *Archaeopteryx lithographica* (Bergmann et al., 2010; Wogelius et al., 2011; Manning, Edwards, et al., 2013), Green River Formation reptile skin (Edwards et al., 2011) and *Confuciusornis sanctus* (Wogelius et al., 2011) have provided great insight to the chemistry and taphonomy of fossilized soft tissues. This powerful, but nondestructive technique offers new insight to the mass transfer of elements and organic compounds within specific sedimentary lithofacies and provides high-resolution maps of chemistry within discrete biological structures. When combined with supporting chemical data derived from the tissues of extant species that phylogenetically bracket the extinct samples being studied, it can provide evidence for potential endogenous compounds and biosynthetic pathways.

SRS-XRF maps are generated by collecting signals from multiple elements that are read out at intervals of only a few milliseconds per pixel during bidirectional scans (Bergmann et al., 2010; Edwards et al., 2011). This dramatically reduces scan times to ~30 seconds for each square centimeter scanned at 100 μm resolution. Because of the intensity of the incident synchrotron X-ray beam, imaging is not only rapid enough to be practical for large objects but also can successfully record spatial variation at even lower concentrations of an element than is typically possible with standard electron beam methods; approximately an order of magnitude better sensitivity can be achieved routinely with the current Stanford Synchrotron Radiation Lightsource (SSRL) configuration (Wogelius et al., 2011). SRS-XRF applied to fossils makes it possible to simultaneously probe elemental distributions of large specimens and their embedding geological matrix, thus resolving the exchange of material between the organism and the surrounding sediment during fossilization (Bergmann et al., 2010; Wogelius et al., 2011).

Samples of fossil skin from MRF-03 were chemically mapped using SRS-XRF on beam line 3–2 at the SSRL (SLAC National Laboratory; Fig. 36.4). In addition to the SRS-XRF, extended X-ray absorption fine structure (EXAFS) analysis was also undertaken with samples from MRF-03 on beam line 2–3 at SSRL. EXAFS spectroscopy is sensitive to the electronic structure of the probed central absorber atom, and is especially able to accurately quantify distances to shells of surrounding atoms, thus providing a great deal of information on biological or geochemical context.

The EXAFS data can be used to diagnose whether trace metals such as $Fe^{2+}$and Mn^{2+} are sequestered within channels of eumelanin- or melanin-derived organic compounds. Given that divalent trace metals are chelated by melanin pigments, this may provide useful information about presence and distribution of original pigment (Wogelius et al., 2011).

RESULTS

The SRS-XRF maps (Fig. 36.4) show elevated levels of iron, potassium, calcium, and manganese, which are constrained within the fossil skin of MRF-03, but attenuated within the surrounding matrix. There is also a low signal for both copper and zinc, again concentrated within the skin (Fig. 36.4). The origin of the elemental inventory can be explained from two sources, or possibly from a combination of the two: geochemical precipitates or endogenically derived compounds from the dinosaurs' skin.

DISCUSSION

Pigment of Our Imagination?

Resolving pigment patterns in extinct species might provide one of the keys to understanding both selection processes and associated functions of pigmentation, such as camouflage, communication, and sexual selection. However, up until four years ago little was known for certain about pigmentation in extinct species (Vinther et al., 2008). Two approaches have recently yielded success. One has used electron microscopy to study the structure of proposed fossilized melanosomes in order to resolve potential color patterning (Vinther et al., 2008, 2010; Li et al., 2010; Zhang et al., 2010). A second approach, focused on chemistry, and has used synchrotron-based imaging such as SRS-XRF and EXAFS (Wogelius et al., 2011) to detect and map trace metals in fossils. Later studies have combined additional multiple analytical techniques (Glass et al., 2012; Lindgren et al., 2012) that confirmed that organometallic chelates derived from original eumelanin can serve as pigment biomarkers. Hence chemical diagnosis and mapping, including metal chelates, can serve to nondestructively provide tremendous details of pigment density in an entire specimen.

The synthesis of melanin (melanogenesis) takes place in the pigment containing organelle, the melanosome, in the presence of the catalytic enzyme tyrosinase for the formation of melanin in vertebrates (Sánchez-Ferrer et al., 1995). This enzyme completes the initial oxidation step to form dopaquinone, which can then combine with cysteine or other amino acids to generate eumelanin or phaeomelanin. Another diagnostic and functional component of the molecular structure of melanins is their carboxyl substitutes. These negatively charged end groups function as cation chelators, selectively binding positively charged particles, such as transition metals (Riley, 1997). Copper is a critical cofactor of tyrosinase, which is remarkably conserved both functionally and genetically across major animal lineages (Sato et al., 2001). When melanin binds transition metals, they can help protect the organism from the toxic effects of high metal accumulation; they have strong interlinking properties and might also strengthen tissues and help them resist breakage, wear, and microbial degradation (McGraw, 2006; Burtt et al., 2011).

Eumelanins are commonly found in all vertebrates (Fox and Vevers, 1960; Needham, 1964; Fox, 1976). However, little is known about the distribution of eumelanin among invertebrates, save the ink of cephalopods (Prota, 1992), jellyfish (Fox and Millot, 1954), the plumose anemone (Fox and Pantin, 1944), sea urchins (Jacobson and Millot, 1953), the brittlestar (Fontaine, 1962), various species of worms (Kennedy, 1969), the black slug, the snail *Limnaca* sp., and other gastropods (Goodwin, 1972).

The detailed chemical analysis of exceptionally preserved fossil skin has already demonstrated that a residue comprising endogenous biomolecules can survive (Manning et al., 2009; Edwards et al., 2011). These earlier studies found traces of hopanes, important to the adjustment of bacteria cell membrane permeability, in matrix surrounding fossil skin. However, that trace disappeared on extraction, indicating it was not associated with the kerogen in the skin. Therefore, a bacterial biofilm was not responsible for the observed chemical difference between the sediment and the fossilized skin (Manning et al., 2009; Edwards et al., 2011). Such exceptional preservation makes it possible to constrain mass transfer processes between the fossil and its embedding sedimentary matrix (Bergmann et al., 2010), and influences of microbial reprocessing that may also be detected (Barden et al., 2011; Edwards et al., 2011).

Chemical analysis improves our understanding of the pathways that lead to exceptional preservation, and may even reveal new biomarkers that may be used in conjunction with structural information (Lindgren et al., 2012). This approach might provide evidence for the presence of other compounds within fossilized integument, including chemically based pigments. Furthermore, given that fossilized microbes may be difficult to distinguish from fossilized melanosomes (Lingham-Soliar and Glab, 2010; Barden et al., 2011), developing a chemical test for pigment residue would be of significant benefit to the structural analysis approach because it provides an independent test for the interpretation of any structures. Finally, because geological deformation may destroy structural information while leaving a chemical signal intact (Wogelius et al., 2011), chemical imaging may be able to resolve pigments in a variety of geological settings and over longer periods of geological time if structural information is absent.

The chemical approach has also achieved something perhaps even more remarkable than resolving the organic residue of 120-million-year-old pigments from feathers (Wogelius et al., 2011); it has provided the first direct ångström-scale measurements of trace metal coordination chemistry for eumelanin within extant species as well as in residue from extinct species via X-ray absorption spectroscopy (both EXAFS and X-ray absorption near edge structure [XANES]) (Wogelius et al., 2011). Despite numerous excellent studies of eumelanin in the biological literature (see Prota, 1992, for review), there are still a number of unknowns concerning this most ubiquitous pigment. Indeed, we have had to routinely image extant soft tissue samples via SRS-XRF so that we had control specimens to compare to our fossil data. Because SRS-XRF has only recently been developed, can resolve extremely low concentrations (at least 10× lower than standard techniques) with excellent spatial resolution (micron scale), and can do this 3000 times faster than conventional methods, we routinely observe elemental patterns in discrete biological structures (e.g., feather barbule pigment pattern) in modern organisms that have not been reported previously. Thus the study of fossils using SRS-XRF has enhanced our knowledge of extant species as well. This underlines the fact that our chemical approach to analyzing fossil material needs to be supported by analyses of comparable extant organisms and that aspects of what we learn will be of interest not only to palaeontologists but also to the wider biological community. The samples from MRF-03 are no exception, as here, too, we can observe discrete elemental inventories constrained within the fossil skin of this ~66-million-year-old hadrosaur, but do they reflect geochemical precipitates or organically derived molecules endogenous to MRF-03?

Trace-Metal Jacket

Previous studies have shown an intimate relationship between trace metals and pigmentation (McGraw, 2006; Burtt et al., 2011; Wogelius et al., 2011). The trace-metal central absorber atom that dominates the melanin molecule can be mapped and quantified using SRS-XRF and EXAFS. The combination of both keratinous skin tissue and pigment provide ideal substrates for the chelation of divalent trace metals. It is worth noting that if the trace metals were keratin derived they would have sulfur in the first shell – and this was clearly not the case (Wogelius et al., 2011) – supporting coordination chemistry that is agreeable with melanin pigment. It is clear from this and prior studies (Barden et al., 2011; Edwards et al., 2011; Wogelius et al., 2011) that divalent trace metals are also intimately associated with preserved biological structures, such as the rachis and barbs of feathers. The SRS-XRF maps (Fig. 36.4) clearly indicate that levels of iron, potassium, calcium, copper, zinc, and manganese are elevated within the skin of MRF-03, but are lower or absent within the matrix. To discern the origin of the elemental inventory we have to review both geochemical and endogenically derived sources for such elements.

The mobility of elements, which control geochemical precipitates within the skin, would have been high in the organic-rich pore waters of the Hell Creek Formation. However, it is interesting to note that many of these elements are present in elevated levels in the skin of MRF-03, but in significantly reduced amounts in the surrounding matrix (Fig. 36.4). The partial pressure of CO_2 is very high in this environment (reducing) because of the microbial degradation of organic matter. The precipitation of iron, calcium, and manganese minerals is likely a function of the carbonate rich pore waters generated by the organic acids liberating CO_2 from the immature Hell Creek Formation sediments. BSE microscopy indicated that the sediments are composed of framework grains comprising relict (skeletal) feldspars, rock fragments, detrital dolomite, quartz, and woody organic matter with significant volumes of both grain dissolution and intergranular porosity (Manning et al., 2009). Some minerals dissolve altogether (e.g., detrital calcites), liberating alkali and alkaline earth elements (Na, K, Ca, Mg) as soluble products. It is interesting to note that a great deal of acid-generating reactions produce carbonate cements, but that Fe(III) reduction and mineral dissolution tends to generate significant alkalinity to buffer such environments. Thus, the decay products of organic matter and of unstable minerals may combine to cause the precipitation of carbonate minerals such as ferroan and manganese-rich calcite (Ca, Fe, Mg). Similar carbonate mineralization of skin has been noted from the lacustrine Las Hoya limestone (Briggs et al., 1997) and a marine carbonate mudstone (Martill et al., 2000). Manning et al. (2009) also concluded that the combination of the intensely reducing pore waters, from the decay of plant material, caused oxidized iron and manganese species to be reduced and feldspars and rock fragments to partially dissolve. The calcium, reduced iron, and reduced manganese in solution were readily available to replace the soft tissue of MRF-03 with carbonate minerals. Crucially, supply of iron to the rotting dinosaur was maintained by Fe-rich ground waters that preferentially flowed through the higher porosity sandy units that were deposited as channel sands rather than through the mud-dominated overbank deposits. Rapid precipitation of carbonates was critical in preservation of the skin envelope. However, it is possible that certain elements, such as Ca, were crucial to the early precipitation of carbonates were already present in the skin (epidermis) of MRF-03 and might have played a role in the high-fidelity preservation observed in the skin of this dinosaur. To explain this, we need to look at the elemental inventory that controls the formation and regulation of vertebrate skin.

The epidermis consists of five major layers: stratum corneum, stratum lucidum, stratum granulosum, stratum spinosum, and stratum basale (Fig. 36.5). The cornified stratum corneum functions as a keratinous barrier that prevents the desiccation of terrestrial organisms. Calcium and potassium are essential elements that help form and maintain the keratinous layers (Lee et al., 1992; Mauro et al., 1997). In undamaged skin, the highest concentrations of calcium occur in the stratum granulosum and the lowest occur in the stratum basale; potassium concentration is highest in the stratum spinosum and lowest in the stratum granulosum (Lee et al., 1992; Mauro et al., 1997; Elias et al., 2002). High concentrations of calcium and low concentrations of potassium in the stratum granulosum triggers calcium-induced

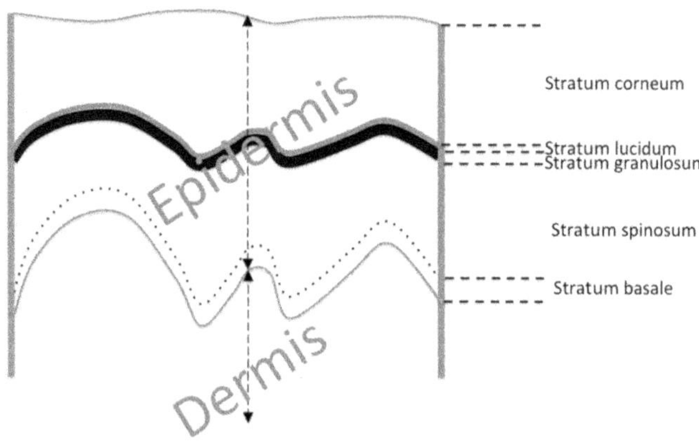

36.5. Cross section of vertebrate epidermis showing gross morphology.

keratinocyte differentiation that leads to the formation of the stratum corneum. If the concentrations of either calcium or potassium decrease in the epidermis due to damage, a series of metabolic responses occur to restore barrier homeostasis (Lee et al., 1992; Mauro et al., 1997; Behne et al., 2011).

The elemental maps of the sectioned skin of MRF-03 (Fig. 36.4) show a strong correlation between the skin and high levels of both potassium and calcium. The cell-like morphology of the skin of MRF-03, seen in polished thin-sections, was previously mapped in calcium by Manning et al. (2009) using an electron microprobe. The carbonates were interpreted as having a geochemical origin, but still recording original structure of the epidermis. It is possible the earlier study overlooked the potential for the soft tissue as being an additional source for the elements required to preserve the skin. It is interesting to speculate that the high concentrations of calcium in the stratum granulosum and potassium in the stratum spinosum (Lee et al., 1992; Mauro et al., 1997; Elias et al., 2002) might provide a high-fidelity framework for the preciptation of minerals that could faithfully reproduce biological cell-like structures, as have been previously described from MRF-03 (Manning et al., 2009). This might well explain some of the detailed preservation we observe in this unique fossil.

The presence of zinc and copper in the skin of MRF-03 is low, but when compared to the surrounding sediment, the skin preserves elevated levels of these divalent trace metals. Such trace-metal loadings in both melanin pigments and keratinous tissues are not uncommon (see Wogelius et al., 2011). The presence of such chelated metals have been suggested to act as effective biocides in extant soft tissues (Goldstein et al., 2004), and have been postulated to likewise contribute to the exceptional preservation of fossils (Wogelius et al., 2011). This biocidal action reduces or slows the local breakdown of soft tissue, enabling such biomaterials to survive long enough to be mineralized.

While we suggest the possible presence of skin pigment in MRF-03 from the elevated levels of zinc and (less so) copper, the color of the skin is a more complex question. Other vectors that have an effect upon coloration of MRF-03, such as carotenoid pigments and even the physical structure of the skin, might also have to be taken into account. Carotenoids are sequestered from compounds as a function of diet and can be pooled into contiguous patches of pigment (McGraw, 2006). Carotenoids are produced by plants, and are acquired by eating plants or by eating something that has eaten a plant. Colors can also result from the refraction, reflection, or absorption of incident light caused by the microscopic structure within the tissue in question, splitting light into rich, component colors (Kinoshita and Yoshioka, 2005).

Skin coloration and patterns are often a function of the selection pressures acting upon a species in distinct habitats. The recognition of trace metals that might signify the presence of melanin does not exclude the occurrence of other brightly colored pigments whose presence might be masked, but would have still contributed to the coloration of MRF-03. While we might infer the presence of a pigment, we are still not in a position to diagnose color, but a melanin pattern might well be possible to map using SRS-XRF on well-preserved dinosaur skin, potentially yielding a monochrome picture of dinosaurs.

CONCLUSION

This study rejects earlier assumptions that mummified dinosaur skin is merely an impression (Carpenter, 2007), but preserves biologically derived material. We suggest that if suitable chemical conditions prevail during burial that organometallic compounds may aid the preservation of dinosaur skin. The presence of trace metal-rich pigment in the skin of MRF-03 (Fig. 36.4) might have acted as a natural biocide (Wogelius et al., 2011), resistant to the bacteria that would consume the rest of the carcass. Thus, the presence of indigestible pigments and the elevated levels of potassium and calcium associated with keratin production and the mineral-rich pore waters were possibly factors that might explain both the rapidity and exceptional preservation of dinosaur skin at the cellular level. When combined with the results from earlier work, the synchrotron-based imaging in this study provides the first evidence for the presence of Cu-O/Cu-N complexation, which is indicative of endogenous pigments in dinosaur skin. If these trace metals were just associated with keratin, there would be sulfur present in the first coordination shell, but this is not the case. The presence of pigment-associated trace metals in skin might well have played a taphonomic role in the spectacular soft tissue preservation shown by MRF-03. It seems possible that the chemistry of life might well have assisted in the preservation of this dinosaur after its death, ultimately being immortalized in stone.

ACKNOWLEDGMENTS

The authors would like to thank the National Geographic Society for supporting much of the preliminary work undertaken on the dinosaur mummy (MRF-03). We also would like to especially thank the Marmarth Research Foundation for their support and access to the unique fossil that is the subject of this chapter. We also wish to thank the staff of the Stanford Synchrotron Radiation Lightsource for their continued support and the Department of Energy for the beam time that made the synchrotron-based imaging possible. Finally, we thank the editors and reviewers Dave Eberth, David Evans, Patty Ralrick, and Jakob Vinther for their constructive reviews that helped improve this chapter.

LITERATURE CITED

Barden, H. E., R. A. Wogelius, P. L. Manning, N. P. Edwards, H. You, and B. E. van Dongen. 2011. Morphological and geochemical evidence of melanin preservation in the feathers of the early Cretaceous bird *Gansus yumenensi*. PLoS ONE 6(10):e25494.

Behne, M., S. Sanchez, N. Barry, N. Kirschner, W. Meyer, T. Mauro, I. Moll, and E. Gratton. 2011. Major translocation of calcium upon epidermal barrier insult: imaging and quantification via FLIM/Fourier vector analysis. Archives of Dermatological Research 303:103–115.

Bergmann, U., R. W. Morton, P. L. Manning, W. I. Sellers, S. Farrar, K. G. Huntley, R. A. Wogelius, and P. Larson. 2010. *Archaeopteryx* feathers and bone chemistry fully revealed via synchrotron imaging. Proceedings of the National Academy of Sciences 107:9060–9065.

Briggs, D. E. G., P. R. Wilby, B. P. Perez-Moreno, J. L. Sandz, and M. Fregenal-Martinez. 1997. The mineralization of dinosaur soft tissue in the lower Cretaceous of Las Hoyas, Spain. Journal of the Geological Society of London 154:587–588.

Buckely, M., A. Walker, S. Y. W. Ho, Y. Yang, C. Smith, P. Ashton, J. T. Oates, E. Cappellini, H. Koon, K. Penkman, B. Elsworth, D. Ashford, C. Solazzo, P. Andrews, J. Strahler, B. Shapiro, P. Ostrom, H. Gandhi, W. Miller, B. Raney, M. I. Zylber, M. T. P. Gilbert, R. V. Prigodich, M. Ryan, K. F. Rijsdijk, A. Janoo, and M. J. Collins. 2008. Comment on "Protein sequences from mastodon and *Tyrannosaurus rex* revealed by mass spectrometry." Science 319:33.

Burtt, E. H., M. R. Schroeder, L. A. Smith, J. E. Sroka, and K. J. McGraw. 2011. Colourful parrot feathers resist bacterial degradation. Biology Letters 7:214–216.

Carpenter, K. 2007. How to make a fossil: part 2: dinosaur mummies and other soft tissue. Journal of Paleontological Science JPS.C.07.0002:1–23.

Edwards, N. P., B. E. Barden, B. E. van Dongen, P. L. Manning, U. Bergman, W. I. Sellers, and R. A. Wogelius. 2011. Infrared mapping resolves soft tissue preservation in 50 million year old reptile skin. Proceedings of the Royal Society B 278:3209–3218.

Elias, P. M., S. K. Ahn, B. E. Brown, D. Crumrine, and K. R. Feingold. 2002. Origin of the epidermal calcium gradient: regulation by barrier status and role of active versus passive mechanisms. Journal of Investigative Dermatology 119:1269–1274.

Evans, D. C., and R. R. Reisz. 2007. Anatomy and relationships of *Lambeosaurus magnicristatus*, a crested hadrosaurid dinosaur (Ornithischia) from the Dinosaur Park Formation, Alberta. Journal of Vertebrate Paleontology 27:373–393.

Fontaine, A. R. 1962. The colours of *Ophiocomina nigra* (Abildgaard): III: carotenoid pigments. Journal of the Marine Biological Association of the United Kingdom 42:33–47.

Fox, D. L. 1976. Animal Biochronics and Structural Colors. University of California Press, Berkeley, California, 450 pp.

Fox, D. L., and N. Millot. 1954. The pigmentation of the jelly fish, *Pelagia noctiluca* (Forskål) var. *panopyra* Péron & Lesueur. Proceedings of the Royal Society of London B 142:392–408.

Fox, D. L., and C. F. A. Pantin. 1944. Pigments in the Coelenterata. Biological Reviews 19:121.

Fox, D. L., and G. Vevers. 1960. The Nature of Animal Colours. Sidgwick and Jackson, London, 246 pp.

Glass, K., S. Ito, P. R. Wilby, T. Sota, A. Nakamura, C. R. Bowers, J. Vinther, S. Dutta, R. Summons, D. E. G. Briggs, K. Wakamatsu, and J. D. Simon. 2012. Direct chemical evidence for eumelanin pigment from Jurassic Period. Proceedings of the National Academy of Sciences 109:10218–10223.

Goldstein, G. I., D. E. Newbury, P. Echlin, D. C. Joy, C. Fiori, and E. Lifshin. 1981. Scanning Electron Microscopy and X-Ray Microanalysis. Plenum Press, New York, 673 pp.

Goldstein, G. K. R., B. A. B. Flory, M. Samia, J. M. Ichida, and E. H. Burtt. 2004. Bacterial degradation of black and white feathers. Auk 121:656–659.

Goodwin, T. W. 1972. Pigments in Mollusca; pp. 187–199 in M. Florkin, and B. T. Scheer (eds.), Chemical Zoology, volume 7, Mollusca. Academic Press, New York.

Gordon, A. H. 1975. Electrophoresis of Proteins in Polyacrylamide and Starch Gels. American Elsevier, New York, 216 pp.

Griffiths, P. R., and J. A. Haseth. 2007. Fourier Transform Infrared Spectrometry. John Wiley & Sons, Hoboken, New Jersey, 529 pp.

Gupta, N. S., D. E. G. Briggs, M. E. Collinson, R. P. Evershed, R. Michels, and R. D. Pancost. 2007. Molecular preservation of plant and insect cuticles from the Oligocene Enspel Formation, Germany: evidence against derivation of aliphatic polymer from sediment. Organic Geochemistry 38:404–418.

Gupta, N. S., R. Michels, D. E. G. Briggs, M. E. Collinson, R. P. Evershed, and R. D. Pancost. 2007. Experimental evidence for land plant lipids as a source of aliphaticrich kerogen. Organic Geochemistry 38:28–36.

Halket, J. M., and V. Z. Zaikin. 2006. Derivatization in mass spectrometry–7: on-line derivatisation/degradation. European Journal of Mass Spectrometry 12(1):1–14.

Jacobson, F. W., and N. Millot. 1953. Phenolases and melanogenesis in the coelomic fluid of the ethinoid Diadema antillarum philippi. Proceedings of the Royal Society of London B 141:231–247.

Kennedy, G. Y. 1969. Pigments of Annelida, Echiuroidea, Sipunculoidea, Priapuloidea, and Phoronidea; pp. 311–376 in M. Florkin and B. T. Scheer (eds.), Chemical Zoology, Vol. IV. Academic Press, New York.

Killops, S., and V. Killops. 2005. Introduction to Organic Geochemistry. Blackwell, Oxford, U.K., 393 pp.

Kinoshita, S., and S. Yoshioka. 2005. Structural colors in nature: the role of regularity and irregularity in the structure. ChemPhysChem 6:1–19.

Lee, S. H., P. M. Elias, E. Proksch, G. K. Menon, M. Mao-Quiang, and K. R. Feingold. 1992. Calcium and potassium are important regulators of barrier homeostasis in murine epidermis. Journal of Clinical Investigation 89:530–538.

Li, Q., K.-Q. Gao, J. Vinther, M. D. Shawkey, J. A. Clarke, L. D'Alba, Q. Meng, D. E. G. Briggs, and R. O. Prum. 2010. Plumage color patterns of an extinct dinosaur. Science 327:1369–1372.

Lindgren, J., P. Uvdal, P. Sjövall, D. E. Nilsson, A. Engdahl, B. P. Schultz, and V. Thiel. 2012. Molecular preservation of the pigment melanin in fossil melanosomes. Nature Communications 3(824) doi:10.1038/ncomms1819.

Lingham-Soliar, T., and J. Glab. 2010. Dehydration: a mechanism for the preservation of fine detail in fossilised soft tissue of ancient terrestrial animals. Palaeogeography, Palaeoclimatology, Palaeoecology 291:481–487.

Manning, P. L. 2008. Grave Secrets of Dinosaurs: Soft Tissue and Hard Science. National Geographic Society, Washington, D.C., 320 pp.

Manning, P. L., P. M. Morris, A. McMahon, E. Jones, A. Gize, J. H. S. Macquaker, J. Marshall, T. Lyson, G. Wolff, M. Buckley, and R. A. Wogelius. 2009. Preserved soft-tissue structures and organic molecules in a mummified hadrosaur dinosaur from the Hell Creek Formation, North Dakota (USA). Proceedings of the Royal Society Series B 276:3429–3437.

Manning, P. L., N. P. Edwards, R. A. Wogelius, H. Barden, P. L. Larson, D. Schwarz-Wings, U. Bergmann, V. M. Egerton, and W. I. Sellers. 2013. Synchrotron-based chemical imaging

reveals plumage patterns in a 150 million year old early bird. Royal Society of Chemistry. Journal of Analytical Atomic Spectrometry 28:1024–1030.
Martill, D. M., D. J. Batten, and D. K. Lydell. 2000. A new specimen of the thyreophoran dinosaur cf. *Scelidosaurus* with soft tissue preservation. Palaeontology 43:549–559.
Mauro, T., D. B. Dixon, L. Komuves, K. Hanley, and P. A. Pappone. 1997. Keratinocyte K+ Channels Mediate Ca2+-Induced Differentiation. Journal of Investigative Dermatology 108:864–870.
McGraw, K. J. 2006. Mechanics of melanin-based coloration; pp. 243–294 in G. E. Hill and K. J. McGraw (eds.), Bird Coloration: Mechanisms and Measurements, volume 1. Harvard University Press, Cambridge, Massachusetts.
Morton, R.W., and K. C. Witherspoon. 1993. Elemental x-ray imaging of fossils. Advances in X-Ray Analysis 36:97–104.
Needham, A. E. 1964. The Significance of Zoochromes. Springer-Verlag, Berlin, 429 pp.
Nelson, D. L., and M. M. Cox. 2008. Lehninger's Principles of Biochemistry, 5th edition. W. H. Freeman and Company, New York, 1263 pp.
Osborn, H. F. 1909. The epidermis of an iguanodont dinosaur. Science 29:793–795.
Osborn, H. F. 1911. A dinosaur mummy. American Museum Journal 11:7–11.
Osborn, H. F. 1912. Integument of the iguanodont dinosaur *Trachodon*. Memoirs of the American Museum of Natural History 1:33–54.
Prota, G. 1992. Melanins and Melanogenesis. Academic Press, New York, 290 pp.
Riley, P. A. 1997. Melanin. International Journal of Biochemistry and Cell Biology 29:1235–1239.
Sánchez-Ferrer, A., J. N. Rodríguez-López, F. García-Cánovas, and F. García-Carmona. 1995. Tyrosinase: a comprehensive review of its mechanism. Biochim Biophys Acta 1247:1–11.
Sato, S., M. Tanaka, H. Miura, K. Ikeo, T. Gojobori, T. Takeuchi, and H. Yamamoto. 2001. Functional conservation of the promoter regions of vertebrate tyrosinase genes. Journal of Investigative Dermatology 6:10–18.
Schweitzer, M. H. 2011. Soft tissue preservation in terrestrial Mesozoic vertebrates. Annual Review of Earth and Planetary Sciences 39:187–216.
Sellers, W., P. L. Manning, T. Lyson, K. Stevens, and L. Margetts. 2009. Virtual palaeontology: gait reconstruction of extinct vertebrates using high performance computing. Palaeontologica Electronica 12.3.13A:1–26.
Tissot, B. P., and T. H. Welte. 1984. Petroleum Formation and Occurrence. Springer-Verlag, New York, 699 pp.
Vinther, J., D. E. G. Briggs, R. O. Prum, and V. Saranathan. 2008. The colour of fossil feathers. Biology Letters 4:522–525.
Vinther, J., D. E. G. Briggs, J. Clarke, G. Mayr, and R. O. Prum. 2010. Structural coloration in a fossil feather. Biology Letters 6:128–131.
Wehmiller, J. F., and G. H. Miller. 2000. Aminostratigraphic dating methods in Quaternary geology; pp. 187–222 in J. S. Noller, J. M. Sowers, and W. R. Lettis (eds.), Quaternary Geochronology: Methods and Applications. American Geophysical Union, Washington, D.C.
Witmer, L. M. 1995. Extant phylogenetic bracket and the importance of reconstructing soft tissue in fossils; pp. 19–33 in J. J. Thomason (ed.), Functional Morphology in Vertebrate Palaeontology. Cambridge University Press, Cambridge, U.K.
Wogelius, R. A., P. L. Manning, P. L. Larson, H. Barden, N. P. Edwards, S. M. Webb, W. I. Sellers, K. G. Taylor, P. Dodson, H. You, L. Daqing, and U. Bergmann. 2011. Trace metals as biomarkers for eumelanin pigment in the fossil record. Science 330:1622–1626.
Zhang, F., S. L. Kearns, P. J. Orr, M. J. Benton, Z. Zhou, D. Johnson, X. Xu, and X. Wang. 2010. Fossilized melanosomes and the colour of Cretaceous dinosaurs and birds. Nature 463:1075–1078.

Afterword

A

John R. Horner

COOL!! A DUCKBILL CELEBRATION TO HONOR OUR FRIEND Dave Weishampel *and* the current word on my favorite group of dinosaurs! Thirty-six chapters covering the origins, anatomy, variation, distribution, function, and traces of the Hadrosauridae – Joseph Leidy and Edward Cope would likely be astonished. I recall many years ago when Dave Weishampel and I, sitting on an outcrop in northern Montana, wondered if we would be able to interest any new students in hadrosaurs, or if they would all be mesmerized by theropods and sauropods. Little did we know!

I am pleasantly surprised by the number of chapters concerning hadrosaur origins (chapters 2–8, and 15). We are now beginning to finally see some tractable evidence of their early Cretaceous Asian origin, followed by the middle Cretaceous North American dispersals, and finally the back-and-forth flow that seemingly confounded our previous biogeographic interpretations. I do, however, agree with Dave Norman (chapter 2), that we must come to some agreement on definitions before a palpable phylogeny is resolved. In my mind the European (chapters 15 and 16) and South American (chapter 18) taxa exemplify not only the need for clear definitions, but also the need for additional specimens.

The six chapters of part 3 (9–14) are interesting not only for their anatomical perspectives, but more so for their discussions on morphological variation, which of course is a result of having multiple specimens, with some showing ontogenetic variation (chapter 14). Nic Campione's description (chapter 13) of the postcranial skeleton of *Edmontosaurus*, the first comprehensive anatomical study of a hadrosaur postcranium, has been in need for more than 150 years, ever since the description of *Hadrosaurus foulkii* (Leidy, 1858).

The continued tradition of relocating old quarries by Darren Tanke and Dave Evans (chapter 21) is nothing short of commendable, for as we know these days, knowledge of geographic and stratigraphic positions, as well as ontogenetic status, is mandatory before specimens can be legitimately placed in any type of phylogram. The stratigraphic distribution of hadrosaurids from the Upper Cretaceous formations of New Mexico by Robert Sullivan and Spencer Lucas (chapter 20) is also very welcome so that northern and southern North American faunas can be better evaluated. The more thorough descriptions of the hadrosaurs from Russia (chapter 17) and Mexico (chapter 19) also provide valuable data.

The models presented in part 5, "Function and Growth" (chapters 22–28) are both interesting and thought provoking, and will certainly evoke future discussions, particularly the study by Don Brinkman (chapter 23), in which he suggests conflicting size-frequency distributions between Montana and Alberta.

The final section (part 6), entitled "Preservation, Tracks and Traces," includes eight informative chapters (29–36) on various subjects including taphonomy (chapters 29 and 30), skin and footprint traces (chapters 32, 34, and 35), pathological elements (chapter 33), and biochemistry (chapter 36). Among these, the size-distribution study by Hone and colleagues (chapter 31) may well provide an answer to a question that has puzzled me for many years concerning the whereabouts of skeletally mature dinosaurs (Horner et al., 2011).

Clearly this volume provides a wealth of new information, but as each of us who study hadrosaurian dinosaurs well knows, a book like this could be published on an annual basis for many more years to come. Hadrosaurian dinosaurs could easily be called the "cows of the Cretaceous" as they are by far the best known of all dinosaurs, being represented by eggs, embryos, and virtually all stages of their ontogeny. We have their nesting grounds, massive bonebeds, coprolites, footprints, and skin impressions. We have cut open their bones to reveal their growth rates, ages at death, proteins, and physiology. We have a fair understanding of their diversity, phylogeny, biogeography, and even some of their behaviors, and yet publications concerning these animals are far from complete. This book is a beginning, with much more to come.

So other than naming new-found species, or arguing over their phylogenetic relationships or other biological attributes, what is left to discover about these bourgeois extinct animals? Plenty! But it is going to take reevaluations of outdated and preconceived ideas, exploration of formations that might be logistically difficult, and much more quantitative analysis. We need more specimens from Europe, eastern North America, and South America to have even a fair understanding of their origins. The specimens are out there, as are incredible finds within the bones themselves. With continued exploration we are sure to discover the very nature of these incredible honking dinosaurs!

REFERENCES CITED

Horner, J. R., M. B. Goodwin, and N. Myhrvold. 2011. Dinosaur census reveals abundant *Tyrannosaurus* and rare ontogenetic stages in the Upper Cretaceous Hell Creek Formation (Maastrichtian), Montana, USA. PLoS ONE 6(2):e16574.

Leidy, J. 1858. *Hadrosaurus foulkii,* a new saurian from the Cretaceous of New Jersey, related to the *Iguanodon.* Academy of Natural Sciences of Philadelphia, Proceedings, 10:213–218.

Subject Index

Locality Index (by country)

Stratigraphy Index (by country)

Taxonomic Index

DAVID A. EBERTH is a senior research scientist at the Royal Tyrrell Museum in Drumheller, Alberta.

DAVID C. EVANS is a curator in vertebrate paleontology at the Royal Ontario Museum and an assistant professor in the Department of Ecology and Evolutionary Biology at the University of Toronto.

This book was designed by Jamison Cockerham at Indiana University Press, set in type by Jamie McKee at MacKey Composition, and printed by Thompson-Shore, Inc.

The fonts are Electra, designed by William A. Dwiggins in 1935, Frutiger, designed by Adrian Frutiger in 1975, and Futura, designed by Paul Renner in 1927. All were published by Adobe Systems Incorporated.

www.ingramcontent.com/pod-product-compliance
Lightning Source LLC
LaVergne TN
LVHW070404060826
844660LV00009B/292
* 9 7 8 0 2 5 3 0 1 3 8 5 9 *